INSTRUCTOR'S
SOLUTIONS MANUAL

JOHN R. MARTIN

Tarrant County College, Northeast Campus

TECHNICAL CALCULUS
WITH ANALYTIC GEOMETRY

FOURTH EDITION

ALLYN J. WASHINGTON
Duchess Community College

Addison
Wesley

Boston San Francisco New York
London Toronto Sydney Tokyo Singapore Madrid
Mexico City Munich Paris Cape Town Hong Kong Montreal

Reproduced by Addison-Wesley from camera-ready copy supplied by the author.

Copyright © 2002 Pearson Education, Inc.

ISBN   0-201-71118-4

2 3 4 5 6 7 8 9 10 PH 04 03 02 01

CONTENTS

CHAPTER 4 APPLICATIONS OF THE DERIVATIVE

CHAPTER 5 INTEGRATION

CHAPTER 6 APPLICATIONS OF INTEGRATION

CHAPTER 7 DIFFERENTIATION OF THE TRIGONOMETRIC AND INVERSE TRIGONOMETRIC FUNCTIONS

CHAPTER 8 DIFFERENTIATION OF THE EXPONENTIAL AND LOGARITHMIC FUNCTIONS

CHAPTER 9 INTEGRATION BY STANDARD FORMS

CHAPTER 10 METHODS OF INTEGRATION

CHAPTER 11 INTRODUCTION TO PARTIAL DERIVATIVES AND DOUBLE INTEGRALS

CHAPTER 12 POLAR AND CYLINDRICAL COORDINATES

CHAPTER 13 EXPANSION OF FUNCTIONS IN SERIES

CHAPTER 14 FIRST-ORDER DIFFERENTIAL EQUATIONS

CHAPTER 15 HIGHER-ORDER DIFFERENTIAL EQUATIONS

CHAPTER 16 OTHER METHODS OF SOLVING DIFFERENTIAL EQUATIONS

APPENDIX A SUPPLEMENTARY TOPICS

APPENDIX C

Chapter 1

FUNCTIONS AND GRAPHS

1.1 Introduction to Functions

1. (a) $A(r) = \pi r^2$

(b) $A(d) = \pi \left(\dfrac{d}{2}\right)^2 = \pi \dfrac{d^2}{4}$

2. From geometry, $c = 2\pi r$ (b) $c = \pi d$

3. From geometry,

$$V = \frac{4}{3}\pi r^3 = \frac{4}{3}\pi \left(\frac{d}{2}\right)^3$$

$$= \frac{4}{3}\pi \frac{d^3}{8} = \frac{1}{3}\pi \frac{d^3}{2} = \frac{1}{6}\pi d^3$$

4. From geometry, $A = 6e^2$; $e^2 = \dfrac{A}{6}$; $e = \sqrt{\dfrac{A}{6}}$

5. $A(l) = lw = 5l$

6. From geometry, $V = \dfrac{1}{3}\pi r^2 h = \dfrac{1}{3}\pi r^2 (8) = \dfrac{8}{3}\pi r^2$

7. From geometry, $\quad A = s^2$
$$\sqrt{A} = \sqrt{s^2}$$
$$s = \sqrt{A}$$

8. From geometry, $\quad p = 4s$
$$\frac{4s}{4} = \frac{p}{4}$$
$$s = \frac{p}{4}$$

9. $f(x) = 2x + 1; f(1) = 2 \cdot 1 + 1 = 3;$

$f(-1) = 2(-1) + 1 = -1$

10. $f(x) = 5x - 9$
$f(2) = 5(2) - 9 = 10 - 9 = 1$
$f(-2) = 5(-2) - 9 = -10 - 9 = -19$

11. $f(x) = 5 - 3x$
$f(-2) = 5 - 3(-2) = 5 + 6 = 11$
$f(0.4) = 5 - 3(0.4) = 5 - 1.2 = 3.8$

12. $f(T) = 7.2 - 2.5T$
$f(2.6) = 7.2 - 2.5(2.6) = 0.7$
$f(-4) = 7.2 - 2.5(-4) = 17.2$

13. $\phi(x) = \dfrac{6 - x^2}{2x}; \phi(1) = \dfrac{6 - 1^2}{2(1)} = \dfrac{5}{2};$

$\phi(-2) = \dfrac{6 - (-2)^2}{2(-2)} = \dfrac{2}{-4} = -\dfrac{1}{2}$

14. $H(q) = \dfrac{8}{q} + 2\sqrt{q}$

$H(4) = \dfrac{8}{4} + 2\sqrt{4} = 2 + 2(2) = 2 + 4 = 6$

$H(0.16) = \dfrac{8}{0.16} + 2\sqrt{0.16} = 50 + 2(0.4)$

$\qquad\qquad = 50 + 0.8 = 50.8$

15. $g(t) = at^2 - a^2 t$

$g\left(-\dfrac{1}{2}\right) = a\left(-\dfrac{1}{2}\right)^2 - a^2\left(-\dfrac{1}{2}\right)$

$\qquad\qquad = a\left(\dfrac{1}{4}\right) + a^2\left(\dfrac{1}{2}\right)$

$\qquad\qquad = \dfrac{1}{4}a + \dfrac{1}{2}a^2$

$g(a) = a(a^2) - a^2(a) = a^3 - a^3 = 0$

16. $s(y) = 6\sqrt{y+1} - 3$
$s(8) = 6\sqrt{8+1} - 3 = 6\sqrt{9} - 3$
$\qquad = 6(3) - 3 = 18 - 3 = 15$
$s(a^2) = 6\sqrt{a^2+1} - 3$

17. $K(s) = 3s^2 - s + 6;$
$K(-s) = 3(-s)^2 - (-s) + 6 = 3s^2 + s + 6$
$K(2s) = 3(2s)^2 - 2s + 6 = 12s^2 - 2s + 6$

18. $T(t) = 5t + 7$
$T(-2t) = 5(-2t) + 7 = -10t + 7$
$T(t+1) = 5(t+1) + 7 = 5t + 5 + 7 = 5t + 12$

19. $\qquad f(x) = 2x + 4$
$f(3x) - 3f(x) = 2(3x) + 4 - 3(2x + 4)$
$\qquad\qquad = 6x + 4 - 6x - 12$
$\qquad\qquad = -8$

20. $\qquad\qquad f(x) = 2x^2 + 1$
$f(x+2) - f(x) - 2 = 2(x+2)^2 + 1 - 2x^2 - 1 - 2$
$\qquad\qquad = 2(x^2 + 4x + 4) - 2x^2 - 2$
$\qquad\qquad = 2x^2 + 8x + 8 - 2x^2 - 2$
$\qquad\qquad = 8x + 6$

21. 3 is an integer (whole number); 3 is rational (may be written as a ratio of integers, 3/1); 3 is real (not the square root of a negative number). $-\pi$ is irrational; it is not a ration of integers; $-\pi$ is real. $-\sqrt{-6}$ is imaginary; $\sqrt{7}/3$ is irrational (not the ratio of integers) and real.

22. $\dfrac{5}{4}$: real, rational

$\sqrt{-4}$: imaginary

$-\dfrac{7}{3}$: real, rational

$\dfrac{\pi}{6}$: real, irrational

23. $|3| = 3$. $\left|\dfrac{7}{2}\right| = \dfrac{7}{2}$

$\left|-\dfrac{6}{7}\right| = \dfrac{6}{7}$; $|-\sqrt{3}| = \sqrt{3}$

24. $|-4| = -(-4) = 4$

$|\sqrt{2}| = \sqrt{2}$

$\left|-\dfrac{\pi}{2}\right| = -\left(-\dfrac{\pi}{2}\right) = \dfrac{\pi}{2}$

$\left|-\dfrac{19}{4}\right| = -\left(-\dfrac{19}{4}\right) = \dfrac{19}{4}$

25. (a) $-4 < 0$

(b) $1 > -\pi$ since $-\pi \approx -3.14$ and $1 > -3.14$

(c) $-\dfrac{1}{3} > -\dfrac{1}{2}$

26. (a) $-3 < -2$

(b) $-\sqrt{2} > -1.42$

(c) $-|-3| = -3 \Rightarrow -4 < -3 = -|-3|$

27. (a) $b - a$; $b > a$, positive integer

(b) $a - b$; $b > a$, negative integer.

(c) $\dfrac{b-a}{b+a}$, positive rational number less than 1

28. (a) $a + b$, positive integer

(b) $\dfrac{a}{b}$, positive rational integer.

(c) $a \times b$, positive integer

29. (a) Let x be a positive integer, then $|x| = x$ which is positive integer and thus an integer. Yes. Let x be a negative integer, then $|x| = -x$ which is a positive integer and thus an integer. Yes.

(b) Let x be a positive or negative integer, then the reciprocal is $\dfrac{1}{x}$ which is the ratio of two integers, 1 and x, and is thus a rational number. Yes.

30. (a) Yes, |positive or negative rational| = positive rational

(b) Yes, $\dfrac{\text{integer}}{\text{integer}}$ has reciprocal $\dfrac{\text{integer}}{\text{integer}}$ which is rational

31. For $x < 0$, $|x| > 0$ which is to the right of zero on number line.

32. (a) $|x| < 1$ describes

$\Leftrightarrow -1 < x < 1$

(b) $|x| > 2$ describes

$\Leftrightarrow x < -2$ or $x > 2$

33. $s = f(t) = 17.5 - 4.9t^2$; $f(1.2) = 17.5 - 4.9(1.2)^2$
$= 10.4$ m

34. $C = 0.014(T - 40)$
$f(T) = 0.014(T - 40)$
$f(15) = 0.014(15 - 40) = -0.35$ in, the change in length at $T = 55°$F.

35. $d = v + 0.05v^2$
$f(v) = v + 0.05v^2$
$f(30) = 30 + 0.05(30) = 75$ ft
$f(2v) = 2v + 0.05(2v)^2 = 2v + 0.2v^2$
$f(60) = 60 + 0.05(60)^2 = 240$ ft
$f[2(30)] = 2(30) + 0.2(30)^2 = 240$ ft

36. $P = \dfrac{200R}{(100 + R)^2}$

$f(R) = \dfrac{200R}{(100 + R)^2}$

$f(R + 10) = \dfrac{200(R + 10)}{[100 + (R + 10)]^2}$

$= \dfrac{200(R + 10)}{(110 + R)^2}$

1.2   Algebraic Functions

1. $F[G(x)] = F[x - 1] = \sqrt{(x - 1)^2 + 4}$
$= \sqrt{x^2 - 2x + 1 + 4}$
$= \sqrt{x^2 - 2x + 5}$

2. $[G(x)]^3 = (x-1)^3 = (x-1)(x-1)(x-1)$
$$= (x^2 - 2x + 1)(x - 1)$$
$$= x^3 - x^2 - 2x^2 + 2x + x - 1$$
$$= x^3 - 3x^2 + 3x - 1$$

3. $\dfrac{G(x)}{F(x)} = \dfrac{x-1}{\sqrt{x^2 + 4}}$

4. $\{G[F(x)]\}^2 = \{G[\sqrt{x^2 + 4}]\}^2$
$$= \{\sqrt{x^2 + 4} - 1\}^2$$
$$= x^2 + 4 - 2\sqrt{x^2 + 4} + 1$$
$$= x^2 - 2\sqrt{x^2 + 4} + 5$$

5. $\sqrt{\sqrt[3]{x^6 + 1}} = ((x^6 + 1)^{1/3})^{1/2} = (x^6 + 1)^{1/6}$
$$= \sqrt[6]{x^6 + 1}$$

6. $(x-1)(\sqrt{x^2 + x + 1}) = \sqrt{(x-1)(x^2 + x + 1)}$
$$= \sqrt{x^3 + x^2 + x - x^2 - x - 1}$$
$$= \sqrt{x^3 - 1}$$

7. $\dfrac{(4x-5)^4}{(4x-5)^{1/2}} = (4x - 5)^{7/2}$

8. $\dfrac{3(x^2 + 1)^0}{(x^2 + 4)^{-1/2}} = 3(1)(x^2 + 4)^{1/2} = 3\sqrt{x^2 + 4}$

9. $(2x+1)^{1/2} + (x+3)(2x+1)^{-1/2}$
$$= \sqrt{2x+1} + \frac{x+3}{\sqrt{2x+1}}$$
$$= \frac{\sqrt{2x+1}\sqrt{2x+1} + x + 3}{\sqrt{2x+1}}$$
$$= \frac{2x + 1 + x + 3}{\sqrt{2x+1}}$$
$$= \frac{3x + 4}{\sqrt{2x+1}}$$

10. $(3x-1)^{-2/3}(1-x) - (3x-1)^{1/3}$
$$= \frac{1-x}{(3x-1)^{2/3}} - (3x-1)^{1/3}$$
$$= \frac{1 - x - (3x-1)^{1/3}(3x-1)^{2/3}}{(3x-1)^{2/3}}$$
$$= \frac{1 - x - (3x-1)}{(3x-1)^{2/3}}$$
$$= \frac{1 - x - 3x + 1}{(3x-1)^{2/3}}$$
$$= \frac{2 - 4x}{(3x-1)^{2/3}}$$

11. $\dfrac{(x^2+1)^{1/2} - x^2(x^2+1)^{-1/2}}{(x^2+1)} \cdot \dfrac{(x^2+1)^{1/2}}{(x^2+1)^{1/2}}$
$$= \frac{x^2 + 1 - x^2}{(x^2+1)^{3/2}}$$
$$= \frac{1}{(x^2+1)^{3/2}}$$

12. $\dfrac{(1-2x^2)^{1/4}(2x) + 4x^3(1-2x^2)^{-3/4}}{(1-2x^2)^{1/2}} \cdot \dfrac{(1-2x^2)^{3/4}}{(1-2x^2)^{3/4}}$
$$= \frac{(1-2x^2)(2x) + 4x^3}{(1-2x^2)^{5/4}}$$
$$= \frac{2x - 4x^3 + 4x^3}{(1-2x^2)^{5/4}}$$
$$= \frac{2x}{(1-2x^2)^{5/4}}$$

13. The domain and range of $f(x) = x + 5$ are each all real numbers.

14. $g(u) = 3 - u^2$; since $3 - u^2$ is defined for all real numbers, the domain is the set of all real numbers. However, the range is all real numbers $g(u) \leq 3$, since u^2 is never negative.

15. $G(R) = \dfrac{3.2}{R}$ is not defined for $R = 0$.

Domain: all real numbers except 0.
Range: all real numbers except 0.

16. $F(r) = \sqrt{r + 4}$ is not defined for real numbers less than -4.
Domain: all real numbers $r \geq -4$ and the range cannot be negative due to the principal square root of $r + 4$.
Range: all real numbers $F(r) \geq 0$.

17. The domain of $f(s) = \dfrac{2}{s^2}$ is all real numbers except zero since it gives a division by zero. The range is all positive real numbers because $\frac{2}{s^2}$ is always positive.

18. $T(t) = 2t^4 + t^2 - 1$; the domain is all real numbers and the range is not defined for any real numbers less than -1. So the range is all real numbers $T(t) \geq -1$.

19. $H(h) = 2h + \sqrt{h} + 1$ where $\sqrt{h}$ is not defined for $h < 0$.
Domain: all real numbers $h \geq 0$.
Range: all real numbers $H(h) \geq 1$.

20. $f(x) = \dfrac{6}{\sqrt{2-x}}$ is not defined for real numbers greater than or equal to 2.

Domain: all real numbers $x < 2$, and the range cannot be negative due to the principal square root of $2 - x$, thus the range will be greater than 0.

21. The domain of $Y(y) = \dfrac{y+1}{\sqrt{y-2}}$ is $y > 2$ because the square root requires $y - 2 \geq 0$ or $y \geq 2$ and to avoid a division by zero, $y > 2$ is required.

22. $f(n) = \dfrac{n}{6-2n}$; since division by zero is undefined, the domain must be restricted to exclude any values when $6 - 2n = 0$. In this case, $n = 3$ must be excluded. So the domain is the set of all real numbers except $n = 3$.

23. $f(D) = \dfrac{D}{D-2} + \dfrac{4}{D+4} - \dfrac{D-3}{D-6}$; since division by zero is undefined, the domain must be restricted to exclude any value for which $D - 2$, $D + 4$, or $D - 6$ are equal to zero. In this case, $D \neq 2, -4, 6$. So the domain is the set of all real numbers except $2, -4, 6$.

24. $g(x) = \dfrac{\sqrt{x-2}}{x-3}$; since division by zero is undefined, the domain must be restricted to exclude any value for which $x - 3 = 0$. In this case $x = 3$ must be excluded. And square roots aren't defined for negative values, so the domain is all real numbers $x \geq 2$, except $x = 3$.

25. $F(t) = 3t - t^2$ for $t \leq 2$; $F(2) = 3 \cdot 2 - 2^2 = 2$; $F(3)$ does not exist.

26. $h(-8) = 2(-8) = -16$ (since $-8 < -1$)

$h\left(-\dfrac{1}{2}\right) = -\dfrac{1}{2} + 1 = \dfrac{1}{2}$ (since $-\dfrac{1}{2} \geq -1$)

27. $f(1) = \sqrt{1+3} = \sqrt{4} = 2$ (since $1 \geq 1$)

$f\left(-\dfrac{1}{4}\right) = -\dfrac{1}{4} + 1 = \dfrac{3}{4}$ (since $-\dfrac{1}{4} < 1$)

28. $g\left(\dfrac{1}{5}\right) = \dfrac{1}{\frac{1}{5}} = 5$ (since $\dfrac{1}{5} \neq 0$)

$g(0) = 0$ (since $0 = 0$)

29. $d(t) = 40(2) + 55t = 80 + 55t$

30. $C = f(r) = 3(2\pi rh + 2\pi r^2) = 6\pi r(2) + 6\pi r^2$
$= 12\pi r + 6\pi r^2$

31. $w = f(t) = 5500 - 2t$

32. $p = f(c) = 100c - 300$

33. $m(h) = 110 + 0.5(h - 1000)$ for $h > 1000$

34. $n = f(x) = 0.5x + 0.7(100) = 0.5x + 70$

35. $C = f(l) = 500 + 5(l - 50)$
$= 5l + 250$

36. $M = f(h) = \dfrac{8}{h}$

37. (a) $0.1x + 0.4y = 1200 \Rightarrow y(x) = \dfrac{1200 - 0.1x}{0.4}$

(b) $y(400) = \dfrac{1200 - 0.1(400)}{0.4} = 2900$ L

38. $18^2 = (18 - d)^2 + r^2$
$r = \sqrt{36d - d^2}$
$c = 2\pi r = 2\pi\sqrt{36d - d^2}$

39. For the square, $p = x$, side $= \dfrac{x}{4}$;

$$A_{\text{square}} = \left(\dfrac{x}{4}\right)^2 = \dfrac{x^2}{16} = \dfrac{p^2}{16}$$

For the circle, $c = 60 - x = 2\pi r$;

$$r = \dfrac{60 - x}{2\pi} = \dfrac{60 - p}{2\pi}$$

$$A_{\text{circle}} = \pi r^2 = \dfrac{\pi(60 - p)^2}{4\pi^2}.$$

Thus, the total

$$A = \dfrac{p^2}{16} + \dfrac{(60 - p)^2}{4\pi}$$

40. $A = f(d)$

This is a circle and a square.

$$A = \pi\left(\dfrac{d}{2}\right)^2 + d^2 = \dfrac{1}{4}\pi d^2 + d^2$$

41. $A = f(x) = \pi(6-x)^2$ with domain $0 \leq x \leq 6$ since x represents the radius and it must be greater than or equal to zero and less than or equal to six. Using the end point values for the radius gives a range $0 \leq A \leq 36\pi$.

42. $d = f(h)$
$120^2 + h^2 = d^2$
$d = \sqrt{14400 + h^2}$

The domain is $h \geq 0$ since the distance above the ground is nonnegative.
The range is $d \geq 120$ m since 120 m is the value of d when $h = 0$.

43. $s = f(t)$

$d = st = 300$

$s = 300/t$ (Note: cannot have negative time or speed)

Domain: can't divide by zero, so all real numbers $t > 0$

Range: all real numbers $s > 0$ (upper limits depend on truck)

44. $l = f(w)$

$A = lw = 8$

$l = \dfrac{8}{w}$ (Note: cannot have negative width or length)

Domain: can't divide by zero, so all real numbers $0 < w < 8$

Range: all real numbers $0 < l < 8$

45. The domain of $f = \dfrac{1}{2\pi\sqrt{C}}$ is $C > 0$ because C must be ≥ 0 to avoid taking the square root of a negative and > 0 to prevent division by zero.

46. $y = f(x) = 550 - x$

Domain: all real numbers greater than zero and less than 550 (since distance cannot be negative)

47. From Ex. 33,

$m = f(h) = 110 + 0.5(h - 1000) = 0.5h - 390$

$m = \begin{cases} 0.5h - 390 & \text{for} \quad h > 1000 \\ 110 & \text{for} \quad 0 \leq h \leq 1000 \end{cases}$

48. From Ex 35, $C = f(l) = 5l + 250$

$C = \begin{cases} 5l + 250 & \text{for} \quad l > 50 \\ 500 & \text{for} \quad 0 \leq l \leq 50 \end{cases}$

1.3   Rectangular Coordinates

1. $A(2, 1)$; $B(-1, 2)$; $C(-2, -3)$

2. $D = (3, -2)$; $E = (-3.5, 0.5) = \left(-\dfrac{7}{2}, \dfrac{1}{2}\right)$;

$F = (0, -4)$

3.

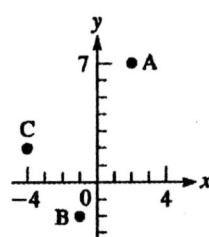

4.

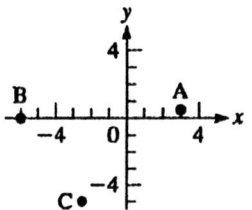

5. Joining the points in the order ABCA form an isosceles triangle.

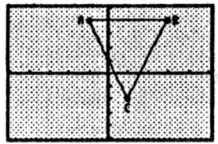

6. Isosceles right triangle

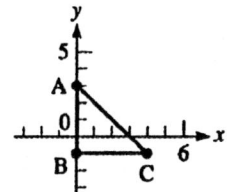

7. Rectangle

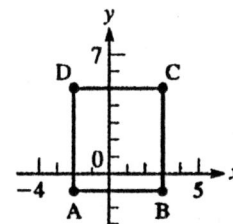

8. Parallelogram

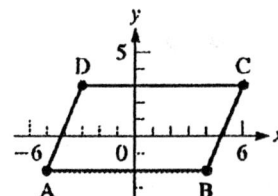

9. The coordinates of the fourth vertex, V, are (5, 4).

10. The abscissa is the x-coordinate, since this is an equilateral triangle and we know the base is $(2,1)$, $(7,1)$ then the third vertex must be equi-distant between 2 and 7. So $\frac{2+7}{2} = \frac{9}{2}$

11. In order for the x-axis to be the perpendicular bisector of the line segment join P and Q, Q must be $(3,-2)$.

12. The line segment joining P and Q are bisected by the origin gives point $Q(4,-1)$.

13. All points with abscissas of 1 are on a vertical line through $(1, 0)$. The equation of this vertical line is $x = 1$.

14. Ordinates are y-coordinates; the points whose ordinates are -3 are all on a line parallel to the x-axis, 3 units below.

15. All points $(x,3)$, where x is any real number, are points on a line parallel to the x-axis, 3 units above it.

16. All points $(-2,y)$, where y is any real number, are points on a line parallel to the y-axis, 2 units to the left.

17. All points whose abscissas equal their ordinates are on a 45° line through the origin. The equation of this line is $y = x$ and it bisects the first and third quadrants.

18. When the abscissa equals the negative of their ordinates, then $x = -y$. These pairs lie on a line formed of points made by varying x. These points form a line bisecting quadrants two and four.

19. Abscissas are x-coordinates; thus the abscissa of all points on the y-axis is zero.

20. Ordinates are y-coordinates; thus the ordinate of all points on the x-axis is zero.

21. All points for which $x > 0$ are to the right of the y-axis.

22. All points below the x-axis.

23. All points which lie to the left of a line that is parallel to the y-axis, one unit to the left have $x < -1$.

24. All points which lie above a line parallel to the x-axis, 4 units above have $y > 4$.

25. The ratio $\frac{y}{x}$ is positive in QI and QIII.

26. The ratio y/x is negative in quadrants two and four.

27. (a) $d = 3 - (-5) = 8$

(b) $d = 4 - (-2) = 6$

28. From exercise 27, the distance between $(-5,-2)$ and $(3,4)$ is 10 because it is the hypotenuse of a right triangle with legs of 6 and 8.

1.4 The Graph of a Function

1. $y = 2x - 4$

x	y
0	-4
1	-2
2	0

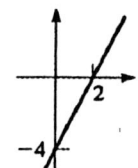

2. $y = 6 - \frac{1}{3}x$

x	y
-1	6.3
0	6
1	$5.\bar{6}$
3	5
18	0

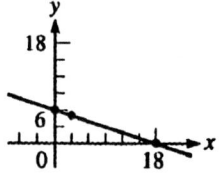

3. $y = 3 - x^2$

x	y
-2	-1
-1	2
0	3
1	2
2	-1

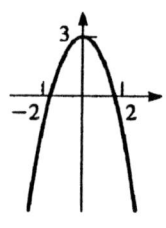

4. $y = 2x^2 + 1$

x	y
-2	9
-1	3
0	1
1	3
2	9

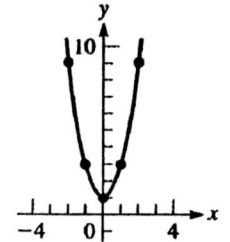

5. $h = 20t - 5t^2$

t	h
0	0
1	15
2	20
3	15
4	0

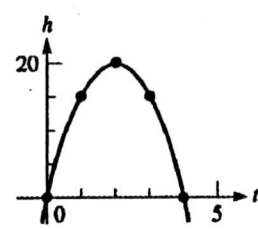

6. $y = 2 + 3x + x^2$

x	y
−3	2
−2	0
−1	0
0	2
1	6
2	12

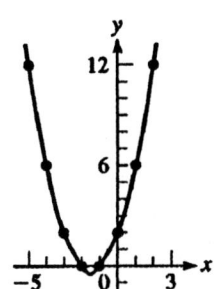

7. $V = e^3$

e	V
−2	−8
−1	−1
0	0
1	1
2	8

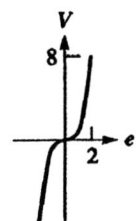

8. $y = 3x - x^3$

x	y
−3	18
−2	2
−1	−2
0	0
1	2
2	−2
3	−18

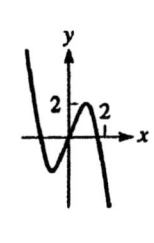

9. $y = \dfrac{2}{x+2}$

x	y
−3	−2
−1	2
0	1
1	0.66
3	0.4

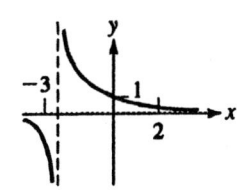

10. $p = \dfrac{1}{n^2 + 0.5}$

p	n
−2	0.22
−1	0.67
0	2
1	0.67
2	0.22

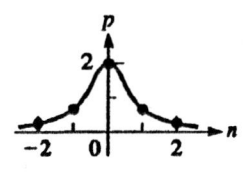

11. $y = \sqrt{4 - x}$

x	y
−5	3
0	2
3	1
4	0

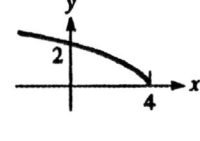

12. $y = \sqrt{x^2 - 16}$

x	y
−6	4.5
−5	3
−4	0
4	0
5	3
6	4.5

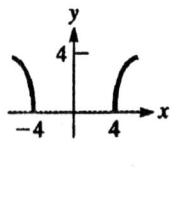

13.

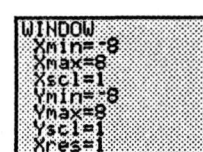 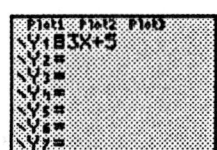

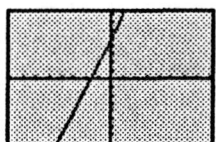

14.

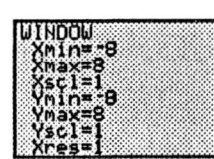 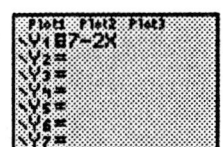

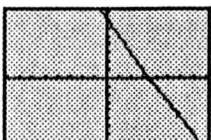

15.

19.

16.

20.

17.

21.

18.

22.

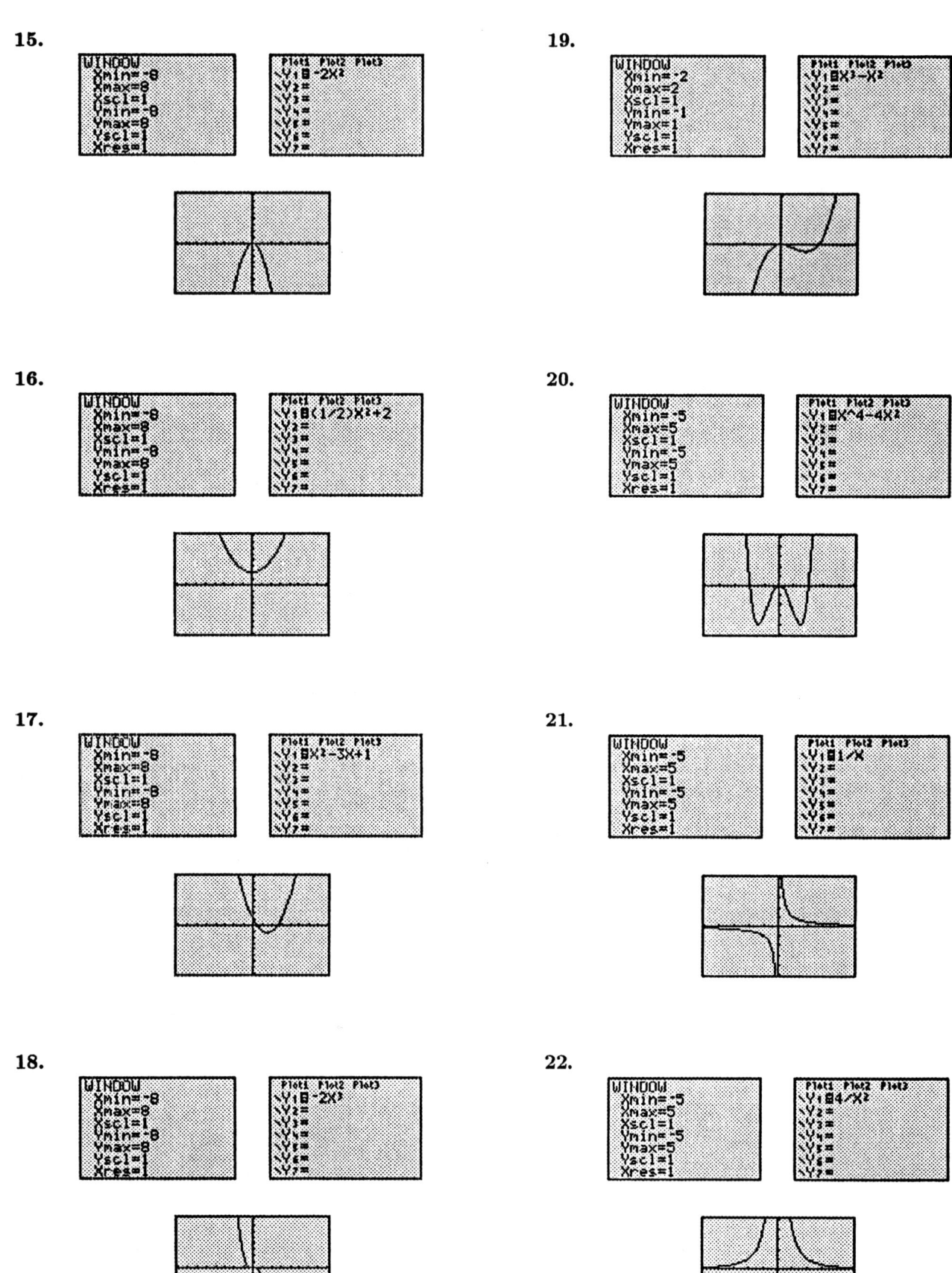

23.

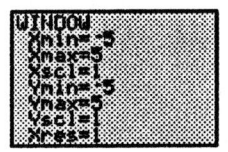 

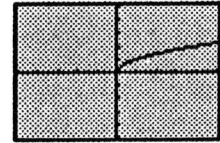

24.

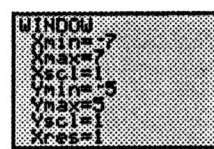 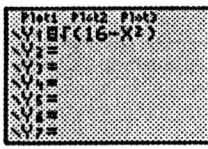

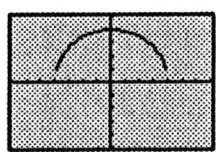

25. To solve $x^2 - 41 = 0$ using a graphing calculator, we let $y = x^2 - 41$. Using the trace, we see that the solutions for $y = 0$ are approximately -6.4 and 6.4. With the zoom feature these results may be read more accurately.

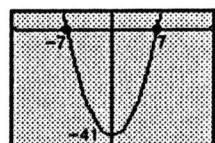

26. $w(w-4) = 9$; $w^2 - 4w = 9$. Collect all terms on the left; $w^2 - 4w - 9 = 0$. Let $y = x^2 - 4x - 9$. Using the trace we see that the solutions are approximately -1.6 and 5.6. With the zoom feature these results may be read more accurately.

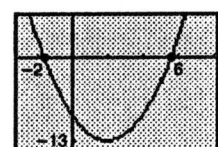

27. $\sqrt{5R+2} = 3 \Rightarrow \sqrt{5R+2} - 3 = 0$

Graph $y = \sqrt{5x+2} - 3$ and use zero feature to solve.

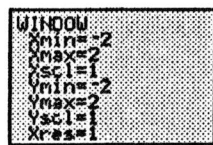 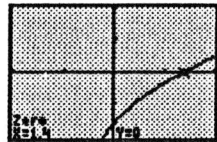 $R = 1.4$

28. $x - 2 = \frac{1}{x}$. Collect all terms on the left; $x - \frac{1}{x} - 2 = 0$. Let $y = x - \frac{1}{x} - 2$. Using the trace, we see that the solutions are approximately -0.4 and 2.4. With the zoom feature, the results may be read more accurately.

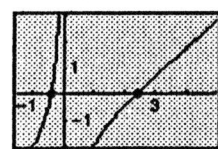

29. From the graph, $y = \dfrac{4}{x^2 - 4}$ has range $y \le -1$ or $y > 0$.

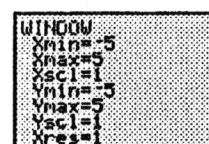 

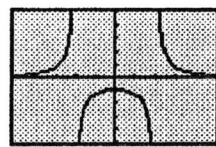

30. Set the range at $x_{min} = -2, x_{max} = 2$, $y_{min} = -2, y_{max} = 2$. From the graph, using the trace, we see that the minimum y-value is approximately -0.25. Therefore, the range is all numbers greater than -0.25.

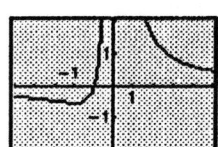

31. Set the range at $x_{\min} = -4$, $x_{\max} = 4$, $y_{\min} = -6$, $y_{\max} = 4$. From the graph, using the trace, we see that the range appears to be $y \geq 0$ or $y \leq -4$.

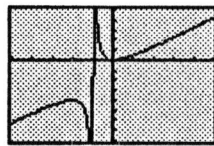

32. Set the range at $x_{\min} = -3$, $x_{\max} = 3$, $y_{\min} = -3$, $y_{\max} = 3$. From the graph the appears to be the set of all real numbers.

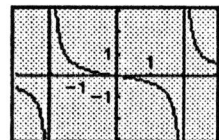

33. Graph $y = \dfrac{x+1}{\sqrt{x-2}}$ on graphing calculator and use the minimum feature, then from the graph, $Y(y) = \dfrac{y+1}{\sqrt{y-2}}$ has range $Y(y) \geq 3.464$.

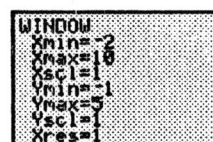 

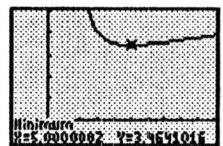

34. Set the range at $x_{\min} = -2$, $x_{\max} = 5$, $y_{\min} = -4$, $y_{\max} = 4$. The range appears to be the set of all real numbers excluding $-\frac{1}{2}$.

35. Graph $y = \dfrac{x}{x-2} + \dfrac{4}{x+4} - \dfrac{x-3}{x-6}$ using a graphing calculator, the range appears to be the set of all real numbers.

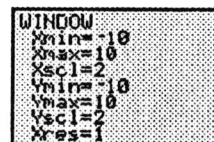 

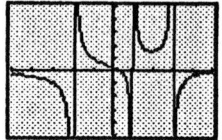

36. Using a graphing calculator with $x_{\min} = 0$, $x_{\max} = 5$, $y_{\min} = -4$, $y_{\max} = 4$, the range appears to be the set of all real numbers.

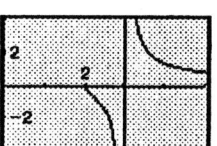

37. $c = 0.011r + 4.0$

r	c
500	9.5
1000	15
2000	26

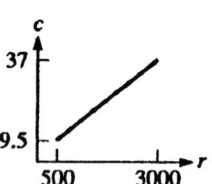

38. $H = 240I^2$

I	H
0	0
0.2	9.6
0.4	38.4
0.6	86.4
0.8	153.6

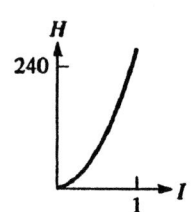

39. $p = \dfrac{0.05(1+m)}{m}$

m	p
0.1	0.55
0.2	0.3
0.3	0.216
0.4	0.175

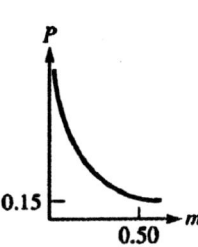

40. $N = \sqrt{n^2 - 1.69}$

n	N
1.3	0
1.5	0.748
1.7	1.095
2.0	1.520

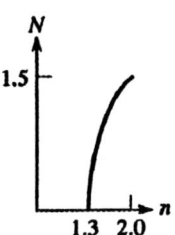

41. $e^3 + (e + 5.00)^3 = 40{,}000$

$$e = 24.4 \text{ cm}$$
$$e + 5.00 = 29.4 \text{ cm}$$

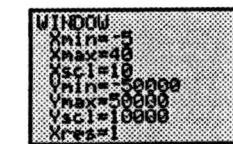 

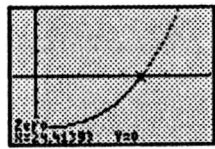

42. $A = 520 = lw = (w + 12)w$
$$= w^2 + 12w \Rightarrow w^2 + 12w - 520 = 0$$

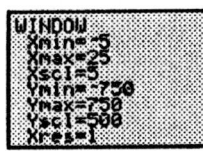 

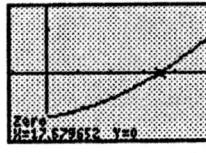

The approximate dimensions, in cm, to 2 significant digits are $w \approx 18$ cm, $l \approx 30$ cm.

43. $h = 15 + 86t - 4.9t^2$; when $h = 0, t = 18$ s

t	h
0	50
10	1250
20	−750

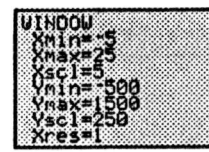 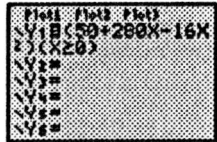

44. Graph $y = 9x^3 - 2400x^2 + 240{,}000x - 8{,}000{,}000$ and use the zero feature to solve.

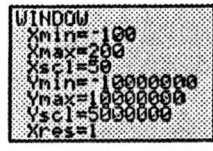 

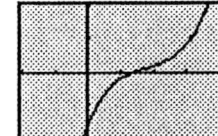 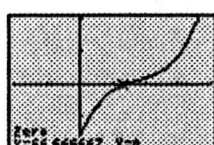

45. $P = 2l + 2w = 200 \Rightarrow l = 100 - w$
$A = lw = (100 - w)w = 100w - w^2$
for $30 \le w \le 70$

w	30	40	50	60	70
A	2100	2400	2500	2400	2100

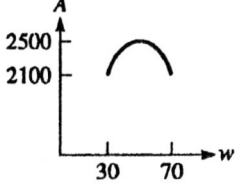

46. $y = x(10 - 2x)(12 - 2x)$
$y = x(120 - 44x + 4x^2)$
$y = 120x - 44x^2 + 4x^3$

x	y
0	0
1	80
2	96
3	72
4	32

From the table $y = 90$ for x between 1 and ⁚
for x between 2 and 3. Graph

$$y_1 = 120x - 44x^2 + 4x^3 - 90$$

and use the zero feature to solve.

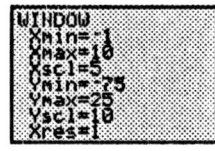 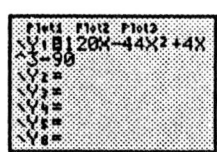

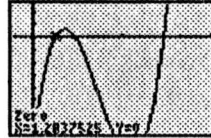 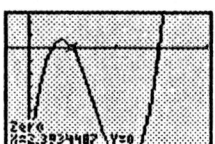

From the graphs, $y = 90$ when $x = 1.3$ in and
$x = 2.4$ in

47. $s = \sqrt{t - 4t^2}$. Graph $y_1 = \sqrt{x - 4x^2}$ and use the
maximum feature to find the maximum cutting
speed.

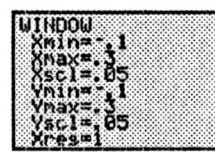 

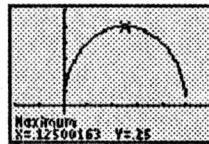

From the graph the maximum cutting speed is
0.25 ft/min.

48. To determine the maximum capacity graph
$y = 120x - 44x^2 + 4x^3$ from Exercise 32 and use
the maximum feature to solve.

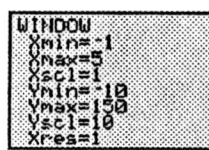 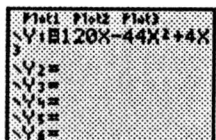

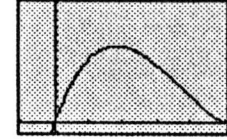 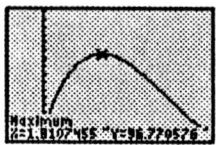

The maximum capacity is 96.8 in³.

49.

x	y
-2	2
-1	1
0	0
1	1
2	2

$y = x$ is the same as
$y = |x|$ for $x \geq 0$.
$y = |x|$ is the same as
$y = -x$ for $x < 0$.

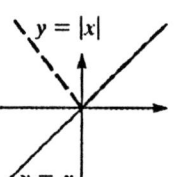

For negative values of x, $y = |x|$ becomes $y = -x$.

50. The graphs differ because the absolute value does
not allow the graph to go below the x-axis.
$y = 2 - x$

x	y
-1	3
0	2
1	1
2	0
3	-1

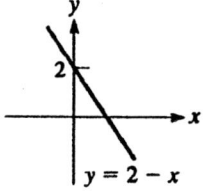

$y = |2 - x|$

x	y
-2	4
1	1
2	0
3	1

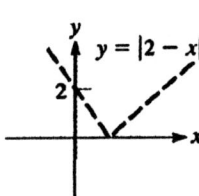

51. $f(x) = \begin{cases} 3 - x & x < 1 \\ x^2 + 1 & x \geq 1 \end{cases}$

x	y
-2	5
-1	4
0	3
1	2
2	5
3	10

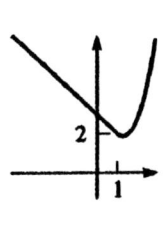

52. $f(x) = \begin{cases} \dfrac{1}{x - 1} & x < 0 \\ \sqrt{x + 1} & x \geq 0 \end{cases}$

x	y
-2	-0.3
-1	-0.5
-0.1	-0.9
0	1
1	1.4
3	2

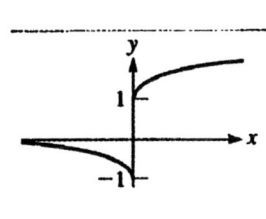

53. The graph passes the vertical line test and is, therefore, a function.

54. Some vertical lines will intercept the graph at multiple points. Graph is that of a relation.

55. No. Some vertical lines will intercept the graph at multiple points.

56. Any vertical line will intercept the graph at only one point. Graph is that of a function.

Chapter 1 Review Exercises

1. $A = \pi r^2 = \pi(2t)^2$
  $A = 4\pi t^2$

2. $A = \pi r s = \pi r \sqrt{h^2 + r^2} = 3\pi \sqrt{h^2 + 9}$

3. $2000(x) + 1800(y) = 50{,}000$

$$= -\frac{10}{9}x + \frac{250}{9}$$

4. $C = 20l + 2w(10) + 10l$
  $= 30l + 20w$
  $= 30(w + 20) + 20w$
  $= 30w + 600 + 20w$
  $= 50w + 600$

5. $f(x) = 7x - 5$
  $f(3) = 7(3) - 5 = 21 - 5 = 16$
  $f(-6) = 7(-6) - 5 = -42 - 5 = -47$

6. $g(I) = 8 - 3I$
  $g\left(\dfrac{1}{6}\right) = 8 - 3\left(\dfrac{1}{6}\right) = \dfrac{15}{2}$
  $g(-4) = 8 - 3(-4) = 20$

7. $H(h) = \sqrt{1 - 2h}$
  $H(-4) = \sqrt{1 - 2(-4)} = 3$
  $H(2h) = \sqrt{1 - 2(2h)} = \sqrt{1 - 4h}$

8. $\phi(v) = \dfrac{3v - 2}{v + 1}$
  $\phi(-2) = \dfrac{3(-2) - 2}{-2 + 1} = 8$
  $\phi(v + 1) = \dfrac{3(v + 1) - 2}{v + 1 + 1} = \dfrac{3v + 1}{v + 2}$

9. $f(x) = 3x^2 - 2x + 4$
  $f(x + h) - f(x)$
  $= 3(x + h)^2 - 2(x + h) + 4 - (3x^2 - 2x + 4)$
  $= 3(x^2 + 2xh + h^2) - 2x - 2h + 4 - 3x^2 + 2x - 4$
  $= 3x^2 + 6xh + 3h^2 - 2x - 2h + 4 - 3x^2 + 2x - 4$
  $= 6xh + 3h^2 - 2h$

10. $F(x) = x^3 + 2x^2 - 3x$
  $F(3 + h) - F(3)$
  $= (3 + h)^3 + 2(3 + h)^2 - 3(3 + h)$
  $\quad - (3^3 + 2(3)^2 - 3(3))$
  $= 3^3 + 3(3)^2h + 3(3)h^2 + h^3$
  $\quad + 2(9 + 6h + h^2) - 9 - 3h - 36$
  $= 27 + 27h + 9h^2 + h^3 + 18 + 12h$
  $\quad + 2h^2 - 9 - 3h - 36$
  $= h^3 + 11h^2 + 36h$

11. $f(x) = 3 - 2x$
  $f(2x) - 2f(x) = 3 - 2(2x) - 2(3 - 2x)$
  $\qquad\qquad\qquad = 3 - 4x - 6 + 4x = -3$

12. $f(x) = 1 - x^2$
  $[f(x)]^2 - f(x^2) = (1 - x^2)^2 - (1 - (x^2)^2)$
  $\qquad\qquad\qquad = 1 - 2x^2 + x^4 - 1 + x^4$
  $\qquad\qquad\qquad = 2x^4 - 2x^2$

13.
$$f(x) = 8.07 - 2x$$
$$f(5.87) = 8.07 - 2(5.87) = -3.67$$
$$f(-4.29) = 8.07 - 2(-4.29) = 16.65 \approx 16.7$$

14.
$$g(x) = 7x - x^2$$
$$g(45.81) = 7(45.81) - (45.81)^2$$
$$= -1778$$
$$g(-21.85) = 7(-21.85) - (-21.85)^2$$
$$= -630.4$$

15.
$$G(S) = \frac{S - 0.087629}{3.0125S}$$
$$G(0.17427) = \frac{0.17427 - 0.087629}{3.0125(0.17427)}$$
$$= 0.16503$$
$$G(0.053206) = \frac{0.053206 - 0.087629}{3.0125(0.053206)}$$
$$= -0.21476$$

16.
$$h(t) = \frac{t^2 - 4t}{t^3 + 564}$$
$$h(8.91) = \frac{8.91^2 - 4(8.91)}{8.91^3 + 564}$$
$$= 0.0344$$
$$h(-4.91) = \frac{(-4.91)^2 - 4(-4.91)}{(-4.91)^3 + 564}$$
$$h(-4.91) = 0.0982$$

17. The domain of $f(x) = x^4 + 1$ is $-\infty < x < \infty$. The range is $f(x) \geq 1$.

18. The domain of $G(z) = \dfrac{4}{z^3}$ is all real numbers except 0.
The range is $y \neq 0$.

19. The domain of $g(t) = \dfrac{2}{\sqrt{t+4}}$ is all real numbers $t > -4$.
The range is all real numbers $g(t) > 0$.

20. The domain of $F(y) = 1 - 2\sqrt{y}$ is all real numbers $y \geq 0$.
The range is all real numbers $F(y) \leq 1$.

21. $y = 4x + 2$

x	y
0	2
$-\frac{1}{2}$	0

22. $y = 5x - 10$

x	y
0	-10
2	0

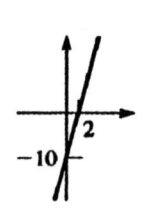

23. $y = 4x - x^2$

x	y
0	0
1	3
2	4
3	3
4	0

24. $y = x^2 - 8x - 5$

x	y
-1	4
0	-5
1	-12
2	-17
3	-20
4	-21
5	-20

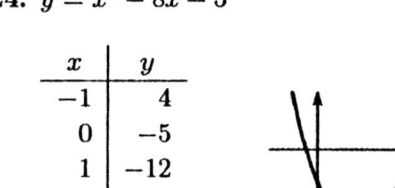

25. $y = 3 - x - 3x^2$

x	y
-2	-3
-1	2
0	3
1	0
2	-7

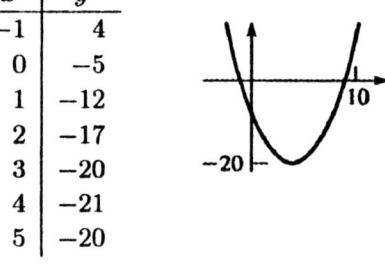

26. $y = 6 + 4x + x^2$

x	y
-3	3
-2	2
-1	3
0	6
1	11
2	18
3	27

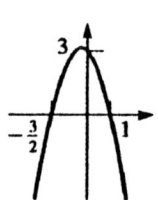

27. $y = x^3 - 6x$

x	y
-3	-9
-2	4
-1	5
0	0
1	-5
2	-4
3	9

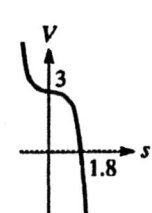

28. $V = 3 - 0.5s^3$

s	V
-2	7
0	3
1	2.5
2	-1

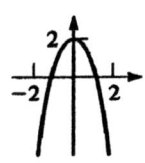

29. $y = 2 - x^4$

x	y
-2	-14
-1	1
0	2
1	1
2	-14

30. $y = x^4 - 4x$

x	y
-2	24
-1	5
0	0
1	-3
2	8

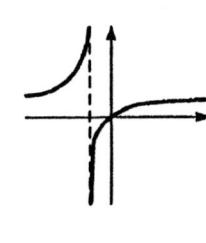

31. $y = \dfrac{x}{x+1}$

x	y
-4	$\frac{4}{3}$
-3	$\frac{3}{2}$
-2	-2
0	0
1	$\frac{1}{2}$
2	$\frac{2}{3}$

32. $Z = \sqrt{25 - 2R^2}$

x	y
-3	2.65
-2	4.12
-1	4.80
0	5.00
1	4.80
2	4.12
3	2.65

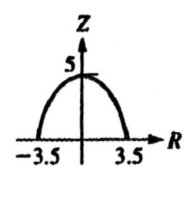

33. $7x - 3 = 0$

Graph $y = 7x - 3$ and use the zero feature to solve.

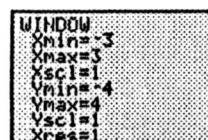 

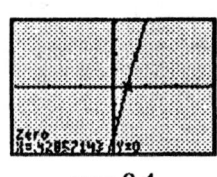

$$x = 0.4$$

34. $3x + 11 = 0$. Graph $y = 3x + 11$ and use zero feature to solve.

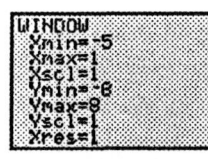 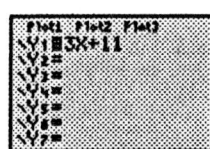

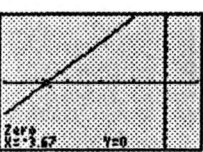

$$x = -3.7$$

35. $x^2 + 1 = 6x$. Graph $y = x^2 + 1 - 6x$ and use zero feature to solve.

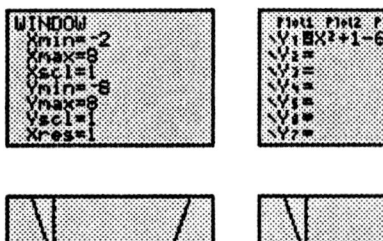 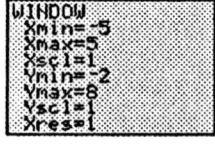

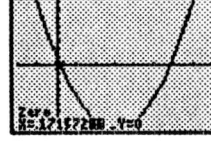 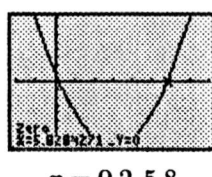

$$x = 0.2, 5.8$$

36. $3t - 2 = t^2$. Graph $y = 3x - 2 - x^2$ and use the zero feature to solve.

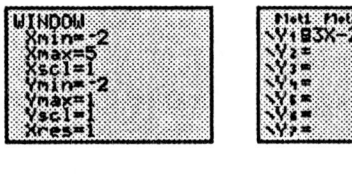

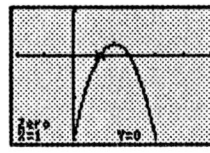 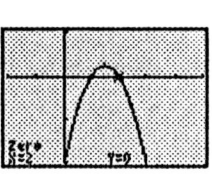

$$t = 1.0, 2.0$$

37. $x^3 - x^2 = 2 - x \Rightarrow x^3 - x^2 + x - 2 = 0$

Graph $y = x^3 - x^2 + x - 2$ use the zero feature to solve.

 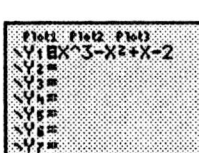 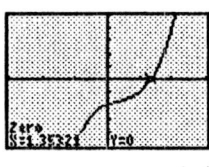

$$x = 1.4$$

38. $5 - x^3 = 2x^2$. Graph $y = 5 - x^3 - 2x^2$ and use the zero feature to solve.

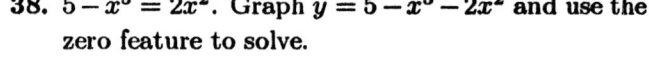

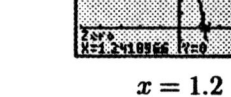

$$x = 1.2$$

39. $\dfrac{1}{x} = 2x$. Graph $y = \dfrac{1}{x} - 2x$ and use the zero feature to solve.

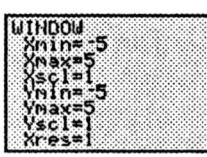 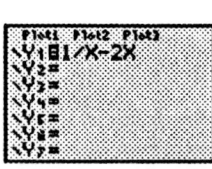

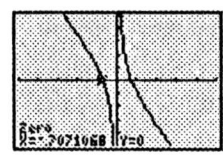 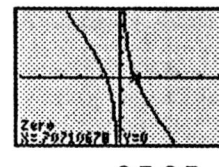

$$x = -0.7, 0.7$$

40. $\sqrt{x} = 2x - 1$. Graph $y = \sqrt{x} - 2x + 1$ and use the zero feature to solve.

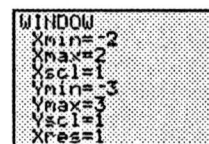 

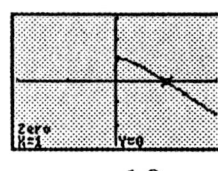

$$x = 1.0$$

41. Graph $y = x^4 - 5x^2$ and use the minimum feature, then from the graph, the range is $y \geq -6.25$.
All real numbers $y \geq -6.25$

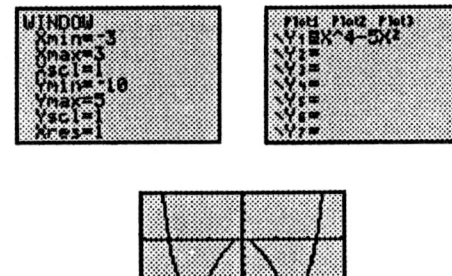

42. Graph $y = x\sqrt{4 - x^2}$ and use the maximum and minimum feature. From the graph the range is $-2 \leq y \leq 2$.

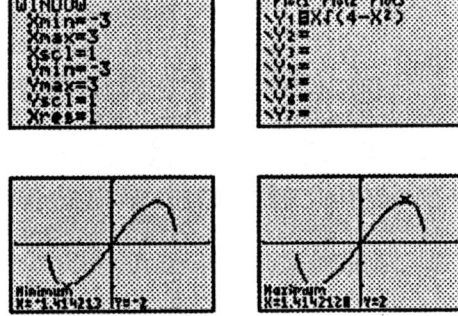

43. $A = w + \dfrac{2}{w}$. Graph $y_1 = x + \dfrac{2}{x}$.

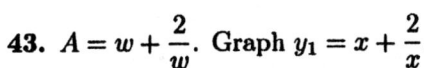

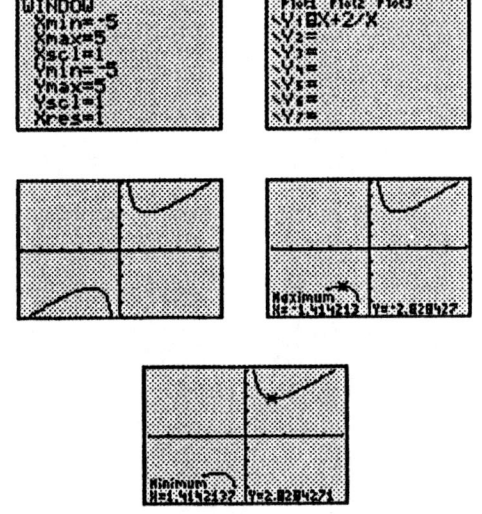

From the graph the range is all real numbers
$w \leq -2.83$ or $w \geq 2.83$

44. Graph $y = 2x + \dfrac{3}{\sqrt{x}}$ and use minimum feature. From the graph the range is all real numbers $y \geq 4.95$

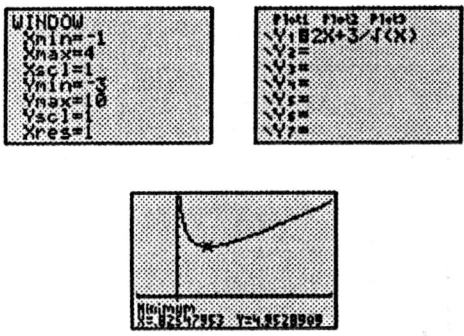

45. $A(a, b) = A(2, -3)$ is in QIV while $B(b, a) = B(-3, 2)$ is in QII.

46. distance from $(0, 0)$ to (a, b)
$$= \sqrt{(a - 0)^2 + (b - 0)^2}$$
$$= \sqrt{a^2 + b^2}$$

47. Let the coordinates of the third vertex be $(1, y)$, then
$$\sqrt{(1 - 0)^2 + (y - 0)^2} = 2$$
$$1 + y^2 = 4$$
$$y^2 = 3$$
$$y = \pm\sqrt{3}$$

The coordinates of the third vertex are $(1, \sqrt{3})$ or $(1, -\sqrt{3})$.

48. The side s of the square is the distance between $(1, 2)$ and $(1, -3)$.
$$s = \sqrt{(1 - 1)^2 + (2 - (-3))^2}$$
$$= 5.$$

The other two vertices are $(6, 2)$ and $(6, -3)$, or $(-4, 2)$ and $(-4, -3)$.

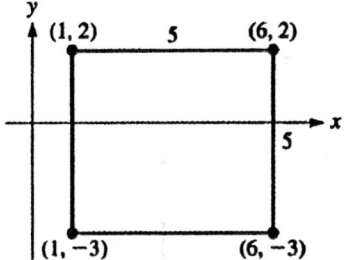

49. $I = f(m) = 12.5\sqrt{1 + 0.5m^2}$
$f(0.55) = 12.5\sqrt{1 + 0.5(0.55)^2} = 13.4$

50. $p = f(t) = \dfrac{100t}{t + 1.5}$

$f(0.4) = \dfrac{100(0.4)}{0.4 + 1.5}$

$= 20$

51. $A = 8.0 + 12t^2 - 2.0t^3$, $0 \le t \le 6$. Graph $y = 8 + 12x^2 - 2x^3$ and use the maximum feature to find the greatest value of $A = 72.0°$

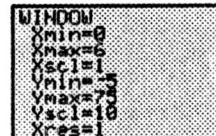 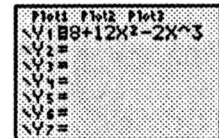

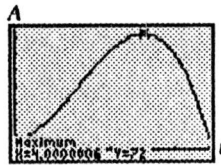

52. $P = \dfrac{24R}{R^2 + 1.40R + 0.49}$

Graph $y = \dfrac{24x}{x^2 + 1.40x + 0.49}$

and use the maximum feature to find the maximum power.

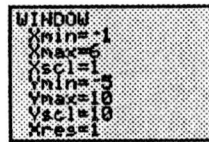 

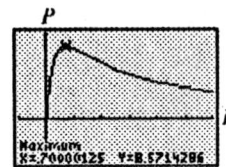

The maximum power is 8.6 W.

53. $T = f(t) = 28.0 + 0.15t$, $0 \le t \le 30$

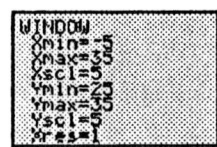 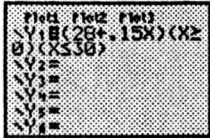

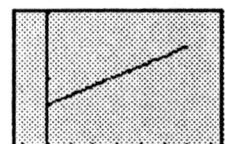

54. $N(t) = 500 + 7000t$. Graph $y = 500 + 7000x$ for

$0 \le x \le \dfrac{199}{14}$ since the tank fills at time

$t = \dfrac{100{,}000 - 500}{7000} = \dfrac{199}{14} \approx 14$ h.

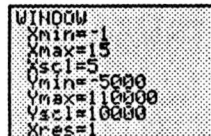 

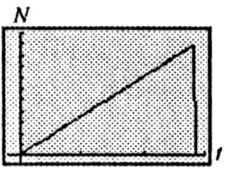

55. $160 = 5F - 9C$
$9C = 5F - 160$

$C = f(F) = \dfrac{5}{9}F - \dfrac{160}{9}$. Graph $y = \dfrac{5}{9}x - \dfrac{160}{9}$.

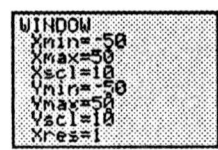 

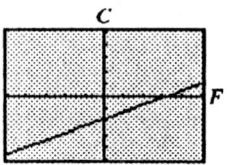

56. $C(n) = 1000 + 10n$. Graph $y = 1000 + 10x$

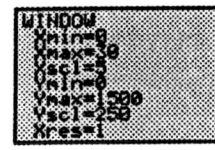 

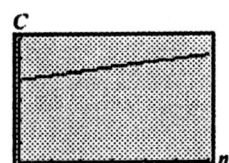

59. $N(t) = \dfrac{1000}{\sqrt{t+1}}$, $t > 0$. Graph $y = \dfrac{1000}{\sqrt{t+1}}$, $x > 0$.

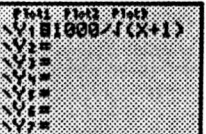

57. $P = f(i) = 1.5 \times 10^{-6} i^3 - 0.77$, $80 \le i \le 140$

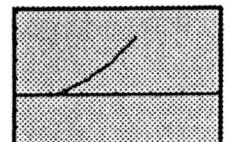

60. $E = f(r) = \dfrac{25}{r^2}$, $0 < r \le 10$. Graph $y = \dfrac{25}{x^2}$, $0 < x \le 10$.

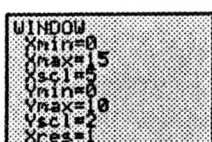 

58. $C(v) = 0.025v^2 - 1.4v + 35$, $10 \le v \le 60$.
Graph $y = 0.025x^2 - 1.4x + 35$, $10 \le x \le 60$.

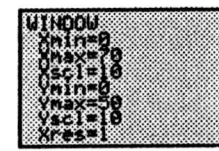 

61. $r(v) = 0.42v^2$. Graph $y = 0.42x^2$.

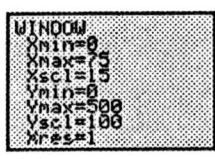 

62. $h(t) = 1500t - 4.9t^2$. Graph $y = 1500x - 4.9t^2$

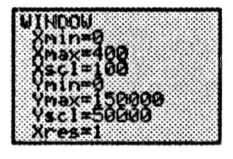 

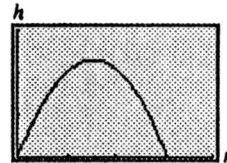

63. $F = f(x) = x^4 - 12x^3 + 46x^2 - 60x + 25, 1 \le x \le 5$
Graph $y = x^4 - 12x^3 + 46x^2 - 60x + 25, 1 \le x \le 5$.

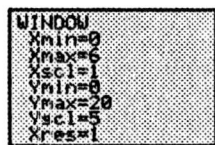 

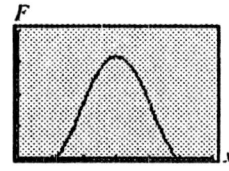

64. $V(x) = x(x)(x - 2.0)$
$= x^2(x - 2.0)$
$= x^3 - 2.0x^2$. Graph $y = x^3 - 2.0x^2$.

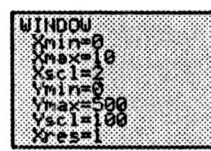 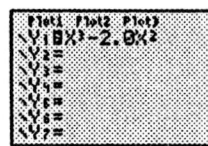

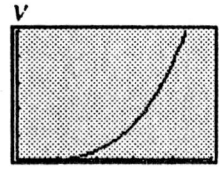

65. $d = f(t) = 250 - 60y, 0 \le t \le 2;$
$d = f(t) = 130 - 40(t - 2), 2 < t < 5.25$

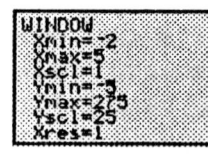 

The person is 70 mi. from home after 3.5 hours.

66. $0.30(120) + 0.10x = 50$
$36 + 0.10x = 50$
$0.10x - 14 = 0$

Graph $y = 0.10x - 14$ and use the zero feature to solve.

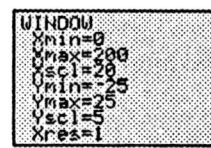 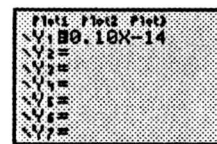

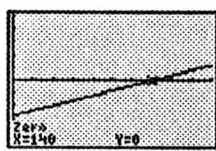

140 gal of the second cleaner must be used.

67. For $s = 500$,
$$s = 135 + 4.9T + 0.19T^2 \text{ is}$$
$$500 = 135 + 4.9T + 0.19T^2$$
$$0.19T^2 + 4.9T - 365 = 0$$

Graph $y = 0.19x^2 + 4.9x - 365$ and use the zero feature to solve.

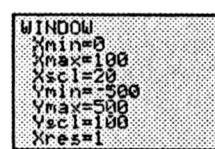 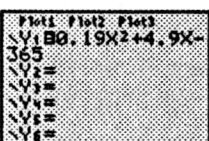

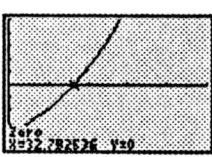

For $s = 500 \text{ kg/m}^3$, $T = 33°C$

68. $V = \pi r^2 h = 2000$

$$h = \frac{2000}{\pi r^2}$$

$$A = \pi r^2 + 2\pi r h$$

$$A = \pi r^2 + 2\pi r\left(\frac{2000}{\pi r^2}\right)$$

$$A(r) = \pi r^2 + \frac{4000}{r}$$

Graph $y = \pi x^2 + \dfrac{4000}{x}$ and use the value feature.

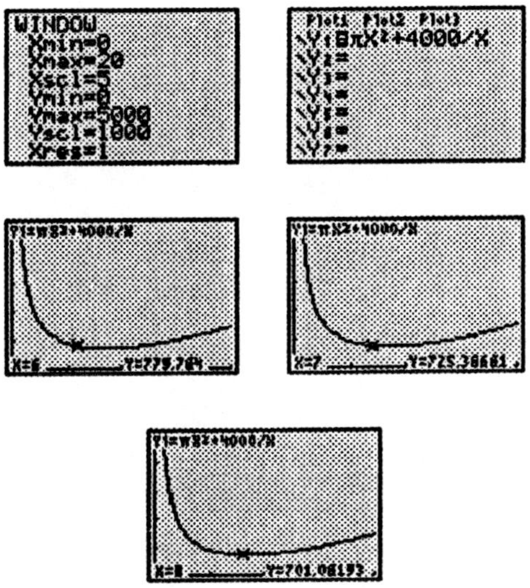

$A(6.00) = 780\ \text{cm}^2$
$A(7.00) = 725\ \text{cm}^2$
$A(8.00) = 701\ \text{cm}^2$

69. $v = 7.6x - 2.1x^2, 0 \le x \le 1.75$

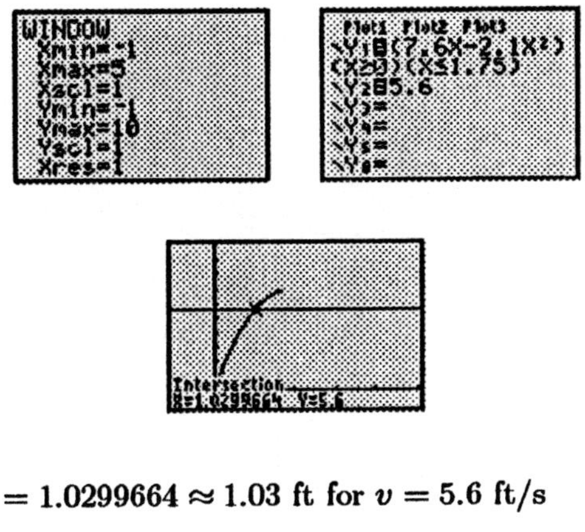

$x = 1.0299664 \approx 1.03$ ft for $v = 5.6$ ft/s

70.

$$V = \frac{4}{3}\pi r^3$$

$$2V = \frac{4}{3}\pi (r + 1.00)^3$$

$$\frac{1}{2} = \frac{r^3}{(r + 1.00)^3}$$

$$r^3 - \frac{1}{2}(r + 1.00)^3 = 0$$

Graph $y = x^3 - \dfrac{1}{2}(x + 1.00)^3$ and use zero feature to solve.

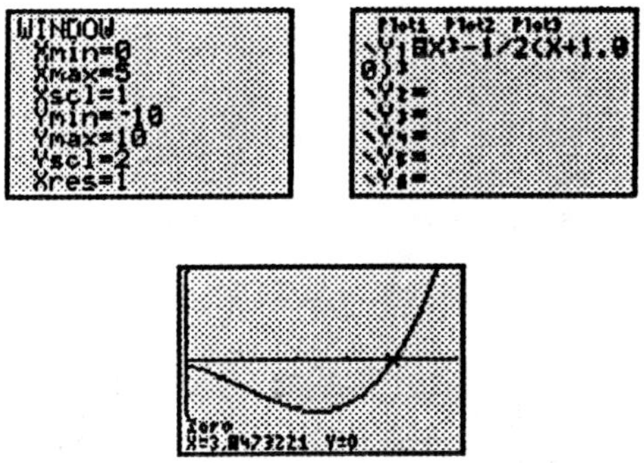

The radii of the ball bearings are 3.85 mm and 4.85 mm.

71. $T = \dfrac{4t^2}{t+2} - 20.$

Graph $y = \dfrac{4x^2}{x+2} - 20$ and use zero feature to solve.

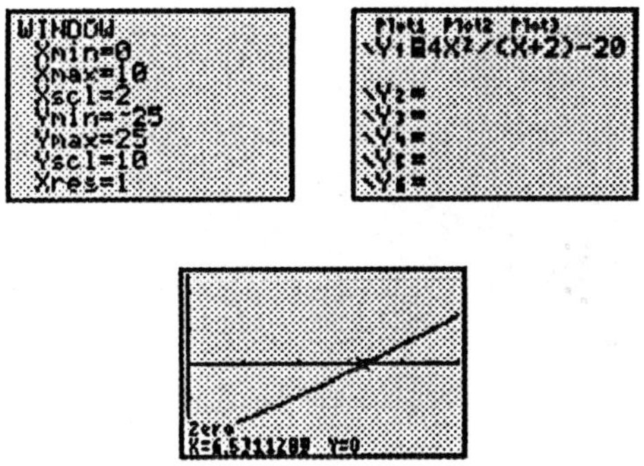

The temperature will reach 0°C after 6.5 h

72. $R_T = \dfrac{R_1 R_2}{R_1 + R_2} = \dfrac{R_1(R_1 + 2.0)}{R_1 + R_1 + 2.0}$

$R_T = 6.0 = \dfrac{R_1^2 + 2.0R_1}{2R_1 + 2.0}$

Graph $y = \dfrac{x^2 + 2.0x}{2x + 2.0} - 6.0$ and use zero feature to solve.

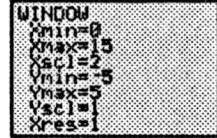 

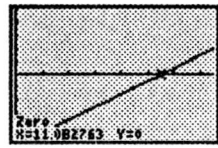

If $R_T = 6.0 \ \Omega, R_1 = 11 \ \Omega$.

73. $V = \pi r^2 h = 250.0 \Rightarrow h = \dfrac{250.0}{\pi r^2}$

$A = A_{\text{base}} + A_{\text{side}} = \pi r^2 + 2\pi r h$

$\qquad\qquad = \pi r^2 + 2\pi r\left(\dfrac{250.0}{\pi r^2}\right)$

$\qquad\qquad = \pi r^2 + \dfrac{500.0}{r}$

$A = 175 = \pi r^2 + \dfrac{500.0}{r}$

Graph $y_1 = \pi x^2 + \dfrac{500.0}{x}$ and $y = 175$ and use the intersection feature to solve.

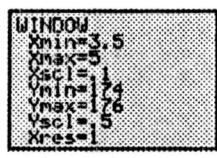 

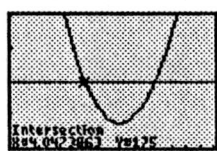 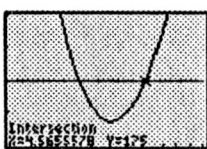

$r = 4.047$ cm or 4.566 cm when $A = 175.0 \text{ cm}^2$

PLANE ANALYTIC GEOMETRY

2.1 Basic Definitions

1. Given: $(x_1, y_1) = (3, 8)$; $(x_2, y_2) = (-1, -2)$

$$\begin{aligned} d &= \sqrt{(x_2 - x_1)^2 + (y_2 - y_1)^2} \\ &= \sqrt{(-1 - 3)^2 + (-2 - 8)^2} \\ &= \sqrt{(-4)^2 + (-10)^2} \\ &= \sqrt{16 + 100} = \sqrt{116} = \sqrt{4 \times 29} = 2\sqrt{29} \end{aligned}$$

2. Given: $(x_1, y_1) = (-1, 3)$; $(x_2, y_2) = (-8, -4)$

$$\begin{aligned} d &= \sqrt{(x_2 - x_1)^2 + (y_2 - y_1)^2} \\ &= \sqrt{(-8 - (-1))^2 + (-4 - 3)^2} \\ &= \sqrt{(-7)^2 + (-7)^2} \\ &= \sqrt{49 + 49} = \sqrt{98} \\ &= \sqrt{2 \times 49} = 7\sqrt{2} \end{aligned}$$

3. Given: $(x_1, y_1) = (4, -5)$; $(x_2, y_2) = (4, -8)$

$$\begin{aligned} d &= \sqrt{(x_2 - x_1)^2 + (y_2 - y_1)^2} \\ &= \sqrt{(4 - 4)^2 + (-8 - (-5))^2} \\ &= \sqrt{0^2 + (-3)^2} \quad \text{vertical line} \\ &= \sqrt{9} = 3 \end{aligned}$$

4. Given: $(x_1, y_1) = (-3, 7)$; $(x_2, y_2) = (2, 10)$

$$\begin{aligned} d &= \sqrt{(x_2 - x_1)^2 + (y_2 - y_1)^2} \\ &= \sqrt{(2 - (-3))^2 + (10 - 7)^2} \\ &= \sqrt{(5)^2 + (3)^2} = \sqrt{25 + 9} \\ &= \sqrt{34} \end{aligned}$$

5. Given: $(x_1, y_1) = (-12, 20)$; $(x_2, y_2) = (32, -13)$

$$\begin{aligned} d &= \sqrt{(x_2 - x_1)^2 + (y_2 - y_1)^2} \\ &= \sqrt{(32 + 12)^2 + (-13 - 20)^2} \\ &= \sqrt{(44)^2 + (-33)^2} \\ &= \sqrt{1936 + 1089} = \sqrt{3025} = 55 \end{aligned}$$

6. Given: $(x_1, y_1) = (23, -9)$; $(x_2, y_2) = (-25, 11)$

$$\begin{aligned} d &= \sqrt{(x_2 - x_1)^2 + (y_2 - y_1)^2} \\ &= \sqrt{(-25 - 23)^2 + (11 - (-9))^2} \\ &= \sqrt{(48)^2 + (20)^2} \\ &= \sqrt{2304 + 400} = \sqrt{2704} \\ &= 52 \end{aligned}$$

7. $d = \sqrt{(\sqrt{32} - (-\sqrt{50}))^2 + (-\sqrt{18} - \sqrt{8})^2}$
$= \sqrt{212} = 2\sqrt{53}$

8. $d = \sqrt{(e - (-2e))^2 + (-\pi - (-\pi))^2}$
$d = \sqrt{(3e)^2}$
$d = 3e$
$d \approx 8.15$

9. Given: $(x_1, y_1) = (1.22, -3.45)$;
$(x_2, y_2) = (-1.07, -5.16)$

$$\begin{aligned} d &= \sqrt{(x_2 - x_1)^2 + (y_2 - y_1)^2} \\ &= \sqrt{(-1.07 - 1.22)^2 + (-5.16 - (-3.45))^2} \\ &= \sqrt{(-2.29)^2 + (-5.16 + 3.45)^2} \\ &= \sqrt{(-2.29)^2 + (-1.71)^2} = \sqrt{8.1682} = 2.86 \end{aligned}$$

10. Given: $(x_1, y_1) = (-5.6, 2.3)$; $(x_2, y_2) = (8.2, -7.5)$

$$\begin{aligned} d &= \sqrt{(x_2 - x_1)^2 + (y_2 - y_1)^2} \\ &= \sqrt{(8.2 - (-5.6))^2 + (-7.5 - 2.3)^2} \\ &= \sqrt{(13.8)^2 + (-9.8)^2} \\ &= \sqrt{190.44 + 96.04} \\ &= \sqrt{286.48} = 16.9 \end{aligned}$$

11. Given: $(x_1, y_1) = (3, 8)$; $(x_2, y_2) = (-1, -2)$

$$m = \frac{y_2 - y_1)}{x_2 - x_1} = \frac{-2 - 8}{-1 - 3} = \frac{-10}{-4} = \frac{10}{4} = \frac{5}{2}$$

12. Given: $(x_1, y_1) = (-1, 3)$; $(x_2, y_2) = (-8, -4)$

$$m = \frac{y_2 - y_1}{x_2 - x_1} = \frac{-4 - 3}{-8 - (-1)} = \frac{-7}{-7} = 1$$

13. Given: $(x_1, y_1) = (4, -5)$; $(x_2, y_2) = (4. - 8)$

$$m = \frac{y_2 - y_1}{x_2 - x_1} = \frac{-8 - (-5)}{4 - 4}$$

Since $x_2 - x_1 = 4 - 4 = 0$, the slope is undefined.

14. Given: $(x_1, y_1) = (-3, 7)$; $(x_2, y_2) = (2, 10)$

$$m = \frac{y_2 - y_1}{x_2 - x_1} = \frac{10 - 7}{2 - (-3)} = \frac{3}{5}$$

15. Given: $(x_1, y_1) = (-12, 20)$; $(x_2, y_2) = (32, -13)$

$$m = \frac{y_2 - y_1}{x_2 - x_1} = \frac{-13 - 20}{32 - (-12)} = \frac{-33}{44} = -\frac{3}{4}$$

16. Given: $(x_1, y_1) = (23, -9)$; $(x_2, y_2) = (-25, 11)$

$$m = \frac{y_2 - y_1}{x_2 - x_1} = \frac{11 - (-9)}{-25 - 23} = \frac{20}{-48} = -\frac{5}{12}$$

23

17. Given: $(x_1, y_1) = (\sqrt{32}, -\sqrt{18})$;
$\quad\quad\quad\quad (x_2, y_2) = (-\sqrt{50}, \sqrt{8})$

$$m = \frac{y_2 - y_1}{x_2 - x_1} = \frac{\sqrt{8} - (-\sqrt{18})}{-\sqrt{50} - \sqrt{32}} = \frac{-5}{9}$$

18. $m = \dfrac{y_2 - y_1}{x_2 - x_1} = \dfrac{-\pi - (-\pi)}{-2e - e}$

$m = 0$

19. Given: $(x_1, y_1) = (1.22, -3.45)$;
$\quad\quad\quad\quad (x_2, y_2) = (-1.07, -5.16)$

$$m = \frac{y_2 - y_1}{x_2 - x_1} = \frac{-5.16 - (-3.45)}{-1.07 - 1.22}$$

$$= \frac{-1.71}{-2.29} = 0.747$$

20. Given: $(x_1, y_1) = (-5.6, 2.3)$; $(x_2 - y_2) = (8.2, -7.5)$

$$m = \frac{y_2 - y_1}{x_2 - x_1} = \frac{-7.5 - 2.3}{8.2 - (-5.6)} = \frac{-9.8}{13.8} = -0.71$$

21. Given: $\alpha = 30°$; $m = \tan \alpha$, $0° < \alpha < 180°$

$$\tan 30° = \frac{\sqrt{3}}{3} \text{ or } \frac{1}{3}\sqrt{3}$$

22. Given: $\alpha = 62.5°$; $\tan \alpha$; $0° < \alpha < 180°$
$m = \tan 62.5° = 1.92$

23. Given: $\alpha = 132.7°$, $m = \tan \alpha$; $0° < \alpha < 180°$
$m = \tan 132.7° = -1.084$

24. Given: $\alpha = 135°$; $m = \tan \alpha$; $0° < \alpha < 180°$
$m = \tan 135° = -1$

25. Given: $m = 0.364$; $m = \tan \alpha$; $0.364 = \tan \alpha$;
$\alpha = 20.0°$

26. Given: $m = 0.824$; $m = \tan \alpha$
$0.824 = \tan \alpha$; $\alpha = 39.5°$

27. Given: $m = -6.691$; $0° < \alpha < 180°$; (negative quadrant two)
$m = \tan \alpha$
$-6.691 = \tan \alpha$; $\alpha = 180° - 81.500°$; $\alpha = 98.50°$

28. Given: $m = -1.428$; $m = \tan \alpha$ (α in quadrant two)
$-1.428 = \tan \alpha$; $\alpha = (180° - 55°) = 125°$

29. Given: $(x, y) = (6, -1)$; $(x_1, y_1) = (4, 3)$
$\quad\quad\quad\quad (x_2, y_2) = (-5, 2)$; $(x_3, y_3) = (-7, 6)$

$$m_1 = \frac{y - y_1}{x - x_1} = \frac{-1 - 3}{6 - 4} = \frac{-4}{2} = -2$$

$$m_2 = \frac{y_2 - y_3}{x_2 - x_3} = \frac{2 - 6}{-5 - (-7)} = \frac{-4}{-5 + 7}$$

$$= \frac{-4}{2} = -2$$

$m_3 = m_2$ for all parallel lines.

30. Given: $(x_1, y_1) = (-3, 9)$; $(x_2, y_2) = (4, 4)$
$\quad\quad\quad\quad (x_3, y_3) = (9, -1)$; $(x_4, y_4) = (4, -8)$

$$m_1 = \frac{y_2 - y_1}{x_2 - x_1} = \frac{4 - 9}{4 - (-3)}$$

$$= \frac{-5}{7} = -\frac{5}{7}$$

$$m_2 = \frac{y_4 - y_3}{x_4 - x_3} = \frac{-8 - (-1)}{4 - 9}$$

$$= \frac{-7}{-5} = \frac{7}{5}$$

m_1 is negative reciprocal of m_2; therefore, lines are perpendicular.

31. Given: $(x_1, y_1) = (-1, -4)$; $(x_2, y_2) = (2, 3)$ line 1
$\quad\quad\quad\quad (x_3, y_3) = (-5, 2)$; $(x_4, y_4) = (-19, 8)$ line 2

$$m_1 = \frac{y_2 - y_1}{x_2 - x_1} = \frac{3 - (-4)}{2 - (-1)} = \frac{7}{3}$$

$$m_2 = \frac{y_4 - y_3}{x_4 - x_3} = \frac{8 - 2}{-19 - (-5)}$$

$$= \frac{6}{-14} = -\frac{3}{7}$$

m_1 is negative reciprocal of m_2; therefore, lines are perpendicular.

32. Given: $(x_1, y_1) = (-1, -2)$; $(x_2, y_2) = (3, 6)$
$\quad\quad\quad\quad (x_3, y_3) = (2, -6)$; $(x_4, y_4) = (5, 0)$

$$m_1 = \frac{y_2 - y_1}{x_2 - x_1} = \frac{6 - (-2)}{3 - (-1)} = \frac{8}{4} = 2$$

$$m_2 = \frac{y_4 - y_3}{x_4 - x_3} = \frac{0 - (-6)}{5 - 2} = \frac{6}{3} = 2$$

$m_1 = m_2$; therefore, lines are parallel.

33. Given: distance between $(-1, 3)$ and $(11, k)$ is 13.

$$d = \sqrt{(x_1 - x_2)^2 + (y_1 - y_2)^2}$$
$$13 = \sqrt{(-1 - 11)^2 + (3 - k)^2}$$
$$= \sqrt{(-12)^2 + (3 - k)^2}$$
$$= \sqrt{144 + (3 - k)^2}$$
$$169 = 144 + (3 - k)^2;$$
$$(3 - k)^2 = 25; \quad 3 - k = \pm 5$$
$$-k = -3 \pm 5; \quad k = -2, 8$$

34. Given: $(x_1, y_1) = (k, 0); \quad (x_2, y_2) = (0, 2k)$

$$d = \sqrt{(x_2 - x_1)^2 + (y_2 - y_1)^2}; \quad d = 10$$
$$10 = \sqrt{(0 - k)^2 + (2k - 0)^2} = \sqrt{(-k)^2 + (2k)^2}$$
$$= \sqrt{k^2 + 4k^2} = \sqrt{5k^2}$$
$$100 = 5k^2; \quad k^2 = 20; \quad k = \pm\sqrt{20}$$
$$k = \pm\sqrt{4 \times 5} = \pm 2\sqrt{5}$$

35. Given $(6, -1), (3, k); (-3, -7)$ on same line, therefore slope constant between points.

$(x_1, y_1) = (6, -1); \quad (x_2, y_2) = (-3, -7)$

Slope between these two points is:

$$m = \frac{y_2 - y_1}{x_2 - x_1} = \frac{-7 - (-1)}{-3 - 6} = \frac{-6}{-9} = \frac{2}{3}$$

Therefore, slope between $(6, -1)$ and $(3, k)$ must be $\frac{2}{3}$.

$(x_1, y_1) = (6, -1); \quad (x_3, y_3) = (3, k)$

$$m = \frac{y_3 - y_1}{x_3 - x_1}; \quad \frac{2}{3} = \frac{k - (-1)}{3 - 6}; \quad \frac{2}{3} = \frac{k + 1}{-3};$$
$$-6 = 3(k + 1) = 3k + 3; \quad 3k = -9; \quad k = -3$$

36. Given: $A(6, -1) = (x_1, y_1); B(3, k) = (x_2, y_2);$
$C(-3 - 7) = (x_3, y_3)$
By Pythagorus:

$$(\text{hypotenuse})^2 = (\text{opposite})^2 + (\text{adjacent})^2$$
Hypotenuse is opposite $(3, k)$.

$$(x_3 - x_1)^2 + (y_3 - y_1)^2$$
$$= (x_2 - x_1)^2 + (y_2 - y_1)^2 + (x_3 - x_2)^2 + (y_3 - y_2)^2$$
$$(-3 - 6)^2 + (-7 - (-1))^2$$
$$= (3 - 6)^2 + (k - (-1))^2 + (-3 - 3)^2 + (-7 - k)^2$$
$$(-9)^2 + (-6)^2 = (-3)^2 + (k + 1)^2 + (-6)^2 + (-7 - k)^2$$
$$81 + 36 = 9 + (k + 1)^2 + 36 + (-1(7 + k))^2$$
$$81 + 36 - 9 - 36 = k^2 + 2k + 1 + 49 + 14k + k^2$$
$$72 = 2k^2 + 16k + 50; \quad 2k^2 + 16k - 22 = 0$$
$$k^2 + 8k - 11 = 0$$

Completing the square.

$$k^2 + 8k + 16 = 11 + 16; \quad (k + 4)^2 = 27; \quad k + 4 = \pm\sqrt{27}$$
$$k = -4 \pm \sqrt{27}$$
$$k = -4 \pm 3\sqrt{3}$$

37. $d_1 = \sqrt{(9 - 7)^2 + [4 - (-2)]^2}$
$$= \sqrt{2^2 + 6^2} = \sqrt{40} = 2\sqrt{10}$$
$d_2 = \sqrt{(9 - 3)^2 + (4 - 2)^2} = \sqrt{6^2 + 2^2}$
$$= \sqrt{40} = 2\sqrt{10}$$
$d_1 = d_2$ so the triangle is isosceles.

38. Given: $A(-1, 3) = (x_1, y_1); B(3, 5) = (x_2, y_2);$
$C(5, 1) = (x_3, y_3)$

Therefore, by inspection $(3, 5)$ appears to be a right angle. Therefore, if slopes for AB and BC are negative reciprocals, we have a right triangle.

$$m_1 = \frac{y_2 - y_1}{x_2 - x_1} = \frac{5 - 3}{3 - (-1)} = \frac{2}{4} = \frac{1}{2}$$

$$m_2 = \frac{y_3 - y_2}{x_3 - x_2} = \frac{1 - 5}{5 - 3} = \frac{-4}{2} = -2$$

m_1 is negative reciprocal of m_2, therefore a right triangle.

39. Given: $A(-5, -4), B(7, 1), C(10, 5), D(-2, 0)$.
By inspection, if slopes of AD and BC are equal and slope of AB and DC are equal, then a parallelogram is formed.
Let $(x_1, y_1) = (-5 - 4); \quad (x_2, y_2) = (7, 1)$
$(x_3, y_3) = (10, 5); \quad (x_4, y_4) = (-2, 0)$

Slope AD $= \dfrac{0 - (-4)}{-2 - (-5)} = \dfrac{4}{3}$

Slope BC $= \dfrac{5 - 1}{10 - 7} = \dfrac{4}{3}; \dfrac{4}{3} = \dfrac{4}{3}$

Slope AB $= \dfrac{1 - (-4)}{7 - (-5)} = \dfrac{5}{12}$

Slope DC $= \dfrac{5 - 0}{10 - (-2)} = \dfrac{5}{12}; \dfrac{5}{12} = \dfrac{5}{12}$

Therefore, a parallelogram is formed.

40. Given: $A(-5, 6) = (x_1, y_1)$
$B(0, 8) = (x_2, y_2)$
$C(-3, 1) = (x_3, y_3)$
$D(2, 3) = (x_4, y_4)$

We have a square if four sides are of equal length and four angles are right angles.

$d_1 = \sqrt{(x_2 - x_1)^2 + (y_2 - y_1)^2}$
$$= \sqrt{(0 - (-5))^2 + (8 - 6)^2}$$
$$= \sqrt{(5)^2 + (2)^2} = \sqrt{25 + 4} = \sqrt{29}$$
$d_2 = \sqrt{(x_3 - x_1)^2 + (y_3 - y_1)^2}$
$$= \sqrt{(-3 - (-5))^2 + (1 - 6)^2}$$
$$= \sqrt{(2)^2 + (-5)^2} = \sqrt{4 + 25}$$
$$= \sqrt{29}$$

$$d_3 = \sqrt{(x_4 - x_2)^2 + (y_4 - y_2)^2}$$
$$= \sqrt{(2-0)^2 + (3-8)^2}$$
$$= \sqrt{(2)^2 + (-5)^2} = \sqrt{4 + 25}$$
$$= \sqrt{29}$$

$$d_4 = \sqrt{(x_4 - x_3)^2 + (y_4 - y_3)^2}$$
$$= \sqrt{2 - (-3))^2 + (3 - 1)^2}$$
$$= \sqrt{(5)^2 + (2)^2} = \sqrt{25 + 4}$$
$$= \sqrt{29}$$

$$m_1 = \frac{y_2 - y_1}{x_2 - x_1} = \frac{2}{5}; \quad m_2 = \frac{y_3 - y_1}{x_3 - x_1} = \frac{-5}{2} = -\frac{5}{2}$$

$$m_3 = \frac{y_4 - y_2}{x_4 - x_2} = \frac{-5}{2} = -\frac{5}{2}; \quad m_4 = \frac{y_4 - y_3}{x_4 - x_3} = \frac{2}{5}$$

Above conditions satisfied; therefore, we have a square.

41. $d_1 = \sqrt{(3 - 5)^2 + (-1 - 3)^2} = \sqrt{(-2)^2 + (-4)^2}$
$$= \sqrt{4 + 16} = \sqrt{20}$$

$$m_1 = \frac{y - y_1}{x - x_1} = \frac{5 - 3}{3 - (-1)} = \frac{5 - 3}{3 + 1} = \frac{2}{4} = \frac{1}{2}$$

$$d_2 = \sqrt{(5 - 1)^2 + (3 - 5)^2} = \sqrt{(4)^2 + (-2)^2}$$
$$= \sqrt{16 + 4} = \sqrt{20}$$

$$m_2 = \frac{y - y_1}{x - x_1} = \frac{5 - 1}{3 - 5} = \frac{4}{-2} = -2$$

$$m_1 = \frac{-1}{m_2}, \quad m_1 \perp m_2$$

$$A = \frac{1}{2}d_1 d_2 = \frac{1}{2}\sqrt{20}\sqrt{20}$$
$$= \frac{1}{2}(20) = 10$$

42. Area of square = (side)^2
Area = $(\sqrt{29})^2 = 29$ square units

43. Given A(2, 3); B(4, 9); C(−2, 7)
The perimeter $p = d_1 + d_2 + d_3$
Let $(x_1, y_1) = (2, 3)$; $(x_2, y_2) = (4, 9)$;
$(x_3, y_3) = (-2, 7)$

$$d_1 = \sqrt{(4 - 2)^2 + (9 - 3)^2}$$
$$= \sqrt{2^2 + 6^2}$$
$$= \sqrt{40} = 2\sqrt{10}$$

$$d_2 = \sqrt{(-2 - 2)^2 + (7 - 3)^2}$$
$$= \sqrt{(-4)^2 + 4^2}$$
$$= \sqrt{32} = 4\sqrt{2}$$

$$d_3 = \sqrt{(-2 - 4)^2 + (7 - 9)^2}$$
$$= \sqrt{(-6)^2 + (-2)^2} = \sqrt{40}$$
$$= 2\sqrt{10}$$

Perimeter = $2\sqrt{10} + 4\sqrt{2} + 2\sqrt{10}$
$p = 4\sqrt{10} + 4\sqrt{2} = 4(\sqrt{10} + \sqrt{2}) = 18.3$

44. Given: A(−5, −4) = (x_1, y_1)
$$B(7, 1) = (x_2, y_2)$$
$$C(10, 5) = (x_3, y_3)$$
$$D(-2, 0) = (x_4, y_4)$$

Perimeter = $2d_1 + 2d_2$

$$d_1 = \sqrt{(x_2 - x_1)^2 + (y_2 - y_1)^2}$$
$$= \sqrt{(7 - (-5))^2 + (1 - (-4))^2}$$
$$= \sqrt{(12)^2 + (5)^2}$$
$$= \sqrt{169}$$
$$= 13$$

$$d_2 = \sqrt{(x_4 - x_1)^2 + (y_4 - y_1)^2}$$
$$= \sqrt{(-2) - (-5))^2 + (0 - (-4))^2}$$
$$= \sqrt{(3)^2 + (4)^2}$$
$$= \sqrt{25}$$
$$= 5$$

Perimeter = $2(13 + 5) = 2(18) = 36$ units

45. $\left(\dfrac{-4 + 6}{2}, \dfrac{9 + 1}{2}\right) = \left(\dfrac{2}{2}, \dfrac{10}{2}\right)$
$$= (1, 5)$$

46. Given: $(x_1, y_1) = (-1, 6)$; $(x_2, y_2) = (-13, -8)$

$$\left(\frac{x_1 + x_2}{2}, \frac{y_1 + y_2}{2}\right) = \left(\frac{-1 + (-13)}{2}, \frac{6 + (-8)}{2}\right)$$
$$= (-7, -1)$$

47. Given: A(−12.4, 25.7); B(6.8, −17.3)
Midpoint A and B

$$x = \frac{x_1 + x_2}{2} = \frac{-12.4 + 6.8}{2} = -2.8$$

$$y = \frac{y_1 + y_2}{2} = \frac{25.7 + (-17.3)}{2} = 4.2$$

$(-2.8, 4.2)$

48. Given: $(x_1 y_1) = (2.6, 5.3)$; $(x_2, y_2) = (-4.2, -2.7)$

$$\left(\frac{x_1 + x_2}{2}, \frac{y_1 + y_2}{2}\right) = \left(\frac{2.6 + (-4.2)}{2}, \frac{5.3 + (-2.7)}{2}\right)$$
$$= (-0.8, 1.3)$$

2.2   The Straight Line

1. Given: $m = 4; (x_1, y_1) = (-3, 8)$

$y - y_1 = m(x - x_1)$
$y - 8 = 4[x - (-3)] = 4(x + 3) = 4x + 12$
$y = 4x + 20$ or $4x - y + 20 = 0$

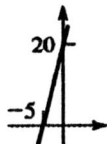

2. Given: $(x_1, y_1) = (-2, -1); m = -2$

$y - y_1 = m(x - x_1)$
$y - (-1) = -2(x - (-2))$
$y + 1 = -2(x + 2)$
$y + 1 = -2x - 4$
$y = -2x - 5$ or $2x + y + 5 = 0$

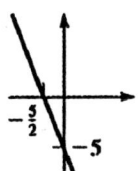

3. Given: $(x_1, y_1) = (2, -5); (x_2, y_2) = (4, 2)$

$$\frac{y - y_1}{x - x_1} = \frac{y_2 - y_1}{x_2 - x_1}$$

$$\frac{y - (-5)}{x - 2} = \frac{2 - (-5)}{4 - 2}$$

$$\frac{y + 5}{x - 2} = \frac{7}{2};$$

$2(y + 5) = 7(x - 2); 2y + 10 = 7x - 14; y = \frac{7}{2}x - 12$

or $7x - 2y - 24 = 0$

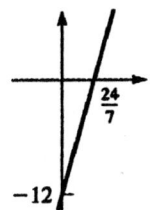

4. Given: $(x_1, y_1) = (-3, 5); (x_2, y_2) = (-2, 3)$

$$\frac{y - y_1}{x - x_1} = \frac{y_2 - y_1}{x_2 - x_1}$$

$$\frac{y - 5}{x - (-3)} = \frac{3 - 5}{-2 - (-3)} = \frac{-2}{1}$$

$y - 5 = -2(x + 3); y - 5 = -2x - 6; y = -2x - 1$
or $2x + y + 1 = 0$

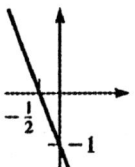

5. Given: $(x_1, y_1) = (1, 3); \alpha = 45°$

$m = \tan \alpha = \tan 45° = 1$
$y - y_1 = m(x - x_1)$
$y - 3 = 1(x - 1) = x - 1; y = x + 2$
or $x - y + 2 = 0$

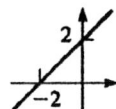

6. $\alpha = 120°; m = \tan \alpha$
$m = \tan 120°$
$m = -\sqrt{3} \qquad (x_1, y_1) = (0, -2)$
$y - y_1 = m(x - x_1)$
$y - (-2) = -\sqrt{3}(x - 0)$
$y + 2 = -\sqrt{3}x$
$y = -\sqrt{3}x - 2$ or $\sqrt{3}x + y + 2 = 0$

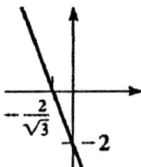

7. $m = 0$
$y - y_1 = m(x - x_1)$
$y - (-2.7) = 0(x - 5.3)$
$y + 2.7 = 0$
$y = -2.7$

8. Slope is undefined
Therefore, $x = -4$

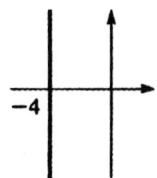

9. Parallel to y-axis and
3 units left of y-axis.
$x = -3$

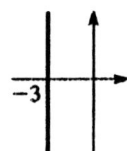

10. $m = 0$; $(x_1, y_1) = (0, -4.1)$
Therefore, $y = -4.1$

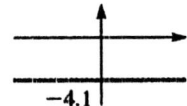

11. $(x_1, y_1) = (0, -6)$; $(x_2, y_2) = (4, 0)$

$$m = \frac{y_2 - y_1}{x_2 - x_1} = \frac{0 - (-6)}{4 - 0}$$
$$= \frac{6}{4} = \frac{3}{2}$$

$$y - y_1 = m(x - x_1)$$

$$y - (-6) = \frac{3}{2}(x - 0)$$

$$y + 6 = \frac{3}{2}x$$

$$y = \frac{3}{2}x - 6 \text{ or } 3x - 2y - 12 = 0$$

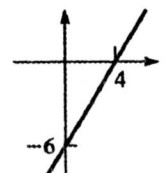

12. $(x_1, y_1) = (-3, 0)$; $m = 2$
$y - y_1 = m(x - x_1)$
$y - 0 = 2(x - (-3))$
$y = 2(x + 3)$
$y = 2x + 6 \text{ or } 2x - y + 6 = 0$

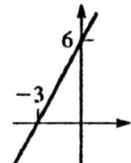

13. Perpendicular to line with slope 3;
$(x_1, y_1) = (1, -2)$

$y - y_1 = (x - x_1); y - (-2) = -\frac{1}{3}(x - 1)$

$y + 2 = -\frac{1}{3}x + \frac{1}{3}; y = -\frac{1}{3}x - \frac{5}{3}$

or $-\frac{1}{3}x - y - \frac{5}{3} = 0; x + 3y + 5 = 0$

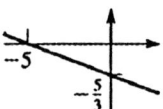

14. Perpendicular to line with $m = -4$
Therefore, slope of line $= \dfrac{1}{4}$
$(x_1, y_1) = (0, 3)$
$y - y_1 = m(x - x_1)$

$y - 3 = \dfrac{1}{4}(x - 0)$

$y - 3 = \dfrac{1}{4}x$

$y = \dfrac{1}{4}x + 3 \text{ or } x - 4y + 12 = 0$

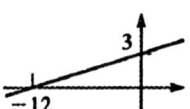

15. Parallel to line with $m = -\dfrac{1}{4}$
Therefore, slope of line $= -\dfrac{1}{4}$
$(x_1, y_1) = (1, 2)$
$y - y_1 = m(x - x_1)$

$y - 2 = -\dfrac{1}{4}(x - 1)$

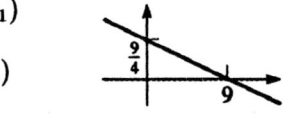

$x + 4y - 9 = 0$

16. Parallel to a line through $(7, -1)$ and $(4, 3)$; y-intercept is -2

$$m = \frac{y_2 - y_1}{x_2 - x_1} = \frac{3 - (-1)}{4 - 7} = \frac{4}{-3} = -\frac{4}{3}$$

y-intercept is $b = -2$

$$y = mx + b; \quad y = -\frac{4}{3}x - 2; \quad 4x + 3y + 6 = 0$$

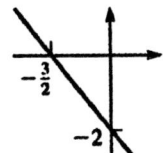

17. Given: $6.0x - 2.4y - 3.9 = 0$
or $y = 2.5x - 1.625$
Perpendicular to this line
Therefore, slope of desired line
$= -\frac{1}{2.5} = -0.4 = m$
$(x_1, y_1) = (7.5, -4.7)$
$y - y_1 = m(x - x_1)$
$y - (-4.7) = -0.4(x - 7.5)$
$y + 4.7 = -0.4x + 3$
$y = -0.4x - 1.7$ or $0.4x + y + 1.7 = 0$

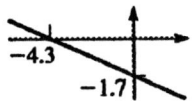

18. Given: $2y - 6x - 5 = 0$ or $y = 3x + \dfrac{5}{2}$

Parallel to this line
Therefore, slope of desired line $= 3 = m$

$(x_1, y_1) = (-4, -5)$
$y - y_1 = m(x - x_1)$
$y - (-5) = 3(x - (-4))$
$y + 5 = 3(x + 4)$
$y + 5 = 3x + 12$
$y = 3x + 7$ or $3x - y + 7 = 0$

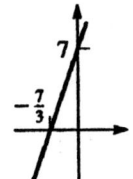

19. $5x - y = 6$ and $x + y = 12$ intersect at $(3, 9)$
$y - y_1 = m(x - x_1)$
$y - 9 = -3(x - 3)$
$y - 9 = -3x + 9$
$3x + y - 18 = 0$

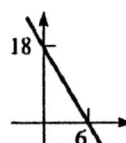

20. $2x + y - 3 = 0$ and $x - y - 3 = 0$ intersect at $(2, -1)$

$$m = \frac{y_2 - y_1}{x_2 - x_1} = \frac{-3 - (-1)}{4 - 2} = -1$$

$$y - y_1 = m(x - x_1)$$
$$y - (-1) = -1(x - 2)$$
$$y + 1 = -x + 2$$
$$x + y - 1 = 0$$

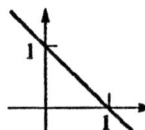

21. Given: $4x - y = 8, y = 4x - 8, m = 4, b = -8$
When $x = 0, y = -8$
$\qquad y = 0, x = 2$

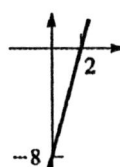

22. Given: $2x - 3y - 6 = 0, y = \dfrac{2}{3}x - 2, m = \dfrac{2}{3}, b = -2$

When $x = 0, y = -2$
$\qquad y = 0, x = 3$

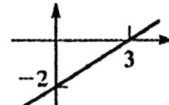

23. Given: $3x + 5y - 10 = 0, y = \dfrac{-3}{5}x + 2,$

$m = \dfrac{-3}{5}, b = 2$

When $x = 0, y = 2$

$\quad\quad y = 0, x = \dfrac{10}{3}$

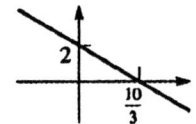

24. Given: $4y = 6x - 9, y = \dfrac{3}{2}x - \dfrac{9}{4}, m = \dfrac{3}{2}, b = \dfrac{-9}{4}$

When $x = 0, y = -\dfrac{9}{4}$

$\quad\quad y = 0, x = \dfrac{9}{6} = \dfrac{3}{2}$

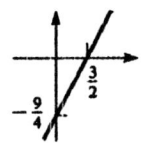

25. Given: $3x - 2y - 1 = 0$

$3x - 2y - 1 = 0; -2y = -3x + 1$

$y = \dfrac{-3}{-2}x + \dfrac{1}{-2}; y = \dfrac{3}{2}x - \dfrac{1}{2}$

Slope $= \dfrac{3}{2} = m;$

y-intercept $= -\dfrac{1}{2} = b$

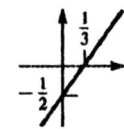

26. Given: $4x + 2y - 5 = 0; 2y = -4x + 5; y = -2x + \dfrac{5}{2}$

Slope $= -2;$ y-intercept $= \dfrac{5}{2}$

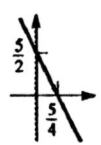

27. Given: $11.2x - 3.2y + 1.6 = 0; 3.2y = 11.2x + 1.6$

$y = 3.5x + 0.5$

$m = 3.5;$ y-intercept $= 0.5$

28. Given: $11.5x + 4.60y - 5.98 = 0$

$4.60y = -11.5x + 5.98; y = -2.50x + 1.30$

$m = -2.50;$ y-intercept $= 1.30$

29. Given: $4x - ky = 6 \parallel 6x + 3y + 2 = 0$

$6x + 3y + 2 = 0; 3y = -6x - 2$

$y = \dfrac{-6}{3}x - \dfrac{2}{3}; y = -2x - \dfrac{2}{3};$ slope is -2

$4x - ky = 6; -ky = -4x + 6$

$y = \dfrac{-4}{-k}x + \dfrac{6}{-k}; y = \dfrac{4}{k}x - \dfrac{6}{k};$ slope is $\dfrac{4}{k}$

Since the lines are parallel, the slopes are equal.

$\dfrac{4}{k} = -2; 4 = -2k; k = -2$

30. Given: $4x - ky = 6 \perp$ to $6x + 3y + 2 = 0$

$6x + 3y + 2 = 0; 3y = -6x - 2; y = -2x - \dfrac{2}{3}$

$m_1 = -2$

Therefore, slope of other line $= \dfrac{1}{2} = m$

$4x - ky = 6; ky = 4x - 6; y = \dfrac{4}{k}x - \dfrac{6}{k}, \dfrac{4}{k} = \dfrac{1}{2}$

Therefore, $k = 8$

31. Given: $3x - y - 9 = 0 \perp$ to $kx + 3y = 5$

Slope of line $3x - y - 9$ is 3

Therefore, slope of line $kx + 3y = 5$ is $-\dfrac{1}{3}$

Slope of line $kx + 3y = 5$ is $-\dfrac{k}{3}$

Therefore, $k = 1$

32. Slope of line $3x - y - 9 = 0$ is 3

Therefore, slope of line $kx + 3y = 5$ is also 3

Therefore, $-\dfrac{k}{3} = 3, k = -9$

33. $3x - 2y + 5 = 0; -2y = -3x - 5;$

$y = \frac{-3}{-2}x + \frac{-5}{-2}; y = \frac{3}{2}x + \frac{5}{2};$

slope $= \frac{3}{2} = m_1$

$4y = 6x - 1; y = \frac{6}{4}x - \frac{1}{4};$

$y = \frac{3}{2}x - \frac{1}{4};$ slope $= \frac{3}{2} = m_2$

$m_1 = m_2$ for all parallel lines.

34. $8x - 4y + 1 = 0 \Rightarrow y = 2x + \dfrac{1}{4}, m_1 = 2$

$4x + 2y - 3 = 0 \Rightarrow y = -2x + \dfrac{3}{2}, m_2 = -2$

Neither perpendicular or parallel

35. Line one: $6x - 3y - 2 = 0;$ $m_1 = 2;$ negative reciprocal

Line two: $x + 2y - 4 = 0;$ $m_2 = -\dfrac{1}{2}$

Therefore, lines are perpendicular.

36. Line one: $3y - 2x = 4;$ $m_1 = \dfrac{2}{3};$ $m_1 = m_2$

Line two: $6x - 9y = 5;$ $m_2 = \dfrac{6}{9} = \dfrac{2}{3}$

Therefore, lines are parallel.

37. $5x + 2y = 3 \Rightarrow y = \frac{-5}{2} \cdot x + \frac{3}{2}$

$10y = 7 - 4x \Rightarrow y = \frac{-4}{10}x + \frac{7}{10}$

$m_1 m_2 = \frac{-5}{2}\left(\frac{-4}{10}\right) = 1 \neq -1$

$m_1 \neq m_2$

Lines are neither perpendicular nor parallel.

38. $48y - 36x = 71 \Rightarrow y = \frac{3}{4}x + \frac{71}{48}, m_1 = \frac{3}{4}$

$52x = 17 - 39y \Rightarrow y = -\frac{4}{3}x + \frac{17}{39}, m_2 = -\frac{4}{3}$

$m_1 m_2 = \frac{3}{4}\left(\frac{-4}{3}\right) = -1,$ perpendicular

39. Line one: $4.5x - 1.8y = 1.7;$ $m_1 = 2.5;$
negative reciprocal
Line two: $2.4x + 6.0y = 0.3;$ $m_2 = -0.4$
Therefore, lines are perpendicular.

40. $3.5y = 4.3 - 1.5x \Rightarrow y = -\frac{3}{7}x + \frac{43}{35}, m_1 = \frac{-3}{7}$

$3.6x + 8.4y = 1.7 \Rightarrow y = -\frac{3}{7}x + \frac{17}{84}, m_2 = -\frac{3}{7}$

Therefore lines are parallel

41. $v = v_0 + at$
$35.4 = 12.2 + a(4.50)$
$4.50a = 35.4 - 12.2$
$a = 5.16 \text{ ft/s}^2$
$v = 12.2 + 5.16t$

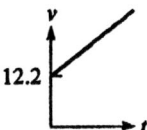

42. $9.17 \text{ mA} = 9.17 \times 10^{-3} \text{ A}$
$V = E - iR$
$4.35 = 6.00 - 9.17 \times 10^{-3} \text{ R}$
$R = 180 \ \Omega$
Therefore, $V = 6.00 - 180i$

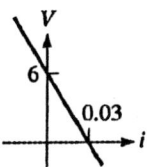

43. $v = mT + b, 343 = 0.607(20°) + b \Rightarrow b = 331$
$v = 0.607T + 331$

44. $0.20x + 0.30y = 20;$ therefore, $2x + 3y = 200$

45. $50x + 60y = 12,200; 5x + 6y = 1220$

46. $x = 0; T = 3 °C$
$T = kx + T_1$
$3 = 0 + T_1$
$T = kx + 3$
$x = 15 \text{ cm}; T = 23 °C$
$23° = k(15) + 3$
$k = \dfrac{20}{15} = \dfrac{4}{3} °C/cm,$ slope = change in
temperature in °C per cm.
Therefore, $T = \dfrac{4}{3}x + 3$

47. $0.72x + 0.90y = 135,000$
$y = \dfrac{135,000 - 0.72x}{0.90}; y = 150,000 - 0.80x$

48. $l = w + 10$
$p = 2w + 2l$
$p = 2w + 2(w + 10)$
$p = 4w + 20,$ slope = 4. Therefore each one cm change in width produces a 4 cm change in perimeter.

49. $m = \tan(180° - 0.0032°)$
$b = 24\mu m = 24 \times 10^{-6}$ m $= 2.4 \times 10^{-5}$ m
$m = -5.6 \times 10^{-5}$
$y = mx + b = -5.6 \times 10^{-5}x + 2.4 \times 10^{-5}$
$y = (-5.6x + 2.4)10^{-5}$

50. $m = \tan\alpha$
$m = \tan(90° - 16.5°)$
$m = \tan 73.5°$
$m = 3.38$

Therefore, $y = 3.38x$

51. Start is 6:30 $= 0$ min
Therefore, $(t_1 n_1) = (30, 45);\ (t_2 n_2) = (90, 115)$
$$m = \frac{n_2 - n_1}{t_2 - t_1} = \frac{115 - 45}{90 - 30} = \frac{7}{6}$$
$$n - n_1 = m(t - t_1)$$
$$n - 45 = \frac{7}{6}(t - 30)$$
$$n - 45 = \frac{7}{6}t - 35;\ n = \frac{7}{6}t + 10$$

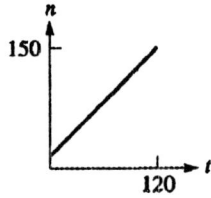

Therefore, n at 6:30 $(t = 0) = 10$; at 8:30
$(t = 120) = 150$

52. $t = 0$ month: $w = 30$ mg
Rate of growth is -2 mg/month
Therefore, $w = -2t + 30$

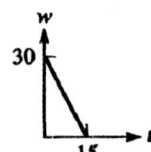

53. $n = 1200\sqrt{t} + 0$
$m = 1200$
$b = 0$

t	$\sqrt{t}$	h
0	0	0
1	1	1200
4	2	2400

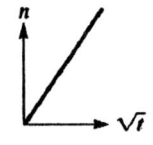

54. $F = \dfrac{40}{d} = 40\left(\dfrac{1}{d}\right)$

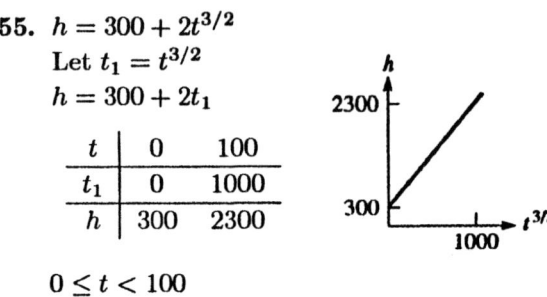

d	1	2	$\frac{1}{2}$
$\frac{1}{d}$	1	$\frac{1}{2}$	2
F	40	20	80

55. $h = 300 + 2t^{3/2}$
Let $t_1 = t^{3/2}$
$h = 300 + 2t_1$

t	0	100
t_1	0	1000
h	300	2300

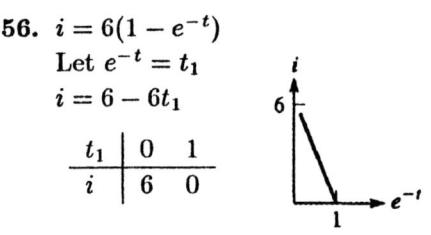

$0 \le t < 100$

56. $i = 6(1 - e^{-t})$
Let $e^{-t} = t_1$
$i = 6 - 6t_1$

t_1	0	1
i	6	0

(e^{-t} never equals zero)

57. Slope is found by measuring between points. The vertical displacement and the horizontal displacement between the extreme points is in a 1 to 2 ratio; $m = \frac{1}{2}$.
Since the graph is linear, the log equation is of the form $\log y = m \log x + \log a$, where a is the intercept $(1, a)$.
$y = ax^n;\ y = 3x^4$
$a = 3, n = 4$

x	y
1.0	3.0
1.1	4.4
1.2	6.2
1.3	8.6

$\log y = \log a + n \log x$
$\log y = \log 3 + 4 \log x$
Verify
(1) Slope is $\dfrac{\log y - \log a}{\log x} = 4$.

Vertical and horizontal measures in millimeters between points are shown. Each slope is 4.

(2) The intercept is $a = 3$.
The line crosses the vertical axis at $x = 1.0$,
$y = 3.0$.

58. $y = a(b^x)$
$\log y = \log a + x \log b$; straight line form
Given: $y = 3(2^x)$
Intercept: $x = 0$
$\log a = \log y$; $a = y$

Slope: $\log b = \dfrac{\log y - \log a}{x}$;

$\log b = m$

x	y
0	3
1	6
2	12
3	24
4	48
5	96

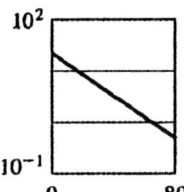

$\log y = \log 3 + x \log 2$
Intercept on graph is $3 = a$.
Slope: use $(3, 24)$

$$m = \log b = \frac{\log 24 - \log 3}{3} = \frac{\log 8}{3} = 0.3010\ldots$$

$b = 2$
$m = 0.3010\ldots$

59. Log paper, therefore $y = ax^n$;
$p = aV^n$; $\log p = \log a + n \log V$
From graph: When $V = 1, a = 0.8$
Slope is negative, therefore, $n < 0$.
Using typical log paper and
measuring horizontal and
vertical distances:

$$m = \frac{-7.2 \text{ cm}}{5.1 \text{ cm}} = -1.4$$

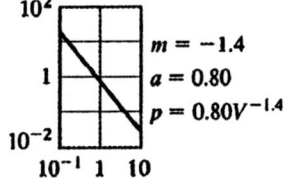

60. Semilog paper, therefore, $y = a(b^x)$
$v = a(b^t)$; $\log v = \log a + t \log b$
$t = 0$; $v = 40$, therefore, $a = 40$
Use point $(20, 15)$ to solve for b.

$$\log 15 = \log 40 + 20 \log b; \log b = \frac{\log 15 - \log 40}{20}$$

$\log b = -0.0213 = m$; $b = 0.952$
Therefore, $v = 40(0.952)^t$

2.3   The Circle

1. $(x - 2)^2 + (y - 1)^2 = 25$
Center at $(2, 1)$, radius is 5.

2. $(x - 3)^2 + (y + 4)^2 = 49$
Center at $(3, -4)$, radius is 7.

3. $(x + 1)^2 + y^2 = 4$
Center at $(-1, 0)$, radius is 2.

4. $x^2 + (y - 6)^2 = 64$
Center at $(0, 6)$, radius is 8.

5. $(x - h)^2 + (y - k)^2 = r^2; C(0, 0), r = 3$
$(x - 0)^2 + (y - 0)^2 = 3^2; x^2 + y^2 = 9$

6. $(x - h)^2 + (y - k)^2 = r^2 \ C(0, 0), r = 1$
$(x - 0)^2 + (y - 0)^2 = 1^2; x^2 + y^2 = 1$

7. $(x - 2)^2 + (y - 2)^2 = 4^2; C(2, 2), r = 4$
$x^2 + y^2 - 4x - 4y - 8 = 0$

8. $(x - 0)^2 + (y - 2)^2 = 2^2; C(0, 2), r = 2$
$x^2 + (y - 2)^2 = 4; x^2 + y^2 - 4y = 0$

9. $(x - h)^2 + (y - k)^2 = r^2; C(-2, 5), r = \sqrt{5}$
$[x - (-2)]^2 + (y - 5)^2 = (\sqrt{5})^2$;
$(x + 2)^2 + (y - 5)^2 = \sqrt{25}$
$x^2 + 4x + 4 + y^2 - 10y + 25 = 5$
$x^2 + y^2 + 4x - 10y + 4 + 25 - 5 = 0$;
$x^2 + y^2 + 4x - 10y + 24 = 0$

10. $(x-(-3))^2 + (y-(-5))^2 = (2\sqrt{3})^2$;
$C(-3,-5)$, $r = 2\sqrt{3}$
$(x+3)^2 + (y+5)^2 = 12$;
$x^2 + y^2 + 6x + 10y + 22 = 0$

11. $(x-12)^2 + (y-(-15))^2 = 18^2$; $C(12,-15)$, $r = 18$
$(x-12)^2 + (y+15)^2 = 324$;
$x^2 + y^2 - 24x + 30y + 45 = 0$

12. $\left(x - \frac{3}{2}\right)^2 + (y-(-2))^2 = \left(\frac{5}{2}\right)^2$;

$C\left(\frac{3}{2}, -2\right)$, $r = \frac{5}{2}$

$\left(x - \frac{3}{2}\right)^2 + (y+2)^2 = \frac{25}{4}$;

$4x^2 + 4y^2 - 12x + 16y = 0$

13. $C(2,1)$, passes through $(4,-1)$
$r^2 = (2-4)^2 + (1+1)^2 = (-2)^2 + (2)^2 = 8$
$(x-h)^2 + (y-k)^2 = r^2$;
$(x-2)^2 + (y-1)^2 = 8$
$x^2 - 4x + 4 + y^2 - 2y + 1 = 8$;
$x^2 + y^2 - 4x - 2y - 3 = 0$

14. $r^2 = (-2-(-1))^2 + (3-4)^2$;
$r^2 = (-1)^2 + (-1)^2 = 2$;
$r = \sqrt{2}$; $C(-1,4)$, passes through $(-2,3)$.
$(x-h)^2 + (y-k)^2 = r^2$; $(x-(-1))^2 + (y-4)^2 = 2$
$(x+1)^2 + (y-4)^2 = 2$
$x^2 + y^2 + 2x - 8y + 15 = 0$

15. Touches x-axis at $(-3,0)$
Therefore, radius is 5.

$(x-(-3))^2 + (y-5)^2 = 5^2$
$(x+3)^2 + (y-5)^2 = 25$
$x^2 + y^2 + 6x - 10y + 9 = 0$

16. Touches x-axis at $(0,-4)$
Therefore, radius is 2.

$(x-2)^2 + (y-(-4))^2 = 2^2$
$(x-2)^2 + (y+4)^2 = 4$
$x^2 + y^2 - 4x + 8y + 16 = 0$

17. The center is $(2,2)$ and radius is 2.
$(x-h)^2 + (y-k)^2 = r^2$
$(x-2)^2 + (y-2)^2 = 2^2$
$x^2 - 4x + 4 + y^2 - 4y + 4 = 4$
$x^2 + y^2 - 4x - 4y + 4 = 0$

18. Center at $(-4,4)$.
$(x-(-4))^2 + (y-4)^2 = 4^2$
$(x+4)^2 + (y-4)^2 = 16$
$x^2 + y^2 + 8x - 8y + 16 = 0$

19. Center on line $5x = 2y$; $y = \frac{5}{2}x$; $r = 5$

Tangent to x-axis; center at $(2,5)$ or $(-2,-5)$
$(x-2)^2 + (y-5)^2 = 5^2$; $x^2 + y^2 - 4x - 10y + 4 = 0$
$(x+2)^2 + (y+5)^2 = 5^2$; $x^2 + y^2 + 4x + 10y + 4 = 0$

20. Ends of diameter: $(3,8),(-3,0)$
Diameter $= \sqrt{6^2 + 8^2} = \sqrt{100} = 10$; $r = 5$
Therefore, halfway point (center) is at $(0,4)$.
$(x-0)^2 + (y-4)^2 = (5)^2$
$x^2 + (y-4)^2 = 25$
$x^2 + y^2 - 8y - 9 = 0$

21. $x^2 + (y-3)^2 = 4$ is the same as
$(x-0)^2 + (y-3)^2 = 2^2$, so
Therefore, $h = 0, k = 3, r = 2$
$C(0,3)$

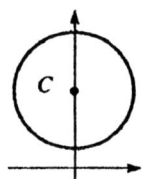

22. $(x-2)^2 + (y+3)^2 = 49$
$(x-h)^2 + (y-k)^2 = r^2$
Therefore, $h = 2, k = -3, r = 7$, $C(2,-3)$

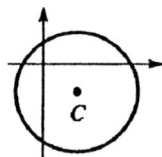

23. $4(x+1)^2 + 4(y-5)^2 = 81$

$(x+1)^2 + (y-5)^2 = \frac{81}{4}$

$h = -1, k = 5, r = \sqrt{\frac{81}{4}} = \frac{9}{2}$; $C(-1,5)$

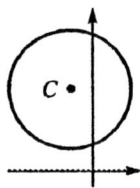

Center is $(2.10, 1.30)$; radius is 3.1.

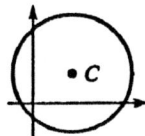

24. $2(x+4)^2 + 2(y+3)^2 = 25$

$(x+4)^2 + (y+3)^2 = \dfrac{25}{2}$

$h = -4, k = -3,$ C$(-4, -3)$;

$r = \dfrac{5}{\sqrt{2}} = \dfrac{5\sqrt{2}}{2}$

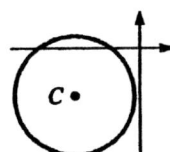

28. $x^2 + y^2 + 22x + 14y = 26$
$x^2 + 22x + 121 + y^2 + 14y + 49 = 26 + 121 + 49$
$(x - (-11))^2 + (y - (-7))^2 = 196 = 14^2$
C$(-11, -7), r = 14$

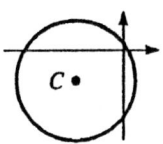

25. $x^2 + y^2 - 2x - 8 = 0$
$x^2 - 2x + 1 + y^2 = 9$
$(x-1)^2 + (y-0)^2 = 9$
$h = 1, k = 0, r = 3$
$C(1, 0)$

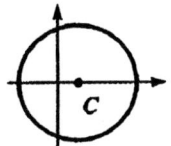

29. $4x^2 + 4y^2 - 16y = 9$
$4(x-0)^2 + 4(y^2 - 4y + 4) = 9 + 16$

$(x-0)^2 + (y-2) = \dfrac{25}{4}$

$h = 0, k = 2, r = \dfrac{5}{2}$

$C(0, 2)$

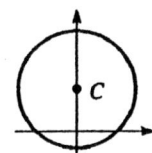

26. $x^2 + y^2 - 4x - 6y - 12 = 0$
$x^2 - 4x + y^2 - 6y = 12$
$x^2 - 4x + 4 + y^2 - 6y + 9 = 12 + 13$
$(x-2)^2 + (y-3)^3 = 25$
$h = 2, k = 3, r = 5;$ C$(2, 3)$

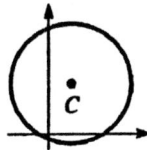

30. $9x^2 + 9y^2 + 18y = 7$
$9 \cdot x^2 + 9(y^2 + 2y + 1) = 7 + 9 = 16$

$(x-0)^2 + (y-(-1))^2 = \dfrac{16}{9} = \left(\dfrac{4}{3}\right)^2$

27. $x^2 + y^2 - 4.20x - 2.60y - 3.51 = 0$

Complete the square by dividing the coefficient of x and y by 2 and squaring.

$x^2 - 4.20x + 4.41 + y^2 - 2.60y + 1.69$
$= 3.51 + 4.41 + 1.69$
$(x - 2.10)^2 + (y - 1.30)^2 = 9.61$
$[x - (-2.10)]^2 + (y - 1.30)^2 = 3.1^2$

C$(0, -1), r = \dfrac{4}{3}$

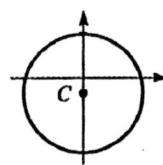

31. $2x^2 - 4x + 2y^2 - 8y = 1$

$$x^2 - 2x + y^2 - 4y = \frac{1}{2}$$

$$x^2 - 2x + 1 + y^2 - 4y + 4 = \frac{1}{2} + 5$$

$$(x-1)^2 + (y-2)^2 = \frac{1}{2} + 5$$

$$(x-1)^2 + (y-2)^2 = \frac{11}{2}$$

$h = 1, k = 2,$ C$(1, 2)$

$$r = \sqrt{\frac{11}{2}} = \frac{\sqrt{22}}{2}$$

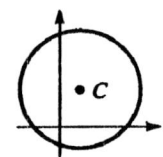

32. $3x^2 - 12x + 3y^2 = -4$

$$x^2 - 4x + y^2 = -\frac{4}{3}$$

$$x^2 - 4x + 4 + y^2 = -\frac{4}{3} + 4$$

$$(x-2)^2 + (y-0)^2 = \frac{8}{3}$$

$h = 2, k = 0,$ C$(2, 0)$

$$r = \sqrt{\frac{8}{3}} = \frac{2\sqrt{6}}{3}$$

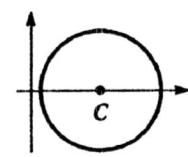

33. $(-x)^2 + y^2 = 100;\ x^2 + y^2 = 100$
Symmetrical to y-axis
$x^2 + (-y)^2 = 100;\ x^2 + y^2 = 100$
Symmetrical to x-axis
$(-x)^2 + (-y)^2 = 100;\ x^2 + y^2 = 100$
Symmetrical to origin

34. $x^2 - 4x + y^2 - 5 = 0;\ (-x)^2 - 4(-x) + (-y)^2 - 5 = 0$
$x^2 + 4x + y^2 - 5 = 0$
$(-y)^2 = y^2$
$4x \neq -4x;$ therefore, symmetrical to x-axis.

35. Replace x with $-x$ and y with $-y$:

$$3(-x)^2 + 3(-y^2) + 24(-y) = 8$$

$3x^2 + 3y^2 - 24y = 8$, not symmetrical with **respect** to origin
Replace x with $-x$:

$$3(-x^2) + 3y^2 + 24y = 8$$

$3x^2 + 3y^2 + 24y = 8$, symmetrical with respect to y-axis
Replace y with $-y$:

$$3x^2 + 3(-y)^2 + 24(-y) = 8$$

$3x^2 + 3y^2 - 24y = 8$, not symmetrical with respect to x-axis.

36. Replace x with $-x$ and y with $-y$:

$$5(-x)^2 + 5(-y)^2 - 10(-x) + 20(-y) = 3$$

$5x^2 + 5y^2 + 10x - 20y = 3$, not symmetrical with respect to origin
Replace x with $-x$:

$$5(-x)^2 + 5y^2 - 10(-x) + 20y = 3$$

$5x^2 + 5y^2 + 10x + 20y = 3$, not symmetrical with respect to y-axis.
Replace y with $-y$:

$$5x^2 + 5(-y)^2 - 10x + 20(-y) = 3$$

$5x^2 + 5y^2 - 10x - 20y = 3$, not symmetrical with respect to x-axis.

37. Find all points for which $y = 0$.
$x^2 - 6x + (0)^2 - 7 = 0;\ x^2 - 6x - 7 = 0 \Rightarrow$
$(x+1)(x-7) = 0;\ x = -1$ or $x = 7;$
$(-1, 0)$ and $(7, 0)$

38. Points of intersection occur at same **coordinates** on both circle and line.
Therefore, substitute $y = x - 1$ into equation of circle.

$x^2 + y^2 - x - 3y = 0;\ x^2 + (x-1)^2 - x - 3(x-1) = 0$
$x^2 + x^2 - 2x + 1 - x - 3x + 3 = 0;\ 2x^2 - 6x + 4 = 0$
$x^2 - 3x + 2 = 0;\ (x-2)(x-1) = 0$

Therefore, $x = 2$ and $x = 1$.
Substituting values into $y = x - 1$ gives:

$x = 2, y = 1;\ x = 1, y = 0$
$(2, 1), (1, 0)$ are intersection points.

39. $d_2 = 2d_1$

By Pythagorus:

$d_2 = \sqrt{(2-x)^2 + (4-y)^2}$
$d_1 = \sqrt{(x-0)^2 + (y-0)^2}$
$d_2 = 2d_1$
$\sqrt{(2-x)^2 + (4-y)^2} = 2\sqrt{x^2 + y^2}$
$(2-x)^2 + (4-y)^2 = 4(x^2 + y^2)$
$4 - 4x + x^2 + 16 - 8y + y^2 = 4x^2 + 4y^2$
$3x^2 + 4x + 3y^2 + 8y - 20 = 0$

This is the equation of a circle.

40. $m_1 = -\dfrac{1}{m_2}$

$m_1 m_2 = -1$

$m_1 = \dfrac{y-0}{x-2} = \dfrac{y}{x-2}$

$m_2 = \dfrac{y-0}{x-(-2)} = \dfrac{y}{x+2}$

$\left(\dfrac{y}{x-2}\right)\left(\dfrac{y}{x+2}\right) = -1$

$y^2 = -(x-2)(x+2)$
$y^2 = -(x^2 - 4) = -x^2 + 4$
$x^2 + y^2 = 4$

This is the equation of a circle.

41. $x^2 + y^2 + 5y - 4 = 0$
$y^2 + 5y + (x^2 - 4) = 0$; solve for y

$y = \dfrac{-5 \pm \sqrt{5^2 - 4(x^2 - 4)}}{2} = \dfrac{-5 \pm \sqrt{41 - 4x^2}}{2}$

$= -2.5 \pm \sqrt{10.25 - x^2}$

Set the range for

$x_{min} = -6, x_{max} = 6, y_{min} = -6, y_{max} = 2$

$y_1 = -2.5 + \sqrt{10.25 - x^2}$
$y_2 = -2.5 - \sqrt{10.25 - x^2}$

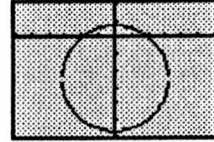   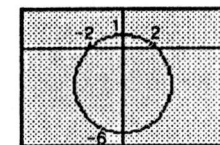

42. $2x^2 + 2y^2 + 2y - x - 1 = 0$; solve for y
$2y^2 + 2y + (2x^2 - x - 1) = 0$

$y = \dfrac{-2 \pm \sqrt{2^2 - 4(2)(2x^2 - x - 1)}}{2(2)}$

$= \dfrac{-2 \pm \sqrt{12 + 8x - 16x^2}}{4}$

$= -0.5 \pm \sqrt{0.75 + 0.5x - x^2}$

Set range:

$x_{min} = -1, x_{max} = 2, y_{min} = -1.5, y_{max} = 0.5$

$y_1 = -0.5 + \sqrt{0.75 + 0.5x - x^2}$
$y_2 = -0.5 - \sqrt{0.75 + 0.5x - x^2}$

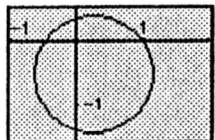

43. $x^2 + y^2 = 14.5$ is a circle with center $(0,0)$,
$r = \sqrt{14.5}$

$x^2 + y^2 - 19.6y + 86 = 0$
$x^2 + y^2 - 19.6y + 96.04 = -86 + 96.04 = 10.04$
$(x - 0)^2 + (y - 9.8)^2 = 10.04$, circle with center
$(0, 9.8)$,
$r = \sqrt{10.04}$
distance between circles $= 9.8 - \sqrt{10.04} - \sqrt{14.5}$
$\qquad\qquad\qquad\qquad = 2.82$ in.

44. $x^2 + y^2 = 42.5$; C$(0, 0)$, $r = \sqrt{42.5} = 6.52$
$x^2 + y^2 + 3.06y - 1.24 = 0$; C$(0, 1.53)$, $r = 1.89$

The least distance is d, the distance between the
two points as the straight line $x = 0$ (y-axis), cutting the two circles at $(0, x_1)$ and $(0, x_2)$.
$x_1 = 6.52$ and $x_2 = 1.53 + 1.89 = 3.42$

Therefore, $d = 6.52 - 3.42 = 3.10$ in

45. 60 Hz = 60 cycles/s = 37.7 m/s; $(h, k) = (0, 0)$
60 cycles = 37.7 m; 1 cycle = 0.628 m
$r = 0.628$ m $\div 2\pi$; $r = 0.10$
$x^2 + y^2 = (0.10)^2$; $x^2 + y^2 = 0.0100$

46. $r_1 = 3960$ mi; $r_2 = r_1 + 22,500$ mi; $r_2 = 26,460$ mi

By Pythagorus:

$$v^2 = v_H^2 + v_V^2$$

$$w = \frac{1 \text{ r}}{24 \text{ h}} = \frac{2\pi \text{ rad}}{24 \text{ h}}$$

$$v = rw = 26,460 \left(\frac{2\pi}{24}\right) = 6927.2 \text{ mi/h (constant)}$$

Therefore, $v_H^2 + v_V^2 = v^2 = (6927.2)^2 = 4.80 \times 10^7$

47.

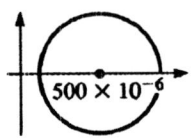

Center of circle is at $(500, 0)$.
Radius of circle is 300.

$$(x - h)^2 + (y - k)^2 = r^2$$
$$(x - 500)^2 + (y - 0)^2 = 400^2$$
$$x^2 - 1000x + 25 \times 10^4 + y^2 = 16 \times 10^4$$
$$x^2 + y^2 - 1000x + 9 \times 10^4 = 0$$
$$x^2 + y^2 - 1000 \times 10^{-6} + 9 \times 10^{-8} = 0$$
$$\text{or } (x - 500 \times 10^{-6})^2 + y^2 = (400 \times 10^{-6})^2$$
$$= 0.16 \times 10^{-6}$$

48. Area of window is made up of area of top semicircle plus area of bottom rectangle.

$$x^2 + y^2 - 3.00y + 1.25 = 0$$
$$(x - 0)^2 + y^2 - 3.00y + (1.500)^2 = -1.25 + (1.500)^2$$
$$(x - 0)^2 + (y - 1.50)^2 = 1$$
$$h = 0, k = 1.50, r = 1$$

Area of semicircle $= \dfrac{\pi r^2}{2} = \dfrac{\pi (1)^2}{2} = \dfrac{\pi}{2}$ m^2

Area of rectangle $= lw = 1.50 \times 2(1) = 3.00$ m^2
Total area is 4.57 m^2.

2.4   The Parabola

1. $y^2 = 4x$
$y^2 = 4px$
$y^2 = 4x = 4(1)x; p = 1$
$F(1, 0)$; directrix $x = -1$

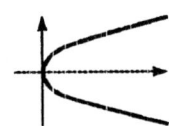

2. $y^2 = 16x; y^2 = 4px$
$4p = 16; p = 4$
$F(4, 0)$; directrix $x = -4$

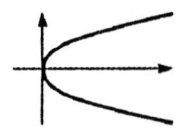

3. $y^2 = -4x; y^2 = 4px$
$4p = -4; p = -1$
$F(-1, 0)$; directrix $x = 1$

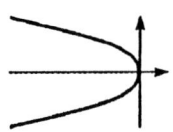

4. $y^2 = -16x$
$4p = -16; p = -4$
$F(-4, 0)$; directrix $x = 4$

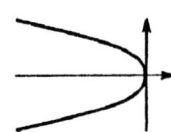

5. $x^2 = 8y$
$x^2 = 4py$
$x^2 = 8y = 4(2)y; p = 2$
$F(0, 2)$; directrix $y = -2$

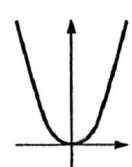

6. $x^2 = 10y$

$x^2 = 4py$

$4p = 10; \; p = \dfrac{10}{4} = \dfrac{5}{2}$

$F\left(0, \dfrac{5}{2}\right)$; directrix $y = -\dfrac{5}{2}$

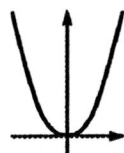

7. $x^2 = -4y$

$4p = -4; \; p = -1$

$F(0, -1)$; directrix $y = 1$

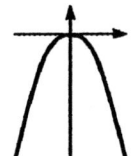

8. $x^2 = -12y$

$4p = -12; \; p = -3$

$F(0, -3)$; directrix $y = 3$

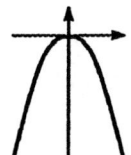

9. $2y^2 = 5x$

$y^2 = \dfrac{5}{2}x = 4px$

$p = \dfrac{5}{8}$

$F(\dfrac{5}{8}, 0)$; directrix $x = -\dfrac{5}{8}$

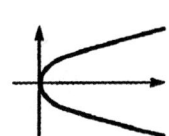

10. $3x^2 = 8y, \; x^2 = \dfrac{8}{3}y$

$4p = \dfrac{8}{3}; \; p = \dfrac{2}{3}$

$F\left(0, \dfrac{2}{3}\right)$; directrix $y = -\dfrac{2}{3}$

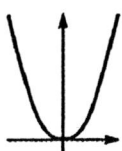

11. $y = 0.48x^2$

$x^2 = \dfrac{y}{0.48} = \dfrac{100}{48}y$

$4p = \dfrac{100}{48}; \; p = \dfrac{25}{48}$

$F\left(0, \dfrac{25}{48}\right)$; directrix $y = -\dfrac{25}{48}$

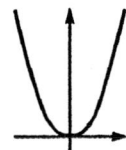

12. $x = 7.6y^2$

$y^2 = \dfrac{x}{7.6} = \dfrac{10}{76}x$

$4p = \dfrac{10}{76}; \; p = \dfrac{10}{4(76)} = \dfrac{5}{152}$

$F\left(\dfrac{5}{152}, 0\right)$;

directric $x = -\dfrac{5}{152}$

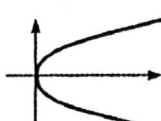

13. $F(3, 0)$; directrix $x = -3; \; p = 3$

$y^2 = 4px$

$y^2 = 4(3)x$;

$y^2 = 12x$

14. $F(-2, 0)$; directrix $x = 2$, therefore $p = -2$

Therefore, $y^2 = -8x$

15. $F(0,4)$; $V(0,0)$, therefore, $p = 4$
Therefore, $x^2 = 16y$

16. $F(-3,0)$; $V(0,0)$, therefore $p = -3$
Therefore, $y^2 = -12x$

17. $V(0,0)$, directrix $y = -0.16$
$F(0, 0.16)$, $p = 0.16$
$x^2 = 4py = 4(0.16)y$
$x^2 = 0.64y$

18. $V(0,0)$; directrix $y = 2.3$, therefore $p = -2.3$
Therefore, $x^2 = -9.2y$

19. $V(0,0)$
Therefore, $x^2 = 4py$

$(-1)^2 = 4p(8); \; 1 = 32p; \; p = \dfrac{1}{32}$

Therefore, $x^2 = \dfrac{1}{8}y$

20. $V(0,0)$

$y^2 = 4px; \; (-1)^2 = 4p(2); \; p = \dfrac{1}{8}$

Therefore, $y^2 = \dfrac{1}{2}x$

21. $F(6,1)$; directrix $x = 0$; $V(3,1)$
$d_1 = d_2$
$d_1 = x$
$d_2 = \sqrt{(x-6)^2 + (y-1)^2}$
$x = \sqrt{(x-6)^2 + (y-1)^2}$
$x^2 = (x-6)^2 + (y-1)^2$
$x^2 = x^2 - 12x + 36 + y^2 - 2y + 1$
$0 = -12x + 36 + y^2 - 2y + 1$
$y^2 - 2y - 12x + 37 = 0$

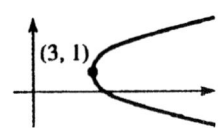

22. $F(1,1)$; directrix $y = 5$
Therefore, V is $(1, 3)$.

$d_1 = d_2$
$d_2 = 5 - y$
$d_1 = \sqrt{(x-1)^2 + (y-1)^2}$
$\sqrt{(x-1)^2 + (y-1)^2} = 5 - y$
$(x-1)^2 + (y-1)^2 = (5-y)^2$

$x^2 - 2x + 1 + y^2 - 2y + 1 = 25 - 10y + y^2$
$x^2 - 2x + 2 + 8y - 25 = 0$
$x^2 - 2x + 8y - 23 = 0$

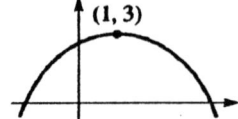

23. $y^2 + 2x + 8y + 13 = 0$; solve for y
$y^2 + 8y + (2x + 13) = 0$

$y = \dfrac{-8 \pm \sqrt{8^2 - 4(2x + 13)}}{2}$

$y = \dfrac{-8 \pm \sqrt{12 - 8x}}{2}$

$y_1 = -4 + \sqrt{3 - 2x}, \; y_2 = -4 - \sqrt{3 - 2x}$

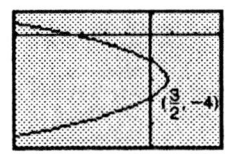

24. $y^2 - 2x - 6y + 19 = 0$; solve for y
$y^2 = -6y + (-2x - 19) = 0$

$y = \dfrac{6 \pm \sqrt{(-6)^2 - 4(-2x + 19)}}{2}$

$= \dfrac{6 \pm \sqrt{8x - 40}}{2}$

$y_1 = 3 + \sqrt{2x - 10}; \; y_2 = 3 - \sqrt{2x - 10}$

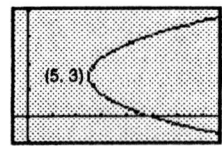

25. Vertex at $(2, -3)$
Focus is 2 units from
vertex at $(4, -3)$.
$(y + 3)^2 = 8(x - 2)$

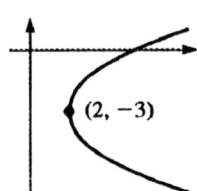

26. $V(h,k)$; $(x-h)^2 = 4p(y-k)$
(h,k) is $(-1,2)$; $p = -3$
$(x+1)^2 = -12(y-2)$

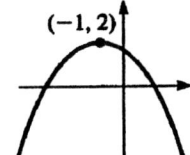

(-1, 2)

27. $y^2 = 4px$

When $x = p$; $y^2 = 4p^2$; $y = 2p$
Therefore, latus rectum intersects parabola at $(p, 2p)$.
Therefore, length of latus rectum is $2(2p) = 4p$.

28. $x^2 = 8y$
$4p = 8$; $p = 2$
$F(0,2)$; directrix $y = -2$
Center is circle of $(0, 1)$.
Radius is 1.

$(x-h)^2 + (y-k)^2 = r^2$
$(x-0)^2 + (y-1)^2 = 1$
$x^2 + y^2 - 2y + 1 = 1$
$x^2 + y^2 - 2y = 0$

29. Let the vertex of the parabola be at the origin.
$x^2 = 4py$. A point on the parabola will be $(2100, 300)$.
Substitute this into the equation and solve for p.

$2100^2 = 4p(300)$; $1200p = 4,410,000$; $p = 3675$
$x^2 = 4(3675)y = 14,700y$

30. Equation of parabolas is $x^2 = -4py$.
Curve passes through $(3.7, -5.6)$.
$(3.7)^2 = -4p(-5.6)$
$p = 0.611$

Therefore, $x^2 = -2.44y$

31. $H = Ri^2$; $R = 6.0\ \Omega$
Therefore, $H = 6.0i^2$; H vs i
of the form $y = 4px$

$4p = 6.0$; $p = \dfrac{3}{2}$

H and $i > 0$

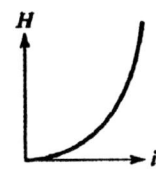

32. Place origin at top.
Therefore, equation of curve is $x^2 = -4py$.
$(2.10, -5)$ lies on curve.
Therefore, $(2.10)^2 = -4p(-5)$
$p = 0.2205$
Therefore, $x^2 = -0.822y$
Substitute $y = -2.50$; $x = 1.485$
Therefore, length of bar $= 2.97$ ft

33. $y^2 = 4px$
$(1.20)^2 = 4p(0.00625)$

$p = 57.6$ m

34. $x = v_0 t$, therefore $t = \dfrac{x}{v_0}$

$y = \dfrac{1}{2}gt^2$

Substitute: $y = \dfrac{1}{2}g\left(\dfrac{x}{v_0}\right) = \dfrac{1}{2}g\dfrac{x^2}{v_0^2} = \dfrac{g}{2v_0^2}x^2$

35. x-value of focus occurs where $y = 0$.
$0 = -12.0x + 3.6$
Therefore, $x = 0.3$
Focus $(0.3, 0)$
Therefore, $p = 0.3$
Equation is $y^2 = 4px$.
Therefore, $y^2 = 4(0.3)x = 1.2x$

36. Equation of curve is $x^2 = 4py$

Substitute $(100, 6.0)$; $100^2 = 4p(6)$; $4p = \dfrac{10^4}{6}$

Therefore, $x^2 = \dfrac{10^4}{6}y$

Substitute $x = 50.0$; $50.0^2 = \dfrac{10^4}{6}y$; $y = 1.5$ ft

Therefore, the wire is $1.5 + 30.0 = 31.5$ ft above
the ground, 50.0 ft from either pole by symmetry.

37. The graph is parabolic since it can be
transformed into the form $f^2 = 4pA$.

$f = 0.065\sqrt{A}$
$\quad = 0.065\sqrt{200}$
$\quad = 0.92$

38. $v = 8\sqrt{h}$; v versus h

$v^2 = 64h$ is a parabola of form $y^2 = 4px$.

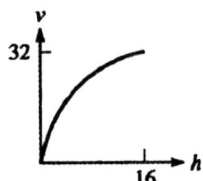

39. Path of ship channel is a parabola with focus at $(0, -2)$ and vertex $(0,0)$ and directrix at $y = 2$. Therefore, $p = -2$. Parabola of the type $x^2 = -4py$; therefore, $x^2 = -8y$. If positions of island and shoreline are interchanged, the equation would be $x^2 = 8y$. If a parabola of the form $y^2 = 4px$ is assumed, the path would be $y^2 = 8x$ or $y^2 = -8x$.

40. $P_{max} = \dfrac{kE_0^2}{R_i}$, $P_{max} = 10$ W; $E_0 = 2.0$ V;

$R_i = 0.10\ \Omega$

$10 = \dfrac{k(2.0)^2}{0.10}$; $k = 0.25$; $P_{max} = \dfrac{0.25E_0^2}{R_i}$

For constant R_i, $P = 2.5E_0^2$

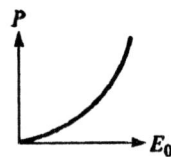

2.5   The Ellipse

1. $\dfrac{x^2}{4} + \dfrac{y^2}{1} = 1$

$a^2 = 4, b^2 = 1$

$\dfrac{x^2}{a^2} + \dfrac{y^2}{b^2} = 1$

$c^2 = a^2 - b^2$

$c^2 = 4 - 1 = 3, c = \sqrt{3}$

$V(\pm 2, 0), F(\pm\sqrt{3}, 0)$,

y-intercepts $(0, \pm 1)$

2. $\dfrac{x^2}{100} + \dfrac{y^2}{64} = 1$

$\dfrac{x^2}{a^2} + \dfrac{y^2}{b^2} = 1$

$a^2 = 100;\ a = 10$

$b^2 = 64;\ b = 8$

$c^2 = a^2 - b^2$

$c^2 = 100 - 64 = 36$

$c = 6$

$V(\pm 10, 0), F(\pm 6, 0)$, y-intercepts $(0, 8)$

3. $\dfrac{x^2}{25} + \dfrac{y^2}{36} = 1$

$a^2 = 36;\ a = 6$

$b^2 = 25;\ b = 5$

$c^2 = 36 - 25 = 11$

$c = \sqrt{11}$

$V(0, \pm 6), F(0, \pm\sqrt{11})$, x-intercepts $(\pm 5, 0)$

4. $\dfrac{x^2}{49} + \dfrac{y^2}{81} = 1$

$a^2 = 81;\ a = 9$

$b^2 = 49;\ b = 7$

$c^2 = 81 - 49 = 32$

$c = \sqrt{32} = 4\sqrt{2}$

$V(0, \pm 9), F(0, \pm 4\sqrt{2})$, x-intercepts $(\pm 7, 0)$

5. $4x^2 + 9y^2 = 36$

$\dfrac{4x^2}{36} + \dfrac{9y^2}{36} = 1$

$\dfrac{x^2}{9} + \dfrac{y^2}{4} = 1$

$a^2 = 9, b^2 = 4$

$c^2 = 9 - 4 = 5; c = \sqrt{5}$

$V(\pm 3, 0), F(\pm\sqrt{5}, 0)$,

y-intercepts $(0, \pm 2)$

6. $x^2 + 36y^2 = 144$

$\dfrac{x^2}{144} + \dfrac{36y^2}{144} = 1$

$a = 12, b = \dfrac{12}{6} = 2$

$c^2 = 144 - \dfrac{144}{36} = \dfrac{144(36 - 1)}{36}$

$c = 2\sqrt{35}$

$V(\pm 12, 0), F(\pm 2\sqrt{35}, 0)$, y-intercepts $(\pm 2, 0)$

7. $49x^2 + 4y^2 = 196$

$$\frac{49}{196}x^2 + \frac{4}{196}y^2 = 1$$

$$\frac{x^2}{196/49} + \frac{y^2}{196/4} = 1$$

$$a^2 = \frac{196}{4}; \ a = 7;$$

$$b^2 = \frac{196}{49}; \ b = 2$$

$$c^2 = \frac{196}{4} - \frac{196}{49}, \ c = 3\sqrt{5}$$

$V(0, \pm 7)$, $F(0, \pm 3\sqrt{5})$, x-intercepts $(\pm 2, 0)$

8. $25x^2 + y^2 = 25$

$$\frac{x^2}{1} + \frac{y^2}{25} = 1$$

$$a^2 = 25; \ a = 5$$
$$b^2 = 1; \ b = 1$$
$$c^2 = 25 - 1 = 24$$
$$c = 2\sqrt{6}$$

$V(0, \pm 5)$, $F(0, \pm 2\sqrt{6})$, x-intercepts $(\pm 1, 0)$

9. $8x^2 + y^2 = -16$

$$\frac{8x^2}{16} + \frac{y^2}{16} = 1$$

$$\frac{x^2}{2} + \frac{y^2}{16} = 1$$

$$\frac{y^2}{16} + \frac{x^2}{2} = 1$$

$$a^2 = 16, b^2 = 2, c^2 = 16 - 2 = 14$$
$V(0, \pm 4)$, $F(0, \pm\sqrt{14})$, x-intercepts $(\pm\sqrt{2}, 0)$

10. $2x^2 + 3y^2 = 600$

$$\frac{x^2}{300} + \frac{y^2}{200} = 1$$

$$a = \sqrt{300}, b = \sqrt{200}, c = 10$$

$V(\pm\sqrt{300}, 0)$, $F(\pm 10, 0)$, y-intercepts $(0, \pm\sqrt{200})$

11. $4x^2 + 25y^2 = 0.25 = \dfrac{1}{4}$

$$16x^2 + 100y^2 = 1$$

$$\frac{x^2}{1/16} + \frac{y^2}{1/100} = 1$$

$$a = \frac{1}{4} = 0.25$$

$$b = \frac{1}{10} = 0.1$$

$$c = \frac{\sqrt{84}}{40} = 0.23$$

$V(\pm 0.25, 0)$, $F(\pm 0.23, 0)$, y-intercepts $(0, \pm 0.1)$

12. $9x^2 + 4y^2 = 0.09 = \dfrac{9}{100}$

$$\frac{900}{9}x^2 + \frac{400}{9}y^2 = 1$$

$$\frac{x^2}{1/100} + \frac{y^2}{9/400} = 1$$

$$a = \frac{3}{20}, b = \frac{1}{10}, c = \frac{\sqrt{5}}{20}$$

$V\left(0, \pm\dfrac{3}{20}\right)$, $F\left(0, \pm\dfrac{\sqrt{5}}{20}\right)$, x-intercepts $\left(\pm\dfrac{1}{10}, 0\right)$

13. $V(15, 0); F(9, 0)$

$a = 15, a^2 = 225;$
$c = 9, c^2 = 81; a^2 - c^2 = b^2$

$$b^2 = 144; \frac{x^2}{a^2} + \frac{y^2}{b^2} = 1;$$

$$\frac{x^2}{225} + \frac{y^2}{144} = 1$$
$$144x^2 + 225y^2 = 32,400$$

14. Minor axis 8, therefore $b = 4$
$V(0, -5)$, therefore $a = 5$
$c^2 = 25 - 16 = 9$
$c = 3$

$$\frac{y^2}{25} + \frac{x^2}{16} = 1 \text{ or } 25x^2 + 16y^2 = 400$$

15. $F(0, 2)$, therefore $c = 2$
Major axis 6, therefore $a = 3$
$c^2 = a^2 - b^2$
$b^2 = 9 - 4 = 5$
$b = \sqrt{5}$

$$\frac{y^2}{9} + \frac{x^2}{5} = 1 \text{ or } 9x^2 + 5y^2 = 45$$

16. $2a + 2b = 18, F(3,0) \Rightarrow c = 3$

$$a + b = 9$$
$$b = 9 - a$$
$$a^2 = b^2 + c^2 = b^2 + 3^2$$
$$a^2 = (9 - a)^2 + 9$$
$$a^2 = 81 - 18a + a^2 + 9$$
$$18a = 90$$
$$a = 5$$
$$b = 9 - 5 = 4$$

$$\frac{x^2}{25} + \frac{y^2}{16} = 1$$

17. Vertex $(8,0); (x,y)$ is $(2,3); a^2 = 8^2 = 64$

$$\frac{x^2}{a^2} + \frac{y^2}{b^2} = 1; \frac{x^2}{64} + \frac{y^2}{b^2} = 1$$

$$\frac{(2)^2}{64} + \frac{(3)^2}{b^2} = 1; \frac{9}{b^2} = \frac{16}{16} - \frac{1}{16};$$

$$15b^2 = 144; b^2 = \frac{144}{15}$$

$$\frac{x^2}{64} + \frac{15y^2}{144} = 1; 144x^2 + 960y^2 = 9216$$

$$3x^2 + 20y^2 = 192$$

18. $F(0,2), c = 2; (x,y) = (-1, \sqrt{3})$

$$a^2 = b^2 + c^2; a^2 = b^2 + 4$$

$$\frac{y^2}{a^2} + \frac{x^2}{b^2} = 1; \frac{(\sqrt{3})^2}{a^2} + \frac{(-1)^2}{b^2} = 1; \frac{3}{a^2} + \frac{1}{b^2} = 1$$

$$\frac{3}{b^2 + 4} + \frac{1}{b^2} = 1; 3b^2 + b^2 + 4 = b^4 + 4b^2$$

$$b^4 = 4, b^2 = 2, a^2 = 2 + 4 = 6$$

Therefore, $\frac{y^2}{6} + \frac{x^2}{2} = 1$ or $3x^2 + y^2 = 6$

19. $(x_1, y_1) = (2,2), (x_2, y_2) = (1,4)$

$$\frac{x^2}{b^2} + \frac{y^2}{a^2} = 1$$

Substitute: $\frac{4}{b^2} + \frac{4}{a^2} = 1$

Therefore, $4a^2 + 4b^2 = a^2b^2$

$$\frac{1}{b^2} + \frac{16}{a^2} = 1$$

Therefore, $a^2 + 16b^2 = a^2b^2$

$$a^2 + 16b^2 = a^2b^2$$
$$16b^2 = a^2b^2 - a^2 = a^2(b^2 - 1)$$

Therefore, $a^2 = \frac{16b^2}{b^2 - 1}$

Substitute:

$$4a^2 + 4b^2 = a^2b^2; \frac{64b^2}{b^2 - 1} + 4b^2 = \frac{16b^4}{b^2 - 1}$$

$$64b^2 + 4b^4 - 4b^2 = 16b^4; -12b^4 + 60b^2 = 0$$
$$12b^2(-b^2 + 5) = 0$$
$$b^2 = 5$$

Therefore, $a^2 = \frac{16(5)}{4} = 20$

Therefore, $\frac{y^2}{20} + \frac{x^2}{5} = 1$

or: $5y^2 + 20x^2 = 100; 4x^2 + y^2 = 20$

20. $(x_1, y_1 0 = (-2, 2), (x_2, y_2) = (1, \sqrt{6})$

$$\frac{x^2}{a^2} + \frac{y^2}{b^2} = 1$$

Substitute: $\frac{4}{a^2} + \frac{4}{b^2} = 1$

Therefore, $4b^2 + 4a^2 = a^2b^2$

$$\frac{1}{a^2} + \frac{6}{b^2} = 1$$

Therefore, $b^2 + 6a^2 = a^2b^2$

$$b^2 + 6a^2 = a^2b^2$$
$$b^2 = a^2b^2 - 6a^2 = a^2(b^2 - 6)$$

Therefore, $a^2 = \frac{b^2}{b^2 - 6}$

Substitute:

$$4b^2 + 4a^2 = a^2b^2;$$

$$4b^2 + \frac{4b^2}{b^2 - 6} = \frac{b^4}{b^2 - 6}$$

$$4b^4 - 24b^2 + 4b^2 = b^4;$$
$$3b^4 - 20b^2 = 0$$

$$b^2(3b^2 - 20) = 0; b^2 = \frac{20}{3}$$

Therefore, $a^2 = \dfrac{\frac{20}{3}}{\frac{20}{3} - 6} = 10$

Therefore, $\dfrac{x^2}{10} + \dfrac{y^2}{\frac{20}{3}} = 1$ or $2x^2 + 3y^2 = 20$

21. $F(-2, 1)$ and $(4, 1)$, major axis 10

$$\sqrt{[x - (-2)]^2 + (y - 1)^2} + \sqrt{(x - 4)^2 + (y - 1)^2} = 10$$

$$\sqrt{(x + 2)^2 + (y - 1)^2} = 10 - \sqrt{(x - 4)^2 + (y - 1)^2}$$

$$(x + 2)^2 + (y - 1)^2 = 100 - 20\sqrt{(x - 4)^2 + (y - 1)^2} + (x - 4)^2 + (y - 1)^2$$

$$x^2 + 4x + 4 + y^2 - 2y + 1 = 100 - 20\sqrt{(x - 4)^2 + (y - 1)^2} + x^2 - 8x + 16 + y^2 - 2y + 1$$

$$x^2 + 4x + 4 + y^2 - 2y + 1 - 100 - x^2 + 8x - 16 - y^2 + 2y - 1 = -20\sqrt{(x - 4)^2 + (y - 1)^2}$$

$$12x - 112 = -20\sqrt{(x - 4)^2 + (y - 1)^2}$$

$$3x - 28 = -5\sqrt{(x - 4)^2 + (y - 1)^2}$$
$$(3x - 28)^2 = 25[(x - 4)^2 + (y - 1)^2]$$
$$9x^2 - 168x + 784 = 25(x^2 - 8x + 16 + y^2 - 2y + 1)$$
$$9x^2 - 168x + 784 = 25x^2 - 200x + 400 + 25y^2 - 50y + 25$$
$$-16x^2 - 25y^2 + 32x + 50y + 359 = 0$$
$$16x^2 + 25y^2 - 32x - 50y - 359 = 0$$

22. $\sqrt{(x - 1)^2 + (y - 4)^2} + \sqrt{(x - 1)^2 + (y - 0)^2} = $ constant. Let $(x, y) = (4, 4)$

$$\sqrt{(4 - 1)^2 + (4 - 4)^2} + \sqrt{(4 - 1)^2 + (4 - 0)^2} = \text{constant}$$

$$\sqrt{3^2 + 0^2} + \sqrt{3^2 + 4^2} = \text{constant}$$

$$3 + 5 = 8 = \text{constant}$$

$$\sqrt{(x - 1)^2 + (y - 4)^2} + \sqrt{(x - 1)^2 + y^2} = 8$$

$$\sqrt{(x - 1)^2 + (y - 4)^2} = 8 - \sqrt{(x - 1)^2 + y^2}$$

$$x^2 - 2x + 1 + y^2 - 8y + 16 = 64 - 16\sqrt{(x - 1)^2 + y^2} + x^2 - 2x + 1 + y^2$$

$$-8y - 48 = -16\sqrt{(x - 1)^2 + y^2}$$

$$y + 6 = 2\sqrt{(x - 1)^2 + y^2}$$

$$y^2 + 12y + 36 = 4(x^2 - 2x + 1 + y^2)$$
$$y^2 + 12y + 36 = 4x^2 - 8x + 4 + 4y^2$$
$$0 = 4x^2 + 3y^2 - 8x - 12y - 32$$

$$4x^2 + 3y^2 - 8x - 12y - 32 = 0$$

x	y_1	y_2
1	6	-2
4	4	0
0	5.83	-1.83
-2.46	2	2
4.46	2	2

To sketch, solve $3y^2 - 12y + 4x^2 - 8x - 32 = 0$ for y using quadratic formula,

$$y_1 = \frac{12 + \sqrt{144 - 4(3)(4x^2 - 8x - 32)}}{2(3)} = \frac{12 + \sqrt{144 - 12(4x^2 - 8x - 32)}}{6}$$

$$y_2 = \frac{12 - \sqrt{144 - 12(4x^2 - 8x - 32)}}{6}$$

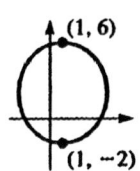

23. $4x^2 + 3y^2 + 16x - 18y + 31 = 0$; solve for y

$3y^2 - 18y + (4x^2 + 16x + 31) = 0$

$$y = \frac{18 \pm \sqrt{(-18)^2 - 4(3)(4x^2 + 16x + 31)}}{2(3)}$$

$$= \frac{18 + \sqrt{-48x^2 - 192x - 48}}{2}$$

$$y_1 = 3 + \frac{\sqrt{-12x^2 - 48x - 12}}{3}$$

$$y_2 = 3 - \frac{\sqrt{-12x^2 - 48x - 12}}{3}$$

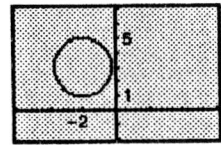

24. $4x^2 + 8y^2 + 4x - 24y + 1 = 0$; solve for y

$8y^2 - 24y + (4x^2 + 4x + 1) = 0$

$$y = \frac{24 \pm \sqrt{(-24)^2 - 4(8)(4x^2 + 4x + 1)}}{2(8)}$$

$$= \frac{24 \pm \sqrt{-128x^2 - 128x + 544}}{16}$$

$$y_1 = 1.5 + \frac{\sqrt{-8x^2 - 8x + 34}}{4}$$

$$y_2 = 1.5 - \frac{\sqrt{-8x^2 - 8x + 34}}{4}$$

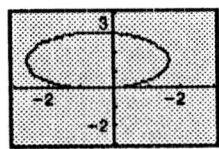

25. $2a = 6$; $a = 3$; $2b = 4$; $b = 2$; $(h, k) = (2, -1)$

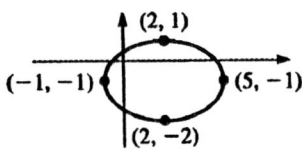

26. Center (h, k); $\dfrac{(y - k)^2}{a^2} + \dfrac{(x - h)^2}{b^2} = 1$

Major axis = 8; minor axis = 6; $(h, k) = (1, 3)$

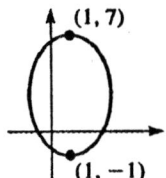

27. $x^2 + y^2 = 1, k > 0$

Therefore, $\dfrac{x^2}{1} + \dfrac{y^2}{\dfrac{1}{k}} = 1$

$\dfrac{1}{k} = a^2 = 1$

Therefore, $\sqrt{\dfrac{1}{k}} = a > 1 \Rightarrow \dfrac{1}{k} > 1$

Therefore, $0 < k < 1$

28. $x^2 + k^2 y^2 = 25$

$\dfrac{x^2}{25} + \dfrac{k^2 y^2}{25} = 1$

Therefore, $\dfrac{x^2}{25} + \dfrac{y^2}{\dfrac{25}{k^2}} = 1$

$F(3, 0)$, therefore, $c = 3$

Equation of form $\dfrac{x^2}{a^2} + \dfrac{y^2}{b^2} = 1$

$a = 5$, therefore $b = 4$

Therefore, $\dfrac{25}{k^2} = 16$; $k^2 = \dfrac{25}{16}$; $k = \pm\dfrac{5}{4}$

29. Given: $2x^2 + 3y^2 - 8x - 4 = 0$

$2x^2 + 3(-y)^2 - 8x - 4 = 2x^2 + 3y^2 - 8x - 4$

30. $5x^2 + y^2 - 3y - 7 = 5(-x)^2 + y^2 - 3y - 7$

31. Eccentricity $e = \dfrac{c}{a}$

$x^2 + 9y^2 = 81$; $\dfrac{x^2}{81} + \dfrac{9y^2}{81} = 1$; $\dfrac{x^2}{81} + \dfrac{y^2}{\dfrac{81}{9}} = 1$

$a^2 = 81$; $a = 9$; $b^2 = \dfrac{81}{9}$; $b = 3$

$c^2 = a^2 - b^2$; $c = \sqrt{81 - 9} = \sqrt{72} = 6\sqrt{2}$

Therefore, $e = \dfrac{6\sqrt{2}}{9}$; $e = \dfrac{2\sqrt{2}}{3} \approx 0.943$

32. $2a = 4.6 + 2.8 = 7.4; a = 3.7$
$2c = 4.6 - 2.8 = 1.8; c = 0.9$

Therefore, $e = \dfrac{c}{a} = \dfrac{0.9}{3.7} = 0.24$

33. If the two vertices of each base are fixed at $(-3,0)$ and $(3,0)$, and the sum of the two leg lengths is also fixed, the third vertex lies on an ellipse. The base is 6 cm, so
$d_1 + d_2 = 14$ cm -6 cm $= 8$ cm
$(-3,0)$ and $(3,0)$ are foci $(-c,0)$ and $(c,0)$
$d_1 + d_2 = 2a = 8; a = 4$
$a^2 - c^2 = b^2$
$4^2 - 3^2 = b^2$
$b^2 = 7, a^2 = 16$

The equation is $\dfrac{x^2}{16} + \dfrac{y^2}{7} = 1$, or $7x^2 + 16y^2 = 112$

34. $P = Ri^2$
$P_T = R_1 i_1^2 + R_2 i_2^2$
$64 = 2i_1^2 + 8i_2^2$
$\dfrac{2}{64}i_1^2 + \dfrac{8}{64}i_2^2 = 1$
$\dfrac{i_1^2}{32} + \dfrac{i_2^2}{8} = 1$
$i_1^2 + 4i_2^2 = 32$
$a^2 = 32; a = \sqrt{32} \approx 5.7$
$b^2 = 8; b = \sqrt{8} \approx 2.8$

35. $36x^2 + 225y^2 = 8100$

Two people must be separated by a distance $= 2c$.

$\dfrac{36x^2}{8100} + \dfrac{225y^2}{8100} = 1$

$\dfrac{x^2}{225} + \dfrac{y^2}{36} = 1$

$a^2 = 225; a = 15$
$b^2 = 36; b = 6$
$c^2 = a^2 - b^2 = 225 - 36 = 189; c = 13.748$
$2c = 27.5$ m 34

36. $a = 4.20, b = 0.60$
$\dfrac{x^2}{a^2} + \dfrac{y^2}{b^2} = 1$, therefore $\dfrac{x^2}{(4.20)^2} + \dfrac{y^2}{(0.60)^2} = 1$

$\dfrac{x^2}{17.64} + \dfrac{y^2}{0.36} = 1$

$0.36x^2 + 17.64y^2 = 6.3504; x^2 + 49y^2 = 17.6$

37. $a = \dfrac{64}{2} = 32, b = 18$

Let the center be at the origin. The equation of the ellipse is

$\dfrac{x^2}{32^2} + \dfrac{y^2}{18^2} = 1; \dfrac{x^2}{1024} + \dfrac{y^2}{324} = 1$

If $x = 22, \dfrac{22^2}{1024} + \dfrac{y^2}{324} = 1$

$y = 13$ ft

38. $a = 2.25; b = 1.60; p = \pi(a+b); p = \pi(2.25+1.60);$
$p = 12.1$ ft

39. $9x^2 + 20y^2 = 180; \dfrac{9x^2}{180} + \dfrac{20y^2}{180} = 1; \dfrac{x^2}{20} + \dfrac{y^2}{9} = 1$

$a = \sqrt{20} = 2\sqrt{5} \approx 4.5; b = 3;$
$V =$ area of end $\times$ length
$V = \pi ab \times 20.0$
$V = \pi(2\sqrt{5})(3)(20.0) = 843$ ft^3

40. $\cos 28° = \dfrac{6.80}{2a}$, therefore $a = \dfrac{3.40}{\cos 28°} = 3.851$ mm

$b = 3.40$ mm

Laser beam is circular in cross section and appears as an ellipse on plane surface when incident at 62°. One diameter is elongated; other diameter, at right angles to first, is in true length.
Ellipse of minor axis $= 6.80$ mm, major axis $= 7.702$ mm.

$A = \pi ab$
$A = \pi(3.851)(3.40)$
$A = 41.1$ mm^2

2.6   The Hyperbola

1. $\dfrac{x^2}{25} - \dfrac{y^2}{144} = 1$
$a^2 = 25; a = 5$
$b^2 = 144; b = 12$
$c^2 = a^2 + b^2$
$c^2 = 169; c = 13$
$V(\pm 5, 0), F(\pm 13, 0)$

2. $\dfrac{x^2}{16} - \dfrac{y^2}{4} = 1$

$a^2 = 16;\ a = 4$
$b^2 = 4;\ b = 2$
$c^2 = a^2 + b^2$
$c^2 = 20;\ c = 2\sqrt{5}$
$V(\pm 4, 0), F(\pm 2\sqrt{5}, 0)$

3. $\dfrac{y^2}{9} - \dfrac{x^2}{1} = 1$

$a^2 = 9;\ a = 3$
$b^2 = 1;\ b = 1$
$c^2 = 10;\ c = \sqrt{10}$
$V(0, \pm 3), F(0, \pm\sqrt{10})$

4. $\dfrac{y^2}{2} - \dfrac{x^2}{2} = 1$

$a^2 = 2;\ a = \sqrt{2}$
$b^2 = 2;\ b = \sqrt{2}$
$c^2 = 4;\ c = 2$
$V(0, \pm\sqrt{2}), F(0, \pm 2)$

5. $4x^2 - y^2 = 4$

$\dfrac{4x^2}{4} - \dfrac{y^2}{4} = 1;$

$\dfrac{x^2}{1} - \dfrac{y^2}{4} = 1$

$a^2 = 1;\ b^2 = 4;\ c^2 = 5$
$V(\pm 1, 0);\ F(\pm\sqrt{5}, 0)$

6. $x^2 - 9y^2 = 81$

$\dfrac{x^2}{81} - \dfrac{9y^2}{81} = 1$

$\dfrac{x^2}{81} - \dfrac{y^2}{9} = 1$

$a^2 = 81;\ a = 9$
$b^2 = 9;\ b = 3$
$c^2 = 90;\ c = 3\sqrt{10}$
$V(\pm 9, 0), F(\pm 3\sqrt{10}, 0)$

7. $2y^2 - 5x^2 = 10$

$\dfrac{2y^2}{10} - \dfrac{5x^2}{10} = 1$

$\dfrac{y^2}{5} - \dfrac{x^2}{2} = 1$

$a^2 = 5;\ a = \sqrt{5}$
$b^2 = 2;\ b = \sqrt{2}$
$c^2 = 7;\ c = \sqrt{7}$
$V(0, \pm\sqrt{5}), F(0, \pm\sqrt{7})$

8. $3y^2 - 2x^2 = 300$

$\dfrac{3y^2}{300} - \dfrac{2x^2}{300} = 1$

$\dfrac{y^2}{100} - \dfrac{x^2}{150} = 1$

$a^2 = 100;\ a = 10$
$b^2 = 150;\ b = 5\sqrt{6}$
$c^2 = 250;\ c = 5\sqrt{10}$
$V(0, \pm 10), F(0, \pm 5\sqrt{10})$

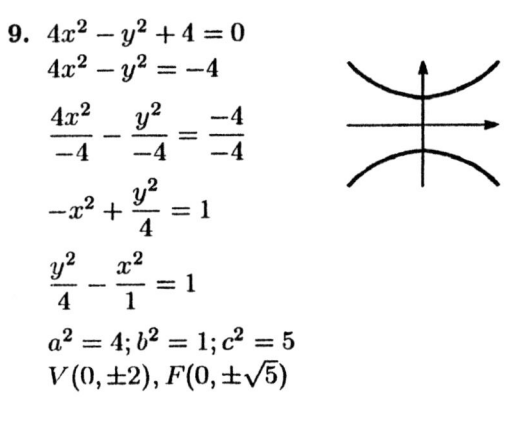

9. $4x^2 - y^2 + 4 = 0$

$4x^2 - y^2 = -4$

$\dfrac{4x^2}{-4} - \dfrac{y^2}{-4} = \dfrac{-4}{-4}$

$-x^2 + \dfrac{y^2}{4} = 1$

$\dfrac{y^2}{4} - \dfrac{x^2}{1} = 1$

$a^2 = 4;\ b^2 = 1;\ c^2 = 5$
$V(0, \pm 2), F(0, \pm\sqrt{5})$

10. $9x^2 - y^2 = 9$

$\dfrac{x^2}{1} - \dfrac{y^2}{9} = 1$

$a = 1;\ b = 3;$
$c^2 = 10;\ c = \sqrt{10}$
$V(\pm 1, 0), F(\pm\sqrt{10}, 0)$

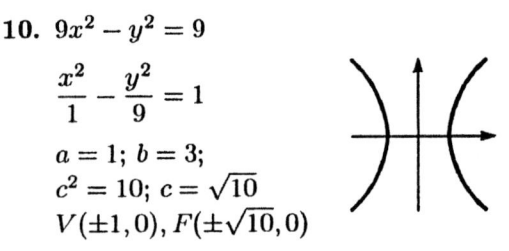

11. $4x^2 - y^2 = 0.64$

$\dfrac{4x^2}{0.64} - \dfrac{y^2}{0.64} = 1$

$\dfrac{x^2}{0.16} - \dfrac{y^2}{0.64} = 1$

$a^2 = 0.16;\ a = 0.40$
$b^2 = 0.64;\ b = 0.80$
$c^2 = 0.80;\ c = 0.89$
$V(\pm 0.4, 0), F(\pm 0.89, 0)$

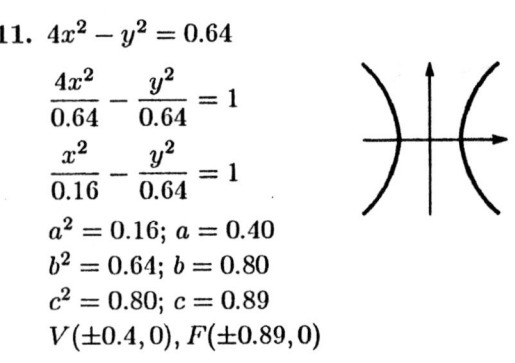

12. $9y^2 - x^2 = 0.36$

$$\frac{9y^2}{0.36} - \frac{x^2}{0.36} = 1$$

$$\frac{y^2}{0.04} - \frac{x^2}{0.36} = 1$$

$a^2 = 0.04;\ a = 0.2$

$b^2 = 0.36;\ b = 0.6$

$c^2 = 0.40;\ c^2 = \dfrac{40}{100}$

$c = \dfrac{2}{10}\sqrt{10};\ c = \dfrac{1}{5}\sqrt{10}$

$V(0, \pm 0.2), F\left(0, \pm\dfrac{\sqrt{10}}{5}\right)$

13. $V(3,0);\ F(5,0)$

$a = 3;\ c = 5;\ a^2 = 9;\ c^2 = 25$

$b^2 = c^2 - a^2 = 25 - 9 = 16$

$\dfrac{x^2}{a^2} - \dfrac{y^2}{b^2} = 1;\ \dfrac{x^2}{9} - \dfrac{y^2}{16} = 1;$

$16x^2 - 9y^2 = 144$

14. $V(0,1);\ F(0, \sqrt{3})$

$a = 1;\ c = \sqrt{3};\ c^2 = a^2 + b^2;\ b^2 = 3 - 1 = 2;$

$b = \sqrt{2}$

$\dfrac{y^2}{a^2} - \dfrac{x^2}{b^2} = 1;\ \dfrac{y^2}{1} - \dfrac{x^2}{2} = 1$

$2y^2 - x^2 = 2$

15. Conjugate axis $= 12;\ V(0, 10)$

$a = 10;\ 2b = 12;\ b = 6$

$\dfrac{y^2}{a^2} - \dfrac{x^2}{b^2} = 1;\ \dfrac{y^2}{100} - \dfrac{x^2}{36} = 1$

$9y^2 - 25x^2 = 900$

16. focus $(10, 0) \Rightarrow c = 10$

$a + b = 14, b = 14 - a$

$$a^2 + b^2 = c^2 = 10^2$$
$$a^2 + (14 - a)^2 = 100$$
$$a^2 + 196 - 28a + a^2 = 100$$
$$2a^2 - 28a + 96 = 0$$
$$a^2 - 14a + 48 = 0$$
$$(a - 6)(a - 8) = 0$$
$$a - 6 = 0\ ,\qquad a - 8 = 0$$
$$a = 6\qquad\qquad a = 8$$
$$b = 14 - 6 = 8\quad b = 14 - 8 = 6$$

$\dfrac{x^2}{36} - \dfrac{y^2}{64} = 1$ or $\dfrac{x^2}{64} - \dfrac{y^2}{36} = 1$

17. (x, y) is $(2, 3);\ F(2, 0), (-2, 0);\ c = \pm 2, c^2 = 4$

$d_1 = \sqrt{(2 - [-2])^2 + (3 - 0)^2}$

$\quad = \sqrt{4^2 + 3^2} = \sqrt{16 + 9}$

$\quad = \sqrt{25} = 5$

$d_2 = \sqrt{(2 - 2)^2 + (3 - 0)^2}$

$\quad = \sqrt{0 + 9} = \sqrt{9} = 3$

$d_1 - d_2 = 2a;\ 5 - 3 = 2a;\ 2 = 2a;\ 1 = a;\ a^2 = 1$

$c^2 = 4;\ b^2 = c^2 - a^2 = 3$

$\dfrac{x^2}{a^2} - \dfrac{y^2}{b^2} = 1;\ \dfrac{x^2}{1} - \dfrac{y^2}{3} = 1$

$3x^2 - y^2 = 3$

18. $(x_1, y_1) = (8, \sqrt{3}), V(4, 0), a = 4$

$\dfrac{x^2}{a^2} - \dfrac{y^2}{b^2} = 1;\ \dfrac{8^2}{16} - \dfrac{3}{b^2} = 1;\ 4 - \dfrac{3}{b^2} = 1;$

$\dfrac{3}{b^2} = 3;\ b^2 = 1;\ b = 1$

Therefore, $\dfrac{x^2}{16} - \dfrac{y^2}{1} = 1;\ x^2 - 16y^2 = 16$

19. $(x_1, y_1) = (5.4),\ (x_2, y_2) = \left(3, \dfrac{4}{5}\sqrt{5}\right)$

Use: $\dfrac{x^2}{a^2} - \dfrac{y^2}{b^2} = 1$

Solve for a^2 and b^2:

$\dfrac{25}{a^2} - \dfrac{16}{b^2} = 1;\ 25b^2 - 16a^2 = a^2 b^2$

$25b^2 - a^2 b^2 = 16a^2;\ b^2(25 - a^2) = 16a^2$

$b^2 = \dfrac{16a^2}{25 - a^2}$

Substitute:

$\dfrac{9}{a^2} - \dfrac{\frac{16}{5}}{b^2} = 1;\ \dfrac{9}{a^2} - \dfrac{\frac{16}{5}}{\frac{16a^2}{25} - a^2} = 1$

$\dfrac{9}{a^2} - \dfrac{25 - a^2}{5a^2} = 1;\ 45 - 25 + a^2 = 5a^2;\ 4a^2 = 20$

$a^2 = 5$

Therefore, $b^2 = \dfrac{16(5)}{25 - 5} = 4$

Therefore, $\dfrac{x^2}{5} - \dfrac{y^2}{4} = 1;\ 4x^2 - 5y^2 = 20$

20. $(x_1, y_1) = (1, 2), (x_2, y_2) = (2, 2\sqrt{2})$

Use: $\dfrac{y^2}{a^2} - \dfrac{x^2}{b^2} = 1$

Solve for a^2 and b^2:

$\dfrac{4}{a^2} - \dfrac{1}{b^2} = 1;\ 4b^2 - a^2 = a^2b^2;\ 4b^2 - a^2b^2 = a^2$

$b^2(4 - a^2) = a^2;\ b^2 = \dfrac{a^2}{4 - a^2}$

Substitute:

$\dfrac{8}{a^2} - \dfrac{4}{b^2} = 1;\ \dfrac{8}{a^2} - \dfrac{4}{\dfrac{a^2}{4} - a^2} = 1$

$\dfrac{8}{a^2} - \dfrac{(4 - a^2)4}{a^2} = 1;\ 8 - 16 + 4a^2 = a^2;\ 3a^2$

$a^2 = \dfrac{3}{8}$

Therefore, $b^2 = \dfrac{\dfrac{8}{3}}{4 - \dfrac{8}{3}};\ b^2 = 2$

Therefore, $\dfrac{y^2}{\dfrac{8}{3}} - \dfrac{x^2}{2} = 1;\ 3y^2 - 4x^2 = 8$

21. $xy = 2;\ y = \dfrac{2}{x}$

x	y
$\frac{1}{2}$	4
1	2
2	1
4	$\frac{1}{2}$
8	$\frac{1}{4}$

x	y
$-\frac{1}{2}$	-4
-1	-2
-2	-1
-4	$-\frac{1}{2}$
-8	$-\frac{1}{4}$

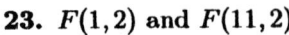

22. $xy = -4;\ y = -\dfrac{4}{x}$

x	y
$\frac{1}{2}$	-8
1	-4
2	-2
4	-1
8	$-\frac{1}{2}$

x	y
$-\frac{1}{2}$	8
-1	4
-2	2
-4	1
-8	$\frac{1}{2}$

23. $F(1, 2)$ and $F(11, 2)$

Transverse axis $= 8 = 2a$

$d_1 = \sqrt{(x - 1)^2 + (y - 2)^2}$
$d_2 = \sqrt{(x - 11)^2 + (y - 2)^2}$
$d_1 - d_2 = 2a$
$\sqrt{(x - 1)^2 + (y - 2)^2} - \sqrt{(x - 11)^2 + (y - 2)^2} = 8$
$\sqrt{(x - 1)^2 + (y - 2)^2} = 8 + \sqrt{(x - 11)^2 + (y - 2)^2}$

Square both sides.

$(x - 1)^2 + (y - 2)^2$
$= 64 + 16\sqrt{(x - 11)^2 + (y - 2)^2} + (x - 11)^2 + (y - 2)^2$

Expand, gather like terms, keep radical to right.

$20x - 184 = 16\sqrt{(x - 11)^2 + (y - 2)^2}$ Factor.
$5x - 46 = 4\sqrt{(x - 11)^2 + (y - 2)^2}$ Square both sides.
$25x^2 - 460x + 2116 = 16[(x - 11)^2 + (y - 2)^2]$

Expand, gather terms.

$25x^2 - 460x + 2116$
$= 16(x^2 - 22x + 121 + y^2 - 4y + 4)$
$25x^2 - 460x + 2116$
$= 16x^2 - 352x + 1936 + 16y^2 - 64y + 64$
$9x^2 - 16y^2 - 108x + 64y + 116 = 0$

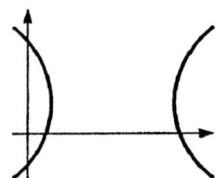

24. $V(-2, 4);\ V(-2, -2)$
Conjugate axis $= 4$
Therefore, $2b = 4;\ b = 2$
Transverse axis $= 6$
$2a = 6;\ a = 3$
Therefore, $c = \sqrt{13}$
Center of transeve axis $(-2, 1)$
$F_1(-2, 1 + \sqrt{13});$
$F_2(-2, 1 - \sqrt{13})$

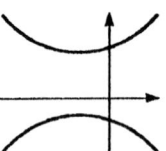

$d_2 - d_1 = 6$
$\sqrt{(x + 2)^2 + (y - 1 + \sqrt{13})^2} - \sqrt{(x + 2)^2 + (y - 1 - \sqrt{13})^2}$
$= 6$
$\sqrt{(x + 2)^2 + (y - 1 + \sqrt{13})^2}$
$= 6 + \sqrt{(x + 2)^2 + (y - 1 - \sqrt{13})^2}$

Square both sides.

$(x + 12)^2 + (y - 1 - \sqrt{13})^2$
$= 36 + 12\sqrt{(x + 2)^2 + (y - 1 - \sqrt{13})^2}$
$+ (x + 2)^2 + (y - 1 - \sqrt{13})^2$

Expand trinomials and cancel like terms.

$$4\sqrt{13}y - 4\sqrt{13} - 36 = 12\sqrt{(x+2)^2 + (y-1-\sqrt{13})^2}$$

Square both sides.

$$52y^2 - 104y - 72\sqrt{13}y + 52 + 72\sqrt{13} + 32y$$
$$= 36(x^2 + 4x + 4 + y^2 - 2y - 2\sqrt{13}y + 2\sqrt{13} + 14);$$
$$52y^2 - 104y - 72\sqrt{13}y + 376 + 72\sqrt{13}$$
$$= 36x^2 + 144x + 144 + 36y^2 - 72y + 72\sqrt{13}y$$
$$+ 72\sqrt{13} + 504$$
$$16y^2 - 36x^2 - 32y - 144x - 272 = 0$$
$$4y^2 - 9x^2 - 8y - 36x - 68 = 0$$

25. $x^2 - 4y^2 + 4x + 32y - 64 = 0$; solve for y
$$4y^2 - 34y + (-x^2 - 4x + 64) = 0$$

$$y = \frac{32 \pm \sqrt{(-32)^2 - 4(4)(-x^2 - 4x + 64)}}{2(4)}$$

$$= \frac{32 \pm \sqrt{16x^2 + 64x}}{8}$$

$$y_1 = 4 + 0.5\sqrt{x^2 + 4x}, \quad y_2 = 4 - 0.5\sqrt{x^2 + 4x}$$

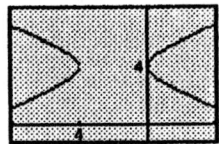

26. $5y^2 - 4x^2 + 8x + 40y + 56 = 0$; solve for y
$$5y^2 + 40y + (-4x^2 + 8x + 56) = 0$$

$$y = \frac{-40 \pm \sqrt{40^2 - 4(5)(-4x^2 + 8x + 56)}}{2(5)}$$

$$= \frac{-40 \pm \sqrt{80x^2 - 160x + 480}}{10}$$

$$y_1 = -4 + 0.4\sqrt{5x^2 - 10x + 30}$$
$$y_2 = -4 - 0.4\sqrt{5x^2 - 10x + 30}$$

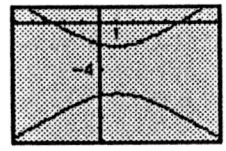

27. Transverse axis $= 4$; conjugate axis $= 6$;
$(h, k) = (-3, 2)$

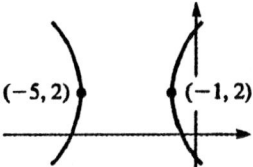

28. Transverse axis $= 2$; conjugate axis $= 8$;
$(h, k) = (5, 0)$

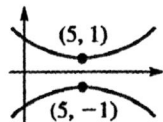

29. $V(0,1), F(0,\sqrt{3}); c^2 = a^2 + b^2$ where $c = \sqrt{3}$ and $a = 1; b^2 = \sqrt{3}^2 - 1^2 = 2$

$$\frac{y^2}{1^2} - \frac{x^2}{\sqrt{2}^2} = 1$$

The transverse axis of the first equation is length $2a = 2\sqrt{1}$ along the y-axis. Its conjugate axis is length $2b = 2\sqrt{2}$ along the x-axis.

The transverse axis of the conjugate hyperbola is length $2\sqrt{2}$ along the x-axis, and its conjugate axis is length $2\sqrt{1}$ along the y-axis.

The equation, then, is $\dfrac{x^2}{\sqrt{2}^2} - \dfrac{y^2}{\sqrt{1}^2} = 1$

$$\frac{x^2}{2} - \frac{y^2}{1} = 1 \text{ or } x^2 - 2y^2 = 2$$

30. $e = \dfrac{c}{a}$

$$2x^2 - 3y^2 = 24; \frac{2}{24}x^2 - \frac{3}{24}y^2 = 1; \frac{x^2}{12} - \frac{y^2}{8} = 1$$

Therefore, $a^2 = 12; a = 2\sqrt{3}; b^2 = 8; b = 2\sqrt{2}$

$$c^2 = a^2 + b^2; c = 12 + 8 = 20; c = 2\sqrt{5}$$

$$e = \frac{2\sqrt{5}}{2\sqrt{3}} = \frac{\sqrt{5}}{\sqrt{3}} = \frac{\sqrt{15}}{3} \approx 1.29$$

31.

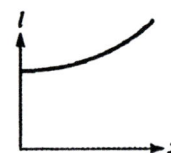

By Pythagorus: $l^2 = 2000^2 + x^2$
Therefore, $l^2 - x^2 = 2000^2$ is the equation of a hyperbola.

$$\frac{l^2}{2000^2} - \frac{x^2}{2000^2} = 1$$

$a = 2000$, therefore transverse axis $= 4000$ m
$b = 2000$, therefore transverse axis $= 4000$ m

32. $\pi R^2 - 2\pi r^2 = 24$

$$\frac{\pi R^2}{24} - \frac{2\pi r^2}{24} = 1$$

$$\frac{R^2}{24/\pi} - \frac{r^2}{12/\pi} = 1 \text{ is the equation of a hyperbola.}$$

$$a^2 = \frac{24}{\pi}; \quad b^2 = 12/\pi$$

33. $V = iR$ (Ohm's law)
$6.00 = iR$

Therefore, $i = \dfrac{6.00}{R}$

R	i
0.5	12
1	6
2	3
3	2
4	1.5
6	1
9	0.7
12	0.5

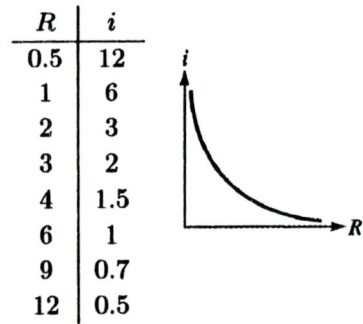

34. Foci at $(-3.5, 0)$; $(3.5, 0)$
Vertices at $(-2.8, 0)$; $(2.8, 0)$
Therefore, $a = 2.8$; $c = 3.5$
$c^2 = a^2 + b^2$; $b^2 = (3.5)^2 - (2.8)^2$; $b^2 = 4.41$

Equation of type:

$$\frac{x^2}{a^2} - \frac{y^2}{b^2} = 1; \quad \frac{x^2}{(2.8)^2} - \frac{y^2}{4.41} = 1; \quad \frac{x^2}{7.84} - \frac{y^2}{4.41} = 1$$

35.

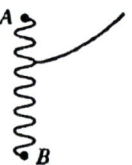

$d_1 - d_2 = $ constant
Let t_1 be time for signal to go from B to ship.
The t_2 is time for signal to go from A to ship.
Where $t_2 = t_1 - 1.20$ ms
$V = \dfrac{s}{t}$, therefore, $s = Vt$

Therefore, $d_1 = 300t_1$ and $d_2 = 300(t_1 - 1.20)$
$d_1 - d_2 = 2a$
$300t_1 - 300(t_1 - 1.20) = $ constant $= 360$ km $= 2a$
Therefore, the ship could lie anywhere on the hyperbolic arc so sketched.
Foci at $(\pm 300, 0)$, therefore $c = 300$
Vertices at $(\pm 180, 0)$, therefore $a = 180$; therefore $b = 240$

36. Foci at $(\pm 2, 0)$, therefore $c = 2$
$d_1 - d_2 = 2 = 2a$, therefore $a = 1$
$b^2 = 4 - 1 = 3$
$$\frac{x^2}{a^2} - \frac{y^2}{b^2} = 1; \quad \frac{x^2}{1} - \frac{y^2}{3} = 1; \quad 3x^2 - y^2 = 3$$

2.7   Translation of Axes

1. $(y - 2)^2 = 4(x + 1)$; parabola
$y - 2 = y'$; $x + 1 = x'$; $y'^2 = 4x'$
Origin O' at $(h, k) = (-1, 2)$
$y'^2 = 4(1)x'$; $p = 1$;
Focus $(-1 + p, 2)$ or $(0, 2)$
Directrix $x' = -p$; $x + 1 = -1$; $x = -2$
Vertex $(-1, 2)$

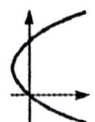

2. $\dfrac{(x + 4)^2}{4} + \dfrac{(y - 1)^2}{1} = 1$; ellipse

Center $(-4, 1)$; $a = 2$; $b = 1$

3. $\dfrac{(x-1)^2}{4} - \dfrac{(y-2)^2}{9} = 1$; hyperbola

Center $(1,2)$; $a = 2$; $b = 3$

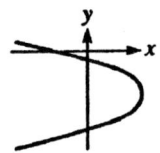

4. $(y+5)^2 = -8(x-2)$; parabola

$y + 5 = y'$; $x - 2 = x'$

$y'^2 = -8x'$

Origin $0'$ at $(h,k) = (2,-5)$

$y'^2 = 4(-2)x'$; $p = -2$

Vertex $(2,-5)$, focus $(0,-5)$, directrix $x = 4$

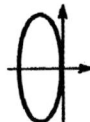

5. $\dfrac{(x+1)^2}{1} + \dfrac{y^2}{9} = 1$; ellipse

$x' = x + 1; x - h = x + 1; h = -1$

$y' = y - 0; y - k = y - 0; k = 0$

$\dfrac{x'^2}{1} - \dfrac{y'^2}{9} = 1; \dfrac{y'^2}{9} + \dfrac{x'^2}{1} = 1$

Center (h,k) at $(-1,0)$

$a^2 = 9; a = 3; b^2 = 1; b = 1;$

$c^2 = 8; c = 2\sqrt{2}$

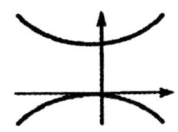

6. $\dfrac{(y-4)^2}{16} - \dfrac{(x+2)^2}{4} = 1;$

eq. (21-33), hyperbola

Center $(-2,4)$

$a = 4$

$b = 2$

7. $(x+3)^2 = -12(y-1)$; eq. (21-29), parabola

$x' = x + 3; y' = y - 1$

$x'^2 = -12y'$

Origin at O' at $(h,k) = (-3,1)$

$x'^2 = 4(3y')$; therefore $p = 3$

Vertex $(-3,1)$, focus $(-3,-2)$, directrix $y = 4$

8. $\dfrac{x^2}{0.16} + \dfrac{(y+1)^2}{0.25} = 1, \dfrac{(x-0)}{0.4^2} + \dfrac{(y-(-1))^2}{0.5^2} = 1,$
ellipse

Center $(0,-1)$, $a = 0.5, b = 0.4$

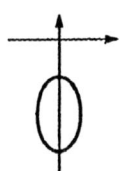

9. Parabola: $V(-1,3); p = 4$; parallel to x-axis;
vertex (h,k) at $(-1,3)$

$y'^2 = 4px'; (y-k)^2 = 4p(x-h)$

$(y-3)^2 = 4(4)[x-(-1)]; (y-3)^2 = 16(x+1)$

$y^2 - 6y + 9 = 16x + 16; y^2 - 6y - 16x + 9 - 16 = 0$

$y^2 - 6y - 16x - 7 = 0$

10. Parabola: $V(2,-1)$, directrix $y = 3$
Focus at $(2,-5); p = 4$

$(x')^2 = -4py'$; parallel to y-axis.

$(x-2)^2 = -16(y+1)$

or $x^2 - 4x + 16y + 20 = 0$

11. $F(12,0), V(6,0), p = 6$

$(y-k)^2 = 4p(x-h)^2$

$(y-0)^2 = 4(6)(x-6)$

$y^2 = 24(x-6)$

12. Parabola: $F(2,4)$, directrix $x = 6$
Therefore, vertex at $V(4,4); p = 2$

$(y')^2 = -4px'$; parallel to x-axis

$(y-4)^2 = -8(x-4)$

or $y^2 - 8y + 8x - 16 = 0$

13. Ellipse: center $(-2,2)$; foci $(-5,2),(1,2)$;
vertices $(-7,2),(3,2)$
(h,k) at $(-2,2)$; $c=3$; $c^2=9$; $a=5$; $a^2=25$;
$b^2=a^2-c^2=16$

$$\frac{x'^2}{a^2}+\frac{y'^2}{b^2}=1;\ \frac{[x-(-2)]^2}{25}+\frac{(y-2)^2}{16}=1$$

$$\frac{(x+2)^2}{25}+\frac{(y-2)^2}{16}=1;\ 16(x+2)^2+25(y-2)^2=400$$

$$16(x^2+4x+4)+25(y^2-4y+4)=400$$
$$16x^2+64x+64+25y^2-100y+100-400=0$$
$$16x^2+25y^2+64x-100y-236=0$$

14. Ellipse: center $(0,3)$, $F(12,3)$, major axis $=26$
$2a=26$; $a=13$; $c=12$;
$a^2=b^2+c^2$; $b^2=13^2-12^2=25$; $b=5$

Therefore, $(h,k)=(0,3)$; major axis parallel to
x-axis.

$$\frac{x^2}{a^2}+\frac{(y-k)^2}{b^2}=1;\ \frac{x^2}{169}+\frac{(y-3)^2}{25}=1$$

or $25x^2+169y^2-1014y-2704=0$

15. Ellipse: $V(-2,-3),V(-2,5)$, end minor axis $(0,1)$
$a=4$; $b=2$; therefore, $c=\sqrt{12}$
$(h-k)=(-2,1)$
Major axis parallel to y-axis.

$$\frac{(y-k)^2}{a^2}+\frac{(x-h)^2}{b^2}=1$$

$$\frac{(y-1)^2}{16}+\frac{(x+2)^2}{4}=1$$

or $4x^2+y^2+16x-2y+1=0$

16. Ellipse: $F(1,-2),F(1,10)$, minor axis $=5$
Therefore, $2c=12$; $c=6$

$b=\dfrac{5}{2}$; center $(1,4)$

$a^2=b^2+c^2$

$$a^2=(2.5)^2+36=\left(\frac{5}{2}\right)^2+36$$

$$a^2=\frac{25}{4}+36=\frac{169}{4}$$

Major axis parallel to y-axis.

$$\frac{(y-k)^2}{a^2}+\frac{(x-h)^2}{b^2}=1;\ (h,k)=(1,4)$$

$$\frac{(y-4)^2}{\dfrac{169}{4}}+\frac{(x-1)^2}{\dfrac{25}{4}}=1$$

$$\frac{4(y-4)^2}{169}+\frac{4(x-1)^2}{25}=1$$

$$676x^2+100y^2-1352x-800y-1949=0$$

17. Hyperbola: center $(h,k)=(-1,2)$; $F_1=(-1,4)$
and
$V_1=(-1,1)$
$c^2=a^2+b^2$ where $c=\sqrt{(-1+1)^2+(4-2)^2}=2$
is the distance between F_1 and (h,k).
Substituting known values:

$$\frac{(y-2)^2}{a^2}-\frac{(x+1)^2}{b^2}=1\quad (h,k)\text{ substituted}$$

$$\frac{(1-2)^2}{a^2}-\frac{(-1+1)^2}{b^2}=1\quad V_1\text{ substituted}$$

$$\frac{1}{a^2}-\frac{0}{b^2}=1;\ a^2=1$$

Since $c^2=a^2+b^2$; $2^2=1^2+b^2$; $b^2=3$

$$\frac{(y-2)^2}{1}-\frac{(x+1)^2}{3}=1;\ 3(y-2)^2-(x+1)^2=3$$
or
$$x^2-3y^2+2x+12y-8=0$$

18. Hyperbola: $F(2,1)$; $F(8,1)$
Conjugate axis $=6$
Therefore, $2c=6$; $c=3$
$2b=6$; $b=3$
$c=b$; center $(h,k)=(5,1)$
$a^2=18$; $a=3\sqrt{2}$
No hyperbola.
c must be $>b$.
$2a$ must be less than 6.
Therefore, no hyperbola.

19. Hyperbola: $V(2,1),V(-4,1),F(-6,1)$
Center: $(h,k)=(-1,1)$
$2a=6$; $a=3$; $c=5$
Therefore, $b^2=c^2-a^2=25-9=16$
Transverse axis parallel to x-axis.

Therefore, $\dfrac{(x-h)^2}{a^2}-\dfrac{(y-k)^2}{b^2}=1$

$$\frac{(x+1)^2}{9}-\frac{(y-1)^2}{16}=1$$

or $16x^2-9y^2+32x+18y-137=0$

20. Hyperbola; $C(1,-4),F(1,1)$
Transverse axis $=8$; $2a=8$; $a=4$
$(h,k)=(1,-4)$; $c=5$
$b^2=25-16=9$
Transverse axis parallel to y-axis.

Therefore, $\dfrac{(y-k)^2}{a^2}-\dfrac{(x-h)^2}{b^2}=1$

$$\frac{(y+4)^2}{16}-\frac{(x-1)^2}{9}=1$$

or $9y^2-16x^2+32x+72y-16=0$

21. $x^2 + 2x - 4y - 3 = 0;\ x^2 + 2x - 3 = 4y$
$x^2 + 2x = 4y + 3;\ x^2 + 2x + 1 = 4y + 3 + 1$
$(x + 1)^2 = 4y + 4;\ (x + 1)^2 = 4(y + 1)$
Parabola, $p = 1;\ x - h = x + 1;\ h = -1;$
$y - k = y + 1;\ k = -1$
Vertex is $(-1, -1)$

22. $y^2 - 2x - 2y - 9 = 0$
$y^2 - 2y - 2x = 9$
$y^2 - 2y + 1 = 2x + 10$
$(y - 1)^2 = 2(x + 5)$
$y'^2 = 2(x');$ parabola
$y - k = y - 1;\ k = 1$
$x - h = x + 5;\ h = -5$

$4p = 2;\ p = \dfrac{1}{2}$

Vertex: $(-5, 1)$

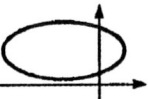

23. $4x^2 + 9y^2 + 24x = 0$
$4x^2 + 24x + 9y^2 = 0$

$x^2 + 6x + \dfrac{9}{4}y^2 = 0$

$x^2 + 6x + 9 + \dfrac{9}{4}y^2 = 9$

$(x + 3)^2 + \dfrac{9}{4}y^2 = 0$

$\dfrac{(x + 3)^2}{9} + \dfrac{y^2}{4} = 1;$ ellipse

$x - h = 3;\ h = -3;\ a = 3$
$y - k = y;\ k = 0;\ b = 2$
Center: $(-3, 0)$

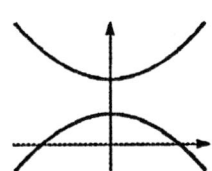

24. $2x^2 + 9y^2 + 8x - 72y + 134 = 0$
$2x^2 + 8x + 9y^2 - 72y = -134$
$2(x^2 + 4x) + 9(y^2 - 8y) = -134$

$\dfrac{(x^2 + 4x + 4)}{\dfrac{1}{2}} + \dfrac{(y^2 - 8y + 16)}{\dfrac{1}{9}} = -134 + 8 + 144$

$\dfrac{(x + 2)^2}{\dfrac{1}{2}} + \dfrac{(y - 4)^2}{\dfrac{1}{9}} = 18$

$\dfrac{(x + 2)^2}{9} + \dfrac{(y - 4)^2}{2} = 1;$ ellipse

Center: $(-2, 4)$

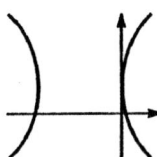

25. $9x^2 - y^2 + 8y - 7 = 0$
$9x^2 - (y^2 - 8y + 7 + 9) = -9$
$9x^2 - (y - 4)^2 = -9$

$\dfrac{9x^2}{-9} + \dfrac{(y - 4)^2}{-9} = 1$

$-x^2 + \dfrac{(y - 4)^2}{9} = 1$

$\dfrac{(y - 4)^2}{9} - \dfrac{x^2}{1} = 1$

Hyperbola, (h, k) is $(0, 4);\ a = 3;\ b = 1$

26. $5x^2 - 4y^2 + 20x + 8y = 4$
$5x^2 + 20x - 4y^2 + 8y = 4$
$5(x^2 + 4x) - 4(y^2 - 2y) = 4$

$\dfrac{(x^2 + 4x + 4)}{\dfrac{1}{5}} - \dfrac{(y^2 - 2y + 1)}{\dfrac{1}{4}} = 4 + 20 - 4$

$\dfrac{(x + 2)^2}{\dfrac{1}{5}} - \dfrac{(y - 1)^2}{\dfrac{1}{4}} = 20$

$\dfrac{(x^2 + 2)^2}{4} - \dfrac{(y - 1)^2}{5} = 1$

Hyperbola, center $(-2, 1)$

27. $2x^2 - 4x = 9y - 2$

$$x^2 - 2x = \frac{9}{2}y - 1$$

$$(x^2 - 2x + 1) = \frac{9}{2}y - 1 + 1$$

$$(x - 1)^2 = \frac{9}{2}y$$

$$(x')^2 = \frac{9}{2}y'$$

A parabola, parallel to y-axis
$x - h = x - 1$; $h = 1$
$y - k = y$; $k = 0$

Vertex $(1,0)$; $4p = \frac{9}{2}$; $p = \frac{9}{8}$

28. $0.04x^2 + 0.16y^2 = 0.0y$; $0.04x^2 + 0.16y^2 - 0.01y = 0$

$$x^2 + 4y^2 - \frac{1}{4}y = 0$$

$$x^2 + 4\left(y^2 - \frac{1}{16}y\right) = 0$$

$$x^2 + 4\left(y^2 - \frac{1}{16}y + \left(\frac{1}{32}\right)^2\right) = \left(\frac{1}{32}\right)^2 = 4$$

$$x^2 + \frac{\left(y - \frac{1}{32}\right)^2}{\frac{1}{4}} = \frac{1}{256};$$

$$256x^2 + 256\frac{\left(y - \frac{1}{32}\right)^2}{\frac{1}{4}} = 1$$

$$\frac{x^2}{\frac{1}{256}} + \frac{\left(y - \frac{1}{32}\right)^2}{\frac{1}{1024}} = 1$$

$$a^2 = \frac{1}{256}; \ b^2 = \frac{1}{1024}$$

An ellipse. Center $\left(0, \frac{1}{32}\right)$ or $(0, 0.03)$

29. $4x^2 - y^2 + 32x + 10y + 35 = 0$
$4(x^2 + 8x) - (y^2 - 10y) = -35$
$4(x^2 + 8x + 16) - (y^2 - 10y + 25) = -35 + 64 - 25$

$$\frac{(x + 4)^2}{1^2} - \frac{(y - 5)^2}{2^2} = 1, \text{ hyperbola}$$

$C(-4, 5)$

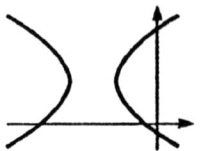

30. $2x^2 + 2y^2 - 24x + 16y + 95 = 0$
$2(x^2 - 12x + 36) + 2(y^2 + 8y + 16) = -95 + 72 + 32 = 9$

$$(x - 6)^2 + (y + 4)^2 = \frac{9}{2}; \text{ circle,}$$

center $(6, -4)$, $r = \dfrac{3}{\sqrt{2}}$

31. $9x^2 + 4y^2 - 12x + 16y + 16 = 0$
$9x^2 - 12x + 4y^2 + 16y + 16 = 0$

$$9\left(x^2 - \frac{4}{3}x + \frac{4}{9}\right) + 4(y^2 + 4y + 4) = -16 + 16 + 4$$

$$\frac{\left(x - \frac{2}{3}\right)^2}{\frac{4}{9}} + \frac{(y - (-2))^2}{1} = 1; \text{ ellipse,}$$

center $\left(\frac{2}{3}, -2\right)$

32. $5x^2 - 3y^2 - 40x + 95 = 0$
$$5(x^2 - 8x + 16) - 3y^2 = -95 + 80$$
$$= -15$$

$$\frac{y^2}{5} - \frac{(x-4)^2}{3} = 1, \text{ hyperbola,}$$

center $(4, 0)$

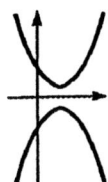

33. $7x^2 - y^2 - 14x - 16y - 64 = 0$
$$7x^2 - 14x - y^2 - 16y - 64 = 0$$
$$7(x^2 - 2x + 1) - (y^2 + 16y + 64) = 64 + 7 - 64$$
$$\frac{(x-1)^2}{1^2} - \frac{(y+8)^2}{\sqrt{7}^2} = 1$$

hyperbola, $C(1, -8)$

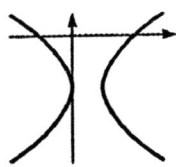

34. $5x^2 - 2y^2 + 12y + 18 = 0$
$$5x^2 - 2(y^2 - 6y + 9) = -18 - 18$$
$$2(y-3)^2 - 5x^2 = 36; \text{ hyperbola,}$$

center $(0, 3)$

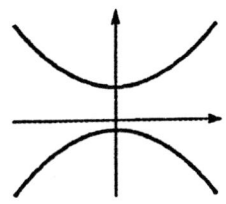

35. $9x^2 + 9y^2 - 6x - 24y + 14 = 0$

$$9\left(x^2 - \frac{2}{3}x + \frac{1}{9}\right) + 9\left(y^2 - \frac{8}{3}y + \frac{16}{9}\right)$$
$$= -14 + 1 + 16 = 3$$

$$\left(x - \frac{1}{3}\right)^2 + \left(y - \frac{4}{3}\right)^2 = \frac{1}{3}; \text{ circle,}$$

center $\left(\frac{1}{3}, \frac{4}{3}\right), r = \frac{1}{\sqrt{3}}$

36. $4y^2 - 15x - 12y + 29 = 0$

$$4\left(y^2 - 3y + \frac{9}{4}\right) = 15x - 29 + 9$$
$$= 15x - 20$$

$$4\left(y - \frac{3}{2}\right)^2 = 15\left(x - \frac{4}{3}\right)$$

$$\left(y - \frac{3}{2}\right)^2 = \frac{15}{4}\left(x - \frac{4}{3}\right); \text{ parabola,}$$

vertex $\left(\frac{4}{3}, \frac{3}{2}\right)$

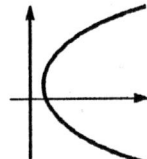

37. Hyperbola: asymptotes: $x - y = -1$ or $x + 1 = y$, and $x + y = -3$ or $y = -x - 3$; vertices $(3, -1)$ and $(-7, -1)$. The center is at the point of inter-action of the asymptotes. The equations for the asymptotes are solved simultaneously by adding, $2y = -2; y = -1; -1 = x + 1; x = -2$. Therefore, the coordinates of the center are $(-2, -1)$. Since the slopes are 1 and $-1, a = b$, where a is the distance from the center $(-2, -1)$ to the vertex $(3, -1); a = 5, b = 5$.

$$\frac{(x-h)^2}{a^2} - \frac{(y-k)^2}{b^2} = 1;$$

$$\frac{[x-(-2)]^2}{25} - \frac{[y-(-1)]^2}{25} = 1$$

$$\frac{(x+2)^2}{25} - \frac{(y+1)^2}{25} = 1$$

$$x^2 + 4x + 4 - (y^2 + 2y + 1) = 25;$$
$$x^2 + 4x + 4 - 2y - 1 = 25$$
$$x^2 - y^2 + 4x - 2y - 22 = 0$$

38. $x^2 + y^2 + 4x - 5 = 0$
$x^2 + 4x + 4 + y^2 = 5 + 4$
$(x+2)^2 + y^2 = 9$
Center $(-2, 0)$; $r = 3$
Ellipse: $c = 3$; $b = 3$
Therefore, $a = \sqrt{18} = 3\sqrt{2}$
Center $(h, k) = (-2, 0)$

$$\frac{(x+2)^2}{18} + \frac{y^2}{9} = 1$$

$$x^2 + 4x + 4 + 2y^2 = 18$$
$$x^2 + 2y^2 + 4x - 14 = 0$$

39. $y^2 = 4x$ (second parabola)
$4p = 4$; $p = 1$
Therefore, focus of second is at $(1, 0)$.
Vertex of second is at $(0, 0)$.
Therefore, focus of first is $(0, 0)$.
Vertex of first is $(1, 0)$; $p = -1$
$y'^2 = 4px'$
$(y-k)^2 = 4p(x-h)$
$(y-0)^2 = 4(-1)(x-1)$
$y^2 = -4x + 4$
$y^2 + 4x - 4 = 0$

40. $4y^2 - x^2 - 6x - 2y - 14 = 0;$ $4y^2 - 2y - x^2 - 6x = 14$

$$4\left(y^2 - \frac{1}{2}y\right) - (x^2 + 6x) = 14$$

$$4\left(y^2 - \frac{1}{2}y + \frac{1}{16}\right) - (x^2 + 6x + 9) = 14 + \frac{1}{4} - 9$$

$$\frac{\left(y - \frac{1}{4}\right)^2}{\frac{1}{4}} - (x+3)^2 = \frac{21}{4};$$

$$\frac{\left(y - \frac{1}{4}\right)^2}{\frac{21}{16}} - \frac{(x+3)^2}{\frac{21}{4}} = 1$$

Hyperbola

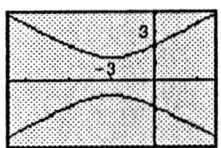

41. $(x-h)^2 = 4p(y-k)$
$(x-95)^2 = 4p(y-60)$
Solve for $4p$ using $(x, y) = (0, 0)$.
$(-95)^2 = 4p(-60)$

$$4p = \frac{95^2}{-60}$$

$$(x-95)^2 = \frac{95^2}{-60}(y-60)$$

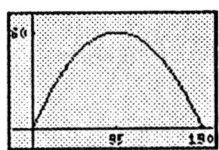

42. $|Z| = \sqrt{R^2 + (X_L - X_c)^2}$; X_c constant, R constant
$Z^2 = R^2 + (X_L - X_c)^2$
$(X_L - X_c)^2 = Z^2 - R^2$; $Z^2 - (X_L - X_c)^2 = R^2$

$$\frac{Z^2}{R^2} - \frac{(X_L - X_c)^2}{R^2} = 1$$

Hyperbola

$$\frac{(x-h)^2}{a^2} - \frac{(y-k)^2}{b^2} = 1$$

$h = 0$, $k = X_c$, center $(0, X_c)$
a and $b = R$

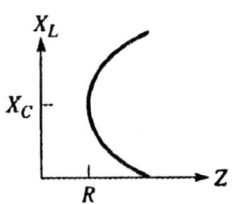

43. First ellipse:
$a = 4, b = 3$, therefore $c = \sqrt{7}$
Center $(0,0)$

$$\frac{y^2}{a^2} + \frac{x^2}{b^2} = 1; \; \frac{y^2}{16} + \frac{x^2}{9.0} = 1$$

Second ellipse:
$a = 4, b = 3$, therefore $c = \sqrt{7}$
Center $(7.0, 0.0)$

$$\frac{(x-h)^2}{a^2} + \frac{(y-k)^2}{b^2} = 1; \; \frac{(x-7)^2}{16} + \frac{y^2}{9.0} = 1$$

44.

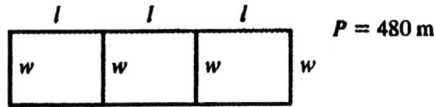

$P = 2l + 2w = 480; \; l + w = 240; \; l = 240 - w$
$A = lw; \; A = (240 - w)(w); \; A = 240w - w^2$ is a parabola, concave down.

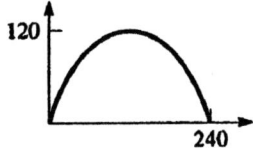

A is greatest at $w = 120$ m (position of axis of parabola).

2.8   The Second-Degree Equation

1. $x^2 + 2y^2 - 2 = 0$
$A \neq C$, they have the same sign, and $B = 0$; ellipse

2. $x^2 - y = 0; \; y = x^2$
$A \neq 0; \; B = 0; \; C = 0$; parabola

3. $2x^2 - y^2 - 1 = 0$
A and C have different signs, $B = 0$; hyperbola

4. $y(y + x^2) = 4$
$y^2 + yx^2 = 4$, not second degree.

5. $2x^2 + 2y^2 - 3y - 1 = 0$
$A = C; \; B = 0$; circle

6. $x^2 - 3x = y - 2y^3$, not second degree.

7. $2.2x^2 - x - y = 1.6$
$A \neq 0; \; C = 0; \; B = 0$; parabola

8. $2x^2 + 4y^2 - y - 2x = 4$
$A \neq C; \; B = 0$; ellipse

9. $x^2 = y^2 - 1; \; x^2 - y^2 + 1 = 0$
A and C have different signs;
$B = 0$; hyperbola

10. $32x^2 = 21y - 47y^2; \; 32x^2 + 47y^2 - 21y = 0$
$A \neq C; \; B = 0$; ellipse

11. $3.6x^2 = 1.1y - 3.6y^2; \; 3.6x^2 + 3.6y^2 - 1.1y = 0$
$A = C; \; B = 0$; circle

12. $y = 3 - 6x^2; \; 6x^2 + y - 3 = 0$
$A \neq 0; \; B = 0; \; C = 0$; parabola

13. $y(3 - 2x) = x(5 - 2y)$
$3y - 2xy = 5x - 2xy$

$$y = \frac{5}{3}x, \text{ line}$$

14. $x(13 - 5x) = 5y^2; \; 13x - 5x^2 = 5y^2;$
$5y^2 + 5x^2 - 13x = 0$
$A = C; \; B = 0$; circle

15. $2xy + x - 3y = 6$
$A = 0; \; B \neq 0; \; C = 0$; hyperbola

16. $(y + 1)^2 = x^2 + y^2 - 1; \; y^2 + 2y + 1 = x^2 + y^2 - 1$
$x^2 - 2y - 2 = 0$
$A \neq 0; \; B = 0; \; C = 0$; parabola

17. $2x(x - y) = y(3 - y - 2x);$
$2x^2 - 2xy = 3y - y^2 - 2xy$
$2x^2 - 2xy + 2xy + y^2 - 3y = 0;$
$2x^2 + y^2 - 3y = 0$
$A \neq 0$, same sign; $B = 0$; ellipse

18. $2x^2 = x(x - 1) + 4y^2; \; 2x^2 = x^2 - x + 4y^2$
$x^2 - 4y^2 + x = 0$
A and C are different signs, $B = 0$; hyperbola

19. $x(y + 3x) = x^2 + xy - y^2 + 1;$
$xy + 3x^2 = x^2 + xy - y^2 + 1$
$2x^2 + y^2 - 1 = 0$
$A \neq C; \; B = 0$; ellipse

20. $4x(x - 1) = 2x^2 - 2y^2 + 3;$
$4x^2 - 4x = 2x^2 - 2y^2 + 32x^2 + 2y^2 - 4x - 3 = 0$
$A = C; \; B = 0$; circle

21. $x^2 = 8(y - x - 2)$

$x^2 = 8y - 8x - 16$

$x^2 + 8x - 8y + 16 = 0$

$A \neq 0; B = 0;$

$C = 0;$ parabola

$x^2 + 8x - 8y + 16 = 0$

$x^2 + 8x + 16 = 8y$

$(x + 4)^2 = 4(2)y; p = 2$

Vertex $(-4, 0)$, focus $(-4, 2)$

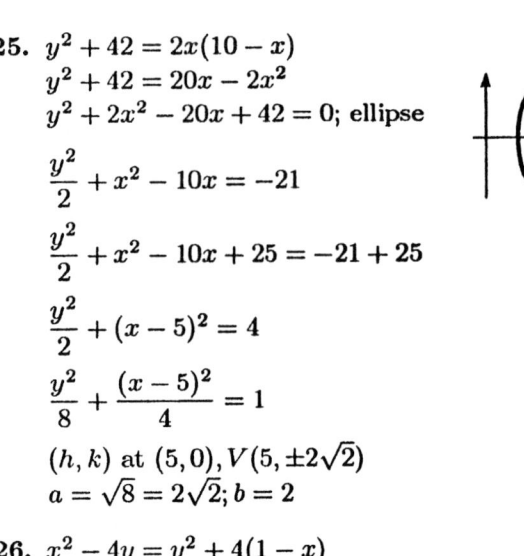

22. $x^2 = 6x - 4y^2 - 1$

$x^2 + 4y^2 - 6x + 1 = 0$

$A \neq C; B = 0;$ ellipse

$x^2 - 6x + 4y^2 = -1$

$x^2 - 6x + 9 + 4y^2 = -1 + 9$

$(x - 3)^2 + 4y^2 = 8$

$\dfrac{(x - 3)^2}{8} + \dfrac{y^2}{2} = 1$

$(h, k) = (3, 0)$

$a = 2\sqrt{2}, b = \sqrt{2}$, therefore $c = \sqrt{6}$

$V(3 \pm 2\sqrt{2}, 0)$

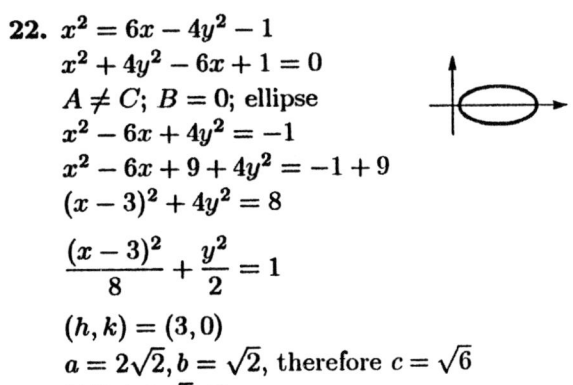

23. $y^2 = 2(x^2 - 2x - 2y)$

$y^2 = 2x^2 - 4x - 4y$

$y^2 - 2x^2 + 4y + 4x = 0$

A and C are different signs.

$B = 0;$ hyperbola

$y^2 + 4y - 2x^2 + 4x = 0$

$y^2 + 4y + 4 - 2(x^2 - 2x + 1) = 4 - 2$

$(y + 2)^2 - 2(x - 1)^2 = 2$

$\dfrac{(y + 2)^2}{2} - (x - 1)^2 = 1$

$(h, k) = (1, -2)$

$a^2 = 2; b^2 = 1; c^2 = 3$

$C(1, -2), V(1, -2 \pm \sqrt{2})$

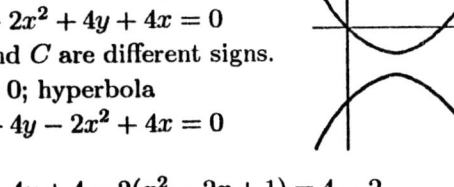

24. $4x^2 + 4 = 9 - 8x - 4y^2$

$4x^2 + 4y^2 + 8x - 5 = 0$

$A = C; B = 0;$ circle

$4x^2 + 8x + 4y^2 = 5$

$x^2 + 2x + y^2 = \dfrac{5}{4}$

$(x + 1)^2 + y^2 = \dfrac{5}{4} + 1$

$(x + 1)^2 + y^2 = \dfrac{9}{4}$

$(h, k) = (-1, 0); r = \dfrac{3}{2}$

25. $y^2 + 42 = 2x(10 - x)$

$y^2 + 42 = 20x - 2x^2$

$y^2 + 2x^2 - 20x + 42 = 0;$ ellipse

$\dfrac{y^2}{2} + x^2 - 10x = -21$

$\dfrac{y^2}{2} + x^2 - 10x + 25 = -21 + 25$

$\dfrac{y^2}{2} + (x - 5)^2 = 4$

$\dfrac{y^2}{8} + \dfrac{(x - 5)^2}{4} = 1$

(h, k) at $(5, 0), V(5, \pm 2\sqrt{2})$

$a = \sqrt{8} = 2\sqrt{2}; b = 2$

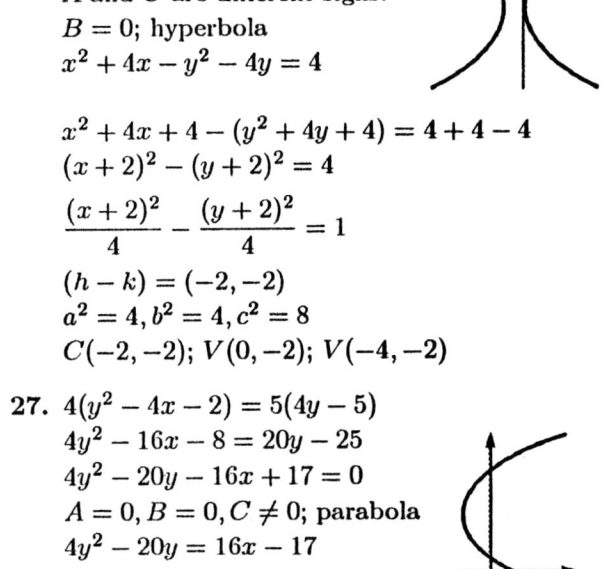

26. $x^2 - 4y = y^2 + 4(1 - x)$

$x^2 - 4y - y^2 - 4 + 4x = 0$

$x^2 - y^2 + 4x - 4y - 4 = 0$

A and C are different signs.

$B = 0;$ hyperbola

$x^2 + 4x - y^2 - 4y = 4$

$x^2 + 4x + 4 - (y^2 + 4y + 4) = 4 + 4 - 4$

$(x + 2)^2 - (y + 2)^2 = 4$

$\dfrac{(x + 2)^2}{4} - \dfrac{(y + 2)^2}{4} = 1$

$(h - k) = (-2, -2)$

$a^2 = 4, b^2 = 4, c^2 = 8$

$C(-2, -2); V(0, -2); V(-4, -2)$

27. $4(y^2 - 4x - 2) = 5(4y - 5)$

$4y^2 - 16x - 8 = 20y - 25$

$4y^2 - 20y - 16x + 17 = 0$

$A = 0, B = 0, C \neq 0;$ parabola

$4y^2 - 20y = 16x - 17$

$y^2 - 5y = 4x - \dfrac{17}{4}$

$\left(y^2 - 5y + \dfrac{25}{4} \right) = 4x - \dfrac{17}{4} + \dfrac{25}{4}$

$\left(y - \dfrac{5}{2} \right)^2 = 4x + 2$

$\left(y - \dfrac{5}{2} \right)^2 = 4\left(x + \dfrac{1}{2} \right)$

$(y')^2 = 4p(x')$

$y - k = y - \dfrac{5}{2}; k = \dfrac{5}{2}$

$x - h = x + \dfrac{1}{2}; h = -\dfrac{1}{2}$

$4p = 4; p = 1$

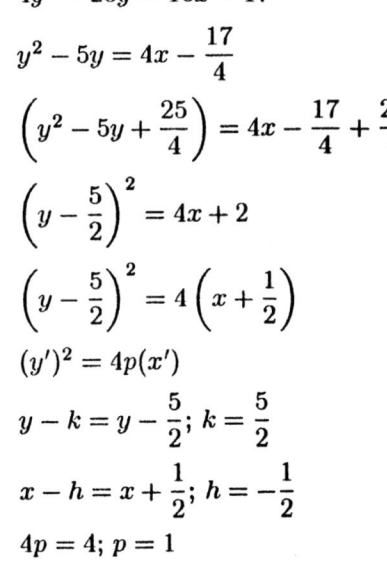

$$V\left(-\frac{1}{2}, \frac{5}{2}\right), F\left(\frac{1}{2}, \frac{5}{2}\right)$$

Directrix $x = -\frac{3}{2}$

28. $2(2x^2 - y) = 8 - y^2$

$4x^2 - 2y - 8 + y^2 = 0$

$4x^2 + y^2 - 2y - 8 = 0$

$A \neq C;\ B = 0;$ ellipse

$4x^2 + y^2 - 2y + 1 = 8 + 1$

$4x^2 + (y-1)^2 = 9$

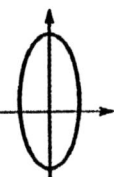

$$\frac{x^2}{\frac{9}{4}} + \frac{(y-1)^2}{9} = 1$$

$$\frac{(y-1)^2}{9} + \frac{x^2}{\frac{9}{4}} = 1$$

$(h, k) = (0, 1)$

$a^2 = 9, b^2 = \frac{9}{4}, c^2 = \frac{27}{4}$

$C(0, 1);\ V(0, 4), V(0, -2)$

29. $x^2 + 2y^2 - 4x + 12y + 14 = 0;$ solve for y

$x^2 + 0xy + 2y^2 - 4x + 12y + 14 = 0$

$A \neq C$, they have the same sign, and $B = 0$; ellipse

Solve for y.

$2y^2 + 12 + (x^2 - 4x + 14) = 0$

$$y = \frac{-12 \pm \sqrt{12^2 - 4(2)(x^2 - 4x + 14)}}{2(2)}$$

$$= \frac{-12 \pm \sqrt{-8x^2 + 32x + 32}}{4}$$

$y_1 = -3 + 0.5\sqrt{-2x^2 + 8x + 8}$

$y_2 = -3 - 0.5\sqrt{-2x^2 + 8x + 8}$

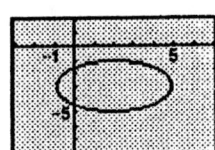

30. $4y^2 - x^2 + 40y - 4x + 60 = 0$

A and C have different signs; $B = 0$; hyperbola;
solve for y.

$4y^2 + 40y + (-x^2 - 4x + 60) = 0$

$$y = \frac{-40 \pm \sqrt{40^2 - 4(4)(-x^2 - 4x + 60)}}{2(4)}$$

$$= \frac{-40 \pm \sqrt{16x^2 + 64x + 640}}{8}$$

$y_1 = -5 + 0.5\sqrt{x^2 + 4x + 10}$

$y_2 = -5 - 0.5\sqrt{x^2 + 4x + 10}$

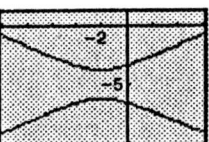

31. $x^2 + 6xy + 9y^2 - 2x + 14y - 10 = 0;$ solve for y.

$9y^2 + (14 + 6x)y + (x^2 - 2x - 10) = 0$

$$y = \frac{-(14 + 6x) \pm \sqrt{(14 + 6x)^2 - 4(9)(x^2 - 2x - 10)}}{2(9)}$$

$$= \frac{-(14 + 6x) \pm \sqrt{240x + 556}}{18}$$

$$y_1 = \frac{-7 - 3x + \sqrt{60x + 139}}{9}$$

$$y_2 = \frac{-7 - 3x - \sqrt{60x + 139}}{9}$$

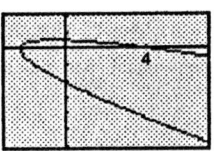

32. $x^2 - xy + y^2 - 6 = 0;$ solve for y

$y^2 - xy + (x^2 - 6) = 0$

$$y = \frac{x \pm \sqrt{(-x)^2 - 4(x^2 - 6)}}{2}$$

$$= \frac{x \pm \sqrt{24 - 3x^2}}{2}$$

$y_1 = 0.5(x + \sqrt{24 - 3x^2})$

$y_2 = 0.5(x - \sqrt{24 - 3x^2})$

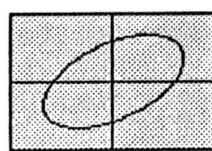

33. (a) If $k = 1$, $x^2 + ky^2 = a^2$; $x^2 + (1)y^2 = a^2$
$x^2 + y^2 = a^2$(circle)

(b) If $k < 0$, $x^2 + ky^2 = a^2$; $x^2 - |k|y^2 = a^2$

$\dfrac{x^2}{a^2} - \dfrac{y^2}{a^2/|k|} = 1$ (hyperbola)

(c) If $k > 0$ ($k \neq 1$), $x^2 + ky^2 = a^2$

$\dfrac{x^2}{a^2} + \dfrac{y^2}{a^2/k} = 1$ (ellipse)

34. $\dfrac{x^2}{4-C} - \dfrac{y^2}{C} = 1$

(a) $C < 0$; $\dfrac{x^2}{4+C} - \dfrac{y^2}{-C} = 1$; $\dfrac{x^2}{4+C} + \dfrac{y^2}{C}$

Let $\dfrac{1}{4+C} = k_1$ and $\dfrac{1}{C} = k_2$

$k_1 x^2 + k_2 y^2 - 1 = 0$
Eq. (21-34): $k_1 \neq k_2$, $B = 0$; therefore ellipse

(b) $0 < C < 4$; $\dfrac{x^2}{4-C} - \dfrac{y^2}{C} = 1$

Therefore $4 - C > 0$; let $\dfrac{1}{4-C} = k_1 > 0$

$C > 0$; let $\dfrac{1}{C} = k_2 > 0$

$k_1 x^2 - k_2 y^2 = 1$
Eq. (21-34): k_1 and k_2 different signs, $B = 0$
hyperbola

35. $Ax^2 + Bxy + Cy^2 + Dx + Ey + F = 0$
$A = C \neq 0$; $B = D = E = F = 0$
Let $A = C = k$
$kx^2 + ky^2 = 0$; $y^2 = -x^2$; $y = \sqrt{-x^2}$
Valid in real number system for $x = 0$ only. For
$x < 0$ and $0 < x$ we have imaginary number.
Solution is $(0,0)$ origin.

36. $\dfrac{x^2}{4-C} - \dfrac{y^2}{C} = 1$; $C > 4$; $4 - C < 0$

$\dfrac{y^2}{C} = \dfrac{x^2}{4-C} - 1$; $y^2 = C\left(\dfrac{x^2}{4-C} - 1\right)$

$y = \sqrt{C\left(\dfrac{x^2}{4-C} - 1\right)}$; $\dfrac{x^2}{4-C} < 0$ for all $x, C > 4$

Therefore, $C\left(\dfrac{x^2}{4-C} - 1\right) < 0$ for all x.

No real solution (root of negative numbers).

37. $x^2 + y^2 = (x+3)^2$
$x^2 + y^2 = x^2 + 6x + 9$

$y^2 = 6x + 9 = 6\left(x + \dfrac{3}{2}\right)$

Therefore a parabola.

38. $A_T = \pi(r-2)^2 + \pi r^2$; $A_T = \pi(r^2 - 2r + 4) + \pi r^2$
$A_T = \pi r^2 - 2\pi r + 4\pi + \pi r^2$; $A_T = 2\pi r^2 - 2\pi r + 4\pi$
$A_T - 2\pi r^2 + 2\pi r - 4\pi = 0$
Coefficient of $A_T^2 = 0$; coefficient of $r^2 \neq 0$; no Ar
term
Therefore a parabola.

39. (a) Beam is perpendicular to floor. We have a
circle.

(b) Beam is not perpendicular to floor. We have
an ellipse.

* See conic section diagrams, Fig. 2-90, p. 63 of
text.

40. Shape of curve on lake is that of a hyperbola.

* See conic section diagrams, Fig. 2-90, p. 63 of
text.

Chapter 2 Review Exercises

1. Given: straight line; (x_1, y_1) is $(1, -7)$; $m = 4$
$y - y_1 = m(x - x_1)$; $y - (-7) = 4(x - 1)$
$y + 7 = 4x - 4$; $y = 4x - 4 - 7$
$y = 4x - 11$ or $4x - y - 11 = 0$

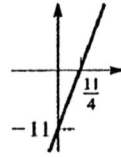

2. Given: straight line; $(x_1, y_1) = (-1, 5)$ and
$(x_2, y_2) = (-2, -3)$

$m = \dfrac{y_2 - y_1}{x_2 - x_1} = \dfrac{-3 - 5}{-2 - (-1)}$

$= 8$

$y - y_1 = m(x - x_1)$

$$y - 5 = 8(x - (-1))$$
$$= 8(x + 1)$$
$$= 8x + 8$$
$$8x - y + 13 = 0$$

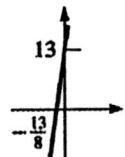

3. Given: straight line; perpendicular to
$3x - 2y + 8 = 0$ with y-intercept of $(0, -1)$

$$3x - 2y + 8 = 0$$
$$2y = 3x + 8$$
$$y = \frac{3}{2}x + 4,$$
$$m = \frac{3}{2} \text{ from which } m = -\frac{2}{3}$$

for given line.

$$y = mx + b = -\frac{2}{3}x + (-1)$$
$$= -\frac{2}{3}x - 1$$
$$3y = -2x - 3$$
$$2x + 3y + 3 = 0$$

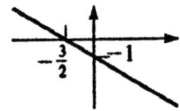

4. Given: straight line; parallel to
$2x - 5y + 1 = 0$ with x-intercept $(2, 0)$.

$$2x - 5y + 1 = 0$$
$$5y = 2x + 1$$
$$y = \frac{2}{5}x + \frac{1}{5} \text{ from which}$$

$m = \frac{2}{5}$ for the given line.

$$y - y_1 = m(x - x_1)$$
$$y - 0 = \frac{2}{5}(x - 2)$$
$$5y = 2x - 4$$
$$2x - 5y - 4 = 0$$

5. Given: circle; (h, k) is $(1, -2)$; (x, y) is $(4, -3)$
$$r = \sqrt{(1 - 4)^2 + [-2 - (-3)]^2}$$
$$= \sqrt{(-3)^2 + (-2 + 3)^2}$$
$$= \sqrt{9 + 1} = \sqrt{10}$$

$$(x - h)^2 + (y - k)^2 = r^2;$$
$$(x - 1)^2 + (y + 2)^2 = \sqrt{10}^2$$
$$x^2 - 2x + 1 + y^2 + 4y + 4 = 10$$
$$x^2 + y^2 - 2x + 4y + 1 + 4 - 10 = 0$$
$$x^2 + y^2 - 2x + 4y + 1 + 4 - 10 = 0;$$
$$x^2 + y^2 - 2x + 4y - 5 = 0$$

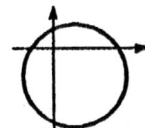

6. Given: circle; tangent to line $x = 3$, center $(5, 1)$
$$(x - h)^2 + (y - k)^2 = r^2 \text{ and letting } r = 2,$$
$$(x - 5)^2 + (y - 1)^2 = 2^2$$
$$x^2 - 10x + 25 + y^2 - 2y + 1 = 4$$
$$x^2 + y^2 - 10x - 2y + 22 = 0$$

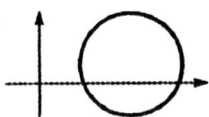

7. Given: parabola; focus $(3, 0)$, vertex $(0, 0)$

$$(y - k)^2 = 4p(x - h)$$
$$(y - 0)^2 = 4(3)(x - 0)$$
$$y^2 = 12x$$

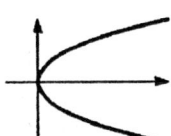

8. Given: parabola; directrix, $y = -5$, vertex $(0, 0)$

$$(x - h)^2 = 4p(y - k)$$
$$(x - 0)^2 = 4(5)(y - 0)$$
$$x^2 = 20y$$

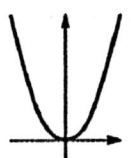

9. Given: ellipse; vertex $(10,0)$; focus $(8,0)$;
(h,k) is $(0,0)$

$$\frac{(x-h)^2}{a^2} + \frac{(y-k)^2}{b^2} = 1$$

$$\frac{x^2}{100} + \frac{y^2}{36} = 1$$

$$9x^2 + 25y^2 = 900$$

$a = 10 \qquad\qquad c = 8$
$b^2 = a^2 - c^2$
$b^2 = 100 - 64$
$b^2 = 36$

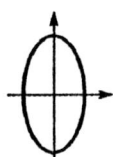

10. Given: ellipse; center $(0,0)$, passes through $(0,3)$ and
$(2,1)$.

$$\frac{(x-h)^2}{a^2} + \frac{(y-k)^2}{b^2} = 1$$

$$\frac{(x-0)^2}{a^2} + \frac{(y-0)^2}{3^2} = 1$$

$$\frac{x^2}{a^2} + \frac{y^2}{9} = 1, \text{ letting } (x,y) = (2,1)$$

$$\frac{2^2}{a^2} + \frac{1^2}{9} = 1$$

$$\frac{1}{a^2} = \frac{2}{9}$$

$$\frac{2x^2}{9} + \frac{y^2}{9} = 1$$

$$2x^2 + y^2 = 9$$

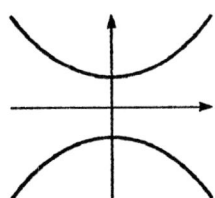

11. Given: hyperbola, $V(0,13), C(0,0)$, conj. axis 24.

$$\frac{(y-h)^2}{a^2} - \frac{(x-k)^2}{b^2} = 1$$

$$\frac{(y-0)^2}{13^2} - \frac{(x-0)^2}{12^2} = 1$$

$$144y^2 - 169x^2 = 24{,}336$$

12. Given: hyperbola; $F(0,10), F(0,-10), V(0,8)$

$$\frac{(y-h)^2}{a^2} - \frac{(x-k)^2}{b^2} = 1$$

$$\frac{(y-0)^2}{8^2} - \frac{x^2}{b^2} = 1$$

$$c^2 = a^2 + b^2$$

$$10^2 = 8^2 + b^2$$

$$36 = b^2$$

$$\frac{y^2}{64} - \frac{x^2}{36} = 1$$

$$36y^2 - 64x^2 = 2304$$

$$9y^2 - 16x^2 = 576$$

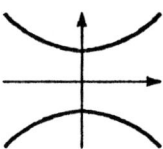

13. Given: $x^2 + y^2 + 6x - 7 = 0$
$(x^2 + 6x) + (y^2) = 7; (x^2 + 6x + 9) + y^2 = 7 + 9$
$(x+3)^2 + (y+0)^2 = 16$
$[x - (-3)]^2 + (y-0)^2 = 4^2$
center $(h,k) = (-3,0)$; radius $r = 4$

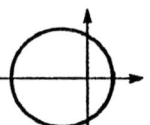

14. Given: $x^2 + y^2 - 4x + 2y - 20 = 0$
$$x^2 - 4x + y^2 + 2y = 20$$
$$x^2 - 4x + 4 + y^2 + 2y + 1 = 25$$
$$(x-2)^2 + (y+1)^2 = 5^2$$

center $(h,k) = (2,-1)$; radius $r = 5$

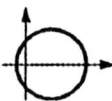

15. Given: $x^2 = -20y$
$$x^2 = 4(-5)y, \ p = -5$$

focus $(0,-5)$. directrix $y = 5$

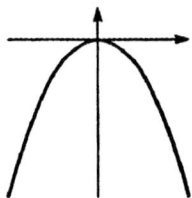

16. Given: $y^2 = 24x = 4(6)x, p = 6$

focus $(6, 0)$, directrix, $x = -6$

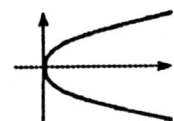

17. Given: $16x^2 + y^2 = 16$

$\dfrac{16x^2}{16} + \dfrac{y^2}{16} = 1; \dfrac{x^2}{1^2} + \dfrac{y^2}{4^2} = 1; \dfrac{y^2}{4^2} + \dfrac{x^2}{1^2} = 1$

$a = 4, b = 1, c = \sqrt{16 - 1} = \sqrt{15}$

vertices $(0, a), (0, -a)$ or $(0, 4), (0, -4)$

foci $(0, c), (0, -c)$ or $(0, \sqrt{15}), (0, -\sqrt{15})$

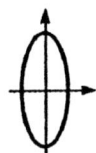

18. Given: $2y^2 - 9x^2 = 18$

$\dfrac{y^2}{9} - \dfrac{x^2}{2} = 1$

$\dfrac{y^2}{3^2} - \dfrac{x^2}{\sqrt{2}^2} = 1$

$c^2 = a^2 + b^2 = 9 + 2$

$\quad = 11$

$c = \sqrt{11}$

vertices: $(0, 3), (0, -3)$.

foci: $(0, \sqrt{11}), (0, -\sqrt{11})$

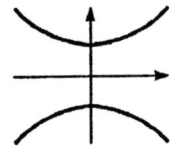

19. Given: $\qquad 2x^2 - 5y^2 = 0.25$

$\dfrac{x^2}{(\frac{1}{2\sqrt{2}})^2} - \dfrac{y^2}{(\frac{1}{2\sqrt{5}})^2} = 1$

$c^2 = a^2 + b^2 = \dfrac{1}{8} + \dfrac{1}{20}$

$\quad = \dfrac{7}{40}$

$c = \dfrac{\sqrt{70}}{20}$

vertices: $\left(\dfrac{1}{2\sqrt{2}}, 0\right), \left(\dfrac{-1}{2\sqrt{2}}, 0\right)$

foci: $\left(\dfrac{\sqrt{70}}{20}, 0\right), \left(\dfrac{-\sqrt{70}}{20}, 0\right)$

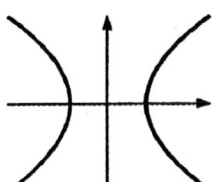

20. Given: $\qquad 2x^2 + 25y^2 = 800$

$\dfrac{x^2}{400} + \dfrac{y^2}{32} = 1$

$\dfrac{x^2}{20^2} + \dfrac{y^2}{(4\sqrt{2})^2} = 1$

$c^2 = a^2 + b^2 = 400 + 32 = 432$

$c = 4\sqrt{23}$

vertices: $(20, 0), (-20, 0)$

foci: $(4\sqrt{23}, 0), (-4\sqrt{23}, 0)$

21. Given: $x^2 - 8x - 4y - 16 = 0$

$x^2 - 8x = 4y + 16; \ x^2 - 8x + 16 = 4y + 16 + 16$

$(x - 4)^2 = 4y + 32; \ (x - 4)^2 = 4(y + 8)$

$(x - 4)^2 = 4(1)(y + 8); \ p = 1$

vertex (h, k) is $(4, -8)$; focus is $(4, -7)$

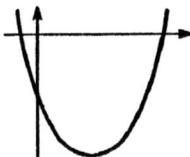

22. Given: $y^2 - 4x + 4y + 24 = 0$

$y^2 + 4y + 4 = 4x - 24 + 4$

$(y + 2)^2 = 4x - 20$

$(y + 2)^2 = 4(1)(x - 5), p = 1$

vertex: $(5, -2)$, directrix: $x = 4$

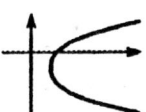

23. Given:

$$4x^2 + y^2 - 16x + 2y + 13 = 0$$
$$4x^2 - 16x + y^2 + 2y = -13$$
$$4(x^2 - 4x + 4) + y^2 + 2y + 1 = -13 + 16 + 1$$
$$4(x - 2)^2 + (y + 1)^2 = 4$$
$$\frac{(x - 2)^2}{1} + \frac{(y + 1)^2}{4} = 1$$

center: $(2, -1)$

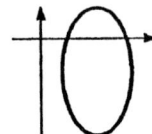

24.
$$x^2 - 2y^2 + 4x + 4y = 6$$
$$x^2 + 4x + 4 - 2(y^2 - 2y + 1) = -4$$
$$\frac{(y - 1)^2}{2} - \frac{(x + 2)^2}{4} = 1, \text{ center } (-2, 1)$$

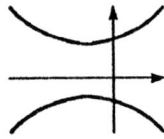

25. $x^2 + y^2 = 9$ circle; center $(0, 0)$; radius 3
$x^2 + y^2 = 3^2$

$4x^2 + y^2 = 16$ ellipse; centered at $(0, 0)$
                             $(\pm 2, 0)$ and $(0, \pm 4)$ vertices

$$\frac{x^2}{4} + \frac{y^2}{16} = 1$$
$$\frac{x^2}{2^2} + \frac{y^2}{4^2} = 1$$

four real solutions

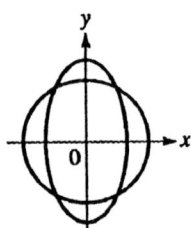

26. $y = |x|$ absolute value function
$y = x, x \geq 0$
$y = -x, x < 0$

$x^2 - y^2 = 1$ hyperbola with $y = \pm 1$ asymptotes.

There are no points of intersection and thus no real solutions.

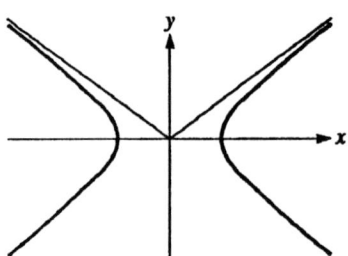

27. $x^2 + y^2 - 4y - 5 = 0$
$x^2 + y^2 - 4y + 4 = 5 + 4$
$x^2 + (y - 2)^2 = 9 = 3^2$

is a circle with center $(0, 2)$ and radius 3

$$y^2 - 4x^2 - 4 = 0$$
$$y^2 - 4x^2 = 4$$
$$\frac{y^2}{4} - \frac{x^2}{1} = 1$$

is an hyperbola with center $(0, 0)$ and asymptotes $y = \pm 2x$

There are 2 points of intersection and thus 2 real solutions.

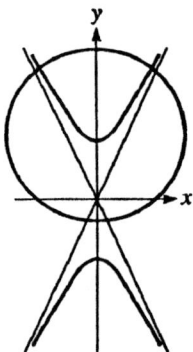

28. $x^2 - 4y^2 + 2x - 3 = 0$

$$x^2 + 2x + 1 - 4y^2 = 3 + 1$$
$$(x + 1)^2 - 4y^2 = 4$$
$$\frac{(x + 1)^2}{4} - \frac{y^2}{1} = 1,$$

hyperbola with center $(-1, 0)$ and asymptotes

$$y = \pm \frac{1}{2}(x + 1)$$

$$y^2 - 4x - 4 = 0$$
$$y^2 = 4x + 4$$
$$y^2 = 4(x + 1),$$

parabola with vertex $(-1, 0)$

From the graphs there are 2 intersection points and thus 2 real solutions.

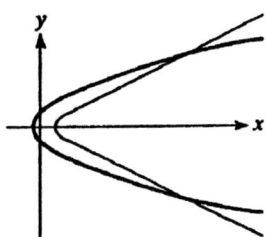

29. $x^2 - 4y^2 + 4x + 24y - 48 = 0$. Solve for y by completing the square.

$$y^2 - 6y = 0.25x^2 + x - 12$$
$$y^2 - 6y + 9 = 0.25x^2 + x - 3$$
$$(y - 3)^2 = 0.25x^2 + x - 3$$
$$y = \pm\sqrt{0.25x^2 + x - 3} + 3$$

 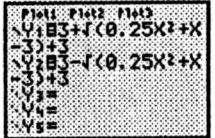

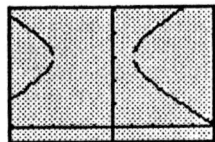

30. $x^2 + 2xy + y^2 - 3x + 8y = 0$. Solve for y with quadratic formula.

$$y^2 + (2x + 8)y + (x^2 - 3x) = 0$$

$$y = \frac{-(2x + 8) \pm \sqrt{(2x + 8)^2 - 4(1)(x^2 + 3x)}}{2(1)}$$

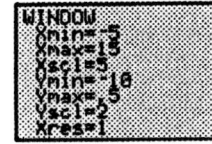 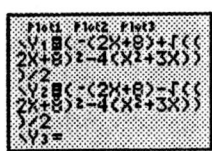

31. $y^2 - 4y + 6x - 8 = 0$. Solve for y with quadratic formula.

$$y = \frac{-(-4) \pm \sqrt{(-4)^2 - 4(6x - 8)}}{2}$$

$$= \frac{4 \pm \sqrt{16 - 4(6x - 8)}}{2}$$

$$y = 2 \pm \sqrt{12 - 6x}$$

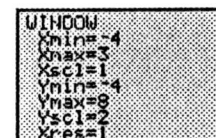 

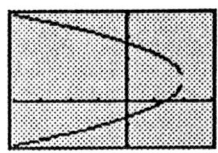

32. $9x^2 + 4y^2 - 72x - 8y + 144 = 0$. Solve for y with quadratic formula.

$$4y^2 - 8y + 9x^2 - 72x + 144 = 0$$

$$y = \frac{-(-8) \pm \sqrt{(-8)^2 - 4(4)(9x^2 - 72x + 144)}}{2(4)}$$

$$y = \frac{8 \pm \sqrt{-144x^2 + 1152x - 2240}}{8}$$

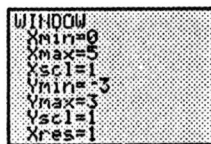 

33. (a) The slope of the line between $(-3, 11)$ and $(2, -1)$ is $-\frac{12}{5}$. The slope of the line between $(14, 4)$ and $(2, -1)$ is $\frac{5}{12}$. The product of the slopes is negative one which shows that the line segments form a right triangle.

(b) The distance between $(-3, 11)$ and $(2, -1)$ is 13. The distance between $(14, 4)$ and $(2, -1)$ is 13. The distance between $(-3, 11)$ and $(14, 4)$ is $\sqrt{338}$. Since $13^2 + 13^2 = 338$ the line segments form a right triangle by the Pythagorean Theorem.

34. Let $A = (-2, 5), B = (3, -1), C = (2, -4)$.

$$m_{AB} = \frac{-1 - 5}{3 - (-2)} = -\frac{6}{5}$$

The equation of the altitude from C to AB is

$$y + 4 = \frac{5}{6}(x - 2)$$

$$6y + 24 = 5x - 10$$

(1) $5x - 6y = 34$

$$m_{BC} = \frac{-4 - (-1)}{2 - 3} = 3$$

The equation of the altitude from A to BC is

$$y - 5 = -\frac{1}{3}(x + 2)$$

$$3y - 15 = -x - 2$$

(2) $x + 3y = 13$

$$m_{AC} = \frac{-4 - 5}{2(-2)} = -\frac{9}{4}$$

The equation of the altitude from B to AC is

$$y + 1 = \frac{4}{9}(x - 3)$$

$$9y + 9 = 4x - 12$$

(3) $4x - 9y = 21$

Solving (1) and (2) gives $\left(\frac{60}{7}, \frac{31}{21}\right)$ and since this is a solution of (3), the altitudes meet at $\left(\frac{60}{7}, \frac{31}{21}\right)$.

35. The line $y = x$ intersects the ellipse $7x^2 + 2y^2 = 18$ when $7x^2 + 2x^2 = 18$ or $x = \pm\sqrt{2}$. In first quadrant the intersection point is $(\sqrt{2}, \sqrt{2})$. The area of the square is

$$A = 4(\sqrt{2})^2 = 8$$

36. The equation of the ellipse is $\frac{x^2}{a^2} + \frac{y^2}{b^2} = 1$ or

$$x^2 = a^2\left(1 - \frac{y^2}{b^2}\right)$$

(slope of PA)(slope of PB)

$$= \left(\frac{y - 0}{x - a}\right)\left(\frac{y - 0}{x + a}\right) = \frac{y^2}{x^2 - a^2}$$

$$= \frac{y^2}{a^2\left(1 - \frac{y^2}{b^2}\right) - a^2}$$

$$= \frac{y^2}{a^2 - \frac{a^2 y^2}{b^2} - a^2}$$

$$= \frac{y^2}{-\frac{a^2 y^2}{b^2}}$$

$$= -\frac{b^2}{a^2}$$

37. Given: focus $(3, 1)$; directrix $y = -3$; vertex $(3, -1)$ is (h, k); $p = 2$
By definition, $d_1 = d_2$ where d_1 is from (x, y) to $(x, -3)$, a point on the directrix, and d_2 is from (x, y) to $(3, 1)$, the focus.

$d_1 = \sqrt{(x - x)^2 + [y - (-3)]^2}$;
$d_2 = \sqrt{(x - 3)^2 + (y - 1)^2}$
$\sqrt{0^2 + (y + 3)^2} = \sqrt{(x - 3)^2 + (y - 1)^2}$;
$0 + (y + 3)^2 = (x - 3)^2 + (y - 1)^2$
$y^2 + 6y + 9 = x^2 - 6x + 9 + y^2 - 2y + 1$
$0 = x^2 + y^2 - y^2 - 6x - 2y - 6y + 9 + 1 - 9$;
$0 = x^2 - 6x - 8y + 1$
By translation, $(x - h)^2 = 4(p)(y - k)$

$(x - 3)^2 = 4(2)[y - (-1)]$; $(x - 3)^2 = 8(y + 1)$
$x^2 - 6x + 9 = 8y + 8$; $x^2 - 6x - 8y + 9 - 8 = 0$
$x^2 - 6x - 8y + 1 = 0$

38. Rewrite $x^2 - ky^2 = 1$ as $\frac{x^2}{1} + \frac{y^2}{-\frac{1}{k}} = 1$.

For an ellipse $-\frac{1}{k} > 0$ or $\frac{1}{k} < 0$ which implies $k < 0$. (1)

For the vertices to be on y-axis $-\frac{1}{k} > 1$ or

$\frac{1}{k} < -1$ and using (1) this may be written as

$1 > -k \iff -1 < k$. (2)
Combining (1) and (2) gives $-1 < k < 0$

39. $R_T = R + 2.5$

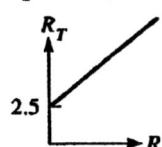

40. $a = \dfrac{v - v_o}{t - t_0} = 20$

$$\dfrac{v - 5}{t - 0} = 20$$

$$v = 20t + 5$$

41. $2500x + 1500y = 37{,}500$

$$1500y = -2500x + 37{,}500$$

$$y = -\dfrac{5}{3}x + 25$$

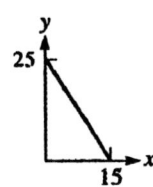

42. $v(t) = 100 - 20{,}000t$

$t(\text{h})$	$v(\text{mi/h})$
0	100
0.001	80
0.002	60
0.004	20
0.005	0

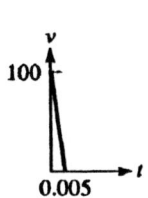

43. $y = 2.010(50.00)(T - 100)$

$y = 100.5T - 10{,}050$

$T(^\circ\text{C})$	$y(\text{kJ})$
100	0
200	10,500
500	40,200
1000	90,450

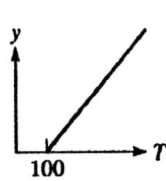

44. $(0, 27)$ and $(2500, 12)$ are points on the graph of T vs. h.

$$m = \dfrac{12 - 27}{2500 - 0} = -0.006$$

The equation is

$$T - 12 = -0.006(h - 2500)$$

$$T - 12 = -0.006h + 15$$

$$T = -0.006h + 27$$

45. $A = \pi r^2 = \pi(490(\tan 7^\circ))^2$

$A = 11{,}000 \text{ ft}^2$

46. $c = 2\pi r = 191$

$$= \dfrac{191}{2\pi}$$

center $\left(0, 2.0 + \dfrac{191}{2\pi}\right)$

$$(x - 0)^2 + \left(y - \left(2.0 + \dfrac{191}{2\pi}\right)\right)^2 = \left(\dfrac{191}{2\pi}\right)^2$$

$$x^2 + y^2 - 2\left(2.0 + \dfrac{191}{2\pi}\right)y + \left(2.0 + \dfrac{191}{2\pi}\right)^2 = \left(\dfrac{191}{2\pi}\right)^2$$

$$x^2 + y^2 - 64.8y + 126 = 0$$

47. The equation of the parabola is $y^2 = 4px$ and $(50, 40)$ is on the graph, $40^2 = 4p(50)$ gives $4p = 32$ from which $y^2 = 32x$.

48. $Q = \dfrac{1}{R}\sqrt{\dfrac{L}{C}} = \dfrac{1}{1000}\sqrt{\dfrac{L}{4.00 \times 10^{-6}}} = \dfrac{1}{2}\sqrt{L}$

$$Q(L) = \dfrac{1}{2.00}\sqrt{L}$$

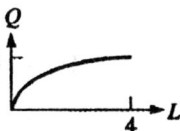

49.

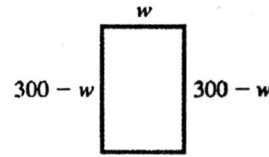
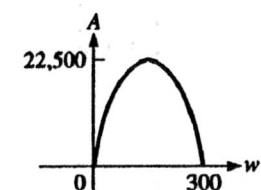

$$A = w(300 - w) = 300w - w^2$$

$$-A = w^2 - 300w$$

$$-A + 150^2 = w^2 - 300w + 150^2 = (w - 150)^2$$

$$-1(A - 150^2) = (w - 150)^2$$

$$(w - 150)^2 = -1(A - 150^2)$$

Parabola with $(h, k) = (150, 150^2)$

$$= (150, 22{,}500)$$

50. $\dfrac{H_T}{H_o} = 1 - \left(\dfrac{T}{T_o}\right)^2$

$\dfrac{T}{T_o}$	$\dfrac{H_T}{H_o}$
0	1
0.25	0.9375
0.50	0.75
0.75	0.4375
1	0

51. $P = 12.0i - 0.500i^2$

i(A)	P(W)
0.00	0.00
6.00	54.0
12.0	72.0
18.0	54.0
24.0	0.00

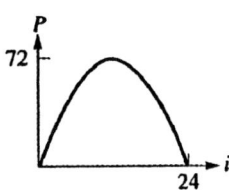

52. $A = \pi ab = \pi(94)(78)$
$\qquad = 2.3 \times 10^4 \text{ m}^2$

53. $a = \dfrac{120}{2} = 60; \; c = 60 - 15 = 45;$

$\quad b = \sqrt{a^2 - b^2} = \sqrt{60^2 - 45^2} = \sqrt{1575}$
$\quad A = \pi ab = \pi(60)\sqrt{1575} = 7500 \text{ ft}^2$

54. $\dfrac{(d-10)^2}{10^2} + \dfrac{(f-0)^2}{1^2} = 1$

$\dfrac{(d-10)^2}{100} + \dfrac{f^2}{1} = 1, 0 \le f \le 1 \text{ and } 0 < d \le 10$

55. If the pins are on the x-axis and centered, the equation of the ellipse is

$$\frac{x^2}{5^2} + \frac{y^2}{3^2} = 1.$$

$a^2 = b^2 + c^2$
$c^2 = a^2 - b^2$
$c^2 = 25 - 9$
$c^2 = 16$
$c = 4$

The coordinates of the pin art $(\pm 4, 0)$, so the pins should be 8 cm apart. The length of the string, l, is

$$l = 2a + 2c$$
$$l = 2(5) + 2(4)$$
$$l = 18 \text{ cm}$$

56. The length of the major axis is $70+2(1080)+190 = 2420$, from which $a = 2420/2 = 1210$. Since $70+1080 = 1150$ is less than 1210, the distance from the center to focus is $1210-1150 = 60$. From

$$a^2 = b^2 + c^2$$
$$1210^2 = b^2 + 60^2$$
$$b^2 = 1210^2 - 60^2.$$

The equation of the ellipse is

$$\frac{x^2}{a^2} + \frac{y^2}{b^2} = 1$$
$$\frac{x^2}{1210^2} + \frac{y^2}{1210^2 - 60^2} = 1$$
$$\frac{x^2}{1.464 \times 10^6} + \frac{y^2}{1.461 \times 10^6} = 1$$

57.

$$\frac{x^2}{a^2} - \frac{y^2}{b^2} = 1$$
$$y = 40 \text{ when } x = 40$$
$$y = 100 \text{ when } x = 50$$

$$\begin{cases} \dfrac{40^2}{a^2} - \dfrac{40^2}{b^2} = 1 \text{ or } 40^2b^2 - 40^2a^2 = a^2b^2 \\ \dfrac{50^2}{a^2} - \dfrac{100^2}{b^2} = 1 \text{ or } 50^2b^2 - 100^2a^2 = a^2b^2 \end{cases}$$

M by 100^2 $\begin{cases} 40^2b^2 - 40^2a^2 = a^2b^2 \\ 50^2b^2 - 100^2a^2 = a^2b^2 \end{cases}$
M by -40^2

$\begin{cases} 100^240^2b^2 - 100^240^2a^2 = 100^2a^2b^2 \\ -40^250^2b^2 + 40^2100^2a^2 = -40^2a^2b^2 \end{cases}$

Add $\begin{cases} 16 \times 10^6b^2 - 1.6 \times 10^7a^2 = 10 \times 10^3a^2b^2 \\ 4 \times 10^6b^2 + 1.6 \times 10^7a^2 = -1.6 \times 10^3a^2b^2 \end{cases}$

$$12 \times 10^6b^2 = 8.4 \times 10^3a^2b^2$$
$$12 \times 10^6 = 8.4 \times 10^3a^2$$

$$a^2 = \frac{12 \times 10^6}{8.4 \times 10^3} = 1.42 \times 10^3$$

$$a = 37.8 \text{ ft}$$

58. Let (x, y) be the point where the earthquake occurred, $(0, p)$ be Palo Alto, $(0, -p)$ be Pasadena where $p = \frac{510}{2}$, then

$$\sqrt{(x-0)^2 + (y+p)^2} = 5.0t, t = \text{time to reach}$$

Caltech

$$\sqrt{(x-0)^2 + (y-p)^2} = 5.0(t + 36)$$
$$= 5.0t + 180$$

$$\sqrt{x^2 + (y-p)^2} - \sqrt{x^2 + (y+p)^2} = 180$$
$$\sqrt{x^2 + (y-p)^2} = 180 + \sqrt{x^2 + (y+p)^2}$$
$$x^2 + y^2 - 2py + p^2$$
$$= 180^2 + 360\sqrt{x^2 + (y+p)^2}$$
$$+ x^2 + y^2 + 2py + p^2$$
$$-4py - 180^2 = 360\sqrt{x^2 + (y+p)^2}$$
$$16p^2y^2 + 8(180^2)py + 180^4$$
$$= 360^2(x^2 + y^2 + 2py + p^2)$$
$$16p^2y^2 + 8(180^2)py + 180^4$$
$$= 360^2x^2 + 360^2y^2 + 2(360^2)py + 360^2p^2$$
$$(16p^2 - 360^2)y^2 - 360^2x^2 = 360^2p^2 - 180^4$$

$$\frac{y^2}{\frac{360^2p^2 - 180^4}{16p^2 - 360^2}} - \frac{x^2}{\frac{360^2p^2 - 180^4}{16p^2 - 360^2}} = 1, \text{ letting } p = \frac{510}{2}$$

$$\frac{y^2}{8100} - \frac{x^2}{57,000} = 1$$

(Palo Alto is at upper focus and Pasadena at lower focus.)

59. Let the coordinates of the top of the hyperbolic arch be $(0, -a)$ where $a > 0$. The coordinates of the top right corner of the building and the arch are $(9, -a - 3)$. The focus coordinates $(0, -a - 3)$. The equation of the arch is

$$\frac{y^2}{a^2} - \frac{x^2}{b^2} = 1 \text{ where}$$

$$b^2 = (-a - 3)^2 - a^2$$

$$\frac{y^2}{a^2} - \frac{x^2}{(-a-3)^2 - a^2} = 1$$

$$\frac{y^2}{a^2} - \frac{x^2}{a^2 + 6a + 9 - a^2} = 1$$

(1) $\qquad \dfrac{y^2}{a^2} - \dfrac{x^2}{6a + 9} = 1$ and letting

$$(x, y) = (9, -a - 3)$$

$$\frac{a^2 + 6a + 9}{a^2} - \frac{81}{6a + 9} = 1$$

$$(a^2 + 6a + 9)(6a + 9) - 81a^2 = a^2(6a + 9)$$
$$6a^3 + 9a^2 + 36a^2 + 54a + 54a + 81 - 81a^2$$
$$= 6a^3 + 9a^2$$

$-45a^2 + 108a + 81 = 0$ from which $a = 3$

(1) is now $\dfrac{y^2}{9} - \dfrac{x^2}{27} = 1$ or $3y^2 - x^2 = 27$.

60. Let PM be the distance from the pulley to the person, then

$$10 - h + PM = 60$$
$$PM = 50 + h \text{ and}$$
$$(50 + h)^2 = 6^2 + x^2.$$

Since $h \geq 0, x \geq \sqrt{2464}$.

$$50 + h = \sqrt{36 + x^2}$$
$$h = \sqrt{36 + x^2} - 50$$

61. Let $P(x, y)$ be the coordinates of the recorder in a coordinate system with origin at the target. Let rifle be at $(0, r)$, then

$$\sqrt{x^2 + (y - r)^2} = v_s(t_0 + t_1) \text{ where } t_0 \text{ is the time}$$
for bullet to reach target and t_1 is the time for sound to reach detector from the target.

$$\sqrt{x^2 + y^2} = v_s t_1$$

$$\sqrt{x^2 + (y - r)^2} - \sqrt{x^2 + y^2} = v_s t_0 = \text{constant}$$
$$= 2a$$

$$\sqrt{x^2 + (y - r)^2} = 2a + \sqrt{x^2 + y^2}$$
$$x^2 + y^2 - 2ry + r^2 = 4a^2 + 4a\sqrt{x^2 + y^2} + x^2 + y^2$$
$$-2ry + (r^2 - 4a^2) = 4a\sqrt{x^2 + y^2}$$
$$4r^2y^2 - 4r(r^2 - 4a^2)y + (r^2 - 4a^2) = 16a^2(x^2 + y^2)$$
$$(4r^2 - 16a^2)y^2 - 4r(r^2 - 4a^2)y - 16a^2x^2 = 4a^2 - r^2$$

which has the form $Ay^2 - By - dx^2 = e$

$$y^2 - \frac{B}{A}y + \frac{B^2}{4A^2} - \frac{d}{A}x^2 = \frac{e}{A} + \frac{B^2}{4A^2}$$

$$\left(y - \frac{B}{2A}\right)^2 - Dx^2 = F$$

$$\frac{\left(y - \frac{B}{2A}\right)^2}{F} - \frac{x^2}{\frac{F}{D}} = 1,$$

which is a hyperbola.

Summary: The distance from the rifle to P minus the distance from the target to P = constant which is related to the distance from the rifle to the target.

THE DERIVATIVE

3.1 Limits

1. $f(x) = 3x - 2$ is continuous for all real x since it is defined for all x, and any small change in x will produce only a small change in $f(x)$.

2. $f(x) = 9 - x^2$ is continuous for all x.

3. $f(x) = \dfrac{2}{x^2 - x} = \dfrac{2}{x(x-1)}$ is not continuous for $x = 0$

or for $x = 1$; division by zero.

4. $f(x) = \dfrac{1}{\sqrt{x}}$ is not defined for $x = 0$ since

$\sqrt{0} = 0$ and division by zero is undefined, and is not defined for $x < 0$ since square roots of negative numbers are not real. The function is continuous for all $x > 0$ since it is defined, and any small change in x will produce only a small change in $f(x)$.

5. $f(x) = \sqrt{\frac{x}{x-2}}$ is continuous for $x \leq 0$ and $x > 2$. The function is not defined for $0 < x \leq 2$. $0 < x < 2$ gives the square root of a negative and $x = 2$ gives division by zero.

6. $f(x) = \dfrac{\sqrt{x+2}}{x}$ is continuous for $-2 \leq x < 0$ and for $x > 0$. Not continuous for $x = 0$, division by zero.

7. Continuous for all x.

8. Continuous for all x.

9. The graph is not continuous at $x = 1$ since $f(0, 9)$ is a value between -1 and -2, and $f(1.1)$ is a value greater than $+2$. This amount of change in y for a 0.2 change in x is not consistent with equivalent changes in x at other values in the function domain. A small change in x does not produce a small change in y at $x = 1$. The function is continuous for $x < 1$, and continuous for $x > 1$.

10. Not continuous at $x = 2$. Function does not exist at $x = 2$.

11. Continuous for $x \leq 2$

12. Continuous for $-2 < x \leq 2$

13. $f(x) = \begin{cases} x^2 & \text{for } x < 2 \\ 2 & \text{for } x \geq 2 \end{cases}$

Not continuous at $x = 2$. Small change in x around $x = 2$ produces a large change in $f(x)$.

14. $f(x) = \begin{cases} \dfrac{x^3 - x^2}{x - 1} & \text{for } x \neq 1 \\ 1 & \text{for } x = 1 \end{cases}$

$\dfrac{x^3 - x^2}{x - 1} = \dfrac{x^2(x - 1)}{(x - 1)}$

For all $x \neq 1$; $\dfrac{(x-1)}{x-1} = 1$

Therefore, $f(x) = x^2$ for $x \neq 1$
Continuous for all x.

15. $f(x) = \begin{cases} \dfrac{2x^2 - 18}{x - 3} & \text{for } x < 3 \text{ or } x > 3 \\ 12 & \text{for } x = 3 \end{cases}$

$\dfrac{2x^2 - 18}{x - 3} = \dfrac{2(x^2 - 9)}{(x - 3)} = \dfrac{2(x - 3)(x + 3)}{(x - 3)}$

For all $x \neq 3$, $\dfrac{x - 3}{x - 3} = 1$

Therefore, $f(x) = 2(x + 3)$ for $x \neq 3$
Continuous for all x.

16. $f(x) = \begin{cases} \dfrac{x + 2}{x^2 - 4} & \text{for } x < -2 \\ \dfrac{x}{8} & \text{for } x > -2 \end{cases}$

$\dfrac{x + 2}{x^2 - 4} = \dfrac{(x + 2)}{(x + 2)(x - 2)}$

For $x \neq -2$, $\dfrac{x + 2}{x + 2} = 1$

Therefore, $f(x) = \dfrac{1}{x - 2}$ for $x < -2$

Not continuous at $x = -2$.
$f(x)$ not defined at $x = -2$.

17. $f(x) = 3x - 2$

x	2.900	2.990	2.999
$f(x)$	6.700	6.970	6.997

x	3.001	3.010	3.100
$f(x)$	7.003	7.030	7.300

Therefore, $\lim\limits_{x \to 3} (3x - 2) = 7$

18. $f(x) = x^2 - 7$

x	3.900	3.990	3.999
$f(x)$	8.210	8.920	8.992

x	4.001	4.010	4.100
$f(x)$	9.008	9.080	9.810

Therefore, $\lim\limits_{x \to 4} (x^2 - 7) = 9$

19. $f(x) = \dfrac{x^3 - x}{x - 1}$

x	0.900	0.990	0.999
$f(x)$	1.7100	1.9701	1.9970

x	1.001	1.010	1.100
$f(x)$	2.0030	2.0301	2.3100

Therefore, $\lim\limits_{x \to 1} f(x) = 2$

20. $f(x) = \dfrac{x^3 + 2x^2 - 2x + 3}{x + 3}$

x	−3.100	−3.010	−3.001
$f(x)$	13.7100	13.0701	13.0070

x	−2.999	−2.990	−2.900
$f(x)$	12.9930	12.9301	12.3100

Therefore, $\lim\limits_{x \to -3} f(x) = 13$

21. $f(x) = \dfrac{2 - \sqrt{x + 2}}{x - 2}$

x	1.900	1.990	1.999
$f(x)$	−0.2516	−0.2502	−0.25002

x	2.001	2.010	2.100
$f(x)$	−0.24998	−0.2498	−0.2485

Therefore, $\lim\limits_{x \to 2} f(x) = -0.25$

22. $f(x) = \dfrac{2 - \sqrt{x}}{4 - x}$

x	3.900	3.990	3.999
$f(x)$	0.2516	0.2502	0.25002

x	4.001	4.010	4.100
$f(x)$	0.24998	0.2498	0.2485

Therefore, $\lim\limits_{x \to 4} f(x) = 0.25$

23. $f(x) = \dfrac{2x + 1}{5x - 3}$

x	10	100	1000
$f(x)$	0.4468	0.4044	0.4004

Therefore, $\lim\limits_{x \to \infty} f(x) = 0.4$

24. $f(x) = \dfrac{1 - x^2}{8x^2 + 5}$

x	10	100	1000
$f(x)$	−0.1230	−0.12498	−0.1249998

Therefore, $\lim\limits_{x \to \infty} f(x) = -0.125$

25. Since the function is continuous at $x = 3$, we may evaluate the limit by substitution. For $f(x) = 3x - 2$, $f(3) = 7$.

26.
$$\lim_{x \to 4} \sqrt{x^2 - 7} = \sqrt{\lim_{x \to 4} (x^2 - 7)}$$
$$= \sqrt{4^2 - 7} = \sqrt{16 - 7}$$
$$= \sqrt{9} = 3$$

27. $\lim\limits_{x \to 2} \dfrac{x^2 - 1}{x + 1}$; $f(2) = \dfrac{3}{3} = 1$ continuous; at $x = 2$

Therefore, $\lim\limits_{x \to 2} \dfrac{x^2 - 1}{x + 1} = \dfrac{3}{3} = 1$

28. $\lim\limits_{x \to 5} \left(\dfrac{3}{x^2 + 2} \right)$; $f(5) = \dfrac{3}{27}$; continuous at $x = 5$

Therefore, $\lim\limits_{x \to 5} \dfrac{3}{x^2 + 2} = \dfrac{1}{9}$

29.
$$\lim_{x \to 0} \frac{x^2 + x}{x} = \lim_{x \to 0} \frac{x(x + 1)}{x}$$
$$= \lim_{x \to 0} (x + 1)$$
$$= 0 + 1 = 1$$

30.
$$\lim_{x \to 2} \frac{4x^2 - 8x}{x - 2} = \lim_{x \to 2} \frac{4(x^2 - 2x)}{x - 2}$$
$$= \lim_{x \to 2} \frac{4x(x - 2)}{(x - 2)}$$
$$= \lim_{x \to 2} 4x = 8$$

31. $\lim\limits_{x \to -1} \dfrac{x^2 - 1}{3x + 3} = \lim\limits_{x \to -1} \dfrac{x^2 - 1}{3(x + 1)}$

$= \lim\limits_{x \to -1} \dfrac{(x - 1)(x + 1)}{3(x + 1)}$

$= \lim\limits_{x \to -1} \dfrac{(x - 1)}{3} = -\dfrac{2}{3}$

32. $\lim\limits_{x \to 3} \dfrac{x^2 - 2x - 3}{(3 - x)} = \lim\limits_{x \to 3} \dfrac{(x - 3)(x + 1)}{-(x - 3)}$

$= \lim\limits_{x \to 3} -(x + 1)$

$= -4$

33. $\lim\limits_{x \to 1} \dfrac{x^3 - x}{x - 1} = \lim\limits_{x \to 1} \dfrac{x(x^2 - 1)}{x - 1}$

$= \lim\limits_{x \to 1} \dfrac{x(x + 1)(x - 1)}{x - 1}$

$= \lim\limits_{x \to 1} x(x + 1) = 1(2) = 2$

34. $\lim\limits_{x \to \frac{1}{3}} \dfrac{3x - 1}{3x^2 + 5x - 2} = \lim\limits_{x \to \frac{1}{3}} \dfrac{(3x - 1)}{(3x - 1)(x + 2)}$

$= \lim\limits_{x \to \frac{1}{3}} \dfrac{1}{x + 2}$

$= \dfrac{3}{7}$

35. $\lim\limits_{x \to 1} \dfrac{(2x - 1)^2 - 1}{2x - 2} = \lim\limits_{x \to 1} \dfrac{4x^2 - 4x}{2(x - 1)}$

$= \lim\limits_{x \to 1} \dfrac{4x(x - 1)}{2(x - 1)}$

$= \lim\limits_{x \to 1} 2x = 2$

36. $\lim\limits_{x \to 0} \dfrac{(2 + x)^2 - 4}{x} = \lim\limits_{x \to 0} \dfrac{4x + x^2}{x}$

$= \lim\limits_{x \to 0} \dfrac{x(4 + x)}{x}$

$= \lim\limits_{x \to 0} (4 + x) = 4$

37. $\lim\limits_{x \to -1} \sqrt{x}(x + 1)$ does not exist since x cannot approach -1 without the function going through imaginary values. $f(0)$, which is real, is the only defined value for the function near $x = -1$.

38. $\lim\limits_{x \to 1} (x - 1)\sqrt{x^2 - 4}$; $\lim\limits_{x \to 1} (x - 1) = 0$; $\lim\limits_{x \to 1} \sqrt{x^2 - 4}$ is imaginary. Limit does not exist.

39. $\lim\limits_{x \to \infty} \dfrac{\frac{2}{x}}{1 - 2x} = \lim\limits_{x \to \infty} \dfrac{\frac{2}{x^2}}{\frac{1}{x} - 2} = \dfrac{0}{0 - 2} = 0$

40. $\lim\limits_{x \to \infty} \dfrac{6}{1 + \frac{2}{x^2}} = \dfrac{6}{1 + 0} = 6$

41. $\lim\limits_{x \to \infty} \dfrac{3x^2 + 5}{x^2 - 2} = \lim\limits_{x \to \infty} \dfrac{(3x^2 + 5) \div x^2}{(x^2 - 2) \div x^2}$

$= \lim\limits_{x \to \infty} \dfrac{3 + \frac{5}{x^2}}{1 - \frac{2}{x^2}}$

$= \dfrac{3 + 0}{1 - 0} = 3$

42. $\lim\limits_{x \to \infty} \dfrac{x - 1}{7x + 4} = \lim\limits_{x \to \infty} \dfrac{1 - \frac{1}{x}}{7 + \frac{4}{x}} = \dfrac{1 - 0}{7 + 0} = \dfrac{1}{7}$

43. $\lim\limits_{x \to \infty} \dfrac{2x - 6}{x^2 - 9} = \lim\limits_{x \to \infty} \dfrac{2(x - 3)}{(x - 3)(x + 3)} = \lim\limits_{x \to \infty} \dfrac{2}{x + 3}$

$= \lim\limits_{x \to \infty} \dfrac{\frac{2}{x}}{1 + \frac{3}{x}} = \dfrac{0}{1 + 0} = 0$

44. $\lim\limits_{x \to \infty} \dfrac{1 - 2x^2}{(4x + 3)^2} = \lim\limits_{x \to \infty} \dfrac{1 - 2x^2}{16x^2 + 24x + 9}$

$= \lim\limits_{x \to \infty} \dfrac{\frac{1}{x^2} - 2}{16 + \frac{24}{x} + \frac{9}{x^2}} = \dfrac{-2}{16}$

$= \dfrac{-1}{8}$

45. $\lim\limits_{x \to 0} \dfrac{x^2 - 3x}{x}$

x	-0.1	-0.01	-0.001
$f(x)$	-3.10	-3.01	-3.001

x	0.001	0.01	0.1
$f(x)$	-2.999	-2.99	-2.9

We see that $f(x) \to -3$ as $x \to 0$. We now find the limit by changing the algebraic form.

$\lim\limits_{x \to 0} \dfrac{x^2 - 3x}{x} = \lim\limits_{x \to 0} \dfrac{x(x - 3)}{x}$

$= \lim\limits_{x \to 0} x - 3$

$= 0 - 3 = -3$

46. $\lim\limits_{x \to 3} \dfrac{2x^2 - 6x}{x - 3} = \lim\limits_{x \to 3} \dfrac{2x(x - 3)}{(x - 3)} = \lim\limits_{x \to 3} 2x = 6$

x	2.9	2.99	2.999	3.001	3.01	3.1
$f(x)$	5.8	5.98	5.998	6.002	6.02	6.2

47. $\lim\limits_{x\to\infty} \dfrac{2x^2+x}{x^2-3} = \lim\limits_{x\to\infty} \dfrac{2+\dfrac{1}{x}}{1-\dfrac{3}{x^2}} = \dfrac{2}{1} = 2$

x	10	100	1000
$f(x)$	2.1649	2.0106	2.0010

48.

x		10	100
$f(x) = \dfrac{x^2+5}{\sqrt{64x^4+1}}$		0.1312498975	0.1250625

x	1000
$f(x) = \dfrac{x^2+5}{\sqrt{64x^4+1}}$	0.125000625

$\lim\limits_{x\to\infty} f(x) = 0.125 = \dfrac{1}{8}$

$\lim\limits_{x\to\infty} \dfrac{x^2+5}{\sqrt{64x^4+1}} = \lim\limits_{x\to\infty} \dfrac{x^2\left(1+\dfrac{5}{x^2}\right)}{x^2\sqrt{64+\dfrac{1}{x^4}}}$

$= \lim\limits_{x\to\infty} \dfrac{\left(1+\dfrac{5}{x^2}\right)}{\sqrt{64+\dfrac{1}{x^4}}}$

$= \dfrac{1}{\sqrt{64}} = \dfrac{1}{8}$

49. $\lim\limits_{t\to 0} v = 3 \text{ cm/s}$

d	0.480000	0.280000	0.029800
t	0.200000	0.100000	0.010000
$v = \frac{d}{t}$	2.40000	2.80000	2.98000

d	0.002998	0.00029998
t	0.001000	0.00010000
$v = \frac{d}{t}$	2.99800	2.99800

50. $P = 2l + 2w$

$A = 520 \text{ cm}^2$

$l \cdot w = 520; \; l = \dfrac{520}{w}$

$P = 2\left(\dfrac{520}{w}\right) + 2w; \; P = \dfrac{1040}{w} + 2w$

w	19.0	19.9	19.99
p	92.737	92.061	92.006

w	20.01	20.1	21
p	91.994	91.941	91.524

$\lim\limits_{w\to 20} \dfrac{1040}{w} + 2w = 92 \text{ cm}$

51.

t	T
0	100
1	90
2	81

etc.

This is a geometric progression

$a_n = a_1 r^{n-1}, \; r = 0.9, a_1 = 100$

Therefore, $T = 100(0.9)^t$

$\lim\limits_{t\to 10} 100(0.9)^t = 34.9°\text{C}$

$\lim\limits_{t\to\infty} 100(0.9)^t = 100 \lim\limits_{t\to\infty} (0.9)^t = 100(0) = 0°\text{C}$

52. $\lim\limits_{R\to\infty} \dfrac{5R}{5+R} = \lim\limits_{R\to\infty} \dfrac{5}{\dfrac{5}{R}+1} = 5\,\Omega$

53. $\lim\limits_{x\to 0} (1+x)^{1/x} = 2.71827 \text{ or } e$

x	0.1	0.01	0.001	0.0001	0.00001
$(1+x)^{1/x}$	2.6	2.70	2.716	2.718	2.71827

54. $\lim\limits_{x\to 0} \dfrac{\sin x}{x}$ (x in radians)

x	1	0.1	0.01
$\frac{\sin x}{x}$	0.841	0.998	0.99998

Clearly, $\lim\limits_{x\to 0} \dfrac{\sin x}{x} = 1$

55. For $x > 0$, $f(x) = \dfrac{x}{|x|} = \dfrac{x}{x} = 1$ from which

$\lim\limits_{x\to 0^+} f(x) = 1$

For $x < 0$, $f(x) = \dfrac{x}{|x|} = \dfrac{x}{-x} = -1$ from which

$\lim\limits_{x\to 0^-} f(x) = -1$

Since $RHL \neq LHL$, limit D.N.E. which means $f(x)$ is not continuous.

56. $\lim\limits_{x\to 0^+} 2^{1/x}; \; x \to 0^+; \dfrac{1}{x}$ will be positive and large

Therefore, $\lim\limits_{x\to 0^+} 2^{1/x} \left(2^{\text{Large+No.}}\right) = \infty$

$\lim\limits_{x\to 0^-} 2^{1/x}; \; x \to 0^-; \dfrac{1}{x}$ will be negative and large

Therefore,

$\lim\limits_{x\to 0^-} 2^{1/x}\left(2^{\text{Large-No.}}\right) = \lim\limits_{x\to\infty} \dfrac{1}{2^{1/x}} = 0$

3.2   The Slope of a Tangent to a Curve

1. $y = x^2$; $P = (2, 4)$

	Q_1	Q_2	Q_3	Q_4	P
x_2	1.5	1.9	1.99	1.99	2
y_2	2.25	3.61	3.9601	3.996001	4
$y_2 - 4$	−1.75	−0.39	−0.0399	−0.003999	
$x_2 - 2$	−0.5	−0.1	−0.01	−0.001	
$m = \frac{y_2 - 4}{x_2 - 2}$	3.5	3.9	3.99	3.999	
$m_{\tan} = 4$					

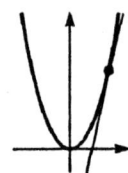

2. $y = 1 - \frac{1}{2}x^2$; $P = (2, -1)$

	Q_1	Q_2	Q_3	Q_4	P
x_2	1.5	1.9	1.99	1.999	2
y_2	−0.125	−0.805	−0.98005	−0.9980005	−1
$y_2 + 1$	0.875	0.195	0.01995	0.0019995	
$x_2 - 2$	−0.5	−0.1	−0.01	−0.001	
$m = \frac{y_2 + 1}{x_2 - 2}$	−1.75	−1.95	−1.995	−1.9995	
$m_{\tan} = -2$					

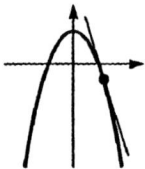

3. $y = 2x^2 + 5x$; $P = (-2, -2)$

	Q_1	Q_2	Q_3	Q_4	P
x_2	−1.5	1.9	1.99	1.999	−2
y_2	−3	−2.28	−2.0298	−2.002998	−2
$y_2 - (-2)$	−1	−0.28	−0.0298	−0.002998	
$x_2 - (-2)$	0.5	0.1	0.01	0.001	
$m = \frac{y_2 - (-2)}{x_2 - (-2)}$	−2	−2.8	−2.98	−2.998	
$m_{\tan} = -3$					

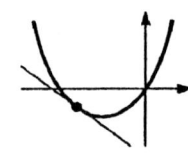

$y = x^3 + 1$; $p = (-1, 0)$

4.

	Q_1	Q_2	Q_3	Q_4	P
x_2	−0.5	−0.9	−0.99	−0.999	−1
y_2	0.875	0.271	0.029701	0.002997001	0
$y_2 - 0$	0.875	0.271	0.029701	0.002997001	
$x_2 + 1$	0.5	0.1	0.01	0.001	
$m = \frac{y_2}{x_2 + 1}$	1.75	2.71	2.9701	2.99700000	
$m_{\tan} = 3$					

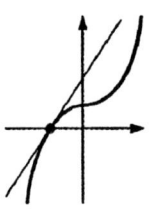

5. $y = x^2$; $P = (2, 4)$

$4 + \Delta y = (2 + \Delta x)^2$; $4 + \Delta y = 4 + 4\Delta x + (\Delta x)^2$

$\Delta y = 4 - 4 + 4\Delta x + (\Delta x)^2$; $\Delta y = 4\Delta x + (\Delta x)^2$

$$m_{PQ} = \frac{\Delta y}{\Delta x} = \frac{4\Delta x + (\Delta x)^2}{\Delta x} = \frac{\Delta x(4 + \Delta x)}{\Delta x}$$

As Δx approaches zero, $m_{\tan} = 4 + 0 = 4$

6. $y = 1 - \frac{1}{2}x^2$; $P = (2, -1)$

$-1 + \Delta y = 1 - \frac{1}{2}(2 + \Delta x)^2$

$-1 + \Delta y = 1 - \frac{1}{2}(4 + 4\Delta x + \Delta x^2)$

$-1 + \Delta y = 1 - 2 - 2\Delta x - \frac{1}{2}\Delta x^2$;

$\Delta y = -2\Delta x - \frac{1}{2}\Delta x^2$

$$m_{PQ} = \frac{\Delta y}{\Delta x} = \frac{-\Delta\left(2 + \frac{1}{2}\Delta x\right)}{\Delta x} = -\left(2 + \frac{1}{2}\Delta x\right)$$

As Δx approaches zero,
$m_{PQ} = -(2 + 0) = -2 = m_{\tan}$

7. $y = 2x^2 + 5x$; $P = (-2, -2)$

$-2 + \Delta y = 2(-2 + \Delta x)^2 + 5(-2 + \Delta x)$

$-2 + \Delta y = 2(4 - 4\Delta x + \Delta x^2) - 10 + 5\Delta x$

$-2 + \Delta y = 8 - 8\Delta x + 2\Delta x^2 - 10 + 5\Delta x$

$\Delta y = 2\Delta x^2 - 3\Delta x$

$$M_{PQ} = \frac{\Delta y}{\Delta x} = 2\Delta x - 3$$

as $\Delta x \to 0$, $m_{\tan} = -3$

8. $y = x^3 + 1$; $P(-1, 0)$

$0 + \Delta y = (-1 + \Delta x)^3 + 1$

$\Delta y = -1 + 3\Delta x - 3\Delta x^2 + \Delta x^3 + 1$

$\Delta y = 3\Delta x - 3\Delta x^2 + \Delta x^3 = \Delta x(3 - 3\Delta x + \Delta x^2)$

$$m_{PQ} = \frac{\Delta y}{\Delta x} = 3 - 3\Delta x + \Delta x^2$$

As Δx approaches zero, $m_{PQ} = m_{\tan} = 3$

9. $y = x^2$; $x = 2$, $x = -1$

$y_1 + \Delta y = (x_1 + \Delta x)^2$;

$y_1 + \Delta y = x_1^2 + 2x_1\Delta x + (\Delta x)^2$

$y_1 + \Delta y - y_1 = x_1^2 + 2x_1\Delta x + (\Delta x)^2 - x_1^2$

$\Delta y = 2x_1\Delta x + (\Delta x)^2$

$$m_{\tan} = \frac{\Delta y}{\Delta x} = \frac{2x_1\Delta x + (\Delta x)^2}{\Delta x}$$

$$= \frac{(\Delta x)(2x_1 + \Delta x)}{\Delta x}$$

$$= 2x_1 + \Delta x$$

As $\Delta x \to 0$, $m_{\tan} = 2x_1$

$x_1 = 2$, $m_{\tan} = 4$

$x_1 = -1$, $m_{\tan} = -2$

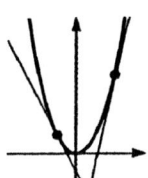

10. $y = 1 - \frac{1}{2}x^2$; $x = 2$; $x = -2$

$y_1 + \Delta y = 1 - \frac{1}{2}(x_1 + \Delta x)^2$

$= 1 - \frac{1}{2}(x_1^2 + 2x_1\Delta x + \Delta x^2)$

$y_1 + \Delta y = 1 - \frac{1}{2}x_1^2 - x_1\Delta x - \frac{1}{2}\Delta x^2$

$\Delta y = 1 - \frac{1}{2}x_1^2 - x_1\Delta x - \frac{1}{2}\Delta x^2 - y_1$

$\Delta y = 1 - \frac{1}{2}x_1^2 - x_1\Delta x - \frac{1}{2}\Delta x^2 - 1 + \frac{1}{2}x_1^2$

$\Delta y = -x_1\Delta x - \frac{1}{2}\Delta x^2 = \Delta x\left(-x_1 - \frac{1}{2}\Delta x\right)$

$$m_{PQ} = \frac{\Delta y}{\Delta x} = -x_1 - \frac{1}{2}\Delta x = m_{PQ}$$

As $\Delta x \to 0$, m_{PQ} becomes $m_{\tan}$

$m_{\tan} = -x_1$

$x_1 = 2$, $m_{\tan} = -2$

$x_1 = -2$, $m_{\tan} = 2$

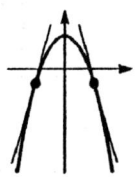

11. Find the slope of tangent line to $y = 2x^2 + 5x$ at (x_1, y_1).

$$y_1 + \Delta y = 2(x_1 + \Delta x)^2 + 5(x_1 + \Delta x)$$
$$y_1 + \Delta y = 2(x_1^2 + 2x_1\Delta x + \Delta x^2) + 5x_1 + 5\Delta x$$
$$y_1 + \Delta y = 2x_1^2 + 4x_1\Delta x + 2\Delta x^2 + 5x_1 + 5\Delta x \quad (1)$$
$$\underline{y_1 \qquad\qquad = 2x_1^2 + 5x_1 \qquad\qquad\qquad\qquad\quad (2)}$$
$$\Delta y = 4x_1\Delta x + 2\Delta x^2 + 5\Delta x, \text{ subtracting}$$

$$\frac{\Delta y}{\Delta x} = 4x_1 + 2\Delta x + 5, \Delta x \to 0$$

$$m_{\tan} = 4x_1 + 5 = 4(-2) + 5 = -3 \text{ for } x_1 = -2$$
$$m_{\tan} = 4(0.5) + 5 = 7 \text{ for } x_1 = 0.5$$

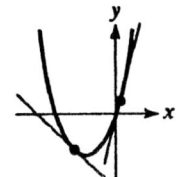

12. $y = 4 - 3x^2$; $x = 0, x = 2$

$$y_1 + \Delta y = 4 - 3(x_1 + \Delta x)^2$$
$$= 4 - 3(x_1^2 + 2x_1\Delta x + \Delta x^2)$$

$$\Delta y = 4 - 3x_1^2 - 6x_1\Delta x - 3\Delta x^2 - 4 + 3x_1^2$$

$$\Delta y = -6x_1\Delta x - 3\Delta x^2 = -\Delta x(6x_1 + 3\Delta x)$$

$$m_{PQ} = \frac{\Delta y}{\Delta x} = -(6x_1 + 3\Delta x)$$

As $\Delta x \to 0$, $m_{PQ} =$ becomes $m_{\tan} = -6x_1$
$$x_1 = 0, m_{\tan} = 0$$
$$x_1 = 2, m_{\tan} = -12$$

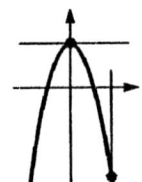

13. $y = x^2 + 4x + 5$; $x = -3, x = 2$; $y_1 = x_1^2 + 4x_1 + 5$
$$y_1 + \Delta y = (x_1 + \Delta x)^2 + 4(x_1 + \Delta x) + 5$$
$$y_1 + \Delta y = x_1^2 + 2x_1\Delta x + (\Delta x)^2 + 4x_1 + 4\Delta x + 5$$
$$y_1 + \Delta y - y_1 = \Delta y = 2x_1\Delta x + (\Delta x)^2 + 4\Delta x$$

$$\frac{\Delta y}{\Delta x} = \frac{\Delta x(2x_1 + \Delta x + 4)}{\Delta x} = 2x_1 + 4 + \Delta x$$

As $\Delta x \to 0$, $m_{\tan} = 2x_1 + 4$
$$x_1 = -3, m_{\tan} = -2$$
$$x_1 = 2, m_{\tan} = 8$$

14. $y = 2x^2 - 4x$; $x = 1, x = 1.5$
$$y_1 + \Delta y = 2(x_1 + \Delta x)^2 - 4(x_1 + \Delta x)$$
$$y_1 + \Delta y = 2(x_1^2 + 2x_1\Delta x + \Delta x^2) - 4x_1 - 4\Delta x_1$$
$$\Delta y = 2x_1^2 + 4x_1\Delta x + 2\Delta x^2 - 4x_1 - 4\Delta x - 2x_1^2 + 4x_1$$
$$\Delta y = 4x_1\Delta x + 2\Delta x^2 - 4\Delta x = \Delta x(4x_1 + 2\Delta x - 4)$$

$$m_{PQ} = \frac{\Delta y}{\Delta x} = 4x_1 + 2\Delta x - 4$$

As $\Delta x \to 0$, $m_{PQ} = \to m_{\tan} = 4x_1 - 4$
$$x_1 = 1, m_{\tan} = 0$$
$$x_1 = 1.5, m_{\tan} = 2$$

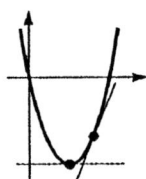

15. $y = 6x - x^2$; $x = -2, x = 3$
$$y_1 + \Delta y = 6(x_1 + \Delta x) - (x_1 + \Delta x)^2$$
$$y_1 + \Delta y = 6x_1 + 6\Delta x - x_1^2 - 2x_1\Delta x^2 - \Delta x^2$$
$$\Delta y = 6x_1 + 6\Delta x - x_1^2 - 2x_1\Delta x - \Delta x^2 - 6x_1 + x_1^2$$
$$\Delta y = 6\Delta x - 2x_1\Delta x^2 - \Delta x^2 = \Delta x(6 - 2x_1 - \Delta x)$$

$$m_{PQ} = \frac{\Delta y}{\Delta x} = 6 - 2x_1 - \Delta x$$

As $\Delta x \to 0$, $m_{PQ} \to m_{\tan} = 6 - 2x_1$
$$x_1 = -2, m_{\tan} = 10$$
$$x_1 = 3, m_{\tan} = 0$$

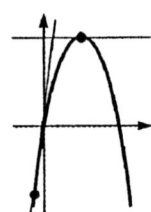

16. $y = x^3 - 2x; \ x = -1, x = 0, x = 1; \ y_1 = x_1^3 - 2x_1$

$y_1 + \Delta y = (x_1 + \Delta x)^3 - 2(x_1 + \Delta x)$

$y_1 + \Delta y = x_1^3 + 3x_1^2 \Delta x + 3x_1(\Delta x)^2 + (\Delta x)^3$
$\qquad - 2x_1 - 2\Delta x$

$y_1 + \Delta y - y_1 = \Delta y; \ \Delta y = 3x_1^2 \Delta x + 3x_1(\Delta x)^2$
$\qquad + (\Delta x)^3 - 2\Delta x$

$\dfrac{\Delta y}{\Delta x} = \dfrac{\Delta x[3x_1^2 + 3x_1\Delta x + (\Delta x)^2 - 2]}{\Delta x}$

$\qquad = 3x_1^2 + 3x_1\Delta x + (\Delta x)^2 - 2$

As $\Delta x \to 0$, $m_{\tan} = 3x_1^2 - 2$

$x_1 = -1, m_{\tan} = 1$

$x_1 = 0, m_{\tan} = -2$

$x_1 = 1, m_{\tan} = 1$

17. $y = x^4; \ x = 0, \ x = 0.5, \ x = 1$

$y_1 + \Delta y = (x_1 + \Delta x)^4$

$y_1 + \Delta y = x_1^4 + 4x_1^3 \Delta x + 6x_1^2 \Delta x^2 + 4x_1\Delta x^3 + x^4$

$\Delta y = 4x_1^3 \Delta x + 6x_1^2 \Delta x^2 + 4x_1\Delta x^3 + \Delta x^4$

$\Delta y = \Delta x(4x_1^3 + 6x_1^2 \Delta x + 4x_1\Delta x^2 + \Delta x^3)$

$m_{PQ} = \dfrac{\Delta y}{\Delta x} = 4x_1^3 + 6x_1^2 \Delta x + 4x_1\Delta x^2 + \Delta x^3$

$\Delta x \to 0, \ m_{PQ} \to m_{\tan} = 4x_1^3$

$x_1 = 0, m_{\tan} = 0$

$x_1 = 0.5, \ m_{\tan} = 0.5$

$x_1 = 1, \ m_{\tan} = 4$

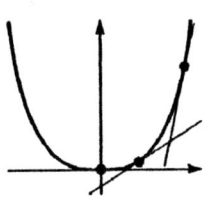

18. $y = 1 - x^4; \ x = 0, x = 1, x = 2$

$y_1 + \Delta y = 1 - (x_1 + \Delta x)^4$

$y_1 + \Delta y = 1 - x_1^4 - 4x_1^3 \Delta x - 6x_1^2 \Delta x^2 - 4x_1\Delta x^3$
$\qquad - \Delta x^4$

$\Delta y = -\Delta x(4x_1^3 + 6x_1^2 \Delta x + 4x_1\Delta x^2 + \Delta x^3)$

$m_{PQ} = \dfrac{\Delta y}{\Delta x} = -(4x_1^3 + 6x_1^2 \Delta x + 4x_1\Delta x^2 + \Delta x^3)$

As $\Delta x \to 0, m_{PQ} \to m_{\tan} = -(4x_1^3) = -4x_1^3$

$x_1 = 0, m_{\tan} = 0$

$x_1 = 1, m_{\tan} = -4$

$x_1 = 2, m_{\tan} = -32$

19. $y = x^5; \ x = 0, x = 0.5, x = 1$

$y + \Delta y = (x + \Delta x)^5$

$\qquad = x^5 + 5x^4\Delta x + 10x^3\Delta x^2 + 10x^2\Delta x^3$
$\qquad + 5x\Delta x^4 + \Delta x^5$

$y + \Delta y - y = 5x^4\Delta x + 10x^3\Delta x^2 + 10x^2\Delta x^3$
$\qquad + 5x\Delta x^4 + \Delta x^5$

$\dfrac{\Delta y}{\Delta x} = 5x^4 + 10x^3\Delta x + 10x^2\Delta x^2 + 5x\Delta x^3 + \Delta x^4$

$\Delta x \to 0, m_{\tan} = 5x^4$

$x_1 = 0, m_{\tan} = 0$

$x_1 = 0.5, m_{\tan} = 0.31$

$x_1 = 1, m_{\tan} = 5$

20. $y = \dfrac{1}{x}; \ x = 0.5, x = 1, x = 2$

$y_1 + \Delta y = \dfrac{1}{x_1 + \Delta x}$

$\Delta y = \dfrac{1}{x_1 + \Delta x} - \dfrac{1}{x_1} = \dfrac{x_1 - (x_1 + \Delta x)}{x_1(x_1 + \Delta x)}$

$\qquad = \dfrac{-\Delta x}{x_1(x_1 + \Delta x)}$

$m_{PQ} = \dfrac{\Delta y}{\Delta x} = \dfrac{-1}{x_1(x_1 + \Delta x)}$

$\Delta x \to 0, m_{PQ} \to m_{\tan} = \dfrac{-1}{x_1^2}$

$x_1 = 0.5, m_{\tan} = -4$

$x_1 = 1, m_{\tan} = -1$

$x_1 = 2, m_{\tan} = -\dfrac{1}{4}$

21.

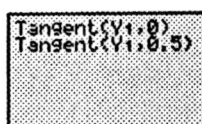

22.

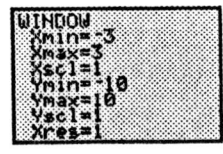
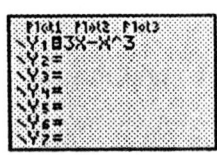

23.

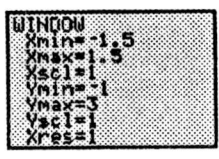
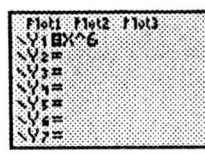

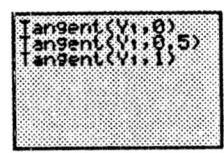

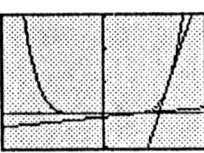

24.

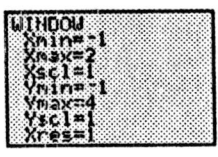
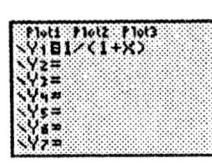

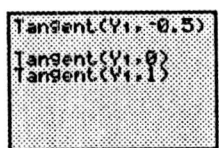

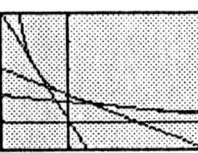

25. $y = x^2 + 2$; $P(2,6), Q(2.1,6.41)$
From P to Q, x changes by 0.1 units and Q by 0.41 units. The average change in y for 1 unit change in x is $\frac{0.41}{0.1} = 4.1$ units.

$$y + \Delta y = (x + \Delta x)^2 + 2$$
$$y + \Delta y = x^2 + 2x\Delta x + \Delta x^2 + 2$$
$$y + \Delta y - y = (x^2 + 2x\Delta x + \Delta x^2 + 2) - (x^2 + 2)$$
$$\Delta y = 2x\Delta x + \Delta x^2$$

$$m_{PQ} = \frac{\Delta y}{\Delta x} = 2x + \Delta x;$$
$$m_{\tan} = \lim_{\Delta x \to 0} 2x + \Delta x = 2x$$
$$x = 2, \ m_{\tan} = 4 \ \text{(Instantaneous rate of change)}$$
$$m_{PQ} = 4.1 \ \text{(Average rate of change)}$$

26. $y = 1 - 2x^2$; $P(1,-1), Q(1.1 - 1.42)$

$\Delta y = -0.42$; $\Delta x = 0.1$; $\dfrac{\Delta y}{\Delta x} = -4.2$ (Average rate of change)
$$y + \Delta y = 1 - 2(x + \Delta x)^2 = 1 - 2(x^2 + 2x\Delta x + \Delta x^2)$$
$$\Delta y = 1 - 2x^2 - 4x\Delta x - 2\Delta x^2 - 1 + 2x^2$$
$$\Delta y = -\Delta x(4x + 2\Delta x)$$

$$m_{PQ} = \frac{\Delta y}{\Delta x} = -1(4 + 2\Delta x) = -1(4(1) + 2(0.1))$$
$$= -4.2$$
$$m_{\tan} = \lim_{\Delta x \to 0} -(4x + 2\Delta x) = -4x$$
$$x = 1, \ m_{\tan} = -4 \ \text{(Instantaneous rate of change)}$$
$$m_{PQ} = -4.2 \ \text{(Average rate of change)}$$

27. $y = 9 - x^3$; $P(2,1), Q(2.1,-0.261)$
$\Delta y = -1.261, \Delta x = 0.1$

$$\frac{\Delta y}{\Delta x} = -12.61 \ \text{(Average rate of change)}$$

$$y + \Delta y = 9 - (x + \Delta x)^3$$
$$= 9 - x^3 - 3x^2\Delta x - 3x\Delta x^2 - \Delta x^3$$
$$\Delta y = -3x^2\Delta - 3x\Delta x^2 - \Delta x^3$$
$$= -\Delta x(3x^2 + 3x\Delta x + \Delta x^2)$$

$$m_{PQ} = \frac{\Delta y}{\Delta x} = -(3x^2 + 3x\Delta x + \Delta x^2)$$
$$= -1(3(4) + 3(2)(0.1) + 0.1^2) = -12.61$$
$$m_{\tan} = \lim_{\Delta x \to 0} -(3x^2 + 3x\Delta x + \Delta x^3) = -3x^2$$
$$x = 2, \ m_{\tan} = -12 \ \text{(instantaneous rate of change)}$$
$$m_{PQ} = -12.61 \ \text{(Average rate of change)}$$

28. $y = x^3 - 6x$; $P(3,9), Q(3.1,11.191)$

$\Delta y = 2.191$; $\Delta x = 0.1$; $\dfrac{\Delta y}{\Delta x} = 21.91$

$$y + \Delta y = (x + \Delta x)^3 - 6(x + \Delta x)$$
$$\Delta y = x^3 + 3x^2\Delta x + 3x\Delta x^2 + \Delta x^3 - 6x - 6\Delta x$$
$$\qquad - x^3 + 6x$$
$$\Delta y = \Delta x(3x^2 + 3x\Delta x + \Delta x^2 - 6)$$

$$m_{PQ} = \frac{\Delta y}{\Delta x} = 3x^2 + 3x\Delta x + \Delta x^2 - 6$$
$$= 3(9) + 3(3)(0.1) + 0.1^2 - 6 = 21.91$$
$$m_{\tan} = \lim_{\Delta x \to 0} (3x^2 + 3x\Delta x + \Delta x^2 - 6) = 3x^2 - 6$$
$$x = 3, \ m_{\tan} = 21 \ \text{(instantaneous rate of change)}$$
$$m_{PQ} = 21.91 \ \text{(Average rate of change)}$$

3.3   The Derivative

1. $y = 3x - 1;\ y + \Delta y = 3(x + \Delta x) - 1$
$y + \Delta y = 3x + 3\Delta x - 1$
$y + \Delta y - y = 3x + 3\Delta x - 1 - (3x - 1)$
$\quad \Delta y = 3\Delta x$

$\quad \dfrac{\Delta y}{\Delta x} = \dfrac{3\Delta x}{\Delta x};$

$\quad \lim_{\Delta x \to 0} = 3;\ \dfrac{dy}{dx} = 3$

2. $y = 6x + 3;\ y + \Delta y = 6(x + \Delta x) + 3 = 6x + 6\Delta x + 3$
$\Delta y = 6x + 6\Delta + 3 - 6x - 3 = 6\Delta x$

$\dfrac{\Delta y}{\Delta x} = 6;\ \lim_{\Delta x \to 0} 6 = 6;\ \dfrac{dy}{dx} = 6$

3. $y = 1 - 2x;\ y + \Delta y = 1 - 2(x + \Delta x) = 1 - 2x - 2\Delta x$
$\Delta y = 1 - 2x - 2\Delta x - 1 + 2x = -2\Delta x$

$\dfrac{\Delta y}{\Delta x} = -2;\ \lim_{\Delta x \to 0} (-2) = -2;\ \dfrac{dy}{dx} = -2$

4. $y = 2 - 5x;\ y + \Delta y = 2 - 5(x + \Delta x) = 2 - 5x - 5\Delta x$
$\Delta y = 2 - 5x - 5\Delta x - 2 + 5x = -5\Delta x$

$\dfrac{\Delta y}{\Delta x} = -5;\ \lim_{\Delta x \to 0} (-5) = -5;\ \dfrac{dy}{dx} = -5$

5. $y = x^2 - 1;\ y + \Delta y = (x + \Delta x)^2 - 1$
$y + \Delta y = x^2 + 2x\Delta x + (\Delta x)^2 - 1$
$y + \Delta y - y = x^2 + 2x\Delta x + (\Delta x)^2 - 1 - (x^2 - 1)$
$\Delta y = 2x\Delta x + (\Delta x)^2;$

$\dfrac{\Delta y}{\Delta x} = \dfrac{\Delta x(2x + \Delta x)}{\Delta x} = 2x + \Delta x$

$\lim_{\Delta x \to 0} (2x + \Delta x) = 2x + 0 = 2x;\ \dfrac{dy}{dx} = 2x$

6. $y = 4 - x^2;\ y + \Delta y = 4 - (x + \Delta x)^2$
$\qquad\qquad = 4 - x^2 - 2x\Delta x - \Delta x^2$
$\Delta y = 4 - x^2 - 2x\Delta x - \Delta x^2 - 4 + x^2 = -2x\Delta x - \Delta x^2$
$\qquad = -\Delta x(2x + \Delta x)$
$\dfrac{\Delta y}{\Delta x} = -2x - \Delta x;\ \lim_{\Delta x \to 0} (-2x - \Delta x) = -2x;$
$\dfrac{dy}{dx} = -2x$

7. $y = 5x^2;$
$y + \Delta y = 5(x + \Delta x)^2 = 5(x^2 + 2x\Delta x + \Delta x^2)$
$\Delta y = 5x^2 + 10x\Delta x + 5\Delta x^2 - 5x^2 = 10x\Delta x + 5\Delta x^2$
$\qquad = 5\Delta x(2x + 5\Delta x)$

$\dfrac{\Delta y}{\Delta x} = 5(2x + 5\Delta x);\ \lim_{\Delta x \to 0} 5(2x + 5\Delta x) = 10x;$
$\dfrac{dy}{dx} = 10x$

8. $y = -6x^2;$
$y + \Delta y = -6(x + \Delta x) = -6(x^2 + 2x\Delta x + \Delta x^2)$
$\Delta y = -6x^2 - 12x\Delta x - 6\Delta x^2 + 6x^2$
$\qquad = -6\Delta x(2x + \Delta x)$

$\dfrac{\Delta y}{\Delta x} = -6(2x + \Delta x);\ \lim_{\Delta x \to 0} -6(2x + \Delta x) = -12x$

$\dfrac{dy}{dx} = -12x$

9. $y = x^2 - 7x;\ y + \Delta y = (x + \Delta x)^2 - 7(x + \Delta x)$
$y + \Delta y = x^2 + 2x\Delta x + (\Delta x)^2 - 7x - 7\Delta x)$
$y + \Delta y - y = x^2 + 2x\Delta x + (\Delta x)^2 - 7x - 7\Delta x - (x^2 - 7x)$
$\Delta y = 2x\Delta x + (\Delta x)^2 - 7\Delta x$

$\dfrac{\Delta y}{\Delta x} = \dfrac{\Delta x(2x + \Delta x - 7)}{\Delta x} = 2x + \Delta x - 7$

$\lim_{\Delta x \to 0} (2x - \Delta x - 7) = 2x - 7;\ \dfrac{dy}{dx} = 2x - 7$

10. $y = x^2 + 4x;$
$y + \Delta y = (x + \Delta x)^2 + 4(x + \Delta x)$
$\qquad = x^2 + 2x\Delta x + \Delta x^2 + 4x + 4\Delta x$
$\Delta y = x^2 + 2x\Delta x + \Delta x^2 + 4x + 4\Delta x - x^2 - 4x$
$\qquad = 2x\Delta x + \Delta x^2 + 4\Delta x$

$\dfrac{\Delta y}{\Delta x} = 2x + \Delta x + 4;\ \lim_{\Delta x \to 0} 2x + \Delta x + 4 = 2x + 4;$

$\dfrac{dy}{dx} = 2x + 4$

11. $y = 8x - 2x^2;$
$y + \Delta y = 8(x + \Delta x) - 2(x + \Delta x)^2$
$\qquad = 8x + 8\Delta x - 2x^2 - 4x\Delta x - 2\Delta x^2$
$\Delta y = 8x + 8\Delta x - 2x^2 - 4x\Delta x - 2\Delta x^2 - 8x + 2x^2$
$\qquad = 8\Delta x - 4x\Delta x - 2\Delta x^2$

$\dfrac{\Delta y}{\Delta x} = 8 - 4x - 2\Delta x;\ \lim_{\Delta x \to 0} 8 - 4x - 2\Delta x = 8 - 4x$

$\dfrac{dy}{dx} = 8 - 4x$

12. $y = 3x - \dfrac{1}{2}x^2;$
$y + \Delta y = 3(x + \Delta x) - \dfrac{1}{2}(x + \Delta x)^2$

$\qquad = 3x + 3\Delta x - \dfrac{1}{2}x^2 - x\Delta x - \dfrac{1}{2}\Delta x^2$

$\Delta y = 3x + 3\Delta x - \dfrac{1}{2}x^2 - x\Delta x - \dfrac{1}{2}\Delta x^2 - 3x + \dfrac{1}{2}x^2$

$\qquad = 3\Delta x - x\Delta x - \dfrac{1}{2}\Delta x^2$

$\dfrac{\Delta y}{\Delta x} = 3 - x - \dfrac{1}{2}\Delta x;\ \lim_{\Delta x \to 0} 3 - x - \dfrac{1}{2}\Delta x = 3 - x;$

$\dfrac{dy}{dx} = 3 - x$

13. $y = x^3 + 4x - 6$;
$$y + \Delta y = (x + \Delta x)^3 + 4(x + \Delta x) - 6$$
$$y + \Delta y = x^3 + 3x^2\Delta x + 3x(\Delta x)^2 + (\Delta x)^3 + 4x + 4\Delta x - 6$$
$$y + \Delta y - y = x^3 + 3x^2\Delta x + 3x(\Delta x)^2 + (\Delta x)^3 + 4x + 4\Delta x$$
$$\qquad\quad - 6 - (x^3 + 4x - 6)$$
$$\Delta y = 3x^2\Delta x + 3x(\Delta x)^2 + (\Delta x)^3 + 4\Delta x$$

$$\frac{\Delta y}{\Delta x} = \frac{\Delta x[3x^2 + 3x\Delta x + (\Delta x)^2 + 4]}{\Delta x} = 3x^2 + 3x\Delta x + (\Delta x)^2 + 4$$

$$\lim_{\Delta x \to 0} [3x^2 + 3x\Delta x + (\Delta x)^2 + 4] = 3x^2 + 4$$

$$\frac{dy}{dx} = 3x^2 + 4$$

14. $y = 2x - 4x^3$; $y + \Delta y = 2(x + \Delta x) - 4(x + \Delta x)^3$
$$\Delta y = 2x + 2\Delta x - 4(x^3 + 3x^2\Delta x + 3x\Delta x^2 + \Delta x^3) - 2x + 4x^3$$
$$\Delta y = 2x + 2\Delta x - 4x^3 - 12x^2\Delta x - 12x\Delta x^2 - 4\Delta x^3 - 2x + 4x^3$$
$$\Delta y = 2\Delta x - 12x^2\Delta x - 12x\Delta x^2 - 4\Delta x^3$$

$$\frac{\Delta y}{\Delta x} = 2 - 12x^2 - 12x\Delta x - 4\Delta x^2;$$

$$\lim_{\Delta x \to 0} \frac{\Delta y}{\Delta x} = 2 - 12x^2$$

$$\frac{dy}{dx} = 2 - 12x^2$$

15. $y = \dfrac{1}{x + 2}$; $y + \Delta y = \dfrac{1}{x + \Delta x + 2}$

$$\Delta y = \frac{1}{x + \Delta x + 2} - \frac{1}{x + 2}$$

$$\Delta y = \frac{x + 2 - (x + \Delta x + 2)}{(x + 2)(x + \Delta x + 2)} = \frac{x + 2 - x - \Delta x - 2}{(x + 2)(x + \Delta x + 2)} = \frac{-\Delta x}{(x + 2)(x + \Delta x + 2)}$$

$$\frac{\Delta y}{\Delta x} = \frac{-1}{(x + 2)(x + \Delta x + 2)};$$

$$\lim_{\Delta x \to 0} \frac{\Delta y}{\Delta x} = \frac{-1}{(x + 2)(x + 2)}$$

$$\frac{dy}{dx} = \frac{-1}{(x + 2)^2}$$

16. $\qquad y = \dfrac{3}{2x + 1}$

$$y + \Delta y = \frac{3}{2(x + \Delta x) + 1}$$

$$\Delta y = \frac{3}{2(x + \Delta x) + 1} - y = \frac{3}{2(x + \Delta x) + 1} - \frac{3}{2x + 1} = \frac{3(2x + 1) - 3(2(x + \Delta x) + 1)}{(2(x + \Delta x) + 1)(2x + 1)}$$

$$\Delta y = \frac{6x + 3 - 3(2x + 2\Delta x + 1)}{(2(x + \Delta x) + 1)(2x + 1)} = \frac{6x + 3 - 6x - 6\Delta x - 3}{(2(x + \Delta x) + 1)(2x + 1)} = \frac{-6\Delta x}{(2(x + \Delta x) + 1)(2x + 1)}$$

$$\lim_{\Delta x \to 0} \frac{\Delta y}{\Delta x} = \lim_{\Delta x \to 0} \frac{-6}{(2(x + \Delta x) + 1)(2x + 1)} = \frac{-6}{(2x + 1)^2}$$

$$\frac{dy}{dx} = \frac{-6}{(2x + 1)^2}$$

17. $y = x + \dfrac{4}{3x}$; $\quad y + \Delta y = x + \Delta x + \dfrac{4}{3(x + \Delta x)}$

$$y + \Delta y - y = x + \Delta x + \frac{4}{3(x + \Delta x)} - x - \frac{4}{3x}$$

$$\Delta y = \Delta x + \frac{4}{3(x + \Delta x)} - \frac{4}{3x}$$

$$\Delta y = \frac{\Delta x(3x(x + \Delta x)) + 4x - 4(x + \Delta x)}{3x(x + \Delta x)}$$

$$\frac{\Delta y}{\Delta x} = \frac{3x^2 + 3x\Delta x - 4}{3x(x + \Delta x)}$$

$$\lim_{\Delta x \to 0} \frac{3x^2 + 3x\Delta x - 4}{3x(x + \Delta x)} = \frac{3x^2 - 4}{3x^2} = 1 - \frac{4}{3x^2}; \quad \frac{dy}{dx} = 1 - \frac{4}{3x^2}$$

18. $y = \dfrac{x}{x - 1}$; $\quad y + \Delta y = \dfrac{x + \Delta x}{x + \Delta x - 1}$

$$\Delta y = \frac{x + \Delta x}{x + \Delta x - 1} - \frac{x}{x - 1}$$

$$\Delta y = \frac{(x + \Delta x)(x - 1) - x(x + \Delta x - 1)}{(x + \Delta x - 1)(x - 1)} = \frac{x^2 + x\Delta x - x - \Delta x - x^2 - x\Delta x + x}{(x + \Delta x - 1)(x - 1)}$$

$$\Delta y = \frac{-\Delta x}{(x + \Delta x - 1)(x - 1)}$$

$$\frac{\Delta y}{\Delta x} = \frac{-1}{(x + \Delta x - 1)(x - 1)}; \quad \lim_{\Delta x \to 0} \frac{\Delta y}{\Delta x} = \frac{-1}{(x - 1)(x - 1)}; \quad \frac{dy}{dx} = \frac{-1}{(x - 1)^2}$$

19. $y = \dfrac{2}{x^2}$; $\quad y + \Delta y = \dfrac{2}{(x + \Delta x)^2}$

$$\Delta y = \frac{2}{(x + \Delta x)^2} - \frac{2}{x^2} = \frac{2x^2 - 2(x + \Delta x)^2}{(x + \Delta x)^2 \cdot x^2}$$

$$\Delta y = \frac{2x^2 - 2x^2 - 4x\Delta x - 2\Delta x^2}{x^2(x + \Delta x)^2} = \frac{-2\Delta x(2x + \Delta x)}{x^2(x + \Delta x)^2}$$

$$\frac{\Delta y}{\Delta x} = \frac{-2(2x + \Delta x)}{x^2(x + \Delta x)^2}; \quad \lim_{\Delta x \to 0} \frac{\Delta y}{\Delta x} = \frac{-4x}{x^4}; \quad \frac{dy}{dx} = \frac{-4}{x^3}$$

20. $y = \dfrac{2}{x^2 + 4}$; $\quad y + \Delta y = \dfrac{2}{(x + \Delta x)^2 + 4}$

$$\Delta y = \frac{2}{(x + \Delta x)^2 + 4} - \frac{2}{x^2 + 4}$$

$$\Delta y = \frac{2(x^2 + 4) - 2(x^2 + 2x\Delta x + \Delta x^2 + 4)}{[(x + \Delta x)^2 + 4](x^2 + 4)} = \frac{2x^2 + 8 - 2x^2 - 4x\Delta x - 2\Delta x^2 - 8}{[(x + \Delta x)^2 + 4](x^2 + 4)}$$

$$\Delta y = \frac{-4x\Delta x - 2\Delta x^2}{[(x + \Delta x)^2 + 4](x^2 + 4)}$$

$$\frac{\Delta y}{\Delta x} = \frac{-4x - 2\Delta x}{[(x + \Delta x)^2 + 4](x^2 + 4)}$$

$$\frac{dy}{dx} = \lim_{\Delta x \to 0} \frac{\Delta y}{\Delta x} = \frac{-4x}{(x^2 + 4)^2}$$

21. $y = x^4 + x^3 + x^2 + x$

$y + \Delta y = (x + \Delta x)^4 + (x + \Delta x)^3 + (x + \Delta x)^2 + (x + \Delta x)$

$\qquad = [x^4 + 4x^3\Delta x + 6x^2(\Delta x)^2 + 4x(\Delta x)^3 + (\Delta x)^4]$

$\qquad\qquad + [x^3 + 3x^2\Delta x + 3x(\Delta x)^2 + (\Delta x)^3] + [x^2 + 2x\Delta x + (\Delta x)^2] + (x + \Delta x)$

$y + \Delta y - y = 4x^3\Delta x + 6x^2(\Delta x)^2 + 4x(\Delta x)^3 + (\Delta x)^4 + 3x^2\Delta x + 3x(\Delta x)^2$

$\qquad\qquad + (\Delta x)^3 + 2x\Delta x + (\Delta x)^2 + \Delta x$

$\Delta y = (\Delta x)[4x^3 + 6x^2\Delta x + 4x(\Delta x)^2 + (\Delta x)^3 + 3x^2 + 3x(\Delta x) + (\Delta x)^2 + 2x + \Delta x + 1]$

$\dfrac{\Delta y}{\Delta x} = (\Delta x)[4x^3 + 6x^2\Delta x + 4x(\Delta x)^2 + (\Delta x)^3 + 3x^2 + 3x(\Delta x) + (\Delta x)^2 + 2x + \Delta x + 1]/\Delta x$

$\dfrac{\Delta y}{\Delta x} = 4x^3 + 6x^2\Delta x + 4x(\Delta x)^2 + (\Delta x)^3 + 3x^2 + 3x(\Delta x) + (\Delta x)^2 + 2x + \Delta x + 1$

$\lim\limits_{\Delta x \to 0} \dfrac{\Delta y}{\Delta x} = 4x^3 + 3x^2 + 2x + 1; \quad \dfrac{dy}{dx} = 4x^3 + 3x^2 + 2x + 1$

22. $y = \dfrac{1}{3}x^3 + \dfrac{1}{2}x^2 + x$

$y + \Delta y = \dfrac{1}{3}(x + \Delta x)^3 + \dfrac{1}{2}(x + \Delta x)^2 + x + \Delta x$

$y + \Delta y = \dfrac{1}{3}(x^3 + 3x^2\Delta x + 3x\Delta x^2 + \Delta x^3) + \dfrac{1}{2}(x^2 + 2x\Delta x + \Delta x^2) + x + \Delta x$

$y + \Delta y = \dfrac{1}{3}x^3 + x^2\Delta x + x\Delta x^2 + \dfrac{1}{3}\Delta x^3 + \dfrac{1}{2}x^2 + x\Delta x + \dfrac{1}{2}\Delta x^2 + x + \Delta x$

$\Delta y = x^2\Delta x + x\Delta x^2 + \dfrac{1}{3}\Delta x^3 + x\Delta x + \dfrac{1}{2}\Delta x^2 + \Delta x$

$\dfrac{\Delta y}{\Delta x} = x^2 + x\Delta x + \dfrac{1}{3}\Delta x^2 + x + \dfrac{1}{2}\Delta x + 1; \quad \dfrac{dy}{dx} = \lim\limits_{\Delta x \to 0} \dfrac{\Delta y}{\Delta x} = x^2 + x + 1$

23. $y = x^4 - \dfrac{2}{x}$

$y + \Delta y = (x + \Delta x)^4 - \dfrac{2}{x + \Delta x} = \dfrac{(x + \Delta x)^5 - 2}{x + \Delta x}$

$\Delta y = \dfrac{x^5 + 5x^4\Delta x + 10x^3\Delta x^2 + 10x^2\Delta x^3 + 5x\Delta x^4 + \Delta x^5 - 2}{x + \Delta x} - \dfrac{(x^5 - 2)}{x}$

$\Delta y = (x^6 + 5x^5\Delta x + 10x^4\Delta x^2 + 10x^3\Delta x^3 + 5x^2\Delta x^4 + x\Delta x^5 - 2x - (x^5 - 2)(x + \Delta x)/(x(x + \Delta x))$

$\Delta y = (x^6 + 5x^5\Delta x + 10x^4\Delta x^2 + 10x^3\Delta x^3 + 5x^2\Delta x^4 + x\Delta x^5 - 2x - x^6 + 2x - x^5\Delta x + 2\Delta x)/(x(x + \Delta x))$

$\dfrac{\Delta y}{\Delta x} = \dfrac{5x^5 + 10x^4\Delta x + 10x^3\Delta x^2 + 5x^2\Delta x^3 + x\Delta x^4 - x^5 + 2}{x(x + \Delta x)}$

$\dfrac{dy}{dx} = \lim\limits_{\Delta x \to 0} \dfrac{\Delta y}{\Delta x} = \dfrac{5x^5 - x^5 + 2}{x^2} = \dfrac{4x^5 + 2}{x^2} = 4x^3 + \dfrac{2}{x^2}$

24. $y = \dfrac{1}{x} + \dfrac{1}{x^2}$

$y + \Delta y = \dfrac{1}{x + \Delta x} + \dfrac{1}{(x + \Delta x)^2} = \dfrac{(x + \Delta x) + 1}{(x + \Delta x)^2}$

$\Delta y = \dfrac{x + \Delta x + 1}{(x + \Delta x)^2} - \left(\dfrac{x + 1}{x^2}\right) = \dfrac{(x + \Delta x + 1)x^2 - (x + 1)(x + \Delta x)^2}{x^2(x + \Delta x)^2}$

$\Delta y = (x^3 + x^2\Delta x + x^2 - x^3 - 2x^2\Delta x - x\Delta x^2 - x^2 - 2x\Delta x - \Delta x^2)/(x^2(x + \Delta x)^2)$

$\Delta y = \dfrac{-x^2\Delta x - x\Delta x^2 - 2x\Delta x - \Delta x^2}{x^2(x + \Delta x)^2}$

$\dfrac{\Delta y}{\Delta x} = \dfrac{-x^2 - x\Delta x - 2x - \Delta x}{x^2(x + \Delta x)^2}$

$\dfrac{dy}{dx} = \lim\limits_{\Delta x \to 0} \dfrac{\Delta y}{\Delta x} = \dfrac{-x^2 - 2x}{x^4} = \dfrac{x(-x - 2)}{x^4} = \dfrac{-x - 2}{x^3}; \dfrac{dy}{dx} = \dfrac{-x}{x^3} - \dfrac{2}{x^3} = \dfrac{-1}{x^2} - \dfrac{2}{x^3}$

25. $y = 3x^2 - 2x; \ (-1, 5)$

$y + \Delta y = 3(x + \Delta x)^2 - 2(x + \Delta x)$

$y + \Delta y = 3x^2 + 6x\Delta x + 3\Delta x^2 - 2x - 2\Delta x$

$y + \Delta y - y = 3x^2 + 6x\Delta x + 3\Delta x^2 - 2x - 2\Delta x - 3x^2 + 2x$

$\Delta y = 6x\Delta x + 3\Delta x^2 - 2\Delta x$

$\dfrac{\Delta y}{\Delta x} = 6x + 3\Delta x - 2; \dfrac{dy}{dx} = \lim\limits_{\Delta x \to 0} \dfrac{\Delta y}{\Delta x} = 6x - 2; \ \left.\dfrac{dy}{dx}\right|_{(-1.5)} = 6(-1) - 2 = -8$

26. $y = 9x - x^3; \ (2, 10)$

$y + \Delta y = 9(x + \Delta x) - (x + \Delta x)^3 = 9x + 9\Delta x - x^3 - 3x^2\Delta x - 3x\Delta x^2 - \Delta x^3$

$\Delta y = 9\Delta x - 3x^2\Delta x - 3x\Delta x^2 - \Delta x^3$

$\dfrac{\Delta y}{\Delta x} = 9 - 3x^2 - 3x\Delta x - \Delta x^2; \dfrac{dy}{dx} = \lim\limits_{\Delta x \to 0} \dfrac{\Delta y}{\Delta x} = 9 - 3x^2; \ \left.\dfrac{dy}{dx}\right|_{(2,10)} = 9 - 3(2^2) = -3$

27. $\qquad y = \dfrac{11}{3x + 2}$

$y + \Delta y = \dfrac{11}{3(x + \Delta x) + 2}$

$\Delta y = \dfrac{11}{3(x + \Delta x) + 2} - y = \dfrac{11}{3(x + \Delta x) + 2} - \dfrac{11}{3x + 2}$

$\Delta y = \dfrac{11(3x + 2) - 11(3(x + \Delta x) + 2)}{(3(x + \Delta x) + 2)(3x + 2)} = \dfrac{33x + 22 - 33x - 33\Delta x - 22}{(3(x + \Delta x) + 2)(3x + 2)} = \dfrac{-33\Delta x}{(3(x + \Delta x) + 2)(3x + 2)}$

$\lim\limits_{\Delta x \to 0} \dfrac{\Delta y}{\Delta x} = \lim\limits_{\Delta x \to 0} \dfrac{-33}{(3(x + \Delta x) + 2)(3x + 2)} = \dfrac{-33}{(3x + 2)^2}; \dfrac{dy}{dx} = \dfrac{-33}{(3x + 2)^2}$

$\left.\dfrac{dy}{dx}\right|_{x=3} = \dfrac{-33}{(3 \cdot 3 + 2)^2} = \dfrac{-3}{11} = -0.\overline{27}$

28. $y = x^2 - \dfrac{2}{x}; \ (-2, 5)$

$$y + \Delta y = (x + \Delta x)^2 - \frac{2}{x + \Delta x}$$

$$\Delta y = x^2 + 2x\Delta x + \Delta x^2 - \frac{2}{x + \Delta x} - x^2 + \frac{2}{x}$$

$$\Delta y = 2x\Delta x + \Delta x^2 - \frac{2}{x + \Delta x} + \frac{2}{x} = 2x\Delta x + \Delta x^2 + \frac{-2x + 2x + 2\Delta x}{x(x + \Delta x)}$$

$$\Delta y = 2x\Delta x + \Delta x^2 + \frac{2\Delta x}{x(x + \Delta x)}$$

$$\frac{\Delta y}{\Delta x} = 2x + \Delta x + \frac{2}{x(x + \Delta x)}; \ \frac{dy}{dx} = \lim_{\Delta x \to 0} \frac{\Delta y}{\Delta x} = 2x + \frac{2}{x^2}; \ \frac{dy}{dx}\bigg|_{(-2,5)} = -4 + \frac{1}{2} = -\frac{7}{2}$$

```
nDeriv(X²-2/X,X,
-2)
          -3.499999875
```

29. $y = 1 + \dfrac{2}{x}$

$$y + \Delta y = 1 + \frac{2}{x + \Delta x} = \frac{x + \Delta x + 2}{x + \Delta x}$$

$$y + \Delta y - y = \frac{x + \Delta x + 2}{x + \Delta x} - 1 - \frac{2}{x} = \frac{x^2 + x\Delta x + 2x - x^2 - x\Delta x - 2x - 2\Delta x}{x(x + \Delta x)}$$

$$\Delta y = \frac{-2\Delta x}{x(x + \Delta x)}; \ \frac{\Delta y}{\Delta x} = \frac{-2}{x(x + \Delta x)}$$

$$\frac{dy}{dx} = \lim_{\Delta x \to 0} \frac{\Delta y}{\Delta x} = -\frac{2}{x^2}$$

Differentiable for all x, except $x = 0$.

30. $y = \dfrac{5x}{x - 4}$

$$y + \Delta y = \frac{5(x + \Delta x)}{x + \Delta x - 4} = \frac{5x + 5\Delta x}{x - 4 + \Delta x}$$

$$\Delta y = \frac{5x + 5\Delta x}{x - 4 + \Delta x} - y = \frac{5x + 5\Delta x}{x - 4 + \Delta x} - \frac{5x}{x - 4} = \frac{(5x + 5\Delta x)(x - 4) - 5x(x - 4 + \Delta x)}{(x - 4 + \Delta x)(x - 4)}$$

$$\Delta y = \frac{5x^2 - 20x + 5x\Delta x - 20\Delta x - 5x^2 + 20x - 5x\Delta x}{(x - 4 + \Delta x)(x - 4)} = \frac{-20\Delta x}{(x - 4 + \Delta x)(x - 4)}$$

$$\lim_{\Delta x \to 0} \frac{\Delta y}{\Delta x} = \lim_{\Delta x \to 0} \frac{-20}{(x - 4 + \Delta x)(x - 4)} = \frac{-20}{(x - 4)^2}; \ \frac{dy}{dx} = \frac{-20}{(x - 4)^2}; \ \text{differentiable for all } x \text{ except } x = 4.$$

31. $y = \dfrac{3}{x^2 - 1}; \ y + \Delta y = \dfrac{3}{(x + \Delta x)^2 - 1}$

$$\Delta y = \frac{3}{(x + \Delta x)^2 - 1} - \frac{3}{x^2 - 1}$$

$$\Delta y = \frac{3(x^2 - 1) - 3(x^2 + 2x\Delta x + \Delta x^2 - 1)}{[(x + \Delta x)^2 - 1](x^2 - 1)}$$

$$\Delta y = \frac{3x^2 - 3 - 3x^2 - 6x\Delta x - 3\Delta x^2 + 3}{[(x + \Delta x)^2 - 1](x^2 - 1)} = \frac{-6x\Delta x - 3\Delta x^2}{[(x + \Delta x)^2 - 1](x^2 - 1)}$$

$$\frac{\Delta y}{\Delta x} = \frac{-6x - 3\Delta x}{[(x + \Delta x)^2 - 1](x^2 - 1)}; \ \frac{dy}{dx} = \lim_{\Delta x \to 0} \frac{\Delta y}{\Delta x} = \frac{-6x}{(x^2 - 1)^2}$$

Differentiable for all x, except $x = \pm 1$.

32. $y = \dfrac{2}{x^2+1}; \; y + \Delta y = \dfrac{2}{(x+\Delta x)^2+1}$

$\Delta y = \dfrac{2}{(x+\Delta x)^2+1} - \dfrac{2}{x^2+1}$

$\Delta y = \dfrac{2(x^2-1)-2(x^2+2x\Delta x+\Delta x^2+1)}{[(x+\Delta x)^2+1](x^2+1)} = \dfrac{2x^2+2-2x^2-4x\Delta x-2\Delta x^2-2}{[(x+\Delta x)^2+1](x^2+1)}$

$\Delta y = \dfrac{-4x\Delta x - 2\Delta x^2}{[(x+\Delta x)^2+1](x^2+1)}$

$\dfrac{\Delta y}{\Delta x} = \dfrac{-4x+2\Delta x}{[(x+\Delta x)^2+1](x^2+1)}; \; \dfrac{dy}{dx} = \lim\limits_{\Delta x \to 0} \dfrac{\Delta y}{\Delta x} = \dfrac{-4x}{(x^2+1)^2}$

Differentiable for all x.

33. $y = \sqrt{x+1}; \; y^2 = x+1; \; (y+\Delta y)^2 = x+\Delta x+1$

$y^2 + 2y\Delta y + \Delta y^2 = x + \Delta x + 1$

$2y\Delta y + \Delta y^2 = x + \Delta x + 1 - x - 1 = \Delta x$

$\Delta y(2y + \Delta y) = \Delta x$

$\Delta y = \dfrac{\Delta x}{2y + \Delta y}; \; \dfrac{\Delta y}{\Delta x} = \dfrac{1}{2y + \Delta y}; \; \begin{pmatrix} \Delta x \to 0 \\ \Delta y \to 0 \end{pmatrix}$

$\dfrac{dy}{dx} = \lim\limits_{\Delta x \to 0} \dfrac{\Delta y}{\Delta x} = \lim\limits_{\Delta y \to 0} \dfrac{1}{2y + \Delta y} = \dfrac{1}{2y}$

$\dfrac{dy}{dx} = \dfrac{1}{2y} = \dfrac{1}{2\sqrt{x+1}}$

34. $y = \sqrt{x^2+3}; \; y^2 = x^2+3; \; (y+\Delta y)^2 = (x+\Delta x)^2+3; \; y^2+2y\Delta y+\Delta y^2 = x^2+2x\Delta x+\Delta x^2+3$

$2y\Delta y + \Delta y^2 = x^2+2x\Delta x+\Delta x^2+3-x^2-3; \; \Delta y(2y+\Delta y) = 2x\Delta x+\Delta x^2$

$\Delta y = \dfrac{\Delta x(2x+\Delta x)}{2y+\Delta y}; \; \dfrac{\Delta y}{\Delta x} = \dfrac{2x+\Delta x}{2y+\Delta y}; \; \dfrac{dy}{dx} = \lim\limits_{\substack{\Delta x \to 0 \\ \Delta y \to 0}} \dfrac{\Delta y}{\Delta x} = \dfrac{2x}{2y} = \dfrac{x}{y} = \dfrac{x}{\sqrt{x^2+3}}$

35. $y = \sqrt{x}; \; y+\Delta y = \sqrt{x+\Delta x}; \; \Delta y = \sqrt{x+\Delta x} - \sqrt{x}$

$\Delta y = \dfrac{(\sqrt{x+\Delta x}-\sqrt{x})(\sqrt{x+\Delta x}+\sqrt{x})}{(\sqrt{x+\Delta x}+\sqrt{x})} = \dfrac{x+\Delta x-x}{\sqrt{x+\Delta x}+\sqrt{x}}$

$\dfrac{\Delta y}{\Delta x} = \dfrac{1}{\sqrt{x+\Delta x}+\sqrt{x}}; \; \dfrac{dy}{dx} = \lim\limits_{\Delta x \to 0} \dfrac{\Delta y}{\Delta x} = \dfrac{1}{\sqrt{x}+\sqrt{x}} = \dfrac{1}{2\sqrt{x}}$

36. $y = \sqrt{x-2}; \; y+\Delta y = \sqrt{x+\Delta x-2}$

$\Delta y = \sqrt{x+\Delta x-2} - \sqrt{x-2}$

$\Delta y = \dfrac{(\sqrt{x+\Delta x-2}-\sqrt{x-2})(\sqrt{x+\Delta x-2}+\sqrt{x-2})}{(\sqrt{x+\Delta x-2}+\sqrt{x-2})}$

$\Delta y = \dfrac{x+\Delta x-2-(x-2)}{\sqrt{x+\Delta x-2}+\sqrt{x-2}} = \dfrac{\Delta x}{\sqrt{x+\Delta x-2}+\sqrt{x-2}}$

$\dfrac{\Delta y}{\Delta x} = \dfrac{1}{\sqrt{x+\Delta x-2}+\sqrt{x-2}}; \; \dfrac{dy}{dx} = \lim\limits_{\Delta x \to 0} \dfrac{\Delta y}{\Delta x} = \dfrac{1}{2\sqrt{x-2}}$

The function is differentiable for $2 < x$. The function, exists at $x = 2$ but the derivative does not exist.

3.4 The Derivative as an Instantaneous Rate of Change

1. $y = x^2 - 1;\ (2,3)$

$y + \Delta y = (x + \Delta x)^2 - 1$

$y + \Delta y = x^2 + 2x\Delta x + (\Delta x)^2 - 1$

$y + \Delta y - y = x^2 + 2x\Delta x + (\Delta x)^2 - 1 - (x^2 - 1)$

$\Delta y = 2x\Delta x + (\Delta x)^2$

$\dfrac{\Delta y}{\Delta x} = 2x + \Delta x$

$\displaystyle\lim_{\Delta x \to 0} \frac{\Delta y}{\Delta x} = 2x;\ m_{\tan(2,3)} = 2(2) = 4$

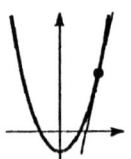

2. $y = 2x - x^2;\ (-1,-3)$

$y + \Delta y = 2(x + \Delta x) - (x + \Delta x)^2$

$y + \Delta y = 2x + 2\Delta x - x^2 - 2x\Delta x - \Delta x^2$

$\Delta y = 2x + 2\Delta x - x^2 - 2x\Delta x - \Delta x^2 - 2x + x^2$

$\Delta y = 2x - 2x\Delta - \Delta x^2 = \Delta x(2 - 2x - \Delta x)$

$\dfrac{\Delta y}{\Delta x} = 2 - 2x - \Delta x;\ \displaystyle\lim_{\Delta x \to 0} \frac{\Delta y}{\Delta x} = 2 - 2x = \frac{dy}{dx}$

$m_{\tan(-1,-3)} = 2 - 2(-1) = 4$

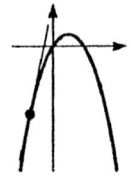

3. $\qquad y = \dfrac{16}{3x + 1} \qquad (-3,-2)$

$y + \Delta y = \dfrac{16}{3(x + \Delta x) + 1}$

$\Delta y = \dfrac{16}{3(x + \Delta x) + 1} - y = \dfrac{16}{3(x + \Delta x) + 1} - \dfrac{16}{3x + 1}$

$\Delta y = \dfrac{16(3x + 1) - 16(3(x + \Delta x) + 1)}{(3(x + \Delta x) + 1)(3x + 1)} = \dfrac{48x + 16 - 48x - 48\Delta x - 16}{(3(x + \Delta x) + 1)(3x + 1)}$

$\dfrac{dy}{dx} = \displaystyle\lim_{\Delta x \to 0} \frac{\Delta y}{\Delta x} = \lim_{\Delta x \to 0} \frac{-48}{(3(x + \Delta x) + 1)(3x + 1)} = \frac{-48}{(3x + 2)^2}$

$\left.\dfrac{dy}{dx}\right|_{(-3,-2)} = \dfrac{-48}{(3(-3) + 1)^2} = \dfrac{-3}{4}$

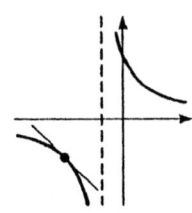

4. $y = 3 - \dfrac{16}{x^2}$; $(2, -1)$

$$y + \Delta y = 3 - \frac{16}{(x + \Delta x)^2}$$

$$\Delta y = 3 - \frac{16}{(x + \Delta x)^2} - 3 + \frac{16}{x^2}$$

$$\Delta y = \frac{-16}{(x + \Delta x)^2} + \frac{16}{x^2}$$

$$\Delta y = \frac{-16(x^2) + 16(x + \Delta x)^2}{(x + \Delta x)^2 (x^2)}$$

$$\Delta y = \frac{-16x^2 + 16x^2 + 32x\Delta x + 16x\Delta x^2}{(x + \Delta x)^2 x^2}$$

$$\Delta y = \frac{\Delta x(32x + 16\Delta x)}{(x + \Delta x)^2 x^2}$$

$$\frac{\Delta y}{\Delta x} = \frac{32x + 16\Delta x}{(x + \Delta x)^2 x^2}$$

$$\lim_{\Delta x \to 0} \frac{\Delta y}{\Delta x} = \frac{32x}{x^4} = \frac{32}{x^3}$$

$$m_{\tan(2, -1)} = \frac{32}{2^3} = 4$$

5. $s = 4t + 10$

	Q_1	Q_2	Q_3	Q_4	Q_5	P
t_2	2.0	2.5	2.9	2.99	2.999	3
s_2	18	20	21.6	21.96	21.996	22
t_{2-3}	-1	-0.5	-0.1	-0.01	-0.001	
$s_2 - 22$	-4	-2	-0.4	-0.04	-0.004	
$v = \frac{s_2 - 22}{t_2 - 3}$	4.00	4.00	4.00	4.00	4.00	

$$\lim_{t \to 3} \frac{\Delta s}{\Delta t} = 4 \text{ ft/s}$$

6. $s = 6 - 3t$; $t = 4$, $s = -6$

t	3.0	3.5	3.9	3.99	3.999
s	-3.0	-4.5	-5.7	-5.97	-5.997
$(-6 - s)\Delta s$	-3.0	-1.5	-0.3	-0.03	-0.003
$(4 - t)\quad \Delta t$	1.0	0.5	0.1	0.01	0.001
$\dfrac{\Delta s}{\Delta t}\qquad v$	-3.0	-3.0	-3.0	-3.0	-3.0

$$\lim_{t \to 4} \frac{\Delta s}{\Delta t} = v = 3 \text{ ft/s}$$

7. $s = 3t^2 - 4t;\ t = 2, s = 4$

t	1.0	1.5	1.9	1.99	1.999
s	-1.0	0.75	3.23	3.9203	2.992003
Δs	5.0	3.25	0.77	0.0797	0.007997
Δt	1.0	0.5	0.1	0.01	0.001
$\dfrac{\Delta s}{\Delta t}$	5.0	6.5	7.7	7.97	7.997

$$\lim_{t \to 2} \frac{\Delta s}{\Delta t} = v = 8 \text{ ft/s}$$

8. $s = 120t - 16t^2;\ t = 0.5, s = 56$

t	0.4	0.45	0.49	0.499	0.4999
s	45.44	50.76	54.9584	55.895984	55.989600
Δs	10.56	5.24	1.0416	0.104016	0.01040016
Δt	0.1	0.05	0.01	0.001	0.0001
$\dfrac{\Delta s}{\Delta t}$	105.6	104.8	104.16	104.016	104.0016

$$\lim_{t \to 0.5} \frac{\Delta s}{\Delta t} = v = 104 \text{ ft/s}$$

9. $s = 4t + 10;\ t = 3$

$s + \Delta s = 4(t + \Delta t) + 10 = 4t + 4\Delta t + 10$

$s + \Delta s - s = 4t + 4\Delta t + 10 - (4t + 10)$

$\Delta s = 4\Delta t;\ \dfrac{\Delta s}{\Delta t} = 4;\ \lim\limits_{\Delta t \to 0} \dfrac{\Delta s}{\Delta t} = \dfrac{ds}{dt} = 4$

$\left.\dfrac{ds}{dt}\right|_{t=3} = 4 \text{ ft/s}$

10. $s = 6 - 3t;\ t = 4$

$s + \Delta s = 6 - 3(t + \Delta t) = 6 - 3t - 3\Delta t$

$\Delta s = 6 - 3t - 3\Delta t - 6 + 3t = -3\Delta t$

$\dfrac{\Delta s}{\Delta t} = -3;\ \lim\limits_{\Delta t \to 0} \dfrac{\Delta s}{\Delta t} = \dfrac{ds}{dt} = -3;$

$\left.\dfrac{ds}{dt}\right|_{t=4} = -3 \text{ ft/s}$

11. $s = 3t^2 - 4t;\ t = 2$

$s + \Delta s = 3(t + \Delta t)^2 - 4(t + \Delta t)$

$\qquad\quad = 3(t^2 + 2t\Delta t + \Delta t^2) - 4t - 4\Delta t$

$\Delta s = 3t^2 + 6t\Delta t + 3\Delta t^2 - 4t - 4\Delta t - 3t^2 + 4t$

$\Delta s = \Delta t(6t + 3t\Delta t - 4);\ \dfrac{\Delta s}{\Delta t} = 6t + 3t\Delta t - 4$

$\lim\limits_{\Delta t \to 0} \dfrac{\Delta s}{\Delta t} = \dfrac{ds}{dt} = 6t - 4;$

$\left.\dfrac{ds}{dt}\right|_{t=2} = 8 \text{ ft/s}$

12. $s = 120t - 16t^2; t = 0.5$

$s + \Delta s = 120(t + \Delta t) - 16(t + \Delta t)^2$

$s + \Delta s = 120t + 120\Delta t - 16t^2 - 32t\Delta t - 16\Delta t^2$

$\Delta s = 120t + 120\Delta t - 16t^2 - 32t\Delta t - 16\Delta t^2 - 120t + 16t^2$

$\Delta s = \Delta t(120 - 32t - 16\Delta t);$

$\dfrac{\Delta s}{\Delta t} = 120 - 32t - 16\Delta t$

$\lim\limits_{\Delta t \to 0} \dfrac{\Delta s}{\Delta t} = \dfrac{ds}{dt} = 120 - 32t;$

$\dfrac{ds}{dt}\bigg|_{t=0.5} = 120 - 16 = 104 \text{ ft/s}$

13. $v = \lim\limits_{\Delta t \to 0} \dfrac{\Delta s}{\Delta t}; \ s + \Delta s = 3(t + \Delta t) - \dfrac{2}{5(t + \Delta t)}$

$s + \Delta s - s = 3t + 3\Delta t - \dfrac{2}{5(t + \Delta t)} - \left(3t - \dfrac{2}{5t}\right)$

$\Delta s = 3\Delta t - \dfrac{2}{5(t + \Delta t)} + \dfrac{2}{5t}$

$= 3\Delta t + \dfrac{-2t + 2(t + \Delta t)}{5(t + \Delta t)(t)}$

$\Delta s = 3\Delta t + \dfrac{2\Delta t}{5(t + \Delta t)(t)};$

$\dfrac{\Delta s}{\Delta t} = \dfrac{3\Delta t}{\Delta t} + \dfrac{2\Delta t}{5(t + \Delta t)(t)} \times \dfrac{1}{\Delta t}$

$\dfrac{\Delta s}{\Delta t} = 3 + \dfrac{2}{5(t + \Delta t)(t)}$

$\lim\limits_{\Delta t \to 0} \dfrac{\Delta s}{\Delta t} = \dfrac{ds}{dt} = 3 + \dfrac{2}{5t^2}$

$= v \text{ (instantaneous velocity)}$

14. $s = \dfrac{2t}{t + 2}; \ s + \Delta s = \dfrac{2(t + \Delta t)}{t + \Delta t + 2}$

$\Delta s = \dfrac{2t + 2\Delta t}{t + \Delta t + 2} - \dfrac{2t}{t + 2}$

$= \dfrac{(2t + 2\Delta t)(t + 2) - 2t(t + \Delta t + 2)}{(t + \Delta t + 2)(t + 2)}$

$\Delta s = \dfrac{2t^2 + 2t\Delta t + 4t + 4\Delta t - 2t^2 - 2t\Delta t - 4t}{(t + \Delta t + 2)(t + 2)}$

$\Delta s = \dfrac{4\Delta t}{(t + \Delta t + 2)(t + 2)}; \ \dfrac{\Delta s}{\Delta t} = \dfrac{4}{(t + \Delta t + 2)(t + 2)}$

$\lim\limits_{\Delta t \to 0} \dfrac{\Delta s}{\Delta t} = \dfrac{ds}{dt} = \dfrac{4}{(t + 2)^2}$

$= v \text{ (instantaneous velocity)}$

15. $s = 3t^2 - 2t^3; \ s + \Delta s = 3(t + \Delta t)^2 - 2(t + \Delta t)^3$

$s + \Delta s = 3t^2 + 6t\Delta t + 3\Delta t^2 - 2(t^3 + 3t^2\Delta t + 3t\Delta t^2 + \Delta t^3)$

$s + \Delta s = 3t^2 + 6t\Delta t + 3\Delta t^2 - 2t^3 - 6t^2\Delta t - 6t\Delta t^2 - 3\Delta t^3$

$\Delta s = 6t\Delta + 3\Delta t^2 - 6t^2\Delta t - 6t\Delta t^2 - 3\Delta t^3$

$\dfrac{\Delta s}{\Delta t} = 6t + 3\Delta t - 6t^2 + 6t\Delta t - 3\Delta t^2$

$\lim\limits_{\Delta t \to 0} \dfrac{\Delta s}{\Delta t} = 6t - 6t^2 = \dfrac{ds}{dt} = v$

16. $s = s_0 + v_0 t - \dfrac{1}{2}at^2;$

$s + \Delta s = s_0 + v_0(t + \Delta t) - \dfrac{1}{2}a(t + \Delta t)^2$

$s + \Delta s = s_0 + v_0 t + v_0\Delta t - \dfrac{1}{2}at^2 - at\Delta t - \dfrac{1}{2}a\Delta t^2$

$\Delta s = s_0 + v_0 t + v_0\Delta t - \dfrac{1}{2}at^2 - at\Delta t - \dfrac{1}{2}a\Delta t^2$

$\qquad - s_0 - v_0 t + \dfrac{1}{2}at^2$

$\Delta s = v_0\Delta t - at\Delta t - \dfrac{1}{2}a\Delta t^2$

$\dfrac{\Delta s}{\Delta t} = v_0 - at - \dfrac{1}{2}a\Delta t; \ \lim\limits_{\Delta t \to 0} \dfrac{\Delta s}{\Delta t} = \dfrac{ds}{dt} = v = v_0 - at$

17. $a = \lim\limits_{\Delta t \to 0} \dfrac{\Delta v}{\Delta t}; \ v + \Delta v = 6(t + \Delta t)^2 - 4(t + \Delta t) + 2$

$v + \Delta v = 6[t^2 + 2t\Delta t + (\Delta t)^2] - 4t - 4\Delta t + 2$

$v + \Delta v - v = 6t^2 + 12t\Delta t + 6(\Delta t)^2 - 4t - 4\Delta t + 2$

$\qquad\qquad\qquad - (6t^2 - 4t + 2)$

$\Delta v = 12t\Delta t + 6(\Delta t)^2 - 4\Delta t$

$\dfrac{\Delta v}{\Delta t} = \dfrac{\Delta t(12t + 6\Delta t - 4)}{\Delta t} = 12t + 6\Delta t - 4$

$\lim\limits_{\Delta t \to 0} \dfrac{\Delta v}{\Delta t} = \dfrac{dv}{dt} = 12t - 4 = a \text{ (instantaneous acceleration)}$

18. $v = \sqrt{2t + 1}; \ \lim\limits_{\Delta t \to 0} \dfrac{\Delta v}{\Delta t} = \dfrac{dv}{dt} = a$

$v^2 = 2t + 1; \ (v + \Delta v)^2 = 2(t + \Delta t) + 1 = 2t + 2\Delta t + 1$

$v^2 + 2v\Delta v + \Delta v^2 = 2t + 2\Delta t + 1$

$\Delta v(2v + \Delta v) = 2\Delta t; \ \Delta v = \dfrac{2\Delta t}{2v + \Delta v}$

$\dfrac{\Delta v}{\Delta t} = \dfrac{2}{2v + \Delta v}; \ \lim\limits_{\substack{\Delta t \to 0 \\ \Delta v \to 0}} \dfrac{\Delta v}{\Delta t} = \dfrac{2}{2v} = \dfrac{1}{v} = \dfrac{1}{\sqrt{2t + 1}}$

19. $s = t^3 + 2t;\quad s + \Delta s = (t + \Delta t)^3 + 2(t + \Delta t)$

$s + \Delta s = t^3 + 3t^2\Delta t + 3t\Delta t^2 + \Delta t^3 + 2t + 2\Delta t$

$\Delta s = 3t^2\Delta t + 3t\Delta t^2 + \Delta t^3 + 2\Delta t$

$\dfrac{\Delta s}{\Delta t} = 3t^2 + 3t\Delta t + \Delta t^2 + 2$

$\lim\limits_{\Delta t \to 0} \dfrac{\Delta v}{\Delta t} = 3t^2 + 2 = v$

$v + \Delta v = 3(t + \Delta t)^2 + 2$

$\qquad\quad = 3(t^2 + 2t\Delta t + \Delta t^2) + 2$

$\Delta v = 3t^2 + 6t\Delta t + 3\Delta t^2 + 2 - 3t^2 - 2$

$\Delta v = \Delta t(6t + 3\Delta t);$

$\dfrac{\Delta v}{\Delta t} = 6t + 3\Delta t$

$\lim\limits_{\Delta t \to 0} \dfrac{\Delta v}{\Delta t} = 6t = \dfrac{dv}{dt} = a$

20. $s = s_0 + v_0 t - \dfrac{1}{2}at^2$ (from problem 16)

$\lim\limits_{\Delta t \to 0} \dfrac{\Delta v}{\Delta t} = v = v_0 - at$

$v + \Delta v = v_0 - a(t + \Delta t)$

$\qquad\quad = v_0 - at - a\Delta t$

$\Delta v = v_0 - at - a\Delta t - v_0 + at = -a\Delta t$

$\dfrac{\Delta v}{\Delta t} = -a;\quad \lim\limits_{\Delta t \to 0} \dfrac{\Delta v}{\Delta t} = \dfrac{dv}{dt} = -a$

21. $q = 30 - 2t;\quad q + \Delta q = 30 - 2(t + \Delta t)$

$q + \Delta q = 30 - 2t - 2\Delta t$

$q + \Delta q - q = (30 - 2t - 2\Delta t) - (30 - 2t)$

$\qquad\qquad\quad = 30 - 2t - 2\Delta t - 30 + 2t$

$\Delta q = -2\Delta t;\quad \dfrac{\Delta q}{\Delta t} = -2$

Therefore, $i = \lim\limits_{\Delta t \to 0} \dfrac{\Delta q}{\Delta t} = -2$

22. $L = 5x - 0.5x^2;\quad L + \Delta L = 5(x + \Delta x) - 0.5(x + \Delta x)^2$

$\Delta L = 5x + 5\Delta x - 0.5x^2 - x\Delta x - 0.5\Delta x^2 - 5x + 0.5x^2$

$\Delta L = 5\Delta x - x\Delta x - 0.5\Delta x^2;$

$\dfrac{\Delta L}{\Delta x} = 5 - x - 0.5\Delta x$

$\lim\limits_{\Delta x \to 0} \dfrac{\Delta L}{\Delta x} = \dfrac{dL}{dx} = 5 - x$

23. $A = lw;\quad l = 3w$

$A = 3w(w) = 3w^2$

$A + \Delta A = 3(w + \Delta w)^2$

$\Delta A = 3w^2 + 6w\Delta w + 3\Delta w^2 - 3w^2$

$\Delta A = 6w\Delta w + 3\Delta w^2;$

$\dfrac{\Delta A}{\Delta w} = 6w + 3\Delta w$

$\lim\limits_{\Delta w \to 0} \dfrac{\Delta A}{\Delta w} = \dfrac{dA}{dw} = 6w$

24. $A = \pi r^2;\quad r = 240$ m

$A + \Delta A = \pi(r + \Delta r)^2$

$A + \Delta A = \pi r^2 + 2\pi r\Delta r + \pi \Delta r^2$

$\Delta A = 2\pi r\Delta r + \pi \Delta r^2$

$\dfrac{\Delta A}{\Delta r} = 2\pi r + \pi \Delta r$

$\lim\limits_{\Delta r \to 0} \dfrac{\Delta A}{\Delta r} = \dfrac{dA}{dr} = 2\pi r$

$\left.\dfrac{dA}{dr}\right|_{r=240} = 2\pi(240) = 1508\dfrac{\text{m}^2}{\text{m}} = 1510$ m

25. $P = 500 + 250m^2$

$P + \Delta P = 500 + 250(m + \Delta m)^2$

$P + \Delta P = 500 + 250(m^2 + 2m\Delta m + (\Delta m)^2)$

$P + \Delta P = 500 + 250m^2 + 500m\Delta m + (\Delta m)^2$

$P + \Delta P - P = 500 + 250m^2 + 500m\Delta m$

$\qquad\qquad\qquad + (\Delta m)^2 - (500 + 250m^2)$

$\Delta P = 500m\Delta m + (\Delta m)^2;\quad \dfrac{\Delta P}{\Delta x} = 500m + \Delta m$

$\lim\limits_{\Delta m \to 0} \dfrac{\Delta P}{\Delta m} = \dfrac{dP}{dm} = 500m$

$\left.\dfrac{dP}{dm}\right|_{m=0.92} = 500(0.92) = 460$ W

26. $V = \dfrac{1}{6}\pi h^3 + 2.00\pi h$

$V + \Delta V = \dfrac{1}{6}\pi(h + \Delta h)^3 + 2.00\pi(h + \Delta h)$

$V + \Delta V = \dfrac{\pi}{6}(h^3 + 3h^2\Delta h + 3h\Delta h^2 + \Delta h^3)$

$\qquad\qquad + 2.00\pi(h + \Delta h)$

$\Delta V = \dfrac{\pi}{6}h^3 + \dfrac{\pi}{2}h^2\Delta h + \dfrac{\pi}{2}h\Delta h^2 + \dfrac{\pi}{6}\Delta h^3$

$\qquad\quad + 2.00\pi h + 2.00\pi\Delta h - \dfrac{\pi}{6}h^3 - 2.00\pi h$

$\Delta V = \dfrac{\pi}{2}h^2\Delta h + \dfrac{\pi}{2}h\Delta h^2 + \dfrac{\pi}{6}\Delta h^3 + 2.00\pi\Delta h$

$\dfrac{\Delta V}{\Delta h} = \dfrac{\pi}{2}h^2 + \dfrac{\pi}{2}h\Delta h + \dfrac{\pi}{6}\Delta h^2 + 2.00\pi$

$\dfrac{dV}{dh} = \lim\limits_{\Delta h \to 0}\left(\dfrac{\pi}{2}h^2 + \dfrac{\pi}{2}h\Delta h + \dfrac{\pi}{6}\Delta h^2 + 2.00\pi\right)$

$\qquad = \dfrac{\pi}{2}h^2 + 2.00\pi$

$\left.\dfrac{dV}{dh}\right|_{h=0.6 \text{ cm}} = \dfrac{\pi}{2}(0.6)^2 + 2.00\pi = 6.8\dfrac{\text{cm}^3}{\text{cm}}$

27. $H = \dfrac{5000}{t^2 + 10}$; $(-6 \leq t \leq 6)$

$t = 3 \text{ p.m.} = +3$

$$H + \Delta H = \frac{5000}{(t + \Delta t)^2 + 10}; \quad \Delta H = \frac{5000}{(t + \Delta t)^2 + 10} - \frac{5000}{t^2 + 10}$$

$$\Delta H = \frac{5000(t^2 + 10) - 5000(t^2 + 2t\Delta t + \Delta t^2 + 10)}{[(t + \Delta t)^2 + 10](t^2 + 10)} = \frac{5000t^2 + 50\,000 - 5000t^2 - 10\,000t\Delta t - 5000\Delta t^2 - 50\,000}{[(t + \Delta t)^2 + 10](t^2 + 10)}$$

$$\frac{\Delta H}{\Delta t} = \frac{-10\,000t - 5000\Delta t}{[(t + \Delta t)^2 + 10](t^2 + 10)}$$

$$\lim_{\Delta t \to 0} \frac{\Delta H}{\Delta t} = \frac{dH}{dt} = \frac{-10\,000t}{(t^2 + 10)^2}$$

$$\left.\frac{dH}{dt}\right|_{t=3} = -83.1 \text{ W/(m}^2 \cdot \text{h)}$$

28. $V = \dfrac{48}{t + 3}$; $t = 3$ years

$$V + \Delta V = \frac{48}{t + \Delta t + 3}; \quad \Delta V = \frac{48}{(t + \Delta t + 3)} - \frac{48}{t + 3}$$

$$\Delta V = \frac{48(t + 3) - 48(t + \Delta t + 3)}{(t + \Delta t + 3)(t + 3)} = \frac{48t + 144 - 48t - 48\Delta t - 144}{(t + \Delta t + 3)(t + 3)}$$

$$\frac{\Delta V}{\Delta t} = \frac{-48}{(t + \Delta t + 3)(t + 3)}; \quad \lim_{\Delta t \to 0} \frac{\Delta V}{\Delta t} = \frac{dV}{dt} = \frac{-48}{(t + 3)^2}$$

$$\left.\frac{dV}{dt}\right|_{t=3} = \frac{-48}{36} = -\$1300/\text{year}$$

29. The volume of a cone $= \frac{1}{3}\pi r^2 h$. For this cone, the radius and height are equal to 4 cm. Due to the similarity of the figures, as the level of the oil decreases, the radius and height will still be equal; $r = h = d$. Therefore,

$$V = \frac{1}{3}\pi d^2(d) = \frac{1}{3}\pi d^3$$

$$V + \Delta V = \frac{1}{3}\pi(d + \Delta d)^3 = \frac{1}{3}\pi(d^3 + 3d\Delta d^2 + 3d^2\Delta d + \Delta d^3)$$

$$V + \Delta V - V = \frac{1}{3}\pi d^3 + \pi d\Delta d^2 + \pi d^2\Delta d + \frac{1}{3}\pi\Delta d^3 - \frac{1}{3}\pi d^3$$

$$\Delta V = \pi d\Delta d^2 + \pi d^2\Delta d + \frac{1}{3}\pi\Delta d^3$$

$$\frac{\Delta V}{\Delta d} = \pi d\Delta d + \pi d^2 + \frac{1}{3}\pi\Delta d^2; \quad \lim_{\Delta d \to 0} \frac{\Delta V}{\Delta d} = \frac{dV}{dd} = \pi d^2$$

30. $t = kn^2$

$n = 6400, t = 25.0$, therefore $k = 6.10 \times 10^{-7}$

$t + \Delta t = k(n + \Delta n)^2$; $t + \Delta t = kn^2 - 2kn\Delta n + k\Delta n^2$

$\Delta t = kn^2 + 2kn\Delta n + k\Delta n^2 - kn^2$

$$\Delta t = 2kn\Delta n + k\Delta n^2; \quad \frac{\Delta t}{\Delta n} = 2kn + k\Delta n$$

$$\lim_{\Delta n \to 0} \frac{\Delta t}{\Delta n} = \frac{dt}{dn} = 2kn$$

$$\left.\frac{dt}{dn}\right|_{n=8000} = 2(6.10 \times 10^{-7})8000 = 0.009\,77$$

$$\frac{dt}{dn} = 9.77 \times 10^{-3} \text{ s/cell}$$

31. $r = k\sqrt{\lambda}$

$r = 3.72 \times 10^{-2}$ m, $\lambda = 59.2 \times 10^{-8}$ m,

$$k = \frac{3.72 \times 10^{-2}}{\sqrt{59.2 \times 10^{-8}}} = 48.35$$

$r^2 = k^2\lambda$

$(r + \Delta r)^2 = k^2(\lambda + \Delta\lambda);$

$r^2 + 2r\Delta r + \Delta r^2 = k^2\lambda + k^2\Delta\lambda$

$2r\Delta r + \Delta r^2 = k^2\lambda + k^2\Delta\lambda - k^2\lambda$

$$\Delta r(2r + \Delta r) = k^2\Delta\lambda; \quad \Delta r = \frac{k^2\Delta\lambda}{2r + \Delta r}$$

$$\frac{\Delta r}{\Delta\lambda} = \frac{k^2}{2r + \Delta r}$$

$$\lim_{\substack{\Delta r \to 0 \\ \Delta\lambda \to 0}} \frac{\Delta r}{\Delta\lambda} = \frac{dr}{d\lambda} = \frac{k^2}{2r} = \frac{k^2}{2k\sqrt{\lambda}} = \frac{k}{2\sqrt{\lambda}};$$

$$\frac{dr}{d\lambda} = \frac{24.2}{\sqrt{\lambda}}$$

32. $F = 0.12$ N, $r = 0.060$ m, $k = 4.32 \times 10^{-4}$ N $\cdot$ m^2

$$F = \frac{k}{r^2}; \quad F + \Delta F = \frac{k}{(r + \Delta r)^2}$$

$$\Delta F = \frac{k}{(r + \Delta r)^2} - \frac{k}{r^2} = \frac{kr^2 - k(r + \Delta r)^2}{(r + \Delta r)^2(r^2)}$$

$$\Delta F = \frac{kr^2 - kr^2 - 2kr\Delta r - k\Delta r^2}{(r + \Delta r)^2(r^2)}$$

$$\Delta F = \frac{-2kr\Delta r - k\Delta r^2}{(r + \Delta r)^2(r^2)}; \quad \frac{\Delta F}{\Delta r} = \frac{-2kr - k\Delta r}{(r + \Delta r)^2(r^2)}$$

$$\lim_{\Delta r \to 0} \frac{\Delta F}{\Delta r} = \frac{dF}{dr} = \frac{-2kr}{r^4} = -\frac{2k}{r^3}$$

$$\left.\frac{dF}{dr}\right|_{r=0.120} = \frac{-2(4.32 \times 10^{-4})}{(0.120)^3} = -0.50 \text{ N/m}$$

3.5   Derivatives of Polynomials

1. $y = x^5; \dfrac{dy}{dx} = 5x^{5-1} = 5x^4$

2. $y = x^{12}; \dfrac{dy}{dx} = 12x^{12-1} = 12x^{11}$

3. $y = -4x^9; \dfrac{dy}{dx} = -4(9x^8) = -36x^8$

4. $y = -7x^6; \dfrac{dy}{dx} = -7(6x^5) = -42x^5$

5. $y = x^4 - 6; \dfrac{dy}{dx} = 4x^{4-1} - 0 = 4x^3$

6. $y = 3t^5 - 1; \dfrac{dy}{dt} = 15t^4 - 0 = 15t^4$

7. $y = x^2 + 2x; \dfrac{dy}{dx} = 2x + 2$

8. $y = x^3 - 2x^2; \dfrac{dy}{dx} = 3x^2 - 4x$

9. $p = 5r^3 - 2r + 1; \dfrac{dp}{dr} = 5(3r^2) - 2 + 0$

$\qquad = 15r^2 - 2$

10. $y = 6x^2 - 6x + 5; \dfrac{dy}{dx} = 12x - 6$

11. $y = x^8 - 4x^7 - x; \dfrac{dy}{dx} = 8x^7 - 28x^6 - 1$

12. $y = 4x^4 - 2x + 9; \dfrac{dy}{dx} = 16x^3 - 2$

13. $f(x) = -6x^7 + 5x^3 + \pi^2$

$\qquad \dfrac{f(x)}{dx} = -6(7x^6) + 5(3x^2) + 0 = -42x^6 + 15x^2$

14. $y = 13x^4 - 6x^3 - x - 1; \dfrac{dy}{dx} = 52x^3 - 18x^2 - 1$

15. $y = \dfrac{1}{3}x^3 + \dfrac{1}{2}x^2; \dfrac{dy}{dx} = x^2 + x$

16. $f(z) = -\dfrac{1}{4}z^8 + \dfrac{1}{2}z^4 - 2^3; f'(z) = -2z^7 + 2z^3$

17. $y = 6x^2 - 8x + 1;$

$$\frac{dy}{dx} = \frac{d(6x^2)}{dx} - \frac{d(8x)}{dx} + \frac{d(1)}{dx} = 12x - 8 + 0$$

Since the derivative is a function of only x, we now evaluate it for $x = 2$.

$$\left.\frac{dy}{dx}\right|_{x=2} = 12(2) - 8 = 24 - 8 = 16$$

18. $s = 2t^3 - 5t^2 + 4, (-1, -3)$

$$\left.\frac{ds}{dt} = 6t^2 - 10t\right|_{(-1,-3)} = 6(-1)^2 - 10(-1) = 16$$

19. $y = 2x^3 + 9x - 7; (-2, -4);$

$$\left.\frac{dy}{dx} = 6x^2 + 9\right|_{x=-2} = 33$$

```
nDeriv(2X³+9X-7,
X,-2)
          33.000002
```

20. $y = x^4 - 9x^2 - 5x$; $(3, -15)$;

$$\frac{dy}{dx} = 4x^3 - 18x - 5\bigg|_{x=3} = 49$$

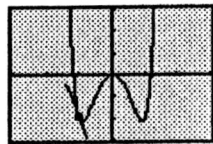

21. $y = 2x^6 - 4x^2$; $m_{tan} = \frac{dy}{dx} = 12x^5 - 8x$

$$\frac{dy}{dx}\bigg|_{x=-1} = m_{tan} = 12(-1)^5 - 8(-1) = -12 + 8$$
$$= -4$$

Move the trace to $x = -1$ and observe that the function is decreasing and that the slope is negative.

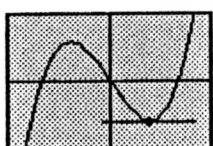

22. $y = 3x^3 - 9x$; $\frac{dy}{dx} = 9x^2 - 9\bigg|_{x=1} = 0$

23. $y = 35x - 2x^4$; $(x = 2)$; $\frac{dy}{dx} = 35 - 8x^3\bigg|_{x=2} = -29$

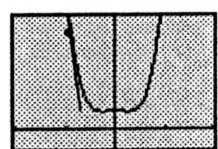

24. $y = x^4 - \frac{1}{2}x^2 + 2$; $(x = -2)$;

$$\frac{dy}{dx} = 4x^3 - x\bigg|_{x=-2} = -30$$

25. $s = 6t^5 - 5t + 2$; $v = \frac{ds}{dt} = 30t^4 - 5$

26. $s = 20 + 60t - 4.9t^2$; $v = \frac{ds}{dt} = 60 - 9.8t$

27. $s = 2 - 6t - 2t^3$; $v = \frac{ds}{dt} = -6 - 6t^2$

28. $s = s_0 + v_0 t + \frac{1}{2}at^2$; $v = \frac{ds}{dt} = v_0 + at$

29. $s = 2t^3 - 4t^2$; $t = 4$

$$v = \frac{ds}{dt} = 2(3t^2) - 4(2t) = 6t^2 - 8t$$
$$v\big|_{t=4} = 6(4^2) - 8(4) = 64$$

30. $s = 120 + 80t - 16t^2$; $t = 2.5$

$$v = \frac{ds}{dt} = 80 - 32t\bigg|_{t=2.5} = 0$$

31. $s = 0.5t^4 - 1.5t^2 + 2.5$; $t = 3$

$$v = \frac{ds}{dt} = 2.0t^3 - 3.0t\bigg|_{t=3} = 45$$

32. $s = 8t^2 - 10t + 6$; $t = 5$

$$v = \frac{ds}{dt} = 16t - 10\bigg|_{t=5} = 70$$

33. $y = 3x^2 - 6x$; $m_{tan} = \frac{dy}{dx} = 6x - 6$

Tangent is parallel where slope is zero. Therefore $6x - 6 = 0$; $x = 1$

34. $y = ax^2 + 2x$; $\frac{dy}{dx} = -4$ for $x = 2$

$$\frac{dy}{dx} = 2ax + 2 = -4; \ 2ax = -6$$

$$a = \frac{-6}{2x} = \frac{-6}{4} = -\frac{3}{2}$$

35. $y = 3x^2 - 4x$; $\frac{dy}{dx} = 8$ at (x, y)?

$$\frac{dy}{dx} = 6x - 4 = 8; \ 6x = 12$$

$x = 2$, therefore $(2, 4)$

36. $y = 5x^3 + 4x - 3$; why is $\frac{dy}{dx} \geq 4$?

$\frac{dy}{dx} = 15x^2 + 4$; since $x^2 \geq 0$, minimum value of $\frac{dy}{dx} = 4$.

Therefore, $m \geq 4$

37. $y = 4x^2 + 3x; m_{\tan} = \dfrac{dy}{dx} = 8x + 3;$

$y = 5 - 2x^2$

$m_{\tan} = \dfrac{dy}{dx} = -4x$

The slopes are equal. Therefore, $8x + 3 = -4x$; $12x = -3$; $x = -\frac{1}{4}$

38. $s_1 = 5t - 2t^2; \; s_2 = 3t^2 + 4; \; \dfrac{ds_1}{dt} = \dfrac{ds_2}{dt}$ at (x, y)?

$\dfrac{ds_1}{dt} = 5 - 4t; \; \dfrac{ds_2}{dt} = 6t$

$5 - 4t = 6t; \; 10t = 5; \; t = \dfrac{1}{2}$

39. $V = \pi r^2 \cdot h$

$V = \pi r^2 \cdot r = r^3$

$\dfrac{dV}{dr} = 3\pi r^2$

40. $A_{\text{one surface}} = e^2$. There are 6 surfaces. Therefore $A_T = 6e^2$

$\dfrac{dA_T}{de} = 12e$

41. $P = 16i^2 + 60i;$

$\dfrac{dP}{di} = 16(2i) + 60 = 32i + 60$

$\dfrac{dP}{di}\bigg|_{i=0.75} = 32(0.75) + 60 = 84 \text{ W/A}$

42. $T \, \alpha \, d^3; \; T = kd^3; \; T = 850; \; d = 0.925$

$k = \dfrac{T}{d^3} = \dfrac{850}{0.925} = 1074$

$\dfrac{dT}{dd} = 3kd^2 = 3(1074)d^2 = 3220d^2$

43. $P = a(c_1 E + c_2 E^2 + c_3 E^3);$

$\dfrac{dP}{dE} = a(c_1 + 2c_2 E + 3c_3 E^2)$

44. $y = kx^2(x^3 + 450x - 3500);$
$y = k(x^5 + 450x^3 - 3500x^2)$

$\dfrac{dy}{dx} = k(5x^4 + 1350x^2 - 7000x)$
$\quad\;\; = kx(5x^3 + 1350x - 7000)$

45. $h = 0.000104x^4 - 0.0417x^3 + 4.21x^2 - 8.33x$

$\dfrac{dh}{dx} = 0.000416x^3 - 0.1251x^2 + 8.42x - 8.33\big|_{x=120}$

$\dfrac{dh}{dx} = -80.5 \text{ m/km}$

46. $F = x^4 - 12x^3 + 46x^2 - 60x + 25$

$\dfrac{dF}{dx} = 4x^3 - 12(3x^2) + 46(2x) - 60$
$\quad\;\; = 4x^3 - 36x^2 + 92x - 60$

$\dfrac{dF}{dx}\bigg|_{x=4} = 4(4)^3 - 36(4^2) + 92(4) - 60$
$\quad\quad\;\; = 256 - 576 + 368 - 60$
$\quad\quad\;\; = -12 \text{ N/cm}$

47. $V_T = V_1 + V_2$

$V_T = \dfrac{4}{3}\pi {}_1^3 + \dfrac{4}{3}\pi r^3$

$V_T = \dfrac{4}{3}\pi[(r + 1.20)^3 + r^3]$

$V_T = \dfrac{4}{3}\pi[r^3 + 3r^2(1.20) + 3r(1.20)^2 + 1.20^3 + r^3]$

$V_T = \dfrac{4}{3}\pi[2r^3 + 3.6r^2 + 4.32r + 1.728]$

$\dfrac{dVr}{dr} = \dfrac{4}{3}\pi[6r^2 + 7.2r + 4.32]\bigg|_{r=3.30} = 391 \text{ mm}^2$

48. $V = l \cdot w \cdot h$
$V = (8 - 2x)(6 - 2x)x$
$V = 48x - 28x^2 + 4x^3$

$\dfrac{dV}{dx} = 48 - 56x + 12x^2$

$\dfrac{dV}{dx}\bigg|_{x=1.75} = -13.3 \text{ in}^2$

3.6 Derivatives of Products and Quotients of Functions

1. $\quad y = x^2(3x + 2); \; u = x^2;$

$\dfrac{du}{dx} = 2x; \; v = 3x + 2; \; \dfrac{dv}{dx} = 3$

$\dfrac{dy}{dx} = x^2(3) + (3x + 2)(2x)$
$\quad\;\; = 3x^2 + 6x^2 + 4x = 9x^2 + 4x$

2. $y = 3x(x^3 + 1)$

$\dfrac{dy}{dx} = 3x(3x^2) + (x^3 + 1)(3) = 9x^3 + 3x^3 + 3$
$\quad\;\; = 12x^3 + 3$

3. $y = 6x(3x^2 - 5x)$

$$\frac{dy}{dx} = 6x(6x - 5) + (3x^2 - 5x)(6)$$
$$= 36x^2 - 30x + 18x^2 - 30x = 54x^2 - 60x$$

4. $y = 2x^3(3x^4 + x)$

$$\frac{dy}{dx} = 2x^3(12x^3 + 1) + (3x^4 + x)(6x^2)$$
$$= 24x^6 + 2x^3 + 18x^6 + 6x^3 = 42x^6 + 8x^3$$

5. $s = (3t + 2)(2t - 5)$
    $u = 3t + 2,\ v = 2t - 5$

$$\frac{du}{dt} = 3 \qquad \frac{dv}{dt} = 2$$

$$\frac{ds}{dt} = (3t + 2)(2) + (2t - 5)(3)$$

$$= 6t + 4 + 6t - 15 = 12t - 11$$

6. $f(x) = (3x - 2)(4x^2 + 3)$
    $f'(x) = (3x - 2)(8x) + (4x^2 + 3)(3)$
    $\qquad = 24x^2 - 16x + 12x^2 + 9$
    $f'(x) = 36x^2 - 16x + 9$

7. $y = (x^4 - 3x^2 + 3)(1 - 2x^3)$

$$\frac{dy}{dx} = (x^4 - 3x^2 + 3)(-6x^2) + (1 - 2x^3)(4x^3 - 6x)$$
$$= -6x^6 + 18x^4 - 18x^2 + 4x^3 - 8x^6 - 6x + 12x^4$$
$$= -14x^6 + 30x^4 + 4x^3 - 18x^2 - 6x$$

8. $y = (x^3 - 6x)(2 - 4x^3)$

$$\frac{dy}{dx} = (x^3 - 6x)(-12x^2) + (2 - 4x^3)(3x^2 - 6)$$
$$= -12x^5 + 72x^3 + 6x^2 - 12x^5 - 12 + 24x^3$$
$$= -24x^5 + 96x^3 + 6x^2 - 12$$

9. $y = (2x - 7)(5 - 2x);\ u = (2x - 7);\ v = (5 - 2x)$

$$\frac{dy}{dx} = (2x - 7)(-2) + (5 - 2x)(2)$$

$$= -4x + 14 + 10 - 4x = -8x + 24$$
$$y = (2x - 7)(5 - 2x)$$
$$= 10x - 4x^2 - 35 + 14x = -4x^2 + 24x - 35$$

$$\frac{dy}{dx} = -8x + 24$$

10. $f(s) = (5s^2 + 2)(2s^2 - 1)$
    $f'(s) = (5s^2 + 2)(4s) + (2s^2 - 1)(10s)$
    $\qquad = 20s^3 + 8s + 20s^3 - 10s$
    $f'(s) = 40s^3 - 2s$
    $f(s) = 10s^4 - 5s^2 + 4s^2 - 2 = 10s^4 - s^2 - 2$
    $f'(s) = 40s^3 - 2s$

11. $y = (x^3 - 1)(2x^2 - x - 1)$

$$\frac{dy}{dx} = (x^3 - 1)(4x - 1) + (2x^2 - x - 1)(3x^2)$$
$$= 4x^4 - x - x^3 + 1 + 6x^4 - 3x^3 - 3x^2$$
$$= 10x^4 - 4x^3 - 3x^2 - 4x + 1$$
$$y = 2x^5 - x^4 - x^3 - 2x^2 + x + 1$$

$$\frac{dy}{dx} = 10x^4 - 4x^3 - 3x^2 - 4x + 1$$

12. $y = (3x^2 - 4x + 1)(5 - 6x^2)$

$$\frac{dy}{dx} = (3x^2 - 4x + 1)(-12x) + (5 - 6x^2)(6x - 4)$$

$$= -36x^3 + 48x^2 - 12x + 30x - 20 - 36x^3 + 24x^2$$
$$= -72x^3 + 72x^2 + 18x - 20$$
$$y = -18x^4 + 24x^3 + 9x^2 - 20x + 5$$

$$\frac{dy}{dx} = -72x^3 + 72x^2 + 18x - 20$$

13. $y = \dfrac{x}{2x + 3};\ u = x;\ \dfrac{du}{dx} = 1;\ v = 2x + 3;\ \dfrac{dv}{dx} = 2$

$$\frac{dy}{dx} = \frac{(2x + 3)(1) - x(2)}{(2x + 3)^2}$$

$$= \frac{2x + 3 - 2x}{(2x + 3)^2} = \frac{3}{(2x + 3)^2}$$

14. $y = \dfrac{2x}{x + 1}$

$$\frac{dy}{dx} = \frac{(x + 1)2 - 2x(1)}{(x + 1)^2} = \frac{2x + 2 - 2x}{(x + 1)^2}$$

$$= \frac{2}{(x + 1)^2}$$

15. $y = \dfrac{1}{x^2 + 1};\ \dfrac{dy}{dx} = \dfrac{(x^2 + 1)(0) - 1(2x)}{(x^2 + 1)^2} = \dfrac{-2x}{(x^2 + 1)^2}$

16. $R = \dfrac{5i + 2}{2i + 3}$

$$\frac{dR}{di} = \frac{(2i + 3)(5) - (5i + 2)(2)}{(2i + 3)^2}$$

$$= \frac{10i + 15 - 10i - 4}{(2i + 3)^2}$$

$$\frac{dR}{di} = \frac{11}{(2i + 3)^2}$$

17. $y = \dfrac{x^2}{3 - 2x};\ u = x^2;$

$$\frac{du}{dx} = 2x;\ v = 3 - 2x;\ \frac{dv}{dx} = -2$$

$$\frac{dy}{dx} = \frac{(3 - 2x)(2x) - (x^2)(-2)}{(3 - 2x)^2}$$

$$= \frac{6x - 4x^2 + 2x^2}{(3 - 2x)^2} = \frac{6 - 2x^2}{(3 - 2x)^2}$$

18. $y = \dfrac{2}{3x^2 - 5x}$

$\dfrac{dy}{dx} = \dfrac{(3x^2 - 5x)(0) - 2(6x - 5)}{(3x^2 - 5x)^2} = \dfrac{-12x + 10}{(3x^2 - 5x)^2}$

19. $y = \dfrac{2x - 1}{3x^2 + 2}; \ \dfrac{dy}{dx} = \dfrac{(3x^2 + 2)2 - (2x - 1)6x}{(3x^2 + 2)^2}$

$\dfrac{dy}{dx} = \dfrac{6x^2 + 4 - 12x^2 + 6x}{(3x^2 + 2)^2} = \dfrac{-6x^2 + 6x + 4}{(3x^2 + 2)^2}$

20. $y = \dfrac{2x^3}{4 - x}; \ \dfrac{dy}{dx} = \dfrac{(4 - x)6x^2 - 2x^3(-1)}{(4 - x)^2}$

$\dfrac{dy}{dx} = \dfrac{24x^2 - 6x^3 + 2x^3}{(4 - x)^2} = \dfrac{-4x^3 + 24x^2}{(4 - x)^2}$

21. $f(x) = \dfrac{3x + 8}{x^2 + 4x + 2}$

$\dfrac{df(x)}{dx} = \dfrac{(x^2 + 4x + 2)(3) - (3x + 8)(2x + 4)}{(x^2 + 4x + 2)^2}$

$= \dfrac{-3x^2 - 16x - 26}{(x^2 + 4x + 2)^2}$

22. $y = \dfrac{3x}{4x^5 - 3x - 4}$

$\dfrac{dy}{dx} = \dfrac{(4x^5 - 3x - 4)3 - 3x(20x^4 - 3)}{(4x^5 - 3x - 4)^2}$

$\dfrac{dy}{dx} = \dfrac{12x^5 - 9x - 12 - 60x^5 + 9x}{(4x^5 - 3x - 4)^2}$

$= \dfrac{-48x^5 - 12}{(4x^5 - 3x - 4)^2}$

23. $y = \dfrac{2x^2 - x - 1}{x^3 + 2x^2}$

$\dfrac{dy}{dx} = \dfrac{(x^3 + 2x^2)(4x - 1) - (2x^2 - x - 1)(3x^2 + 4x)}{(x^3 + 2x^2)^2}$

$= \dfrac{4x^4 + 8x^3 - x^3 - 2x^2 - 6x^4 + 3x^3 + 3x^2 - 8x^3 + 4x^2 + 4x}{(x^3 + 2x^2)^2}$

$= \dfrac{-2x^4 + 2x^3 + 5x^2 + 4x}{(x^3 + 2x^2)^2} = \dfrac{-2x^3 + 2x^2 + 5x + 4}{x^3(x + 2)^2}$

24. $y = \dfrac{3x^3 - x}{2x^2 - 5x + 4}$

$\dfrac{dy}{dx} = \dfrac{(2x^2 - 5x + 4)(9x^2 - 1) - (3x^3 - x)(4x - 5)}{(2x^2 - 5x + 4)^2}$

$= \dfrac{18x^4 - 45x^3 + 36x^2 - 2x^2 + 5x - 4 - 12x^4 + 4x^2 + 15x^3 - 5x}{(2x^2 - 5x + 4)^2}$

$= \dfrac{6x^4 - 30x^3 + 38x^2 - 4}{(2x^2 - 5x + 4)^2}$

25. $y = (3x - 1)(4 - 7x)$

$\dfrac{dy}{dx} = (3x - 1)(-7) + (4 - 7x)(3) = -21x + 7 + 12 - 21x = -42x + 19$

$\dfrac{dy}{dx}\bigg|_{x=3} = -42(3) + 19 = -126 + 19 = -107$

26. $y = (3x^2 - 5)(2x^2 - 1);\ x = -1$

$$\frac{dy}{dx} = (3x^2 - 5)(4x) + (2x^2 - 1)(6x) = 12x^3 - 20x + 12x^3 - 6x = 24x^3 - 26x$$

$$\left.\frac{dy}{dx}\right|_{x=-1} = 24(-1)^3 - 26(-1) = 2$$

27. $y = (2x^2 - x + 1)(4 - 2x - x^2);\ x = -3$

$$\frac{dy}{dx} = (2x^2 - x + 1)(-2 - 2x) + (4 - 2x - x^2)(4x - 1)$$

$$\frac{dy}{dx} = -4x^2 + 2x - 2 - 4x^3 + 2x^2 - 2x + 16x - 8x^2 - 4x^3 - 4 + 2x + x^2 + x^2 = -8x^3 - 9x^2 + 18x - 6$$

$$\left.\frac{dy}{dx}\right|_{x=-3} = -8(-3)^3 - 9(-3)^2 + 18(-3) - 6 = 75$$

28. $y = (4x^4 + 0.5x^2 + 1)(3x - 2x^2);\ x = 0.5$

$$\frac{dy}{dx} = (4x^4 + 0.5x^2 + 1)(3 - 4x) + (3x - 2x^2)(16x^3 + 1.0x)$$

$$\frac{dy}{dx} = 12x^4 + 1.5x^2 + 3 - 16x^5 - 2.0x^3 - 4x + 48x^4 + 3.0x^2 - 32x^5 - 2.0x^3$$

$$= -48x^5 + 60x^4 - 4x^3 + 4.5x^2 - 4x + 3$$

$$\left.\frac{dy}{dx}\right|_{x=\frac{1}{2}} = -48\left(\frac{1}{2}\right)^5 + 60\left(\frac{1}{2}\right)^4 - 4\left(\frac{1}{2}\right)^3 + 4.5\left(\frac{1}{2}\right)^2 - 4\left(\frac{1}{2}\right) + 3 = \frac{31}{8} = 3.875$$

29. $y = \dfrac{3x - 5}{2x + 3};\ y = 3x - 5;\ v = 2x + 3;\ du = 3dx;\ dv = 2dx$

$$\frac{dy}{dx} = \frac{(2x + 3)(3) - (3x - 5)(2)}{(2x + 3)^2}$$

$$= \frac{6x + 9 - 6x + 10}{(2x + 3)^2} = \frac{19}{(2x + 3)^2}$$

$$\left.\frac{dy}{dx}\right|_{x=-2} = \frac{19}{(2(-2) + 3)^2} = \frac{19}{1} = 19$$

30. $\quad y = \dfrac{2x^2 - 5x}{3x + 2};$

$$\frac{dy}{dx} = \frac{(3x + 2)(4x - 5) - (2x^2 - 5x)(3)}{(3x + 2)^2}$$

$$\frac{dy}{dx} = \frac{12x^2 - 15x + 8x - 10 - 6x^2 + 15x}{(3x + 2)^2} = \frac{6x^2 + 8x - 10}{(3x + 2)^2}$$

$$\left.\frac{dy}{dx}\right|_{x=2} = \frac{6(2)^2 + 8(2) - 10}{(3(2) + 2)^2} = \frac{30}{64} = \frac{15}{32}$$

31. $\quad S = \dfrac{2n^3 - 3n + 8}{2n - 3n^4}, \, n = -1$

$$\frac{dS}{dn} = \left.\frac{(2n - 3n^4)(6n^2 - 3) - (2n^3 - 3n + 8)(2 - 12n^3)}{(2n - 3n^4)^2}\right|_{n=-1}$$

$$\frac{dS}{dn} = \frac{(2(-1) - 3(-1)^4)(6(-1)^2 - 3) - (2(-1)^3 - 3(-1) + 8)(2 - 12(-1)^3)}{(2(-1) - 3(-1)^4)^2}$$

$$\frac{dS}{dn} = -5.64$$

```
nDeriv((2N³-3N+8
)/(2N-3N^4),N,⁻1
)
        ⁻5.64000713
```

32. $\quad y = \dfrac{2x^3 - x^2 - 2}{4x + 3}$

$$\frac{dy}{dx} = \frac{(4x + 3)(6x^2 - 2x) - (2x^3 - x^2 - 2)(4)}{(4x + 3)^2}$$

$$\frac{dy}{dx} = \frac{24x^3 - 8x^2 + 18x^2 - 6x - 8x^3 + 4x^2 + 8}{(4x + 3)^2}$$

```
nDeriv((2X³-X²-2
)/(4X+3),X,0.5)
      .4200003488
```

$$\frac{dy}{dx} = \frac{16x^3 + 14x^2 - 6x + 8}{(4x + 3)^2}; \quad \left.\frac{dy}{dx}\right|_{x=\frac{1}{2}} = \frac{\frac{21}{2}}{25} = \frac{21}{50} = 0.42$$

33. **(1)** $\quad y = \dfrac{x^2(1 - 2x)}{3x - 7}$

$$\frac{dy}{dx} = \frac{(3x - 7)[x^2(-2) + (1 - 2x)(2x)] - x^2(1 - 2x)(3)}{(3x - 7)^2} = \frac{(3x - 7)(-6x^2 + 2x) - 3x^2 + 6x^3}{(3x - 7)^2}$$

$$= \frac{-18x^3 + 6x^2 + 42x^2 - 14x - 3x^2 + 6x^3}{(3x - 7)^2} = \frac{-12x^3 + 45x^2 - 14x}{(3x - 7)^2}$$

    (2) $\quad y = \dfrac{x^2 - 2x^3}{3x - 7}$

$$\frac{dy}{dx} = \frac{(3x - 7)(2x - 6x^2) - (x^2 - 2x^3)(3)}{(3x - 7)^2} = \frac{-12x^3 + 45x^2 - 14x}{(3x - 7)^2}$$

34. **(1)** $\quad y = 4x^2 - \dfrac{1}{x - 1}; \quad \dfrac{dy}{dx} = 8x - \left\{\dfrac{(x - 1)0 - 1(1)}{(x - 1)^2}\right\};$

$$\frac{dy}{dx} = 8x + \frac{1}{(x - 1)^2} = \frac{8x(x - 1)^2 + 1}{(x - 1)^2}$$

$$\frac{dy}{dx} = \frac{8x(x^2 - 2x + 1) + 1}{(x - 1)^2} = \frac{8x^3 - 16x^2 + 8x + 1}{(x - 1)^2}$$

(2) $y = 4x^2 - \dfrac{1}{x-1} = \dfrac{4x^2(x-1)-1}{(x-1)}$

$\qquad = \dfrac{4x^3 - 4x^2 - 1}{x-1}$

$\dfrac{dy}{dx} = \dfrac{(x-1)(12x^2 - 8x) - (4x^3 - 4x^2 - 1)(1)}{(x-1)^2}$

$\dfrac{dy}{dx} = \dfrac{12x^3 - 8x^2 - 12x^2 + 8x - 4x^3 + 4x^2 + 1}{(x-1)^2}$

$\dfrac{dy}{dx} = \dfrac{8x^3 - 16x^2 + 8x + 1}{(x-1)^2}$

35. $y = (4x+1)(x^4 - 1); \; (-1, 0)$

$\dfrac{dy}{dx} = (4x+1)(4x^3) + (x^4 - 1)4$

$\qquad = 16x^4 + 4x^3 + 4x^4 - 4$

$\left. \dfrac{dy}{dx} \right|_{x=-1} = 20x^4 + 4x^3 - 4 \big|_{x=-1} = 12$

36. $y = (3x+4)(1-4x); \; (2, -70)$

$\dfrac{dy}{dx} = (3x+4)(-4) + (1-4x)(3)$

$\qquad = -12x - 16 + 3 - 12x$

$\dfrac{dy}{dx} = -24x - 13$

$\left. \dfrac{dy}{dx} \right|_{x=2} = -24x - 13 \big|_{x=2} = -61$

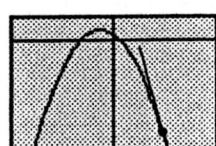

37. $y = \dfrac{x}{x^2+1}; \; y = x; \; \dfrac{du}{dx} = 1;$

$v = x^2 + 1; \; \dfrac{dv}{dx} = 2x$

$\dfrac{dy}{dx} = \dfrac{(x^2+1)(1) - (x)(2x)}{(x^2+1)^2}$

$\qquad = \dfrac{x^2 + 1 - 2x^2}{(x^2+1)^2} = \dfrac{-x^2+1}{(x^2+1)^2}$

Therefore, $m_{\text{tan}} = 0$ when $\dfrac{-x^2+1}{(x^2+1)^2} = 0$;

$-x^2 + 1 = 0; \; x^2 = 1; \; x = 1, -1$

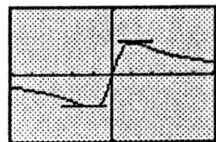

38. $y = \dfrac{2x-1}{1-x^2}; \; x = -2, -1, 0, 1, 2$

$\dfrac{dy}{dx} = \dfrac{(1-x^2)(2) - (2x-1)(-2x)}{(1-x^2)^2}$

$\dfrac{dy}{dx} = \dfrac{2 - 2x^2 + 4x^2 - 2x}{(1-x^2)^2} = \dfrac{2x^2 - 2x + 2}{(1-x^2)^2}$

$x = -2, \; \dfrac{dy}{dx} > 0; \; x = -1, \; \dfrac{dy}{dx}$ undefined (vert. asymptote);

$x = 0, \; \dfrac{dy}{dx} > 0; \; x = 1, \; \dfrac{dy}{dx}$ undefined (vert. asymptote);

$x = 2, \; \dfrac{dy}{dx} > 0$

$m_{\text{tan}} = \dfrac{2(x^2 - x + 1)}{(1-x^2)^2}$

By completing the square:

$m_{\text{tan}} = \dfrac{2\left\{ \left(x - \dfrac{1}{2} \right)^2 + \dfrac{3}{4} \right\}}{(1-x^2)^2}$

$\left(x - \dfrac{1}{2} \right)^2 > 0$ for all x; $(1-x^2)^2 > 0$ for all x

Therefore, m_{tan} is never negative.

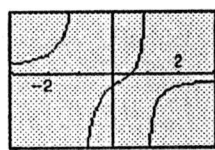

39. $s = (t^2 - 8t)(2t^2 + t + 1)$

$\dfrac{ds}{dt} = (t^2 - 8t)(4t+1) + (2t^2 + t + 1)(2t - 8)$

$v = 4t^3 + t^2 - 32t^2 - 8t + 4t^3$

$\qquad + 2t^2 + 2t - 16t^2 - 8t - 8$

$v = 8t^3 - 45t^2 - 14t - 8$

40. $n = \dfrac{75{,}000}{e(e+6)} = \dfrac{75{,}000}{e^2 + 6e}$

$\dfrac{dn}{de} = \dfrac{(e^2 + 6e)0 - 75{,}000(2e+6)}{(e^2 + 6e)^2}$

$\dfrac{dn}{de} = \dfrac{-150{,}000e - 450{,}000}{(e^2 + 6e)^2}$

$\dfrac{dn}{de} = \dfrac{-150{,}000(e+3)}{[e(e+6)]^2} = \dfrac{-150{,}000(e+3)}{e^2(e+6)^2}$

41. $V = \dfrac{6R + 25}{R + 3}$

$\dfrac{dV}{dR} = \dfrac{6(R+3) - (6R+25)}{(R+3)^2}$

$= \dfrac{6R + 18 - 6R - 25}{(R+3)^2}$

$= \dfrac{-7}{(R+3)^2}$

$\left.\dfrac{dV}{dR}\right|_{R=7} = \dfrac{-7}{(7+3)^2} = \dfrac{-7}{100} = -0.07 \text{ V}/\Omega$

42. $n = \dfrac{6t^2}{2t^2 + 3}; \quad \dfrac{dn}{dt} = \dfrac{(2t^2+3)12 - 6t^2(4t)}{(2t^2+3)^2}$

$= \dfrac{24t^3 + 36t - 24t^3}{(2t^2+3)^2}$

$\left.\dfrac{dn}{dt}\right|_{t=3.0\text{ s}} = \left.\dfrac{36t}{(2t^2+3)^2}\right|_{t=3.0\text{ s}} = 0.24 \text{ g/s}$

43. $T = \dfrac{2t}{0.05t + 1} - 20; \quad t = 6 \text{ h}$

$\dfrac{dT}{dt} = \dfrac{(0.05t+1)2 - 2t(0.05)}{(0.05t+1)^2} - 0$

$= \dfrac{0.1t + 2 - 0.1t}{(0.05t+1)^2}$

$\left.\dfrac{dT}{dt}\right|_{t=6.0\text{ h}} = 1.2°\text{C/h}$

44. $I = 5.00 + 0.01t^2; \quad R = 15.00 - 0.10t$

$V = IR = (5.00 + 0.01t^2)(15.00 - 0.10t)$

$\dfrac{dV}{dt} = (5.00 + 0.01t^2)(-0.10) + (15.00 - 0.010t)(0.02t)$

$\dfrac{dV}{dt} = -0.500 - 0.001t^2 + 0.300t - 0.002t^2$

$\dfrac{dV}{dt} = -0.003t^2 + 0.300t - 0.500$

$\left.\dfrac{dV}{dt}\right|_{t=5.00\text{ s}} = 0.925 \text{ V/s}$

45. $r_f = \dfrac{2(R^2 + Rr + r^2)}{3(R+r)}$

$\dfrac{dr_f}{dR} = \dfrac{2(2R+r)(3)(R+r) - 2(R^2 + Rr + r^2)(3)}{9(R+r)^2}$

$= \dfrac{6(2R+r)(R+r) - 6(R^2 + Rr + r^2)}{9(R+r)^2}$

$= \dfrac{12R^2 + 18Rr + 6r^2 - 6R^2 - 6Rr - 6r^2}{9(R+r)^2}$

$= \dfrac{6R^2 + 12Rr}{9(R+r)^2} = \dfrac{6R(R+2)}{9(R+r)^2} = \dfrac{2R(R+2r)}{3(R+r)^2}$

46. $T = \left(p + \dfrac{a}{V^2}\right)\left(\dfrac{V-b}{R}\right); \quad a, b, R$ are constants, assume p constant.

$T = \left(p + \dfrac{a}{V^2}\right)\dfrac{1}{R}(V - b)$

$\dfrac{dT}{dV} = \dfrac{1}{R}\left\{\left(p + \dfrac{a}{V^2}\right)(1) + (V-b)\left(0 + \dfrac{V^2(0) - a2V}{V^4}\right)\right\}$

$\dfrac{dT}{dV} = \dfrac{1}{R}\left\{\left(p + \dfrac{a}{V^2}\right) + (V-b)\left(\dfrac{-2aV}{V^4}\right)\right\}$

$\dfrac{dT}{dV} = \dfrac{1}{R}\left\{\left(\dfrac{pV^2 + a}{V^2}\right) - \dfrac{2a(V-b)}{V^3}\right\}$

$\dfrac{dT}{dV} = \dfrac{1}{R}\left\{\dfrac{pV^3 - aV - 2aV + 2ab}{V^3}\right\}$

$= \dfrac{pV^3 - aV + 2ab}{RV^3}$

47. $P = \dfrac{E^2 r}{R^2 + 2Rr + r^2}$

$\dfrac{dP}{dr} = \dfrac{(R^2 + 2Rr + r^2)E^2 - E^2 r(0 + 2R + 2r)}{(R^2 + 2Rr + r^2)^2}$

$\dfrac{dP}{dr} = \dfrac{E^2 R^2 + 2E^2 Rr + E^2 r^2 - 2E^2 Rr - 2E^2 r^2}{(R^2 + 2Rr + r^2)^2}$

$\dfrac{dP}{dr} = \dfrac{E^2 R^2 - E^2 r^2}{(R^2 + 2Rr + r^2)^2} = \dfrac{E^2(R^2 - r^2)}{[(R+r)^2]^2}$

$\dfrac{dP}{dr} = \dfrac{E^2(R-r)(R+r)}{(R+r)^4} = \dfrac{E^2(R-r)}{(R+r)^3}$

48. $P = \dfrac{kf^2}{w^2 - 2wf + f^2 + a^2}$

$\dfrac{dP}{df} = \dfrac{(w^2 + 2wf + f^2 + a^2)2kf - kf^2(-2w + 2f)}{(w^2 - 2wf + f^2 + a^2)^2}$

$\dfrac{dP}{df} = \dfrac{2kw^2 f - 4kwf^2 + 2kf^3 + 2ka^2 f - 2kwf^2 - 2kf^3}{(w^2 - 2wf + f^2 + a^2)^2}$

$\dfrac{dP}{df} = \dfrac{2kw^2 f - 2kwf^2 + 2ka^2 f}{(w^2 - 2wf + f^2 + a^2)^2}$

$\dfrac{dP}{df} = \dfrac{2kf(w^2 - wf + a^2)}{(w^2 - 2wf + f^2 + a^2)^2}$

3.7   The Derivative of a Power of a Function

1. $y = \sqrt{x} = x^{1/2}$

$\dfrac{dy}{dx} = \dfrac{1}{2}x^{1/2-1} = \dfrac{1}{2}x^{-1/2} = \dfrac{1}{2}\left(\dfrac{1}{x^{1/2}}\right) = \dfrac{1}{2x^{1/2}}$

2. $y = \sqrt[4]{x^3} = x^{3/4}$

$\dfrac{dy}{dx} = \dfrac{3}{4}x^{3/4-1} = \dfrac{3}{4}x^{-1/4} = \dfrac{3}{4\sqrt[4]{x}}$

3. $v = \dfrac{3}{t^2} = 3t^{-2}$

$\dfrac{dv}{dt} = -6t^{-2-1} = -6t^{-3} = \dfrac{-6}{t^3}$

4. $y = \dfrac{2}{x^4} = 2x^{-4};\ \dfrac{dy}{dx} = 2(-4x^{-5}) = \dfrac{-8}{x^5}$

5. $y = \dfrac{3}{\sqrt[3]{x}} = \dfrac{3}{x^{1/3}} = 3x^{-1/3};\ \dfrac{dy}{dx} = 3\left(-\dfrac{1}{3}x^{-1/3-1}\right)$

$= -1x^{-4/3} = -1\left(\dfrac{1}{x^{4/3}}\right) = -\dfrac{1}{x^{4/3}}$

6. $y = \dfrac{1}{\sqrt[5]{x^2}} = x^{-2/5};\ \dfrac{dy}{dx} = -\dfrac{2}{5}x^{-7/5} = \dfrac{-2}{5\sqrt[5]{x^7}}$

7. $y = x\sqrt{x} - \dfrac{1}{x} = x \cdot x^{1/2} - x^{-1} = x^{3/2} - x^{-1}$

$\dfrac{dy}{dx} = \dfrac{3}{2}x^{1/2} - (-1x^{-2}) = \dfrac{3\sqrt{x}}{2} + \dfrac{1}{x^2}$

8. $f(x) = 2x^{-3} - 3x^{-2}$
$f'(x) = -6x^{-4} + 6x^{-3}$

9. $y = (x^2 + 1)^5;$

$\dfrac{dy}{dx} = 5(x^2+1)^4(2x) = 10x(x^2+1)^4$

10. $y = (1-2x)^4;\ \dfrac{dy}{dx} = 4(1-2x)^3(-2) = -8(1-2x)^3$

11. $y = 2(7 - 4x^3)^8$

$\dfrac{dy}{dx} = 16(7-4x^3)^7(-12x^2) = -192x^2(7-4x^3)^7$

12. $y = 3(8x^2 - 1)^6$

$\dfrac{dy}{dx} = 18(8x^2-1)^5(16x) = 288x(8x^2-1)^5$

13. $y = (2x^3 - 3)^{1/3};$

$\dfrac{dy}{dx} = \dfrac{1}{3}(2x^3-3)^{-2/3}(6x^2) = \dfrac{2x^2}{(2x^3-3)^{2/3}}$

14. $y = (1 - 6x)^{3/2}$

$\dfrac{dy}{dx} = \dfrac{3}{2}(1-6x)^{1/2}(-6) = -9(1-6x)^{1/2}$

15. $f(y) = \dfrac{3}{(4-y^2)^4} = 3(4-y^2)^{-4}$

$f'(y) = -12(4-y^2)^{-5}(-2y) = \dfrac{24y}{(4-y^2)^5}$

16. $y = \dfrac{4}{\sqrt{1-3x}} = \dfrac{4}{(1-3x)^{1/2}} = 4(1-3x)^{-1/2}$

$\dfrac{dy}{dx} = -2(1-3x)^{-3/2}(-3) = 6(1-3x)^{-3/2}$

$= \dfrac{6}{(1-3x)^{3/2}}$

17. $y = 4(2x^4 - 5)^{3/4};\ u = 2x^4 - 5;\ \dfrac{du}{dx} = 8x^3$

$\dfrac{dy}{dx} = 4\left[\dfrac{3}{4}(2x^4-5)^{-1/4}(8x^3)\right] = \dfrac{24x^3}{(2x^4-5)^{1/4}}$

18. $r = 5(3\theta^6 - 4)^{2/3}$

$\dfrac{dr}{d\theta} = \dfrac{10}{3}(3\theta^6-4)^{-1/3} \cdot (18\theta^5)$

$\dfrac{dr}{d\theta} = \dfrac{60\theta^5}{(3\theta^6-4)^{1/3}}$

19. $y = \sqrt[4]{1-8x^2} = (1-8x^2)^{1/4}$

$\dfrac{dy}{dx} = \dfrac{1}{4}(1-8x^2)^{-3/4}(-16x)$

$\dfrac{dy}{dx} = \dfrac{-4x}{(1-8x^2)^{3/4}}$

20. $y = \sqrt[3]{4x^6 + 2} = (4x^6 + 2)^{1/3}$

$\dfrac{dy}{dx} = \dfrac{1}{3}(4x^6+2)^{-2/3}(24x^5)$

$\dfrac{dy}{dx} = \dfrac{8x^5}{(4x^6+2)^{2/3}}$

21. $y = x\sqrt{8x+5} = x(8x+5)^{1/2}$

$\dfrac{dy}{dx} = x\left(\dfrac{1}{2}\right)(8x+5)^{-1/2}(8) + (8x+5)^{1/2}(1)$

$= 4x(8x+5)^{-1/2} + (8x+5)^{1/2}(1)$

$= \dfrac{4x}{(8x+5)^{1/2}} + \dfrac{(8x+5)}{(8x+5)^{1/2}} = \dfrac{12x+5}{(8x+5)^{1/2}}$

22. $y = x^2(1 - 3x)^5$

$$\frac{dy}{dx} = x^2(5)(1 - 3x)^4(-3) + (1 - 3x)^5(2x)$$
$$= -15x^2(1 - 3x)^4 + 2x(1 - 3x)^5$$
$$= x(1 - 3x)^4(-15x + 2(1 - 3x))$$
$$= x(1 - 3x)^4(-15x + 2 - 6x)$$
$$= x(1 - 3x)^4(2 - 21x)$$

23. $f(R) = \sqrt{\dfrac{2R + 1}{4R + 1}} = \left(\dfrac{2R + 1}{4R + 1}\right)^{1/2}$

$$f'(R) = \frac{1}{2}\left(\frac{2R + 1}{4R + 1}\right)^{-1/2} \cdot \frac{(4R + 1)(2) - (2R + 1)(4)}{(4R + 1)^2}$$

$$f'(R) = \frac{1}{2\sqrt{\dfrac{2R + 1}{4R + 1}}} \cdot \frac{8R + 2 - 8R - 4}{(4R + 1)^2}$$

$$= \frac{-1}{\sqrt{2R + 1}(4R + 1)^{3/2}}$$

24. $y = \left(\dfrac{2x + 1}{3x - 2}\right)^2$

$$\frac{dy}{dx} = 2\left(\frac{2x + 1}{3x - 2}\right) \cdot \frac{(3x - 2)(2) - (2x + 1)(3)}{(3x - 2)^2}$$

$$= 2\left(\frac{2x + 1}{3x - 2}\right) \cdot \frac{6x - 4 - 6x - 3}{(3x - 2)^2}$$

$$\frac{dy}{dx} = \frac{2(2x + 1)(-7)}{(3x - 2)^3} = \frac{-14(2x + 1)}{(3x - 2)^3}$$

25. $y = \sqrt{3x + 4}; \; x = 7$

$$y = (3x + 4)^{1/2}; \; u = 3x + 4; \; n = \frac{1}{2}; \; \frac{du}{dx} = 3$$

$$\frac{dy}{dx} = \frac{1}{2}(3x + 4)^{-1/2}(3) = \frac{3}{2}(3x + 4)^{-1/2}$$

$$= \frac{3}{2\sqrt{3x + 4}}$$

$$\frac{dy}{dx}\bigg|_{x=7} = \frac{3}{2\sqrt{3(7) + 4}} = \frac{3}{2\sqrt{25}} = \frac{3}{2(5)} = \frac{3}{10}$$

26. $y = (4 - x^2)^{-1}; \; x = -1$

$$\frac{dy}{dx} = -1(4 - x^2)^{-2} - (2x) = \frac{2x}{(4 - x^2)^2}$$

$$\frac{dy}{dx}\bigg|_{x=-1} = -\frac{2}{9}$$

27. $y = \dfrac{\sqrt{x}}{1 - x} = \dfrac{x^{1/2}}{1 - x}; \; x = 4$

$$\frac{dy}{dx} = \frac{(1 - x)\dfrac{1}{2}x^{-1/2} - x^{1/2}(-1)}{(1 - x)^2}$$

$$= \frac{\dfrac{(1 - x)}{2x^{1/2}} + \dfrac{x^{1/2}}{1}}{(1 - x)^2}$$

$$\frac{dy}{dx} = \frac{\left(\dfrac{1 - x + 2x}{2x^{1/2}}\right)}{(1 - x)^2}$$

$$= \frac{(1 + x)}{2x^{1/2}(1 - x)^2}$$

$$\frac{dy}{dx}\bigg|_{x=4} = \frac{5}{36}$$

```
nDeriv(f(X)/(1-X
),X,4)
         .1389009019
5/36
         .1388888889
```

28. $y = x^2\sqrt[3]{3x + 2} = x^2(3x + 2)^{1/3}$

$$\frac{dy}{dx} = x^2\frac{1}{2}(3x + 2)^{-2/3}(3) + (3x + 2)^{1/3}(2x)$$

$$\frac{dy}{dx} = \frac{x^2}{(3x + 2)^{2/3}} + 2x(3x + 2)^{1/3}$$

$$\frac{dy}{dx} = \frac{x^2 + 2x(3x + 2)}{(3x + 2)^{2/3}} = \frac{x^2 + 6x^2 + 4x}{(3x + 2)^{2/3}}$$

$$\frac{dy}{dx} = \frac{7x^2 + 4x}{(3x + 2)^{2/3}}$$

$$\frac{dy}{dx}\bigg|_{x=2} = 9$$

```
nDeriv(X²√(3X+2
),X,2)
         9.000000151
```

29. (a) $y = \dfrac{1}{x^3}; \; u = 1; \; \dfrac{du}{dx} = 0$

$$= v = x^3; \; \frac{dv}{dx} = 3x^2$$

$$\frac{dy}{dx} = \frac{x^3(0) - 1(3x^2)}{x^6} = \frac{-3x^2}{x^6} = \frac{-3}{x^4}$$

(b) $y = x^{-3}; \; \dfrac{dy}{dx} = -3x^{-3-1} = -3x^{-4} = \dfrac{-3}{x^4}$

30. (a) $y = \dfrac{2}{4x + 3}$

$$\frac{dy}{dx} = \frac{(4x + 3)(0) - 2(4)}{(4x + 3)^2} = \frac{-8}{(4x + 3)^2}$$

(b) $y = 2(4x + 3)^{-1}$

$$\frac{dy}{dx} = -2(4x + 3)^{-2}(4) = \frac{-8}{(4x + 3)^2}$$

31. $y = \dfrac{x^2}{\sqrt{x^2+1}}$; $\dfrac{dy}{dx} = 0$ for (x,y); $y = \dfrac{x^2}{(x^2+1)^{1/2}}$

$$\frac{dy}{dx} = \frac{(x^2+1)^{1/2}2x - x^2\frac{1}{2}(x^2+1)^{-1/2}(2x)}{(x^2+1)}$$

$$\frac{dy}{dx} = \frac{2x(x^2+1)^{1/2} - \dfrac{x^3}{(x^2+1)^{1/2}}}{(x^2+1)}$$

$$= \frac{\dfrac{2x(x^2+1) - x^3}{(x^2+1)^{1/2}}}{(x^2+1)}$$

$$\frac{dy}{dx} = \frac{x^3+2x}{(x^2+1)^{3/2}}; \quad \frac{dy}{dx} = 0$$

$$x^3 + 2x = 0; \quad x(x^2+2) = 0$$

$x = 0$ or $x = \sqrt{-2}$ (imaginary). Therefore,

$$\frac{dy}{dx} = 0 \text{ where } x = 0$$

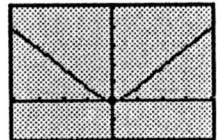

32. $y = \dfrac{x}{\sqrt{4x-1}}$; $\dfrac{dy}{dx} = 0$ for (x,y);

$$y = \frac{x}{(4x-1)^{1/2}}$$

$$\frac{dy}{dx} = \frac{(4x-1)^{1/2}(1) - x\frac{1}{2}(4x-1)^{-1/2}(4)}{(4x-1)}$$

$$\frac{dy}{dx} = \frac{(4x-1)^{1/2} - \dfrac{2x}{(4x-1)^{1/2}}}{(4x-1)}$$

$$\frac{dy}{dx} = \frac{(4x-1) - 2x}{(4x-1)^{3/2}} = \frac{2x-1}{(4x-1)^{3/2}}$$

$$\frac{dy}{dx} = 0; \quad 2x - 1 = 0, \quad x = \frac{1}{2}$$

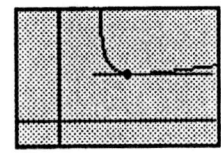

33. $y^2 = 4x$; $y = \sqrt{4x} = 2\sqrt{x} = 2x^{1/2}$

$$m_{\text{tan}} = \frac{dy}{dx} = \frac{d(2x^{1/2})}{dx} = 2\left(\frac{1}{2}\right)x^{-1/2} = \frac{1}{\sqrt{x}}$$

$$m_{\text{tan}}\big|_{x=1} = \frac{1}{\sqrt{1}} = \frac{1}{1} = 1$$

   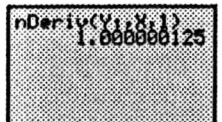

34. $x^2 + y^2 = 25$; $(4, 3)$

$$y = \sqrt{25-x^2} = (25-x^2)^{1/2}$$

$$\frac{dy}{dx} = \frac{1}{2}(25-x^2)^{-1/2}(-2x)$$

$$= \frac{-x}{\sqrt{25-x^2}}$$

$$\frac{dy}{dx}\bigg|_{x=4} = \frac{-4}{\sqrt{9}}$$

$$= -\frac{4}{3}$$

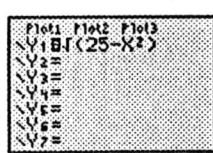

35. $s = (8t - t^2)^{2/3}$; $t = 6.25$ s

$$\frac{ds}{dt} = \frac{2}{3}(8t - t^2)^{-1/3}(8 - 2t)$$

$$v = \frac{2(8 - 2t)}{3\sqrt[3]{8t - t^2}}; \quad v\big|_{t=6.25}$$

$$= -1.35 \text{ cm/s}$$

36. $w = \sqrt{2rh - h^2} = (2rh - h^2)^{1/2}$

$$\frac{dw}{dh} = \frac{1}{2}(2rh - h^2)^{-1/2}(2r - 2h)$$

$$\frac{dw}{dh} = (2rh - h^2)^{-1/2}(r - h)\big|_{h=2.25, r=6.00}$$

$$\frac{dw}{dh} = (2(6)(2.25) - (2.25)^2)^{-1/2}(6 - 2.25)$$

$$\frac{dw}{dh} = 0.801$$

37. $P = \dfrac{k}{V^{3/2}}$; $P = 300$ kPa when $V = 100$ cm^3

$$300 = \frac{k}{100^{3/2}}$$

$$k = 300(100)^{3/2} = 300(10^3)$$
$$= 300(1000) = 300,000$$

$$P = \frac{300,000}{V^{3/2}} = 300,000V^{-3/2}$$

$$\frac{dP}{dV} = 300,000\left(-\frac{3}{2}\right)V^{-5/2}$$

$$\frac{dP}{dV} = -450,000V^{-5/2} = \frac{-450,000}{V^{5/2}}$$

$$\left.\frac{dP}{dV}\right|_{V=100} = \frac{-450,000}{100^{5/2}} = \frac{-450,000}{100,000}$$

$$= -4.50 \text{ kPa/cm}^3$$

38. $G = \dfrac{k}{\lambda^2}$; $G = 5.0 \times 10^4$; $\lambda = 0.35$ ft; $k = G\lambda^2$

$$G = k\lambda^{-2}; \quad \frac{dG}{d\lambda} = -2k\lambda^{-3} = \frac{-2k}{\lambda^3} = \frac{-2G\lambda^2}{\lambda^3}$$

$$\frac{dG}{d\lambda} = \frac{-2G}{\lambda}$$

$$\left.\frac{dG}{d\lambda}\right|_{\lambda=0.35} = \frac{-2(5.0 \times 10^4)}{(0.35)} = -2.9 \times 10^5/\text{ft}$$

39. $H = \dfrac{4000}{\sqrt{t^6 + 100}}$; $(-6 < t < 6)$

$$\frac{dH}{dt} = 4000(t^6 + 100)^{-1/2}$$

$$\frac{dH}{dt} = 4000\left(-\frac{1}{2}\right)(t^6 + 100)^{-3/2}(6t^5)$$

$$\frac{dH}{dt} = \frac{-12\,000t^5}{(t^6 + 100)^{3/2}}$$

$$\left.\frac{dH}{dt}\right|_{t=4.0} = -45.2 \text{ W/(m}^2 \cdot \text{h)}$$

40. $t = \dfrac{\sqrt{a^2 + x^2}}{v_1} + \dfrac{\sqrt{b^2 + (c-x)^2}}{v_2}$; $\dfrac{dt}{dx} = ?$

$$t = \frac{1}{v_1}(a^2 + x^2)^{1/2} + \frac{1}{v_2}(b^2 + (c-x)^2)^{1/2}$$

$$\frac{dt}{dx} = \frac{1}{v_1} \cdot \frac{1}{2}(a^2 + x^2)^{-1/2}(2x) + \frac{1}{v_2}$$

$$\cdot \frac{1}{2}[b^2 + (c-x)^2]^{-1/2} \times 2(c-x)(-1)$$

$$\frac{dt}{dx} = \frac{x}{v_1(a^2 + x^2)^{1/2}} + \frac{x-c}{v_2[b^2 + (c-x)^2]^{1/2}}$$

41. $\lambda_r = \dfrac{2a\lambda}{\sqrt{4a^2 - \lambda^2}} = \dfrac{2a\lambda}{(4a^2 - \lambda^2)^{1/2}}$

$$\frac{d\lambda_r}{d\lambda} = \frac{(4a^2 - \lambda^2)^{1/2}(2a) - 2a\lambda\left(\frac{1}{2}\right)(4a^2 - \lambda^2)^{-1/2}(-2\lambda)}{(4a^2 - \lambda^2)}$$

$$= \frac{2a(4a^2 - \lambda^2)^{1/2} + 2a\lambda^2(4a^2 - \lambda^2)^{-1/2}}{(4a^2 - \lambda^2)}$$

$$= \frac{(4a^2 - \lambda^2)^{-1/2}[(2a)(4a^2 - \lambda^2) + 2a\lambda^2]}{(4a^2 - \lambda^2)}$$

$$= \frac{8a^3}{(4a^2 - \lambda^2)^{3/2}}$$

42. $I = \dfrac{V}{\sqrt{R^2 + (\omega L)^2}}$

$$= \frac{V}{[R^2 + (\omega L)^2]^{1/2}}$$

$$= V[R^2 + (\omega L)^2]^{-1/2}$$

$$\frac{dI}{dL} = V\left(-\frac{1}{2}\right)[R^2 + (\omega L)^2]^{-3/2}2(\omega L)\omega$$

$$\frac{dI}{dL} = \frac{-\omega^2 LV}{[R^2 + (\omega L)^2]^{3/2}}$$

43. $D^2 = w^2 + (w + 2)^2 = w^2 + w^2 + 4w + 4$
$D = \sqrt{2w^2 + 4w + 4}$
$D = (2w^2 + 4w + 4)^{1/2}$

$$\frac{dD}{dw} = \frac{1}{2}(2w^2 + 4w + 4)^{-1/2}(4w + 4)$$

$$\frac{dD}{dw} = \frac{2w + 2}{(2w^2 + 4w + 4)^{1/2}}$$

$$= \frac{2(w + 1)}{(2w^2 + 4w + 4)^{1/2}}$$

$l = w + 2$

44. $l^2 = 5^2 + (x - 3)^2$
$l = \sqrt{25 + (x - 3)^2}$
$l = [25 + (x - 3)^2]^{1/2}$

$$\frac{dl}{dx} = \frac{1}{2}[25 + (x - 3)^2]^{-1/2} \cdot 2(x - 3)^1$$

$$\frac{dl}{dx} = \frac{x - 3}{[25 + (x - 3)^2]^{1/2}}$$

$$= \frac{x - 3}{\sqrt{x^2 - 6x + 34}}$$

3.8   Differentiation of Implicit Functions

1. $3x + 2y = 5$; $\dfrac{d(3x)}{dx} + \dfrac{d(2y)}{dx} = \dfrac{d(5)}{dx}$

$3 + \dfrac{2dy}{dx} = 0$; $\dfrac{2dy}{dx} = -3$; $\dfrac{dy}{dx} = -\dfrac{3}{2}$

2. $6x - 3y = 4$; $\dfrac{d}{dx}(6x) - \dfrac{d}{dx}(3y) = \dfrac{d}{dx}(4)$

$6\dfrac{d}{dx}(x) - 3\dfrac{d}{dx}(y) = 0$; $6(1) - 3\dfrac{dy}{dx} = 0$;

$3\dfrac{dy}{dx} = 6$; $\dfrac{dy}{dx} = 2$

3. $4y - 3x^2 = x$; $\dfrac{d}{dx}(4y) - \dfrac{d}{dx}(3x^2) = \dfrac{d}{dx}(x)$

$4\dfrac{dy}{dx} - 6x = 1$; $4\dfrac{dy}{dx} = 1 + 6x$; $\dfrac{dy}{dx} = \dfrac{1 + 6x}{4}$

4. $x^5 - 5y = 6 - x$; $\dfrac{d}{dx}(x^5) - \dfrac{d}{dx}(5y) = \dfrac{d}{dx}(6 - x)$

$5x^4 - 5\dfrac{dy}{dx} = -1$; $5\dfrac{dy}{dx} = 5x^4 + 1$; $\dfrac{dy}{dx} = \dfrac{5x^4 + 1}{5}$

5. $x^2 - 4y^2 - 9 = 0$; $\dfrac{d(x^2)}{dx} - \dfrac{d(4y^2)}{dx} - \dfrac{d(9)}{dx} = \dfrac{d(0)}{dx}$

$2x - \dfrac{8y\,dy}{dx} - 0 = 0$; $\dfrac{-8y\,dy}{dx} = -2x$; $\dfrac{dy}{dx} = \dfrac{x}{4y}$

6. $x^2 + 2y^2 - 11 = 0$; $\dfrac{d}{dx}(x^2) + \dfrac{d}{dx}(2y^2) - \dfrac{d}{dx}(11) = 0$

$2x + 4y\dfrac{dy}{dx} = 0$; $4y\dfrac{dy}{dx} = -2x$; $\dfrac{dy}{dx} = \dfrac{-2x}{4y} = \dfrac{-x}{2y}$

7. $y^5 = x^2 - 1$; $\dfrac{d}{dx}(y^5) = 2x$; $5x\dfrac{dy}{dx} = 2x$; $\dfrac{dy}{dx} = \dfrac{2x}{5y^4}$

8. $y^4 = 3x^3 - x$; $4y^3\dfrac{dy}{dx} = 9x^2 - 1$; $\dfrac{dy}{dx} = \dfrac{9x^2 - 1}{4y^3}$

9. $y^2 + y = x^2 - 4$; $\dfrac{d(y^2)}{dx} + \dfrac{d(y)}{dx} = \dfrac{d(x^2)}{dx} - \dfrac{d(4)}{dx}$

$\dfrac{2y\,dy}{dx} + \dfrac{dy}{dx} = 2x - 0$

$\dfrac{dy}{dx}(2y + 1) = 2x$; $\dfrac{dy}{dx} = \dfrac{2x}{2y + 1}$

10. $2y^3 - y = 7 - x^4$; $6y^2\dfrac{dy}{dx} - \dfrac{dy}{dx} = -4x^3$

$\dfrac{dy}{dx}(6y^2 - 1) = -4x^3$; $\dfrac{dy}{dx} = \dfrac{-4x^3}{6y^2 - 1} = \dfrac{4x^3}{1 - 6y^2}$

11. $y + 3xy - 4 = 0$

$\dfrac{dy}{dx} + 3\dfrac{d}{dx}(xy) = 0$; $\dfrac{dy}{dx} + 3\left(x\dfrac{dy}{dx} + y(1)\right) = 0$

$\dfrac{dy}{dx} + 3x\dfrac{dy}{dx} + 3y = 0$; $\dfrac{dy}{dx}(1 + 3x) = -3y$

$\dfrac{dy}{dx} = \dfrac{-3y}{1 + 3x}$

12. $8y - xy - 7 = 0$; $8\dfrac{dy}{dx} - \dfrac{d}{dx}(xy) = 0$

$8\dfrac{dy}{dx} - \left(x\dfrac{dy}{dx} + y\right) = 0$; $8\dfrac{dy}{dx} - x\dfrac{dy}{dx} - y = 0$

$\dfrac{dy}{dx}(8 - x) = y$; $\dfrac{dy}{dx} = \dfrac{y}{8 - x}$

13. $xy^3 + 3y + x^2 = 9$; $\dfrac{d(xy^3)}{dx} + \dfrac{d(3y)}{dx} + \dfrac{d(x^2)}{dx} = \dfrac{d(9)}{dx}$

$\dfrac{x\,dy^3}{dx} + \dfrac{y^3\,dx}{dx} + \dfrac{3\,dy}{dx} + 2x = 0$

$x(3y^2)\dfrac{dy}{dx} + y^3(1) + \dfrac{3\,dy}{dx} + 2x = 0$

$3xy^2\dfrac{dy}{dx} + y^3 + \dfrac{3\,dy}{dx} + 2x = 0$

$3xy^2\dfrac{dy}{dx} + \dfrac{3\,dy}{dx} = -y^3 - 2x$

$\dfrac{dy}{dx}(3xy^2 + 3) = -y^3 - 2x$; $\dfrac{dy}{dx} = \dfrac{-2x - y^3}{3xy^2 + 3}$

14. $y^2x - \dfrac{5y}{x + 1} + 3x = 4$

$y^2(1) + x2y\dfrac{dy}{dx} - \left\{\dfrac{5(x + 1)\dfrac{dy}{dx} - 5y(1)}{(x + 1)^2}\right\} + 3 = 0$

$2xy\dfrac{dy}{dx} - \dfrac{5}{x + 1}\cdot\dfrac{dy}{dx} + \dfrac{5y}{(x + 1)^2} = -3 - y^2$

$\dfrac{dy}{dx}\left(2xy - \dfrac{5}{x + 1}\right) = -3 - y^2 - \dfrac{5y}{(x + 1)^2}$

$\qquad\qquad = \dfrac{-3(x + 1)^2 - y^2(x + 1)^2 - 5y}{(x + 1)^2}$

$\dfrac{dy}{dx}\left(\dfrac{2xy(x + 1) - 5}{x + 1}\right) = \dfrac{-3(x + 1)^2 - y^2(x + 1)^2 - 5y}{(x + 1)^2}$

$\qquad\qquad = \dfrac{-\{(x + 1)^2(3 + y^2) + 5y\}}{(x + 1)^2}$

$\dfrac{dy}{dx} = \dfrac{-\{(x + 1)^2(3 + y^2) + 5y\}}{(x + 1)^2}\cdot\dfrac{(x + 1)}{(2x^2y + 2xy - 5)}$

$\dfrac{dy}{dx} = \dfrac{(x + 1)^2(3 + y^2) + 5y}{(5 - 2xy - 2x^2y)(x + 1)}$

15. $\dfrac{3x^2}{y^2+1} + y = 3x + 1$

$$\dfrac{(y^2+1)6x - 3x^2\left(2\dfrac{dy}{dx}\right)}{(y^2+1)^2} + \dfrac{dy}{dx} = 3$$

$$\dfrac{6x(y^2+1) - 6x^2y\dfrac{dy}{dx} + (y^2+1)^2\dfrac{dy}{dx}}{(y^2+1)^2} = 3$$

$$(y^2+1)^2\dfrac{dy}{dx} - 6x^2y\dfrac{dy}{dx} = 3(y^2+1)^2 - 6x(y^2+1)$$
$$= 3(y^2+1)(y^2+1-2x)$$

$$\dfrac{dy}{dx}[(y^2+1)^2 - 6x^2y] = 3(y^2+1)(y^2-2x+1)$$

$$\dfrac{dy}{dx} = \dfrac{3(y^2+1)(y^2-2x+1)}{(y^2+1)^2 - 6x^2y}$$

16. $2x - x^3y^2 = y - x^2 - 1$

$$2 - \left(x^3 2y\dfrac{dy}{dx} + y^2 3x^2\right) = \dfrac{dy}{dx} - 2x$$

$$2 - 2x^3y\dfrac{dy}{dx} - 3x^2y^2 = \dfrac{dy}{dx} - 2x$$

$$\dfrac{dy}{dx} + 2x^3y\dfrac{dy}{dx} = 2 + 2x - 3x^2y^2$$

$$\dfrac{dy}{dx}(1 + 2x^3y) = 2 + 2x - 3x^2y^2$$

$$\dfrac{dy}{dx} = \dfrac{2 + 2x - 3x^2y^2}{1 + 2x^3y}$$

17. $(2y - x)^4 + x^2 = y + 3$

$$4(2y-x)^3\left(2\dfrac{dy}{dx} - 1\right) + 2x = \dfrac{dy}{dx} + 0$$

$$4(2y-x)^3\left(2\dfrac{dy}{dx}\right) - 4(2y-x)^3 + 2x = \dfrac{dy}{dx}$$

$$8\dfrac{dy}{dx}(2y-x)^3 - \dfrac{dy}{dx} = 4(2y-x)^3 - 2x$$

$$\dfrac{dy}{dx}[8(2y-x)^3 - 1] = 4(2y-x)^3 - 2x$$

$$\dfrac{dy}{dx} = \dfrac{4(2y-x)^3 - 2x}{8(2y-x)^3 - 1}$$

18. $(y^2+2)^3 = x^4y + 11$

$$3(y^2+2)^2 2y\dfrac{dy}{dx} = x^4\dfrac{dy}{dx} + y4x^3 + 0$$

$$6y(y^2+2)^2\dfrac{dy}{dx} - x^4\dfrac{dy}{dx} = 4x^3y$$

$$\dfrac{dy}{dx}[6y(y^2+2)^2 - x^4] = 4x^3y$$

$$\dfrac{dy}{dx} = \dfrac{4x^3y}{6y(y^2+2)^2 - x^4}$$

19. $2(x^2+1)^3 + (y^2+1)^2 = 17$

$$6(x^2+1)^2(2x) + 2(y^2+1)^1 2y\dfrac{dy}{dx} = 0$$

$$4y(y^2+1)\dfrac{dy}{dx} = -12x(x^2+1)^2$$

$$\dfrac{dy}{dx} = \dfrac{-12x(x^2+1)^2}{4y(y^2+1)} = \dfrac{-3x(x^2+1)^2}{y(y^2+1)}$$

20. $(2x+1)(1-3y) + y^2 = 13$

$$(2x+1)\left(-3\dfrac{dy}{dx}\right) + (1-3y)2 + 2y\dfrac{dy}{dx} = 0$$

$$3(2x+1)\dfrac{dy}{dx} - 2y\dfrac{dy}{dx} = 2(1-3y)$$

$$\dfrac{dy}{dx}[6x+3-2y] = 2(1-3y)$$

$$\dfrac{dy}{dx} = \dfrac{2(1-3y)}{6x-2y+3} = \dfrac{2(3y-1)}{2y-6x-3}$$

21. $3x^3y^2 - 2y^3 = -4;\ (1,2)$

$$3x^3(2y)\dfrac{dy}{dx} + y^2(9x^2) - 6y^2\dfrac{dy}{dx} = 0$$

$$6x^3y\dfrac{dy}{dx} + 9x^2y^2 - 6y^2\dfrac{dy}{dx} = 0$$

$$6x^3y\dfrac{dy}{dx} - 6y^2\dfrac{dy}{dx} = -9x^2y^2$$

$$\dfrac{dy}{dx}(6x^3y - 6y^2) = -9x^2y^2$$

$$\dfrac{dy}{dx} = \dfrac{-9x^2y^2}{6x^3y - 6y^2} = \dfrac{-9x^2y^2}{3(2x^3y - 2y^2)}$$

$$\dfrac{dy}{dx} = \dfrac{-3x^2y^2}{2x^3y - 2y^2}$$

$$\left.\dfrac{dy}{dx}\right|_{(1,2)} = \dfrac{-3(1^2)(2^2)}{2(1^3)(2) - 2(2^2)} = \dfrac{-12}{-4} = 3$$

22. $2y + 5 - x^2 - y^3 = 0;\ (2,1)$

$$2\dfrac{dy}{dx} - 2x - 3y^2\dfrac{dy}{dx} = 0$$

$$\dfrac{dy}{dx}(2 - 3y^2) = 2x;\quad \dfrac{dy}{dx} = \dfrac{2x}{2 - 3y^2};\quad \left.\dfrac{dy}{dx}\right|_{(2,-1)} = -4$$

23. $5y^4 + 7 = x^4 - 3y;\ (3,-2)$

$$20y^3\dfrac{dy}{dx} = 4x^3 - 3\dfrac{dy}{dx};\quad 20y^3\dfrac{dy}{dx} + 3\dfrac{dy}{dx} = 4x^3$$

$$\dfrac{dy}{dx}(20y^3 + 3) = 4x^3;\quad \dfrac{dy}{dx} = \dfrac{4x^3}{20y^3 + 3};$$

$$\left.\dfrac{dy}{dx}\right|_{(3,-2)} = -\dfrac{108}{157}$$

24. $(xy - y^2)^3 = 5y^2 + 22$; $(4, 1)$

$$3(xy - x^2)^2 \left(x\frac{dy}{dx} + y - 2y\frac{dy}{dx} \right) = 10y\frac{dy}{dx}$$

$$3x(xy - y^2)^2\frac{dy}{dx} + 3y(xy - y^2)^2 - 6y(xy - y^2)^2\frac{dy}{dx}$$

$$= 10y\frac{dy}{dx}$$

$$10y\frac{dy}{dx} + 6y(xy - y^2)^2\frac{dy}{dx} - 3x(xy - y^2)^2\frac{dy}{dx}$$

$$= 3y(xy - y^2)^2$$

$$\frac{dy}{dx} = \frac{3y(xy - y^2)^2}{10y + 6y(xy - y^2)^2 - 3x(xy - y^2)^2};$$

$$\frac{dy}{dx}\bigg|_{(4,1)} = -\frac{27}{44}$$

25. $xy + y^2 + 2 = 0$; $\dfrac{d(xy)}{dx} + \dfrac{d(y^2)}{dx} + \dfrac{d(2)}{dx} = \dfrac{d(0)}{dx}$

$$\frac{x\,dy}{dx} + \frac{y\,dx}{dx} + \frac{2y\,dy}{dx} + 0 = 0$$

$$\frac{x\,dy}{dx} + y(1) + \frac{2y\,dy}{dx} = 0; \quad \frac{x\,dy}{dx} + \frac{2y\,dy}{dx} = -y$$

$$\frac{dy}{dx}(x + 2y) = -y; \quad \frac{dy}{dx} = \frac{-y}{x + 2y}$$

$$\frac{dy}{dx}\bigg|_{(-3,1)} = \frac{-1}{-3 + 2(1)} = \frac{-1}{-1} = 1$$

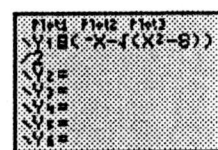 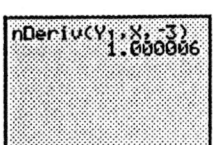

26. $s^3 - t^2 = 7t$; $\dfrac{ds}{dx}$ at $(4.01, 5.25)$

$$3s^2\frac{ds}{dx} - 2t = 7$$

$$3s^2\frac{ds}{dx} = 7 + 2t;$$

$$\frac{ds}{dx} = \frac{7 + 2t}{3s^2}$$

$$\frac{ds}{dx}\bigg|_{(s=4.01, t=5.25)} = 0.363 \text{ m/s}$$

27. $x^2 + y^2 = 2.38^2$; $2x + 2y\dfrac{dy}{dx} = 0$; $\dfrac{dy}{dx} = \dfrac{-x}{y}$

28. $A_T = \pi r^2 + 2\pi rh = 940$

$\pi r^2 + 2\pi rh = 940$

$$2\pi r\frac{dr}{dh} + 2\pi \left(r + h\frac{dr}{dh} \right) = 0$$

$$2\pi r\frac{dr}{dh} + 2\pi r + 2\pi h\frac{dr}{dh} = 0$$

$$2\pi r\frac{dr}{dh} + 2\pi h\frac{dr}{dh} = -2\pi r$$

$$\frac{dr}{dh}(2\pi r + 2\pi h) = -2\pi r; \quad \frac{dr}{dh} = \frac{-2\pi r}{2\pi(r + h)}$$

$$\frac{dr}{dh} = \frac{-r}{r + h}$$

29. $r^2 = 2rR + 2R - 2r$; $2r = 2R + \dfrac{dR}{dr}(2r) + 2\dfrac{dR}{dr} - 2$

$$2r - 2R + 2 = 2r\frac{dR}{dr} + 2\frac{dR}{dr} = \frac{dR}{dr}(2r + 2)$$

$$\frac{dR}{dr} = \frac{2(r - R + 1)}{2(r + 1)} = \frac{r - R + 1}{r + 1}$$

30. $I = \dfrac{1}{12}(b^3h + bh^3)$; $\dfrac{db}{dh} = ?$

$$0 = \frac{1}{12}\left(b^3 + h3b^2\frac{db}{dh} + b3h^2 + h^3\frac{db}{dh} \right)$$

$$3b^2h\frac{db}{dh} + h^3\frac{db}{dh} = -b^3 - 3bh^2$$

$$\frac{db}{dh}(3b^2h + h^3) = -b(b^2 + 3h^2)$$

$$\frac{db}{dh} = \frac{-b(b^2 + 3h^2)}{h(3b^2 + h^2)} = -\frac{b^3 + 3bh^2}{h^3 + 3b^2h}$$

31. $24C^3Sr^3 = 40C^3r^3 + 9LC^2r^2 - 3L^3$

$$24C^3S3r^2 = 40C^33r^2 + 9C^2\left(L2r + r^2\frac{dL}{dr} \right) - 9L^2\frac{dL}{dr}$$

$$72C^3Sr^2 - 120C^3r^2 = 18C^2Lr + 9C^2r^2\frac{dL}{dr} - 9L^2\frac{dL}{dr}$$

$$\frac{dL}{dr}(9C^2r^2 - 9L^2) = 72C^3Sr^2 - 120C^3r^2 - 18C^2Lr$$

$$\frac{dL}{dr} = \frac{6C^2r(12CSr - 20Cr - 3L)}{9(C^2r^2 - L^2)}$$

$$\frac{dL}{dr} = \frac{2C^2r(12CSr - 20Cr - 3L)}{3(C^2r^2 - L^2)}$$

32. $(x^2+y^2)^3 = 64x^2y^2$; $m_{\tan}$ at $(2,00,0.56)$, $(2.00,3.07)$

$$3(x^2+y^2)^2\left(2x+2y\frac{dy}{dx}\right) = 64\left(x^22y\frac{dy}{dx}+y^22x\right)$$

$$6x(x^2+y^2)^2 + 6y(x^2+y^2)^2\frac{dy}{dx}$$

$$= 128x^2y\frac{dy}{dx} + 128xy^2$$

$$6y(x^2+y^2)^2\frac{dy}{dx} - 128x^2y\frac{dy}{dx}$$

$$= 128xy^2 - 6x(x^2+y^2)^2$$

$$\frac{dy}{dx} = \frac{128xy^2 - 6x(x^2+y^2)^2}{6y(x^2+y^2)^2 - 128x^2y}$$

$$m_{\tan(2.00,0.56)} = \frac{-143.004}{-224.200} = 0.638$$

$$m_{\tan(2.00,3.07)} = \frac{250.039}{1747.959} = 0.143$$

3.9   Higher Derivatives

1. $y = x^3 + x^2$; $y' = 3x^2 + 2x$; $y'' = 6x + 2$; $y''' = 6$;
$y^{(n)} = 0$; $n \geq 4$

2. $f(x) = 3x - x^4$; $f'(x) = 3 - 4x^3$; $f''(x) = -12x^2$;
$f'''(x) = -24x$; $f^{(4)}(x) = -24$; $f^{(n)}(x) = 0$, $n \geq 5$

3. $f(x) = x^3 - 6x^4$; $f'(x) = 3x^2 - 24x^3$;
$f''(x) = 6x - 72x^2$; $f'''(x) = 6 - 144x$;
$f^{(4)}(x) = -144$; $f^{(n)}(x) = 0$, $n \geq 5$

4. $s = 2t^5 + 5t^4$

$$\frac{ds}{dt} = 10t^4 + 20t^3$$

$$\frac{d^2s}{dt^2} = 40t^3 + 60t^2$$

$$\frac{d^3s}{dt^3} = 120t^2 + 120t$$

$$\frac{d^4s}{dt^4} = 240t + 120$$

$$\frac{d^5s}{dt^5} = 240$$

$$\frac{d^ns}{dt^n} = 0, \; n \geq 6$$

5. $y = (1 - 2x)^4$; $y' = 4(1 - 2x)^3(-2) = -8(1 - 2x)^3$
$y'' = -24(1 - 2x)^2(-2) = 48(1 - 2x)^2$;
$y''' = 96(1 - 2x)(-2) = -192(1 - 2x)$
$y^{(4)} = -2(-192) = 384$; $y^{(n)} = 0$, $n \geq 5$

6. $f(x) = (3x + 2)^3$; $f'(x) = 3(3x + 2)^2(3)$
      $= 9(3x + 2)^2$;
$f''(x) = 18(3x + 2)(3) = 54(3x + 2)$;
$f'''(x) = 54(3) = 162$; $f^{(n)} = 0$, $n \geq 4$

7. $f(r) = r(4r + 1)^3 = r(64r^3 + 48r^2 + 12r + 1)$
   $f(r) = 64r^4 + 48r^3 + 12r^2 + r$
   $f'(r) = 256r^3 + 144r^2 + 24r + 1$
   $f''(r) = 768r^2 + 288r + 24$
   $f'''(r) = 1536r + 288$
   $f^{(4)}(r) = 1536$
   $f^{(n)}(r) = 0$, $n \geq 5$

8. $y = x(x - 1)^3$; $y' = x(3)(x - 1)^2 + (x - 1)^3$
   $= (x - 1)^2(3x + x - 1) = (x - 1)^2(4x - 1)$
   $y'' = (x - 1)^2(4) + (4x - 1)(2)(x - 1)$
   $y'' = 2(x - 1)[2(x - 1) + 4x - 1]$
   $= 2(x - 1)(6x - 3)$
   $y'' = 12x^2 - 18x + 6$; $y''' = 24x - 18$
   $y^{(4)} = 24$; $y^{(n)} = 0$, $n \geq 5$

9. $y = 2x^7 - x^6 - 3x$; $y' = 14x^6 - 6x^5 - 3$;
$y'' = 84x^5 - 30x^4$

10. $y = 6x - 2x^5$; $y' = 6 - 10x^4$; $y'' = -40x^3$

11. $y = 2x + \sqrt{x} = 2x + x^{1/2}$; $y' = 2 + \frac{1}{2}x^{-1/2}$

$$y'' = \frac{1}{2}\left(-\frac{1}{2}\right)x^{-3/2} = \frac{-1}{4x^{3/2}}$$

12. $r = 3\theta^2 - \frac{1}{2\sqrt{\theta}} = 3\theta^2 - \frac{1}{2}\theta^{-1/2}$

$$\frac{dr}{d\theta} = 6\theta + \frac{1}{2 \cdot 2}\theta^{-3/2}$$

$$= 6\theta + \frac{1}{4\theta^{3/2}}$$

$$\frac{d^2r}{d\theta^2} = 6 - \frac{1}{2 \cdot 2 \cdot 2}\theta^{-5/2}$$

$$= 6 - \frac{3}{8\theta^{5/2}}$$

13. $f(x) = \sqrt[4]{8x - 3} = (8x - 3)^{1/4}$

$$f'(x) = \frac{1}{4}(8x - 3)^{-3/4}(8) = -12(8x - 3)^{-3/4}$$

$$f''(x) = -\frac{3}{2}(8x - 3)^{-7/4}(8) = -12(8x - 3)^{-7/4}$$

$$= \frac{-12}{(8x - 3)^{7/4}}$$

14. $f(x) = \sqrt[3]{6x + 5} = (6x + 5)^{1/3};$

$f'(x) = \frac{1}{3}(6x + 5)^{-2/3}(6);$

$f'(x) = 2(6x + 5)^{-2/3}$

$f''(x) = -\frac{4}{3}(6x + 5)^{-5/3}(6) = \dfrac{-8}{(6x+5)^{5/3}}$

15. $f(p) = \dfrac{4}{\sqrt{1 + 2p}} = 4(1 + 2p)^{-1/2}$

$f'(p) = -2(1 + 2p)^{-3/2}(2) = -4(1 - 2p)^{-3/2}$

$f''(p) = 6(1 + 2p)^{-5/2}(2) = \dfrac{12}{(1 + 2p)^{5/2}}$

16. $f(x) = \dfrac{5}{\sqrt{3 - 4x}} = 5(3 - 4x)^{-1/2}$

$f'(x) = -\frac{5}{2}(3 - 4x)^{-3/2}(-4)$

$f'(x) = 10(3 - 4x)^{-3/2}$

$f''(x) = -15(3 - 4x)^{-5/2}(-4) = \dfrac{60}{(3 - 4x)^{5/2}}$

17. $y = 2(2 - 5x)^4; \ y' = 8(2 - 5x)^3(-5)$
$\quad = -40(2 - 5x)^3;$
$y'' = -120(2 - 5x)^2(-5) = 600(2 - 5x)^2$

18. $y = (4x + 1)^6; \ y' = 6(4x + 1)^5(4) = 24(4x + 1)^5;$
$y'' = 120(4x + 1)^4(4) = 480(4x + 1)^4$

19. $y = (3x^2 - 1)^5; \ y' = 5(3x^2 - 1)^4(6x) = 30x(3x^2 - 1)^4;$
$y'' = 30x(4)(3x^2 - 1)^3(6x) + (3x^2 - 1)^4(30)$
$\quad = 30(3x^2 - 1)^3(27x^2 - 1)$

20. $y = 3(2x^3 + 3)^4; \ y' = 12(2x^3 + 3)^3(6x^2)$
$\quad = 72x^2(2x^3 + 3)^3;$
$y'' = 72[x^2 \cdot 3(2x^3 + 3)^2(6x^2) + (2x^3 + 3)^3 2x]$
$y'' = 72(2x)(2x^3 + 3)^2[9x^3 + 2x^3 + 3]$
$\quad = 144x(2x^3 + 3)^2(11x^3 + 3)$

21. $f(x) = \dfrac{2x}{1 - x}$

$f'(x) = \dfrac{(1 - x)(2) - (2x)(-1)}{(1 - x)^2}$

$\quad = \dfrac{2}{(1 - x)^2}$

$f''(x) = \dfrac{(1 - x)^2(0) - 2(2)(1 - x)(-1)}{(1 - x)^4}$

$\quad = \dfrac{4(1 - x)}{(1 - x)^4}$

$\quad = \dfrac{4}{(1 - x)^3}$

22. $fR) = \dfrac{1 - R}{1 + R}$

$f'(R) = \dfrac{(1 + R)(-1) - (1 - R)1}{(1 + R)^2} = \dfrac{-2}{(1 + R)^2}$

$f'(R) = -2(1 + R)^{-2}$

$f''(R) = 4(1 + R)^{-3} = \dfrac{4}{(1 + R)^3}$

23. $y = \dfrac{x^2}{x + 1}$

$y' = \dfrac{(x + 1)2x - x^2}{(x + 1)^2} = \dfrac{x^2 + 2x}{(x + 1)^2}$

$y'' = \dfrac{(x + 1)^2(2x + 2) - (x^2 + 2x)(2)(x + 1)^1}{(x + 1)^4}$

$\quad = \dfrac{2(x + 1)[(x + 1)^2 - (x^2 + 2x)]}{(x + 1)^4}$

$y'' = \dfrac{2(1)}{(x + 1)^3} = \dfrac{2}{(x + 1)^3}$

24. $y = \dfrac{x}{\sqrt{1 - x^2}} = \dfrac{x}{(1 - x^2)^{1/2}}$

$y' = \dfrac{(1 - x^2)^{1/2} - x \cdot \frac{1}{2}(1 - x^2)^{-1/2}(-2x)}{1 - x^2}$

$y' = \dfrac{(1 - x^2)^{1/2} + \frac{x^2}{(1 - x^2)^{1/2}}}{1 - x^2}$

$\quad = \dfrac{1 - x^2 + x^2}{(1 - x^2)^{3/2}}$

$y' = \dfrac{1}{(1 - x^2)^{3/2}} = (1 - x^2)^{-3/2}$

$y'' = -\frac{3}{2}(1 - x^2)^{-5/2}(-2x)$

$\quad = \dfrac{3x}{(1 - x^2)^{5/2}}$

25. $x^2 - y^2 = 9; \ 2x - 2yy' = 0; \ 2x = 2yy'; \ \dfrac{2x}{2y} = \dfrac{x}{y} = y'$

$y'' = \dfrac{y(1) - xy'}{y^2} = \dfrac{y - x\left(\frac{x}{y}\right)}{y^2}$

$\quad = \dfrac{\frac{y^2 - x^2}{y}}{y^2} = \dfrac{y^2 - x^2}{y^3}$

$\quad = \dfrac{-9}{y^3}$

26. $xy + y^2 = 4$; $xy' + y + 2yy' = 0$; $xy' + 2yy' = -y$

$$y'(x + 2y) = -y; \quad y' = \frac{-y}{x + 2y}$$

$$y'' = \frac{(x + y)(-y') - (-y)(1 + 2y')}{(x + 2y)^2}$$

$$y'' = \frac{(x + 2y)\frac{y}{(x+2y)} + y\left(1 - \frac{2y}{x+2y}\right)}{(x + 2y)^2}$$

$$y'' = \frac{y + y - \frac{2y^2}{x+2y}}{(x + 2y)^2} = \frac{2y(x + 2y) - 2y^2}{(x + 2y)^3}$$

$$y'' = \frac{2xy + 2y^2}{(x + 2y)^3} = \frac{2(xy + y^2)}{(x + 2y)^3} = \frac{2(4)}{(x + 2y)^3}$$

$$= \frac{8}{(x + 2y)^3}$$

27. $x^2 - xy = 1 - y^2$; $2x - (xy' + y) = -2yy'$
$2x - xy' - y = -2yy'$; $2yy' - xy' = y - 2x$

$$y' = \frac{y - 2x}{2y - x};$$

$$y'' = \frac{(2y - x)(y' - 2) - (y - 2x)(2y' - 1)}{(2y - x)^2}$$

$$y'' = \left[(2y-x)\left(\frac{y-2x}{2y-x} - 2\right) - (y-2x)\left(2\frac{y-2x}{2y-x} - 1\right)\right]$$
$$\div (2y - x)^2$$

$$y'' = \left[y - 2x - 2(2 - y) - \frac{2(y-2x)^2}{2y-x} + y - 2x\right]$$
$$\div (2y-x)^2$$

$$y'' = \left[-2(y + x) - \frac{2(y - 2x)^2}{(2y - x)}\right] \div (2y - x)^2$$

$$y'' = (-4y^2 - 2xy + 2x^2 - 2y^2 + 8xy - 8x^2) \div (2y - x)^3$$

$$y'' = \frac{-6(y^2 - xy + x^2)}{(2y - x)^3}$$

28. $xy = y^2 + 1$; $xy' + y = 2yy'$; $2yy' - xy' = y$;
$y'(2y - x) = y$

$$y' = \frac{y}{2y - x}; \quad y'' = \frac{(2y - x)y' - y(2y' - 1)}{(2y - x)^2}$$

$$y'' = \frac{(2y - x)\frac{y}{2y-x} - y\left[\frac{2y}{2y-x} - 1\right]}{(2y - x)^2}$$

$$y'' = \frac{y - \frac{2y^2}{2y-x} + y}{(2y - x)^2} = \frac{\frac{2y(2y-x) - 2y^2}{2y-x}}{(2y - x)^2}$$

$$= \frac{2y^2 - 2xy}{(2y - x)^3} = \frac{-2y(x - y)}{-(x - 2y)^3} = \frac{2y(x - y)}{(x - 2y)^3}$$

29. $f(x) = \sqrt{x^2 + 9} = (x^2 + 9)^{1/2}$

$$f'(x) = \frac{1}{2}(x^2 + 9)^{-1/2}(2x) = x(x^2 + 9)^{-1/2}$$

$$f''(x) = x\left[-\frac{1}{2}(x^2 + 9)^{-3/2}(2x)\right] + (x^2 + 9)^{-1/2}$$

$$= -x^2(x^2 + 9)^{-3/2} + (x^2 + 9)^{-1/2}$$

$$= \frac{-x^2}{\sqrt{(x^2 + 9)^3}} + \frac{1}{\sqrt{x^2 + 9}}$$

$$f''(4) = \frac{-16}{(\sqrt{25})^3} + \frac{1}{\sqrt{25}} = \frac{-16}{5^3} + \frac{1}{5}$$

$$= \frac{-16}{125} + \frac{25}{125} = \frac{9}{125}$$

30. $f(x) = x - \frac{2}{x^3} = x - 2x^{-3}$; $f'(x) = 1 + 6x^{-4}$

$$f''(x) = -24x^{-5} = \frac{-24}{x^5}; \quad f''(-1) = 24$$

31. $y = 3x^{2/3} - \frac{2}{x} = 3x^{2/3} - 2x^{-1}$

$$y' = 2x^{-1/3} + 2x^{-2}$$

$$y'' = -\frac{2}{3}x^{-4/3} - 4x^{-3} = \frac{-2}{3x^{4/3}} - \frac{4}{x^3};$$

$$y''|_{x=-8} = \frac{-2}{48} + \frac{4}{512}$$

$$y''|_{x=-8} = \frac{-2}{6 \times 8} + \frac{4}{8 \times 64} = -\frac{13}{384}$$

32.
$$y = 3(1 + 2x)^4$$
$$y' = 12(1 + 2x)^3(2) = 24(1 + 2x)^3$$
$$y'' = 72(1 + 2x)^2(2) = 144(1 + 2x)^2$$
$$y''|_{x=1/2} = 576$$

33.
$$y = x(1 - x)^5$$
$$y' = x[5(1 - x)^4(-1)] + (1 - x)^5(1)$$
$$= -5x(1 - x)^4 + (1 - x)^5$$
$$y'' = (-5x)[4(1 - x)^3(-1)] + (1 - x)^4(-5)$$
$$+ 5(1 - x)^4(-1)$$
$$= 20x(1 - x)^3 - 10(1 - x)^4$$
$$y''|_{x=2} = 20(2)(1 - 2)^3 - 10(1 - 2)^4$$
$$= -40 - 10 = -50$$

34.
$$y = \frac{x}{2 - 3x}$$

$$y' = \frac{(2 - 3x) - x(-3)}{(2 - 3x)^2} = \frac{2}{(2 - 3x)^2}$$

$$= 2(2 - 3x)^{-2}$$

$$y'' = -4(2 - 3x)^{-3}(-3) = \frac{12}{(2 - 3x^3)}$$

$$y''|_{x=-1/3} = \frac{4}{9}$$

35. $\frac{d}{dx}$ of $\frac{dy}{dx} = y''$; $x = 1$

$$y = (1 - 2x)^4$$
$$y' = 4(1 - 2x)^3(-2) = -8(1 - 2x)^2$$
$$y'' = -24(1 - 2x)^2(-2) = 48(1 - 2x)^2$$
$$y''|_{x=1} = 48$$

36. $\quad y = 0.0001(x^5 - 25x^2)$

$$\frac{dy}{dx} = 0.0001(5x^4 - 50x)$$

$$\frac{d^2y}{dx} = 0.0001(20x^3 - 50)$$

$$\frac{d^2y}{dx^2} = 0.0001(20(3.00)^3 - 50)$$

$$= 540$$

37. $s = 2250t - 16.1t^2$; $s' = 2250 - 32.2t$

$s'' = a = -32.2$ ft/s^2

38. $s = 57.6t - 1.20t^3$

$$v = \frac{ds}{dt} = 57.6 - 3.60t^2; \quad v|_{t=4.00} = 0 \text{ ft/s}$$

$$\frac{dv}{dt} = a = -7.20t; \quad a|_{t=4.00} = -28.8 \text{ ft/s}^2$$

39. $V = L\left(\frac{d^2q}{dt^2}\right)$; $L = 1.60$ H

$$q = \sqrt{2t+1} - 1 = (2t+1)^{1/2} - 1$$

$$\frac{dq}{dt} = \frac{1}{2}(2t+1)^{-1/2}(2) = (2t+1)^{-1/2}$$

$$\frac{d^2q}{dt^2} = -\frac{1}{2}(2t+1)^{-3/2}(2) = \frac{-1}{(2t+1)^{3/2}}$$

$$V = \frac{-1.60}{(2t+1)^{3/2}}$$

40. $H = \frac{5000}{t^2 + 10}$; $\frac{d^2H}{dt^2} = ?, t = 3$

$$H = 500(t^2 + 10)^{-1}$$

$$\frac{dH}{dt} = -5000(t^2 + 10)^{-2}(2t) = \frac{-10\,000t}{(t^2 + 10)^2}$$

$$\frac{d^2H}{dt^2} = -\frac{(t^2 + 10)^2 10\,000 - (10\,000t)(2)(t^2 + 10)^1(2t)}{(t^2 + 10)^4}$$

$$\frac{d^2H}{dt^2} = -\frac{10\,000(t^2 + 10)(t^2 + 10 - 4t^2)}{(t^2 + 10)^4}$$

$$\frac{d^2H}{dt^2} = \frac{-10\,000(10 - 3t^2)}{(t^2 + 10)^3}$$

$$\left.\frac{d^2H}{dt^2}\right|_{t=3} = 24.8 \text{ W/(m}^2 \cdot \text{h}^2)$$

Chapter 3 Review Exercises

1. $\lim_{x \to 4} (8 - 3x) = 8 - 3(4) = -4$

2. $\lim_{x \to 3} (2x^2 - 10) = 2(3)^2 - 10 = 8$

3. $\lim_{x \to -3} \frac{2x + 5}{x - 1} = \frac{2(-3) + 5}{-3 - 1} = \frac{1}{4}$

4. $\lim_{x \to 1} (x - 1)\sqrt{x^2 + 9} = (1 - 1)\sqrt{1^2 + 9} = 0$

5. $\lim_{x \to 2} \frac{4x - 8}{x^2 - 4} = \lim_{x \to 2} \frac{4(x - 2)}{(x - 2)(x + 2)}$

$$= \lim_{x \to 2} \frac{4}{x + 2}$$

$$= \frac{4}{2 + 2} = 1$$

6. $\lim_{x \to 5} \frac{x^2 - 25}{3x - 15} = \lim_{x \to 5} \frac{(x + 5)(x - 5)}{3(x - 5)}$

$$= \lim_{x \to 5} \frac{x + 5}{3}$$

$$= \frac{5 + 5}{3} = \frac{10}{3}$$

7. $\lim_{x \to 2} \frac{x^2 + 3x - 10}{x^2 - x - 2} = \lim_{x \to 2} \frac{(x + 5)(x - 2)}{(x + 1)(x - 2)}$

$$= \lim_{x \to 2} \frac{x + 5}{x + 1}$$

$$= \frac{2 + 5}{2 + 1} = \frac{7}{3}$$

8. $\lim_{x \to 0} \frac{(x - 3)^2 - 9}{x} = \lim_{x \to 0} \frac{x^2 - 6x + 9 - 9}{x}$

$$= \lim_{x \to 0} \frac{x(x - 6)}{x}$$

$$= \lim_{x \to 0} (x - 6) = -6$$

9. $\lim_{x \to \infty} \frac{2 + \frac{1}{x+4}}{3 - \frac{1}{x^2}} = \frac{2 + 0}{3 - 0} = \frac{2}{3}$

10. $\lim_{x \to \infty} \left(7 - \frac{1}{x + 1}\right) = \lim_{x \to \infty} 7 - \lim_{x \to \infty} \frac{1}{x + 1} = 7$

11. $\lim_{x \to \infty} \frac{x - 2x^3}{(1 + x)^3} = \lim_{x \to \infty} \frac{x - 2x^3}{1 + 3x^2 + 3x + x^3}$

$$= \lim_{x \to \infty} \frac{\frac{1}{x^2} - 2}{\frac{1}{x^3} + \frac{3}{x} + \frac{3}{x^2} + 1}$$

$$= -2$$

12. $\lim\limits_{x\to\infty} \dfrac{2x+5}{3x^3-2x} = \lim\limits_{x\to\infty} \dfrac{\frac{2}{x^2}+\frac{5}{x^3}}{3-\frac{2}{x^2}}$

$\qquad\qquad\qquad = \dfrac{0}{3}$

$\qquad\qquad\qquad = 0$

13. $\qquad y = 7+5x$

$\qquad y+\Delta y = 7+5(x+\Delta x) = 7+5x+5\Delta x$

$\qquad\qquad \Delta y = 7+5x+5\Delta x - (7+5x) = 5\Delta x$

$\qquad\qquad \dfrac{\Delta y}{\Delta x} = \dfrac{5\Delta x}{\Delta x}$

$\qquad\qquad\qquad = 5$

$\qquad \lim\limits_{\Delta x\to 0} 5 = 5$

14. $\qquad y = 6x-2$

$\qquad y+\Delta y = 6(x+\Delta x)-2$

$\quad 6x-2+\Delta y = 6x+6\Delta x-2$

$\qquad\qquad \Delta y = 6\Delta x$

$\qquad\qquad \dfrac{\Delta y}{\Delta x} = 6$

$\qquad \lim\limits_{\Delta x\to 0} \dfrac{\Delta y}{\Delta x} = \lim\limits_{\Delta x\to 0} 6$

$\qquad\qquad\qquad = 6$

15. $\qquad y = 6-2x^2$

$\qquad y+\Delta y = 6-2(x+\Delta x)^2$

$\qquad\qquad = 6-2x^2-4x\Delta x-2\Delta x^2$

$\quad 6-2x^2+\Delta y = 6-2x^2-4x\Delta x-2\Delta x^2$

$\qquad\qquad \Delta y = -4x\Delta x-2\Delta x^2$

$\qquad\qquad \dfrac{\Delta y}{\Delta x} = -4x-2\Delta x$

$\qquad \lim\limits_{\Delta x\to 0} \dfrac{\Delta y}{\Delta x} = \lim\limits_{\Delta x\to 0}(-4x-2\Delta x)$

$\qquad\qquad\qquad = -4x$

16. $\qquad y = 2x^2-x^3$

$\qquad y+\Delta y = 2(x+\Delta x)^2-(x+\Delta x)^3$

$\qquad\qquad = 2x^2+4x\Delta x+2\Delta x^2-x^3$

$\qquad\qquad\quad -3x^2\Delta x-3x\Delta x^2-\Delta x^3$

$\quad 2x^2-x^3+\Delta y = 2x^2+4x\Delta x+2\Delta x^2-x^3$

$\qquad\qquad\quad -3x^2\Delta x-3x\Delta x^2-\Delta x^3$

$\qquad\qquad \Delta y = 4x\Delta x+2\Delta x^2-3x^2\Delta x$

$\qquad\qquad\quad -3x\Delta x^2-\Delta x^3$

$\qquad\qquad \dfrac{\Delta y}{\Delta x} = 4x+2\Delta x-3x^2-3x\Delta x-\Delta x^2$

$\qquad \lim\limits_{\Delta x\to 0} \dfrac{\Delta y}{\Delta x} = 4x-3x^2$

17. $\qquad y = \dfrac{2}{x^2}$

$\qquad y+\Delta y = \dfrac{2}{(x+\Delta x)^2}$

$\qquad\qquad \Delta y = \dfrac{2}{(x+\Delta x)^2} - \dfrac{2}{x^2} = \dfrac{2x^2-2(x+\Delta x)^2}{x^2(x+\Delta x)^2}$

$\qquad\qquad = \dfrac{2x^2-2x^2-4x\Delta x-2(\Delta x)^2}{x^2(x+\Delta x)^2}$

$\qquad\qquad = \dfrac{\Delta x(-4x-2\Delta x)}{x^2(x+\Delta x)^2}$

$\qquad\qquad \dfrac{\Delta y}{\Delta x} = \dfrac{-4x-2\Delta x}{x^2(x+\Delta x)^2}$

$\qquad \lim\limits_{\Delta x\to 0} \dfrac{-4x-2\Delta x}{x^2(x+\Delta x)^2} = \dfrac{-4x-0}{x^2(x+0)^2} = \dfrac{-4x}{x^2(x^2)}$

$\qquad\qquad\qquad = \dfrac{-4}{x^3}$

18. $\qquad y = \dfrac{1}{1-4x}$

$\qquad y+\Delta y = \dfrac{1}{1-4(x+\Delta x)}$

$\quad \dfrac{1}{1-4x}+\Delta y = \dfrac{1}{1-4(x+\Delta x)}$

$\qquad\qquad \Delta y = \dfrac{1}{1-4(x+\Delta x)} - \dfrac{1}{1-4x}$

$\qquad\qquad = \dfrac{1-4x-1+4x+4\Delta x}{(1-4(x+\Delta x))(1-4x)}$

$\qquad\qquad \dfrac{\Delta y}{\Delta x} = \dfrac{4}{(1-4(x+\Delta x))(1-4x)}$

$\qquad \lim\limits_{\Delta x\to 0} \dfrac{\Delta y}{\Delta x} = \dfrac{4}{(1-4x)^2}$

19. $\qquad y = \sqrt{x+5}$

$\qquad y+\Delta y = \sqrt{x+\Delta x+5}$

$\quad \sqrt{x+5}+\Delta y = \sqrt{x+\Delta x+5}$

$\qquad\qquad \Delta y = \left(\sqrt{x+\Delta x+5}-\sqrt{x+5}\right)$

$\qquad\qquad\quad \cdot\left(\dfrac{\sqrt{x+\Delta x+5}+\sqrt{x+5}}{\sqrt{x+\Delta x+5}+\sqrt{x+5}}\right)$

$\qquad\qquad \Delta y = \dfrac{x+\Delta x+5-x-5}{\sqrt{x+\Delta x+5}+\sqrt{x+5}}$

$\qquad\qquad = \dfrac{\Delta x}{\sqrt{x+\Delta x+5}+\sqrt{x+5}}$

$\qquad \lim\limits_{\Delta x\to 0} \dfrac{\Delta y}{\Delta x} = \lim\limits_{\Delta x\to 0} \dfrac{1}{\sqrt{x+\Delta x+5}+\sqrt{x+5}}$

$\qquad\qquad\qquad = \dfrac{1}{2\sqrt{x+5}}$

20.
$$y = \frac{1}{\sqrt{x}}$$

$$y + \Delta y = \frac{1}{\sqrt{x + \Delta x}}$$

$$\frac{1}{\sqrt{x}} + \Delta y = \frac{1}{\sqrt{x + \Delta x}}$$

$$\Delta y = \frac{1}{\sqrt{x + \Delta x}} - \frac{1}{\sqrt{x}}$$

$$= \frac{\sqrt{x} - \sqrt{x + \Delta x}}{\sqrt{x + \Delta x}\sqrt{x}} \cdot \frac{\sqrt{x} + \sqrt{x + \Delta x}}{\sqrt{x + \Delta x}\sqrt{x}}$$

$$\Delta y = \frac{x - x - \Delta x}{\sqrt{x + \Delta x}\sqrt{x}(\sqrt{x} + \sqrt{x + \Delta x})}$$

$$\lim_{\Delta x \to 0} \frac{\Delta y}{\Delta x} = \lim_{\Delta x \to 0} \frac{-1}{\sqrt{x + \Delta x}\sqrt{x}(\sqrt{x} + \sqrt{x + \Delta x})}$$

$$= \frac{-1}{2x^{3/2}}$$

21. $y = 2x^7 - 3x^2 + 5$

$$\frac{dy}{dx} = 2(7x^6) - 3(2x) + 0 = 14x^6 - 6x$$

22. $y = 8x^7 - 2^5 - x$

$$\frac{dy}{dx} = 56x^6 - 1$$

23. $y = 4\sqrt{x} - \frac{3}{x} + \sqrt{3}$

$$\frac{dy}{dx} = \frac{2}{x^{1/2}} + \frac{3}{x^2}$$

24. $y = \frac{3}{x^2} - 8\sqrt[4]{x} = 3x^{-2} - 8x^{1/4}$

$$\frac{dy}{dx} = -6x^{-3} - 2x^{-3/4} = \frac{6}{x^3} - \frac{2}{x^{3/4}}$$

25. $f(y) = \frac{3y}{1 - 5y}$

$$\frac{df(y)}{dy} = \frac{(1 - 5y)(3) - 3y(-5)}{(1 - 5y)^2}$$

$$= \frac{3 - 15y + 15y}{(1 - 5y)^2}$$

$$\frac{df(y)}{dy} = \frac{3}{(1 - 5y)^2}$$

26. $y = \frac{2x - 1}{x^2 + 1}$

$$\frac{dy}{dx} = \frac{(x^2 + 1)(2) - (2x - 1)(2x)}{(x^2 + 1)^2}$$

$$= \frac{2x^2 + 2 - 4x^2 + 2x}{(x^2 + 1)^2}$$

$$\frac{dy}{dx} = \frac{-2x^2 + 2x + 2}{(x^2 + 1)^2}$$

$$= \frac{-2(x^2 - x - 1)}{(x^2 + 1)^2}$$

27. $y = (2 - 3x)^4$

$$\frac{dy}{dx} = 4(2 - 3x)^3(-3)$$

$$= -12(2 - 3x)^3$$

28. $y = (2x^2 - 3)^6$

$$\frac{dy}{dx} = 6(2x^2 - 3)^5(4x)$$

$$= 24x(2x^2 - 3)^5$$

29. $y = \frac{3}{(5 - 2x^2)^{3/4}}; u = 3, \frac{du}{dx} = 0;$

$$v = (5 - 2x^2)^{3/4}$$

$$\frac{dy}{dx} = \frac{(5 - 2x^2)^{3/4}(0) - 3[-3x(5 - 2x^2)^{-1/4}]}{(5 - 2x^2)^{3/2}}$$

$$\frac{dv}{dx} = \frac{3}{4}(5 - 2x^2)^{-1/4}(-4x)$$

$$= -3x(5 - 2x^2)^{-1/4}$$

$$= \frac{9x(5 - 2x^2)^{-1/4}}{(5 - 2x^2)^{3/2}}$$

$$= \frac{9x}{(5 - 2x^2)^{3/2}(5 - 2x^2)^{1/4}}$$

$$= \frac{9x}{(5 - 2x^2)^{7/4}}$$

30. $y = \frac{3}{(5 - 2x^2)^{3/4}}$

$$= 3(5 - 2x^2)^{-3/4}$$

$$\frac{dy}{dx} = -\frac{9}{4}(5 - 2x^2)^{-3/4}(-4x)$$

$$= \frac{9x}{(5 - 2x^2)^{7/4}}$$

31. $v = \sqrt{1 + \sqrt{1 + \sqrt{1 + 8s}}}$ let $u = \sqrt{1 + 8s}$

$$\frac{du}{ds} = \frac{4}{\sqrt{1 + 8s}}$$

$$v = \sqrt{1 + \sqrt{1 + u}}$$ let $w = \sqrt{1 + u}$

$$\frac{dw}{du} = \frac{1}{2 + \sqrt{1 + u}}$$

$$v = \sqrt{1 + w}, \ \frac{dv}{dw} = \frac{1}{2\sqrt{1 + w}}$$

$$\frac{dv}{ds} = \frac{dv}{dw}\frac{dw}{du}\frac{du}{ds}$$

$$= \frac{1}{2\sqrt{1 + w}} \cdot \frac{1}{2\sqrt{1 + u}} \cdot \frac{4}{\sqrt{1 + 8s}}$$

$$\frac{dv}{ds} = \frac{1}{\sqrt{(1 + w)(1 + u)(1 + 8s)}}$$

$$= \frac{1}{\sqrt{(1 + \sqrt{1 + u})(1 + u)(1 + 8s)}}$$

$$\frac{dv}{ds} = \frac{1}{\sqrt{(1 + \sqrt{1 + u})}\sqrt{1 + u}\sqrt{1 + 8s}}$$

$$\frac{dv}{ds} = \frac{1}{\sqrt{1 + \sqrt{1 + \sqrt{1 + 8s}}}\sqrt{1 + \sqrt{1 + 8s}}\sqrt{1 + 8s}}$$

32. $y = (x - 1)^3(x^2 - 2)^2$

$$\frac{dy}{ds} = (x - 1)^3(2(x^2 - 2)(2x)) + (x^2 - 2)^2(3(x - 1)^2)$$

$$\frac{dy}{ds} = (x - 1)^2(x^2 - 2)(4x(x - 1) + (x^2 - 2)(3))$$

$$\frac{dy}{ds} = (x - 1)^2(x^2 - 2)(4x^2 - 4x + 3x^2 - 6)$$

$$\frac{dy}{ds} = (x - 1)^2(x^2 - 2)(7x^2 - 4x - 6)$$

33. $y = \frac{\sqrt{4x + 3}}{2x}; \ u = \sqrt{4x + 3} = (4x + 3)^{1/2};$

$$\frac{du}{dx} = \frac{1}{2}(4x + 3)^{-1/2}(4)$$

$$= 2(4x + 3)^{-1/2}$$

$$v = 2x, \ \frac{dv}{dx} = 2$$

$$\frac{dy}{dx} = \frac{2x(2)(4x + 3)^{-1/2} - (4x + 3)^{1/2}(2)}{(2x)^2}$$

$$= \frac{(4x + 3)^{-1/2}[4x - (4x + 3)(2)]}{4x^2}$$

$$= \frac{-4x - 6}{(4x + 3)^{1/2}(4x^2)}$$

$$= \frac{2(-2x - 3)}{2(2x^2)(4x + 3)^{1/2}}$$

$$= \frac{-2x - 3}{2x^2(4x + 3)^{1/2}}$$

34. $R = \frac{\sqrt{t} + 1}{\sqrt{t} - 1}$

$$\frac{dR}{dt} = \frac{(\sqrt{t} - 1)(\frac{1}{2\sqrt{t}}) - (\sqrt{t} + 1)(\frac{1}{2\sqrt{t}})}{(\sqrt{t} - 1)^2}$$

$$= \frac{\frac{1}{2} - \frac{1}{2\sqrt{t}} - \frac{1}{2} - \frac{1}{2\sqrt{t}}}{(\sqrt{t} - 1)^2}$$

$$\frac{dR}{dt} = \frac{-1}{\sqrt{t}(\sqrt{t} - 1)^2}$$

35. $(2x - 3y)^3 = x^2 - y$

$$3(2x - 3y)^2\left(2 - 3\frac{dy}{dx}\right) = 2x - \frac{dy}{dx}$$

$$6(2x - 3y)^2 - 9(2x - 3y)^2\frac{dy}{dx} = 2x - \frac{dy}{dx}$$

$$(1 - 9(2x - 3y)^2)\frac{dy}{dx} = 2x - 6(2x - 3y)^2$$

$$\frac{dy}{dx} = \frac{2x - 6(2x - 3y)^2}{1 - 9(2x - 3y)^2}$$

36. $x^2y^2 = x^2 + y^2$

$$x^2\left(2y\frac{dy}{dx}\right) + y^2(2x) = 2x + 2y\frac{dy}{dx}$$

$$(2x^2y - 2y)\frac{dy}{dx} = 2x - 2xy^2$$

$$\frac{dy}{dx} = \frac{x - xy^2}{x^2y - y}$$

$$= \frac{x(1 - y^2)}{y(x^2 - 1)}$$

37. $y = \frac{4}{x} + 2\sqrt[3]{x}, \ x = 8$

$$y = 4x^{-1} + 2x^{1/3};$$

$$\frac{dy}{dx} = 4(-1)x^{-2} + 2\left(\frac{1}{3}x^{-2/3}\right) = \frac{-4}{x^2} + \frac{2}{3x^{2/3}}$$

$$\left.\frac{dy}{dx}\right|_{x=8} = \frac{-4}{8^2} + \frac{2}{3(8)^{2/3}}$$

$$= \frac{-4}{64} + \frac{2}{3(4)} = \frac{-1}{16} + \frac{1}{6}$$

$$= \frac{-3}{48} + \frac{8}{48} = \frac{5}{48}$$

38. $y = (3x - 5)^4, \ x = -2$

$$\left.\frac{dy}{dx}\right|_{x=-2} = 4(3x - 5)^3(3)\big|_{x=-2}$$

$$= 12(3(-2) - 5)^3$$

$$= -15{,}972$$

39. $y = 2x\sqrt{4x+1}, x = 6$

$$\frac{dy}{dx} = 2x\left(\frac{1}{2\sqrt{4x+1}}\right)(4) + \sqrt{4x+1}(2)$$

$$\frac{dy}{dx}\Big|_{x=6} = \frac{4(6)}{\sqrt{4(6)+1}} + 2\sqrt{4(6)+1} = \frac{74}{5}$$

40. $y = \dfrac{\sqrt{2x^2+1}}{3x}, x = 2$

$$\frac{dy}{dx} = \frac{3x\left(\frac{4x}{2\sqrt{2x^2+1}}\right) - 3\sqrt{2x^2+1}}{9x^2}$$

$$\frac{dy}{dx}\Big|_{x=2} = \frac{3(2)\left(\frac{4(2)}{2\sqrt{2(2)^2+1}}\right) - 3\sqrt{2(2)^2+1}}{9(2)^2}$$

$$= -\frac{1}{36}$$

41. $y = 3x^4 - \dfrac{1}{x} = 3x^4 - x^{-1}$

$$y' = 12x^3 + x^{-2}$$
$$y'' = 36x^2 - 2x^{-3}$$

42. $y = \sqrt{1-8x}$

$$\frac{dy}{dx} = \frac{1}{2\sqrt{1-8x}}(-8)$$

$$\frac{dy}{dx} = \frac{-4}{\sqrt{1-8x}}$$

$$\frac{d^2y}{dx^2} = \frac{\sqrt{1-8x}(0) - (-4)\frac{1}{2\sqrt{1-8x}}(-8)}{(\sqrt{1-8x})^2}$$

$$\frac{d^2y}{dx^2} = \frac{-16}{(1-8x)^{3/2}}$$

43. $y = \dfrac{1-3x}{1+4x}$

$$\frac{dy}{dx} = \frac{(1+4x)(-3) - (1-3x)(4)}{(1+4x)^2}$$

$$= \frac{-3 - 12x - 4 + 12x}{(1+4x)^2}$$

$$= \frac{-7}{(1+4x)^2}$$

$$\frac{d^2y}{dx^2} = \frac{(1+4x)^2(0) - (-7)(2(1+4x)(4))}{(1+4x)^4}$$

$$= \frac{56}{(1+4x)^3}$$

44. $y = 2x(6x+5)^4$

$$\frac{dy}{dx} = 2x(4(6x+5)^3(6)) + (6x+5)^4(2)$$

$$= 48x(6x+5)^3 + 2(6x+5)^4$$

$$\frac{d^2y}{dx^2} = 48x(3(6x+5)^2(6))$$

$$\qquad + (6x+5)^3(48) + 8(6x+5)^3(6)$$

$$\frac{d^2y}{dx^2} = (6x+5)^2(1440x + 48(6x+5) + 48(6x+5))$$

$$\frac{d^2y}{dx^2} = (6x+5)^2(1440x + 480)$$

$$= 480(6x+5)^2(3x+1)$$

45. $y = \dfrac{2(x^2-4)}{x-2}, x \neq 2$

$$y = \frac{2(x+2)(x-2)}{(x-2)}$$

$$= 2(x+2)\big|_{x=2}$$
$$= 2(2+2) = 8$$

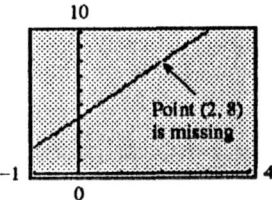

Just to the left of $x = 2$, the trace feature gives $x = 1.9787234, y = 7.9574468$ and just to the right of $x = 2$, the trace feature gives $x = 2.0319149, y = 8.0638298$ which would appear to give $y = 8$ for $x = 2$. However, using the value feature shows there is no y-value for $x = 2$.

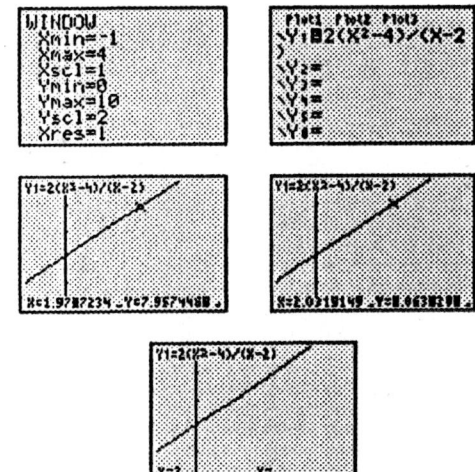

46. Since the graph of the function must touch or cross the x-axis an odd number of times between $x = 0$ and $x = 1$, $f(x) = 0$ will have an odd number of solutions.

47. (a) The initial velocity is

$$v(0) = \frac{6(0+5)}{0+1} = 30 \text{ ft/s}$$

(b) $v = \dfrac{6(t+5)}{t+1} = \dfrac{6t+30}{t+1}$

$$= \frac{6 + \frac{30}{t}}{1 + \frac{1}{t}}$$

from which $v \to 6$ as $t \to \infty$, thus the terminal velocity is 6 ft/s

48. $f = \dfrac{f_1 f_2}{f_1 + f_2 - d}$

$$= \frac{f_2}{1 + \frac{f_2}{f_1} - \frac{d}{f_1}} \to f_2 \text{ as } f_1 \to \infty.$$

49. $y = 7x^4 - x^3$

$$\frac{dy}{dx} = 28x^3 - 3x^2 \Big|_{(-1,8)}$$

$$= 28(-1)^3 - 3(-1)^2$$

$$= -31 = m_{\text{tangent}} \text{ at } (-1,8)$$

The nDeriv feature of calculator gives -31.000029

50. $y = \sqrt[3]{3 - 8x} = (3 - 8x)^{1/3}$

$$\frac{dy}{dx} = \frac{1}{3}(3 - 8x)^{-2/3}(-8) \Big|_{(-3,3)}$$

$$= -\frac{8}{3}(3 - 8(-3))^{-2/3}$$

$$= -\frac{8}{27} = m_{\text{tangent}} \text{ at } (-3, 3).$$

$$-\frac{8}{27} = -0.2962962963$$

Using the nDeriv ffeature of calculator gives -0.2962963012.

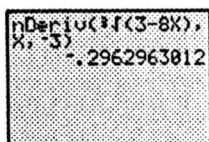

51. $y = 0.0015x^2 + C$

$$\frac{dy}{dx} = 2(0.0015)x = 0.3$$

$$x = \frac{0.3}{2(0.0015)}$$

$$y = 0.3\left(\frac{0.3}{2(0.0015)}\right) - 10$$

$$= 0.0015\left(\frac{0.3}{2(0.0015)}\right)^2 + C$$

$$C = 5$$

52. $s = 8t - t^2$

$$v = \frac{ds}{dt} = 8 - 2t = 4$$

$$2t = 4$$

$$t = 2$$

when velocity of piston is 4 cm/s

53. $R = 1 - kt + \dfrac{k^2 t^2}{2} - \dfrac{k^3 t^3}{6}$

$$= 1 - kt + \frac{1}{2}(k^2 t^2) - \frac{1}{6}(k^3 t^3)$$

$$R' = -k + k^2 t - \frac{1}{2}k^3 t^2$$

54. $s = 40t - 5t^2$

$$v = \frac{ds}{dt} = 40 - 10t = 0 \Rightarrow t = 4$$

$$s = 40(4) - 5(4)^2 = 80 \text{ ft}$$

55. $E = \dfrac{k}{r^2}$

$$\frac{dE}{dr} = \frac{-2k}{r^3}$$

56. $v = \sqrt{v_0^2 + 2as}$

$$\frac{dv}{ds} = \frac{1}{2\sqrt{v_0^2 + 2as}} \cdot 2a$$

$$= \frac{a}{\sqrt{v_0^2 + 2as}}$$

57. $E = \dfrac{L dI}{dt}$; $I = t(0.01t + 1)^3$;

$$\frac{dI}{dt} = (0.01t + 1)^2(0.04t + 1); \quad L = 0.4H$$

$E = 0.4\dfrac{dI}{dt}$; substituting the value for $\dfrac{dI}{dt}$,

$$E = 0.04(0.01t + 1)^2(0.04t + 1)$$

58. $v = \dfrac{z}{\alpha(1-z^2) - \beta}$

$\dfrac{dv}{dz} = \dfrac{\alpha(1-z^2) - \beta - z(-2\alpha z)}{(\alpha(1-z^2) - \beta)^2}$

$\quad = \dfrac{\alpha - \alpha z^2 - \beta + 2\alpha z^2}{(\alpha(1-z^2) - \beta)^2}$

$\dfrac{dv}{dz} = \dfrac{\alpha z^2 + \alpha - \beta}{(\alpha(1-z^2) - \beta)^2}$

59. $r_f = \dfrac{2(R^3 - r^3)}{3(R^2 - r^2)}$

$\dfrac{dr_f}{dR} = \dfrac{3(R^2 - r^2)(6R^2) - 2(R^3 - r^3)(6R)}{9(R^2 - r^2)^2}$

$\dfrac{dr_f}{dR} = \dfrac{18R^2(R+r)(R-r) - 12R(R-r)(R^2 + Rr + r^2)}{9(R+r)^2(R-r)^2}$

$\dfrac{dr_f}{dR} = \dfrac{6R^2(R+r) - 4R(R^2 + Rr + r^2)}{3(R+r)^2(R-r)}$

$\dfrac{dr_f}{dR} = \dfrac{6R^3 + 6R^2r - 4R^3 - 4R^2r - 4Rr^2}{3(R+r)^2(R-r)}$

$\dfrac{dr_f}{dR} = \dfrac{2R^3 + 2R^2r - 4Rr^2}{3(R+r)^2(R-r)}$

$\quad = \dfrac{2R(R^2 + Rr - 2r^2)}{3(R+r)^2(R-r)}$

$\dfrac{dr_f}{dR} = \dfrac{2R(R+2r)(R-r)}{3(R+r)^2(R-r)}$

$\quad = \dfrac{2R(R+2r)}{3(R+r)^2}$

60. $V = 5000(60 - t)^2$

$\dfrac{dV}{dt} = 5000(2)(60 - t)(-1)\big|_{t=4.00}$

$\quad = -560{,}000$

After 4.00 h the pond is being drained at a rate of 560,000 m^3/h.

61. $f = \dfrac{1}{2\pi\sqrt{C(L+2)}}$

$\quad = \dfrac{1}{2\pi}(CL + 2C)^{-1/2}$

$\dfrac{df}{dL} = -\dfrac{1}{2}\left(\dfrac{1}{2\pi}\right)(CL + 2C)^{-3/2}(C)$

$\quad = -\dfrac{C}{4\pi}(CL + 2C)^{-3/2}$

$\quad = -\dfrac{C}{4\pi\sqrt{C^3(L+2)^3}}$

$\quad = -\dfrac{C}{4\pi C\sqrt{C(L+2)^3}}$

$\quad = -\dfrac{1}{4\pi\sqrt{C(L+2)^3}}$

$\quad = -\dfrac{1}{4\pi\sqrt{C}(L+2)^{3/2}}$

62. $V = \dfrac{aI^2}{b - I}$

$\dfrac{dV}{dI} = \dfrac{(b-I)(2aI) - aI^2(-1)}{(b-I)^2}$

$\quad = \dfrac{aI(2b - 2I + I)}{(b-I)^2}$

$\dfrac{dV}{dI} = \dfrac{aI(2b - I)}{(b-I)^2}$

63. $T = \dfrac{10(1-t)}{0.5t + 1}$

$\dfrac{dT}{dt} = \dfrac{(0.5t+1)(-10) - 10(1-t)(0.5)}{(0.5t+1)^2}$

$\quad = \dfrac{-5t - 10 - 5 + 5t}{(0.5t+1)^2}$

$\dfrac{dT}{dt} = \dfrac{-15}{(0.5t+1)^2}$

64. $e = 100\left(1 - \dfrac{1}{(V_1/V_2)^{0.4}}\right)$

$\quad = 100 - 100\left(\dfrac{V_1}{V_2}\right)^{-0.4}$

$\dfrac{de}{dV} = 40\left(\dfrac{V_1}{V_2}\right)^{-1.4}\left(\dfrac{1}{V_2}\right)$

$\quad = \dfrac{40V_2^{0.4}}{V_1^{1.4}}$

65. $y = \dfrac{w}{24EI}(6L^2x^2 - 4Lx^3 + x^4)$

$y' = \dfrac{w}{24EI}(12L^2x - 12Lx^2 + 4x^3)$

$\quad = \dfrac{w}{6EI}(3L^2x - 3Lx^2 + x^3)$

$y'' = \dfrac{w}{6EI}(3L^2 - 6Lx + 3x^2) = \dfrac{w}{2EI}(L - x)^2$

$y''' = \dfrac{w}{2EI}(-2L + 2x) = \dfrac{w}{EI}(x - L)$

$y^{iv} = \dfrac{w}{EI}$

66. $n = \dfrac{2t}{t + 1}$

$\dfrac{dn}{dt} = \dfrac{(t + 1)(2) - 2t(1)}{(t + 1)^2}$

$\quad = \dfrac{2t + 2 - 2t}{(t + 1)^2} = \dfrac{2}{(t + 1)^2}$

$\dfrac{d^2n}{dt^2} = \dfrac{-4}{(t + 1)^3}\Big|_{t=4.00}$

$\quad = \dfrac{-4}{(4.00 + 1)^3}$

$\quad = -0.0320 \text{ g/min}^2$

67. $A = lw = 75 \Rightarrow l = \dfrac{75}{w}$

$p = 2l + 2w = \dfrac{150}{w} + 2w$

$\dfrac{dp}{dw} = \dfrac{-150}{w^2} + 2$

68. $V = \pi r^2 h = 100 \Rightarrow h = \dfrac{100}{\pi r^2}$

$A = 2\pi r^2 + 2\pi rh = 2\pi r^2 + \dfrac{200}{r}$

$\dfrac{dA}{dr} = 4\pi r - \dfrac{200}{r^2}$

69. $A = xy = x(4 - x^2) = 4x - x^3$

$\dfrac{dA}{dx} = 4 - 3x^2$

70. $x^2 = h^2 + (vt)^2, h = \dfrac{2640}{5280}, v = 400$

$2x\dfrac{dx}{dt} = 2v^2t$

$\dfrac{dx}{dt} = \dfrac{v^2t}{x}\Big|_{v=400, x=\sqrt{(\frac{2640}{5280})^2+(400(\frac{0.600}{60}))^2}, t=\frac{0.600}{60}}$

$\dfrac{dx}{dt} = 397 \text{ mi/h}$

71. $V = \dfrac{1{,}500{,}000}{2t + 10}$

$\dfrac{dV}{dt} = \dfrac{-3{,}000{,}000}{(2t + 10)^2}\Big|_{t=5} = -7500$

$\dfrac{d^2V}{dt^2} = \dfrac{12{,}000{,}000}{(2t + 10)^3}\Big|_{t=5} = 1500$

At $t = 5$ years, $dV/dt = -\$7500/\text{year}$ (rate of appreciation is decreasing); $\frac{d^2V}{dt^2} = \$1500/\text{year}^2$ (rate at which appreciation changes is increasing), (Machinery is depreciating, but depreciation is lessening.)

72. $y = 1.2x - x^2, 0 < x < 1.2$

$y' = 1.2 - 2x$

$y'' = -2$

$R = \dfrac{[1 + (y')^2]^{3/2}}{|y''|}$

$\quad = \dfrac{[1 + (1.2 - 2x)^2]^{3/2}}{|-2|}\Big|_{x=0.2}$

$\quad = 1.05 \text{ mi}$

$R = \dfrac{[1 + (1.2 - 2x)^2]^{3/2}}{2}\Big|_{x=0.6}$

$\quad = 0.50 \text{ mi}$

73. Take derivative of $f(x)$ and $g(x)$. Evaluate at intersection to find inclinations. Difference in inclinations is angle at which paths cross.

APPLICATIONS OF THE DERIVATIVE

4.1 Tangents and Normals

1. $y = x^2 + 2$ at $(2,6)$

$\dfrac{dy}{dx} = 2x;\ \dfrac{dy}{dx}\bigg|_{(2,6)} = 4$

$m = 4;\ (x_1, y_1) = (2,6)$
Eq. T.L.: $y - y_1 = m(x - x_1)$
$y - 6 = 4(x - 2) = 4x - 8$
$y = 4x - 8 + 6 = 4x - 2;$
$4x - y - 2 = 0$

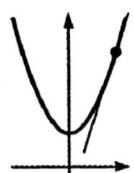

2. $y = \dfrac{1}{3}x^3 - 5x;\ (3, -6);\ \dfrac{dy}{dx} = m_{\tan} = x^2 - 5$

$m_{\tan_{x=3}} = 4$
Eq. T.L.: $y - y_1 = m(x - x_1)$
$y + 6 = 4(x - 3);\ y + 6 = 4x - 12;\ y = 4x - 18$
Therefore, $4x - y - 18 = 0$

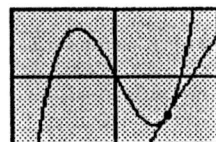

3. $y = \dfrac{1}{x^2 + 1};\ \left(1, \dfrac{1}{2}\right);\ y = (x^2 + 1)^{-1}$

$\dfrac{dy}{dx} = m_{\tan} = -(x^2 + 1)^{-2}(2x);\ m_{\tan} = \dfrac{-2x}{(x^2 + 1)^2};$

$m_{\tan_{x=1}} = -\dfrac{1}{2}$

Eq. of T.L: $y - \dfrac{1}{2} = -\dfrac{1}{2}(x - 1);\ 2y - 1 = -x + 1$

$y = -\dfrac{1}{2}x + 1$

Therefore, $x + 2y - 2 = 0$

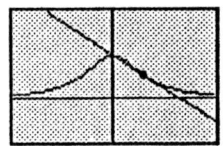

4. $x^2 + y^2 = 25;\ (3,4)$

$2x + 2y\dfrac{dy}{dx} = 0$

$\dfrac{dy}{dx} = m_{\tan} = -\dfrac{x}{y}$

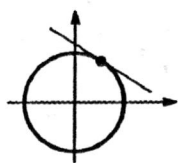

$m_{\tan_{(3,4)}} = -\dfrac{3}{4}$

Eq. of T.L.: $y - 4 = -\dfrac{3}{4}(x - 3);\ 4y - 16 = -3x + 9;$

$3x + 4y - 25 = 0;\ y = -\dfrac{3}{4}x + \dfrac{25}{4}$

5. $y = 6x - 2x^2$ at $(2,4)$

$\dfrac{dy}{dx} = 6 - 4x;\ m_{\tan(2.4)} = 2$

Eq. of normal: $m_{\text{normal}} = \dfrac{1}{2}$

$y - y_1 = m(x - x_1)$

$y - 4 = \dfrac{1}{2}(x - 2);\ 2y - 8 = x - 2$

$x - 2y + 6 = 0$ or $y = \dfrac{1}{2}x + 3$

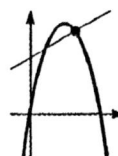

6. $y = 8 - x^3;\ (-1, 9);\ \dfrac{dy}{dx} = m_{\tan} = -3x^2;$

$m_{\tan_{(-1,9)}} = -3;\ m_{\text{normal}} = \dfrac{1}{3}$

Eq. of normal: $y - 9 = \dfrac{1}{3}(x + 1);\ y = \dfrac{1}{3}x + \dfrac{28}{3}$

$3y - 27 = x + 1$

Therefore, $x - 3y + 28 = 0$.

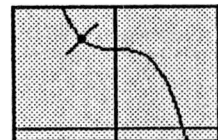

7. $y = \dfrac{6}{(x^2 + 1)^2}$; $\left(1, \dfrac{3}{2}\right)$; $y = 6(x^2 + 1)^{-2}$

$\dfrac{dy}{dx} = m_{\tan} = -12(x^2 + 1)^{-3}(2x)$

$m_{\tan} = \dfrac{-24x}{(x^2 + 1)^3}$; $m_{\tan(1,3/2)}\big|_{x=1} = -3$;

$m_{\text{normal}} = \dfrac{1}{3}$

Eq. of normal: $y - \dfrac{3}{2} = \dfrac{1}{3}(x - 1)$

Therefore, $2x - 6y + 7 = 0$

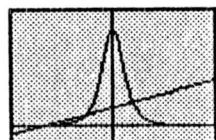

8. $x^2 - y^2 = 8$; $(3, 1)$

$2x - 2y\dfrac{dy}{dx} = 0$

$\dfrac{dy}{dx} = m_{\tan} = \dfrac{x}{y}$;

$m_{\tan(3,1)} = 3$

$m_{\text{normal}} = -\dfrac{1}{3}$

Eq. of normal:

$y - 1 = -\dfrac{1}{3}(x - 3)$; $3y - 3 = -x + 3$; $x + 3y - 6 = 0$

9. $y = \dfrac{1}{\sqrt{x^2 + 1}}$ where $x = \sqrt{3}$, $y = \dfrac{1}{2}$

$y = (x^2 + 1)^{-1/2}$;

$\dfrac{dy}{dx} = -\dfrac{1}{2}(x^2 + 1)^{-3/2}(2x) = -\dfrac{x}{(x^2 + 1)^{3/2}}$

$m_{\tan(\sqrt{3},1/2)} = \dfrac{-\sqrt{3}}{8}$; $m_{\text{normal}} = \dfrac{8}{\sqrt{3}}$

Eq. of T.L.:

$y - \dfrac{1}{2} = -\dfrac{\sqrt{3}}{8}(x - \sqrt{3})$; $8y - 4 = -\sqrt{3x} + 3$

$\sqrt{3}x + 8y - 7 = 0$

Eq. of N.L.:

$y - \dfrac{1}{2} = \dfrac{8}{\sqrt{3}}(x - \sqrt{3})$; $2\sqrt{3}y - \sqrt{3} = 16x - 16\sqrt{3}$

$16x - 2\sqrt{3}y - 15\sqrt{3} = 0$

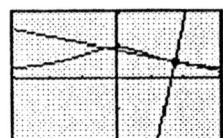

10. $y = \dfrac{4}{(5 - 2x)^2}$; $x = 2, y = 4$

$y = 4(5 - 2x)^{-2}$; $\dfrac{dy}{dx} = -8(5 - 2x)^{-3}(-2)$

$m_{\tan} = \dfrac{16}{(5 - 2x)^3}$; $m_{\tan(2,4)} = 16$; $m_{\text{normal}} = -\dfrac{1}{16}$

Eq. of T.L.:
$y - 4 = 16(x - 2)$; $y - 4 = 16x - 32$, therefore
$16x - y - 28 = 0$
Eq. of N.L.:

$y - 4 = -\dfrac{1}{16}(x - 2)$; $16y - 64 = -x + 2$, therefore

$x + 16y - 66 = 0$

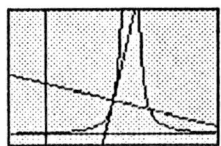

11. $V(0, 3), F(0, 0)$ at $x = -1$
$x^2 = -4p(y - k)$; $p = 3, k = 3$
$x^2 = -12(y - 3)$
$y = -\dfrac{x^2}{12} + 3$

$\dfrac{dy}{dx} = -\dfrac{1}{6}x$, $m_{\tan}\big|_{x=-1} = \dfrac{1}{6}$, $m_{\text{normal}} = -6$

Eq. of T.L: $\left(x = -1, y = \dfrac{35}{12}\right)$

$y - \dfrac{35}{12} = \dfrac{1}{6}(x + 1)$; $12y - 35 = 2x + 2$;

$2x - 12y + 37 = 0$

Eq. of N.L:

$$y - \frac{35}{12} = -6(x+1); \quad 12y - 35 = -72x - 72$$

$$72x + 12y + 37 = 0$$

12. $F(4,0), V(5,0), C(0,0), x = 2$
$a = 5, c = 4$
Therefore, $b^2 = a^2 - c^2$
$b^2 = 25 - 16 = 9; \ b = 3$

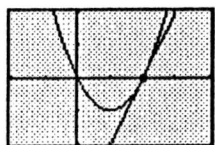

$$\frac{x^2}{a^2} + \frac{y^2}{b^2} = 1; \quad \frac{x^2}{25} + \frac{y^2}{9} = 1$$

$$\frac{2}{25}x + \frac{2}{9}y\frac{dy}{dx} = 0; \quad \frac{dy}{dx} = \frac{-9x}{25y}$$

$$x = 2, y_1 = \frac{3\sqrt{21}}{5}, \ y_2 = -\frac{3\sqrt{21}}{5}$$

$$m_{\text{TL}} = \pm\frac{6}{5\sqrt{21}}; \quad m_{\text{NL}} = \pm\frac{5\sqrt{21}}{6}$$

For $x = 2, y = \frac{3\sqrt{21}}{5}$

E. of T.L:

$$y - \frac{3\sqrt{21}}{5} = -\frac{6}{5\sqrt{21}}(x-2); \quad 6x + 5\sqrt{21}y - 75 = 0$$

Eq of N.L:

$$y - \frac{3\sqrt{21}}{5} = \frac{5\sqrt{21}}{6}(x-2)$$

$$30y - 18\sqrt{21} = 25\sqrt{21}x - 50\sqrt{21}$$

$$25\sqrt{21}x - 30y - 32\sqrt{21} = 0$$

For $x = 2, y = -\frac{3\sqrt{21}}{5}$

E. of T.L: $y + \frac{3\sqrt{21}}{5} = \frac{6}{5\sqrt{21}}(x-2)$

$$5\sqrt{21}y + 63 = 6x - 12; \quad 6x - 5\sqrt{21}y - 75 = 0$$

Eq of N.L: $y + \frac{3\sqrt{21}}{5} = -\frac{5\sqrt{21}}{6}(x-2)$

$$30y + 18\sqrt{21} = -25\sqrt{21}x + 50\sqrt{21}$$
$$25\sqrt{21}x + 30y - 32\sqrt{21} = 0$$

13. $y = x^2 - 2x$; tangent line with slope of 2

$$m_{\text{tan}} = \frac{dy}{dx} = 2x - 2; \ 2 = 2x - 2; \ x = 2;$$

$$y = 2^2 - 2(2) = 0$$

Therefore, the point at which the slope is 2 is $(2, 0)$, Using the point slope formula for the equation of a line, $y - 0 = 2(x - 2); \ y = 2x - 4$

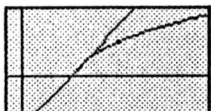

14. $y = \sqrt{2x - 9}$; tangent line $m = 1$

$$y = (2x - 9)^{1/2};$$

$$m_{\text{tan}} = \frac{1}{2}(2x-9)^{-1/2}(2) = \frac{1}{\sqrt{2x-9}}$$

$$\frac{1}{\sqrt{2x-9}} = 1; \ \sqrt{2x-9} = 1; \ 2x - 9 = 1; \ 2x = 10;$$

$x = 5, y = 1$
Eq. of T.L.: $y - 1 = 1(x - 5); \ y - 1 = x - 5$
Therefore, $x - y - 4 = 0; \ y = x - 4$

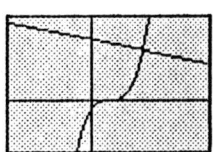

15. $y = (2x - 1)^3$; normal line $m = -\frac{1}{24}, x > 0$

Therefore, $m_{\text{tan}} = 24$
$m_{\text{tan}} = 3(2x-1)^2(2) = 6(2x-1)^2$
$6(2x-1)^2 = 24; \ (2x-1)^2 = 4; \ 2x - 1 = \pm 2$
$2x = \pm 2 + 1 = 3$ and -1

$$x = \frac{3}{2}, y = 8$$

Eq. of N.L.: $y - 8 = -\frac{1}{24}\left(x - \frac{3}{2}\right);$

$$24y - 192 = -x + \frac{3}{2}$$

$$48y - 384 = -2x + 3; \ 2x + 48y - 387 = 0$$

16. $y = \frac{1}{2}x^4 + 1$; normal line $m = 4, m_{\tan} = -\frac{1}{4}$

$m_{\tan} = 2x^3$; $2x^3 = -\frac{1}{4}$; $x^3 = -\frac{1}{8}$; $x = -\frac{1}{2}$,

$y = \frac{33}{32}$

Eq. of N.L.: $y - \frac{33}{32} = 4\left(x + \frac{1}{2}\right)$;

$32y - 33 = 128x + 64$

$128x - 32y + 97 = 0$

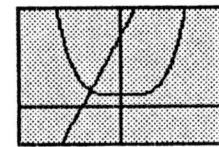

17. For the parabola $y^2 = 4x$, $2yy' = 4$,

$y' = \frac{2}{y}\Big|_{(a,b)} = \frac{2}{b}$

$m_{\text{parabola}}(a, b) = \frac{2}{b}$

For the ellipse, $2x^2 + y^2 = 6$, $4x + 2yy' = 0$,

$y' = \frac{-2x}{y}\Big|_{(a,b)} = \frac{-2a}{b}$

$m_{ellipse}(a, b) = \frac{-2a}{b}$

$m_{\text{parabola}}(a, b) \cdot m_{ellipse}(a, b) = \frac{2}{b} \cdot \frac{-2a}{b} = \frac{-4a}{b^2}$ and since $b^2 = 4a$

$m_{\text{parabola}}(a, b) \cdot m_{ellipse}(a, b) = \frac{-4a}{-4a} = -1$ which

implies the TL's are perpendicular; they intersect at right angles.

18. $y = x^{3/4}$; y-intercept of N.L. where $x = 16, y = 8$

$\frac{dy}{dx} = \frac{3}{4}x^{-1/4} = \frac{3}{4x^{1/4}} = m_{TL}\big|_{x=16} = \frac{3}{8}$; therefore,

$m_{NL} = -\frac{8}{3}$

Eq. of N.L.: $y - 8 = -\frac{8}{3}(x - 16)$;

$3y - 24 = -8x + 128$

$8x + 3y - 152 = 0$

y int. $x = 0$; $0 + 3y - 152 = 0$; $b = \frac{152}{3}$

19. Equation of curve:

$x^2 = 4py$

$p = \frac{x^2}{4y} = \frac{(100)^2}{4(30)}$

$p = \frac{250}{3}$

Therefore, $x^2 = \frac{1000}{3}y$

Eq. of T.L. at $(100, 30)$

$2x = \frac{1000}{3}\frac{dy}{dx}$; $\frac{dy}{dx} = \frac{6x}{1000}$; $m_{TL}\big|_{x=100} = \frac{6}{10} = \frac{3}{5}$

$y - 30 = \frac{3}{5}(x - 100)$; $5y - 150 = 3x - 300$

$3x - 5y - 150 = 0$

20. Equation of circle: $x^2 + y^2 = 9$

$2x + 2y\frac{dy}{dx} = 0$; $m_{\tan} = -\frac{x}{y}$

Therefore, $l = \sqrt{25 - 9} = 4$

Using distance formula:

$9 = y_1^2 + x_1^2$; $16 = y_1^2 + (x_1 - 5)^2$

Solve: $x_1 = \frac{9}{5}, y = \frac{12}{5}$, therefore

$m_{\tan} = -\frac{9}{12} = -\frac{3}{4}$

Eq. of T.L.: $y - \frac{12}{5} = -\frac{3}{4}\left(x - \frac{9}{5}\right)$

$20y - 48 = -15x + 27$; $15x + 20y - 75 = 0$

$3x + 4y - 15 = 0$

21. $y = \sqrt{2x^2 + 8} = (2x^2 + 8)^{1/2}$; $m = \tan 135° = -1$

$\frac{dy}{dx} = \frac{1}{2}(2x^2 + 8)^{-1/2}(4x) = \frac{2x}{\sqrt{2x^2 + 8}} = 1$

$2x = \sqrt{2x^2 + 8}$; $4x^2 = 2x^2 + 8$; $2x^2 = 8$; $x^2 = 4$

$x = \pm 2$; $y = \sqrt{2(4) + 8} = \sqrt{16} = 4$

$y - y_1 = m(x - x_1)$

$y - 4 = -1(x - 2)$

$y + x - 6 = 0$

22. $y^2 = 8x$; reflected line passes through the focus (p).

$4p = 8$; $p = 2$

Eq. of reflected line:

$m = \tan(43.60°)$

$\quad = 0.9522$

$y - 0 = 0.952(x - 2)$

$y = 0.952x - 1.906$

$(0.952x - 1.906)^2 = 8x$

$0.906x^2 - 3.629x + 3.633 = 8x$

$0.906x^2 - 11.629x + 3.633 = 0$

$x^2 - 12.836x + 4.010 = 0$

$x^2 - 12.836x + 41.191 = -4.010 + 41.191$

$(x - 6.418)^2 = 37.181;$

$x - 6.418 = \pm 6.098$

$x = 6.418 \pm 6.098 = 12.516$ and 0.32

$x = 0.32, y = 1.60$

$2y \dfrac{dy}{dx} = 8$

$\dfrac{dy}{dx} = \dfrac{8}{2y} = m_{TL}\big|_{(0.32, 1.6)}$

$\quad = \dfrac{8}{3.2} = 2.5;$

$m_{NL} = \dfrac{1}{2.5} = -0.4$

Eq. of N.L.:

$y - 1.60 = -0.40(x - 0.32);$

$y - 1.60 = -0.40x + 0.128$

$0.40x + y - 1.728 = 0$

23. $y = \dfrac{4}{x^2 + 1}$; $(-2 < x < 2)$

Supports at $(-1, 2), (0, 4), (1, 2)$

$y = 4(x^2 + 1)^{-1}$

$\dfrac{dy}{dx} = m_{\tan} = -4(x^2 + 1)^{-2}(2x)$

$m_{\tan} = \dfrac{-8x}{(x^2 + 1)^2}$

$m_{\tan}\big|_{x=-1} = 2;\ m_{\tan}\big|_{x=0} = 0;$

$m_{\tan}\big|_{x=1} = -2$

Eq. of N.L.: at $(-1, 2)$:

$y - 2 = -\dfrac{1}{2}(x + 1);\ 2y - 4 = -x + 1;$

$x + 2y - 3 = 0$

$m_{NL} = -\dfrac{1}{2}$

Eq. of N.L.: at $(0, 4)$:

$x = 0$; m_{NL} is undefined (vertical).

Eq. of N.L.: at $(1, 2)$:

$y - 2 = \dfrac{1}{2}(x - 1);\ 2y - 4 = x - 1;$

$x - 2y + 3 = 0$

$m_{NL} = \dfrac{1}{2}$

24. $y = -10$

$4y - 3x - 50 = 0$

$4y = 3x + 50$

$y = \dfrac{3}{4}x + \dfrac{50}{4}$

$m_{\tan \text{ line}} = \dfrac{3}{4}$

$x^2 + y^2 = 100$

$2x + 2y \dfrac{dy}{dx} = 0$

$\dfrac{dy}{dx} = m_{TL} = -\dfrac{x}{y}$ for curve

$-\dfrac{x}{y} = \dfrac{3}{4}, y = -\dfrac{4x}{3}$

$x^2 + \left(-\dfrac{4x}{3}\right)^2 = 100;\ 9x^2 + 16x^2 = 900$

$25x^2 = 900, x = \pm 6, y = \pm 8$

Only $(-6, 8)$ satisfies $4y - 3x - 50 = 0$.

Points of contact: $(-6, 8)$ and $(0, -10)$.

4.2   Newtons Method for Solving Equations

1. $x^2 - 2x - 5 = 0$ (between 3 and 4)

$f(x) = x^2 - 2x - 5;\ f'(x) = 2x - 2$

Let $x_1 = 3.3$, then

$x_2 = x_1 - \dfrac{f(x_1)}{f'(x_1)} = 3.3 - \dfrac{(-0.71)}{4.6}$

$\quad = 3.454347826$

$x_3 = x_2 - \dfrac{f(x_2)}{f'(x_2)}$

$x_3 = 3.454347826 - \dfrac{f(3.454347826)}{f'(3.45347826)}$

$x_3 = 3.449494551$

Using the quadratic formula,

$$x = \frac{-(-2) \pm \sqrt{(-2)^2 - 4(1)(-5)}}{2} = \frac{2 \pm \sqrt{24}}{2}$$

The positive root is the one between 3 and 4

$$x = \frac{2 + \sqrt{24}}{2} = 3.449489743$$

The results agree to four (rounded off) decimal places.

2. $f(x) = 2x^2 - x - 2$ (between 1 and 2)
$f(1) = -1; \; f(2) = 4$

Closer to $(1, -1)$
Let $x_1 = 1.2$

$$x_2 = x_1 - \frac{f(x_1)}{f'(x_1)};$$

$$f'(x) = 4x - 1$$

$$x_2 = 1.2 - \frac{f(1.2)}{f'(1.2)}$$

$$= 1.2 - \frac{(-0.32)}{3.8}$$

$$= 1.284210526$$

$$x_3 = 1.2842105 - \frac{f(1.284211)}{f'(1.284211)}$$

$$= 1.284210526 - \frac{(0.014182717)}{(4.136842)}$$

$$= 1.2807821$$

Quadratic formula:

$$x = \frac{-b \pm \sqrt{b^2 - 4ac}}{2a}$$

$$= \frac{1 + \sqrt{1^2 + 16}}{4}$$

$$= 1.2807764$$

3. $f(x) = 3x^2 - 5x - 1$ (between -1 and 0)
$f'(x) = 6x - 5$
$f(-1) = 7; \; f(0) = -1$

Closer to $x = 0$
Let $x_1 = -0.2$

$$x_2 = -0.2 - \frac{f(-0.2)}{f'(-0.2)}$$

$$= -0.2 - \frac{0.12}{-6.2}$$

$$= -0.180645161$$

$$x_3 = -0.180645161 - \frac{(0.001123829)}{(-6.0838710)}$$

$$= -0.1804604$$

Quadratic formula:

$$x = \frac{5 - \sqrt{25 + 12}}{6}$$

$$= -0.1804604$$

4. $f(x) = x^2 + 4x + 2$ (between -4 and -3)
$f'(x) = 2x + 4$
$f(-4) = 2; \; f(-3) = -1$

Closer to -3
Let $x_1 = -3.4$

$$x_2 = -3.4 - \frac{f(-3.4)}{f'(-3.4)}$$

$$= -3.4 - \frac{-0.04}{-2.8}$$

$$= -3.4142857$$

$$x_3 = -3.4142857 - \frac{f(x_2)}{f'(x_2)}$$

$$= -3.4142857 - \frac{(0.00020408)}{(-2.8285714)}$$

$$= -3.4142136$$

Quadratic formula:

$$x = \frac{-4 - \sqrt{16 - 8}}{2}$$

$$= -3.4142136$$

5. $x^3 - 6x^2 + 10x - 4 = 0$ (between 0 and 1)

$f(x) = x^3 - 6x^2 + 10x - 4; \; f'(x) = 3x^2 - 12x + 10;$

$f(0) = -4; \; f(1) = 1$

The root is probably closer to 1. Let $x_1 = 0.7$

n	x_n	$f(x_n)$	$f'(x_n)$	$x_n - \frac{f(x_n)}{f'(x_n)}$
1	0.7	0.403	3.07	0.5687296
2	0.5687296	−0.0694666	4.1456049	0.5854863
3	0.5854863	−0.0012009	4.002547	0.5857863
4	0.5857863	−0.0000005	4.0000012	0.5857864

$x_4 = x_3 = 0.5857864$ to seven decimal places

6. $f(x) = x^3 - 3x^2 - 2x + 3; \; f'(x) = 3x^2 - 6x - 2$

$f(0) = 3; \; f(1) = -1;$ closer to $x = 1;$ let $x_1 = 0.8$

n	x_n	$f(x_n)$	$f'(x_n)$	$x_n - \frac{f(x_n)}{f'(x_n)}$
1	0.8	0.008	−4.88	0.7983606
2	0.7983606	−0.000001617	−4.8780247	0.7983603
3	0.7983603	0.0000001		0.7983603

$x_3 = 0.7983603$

7. $f(x) = x^3 + 5x^2 + x - 1 = 0; \; f'(x) = 3x^2 + 10x + 1$

$f(0) = -1; \; f(1) = 6;$ let $x_1 = 0.3$

n	x_n	$f(x_n)$	$f'(x_n)$	$x_n - \frac{f(x_n)}{f'(x_n)}$
1	0.3	−0.223	4.27	0.3522248
2	0.3522248	0.0162343	4.8944352	0.3489079
3	0.3489079	0.000066588	4.8542896	0.3488942
4	0.3488942	ϕ		0.3488942

Therefore, $x_4 = 0.3488942$

8. $f(x) = 2x^3 + 2x^2 - 11x + 3 = 0; \; f'(x) = 6x^2 + 4x - 11$

$f(0) = 3; \; f(1) = -4$ (between 0 and 1)

$f(1) = -4; \; f(2) = 5$ (between 1 and 2)

Let $x_1 = 1.4$

n	x_n	$f(x_n)$	$f'(x_n)$	$x_n - \frac{f(x_n)}{f'(x_n)}$
1	1.4	−2.992	6.36	1.87044025
2	1.87044025	2.5098960	17.4730414	1.7267963
3	1.7267963	0.2669025	13.7981390	1.7074530
4	1.7074530	0.00461055	13.3321858	1.7071069
5	1.7071069	0.0000015	13.3137112	1.7071068

$x_5 = 1.7071068$

9. $x^4 - x^3 - 3x^2 - x - 4 = 0$; (between 2 and 3)

$f(x) = x^4 - x^3 - 3x^2 - x - 4$; $f'(x) = 4x^3 - 3x^2 - 6x - 1$

$f(2) = -10$; $f(3) = 20$; the root is possibly closer to 2.

Let $x_1 = 2.3$

n	x_n	$f(x_n)$	$f'(x_n)$	$x_n - \frac{f(x_n)}{f'(x_n)}$
1	2.3	−6.3529	17.998	2.6529781
2	2.6529781	3.0972725	36.657001	2.5684848
3	2.5684848	0.2175007	31.576097	2.5615967
4	2.5615967	0.0013683	−31.179599	2.5615528

$x_3 = x_4 = 2.5615528$ to seven decimal places.

10. $f(x) = 2x^4 - 2x^3 - 5x^2 - x - 3 = 0$

$f'(x) = 8x^3 - 6x^2 - 10x - 1$; $f(2) = -9$; $f(3) = 57$

Let $x_1 = 2.2$

n	x_n	$f(x_n)$	$f'(x_n)$	$x_n - \frac{f(x_n)}{f'(x_n)}$
1	2.2	−3.8448	33.144	2.31 0029
2	2.3160029	0.5613660	43.0386513	2.3029596
3	3.3029596	0.0076992	41.8609008	2.3027757
4	2.3027757	0.0000015	41.8444135	2.3027756

Therefore, $x_4 = 2.3027756$

11. $f(x) = x^4 - 2x^3 - 8x - 16 = 0$; $f'(x) = 4x^3 - 6x^2 - 8$

$f(0) = -16$; $f(-1) = -5$; $f(-2) = 32$; let $x_1 = -1.2$

n	x_n	$f(x_n)$	$f'(x_n)$	$x_n - \frac{f(x_n)}{f'(x_n)}$
1	−1.2	−0.8704	−23.552	−1.2369565
2	−1.2369565	0.0219791	−24.7508465	−1.2360685
3	−1.2360685	0.0000131	−24.7213772	−1.2360680
4	−1.2360680	ϕ		

$x_4 = -1.2360680$

12. $f(x) = 3x^4 - 3x^3 - 11x^2 - x - 4 = 0$

$f'(x) = 12x^3 - 9x^2 - 22x - 1$

$f(0) = -4$; $f(-1) = -8$; $f(-2) = 26$; let $x_1 = -1.3$

n	x_n	$f(x_n)$	$f'(x_n)$	$x_n - \frac{f(x_n)}{f'(x_n)}$
1	−1.3	−6.1307	−13.974	1.7387219
2	−1.7387219	7.6716792	−53.0335898	1.5940649
3	−1.5940649	1.1650530	−37.4070087	1.5629196
4	−1.5629196	0.0469486	−34.4134852	1.5615554
5	−1.5615554	0.0000875	−34.2852711	1.5615528
6	−1.5615528	ϕ		

$x_6 = -1.5615528$

13. $2x^2 = \sqrt{2x+1}$ or $2x^2 - \sqrt{2x+1} = 0$

$4x^4 - 2x - 1 = 0$ (Square both sides.)

$f(x) = 4x^4 - 2x - 1$

From sketch, the intersections lie between 0 and 1, and between 0 and -1. Approximate the positive root at 0.8.

n	x_n	$f(x_n)$	$f'(x_n)$	$x_n - \frac{f(x_n)}{f'(x_n)}$
1	0.8	-0.9616	6.192	0.9552972
2	0.955 2972	0.4207071	11.948757	0.9200879
3	0.920 0879	0.0264914	10.462579	0.9175559
4	0.917 5559	0.0001301	10.3601	0.9175433

The positive root is approximately 0.9175433.

14. $f(x) = x^3 - \sqrt{x+1}$; $f'(x) = 3x^2 - \dfrac{1}{2\sqrt{x+1}}$

$f(1) = -0.414$; $f(2) = 6.268$; let $x = 1.1$

n	x_n	$f(x_n)$	$f'(x_n)$	$x_n - \frac{f(x_n)}{f'(x_n)}$
1	1.1	-0.1181377	3.2849672	1.1359631
2	1.1359631	0.0043673	3.5291209	1.1347256
3	1.1347256	0.0000052	3.5205918	1.1347241
4	1.1347241	ϕ		

$x_4 = 1.1347241$

15. $f(x) = x - \dfrac{1}{\sqrt{x+2}}$; $f'(x) = 1 + \dfrac{1}{2(x+2)^{3/2}}$

$f(0) = -0.707$; $f(1) = 0.423$; let $x_1 = 0.5$

n	x_n	$f(x_n)$	$f'(x_n)$	$x_n - \frac{f(x_n)}{f'(x_n)}$
1	0.5	-0.1324555	1.1264911	0.6175824
2	0.6175824	-0.0005049	1.1180645	0.0.6180340
3	0.6180340	ϕ		

$x_3 = 0.6180340$

16. $f(x) = x^{3/2} - \dfrac{1}{(2x+1)}$; $f'(x) = \dfrac{3}{2}x^{1/2} + \dfrac{2}{(2x+1)^2}$

$f(0) = -1$; $f(1) = 0.667$; let $x_1 = 0.5$

n	x_n	$f(x_n)$	$f'(x_n)$	$x_n - \frac{f(x_n)}{f'(x_n)}$
1	0.5	-0.1464466	1.5606602	0.5938363
2	0.5938363	0.0005081	1.5738049	0.5935135
3	0.5935135	ϕ		

$x_3 = 0.5935135$

17. $f(x) = x^3 - 2x^2 - 5x + 4$. From sketch, one root lies between -1 and -2, and the other between 0 and 1.

$f'(x) = 3x^2 - 4x - 5$
$f(-1) = 6$ and $f(-2) = -2$

One root is possibly closer to -2.
Let $x_1 = -1.7$
$f(0) = 4$; $f(1) = -2$. The second root is possibly closer to 1. Let $x_1 = 0.7$
$f(3) = -2$; $f(4) = 16$. The third root is closer to 3. Let $x_1 = 3.1$

n	x_n	$f(x_n)$	$f'(x_n)$	$x_n - \frac{f(x_n)}{f'(x_n)}$
1	-1.7	1.807	10.47	-1.8725883
2	-1.8725883	-0.2166261	13.010114	-1.8559377
3	-1.8559377	-0.0021072	12.757265	-1.8557725
4	-1.8557725	-0.000000057	11.562575	-1.8557725

1	0.7	-0.137	-6.33	0.678357
2	0.678357	0.0000369	-6.332923	$0.678362\ 8$

1	3.1	-0.929	11.43	3.1812773
2	3.11812773	0.0487617	12.636467	3.1774185
3	3.1774185	0.000111	12.578291	3.1774097

The roots are $-1.8557725, 0.6783628$, and 3.1774097.

18. $f(x) = x^3 - 2x^2 - 2x - 7$
$f'(x) = 3x^2 - 4x - 2$
From sketch, one root between $x = 3$ and $x = 4$
$f(3) = -4$; $f(4) = 17$
Let $x_1 = 3.2$

n	x_n	$f(x_n)$	$f'(x_n)$	$x_n - \frac{f(x_n)}{f'(x_n)}$
1	3.2	-1.112	15.92	3.2698492
2	3.2698492	0.0374206	16.9963453	3.2676476
3	3.2676476	0.0000378	16.9619715	3.2676453
4	3.2676453	ϕ		

$x_4 = 3.2676453$

19. $\sqrt[3]{4}$; $x = \sqrt[3]{4}$; $x^3 = 4$, therefore, $f(x) = x^3 - 4$; $f'(x) = 3x^2$

n	x_n	$f(x_n)$	$f'(x_n)$	$x_n - \frac{f(x_n)}{f'(x_n)}$
1	1.5	-0.625	6.75	1.5925926
2	1.5925926	0.0393741	7.6090535	1.5874182
3	1.5874182	0.0001292	7.5596892	1.5874011

Therefore, $\sqrt[3]{4} = 1.5874011$

20. 0.00962518 should be 0.00961518, possibly a typographical or computational error.

21. $V = \frac{1}{6}\pi h(h^2 + 3r^2) = \frac{1}{6}\pi h^3 + \frac{1}{2}\pi r^2 h$

$180000 = \frac{1}{6}\pi h[h^2 + 3(60)^2] = \frac{1}{6}\pi h^3 + 1800\pi h$

$f(h) = \frac{1}{6}\pi h^3 + 1800\pi h - 180000$

$f'(h) = \frac{1}{2}\pi h^2 + \frac{1}{2}\pi r^2 = \frac{1}{2}\pi h^2 + 1800\pi$

n	h_n	$f(h_n)$	$f'(h_n)$	$h_n - \frac{f(h_n)}{f'(h_n)}$
1	29	−3238.813	6975.906	29.464286
2	29.464286	9.874670	7018.544	29.462879
3	29.462879			

$h_3 = 29.462879$; $h = 29.5$ m

22. Simplify $\frac{1}{C} + \frac{1}{C+1} + \frac{1}{C+2} = 1$ gives $C^3 - 4C - 2 = 0$

$f(C) = C^3 - 4C - 2$; $f'(C) = 3C^2 - 4$
$f(2) = -2$; $f(3) = 13$; let $C_1 = 2.2$

n	h_n	$f(h_n)$	$f'(h_n)$	$h_n - \frac{f(h_n)}{f'(h_n)}$
1	2.2	−0.152	10.52	2.2144487
2	0.214448 7	0.001380 86	10.7113487	2.2143198
3	2.214319 8	ϕ		

$C_3 = 2.214\ 319\ 8$
Therefore, $C = 2.2143\ \mu\mathrm{F}$, $C + 1 = 3.2143\ \mu\mathrm{F}$,
$C + 2 = 4.2143\ \mu\mathrm{F}$

23. Volume = 2(half spheres) + cylinder

$V = \frac{4}{3}\pi r^3 + \pi r^2 h$; $1500 = \frac{4}{3}\pi r^3 + 12.0\pi r^2$

$f(r) = \frac{4}{3}\pi r^3 + 12.0\pi r^2 - 1500$; $f'(r) = 4\pi r^2 + 24.0\pi r$

$f(5) < 0$; $f(6) > 0$; let $r_1 = 5.1$

n	x_n	$f(x_n)$	$f'(x_n)$	$x_n - \frac{f(x_n)}{f'(x_n)}$
1	5.1	36.2011085	711.382405	5.0491116
2	5.0491116	0.2630481	701.0551673	5.0487364
3	5.0487364	ϕ		

$r_3 = 5.048\ 736\ 4$; $r = 5.05$ ft

24. $V_1 = 16.00$ cm^3
$V_2 = 32.00$ cm^3
Let each dimension increase by x cm as a result of the expansion.
Therefore, $32 = (4.00 + x)(2.00 + x)^2$
$32.00 = 16.00 + 20.00x + 8.00x^2 + x^3$
$f(x) = x^3 + 8.00x^2 + 20.00x - 16.00$
$f'(x) = 3x^2 + 16.00x + 20.00$

$f(0) < 0; f(1) > 0;$ let $x_1 = 0.5$

n	x_n	$f(x_n)$	$f'(x_n)$	$x_n - \dfrac{f(x_n)}{f'(x_n)}$
1	0.5	-3.875	28.75	0.6347826
2	0.6347826	0.1750288	31.3653686	0.6292022
3	0.6292022	0.0003083	31.2549232	0.6291924
4	0.6291924	ϕ		

$x_3 = 0.6291924$

Therefore, each edge increases by 0.629 cm.

4.3 Curvilinear Motion

1. $x = 3t; y = 1 - t; t = 4$

$$v_x = \frac{dx}{dt} = 3; \quad v_y = \frac{dy}{dt} = -1$$

$$v_x|_{t=4} = 3; \quad v_y|_{t=4} = -1$$

$$v|_{t=4} = \sqrt{3^2 + (-1)^2} = \sqrt{9 + 1} = \sqrt{10} = 3.16$$

$$\tan\theta = \frac{-1}{3} = -0.3333; \quad \theta = -18.4° = 341.6°$$

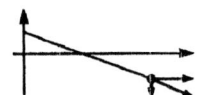

2. $x = \dfrac{5t}{2t + 1}; \dfrac{dx}{dt} = \dfrac{(2t + 1)5 - 5t(2)}{(2t + 1)^2} = \dfrac{5}{(2t + 1)^2} = v_x$

$$y = 0.1(t^2 + 1); \quad \frac{dy}{dx} = 0.1(2t + 1) = 0.2t + 0.1 = v_y$$

$$v_x|_{t=2} = \frac{5}{25} = \frac{1}{5} = 0.2$$

$$v_y|_{t=2} = 0.5$$

$$v = \sqrt{V_x^2 + v_y^2}$$

$$v = \sqrt{(0.2)^2 + (0.5)^2}$$

$$= 0.5385$$

t	x	y	
0	0	0	
1	$\frac{5}{3}$	0.2	
2	2	0.6	←
3	$\frac{15}{7}$	1.2	

$$\theta = \tan^{-1}\frac{v_y}{v_x} = \tan^{-1}\frac{0.5}{0.2}$$

$$\theta = 68.2°$$

Therefore, v is 0.54 at $\theta = 68.2°$.

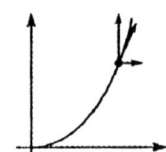

3. $x = t(2t + 1)^2$

$$\frac{dx}{dt} = t(2)(2t + 1)(2) + (2t + 1)^2(1)$$
$$= 12t^2 + 8t + 1 = v_x$$

$$y = 6(4t + 3)^{-1/2}$$

$$\frac{dy}{dt} = 6\left(-\frac{1}{2}\right)(4t + 3)^{-3/2}(4) = \frac{-12}{(4t + 3)^{3/2}} = v_y$$

$$v_x|_{t=0.5} = 8$$
$$v_y|_{t=0.5} = -1.0733$$

$$v = \sqrt{8^2 + (-1.0773)^2}$$
$$v = 8.07$$

$$\alpha = \tan^{-1}\frac{1.0733}{8} = 7.641°$$

t	x	y	
0	0	3.464	
0.5	2	2.683	←
1	9	2.268	

Therefore, $\theta = 360° - \alpha = 352.4°$

Therefore, v is 8.07 at $\theta = 352.4°$

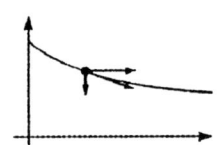

4. $x = (1 + 2t)^{1/2}$

$$\frac{dx}{dt} = \frac{1}{2}(1 + 2t)^{1/2}(2) = \frac{1}{\sqrt{1 + 2t}} = v_x$$

$$y = t - t^2; \quad \frac{dy}{dt} = 1 - 2t = v_y$$

$v_x|_{t=4} = \dfrac{1}{3}$

$v_y|_{t=4} = -7$

$$v = \sqrt{\left(\dfrac{1}{3}\right)^2 + (-7)^2}$$

$$= 7.008$$

$$\alpha = \tan^{-1}\left(\dfrac{7}{\frac{1}{3}}\right)$$

$$= 87.274°$$

$$\theta = 272.7° \text{ or } -87.3°$$

t	x	y
3	2.646	−6
4	3	−12 ←
5	3.317	−20

Therefore, v is 7.01 at $\theta = -87.3°$.

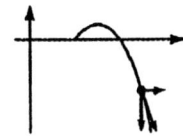

5. $x = 3t; \; y = 1 - t;$

$$t = 4; \; v_x = \dfrac{dx}{dt} = 3$$

$$a_x = \dfrac{d^2x}{dt^2} = \dfrac{d(3)}{dt} = 0;$$

$a_x|_{t=4} = 0$

$$v_y = \dfrac{dy}{dt} = -1; \; a_x = \dfrac{d^2y}{dt^2}$$

$$= \dfrac{d}{dt}(-1) = 0$$

$a_y|_{t=4} = 0$

The particle is not accelerating since
$a = \sqrt{0^2 + 0^2} = 0.$

6. $x = \dfrac{5t}{2t+1}; \; v_x = 5(2t+1)^{-2}; \; a_x = \dfrac{-20}{(2t+1)^3}$

$a_x|_{t=2} = -0.16$

$y = 0.1(t^2 + 1); \; v_y = 0.2t + 0.1; \; a_y = 0.2;$
$a_y|_{t=2} = 0.2$

$$a = \sqrt{a_x^2 + a_y^2}$$

$$a = \sqrt{(0.16)^2 + (0.2)^2}$$

$$a = 0.256$$

$$\alpha = \tan^{-1}\dfrac{0.2}{0.16} = 51.3°$$

$\theta = 128.7°$
Therefore, a is 0.26 at $\theta = 128.7°$.

7. $x = t(2t+1)^2; \; v_x = 12t^2 + 8t + 1; \; a_x = 24t + 8;$
   $a_x|_{t=0.5} = 20.0$

$$y = \dfrac{6}{\sqrt{4t+3}}; \; v_y = -12(4t+3)^{-3/2}$$

$$a_y = \dfrac{72}{(4t+3)^{5/2}}; \; a_y|_{t=0.5} = 1.288$$

$$a = \sqrt{20.0^2 + 1.288^2}$$
$$a = 20.041$$

$$\theta = \tan^{-1}\dfrac{1.288}{20.0}$$

$\theta = 3.68°$
Therefore, a is 20.0 at $\theta = 3.7°$.

8. $x = (1+2t)^{1/2}; \; v_x = (1+2t)^{-1/2}$

$$a_x = \dfrac{-1}{(1+2t)^{3/2}}; \; a_x|_{t=4} = -0.037$$

$y = t - t^2; \; v_y = 1 - 2t; \; a_y = -2; \; a_y|_{t=4} = -2.00$

$$a = \sqrt{(0.037)^2 + (2.00)^2}$$
$$a = 2.000$$

$$\alpha = \tan^{-1}\dfrac{2}{0.037} = 88.9°$$

$\theta = 268.9°$
Therefore, a is 2.00 at $\theta = 268.9°$

9. $y = 4.0 - 0.20x^2;$

$$v_x = \dfrac{dx}{dt} = 5.0 \text{ m/s}$$

$$v_y = \dfrac{dy}{dt} = -0.20\left(2x\dfrac{dx}{dt}\right)$$

$$= -0.40x\dfrac{dx}{dt}$$

$v_y|_{x=4.0} = -0.40(4.0)(5.0) = -8.0 \text{ m/s}$

$$v = \sqrt{(5.0)^2 + (-8.0)^2}$$
$$= \sqrt{25 + 64} = \sqrt{89} = 9.4 \text{ m/s}$$

$$\tan\theta = \dfrac{v_y}{v_x} = \dfrac{-8.0}{5.0} = -1.60;$$

$$\theta = -58° = 302°$$

10. $y = \sqrt{4x+1}$; $v_x = 2x$

Find v at $(2.0, 3.0)$

$y = (4x+1)^{1/2}$;

$\dfrac{dy}{dt} = v_y = \dfrac{1}{2}(4x+1)^{-1/2}(4)\dfrac{dx}{dt} = \dfrac{2v_x}{\sqrt{4x+1}}$

$v_x = 2x$; $v_x|_{x=2.0} = 4.0$

$v_y = \dfrac{4x}{\sqrt{4x+1}}$; $v_y|_{x=2.0} = \dfrac{8}{3}$

$v = \sqrt{(4.0)^2 + \left(\dfrac{8}{3}\right)^2} = 4.8$

$\theta = \tan^{-1}\dfrac{\frac{8}{3}}{4} = 34°$

Therefore, v is 4.8 ft/s at $\theta = 34°$

11. $x = 0.2t^2$; $v_x = 0.4t$; $a_x = 0.4$; $a_x|_{t=2.0} = 0.4$

$y = -0.1t^3$; $v_y = -0.3t^2$; $a_y = -0.6t$;
$a_y|_{t=2.0} = -1.2$

$a = \sqrt{(0.4)^2 + (1.2)^2} = 1.3$

$\alpha = \tan^{-1}\left(\dfrac{1.2}{0.4}\right) = 71.6°$

$\theta = 288.4°$
Therefore, a is 1.3 ft/min² at $\theta = 288°$.

12. $x = 0.2t^3$; $v_x = 0.6t^2$; $a_x = 1.2t$;
$a_x|_{t=3.0} = 3.6$ m/s²

$y = 20t - 2t^2$; $v_y = 20 - 4t$; $a_y = -4$;
$a_y|_{t=3.0} = -4.0$ m/s²

$a = \sqrt{(3.6) + (4.0)^2} = 5.38$ m/s²

$\alpha = \tan^{-1}\left(\dfrac{4.0}{3.6}\right) = 48.0°$

$\theta = 312°$
Therefore, a is 5.4 m/s² at $\theta = 312°$.

13. $x = 96t$; $\dfrac{dx}{dt} = 96$ ft/s $= v_x$;

$y = 120t - 16t^2$

$\dfrac{dy}{dt} = 120 - 32t$;

$\dfrac{dy}{dt}\Big|_{t=6.0} = -72$ ft/s $= v_y$

$v = \sqrt{96^2 + (-72)^2} = 120$ ft/s

$\tan\theta = \dfrac{-72}{96}$; $\theta = 323°$

$\dfrac{d^2x}{dt^2} = 0 = a_x$; $\dfrac{d^2y}{dt^2} = -32 = a_y$;

$a = \sqrt{0^2 + (-32)^2}$
$= 32$ ft/s²

$\tan\theta$ is undefined; $\theta = 270°$

14. $x = 45t$; $v_x = 45$ m/s

$y = -4.9t^2$; $v_y = -9.8t$; $v_y|_{t=3.0} = -29.4$ m/s

$v = \sqrt{(45)^2 + (29.4)^2} = 53.8$ m/s

$\alpha = \tan^{-1}\dfrac{29.4}{45.0} = 33.2°$

$\theta = 326.8°$
Therefore, v is 54 m/s at $\theta = 327°$.

$a_x = 0|_{t=3.0} = 0$ m/s²

$a_y = -9.8|_{t=3.0} = -9.8$ m/s²

Therefore, $a = \sqrt{0^2 + (9.8)^2} = 9.8$ m/s².
$\theta = 270°$
Therefore, a is 9.8 m/s² at $\theta = 270°$.

15. $x = 10(\sqrt{1+t^4} - 1)$; $v_x = \dfrac{20t^3}{\sqrt{1+t^4}}$

$y = 40t^{3/2}$; $v_y = 60t^{1/2}$; $(0 \le t \le 100$ s$)$

For $t = 10$ s:

$v_x = \dfrac{20t^3}{\sqrt{1+t^4}}\Big|_{t=10.0\text{ s}} = 200.0$ m/s

$v_y = 60t^{1/2}|_{t=10.0\text{ s}} = 189.7$ m/s

$v = \sqrt{(200)^2 + (189.7)^2}$
$v = 276$ m/s

$\theta = \tan^{-1}\dfrac{189.7}{200} = 43.5°$

For $t = 100$ s:

$v_x|_{t=100\text{ s}} = 2000.0$ m/s

$v_y =|_{t=100\text{ s}} = 600$ m/s

$$v = \sqrt{(2000)^2 + (600)^2}$$

$$v = 2088 \text{ m/s}: \quad \theta = \tan^{-1}\frac{600}{2000} = 16.7°$$

At 10 s v is 276 m/s at $\theta = 43.5°$, and at 100 s v is

2090 m/s at $\theta = 16.7°$.

16. $x = 3.50t^2$, $y = 20.0 + 0.120t^4 - 3.00\sqrt{t^4 + 1}$,

$0 \le t \le 4.00 \text{ s}$

$$\frac{dx}{dt} = 7.00t$$

$$\frac{dy}{dt} = 0.480t^3 - \frac{3.00}{2}(t^4 + 1)^{-1/2}(4t^3)$$

$$\frac{dy}{dt} = 0.480t^3 - \frac{6.00t^3}{\sqrt{t^4 + 1}}$$

$$v_x = \frac{dx}{dt}\bigg|_{t=4.00} = 7.00(4.00) = 28.0$$

$$v_y = \frac{dy}{dt}\bigg|_{t=4.00} = 0.480(4.00)^3 - \frac{6(4.00)^3}{\sqrt{4.00^4 + 1}}$$

$$= 29.2$$

$$v = \sqrt{v_x^2 + v_y^2} = \sqrt{28.0^2 + 29.2^2} = 40.5 \text{ m/s}$$

$$\tan\theta = \frac{v_y}{v_x} = \frac{29.2}{28.0}$$

$$\theta = 46.2°$$

17. $x = 10(\sqrt{1 + t^4} - 1) = 10(1 + t^4)^{1/2} - 10$;

$y = 40t^{3/2}$

$$\frac{dx}{dt} = 5(1 + t^4)^{-1/2}(4t^3) = 20t^3(1 + t^4)^{-1/2}$$

$$= \frac{20t^3}{\sqrt{1 + t^4}}$$

$$\frac{dy}{dt} = 60t^{1/2} = 60\sqrt{t}$$

$$ax = \frac{(1 + t^4)^{1/2}(60t^2) - (20t^3)\left(\frac{1}{2}\right)(1 + t^4)^{-1/2}(4t^3)}{1 + t^4}$$

$$= \frac{(1 + t^4)^{-1/2}[60t^2(1 + t^4) - 40t^6]}{(1 + t^4)}$$

$$= \frac{60t^2(1 + t^4) - 40t^6}{(1 + t^4)^{3/2}}$$

$$ay = 30t^{-1/2}$$

$$a_x\big|_{t=10.0} = \frac{6000(10,000) - 40,000,000}{1,000,000}$$

$$= 20.0$$

$$a_y\big|_{t=10.0} = 30(10)^{-1/2} = \frac{30}{\sqrt{10}} = 9.5$$

$$a = \sqrt{(20)^2 + (9.5)^2} = 22.1 \text{ m/s}^2$$

$$\tan\theta = \frac{9.5}{20.0} = 0.475; \quad \theta = 25.4°$$

$$a_x\big|_{t=100} = \frac{60(10^4)(10^8) - 40(10^{12})}{(10^8)^{3/2}}$$

$$= \frac{6 \times 10^{13} - 4 \times 10^{13}}{10^{12}}$$

$$= \frac{2 \times 10^{13}}{10^{12}} = 20.0$$

$$a_y\big|_{t=100} = 30(10^2)^{-1/2} = 30(10^{-1}) = 3.0$$

$$a = \sqrt{(20.0)^2 + (3.0)^2} = 20.2 \text{ m/s}^2$$

$$\tan\theta = \frac{3.0}{20.0} = 0.150; \quad \theta = 8.5°$$

18. $a_x = \dfrac{d^2x}{dt^2} = 7.00\big|_{t=4.00} = 7.00$

$$a_y = \frac{d^2y}{dt^2} = 1.44t^2 - \left[\frac{-12.0t^6}{\sqrt{(t^4 + 1)^3}} + \frac{18.0t^2}{\sqrt{t^4 + 1}}\right]\bigg|_{4.00}$$

$$= 17.0$$

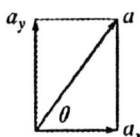

$$a = \sqrt{a_x^2 + a_y^2}$$

$$= \sqrt{7.00^2 + 17.0^2}$$

$$a = 18.4 \text{ m/s}^2$$

$$\tan\theta = \frac{a_y}{a_x} = \frac{17.0}{7.0}$$

$$\theta = 67.6°$$

19. $y = x - \dfrac{1}{90}x^3$; $v_x = x$; rocket hits ground when $y = 0$.

$$v_y = v_x - \frac{1}{30}x^2 v_x; \quad v_y = x - \frac{1}{30}x^2(x) = x - \frac{1}{30}x^3$$

$$y = 0 = x - \frac{1}{90}x^3 = x\left(x - \frac{1}{90}x^2\right)$$

$x = 0, x = \sqrt{90} = 9.487$

$x = 0$; rocket leaves ground

$x = 9.487$; rocket returns (crashed) to ground

$$v_x = x; \quad v_x\big|_{x=9.487} = 9.487$$

$$v_y = x - \frac{1}{30}x^3; \quad v_y\big|_{x=9.487} = -18.97$$

$v = \sqrt{90.0 + (18.97)^2} = 21.2$

$\alpha = \tan^{-1} \dfrac{18.97}{\sqrt{9.487}} = 63.4°$

Therefore, $\theta = 296.6°$
Therefore, v is 21.2 mi/min at $\theta = 296.6°$.

20. $y = 3x^2 - 0.2x^3;\ v_x = 1.2$ km/h; $x = 3.5$ km
$v_y = 6xv_x - 0.6x^2 v_x$

$v_y|_{x=3.5} = 6(3.5)(1.2) - 0.6(3.5)^2(1.2)$
$\qquad = 16.38$ km/h

$v = \sqrt{(1.2)^2 + (16.38)^2} = 16.42$ km/h

$\theta = \tan^{-1} \dfrac{16.38}{1.2} = 85.8°$
Therefore, v is 16.4 km/h at $\theta = 85.8°$.

21. $d = 3.50$ in; $r = 1.75$ in; $x^2 + y^2 = 1.75^2$, $\dfrac{dy}{dx} = -\dfrac{x}{y}$

3600 r/min $= 7200\pi$ rad/min $= \omega$
$v = \omega r = 7200\pi(1.75) = 12,600\pi$ in/min
$x^2 + y^2 = 1.75^2$
$\quad y^2 = 1.75^2 - x^2 = 1.75^2 - 1.20^2$
$\qquad = 3.062 - 1.44 = 1.622$
$\quad y = 1.274$ in

$\dfrac{dy}{dx} = -\dfrac{x}{y} = -\dfrac{1.20}{1.274} = -0.942 = \dfrac{v_y}{v_x}$;

$v_y = -0.942 v_x$

$v = 12,600\pi = \sqrt{v_x^2 + v_y^2}$
$\qquad = \sqrt{(-0.942 v_x)^2 + v_x^2}$
$\qquad = \sqrt{1.893 v_x^2} = 1.376 v_x$
$v_x = 9158\pi = 28,800$ in/min
$v_y = -0.942 v_x = 8654\pi = -27,100$ in/min

22. $2a = 8.0;\ a = 4.0;\ 2b = 4.0;\ b = 2.0;\ \dfrac{x^2}{16} + \dfrac{y^2}{4} = 1$

$v_x = 2.5$ cm/s; $x = -1.5$ cm, $y = 1.854$ cm

$\dfrac{1}{16} 2x v_x + \dfrac{1}{4} 2y v_y = 0$

$x v_x + 4 y v_y = 0$

$v_y = \dfrac{-x v_x}{4y};\ v_y|_{x=-1.5} = \dfrac{-(-1.5)(2.5)}{4(1.854)} = 0.506$

$v = \sqrt{(2.5)^2 + (0.506)^2} = 2.6$ cm/s

$\theta = \tan^{-1} \dfrac{0.506}{2.5} = 11.4°$

Therefore, v is 2.6 cm/s at $\theta = 11°$ (clockwise motion).

23. $h = k\sqrt{x} = kx^{1/2}$

$k = \dfrac{h}{\sqrt{x}} = \dfrac{280}{\sqrt{400}};\ v_x = 350$ m/s

$\dfrac{dh}{dt} = k \cdot \dfrac{1}{2} x^{-1/2} v_x$

$\dfrac{dh}{dt} = \dfrac{k v_x}{2\sqrt{x}};\ \dfrac{dh}{dt}\Big|_{x=400} = \dfrac{\frac{280}{\sqrt{400}}(350)}{2\sqrt{400}} = 122.5$ m/s

$v = \sqrt{350^2 + 122.5^2}$
$v = 370.8$ m/s

$\theta = \tan^{-1} \left(\dfrac{122.5}{350} \right) = 19.3°$

Therefore, v is 370 m/s at $\theta = 19°$.

24. $v = kd^{-1/2}$

$a = \dfrac{dv}{dt} = -\dfrac{1}{2} kd^{-3/2} \dfrac{dd}{dt}$

$a = -\dfrac{kv}{2d^{3/2}} = \dfrac{-k(kd^{-1/2})}{2d^{3/2}}$

$\quad = \dfrac{-k^2}{2d^2} = \dfrac{(-k^2/2)}{d^2}$

Acceleration is inversely proportional to the square of the distance to the center of earth.

4.4   Related Rates

1. $R = 4.000 + 0.003 T^2;\ \dfrac{dT}{dt} = 0.100°C/s$

$\dfrac{dR}{dt} = 0 + 0.006 T \dfrac{dT}{dt}$

$\dfrac{dR}{dt}\Big|_{T=150°C} = 0 + 0.006(150)(0.1)$
$\qquad = 0.0900\ \Omega/s$

2. $K = \dfrac{1}{2} mv^2,\ m = 250$ kg, $a = 5.00$ m/s^2

$\dfrac{dK}{dt} = \dfrac{1}{2} m \left(2v \dfrac{dv}{dt} \right) = mva$

$\dfrac{dK}{dt}\Big|_{v=30.0\ m/s} = (250\ \text{kg})(30.0\ \text{m/s})(5.00\ \text{m/s}^2)$
$\qquad = 37,500$ J/s

3. $D = \sqrt{4.0 + x^2}$

$\dfrac{dD}{dt} = \dfrac{dD}{dx} \cdot \dfrac{dx}{dt} = \dfrac{x}{\sqrt{4.0 + x^2}} \cdot \dfrac{dx}{dt}\Big|_{\substack{x=6.2 \\ \frac{dx}{dt}=350}}$

$\dfrac{dD}{dt} = \dfrac{6.2}{\sqrt{4 + 6.2^2}}(350) = 330$ mi/h

4. $R_T = \dfrac{8R}{8+R}$; $\dfrac{dR_T}{dt} = \dfrac{(8+R)8\frac{dR}{dt} - 8R\left(\frac{dR}{dt}\right)}{(8+R)^2}$

$\dfrac{dR}{dt} = 0.30\ \Omega/\text{min}$; $R = 6.0\ \Omega$

$\dfrac{dR}{dt} = \dfrac{64\frac{dR}{dt}}{(8+R)^2}$; $\dfrac{dR}{dt}\Big|_{R=6.0} = 0.098\ \Omega/\text{min}$

5. $r = \sqrt{0.4\lambda}$; $\dfrac{d\lambda}{dt} = 0.10 \times 10^{-7}$; $r = (0.4\lambda)^{1/2}$

$\dfrac{dr}{dt} = \dfrac{1}{2}(0.4\lambda)^{-1/2}(0.4)\dfrac{d\lambda}{dt}$

$\qquad = 0.2(0.4\lambda)^{-1/2}\dfrac{d\lambda}{dt}$

$\dfrac{dr}{dt}\Big|_{\lambda=6.0\times10^{-7}} = 0.2[0.4(6.0\times10^{-7})]^{-1/2}(0.10\times10^{-7})$

$\qquad = \dfrac{2\times10^{-9}}{\sqrt{24\times10^{-4}}} = 4.1\times10^{-6}\ \text{m/s}$

6. $\dfrac{x^2}{28.0} + \dfrac{y^2}{27.6} = 1$; $\dfrac{dx}{dt} = 7750\ \text{mi/h}$

$x = 2020\ \text{mi}$, $y > 0$, $y = 4856$

$\left(\dfrac{1}{28.0}\right)2x\dfrac{dx}{dt} + \left(\dfrac{1}{27.6}\right)2y\dfrac{dy}{dt} = 0$

$\dfrac{2}{27.6}y\dfrac{dy}{dt} = -\dfrac{2}{28.0}x\dfrac{dx}{dt}$

$\dfrac{dy}{dt} = -\dfrac{x}{y}\cdot\dfrac{27.6}{28.0}\cdot\dfrac{dx}{dt} = -\dfrac{2020}{4856}\cdot\dfrac{27.6}{28.0}(7750)$

$\qquad = -3180\ \text{mi/h}$

7. $B = \dfrac{k}{\left[r^2+\left(\frac{\ell}{2}\right)^2\right]^{3/2}} = k\left[r^2+\left(\dfrac{\ell}{2}\right)^2\right]^{-3/2}$

$\dfrac{dB}{dt} = -\dfrac{3}{2}k\left[r^2+\left(\dfrac{\ell}{2}\right)^2\right]^{-5/2}\left(2r\dfrac{dr}{dt}\right)$

$\dfrac{dB}{dt} = \dfrac{-3kr\frac{dr}{dt}}{\left[r^2+\left(\frac{\ell}{2}\right)^2\right]^{5/2}}$

8. $pv^{1.4} = k$; $p = 4200\ \text{kPa}$; $v = 75\ \text{cm}^3$;

$\dfrac{dv}{dt} = +850\ \text{cm}^3/s$

$k = 4200(75^{1.4}) = 1.77\times10^6$

$p\dfrac{d}{dt}v^{1.4} + v^{1/4}\dfrac{dp}{dt} = 0$; $p1.4v^{0.4}\dfrac{dv}{dt} + v^{1/4}\dfrac{dp}{dt} = 0$

$v^{1.4}\dfrac{dp}{dt} = -1.4pv^{0.4}\dfrac{dv}{dt}$

$\dfrac{dp}{dt} = \dfrac{-1.4pv^{0.4}}{v^{1.4}}\dfrac{dv}{dt} = \dfrac{-1.4p\frac{dv}{dt}}{v}$

$\dfrac{dp}{dt}\Big|_{v=75} = \dfrac{-1.4(4200)(850)}{75} = -67{,}000\ \text{kPa/s}$

9. $A = \pi r^2$; $\dfrac{dr}{dt} = 0.020\ \text{mm/mo}$; $\dfrac{dA}{dt} = 2\pi r\dfrac{dr}{dt}$

$\dfrac{dA}{dt}\Big|_{r=1.2} = 2\pi(1.2)(0.020) = 0.15\ \text{mm}^2/\text{month}$

10. $A = x^2$; $\dfrac{dx}{dt} = 0.25$

$\dfrac{dA}{dt} = 2x\dfrac{dx}{dt}$; $\dfrac{dA}{dt}\Big|_{x=6.50} = 2(6.50)(0.25)$

$\qquad = 3.25\ \text{in}^2/\text{s}$

11. $V = x^3$; $\dfrac{dx}{dt} = -0.50\ \text{mm/min}$

$\dfrac{dV}{dt} = 3x^2\dfrac{dx}{dt}$; $\dfrac{dV}{dt}\Big|_{x=8.20} = 3(8.20)^2(-0.50)$

$\qquad = -101\ \text{mm}^3/\text{min}$

12. By similar triangles:

Let x be distance from door opening to start of door's shadow.

$\dfrac{9.50}{x+12} = \dfrac{y}{x}$

$y = \dfrac{9.50\ \mathsf{X}}{x+12}$

$y = 2.00\ \text{ft}$, $x = 3.20\ \text{ft}$

$\dfrac{dy}{dt} = \dfrac{(x+12)9.50\frac{dx}{dt} - 9.40x\frac{dx}{dt}}{(x+12)^2} = -1.50\ \text{ft/s}$

$\dfrac{dx}{dt} = \dfrac{(x+12)^2\frac{dy}{dt}}{114}$; $\dfrac{dx}{dt}\Big|_{x=3.20}$

$\qquad = \dfrac{(3.2+12)^2(-1.50)}{114}$

$\qquad = -3.04\ \text{ft/s}.$

The negative sign indicates x is decreasing in length; i.e. the shadow is moving toward the door.

13. $p = \dfrac{k}{v}$; $\dfrac{dv}{dt} = 20\ \text{cm}^3/\text{min}$; $v = 810\ \text{cm}^3$

$230 = \dfrac{k}{650}$; $k = 1.495\times10^5\ \text{kPa}\times\text{cm}^3$

$p = \dfrac{149\,500}{v} = 149\,500v^{-1}$; $\dfrac{dp}{dt} = -149\,500v^{-2}\dfrac{dv}{dt}$

$\dfrac{dp}{dt}\Big|_{v=810} = -149\,500(810)^{-2}(20)$

$\qquad = -4.6\ \text{kPa/min}$

14. $f \propto \dfrac{1}{\sqrt{C}}$; therefore $f = \dfrac{k}{\sqrt{C}}$

$f = 920$ kHz, $C = 3.5$ pF; $\dfrac{dC}{dt} = 0.3$ pF/s

$f = kC^{-1/2}$; $\dfrac{df}{dt} = -\dfrac{1}{2}kC^{-3/2}\dfrac{dC}{dt}$; $k = f\sqrt{C}$

$\dfrac{df}{dt} = \dfrac{-k\frac{dC}{dt}}{2C^{3/2}} = \dfrac{-f\sqrt{C}\frac{dC}{dt}}{2c^{3/2}} = \dfrac{-f\frac{dC}{dt}}{2C}$

$\left.\dfrac{df}{dt}\right|_{f=920} = \dfrac{-920(0.3)}{2(3.5)} = -39.4$ kHz/s

$\qquad\qquad = -39$ kHz/s

15. $V = \dfrac{4}{3}\pi r^3$; $\dfrac{dr}{dt} = 5.00$ mm/s; $r = 225$ mm

$\dfrac{dV}{dt} = 4\pi r^2 \dfrac{dr}{dt}$

$\left.\dfrac{dV}{dt}\right|_{r=225} = 4\pi(225)^2(5.00) = 3.18 \times 10^6$ mm³/s

16. $g = \dfrac{k}{d}$; $d = r + h$

$r = 3960$ mi

$h = 25,500$ mi $= 25,500(5280)$ ft

$g = 32.2$ ft/s²; $h = 0$

$\dfrac{dh}{dt} = -4500$ ft/s

$g = \dfrac{k}{3960 + h}$

$g = \dfrac{k}{3960(5280) + h}$

$-32.2 = \dfrac{k}{3960(5280)}$; $k = -6.73 \times 10^8$

$g = k[3960(5280) + h]^{-1}$

$\dfrac{dg}{dt} = -k[3960(5280) + h]^{-2}\dfrac{dh}{dt}$

$\qquad = \dfrac{-k\frac{dh}{dt}}{[3960(5280) + h]^2}$

$\left.\dfrac{dg}{dt}\right|_{h=25\,500} = \dfrac{-32.2(3960)5280(-4500)}{[3960(5280) + 25,500(5280)]^2}$

$\dfrac{dg}{dt} = 1.3 \times 10^{-4}$ ft/s³

17. $\dfrac{x}{h} = \dfrac{1.15}{3.6}$

$V = \dfrac{1}{3}\pi r^2 h = \dfrac{1}{3}\pi\left(\dfrac{1.15}{3.6}\right)^2 h^3$

$\dfrac{dV}{dt} = \dfrac{dV}{dh}\cdot\dfrac{dh}{dt}$

$0.50 = \pi\cdot\left(\dfrac{1.15}{3.6}\right)^2\cdot h^2\cdot\dfrac{dh}{dt}$

$0.50 = \pi\left(\dfrac{1.15}{3.6}\right)^2(1.8)^2\cdot\dfrac{dh}{dt}$

$\dfrac{dh}{dt} = 0.48$ m/min

18. $x^2 + y^2 = 100$; $\dfrac{dy}{dt} = -10.0$ ft/s; $x = 6.00$ ft, $y = 8.00$ ft

$2x\dfrac{dx}{dt} + 2y\dfrac{dy}{dt} = 0$

$\dfrac{dx}{dt} = \dfrac{-y}{x}\dfrac{dy}{dt}$

$\left.\dfrac{dx}{dt}\right|_{x=6.00} = \dfrac{-8.00}{6.00}(-10.0)$

$\dfrac{dx}{dt} = 13.3$ ft/s

19. $z^2 = 20^2 + x^2$; $\dfrac{dz}{dt} = -10.0$ ft/s; $z = 36.0$ ft, $x = 29.9$ ft

$2z\dfrac{dz}{dt} = 2x\dfrac{dx}{dt}$

$\dfrac{dx}{dt} = \dfrac{z}{x}\dfrac{dz}{dt}$

$\left.\dfrac{dx}{dt}\right|_{z=36.0} = \dfrac{36.0}{29.9}(-10.0)$

$\dfrac{dx}{dt} = -12.0$ ft/s

The negative sign indicates the boat is approaching the wharf.

20. $z^2 = h^2 + 350^2$; $\dfrac{dh}{dt} = 12$ ft/s; $h = 250$ ft, $z = 430$ ft

$2z\dfrac{dz}{dt} = 2h\dfrac{dh}{dt}$;

$\dfrac{dz}{dt} = \dfrac{h}{z}\dfrac{dh}{dt}$

$\left.\dfrac{dz}{dt}\right|_{h=250} = \dfrac{250}{430}(12) = 6.97$ ft/s

$\dfrac{dz}{dt} = 7.0$ ft/s

21. Let x be the distance traveled by the jet going due east, and y be the distance traveled by the jet going north of east.

Since the second jet remains due north of the first jet, we have a right triangle and can use the Pythagorean theorem. $x^2 + z^2 = y^2$

Taking the derivative of this expression,

$$2x\frac{dx}{dt} + 2z\frac{dz}{dt} = 2y\frac{dy}{dt}$$

$$x\big|_{t=(1/2)} = 1600\left(\frac{1}{2}\right) = 800 \text{ mi};$$

$$y\big|_{t=(1/2)} = 1800\left(\frac{1}{2}\right) = 900 \text{ mi}$$

$$z = \sqrt{y^2 - x^2} = \sqrt{900^2 - 800^2} = 412.3 \text{ mi}$$

$$\frac{dx}{dt} = 1600; \; \frac{dy}{dt} = 1800 \text{ mi/h}$$

Substituting,

$$2(800)(1600) + 2(412.3)\frac{dz}{dt} = 2(900)(1800)$$

$$\frac{dz}{dt} = 820 \text{ mi/h}$$

22. Car: $v = 15.0 \text{ m/s} = \dfrac{dx}{dt}$

Boat: $v = 4.0 \text{ m/s} = \dfrac{dz}{dt}$

5 seconds: car travels 75.0 m $= x$
boat travels 20.0 m $= z$

$$y^2 + x^2 + z^2$$
$$d^2 = y^2 + (10.5)^2 = x^2 + z^2 + (10.5)^2$$
$$x = 75.0 \text{ m}, z = 20.0 \text{ m}, d = 78.3 \text{ m}$$

$$2d\frac{dd}{dt} = 2x\frac{dx}{dt} + 2z\frac{dz}{dt} + 0$$

$$\frac{dd}{dt} = \frac{x\frac{dx}{dt} + z\frac{dz}{dt}}{d}$$

$$\frac{dd}{dt}\bigg|_{x=75.0} = \frac{(75.0)(15.0) + (20.0)(4.0)}{78.3} = 15.4 \text{ m/s}$$

23. By similar triangles;

$$\frac{6}{d-x} = \frac{15}{d}; \; \frac{dx}{dt} = -500 \text{ ft/s}; \; x = 10.0 \text{ ft}$$

$$6d = 15(d - x)$$
$$6d = 15d - 15x$$
$$9d = 15x$$

$$9\frac{dd}{dt} = 15\frac{dx}{dt}$$

$$\frac{dd}{dt} = \frac{15}{9}\frac{dx}{dt}; \; \frac{dd}{dt}\bigg|_{x=10.0} = -8.33 \text{ ft/s}$$

The negative sign means the end of man's shadow is approaching the light post.

24. *Law of cosines*

$$a^2 = b^2 + c^2 - 2bc \; \cos A; \; \frac{dx}{dt} = 1.50 \text{ cm/s};$$

$$x = 10.0 \text{ cm}$$

$$50^2 = x^2 + y^2 - 2xy \cos 120°$$
$$50^2 = x^2 + y^2 + xy$$

$$0 = 2x\frac{dx}{dt} + 2y\frac{dy}{dt} + \left(x\frac{dy}{dt} + y\frac{dx}{dt}\right)$$

$$2y\frac{dy}{dt} + x\frac{dy}{dt} = -y\frac{dx}{dt} - 2x\frac{dx}{dt}$$

$$x = 10.0 \text{ cm}, y = 44.24 \text{ cm}$$

$$\frac{dy}{dt} = \frac{-(y + 2x)\frac{dx}{dt}}{(2y + x)}$$

$$\frac{dy}{dt}\bigg|_{x=10.0} = -\frac{(44.24 + 20.0)(1.50)}{2(44.24) + 10}$$

$$= -0.978 \text{ cm/s}$$

4.5   Using Derivatives in Curve Sketching

1. $y = x^2 + 2x; \; y' = 2x + 2; \; 2x + 2 > 0$
$2x > -2; \; x > -1; \; f(x)$ increases.
$2x + 2 < 0; \; 2x < -2; \; x < -1; \; f(x)$ decreases.

2. $y = 2 + 6x - 3x^2$
$y' = 6 - 6x$
$6 - 6x > 0$
$-6x > -6$
for $x < 1, \; f(x)$ increases
$6 - 6x < 0$
$-6x < -6$
for $x > 1, \; f(x)$ decreases

3. $y = 12x - x^3; \; \dfrac{dy}{dx} = 12 - 3x^2; \; 12 - 3x^2 = 0; \; x = \pm 2$

$\dfrac{dy}{dx} < 0$ for $x < -2$, therefore $f(x)$ decreases for $x < -2$

$\dfrac{dy}{dx} < 0$ for $x > 2$, therefore $f(x)$ decreases for $x > 2$

$\dfrac{dy}{dx} > 0$ for $-2 < x < 2$, therefore $f(x)$ increases for

$-2 < x < 2$

4. $y = x^4 - 6x^2$

$\dfrac{dy}{dx} = 4x^3 - 12x = 4x(x^2 - 3);\ x = 0, x = \pm\sqrt{3}$

$\dfrac{dy}{dx} < 0$ for $x < -\sqrt{3};\ \dfrac{dy}{dx} < 0$ for $0 < x < \sqrt{3}$

$\dfrac{dy}{dx} > 0$ for $-\sqrt{3} < x < 0;\ \dfrac{dy}{dx} > 0$ for $x > \sqrt{3}$

Therefore, $f(x)$ increasing for $-\sqrt{3} < x < 0$ and for

$x > \sqrt{3}$

$f(x)$ decreasing for $0 < x < \sqrt{3}$ and for $x < -\sqrt{3}$

5. $y = x^2 + 2x;\ y' = 2x + 2;\ y' = 0$ at $x = 1$

$y'' = 2 > 0$ at $x = -1$ and $(-1, -1)$

is a relative minimum.

6. $y = 2 + 6x - 3x^2$

$y' = 6 - 6x$

$6 - 6x = 0$

$\quad x = 1$

$y'' = -6x$

$y''(1) = -6(1) = -6 < 0$

$f(x)$ has max @ $(1, 5)$

7. $y = 12 - x^3;\ f'(x) = 12 - 3x^2 = 0;\ x = -2;\ x = 2$

$f''(x) = -6x;\ f''(-2) > 0$, concave up

$\qquad\qquad\qquad f''(2) < 0$, concave down

$f(-2) = -16$, therefore $(-2, -16)$ rel. min.

$f(2) = 16$, therefore $(2, 16)$ rel. max.

8. $y = x^4 - 6x^2;\ f(x) = 4x^3 - 12x = 4x(x^2 - 3) = 0;$

therefore $x = 0, x = \pm\sqrt{3}$

$f''(x) = 12x^2 - 12x;$

$f''(0) = 0;\ f'(-1) < 0$, concave down; $f(0) = 0$

$f''(-\sqrt{3}) > 0$; concave up, therefore, rel. min.

$f(-\sqrt{3}) = -9$

$f''(\sqrt{3}) > 0$ concave up, therefore rel. min.

$f(\sqrt{3}) = -9$

Therefore, $(0, 0)$ rel. max, $(-\sqrt{3}, -9)$ rel. min.,

$(\sqrt{3}, -9)$ rel min.

9. $y = x^2 + 2x;\ y' = 2x + 2;\ y'' = 2$

Thus, $y'' > 0$ for all x. The graph is concave

up for all x and has no points of inflection.

10. $\quad y = 2 + 6x - 3x^2$

$\quad\ y' = 6 - 6x$

$\quad\ y'' = -6 < 0$

$f(x)$ is concave down everywhere

$f(x)$ has no inflection points

11. $y = 12x - x^3;$

$f'(x) = 12 - 3x^2;$

$f''(x) = -6x = 0;\ x = 0$

$f''(x) < 0$ for all $x > 0;$

$f''(x) > 0$ for $x < 0$

Therefore, $f(x)$ is concave down for $x > 0;$

$f(0) = 0$

$f(x)$ is concave up for $x < 0$

$(0, 0)$ is infl. point

12. $y = x^4 - 6x^2;\ f'(x) = 4x^3 - 12x;$

$f''(x) = 12x^2 - 12 = 0$

$12(x^2 - 1) = 0;\ x = \pm 1$

$x < -1,\ f''(x) > 0,$

concave up, $f(-1) = -5$

$-1 < x < 1,\ f''(x) < 0,$

concave down, $f(1) = -5$

$x > 1,\ f''(x) > 0$, concave up

$(-1, -5)$ infl. point, $(1, -5)$ infl. point

13. $y = x^2 + 2x$

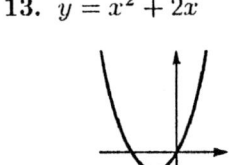

14. $y = 2 + 6x - 3x^2$

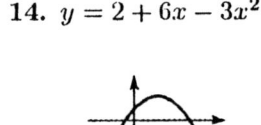

15. $y = 12x - x^3$

16. $y = x^4 - 6x^2$

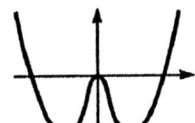

17. $y = 12x - 2x^2$; $y' = 12 - 4x$
$y' = 0$ at $x = 3$; for $x = 3$,
$y = 12(3) - 2(3)^2 = 18$
and $(3, 18)$ is a critical point. $12 - 4x < 0$ for $x > 3$
and the function decreases; $12 - 4x > 0$ for $x < 3$
and the function increases; $y'' = -4$; thus $y'' < 0$
for all x. There are no inflections; the graph is
concave down for all x, and $(3, 18)$ is a maximum
point.

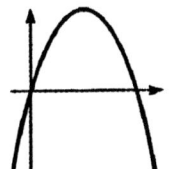

18. $f(x) = 3x^2 - 1$; $f'(x) = 6x = 0$; $x = 0$; $f(0) = -1$
$f''(x) = 6 > 0$, therefore $f(x)$ is concave up for all
x.
Therefore, $(0, -1)$ rel. minimum.
$f(x) = 0$; $3x^2 - 1 = 0$

$$x = \pm\sqrt{\frac{1}{3}}$$

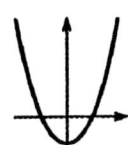

19. $f(x) = 2x^3 + 6x^2$; $f'(x) = 6x^2 + 12x$;
$f''(x) = 12x + 12$
$f'(x) = 0$; $6x(x + 2) = 0$
$x = 0, x = -2$
$f''(0) > 0$, concave up
$f''(-2) < 0$, concave down
$f''(x) = 0, x = -1$ infl. point
$f(0) = 0, (0, 0)$ min. point

$f(-2) = 8, (-2, 8)$ max. point
$f(-1) = 4, (-1, 4)$ infl. point

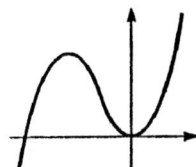

20. $f(x) = x^3 - 9x^2 + 15x + 1$
$f'(x) = 3x^2 - 18x + 15 = 0$; $3(x - 5)(x - 1) = 0$;
$x = 5, x = 1$
$f''(x) = 6x - 18 = 0$; $x = 3$; $f(3) = -8$
$f''(1) < 0$, concave down, $f(1) = 8$
$f''(5) > 0$, concave up, $f(5) = -24$
$(1, 8)$ rel. max.
$(5, -24)$ rel. min.
$(3, -8)$ infl. point
$f(0) = 1$
$(0, 1)$ y-intercept

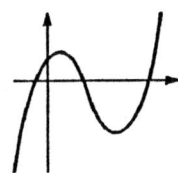

21. $y = x^3 + 3x^2 + 3x + 2$
$y' = 3x^2 + 6x = 3(x^2 + 2x + 1)$
$\quad = 3(x + 1)(x + 1)$
$3(x + 1)(x + 1) = 0$ for $x = -1$
$(-1, 1)$ is a critical point.
$3(x + 1)(x + 1) > 0$ for $x < -1$ and the slope is
positive.
$3(x + 1)(x + 1) > 0$ for $x > -1$ and the slope is
positive.
$y'' = 6x + 6$; $6x + 6 = 0$ for $x = 1$, and $(-1, 1)$ is
an inflection point.
$6x + 6 < 0$ for $x < -1$ and the graph is concave
down.
$6x + 6 > 0$ for $x > -1$ and the graph is concave
up.

Since there is no change in slope from positive to
negative or vice versa, there are no maximum or
minimum points.

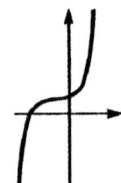

22. $f(x) = x^3 - 12x + 12$
$f'(x) = 3x^2 - 12 = 0;\ 3(x^2 - 4) = 0;\ x = \pm 2$
$f''(x) = 6x = 0;\ x = 0;\ f(0) = 12$
$f''(-2) < 0$, concave down, $f(-2) = 28$
$f''(2) > 0$, concave up, $f(2) = -4$
$(0, 12)$ infl. point
$(-2, 28)$ rel. max.
$(2, -4)$ rel. min.

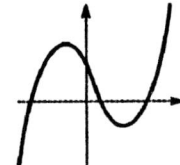

23. $f(x) = 4x^3 - 24x^2 + 36x$
$f'(x) = 12x^2 - 48x + 36 = 0;\ 12(x^3 - 4x + 3) = 0;$
$f'(x) = 12(x-3)(x-1) = 0$; therefore $x = 1, x = 3$
$f''(x) = 24x - 48;$
$x < 1, f(x) > 0$, therefore $f(x)$ increasing
$x > 2, f(x) > 0$, therefore $f(x)$ increasing
$1 < x < 3, f(x) < 0$, therefore $f(x)$ decreasing
$f''(1) < 0$, concave down, $f(1) = 16$
$f''(3) > 0$, concave up, $f(3) = 0$
$f''(x) = 0;\ 24(x - 2) = 0;\ x = 2;\ f(2) = 8$
$(2, 8)$ infl. point
$(1, 16)$ max. point
$(3, 0)$ min. point
$f(0) = 0$
$(0, 0)$ origin

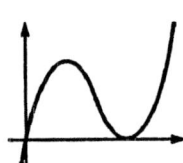

24. $y = x(x - 4)^3 = x(x^3 - 12x^2 + 48x - 64)$
$\quad = x^4 - 12x^3 + 48x^2 - 64x$
$y' = 4x^3 - 36x^2 + 96x - 64$
$y'' = 12x^2 - 72x + 96$

$y' = 0$ @ $x = 1$ and $x = 4$, critical points

$y''(1) = 36 > 0 \Rightarrow$ relative min @ $(1, -27)$
$y''(4) = 0 \Rightarrow$ first derivative test inconclusive
$y' < 0$ for $x < 1$
$y' > 0$ for $x > 1 \Rightarrow (4, 0)$ is HTL but neither
max nor min
$y'' = 0$ for $x = 2$ and $x = 4$, possible
inflection points
$y'' > 0$ for $x < 2$ and $x > 1$
$y'' < 0$ for $2 < x < 4 \Rightarrow (2, -16)$ and $(4, 0)$
are inflection points

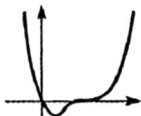

25. $y = 4x^3 - 3x^4;$
$y' = 12x^2 - 12x^3 = 12x^2(1 - x) = 0$
$12x^2(1 - x) = 0$ for $x = 0$ and $x = 1$
$(0, 0)$ and $(1, 1)$ are critical points.
$12x^2 - 12x^3 > 0$ for $x < 0$ and the slope is positive.
$12x - 12x^3 > 0$ for $0 < x < 1$ and the slope is
positive.
$12x - 12x^3 < 0$ for $x > 1$ and the slope is negative.
$y'' = 24x - 36x^2;\ 24x - 36x^2 = 12x(2 - 3x) = 0$
for $x = 0,\ x = \frac{2}{3}$
$(0, 0)$ and $\left(\frac{2}{3}, \frac{16}{27}\right)$ are possible inflection points.
$24x - 36x^2 < 0$ for $x < 0$ and the graph is concave
down.
$24x - 36x^2 > 0$ for $0 < x < \frac{2}{3}$ and the graph is
concave up.
$24x - 36x^2 < 0$ for $x > \frac{2}{3}$ and the graph is concave
down.
$(1, 1)$ is a relative maximum point since $y' = 0$
at $(1, 1)$ and the slope is positive for $x < 1$ and
negative for $x > 1$. $(0, 0)$ and $\left(\frac{2}{3}, \frac{16}{27}\right)$ are inflection
points since there is a concavity change.

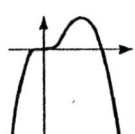

26. $f(x) = x^5 - 20x^2$; $f'(x) = 5x^4 - 40x$;
$f''(x) = 20x^3 - 40$
$f'(x) = 0$; $5x(x^3 - 8) = 0$; $x = 0, x = 2$
$f''(0) < 0$, concave down, $f(0) = 0$
$f''(2) > 0$, concave up, $f(2) = -48$
$f''(x) = 0$; $20(x^3 - 2) = 0$; $x = \sqrt[3]{2}$;
$f(\sqrt[3]{2}) = -18\sqrt[3]{4}$
$(\sqrt[3]{2}, -18\sqrt[3]{4})$ infl. point
$(0, 0)$ max. point
$(2, -48)$ min. point

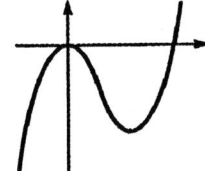

27. $f(x) = x^5 - 5x$; $f'(x) = 5x^4 - 5$; $f''(x) = 20x^3$
$f'(x) = 0$; $5(x^4 - 1) = 0$; $x = \pm 1$
$f''(-1) < 0$, concave down, $f(-1) = 4$
$f''(1) > 0$, concave up $f(1) = -4$
$f''(x) = 0$; $2x^3 = 0$; $x = 0$; $f(0) = 0$
$(0, 0)$ infl. point
$(-1, 4)$ max. point
$(1, -4)$ min. point

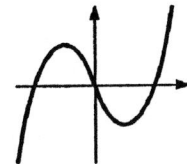

28. $f(x) = x^4 + 8x + 2$; $f'(x) = 4x^3 + 8$; $f''(x) = 12x^2$
$f'(x) = 0$; $4x^3 + 8 = 0$; $x^3 = -2$; $x = \sqrt[3]{-2} = -\sqrt[3]{2}$
$f''(-\sqrt[3]{2}) > 0$, concave up, $f(-\sqrt[3]{2}) = 2 - 6\sqrt[3]{2}$
$f''(x) = 0$; $12x^2 = 0$; $x = 0$
$f''(x) > 0$ for all x. $f(x)$ is concave up for all x.
Rel. minimum
$(-\sqrt[3]{2}, 2 - 6\sqrt[3]{2})$
$f(0) = 2$
$(0, 2)$ y-intercept

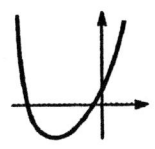

29. $y = x^3 - 12x$; $y' = 3x^2 - 12$; $y'' = 6x$. On graphing calculator with $x_{min} = -5$, $x_{max} = 5$, $y_{min} = -20$, $y_{max} = 20$, enter $y_1 = x^3 - 12x$; $y_2 = 3x^2 - 12$; $y_3 = 6x$. From the graph is observed that the maximum and minimum values of y occur when y' is zero. A maximum value for y occurs when $x = -2$, and a minimum value occurs when $x = 2$. An inflection point (change in curvature) occurs when y'' is zero. x is also zero at this point. Where $y' > 0$, y inc.; $y' < 0$, y dec. $y'' > 0$, y conc. up; $y'' < 0$, y conc. down, $y'' = 0$, y has infl.

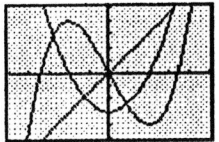

30. $y = 24x - 9x^2 - 2x^3$; $y' = 24 - 18x - 6x^2$; $y'' = -18 - 12x$. On graphing calculator with $x_{min} = -8$, $x_{max} = 4$, $y_{min} = -120$, $y_{max} = 60$, enter $y_1 = 24x - 9x^2 - 2x^3$; $y_2 = 24 - 18x - 6x^2$; $y_3 = -18 - 12x$. From graph, it is observed that y_1 has a maximum and a minimum when $y_2 = 0$. Also, y_1 has a point of inflection when $y_3 = 0$. y_1 increasing when $y_2 > 0$, and y_1 is decreasing when $y_2 < 0$. y_1 is concave up when $y_3 > 0$, and y_1 is concave down when $y_3 < 0$.

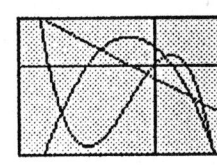

Where $y' > 0$, y inc.; $y' = 0$ y has maximum or minimum $y' < 0$, y dec. $y'' > 0$, y conc. up $y'' = 0$, y has infl. $y'' < 0$, y conc. down

31. $y = x - 0.0025x^2$
$f'(x) = 1 - 0.0050x$
$f''(x) = -0.0050$
$f'(x) = 0 - 1 - 0.0050x$
$x = 200$
$f''(200) < 0$, concave down
$f''(x) < 0$ for all x
$f'(x)$ concave down for all x.
$(200, 100)$ max.
$f(0) = 0$, $(0, 0)$ origin

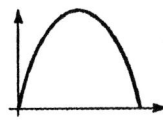

32. $\theta = 10 + 12t^2 - 2t^3; \; 0 \le t \le 6$ s

$f'(t) = 24t - 6t^2$

$f''(t) = 24 - 12t$

$f'(t) = 0 = 6t(4 - t)$

$t = 0, t = 4$

$f''(0) > 0$, concave up

$f''(4) < 0$, concave down

$f''(x) = 0 = 24 - 12t$

$t = 2$, infl. point

$f(0) = 10, (0, 10)$ min. point

$f(4) = 74, (4, 74)$ max. point

$f(2) = 42, (2, 42)$ infl. point

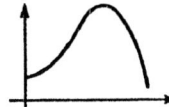

33. $R = 75 - 18i^2 + 8i^3 - i^4$

$R' = -36i + 24i^2 - 4i^3 = -4i(i^2 - 6i + 9)$

$\quad = -4i(i - 3)^2$

$R' = 0$ for $i = 0$ and $i = 3$

$(0, 75)$ and $(3, 48)$ are critical points.

$R' > 0$ for $i < 0$, $R' < 0$ for $0 < i < 3$

$R' < 0$ for $i > 3$

Max. at $(0, 75)$, no max. or min. at $(3, 48)$

$R'' = -36 + 48i - 12i^2 = -12(i - 1)(i - 3)$

$(1, 64)$ and $(3, 48)$ are possible inflection points.

$R'' < 0$ for $i < 1$, concave down

$R'' > 0$ for $1 < i < 3$, concave up

$R'' < 0$ for $i > 3$, concave down

$(1, 64)$ and $(3, 48)$ are inflection points.

(From calculator graph, $R = 0$ for $i = -1.5$ and $i = 5.0$)

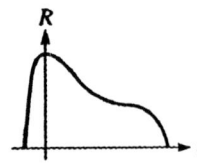

34. $d = 0.00181x^3 - 0.289x^2 + 12.2x + 30.4,$

    $20 < x < 80$

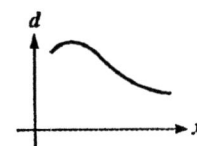

35. $V = x(8 - 2x)(12 - 2x)$

$V = 4(24x - 10x^2 + x^3)$

$\quad = 4x^3 - 40x^2 + 96x$

$f'(x) = 4(24 - 20x + 3x^2) = 0$

By quadratic solution:

$x = 1.57$ and $x = 5.10$ (reject)

$f''(x) = 4(-20 + 6x)$

$f''(1.57) < 0$, rel. max. $(1.57, 67.6)$

$f''(5.10) > 0$, rel. min. $(5.10, -20.2)$

$f''(x) = 0; \; x = \dfrac{20}{6}$, infl. $\left(\dfrac{10}{3}, 23.7 \right)$

$(0, 0)$ origin

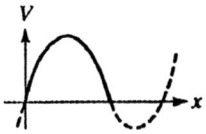

36. $A_T = 2$ ends $+ \, 2$ sides $+$ bottom

$A = 64$ ft^2

$64 = 2(x^2) + 2(xy) + xy$

$64 = 2x^2 + 3xy$

Therefore, $y = \dfrac{64 - 2x^2}{3x}$

$V = x^2 y = x^2 \left(\dfrac{64 - 2x^2}{3x} \right)$

$V = \dfrac{1}{3} x(64 - 2x^2)$

$\quad = \dfrac{2}{3}(32x - x^3)$

$\quad = \dfrac{64}{3}x - \dfrac{2}{3}x^3$

$f'(x) = \dfrac{2}{3}(32 - 3x^2) = 0$

$x = 3.27$ and -3.27

$f''(x) = \dfrac{2}{3}(-6x)$

$f''(3.27) < 0$, max. $(3.27, 46.4)$

$f''(-3.27) > 0$, min. $(-3.27, -46.4)$

$$f''(x) = 0 = \frac{2}{3}(-6x) = 0$$

$x = 0, (0,0)$ infl. point

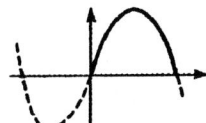

37. $f(1) = 0$; therefore $(1,0)$ is an x-intercept
$f'(x) > 0$ for all x; therefore curve rises left to right
$f''(x) < 0$ for all x; therefore concave down

38. $f(0) = 1$, $f(x)$ has y-intercept of $(0,1)$
$f'(x) < 0$ for all x.
Therefore, $f(x)$ is decreasing for all x.
$f''(x) < 0$ for $x < 0$
Therefore, $f(x)$ is concave down for $x < 0$.
$f''(x) > 0$ for $x > 0$
Therefore, $f(x)$ is concave up for $x > 0$.
Inflection point at $(0,1)$

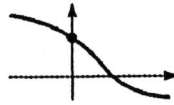

39. $f(-1) = 0$, root at $(-1,0)$
$f(2) = 2$, point on curve $(2,2)$
$f'(x) < 0$ for $x < -1$
Therefore, $f(x)$ decreasing for $x < -1$.
$f'(x) > 0$ for $x > -1$
Therefore, $f(x)$ increasing for $x > -1$.
$f''(x) < 0$ for $0 < x < 2$
Therefore, $f(x)$ concave down for $0 < x < 2$.
$f''(x) > 0$ for $x < 0$ or $x > 2$
Therefore, $f(x)$ concave up for $x < 0$.
Therefore, $f(x)$ concave up for $x > 2$.
Summary: $(-1,0)$ min. point
Inflection point at y-intercept
$(2,2)$ inflection point

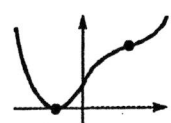

40. $f(x) = x^{2/3}, -2 < x < 2$

$$f'(x) = \frac{2}{3}x^{-1/3}$$
$$= \frac{2}{3x^{1/3}}$$
$$f''(x) = -\frac{2}{9}x^{-4/3}$$
$$= \frac{-2}{9x^{4/3}}$$

$f(x)$ is continuous everywhere
$f'(x)$ is continuous everywhere except $x = 0$
$f''(x)$ is continuous everywhere except $x = 0$

$f(x)$ is concave down everywhere, except $x = 0$
$f'(x)$ is concave down for $x < 0$ and concave up for $x > 0$
$f''(x)$ is concave down everywhere
$x = 0$

$f(x)$ has a minimum at $x = 0$ but $x = 0$ is not an inflection point
$f'(x)$ changes concavity on either side of $x = 0$ but $x = 0$ is not an inflection point since $f'(x)$ is not defined at $x = 0$
$f''(x)$ does not change concavity on either side of $x = 0$

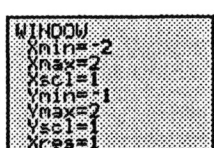

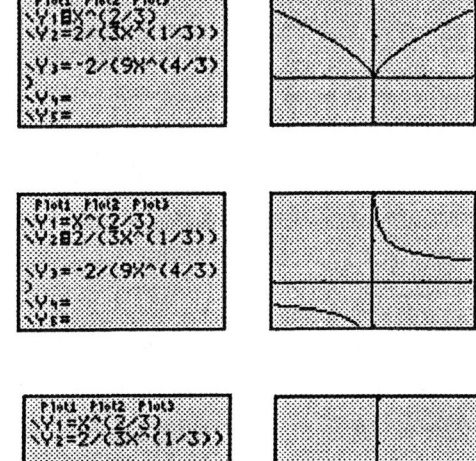

4.6 More on Curve Sketching

1. $y = \dfrac{4}{x^2}$

Intercepts:

(1) y is defined for $x = 0$, so the graph is not continuous at $x = 0$ and cannot cross the y-axis.

(2) Since 4 is positive and x^2 is positive for all non-zero x, $\dfrac{4}{x^2}$ is always positive, and the graph does not cross into quadrants III or IV.

(3) Since $\dfrac{4}{x^2} > 0$ for any x, $y \neq 0$ and the graph does not intersect the x-axis at any point.

Symmetry:

(4) Symmetrical about y-axis since replacing x with $-x$ produces no change. Not symmetrical about x-axis.

Behavior as x becomes large:

(5) As $x \to -\infty$, $\dfrac{4}{x^2}$ approaches 0, and the negative x-axis is an asymptote.

(6) As $x \to +\infty$, $\dfrac{4}{x^2}$ approaches 0, and the positive x-axis is an asymptote.

Derivatives:

(7) $y' = -8x^{-3} = \dfrac{-8}{x^3}$. For negative x, $\dfrac{-8}{x^3}$ is positive and the graph rises. For positive x, $\dfrac{-8}{x^3}$ is negative and the graph falls.

(8) $y'' = 24x^{-4} = \dfrac{24}{x^4}$. y'' is positive for all x. There are no inflection points.
Inc. $x < 0$, dec. $x > 0$
Concave up $x < 0$, $x > 0$
Asym. $x = 0$, $y = 0$

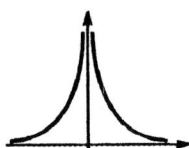

2. $y = \dfrac{2}{x^3}$

(1) Intercepts; $x = 0$, $f(x)$ is undefined; $y \neq 0$

(2) Symmetry: Substitute $-x$ for x:

$$y = \frac{2}{(-x)^3} = \frac{-2}{x^3} \neq \frac{2}{x^3}$$

Substitute $-y$ for y: $-y = \dfrac{2}{x^3}$, therefore
$y = \dfrac{-2}{x^3} \neq \dfrac{2}{x^3}$
Therefore, no symmetry WRT (with respect to) either axis.

Substitute $(-x, -y)$ for (x, y);

$$-y = \frac{2}{(-x)^3} = \frac{-2}{x^3}, \text{ therefore } y = \frac{2}{x^3}$$

Therefore, there is symmetry WRT the origin.

(3) Behavior as x becomes large: $\displaystyle\lim_{x \to -\infty} \frac{2}{x^3} = 0$, $f(x)$ approaches the y-axis from below.

$\displaystyle\lim_{x \to +\infty} \frac{2}{x^3} = 0$, $f(x)$ approaches the y-axis from above.

Therefore, x-axis is a horizontal asymptote.

(4) Vertical asymptotes: $f(x)$ is undefined for $x = 0$, therefore, y-axis is a vertical asymptote.

(5) Domain and range: Domain is all x except $x = 0$.
Range is $y < 0$ and $y > 0$.

(6) Derivatives: $f'(x) = \dfrac{-6}{x^4}$

$f''(x) = \dfrac{24}{x^5}$

$f'(x) < 0$ for all x
Therefore, $f(x)$ decreasing for all x, except $x = 0$
$f''(x) > 0$ for $x > 0, x \neq 0$
Therefore, $f(x)$ concave up for $x > 0$.
$f''(x) < 0$ for $x < 0$
Therefore, $f(x)$ concave down for $x < 0$.

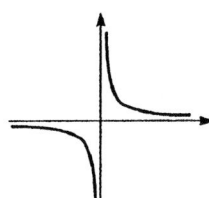

3. $f(x) = \dfrac{2}{x + 1}$

(1) Intercepts: $x = 0, y = 2, (0, 2); y \neq 0$

(2) Symmetry: No symmetry WRT either axis or origin.

(3) Behavior as x becomes large.

$\displaystyle\lim_{x \to -\infty} \frac{2}{x + 1} = 0$ from below.

$$\lim_{x \to +\infty} \frac{2}{x+1} = 0 \text{ from above.}$$

Therefore, x-axis is a horizontal asymptote.

(4) Vertical asymptotes: $f(-1) = \frac{2}{0}$, undefined, therefore $x = -1$ is a vertical asymptote.

(5) Domain and range: Domain is all x except $x = -1$.
Range is $y > 0$ and $y < 0$.

(6) Derivatives: $f'(x) = \dfrac{-2}{(x+1)^2} \neq 0$

$f''(x) = \dfrac{4}{(x+1)^3} \neq 0$

$f'(x) < 0$ for all x
Therefore, $f(x)$ decreasing for all $x, x \neq -1$.
$f''(x) < 0$ for $x < -1$
Therefore, $f(x)$ concave down for $x < -1$.
$f''(x) > 0$ for $x > -1$
Therefore, $f(x)$ concave up for $x > -1$.

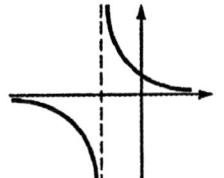

4. $f(x) = \dfrac{x}{x-2}$

(1) Intercepts: $x = 0, y = 0, (0,0)$ origin;
$y = 0, x = 0$

(2) Symmetry: No symmetry WRT either axis or origin.

(3) Behavior as x becomes large:

$$\lim_{x \to -\infty} \frac{x}{x-2} = \lim_{x \to -\infty} \frac{1}{1 - \frac{2}{x}} = 1$$

$$\lim_{x \to -\infty} \frac{1}{1 - \frac{2}{x}} = 1, \text{ therefore, } y = 1 \text{ is a horizontal}$$
asymptote.

(4) Vertical asymptotes: $f(2) = \frac{2}{0}$ undefined, therefore vertical asymptote at $x = 2$.

(5) Domain and range: Domain is all x except $x = 2$.
Range is all y except $y = 1$.

(6) Derivatives: $f'(x) = \dfrac{-2}{(x-2)^2} \neq 0$

$f''(x) = \dfrac{4}{(x-2)^3} \neq 0$

$f'(x) < 0$ for all $x, x \neq 2$
Therefore, $f(x)$ is decreasing all for x.
$f''(x) < 0$ for $x < 2$
Therefore, $f(x)$ concave down for $x < 2$.
$f''(x) > 0$ for $x > 2$
Therefore, $f(x)$ concave up for $x > 2$.

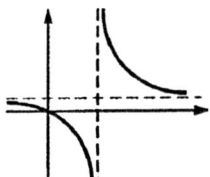

5. $y = x^2 + \dfrac{2}{x} = \dfrac{x^3 + 2}{x}$

(1) $\frac{2}{x}$ is undefined for $x = 0$, so the graph is not continuous at the y-axis; i.e., no y-intercept exists.

(2) $\frac{x^3+2}{x} = 0$ at $x = \sqrt[3]{-2} = -\sqrt[3]{2}$. There is an x-intercept at $(-\sqrt[3]{2}, 0)$.

(3) As $x \to \infty$, $x^2 \to \infty$ and $\frac{2}{x} \to 0$, so $x^2 + \frac{2}{x} \to \infty$.

(4) As $x \to 0$ through positive x, $x^2 \to 0$ and $\frac{2}{x} \to \infty$, so $x^2 + \frac{2}{x} \to \infty$.

(5) As $x \to -\infty$, $x^2 \to \infty$ and $\frac{2}{x} \to 0$ so $x^2 + \frac{2}{x} \to \infty$.

$x = 0$ is a vertical asymptote

(6) As $x \to 0$ through negative numbers, $x^2 \to 0$ and $\frac{2}{x} \to -\infty$, so $x^2 + \frac{2}{x} \to -\infty$.

(7) $y' = 2x - 2x^{-2} = 0$ at $x = 1$ and the slope is zero at $(1,3)$.

(8) $y'' = 2 + 4x^{-3} = 0$ at $x = -\sqrt[3]{2}$ and $(-\sqrt[3]{2}, 0)$ is an inflection point.

(9) $y'' > 0$ at $x = 1$, so the graph is concave up and $(1,3)$ is a relative minimum.

(10) Since $(-\sqrt[3]{2}, 0)$ is an inflection, $f''(-1) < 0$ and the graph is concave down. $f''(-2) > 0$ and the graph is concave up.

(11) Not symmetrical about the x- or y-axis.

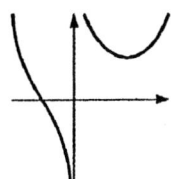

6. $f(x) = x + \dfrac{4}{x^2}$

(1) Intercepts: $x = 0, \frac{4}{0}$ undefined

$y = 0, x = -\sqrt[3]{4}, (-\sqrt[3]{4}, 0)$ a root

(2) Symmetry: No symmetry WRT either axis or origin.

(3) Behavior as x becomes large: $\displaystyle\lim_{x \to \pm\infty} x + \dfrac{4}{x^2}$;

for large values of x, $\dfrac{4}{x^2} \to 0$ and $f(x)$ is approximately equal to x.

Therefore, $\displaystyle\lim_{x \to -\infty} x + \dfrac{4}{x^2}$; $f(x)$ approaches x negatively.

$\displaystyle\lim_{x \to \infty} x + \dfrac{4}{x^2}$; $f(x)$ approaches x positively.

$y = x$ is an asymptote.

(4) Vertical asymptote: $f(0) = 0 + \frac{4}{0}$, undefined, therefore $x = 0$ is a vertical asymptote.

(5) Domain and range: Domain is all x except $x = 0$.
Range is all y.

(6) Derivatives: $f'(x) = 1 - \dfrac{8}{x^3}$

$f''(x) = \dfrac{24}{x^4} \neq 0$

$f'(x) = 0, x = 2$
$f''(2) > 0$, minimum $(2, 3)$
$f''(x) > 0$ for all x
Therefore, $f(x)$ is concave up for all x.
$f'(x) > 0$ for $x < 0$
Therefore, $f(x)$ increases for $x < 0$.
$f'(x) < 0$ for $0 < x > 2$
Therefore, $f(x)$ decreases for $0 < x < 2$.
$f'(x) > 0$ for $x > 2$
Therefore, $f(x)$ increases for $x > 2$.

7. $f(x) = x - \dfrac{1}{x}$

(1) Intercepts: $x = 0, \frac{1}{0}$ undefined
$y = 0, x = -1$ and $x = 1$; $(-1, 0)$ and $(1, 0)$ are roots.

(2) Symmetry: No symmetry WRT either axis or origin.

(3) Behavior as x becomes large: $\displaystyle\lim_{x \to -\infty} \left(x - \tfrac{1}{x}\right)$;

for large values of x. $\frac{1}{x} \to 0$ and $f(x)$ is approximately equal to x.

$\displaystyle\lim_{x \to -\infty} \left(x - \tfrac{1}{x}\right)$; $f(x)$ approaches x negatively.

$\displaystyle\lim_{x \to +\infty} \left(x - \tfrac{1}{x}\right)$; $f(x)$ approaches x positively.

Therefore, $y = x$ is an asymptote.

(4) Vertical asymptotes: $f(0) = 0 - \frac{1}{0}$ undefined,

therefore $x = 0$ is a vertical asymptote.

(5) Domain and range: Domain is all x except $x = 0$.
Range is all y.

(6) Derivatives: $f'(x) = 1 + \dfrac{1}{x^2}$

$f''(x) = -\dfrac{2}{x^3} \neq 0$

$f'(x) = 0, x = \sqrt{-1}$, not real
Therefore, $f'(x) \neq$ for real values.
$f''(x) > 0$ for $x < 0$
Therefore, $f(x)$ concave up for $x < 0$.
$f''(x) < 0$ for $x > 0$
Therefore, $f(x)$ concave down for $x > 0$.
$f'(x) > 0$ for all $x, x \neq 0$
Therefore, $f(x)$ increasing for all $x, x \neq 0$.

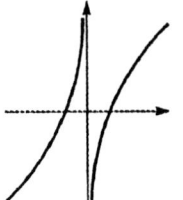

8. $f(x) = 3x + \dfrac{1}{x^3}$

(1) Intercepts: $x = 0$, $\dfrac{1}{x^3}$ undefined; $y = 0$, x not real values; therefore no intercepts.

(2) Symmetry: No symmetry WRT either axis,

however $-y = -3x - \dfrac{1}{x^3}$ is $y = 2x + \dfrac{1}{x^3}$.

Therefore, there is symmetry WRT the origin.

(3) Behavior as x becomes large: As $x \to \infty$, $f(x)$ is approximated by $3x$.

$\lim\limits_{x \to -\infty} \left(3x + \dfrac{1}{x^3}\right)$: $f(x)$ approaches $3x$ negatively.

$\lim\limits_{x \to \infty} \left(3x + \dfrac{1}{x^3}\right)$: $f(x)$ approaches $3x$ positively.

Therefore, $y = 3x$ is an asymptote.

(4) Vertical asymptotes: $f(0)$ is undefined, therefore $x = 0$ is a vertical asymptote.

(5) Domain and range: Domain is all x except $x = 0$.
Range: to be determined

(6) Derivatives: $f'(x) = 3 - \dfrac{3}{x^4}$

$f''(x) = \dfrac{12}{x^5} \neq 0$

$f'(x) = 0 = 3 - \dfrac{3}{x^4}$

$x = -1$ and $x = 1$
$f''(-1) < 0$, max. $(-1, -4)$
$f''(1) > 0$, min. $(1, 4)$
Therefore, range is all y except $-4 < y < 4$.
$f''(x) > 0$ for $x > 0$
Therefore, $f(x)$ concave up for $x > 0$.
$f''(x) < 0$ for $x < 0$
Therefore, $f(x)$ concave down for $x < 0$.

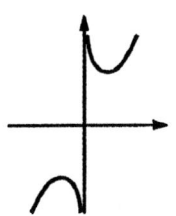

9. $y = \dfrac{x^2}{x+1}$

Intercepts:

(1) Function undefined at $x = -1$; not continuous at $x = -1$.

(2) At $x = 0$, $y = 0$. The origin is the only intercept. Behavior as x becomes large:

(3) As $x \to \infty$, $y \to x$, so $y = x$ is an asymptote. As $x \to \infty$, $y = -\infty$.

Vertical asymptotes:

(4) As $x \to -1$ from the left, $x + 1 \to 0$ through negative values and $\frac{x^2}{x+1} \to -\infty$ since $x^2 > 0$ for all x. As $x \to -1$ from the right, $x + 1 \to 0$ through positive values and $\frac{x^2}{x+1} \to +\infty$. $x = -1$ is an asymptote.

Symmetry:

(5) The graph is not symmetrical about the y-axis or the x-axis.

Derivatives:

(6) $y' = \frac{x^2+2x}{(x+1)^2}$; $y' = 0$ at $x = -2$, $x = 0$. $(-2, -4)$ and $(0, 0)$ are critical points. Checking the derivative at $x = -3$, the slope is positive, and at $x = -1.5$ the slope is negative. $(-2, -4)$ is a relative maximum point. Checking the derivative at $x = -0.5$, the slope is negative, and at $x = 1$ the slope is positive, so $(0, 0)$ is a relative minimum point.
Int. $(0, 0)$, max. $(-2, -4)$, min $(0, 0)$, asym. $x = -1$

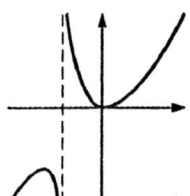

10. $f(x) = \dfrac{9x}{x^2 + 9}$

(1) Intercepts: $x = 0, y = 0, (0, 0)$ origin $y = 0$, $x = 0$

(2) Symmetry: No symmetry WRT either axis,

however $-f(x) = \dfrac{-9x}{(-x)^2 + 9}$ is $f(x)$, therefore symmetry WRT origin.

(3) Behavior as x becomes large:

$$\lim_{x \to -\infty} \frac{9x}{x^2+9} = \lim_{x \to -\infty} \frac{(9/x)}{1+\frac{9}{x^2}}$$

$$\lim_{x \to -\infty} = 0, \text{ negatively: } \lim_{x \to \infty} \frac{9x}{x^2+9} = 0, \text{ positively}$$

Therefore, x-axis is a horizontal asymptote.

(4) Vertical asymptotes: None; no values of x for which $f(x)$ is undefined.

(5) Domain and range: Domain is all x.
Range to be determined.

(6) Derivatives: $f'(x) = \dfrac{81-9x^2}{(x^2+9)^2}$

$$f''(x) = \frac{2x(9x^2-243)}{(x^2+9)^3}$$

$$f'(x) = 0 = \frac{81-9x^2}{(x^2+9)^2}; \ x=-3, x=3$$

$f''(-3) > 0$, min. $(-3, -3/2)$
$f''(3) < 0$, max. $(3, 3/2)$

$$f''(x) = 0 = \frac{2x(9x^2-243)}{(x^2+9)^3}$$

$$x = 0, x = \pm 3\sqrt{3}$$

Summary: $\left(-3, -\dfrac{3}{2}\right)$ min. point

$\left(3, \dfrac{3}{2}\right)$ max. point

Therefore, range is $-\dfrac{3}{2} \le y \le \dfrac{3}{2}$.

$(0,0)$ infl. point

$\left(-3\sqrt{3}, -\dfrac{3\sqrt{3}}{4}\right)$ inflf. point

$\left(3\sqrt{3}, \dfrac{3\sqrt{3}}{4}\right)$ infl. point

11. $f(x) = \dfrac{1}{x^2-1}$

(1) Intercepts: $x = 0, y = -1, (0,-1); y \ne 0$

(2) Symmetry: $f(-x) = f(x)$, therefore symmetry WRT y-axis. No symmetry WRT x-axis or origin.

(3) Behavior as x becomes large:

$$\lim_{x \to \pm\infty} \frac{1}{x^2-1} = 0, \text{ from above (positively)}.$$

$y = 0$ is a horizontal asymptote

(4) Vertical asymptotes: $f(-1)$ and $f(1)$ give $\frac{1}{0}$ undefined. Therefore, vertical asymptotes at $x = -1$ and $x = 1$.

(5) Domain and range: Domain is all x except $x = -1$ and $x = 1$. Range to be determined.

(6) Derivatives: $f'(x) = \dfrac{-2x}{(x^2-1)^2}$

$f''(x) = \dfrac{2(3x^2+1)}{(x^2-1)^3} \ne 0$, no real solutions
$f'(x) = 0; \ -2x = 0; \ x = 0$
$f''(0) < 0$, max. $(0,-1)$

Therefore, range is $-1 \le y < 0$.
$f''(x) < 0$ for $-1 < x < 1$
Therefore, $f(x)$ concave down for $-1 < x < 1$.
$f''(x) > 0$ for $x < -1$ and for $x > 1$
Therefore, $f(x)$ concave up for $x < -1$ and $x > 1$.
$f'(x) > 0$ for $x < 0$
Therefore, $f(x)$ increasing for $x < 0$.
$f'(x) < 0$ for $x > 0$
Therefore, $f(x)$ decreasing for $x > 0$.

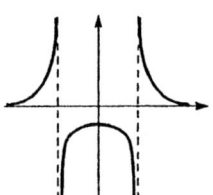

12. $y = \dfrac{x^2-1}{x^3}$

(1) intercepts $(-1,0), (1,0)$

(2) symmetry WRT origin

(3) $f(x) \to 0$ as $x \pm \infty$

(4) $x = 0$ is vertical asymptote

(5) domain all x except 0; range is all y

(6) $f'(x) = \dfrac{3 - x^2}{x^4}$

$\pm\sqrt{3}$ are critical points. $(-\sqrt{3}, -0.385)$ is a min and $(\sqrt{3}, 0.385)$ is a max.

(7) $f''(x) = \dfrac{2(x^2 - 6)}{x} = 0$ for $x = \pm 6$, $(-\sqrt{6}, -0.340)$ and $(\sqrt{6}, 0.340)$ are inflection points.

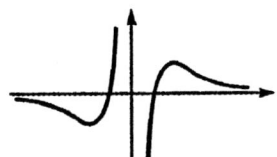

13. $y = \dfrac{4}{x} - \dfrac{4}{x^2}$

Intercepts:

(1) There are no y intercepts since $x = 0$ is undefined.

(2) $y = 0$ when $x = 1$ so $(1, 0)$ is an x-intercept.

Asymptotes:

(3) $x = 0$ is an asymptote; the denominator is 0.
$y = 0$ is an asymptote

Symmetry:

(4) Not symmetrical about the y-axis since $\frac{4}{x} - \frac{4}{x^2}$ is different from $\frac{4}{(-x)} - \frac{4}{(-x)^2}$

(5) Not symmetrical about the x-axis since

$y = \frac{4}{x} - \frac{4}{x^2}$ is different from $-y = \frac{4}{x} - \frac{4}{x^2}$

(6) Not symmetrical about the origin since

$y = \frac{4}{x} - \frac{4}{x^2}$ is different from $-y = \frac{4}{-x} - \frac{4}{(-x)^2}$

Derivatives:

(7) $y' = -4x^{-2} + 8x^{-3} = 0$ at $x = 2$; $(2, 1)$ is a relative maximum.

(8) $y'' = 8x^{-3} - 24x^{-4} = 0$ at $x = 3$ so $\left(3, \frac{8}{9}\right)$ is a possible inflection.
$y'' < 0$ (concave down) for $x < 3$ and 0 (concave up) for $x > 3$ so $\left(3, \frac{8}{9}\right)$ is an inflection.

Behavior as x becomes large:

(9) As $x \to \infty$ or $-\infty$, $\frac{4}{x}$ and $-\frac{4}{x^2}$ each approach 0.

As $x \to 0$, $\frac{4}{x} - \frac{4}{x^2} = \frac{4x - 4}{x^2}$ approaches $-\infty$, through positive or negative values of x.

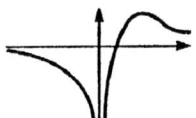

14. $f(x) = 4x + \dfrac{1}{\sqrt{x}}$

(1) Intercepts: $x \neq 0$, division by zero; $y \neq 0$, imaginary values

(2) Symmetry: $f(-x) \neq f(x)$; $-f(x) \neq f(x)$; $-f(-x) \neq f(x)$

Therefore, no symmetry WRT either axis or WRT origin.

(3) Behavior as x becomes large: $f(x)$ does not exist for $x \leq 0$.

$\displaystyle\lim_{x \to +\infty}\left(4x + \dfrac{1}{\sqrt{x}}\right) = \infty + 0$, $f(x)$ increases without limit as $x \to +\infty$.
As $x \to +\infty$, $f(x)$ is approximated by $4x$.

(4) Vertical asymptotes: $f(0) = 0 + \frac{1}{0}$, undefined Therefore, $x = 0$ is a vertical asymptote.

(5) Domain and range: Domain is all $x > 0$. Range to be determined.

(6) Derivatives: $f'(x) = 4 - \dfrac{1}{2x^{3/2}}$

$f''(x) = \dfrac{3}{4x^{5/2}} \neq 0$

$f'(x) = 0 = 4 - \dfrac{1}{2x^{3/2}}$

$8x^{3/2} = 1$

$x = \dfrac{1}{4}$

$f''\left(\dfrac{1}{16\sqrt{2}}\right) > 0$, therefore, min. $\left(\dfrac{1}{4}, 3\right)$

Therefore, range is $y \geq 3$.
$f''(x) > 0$ for $x > 0$

Therefore, $f(x)$ is concave up for all x.

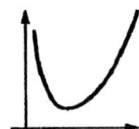

15. $f(x) = x\sqrt{1 - x^2}$

(1) Intercepts: $x = 0, y = 0$; $y = 0, x = -1$ and $x = 1$;
$(0,0), (-1,0), (1,0)$

(2) Symmetry: $-f(-x) = f(-x)$; symmetry to origin. No symmetry WRT either axis.

(3) Behavior as x becomes large: x cannot become 'large'. x can have values $(-1 \le x \le 1)$.

(4) Vertical asymptotes: None. There are no values of x for which $f(x)$ is undefined (division by zero).

(5) Domain and range: Domain is $-1 \le x \le 1$. Range to be determined.

(6) Derivatives: $f'(x) = \dfrac{1 - 2x^2}{(1 - x^2)^{1/2}}$

$f''(x) = \dfrac{x^3 - 3x}{(1 - x^2)^{3/2}}$

$f'(x) = 0, 1 - 2x^2 = 0$

$x = -\dfrac{\sqrt{2}}{2}, x = \dfrac{\sqrt{2}}{2}$

$f''\left(-\dfrac{\sqrt{2}}{2}\right) > 0,\ \min\ \left(-\dfrac{\sqrt{2}}{2}, -\dfrac{1}{2}\right)$

$f''\left(\dfrac{\sqrt{2}}{2}\right) < 0,\ \max.\ \left(\dfrac{\sqrt{2}}{2}, \dfrac{1}{2}\right)$

Therefore, range is $-\dfrac{\sqrt{2}}{2} \le y \le \dfrac{\sqrt{2}}{2}$.

$f''(x) = 0 = x^3 - 3x$
$x = 0, x = \pm\sqrt{3}$ (not in domain)
Infl. point $(0,0)$
No curve in second and fourth quadrants.

16. $f(x) = \dfrac{x - 1}{x^2 - 2x}$

(1) Intercepts: $x = 0, f(x)$ is undefined
$y = 0, x = 1, (1,0)$ a root

(2) Symmetry: No symmetry WRT either axis or origin.

(3) Behavior as x becomes large:

$$\lim_{x \to \infty} f(x) = \lim_{x \to \infty} \dfrac{\frac{1}{x} - \frac{1}{x^2}}{1 - \frac{2}{x}}$$

$\lim_{x \to \infty} f(x) = 0$, positively: $\lim_{x \to -\infty} f(x) = 0$, negatively

Therefore, x-axis is a horizontal asymptote.

(4) Vertical asymptotes: $\dfrac{x - 1}{x(x - 2)} = f(x)$; $f(0)$ undefined, $f(2)$ undefined
Therefore, $x = 0$ and $x = 2$ are vertical asymptotes.

(5) Domain and range: Domain is all x except $x = 0$, $x = 2$.

(6) Derivatives: $f'(x) = \dfrac{-x^2 + 2x - 2}{(x^2 - 2x)^2}$

$f''(x) = \dfrac{(2x - 2)(x^2 - 2x + 4)}{(x^2 - 2x)^3}$

$f'(x) = 0$; $x^2 - 2x + 2 = 0$, no real solution
Consider $x = 0, x = 2$
$f'(x) < 0$ for $x < 0$
Therefore, $f(x)$ decreasing for $x < 0$.
$f'(x) < 0$ for $0 < x < 2$
Therefore, $f(x)$ decreasing for $0 < x < 2$.
$f'(x) < 0$ for $x > 2$
Therefore, $f(x)$ decreasing for $x > 2$.
Therefore, range is all y.
$f''(x) = 0 = 2x - 2$, $x = 1$; $(1,0)$ infl. point

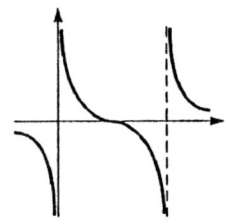

17. $y = \dfrac{9x}{9 - x^2}$

Intercept:

(1) Intercept at $x = 0, y = 0$ only

(2) Asymptotes at $x = -3, x = 3$

Derivatives:

(3) $y' = \dfrac{(9 - x^2)(9) - (9x)(-2x)}{(9 - x^2)^2} = \dfrac{81 + 9x^2}{(9 - x^2)^2}$

$81 + 9x^2 = 0$; $9x^2 = -81$; $x^2 = \sqrt{-9}$ (imaginary)
No real value max. or min., ± 3 are critical values.

(4) $y'' = \dfrac{(9 - x^2)^2(18x) - (81 + 9x^2)(2)(9 - x^2)(-2x)}{(9 - x^2)^2}$

$\quad = \dfrac{-18x^5 - 324x^3 + 4374x}{(9 - x^2)^4}$

$-18x^5 - 324x^3 + 4374x = 0$;
$-18x(x^4 + 18x^2 - 243) = 0$
$-18x = 0$; $x = 0$
$x^4 + 18x^2 - 243 = 0$; $(x^2 + 27)(x^2 - 9) = 0$
(imaginary) $x^2 = 9$; $x = \pm 3$ (these are asymptotes)

$y\big|_{x=0} = 0$; a possible inflection is $(0,0)$

at $\left(-1, -\frac{9}{8}\right)$, $y'' = -4032$; concave down at $\left(1, \frac{9}{8}\right)$,
$y'' = 4032$; concave up, and $(0,0)$ is an inflection
point

Symmetry:

(5) There is symmetry to the origin.

(6) As $x \to +\infty$ and as $x \to -\infty$, $y \to 0$. Therefore, $y = 0$ is an asymptote.

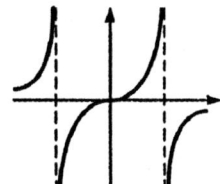

18. $f(x) = \dfrac{x^2 - 4}{x^2 + 4}$

(1) Intercepts: $x = 0, y = -1, (0, -1)$
$y = 0, x = -2, x = 2, (-2, 0), (2, 0)$

(2) Symmetry: $f(-x) = f(x)$, therefore symmetry WRT y-axis.
No symmetry WRT x-axis or origin.

(3) Behavior as x becomes large:

$$\lim_{x \to \pm\infty} \frac{x^2 - 4}{x^2 + 4} = \lim_{x \to \pm\infty} \frac{1 - \frac{4}{x^2}}{1 + \frac{4}{x^2}} = 1$$

Therefore, $y = 1$ is a horizontal asymptote.

(4) Vertical asymptotes: None; no value of x for
which $f(x)$ is undefined.

(5) Domain and range: Domain is all x. Range to
be determined.

(6) Derivatives: $f'(x) = \dfrac{16x}{(x^2 + 4)^2}$

$f''(x) = \dfrac{16(4 - 3x^2)}{(x^2 + 4)^3}$

$f'(x) = 0 = 16x, x = 0$
$f''(0) > 0$, min. $(0, -1)$
$f''(x) = 0 = 4 - 3x^2$

$x = \dfrac{2\sqrt{3}}{3}, x = -\dfrac{2\sqrt{3}}{3}$

Infl. point $\left(-\dfrac{2\sqrt{3}}{3}, -\dfrac{1}{2}\right)$

Infl. point $\left(\dfrac{2\sqrt{3}}{3}, -\dfrac{1}{2}\right)$

Therefore, range is $-1 \leq y < 1$.

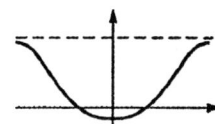

19. $C_T = \dfrac{6C}{6 + C}$; $\dfrac{dC_T}{dC} = \dfrac{36}{(6 + C)^2}$; $\dfrac{d^2C_T}{dC^2} = \dfrac{-72}{(6 + C)^3}$

$f'(C) \neq 0$; $f''(C) = 0$, therefore no max, no min,
no infl. points
$f'(C) > 0$ for all C, therefore C_T is increasing for
all C.
$f''(C) < 0$ for $C > -6$, therefore C_T is concave
down for $C > -6$ (only values of $C \geq 0$ have phys-
ical significance.)
$C = 0, C_T = 0, (0,0)$ is the only intercept

No symmetry WRT axes or origin.

$$\lim_{x \to +\infty} \frac{6C}{6+C} = \lim_{x \to +\infty} \frac{6}{\frac{6}{C}+1} = 6, \text{ therefore, hor-}$$

izontal asymptote at $C_T = 6$.

Vertical asymptote at $C = -6$ (capacitance > 0)

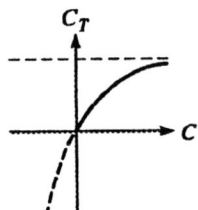

20. $n = \dfrac{75,000}{e(e+6)}$

(1) Intercepts: $e \neq 0$, undefined, $n \neq 0$

(2) Symmetry: No symmetry, $e > 0$

(3) Behavior as e becomes large:

$$\lim_{e \to \infty} \frac{75,000}{e^2 + 6e} = 0, \text{ therefore } e\text{-axis is a horizontal}$$

asymptote (positively).

(4) Vertical asymptotes: $f(0) = \dfrac{75,000}{0}$ is unde-

fined, therefore $e = 0$ is a vertical asymptote.

(5) Domain and range: Domain is $e > 0$, range is $n > 0$.

(6) Derivatives: $f'(e) = \dfrac{-75,000(2e+6)}{(e^2 + 6e)^2}$

$$f''(e) = \frac{450,000(e^2 + 6e + 12)}{(e^2 + 6e)^3}$$

$f'(e) = 0 = 2e + 6$

Therefore, $e = -3$, reject $(e > 0)$

$f'(e) < 0$ for $e > 0$

Therefore, $f(e)$ decreases for $e > 0$.

$f'(e) > 0$ for $e > 0$

Therefore, $f(e)$ is concave up for $e > 0$.

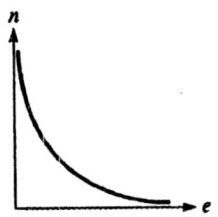

21. $R = \dfrac{200}{\sqrt{t^2 + 40,000}}$

Intercepts:

(1) Not continuous at $t = 0$ (not defined at $t < 0$).

(2) No t-intercept; R-intercept at $(0,1)$

Symmetry:

(3) No symmetry about either axis (R is undefined for $t < 0$).

Derivatives:

(4) $\quad R = 200(t^2 + 40,000)^{-1/2}$
$\quad\quad R' = -100(t^2 + 40,000)^{-3/2}(2t)$
$\quad\quad\quad = -200t(t^2 + 40,000)^{-3/2};$

$$\frac{-200t}{(t^2 + 40,000)^{3/2}} = 0; \; -200t = 0; \; t = 0 \text{ is a max,}$$
$(0,1)$.

since

$R|_{t=0} = 1$ and $R|_{t=1} < 1$ (R is undefined for $t < 0$)

(5)

$$R'' = \frac{(t^2+40,000)^{3/2}(-200)-(-200t)\left(\frac{3}{2}\right)(t^2+40,000)^{1/2}(2}{[(t^2 + 40,000)^{3/2}]^2}$$

$\quad = -200(t^2+40,000)^{3/2}+600t^2(t^2+40,000)^{1/2} = 0$

$(t^2 + 40,000)^{1/2}[-200(t^2 + 40,000) + 600t^2] = 0$
$(t^2 + 40,000)^{1/2} = 0; \; t^2 + 40,000 = 0$
$t^2 = -40,000$ (imaginary)
$-200(t^2 + 40,000) + 600t^2 = 0$
$-200[(t^2+40,000) - 3t^2] = 0; \; t^2+40,000-3t^2 = 0$
$-2t^2 = -40,000; \; t^2 = 20,000$
$t = 141$ possible inflection

$R''|_{t=140} \leq 0; \; R''|_{t=142} > 0; \; R|_{t=141} = 0.82$ is an inflection, $(141, 0.82)$

As x becomes large:

(6) As $x \to \infty$, $\sqrt{t^2 + 40,000}$ becomes infinitely large and $\dfrac{200}{\sqrt{t^2+40,000}}$ is a positive value that becomes infinitely small but never zero.

$R = 0$ is a horizontal asymptote

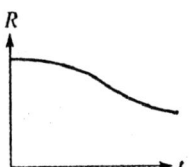

22. $P = \dfrac{36R}{R^2 + 2R + 1}$

(1) Intercepts: $R = 0, P = 0, (0,0)$; $P = 0, R = 0$

(2) Symmetry: No symmetry

(3) Behavior as R becomes large:

$$\lim_{R \to \infty} \frac{36R}{R^2 + 2R + 1} = \lim_{R \to \infty} \frac{\frac{36}{R}}{1 + \frac{2}{R} + \frac{1}{R^2}} = 0$$

Therefore, $P = 0$ is a horizontal asymptote.

(4) Vertical asymptotes: None; $R \geq 0$.

(5) Domain and range: Domain is $R \geq 0$. Range to be determined $(P > 0)$.

(6) Derivatives: $f'(R) = \dfrac{36(1 - R)}{(R + 1)^3}$

$f''(R) = \dfrac{72(R - 2)}{(R + 1)^4}$

$f'(R) = 0; \; R = 1$
$f''(1) < 0; \; \text{max } (1, 9)$
$f''(R) = 0, R = 2$
Infl. point $(2, 8)$
Therefore, range: $0 \leq P \leq 9$.

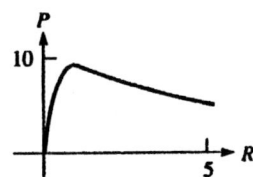

23. $V = \pi r^2 h = 20$

Therefore, $h = \dfrac{20}{\pi r^2}$

$A_T = \text{top} + \text{bottom} + \text{wraparound}$
$ = \pi r^2 + \pi r^2 + 2\pi r h$

$ = 2\pi r^2 + 2\pi r \left(\dfrac{20}{\pi r^2} \right) = 2\pi r^2 + \dfrac{40}{r}$

Vertical asymptote at $r = 0$.

$\dfrac{dA_T}{dr} = 4\pi r - \dfrac{40}{r^2}$

$\dfrac{d^2 A}{dr^2} = 4\pi + \dfrac{80}{r^3}$

$\dfrac{dA_T}{dr} = 0; \; r = \sqrt[3]{\dfrac{10}{\pi}} = 1.47$

$f''(1.47) > 0 \; \text{min.} \; (1.47, 40.8)$

$\dfrac{d^2 A_T}{dr^2} = 0; \; r = -\sqrt[3]{\dfrac{20}{\pi}} \; (\text{reject})$

r must be > 0

$\dfrac{d^2 A}{dr^2} > 0$ for $r > 0$

Therefore, A_T is concave up for $r > 0$.

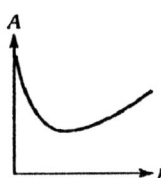

24. $A = xy = 20,000 \text{ m}^2$

$y = \dfrac{20,000}{x}$

$h = x + 2y$

$h = x + \dfrac{40,000}{x} = \dfrac{40,000 + x^2}{x}$

(1) Intercepts: $x = 0$ is undefined, $y = 0$ has no real solution.

(2) Symmetry: No symmetry.

(3) Behavior as $x \to \infty$; $\lim\limits_{x \to \infty} x + \dfrac{40,000}{x} = \infty$, grows without bound.

(4) Vertical asymptotes: $f(0)$ gives $\dfrac{40,000}{0}$, undefined, therefore $x = 0$ is a vertical asymptote.

(5) Domain and range: Domain: $x > 0$
Range to be determined $(y > 0)$.

(6) Derivatives: $f'(x) = 1 - \dfrac{40,000}{x^2}$

$f''(x) = \dfrac{80,000}{x^3} \neq 0$

$f'(x) = 0, x = 200$
$y = 100, l = 400$
$f''(200) > 0, \text{min } (200, 400)$
$f''(x) > 0$ for all x
Therefore, $f(x)$ concave up for all x.
Therefore, range is $l \geq 400$.
Minimum length of fence to use for fixed area is 400 m.

4.7   Applied Maximum and Minimum Problems

1. $s = 112t - 16.0t^2$; find maximum s.
$s' = 112 - 32.0t = 0$; $-32.0t = -112$; $t = 3.50\ s$
$s'' = -32.0 < 0$ for all t, so the graph is concave down and $t = 3.50$ is a maximum.
$s = 112(3.50) - 16.0(3.50)^2 = 196$ ft

2. $P = 8x - 0.02x^2$; max. profit?

$\dfrac{dP}{dx} = 8 - 0.04x = 0$; $x = 200$ barrels

$\dfrac{d^2P}{dx^2} = -0.04 < 0$ for all x, therefore max. $(200, 800)$

Max. profit is for 200 barrels, profit $800.

3. $P = EI - RI^2$; $\dfrac{dP}{dI} = E - 2RI = 0$; $I = \dfrac{E}{2R}$

$\dfrac{d^2P}{dI^2} = -2R < 0$ for all I, therefore max. power
at $I = \dfrac{E}{2R}$

4. $S = -0.00074t^3 + 0.020t^2 + 0.8$

$\dfrac{dS}{dt} = -0.00222t^2 + 0.04t = 0$ for maximum
$t = 18$

$S = -0.00074(18)^3 + 0.020(18)^2 + 0.8 = 2.96$

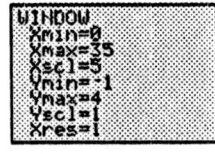

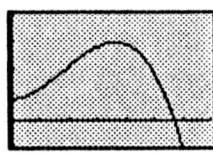
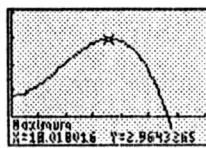

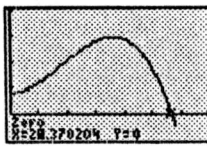

The maximum value of the fund will be about
$\$2.96 \times 10^{12}$ after 18 years, sometime early in 2018.
The fund will run out of money sometime in April
of 2028

5. $S = 360A - 0.1A^2$, find maximum S.
$S' = 360 - 0.3A^2$; $A^2 = 1200$; $A = 35\ \text{m}^2$
$S'' = -0.6A < 0$ for all valid (positive) A so the
graph is concave down and $A = 35\ \text{m}^2$ is a max.
Maximum savings are
$S = 360(35) - 0.1(35)^2 = \$8300$.

6. $h = 16t^3 - 240t^2 + 10\,000$

$\dfrac{dh}{dt} = 48t^2 - 480t = 0$

$48t(t - 10) = 0$
$t = 0, t = 10$

$\dfrac{d^2h}{dt^2} = 96t - 480$

$\left.\dfrac{d^2h}{dt^2}\right|_{t=0} < 0$, max $(0, 10\,000)$

$\left.\dfrac{d^2h}{dt^2}\right|_{t=0} > 0$, min $(10, 2000)$

Therefore, altitude of jet turning out of dive is
2000 ft.

7. $Z = \sqrt{R^2 + (X_L + X_C)^2}$; $R = 2500\ \Omega$;
$X_L = 1500\ \Omega$
$Z = [R^2 + (X_L - X_C)^2]^{1/2}$

$\dfrac{dZ}{dX_C} = \dfrac{1}{2}[R^2 + (X_L - X_C)]^{-1/2}2(X_L - X_C)(-1)$

$= \dfrac{-(X_L - X_C)}{\sqrt{R^2 + (X_L - X_C)^2}}$

$= \dfrac{X_C - 1500}{\sqrt{2500^2 + (1500 - X_C)^2}} = 0$

$X_C = 1500\ \Omega$

Test:

$\left.\dfrac{dZ}{dX_C}\right|_{X_C=1400} < 0$; $\left.\dfrac{dZ}{dX_C}\right|_{X_C=1600} > 0$.

Therefore, concave up, **minimum**.
$X_C = 1500\ \Omega$ gives minimum impedance.

8. $3x + 2y = 6$; $3 + 2\dfrac{dy}{dx} = 0$

$\dfrac{dy}{dx} = -\dfrac{3}{2}$; slope of line

$V = 3x^2 + 2y^2$; $\dfrac{dV}{dx} = 6x + 4y\dfrac{dy}{dx} = 6x + 4y\left(-\dfrac{3}{2}\right)$

$= 6x - 6y = 0$; $x = y$

$3x + 2x = 6; \ 5x = 6; \ x = \dfrac{6}{5} = 1.2$

Test $x = y = 1.2$

$\left.\dfrac{dV}{dx}\right|_{(1.1,1.35)} < 0; \ \left.\dfrac{dV}{dx}\right|_{(1.3,1.05)} > 0$

Concave up, therefore minimum $(1.2, 1.2)$.
Therefore, potential on the line $3x + 2y = 6$ is a minimum at $(1.2, 1.2)$.

9.

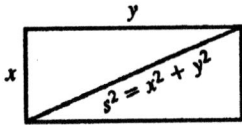

$P = 2x + 2y = 48$

$x + y = 24$

Diagonal will be a minimum if $l = s^2$ is a minimum.

$l = x^2 + y^2$

$l = x^2 + (24 - x)^2$

$l = x^2 + 24x^2 - 48x + x^2$

$l = 2x^2 - 48x + 24^2$

$\dfrac{dl}{dx} = 4x - 48 = 0$

$x = 12$

from which $y = 12$

Dimensions are 12 in by 12 in, a square will minimize the diagonal.

10. $R_1 + R_2 = 32 = R_T$ in series. For parallel,

$\dfrac{1}{R_T} = \dfrac{1}{R_1} + \dfrac{1}{R_2} = \dfrac{1}{R_1} + \dfrac{1}{32 - R_1} = \dfrac{32 - R_1 + R_1}{R_1(32 - R_1)}$

$R_T = \dfrac{R_1(32 - R_1)}{32} = R_1 - \dfrac{R_1^2}{32}$

$\dfrac{dR_T}{dR_1} = 1 - \dfrac{2R_1}{32} = 0, \ R_1 = 16 \ \Omega$

$R_T = 16 - \dfrac{16^2}{32} = 8 \ \Omega$

The maximum value for the total resistance in parallel is 8 Ω when each resistor is 16 Ω.

11. $P = 2x + 2y; \ xy = 25; \ y = \dfrac{25}{x}$

$P = 2x + 2\left(\dfrac{25}{x}\right) = 2x + \dfrac{50}{x}$

$\dfrac{dP}{dx} = 2 - \dfrac{50}{x^2} = 0$

$2x^2 - 50 = 0; \ x = 5 \ \text{mm}$

$\left.\dfrac{d^2P}{dx^2} = \dfrac{100}{x^3}\right|_{x=5} > 0, \ \text{min.} \ (5, 5)$

The dimensions of the chip are 5 mm by 5 mm for min. perimeter.

12. $2x + y = 800$

$y = 800 - 2x$

$A = xy = x(800 - 2x)$

$\quad = 800x - 2x^2$

$\dfrac{dA}{dx} = 800 - 4x = 0$

$x = 200$

$\dfrac{d^2A}{dx^2} = -4 < 0, \ \text{max.} \ (200, 400)$

Therefore, maximum area $= 200 \times 400$

$\quad\quad\quad\quad\quad\quad\quad = 80,000 \ \text{ft}^2$

13. Distance traveled from B is 16.0t; $40.0 - 16.0t$ is side of a right triangle. Distance traveled from A is 18.0t.

$d = \sqrt{(40.0 - 16.0t)^2 + (18.0t)^2}$

$\quad = \sqrt{1600 - 1280t + 580t^2}$

$y = d^2 = 1600 - 1280t + 580t^2$

$\dfrac{dy}{dt} = -1280 + 106t$

Minimum will occur when $-1280 + 1060t = 0$; $t = 1.1 \ \text{h}$

14. Let $\$C$ be the cost/linear meter of two sides and rear. The $2\$C$ is the cost of front/linear meter.

$A = 1350 \ \text{m}^2; \ 1350 = xy; \ y = \dfrac{1350}{x}$

Total cost $= \$Cx + \$Cy + \$Cy + 2\$Cx$

$C_T = 3\$Cx + 2\$Cy = \$C(3x + 2y)$

$\quad = \$C\left(3x + 2 \cdot \dfrac{1350}{x}\right) = \$C\left(3x + \dfrac{2700}{x}\right)$

$\dfrac{dC_T}{dx} = \$C\left(3 - \dfrac{2700}{x^2}\right) = 0$

$3x^2 = 2700; \ x = 30 \ \text{m}$

$\left.\dfrac{d^2C_T}{dx^2} = \$C\left(\dfrac{5400}{x^3}\right)\right|_{x=30} > 0$

Therefore minimum is $(30, 45)$.
Therefore, dimensions of building to give minimum wall cost are 30 m × 45 m; 30 m along front and rear, 45 m deep on sides.

15. $A = \frac{1}{2}xy$; $12.0^2 = x^2 + y^2$

$y^2 = 144 - x^2$; $y = \sqrt{144 - x^2}$

$A = \frac{1}{2}x\sqrt{144 - x^2}$

$\frac{dA}{dx} = \frac{1}{2}\left[x \cdot \frac{1}{2}(144 - x^2)^{-1/2}(-2x) + \sqrt{144 - x^2} \right]$

$\qquad = \frac{1}{2}\left[\frac{-x^2}{\sqrt{144 - x^2}} + \sqrt{144 - x^2} \right]$

$\qquad = \frac{1}{2}\left[\frac{-x^2 + 144 - x^2}{\sqrt{144 - x^2}} \right] = \frac{1}{2}\frac{(144 - 2x^2)}{\sqrt{144 - x^2}}$

$\frac{dA}{dx} = 0 = 144 - 2x^2$; $x = \sqrt{72} = 8.49$

Test $\sqrt{72}$: $f'\sqrt{71} > 0$; $f'\sqrt{73} < 0$
Therefore, max. $(\sqrt{72}, \sqrt{72})$.
Therefore, legs of triangle will be equal at 8.49 cm
for max. area.

16. $l + 2x + 2y \le 108$ in
Square ends: $x = y$
$l + 2x + 2x = 108$; $l = 108 - 4x$
$V = x^2 l$
$V = x^2(108 - 4x)$
$V = 108x^2 - 4x^3$

$\frac{dV}{dx} = 216x - 12x^2 = 0$

$12x(18 - x) = 0$; $x = 0$; reject; $x = 18$ in

$\frac{d^2V}{dx^2} = 216 - 24x\big|_{x=18} = 0$, max. at $x = 18$ in,
therefore $y = 18, l = 36$
**Largest box with square ends will be
18 in ×18 in ×36 in.**

17. $A = \frac{1}{2}(a + b)h$; $a = 6.00$ ft

$A = \frac{1}{2}(6 + 2x)(\sqrt{9 - x^2})$

$\qquad = (3 + x)(9 - x^2)^{1/2}$

$A' = (3 + x)\left(\frac{1}{2}\right)(9 - x^2)^{-1/2}(-2x) + (1)(9 - x^2)^{1/2}$

$\qquad = (-3x - x^2)(9 - x^2)^{-1/2} + (9 - x^2)^{1/2}$

$\qquad = \frac{-3x - x^2}{(9 - x^2)^{1/2}} + \frac{9 - x^2}{(9 - x^2)^{1/2}}$

$\qquad = \frac{9 - x^2 - 3x - x^2}{(9 - x^2)^{1/2}} = \frac{-2x^2 - 3x + 9}{(9 - x^2)^{1/2}}$

$-2x^2 - 3x + 9 = 0$; $2x^2 + 3x - 9 = 0$;
$(2x - 3)(x + 3) = 0$
$x + 3 = 0$; $x = -3$ (not valid); $2x - 3 = 0$; $x = 1.50$
$b = 2x = 3.00$ ft

18. $V = 7x^2 y$
$A_T = 2$ ends $+ 2$ sides $+$ bottom
$A_T = 2xy + 2(7xy) + 7x^2$
$980 = 2xy + 14xy + 7x^2 = 7x^2 + 16xy$

$y = \frac{(980 - 7x^2)}{16x}$

Therefore,

$V = 7x^2 \cdot \frac{980 - 7x^2}{16x} = \frac{7}{16}x(980 - 7x^2)$

$\qquad = \frac{49}{16}(140x - x^3)$

$\frac{dV}{dx} = \frac{49}{16}(140 - 3x^2) = 0$; $x = \sqrt{\frac{140}{3}} = 6.83$ ft

$\frac{d^2V}{dx^2} = \frac{49}{16}(-6x)\Big|_{6.83} < 0$, max. $x = 6.83$ ft

Therefore, $7x = 47.82$ ft., $y = 5.98$ ft
Therefore, dimensions of pool for maximum volume are
$l = 48$ ft, $w = 6.8$ ft, $d = 6.0$ ft.

19. $y = 6x^2 - x^3$; maximum slope? What value of x
gives a maximum $\frac{dy}{dx}$?

$\frac{dy}{dx} = 12x - 3x^2$; $y' = 12x - 3x^2$

$\frac{dy'}{dx} = 12 - 6x = 0$; $x = 2$

$\frac{d^2y'}{dx^2} = -6 < 0$ for all x, therefore $x = 2$,

max. $(2, 16)$

Maximum slope at $x = 2$ is
$12(2) - 3(4) = 24 - 12 = 12$.

20. $y = x^5 - 10x^2$; minimum slope? Minimum y'.

$y' = 5x^4 - 20x$; $\frac{dy'}{dx} = 20x^3 - 20 = 0$; $x = 1$

$\frac{d^2y'}{dx^2} = 60x^2 > 0$ for all x, therefore, $x = 1$ min.
$(1, -9)$

Minimum slope at $x = 1$ at $5(14) - 20(1) = -15$

21. $y = k(2x^4 - 5Lx^3 + 3L^2x^2)$
$\quad = 2kx^4 - 5kLx^3 + 3kL^2x^2$
$y' = 8kx^3 - 15kLx^2 + 6kL^2x = 0$
$kx(8x^2 - 15Lx + 6L^2) = 0$
$kx = 0; \; x = 0$
$8x^2 - 15Lx + 6L^2 = 0$

$$x = \frac{-(-15) \pm \sqrt{(-15)^2 - 4(8)(6)}}{2(8)}$$

$$= \frac{15 \pm \sqrt{33}}{16}$$

$$= \frac{15 \pm 5.75}{16}$$

$x = 0.58L, \, 1.30L$ (not valid—this distance is greater than L, the length of the beam)

22. $E_1 = \dfrac{kq_1}{r_1}$

$E_2 = \dfrac{kq_2}{r_2}$

$E_T = E_1 + E_2 = k\dfrac{2.00}{(10-x)} + k\dfrac{1.00}{x}$

$\dfrac{dE_T}{dx} = k(2.00)(-1)(10-x)^{-2}(-1) + k(1.00)(-1)x^{-2}$

$\qquad = k\left[\dfrac{2.00}{(10-x)^2} - \dfrac{1.00}{x^2}\right] = 0$

$2x^2 - (10-x)^2 = 0; \; 2x^2 - 100 + 20x - x^2 = 0$
$x^2 + 20x - 100 = 0; \; x = -10 \pm \sqrt{200} = 4.14$ mm
Test $x = 4.14$
$f'(4.13) < 0$
$f'(4.15) > 0$
Therefore, minimum total potential energy occurs 4.14 m from 1.00 nC charge and 5.86 mm from 2.00 nC charge.
Therefore, 5.9 mm from 2.00 nC charge.

23. $V = (8.00 - 2x)(8.00 - 2x)x$
$\quad = 64x - 32x^2 + 4x^3$

$\dfrac{dV}{dx} = 64 - 64x + 12x^2 = 0$

$x = 4$ (reject); $x = \dfrac{4}{3}$

$\dfrac{d^2V}{dx^2} = -64 + 24x \Big|_{x=4/3} < 0$, max.

Therefore, the side of the square to be cut out is

$\dfrac{4}{3} = 1.33$ in.

24. $A = \pi rs + \pi r^2$ (minimum)

$\qquad = \dfrac{\pi r\sqrt{\pi^2 r^6 + 300^2}}{\pi r^2} + \pi r^2$

$\qquad = \dfrac{\sqrt{\pi^2 r^6 + 300^2}}{r} + \pi r^2$

$V = 100 \text{ cm}^3; \; V = \dfrac{1}{3}\pi r^2 h = 100; \; h = \dfrac{300}{\pi r^2}$

$s^2 = r^2 + h^2; \; s^2 = r^2 + \dfrac{300^2}{\pi^2 r^4}; \; s = \sqrt{\dfrac{\pi^2 r^6 + 300^2}{\pi^2 r^4}}$

$\dfrac{dA}{dr} = \dfrac{\dfrac{3\pi^2 r^6}{\sqrt{\pi^2 r^6 + 300^2}} - \sqrt{\pi^2 r^6 + 300^2}}{r^2}$

$\qquad = \dfrac{3\pi^2 r^6 - \pi^2 r^6 - 300^2}{r^2\sqrt{\pi^2 r^6 + 300^2}} = 0$

$2\pi^2 r^6 = 300^2; \; r = \sqrt[6]{\dfrac{300^2}{2\pi^2}} = 4.07$

Test $r = 4.07$
$f'(4.00) < 0$
$f'(4.10) > 0$
Minimum at $x = 4.07$
Therefore, $h = 5.76$
Therefore, $h = 5.76$ cm and $r = 4.07$ cm.

25. $2x + \pi d = 400; \; \pi d = 400 - 2x; \; d = \dfrac{400 - 2x}{\pi}$

$A = x(d) = x\left(\dfrac{400 - 2x}{\pi}\right) = \dfrac{400x - 2x^2}{\pi}$

$A' = \dfrac{400 - 4x}{\pi} = 0; \; 400 - 4x = 0; \; x = 100$ m

26. Let x be the number sold in excess of 1000.
Profit on first 1000 units at \$10.00/unit = \$10 000
Profit on units over 1000 at

$\$[10.00 - x(0.2)]/\text{unit} = \$x[10.00 - 0.02x]$

$P_T = 10\,000 - x[10.00 - 0.02x]$
$\quad = 10\,000 - 10.00x - 0.02x^2$

$\dfrac{dP_T}{dx} = -10.00 - 0.04x = 0$

$x = 250$ units in excess of 1000

$\dfrac{d^2 P_T}{dx^2} = -0.04 < 0$ for all x, therefore max. at $x = 250$.

Therefore, company should produce 1250 units per week to maximize profit.

27. $S = kwd^3$; $2.00^2 = d^2 + w^2$; $w = \sqrt{4.00 - d^2}$
$S = kd^3\sqrt{4.00 - d^2}$

$$\frac{ds}{dd} = k\left[d^3\frac{1}{2}(4.00 - d^2)^{-1/2}(-2d)\right.$$
$$\left. + \sqrt{4.00 - d^2}(3d^2)\right]$$

$$\frac{ds}{dd} = k\left[\frac{-d^4}{\sqrt{4.00 - d^2}} + 3d^2\sqrt{4.00 - d^2}\right]$$

$$= k\left[\frac{-d^4 + 3d^2(4.00 - d^2)}{\sqrt{4.00 - d^2}}\right]$$

$$= \frac{k(12d^2 - 4.00d^4)}{\sqrt{4.00 - d^2}} = 0$$

$12d^2 - 4d^4 = 4d^2(3 - d^2) = 0$; $d = 0$, $d = \sqrt{3} = 1.73$
Test $d = 1.73$
$f'(1.72) > 0$
$f'(1.74) < 0$
Max. at $d = 1.73$
Therefore, $w = 1.00$ ft
**Maximum stiffness will occur where $w = 1.00$ ft,
$d = 1.73$ ft**

28. Let s be the distance from any point (x,y) on the curve $y = 8.00 - 2.00x^2$ to the point $(1.20, 7.00)$, then

$$s = \sqrt{(x - 1.20)^2 + (y - 7.00)^2}$$
$$= \sqrt{(x - 1.20)^2 + (8.00 - 2.00x^2 - 7.00)^2}$$
$$s = \sqrt{(x - 1.20)^2 + (1.00 - 2.00x^2)^2}$$

$$s^2 = 4.00x^4 - 3.00x^2 - 2.40x + 2.44$$

$$\frac{ds^2}{dx} = 16.0x^3 - 6.00x - 2.40$$

for $x = 0.757$. From $y = 8.00 - 2.00x^2$;
$y = 8.00 - 2.00(0.757)^2 = 6.85$. $(0.757, 6.85)$
is the point on $y = 8.00 - 2.00x^2$ closest to $(1.20, 7.00)$.
$s = \sqrt{(0.757 - 1.20)^2 + (1.00 - 2.00(0.757)^2)^2}$
$s = 0.466$
The rocket passes 0.466 km from target.

29. Let $C = $ total cost

$$C = 50,000(10 - x) + 80,000(\sqrt{x^2 + 2.5^2})$$
$$= 500,000 - 50,000x + 80,000(x^2 + 6.25)^{1/2}$$
$$C' = -50,000 + 40,000(x^2 + 6.25)^{-1/2}(2x)$$
$$= -50,000 + 80,000x(x^2 + 6.25)^{-1/2}$$

$$= -50,000 + \frac{80,000x}{\sqrt{x^2 + 6.25}}$$

$$= \frac{-50,000\sqrt{x^2 + 6.25} + 80,000x}{\sqrt{x^2 + 6.25}} = 0$$

$$-50,000\sqrt{x^2 + 6.25} + 80,000x = 0$$

$$\sqrt{x^2 + 6.25} = \frac{-80,000x}{-50,000} = \frac{8x}{5}; \; x^2 + 6.25 = \frac{64}{25}x^2$$

$$6.25 = \frac{64}{25}x^2 = \frac{39}{25}x^2; \; x^2 = 6.25\left(\frac{25}{39}\right) = 4.00$$

$x = 2.00$ mi; $10 - x = 8.00$ mi

30. **Total distance travelled by light ray from A to B.**

$$S = AS + SB$$
$$= \sqrt{a^2 + x^2} + \sqrt{b^2 + (c - x)^2}$$

$$\frac{dS}{dx} = \frac{1}{2}(a^2 + x^2)^{-1/2}(2x)$$
$$+ \frac{1}{2}[b^2 + (c - x)^2]^{-1/2}(2)(c - x)(-1)$$
$$= \frac{x}{\sqrt{a^2 + x^2}} - \frac{c - x}{\sqrt{b^2 + (c - x)^2}} = 0$$

$\sin\alpha - \sin\beta = 0$; $\sin\alpha = \sin\beta$; $\alpha = \beta$

Least time from A to B implies distance from A to B is a minimum.

31. $xy = 7000 = A_B$ of building
$(x + 20.0)(y + 40.0) = A_L$ of lot $= A_L$
$A_L = (x + 20.0)(y + 40.0)$

$$y = \frac{7000}{x}$$

Therefore, $A_L = (x + 20.0)\left(\frac{7000}{x}\right)$

$$\frac{dA_L}{dx} = (x + 20.0)\left(\frac{-7000}{x^2} + 40.0\right) + \left(\frac{7000}{x} + 40.0\right)1 = 0$$

$$= \frac{-7000(x + 20.0)}{x^2} + \frac{7000 + 40.0x}{x} = 0$$

$$-7000(x + 20.0) + x(7000 + 40.0x) = 0$$
$$-7000x - 140\,000 + 7000x + 40.0x^2 = 0$$

$$x = \sqrt{\frac{140\,000}{40.0}} = 59.2 \text{ m}$$

Test $x = 59.2$
$f'(59.1) < 0$
$f'(59.3) > 0$
Min. at $x = 59.2$ m
Therefore, $x = 59.2$ m
Therefore, $y = 118$ m
Therefore, **dimensions of building are 59.2 m by 118 m.**

32. $V = \pi r^2 h = 375 \Rightarrow h = \dfrac{375}{\pi r^2}$

$A = (2r)^2 + 2\pi rh = 4r^2 + 2\pi r \cdot \dfrac{375}{\pi r^2}$

$A = 4r^2 + \dfrac{750}{r}$ which is a minimum for

$\dfrac{dA}{dr} = 8r - \dfrac{750}{r^2} = 0$

$r = \sqrt[3]{\dfrac{750}{8}}$

$r = 4.54$

$h = \dfrac{375}{\pi r^2} = 5.79$

The most economical dimensions are $r = 4.54$ cm, $h = 5.79$ cm

---

4.8   Differentials and Linear Approximations

1. $y = x^5 + x;$

$\dfrac{dy}{dx} = 5x^4 + 1$

$dy = (5x^4 + 1)dx$

2. $y = 3x^2 + 6;\ \dfrac{dy}{dx} = 6x;$

$dy = 6x\,dx$

3. $V = \dfrac{2}{r^5} + 3\pi^2;\ \dfrac{dV}{dr} = \dfrac{-10}{r^6};$

$dV = \dfrac{-10\,dr}{r^6}$

4. $y = 2\sqrt{x} - \dfrac{1}{x};$

$\dfrac{dy}{dx} = 2 \cdot \dfrac{1}{2}x^{-1/2} + \dfrac{1}{x^2} = \dfrac{1}{x^{1/2}} + \dfrac{1}{x^2}$

$dy = \left(\dfrac{1}{\sqrt{x}} + \dfrac{1}{x^2}\right)dx$

5. $s = 2(3t^2 - 5)^4;$

$\dfrac{ds}{dt} = 8(3t^2 - 5)^3(6t)$

$ds = 8(3t^2 - 5)^3(6t)dt$

$ds = 48t(3t^2 - 5)^3\,dt$

6. $y = 5(4 + 3x)^{1/3};$

$\dfrac{dy}{dx} = 5 \cdot \dfrac{1}{3}(4 + 3x)^{-2/3}(3)$

$= \dfrac{5}{(4 + 3x)^{2/3}}$

$dy = \dfrac{5\,dx}{(4 + 3x)^{2/3}}$

7. $y = \dfrac{2}{3x^2 + 1} = 2(3x^2 + 1)^{-1}$

$\dfrac{dy}{dx} = -2(3x^2 + 1)^{-2}(6x) = \dfrac{-12x}{(3x^2 + 1)^2}$

$dy = \dfrac{-12x\,dx}{(3x^2 + 1)^2}$

8. $R = \sqrt{\dfrac{u}{1 + 2u}} = \left(\dfrac{u}{1 + 2u}\right)^{1/2}$

$\dfrac{dR}{du} = \dfrac{1}{2}\left(\dfrac{u}{1 + 2u}\right)^{-1/2} \cdot \dfrac{(1 + 2u)(1) - u(2)}{(1 + 2u)^2}$

$= \dfrac{1}{2}\sqrt{\dfrac{1 + 2u}{u}} \cdot \dfrac{1}{(1 + 2u)^2}$

$dR = \dfrac{1}{2}\sqrt{\dfrac{1 + 2u}{u}} \cdot \dfrac{1}{(1 + 2u)^2}\,du$

9. $y = x^2(1 - x)^3$

$dy = [x^2 \cdot 3(1 - x)^2(-1) + (1 - x)^3 \cdot 2x]dx$

$dy = (-3x^2(1 - x)^2 + 2x(1 - x)^3)dx$

$dy = (1 - x)^2(-3x^2 + 2x(1 - x))dx$

$dy = (1 - x)^2(-5x^2 + 2x)dx$

$dy = x(1 - x)^2(-5x + 2)dx$

10. $y = 6x\sqrt{1 - 4x}$

$\dfrac{dy}{dx} = 6x \cdot \dfrac{1}{2}(1 - 4x)^{-1/2}(-4) + \sqrt{1 - 4x}(6)$

$\dfrac{dy}{dx} = \dfrac{-12x}{\sqrt{1 - 4x}} + 6\sqrt{1 - 4x} = \dfrac{-12x + 6(1 - 4x)}{\sqrt{1 - 4x}}$

$dy = \dfrac{6 - 36x}{\sqrt{1 - 4x}}\,dx = \dfrac{6(1 - 6x)dx}{\sqrt{1 - 4x}}$

11. $y = \dfrac{x}{5x + 2};\ \dfrac{dy}{dx} = \dfrac{(5x + 2) - x(5)}{(5x + 2)^2} = \dfrac{2}{(5x + 2)^2}$

$dy = \dfrac{2\,dx}{(5x + 2)^2}$

12. $y = \dfrac{3x + 1}{\sqrt{2x - 1}}$

$\dfrac{dy}{dx} = \dfrac{\sqrt{2x - 1}(3) - (3x + 1)\frac{1}{2}(2x - 1)^{-1/2}(2)}{2x - 1}$

$= \dfrac{3\sqrt{2x - 1} - \frac{(3x+1)}{\sqrt{2x-1}}}{2x - 1}$

$= \dfrac{3(2x - 1) - (3x + 1)}{(2x - 1)^{3/2}}$

$dy = \dfrac{(3x - 4)dx}{(2x - 1)^{3/2}}$

13. $y = f(x) = 7x^2 + 4x, dy = f'(x)dx$
$= (14x + 4)dx$
$\Delta y = f(x + \Delta x) - f(x)$
$= 7(4.2)^2 + 4(4.2) - (7 \cdot 4^2 + 4 \cdot 4)$
$= 12.28$
$dy = (14(4) + 4)(0.2) = 12$

14. $y = 2x^2 - 3x + 1$; $x = 5$; $\Delta x = 0.15 = dx$
$y + \Delta y = 2(x + \Delta x)^2 - 3(x + \Delta x) + 1$
$\Delta y = 2(x^2 + 2x\Delta x + \Delta x^2) - 3x - 3\Delta x + 1$
$\quad - 2x^2 + 3x - 1$
$= 2x^2 + 4x\Delta x + 2\Delta x^2 - 3x - 3\Delta x + 1$
$\quad - 2x^2 + 3x - 1$
$= 4x\Delta x + 2\Delta x^2 - 3\Delta x$

$\Delta y \Big|_{\substack{x=5 \\ \Delta x=0.15}} = 4(5)(0.15) + 2(0.15^2) - 3(0.15)$
$= 2.595$

$dy = (4x - 3)dx; dy \Big|_{\substack{x=5 \\ dx=0.15}} = [4(5) - 3]0.15 = 2.55$

15. $y = 2x^3 - 4x$; $x = 2.5$; $\Delta x = 0.05 = dx$
$y + \Delta y = 2(x + \Delta x)^3 - 4(x + \Delta x)$
$\Delta y = 2(x^3 + 3x^2\Delta x + 3x\Delta x^2 + \Delta x^3) - 4x - 4\Delta x$
$\quad - 2x^3 + 4x$
$= 2x^3 + 6x^2\Delta x + 6x\Delta x^2 + 2\Delta x^3 - 4x - 4\Delta x$
$\quad - 2x^3 + 4x$
$= 6x^2\Delta x + 6x\Delta x^2 + 2\Delta x^3 - 4\Delta x$

$\Delta y \Big|_{\substack{x=2.5 \\ \Delta x=0.05}} = 1.71275$

$\dfrac{dy}{dx} = 6x^2 - 4; dy = (6x^2 - 4)dx;$

$dy \Big|_{\substack{x=2.5 \\ dx=0.05}} = 1.675$

16. $y = x - x^4$; $x = 3.2$; $\Delta x = 0.08$
$y + \Delta y = (x + \Delta x) - (x + \Delta x)^4$
$= x + \Delta x - x^4 - 4x^3\Delta x - 6x^2\Delta x^2 - 4x\Delta x^3$
$\quad - \Delta x^4$
$\Delta y = x + \Delta x - x^4 - 4x^3\Delta x - 6x^2\Delta x^2 - 4x\Delta x^3 - \Delta x^4$
$\quad - x + x^4$
$= \Delta x - 4x^3\Delta x - 6x^2\Delta x^2 - 4x\Delta x^3 - 4\Delta x^4$

$\Delta y \Big|_{\substack{x=3.2 \\ \Delta x=0.08}} = -10.805571$

$dy = (1 - 4x^3)dx; dy \Big|_{\substack{x=3.2 \\ dx=0.08}} = -10.40576$

17. $y = f(x) = (1 - 3x)^5$
$dy = f'(x)dx = 5(1 - 3x)^4(-3)dx = -15(1 - 3x)^4 dx$
$dy = -15(1 - 3(1))^4(0.01) = -2.4$
$f(x + \Delta x) - f(x) = f(1.01) - f(1)$
$= (1 - 3(1.01))^5 - (1 - 3(1))^5$
$= -2.4730881$

18. $y = (x^2 + 2x)^3$; $x = 7$; $\Delta x = 0.02$
$dy = 3(x^2 + 2x)^2(2x + 2)dx$;
$dy \Big|_{\substack{x=7 \\ dx=0.02}} = 3810$

$f(x + \Delta x) = f(7.02) = 253\,881$
$f(x) = f(7) = 250\,047$
$f(x + \Delta x) - f(x) = 3834$

19. $y = x\sqrt{1 + 4x}$; $x = 12$; $\Delta x = 0.06$
$\dfrac{dy}{dx} = x\dfrac{1}{2}(1 + 4x)^{-1/2}(4) + \sqrt{1 + 4x}$
$= \dfrac{2x}{\sqrt{1 + 4x}} + \sqrt{1 + 4x}$
$= \dfrac{2x + 1 + 4x}{\sqrt{1 + 4x}}$
$= \dfrac{1 + 6x}{\sqrt{1 + 4x}}$

$dy = \dfrac{(1 + 6x)dx}{\sqrt{1 + 4x}}; dy \Big|_{\substack{x=12 \\ \Delta x=0.06}} = 0.6257$

$f(x + \Delta x) = f(12.06) = 84.62649$
$f(x) = f(12) = 84; \Delta y = 0.6264903$
$f(x + \Delta x) - f(x) = 0.6264903$

20. $y = \dfrac{x}{\sqrt{6x - 1}}$; $x = 3.5$; $\Delta x = 0.025$

$\dfrac{dy}{dx} = \dfrac{\sqrt{6x - 1} - x\frac{1}{2}(6x - 1)^{-1/2}(6)}{6x - 1}$
$= \dfrac{\sqrt{6x - 1} - \frac{3x}{(6x-1)^{1/2}}}{6x - 1}$
$= \dfrac{6x - 1 - 3x}{(6x - 1)^{3/2}}$
$= \dfrac{3x - 1}{(6x - 1)^{3/2}}$

$dy = \dfrac{(3x - 1)dx}{(6x - 1)^{3/2}}; dy \Big|_{\substack{x=3.5 \\ \Delta x=0.025}} = 0.0026553$

$f(x + \Delta x) = f(3.525) = 0.7852747$
$f(x) = f(3.5) = 0.7826238$
$f(x + \Delta x) - f(x) = 0.0026509$

21. $f(x) = x^2 + 2x; \ f'(x) = 2x + 2$

$$L(x) = f(a) + f'(a)(x - a)$$
$$= f(0) + f'(0)(x - 0)$$
$$L(x) = 0^2 + 2 \cdot 0 + (2 \cdot 0 + 2)(x - 0)$$
$$L(x) = 2x$$

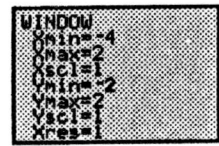 

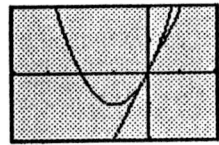

22. $f(x) = 2\sqrt[3]{x}, \ f(8) = 4$

$$f'(x) = \frac{2}{3\sqrt[3]{x^2}}, \ f'(8) = \frac{1}{6}$$

$$L(x) = \frac{1}{6}(x - 8) + 4$$

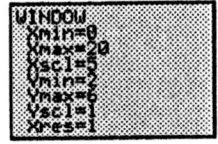 

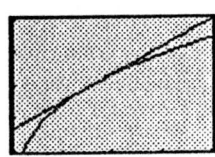

23. $f(x) = \frac{1}{2x + 1}, \ f(-1) = -1$

$$f](x) = \frac{-2}{(2x + 1)^2}, \ f'(-1) = -2$$

$$L(x) = -2(x + 1) - 1 = -2x - 3$$

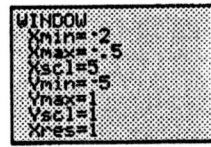 

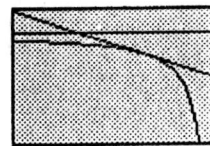

24. $g(x) = x\sqrt{2x + 8}, \ g(-2) = -4$

$$g'(x) = \frac{x}{\sqrt{2x + 8}} + \sqrt{2x + 8}, \ g'(-2) = 1$$

$$L(x) = 1 \cdot (x + 2) - 4$$

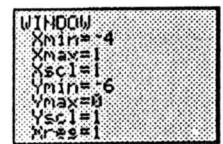 

25. $A = f(x) = x^2$
$dA = f'(x)dx = 2x\,dx$
$dA = 2(0.950)(0.002) = 0.0038 \text{ cm}^2$

26. $A = \pi r^2; \ r = 12.00 \text{ cm}; \ \Delta r = 0.05 \text{ cm}$
$r = 12.00 \pm 0.05 \text{ cm}$
$dA = 2\pi r \, dr$
$\quad = 2\pi(12.00)(0.05) = 3.7699 \text{ cm}^2$

$$\text{Rel. error} = \frac{\text{abs. error}}{\text{true value}}$$

$$\text{Rel. error in area} = \frac{3.7699}{(\pi \cdot 12.00^2)} = 0.0083$$

Rel. error in area as percentage = 0.8%

27. $\lambda = \frac{k}{f}, \ 685 = \frac{k}{4.38 \times 10^{14}};$

$k = 3.00 \times 10^{17} \text{ mm} - \text{H}_z$

$$\frac{d\lambda}{df} = \frac{-k}{f^2}$$

$$d\lambda = \frac{-k}{f^2} df$$

$$d\lambda = \frac{-3.00 \times 10^{17}}{(4.38 \times 10^{14})^2} \cdot (0.20 \times 10^{14})$$

$$d\lambda = -31 \text{ nm}$$

28. $v = \sqrt{100 + 16h}$; $h_1 = 20.0$ ft; $h_2 = 20.5$ ft;
$\Delta h = 0.5$ ft

$$v = (100 + 16h)^{1/2}; \frac{dv}{dh} = \frac{1}{2}(100 + 16h)^{-1/2}(16)$$

$$dv = \frac{8\, dh}{\sqrt{100 + 16h}}; dv\Big|_{\substack{h=20.0 \\ \Delta h=0.5}} = 0.195 \text{ ft/s}$$

Rel. change in v (in %)

$$= \frac{dv}{v}(100\%) = \frac{0.195}{20.494} = 0.95\%$$

29. $r = k\sqrt{\lambda}$

$$\frac{dr}{d\lambda} = \frac{k}{2\sqrt{\lambda}}; \frac{dr}{r} = \frac{k}{r} \cdot \frac{d\lambda}{2\sqrt{\lambda}}$$

$$= \frac{1}{2} \cdot \frac{k}{k\sqrt{\lambda}} \cdot \frac{d\lambda}{\sqrt{\lambda}}$$

$$\frac{dr}{r} = \frac{1}{2} \cdot \frac{d\lambda}{\lambda}$$

30. $F = \dfrac{k}{r^2}$

$$\frac{dF}{dr} = \frac{-2k}{r^3} = \frac{-2}{r} \cdot \frac{k}{r^2} = \frac{-2}{r} \cdot F$$

$$\frac{dF}{F} = \frac{-2\, dr}{r}$$

31. $A = s^2$; $dA = 2s\, ds$; $\dfrac{ds}{s} = 2\% = 0.02$

$$\frac{dA}{A} = \frac{2s\, ds}{s^2} = 2\frac{ds}{s}; \frac{dA}{A}\Big|_{\frac{ds}{s}=0.02} = 2(0.02) = 4\%$$

32. $(2.03)^4$; let $y = x^4$; $x = 2.00$; $\Delta x = 0.03$
$dy = 4x^3 dx; y\big|_{x=2.00} = 16.00$

$$dy\Big|_{\substack{x=2.00 \\ \Delta x=0.03}} = 4(2.00)^3(0.03) = 0.96$$

$$y + dy = y + 0.96 = 16.96$$

33. $f(x) = \sqrt{2-x}$; $f'(x) = \dfrac{-1}{2\sqrt{2-x}}$

$$L(x) = f(a) + f'(a)(x - a)$$

$$L(x) = f(1) + f'(1)(x - 1)$$

$$= \sqrt{2-1} + \frac{-1}{2\sqrt{2-1}}(x-1)$$

$$L(x) = 1 - \frac{1}{2}(x-1) = -\frac{1}{2}x + \frac{3}{2}$$

$$\sqrt{1.9} = f(0.1) \approx L(0.1)$$

$$= 1 - \frac{1}{2}(0.1 - 1) = 1.45$$

34. Let $f(x) = \sqrt[3]{x}$, $f(8) = 2$

$$f'(x) = \frac{1}{3\sqrt[3]{x^2}},$$

$$f'(8) = \frac{1}{12}$$

$$L(x) = \frac{1}{12}(x - 8) + 2$$

$$L(8.03) = \frac{1}{12}(8.03 - 8) + 2$$

$$= 2.0025$$

$\sqrt[3]{8.03} \approx 2.0025$ using **linearization.**

35. $C(V) = \dfrac{3.6}{\sqrt{1 + 2V}}$, $C(4) = 1.2$

$$C'(V) = \frac{-3.6}{(1 + 2V)^{3/2}}, C'(4) = \frac{-2}{15}$$

$$L(V) = -\frac{2}{15}(V - 4) + 1.2$$

$$= -0.13V + 1.73$$

36. $R_T(R) = \dfrac{16R}{16 + R}$, $R_T(4) = 3.2$

$$R_T'(R) = \frac{256}{(16 + R)^2}, R_T'(4) = \frac{16}{25}$$

$$L(R) = \frac{16}{25}(R - 3.2) + 4$$

$$\frac{16}{25}(R-4)+3.2$$

Chapter 4 Review Exercises

1. $y = 3x - x^2$ at $(-1, -4)$; $y' = 3 - 2x$

$$y'\big|_{x=-1} = 3 - 2(-1) = 5$$

$m = 5$ for tangent line
$y - y_1 = 5(x - x_1)$
$y - (-4) = 5[x - (-1)]$;
$y + 4 = 5x + 5$
$5x - y + 1 = 0$

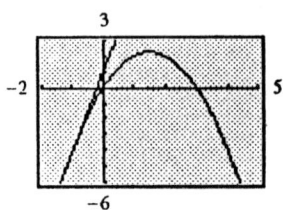

2. $y = x^2 - \dfrac{6}{x}, (2,1)$

$y' = 2x + \dfrac{6}{x^2}\Big|_{x=2}$

$= \dfrac{11}{2} = m$ for tangent line.

$y - y_1 = m(x - x_1)$

$y - 1 = \dfrac{11}{2}(x - 2)$

$11x - 2y - 20 = 0$

3. $x^2 - 4y^2 = 9, (5,2)$

$2x - 8y \cdot y' = 0$

$y' = \dfrac{x}{4y}\Big|_{(5,2)}$

$= \dfrac{5}{4(2)}$

$= \dfrac{5}{8}$

$= m_T$ for tangent line

$m_N = -\dfrac{8}{5}$ for the normal line

$y - y_1 = m(x - x_1)$

$y - 2 = -\dfrac{8}{5}(x - 5)$

$8x + 5y - 50 = 0$

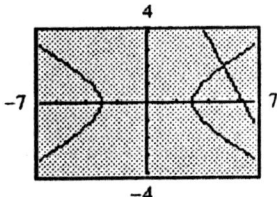

4. $y = \dfrac{1}{\sqrt{x-2}}, \left(6, \dfrac{1}{2}\right)$

$y' = \dfrac{-1}{2(x-2)^{3/2}}\Big|_{x=6}$

$= \dfrac{-1}{16}$

$= m_T$ for tangent line

$m_N = 16$ for normal line

$y - y_1 = m(x - x_1)$

$y - \dfrac{1}{2} = 16(x - 6)$

$2y - 1 = 32(x - 6)$

$32x - 2y - 191 = 0$

5. $y = \sqrt{x^2 + 3}; \; m = \dfrac{1}{2}$

$y = (x^2 + 3)^{1/2};$

$\dfrac{dy}{dx} = \dfrac{1}{2}(x^2 + 3)^{-1/2}(2x)$

$= \dfrac{x}{\sqrt{x^2 + 3}}$

$m_{\tan} = \dfrac{dy}{dx}$

$2x = \dfrac{x}{\sqrt{x^2 + 3}}$

$= \dfrac{1}{2}$

Squaring both sides, $4x^2 = x^2 + 3$; $3x^2 = 3$; $x^2 = 1$; $x = 1$ and $x = -1$. Therefore, the abscissa of the point at which $m = \frac{1}{2}$ is 1 or -1. If $x = 1, y = \sqrt{1^2 + 3} = \sqrt{4} = 2$ or -2. If $x = -1, y = \sqrt{(-1)^2 + 3} = 2$ or -2. The possible points where the slope of the tangent line is $\frac{1}{2}$ are $(1,2)$, $(1,-2)$, $(-1,2)$, $(-1,-2)$. A sketch of the curve shows that the only relative maximum or minimum point is at m(0,1.7). Therefore the point is $(1,2)$.

$y - 2 = \dfrac{1}{2}(x - 1); y = \dfrac{1}{2}x + \dfrac{3}{2}$ or $x - 2y + 3 = 0$ is the equation of the tangent line.

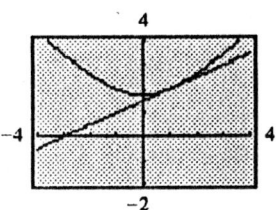

6. $y = \dfrac{1}{2x+1}$

$y' = \dfrac{-2}{(2x+1)^2} = -2$

$= m_T$ since $m_N = \dfrac{1}{2}$

$(2x+1)^2 = 1$

$x = 0, -1$ choose $x = 0$ since $x \geq 0$

when $x = 0, y = \dfrac{1}{2(0)+1} = 1$

$y - y_1 = m(x - x_1)$

$y - 1 = \dfrac{1}{2}(x - 0)$

$x - 2y + 2 = 0$

7. $x = \sqrt{t} + t$

$v_x = \dfrac{dx}{dt} = \dfrac{1}{2\sqrt{t}} + 1$

$v_x = \dfrac{5}{4}$

$y = \dfrac{1}{12}t^3$

$v_y = \dfrac{dy}{dt} = \dfrac{1}{4}t^2$ and at $t = 4$

$v_y = 4$

$v = \sqrt{v_x^2 + v_y^2} = 4.19$

$\theta = \tan^{-1}\dfrac{v_y}{v_x} = 72.6°$

8. $x = 0.1t^2 + 1$

$v_x = \dfrac{dx}{dt} = 0.2t$

$y = \sqrt{4t+1}$

$v_y = \dfrac{dy}{dt} = \dfrac{2}{\sqrt{4t+1}}$ and at $t = 6$

$v = \sqrt{(0.2(6))^2 + (\dfrac{2}{\sqrt{4(6)+1}})^2} = 1.26$

$\theta = \tan^{-1}\dfrac{v_y}{v_x} = \tan^{-1}\dfrac{\frac{2}{\sqrt{4t+1}}}{0.2t}\bigg|_{t=6} = 18.4°$

9. $y = 0.5x^2 + x;\ v_y = \dfrac{dy}{dt} = \dfrac{x\,dx}{dt} + \dfrac{dx}{dt};\ v = 0.5\sqrt{x}$

Substituting, $v_y = x(0.5\sqrt{x}) + 0.5\sqrt{x}$

Find v_y at $(2, 4)$:

$v_y|_{x=2} = 2(0.5\sqrt{2}) + 0.5\sqrt{2} = \sqrt{2} + 0.5\sqrt{2} = 1.5\sqrt{2} = 2.12$

10. $y = \dfrac{1}{x+2}$

$v_y = \dfrac{dy}{dt} = \dfrac{dy}{dx}\dfrac{dx}{dt}$

$v_y = \dfrac{-1}{(x+2)^2}(4)\bigg|_{x=2}$

$v_y = \dfrac{-4}{(2+2)^2} = -0.250$ cm/s

11. $\dfrac{dx}{dt} = \dfrac{1}{2\sqrt{t}} + 1 \qquad \dfrac{dy}{dt} = \dfrac{1}{4}t^2$

$a_x = \dfrac{d^2x}{dt^2} = \dfrac{-1}{4t^3} \qquad a_y = \dfrac{d^2y}{dt^2} = \dfrac{t}{2}$

$a = \sqrt{a_x^2 + a_y^2}\bigg|_{t=4} = 2.00$

$\theta = \tan^{-1}\dfrac{a_y}{a_x}\bigg|_{t=4} + 180° = 90.9°$

12. $v_y = \dfrac{-4}{(x+2)^2}$

$a_y = \dfrac{dv_y}{dt} = \dfrac{dv_y}{dx} \cdot \dfrac{dx}{dt}$

$a_y = \dfrac{8}{(x+2)^3}(4) = \dfrac{32}{(x+2)^3}\bigg|_{x=2} = \dfrac{32}{(2+2)^3} = \dfrac{1}{2}$

$a_x = \dfrac{dv_x}{dt} = \dfrac{d(4)}{dt} = 0$

$\theta = 0.500$ cm/s^2

$\theta = 90°$

13. $x^3 - 3x^2 - x + 2 = 0$ (between 0 and 1)

$f(x) = x^3 - 3x^2 - x + 2;\ f'(x) = 3x^2 - 6x - 1$

$f(0) = 0^3 - 3(0^2) - 0 + 2 = 2;\ f(1) = 1^3 - 3(1^2) - 1 + 2 = -1$

The root is possibly closer to 1 than 0. Let $x_1 = 0.6$:

n	x_n	$f(x_n)$	$f'(x_n)$	$x_n - \dfrac{f(x_n)}{f'(x_n)}$
1	0.6	0.536	−3.52	0.7522727
2	0.7522727	−0.0242935	−3.8158936	0.7459063
3	0.7459063	−0.0000304	−3.8063092	0.7458983

$x_4 = x_3 = 0.7458983$

14. $2x^3 - 4x^2 - 9 = 0$ (between 2 and 3)

$f(x) = 2x^3 - 4x^2 - 9$

$f'(x) = 6x^2 - 8x$, let $x_1 = 2.5$

n	x_n	$f(x_n)$	$f'(x_n)$	$x_n - \dfrac{f(x_n)}{f'(x_n)}$
1	2.5	−2.75	17.5	2.657142839
2	2.657142839	0.279393207	21.10530569	2.643904785
3	2.643904785	0.0020883255	20.79015679	2.643804337
4	2.643804337	1.19897×10^{-7}	20.78777354	2.643804331
5	2.643804331	1.0×10^{-12}	20.78777341	2.643804331

$x_4 = x_5 = 2.643804331$

15. $3x^3 - x^2 - 8x - 2 = 0$ (between 1 and 20

$f(x) = 3x^3 - x^2 - 8x - 2$

$f'(x) = 9x^2 - 2x - 8$ let $x_1 = 1.5$

n	x_n	$f(x_n)$	$f'(x_n)$	$x_n - \dfrac{f(x_n)}{f'(x_n)}$
1	1.5	−6.125	1.5	2.162161947
2	2.162161947	6.351720064	10.75237014	1.948660027
3	1.948660027	0.8122431653	7.194375192	1.91220087
4	1.91220087	0.0218381077	6.641466038	1.911165113
5	1.911165113	1.7389519×10^{-5}	6.625991634	1.911164287
6	1.911164287	1.3×10^{-11}	6.625979297	1.911164287

$x_5 = x_6 = 1.911164287$

16. $x^4 + 3x^3 + 6x + 4 = 0$ (between −1 and 0)

$f(x) = x^4 + 3x^3 + 6x + 4$

$f'(x) = 4x^3 + 9x^2 + 6$ let $x_1 = -0.5$

n	x_n	$f(x_n)$	$f'(x_n)$	$x_n - \dfrac{f(x_n)}{f'(x_n)}$
1	−0.5	0.6875	7.75	−0.588709666
2	−0.588709666	−0.0242442905	8.303073841	−0.5857897487
3	−0.5857897487	$-2.74299308 \times 10^{-5}$	8.284292526	−0.5857864376
4	−0.5857864376	-3.74×10^{-11}	8.284271247	−0.5857864376

$x_3 = x_4 = -0.5857864376$

17. $y = 4x^2 + 16x$

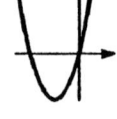

   (1)   The graph is continuous for all x.

   (2)   The intercepts are $(0,0)$ and $(-4,0)$.

   (3)   As $x \to +\infty$ and $-\infty, y \to +\infty$.

   (4)   The graph is not symmetrical about either axis or the origin.

   (5) $y' = 8x + 16$; $y' = 0$ at $x = -2$.
       $(-2, -16)$ is a critical point.

   (6) $y'' = 8 > 0$ for all x; the graph is concave up and $(-2, -16)$ is a minimum.

18. $y = x^3 + 2x^2 + x + 1$

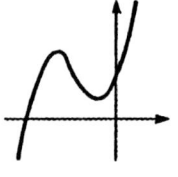

   (1)   The graph is continuous for all x

   (2)   The intercepts are $(0,1)$ and (-1.75). Use Newton's method or zero feature of calculator for x-intercept.

   (3)   As $x \to -\infty, y \to -\infty$; as $x \to \infty, y \to \infty$

   (4)   the graph is not symmetrical about origin or either axis

   (5) $f'(x) = 3x^2 + 4x + 1 = 0$ for $x = -1$ and $x = -\dfrac{1}{3}$

       $(-1, 1)$ and $\left(-\dfrac{1}{3}, \dfrac{23}{27}\right)$ are critical points

   (6) $f''(x) = 6x + 4 = 0$ for $x = -\dfrac{2}{3}$; $\left(-\dfrac{2}{3}, \dfrac{25}{27}\right)$ is a possible inflection point

   (7) $f''(-1) = -2, (-1, 1)$ is a local maximum

       $f''\left(-\dfrac{1}{3}\right) = 2, \left(-\dfrac{1}{3}, \dfrac{23}{27}\right)$ is a local minimum

       $f''(x) < 0$ for $x < -\dfrac{2}{3}, f''(x) > 0$ for $x > -\dfrac{2}{3}, \left(-\dfrac{2}{3}, \dfrac{25}{27}\right)$ is an inflection point.

19. $y = 27x - x^3$

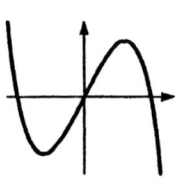

   (1)   The graph is continuous for all x

   (2)   the intercepts are $(\pm 3\sqrt{3}, 0)$ and $(0, 0)$

   (3)   As $x \to -\infty, y \to \infty$; $x \to \infty, y \to -\infty$

   (4)   the graph has symmetry with respect to the origin

   (5) $f'(x) = 27 - 3x^2 = 0$ for $x = \pm 3$; $(\pm 3, \pm 54)$ are critical points

   (6) $f''(x) = -6x = 0$ for $x = 0$; $(0, 0)$ is a possible inflection point

   (7) $f''(3) = -18, (3, 54)$ is a local maximum; $f''(-3) = 18, (-3, -54)$ is a local minimum
       $f''(x) > 0$ for $x < 0$ and $f''(x) < 0$ for $x > 0$, $(0, 0)$ is an inflection point

20. $y = x(6-x)^3$

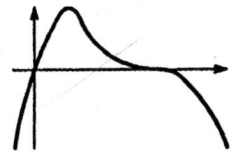

(1) The graph is continuous for all x

(2) the intercepts are $(0,0)$ and (6.0)

(3) As $x \to -\infty, y \to -\infty$; $x \to \infty, y \to -\infty$

(4) the graph is not symmetrical with respect to origin or either axis.

(5) $f'(x) = 2(x-6)^2(3-2x) = 0$ for $x = 6, x = \dfrac{3}{2}$

$(6,0)$ and $\left(\dfrac{3}{2}, \dfrac{2187}{16}\right)$ are critical points

(6) $f''(x) = 12(3-x)(x-6) = 0$ for $x = 3$ and $x = 6$
$(3,81)$ and $(6,0)$ are possible inflection points

(7) $f''\left(\dfrac{3}{2}\right) = -81, \left(\dfrac{3}{2}, \dfrac{2187}{16}\right)$ is a local maximum

$f''(6) = 0$, second derivative test inconclusive

$f'(x) < 0$ for $x > 6$; $f'(x) < 0$ for $\dfrac{3}{2} < x < 6$

$(6,0)$ is neither a local maximum nor a local minimum $f'' < 0$ for $x < 3$
or $x > 6$; $f''(x) > 0$ for $3 < x < 6$
$(3,81)$ and $(6,0)$ are inflection points

21. $y = x^4 - 32x$

(1) The graph is continuous for all x.

(2) The intercepts are $(0,0)$ and $(2\sqrt[3]{4}, 0)$.

(3) As $x \to -\infty$, $y \to +\infty$; as $x \to +\infty$, $y \to +\infty$.

(4) The graph is not symmetrical about either axis or the origin.

(5) $y' = 4x^3 - 32 = 0$ for $x = 2$

(6) $y'' = 12x^2$; $y'' = 0$ at $x = 0$; $(0,0)$ is a possible point of inflection. Since $f''(x) > 0$; the graph is concave up everywhere (except 0) and $(0,0)$ is not an inflection point. $(2, -48)$ is a minimum.

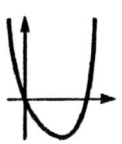

22. $y = x^4 + 4x^3 - 16x$

(1) The graph is continuous for all x.

(2) The intercepts are $(0,0)$ and $(1.68, 0)$.

Use Newton's method or zero feature of calculator for x-intercept

(3) As $x \to \pm\infty$, $y \to \infty$

(4)   The graph is not symmetrical with respect to origin or either axis

(5) $f'(x) = 4x^3 + 12x^2 - 16 = 0$ for $x = -2$ and $x = 1$

       $(-2, 16)$ and $(1, -11)$ are critical points

(6) $f''(x) = 12x^2 + 24x = 0$; for $x = -2$ and $x = 0$
       $(-2, 16)$ and $(0,0)$ are possible inflection points

(7) $f''(-2) = 0$, second derivative test is inconclusive
       $f'(x) < 0$ for $x < -2$ and for $-2 < x < 1$, $(-2, 16)$ is neither a local
       maximum nor a local minimum
       $f''(1) = 0$, second derivative test is inconclusive
       $f'(x) < 0$ for $-2 < x < 1$ and $f'(x) > 0$ for $x > 1$
       $(1, -11)$ is a local minimum
       $f''(x) > 0$ for $x < -2$ or $x > 0$ and $f''(x) < 0$ for $-2 < x < 0$
       $(-2, 16)$ and $(0,0)$ are inflection points

23. $y = \dfrac{x^2}{\sqrt{x^2 - 1}}$

The domain is all x for which $x^2 - 1 > 0$
$$x^2 > 1$$
$$\sqrt{x^2} > 1$$
$$|x| > 1$$
$$x < -1 \quad \text{or } x > 1$$

(1)   The graph is continuous for all x in domain.

(2)   There are no intercepts

(3)   As $x \to \pm\infty$, $y \to \infty$; as $x \to \pm 1, y \to \infty$

       $x = \pm 1$ are vertical asymptotes

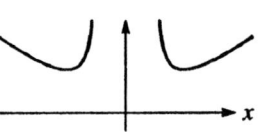

(4)   The graph is symmetrical with respect to the
       y-axis

(5) $f'(x) = \dfrac{x(x^2 - 2)}{(x^2 - 1)^{3/2}}$ for $x = \pm\sqrt{2}$

       $(\pm\sqrt{2}, 2)$ are critical points

(6) $f''(x) = \dfrac{x^2 + 2}{(x^2 - 1)^{5/2}}$, there is no x for which
       $f''(x) = 0$ and thus no inflection points

(7) $f''(\pm\sqrt{2}) = 4 > 0$, $(\pm\sqrt{2}, 2)$ are local minimums
       $f''(x) > 0$ for all x in domain of $f(x)$ concave up
       for $x < -1$ or $x > 1$

24. $y = x^3 + \dfrac{3}{x}$

domain: all x except 0

(1) The graph is continuous for all x in domain.

(2) There are no intercepts

(3) As $x \to -\infty$, $y \to -\infty$; as $x \to \infty$, $y \to \infty$

as $x \to 0^+$, $y \to \infty$; as $x \to 0^-$, $y \to -\infty$

$x = 0$ is a vertical asymptote

(4) The graph is symmetrical with respect to the origin

(5) $f'(x) = 3x^2 - \dfrac{3}{x^2} = 0$ for $x = \pm 1$

$(-1, -4)$ and $(1, 4)$ are critical points

(6) $f''(x) = 6x + \dfrac{6}{x^3}$, there is no x for which $f''(x) = 0$
and thus no inflection points

(7) $f''(-1) = -12 < 0$, $(-1, -4)$ is a local maximum
$f''(x) = 12 > 0$, $(1, 4)$ is a local minimum

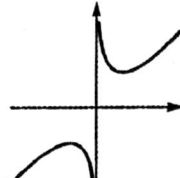

25. $y = f(x)$

$= 4x^3 + \dfrac{1}{x}$

$dy = f'(x)dx$

$= \left(12x^2 - \dfrac{1}{x^2}\right) dx$

26. $y = f(x) = \dfrac{1}{(2x - 1)^2}$

$= (2x - 1)^2$

$f'(x) = -2(2x - 1)^{-3}(2)$

$= \dfrac{-4}{(2x - 1)^3}$

$dy = f'(x)dx = \dfrac{-4\,dx}{(2x - 1)^3}$

27. $y = f(x) = x\sqrt[3]{1 - 3x} = x(1 - 3x)^{1/3}$

$f'(x) = \dfrac{1}{3}x(1 - 3x)^{-2/3}(-3) + (1 - 3x)^{1/3} = \dfrac{-x}{(1 - 3x)^{2/3}} + \dfrac{(1 - 3x)^{1/3}}{1}$

$= \dfrac{-x + (1 - 3x)^{2/3}(1 - 3x)^{1/3}}{(1 - 3x)^{2/3}}$

$= \dfrac{1 - 4x}{(1 - 3x)^{2/3}}$

$dy = f'(x)dx = \dfrac{(1 - 4x)dx}{(1 - 3x)^{2/3}}$

28. $s = f(t) = \sqrt{\dfrac{2+t}{2-t}} = (2+t)^{1/2}(2-t)^{-1/2}$

$f'(t) = (2+t)^{1/2}\left(-\dfrac{1}{2}\right)(2-t)^{-3/2}(-1) + (2-t)^{-1/2}\left(\dfrac{1}{2}\right)(2+t)^{-1/2}(1)$

$f'(t) = \dfrac{(2+t)^{1/2}}{2(2-t)^{3/2}} + \dfrac{1}{2(2-t)^{1/2}(2+t)^{1/2}}$

$f'(t) = \dfrac{2(2+t)^{1/2}(2-t)^{1/2}(2+t)^{1/2} + 2(2-t)^{3/2}}{4(2-t)^2(2+t)^{1/2}}$

$f'(t) = \dfrac{(2+t)(2-t)^{1/2} + (2-t)^{3/2}}{2(2-t)^2(2+t)^{1/2}}$

$f'(t) = \dfrac{(2-t)^{1/2}(2+t+2-t)}{2(2-t)^2(2+t)^{1/2}} = \dfrac{4}{2(2-t)^{3/2}(2+t)^{1/2}}$

$f'(t) = 2\sqrt{\dfrac{1}{(2-t)^3(2+t)}}$

$ds = f'(t)dt = 2\sqrt{\dfrac{1}{(2-t)^3(2+t)}}\,dt$

29. $y = f(x) = x^3, x = 2, \Delta x = 0.1$
$\Delta y - dy = f(x+\Delta x) - f(x) - f'(x)dx = (x+\Delta x)^3 - x^3 - 3x^2\,dx = 2.1^3 - 2^3 - 3\cdot 2^2(0.1)$
$\qquad = 0.061$

30. $y = f(x) = 6x^2 - x, x = 3, \Delta x = 0.2$
$\Delta y - dy = f(x+\Delta x) - f(x) - f'(x)dx$
$\qquad = 6(x+\Delta x)^2 - (x+\Delta x) - [6x^2 - x] - (12x-1)\Delta x$
$\qquad = 6(3.2)^2 - (3.2) - [6(3)^2 - 3] - (12(3)-1)(0.2) = 0.24$

31. $f(x) = \sqrt{x^4 + 3x^2 + 8} = (x^4 + 3x^2 + 8)^{1/2}, a = 2$
$f(2) = (2^4 + 3(2)^2 + 8)^{1/2} = 6, (2,6)$ is on a graph

$f'(x) = \dfrac{1}{2}(x^4 + 3x^2 + 8)^{-1/2}(4x^3 + 6x)$

$f'(2) = \dfrac{1}{2}(2^4 + 3(2)^2 + 8)^{-1/2}(4(2)^3 + 6(2)) = \dfrac{11}{3}$

$y - y_1 = m(x - x_1)$

$y - 6 = \dfrac{11}{3}(x - 2)$

$y = L(x) = \dfrac{11}{3}x - \dfrac{22}{3} + 6 = \dfrac{11}{3}x - \dfrac{22}{3} + \dfrac{18}{3} = \dfrac{11}{3}x - \dfrac{4}{3}$

$L(x) = \dfrac{1}{3}(11x - 4)$

32. $f(x) = x^2(x+1)^4, a = -2$
$f(-2) = (-2)^2(-2+1)^4 = 4, (-2,4)$ is on graph
$f'(x) = x^2(4)(x+1)^3 + 2x(x+1)^4$
$f'(-2) = (-2)^2(4)(-2+1)^3 + 2(-2)(-2+1)^4 = -20$
$L(x) = -20(x+4) + 4 = -20x - 36$

33. $V = f(r) = \dfrac{4}{3}\pi r^3, r = 3.500, \Delta r = 0.012$

$dV = f'(r)dr = 4\pi r^2 dr = 4\pi(3.500)^2(0.012)$

$dV = 1.85 \text{ m}^3$

34. $P = 460 + 230 \, m^2$

$dP = 460 \, m \, dm$

$ = 460(0.86)(0.03)$

$ = 12 \text{ W}$

35. $Z = \sqrt{R^2 + X^2} = (R^2 + X^2)^{1/2}$

$dZ = \dfrac{1}{2}(R^2 + X^2)^{-1/2}(2R)dR$

$dZ = \dfrac{R\,dR}{\sqrt{R^2 + X^2}}$

relative error $= \dfrac{dZ}{Z} = \dfrac{R\,dR}{R^2 + X^2}$

36. $V = \dfrac{4}{3}\pi r^3, \, dV = 4\pi r^2 dr$

relative error in volume $= \dfrac{dV}{V}$

$ = \dfrac{4\pi r^2 dt}{\frac{4}{3}\pi r^3}$

$ = 3\,\dfrac{dr}{r}$

relative error in volume = 3(relative error in radius)

37. $y = x^2 + 2$ and $y = 4x - x^2$

$y' = 2x; \, y' = 4 - 2x$

$2x = 4 - 2x; \, 4x = 4; \, x = 1$

The point $(1,3)$ belongs to both graphs; the slope of the tangent line is 2.

$y - y_1 = 2(x - x_1); \, y - 3 = 2(x - 1); \, y - 3 = 2x - 2$

$2x - y + 1 = 0$ is the equation of the tangent line.

38. $y = x^4 - 8x$

$y' = 4x^3 - 8 = -4$

since TL is perpendicular to $4y - x + 5 = 0$ which has $m = \dfrac{1}{4}$

$4x^3 = 4$

$x^3 = 1$

$x = 1$

$y = 1^4 - 8(1)$

$ = -7, \, (1, -7)$ is point of tangency.

$y - y_1 = m(x - x_1)$

$y + 7 = -4(x - 1) = -4x + 4$

$y = -4x - 3$

39. $y = k(x^4 - 30x^3 + 1000x) = 0$. By inspection $x = 0.0$ m is the first value of x where the deflection is zero. From the graph of $y = x^4 - 30x^3 + 1000x$ there is a zero between $x = 6$ and $x = 7$. Letting $x_1 = 6$, successive iterations of Newton's method give

$x_1 = 6$

$x_2 = 6.593023253$

$x_3 = 6.527855923$

$x_4 = 6.527036576$

$x_5 = 6.527036447$

$x_6 = 6.527036447$

$x = 6.527$ m is the second value where the defection is zero.

40. Graph $y = (3.00 + x)(5.00 + x)(8.00 + x) - 2(3.00)(5.00)(8.00)$ and find the zero using Newton's method. From the graph, the zero is between $x = 1$ and $x = 2$. Let $x_1 = 1$. Successive iterations of Newton's method give

$x_1 = 1$

$x_2 = 1.210526314$

$x_3 = 1.20355493$

$x_4 = 1.203547103$

$x_5 = 1.203547103$

Each edge should be increased 1.20 ft to double the volume.

41. $x = 8t \qquad y = -0.15t^2 \qquad v = \sqrt{8^2 + (-3.6)^2}$

$\dfrac{dx}{dt} = 8 \qquad \dfrac{dy}{dt} = -0.30t \qquad v = \sqrt{64 + 12.96}$

$v_x|_{t=12} = 8 \quad v_y|_{t=12} = -3.6 \quad v|_{t=12} = \sqrt{76.96}$

$\phantom{v_x|_{t=12} = 8 \quad v_y|_{t=12} = -3.6 \quad v|_{t=12}} = 8.8 \text{ m/s}$

$\tan\theta = \dfrac{-3.6}{8} = -0.45; \, \theta = 336°$

42. $y = \dfrac{225}{x}, (10.0, 22.5)$

$$v_y = \frac{dy}{dt} = \frac{dy}{dx}\frac{dx}{dt} = \frac{-225}{x^2}v_x\bigg|_{x=10.0} = \frac{-225}{10.0^2}v_x$$

$$v_x^2 + v_y^2 = 120^2$$

$$v_x^2 + \left(\frac{-225}{10.0^2}\right)^2 v_x^2 = 120^2$$

$$v_x = \sqrt{\frac{120^2}{1 + \left(\frac{-225}{10.0^2}\right)^2}}$$

$$v_x = 48.7 \text{ km/h}$$

$$v_y = \frac{-225}{10.0^2}v_x = -109.7 \text{ km/h}$$

43. $d = \dfrac{1000}{\sqrt{T^2 - 400}}$

$$\frac{dd}{dt} = \frac{dd}{dT}\frac{dT}{dt} = -\frac{-1000T}{(T^2 - 400)^{3/2}}\frac{dT}{dt}\bigg|_{T=28.0, \frac{dT}{dt}=2.00}$$

$$\frac{dd}{dt} = \frac{-1000(28.0)}{(28.0^2 - 400)^{3/2}}(2.00) = -7.44 \text{ cm/s}$$

44. $Z = \sqrt{48 + R^2}$

$$\frac{dZ}{dt} = \frac{dZ}{dR}\frac{dR}{dt} = \frac{R}{\sqrt{48 + R^2}}(0.45)\bigg|_{R=6.5}$$

$$= \frac{6.5}{\sqrt{48 + 6.5^2}}(0.45)$$

$$\frac{dZ}{dt} = 0.31 \ \Omega/\text{min}$$

45. $P = 0.030r^3 - 2.6r^2 + 71r - 200, 6 \leq r \leq 30 \text{ m}^3/\text{s}$

$$\frac{dP}{dr} = 0.09r^2 - 5.2r + 71 = 0$$

$$r = 22.1, r = 35.6 \text{ (reject, } 6 \leq r \leq 30)$$

$$\frac{d^2P}{dr^2} = 0.18r - 5.2\big|_{22.1} = -1.222 < 0, \text{ maximum}$$

P is a maximum when rate is 22.1 m^3/s

46. $h = 1600t - 16t^2$

$$\frac{dh}{dt} = 1600 - 32t = 0, t = 50, (50, 40,000) \text{ is}$$
a critical point and since

$$\frac{d^2h}{dt^2} = -32 < 0, (50, 40,000) \text{ is a maximum}$$

The maximum altitude the rocket attains is 40,000 ft.

47. $f(0) = 2$ implies y-intercept is $(0, 2)$

$f'(x) < 0$ for $x < 0$ implies f is decreasing for $x < 0$

$f'(x) > 0$ for $x > 0$ implies f is increasing for $x > 0$

These three conditions imply $(0, 2)$ is a minimum.

$f''(x) > 0$ for all x implies f is concave up everywhere.

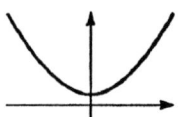

48. $f(0) = 1$ implies the y-intercept is $(0, 1)$

$f'(0) = 0$ implies a horizontal tangent line at $(0, 1)$

$f'(x) > 0$ for $|x| > 0$ implies f is increasing for $|x| > 0$

$f''(x) < 0$ for $x < 0$ implies f is concave down for $x < 0$

$f''(x) > 0$ for $x > 0$ implies f is concave upward for $x > 0$

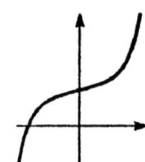

49.

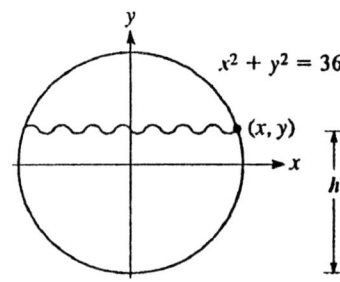

$$w = 2x, h = 6 + y$$

$$w = 2\sqrt{36 - y^2} = 2 \cdot \sqrt{36 - (h - 6)^2}$$

$$\frac{dw}{dt} = \frac{dw}{dh} \cdot \frac{dh}{dt} = \frac{-2(h - 6)}{\sqrt{36 - (h - 6)^2}} \cdot \frac{dh}{dt}$$

$$\frac{dw}{dt} = \frac{-2(1.5 - 6)}{\sqrt{36 - (1.5 - 6)^2}} \cdot (-0.250)$$

$$= -0.567 \text{ ft/s}$$

50. $I = \dfrac{E}{R+r}$

$\dfrac{dI}{dt} = \dfrac{-E}{(R+r)^2}(0.250)\bigg|_{E=3.10,R=6.25,r=0.230}$

$\dfrac{dI}{dt} = \dfrac{-3.10}{(6.25+0.230)^2}(0.250)$

$= -0.0185$ A/min

The current is decreasing at a rate of 18.5 mA/min

51. $A = \pi r^2$

$\dfrac{dA}{dt} = \dfrac{dA}{dt}\dfrac{dr}{dt}$

$= 2\pi r\ \dfrac{dr}{dt} = 2\pi r(15)\bigg|_{r=400}$

$= 2\pi(400)(15)$

$\dfrac{dA}{dt} = 38{,}000$ m^2/min

52. Let $x =$ distance from home plate to player, then

$x^2 = 90.0^2 + (18.0t)^2 = 90.0^2 + 18.0^2 t^2$

$2x\dfrac{dx}{dt} = 2(18.0)^2 t$

$\dfrac{dx}{dt} = \dfrac{18.0^2 t}{x}\bigg|_{x=\sqrt{90^2+(18.0t)^2},\,t=\frac{40.0}{18.0}}$

$= \dfrac{18.0^2\left(\frac{40.0}{18.0}\right)}{\sqrt{90^2 + 18.0^2\left(\frac{40.0}{18.0}\right)^2}}$

$= 7.31$ ft/s

53. $y + 2x = 200;\ y = -2x + 200$

$A = xy = x(-2x + 200) = -2x^2 + 200x$

$A' = -4x + 200 = 0$

$x = \dfrac{-200}{-4} = 50;\ y = 200 - 2(50) = 100$

$A = 50(100) = 5000$ cm^2

54. $A = (l + 24.0)(w + 16.0)$ $\qquad\qquad A = lw = 1200$

$A = lw + 16.0l + 24.0w + 24.0(16.0)$ $\qquad l = \dfrac{1200}{w}$

$A = 1200 + 16.0\dfrac{1200}{w} + 24.0w + 24.0(16.0)$

$\dfrac{dA}{dw} = \dfrac{-16.0(1200)}{w^2} + 24.0 = 0$

$w = \sqrt{\dfrac{16.0(1200)}{24.0}} = 28.3$ ft

$l = \dfrac{1200}{w} = 42.4$ ft

which gives the minimum area since

$\dfrac{d^2A}{dt^2} = \dfrac{16.0(1200)(2)}{w^3} > 0$ for $w > 0$.

55. $y = \dfrac{300}{0.0005x^2 + 2} - 50,\ a = 50$

$\dfrac{dy}{dx} = \dfrac{(0.0005x^2 + 2)(0) - 300(2(0.0005x))}{(0.0005x^2 + 2)^2}$

$\dfrac{dy}{dx} = \dfrac{-300(2(0.0005x))}{(0.0005x^2 + 2)^2}\bigg|_{x=50} = \dfrac{-300(2(0.0005(50)))}{(0.0005(50)^2 + 2)^2}$

$\dfrac{dy}{dx} = -1.42$

$L(x) = -1.42(x - 50) + 423$

$y = 42.3$ for $x = 50$

$y = L(x) = -1.42x + 113$

$\dfrac{dy}{dx} = \dfrac{-1,200,000x}{(x^2+4000)^2} < 0$ for all $x > 0$, thus graph is decreasing everywhere which implies $(0,100)$ is the maximum.

$$\frac{d^2y}{dx^2} = \frac{1,200,000(3x^2 - 4000)}{(x^2+4000)^3} = \begin{cases} <0 & \text{for } 0 < x < 37 \\ 0 & \text{for } x = 37 \\ >0 & \text{for } x > 37 \end{cases}$$

which implies $(37, 63)$ is an inflection point; intercepts are $(0, 100)$ and $(89, 0)$ using **Newton's method**

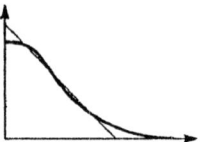

56. $S(t) = \dfrac{500t}{(t+4)^2}, t \geq 0$ since t is months

   (1)   Graph is continuous for all t in domain

   (2) $(0,0)$ is the only intercept

   (3)   As $x \to \infty, S \to 0$ which implies the x-axis is a horizontal asymptote

   (4)   the graph is not symmetrical with respect to the origin or either axis

   (5) $S'(t) = \dfrac{-5000(t-4)}{(t+4)^2} = 0$ for $t = 4$ so $(4, 313)$ is a critical point

   (6) $S''(t) = \dfrac{10,000(t-8)}{(t+4)^4} = 0$ for $t = 8$ so $(8, 278)$ is a possible

   inflection point

   (7) $S''(4) < 0$ so $(4, 313)$ is a local maximum
   $S''(t) < 0$ for $t > 8$ and $S''(t) < 0$ for $t < 8$ so $(8, 278)$ is an inflection point

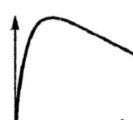

57. let $y = $ altitude of the plane
   $z = $ distance from plane to radar station
   $x = $ distance from a point on the ground directly below the plane
   to the radar station

$$z^2 = x^2 + y^2$$

$$2z \frac{dz}{dt} = 2x \frac{dx}{dt} + 2y \frac{dy}{dt}$$

$$z \frac{dz}{dt} = x \frac{dx}{dt} + y(0)$$

$$\sqrt{x^2 + y^2} \frac{dz}{dt} = x \frac{dx}{dt}$$

$$\sqrt{5.00^2 + \left(\frac{8000}{5280}\right)^2}\,(-680) = 5.00\frac{dx}{dt}$$

$$\frac{dx}{dt} = -710$$

Actual speed of plane $= |-710| = 710$ mi/h

58.
$$\frac{r}{h} = \frac{1.0}{3.0}; \frac{dr}{dt} = \frac{1.0}{3.0}\frac{dh}{dt}$$

$$V = \frac{1}{3}\pi r^2 h$$

$$\frac{dV}{dt} = \frac{\pi}{3}\left[r^2\frac{dh}{dt} + 2rh\frac{dr}{dt}\right]$$

$$\frac{dV}{dt} = \frac{\pi}{3}\left[\left(\frac{(1.0)(2.8)}{3.0}\right)^2(-0.050) + 2\left(\frac{1.0(2.8)}{3.0}\right)(2.8)\left(\frac{1.0}{3.0}(-0.050)\right)\right]$$

$$\frac{dV}{dt} = -0.14 \text{ cm}^3/\text{min}$$

The volume is decreasing 0.14 cm^3/min when the height is 2.8 cm.

59.
$$\frac{1}{C} = \frac{1}{C_1} + \frac{1}{C_2}$$

$$C_1 + C_2 = 12$$

$$C_T = \frac{C_1 + C_2}{C_1 C_2} = \frac{12}{C_1(12 - C_1)} = \frac{12}{12C_1 - C_1^2}$$

$$\frac{dC_T}{dC_1} = \frac{-12(12 - 2C_1)}{(12C_1 - C_1^2)^2} = 0, C_1 = 6$$

which is a minium since

$$\left.\frac{d^2 C_T}{dC_1^2}\right|_{C_1=6} > 0.$$

Each capacitor is 6 μF.

60. Let x = distance from the point directly below A to B

L = total length of cable

$$L = \sqrt{108^2 + x^2} + \sqrt{75^2 + (270 - x)^2}, 0 < x < 270$$

$$\frac{dL}{dx} = \frac{x}{\sqrt{108^2 + x^2}} - \frac{270 - x}{\sqrt{75^2 + (270 - x)^2}} = 0$$

$$x\sqrt{75^2 + (270 - x)^2} = (270 - x)\sqrt{108^2 + x^2}$$
$$x^2(75^2 + (270 - x)^2) = (270 - x)^2(108^2 + x^2)$$
$$75^2 x^2 + x^2(270 - x)^2 = (270 - x)^2 x^2 + 108^2(270 - x)^2$$
$$75^2 x^2 = 108^2(270 - x)^2$$
$$75^2 x^2 = 108^2(270)^2 - 108^2(2)(270)x + 108^2 x^2$$
$(108^2 - 75^2)x^2 - 108^2(2)(270)x + 108^2(270)^2 = 0$ and using quadratic formula $x = 159$, $x = 884$ which must be rejected since $0 < x < 270$. $x = 159$ is a minimum since $\left.\dfrac{d^2 L}{dx^2}\right|_{x=159} > 0.$

The length of rope will be a minimum if B is located 159 ft from base of perpendicular from A.

61. $V = x^2 y;\ x^2 + 4xy = 27, y = \dfrac{27 - x^2}{4x}$

$V = x^2 \cdot \dfrac{27 - x^2}{4x} = \dfrac{27x - x^3}{4}$

$V' = \dfrac{27}{4} - \dfrac{3x^2}{4} = 0$ for $x = 3$

$V'' = \dfrac{-64}{4} < 0$ for $x > 0, V$ is concave down

$V = \dfrac{27(3) - 3^2}{4} = 13.5$ ft^3 is the maximum volume.

62. $A = \dfrac{1}{2} r^2 \theta = 1, \theta = \dfrac{2}{r^2}$

$p = r\theta + 2r = r\left(\dfrac{2}{r^2}\right) + 2r = \dfrac{2}{r} + 2r$

$\dfrac{dp}{dr} = \dfrac{-2}{r^2} + 2 = 0, r = -1$ which is a minimum since $\dfrac{d^2 p}{dr^2} = \dfrac{4}{r^3} > 0$ for $r = -1$.

The perimeter is a minimum for $r = 1, \theta = 2$

63. $V = (12 - x)(10 - 2x)(x), 0 < x < 5$
$V(x) = 120x - 34x^2 + 2x^3$
$V'(x) = 120 - 68x + 6x^2 = 0$ for $x = 2.2, 9.2$

using the quadratic formula. 9.2 is rejected since $0 < x < 5$.
Since $V''(x) = -68 + 12x^2 \big|_{x=2.2} < 0$ the volume of the drawer is a maximum for $x = 2.2$ in.

64. $p = \dfrac{1}{2}(2\pi r) + 2r + 2h = 12$

$h = \dfrac{12 - (\pi + 2)r}{2}$

$A = \dfrac{1}{2}\pi r^2 + 2rh$

$A(r) = \dfrac{1}{2}\pi r^2 + 2r\dfrac{12 - (\pi + 2)r}{2}$

$A(r) = \dfrac{1}{2}\pi r^2 + 12r - (\pi + 2)r^2$

$A'(r) = \pi r + 12 - 2(\pi + 2)r = 0$ for $r = 1.68$

which is a minimum since

$A''(r) = \pi - 2(\pi + 2) < 0.$

$h = \dfrac{12 - (\pi + 2)r}{2} = 1.68$

The dimensions that will admit the most light are $r = h = 1.68$ ft.

65. $V = \dfrac{1}{3}\pi r^2 h = \dfrac{1}{3}\pi r^2 \cdot r = \dfrac{1}{3}\pi r^3$

$\dfrac{dV}{dt} = \pi r^2 \dfrac{dr}{dt}$

$100 = \pi(10.0)^2 \dfrac{dr}{dt}$

$\dfrac{dr}{dt} = 0.318$ ft/min

66. Let x = distance along shore form P toward A

$$t(x) = \frac{\sqrt{4.0^2 + x^2}}{3.0} + \frac{5.0x - x}{5.0}, 0 < x < 5.0$$

$$t'(x) = \frac{x}{3.0\sqrt{4.0^2 + x^2}} - \frac{1}{5.0} = 0$$

$$x = \frac{3.0\sqrt{4.0^2 + x^2}}{5.0}$$

$$x^2 = \left(\frac{3.0}{5.0}\right)^2 (4.0^2 + x^2)$$

$$x^2 \left(1 - \left(\frac{3.0}{5.0}\right)^2\right) = 4.0^2 \left(\frac{3.0}{5.0}\right)^2$$

$$x = 3.0 \text{ which is a minimum}$$

since
$$t''(x) = \frac{3.0\sqrt{4.0^2 + x^2} - x\frac{3.0x}{\sqrt{4.0^2+x^2}}}{(3.0\sqrt{4.0^2 + x^2})^2}\Bigg|_{x=3.0} > 0.$$

The person can reach A in the least time by reaching the shore 3.0 km from P toward A.

67. Let s = cost per cm^2 of stainless steel
$10s$ = cost per cm^2 of silver

$$V = \pi r^2 h = 314$$

$$r = \frac{314}{\pi h} \text{ and } h = \frac{314}{\pi r^2}$$

$$\text{Cost} = c = s(\pi r^2 + 2\pi rh) + 10s(\pi r^2)$$
$$c = 11s\pi r^2 + 2\pi srh, r > 0$$

$$c = 11s\pi r^2 + 2\pi sr\left(\frac{314}{\pi r^2}\right)$$

$$c(r) = 11s\pi r^2 + \frac{2s(314)}{r} > 0$$

$$c'(r) = 22s\pi r - \frac{2s(314)}{r^2} = 0 \text{ for } r = 2.09$$

which is a minimum since
$$c''(r) = 22s\pi + \frac{4s(314)}{r^3} > 0$$

$$h = \frac{314}{\pi r^2} = 23.0$$

The most economical dimensions of the container are $r = 2.09$ cm, $h = 23.0$ cm

68. When the string breaks the object has travelled counter clockwise through an angle of $2.00(1.05)$ rad and has coordinates $x_0 = 2.00\cos(2.00(1.05))$, $y_0 = 2.00\sin(2.00(1.05))$. The slope of the tangent line is $\frac{-x_0}{y_0}$ and the equation of the tangent line is

$$y - y_0 = \frac{-x_0}{y_0}(x - x_0)$$

$$y = \frac{-x_0}{y_0}x + \frac{x_0^2}{y_0^2} + y_0$$

$$y = 0.585x + 2.32$$

69.

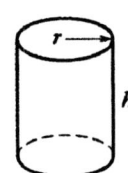

The amount of plastic used is determined by the surface area $S = 2\pi r^2 + 2\pi rh$. $V = \pi r^2 h = $ constant from which $h = \frac{\text{constant}}{\pi r^2}$.

$$S = 2\pi r^2 + 2\pi r \frac{\text{constant}}{\pi r^2}$$

$$S(r) = 2\pi r^2 + \frac{2(\text{constant})}{r}$$

$$\frac{dS}{dr} = 4\pi r - \frac{2(\text{constant})}{r^2} = 0 \text{ for minimum}$$

$$4\pi r^3 = 2(\text{constant})$$

$$r^3 = \frac{\text{constant}}{2\pi}$$

$$\frac{h}{r} = \frac{\text{constant}}{\pi r^3} = \frac{\text{constant}}{\pi \frac{\text{constant}}{2\pi}} = 2$$

The height should be twice the radius to minimize the surface area. In finding the $\frac{h}{r}$ ratio, the constant volume divides out so it is not necessary to specify the volume.

INTEGRATION

5.1 Antiderivatives

1. $3x^2$; the power of x required in the antiderivative is 3. Therefore, we must multiply by $\frac{1}{3}$. The antiderivative of $3x^2$ is $\frac{1}{3}(3x^3) = x^3$. $a = 1$.

2. $f(x) = 5x^4$; power of x required is 5, therefore x^5.

   $\frac{d}{dx}ax^5 = a5x^4 = 5x^4$, therefore $a = 1$

   Therefore, antiderivative of $5x^4$ is x^5, $a = 1$.

3. $f(x) = 18x^5$; power of x required is 6, therefore x^6.

   $\frac{d}{dx}ax^6 = a6x^5 = 18x^5$

   Therefore $6a = 18$; $a = 3$, therefore $3x^6$
   Therefore, antiderivative of $18x^5$ is $3x^6, a = 3$.

4. $f(x) = 40x^7$; power of x required is 8, therefore x^8.

   $\frac{d}{dx}ax^8 = a8x^7 = 40x^7$

   $a8 = 40$; $a = 5$, therefore $5x^8$
   Therefore, antiderivative of $40x^7$ is $5x^8, a = 5$.

5. The power of x required in the antiderivative of $f(x) = 9\sqrt{x}$ is $\frac{3}{2}$. Multiply by $\frac{2}{3}$. The antiderivative of $9\sqrt{x}$ is $\frac{2}{3}\cdot 9x^{3/2}$. $a = 6$.

6. $f(x) = 10x^{1/4}$, $F(x) = ax^{5/4}$

   $\frac{dF(x)}{dx} = \frac{5}{4}\cdot ax^{1/4} = 10x^{1/4} \Rightarrow a = 8$

   $F(x) = 8x^{5/4}$

7. $f(x) = \frac{1}{x^2} = x^{-2}$, $F(x) = \frac{a}{x}$

   $\frac{dF(x)}{dx} = \frac{-a}{x^2} = \frac{1}{x^2} \Rightarrow a = -1$

   $F(x) = \frac{-1}{x}$

8. $f(x) = \frac{6}{x^4}$; $f(x) = 6x^{-4}$; power required is -3;

   $\frac{d}{dx}(ax^{-3}) = -3ax^{-4}$; $-3a = 6$; $a = -2$,

   therefore $-2x^3$

   Therefore, antiderivative is $\frac{-2}{x^3}, a = -2$.

9. The power of x required in the antiderivative of $\frac{5}{2}x^{3/2}$ is $\frac{5}{2}$. Multiply by $\frac{2}{5}$. The antiderivative of $\frac{5}{2}x^{3/2}$ is $\frac{2}{5}\left(\frac{5}{2}\right)x^{5/2} = x^{5/2}$.

10. $f(x) = \frac{4}{3}x^{1/3}$; power of x required is $\frac{1}{3} + 1 = \frac{4}{3}$, therefore $x^{4/3}$

    $\frac{d}{dx}(kx^{4/3}) = k\frac{4}{3}x^{1/3}$; $\frac{4}{3}k = \frac{4}{3}$; $k = 1$

    Therefore, antiderivative of $\frac{4}{3}x^{1/3}$ is $x^{4/3}$.

11. $f(t) = 6t^3 + 2$

    $F(t) = \frac{6t^4}{4} + 2t = \frac{3}{2}t^4 + 2t$

12. $f(x) = 12x^5 + 2x$; $12x^5 \rightarrow ax^6$;

    $\frac{d}{dx}ax^6 = a6x^5 = 12x^5$

    $6a = 12$; $a = 2$, therefore $2x^6$

    $2x \rightarrow ax^2$; $\frac{d}{dx}ax^2 = a2x = 2x$

    $2a = 2$; $a = 1$, therefore x^2
    Therefore, antiderivative of $12x^5 + 2x$ is $2x^6 + x^2$.

13. $f(x) = 2x^2 - x$; $2x^2 \rightarrow ax^3$; $\frac{d}{dx}(ax^3) = a3x^2$

    $3a = 2$; $a = \frac{2}{3}$, therefore $\frac{2}{3}x^3$; $x \rightarrow ax^2$;

    $\frac{d}{dx}(ax^2) = a2x$

    $2a = 1$; $a = \frac{1}{2}$, therefore $\frac{1}{2}x^2$

    Antiderivative of $2x^2 - x$ is $\frac{2}{3}x^3 - \frac{1}{2}x^2$.

14. $f(x) = x^2 - 5$; $x^2 \rightarrow ax^3$; $\frac{d}{dx}(ax^3) = a3x^2 = x^2$

    $3a = 1$; $a = \frac{1}{3}$, therefore $\frac{1}{3}x^3$

    $5 \rightarrow 5x$
    Therefore, antiderivative of $x^2 - 5$ is $\frac{1}{3}x^3 - 5x$.

15. $f(x) = 2\sqrt{x} + 3$; $f(x) = 2x^{1/2} + 3$; $x^{1/2} \to x^{3/2}$;

$\dfrac{d}{dx}(ax^{3/2}) = a\dfrac{3}{2}x^{1/2}$; $\dfrac{3}{2}a = 2$; $a = \dfrac{4}{3}$, therefore

$\dfrac{4}{3}x^{3/2}$

Antiderivative of $2\sqrt{x} + 3$ is $\dfrac{4}{3}x^{3/2} + 3x$.

16. $f(s) = 9\sqrt[3]{s} - 3$; $f(s) = 9s^{1/3} - 3$; $s^{1/3} \to s^{4/3}$;

$\dfrac{d}{ds}(as^{4/3}) = a\dfrac{4}{3}s^{1/3}$; $\dfrac{4}{3}a = 9$; $a = \dfrac{27}{4}$

Therefore, antiderivative is $\dfrac{27}{4}s^{4/3} - 3s$.

17. $f(x) = \dfrac{-7}{x^6}$; $f(x) = -7x^{-6}$; power required is -5, therefore x^{-5};

$\dfrac{d}{dx}ax^{-5} = -5ax^{-6}$; $-5a = -7$; $a = \dfrac{7}{5}$, therefore $\dfrac{7}{5}x^{-5}$

Therefore, antiderivative is $\dfrac{7}{5x^5}$.

18. $f(x) = \dfrac{8}{x^5}$; $f(x) = 8x^{-5}$; $x^{-5} \to x^{-4}$;

$\dfrac{d}{dx}ax^{-4} = -4a^{-5}$; $-4a = 8$; $a = -2$, therefore

$-2x^{-4}$

Therefore, antiderivative is $\dfrac{-2}{x^4}$.

19. $f(v) = 4v + 3\pi^2$, $\dfrac{d(av^2)}{dv} = a2v = 4v$, $a = 2$

$F(v) = 2v^2 + 3\pi^2 v$

20. $f(x) = \dfrac{1}{2\sqrt{x}} + 4$; $f(x) = \dfrac{1}{2}x^{-1/2} + 4$;

$x^{-1/2} \to ax^{1/2}$;

$\dfrac{d}{dx}ax^{1/2} = \dfrac{1}{2}ax^{-1/2}$; $\dfrac{1}{2}a = \dfrac{1}{2}$; $a = 1$, therefore,

$1x^{1/2} = \sqrt{x}$

Antiderivative is $\sqrt{x} + 4x$.

21. The power of x required for the antiderivative of x^2 is 3, so it will be multiplied by $\frac{1}{3}$, the power of x required for $2 = 2x^0$ is 1, and it will be multiplied by 1. The power of x required for x^{-2} is -1, and it will be multiplied by -1. The antiderivative of $x^2 + 2 + x^{-2}$ is $\frac{1}{3}x^3 + 2x - \frac{1}{x}$.

22. $f(x) = x\sqrt{x} - x^{-3}$; $f(x) = x^{3/2} - x^{-3}$;

$x^{3/2} \to ax^{5/2}$

$\dfrac{d}{dx}ax^{5/2} = a\dfrac{5}{2}x^{3/2}$; $\dfrac{5}{2}a = 1$; $a = \dfrac{2}{5}$, therefore

$\dfrac{2}{5}x^{5/2}$

$x^{-3} \to ax^{-2}$, $\dfrac{d}{dx}ax^{-2} = -a2x^{-3}$

$-2a = 1$; $a = -\dfrac{1}{2}$, therefore $-\dfrac{1}{2}x^{-2}$

Antiderivative is $\dfrac{2}{5}x^{5/2} - \left(-\dfrac{1}{2}x^{-2}\right) = \dfrac{2}{5}x^{5/2} + \dfrac{1}{2x^2}$.

23. $f(x) = 6(2x + 1)^5(2)$; $u^n = (2x + 1)^5$, therefore power required is 6.

$u = 2x+1$; $\dfrac{du}{dx} = 2$; $\dfrac{d}{dx}a(2x+1)^6 = 6a(2x+1)^5(2)$,

therefore $a = 1$

Therefore, antiderivative is $(2x + 1)^6$.

24. $f(R) = 3(R^2 + 1)^2(2R)$; $u^n = (R^2 + 1)^2$, power needed is 3.

$u = R^2 + 1$; $\dfrac{du}{dR} = 2R$;

$\dfrac{d}{dR}a(R^2 + 1)^3 = a3(R^2 + 1)^2(2R)$, therefore,

$a = 1$

Therefore, antiderivative is $(R^2 + 1)^3$.

25. The antiderivative requires $(p^2 - 1)^4$. We multiply by $\frac{1}{4}$. Thus, we have $\frac{1}{4}[4(p^2 - 1)^4]$. The derivative of $(p^2 - 1)^4$ is $4(p^2 - 1)^3(2p)$. The antiderivative of $4(p^2 - 1)^3(2p)$ is $(p^2 - 1)^4$.

26. $f(x) = 5(2x^4 + 1)^4(8x^3)$; $u^n = (2x^4 + 1)^4$;

$u = 2x^4 + 1$.

$\dfrac{du}{dx} = 8x^3$

Power needed is 5;

$\dfrac{d}{dx}a(2x^4 + 1)^5 = 5a(2x^4 + 1)^4(8x^3)$, $a = 1$

Therefore, antiderivative is $(2x^4 + 1)^5$.

27. $f(x) = x^3(2x^4 + 1)^4 - (2x^4 + 1)^4$; $u^n = (2x^4 + 1)^4$

$u = 2x^4 + 1$; $\dfrac{du}{dx} = 8x^3$, therefore $\dfrac{1}{8}$ needed with x^3

Power needed is 5.

$\dfrac{d}{dx}a(2x^4 + 1)^5 = 5a(2x^4 + 1)^4 8x^3$; $5a = 1$; $a = \dfrac{1}{5}$

Therefore, $(2x^4 + 1)^4 x^3$ becomes

$$\frac{1}{5}(2x^4 + 1)^5 \frac{1}{8}x^3 = \frac{1}{40}(2x^4 + 5)^5.$$

Check:

$$\frac{d}{dx}\frac{1}{40}(2x^4 + 1)^5 = \frac{1}{40}[5(2x^4 + 1)^4 8x^3]$$

$$= \frac{1}{40}[40(2x^4 + 1)^4 x^3]$$

$$= (2x^4 + 1)^4 x^3$$

Therefore, the antiderivative is $\frac{1}{40}(2x^4 + 1)^5$

28. $f(x) = x(1 - x^2)^7 = (1 - x^2)^7 x;\ u^n = (1 - x^2)^7$

$u = 1 - x^2;\ \dfrac{du}{dx} = -2x$, therefore $-\dfrac{1}{2}$ needed with the x.

Power of u needed is 8.

$$\frac{d}{dx}a(1 - x^2)^8 = 8a(1 - x^2)^7(-2x),\ 8a = 1;\ a = \frac{1}{8}$$

Therefore, $\dfrac{1}{8}(1 - x^2)^8\left(-\dfrac{1}{2}\right) = -\dfrac{1}{16}(1 - x^2)^8$

Check:

$$\frac{d}{dx}\left[-\frac{1}{16}(1 - x^2)^8\right] = -\frac{8}{16}(1 - x^2)^7(-2x)$$

$$= (1 - x^2)^7(x)$$

Therefore, the antiderivative is $-\dfrac{1}{16}(1 - x^2)^8$.

29. The antiderivative requires $(6x + 1)^{3/2}$. We multiply by $\frac{2}{3}$. Thus we have $\frac{2}{3}\left(\frac{3}{2}\right)(6x + 1)^{3/2}$. The derivative of $(6x + 1)^{3/2}$ is $\frac{3}{2}(6x + 1)^{1/2}(6)$. The antiderivative of $\frac{3}{2}(6x + 1)^{1/2}(6)$ is $(6x + 1)^{3/2}$.

30. $f(y) = \dfrac{5}{4}(1 - y)^{1/4}(-1);\ u^n = (1 - y)^{1/4};\ u = 1 - y$

$$\frac{du}{dy} = -1$$

Power of u needed is $\dfrac{5}{4}$.

$$\frac{d}{dy}a(1 - y)^{5/4} = a\frac{5}{4}(1 - y)^{1/4}(-1),\ \text{therefore } a = 1$$

Check: $\dfrac{d}{dy}(1 - y)^{5/4} = \dfrac{5}{4}(1 - y)^{1/4}(-1)$

Therefore, the antiderivative is $(1 - y)^{5/4}$.

31. $f(x) = (3x + 1)^{1/3};\ u^n = (3x + 1)^{1/3};\ u = 3x + 1$

$\dfrac{du}{dx} = 3$, therefore $\dfrac{1}{3}$ needed.

Power of u needed is $\dfrac{4}{3}$.

$$\frac{d}{dx}a(3x+1)^{4/3} = a\frac{4}{3}(3x+1)^{1/3}(3);\ \frac{4}{3}a = 1;\ a = \frac{3}{4}$$

Therefore, $\dfrac{3}{4}(3x + 1)^{4/3}\dfrac{1}{3} = \dfrac{1}{4}(3x + 1)^{4/3}$

Check:

$$\frac{d}{dx}\frac{1}{4}(3x + 1)^{4/3} = \frac{1}{4}\cdot\frac{4}{3}(3x + 1)^{1/3}(3)$$

$$= (3x + 1)^{1/3}$$

Therefore, the antiderivative is $\dfrac{1}{4}(3x + 1)^{4/3}$.

32. $f(x) = (4x + 3)^{1/2};\ u^n = (4x + 3)^{1/3};\ u = 4x + 3$

$\dfrac{du}{dx} = 4$, therefore $\dfrac{1}{4}$ needed

Power of u needed is $\dfrac{3}{2}$.

$$\frac{d}{dx}a(4x+3)^{3/2} = a\frac{3}{2}(4x+3)^{1/2}(4);\ \frac{3}{2}a = 1;\ a = \frac{2}{3}$$

Therefore, $\dfrac{2}{3}(4x + 3)^{3/2}\left(\dfrac{1}{4}\right) = \dfrac{1}{6}(4x + 3)^{3/2}$

Check:

$$\frac{d}{dx}\frac{1}{6}(4x + 3)^{3/2} = \frac{1}{6}\cdot\frac{3}{2}(4x + 3)^{1/2}(4)$$

$$= (4x + 3)^{1/2}$$

Therefore, the antiderivative is $\dfrac{1}{6}(4x + 3)^{3/2}$.

5.2   The Indefinite Integral

1. $\displaystyle\int 2x\,dx = 2\int x\,dx;\ u = x;\ du = dx;\ n = 1$

$$2\int x\,dx = 2\left(\frac{x^{1+1}}{1+1}\right) + C = x^2 + C$$

2. $\displaystyle\int 5x^4\,dx = \int x^4\,dx;\ u = x;\ du = dx;\ n = 4$

$$5\int x^4\,dx = 5\left[\frac{x^5}{5}\right] + C = x^5 + C$$

3. $\int x^7 dx$; $u = x$; $du = dx$; $n = 7$

$$\int x^7 dx = \frac{x^8}{8} + C = \frac{1}{8}x^8 + C$$

4. $\int 0.6y^5 dy = 0.6\frac{y^6}{6} + C = 0.1y^6 + C$

5. $\int 2x^{3/2} dx$; $u = x$; $du = dx$; $n = \frac{3}{2}$

$$\int 2x^{3/2} dx = \frac{2x^{(3/2)+1}}{\frac{3}{2}+1} + C$$

$$= \frac{2x^{5/2}}{\frac{5}{2}} + C$$

$$= \frac{4}{5}x^{5/2} + C$$

6. $\int 6\sqrt[3]{x}\,dx = 6\int x^{1/3} dx$; $u = x$; $du = dx$; $n = \frac{1}{3}$

$$6\int x^{1/3} dx = 6\left[\frac{x^{4/3}}{\frac{4}{3}}\right] + C = 6 \cdot \frac{3}{4}x^{4/3} + C =$$

$$\frac{9}{2}x^{4/3} + C$$

7. $\int x^{-4} dx$; $u = x$; $du = dx$; $n = -4$

$$\int x^{-4} dx = \frac{x^{-3}}{-3} + C = -\frac{1}{3x^3} + C$$

8. $\int \frac{4dx}{\sqrt{x}} = 4\int x^{-1/2} dx$; $u = x$; $du = dx$; $n = -\frac{1}{2}$

$$4\int x^{-1/2} dx = 4\left[\frac{x^{1/2}}{\frac{1}{2}}\right] + C = 8\sqrt{x} + C$$

9. $\int (x^2 - x^5) dx = \int x^2 d - x\int x^5 dx$

$$= \frac{x^3}{3} - \frac{x^6}{6} + C$$

$$= \frac{1}{3}x^3 - \frac{1}{6}x^6 + C$$

10. $\int (1 - 3x) dx$; $u = 1 - 3x$; $du = -3dx$; $n = 1$

$$\int (1 - 3x) dx = -\frac{1}{3}\int (1 - 3x)^1(-3dx)$$

$$-\frac{1}{3}\left[\int (1 - 3x)^1(-3dx)\right] = -\frac{1}{3}\left[\frac{(1 - 3x)^2}{2}\right] + C_1$$

$$= -\frac{1}{6}(1 - 3x)^2 + C_1$$

OR:

$$\int (1 - 3x) dx = \int dx - 3\int x\,dx = x - \frac{3x^2}{2} + C_2$$

Both results are correct. $C_2 = C_1 - \frac{1}{6}$

11. $\int (9x^2 + x + 3) dx = 9\int x^2 dx + \int xdx + 3\int dx$

$$= 9\frac{x^3}{3} + \frac{x^2}{2} + 3x + C$$

$$= 3x^3 + \frac{1}{2}x^2 + 3x + C$$

12. $\int x(x - 2)^2 dx = \int x(x^2 - 4x + 4) dx$

$$= \int (x^3 - 4x^2 + 4x) dx$$

$$= \frac{x^4}{4} - \frac{4x^3}{3} + 2x^2 + C$$

13. $\int \left(\frac{t^2}{2} - \frac{2}{t^2}\right) dt = \frac{t^{2+1}}{2(2+1)} - \frac{2t^{-2+1}}{-2+1} + C$

$$= \frac{t^3}{6} + \frac{2}{t} + C$$

14. $\int \frac{3x^2 - 4}{x^2} dx = \int \left(3 - \frac{4}{x^2}\right) dx$

$$= 3x + \frac{4}{x} + C$$

15. $\int \sqrt{x}(x^2 - x) dx = \int x^{5/2} dx - \int x^{3/2} dx$

$$= \frac{x^{7/2}}{\frac{7}{2}} - \frac{x^{5/2}}{\frac{5}{2}} + C$$

$$= \frac{2}{7}x^{7/2} - \frac{2}{5}x^{5/2} + C$$

16. $\int (3R\sqrt{R} - 5R^2) dR = \int (3R^{3/2} - 5R^2) dR$

$$= \frac{3R^{3/2+1}}{3/2+1} - \frac{5R^3}{3} + C$$

$$= \frac{6}{5}R^{5/2} - \frac{5}{3}R^3 + C$$

17. $\int (2x^{-2/3} + 3^{-2}) dx = \int 2x^{-2/3} dx + \int 3^{-2}x^0 dx$

$$= 2\int x^{-2/3} dx + 3^{-2}\int x^0 dx$$

$$= 2 + \frac{1}{3}(3x^{1/3}) + 3^{-2}(x^1)$$

$$= 6x^{1/3} + \frac{1}{9}x + C$$

18. $\int (x^{1/3} + x^{1/5} + x^{-1/7})dx$

$$= \int x^{1/3}dx + \int x^{1/5}dx + \int x^{-1/7}dx$$

$$= \frac{x^{4/3}}{\frac{4}{3}} + \frac{x^{6/5}}{\frac{6}{5}} + \frac{x^{6/7}}{\frac{6}{7}} + C$$

$$= \frac{3}{4}x^{4/3} + \frac{5}{6}x^{6/5} + \frac{7}{6}x^{6/7} + C$$

19. $\int (1 + 2s^2)^2 ds = \int (1 + 4s^2 + 4s^4)ds$

$$= s + \frac{4s^3}{3} + \frac{4s^5}{5} + C$$

20. $\int (x^2 + 4x + 4)^{1/3}dx = \int [(x+2)^2]^{1/3}dx$

$$= \int (x+2)^{2/3}dx$$

$$u = x + 2; \; du = dx; \; n = \frac{2}{3}$$

$$\int (x^2 + 4x + 4)^{1/3}dx = \frac{(x+2)^{5/3}}{\frac{5}{3}} + C$$

$$= \frac{3}{5}(x+2)^{5/3} + C$$

21. $\int (x^2 - 1)^5 (2x\,dx); \; u = x^2 - 1; \; du = 2x\,dx; \; n = 5$

$$\int (x^2 - 1)^5 (2x\,dx) = \frac{(x^2-1)^6}{6} + C$$

$$= \frac{1}{6}(x^2 - 1)^6 + C$$

22. $\int (x^3 - 2)^6 (3x^2 dx) = \int u^6 du$

let $u = x^3 - 2$ $\quad = \frac{u^7}{7} + C$

$du = 3x^2 dx$

$$= \frac{(x^3 - 2)^7}{7} + C$$

23. $\int (x^4 + 3)^4 (4x^3 dx) = \frac{(x^4 + 3)^5}{5} + C$

$$= \frac{1}{5}(x^4 + 3)^5 + C$$

$$u = x^4 + 3; \; du = 4x^3 dx; \; n = 4$$

24. $\int (1 - 2x)^{1/3}(-2dx) = \frac{(1-2x)^{4/3}}{\frac{4}{3}} + C$

$$= \frac{3}{4}(1 - 2x)^{4/3} + C$$

$$u = 1 - 2x; \; du = -2dx; \; n = \frac{1}{3}$$

25. $\int (2\theta^5 + 5)^7 \theta^4 d\theta = \frac{1}{10}\int (2\theta^5 + 5)^7 \cdot (10\theta^4)d\theta$

$$= \frac{1}{10} \cdot \frac{(2\theta^5 + 5)^8}{8} + C$$

$$= \frac{(2\theta^5 + 5)^8}{80} + C$$

26. $\int 6x^2 (1 - x^3)^{4/3} dx; \; u = 1 - x^3; \; du = -3x^2 dx$

$$n = \frac{4}{3}; \; 6 = (-2)(-3)$$

$$\int 6x^2 (1 - x^3)^{4/3} dx = -2\int (1 - x^3)^{4/3}(-3x^2 dx)$$

$$= -2\frac{(1-x^3)^{7/3}}{\frac{7}{3}} + C$$

$$= -\frac{6}{7}(1 - x^3)^{7/3} + C$$

27. $\int \sqrt{8x + 1}\,dx = \int (8x + 1)^{1/2}dx; \; u = 8x + 1;$

$$du = 8dx; \; n = \frac{1}{2}$$

$$\int (8x + 1)^{1/2}dx = \frac{1}{8}\int (8x + 1)^{1/2}(8)dx$$

$$= \frac{1}{8}\frac{(8x+1)^{3/2}}{\frac{3}{2}} + C$$

$$= \frac{1}{12}(8x + 1)^{3/2} + C$$

28. $\int \frac{dV}{(0.3 + 2V)^3} = \frac{1}{2}\int (0.3 + 2V)^{-3}(2dV)$

$$= \frac{1}{2} \cdot \frac{(0.3 + 2V)^{-3+1}}{-3 + 1} + C$$

$$= \frac{-1}{4(0.3 + 2V)^2} + C$$

29. $\displaystyle\int \frac{x\,dx}{\sqrt{6x^2+1}} = \int (6x^2+1)^{-1/2}x\,dx$

$u = 6x^2+1;\ du = 12x;\ n = -\dfrac{1}{2}$

$\displaystyle\int (6x^2+1)^{-1/2}x\,dx = \frac{1}{12}\int (6x^2+1)^{-1/2}(12x\,dx)$

$\displaystyle\qquad\qquad = \frac{1}{12}\frac{(6x^2+1)^{1/2}}{\frac{1}{2}}+C$

$\displaystyle\qquad\qquad = \frac{1}{6}\sqrt{6x^2+1}+C$

30. $\displaystyle\int \frac{2x^2\,dx}{\sqrt{2x^3+1}} = 2\int (2x^3+1)^{-1/2}x^2\,dx$

$u = 2x^3+1;\ du = 6x^2\,dx;\ n = -\dfrac{1}{2}$

$\displaystyle 2\int (2x^3+1)^{-1/2}x^2\,dx = \frac{2}{6}\int (2x^3+1)^{-1/2}(6x^2)\,dx$

$\displaystyle\qquad\qquad = \frac{1}{3}\frac{(2x^3+1)^{1/2}}{\frac{1}{2}}+C$

$\displaystyle\qquad\qquad = \frac{2}{3}(2x^3+1)^{1/2}+C$

31. $\displaystyle\int \frac{x-1}{\sqrt{x^2-2x}}\,dx = \int (x^2-2x)^{-1/2}(x-1)\,dx$

$u = x^2-2x;\ du = (2x-2)\,dx = 2(x-1)\,dx;\ n = -\dfrac{1}{2}$

$\displaystyle\int (x^2-2x)^{-1/2}(x-1)\,dx = \frac{1}{2}\int (x^2-2x)^{-1/2}(2)(x-1)\,dx$

$\displaystyle\qquad\qquad = \frac{1}{2}\frac{(x^2-2x)^{1/2}}{\frac{1}{2}}+C$

$\displaystyle\qquad\qquad = \sqrt{x^2-2x}+C$

32. $(x^2-x)\left(x^3-\dfrac{3}{2}x^2\right)^8 = dx;\ u = x^3-\dfrac{3}{2}x^2$

$du = (3x^2-3x)\,dx = 3(x^2-x)\,dx;\ n = 8$

$\displaystyle\int \left(x^3-\frac{3}{2}x^2\right)^8 (x^2-x)\,dx$

$\displaystyle\qquad = \frac{1}{3}\int \left(x^3-\frac{3}{2}x^2\right)^8 (3)(x^2-x)\,dx$

$\displaystyle\qquad = \frac{1}{3}\frac{\left(x^3-\frac{3}{2}x^2\right)^9}{9}+C$

$\displaystyle\qquad = \frac{1}{27}\left(x^3-\frac{3}{2}x^2\right)^9+C$

33. $\dfrac{dy}{dx} = 6x^2;\ dy = 6x^2\,dx$

$\displaystyle y = \int 6x^2\,dx = 6\int x^2\,dx = \frac{6x^3}{3}+C = 2x^3+C$

The curve passes through $(0,2)$.　$2 = 2(0^3)+C$,
$C = 2;\ y = 2x^3+2$

34. $\dfrac{dy}{dx} = 8x+1;\ dy = (8x+1)\,dx;\ y = \displaystyle\int (8x+1)\,dx$

$y = 4x^2+x+C$; curve passes through $(-1,4)$
$4 = 4(-1)^2+(-1)+C;\ C = 1$
$y = 4x^2+x+1$

35. $\dfrac{dy}{dx} = x^2(1-x^3)^5;\ x^2(1-x^3)^5\,dx$

$\displaystyle y = \int x^2(1-x^3)^5\,dx = \frac{1}{3}\int (1-x^3)^5(-3x^2\,dx)$

$\displaystyle y = \frac{1}{18}(1-x^3)^6+C$; curve passes through $(1,5)$

$5 = \dfrac{1}{18}(1-1^3)^6+C;\ C = 5$

$\displaystyle y = 5 - \frac{1}{18}(1-x^3)^6$

36. $\dfrac{dy}{dx} = 2x^3(x^4-6)^4;\ dy = 2x^3(x^4-6)^4\,dx$

$\displaystyle y = \int 2x^3(x^4-6)^4\,dx = \frac{1}{2}\int (x^4-6)^4(4x^3\,dx)$

$\displaystyle y = \frac{1}{10}(x^4-6)^5+C$; curve passes through $(2,10)$

$10 = \dfrac{1}{10}(2^4-6)^5+C;\ C = -9990$

$\displaystyle y = \frac{1}{10}(x^4-6)^5 - 9990$

37. Slope: $\dfrac{dy}{dx} = -x\sqrt{1-4x^2}$

$dy = -x\sqrt{1-4x^2}\,dx$

$\displaystyle y = \int -x\sqrt{1-4x^2}\,dx = \int (1-4x^2)^{1/2}(-x\,dx);$

$u = 1-4x^2;\ du = -8x;\ n = \dfrac{1}{2}$

$\displaystyle y = \frac{1}{8}\int (1-4x^2)^{1/2}(-8x\,dx)$

$\displaystyle\qquad = \frac{1}{8}\frac{(1-4x^2)^{3/2}}{\frac{3}{2}}+C$

$\displaystyle\qquad = \frac{1}{12}(1-4x^2)^{3/2}+C$

The curve passes through $(0,7)$.

$7 = \frac{1}{12}[1 - 4(0^2)]^{3/2} + C$

$7 = \frac{1}{12}(1)^{3/2} + C; \ 7 = \frac{1}{12} + C; \ C = \frac{83}{12}$

$y = \frac{1}{12}(1 - 4x^2)^{3/2} + \frac{83}{12}; \ 12y = 83 + (1 - 4x^2)^{3/2}$

38. $\int (x^3 - 1)dx$ cannot be integrated by letting $u = x^3 - 1$ and $du = 3x^2 dx$ because there is no $3x^2$ with dx for proper du.

39. $\frac{di}{dt} = 4t - 0.6t^2; \ di = (4t - 0.6t^2)dt$

$i = \int (4t - 0.6t^2)dt = 2t^2 - 0.2t^3 + C$

$i = 2A$ when $t = 0$ s; $2 = 2(0)^2 - 0.2(0)^3 + C$; $C = 2$
$i = 2t^2 - 0.2t^3 + 2$

40. $\frac{df}{dL} = 80(4 + L)^{-3/2}; \ df = 80(4 + L)^{-3/2}dL$

$f = \int 80(4 + L)^{-3/2}dL = -160(4 + L)^{-1/2} + C$

$f = 80$ HZ when $L = 0$ H: $80 = -160(4)^{-1/2} + C$; $C = 160$

$f = 160 - \frac{160}{(4 + L)^{1/2}}$

41. $\frac{df}{dA} = \frac{0.005}{\sqrt{0.01A + 1}} = 0.005(0.01A + 1)^{-1/2}$

$f(A) = \int 0.005(0.01A + 1)^{-1/2}dA + C$

$\quad = \frac{1}{2} \int (0.01A + 1)^{=1/2}(0.01dA)$

$\quad = (0.01A + 1)^{1/2} + C$

$f = 0$ for $A = 0$ m^2
$f(0) = 0 = (0.01(0) + 1)^{1/2} + C$; $C = -1$
$f(A) = (0.01A + 1)^{1/2} - 1 = \sqrt{0.01A + 1} - 1$

42. $\frac{dp}{dx} = \frac{600(30 - x)}{\sqrt{60x - x^2}} = \frac{300(60 - 2x)}{\sqrt{60x - x^2}}$

$p = 300 \int (60x - x^2)^{-1/2}(60 - 2x)dx$

$p = 300 \cdot 2(60x - x^2)^{1/2} + c$
$p = 600\sqrt{60x - x^2} + c$

When $x = 0, p = -5000 \Rightarrow c = -5000$

$p = 600\sqrt{60x - x^2} - 5000$

43. $\frac{d^2y}{dx^2} = \frac{d(y')}{dx} = 6; \ dy' = 6\,dx$

$y' = 6x + C; \ y' = 8$ for $x = 1$
$8 = 6(1) + C; \ C = 2$

$y' = \frac{dy}{dx} = 6x + 2; \ dy = (6x + 2)dx$

$y = 3x^2 + 2x + C_1$; curve passed through $(1, 2)$
$2 = 3(1)^2 + 2(1) + C_1; \ C_1 = -3$
$y = 3x^2 + 2x - 3$

44. $\frac{d^2y}{dx^2} = \frac{d(y')}{dx} = 12x^2; \ dy' = 12x^2 dx$

$y' = \frac{dy}{dx} = 4x^3 + C; \ y = \int (4x^3 + C)dx$

$y = x^4 + Cx + C_1$

Curve passed through $(1, 6)$ and $(2, 21)$.

$6 = (1)^4 + (1)C + C_1$ and $21 = (2)^4 + (2)C + C_1$
$C + C_1 = 5$ and $2C + C_1 = 5$; $C = 0, C_1 = 5$
$y = x^4 + 5$

5.3   The Area Under a Curve

1. $y = 3x$, between $x = 0$ and $x = 3$

(a)

x	y
1	3
2	6
3	9

$n = 3$
$\Delta x = 1$

$A = 1(0 + 3 + 6) = 9$; (first rectangle has 0 height)

(b)

x	y
0	0
0.3	0.9
0.6	1.8
0.9	2.7
1.2	3.6
1.5	4.5
1.8	5.4
2.1	6.3
2.4	7.2
2.7	8.1
3.0	9.0

$n = 10$
$\Delta x = 0.3$

$A = 0.3(0 + 0.9 + 1.8 + 2.7 + 3.6 + 4.5 + 5.4$
$\quad + 6.3 + 7.2 + 8.1)$
$\quad = 0.3(40.5) = 12.15$

2. $y = 2x + 1$, between $x = 0$ and $x = 2$

    (a) $n = 4 \ (\Delta x = 0.5)$

        $A = (A_1 + A_2 + A_3 + A_4)$

        $A = (y_1 + y_2 + y_3 + y_4)x$

        $A = (1 + 2 + 3 + 4)0.5$

        $A = (10)(0.5) = 5$

x	y
0	1
$\frac{1}{2}$	2
1	3
$\frac{3}{2}$	4
2	5

    (b) $n = 10 \ (\Delta x = 0.2)$

        $A = \sum_{i=1}^{10} A_i$

           $= (\sum_{i=1}^{10} y_i)\Delta x$

        $A = (1 + 1.4 + 1.8 + \cdots + 4.6)0.2$

        $A = 28(0.2) = 5.6$

x	y
0	1
0.2	1.4
0.4	1.8
0.6	2.2
0.8	2.6
1	3
1.2	3.4
1.4	3.8
1.6	4.2
1.8	4.6
2	5

3. $y = x^2$, between $x = 0$ and $x = 2$

    (a) $n = 5 \ (\Delta x = 0.4)$

        $A = \sum_{i=1}^{5} A_i$

           $= \sum_{i=1}^{5} y_i \Delta x$

        $y_1 = f(0) = 0$

        $A = 0.4(0 + 0.16 + 0.64 + 1.44 + 2.56)$

        $A = 0.4(4.8)$

        $A = 1.92$

x	y
0	0
0.4	0.16
0.8	0.64
1.2	1.44
1.6	2.56
2	4

(b) $n = 10 \ (\Delta x = 0.2)$

    $A = \sum_{i=1}^{i=10} A_i$

        $= \sum_{i=1}^{i=10} y_i \Delta x$

    $y_i = f(0) = 0$

    $A = (0.2)(0 + 0.4 + 0.16 + 0.36 + 0.64 + 1$

        $+ 1.44 + 1.96 + 2.56 + 3.24)$

    $A = 0.2(11.4)$

    $A = 2.28$

x	y
0	0
0.2	0.04
0.4	0.16
0.6	0.36
0.8	0.64
1	1
1.2	1.44
1.4	1.96
1.6	2.56
1.8	3.24
2	4

4. $y = x^2 + 2$, between $x = 0$ and $x = 3$

    (a) $n = 3 \ (\Delta x = 1)$

        $A = \sum_{i=1}^{3} A_i = \sum_{i=1}^{3} y_i x$

        $A = (2 + 3 + 6)1$

        $A = 11$

x	y
0	2
1	3
2	6
3	9

    (b) $A = 10 \ (\Delta x = 0.3)$

           $= \sum_{i=1}^{10} A_i$

        $y_1 = \sum_{i=1}^{10} y_i \Delta x$

        $A = f(0) = 2$

        $A = [2 + 2.09 + \cdots 9.29]$

        $A = (45.65)0.3$

        $A = 13.695$

x	y
0	2
0.3	2.09
0.6	2.36
0.9	2.81
1.2	3.44
1.5	4.25
1.8	5.24
2.1	6.41
2.4	7.76
2.7	9.29
3	11

5. $y = 4x - x^2$, between $x = 1$ and $x = 4$

    (a) $n = 6, \Delta x = 0.5$

        $A = 0.5(3.00 + 3.75 + 3.75 + 3.00 + 1.75 + 0.00)$

        $A = 7.625$

x	y
1.0	3.00
1.5	3.75
2.0	4.00
2.5	3.75
3.0	3.00
3.5	1.75
4.0	0.00

($y = 4.00$ is not the height of any inscribed rectangle)

    (b) $n = 10, \Delta x = 0.3$

        $A = 0.3(3.00 + 3.51 + 3.84 + 3.96 + 3.75$

            $+ 3.36 + 2.79 + 2.04 + 1.11)$

        $A = 8.208$

x	y
1.0	3.00
1.3	3.51
1.6	3.84
1.9	3.99
2.2	3.96
2.5	3.75
2.8	3.36
3.1	2.79
3.4	2.04
3.7	1.11
4.0	0.00

($y = 3.99$ is not the height of any inscribed rectangle)

6. $y = 1 - x^2$, between $x = 0.5$ and $x = 1$

    (a) $n = 5, x = \dfrac{1 - 0.5}{5} = 0.1$

        $A = \sum_{i=1}^{5} A_i = \sum_{i=1}^{5} y_1 \Delta x$

    $y_i = f(0.6)$

    $A = (0.64 + 0.51 + 0.36 + 0.19 + 0)(0.1)$

    $A = 1.7(0.1)$

    $A = 0.17$

x	y
0.5	0.75
0.6	0.64
0.7	0.51
0.8	0.36
0.9	0.19
1	0

    (b) $n = 10, \Delta x = \dfrac{0.5}{10} = 0.05$

        $A = \sum_{i=1}^{10} A_i$

            $= \sum_{i=1}^{10} y_i \Delta x$

    $y_1 = f(0.55)$

    $A = (0.6975 + 0.64 + 0.19 + 0.9755 + 0)0.05$

    $A = 3.7875(0.050)$

    $A = 0.189375$

x	y
0.5	0.75
0.55	0.6975
0.6	0.64
0.65	0.5775
0.7	0.51
0.75	0.4375
0.8	0.36
0.85	0.2775
0.9	0.19
0.95	0.0975
1	0

7. $y = \dfrac{1}{x^2}$, between $x = 1$ and $x = 5$

    (a) $n = 4, \Delta x = \dfrac{5 - 1}{4} = 1$

        $A = \sum_{i=1}^{4} A_i = \sum_{i=1}^{4} y_i \Delta x$

    $y_1 = f(2)$

    $A = \left(\dfrac{1}{4} + \dfrac{1}{9} + \dfrac{1}{16} + \dfrac{1}{25}\right)1$

    $A = 0.464$

x	y
1	1
2	$\frac{1}{4}$
3	$\frac{1}{9}$
4	$\frac{1}{16}$
5	$\frac{1}{25}$

    (b) $n = 8, \Delta x - 0.5$

        $A = \sum_{i=1}^{8} A_i$

            $= \sum_{i=1}^{8} y_i \Delta x$

    $y_1 = f(1.5)$

    $A = (0.444 + 0.25 + \cdots + 0.0494 + 0.04)0.5$

    $A = 1.199070925(0.5)$

    $A = 0.5995$

x	y
1	1
1.5	$0.444\ldots$
2	0.25
2.5	0.16
3	$0.111\ldots$
3.5	0.0816
4	0.0625
4.5	0.0494
5	0.04

8. $y = \sqrt{x}$, between $x = 1$ and $x = 4$

(a) $n = 3$, $\Delta x = \dfrac{4-1}{3} = 1$

x	y
1	1
2	$\sqrt{2}$
3	$\sqrt{3}$
4	2

$$A = \sum_{i=1}^{3} A_i$$
$$= \sum_{i=1}^{3} y_i \Delta x$$

$y_1 = f(1)$
$A = (1 + \sqrt{2} + \sqrt{3})1$
$A = 4.146$

(b) $n = 12$, $\Delta x = \dfrac{3}{12} = \dfrac{1}{4}$

$$A = \sum_{i=1}^{12} A_i$$
$$= \sum_{i=1}^{12} y_i \Delta x$$

$y_1 = f(1)$

$$A = (1 + \sqrt{1.25} + \sqrt{1.50} + \cdots + \sqrt{3.75})\frac{1}{4}$$

$$A = (18.161\ 466)\frac{1}{4}$$

$$A = 4.540$$

x	y
1	1
1.25	$\sqrt{1.25}$
1.50	1.22
1.75	$\sqrt{1.75}$
2	$\sqrt{2}$
2.25	$\sqrt{2.25}$
2.5	1.58
2.75	$\sqrt{2.75}$
3	$\sqrt{3}$
3.25	$\sqrt{3.25}$
3.50	1.87
3.75	$\sqrt{3.75}$
4	2

9. $y = \dfrac{1}{\sqrt{x+1}}$, between $x = 3$ and $x = 8$

(a) $n = 5$, $\Delta x = \dfrac{8-3}{5} = 1$

$$A = \sum_{i=1}^{5} A_i = \sum_{i=1}^{5} y_i \Delta x$$

$y_1 = f(4)$
$A = (0.447 + 0.408 + \cdots + 0.354 + 0.333)(1)$
$A = 1.92$

x	y
3	0.5
4	0.447
5	0.408
6	0.378
7	0.355
8	0.333

(b) $n = 10$, $\Delta x = \dfrac{8-3}{10} = 0.5$

$$A = \sum_{i=1}^{10} A_i = \sum_{i=1}^{10} y_i \Delta x$$

$y_1 = f(3.5)$
$A = (0.471 + 0.447 + \cdots + 0.343 + 0.333)(0.5)$
$A = 1.96$

x	y
3	0.5
3.5	0.471
4	0.447
4.5	0.426
5	0.408
5.5	0.392
6	0.378
6.5	0.365
7	0.354
7.5	0.343
8	0.333

10. $y = 2x\sqrt{x^2 + 1}$, between $x = 0$ and $x = 6$

(a) $n = 6$, $\Delta x = \dfrac{6}{6} = 1$

$$A = \sum_{i=1}^{5} A_i = \sum_{i=1}^{5} y_i \Delta x$$

$y_1 = f(1)$
$A = (0 + 2.83 + 8.94 + \cdots + 50.99)1$
$A = (114.721)1 = 114.7$
$A = 115$

x	y
0	0
1	2.83
2	8.94
3	18.97
4	32.98
5	50.99
6	72.99

(b) $n = 12,\ \Delta x = \dfrac{1}{2}$

$$A = \sum_{i=1}^{11} A_i = \sum_{i=1}^{11} y_i \Delta x$$

$$y_1 = f(0.5)$$

$$A = (0 + 1.12 + 2.83 + \cdots + 50.99 + 61.49)\frac{1}{2}$$

$$A = (263.170\,88)\frac{1}{2}$$

$$A = 132$$

x	y
0	0
0.5	1.12
1	2.83
1.5	5.41
2	8.94
2.5	13.46
3	18.97
3.5	25.48
4	32.98
4.5	41.49
5	50.99
5.5	61.49
6	72.99

11. $y = 3x$, between $x = 0$ and $x = 3$

$$A_{0,3} = \left[\int 3x\,dx\right]_0^3 = \frac{3x^2}{2}\bigg|_0^3 = \frac{3}{2}(3^2) - \frac{3}{2}(0)$$

$$= \frac{27}{2} = 13.5$$

12. $y = 2x + 1$, between $x = 0$ and $x = 2$

$$A_{0,2} = \left[\int (2x+1)\,dx\right]_0^2 = (x^2 + x)\big|_0^2$$

$$= 4 + 2 - 0 = 6$$

13. $y = x^2$, between $x = 0$ and $x = 2$

$$A_{0,2} = \left[\int x^2\,dx\right]_0^2 = \frac{x^3}{3}\bigg|_0^2 = \frac{8}{3} - 0 = \frac{8}{3}$$

14. $y = x^2 + 2$, between $x = 0$ and $x = 3$

$$A_{0,3} = \left[\int (x^2 + 2)\,dx\right]_0^3 = \left(\frac{x^3}{3} + 2x\right)\bigg|_0^3$$

$$= 9 + 6 - 0 = 15$$

15. $y = 4x - x^2$, between $x = 1$ and $x = 4$

$$A_{1,4} = \left[\int (4x - x^2)\,dx\right]_1^4 = 2x^2 - \frac{x^3}{3}\bigg|_1^4$$

$$= \left(32 - \frac{64}{3}\right) - \left(2 - \frac{1}{3}\right) = 9$$

16. $y = 1 - x^2$, between $x = 0.5$ and $x = 1$

$$A_{0.5,1} = \left[\int (1 - x^2)\,dx\right]_{0.5}^1 = x - \frac{x^3}{3}\bigg|_{0.5}^1$$

$$= \left(1 - \frac{1}{3}\right) - \left(\frac{1}{2} - \frac{1}{24}\right) = \frac{5}{24}$$

17. $y = \dfrac{1}{x^2} = x^{-2}$, between $x = 1$ and $x = 5$

$$A_{1,5} = \left[\int x^{-2}\,dx\right]_1^5 = \frac{x^{-1}}{-1}\bigg|_1^5 = \frac{-1}{x}\bigg|_1^5$$

$$= -\frac{1}{5} - (-1) = \frac{4}{5} = 0.8$$

18. $y = \sqrt{x}$, between $x = 1$ and $x = 5$

$$A_{1,4} = \left[\int \sqrt{x}\,dx\right]_1^4 = \frac{2}{3}x^{3/2}\bigg|_1^4 = \frac{2}{3}x\sqrt{x}\bigg|_1^4$$

$$= \frac{16}{3} - \frac{2}{3} = \frac{14}{3}$$

19. $y = \dfrac{1}{\sqrt{x+1}}$, between $x = 3$ and $x = 8$

$$A_{3,8} = \left[\int (x+1)^{-1/2}\,dx\right]_3^8 = 2(x+1)^{1/2}\big|_3^8$$

$$= 6 - 4 = 2$$

20. $y = 2x\sqrt{x^2 + 1}$, between $x = 0$ and $x = 6$;
$u = x^2 + 1;\ du = 2x\,dx$

$$A_{0,6} = \left[\int (x^2 + 1)2x\,dx\right]_0^6 = \frac{2}{3}(x^2 + 1)^{3/2}\bigg|_0^6$$

$$= \frac{2}{3}(37^{3/2} - 1) = 149$$

5.4   The Definite Integral

1. $\displaystyle\int_0^1 2x\,dx = \frac{2x^2}{2}\bigg|_0^1 = x^2\bigg|_0^1 = 1^2 - 0^2 = 1$

2. $\int_0^2 3x^2\,dx = \dfrac{3x^3}{3} = x^3\Big|_0^2 = 8 - 0 = 8$

3. $\int_1^4 x^{5/2}\,dx = \dfrac{2}{7}x^{7/2}\Big|_1^4 = \dfrac{256}{7} - \dfrac{2}{7} = \dfrac{254}{7}$

4. $\int_4^9 (p^{3/2} - 3)\,dp = \left(\dfrac{2}{5}p^{5/2} - 3p\right)\Big|_4^9$

$\qquad\qquad = \left(\dfrac{486}{5} - 27\right) - \left(\dfrac{64}{5} - 12\right)$

$\qquad\qquad = \dfrac{347}{5}$

5. $\int_3^6 \left(\dfrac{1}{\sqrt{x}} + 2\right) dx = \int_3^6 \left(\dfrac{1}{\sqrt{x}}\right)dx + \int_3^6 2\,dx$

$\qquad\qquad = \int_3^6 x^{-1/2}\,dx + \int_3^6 2x^0\,dx$

$\qquad\qquad = 2x^{1/2}\Big|_3^6 + 2x\Big|_3^6$

$\qquad\qquad = [2(6)^{1/2} - 2(3)^{1/2}] + [2(6) - 2(3)]$

$\qquad\qquad = 6 + 2\sqrt{6} - 2\sqrt{3}$

6. $\int_{1.2}^{1.6} \left(5 + \dfrac{6}{x^4}\right) dx = 5x \int 6x^{-4}\,dx = 5x + \dfrac{6x^{-3}}{-3}$

$\qquad\qquad = \left(5x - \dfrac{2}{x^3}\right)\Big|_{1.2}^{1.6}$

$\qquad\qquad = 7.5117 - 4.8426$

$\qquad\qquad = 2.67$

7. $u = 1 - x;\ du = -dx$

$\int_{-1.6}^{0.7} (1-x)^{1/3}\,dx = -\int_{-1.6}^{0.7} (1-x)^{1/3}(-dx)$

$\qquad\qquad = -\dfrac{3}{4}(1-x)^{4/3}\Big|_{-1.6}^{0.7}$

$\qquad\qquad = -\dfrac{3}{4}(0.2008 - 3.5752)$

$\qquad\qquad = 2.53$

8. $u = 2x - 1;\ du = 2\,dx$

$\int_1^5 \sqrt{2x-1}\,dx = \dfrac{1}{2}\int_1^5 (2x-1)^{1/2}(2\,dx)$

$\qquad\qquad = \dfrac{2}{3}\cdot\dfrac{1}{2}(2x-1)^{3/2}\Big|_1^5$

$\qquad\qquad = \dfrac{1}{3}(27 - 1) = \dfrac{26}{3}$

9. $\int_{-2}^2 (T-2)(T+2)\,dT$

$= \int_{-2}^2 (T^2 - 4)\,dT = \dfrac{1}{3}T^3 - 4T\Big|_{-2}^2$

$= \dfrac{1}{3}(2)^3 - 4(2) - \left(\dfrac{1}{3}(-2)^3 - 4(-2)\right)$

$= -\dfrac{32}{3}$

10. $\int_1^2 (3x^5 - 2x^3) = \dfrac{3x^6}{6} - \dfrac{2x^4}{4}$

$\qquad\qquad = \left(\dfrac{1}{2}x^6 - \dfrac{1}{2}x^4\right)\Big|_1^2$

$\qquad\qquad = 24 - 0 = 24$

11. $\int_{0.5}^{2.2} (\sqrt[3]{x} - 2)\,dx = \int_{0.5}^{2.2} x^{1/3}\,dx - 2\int_{0.5}^{2.2} dx$

$\qquad\qquad = \left(\dfrac{3}{4}x^{4/3} - 2x\right)\Big|_{0.5}^{2.2}$

$\qquad\qquad = (2.1460 - 4.4) - (0.2976 - 1)$

$\qquad\qquad = -1.5516 = -1.552$

12. $\int_{2.7}^{5.3} \left(\dfrac{1}{x\sqrt{x}} + 4\right) dx = \int (x^{-3/2} + 4)\,dx$

$\qquad\qquad = (-2x^{-1/2} + 4x)\Big|_{2.7}^{5.3}$

$\qquad\qquad = \left(\dfrac{-2}{\sqrt{x}} + 4x\right)\Big|_{2.7}^{5.3}$

$\qquad\qquad = 20.3313 - 9.5828$

$\qquad\qquad = 10.7$

13. $\int_0^4 (1 - \sqrt{x})^2\,dx = \int_0^4 (1 - 2\sqrt{x} + x)\,dx$

$\qquad\qquad = \int_0^4 1\,dx - \int_0^4 2x^{1/2}\,dx + \int_0^4 x\,dx$

$\qquad\qquad = x - \dfrac{2x^{3/2}}{\frac{3}{2}} + \dfrac{x^2}{2}\Big|_0^4$

$\qquad\qquad = x - \dfrac{4}{3}x^{3/2} + \dfrac{x^2}{2}\Big|_0^4$

$\qquad\qquad = \left[4 - \dfrac{4}{3}(4^{3/2}) + \dfrac{4^2}{2}\right] - 0$

$\qquad\qquad = 4 - \dfrac{4}{3}(8) + 8 - 0$

$\qquad\qquad = \dfrac{12}{3} - \dfrac{32}{3} + \dfrac{24}{3} = \dfrac{4}{3}$

14. $\displaystyle\int_1^4 \frac{y+4}{\sqrt{y}}\,dy = \int_1^4 (y^{1/2} + 4y^{-1/2})\,dy$

$\displaystyle \qquad = \frac{2}{3}y^{3/2} + 8y^{1/2}\,\Big|_1^4$

$\displaystyle \qquad = \frac{38}{3}$

15. $u = 4 - x^2;\ du = -2x\,dx$

$\displaystyle \int_{-2}^{-1} 2x(4-x^2)^3\,dx = -\int_{-2}^{-1}(11-x^2)^3(-2x\,dx)$

$\displaystyle \qquad = -\frac{(4-x^2)^4}{4}\,\Big|_{-2}^{-1}$

$\displaystyle \qquad = -\left(\frac{81}{4} - 0\right)$

$\displaystyle \qquad = -\frac{81}{4}$

16. $u = 3x^2 - 1;\ du = 6x\,dx$

$\displaystyle \int_0^1 x(3x^2-1)^3\,dx = \frac{1}{6}\int_0^1 (3x^2-1)^3(6x\,dx)$

$\displaystyle \qquad = \frac{1}{6}\frac{(3x^2-1)^4}{4} = \frac{1}{24}(3x^2-1)^4\,\Big|_0^1$

$\displaystyle \qquad = \frac{1}{24}[16-(1)] = \frac{15}{24} = \frac{5}{8}$

17. $\displaystyle\int_0^4 \frac{x\,dx}{\sqrt{x^2+9}} = \int_0^4 (x^2+9)^{-1/2}x\,dx$

$\displaystyle \qquad = \frac{1}{2}\int_0^4 (x^2+9)^{-1/2}2x\,dx$

$\displaystyle \qquad = \frac{1}{2}\times\frac{(x^2+9)^{1/2}}{\frac{1}{2}}\,\Big|_0^4$

$\displaystyle \qquad = (x^2+9)^{1/2}\big|_0^4$

$\displaystyle \qquad = (4^2+9)^{1/2} - (0^2+9)^{1/2}$

$\displaystyle \qquad = 25^{1/2} - 9^{1/2} = 5 - 3 = 2$

18. $u = x^3 + 2;\ du = 3x^2\,dx$

$\displaystyle \int_{0.2}^{0.7} x^2(x^3+2)^{3/2}\,dx = \frac{1}{3}\int_{0.2}^{0.7}(x^3+2)^{3/2}(3x^2\,dx)$

$\displaystyle \qquad = \frac{1}{3}\frac{(x^3+2)^{5/2}}{\frac{5}{2}}\,\Big|_{0.2}^{0.7}$

$\displaystyle \qquad = \frac{2}{15}(x^3+2)^{5/2}\,\Big|_{0.2}^{0.7}$

$\displaystyle \qquad = \frac{2}{15}(8.4029 - 5.7136)$

$\displaystyle \qquad = 0.359$

19. $u = 6x + 1;\ du = 6\,dx$

$\displaystyle \int_{2.75}^{3.25} \frac{dx}{\sqrt[3]{6x+1}} = \int (6x+1)^{-1/3}dx$

$\displaystyle \qquad = \frac{1}{6}\int (6x+1)^{-1/3}(6\,dx)$

$\displaystyle \qquad = \frac{1}{6}\cdot\frac{3}{2}(6x+1)^{2/3}\,\Big|_{2.75}^{3.25}$

$\displaystyle \qquad = \frac{1}{4}(6x+1)^{2/3}\,\Big|_{2.75}^{3.25}$

$\displaystyle \qquad = \frac{1}{4}(7.4904 - 6.7405)$

$\displaystyle \qquad = 0.1875$

20. $u = 4x + 1;\ du = 4\,dx$

$\displaystyle \int_2^6 \frac{2\,dx}{\sqrt{4x+1}} = 2\int_2^6 (4x+1)^{-1/2}dx$

$\displaystyle \qquad = \frac{2}{4}\int (4x+1)^{-1/2}(4\,dx)$

$\displaystyle \qquad = \frac{1}{2}\cdot\frac{2}{1}(4x+1)^{1/2}\,\Big|_2^6$

$\displaystyle \qquad = \sqrt{4x+1}\,\Big|_2^6$

$\displaystyle \qquad = 5 - 3$

$\displaystyle \qquad = 2$

21. If $u = (2x^2 + 1)$, $n = 3$, and $du = 4x\,dx$, then

$\displaystyle \int_1^3 \frac{2x\,dx}{(2x^2+1)^3} = \frac{1}{2}\int_1^3 \frac{4x\,dx}{(2x^2+1)^3}$

$\displaystyle \qquad = \frac{1}{2}\left[-\frac{1}{2}(2x^2+1)^{-2}\right]\Big|_1^3$

$\displaystyle \qquad = \frac{1}{2}\left[-\frac{1}{2}(2(3)^2+1)^{-2}\right]$

$\displaystyle \qquad\quad -\frac{1}{2}\left[-\frac{1}{2}(2(1)^2+1)^{-2}\right]$

$\displaystyle \qquad = -\frac{1}{4}\left(\frac{1}{19^2}\right) + \frac{1}{4}\left(\frac{1}{3^2}\right)$

$\displaystyle \qquad = -\frac{1}{1444} + \frac{1}{36}$

$\displaystyle \qquad = \frac{88}{3249}$

$\displaystyle \qquad = 0.0271$

22. $u = 6x - 1$; $du = 6\,dx$

$$\int_{12.6}^{17.2} \frac{3\,dx}{(6x-1)^2} = \frac{1}{2}\int_{12.6}^{17.2}(6x-1)^{-2}(2\times 3)\,dx$$

$$= \frac{1}{2}\frac{(6x-1)^{-1}}{-1}$$

$$= \frac{-1}{2(6x-1)}\Big|_{12.6}^{17.2}$$

$$= -0.00489 - (-0.00670)$$

$$= 0.00181$$

23. $u = 4t + 1$; $du = 4\,dt$

$$\int_3^7 \sqrt{16t^2 + 8t + 1}\,dt = \int_3^7 \sqrt{(4t+1)^2}\,dt$$

$$= \int_3^7 \sqrt{(4t+1)}\,dt$$

$$= \frac{1}{4}\int_3^7 (4t+1)^1 4\,dt$$

$$= \frac{1}{4}\frac{(4t+1)^2}{2}$$

$$= \frac{1}{8}(4t+1)^2\Big|_3^7$$

$$= 84$$

24. $u = 6 - 2x$; $du = -2\,dx$

$$\int_{-5}^1 \sqrt{6-2x}\,dx = \int_{-5}^1 (6-2x)^{1/2}$$

$$= -\frac{1}{2}\int_{-5}^1 (6-2x)^{1/2}(-2\,dx)$$

$$= -\frac{1}{2}\left(\frac{6-2x}{\frac{3}{2}}\right)^{3/2}$$

$$= -\frac{1}{2}\cdot\frac{2}{3}(6-2x)^{3/2}$$

$$= -\frac{1}{3}(6-2x)^{3/2}\Big|_{-5}^1$$

$$= -\frac{1}{3}(8-64)$$

$$= \frac{1}{3}(-56)$$

$$= \frac{56}{3}$$

25. $\displaystyle\int_0^2 2x(9-2x^2)^2\,dx$; $u = (9-2x^2)$; $du = -4x$; $n = 2$

$$-\frac{1}{2}\int_0^2 (9-2x^2)^2(-2)(2x\,dx)$$

$$= -\frac{1}{2}\int_0^2 (9-2x^2)^2(-4x\,dx)$$

$$= -\frac{1}{2}\frac{(9-2x^2)^3}{3}\Big|_0^2 = -\frac{1}{6}(9-2x^2)^3\Big|_0^2$$

$$= -\frac{1}{6}[9-2(2)^2]^3 - \left(\frac{1}{6}\right)[9-2(0)^2]^3$$

$$= -\frac{1}{6}(1)^3 - \left(-\frac{1}{6}\right)(9)^3$$

$$= -\frac{1}{6}(1) - \left(-\frac{1}{6}\right)(729)$$

$$= -\frac{1}{6} + \frac{729}{6} = \frac{364}{3}$$

26. $\displaystyle\int_{-1}^2 V(V^3 + 1)\,dV = \int_{-1}^2 (V^4 + V)\,dV$

$$= \frac{V^5}{5} + \frac{V^2}{2}\Big|_{-1}^2$$

$$= 8.1$$

27. $u = x^3 + 9x + 6$; $du = (3x^2 + 9)\,dx = 3(x^2 + 3)\,dx$

$$\int_0^1 (x^2 + 3)(x^3 + 9x + 6)^2\,dx$$

$$= \frac{1}{3}\int_0^1 (x^3 + 9x + 6)^2(x^2 + 3)\,dx$$

$$= \frac{1}{3}\frac{(x^3 + 9x + 6)^3}{3} = \frac{1}{9}(x^3 + 9x + 6)^3\Big|_0^1$$

$$= \frac{1}{9}(16^3 - 6^3) = \frac{3880}{9}$$

28. $u = x^3 + 3x$; $du = (3x^2 + 3)\,dx = 3(x^2 + 1)\,dx$

$$\int_2^3 \frac{x^2 + 1}{(x^3 + 3x)^2}\,dx = \frac{1}{3}\int_2^3 (x^3 + 3x)^{-2}3(x^2 + 1)\,dx$$

$$= \frac{1}{3}\frac{(x^3 + 3x)^{-1}}{-1}$$

$$= \frac{-1}{3(x^3 + 3x)}\Big|_2^3$$

$$= \frac{-1}{3(36)} - \left(\frac{-1}{3(14)}\right)$$

$$= \frac{-7 + 18}{756} = \frac{11}{756}$$

29. $\displaystyle\int_{-1}^{2} \frac{8x-2}{(2x^2-x+1)^3}\,dx$

$\displaystyle = \int_{-1}^{2} (8x-2)(2x^2-x+1)^{-1}\,dx$

$\displaystyle = 2\int_{-1}^{2} (4x-1)(2x^2-x+1)^{-3}\,dx$

$\displaystyle = \left.\frac{2(2x^2-x+1)^{-2}}{-2}\right|_{-1}^{2}$

$\displaystyle = -(2x^2-x+1)^{-2}\big|_{-1}^{2}$

$\displaystyle = -\frac{1}{7^2} - \left(-\frac{1}{4^2}\right)$

$\displaystyle = 0.0421$

30. $u = 2x^3 - 4x + 1;\ du = (6x^2-4)\,dx = 2(3x^2-2)\,dx$

$\displaystyle\int_{-3}^{-2} (3x^2-2)\sqrt[3]{2x^3-4x+1}\,dx$

$\displaystyle = \frac{1}{2}\int_{-3}^{-2} (2x^3-4x+1)^{1/3} 2(3x^2-2)\,dx$

$\displaystyle = \frac{1}{2}\frac{(2x^3-4x+1)^{4/3}}{\frac{4}{3}}$

$\displaystyle = \frac{1}{2}\cdot\frac{3}{4}(2x^3-4x+1)^{4/3}$

$\displaystyle = \frac{3}{8}(2x^3-4x+1)^{4/3}\Big|_{-3}^{-2}$

$\displaystyle = \frac{3}{8}[(-7)^{4/3}-(-41)^{4/3}]$

$\displaystyle = \frac{3}{8}(13.391 - 141.377)$

$\displaystyle = -47.99$

31. $\displaystyle\int_{\sqrt{5}}^{3} 2z\sqrt[4]{z^4+8z^2+16}\,dz$

$\displaystyle = \int_{\sqrt{5}}^{3} 2z((z^2+4)^2)^{1/4}\,dz$

$\displaystyle = \int_{\sqrt{5}}^{3} 2z(z^2+4)^{1/2}\,dz$

$\displaystyle = \frac{2}{3}(z^2+4)^{3/2}\Big|_{\sqrt{5}}^{3}$

$\displaystyle = 13.25$

32. $u = 2x+4,\ du = 2\,dx;\ u = 3x+8,\ du = 3\,dx$

$\displaystyle\int_{-2}^{0} (\sqrt{2x+4} - \sqrt[3]{3x+8})\,dx$

$\displaystyle = \int_{-2}^{0} (2x+4)^{1/2}\,dx - \int_{-2}^{0} (3x+8)^{1/3}\,dx$

$\displaystyle = \frac{1}{2}\int_{-2}^{0} (2x+4)^{1/2} 2\,dx - \frac{1}{3}\int_{-2}^{0} (3x+8)^{1/3} 3\,dx$

$\displaystyle = \left[\frac{1}{2}\cdot\frac{2}{3}(2x+4)^{3/2} - \frac{1}{3}\cdot\frac{3}{4}(3x+8)^{4/3}\right]_{-2}^{0}$

$\displaystyle = \frac{1}{3}(2x+4)^{3/2} - \frac{1}{4}(3x+8)^{4/3}\Big|_{-2}^{0}$

$\displaystyle = \frac{1}{3}(8) - \frac{1}{4}(16) - \left[0 - \frac{1}{4}(2^{4/3})\right] = -0.7034$

33. $\displaystyle W = \int_{0}^{80} (1000 - 5x)\,dx$

$\displaystyle = \left(1000x - \frac{5}{2}x^2\right)\Big|_{0}^{80}$

$\displaystyle = 1000(80) - \frac{5}{2}(80)^2 - [0-0]$

$\displaystyle = 80,000 - 16,000$

$\displaystyle = 64,000\ \text{ft}\cdot\text{lb}$

34. $\displaystyle V = k\left(R^2\int_{0}^{R} r\,dr - \int_{0}^{R} r^3\,dr\right)$

$\displaystyle = kR^2\int_{0}^{R} r\,dr - k\int_{0}^{R} r^3\,dr$

R is a constant, r is a variable.

$\displaystyle V = kR^2\frac{r^2}{2}\Big|_{0}^{R} - k\frac{r^4}{4}\Big|_{0}^{R}$

$\displaystyle = kR^2\left[\frac{R^2}{2} - 0\right] - k\left[\frac{R^4}{4} - 0\right]$

$\displaystyle = \frac{kR^4}{2} - \frac{kR^4}{4} = \frac{2kR^4 - kR^4}{4}$

$\displaystyle = \frac{kR^4}{4}$

35. $u = 3x+9;\ du = 3\,dx$

$\displaystyle A = 4\pi\int_{0}^{2} \sqrt{3x+9}\,dx = 4\pi\cdot\frac{1}{3}\int_{0}^{2} (3x+9)^{1/2} 3\,dx$

$\displaystyle = \frac{4\pi}{3}\cdot\frac{2}{3}(3x+9)^{3/2}\Big|_{0}^{2} = \frac{8\pi}{9}(3x+9)^{3/2}\Big|_{0}^{2}$

$\displaystyle = 86.8\ \text{m}^2$

36. $u = 25 - y^2;\ du = -2y\,dy$

$$F = 19,600 \int_0^5 y\sqrt{25 - y^2}\,dy$$

$$= \frac{-19,600}{2} \int_0^5 (25 - y^2)^{1/2}(-2y\,dy)$$

$$= \frac{-19,600}{2} \cdot \frac{2}{3}(25 - y^2)^{3/2}\Big|_0^5$$

$$= \frac{-19,600}{2}(25 - y^2)^{3/2}\Big|_0^5$$

$$= 817\ \text{kN}$$

5.5 Numerical Integration: The Trapezoidal Rule

1. $\displaystyle\int_0^2 2x^2\,dx;\ n = 4;\ \Delta x = \frac{2-0}{4} = \frac{1}{2};\ \frac{\Delta x}{2} = \frac{1}{4}$

n	x_n	y_n
0	0	0
1	$\frac{1}{2}$	$\frac{1}{2}$
2	1	2
3	$\frac{3}{2}$	$\frac{9}{2}$
4	2	8

$$A_T = \frac{1}{4}\left[0 + 2\left(\frac{1}{2}\right) + 2(2) + 2\left(\frac{9}{2}\right) + 8\right]$$

$$= \frac{11}{2} = 5.50$$

$$\int_0^2 2x^2\,dx = \frac{2x^3}{3}\Big|_0^2 = \frac{16}{3} - 0 = \frac{16}{3} = 5.33$$

2. $\displaystyle\int_0^1 (1 - x^2)\,dx;\ n = 3;\ \Delta x = \frac{1-0}{3} = \frac{1}{3};\ \frac{\Delta x}{2} = \frac{1}{6}$

n	x_n	y_n
0	0	1
1	$\frac{1}{3}$	$\frac{8}{9}$
2	$\frac{2}{3}$	$\frac{5}{9}$
3	1	

$$A_T = \frac{1}{6}\left[1 + 2\left(\frac{8}{9}\right) + 2\left(\frac{5}{9}\right) + 0\right] = \frac{35}{54} = 0.6481$$

$$A = \int_0^1 (1 - x^2)\,dx = \int_0^1 dx - \int_0^1 x^2\,dx$$

$$= \left(x - \frac{x^3}{3}\right)\Big|_0^1 = 1 - \frac{1}{3} = \frac{2}{3} \approx 0.6667$$

3. $\displaystyle\int_1^4 (1 + \sqrt{x})\,dx;\ n = 6;\ \Delta x = \frac{4-1}{6} = \frac{1}{2};\ \frac{\Delta x}{2} = \frac{1}{4}$

n	x_n	y_n
0	1	2
1	1.5	2.22
2	2	2.41
3	2.5	2.58
4	3	2.73
5	3.5	2.87
6	4	3

$$A_T = \frac{1}{4}[2 + 2(2.22) + 2(2.41) + 2(2.58)$$
$$+ 2(2.73) + 2(2.87) + 3]$$

$$A = \frac{1}{4}(30.646) = 7.661$$

$$A = \int_1^4 (1 + x^{1/2})\,dx = \left(x + \frac{2}{3}x^{3/2}\right)\Big|_1^4$$

$$A = 4 + \frac{16}{3} - \left(1 + \frac{2}{3}\right) = \frac{23}{3} = 7.667$$

4. $\displaystyle\int_3^8 \sqrt{1 + x}\,dx;\ n = 5;\ \Delta a = \frac{8-3}{5} = 1;\ \frac{\Delta x}{2} = \frac{1}{2}$

n	x_n	y_n
0	3	2
1	4	2.24
2	5	2.45
3	6	2.65
4	7	2.83
5	8	3

$$A_T = \frac{1}{2}[2 + 2(2.24) + 2(2.45) + 2(2.65) + 2(2.83) + 3$$

$$A_T = 12.66$$

$$u = 1 + x;\ du = dx$$

$$A = \int_3^8 (1 + x)^{1/2}\,dx = \frac{2}{3}(1 + x)^{3/2}\Big|_3^8$$

$$= \frac{2}{3}(27 - 8) = \frac{38}{3} = 12.67$$

5. $\displaystyle\int_2^3 \frac{1}{2x}\,dx;\ n = 2;\ \Delta x = \frac{3-2}{2} = \frac{1}{2};\ \frac{\Delta x}{2} = \frac{1}{4}$

n	x_n	y_n
0	2	$\frac{1}{4}$
1	$\frac{5}{2}$	$\frac{1}{5}$
2	3	$\frac{1}{6}$

$$A_T = \frac{1}{4}\left[\frac{1}{4} + 2\left(\frac{1}{5}\right) + \frac{1}{6}\right] = \frac{1}{4}\left(\frac{49}{60}\right)$$

$$= \frac{49}{240} = 0.2042$$

6. $\displaystyle\int_2^6 \frac{dx}{x+3};\ n = 4;\ \Delta x = \frac{6-2}{4} = 1;\ \frac{\Delta x}{2} = \frac{1}{2}$

n	x_n	y_n
0	2	$\frac{1}{5}$
1	3	$\frac{1}{6}$
2	4	$\frac{1}{7}$
3	5	$\frac{1}{8}$
4	6	$\frac{1}{9}$

$$A_T = \frac{1}{2}\left[\frac{1}{5} + 2\left(\frac{1}{6}\right) + 2\left(\frac{1}{7}\right) + 2\left(\frac{1}{8}\right) + \frac{1}{9}\right]$$

$$A_T = 0.590$$

7. $\displaystyle\int_0^5 \sqrt{25 - x^2}\,dx;\ n = 5;\ x = \frac{5}{5} = 1;\ \frac{\Delta x}{2} = \frac{1}{2}$

n	x_n	y_n
0	0	5
1	1	4.90
2	2	4.58
3	3	4
4	4	3
5	5	0

$$A_T = \frac{1}{2}[5 + 2(4.90) + 2(4.58) + 2(4) + 2(3) + 0]$$

$$= 18.98$$

8. $\displaystyle\int_0^2 \sqrt{x^3 + 1}\,dx;\ n = 4;\ \Delta x = \frac{2}{4} = \frac{1}{2};\ \frac{\Delta x}{2} = \frac{1}{4}$

n	x_n	y_n
0	0	1
1	$\frac{1}{2}$	1.061
2	1	1.414
3	$1\frac{1}{2}$	2.092
4	2	3

$$A_T = \frac{1}{4}[1 + 2(1.061) + 2(1.414) + 2(2.092) + 3]$$

$$A_T = 3.28$$

9. $\displaystyle\int_1^5 \frac{1}{x^2 + x}\,dx;\ n = 10;\ \Delta x = \frac{5-1}{10} = 0.4;$

$$\frac{\Delta x}{2} = 0.2$$

n	x_n	y_n
0	1	0.5000
1	1.4	0.2976
2	1.8	0.1984
3	2.2	0.1420
4	2.6	0.1068
5	3	0.0833
6	3.4	0.0668
7	3.8	0.0548
8	4.2	0.0458
9	4.6	0.0388
10	5	0.3333

$$\begin{aligned}
A = 0.2[&0.5000 + 2(0.2976) + 2(0.1984) \\
&+ 2(0.1420) + 2(0.1068) + 2(0.0833) \\
&+ 2(0.0668) + 2(0.0548) + 2(0.0458) \\
&+ 2(0.0388) + 0.0333] \\
= 0.5205
\end{aligned}$$

10. $\displaystyle\int_2^4 \frac{1}{x^2 + 1}\,dx;\ n = 10;\ \Delta x = \frac{2}{10} = 0.2;\ \frac{\Delta x}{2} = 0.1$

x	y
2	0.2
2.2	0.171
2.4	0.148
2.6	0.129
2.8	0.113
3	0.1
3.2	0.089
3.4	0.080
3.6	0.072
3.8	0.065
4	$\frac{1}{17}$

$$\begin{aligned}
A = 0.1\bigg[&\frac{1}{5} + 2(0.171) + 2(0.148) + 2(0.129) \\
&+ 2(0.113) + 2(0.1) + 2(0.089) + 2(0.080) \\
&+ 2(0.072) + 2(0.065) + \frac{1}{17}\bigg]
\end{aligned}$$

$$A = 0.219$$

11. $\int_0^4 2^x dx$; $n = 12$; $\Delta x = \dfrac{4}{12} = \dfrac{1}{3}$; $\dfrac{\Delta x}{2} = \dfrac{1}{6}$

x	y
0	1
$\frac{1}{3}$	1.260
$\frac{2}{3}$	1.587
1	2
$1\frac{1}{3}$	2.520
$1\frac{2}{3}$	3.175
2	4
$2\frac{1}{3}$	5.040
$2\frac{2}{3}$	6.350
3	8
$3\frac{1}{3}$	10.080
$3\frac{2}{3}$	12.699
4	16

$A = \dfrac{1}{6}[1 + 2(1.260) + 2(1.587) + 2(2) + 2(2.520)$

$\qquad + 2(3.175) + 2(4) + 2(5.040) + 2(6.350)$

$\qquad + 2(8) + 2(10.080) + 2(12.699) + 16]$

$A = 21.74$

12. $\int_0^{1.5} 10^x dx$; $n = 15$; $\Delta x = \dfrac{1.5}{15} = 0.1$; $\dfrac{\Delta x}{2} = 0.05$

x	y
0	1
0.1	1.259
0.2	1.585
0.3	1.995
0.4	2.512
0.5	3.162
0.6	3.981
0.7	5.012
0.8	6.310
0.9	7.943
1	10
1.1	12.589
1.2	15.849
1.3	19.953
1.4	25.119
1.5	31.623

$A = 0.05[1 + 2(1.259) + 2(1.585) + 2(1.995) + 2(2.512)$

$\qquad + 2(3.162) + 2(3.981) + 2(5.012) + 2(6.310)$

$\qquad + 2(7.943) + 2(10) + 2(12.589) + 2(15.849)$

$\qquad + 2(19.953) + 2(25.119) + 31.623]$

$A = 13.36$

13. $\int_2^{14} y dx$; $\Delta x = 2$; $\dfrac{\Delta x}{2} = 1$

x	y
2	0.67
4	2.34
6	4.56
8	3.67
10	3.56
12	4.78
14	6.87

$A = 1[0.67 + 2(2.34) + 2(4.56) + 2(3.67)$

$\qquad + 2(3.56) + 2(4.78) + 6.87]$

$\quad = 45.36$

14. $\int_{1.4}^{3.2} y\, dx$

x	1.4	1.7	2.0	2.3	2.6	2.9	3.2
y	0.18	7.87	18.23	23.53	24.62	20.93	20.76

$\Delta x = 0.3$; $\dfrac{\Delta x}{2} = 0.15$

$A = 0.15[0.18 + 2(7.87) + 2(18.23) + 2(23.53)$

$\qquad + 2(24.62) + 2(20.93) + 20.76]$

$\quad = 0.15(211.3) = 31.695 = 31.70$

15. $F = k \int_0^2 \dfrac{dx}{(4 + x^2)^{3/2}}$; $n = 8$; $\Delta x = \dfrac{2}{8} = \dfrac{1}{4}$;

$\dfrac{\Delta x}{2} = \dfrac{1}{8}$

x	y
0	0.125
0.25	0.122
0.50	0.114
0.75	0.103
1.00	0.089
1.25	0.076
1.50	0.064
1.75	0.053
2.00	0.044

$F = k\dfrac{1}{8}[0.125 + 2(0.122 + 0.114 + 0.103 + 0.089$

$\qquad + 0.076 + 0.064 + 0.053) + 0.044]$

$\quad = k\dfrac{1}{8}(1.4128) = 0.177k$

16. $L = 2\int_0^{100} \sqrt{1.6 \times 10^{-7}x^2 + 1}\,dx;\ n = 10;$

$$\Delta x = \frac{100}{10} = 10;\ \frac{\Delta x}{2} = 5$$

x	y
0	2
10	2.000016
20	2.000064
30	2.000144
40	2.000256
50	2.000400
60	2.000576
70	2.000784
80	2.001024
90	2.001296
100	2.001600

$L = 5[2 + 2(2.000016 + 2.000064 + 2.000144$
$\quad + 2.000256 + 2.000400 + 2.000576$
$\quad + 2.000784 + 2.001024 + 2.001296)$
$\quad + 2.001600]$
$L = 200.054$ ft

5.6   Simpson's Rule

1. $\int_0^2 (1 + x^3)\,dx;\ n = 2;\ \Delta x = 1;\ \frac{\Delta x}{3} = \frac{1}{3}$

$$A_s = \frac{1}{3}[1 + 4(2) + 9]$$

$$A_s = 6$$

$$A = \int_0^2 (1 + x^3)\,dx$$

$$= x + \frac{1}{4}x^4 \Big|_0^2$$

$$= 2 + \frac{1}{4}(16) = 6$$

n	x_n	y_n
0	0	1
1	1	2
2	2	9

2. $\int_0^8 x^{1/3}\,dx;\ n = 2;\ \Delta x = \frac{8}{2} = 4;\ \frac{\Delta x}{3} = \frac{4}{3}$

$$A_S = \frac{4}{3}[0 + 4\sqrt[3]{4} + 2]$$

$$A_S = 11.1$$

n	x_n	y_n
1	0	0
2	4	$\sqrt[3]{4}$
3	8	2

$$A = \int_0^8 x^{1/3}\,dx = \frac{3}{4}x^{4/3}\Big|_0^8 = \frac{3}{4}(8)^{4/3} = 12$$

3. $\int_1^4 (2x + \sqrt{x})\,dx;\ n = 6;\ \Delta x = \frac{4-1}{6} = \frac{1}{2};$

$$\frac{\Delta x}{3} = \frac{1}{6}$$

$$A_S = \frac{1}{6}[3 + 4(4.22) + 2(5.41) + 4(6.58)$$
$$\quad + 2(7.73) + 4(8.87) + 10]$$

$$A_S = \frac{1}{6}(117.999) = 19.67$$

n	x_n	y_n
1	1	3
2	1.5	4.22
3	2	5.41
4	2.5	6.58
5	3	7.73
6	3.5	8.87
7	4	10

$$A = \int_1^4 (2x + \sqrt{x})\,dx = 2\int_1^4 x\,dx + \int x^{1/2}\,dx$$

$$= \left(x^2 + \frac{2}{3}x^{3/2}\right)\Big|_1^4 = 16 + \frac{16}{3} + -\left(1 + \frac{2}{3}\right)$$

$$= \frac{59}{3} = 19.67$$

4. $\int_0^2 x\sqrt{x^2 + 1}\,dx;\ n = 4;\ \Delta x = \frac{2}{4} = \frac{1}{2};\ \frac{\Delta x}{3} = \frac{1}{6}$

$$A_S = \frac{1}{6}[0 + 4(0.559) + 2(1.414) + 4(2.704) + 4.472]$$

$$= \frac{1}{6}(20.353) = 3.3922$$

n	x_n	y_n
1	0	0
2	0.5	0.559
3	1	1.414
4	1.5	2.704
5	2	4.472

$$A = \frac{1}{2}\int_0^2 (x^2 + 1)^{1/2} 2x\,dx$$

$$= \frac{2}{3}\cdot\frac{1}{2}(x^2 + 1)^{3/2}$$

$$= \frac{1}{3}(x^2 + 1)^{3/2}\Big|_0^2$$

$$= \frac{1}{3}[5^{3/2} - 1] = 3.3934$$

5. $\int_2^3 \frac{1}{2x}dx$; $n = 2$; $\Delta x = \frac{1}{2}$; $\frac{\Delta x}{3} = \frac{1}{6}$

$$A_s = \frac{1}{6}\left[\frac{1}{4} + 4\left(\frac{1}{5}\right) + \frac{1}{6}\right]$$

$$= \frac{1}{6}\left(\frac{73}{60}\right) = \frac{73}{360} = 0.2028$$

n	x_n	y_n
0	2	$\frac{1}{4}$
1	$\frac{5}{2}$	$\frac{1}{5}$
2	3	$\frac{1}{6}$

6. $\int_2^6 \frac{dx}{x+3}$; $n = 4$; $\Delta x = \frac{6-2}{4} = 1$; $\frac{\Delta x}{3} = \frac{1}{3}$

$$A_S = \frac{1}{3}\left[\frac{1}{5} + 4\left(\frac{1}{6}\right) + 2\left(\frac{1}{7}\right) + 4\left(\frac{1}{8}\right) + \frac{1}{9}\right]$$

$$= 0.5878$$

n	x_n	y_n
1	2	$\frac{1}{5}$
2	3	$\frac{1}{6}$
3	4	$\frac{1}{7}$
4	5	$\frac{1}{8}$
5	6	$\frac{1}{9}$

7. $\int_0^5 \sqrt{25 - x^2}dx$; $n = 4$; $\Delta x = \frac{5}{4}$; $\frac{\Delta x}{3} = \frac{5}{12}$

$$A_S = \frac{5}{12}[5 + 4(4.841) + 2(4.330) + 4(3.307) + 0]$$

$$= \frac{5}{12}(46.2539) = 19.27$$

n	x_n	y_n
1	0	5
2	1.25	4.841
3	2.50	4.330
4	3.75	3.307
5	5	0

8. $\int_0^2 \sqrt{x^3 + 1}dx$; $n = 4$; $\Delta x = \frac{1}{2}$; $\frac{\Delta x}{3} = \frac{1}{6}$

$$A_S = \frac{1}{6}[1 + 4(1.061) + 2(1.414) + 4(2.092) + 3]$$

$$A_S = \frac{1}{6}[19.438]$$

$$= 3.2396$$

n	x_n	y_n
1	0	1
2	0.5	1.061
3	1	1.414
4	1.5	2.092
5	2	3

9. $\int_1^5 \frac{dx}{x^2 + x}$; $n = 10$; $\Delta x = 0.4$; $\frac{\Delta x}{3} = \frac{0.4}{3}$

$$A_s = \frac{0.4}{3}[0.5000 + 4(0.2976) + 2(0.1984)$$
$$+ 4(0.1420) + 2(0.1068) + 4(0.0833)$$
$$+ 2(0.0668) + 4(0.0548) + 2(0.0458)$$
$$+ 4(0.0388) + 0.0333]$$

$$= \frac{0.4}{3}(3.8349)$$

$$= 0.5114$$

10. $\int_2^4 \frac{dx}{x^2 + 1}$; $1 = 10$; $\Delta x = \frac{2}{10} = \frac{1}{5}$; $\frac{\Delta x}{3} = \frac{1}{15}$

n	x_n	y_n	n	x_n	y_n
1	2	0.2	7	3.2	0.089
2	2.2	0.171	8	3.4	0.080
3	2.4	0.148	9	3.6	0.072
4	2.6	0.129	10	3.8	0.065
5	2.8	0.113	11	4	0.059
6	3	0.1			

$$A_S = [0.2 + 4(0.171) + 2(0.148) + 4(0.129) + 2(0.113)$$
$$+ 4(0.1) + 2(0.089) + 4(0.080) + 2(0.072)$$
$$+ 4(0.065) + 0.059]\left(\frac{1}{15}\right)$$

$$= \frac{1}{15}(3.280)$$

$$= 0.2187$$

11. $\int_{-4}^{5}(2x^4+1)^{0.1}dx$; $n=6$; $\Delta x=\dfrac{9}{6}=\dfrac{3}{2}$; $\dfrac{\Delta x}{3}=\dfrac{1}{2}$

$A_S = [1.866 + 4(1.548) + 2(1.116) + 4(1.012)$

$\qquad + 2(1.419) + 4(1.770) + 2.040] \cdot \dfrac{1}{2}$

$A_S = \dfrac{1}{2}(26.2949)$

$\quad = 13.147$

n	x_n	y_n
1	−4	1.866
2	−2.5	1.548
3	−1	1.116
4	0.5	1.012
5	2	1.419
6	3.5	1.770
7	5	2.040

12. $\int_{0}^{2.4}\dfrac{dx}{(4+\sqrt{x})^{3/2}}$; $n=8$; $\Delta x=\dfrac{2.4}{8}=0.3$;

$\dfrac{\Delta x}{3}=0.1$

n	x_n	y_n	n	x_n	y_n
1	0	0.125	6	1.5	0.084
2	0.3	0.103	7	1.8	0.081
3	0.6	0.096	8	2.1	0.079
4	0.9	0.091	9	2.4	0.076
5	1.2	0.087			

$A_S = (0.1)[0.125 + 4(0.103) + 2(0.096) + 4(0.091)$

$\qquad + 2(0.087) + 4(0.084) + 2(0.081)$

$\qquad + 4(0.079) + 0.076]$

$A_S = 0.1(2.1543) = 0.2154$

13. $\Delta x = 2$; $\dfrac{\Delta x}{3}=\dfrac{2}{3}$

x	y
2	0.67
4	2.34
6	4.56
8	3.67
10	3.56
12	4.78
14	6.87

$\int_{2}^{14} y\,dx = \dfrac{2}{3}[0.67 + 4(2.34) + 2(4.56) + 4(3.67)$

$\qquad + 2(3.56) + 4(4.78) + 6.87]$

$\qquad = 44.63$

14. $\int_{1.4}^{3.2} y\,dx$; $\Delta x = 0.3$; $\dfrac{\Delta x}{3}=0.1$

n	x_n	y_n
1	1.4	0.18
2	1.7	7.87
3	2.0	18.23
4	2.3	23.53
5	2.6	24.62
6	2.9	20.93
7	3.2	20.76

$A_S = 0.1[0.18 + 4(7.87) + 2(18.23) + 4(23.53)$

$\qquad + 2(24.62) + 4(20.93) + 20.76]$

$A_S = 0.1(315.96) = 31.596 = 31.60$

15. $\bar{x} = 0.9129\displaystyle\int_{0}^{3} x\sqrt{0.3-0.1x}\,dx$; $n=12$;

$\Delta x = \dfrac{3}{12}=\dfrac{1}{4}$; $\dfrac{\Delta x}{3}=\dfrac{1}{12}$

n	x_n	y_n	n	x_n	y_n
1	0	0	8	1.75	0.619
2	0.25	0.131	9	2	0.632
3	0.50	0.25	10	2.25	0.616
4	0.75	0.356	11	2.5	0.559
5	1	0.447	12	2.75	0.435
6	1.25	0.523	13	3	0
7	1.50	0.581			

$\bar{x} = A_S = \dfrac{1}{12}[0 + 4(0.131) + 2(0.25) + 4(0.356)$

$\qquad + 2(0.447) + 4(0.523) + 2(0.581)$

$\qquad + 4(0.619) + 2(0.632) + 4(0.616)$

$\qquad + 2(0.559) + 4(0.435) + (0)](0.9129)$

$\qquad = \dfrac{1}{12}(15.6572)(0.9129) = 1.191$

$\bar{x} = 1.191$ in

16. $i_{av} = \dfrac{1}{4}\displaystyle\int_{0}^{4}(4t-t^2)^{0.2}dt$; $n=10$;

$\Delta x = \dfrac{4}{10}=\dfrac{2}{5}=0.4$; $\dfrac{\Delta x}{3}=\dfrac{2}{15}$

n	x_n	y_n	n	x_n	y_n
1	0	0	7	2.4	1.309
2	0.4	1.076	8	2.8	1.274
3	0.8	1.207	9	3.2	1.207
4	1.2	1.274	10	3.6	1.076
5	1.6	1.309	11	4	0
6	2	1.320			

$$i_{av} = A_s = \frac{1}{4} \cdot \frac{2}{15}[0 + 4(1.076) + 2(1.207)$$

$$+ +4(1.274) + 2(1.309) + 4(1.320)$$
$$+ 2(1.309) + 4(1.274) + 2(1.207)$$
$$+ 4(1.076) + 0]$$

$$= \frac{1}{30}(34.1400)$$

$$i_{av} = 1.1380 \text{ A}$$

Chapter 5 Review Exercises

1. $\int (4x^3 - x)dx = \int 4x^3 dx - \int x dx = \frac{4x^4}{4} - \frac{x^2}{2} + C$

$$= x^4 - \frac{1}{2}x^2 + C$$

2. $\int (5 + 3x^2)dx = \int 5\,dx + \int 3x^2 dx$

$$= 5x + x^3 + C$$

3. $\int \sqrt{u}(u^2 + 2)du = \int (u^{5/2} + 2u^{1/2})du$

$$= \frac{2}{7}u^{7/2} + \frac{4}{3}u^{3/2} + C$$

4. $\int x(x - 3x^4)dx = \int (x^2 - 3x^5)dx$

$$= \frac{x^3}{3} - \frac{x^6}{2} + C$$

5. $\int_1^4 \left(\frac{\sqrt{x}}{2} + \frac{2}{\sqrt{x}}\right)dx$

$$= \frac{1}{2}\int_1^4 x^{1/2}dx + 2\int_1^4 x^{-1/2}dx$$

$$= \frac{1}{2}\frac{x^{3/2}}{\frac{3}{2}} + \frac{2x^{1/2}}{\frac{1}{2}}\Big|_1^4$$

$$= \frac{1}{3}x^{3/2} + 4x^{1/2}\Big|_1^4$$

$$= \left[\frac{1}{3}(4)^{3/2} + 4(4)^{1/2}\right] - \left[\frac{1}{3}(1)^{3/2} + 4(1)^{1/2}\right]$$

$$= \frac{19}{3}$$

6. $\int_1^2 \left(x + \frac{1}{x^2}\right)dx = \frac{x^2}{2} - \frac{1}{x}\Big|_1^2$

$$= \frac{2^2}{2} - \frac{1}{2} - \left(\frac{1^2}{2} - \frac{1}{1}\right)$$

$$= 2$$

7. $\int_0^2 x(4 - x)dx = \int_0^2 (4x - x^2)dx$

$$= 2x^2 - \frac{x^3}{3}\Big|_0^2$$

$$= 2(2)^2 - \frac{2^3}{3} = \frac{16}{3}$$

8. $\int_0^1 2t(2t + 1)^2 dt = \int_0^1 2t(4t^2 + 4t + 1)dt$

$$= \int_0^1 (8t^3 + 8t^2 + 1t)dt$$

$$= 2t^4 + \frac{8}{3}t^3 + t^2\Big|_0^1$$

$$= 2(1)^4 + \frac{8}{3}(1)^3 + 1^2 = \frac{17}{3}$$

9. $\int \left(3 + \frac{2}{x^3}\right)dx = \int 3\,dx + \int \frac{2}{x^3}dx$

$$= \int 3\,dx + \int 2x^{-3}dx$$

$$= (3x) + \left(-\frac{2}{2}x^{-2}\right)$$

$$= 3x - x^{-2} = 3x - \frac{1}{x^2} + C$$

10. $\int \left(3\sqrt{x} + \frac{1}{2\sqrt{x}} - \frac{1}{4}\right)dx$

$$= \int 3\sqrt{x}\,dx + \int \frac{1}{2\sqrt{x}}dx - \int \frac{1}{4}dx$$

$$= 3\frac{x^{3/2}}{\frac{3}{2}} + \frac{1}{2}\frac{x^{1/2}}{\frac{1}{2}} - \frac{1}{4}x + C$$

$$= 2x^{3/2} + x^{1/2} - \frac{x}{4} + C$$

11. $\int_{-2}^5 \frac{dx}{\sqrt[3]{x^2 + 6x + 9}}$

$$= \int_{-2}^5 \frac{dx}{\sqrt[3]{(x + 3)^2}}$$

$$= \int_{-2}^5 (x + 3)^{-2/3}dx$$

$$= 3(x + 3)^{1/3}\Big|_{-2}^5$$

$$= 3\sqrt[3]{5 + 3} - 3\sqrt[3]{-2 + 3}$$

$$= 3\sqrt[3]{8} - 3\sqrt[3]{1} = 3(2) - 3$$

$$= 3$$

12. $\displaystyle\int_{0.35}^{0.85} x(\sqrt{1-x^2}+1)\,dx$

$\displaystyle = \int_{1-0.35^2}^{1-0.85^2} (u^{1/2}+1)\left(-\frac{1}{2}\,du\right)$ where

$u = 1 - x^2,\ du = -2x\,dx$

$\displaystyle \int_{0.35}^{0.85} x(\sqrt{1-x^2}+1)\,dx$

$\displaystyle = -\frac{1}{2}\left[\frac{2}{3}u^{2/3}+u\right]_{1-0.35^2}^{1-0.85^2}$

$\displaystyle = -\frac{1}{2}\left[\frac{2}{3}(1-0.85^2)^{3/2}+(1-0.85^2)\right.$

$\displaystyle \left. -\frac{2}{3}(1-0.35^2)^{3/2}-(1-0.35^2)\right]$

$= 0.525$

13. $\displaystyle\int \frac{dn}{(2-5n)^3} = \int (2-5n)^{-3}\,dn$

$\displaystyle -\frac{1}{5}\int (2-5n)^{-3}(-5\ dn) = -\frac{1}{5}\times\frac{(2-5n)^{-2}}{-2}+C$

$\displaystyle = \frac{1}{10}\times\frac{1}{(2-5n)^2}+C$

$\displaystyle = \frac{1}{10(2-5n)^2}+C$

14. $\displaystyle\int \frac{1}{x^2}\sqrt{1+\frac{1}{x}}\,dx = -\int u^{1/2}\,du = -\frac{2}{3}u^{2/3}+C$

$\displaystyle u = 1+\frac{1}{x},\ du = \frac{-1}{x^2}\,dx$

$\displaystyle \int \frac{1}{x^2}\sqrt{1+\frac{1}{x}}\,dx = -\frac{2}{3}\left(1+\frac{1}{x}\right)^{3/2}+C$

15. $\displaystyle\int 3(7-2x)^{3/4}\,dx = -\frac{3}{2}\int u^{3/4}\,du$

$\displaystyle = -\frac{3}{2}\left(\frac{4}{7}\right)u^{7/4}+C$

$u = 7-2x,\ du = -2\,dx$

$\displaystyle \int 3(7-2x)^{3/4}\,dx = -\frac{6}{7}(7-2x)^{7/4}+C$

16. $\displaystyle\int (y^3+3y^2+3y+1)^{2/3}\,dy = \int \left[(y+1)^3\right]^{2/3}\,dy$

$\displaystyle = \int (y+1)^2\,dy$

$\displaystyle = \frac{(y+1)^3}{3}+C$

17. $\displaystyle\int_0^2 \frac{3x\ dx}{\sqrt[3]{1+2x^2}} = \int_0^2 (1+2x^2)^{-1/3}(3x)\,dx$

$\displaystyle u = 1+2x^2;\ du = 4x\ dx;\ n = -\frac{1}{3}$

$\displaystyle \frac{3}{4}\int_0^2 (1+2x^2)^{-1/3}(4x)\,dx$

$\displaystyle = \frac{3}{4}\times\frac{(1+2x^2)^{2/3}}{\frac{2}{3}}\bigg|_0^2$

$\displaystyle = \frac{9}{8}(1+2x^2)^{2/3}\bigg|_0^2$

$\displaystyle = \frac{9}{8}[1+2(2^2)]^{2/3}-\frac{9}{8}[1+2(0)^2]^{2/3}$

$\displaystyle = \frac{9}{8}(9)^{2/3}-\frac{9}{8}(1)^{2/3}$

$\displaystyle = \frac{9}{8}(\sqrt[3]{81}-1) = \frac{9}{8}(3\sqrt[3]{3}-1)$

18. $\displaystyle\int_1^6 \frac{2\ dx}{(3x-2)^{3/4}} = \frac{2}{3}\int_1^{16}\frac{du}{u^{3/4}}$

$\displaystyle = \frac{2}{3}(4)u^{1/4}\bigg|_1^{16}$

$\displaystyle = \frac{8}{3}\left[16^{1/4}-1^{1/4}\right]$

$u = 3x-2,\ du = 3\,dx$

$\displaystyle \int_1^6 \frac{2\ dx}{(3x-2)^{3/4}} = \frac{8}{3}$

19. $\displaystyle\int x^2(1-2x^3)^4\,dx = -\frac{1}{6}\int u^4\,du$

$\displaystyle = -\frac{1}{6}\frac{u^5}{5}+C$

$u = 1-2x^3,\ du = -6x^2\,dx$

$\displaystyle \int x^2(1-2x^3)^4\,dx = -\frac{1}{30}(1-2x^3)^5+C$

20. $\displaystyle\int 3x^2(1-5x^4)^{1/3}\,dx = -\frac{3}{20}\int u^{1/3}\,du$

$\displaystyle = -\frac{3}{20}\left(\frac{3}{4}\right)u^{4/3}+C$

$u = 1-5x^4,\ du = -20x^3\,dx$

$\displaystyle \int 3x^2(1-5x^4)^{1/3}\,dx = -\frac{9}{80}(1-5x^4)^{4/3}+C$

21. $\int \dfrac{(2-3x^2)dx}{(2x-x^3)^2} = \int \dfrac{du}{u^2} = -\dfrac{1}{u} + C$

$u = 2x - x^3,\, du = (2 - 3x^2)dx$

$\int \dfrac{(2-3x^2)dx}{(2x-x^3)^2} = -\dfrac{1}{(2x-x^3)} + C$

22. $\int \dfrac{x^2-3}{\sqrt{6+9x-x^3}}\,dx = -\dfrac{1}{3}\int \dfrac{du}{\sqrt{u}}$

$\qquad\qquad = -\dfrac{1}{3}(2)u^{1/2} + C$

$u = 6 + 9x - x^3,\, du = (9 - 3x^2)dx = -3(x^2 - 3)dx$

$\int \dfrac{x^2-3}{\sqrt{6+9x-x^3}}\,dx = -\dfrac{2}{3}(6+9x-x^3)^{1/2} + C$

23. $\displaystyle\int_1^3 (x^2+x+2)(2x^3+3x^2+12x)dx = \dfrac{1}{6}\int_{17}^{117} u\,du$

$u = 2x^3 + 3x^2 + 12x,$

$du = (6x^2 + 6x + 12)dx = 6(x^2 + x + 2)dx$

$\displaystyle\int_1^3 (x^2+x+2)(2x^3+3x^2+12x)dx$

$= \dfrac{1}{6}\dfrac{u^2}{2}\Big|_{17}^{117}$

$= \dfrac{1}{12}[117^2 - 17^2]$

$= \dfrac{3350}{3}$

24. $\displaystyle\int_0^2 (4x+18x^2)(x^2+3x^3)^2 dx = 2\int_0^{28} u^2 du$

$\qquad\qquad = 2\dfrac{u^3}{3}\Big|_0^{28}$

$u = x^2 + 3x^3,\, du = (2x + 9x^2)dx,$
$2\,du = (4x + 18x^2)dx$

$\displaystyle\int_0^2 (4x+18x^2)(x^2+3x^3)^2 dx = \dfrac{2}{3}(28^3 - 0^3)$

$\qquad\qquad = \dfrac{43,904}{3}$

25. $\dfrac{dy}{dx} = 3 - x^2$

$y = 3x - \dfrac{x^3}{3} + C$

$3 = 3(-1) - \dfrac{(-1)^3}{3} + C$

$C = \dfrac{17}{3}$

$y = 3x - \dfrac{x^3}{3} + \dfrac{17}{3}$

26. $\dfrac{dy}{dx} = x(x^2+1)^2$

$y = \displaystyle\int x(x^2+1)^2 dx, \quad u = x^2+1, du = 2x\,dx$

$y = \dfrac{1}{6}(x^2+1)^3 + C,$

$-2 = \dfrac{1}{6}(1^2+1)^3 + C$

$C = -\dfrac{10}{3}$

$y = \dfrac{1}{6}(x^2+1)^3 - \dfrac{10}{3}$

27. (a) $\displaystyle\int (1-2x)dx = \int dx - 2\int x\,dx$

$\qquad\qquad = x - x^2 + C_1$

(b) $\displaystyle\int (1-2x)dx = -\dfrac{1}{2}\int u\,du = -\dfrac{1}{2}\dfrac{u^2}{2} + C_2$

$u = 1 - 2x,\, du = -2\,dx$

$\displaystyle\int (1-2x)dx = -\dfrac{1}{4}(1-2x)^2 + C_2$

$\qquad\qquad = -\dfrac{1}{4}(1 - 4x + 4x^2) + C_2$

$\qquad\qquad = -\dfrac{1}{4} + x - x^2 + C_2$

$\qquad\qquad = x - x^2 + C_2 - \dfrac{1}{4}$

C_1 and C_2 are not equal, $C_1 = C_2 - \dfrac{1}{4}$. **Indefinite integrals with the same integrand are only equal to within an arbitrary constant.**

28. (a) $\displaystyle\int (3x+2)dx = \int 3x\,dx + \int 2\,dx$

$\qquad\qquad = \dfrac{3}{2}x^2 + 2x + C_1$

(b) $\displaystyle\int (3x+2)dx = \dfrac{1}{3}\int u\,du = \dfrac{1}{6}u^2 + C_2$

$u = 3x + 2,\, du = 3\,dx$

$\displaystyle\int (3x+2)dx = \dfrac{1}{6}(3x+2)^2 + C_2$

$\qquad\qquad = \dfrac{1}{6}(9x^2 + 12x + 4) + C_2$

$\qquad\qquad = \dfrac{9}{6}x^2 + 2x + \dfrac{4}{6} + C_2$

$\qquad\qquad = \dfrac{3}{2}x^2 + 2x + \dfrac{2}{3} + C_2$

from which $C_1 = \dfrac{2}{3} + C_2$. C_1 and C_2 **are not equal.** See exercise 27.

29. $A = \displaystyle\int_{1}^{3} (6x - 1)\,dx$

$A = 3x^2 - x \Big|_{1}^{3}$

$A = 3(3^2) - 3 - [3(1^2) - 1]$

$A = 22$

30. $y = 8x - x^4 = x(8 - x^3)$ has $(0,0)$ and $(0,2)$ as intercepts

$A = \displaystyle\int_{0}^{2} (8x - x^4)\,dx = 4x^2 - \dfrac{x^5}{5}\bigg|_{0}^{2} = 4(2)^2 - \dfrac{2^5}{5}$

$A = \dfrac{48}{5}$

31. $\Delta x = \dfrac{b - a}{n} = \dfrac{3 - 1}{4} = \dfrac{1}{2}, \dfrac{\Delta x}{2} = \dfrac{1}{4}$

$y_0 = 1, f(1) = \dfrac{1}{2(1) - 1} = 1, \; y_1 = 1.5, f(1.5) = \dfrac{1}{2(1.5) - 1} = \dfrac{1}{2}, \; y_2 = 2.0, f(2.0) = \dfrac{1}{2(2.0) - 1} = \dfrac{1}{3},$

$y_3 = 2.5, f(2.5) = \dfrac{1}{2(2.5) - 1} = \dfrac{1}{4}, \; y_4 = 3.0, f(3.0) = \dfrac{1}{2(3.0) - 1} = \dfrac{1}{5}$

$\displaystyle\int_{1}^{3} \dfrac{dx}{2x - 1} \approx \dfrac{1}{4}\left[1 + 2\left(\dfrac{1}{2}\right) + 2\left(\dfrac{1}{3}\right) + 2\left(\dfrac{1}{4}\right) + \dfrac{1}{5}\right] = 0.842$

32. $\Delta x = \dfrac{b - a}{n} = \dfrac{21 - 6.0}{5}, \dfrac{\Delta x}{2} = \dfrac{1}{2}\left(\dfrac{21 - 6.0}{5}\right)$

$\displaystyle\int_{6.0}^{21} f(x)\,dx \approx \dfrac{1}{2}\left(\dfrac{21 - 6.0}{5}\right)[2.0 + 2(1.2) + 2(0.2) + 2(1.0) + 2(6.0) + 12] \approx 46.2$

33. $\Delta x = \dfrac{3 - 1}{4} = \dfrac{1}{2}; \; x_0 = 1, x_1 = 1.5, x_2 = 2, x_3 = 2.5, x_4 = 3.0$

$\displaystyle\int_{1}^{3} \dfrac{dx}{2x - 1} \approx \dfrac{\Delta x}{3}[y_0 + 4y_1 + 2y_2 + 4y_3 + y_4] \approx \dfrac{\frac{1}{2}}{3}\left[\dfrac{1}{2.1 - 1} + \dfrac{4}{2(1.5) - 1} + \dfrac{2}{2(2) - 1} + \dfrac{4}{2(2.5) - 1} + \dfrac{1}{2(3) - 1}\right]$

$\approx \dfrac{73}{90} = 0.811$

34. $\Delta x = \dfrac{b - a}{n} = \dfrac{3.4 - 1.0}{6}, \dfrac{\Delta x}{3} = \dfrac{1}{3}\left(\dfrac{3.4 - 1.0}{6}\right)$

$\displaystyle\int_{1.0}^{3.4} f(x)\,dx \approx \dfrac{1}{3}\left(\dfrac{3.4 - 1.0}{6}\right)[1.45 + 4(1.89) + 2(2.66) + 4(3.50) + 2(3.22) + 4(3.04) + 2.44]$

$\displaystyle\int_{1.0}^{3.4} f(x)\,dx \approx 6.58$

35. $\Delta x = \dfrac{b - a}{n} = \dfrac{4 - 1}{3} = 1$

$\displaystyle\int_{1}^{4} x\sqrt[3]{2x^2 + 1}\,dx \approx \Delta x[f(1) + f(2) + f(3)] \approx 1[1\sqrt[3]{2(2)^2 + 1} + 2\sqrt[3]{2(2)^2 + 1} + 3\sqrt[3]{2(3)^2 + 1}] \approx 13.6$

36. $\Delta x = \dfrac{b-a}{n} = \dfrac{4-1}{6} = \dfrac{1}{2}$

$$\int_1^4 x\sqrt[3]{2x^2+1}\,dx \approx \Delta x[f(1) + f(1.5) + f(2.0) + f(2.5) + f(3.0) + f(3.5)]$$

$$\int_1^4 x\sqrt[3]{2x^2+1}\,dx \approx \frac{1}{2}[1\sqrt[3]{2(1)^2+1} + 1.5\sqrt[3]{2(1.5)^2+1} + 2.0\sqrt[3]{2(2.0)^2+1}$$
$$+ 2.5\sqrt[3]{2(2.5)^2+1} + 3.0\sqrt[3]{2(3.0)^2+1} + 3.5\sqrt[3]{2(3.5)^2+1}]$$

$$\int_1^4 x\sqrt[3]{2x^2+1}\,dx \approx 16.255$$

37. $y = x\sqrt[3]{2x^2+1}, a = 1, b = 4, n = 3.$

$\Delta x = \dfrac{b-a}{n} = \dfrac{4-1}{3} = 1$

$x_0 = 1, y_0 = 1\sqrt[3]{2\cdot 1^2 + 1} = \sqrt[3]{3},\ x_1 = 2, y_1 = 2\sqrt[3]{2\cdot 2^2 + 1} = 2\sqrt[3]{9},$
$x_3 = 3, y_2 = 3\sqrt[3]{2\cdot 3^2 + 1} = 3\sqrt[3]{19},\ x_3 = 4, y_3 = 4\sqrt[3]{2\cdot 4^2 + 1} = 4\sqrt[3]{33}$

$$\int_1^4 x\sqrt[3]{2x^2+1}\,dx \approx \frac{\Delta x}{2}[y_0 + 2y_1 + 2y_2 + y_3] \approx \frac{1}{2}[\sqrt[3]{3} + 4\sqrt[3]{9} + 6\sqrt[3]{19} + 4\sqrt[3]{33}] \approx 19.3016$$

38. $y = x\sqrt[3]{2x^2+1}, a = 1, b = 4, n = 6$

$\Delta x = \dfrac{b-a}{n} = \dfrac{4-1}{6} = \dfrac{1}{2},\ \dfrac{\Delta x}{2} = \dfrac{1}{4}$

$x_0 = 1, y_0 = 1\sqrt[3]{2(1)^2+1} = \sqrt[3]{3}, x_1 = 1.5, y_1 = 1.5\sqrt[3]{2(1.5)^2+1} = 1.5\sqrt[3]{5.5}$
$x_2 = 2.0, y_2 = 2.0\sqrt[3]{2(2.0)^2+1} = 2.0\sqrt[3]{9}, x_3 = 2.5, y_3 = 2.5\sqrt[3]{2(2.5)^2+1} = 2.5\sqrt[3]{13.5}$
$x_4 = 3.0, y_4 = 3.0\sqrt[3]{2(3.0)^2+1} = 3.0\sqrt[3]{19}, x_5 = 3.5, y_5 = 3.5\sqrt[3]{2(3.5)^2+1} = 3.5\sqrt[3]{25.5}$
$x_6 = 4.0, y_6 = 4.0\sqrt[3]{2(4.0)^2+1} = 4.0\sqrt[3]{33}$

$$\int_1^4 x\sqrt[3]{2x^2+1}\,dx \approx \frac{\Delta x}{2}[y_0 + 2y_1 + 2y_2 + 2y_3 + 2y_4 + 2y_5 + y_6]$$

$$\int_1^4 x\sqrt[3]{2x^2+1}\,dx \approx \frac{1}{4}[\sqrt[3]{3} + 2(1.5)\sqrt[3]{5.5} + 2(2.0)\sqrt[3]{9} + 2(2.5)\sqrt[3]{13.5} + 2(3.0)\sqrt[3]{19} + 2(3.5\sqrt[3]{25.5} + 4.0\sqrt[3]{33}]$$

$$\int_1^4 x\sqrt[3]{2x^2+1}\,dx \approx 19.1020$$

39. $y = x\sqrt[3]{2x^2+1}, a = 1, b = 4, n = 6$

$\Delta x = \dfrac{b-a}{n} = \dfrac{4-1}{6} = \dfrac{1}{2},\ \dfrac{\Delta x}{3} = \dfrac{1}{6}$

For $x_0 \ldots x_6$ and $y_0 \ldots y_6$ see exercise 38.

$$\int_1^4 x\sqrt[3]{2x^2+1}\,dx \approx \frac{\Delta x}{3}[y_0 + 4y_1 + 2y_2 + 4y_3 + 2y_4 + 4y_5 + y_6]$$

$$\int_1^4 x\sqrt[3]{2x^2+1}\,dx \approx \frac{1}{6}[\sqrt[3]{3} + 4(1.5)\sqrt[3]{5.5} + 2(2.0)\sqrt[3]{9} + 4(2.5)\sqrt[3]{13.5} + 2(3.0)\sqrt[3]{19} + 4(3.5)\sqrt[3]{25.5} + 4.0\sqrt[3]{33}]$$

$$\int_1^4 x\sqrt[3]{2x^2+1}\,dx \approx 19.0354$$

40. $\displaystyle\int_1^4 x\sqrt[3]{2x^2+1}\,dx = \frac{1}{4}\int_3^{33} \sqrt[3]{u}\,du = \frac{1}{4}\left(\frac{3}{4}\right)u^{4/3}\bigg|_3^{33}$

$u = 2x^2+1, du = 4x\,dx$

$\displaystyle\int_1^4 x\sqrt[3]{2x^2+1}\,dx = \frac{3}{16}[33^{4/3} - 3^{4/3}] \approx 19.0354$

41. $\Delta x = \dfrac{b-a}{n} = \dfrac{4-0}{8} = \dfrac{1}{2}, y = 4 + \sqrt{1 + 8x - 2x^2}$

$x_0 = 0, y_0 = 4 + \sqrt{1 + 8(0) - 2(0)^2} = 5, x_1 = 0.5, y_1 = 4 + \sqrt{4.5}, x_2 = 1.0, y_2 = 4 + \sqrt{7}$
$x_3 = 1.5, y_3 = 4 + \sqrt{8.5}, x_4 = 2.0, y_4 = 4 + \sqrt{9} = 7, x_5 = 2.5, y_5 = 4 + \sqrt{8.5}$
$x_6 = 3.0, y_6 = 4 + \sqrt{7}, x_7 = 3.5, y_7 = 4 + \sqrt{4.5}, x_8 = 4.0, y_8 = 4 + \sqrt{1} = 5$

$A = \Delta x[y_0 + y_1 + y_2 + y_3 + y_4 + y_5 + y_6 + y_7 + y_8]$
$A = \frac{1}{2}[5 + 4 + \sqrt{4.5} + 4 + \sqrt{7} + 4 + \sqrt{8.5} + 4 + \sqrt{8.5} + 4 + \sqrt{7} + 4 + \sqrt{4.5} + 5]$
$A = 24.68 \text{ m}^2$

42. $A_T = \dfrac{\Delta x}{2}[y_0 + 2y_1 + 2y_2 + 2y_3 + 2y_4 + 2y_5 + 2y_6 + 2y_7 + y_8]$

See exercise 41 and Δx and y_n.

$A_T = \frac{1}{4}[5 + 2(4 + \sqrt{4.5}) + 2(4 + \sqrt{7}) + 2(4 + \sqrt{8.5}) + 2(7) + 2(4 + \sqrt{8.5}) + 2(4 + \sqrt{7}) + 2(4 + \sqrt{4.5}) + 5]$

$A_T = 24.68 \text{ m}^2$

43. $A_S = \dfrac{\Delta x}{3}[y_0 + 4y_1 + 2y_2 + 4y_3 + 2y_4 + 4y_5 + 2y_6 + 4y_7 + y_8]$

See exercise 41 and Δx and y_n.

$A_S = \frac{1}{6}[5 + 4(4 + \sqrt{4.5}) + 2(4 + \sqrt{7}) + 4(4 + \sqrt{8.5}) + 2(7) + 4(4 + \sqrt{8.5}) + 2(4 + \sqrt{7}) + 4(4 + \sqrt{4.5}) + 5]$

$A_S = 25.81 \text{ m}^2$

44.

$A = 25.83 \text{ m}^2$

45. $\dfrac{dy}{dx} = k(2L^3 - 12Lx + 2x^4)$
$dy = k(2L^3 x^0 - 12Lx + 2x^4)\,dx$

$y = \displaystyle\int k(2L^3 x^0 - 12Lx + 2x^4)\,dx = k\int (2L^3 x^0 - 12Lx + 2x^4)\,dx = k\left(2L^3 x^0 - \frac{12Lx}{2} + \frac{2x^5}{5}\right) + C$

$y = 0$ for $x = 0; 0 = k(0 - 0 + 0) + C; C = 0; y = k\left(2L^3 x - 6Lx^2 + \frac{2}{5}x^5\right)$

46. $Q = k \int \left(r^2 - \dfrac{r^3}{R} \right) dr = k\dfrac{r^3}{3} - k\dfrac{r^4}{4R} + C$

$Q_0 = k\dfrac{R^3}{3} - k\dfrac{R^4}{4R} + C, \ \ C = -k\dfrac{R^3}{3} + k\dfrac{R^3}{4} + Q_0$

$$C = -k\dfrac{R^3}{12} + Q_0$$

$Q = k\dfrac{r^3}{3} - k\dfrac{R^4}{4R} - k\dfrac{R^3}{12} + Q_0$

$Q = \dfrac{k}{12R}(4r^3R - 3r^4 - R^4) + Q_0$

47. $A = 2 \displaystyle\int_0^5 \sqrt{5-y}\,dy = -2 \int_5^0 \sqrt{u}\,du$

$= -2 \left(\dfrac{2}{3} \right) u^{3/2} \Big|_5^0$

$u = 5 - y, du = -dy$

$A = -\dfrac{4}{3}[0 - 5^{3/2}]$

$A = 14.9 \text{ m}^2$

48. $s = \displaystyle\int_0^4 t\sqrt{4+9t^2}\,dt = \dfrac{1}{18} \int_4^{148} \sqrt{u}\,du = \dfrac{1}{18}\left(\dfrac{2}{3} \right) u^{3/2} \Big|_4^{148}$

$u = 4 + 9t^2, du = 18t\,dt$

$s = \dfrac{1}{27}(148^{3/2} - 4^{3/2})$

$s = 66.4 \text{ in}$

49. The area of a quarter circle is $\frac{1}{4} \cdot \pi r^2$, thus $\pi = \frac{4A}{r^2}$. The area of a quarter circle of radius one may be found from $\int_0^1 \sqrt{1-x^2}\,dx$. This may be evaluated using rectangles. See problem 41.

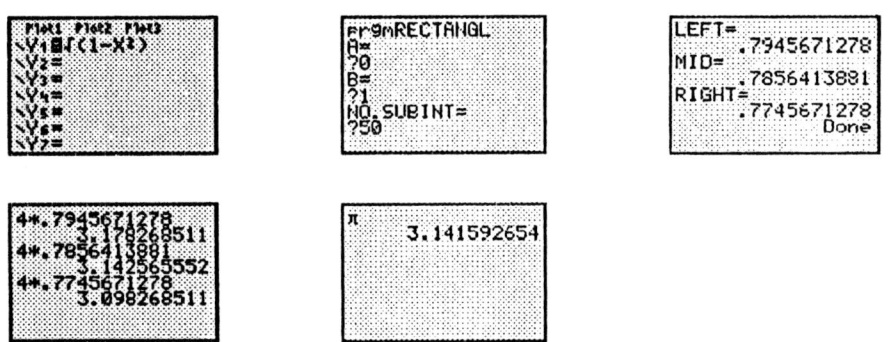

Multiplying each of these values by four gives an approximation for π. Increasing the number of subintervals will increase the accuracy of the approximation.

APPLICATIONS OF INTEGRATION

6.1 Applications of the Indefinite Integral

1. $v = \int a\,dt = \int -32\,dt = -32t + C$;
   $v = 0, t = 0$; $= 0 - 32(0) + C$; $C = 0$
   $v = -32t$; find v for $t = 2.5$ s;
   $v = -32(2.5) = -80$ ft/s; 80 ft/s downward

2. $v = \dfrac{ds}{dt} = 16$ ft/s, $t = 0$

   $a = \dfrac{d^2 s}{dt^2} = -5.0$ ft/s^2

   $\dfrac{dv}{dt} = -5.0$

   $dv = -5.0\,dt$, $\int dv = -5.0 \int dt$

   $v = -5.0t + C_1$; $t = 0, v = 16.0$ ft/s
   $16 = -5.0(0) + C_1$; therefore, $C_1 = 16$

   Therefore,

   $v = -5.0t + 16.0$;
   $v|_{t=6.0} = -5.0(6) + 16.0 = -14.0$ ft/s

   The hoop is on its way down the incline.

3. $\dfrac{ds}{dt} = -0.25$ m/s

   $ds = -0.25\,dt$
   $s = -0.25 \int dt = -0.25t + C_1$
   $t = 0, s = 8$; therefore, $C_1 = 8$
   $s = -0.25t + 8.00 = 8.00 - 0.25t$

4. $v = 6t - 6t^2, t = 0, s = 0$

   $\dfrac{ds}{dt} = 6t - 6t^2$; $ds = (6t - 6t^2)dt$; $s = \int(6t - 6t^2)dt$

   $s = 3t^2 - 2t^3 + C, t = 0, s = 0$; therefore, $C = 0$
   $s = 3t^2 - 2t^3$; $s|_{t=0.75} = 0.84$ mm

5. $a = 90(1 - 4t)^{1/2}$;

   $v = \int 90(1 - 4t)^{1/2}dt = 90\int(1 - 4t)^{1/2}dt$
   $= -\dfrac{1}{4}(90)\int(1 - 4t)^{1/2}(-4\,dt)$

   $= \dfrac{-45}{2}\dfrac{(1-4t)^{3/2}}{\frac{3}{2}} + C = -15(1 - 4t)^{3/2} + C$;

   $v = 0$ for $t = 0$

$0 = -15(1 - 0)^{3/2} + C$; $0 = -15 + C$; $C = 15$
$v = 15(1 - 4t)^{3/2} + 15$; for $t = 0.25$ s;
$v = -15(1 - 1)^{3/2} + 15$; $v = 15$ ft/s

6. $a = \dfrac{dv}{dt} = \dfrac{600t}{(60 + 0.5t^2)^2}$

   $\int dv = \int \dfrac{600t}{(60 + 0.5t^2)^2}$

   $v = \dfrac{-1200}{t^2 + 120} + C$

   @ $t = 0, v = 0$

   $0 = \dfrac{-1200}{0^2 + 120} + C$

   $c = 10$

   $v = \dfrac{-1200}{t^2 + 120} + 10$

7. $a = -9.80$
   $dv = -9.80\,dt$

   $v = -9.80t + C_1 = \dfrac{ds}{dt}$

   Let $t = 0$ be at altitude 16,500 m.
   Therefore, $450 = -9.80(0) + C_1$; therefore,
   $C_1 = 450$
   $v = -9.8t + 450 = \dfrac{ds}{dt}$

   $ds = (-9.8t + 450)dt$; $\int ds = \int(-9.8t + 450)dt$
   $s = -4.9t^2 + 450t + C_2$; $t = 0, s = 16,500$ m
   $16,500 = 0 + 0 + C_2$; therefore, $C_2 = 16,500$
   $s = -4.9t^2 + 450t + 16,500, t = 3$
   $s = -4.9(9) + 450(3) + 16,500 = 17,800$ m

8. $v = -32t + C_1, t = 0, v = 50$; therefore, $C_1 = 50$

   $v = -32t + 50 = \dfrac{ds}{dt}$

   $s = \int(-32t + 50)dt = -16t^2 + 50t + C_2$
   $t = 0, s = 0$; therefore, $C_2 = 0$
   $s = -16t^2 + 50t$; $s|_{t=2.5} = 25$ ft

9. $v = \int -32\, dt = -32t + C$; $v = v_0$; $t = 0$,
$C = v_0$; $v = -32t + v_0$;

$$s = \int (-32t + v_0)\, dt = -16t^2 + v_0 t + C_1$$

$s = 0, t = 0$; $C_1 = 0$; $s = -16t^2 + v_0)$;
$s = 90$ ft when $v = 0$

$$90 = -16t^2 + v_0 t; \quad 0 = -32t + v_0; \quad t = \frac{v_0}{32};$$

$$90 = -16\left(\frac{v_0}{32}\right)^2 + v_0\left(\frac{v_0}{32}\right); \quad \frac{v_0^2}{64} = 90$$

$$v_0 = \sqrt{64(90)} = 76 \text{ ft/s}$$

10. $v = -32t + C_1$; $t = 0, v = 120$ ft/s
Therefore, $C_1 = 120$
$v = -32t + 120$
Let $v = 0$ to get time to maximum height.
$0 = -32t + 120$; $t = 3.75$ s

It will take another 3.75 s for arrow to pass edge
of cliff on way down.

$9 - 2(3.75) = 1.5$ s for arrow to travel down from
cliff's edge downward to ground.

$$\frac{ds}{dt} = -32t + 120, \quad s = \int (-32t + 120)\, dt$$

$s = -16t^2 + 120t + C_2$, $t = 0, , s = 0$; therefore,
$C_2 = 0$
$s = -16t^2 + 120t\big|_{t=3.75} = 225$ ft

$s_{t=9} = -216$ ft. Height of cliff is 216 ft. The minus sign indicates a distance below cliff's edge.
Total distance travelled by arrow is: up 225 ft,
down
$225 + 216$ ft; $s_{\text{total}} = 666$ ft

OR

$v = -32t + C_1$
$t = 0, , v = 0$; therefore, $C_1 = 0$
$v = -32t$

How long does it take the arrow to reach a velocity of -120 ft/s?

$-120 = -32t$, $t = 3.75$ s

Therefore, it will take the arrow $(9.0 - 3.75)$ s
to travel from highest altitude to ground. $9.0 -
3.75 = 5.25$ s

$s = \int -32t\, dt = -16t^2 + C_2$, $t = 0, s = 0$; therefore, $C_2 = 0$
$s = -16t^2$; $s_{t=3.75} = -225$ ft; $s_{t=5.25} = -441$ ft
Height of cliff is $441 - 225 = 216$ ft.

11. $a = -12t, v = 96.0$ ft/s
$dv = -12t\, dt$
$v = \int -12t\, dt = -6t^2 + C_1$
$t = 0, v = 96.0$; therefore, $C_1 = 96.0$

$$v = -6t^2 + 96.0 = \frac{ds}{dt}$$

Car stops when $v = 0, t = \sqrt{\dfrac{96.0}{6}} = 4.00$ s

$s = \int (-6t^2 + 96.0)\, dt = -2t^3 + 96.0t + C_2$
$t = 0, s = 0$; therefore, $C_2 = 0$
$s = -2t^3 + 96.0t$; $s_{t=16} = 256$ ft

12. $a = \sqrt{1 + 0.2t}, v = 0$ m/s, $s = 2$ m, $t = 0$ s
$dv = (1 + 0.2t)^{1/2}\, dt$
$v = \int (1 + 0.2t)^{1/2}\, dt = 5 \int (1 + 0.2t)^{1/2} 0.2\, dt$

$$v = 5 \cdot \frac{2}{3}(1 + 0.2t)^{3/2} + C_1$$

$$u = \frac{10}{3}(1 + 0.2t)^{3/2} + C_1$$

$$u = 1 + 0.2t$$

$$du = 0.2\, dt = \frac{1}{5}\, dt$$

$$0 = \frac{10}{3}(1) + C_1; \text{ therefore, } C_1 = -\frac{10}{3}$$

$$v = \frac{10}{3}(1 + 0.2t)^{3/2} - \frac{10}{3} = \frac{ds}{dt}$$

$$s = \frac{10}{3}\int (1 + 0.2t)^{3/2} dt - \frac{10}{3}\int dt$$

$$= \frac{10}{3}\cdot 5 \int (1 + 0.2t)^{3/2} 0.2\, dt - \frac{10}{3}\int dt$$

$$= \frac{50}{3}\cdot \frac{2}{5}(1 + 0.2t)^{5/2} - \frac{10}{3}t + C_2$$

$$= \frac{20}{3}(1 + 0.2t)^{5/2} - \frac{10}{3}t + C_2; \quad t = 0, s = 2$$

$$2 = \frac{20}{3}(1) - 0 + C_2; \text{ therefore, } C_2 = -\frac{14}{3}$$

$$s = \frac{20}{3}(1 + 0.2t)^{5/2} - \frac{10}{3}t - \frac{14}{3}$$

$$= \frac{2}{3}[10(1 + 0.2t)^{5/2} - 5t - 7]$$

13. $q = \int i\, dt = \int 0.230 \times 10^{-6} dt = 0.230 \times 10^{-6}t + C$;
$q = 0, t = 0$; $C = 0$; $q = 0.230 \times 10^{-6}t$

Find q for $t = 1.50 \times 10^{-3}$s:

$q = 0.230 \times 10^{-6}(1.50 \times 10^{-3})$
$= 0.345 \times 10^{-9} = 0.345$ nC

14. $i = 0.3 - 0.2t - \dfrac{dq}{dt}$

$q = \int (0.3 - 0.2t)dt = 0.3t - 0.1t^2 + C$

$t = 0, q = 0$; therefore $C = 0$

$q = 0.3t - 0.1t^2, q_{t=0.050} = 0.015$ C

15. $i = 0.06t\sqrt{1+t^2}, q = 0.015$ C, $t = 0$

$q = 0.06 \int (1 + t^2)t\, dt, u = 1 + t^2, du = 2t\, dt$

$q = 0.03 \int (1 + t^2)^{1/2} 2t\, dt = 0.03\left(\dfrac{2}{3}\right)(1 + t^2)^{3/2} + C_1$

$t = 0, q = 0.015$ C

$0.015 = 0.02(1) + C_1$; therefore, $C_1 = -0.005$

Therefore, $q = 0.02(1 + t^2)^{3/2} - 0.005$

$q_{t=0.25} = 0.017$ C

16. $i = 8 - t, 0 \le t \le 20$ μs, $q_0 = 0, t_0 = t = 0$

$dq = (8 - t)dt; q = -\int (8 - t) - dt; u = 8 - t,$

$du = -dt$

$q = -\dfrac{1}{2}(8 - t)^2 + C, t = 0, q = 0$

$0 = -\dfrac{1}{2}(8)^2 + C$; therefore, $C = 32$

$q = -\dfrac{1}{2}(8 - t)^2 + 32$

$q = 0$ for what t?

$0 = -\dfrac{1}{2}(8 - t)^2 + 32; \dfrac{1}{2}(8 - t)^2 = 32$

$(8 - t)^2 = 64; 8 - t = \pm 8; t = 8 \pm 8$

$t = 0$ μs, $i = 8; t = 16$ μs, $i = -8$ (change of sign)

Therefore, $t = 16$ μs

The direction of i changed between $t = 0$ and $t = 16$ μs.

17. $V_c = \dfrac{1}{C} \int i\, dt = \dfrac{1}{2.5 \times 10^{-6}} \int 0.025\, dt$

$= \dfrac{1}{2.5 \times 10^{-6}}(0.025)t = 1.0 \times 10^4 t + C$

$v_c = 0, t = 0; C = 0; v_c = 1.0 \times 10^4 t$

Find v_c for $t = 0.012$ s: $v_c = 1.0 \times 10^4 (0.012)$
$= 120$ V

18. $i = 0.042t$ (in mA) $= 0.042 \times 10^{-3}$ A

$V_c = \dfrac{10^{-3}}{C} \int i\, dt$ for voltage across a capacitor

(Eq. 26-6)

$V_c = \dfrac{10^{-3}}{8.50 \times 10^{-9}} \int 0.042t\, dt$

$= \dfrac{10^{-3}}{8.50 \times 10^{-9}} 0.021t^2 + C_1$

$t = 0, V_c = 0$; therefore, $C_1 = 0$

$V_c = \dfrac{0.021t^2}{8.50 \times 10^{-6}}; V_c|_{t=2 \times 10^{-6}} = \dfrac{0.021(2 \times 10^{-6})^2}{8.50 \times 10^{-6}}$

$V = 9.9 \times 10^{-9} V = 9.9$ nV

19. $i = \sqrt[3]{1 + 6t}(\mu A)$

$C = 3.75\ \mu F = 3.75 \times 10^{-6} F$

$V = 4.50$ mV $= 4.50 \times 10^{-3}$ V

$i = \sqrt[3]{1 + 6t} \times 10^{-6}$ A

$V_c = \dfrac{10^{-6}}{3.75 \times 10^{-6}} \int (1 + 6t)^{1/3} dt, u = 1 + 6t, du = 6\, dt$

$V_c = \dfrac{1}{6(3.75)} \int (1 + 6t)^{1/3} 6\, dt$

$= \dfrac{1}{6(3.75)} \dfrac{3}{4}(1 + 6t)^{4/3} + C; t = 0, V = 4.50$ mV

$4.50 \times 10^{-3} = \dfrac{1}{8(3.75)}(1) + C; C = -2.883 \times 10^{-2}$

$V_c = \dfrac{1}{8(3.75)}(1 + 6t)^{4/3} - 2.883 \times 10^{-2};$

$V_c|_{t=0.565 \times 10^{-3}}$

$V_c = 4.65 \times 10^{-3}$ V $= 4.65$ mV

20. $i = \dfrac{t}{\sqrt{t^2 + 1}}$ A, $C = 2.0\ \mu F$

$V_c = \dfrac{1}{2.0 \times 10^{-6}} \int \dfrac{t\, dt}{\sqrt{t^2 + 1}}$

$= \dfrac{1}{2.0 \times 10^{-6}} \int (t^2 + 1)^{-1/2}t\, dt$

$u = t^2 + 1, du = 2t\, dt$

$V_c = \dfrac{1}{4.0 \times 10^{-6}} \int (t^2 + 1)^{-1/2} 2t\, dt$

$= \dfrac{1(2)}{4.0 \times 10^{-6}}(t^2 + 1)^{1/2} + C; t = 0, V = 0$

$0 = \dfrac{2}{4.0 \times 10^{-6}}(1) + C$; therefore,

$C = -\dfrac{1}{2.0 \times 10^{-6}}$

$V_c = \dfrac{1}{2.0 \times 10^{-6}}(t^2 + 1)^{-1/2} - \dfrac{1}{2.0 \times 10^{-6}},$

$V_c = 120$ V

$120 = \dfrac{1}{2.0 \times 10^{-6}}[\sqrt{t^2 + 1} - 1]$

$\sqrt{t^2 + 1} = 1 + 120(2.0 \times 10^{-6})$

$t^2 + 1 = [1 + 120(2.0 \times 10^{-6})]^2$

$t^2 = [1 + 120(2.0 \times 10^{-6})]^2 - 1$

$t = \sqrt{[1 + 120(2.0 \times 10^{-6})]^2 - 1}$

$t = 2.19 \times 10^{-3}$ s $= 21.9$ ms

21. $\omega = \dfrac{d\theta}{dt} = 16t + 0.5t^2$; $d\theta = (16t + 0.50t^2)\, dt$

$\qquad \theta = 8t^2 + \dfrac{0.50t^3}{3} + C$; $\theta = 0, t = 0$; $C = 0$;

$\qquad \theta = 8t^2 + \dfrac{0.50t^3}{3}$; find θ for $t = 10.0$ s

$\qquad \theta = 8(10.0)^2 + \dfrac{0.50}{3}(10.0)^3 = 970$ rad

22. Refer to Fig. 26-3 in the text.

$\qquad \alpha = \sqrt{8t+1}$, $\omega = 0, \theta = 0$ (for $t = 0$)

$\qquad \dfrac{d\omega}{dt} = \sqrt{8t+1}$; $\alpha = \dfrac{d\omega}{dt}$

$\qquad \dfrac{d}{dt} \cdot \dfrac{d\theta}{dt} = \sqrt{8t+1}$; $\omega = \dfrac{d\theta}{dt}$

$\qquad \dfrac{d^2\theta}{dt^2} = \sqrt{8t+1}$; $d\dfrac{d\theta}{dt} = \sqrt{8t+1}\ dt$

$\qquad \omega = \dfrac{d\theta}{dt} = \dfrac{1}{8} \int (8t+1)^{1/2} 8\, dt$

$\qquad = \dfrac{1}{8} \cdot \dfrac{2}{3}(8t+1)^{3/2} + C$

$\qquad = \dfrac{1}{12}(8t+1)^{3/2} + C$; $t = 0, \omega = 0$

$\qquad 0 = \dfrac{1}{12}(1) + C$, therefore $C = -\dfrac{1}{12}$

$\qquad \dfrac{d\theta}{dt} = \dfrac{1}{12}(8t+1)^{3/2} - \dfrac{1}{12}$

$\qquad \theta = \dfrac{1}{12} \int (8t+1)^{3/2} - \dfrac{1}{12} \int dt$

$\qquad = \dfrac{1}{96} \int (8t+1)^{3/2} dt - \dfrac{1}{12} \int dt$

$\qquad = \dfrac{1}{96} \cdot \dfrac{2}{5}(8t+1)^{5/2} - \dfrac{1}{12}t + C$, $\theta = 0, t = 0$

$\qquad 0 = \dfrac{1}{240} + C$; therefore, $C = -\dfrac{1}{240}$

$\qquad \theta = \dfrac{1}{240}(8t+1)^{5/2} - \dfrac{1}{12}t - \dfrac{1}{240}$

$\qquad = \dfrac{1}{240}[(8t+1)^{5/2} - 20t - 1]$

23. $V_L = L\left(\dfrac{di}{dt}\right) = 12.0 - 0.2t, 3.0$ H

$\qquad \dfrac{di}{dt} = \dfrac{1}{L}V_L = \dfrac{1}{3.0}(12.0 - 0.2t)$

$\qquad di = \dfrac{1}{3.0}(12.0 - 0.2t)dt$;

$\qquad i = \dfrac{1}{3.0} \int (12.0 - 0.2t)^1 dt$

$\qquad u = 12.0 - 0.2t, du = -0.2\, dt$

$i = \dfrac{-1}{0.6} \int (12.0 - 0.2t)^1 (-0.2\, dt)$

$\quad = \dfrac{-1}{0.6}\dfrac{(12.0 - 0.2t)^2}{2} + C$

$\quad = \dfrac{-1}{1.2}(12.0 - 0.2t)^2 + C$; $t - 0, i = 0$

$\quad 0 = \dfrac{-1}{1.2}(12.0)^2 + C$; therefore, $C = 120$

$\quad i = \dfrac{-1}{1.2}(12.0 - 0.2t)^2 + 120$

$\quad i_{t=20} = \dfrac{-1}{1.2}(12.0 - 4)^2 + 120 = 66.7$ A

24. $\dfrac{dT}{dx} = f(x)$, $\dfrac{dT}{dx} = 72x^2$

$\qquad dT = 72x^2\, dx$

$\qquad T = 72 \int x^2 dx = 24x^3 + C$

$\qquad x = 0, T = 20°$C

$\qquad 20 = 24(0) + C$; therefore, $C = 20$

$\qquad T = 24x^3 + 20$; $T_{\text{outerwall}} = 20°$ C; $T_{\text{innerwall}} = 23°$C
$\qquad \qquad \qquad \qquad \quad {\scriptstyle (x=0)} \qquad\qquad\quad {\scriptstyle (x=0.5)}$

25. $\dfrac{dV}{dx} = \dfrac{-k}{x^2}$; $V = -\int \dfrac{k}{x^2}\, dx = kx^{-1} + C = \dfrac{k}{x} + C$

$\qquad \lim_{V \to 0} V = \lim_{x \to \infty} \dfrac{k}{x} + C$; $0 = 0 + C$; $C = 0$;

$\qquad$ therefore, $V\big|_{x=x_1} = \dfrac{k}{x_1}$

26. $\dfrac{dy}{dx} = k(x^5 + 1350x^3 - 7000x^2)$

$\qquad y = k\int (x^5 + 1350x^3 - 7000x^2)dx$

$\qquad = k\left(\dfrac{x^6}{6} + 1350\dfrac{x^4}{4} - 7000\dfrac{x^3}{3}\right) + C$

$\qquad x = 0, y = 0$; therefore, $C = 0$

$\qquad y = \dfrac{k}{6}(x^6 + 2025x^4 - 14,000x^3)$

27. $\dfrac{dm}{dt} = \dfrac{-1}{\sqrt{t+1}}$

$\qquad t = 0, m = 1000$ g

$\qquad dm = \dfrac{-dt}{\sqrt{t+1}}$

$\qquad m = -\int (t+1)^{-1/2} dt$

$\qquad u = t + 1, du = dt$

$\qquad m = -2(t+1)^{1/2} + C$; $t = 0, m = 1000$ g

$\qquad 1000 = -2(1) + C$; therefore, $C = 1002$

$\qquad m = -2\sqrt{t+1} + 1002$

To remove all salt, $m = 0$

$$0 = -2\sqrt{t+1} + 1002$$

$$2\sqrt{t+1} = 1002; \; 4(t+1) = (1002)^2;$$

$$t + 1 = \frac{(1002)^2}{4}$$

$$t = \frac{(1002)^2}{4} - 1$$

$$t = 251,000 = 2.51 \times 10^5 \text{ min} \approx 174^+ \text{ days}$$

28. $\dfrac{dr}{d\lambda} = \dfrac{k}{\sqrt{\lambda}}$

$$dr = \frac{kd\lambda}{\sqrt{\lambda}}$$

$$\frac{dr}{d\lambda} = 3.55 \times 10^4, r = 4.08 \text{ cm} = 4.08 \times 10^{-2} \text{ m}$$

$$\lambda = 574 \text{ nm} = 574 \times 10^{-9} \text{ m}$$

$$k = \sqrt{\lambda}\frac{dr}{d\lambda}; \; k = \sqrt{574 \times 10^{-9}} \cdot 3.55 \times 10^4$$

$$= 26.896 \text{ m}^{1/2}$$

$$r = k \int \lambda^{-1/2} d\lambda = k2\lambda^{1/2} + C$$
$$= 2k\sqrt{\lambda} + C = 53.8\sqrt{\lambda} + C$$
$$r = 4.08 \text{ cm}, \lambda = 574 \text{ mm}$$

$$4.08 \times 10^{-2} = 53.8\sqrt{574 \times 10^{-9}} + C$$
Therefore, $C = 0$
Therefore, $r = 53.8\sqrt{\lambda}$

6.2   Areas by Integration

1. $y = 4x; \; y = 0, x = 1$

Using vertical elements,

$$A = \int_0^1 y \, dx = \int_0^1 4x \, dx = 2x^2 \Big|_0^1$$

$$= 2(1)^2 - 2(0) = 2$$

2. $y = 4 - 2x; \; y = 0, x = 0$

Using vertical elements,

$$dA = y \, dx, A = \int y \, dx$$

$$A = \int_0^2 y \, dx = \int_0^2 (4 - 2x) dx$$

$$= (4x - x^2)\Big|_0^2 = 4 - 0 = 4$$

3. $y = x^2; \; y = 0, x = 2$
$$dA = y \, dx$$

$$A = \int_0^2 x^2 dx$$

$$= \frac{x^3}{3}\Big|_0^2 = \frac{8}{3} - 0 = \frac{8}{3}$$

4. $y = 3x^2; \; y = 0, x = 3$

$$A = \int y \, dx$$

$$A = 3\int_0^3 x^2 dx = x^3 \Big|_0^3$$

$$= 27 - 0 = 27$$

5. $y = 6 - 4x; \; x = 0, y = 0, y = 3$

$$y - 6 = -4x, x = -\frac{1}{4}y + \frac{3}{2}$$

$$A = \int_0^3 x \, dy = \int_0^3 \left(-\frac{1}{4}y + \frac{3}{2}\right) dy$$

$$= -\frac{1}{8}y^2 + \frac{3}{2}y \Big|_0^3 = -\frac{1}{8}(3)^2 + \frac{3}{2}(3) + \frac{1}{8}(0)^2 - \frac{3}{2}(0)$$

$$= -\frac{9}{8} + \frac{9}{2} = -\frac{9}{8} + \frac{36}{8} = \frac{27}{8}$$

6. $y = x^2 + 2; \; x = 0, y = 4(x > 0)$
$$x^2 + 2 = 4; \; x = \sqrt{2}$$
$$y = x^2 + 2, x^2 = y - 2; \; x = \sqrt{y - 2}$$
$$dA = x \, dy, A = \int x \, dy$$

$$A = \int_2^4 (y - 2)^{1/2} dy = \frac{2}{3}(y - 2)^{3/2}\Big|_2^4$$

$$u = y - 2, du = dy$$

$$A = \frac{2}{3}(2^{3/2} - 0) = \frac{2}{3} \cdot 2\sqrt{2} = \frac{4}{3}\sqrt{2} = \frac{4}{3}\sqrt{2}$$

7. $y = x^2 - 4, y = 0, x = 4$

$$A = \int_2^4 (x^2 - 4) dx = \frac{1}{3}x^3 - 4x \Big|_2^4$$

$$= \left[\frac{1}{3}(4^3) - 4(4)\right] - \left[\frac{1}{3}(2^3) - 2(4)\right] = \frac{32}{3}$$

8. $y = x^2 - 2x$; $y = 0, dA = l\,dx$

$$A = \int l\,dx = \int_0^2 (0 - y)dx$$

$$= -\int_0^2 y\,dx = -\int_0^2 (x^2 - 2x)dx$$

$$= -\left(\frac{x^3}{3} - x^2\right)\Big|_0^2$$

$$= -\left[\left(\frac{8}{3} - 4\right) - (0)\right] = -\left(-\frac{4}{3}\right) = \frac{4}{3}$$

9. $y = x^{-2}$; $y = 0, x = 2, x = 3$

$$A = \int_2^3 x^{-2}dy = -x^{-1}\Big|_2^3 = -\frac{1}{x}\Big|_2^3$$

$$= -\frac{1}{3} - \left(-\frac{1}{2}\right) = \frac{1}{6}$$

10. $y = 16 - x^2$; $y = 0, x = 1, x = 2$

$$A = \int_1^2 (16 - x^2)dx = \left(16x - \frac{1}{3}x^3\right)\Big|_1^2$$

$$= 32 - \frac{8}{3} - \left(16 - \frac{1}{3}\right) = \frac{41}{3}$$

11. $y = \sqrt{x}$; $x = 0, , y = 1, y = 3$
$y^2 = x$

$$A = \int_1^3 x\,dy = \int_1^3 y^2\,dy = \frac{1}{3}y^3\Big|_1^3$$

$$= \frac{1}{3}(3)^3 - \frac{1}{3}(1)^3$$

$$= \frac{27}{3} - \frac{1}{3} = \frac{26}{3}$$

12. $y = 2\sqrt{x+1}$; $x = 0, y = 4$
$y = 2\sqrt{x+1}$; $y^2 = 4(x+1)$

$$x = \frac{y^2}{4} - 1$$

$$A = \int_2^4 x\,dy$$

$$A = \int_2^4 \left(\frac{y^2}{4} - 1\right)dy = \left(\frac{1}{12}y^3 - y\right)\Big|_2^4$$

$$= \frac{64}{12} - 4 - \left(\frac{8}{12} - 2\right) = \frac{32}{12} = \frac{8}{3}$$

13. $y = \frac{2}{\sqrt{x}}$; $x = 0, y = 1, y = 4$

$$\sqrt{x} = \frac{2}{y}; \quad x = \frac{4}{y^2}$$

$$A = \int_1^4 x\,dy; \quad A = 4\int_1^4 y^{-2}dy = -4y^{-1}\Big|_1^4$$

$$= -\frac{4}{y}\Big|_1^4 = -\frac{4}{4} - \left(-\frac{4}{1}\right)$$

$$= -1 + 4 = 3$$

14. $x = y^2 - y$; $x = 0$
$dA = l\,dy; \quad l = 0 - x = -x$

$$A = \int_0^1 -x\,dy = -\int_0^1 (y^2 - y)dy$$

$$= -\left(\frac{y^3}{3} - \frac{y^2}{2}\right)\Big|_0^1$$

$$= -\left(\frac{1}{3} - \frac{1}{2}\right) - 0 = \frac{1}{6}$$

15. $y = 4 - 2x$; $x = 0, y = 0, y = 3$

$$A = \int x\,dy; \quad x = -\frac{1}{2}y + 2$$

$$A = \int_0^3 \left(-\frac{1}{2}y + 2\right)dy = -\frac{1}{4}y^2 + 2y\Big|_0^3$$

$$= -\frac{9}{4} + 6 = \frac{15}{4}$$

16. $y_2 = x, y_1 = 2 - x, x = 0$
$dA = l\,dx$
$l = y_1 - y_2 = 2 - x - x = 2 - 2x$
$y_1 = y_2; \quad 2 - x = x; \quad x = 1$

$$A = \int_0^1 (2 - 2x)dx = 2\int_0^1 (1 - x)dx$$

$$A = 2\left(x - \frac{x^2}{2}\right)\Big|_0^1 = 2\left[1 - \frac{1}{2} - 0\right] = 1$$

17. $y_2 = x^2$; $y_1 = 2 - x, x = 0 \ (x \geq 0)$
$dA = l\,dx; \quad l = y_1 - y_2 = 2 - x - x^2$
$y_1 = y_2; \quad 2 - x = x^2; \quad x^2 + x - 2 = 0; \quad (x + 2)(x - 1) = 0;$
$x = 1$

$$A = \int_0^1 (2 - x - x^2)\,dx$$

$$= \left(2x - \frac{1}{2}x^2 - \frac{1}{3}x^3\right)\Big|_0^1;$$

$$A = 2 - \frac{1}{2} - \frac{1}{3} - 0 = \frac{7}{6}$$

18. $y_1 = x_1^2$; $y_1 = 2 - x_2, y = 1$

$dA = l\,dy$; $l = x_2 - x_1$

$l = 2 - y - (-\sqrt{y})^*$

$x^2 = 2 - x$; $(x+2)(x-1) = 0$; $x = -2, x = 1$

$y = 2 - x$; therefore, $x = 2 - y$

$$A = \int_1^4 (2 - y + \sqrt{y})dy$$

$$= \left(2y - \frac{1}{2}y^2 + \frac{2}{3}y^{3/2}\right)\Big|_1^4$$

$$A = \left(8 - 8 + \frac{16}{3}\right) - \left(2 - \frac{1}{2} + \frac{2}{3}\right)$$

$$= \frac{19}{6}$$

*$y = x^2$; therefore, $x = \pm\sqrt{y}$. There are two branches of the curve; we want the part in Quad II, namely $-\sqrt{y}$.

19. $y_1 = x^4 - 8x^2 + 16, y_2 = 16 - x^4$

$y_1 = (x^2 - 4)^2$

$y_2 = (4 - x^2)(4 + x^2)$

$dA = l\,dx$; $l = y_2 - y_1$

$l = 16 - x^4 - x^4 + 8x^2 - 16$

$ = 8x^2 - 2x^4$

$$A = 2\int_0^2 (8x^2 - 2x^4)dx = 2\left(\frac{8}{3}x^3 - \frac{2}{5}x^5\right)\Big|_0^2$$

$$= \frac{256}{15}$$

We could integrate from $x = -2$ to $x = 2$ or do as above because of symmetry with respect to y-axis.

20. $y_1 = \sqrt{x-1}, y_2 = 3 - x, y = 0$

$dA = l\,dy$

$l = x_2 - x_1 = 3 - y - (y^2 + 1)$

$l = 3 - y - y^2 - 1 = 2 - y - y^2$

$y^2 = x - 1$; $x = y^2 + 1$

$y = 3 - x$; $x = 3 - y$

$y_1 = y_2$

$\sqrt{x-1} = 3 - x$; $x - 1 = (3-x)^2$

$x - 1 = 9 - 6x + x^2$; $x^2 - 7x + 10 = 0$

$(x-5)(x-2) = 0$; $x = 2$; therefore, $(2,1)$

$$A = \int_0^1 (2 - y - y^2)dy$$

$$A = 2y - \frac{y^2}{2} - \frac{y^3}{3}\Big|_0^1$$

$$A = 2 - \frac{1}{2} - \frac{1}{3} = \frac{7}{6}$$

21. $y_1 = x^2 + 5x, y_2 = 3 - x^2$

$dA = l\,dx$; $l = y_2 - y_1$

$l = 3 - x^2 - (x^2 + 5x) = 3 - x^2 - x^2 - 5x$

$ = 3 - 5x - 2x^2$

$y_1 = y_2$

$x^2 + 5x = 3 - x^2$; $2x^2 + 5x - 3 = 0$

$(2x - 1)(x + 3) = 0$; $x = \frac{1}{2}, x = -3$

$$A = \int_{-3}^{1/2} (3 - 5x - 2x^2)\,dx$$

$$= \left(3x - \frac{5}{2}x^2 - \frac{2}{3}x^3\right)\Big|_{-3}^{1/2}$$

$$A = \left(\frac{3}{2} - \frac{5}{2}\cdot\frac{1}{4} - \frac{2}{3}\cdot\frac{1}{8}\right) - \left(-9 - \frac{45}{2} + 18\right)$$

$$= \frac{343}{24}$$

22. $y = x^2 + 4$ lies above $y_1 = x^3$. Using the graphing calculator, the point of intersection was found to be $x = 2$. Therefore, we integrate from -1 to 2.

$$A = \int_{-1}^2 (x^2 + 4 - x^3)dx = \frac{x^3}{3} + 4x - \frac{x^4}{4}\Big|_{-1}^2$$

$$= \left(\frac{8}{3} + 8 - 4\right) - \left(-\frac{1}{3} - 4 - \frac{1}{4}\right) = \frac{45}{4}$$

23. $y = x^5$; $x = -1, x = 2, y = 0$

$A_T = A_{(-1 \to 0)} + A_{(0 \to 2)}$

$ = \text{negative} + \text{positive}$

$ = -\left(A\big|_{-1}^0\right) + A\big|_0^2$

$l = y$

$$A = -\int_{-1}^0 y\,dx + \int_0^2 y\,dx$$

$$= -\int_{-1}^0 x^5 dx + \int_0^2 x^5 dx = -\frac{x^6}{6}\Big|_{-1}^0 + \frac{x^6}{6}\Big|_0^2$$

$$= 0 - \left(-\frac{1}{6}\right) + \frac{64}{6} - 0 = \frac{65}{6}$$

24. $y_1 = x^2 + 2x - 8, y_2 = x + 4$

$dA = l\,dx$

$l = y_1 - y_2$

$y_1 = y_2$

$x^2 + 2x - 8 = x + 4$; $x^2 + x - 12 = 0$

$(x+3)(x-3) = 0$; $x = -4, x = 3$

$$A = \int_{-4}^{3} [(x+4) - (x^2 + 2x - 8)]dx$$

$$= \int_{-4}^{3} (12 - x - x^2)dx$$

$$= \left(12x - \frac{1}{2}x^2 - \frac{1}{3}x^3\right)\Big|_{-4}^{3}$$

$$A = \left(36 - \frac{9}{2} - 9\right) - \left(-48 - 8 + \frac{64}{3}\right) = \frac{343}{6}$$

25. $y = 8x; x = 0, y = 4$

(a) Using horizontal elements,

$$dA = x\,dy, y = 8x, x = \frac{1}{8}y$$

$$A = \int_{0}^{4} x\,dy = \frac{1}{8}\int_{0}^{4} y\,dy = \frac{1}{8}\frac{y^2}{2}\Big|_{0}^{4}$$

$$= \frac{1}{16}y^2\Big|_{0}^{4} = 1 - 0 = 1$$

(b) Using vertical elements,

$$A = \int_{0}^{1/2} (4 - 8x)\,dx = 4x - 4x^2\Big|_{0}^{1/2} = 1$$

26. (a) $\int_{0}^{\sqrt[3]{3}} (3 - x^3)dx = 3.25$

(b) $\int_{0}^{3} \sqrt[3]{y}\,dy = 3.25$

27. (a) $\int_{0}^{2} (8x - x^4)dx = 9.6 = \frac{48}{5}$

(b) $\int_{0}^{16} \left(\sqrt[4]{y} - \frac{y}{8}\right) dy = 9.6 = \frac{48}{5}$

28. (a) $A = 2\int_{0}^{2} (4x - x^3)dx = 8$

(b) $A = 2\int_{0}^{8} \left(\sqrt[3]{y} - \frac{y}{4}\right) dy = 8$

29. $dw = p\,dt; p = 12t - 4t^2$

$$w = \int_{0}^{3} (12t - 4t^2)\,dt = 6t^2 - \frac{4}{3}t^3\Big|_{0}^{3}$$

$$= 6(3)^2 - \frac{4}{3}(3)^3 - 0 = 54 - 36 = 18.0 \text{ J}$$

30. $Q = \int_{t_1}^{t_2} i\,dt; t_1 = 1 \text{ s}, t_2 = 4 \text{ s}, i = 0.0032t\sqrt{t^2 + 1}$

$$Q = \int_{1}^{4} 0.0032t\sqrt{t^2 + 1}\,dt$$

$$= 0.0032 \cdot \frac{1}{2}\int_{1}^{4} (t^2 + 1)^{1/2}2t\,dt$$

$$u = t^2 + 1, du = 2t\,dt$$

$$Q = 0.0016 \cdot \frac{2}{3}(t^2 + 1)^{3/2}\Big|_{1}^{4}$$

$$= \frac{0.0032}{3}(17^{3/2} - 2^{3/2})$$

$$= 0.072 \text{ C}$$

31. $v = 1 - 0.01\sqrt{2t + 1}; t = 10 \text{ s to } t = 100 \text{ s}$

$$s = \int_{10}^{100} (1 - 0.01\sqrt{2t + 1})\,dt$$

$$= \int_{10}^{100} dt - 0.01\int_{10}^{100} (2t + 1)^{1/2}dt$$

$$u = 2t + 1, du = 2\,dt$$

$$s = \int_{10}^{100} dt - \frac{0.01}{2}\int (2t + 1)^{1/2}2\,dt$$

$$= t - \frac{0.01}{2} \cdot \frac{2}{3}(2t + 1)^{3/2}\Big|_{10}^{100}$$

$$= \left[t - \frac{0.01}{3}(2t + 1)^{3/2}\right]\Big|_{10}^{100}$$

$$= 90.501 - 9.679$$

$$= 80.8 \text{ km}$$

32. $C' = \dfrac{dC}{dx} = \dfrac{100}{(0.01x + 1)^2}$

$n = 100$ units means $x = 0$ to $x = 100$

$$dC = \frac{100\,dx}{(0.01x + 1)^2}$$

$$C = 100\int_{0}^{100} \frac{dx}{(0.01x + 1)^2}$$

$$= \frac{100}{0.01}\int_{0}^{100} (0.01x + 1)^{-2}0.01\,dx$$

$$= 10,000\frac{(0.01x + 1)^{-1}}{-1} = \frac{-10,000}{(0.01x + 1)}\Big|_{0}^{100}$$

$$= -10,000\left[\frac{1}{2} - \frac{1}{1}\right] = -10,000\left[-\frac{1}{2}\right]$$

$$= \$5000$$

33. $y = x^3 - 2x^2 - x + 2$ and $y = x^2 - 1$

Find the points of intersection:

$$x^3 - 2x^2 - x + 2 = x^2 - 1$$

$$x^3 - 3x^2 - x + 3 = 0; \quad (x^2 - 1)(x - 3) = 0$$
$$x^2 = 1; \quad x = 3; \quad x = \pm 1$$

$$A_1 = \int_{-1}^{1} [(x^3 - 2x^2 - x + 2) - (x^2 - 1)] \, dx$$

$$= \int_{-1}^{1} (x^3 - 3x^2 - x + 3) \, dx$$

$$= \frac{1}{4}x^4 - x^3 - \frac{1}{2}x^2 + 3x \Big|_{-1}^{1}$$

$$= \left(\frac{1}{4} - 1 - \frac{1}{2} + 3\right) - \left(\frac{1}{4} + 1 - \frac{1}{2} - 3\right)$$

$$= 4 \text{ cm}^2$$

34. $y_1 = \dfrac{800x}{(x^2 + 10)^2}$, $y_2 = 0.5x^2 - 4x$; $x = 8$

$$A = \int_0^8 l \, dx; \quad l = y_1 - y_2$$

$$A = \int_0^8 \left[\frac{800x}{(x^2 + 10)^2} - (0.5x^2 - 4x)\right] dx$$

$$= 400 \int_0^8 (x^2 + 10)^{-2} 2x \, dx - \int_0^8 (0.5x^2 - 4x) dx$$

$$= \left[-400(x^2 + 10)^{-1} - \frac{0.5x^3}{3} + 2x^2\right]\Big|_0^8$$

$$= -400\left(\frac{1}{74}\right) - \frac{1}{6}(512) + 128$$

$$\quad - \left[-400\left(\frac{1}{10}\right) - 0 + 0\right]$$

$$= 37.261 + 40 = 77.3 \text{ m}^2$$

35. Equation of parabola is $y = -kx^2 + 0.640$.
Substitute $(0.8, 0)$; $0 = -k(0.8)^2 + 0.640$; $k = 1$
Therefore, $y = -x^2 + 0.640$

$$A = \int_0^{0.8} (-x^2 + 0.640) \, dx$$

$$= 2\left(-\frac{x^3}{3} + 0.640x\right)\Big|_0^{0.8}$$

$$= 2\left(-\frac{(0.8)^3}{3} + 0.640(0.8)\right) - 0$$

$$= 0.683 \text{ m}^2$$

36. Equation of line, Quad I: $y = 2.00 - x$
Equation of line, Quad II: $y = 2.00 + x$
Equation of parabola: $y = kx^2 - 4.00$
Substitute $(2, 0)$: $0 = 4k - 4$, $k = 1$
$y = x^2 - 4.00$
$y_1 = 2.00 - x$; $y_2 = x^2 - 4.00$

$$A = 2 \int_0^{2.00} l \, dx$$

$$= 2 \int_0^{2.00} [(2.00 - x) - (x^2 - 4.00)] dx$$

$$= 2 \int_0^{2.00} (6.00 - x - x^2) dx$$

$$= 2\left(6.00 - \frac{x^2}{2} - \frac{x^3}{3}\right)\Big|_0^{2.00}$$

$$= 2(12.00 - 2.00 - 2.67) - 0$$

$$= 14.67$$

Therefore,

volume $= A_{\text{end}} \times \text{depth} = 14.67 \times 6.00 = 88.0 \text{ ft}^3$

6.3   Volumes by Integration

1. $y = 2 - x, x = 0, y = 0$

Disk: $dV = \pi y^2 dx$

$$V = \pi \int_0^2 y^2 dx = \pi \int_0^2 (2 - x)^2 \, dx$$

$$= \pi\left[-\frac{1}{3}(2 - 2)^3 + \frac{1}{3}(2 - 0)^3\right]$$

$$= \frac{8}{3}\pi$$

2. $V = \int_0^2 \pi(2 - y)^2 dy = \dfrac{8\pi}{3}$

3. $V = \int_0^2 2\pi x(2 - x) dx = \dfrac{8\pi}{3}$

4. $V = \int_0^2 2\pi y(2 - y) dy = \dfrac{8\pi}{3}$

5. $y = x, y = 0, x = 2$

Disk: $dV = \pi y^2 dx$

$$V = \pi \int_0^2 x^2 dx = \frac{\pi}{3}x^3 \Big|_0^2 = \frac{8\pi}{3}$$

6. $y = \sqrt{x}, x = 0, y = 2$

$y = \sqrt{x}$; therefore, $x = y^2$

Shell: $dV = 2\pi y x\, dy$

$$V = 2\pi \int_0^2 y y^2\, dy$$

$$= 2\pi \int_0^2 y^3\, dy$$

$$= 2\pi \frac{y^4}{4} = \frac{\pi}{2} y^4 \Big|_0^2 = 8\pi$$

7. $y = 3\sqrt{x}, y = 0, x = 4$

Disk: $dV = \pi y^2 dx$

$$V = \pi \int_0^4 y^2\, dx = \int_0^4 9x\, dx$$

$$= \pi \left(\frac{9}{2} x^2 \right) \Big|_0^4$$

$$= \pi \left[\frac{9}{2}(4)^2 - 0 \right] = 72\pi$$

8. $y = 2x - x^2, y = 0$

Disk: $dV = \pi y^2 dx$

$$V = \pi \int_0^2 (2x - x^2)^2 dx$$

$$= \pi \int_0^2 (4x^2 - 4x^3 + x^4) dx$$

$$= \pi \left(\frac{4}{3} x - x^4 + \frac{1}{5} x^5 \right) \Big|_0^2 = \frac{16\pi}{15}$$

9. $y = x^3, y = 8, x = 0$

$y = x^3$; therefore $x = \sqrt[3]{y}$

Shell: $dV = 2\pi y x\, dy$

$$V = 2\pi \int_0^8 y \sqrt[3]{y}\, dy = 2\pi \int_0^8 y^{4/3} dy$$

$$= 2\pi \frac{3}{7} y^{7/3} = \frac{6\pi}{7} y^{7/3} \Big|_0^8 = \frac{768\pi}{7}$$

10. $y_2 = x^2, y_1 = x, h = x_2 - x_1$

Shell: $dV = 2\pi y h\, dy = 2\pi(x_2 - x_1) dy$

$$V = 2\pi \int_0^1 y(\sqrt{y} - y) dy$$

$$= 2\pi \int_0^1 (y^{3/2} - y^2) dy$$

$$= 2\pi \left(\frac{2}{5} y^{5/2} - \frac{1}{3} y^3 \right) \Big|_0^1$$

$$= 2\pi \left(\frac{2}{5} - \frac{1}{3} \right) = \frac{2\pi}{15}$$

11. $y = x^2 + 1, x = 0,, x = 3, y = 0$

Disk: $dV = \pi y^2 dx$

$$V = \pi \int_0^3 (x^2 + 1)^2 dx$$

$$= \pi \int_0^3 (x^4 + 2x^2 + 1) dx$$

$$= \pi \left(\frac{1}{5} x^5 + \frac{2}{3} x^3 + x \right) \Big|_0^3$$

$$= \pi \left[\frac{1}{5}(3)^5 + \frac{2}{3}(3)^3 + 3 - 0 \right]$$

$$= \frac{348}{5} \pi$$

12. $y = 6 - x - x^2, x = 0, y = 0$ (Quad I)

Disk: $dV = \pi y^2 dx$

$$V = \pi \int_0^2 (6 - x - x^2)^2 dx$$

$$= \pi \int_0^2 (36 - 12x - 11x^2 + 2x^3 + x^4) dx$$

$$= \pi \left(36x - 6x^2 - \frac{11x^3}{3} + \frac{1}{2} x^4 + \frac{1}{5} x^5 \right) \Big|_0^2$$

$$= \frac{496\pi}{15}$$

13. $x = 4y - y^2 - 3, x = 0$

Shell: $dV = 2\pi y x\, dy$

$$V = 2\pi \int_1^3 y(4y - y^2 - 3)\, dy$$

$$= 2\pi \int_1^3 (4y^2 - y^3 - 3y)\, dy$$

$$= 2\pi \left(\frac{4}{3} y^3 - \frac{1}{4} y^4 - \frac{3}{2} y^2 \right) \Big|_1^3$$

$$= 2\pi \left[36 - \frac{81}{4} - \frac{27}{2} - \left(\frac{4}{3} - \frac{1}{4} - \frac{3}{2} \right) \right]$$

$$= \frac{16\pi}{3}$$

14. $y = x^4, x = 0, y = 1, y = 2$

$y = 1, x = 1;\ y = 2, x = \sqrt[4]{2}$

Shell: $dV = 2\pi y x\, dy$

$$V = 2\pi \int_1^2 y \sqrt[4]{y}\, dy$$

$$= 2\pi \int_1^2 y^{5/4} dy$$

$$= 2\pi \cdot \frac{4}{9} y^{9/4}$$

$$= \frac{8\pi}{9} y^{9/4} \Big|_1^2 = \frac{8\pi}{9}(4\sqrt[4]{2} - 1)$$

15. $y = x^{1/3}, x = 0, y = 2$

$x = y^3; x^2 = y^6$

Disk: $dV = \pi x^2 \, dy$

$$V = \pi \int_0^2 y^6 \, dy$$

$$= \frac{\pi}{7} y^7 \Big|_0^2$$

$$= \frac{128\pi}{7}$$

16. $y = \sqrt{x^2 - 1}, y = 0, x = 3$

Shell: $dV = 2\pi xy \, dx$

$$V = 2\pi \int_1^3 x\sqrt{x^2 - 1} \, dx$$

Let $u = x^2 - 1$; then

$du = 2x \, dx$

$$V = \pi \int_0^8 u^{1/2} \, du$$

$$= \pi \left(\frac{2u^{3/2}}{3} \right) \Big|_0^8$$

$$V = \frac{16\pi\sqrt{8}}{3}$$

17. $y = 2\sqrt{x}, x = 0, y = 2$

Disk: $dV = \pi x^2 \, dy$; $x = \frac{1}{4} y^2$

$$V = \pi \int_0^2 \left(\frac{1}{4} y^2 \right) dy$$

$$= \pi \int_0^2 \frac{1}{16} y^4 \, dy$$

$$= \pi \left(\frac{1}{80} y^5 \right) \Big|_0^2$$

$$= \pi \left(\frac{32}{80} - 0 \right)$$

$$= \frac{2}{5} \pi$$

18. $y^2 = x, y = 4, x = 0$

Disk: $dV = \pi x^2 \, dy$

$$V = \pi \int_0^4 y^4 \, dy$$

$$= \frac{\pi}{5} y^5 \Big|_0^4$$

$$= \frac{1024\pi}{15}$$

19. $x^2 - 4y^2 = 4, x = 3, h = 2y$

Shell: $dV = 2\pi x(2y) \, dx$

$$V = 4\pi \int_2^3 x\sqrt{\frac{x^2 - 4}{4}} \, dx$$

$$= \frac{2\pi}{2} \int_2^3 (x^2 - 4)^{1/2} 2x \, dx$$

$u = x^2 - 4, du = 2x \, dx$

$$V = \pi \cdot \frac{2}{3} (x^2 - 4)^{3/2} \Big|_2^3 = \frac{2\pi}{3} (5^{3/2}) - 0$$

$$= \frac{10\sqrt{5}}{3} \pi$$

20. $y = 3x^2 - x^3, y = 0$

Shell: $dV = 2\pi xy \, dx$

$$V = 2\pi \int_0^3 x(3x^2 - x^3) \, dx$$

$$= 2\pi \int_0^3 (3x^3 - x^4) \, dx$$

$$= 2\pi \left(\frac{3}{4} x^4 - \frac{1}{5} x^5 \right) \Big|_0^3 = \frac{243\pi}{10}$$

21. $x = 6y - y^2, x = 0$

Disk: $dV = \pi x^2 \, dy$

$$V = \pi \int_0^6 (36y^2 - 12y^3 + y^4) \, dy$$

$$= \pi \left(12y^3 - 3y^4 + \frac{y^5}{5} \right) \Big|_0^6$$

$$= \pi \left(2592 - 3888 + \frac{7776}{5} \right)$$

$$= \frac{1296\pi}{5}$$

22. $x^2 + 4y^2 = 4$, Quad I

$$y^2 = \frac{1}{4}(4 - x^2)$$

$$y = \frac{1}{2}\sqrt{4 - x^2}$$

Disk: $dV = \pi x^2 \, dy$

$$V = \pi \int_0^1 (4 - 4y^2) \, dy$$

$$= \pi \left(4y - \frac{4}{3} y^3 \right) \Big|_0^1$$

$$= \pi \left(4 - \frac{4}{3} \right) = \frac{8\pi}{3}$$

23. $y = \sqrt{4 - x^2}$, Quad I

Shell: $dV = 2\pi xy\,dx$

$$V = 2\pi \int_0^2 x\sqrt{4 - x^2}\,dx$$

$$u = 4 - x^2, du = -2x\,dx$$

$$V = -\pi \int_0^2 (4 - x^2)^{1/2}(-2x\,dx)$$

$$= -\pi \frac{2}{3}(4 - x^2)^{3/2}\Big|_0^2$$

$$= -\frac{2\pi}{3}(0 - 8) = \frac{16\pi}{3}$$

24. $y = 8 - x^3, x = 0, y = 0$

Shell: $dV = 2\pi xy\,dx$

$$V = 2\pi \int_0^2 x(8 - x^3)\,dx$$

$$= 2\pi \int_0^2 (8x - x^4)\,dx$$

$$= 2\pi \left(4x^2 - \frac{1}{5}x^5\right)\Big|_0^2$$

$$= 2\left(16 - \frac{32}{5}\right) = \frac{96\pi}{5}$$

25. $y = 2x - x^2$, $y = 0$, rotated around

$x = 2$, using shells $r = 2 - x$

$h = y, t = dx$

$dV = 2\pi(2 - x)y\,dx$

$$V = 2\pi \int_0^2 (2 - x)(2 - x^2)\,dx$$

$$= 2\pi \int_0^2 (4x - 4x^2 + x^3)\,dx$$

$$= 2\pi \left[2x^2 - \frac{4}{3}x^3 + \frac{1}{4}x^4\right]\Big|_0^2$$

$$= 2\pi \left[8 - \frac{32}{3} + 4\right] = \frac{8}{3}\pi$$

26. $y_2 = \sqrt{x}$, $y_1 = \dfrac{x}{2}$, rotated around $y = 4$

$r_1 = 4 - y_1, r_2 = 4 - y_2$

Disk: $dV = \pi(4 - y_1)^2 dx - \pi(4 - y_2)^2 dx$

$$V = \pi \int_0^4 \left[\left(4 - \frac{x}{2}\right)^2 - (4 - \sqrt{x})^2 dx\right]$$

$$= \pi \int_0^4 \left(-5x + \frac{1}{4}x^2 + 8\sqrt{x}\right) dx$$

$$= \pi \left(\frac{-5x^2}{2} + \frac{1}{12}x^3 + \frac{16}{3}x^{3/2}\right)\Big|_0^4$$

$$= 8\pi$$

27. $y = \dfrac{r}{h}x; \; y = 0, x = h$

Disk: $dV = \pi y^2 dx$

$$V = \pi \int_0^h \left(\frac{r}{h}x\right)^2 dx$$

$$= \pi \frac{r^2}{h^2} \int_0^h x^2 dx$$

$$= \frac{\pi r^2}{h^2} \cdot \frac{x^3}{3}\Big|_0^h$$

$$= \frac{\pi r^2}{h^2}\left[\frac{h^3}{3}\right] = \frac{1}{3}\pi r^2 h$$

28. Use a semi-circle of radius r, centered at origin.

$x^2 + y^2 = r^2$

$y = \pm\sqrt{r^2 - x^2}$

Use the upper branch of the curve, $y = +\sqrt{r^2 - x^2}$.

Rotate half circle about x-axis.

29. $y = x^4 + 1.5; \; x = 0, x = 1.1$

Shell: $dV = 2\pi xy\,dx$

$$V = 2\pi \int_0^{1.1} xy\,dx = 2\pi \int_0^{1.1} x(x^4 + 1.5)\,dx$$

$$= 2\pi \int_0^{1.1} (x^5 + 1.5x)\,dx$$

$$= 2\pi \left(\frac{x^6}{6} + \frac{3x^2}{4}\right)\Big|_0^{1.1}$$

$$= 2\pi \left(\frac{1.1^6}{6} + \frac{3(1.1)^2}{4}\right) = 7.56 \text{ mm}^3$$

30. Equation of ellipse:

$$\frac{x^2}{a^2} + \frac{y^2}{b^2} = 1$$

$$\frac{x^2}{62^2} + \frac{y^2}{18^2} = 1$$

$$18^2 x^2 + 62^2 y^2 = (62^2)(18^2)$$

$$y^2 = \frac{62^2 \cdot 18^2 - 18^2 x^2}{62^2}$$

Disk: $dV = \pi y^2 dx$

$$V = 2 \left\{ \frac{\pi}{62^2} \int_0^{62} (62^2 \cdot 18^2 - 18^2 x^2) dx \right\}$$

$$= \frac{2\pi}{62^2} 18^2 \int_0^{62} (62^2 - x^2) dx$$

$$= 2\pi \frac{18^2}{62^2} \left(62^2 x - \frac{1}{3} x^3 \right) \Big|_0^{62}$$

$$= 2\pi \frac{18^2}{62^2} \left(62^3 - \frac{62^3}{3} \right)$$

$$= \frac{2\pi 18^2}{62^2} \left(\frac{2(62^3)}{3} \right)$$

$$= \frac{4\pi}{3} (18^2)(62) = 84,100 \text{ ft}^3$$

31. $x^2 + y_2^2 = 9; y_1 = 1$

Volume of lead removed:

$V =$ volume of cylinder + spherical caps

$dV = \pi y_1^2 dx + \pi y_2^2 dx$

$$V = 2 \left[\pi \int_0^{2\sqrt{2}} (1) dx \right] + 2\pi \int_{2\sqrt{2}}^3 y^2 dx$$

$$= 2\pi x \Big|_0^{2\sqrt{2}} + 2\pi \int_{2\sqrt{2}}^3 (9 - x^2) dx$$

$$= 2\pi (2\sqrt{2}) + 2\pi \left(9x - \frac{x^3}{3} \right) \Big|_{2\sqrt{2}}^3$$

$$= 2\pi (2\sqrt{2}) + 2\pi \left(27 - 9 - \left[18\sqrt{2} - \frac{16\sqrt{2}}{3} \right] \right)$$

$$= 2\pi \left[2\sqrt{2} + 18 - 18\sqrt{2} + 16\frac{\sqrt{2}}{3} \right]$$

$V = 18.3 \text{ cm}^3$

32. Place problem in rectangular coordinate system. Equation of parabola is of the form

$$y = -kx^2 + 1.5 = -kx^2 + \frac{3}{2}.$$

Substitute $(2, 1)$

$$1 = -k(4) + 1.5; \text{ therefore, } k = \frac{1}{8}$$

Therefore, $y = -\frac{1}{8} x^2 + \frac{3}{2}$

Disk: $dV = \pi y^2 dx$

$$V = \pi \int_{-2}^2 y^2 dx$$

$$= 2\pi \int_0^2 \left(-\frac{1}{8} x^2 + \frac{3}{2} \right)^2 dx$$

$$= 2\pi \int_0^2 \left(\frac{1}{64} x^4 - \frac{3}{8} x^2 + \frac{9}{4} \right) dx$$

$$= 2\pi \left[\frac{1}{320} x^5 - \frac{1}{8} x^3 + \frac{9}{4} x \right] \Big|_0^2$$

$$= 2\pi \left[\frac{32}{320} - 1 + \frac{9}{2} \right] - 0$$

$$= 2\pi \left(\frac{36}{10} \right) = 22.6 \text{ ft}^3$$

6.4   Centroids

1. $M\bar{x} = m_1 x_1 + m_2 x_2 + m_3 x_3$
$(5.0 + 8.5 + 3.6)\bar{x} = 5.0(1.0) + 8.5(4.2) + 3.6(2.5)$
$\bar{x} = 2.9 \text{ cm}$

2. $M\bar{x} = m_1 x_1 + m_2 x_2 + m_3 x_3$

$(2.3 + 6.5 + 1.2)\bar{x} = 2.3(1.3) + 6.5(5.8) + 1.2(9.5)$
$\bar{x} = 5.2 \text{ cm}$

3. $M\bar{x} = m_1 x_1 + m_2 x_2 + m_3 x_3 + m_4 x_4$

$(42 + 24 + 15 + 84)\bar{x} = 42(-3.5) + 24(0) + 15(2.6)$
$\qquad\qquad\qquad\qquad\qquad + 84(3.7)$
$\bar{x} = 1.2 \text{ cm}$

4. $M\bar{x} = m_1 x_1 + m_2 x_2 + m_3 x_3 + m_4 x_4$

$(550 + 230 + 470 + 120)\bar{x} = 550(-42) + 230(-27)$
$\qquad\qquad\qquad\qquad\qquad\qquad + 470(16) + 120(22)$
$\bar{x} = -14 \text{ cm}$

5. The area of the left rectangle is $4(2) = 8$.
The center is $(-2, 0)$. The area of the right rectangle is $2(4) = 8$. The center is $(1, 1)$;
$8(-2) + 8(1) = (8 + 8)\bar{x}$

$-8 = 16\bar{x}; \bar{x} = -0.5$

$8(0) + 8(1) = (8 + 8)\bar{y}; 8 = 16\bar{y}, \bar{y} = 0.5$

The centroid is $(\bar{x}, \bar{y}) = (-0.5 \text{ in}, 0.5 \text{ in})$.

6. Break area into two rectangles.

One center $(-0.50, 1.50)$; $A_L = 15.00$

Other center $(2.00, 0)$; $A_R = 4.00$

Therefore,

$15.00(-0.50) + 4.00(2.00) = (15.00 + 4.00)\bar{x}$

$\bar{x} = 0.03$

Therefore, $15.00(1.50) + 4.00(0) = (15.00 + 4.00)\bar{y}$

$\bar{y} = 1.18$

Therefore, center of mass is $(0.03 \text{ in}, 1.18 \text{ in})$.

7. Break area into three rectangles.

First: center $(-1.00, -1.00)$; $A_1 = 4.00$

Second: center $(0, 0.50)$; $A_2 = 4.00$

Third: center $(2.50, 1.50)$; $A_3 = 3.00$

Therefore,

$4.00(-1.00) + 4.00(0) + 3.00(2.50)$
$\quad = (4.00 + 4.00 + 3.00)\bar{x}$

$\bar{x} = 0.32$

Therefore,

$4.00(-1.00) + 4.00(0.50) + 3.00(1.50) = 11.00\bar{y}$

$\bar{y} = 0.23$

Therefore, $(0.32 \text{ in}, 0.23 \text{ in})$ is the center of mass.

8. Break area into 5 rectangles.

First: center $(0.50, 0)$; $A_1 = 6.00$

Second: center $(-1.50, 0.50)$; $A_2 = 3.00$

Third: center $(-0.50, 2.50)$; $A_3 = 3.00$

Fourth: center $(3.00, 3.00)$; $A_4 = 8.00$

Fifth: center $(4.00, 1.50)$; $A_5 = 5.00$

Therefore,

$6.00(0.50) + 3.00(-1.50) + 3.00(-0.50)$
$\quad + 8.00(3.00) + 2.00(4.00)$

$\quad = (6.00 + 3.00 + 3.00 + 8.00 + 2.00)\bar{x}$

$\bar{x} = 1.32$

Therefore,

$6.00(0) + 3.00(0.50) + 3.00(2.50) + 8.00(3.00)$
$\quad + 2.00(1.50) = 22.00\bar{y}$

$\bar{y} = 1.64$

Therefore, $(1.32 \text{ in}, 1.64 \text{ in})$ is center of mass.

9. $y = x^2, y = 2$

The curve is symmetrical to the y-axis.
Therefore, $\bar{x} = 0$

$$\bar{y} = \frac{\int_0^2 y(2x)\,dy}{\int_0^2 2x\,dy} = \frac{\int_0^2 y(2\sqrt{y})\,dy}{\int_0^2 2\sqrt{y}\,dy}$$

$$= \frac{\int_0^2 2y^{3/2}\,dy}{\int_0^2 2y^{1/2}\,dy} = \frac{\frac{4}{5}y^{5/2}\big|_0^2}{\frac{4}{3}y^{3/2}\big|_0^2}$$

$$= \frac{\frac{4}{5}\sqrt{32}}{\frac{4}{3}\sqrt{8}}$$

$$= \frac{6}{5}$$

The centroid is $\left(0, \frac{6}{5}\right)$.

10. $x^2 + y^2 = a^2, x = \sqrt{a^2 - y^2}$

Symmetry with respect to y-axis; therefore, $\bar{x} = 0$.

$$\bar{y} = \frac{\int_0^a y(2x)\,dy}{\frac{1}{2}\pi a^2} = \frac{2\int_0^a y\sqrt{a^2 - y^2}\,dy}{\frac{1}{2}\pi a^2}$$

$$= \frac{-\frac{2}{3}\int_0^a (a^2 - y^2)^{1/2}(-2y\,dy)}{\frac{1}{2}\pi a^2}$$

$$= \frac{-\frac{2}{3}(a^2 - y^2)^{3/2}\big|_0^a}{\frac{1}{2}\pi a^2}$$

$$= \frac{-\frac{2}{3}(0 - a^3)}{\frac{1}{2}\pi a^2}$$

$$= \frac{\frac{2}{3}a^3}{\frac{1}{2}\pi a^2} = \frac{4a}{3\pi}; \quad \left(0, \frac{4a}{3\pi}\right)$$

11. $y = 4 - x$, and axes

$$\bar{x} = \frac{\int_0^4 xy\,dx}{\int_0^4 y\,dx} = \frac{\int_0^4 x(4-x)\,dx}{\int_0^4 (4-x)\,dx} = \frac{\int_0^4 (4x - x^2)\,dx}{\int_0^4 (4-x)\,dx}$$

$$= \frac{\left(2x^2 - \frac{1}{3}x^3\right)\big|_0^4}{\left(4x - \frac{1}{2}x^2\right)\big|_0^4} = \frac{\frac{32}{3}}{8} = \frac{32}{24} = \frac{4}{3}$$

$$\bar{y} = \frac{\int_0^4 y(x)\,dy}{\int_0^4 x\,dy} = \frac{\int_0^4 y(4-y)\,dy}{\int_0^4 (4-y)\,dy}$$

$$= \frac{\int_0^4 (4y - y^2)\,dy}{\int_0^4 y(4-y)\,dy}$$

$$= \frac{\left(2y^2 - \frac{1}{3}y^3\right)\big|_0^4}{\left(4y - \frac{1}{2}y^2\right)\big|_0^4} = \frac{4}{3}$$

Therefore, centroid is $\left(\frac{4}{3}, \frac{4}{3}\right)$.

12. $y = x^3, x = 2, x\text{-axis}; \; x = \sqrt[3]{y}$

$$\bar{x} = \frac{\int_0^2 xy\,dx}{\int_0^2 y\,dx} = \frac{\int_0^2 x \cdot x^3 dx}{\int_0^2 x^3 dx} = \frac{\int_0^2 x^4 dx}{\int_0^2 x^3 dx}$$

$$= \frac{\frac{1}{5}x^5\big|_0^2}{\frac{1}{4}x^4\big|_0^2} = \frac{\frac{32}{5}}{4} = \frac{32}{20} = \frac{8}{5}$$

$$\bar{y} = \frac{\int_0^8 y(2-x)dy}{\int_0^8 (2-x)dy}$$

$$= \frac{\int_0^8 y(2-\sqrt[3]{y})dy}{\int_0^8 (2-\sqrt[3]{y})dy}$$

$$= \frac{\int_0^8 (2y - y^{4/3})dy}{\int_0^8 y(2-y^{1/3})dy}$$

$$= \frac{(y^2 - \frac{3}{7}y^{7/3})\big|_0^8}{(2y - \frac{3}{4}y^{4/3})\big|_0^8}$$

$$= \frac{64 - \frac{3}{7}(128)}{16 - \frac{3}{4}(16)}$$

$$= \frac{16}{7}$$

Centroid is $\left(\dfrac{8}{5}, \dfrac{16}{7}\right)$.

13. $y = x^2, y = x^3$

$$\bar{x} = \frac{\int_0^1 x(x^2 - x^3)\,dx}{\int_0^1 (x^2 - x^3)\,dx}$$

$$= \frac{\int_0^1 (x^3 - x^4)\,dx}{\int_0^1 (x^2 - x^3)\,dx}$$

$$= \frac{\frac{1}{4}x^4 - \frac{1}{5}x^5\big|_0^1}{\frac{1}{3}x^3 - \frac{1}{4}x^4\big|_0^1}$$

$$= \frac{\frac{1}{4} - \frac{1}{5}}{\frac{1}{3} - \frac{1}{4}} = \frac{\frac{1}{20}}{\frac{1}{12}} = \frac{3}{5}$$

$$\bar{y} = \frac{\int_0^1 y(y^{1/3} - y^{1/2})\,dy}{\int_0^1 (y^{1/3} - y^{1/2})\,dy}$$

$$= \frac{\int_0^1 (y^{4/3} - y^{3/2})\,dy}{\int_0^1 (y^{1/3} - y^{1/2})\,dy}$$

$$= \frac{\frac{3}{7}y^{7/3} - \frac{2}{5}y^{5/2}\big|_0^1}{\frac{3}{4}y^{4/3} - \frac{2}{5}y^{3/2}\big|_0^1}$$

$$= \frac{\frac{3}{7} - \frac{2}{5}}{\frac{3}{4} - \frac{2}{3}} = \frac{\frac{1}{35}}{\frac{1}{12}} = \frac{12}{35}$$

The centroid is $\left(\frac{3}{5}, \frac{12}{35}\right)$.

14. $y^2 = x, y = 2, x = 0.$ Therefore, $y = \sqrt{x}$

$$\bar{x} = \frac{\int_0^4 x(2-y)dx}{\int_0^4 (2-y)dx} = \frac{\int_0^4 x(2-\sqrt{x})dx}{\int_0^4 (2-\sqrt{x})dx}$$

$$= \frac{\int_0^4 (2x - x^{3/2})dx}{\int_0^4 (2 - x^{1/2})dx} = \frac{(x^2 - \frac{2}{5}x^{5/2})\big|_0^4}{(2x - \frac{2}{3}x^{3/2})\big|_0^4}$$

$$= \frac{\frac{16}{5}}{\frac{8}{3}} = \frac{6}{5}$$

$$\bar{y} = \frac{\int_0^2 y(x)dy}{\int_0^2 x\,dy} = \frac{\int_0^2 yy^2 dy}{\int_0^2 y^2 dy} = \frac{\int_0^2 y^3 dy}{\int_0^2 y^2 dy}$$

$$= \frac{\frac{1}{4}y^4\big|_0^2}{\frac{1}{3}y^3\big|_0^2} = \frac{4}{\frac{8}{3}} = \frac{12}{8} = \frac{3}{2}$$

Centroid is $\left(\dfrac{6}{5}, \dfrac{3}{2}\right)$.

15. $A = \int_0^2 x\,dx + \int_2^3 (6-2x)dx = 3$

$$\bar{x} = \frac{\int_0^2 x(3x - 2x)dx + \int_2^3 x(6-2x)dx}{3} = \frac{5}{3}$$

$$\bar{y} = \frac{\int_0^6 y\left(\frac{y}{2} - \frac{y}{3}\right)dy}{3} = \frac{12}{3} = 4$$

Centroid is $\left(\dfrac{5}{3}, 4\right)$.

16. $y = x^{2/3}, x = 8, y = 0$
Therefore, $x = y^{3/2}$

$$\bar{x} = \frac{\int_0^8 xy\,dx}{\int_0^8 y\,dx} = \frac{\int_0^8 xx^{2/3}dx}{\int_0^8 x^{2/3}dx}$$

$$= \frac{\int x^{5/3}dx}{\int x^{2/3}dx} = \frac{\frac{3}{8}x^{8/3}\big|_0^8}{\frac{3}{5}x^{5/3}\big|_0^8}$$

$$= \frac{\frac{3}{8}(256)}{\frac{3}{5}(32)} = \frac{96}{\frac{96}{5}} = 5$$

$$\bar{y} = \frac{\int_0^4 y(8-x)dy}{\int_0^4 (8-x)dy} = \frac{\int_0^4 y(8-y^{3/2})dy}{\int_0^4 (8-y^{3/2})dy}$$

$$= \frac{\int_0^4 (8y - y^{5/2})dy}{\int_0^4 (8 - y^{3/2})dy}$$

$$= \frac{(4y^2 - \frac{2}{7}y^{7/2})\big|_0^4}{(8y - \frac{2}{5}y^{5/2})\big|_0^4}$$

$$= \frac{64 - \frac{2}{7}(128)}{32 - \frac{2}{5}(32)} = \frac{10}{7}$$

Centroid is $\left(5, \dfrac{10}{7}\right)$

17. $y = x^3, y = 0, x = 1$

Rotated about x-axis, $\overline{y} = 0$

$$\overline{x} = \frac{\int_a^b xy^2\,dx}{\int_a^b y^2\,dx} = \frac{\int_0^1 x(x^3)^2\,dx}{\int_0^1 (x^3)^2\,dx}$$

$$= \frac{\int_0^1 x^7\,dx}{\int_0^1 x^6\,dx} = \frac{\frac{x^8}{8}}{\frac{x^7}{7}}\Big|_0^1$$

$$= \frac{\frac{1}{8}}{\frac{1}{7}} = \frac{7}{8}$$

Centroid is $\left(\frac{7}{8}, 0\right)$.

18. $y = 2 - 2x, x = 0, y = 0$

Rotated about y-axis, $\overline{x} = 0$

$$y = 2 - 2x, 2x = 2 - y, x = \frac{2-y}{2} = 1 - \frac{1}{2}y$$

$$\overline{y} = \frac{\int_0^2 yx^2\,dy}{\int_0^2 x^2\,dy} = \frac{\int_0^2 y\left(1 - \frac{1}{2}y\right)^2\,dy}{\int_0^2 \left(1 - \frac{1}{2}y\right)^2\,dy}$$

$$= \frac{\int_0^2 y\left(1 - y + \frac{1}{4}y^2\right)\,dy}{\int_0^2 \left((1 - y) + \frac{1}{4}y^2\right)\,dy}$$

$$= \frac{\int_0^2 \left(y - y^2 + \frac{1}{4}y^3\right)\,dy}{\int_0^2 \left(1 - y + \frac{1}{4}y^2\right)\,dy}$$

$$= \frac{\left(\frac{1}{2}y^2 - \frac{1}{3}y^3 + \frac{1}{16}y^4\right)\big|_0^2}{\left(y - \frac{1}{2}y^2 + \frac{1}{12}y^3\right)\big|_0^2}$$

$$= \frac{2 - \frac{8}{3} + 1}{2 - 2 + \frac{8}{12}} = \frac{1}{2}$$

Therefore, centroid is $\left(0, \frac{1}{2}\right)$.

19. $y^2 = 4x, y = 0, x = 1$

Rotated about y-axis, $\overline{x} = 0$

$$\overline{y} = \frac{\int_0^2 y(x_1^2 - 1^2)\,dy}{\int_0^2 (x_1^2 - 1^2)\,dy} = \frac{\int_0^2 y\left\{\left(\frac{y^2}{4}\right) - 1\right\}\,dy}{\int_0^2 \left\{\left(\frac{y^2}{4}\right) - 1\right\}\,dy}$$

$$r_1 = 1, r_2 = x; \; y^2 = 4x, x = \frac{y^2}{4}$$

$$\overline{y} = \frac{\int_0^2 \left(\frac{1}{16}y^5 - y\right)\,dy}{\int_0^2 \left(\frac{1}{16}y^4 - 1\right)\,dy} = \frac{\frac{1}{96}y^6 - \frac{1}{2}y^2\big|_0^2}{\frac{1}{80}y^5 - y\big|_0^2}$$

$$= \frac{\frac{64}{96} - 2}{\frac{32}{80} - 2} = \frac{\frac{128}{96}}{\frac{128}{80}} = \frac{5}{6}$$

Therefore, centroid is $\left(0, \frac{5}{6}\right)$

20. $y = x^2, x = 2, x$-axis, $\overline{y} = 0$

$$\overline{x} = \frac{\int_0^2 xy^2\,dx}{\int_0^2 y^2\,dx} = \frac{\int_0^2 x(x^2)^2\,dx}{\int_0^2 (x^2)^2\,dx}$$

$$= \frac{\int_0^2 x^5\,dx}{\int_0^2 x^4\,dx} = \frac{\frac{1}{6}x^6\big|_0^2}{\frac{1}{5}x^5\big|_0^2} = \frac{\frac{64}{6}}{\frac{32}{5}} = \frac{5}{3}$$

Centroid is $\left(\frac{5}{3}, 0\right)$.

21. $y^2 = 4x; x = 1$

Rotated about x-axis, $\overline{y} = 0$

$$\overline{x} = \frac{\int_a^b xy^2\,dx}{\int_a^b y^2\,dx} = \frac{\int_0^1 x(4x)\,dx}{\int_0^1 4x\,dx}$$

$$= \frac{\frac{4x^3}{3}\big|_0^1}{2x^2} = \frac{\frac{4}{3}}{2} = \frac{2}{3}$$

Centroid is $\left(\frac{2}{3}, 0\right)$.

22. $x^2 - y^2 = 9, y = 4, x$-axis

Rotated about y-axis, $\overline{x} = 0$

$$\overline{y} = \frac{\int_0^4 yx^2\,dy}{\int_0^4 x^2\,dy} = \frac{\int_0^4 y(9 + y^2)\,dy}{\int_0^4 (9 + y^2)\,dy}$$

$$= \frac{\int_0^4 (9y + y^3)\,dy}{\int_0^4 (9 + y^2)\,dy} = \frac{\frac{9y^2}{2} + \frac{y^4}{4}\big|_0^4}{9y + \frac{y^3}{3}\big|_0^4} = \frac{102}{43}$$

Therefore, centroid is $\left(0, \frac{102}{43}\right)$.

23. Triangle is area bounded by $y = \frac{a}{b}x$, x-axis, y-axis.

$$\overline{x} = \frac{\int_0^b xy\,dx}{\int_0^b y\,dx} = \frac{\int_0^b x\frac{a}{b}x\,dx}{\int_0^b \frac{a}{b}x\,dx}$$

$$= \frac{\frac{a}{b}\int_0^b x^2\,dx}{\frac{a}{b}\int_0^b x\,dx} = \frac{\frac{x^3}{3}\big|_0^b}{\frac{x^2}{2}\big|_0^b} = \frac{\frac{b^3}{3}}{\frac{b^2}{2}} = \frac{2}{3}b$$

$$\overline{y} = \frac{\int_0^b y(b - x)\,dy}{\int_0^b (b - x)\,dy} = \frac{\int_0^b y\left(b - \frac{b}{a}y\right)\,dy}{\int_0^b \left(b - \frac{b}{a}y\right)\,dy}$$

$$= \frac{\int_0^b \left(by - \frac{b}{a}y^2\right)\,dy}{\int_0^b \left(b - \frac{b}{a}y\right)\,dy} = \frac{\frac{b}{2}y^2 - \frac{b}{3a}y^3\big|_0^a}{by - \frac{b}{2a}y^2\big|_0^a}$$

$$= \frac{\frac{a^2b}{2} - \frac{a^2b}{3}}{ab - \frac{ab}{2}} = \frac{a}{3}$$

Therefore, centroid is $\left(\frac{2}{3}b, \frac{1}{3}a\right)$.

24. Hempisphere is generated by rotating a quarter circle about the y-axis.

Eq. of circle: $x^2 + y^2 = a^2$

$\bar{x} = 0$

$$\bar{y} = \frac{\int_0^a y(x^2)\,dy}{\int_0^a x^2\,dy} = \frac{\int_0^a y(a^2 - y^2)\,dy}{\int_0^a (a^2 - y^2)\,dy}$$

$$= \frac{\int_0^a (a^2 y - y^3)\,dy}{\int_0^a (a^2 - y^2)\,dy} = \frac{\frac{a^2}{2}y^2 - \frac{1}{4}y^4 \big|_0^a}{a^2 y - \frac{1}{3}y^3 \big|_0^a}$$

$$= \frac{\frac{a^4}{2} - \frac{a^4}{4}}{a^3 - \frac{a^3}{3}} = \frac{3}{8}a$$

Therefore, centroid is $\left(0, \frac{3}{8}a\right)$. (If rotated x-axis, centroid is $\left(\frac{3}{8}a, 0\right)$.)

25. $\dfrac{x^2}{5.00^2} + \dfrac{y^2}{1.00^2} = 1;\ y^2 = 1 - \dfrac{x^2}{25.0};$

$x^2 = 25.0 - 25.0y^2$

$$\int_0^{1.00} (yx^2)\,dy = \int_0^{1.00} y(25.0 - 25.0y^2)\,dy$$

$$= \int_0^{1.00} (25.0y - 25.0y^3)\,dy$$

$$= \left(\frac{25.0}{2}y^2 - \frac{25.0}{4}y^4\right)\bigg|_0^{1.00}$$

$$= \frac{25.0}{2} - \frac{25.0}{4} = \frac{25.0}{4}$$

$$\int_0^{1.00} x^2\,dy = \int_0^{1.00} (25.0 - 25.0y^2)\,dy$$

$$= \left(25.0y - \frac{25.0}{3}y^3\right)dy\bigg|_0^{1.00}$$

$$= \left(25.0 - \frac{25.0}{3}\right) = \frac{50.0}{3}$$

$$\bar{y} = \frac{25.0}{4} \div \frac{50.0}{3} = \frac{25.0}{4} \cdot \frac{3}{50.0}$$

$$= 0.375 \text{ cm above center of base}$$

26. $y^2 = \dfrac{4}{x}, y = 1, y = 2, y$-axis

Rotated about y-axis, $\bar{x} = 0$

$$y^2 = \frac{4}{x}, x = \frac{4}{y^2}$$

$$\bar{y} = \frac{\int_1^2 y(x^2)\,dy}{\int_1^2 x^2\,dy}$$

$$= \frac{\int_1^2 y\left(\frac{4}{y^2}\right)^2 dy}{\int_1^2 \left(\frac{4}{y^2}\right)^2 dy}$$

$$\bar{y} = \frac{\int_1^2 \frac{16}{y^3}\,dy}{\int_1^2 \frac{16}{y^4}\,dy} = \frac{\int_1^2 y^{-3}\,dy}{\int_1^2 y^{-4}\,dy} = \frac{-\frac{1}{2}y^{-2}\big|_1^2}{-\frac{1}{3}y^{-3}\big|_1^2}$$

$$= \frac{-\frac{1}{2y^2}\big|_1^2}{-\frac{1}{3y^3}\big|_1^2} = \frac{\frac{1}{8} - \frac{1}{2}}{\frac{1}{24} - \frac{1}{3}} = \frac{9}{7}$$

$$= 1.286 \text{ in}$$

Therefore, centroid of disc (in isolation) is $\left(0, \frac{2}{7}\right)$ on the axis of disc, 0.29 in from larger base.

27. Bounded area: $y = -4x + 80, y = 60, y = 0,$ $x = 0, \bar{x} = 0$

$y = -4x + 80;\ 4x = 80 - y;\ x = 20 - \dfrac{1}{4}y$

$$\bar{y} = \frac{\int_0^{60} yx^2\,dy}{\int_0^{60} x^2\,dy} = \frac{\int_0^{60} y\left(20 - \frac{1}{4}y\right)^2 dy}{\int_0^{60} \left(20 - \frac{1}{4}y\right)^2 dy}$$

$$\bar{y} = \frac{\int_0^{60} \left(400y - 10y^2 + \frac{1}{16}y^3\right)dy}{\int_0^{60} \left(400 - 10y + \frac{1}{16}y^2\right)dy}$$

$$= \frac{200y^2 - \frac{10}{3}y^3 + \frac{1}{64}y^4\big|_0^{60}}{400y - 5y^2 + \frac{1}{48}y^3\big|_0^{60}}$$

$$= \frac{720\,000 - 720\,000 + 202\,500}{24\,000 - 18\,000 + 4\,500}$$

$$= 19.3 \text{ cm from larger base}$$

Centroid is $(0, 19.3)$.

28. Bounded area: $y = \dfrac{3}{2}x - 9, y = 6, y = 0, x = 0$

$y = \dfrac{3}{2}x - 9;\ \dfrac{3}{2}x = y + 9;\ x = \dfrac{2}{3}y + 6$

$$\bar{y} = \frac{\int_0^6 y(2x)\,dy}{\int_0^6 2x\,dy} = \frac{\int_0^6 y\left(\frac{2}{3}y + 6\right)dy}{\int_0^6 \left(\frac{2}{3}y + 6\right)d}$$

$$\bar{y} = \frac{\int_0^6 \left(\frac{2}{3}y^2 + 6y\right)dy}{\int_0^6 \left(\frac{2}{3}y + 6\right)dy}$$

$$= \frac{\frac{2}{9}y^3 + 3y^2\big|_0^6}{\frac{1}{3}y^2 + 6y\big|_0^6} = \frac{48 + 108}{12 + 36}$$

$$= 3.25 \text{ m above the lower base}$$

Centroid is $(0, 3.25)$.

6.5   Moments of Inertia

1. $5.0(2.4)^2 + 3.2(3.5)^2 = 68 \text{ g} \cdot \text{cm}^2$

$(5.0 + 3.2)R^2 = 68;\ R = 2.9 \text{ cm}$

2. $I = m_1 x_1^2 + m_2 x_2^2 + m_3 x_3^2$
$ I = 3.4(-1.5)^2 + 6.0(2.1)^2 + 2.6(3.8)^2$
$ I = 72 \text{ g} \cdot \text{cm}^2$

$I = MR^2$
$72 \text{ g} \cdot \text{cm}^2 = (3.4 + 6.0 + 2.6)R^2$
$\phantom{72 \text{ g} \cdot \text{cm}^2 =} R = 2.4 \text{ cm}$

3. $I = m_1 x_1^2 + m_2 x_2^2 + m_3 x_3^2$
$ I = 45.0(-3.80)^2 + 90.0(0.00)^2 + 62.0(5.50)^2$
$ I = 2530 \text{ g} \cdot \text{cm}^2$

$I = MR^2$
$2530 = (45.0 + 9.0 + 62.0)R^2$
$ R = 3.58 \text{ cm}$

4. $I = m_1 x_1^2 + m_2 x_2^2 + m_3 x_3^2 + m_4 x_4^2$
$ I = 564(-45.0)^2 + 326(-22.5)^2 + 720(15.4)^2$
$ + 205(64.0)^2$
$ I = 2,320,000 \text{ g} \cdot \text{cm}^2$

$I = MR^2$
$2,320,000 = (564 + 326 + 720 + 205)R^2$
$ R = 35.8 \text{ cm}$

5. $y^2 = x, x = 4, x\text{-axis}$, with respect to the x-axis
$$I_x = k \int_0^2 y^2(4 - y^2)\,dy = k \int_0^2 (4y^2 - y^4)\,dy$$
$$= k \left(\frac{4}{3}y^3 - \frac{1}{5}y^5 \right) \Big|_0^2$$
$$= k \left(\frac{32}{3} - \frac{32}{5} \right) = \frac{64}{15}k$$

6. $y = 2x, x = 1, x = 2, x\text{-axis}$, with respect to the y-axis
$$I_y = k \int_1^2 x^2(y)\,dx$$
$$= k \int_1^2 x^2(2x)\,dx = k \int_1^2 2x^3\,dx$$
$$= 2k \int_1^2 x^3\,dx = 2k \frac{x^4}{4} \Big|_1^2$$
$$= 2k \left(\frac{16}{4} - \frac{1}{4} \right) = 2k\frac{15}{4} = \frac{15}{2}k$$

7. $y = x^3, x = 2, x\text{-axis}$, with respect to the y-axis
$$I_y = k \int_0^2 x^2 y\,dx = k \int_0^2 x^2 x^3\,dx = k \int_0^2 x^5\,dx$$
$$= \frac{k}{6}x^6 \Big|_0^2 = k\frac{64}{6} = k\frac{32}{3}$$
$$m = k \int_0^2 y\,dx = k \int_0^2 x^3\,dx = k\frac{x^4}{4} \Big|_0^2 = k4$$
$$R_y^2 = \frac{I_y}{m} = \frac{k\frac{32}{3}}{k4} = \frac{32}{12} = \frac{8}{3}$$
$$R_y = \frac{\sqrt{8}}{\sqrt{3}} = \frac{2\sqrt{2}}{\sqrt{3}} = \frac{2\sqrt{6}}{3}$$

8. $y^2 = 1 - x$, with respect to the x-axis
Quad I
$$I_x = k \int_0^1 y^2(x)\,dy = k \int_0^1 y^2(1 - y^2)\,dy$$
$$= k \int_0^1 (y^2 - y^4)\,dy = k \left(\frac{1}{3}y^3 - \frac{1}{5}y^5 \right) \Big|_0^1 = k\frac{2}{15}$$
$$m = k \int_0^1 x\,dy = k \int_0^1 (1 - y^2)\,dy$$
$$= k \left(y - \frac{1}{3}y^3 \right) \Big|_0^1 = k\frac{2}{3}$$
$$R_x^2 = \frac{I_x}{m} = \frac{k\frac{2}{15}}{k\frac{2}{3}} = \frac{1}{5}; \; R_x = \sqrt{\frac{1}{5}} = \frac{\sqrt{5}}{5}$$

9. $y = \frac{b}{a}x; \; x = a, y = 0$
$$I_x = k \int_0^b y^2(a - x)\,dy = k \int_0^b y^2 \left(a - \frac{ay}{b} \right) dy$$
$$= ka \int_0^b \left(y^2 - \frac{y^3}{b} \right) dy = ka \left(\frac{1}{3}y^3 - \frac{1}{4b}y^4 \right) \Big|_0^b$$
$$= \frac{kab^3}{12}$$

For $k = 1$; $m = \frac{1}{2}ab$; $I_x = ab\left(\frac{b^2}{12} \right) = 2m\left(\frac{b^2}{12} \right) = \frac{1}{6}mb^2$

10. $y = a, x = b$
$$I_y = k \int_0^b x^2 y\,dx = k \int_0^b x^2 a\,dx$$
$$= ka\frac{x^3}{3} = \frac{kab^3}{3} = (ka)\frac{b^3}{3}$$
$$m = k \int_0^b a\,dx = kax \Big|_0^b = \frac{m}{b}\frac{b^3}{3} = \frac{mb^2}{3}$$

11. $y = x^2, x = 2, x$-axis, with respect to the x-axis

$$I_x = k \int_0^4 y^2(2-x)dy = k \int_0^4 y^2(2-\sqrt{y})dy$$

$$= k \int_0^4 (2y^2 - y^{5/2})dy = k\left(\frac{2}{3}y^3 - \frac{2}{7}y^{7/2}\right)\Big|_0^4$$

$$= \frac{128}{21}k$$

$$m = k \int_0^4 (2-x)dy = k \int_0^4 (2-\sqrt{y})dy$$

$$= k\left(2y - \frac{2}{3}y^{3/2}\right)\Big|_0^4$$

$$y = x^2, x = \sqrt{y}, m = \frac{8}{3}k$$

$$R_x^2 = \frac{\frac{128}{21}k}{\frac{8}{3}k}; \; R_x = \frac{4\sqrt{7}}{7}$$

12. $y^2 = x^3, y = 8, y$-axis, with respect to the y-axis

$$I_y = k \int_0^4 x^2(8-y)dx = k \int_0^4 x^2(8 - x^{3/2})dx$$

$$= k \int_0^4 (8x^2 - x^{7/2})dx = k\left(\frac{8}{3}x^3 - \frac{2}{9}x^{9/2}\right)\Big|_0^4$$

$$= k\left(\frac{8}{3}[64] - \frac{2}{9}[512]\right) = k\frac{512}{9}$$

$$y^2 = x^3, y = x^{3/2}$$

$$m = k \int_0^4 (8-y)dx = \int_0^4 (8 - x^{3/2})dx$$

$$= k\left(8x - \frac{2}{5}x^{5/2}\right)\Big|_0^4$$

$$= \frac{96}{5}k$$

$$R_y^2 = \frac{k\frac{512}{9}}{k\frac{96}{5}} = \frac{80}{27}; \; R_y = \frac{4\sqrt{15}}{9}$$

13. $y^2 = x^3, y = 8, y$-axis, with respect to the x-axis

$$I_x = k \int_0^8 y^2 x \, dy = k \int_0^8 y^2(y^{2/3}) \, dy$$

$$= k \int_0^8 y^{8/3} dy = k\frac{3}{11}y^{11/3}\Big|_0^8$$

$$= \frac{3}{11}(8)^{11/3}k = \frac{3}{11}(2)^{11}k = \frac{6144}{11}k$$

$$m = k \int_0^8 x \, dy = k \int_0^8 y^{2/3}dy = k\left(\frac{3}{5}y^{5/3}\right)\Big|_0^8$$

$$= \frac{3}{5}(8)^{5/3}k = \frac{3}{5}(2)^5 k = \frac{96k}{5}$$

$$R^2 = \frac{I_x}{m} = \frac{6144}{11} \div \frac{96k}{5} = \frac{64(5)}{11}$$

$$R = \sqrt{\frac{64(5)}{11}} = \frac{8}{11}\sqrt{55}$$

14. $x = 1, y = 2 - x, y$-axis, with respect to the y-axis

$$I_y = k \int_0^1 x^2(2-x)dx = k \int_0^1 (2x^2 - x^3)dx$$

$$= k\left(\frac{2x^3}{3} - \frac{x^4}{4}\right)\Big|_0^1 = k\frac{5}{12}$$

$$m = k \int_0^1 y \, dx = k \int_0^1 (2-x)dx = k\left(2x - \frac{x^2}{2}\right)\Big|_0^1$$

$$= k\frac{3}{2}$$

$$R_y^2 = \frac{k\frac{5}{12}}{k\frac{3}{2}} = \frac{5}{18}; \; R_y = \frac{\sqrt{10}}{6}$$

15. $y^2 = x, y = 2, y$-axis, rotated about x-axis
$y^2 = x, y = \sqrt{x}$

$$I_x = 2\pi k \int_0^2 xy^3 dy = 2\pi k \int_0^2 y^2 y^3 dy = 2\pi k \int_0^2 y^5 dy$$

$$= 2\pi k \frac{y^6}{6}\Big|_0^2 = \frac{64\pi k}{3}$$

16. $y = 4 - x^2$, rotated about y-axis

$$I_y = 2\pi k \int_0^2 yx^3 dx$$

$$= 2\pi k \int_0^2 (4 - x^2)x^3 dx = 2\pi k \int_0^2 (4x^3 - x^5)dx$$

$$= 2\pi k\left(x^4 - \frac{1}{6}x^6\right)\Big|_0^2 = 2\pi k\left(\frac{16}{3}\right) = \frac{32\pi k}{3}$$

$$m = 2\pi k \int_0^2 xy \, dx = 2\pi k \int_0^2 x(4 - x^2)dx$$

$$= 2\pi k \int_0^2 (4x - x^3)dx = 2\pi k\left(2x^2 - \frac{1}{4}x^4\right)\Big|_0^2$$

$$= 8\pi k$$

$$R_y^2 = \frac{\frac{32\pi k}{3}}{8\pi k} = \frac{4}{3}; \; R_y = \frac{2\sqrt{3}}{3}$$

17. $y = 2x - x^2, y = 0$, rotated about y-axis

$$I_y = 2\pi k \int_0^2 (2x - x^2)(x^3)\,dx$$

$$= 2\pi k \int_0^2 (2x^4 - x^5)\,dx$$

$$= 2\pi k \left[\frac{2}{5}x^5 - \frac{1}{6}x^6 \right]\Big|_0^2$$

$$= 2\pi k \left[\frac{2}{5}(2)^5 - \frac{1}{6}(2)^6 \right]$$

$$= 2\pi k \left[\frac{64}{5} - \frac{64}{6} \right] = \frac{64\pi k}{15}$$

$$m = 2\pi k \int_0^2 (2x - x^2)(x)\,dx$$

$$= 2\pi k \int_0^2 (2x^2 - x^3)\,dx$$

$$= 2\pi k \left[\frac{2}{3}x^3 - \frac{1}{4}x^4 \right]\Big|_0^2$$

$$= 2\pi k \left[\frac{2}{3}(2)^3 - \frac{1}{4}(2)^4 \right]$$

$$= 2\pi k \left[\frac{16}{3} - \frac{16}{4} \right] = \frac{8\pi k}{3}$$

$$R_y^2 = \frac{I_y}{m} = \frac{65\pi k}{15} \div \frac{8\pi k}{3} = \frac{8}{5};$$

$$R_y = \sqrt{\frac{8}{5}\left(\frac{5}{5}\right)}$$

$$= \frac{2}{5}\sqrt{10}$$

18. $y = 2x, y = x^2$, rotated about y-axis

$$I_y = 2\pi k \int_0^2 (y_1 - y_2)x^3\,dx = 2\pi k \int_0^2 (2 - x^2)x^3\,dx$$

$$= 2\pi k \int_0^2 (2x^4 - x^5)$$

$$= 2\pi k \left(\frac{2}{5}x^5 - \frac{1}{6}x^6 \right)\Big|_0^2 = 2\pi k \left(\frac{64}{30} \right) = \frac{64}{15}\pi k$$

$$m = 2\pi k \int_0^2 x(y_1 - y_2)\,dx = 2\pi k \int_0^2 x(2x - x^2)\,dx$$

$$= 2\pi k \int_0^2 (2x^2 - x^3)\,dx = 2\pi k \left(\frac{2}{3}x^3 - \frac{1}{4}x^4 \right)\Big|_0^2$$

$$= 2\pi k \left(\frac{4}{3} \right) = \frac{8\pi k}{3}$$

$$R_y^2 = \frac{\frac{64}{15}\pi k}{\frac{8}{3}\pi k} = \frac{8}{5}; \; R_y = \frac{2\sqrt{10}}{5}$$

19. Eq. of line: $y = \frac{r}{h}x$; therefore, $x = \frac{h}{r}y$

$$I_x = 2\pi k \int_0^r (h - x)y^3\,dy$$

$$= 2\pi k \int_0^r \left(h - \frac{h}{r}y \right) y^3\,dy$$

$$= 2\pi k \int_0^r \left(hy^3 - \frac{h}{r}y^4 \right)\,dy$$

$$= 2\pi k \left(\frac{h}{4}y^4 - \frac{h}{5r}y^5 \right)\Big|_0^r = \pi k \left(\frac{hr^4}{10} \right)$$

$$m = 2\pi k \int_0^r y \left(h - \frac{h}{r}y \right)\,dy$$

$$= 2\pi k \int_0^r \left(hy - \frac{h}{r}y^2 \right)\,dy$$

$$= 2\pi k \left(\frac{hy^2}{2} - \frac{hy^3}{3r} \right)\Big|_0^r = 2\pi k \left(\frac{hr^2}{2} - \frac{hr^2}{3} \right)$$

$$= \pi k \left(\frac{hr^2}{3} \right)$$

$$I_x = \pi k \left(\frac{hr^4}{10} \right) = \frac{\pi k hr^2}{3}\left(\frac{3r^2}{10} \right) = \frac{3mr^2}{10}$$

20. Since thickness is negligible, all mass can be assumed at a distance r from center. Let k = linear density along circumference. Integrate along circumference (as if it were a straight line).

$$I = kr^2 \int_0^{2\pi r} dx = kr^2 x\Big|_0^{2\pi r} = 2\pi k r^3$$

$$m = 2\pi k r; \; I = 2\pi k r(r^2) = mr^2$$

(Since all mass is assumed at a distance r from center, $I = mr^2$ from the meaning of moment of inertia.)

21. $r = 0.600$ cm, $h = 0.800$ cm, $m = 3.00$ g

$$y = \frac{0.600}{0.800}x = 0.750x; \; x = 1.333y$$

$$I_x = 2\pi k \int_0^{0.600} (0.800 - 1.333y)y^3\,dy$$

$$= 2\pi k(0.200y^4 - 0.2667y^5)\Big|_0^{0.600}$$

$$= 2\pi k(0.005\,181)$$

$$m = \frac{k}{3}\pi r^2 h, 2\pi k = \frac{6m}{r^2 h} = \frac{6(3.00)}{(0.600^2)(0.800)}$$

$$= 62.5 \text{ g/cm}^3$$

$$I_x = (62.5)(0.005\,181) = 0.324 \text{ g} \cdot \text{cm}^2$$

22. Equations are: $y = 2, x = 1$
$k = 3 \text{ kg} \cdot \text{m}^2$

$$I_y = k \int_0^1 x^2 y \, dx = k \int_0^1 x^2 (2) \, dx$$

$$= 2k \int_0^1 x^2 \, dx = 2k \frac{x^3}{3} \bigg|_0^1 = \frac{2k}{3}$$

$$I_y = \frac{2}{3}(3) = 2 \text{ kg} \cdot \text{m}^2$$

23. Let a be the thickness.
$m = kV, 1.2 = kV$

$$I_y = 2\pi k \int_{4.0}^{06.0} x^3 (a) \, dx$$

$$I_y = \frac{\pi k a}{2} x^4 \bigg|_{4.0}^{6.0} = \frac{\pi k a}{2} (6.0^4 - 4.0^4)$$

$$= \frac{\pi k}{2} (1298 - 256) = \frac{\pi k a}{2} (1040)$$

$$m = 2\pi k \int_{4.0}^{6.0} x(a) \, dx = \pi k a x^2 \bigg|_{4.0}^{6.0}$$

$$= \pi k a (36 - 16) = 20\pi k a$$

$$2\pi k a = 1.2, \pi k a = \frac{1.2}{20} = 0.06$$

$$I_y = \frac{0.060}{2} (1040) = 31.2 \text{ kg} \cdot \text{cm}^2$$

24. $I_y = k \int x^2 \, dx = k \int_0^L x^2 \, dx = \frac{1}{3} k x^3 \bigg|_0^L = \frac{1}{3} k L^3$

$$m = k \int_0^L dx = k x \big|_0^L = kL$$

$$I_y = kL \left(\frac{1}{3} L^2 \right) = \frac{1}{3} m L^2$$

6.6   Work by A Variable Force

1. $f(x) = kx; \; 6.0 = k(1.5); \; k = 4.0 \text{ lb/in}$

$$W = \int_0^{2.0} 4.0x \, dx = 2.0 x^2 \big|_0^{2.0}$$

$$= 2.0(2.0^2) - 0$$

$$= 8.0 \text{ lb} \cdot \text{in}$$

2. $W = \int_{2.0}^{4.0} 4.0x \, dx = 4.0 \frac{x^2}{2} \bigg|_{2.0}^{4.0}$

$$W = 4.0 \left[\frac{4.0^2}{2} - \frac{2.0^2}{2} \right]$$

$$W = 24 \text{ lb} \cdot \text{in}$$

3. $F = hx; \; 25 = h(16), \; h = \frac{25}{16}$

$$W = \int_0^{16} \frac{25}{16} x \, dx = \frac{25}{16} \frac{x^2}{2} \bigg|_0^{16}$$

$$= \frac{25}{16} \left[\frac{16^2}{2} - \frac{0^2}{2} \right]$$

$$W = 200 \text{ N} \cdot \text{mm}$$

4. $F = kx$
$1200 = k(2); \text{ therefore, } k = 600 \text{ lb/in}$

$$W = \int_2^4 600x \, dx = 300 x^2 \big|_2^4$$

$$= 300(16 - 4) = 300(12)$$

$$= 3600 \text{ lb} \cdot \text{in}$$

5. $F = kx = \frac{25}{16} x$

$$25 = \frac{25}{16} x$$

$$x = 16$$

$$W = \int_{16}^{32} \frac{25}{16} x \, dx = \frac{25}{16} \frac{x^2}{2} \bigg|_{16}^{32}$$

$$W = \frac{25}{16} \left[\frac{32^2}{2} - \frac{16^2}{2} \right]$$

$$W = 600 \text{ N} \cdot \text{mm}$$

6. $f(x) = \frac{kq_1 q_2}{x^2}, 1.0 \text{ pm} = 1.0 \times 10^{-12},$

$4.0 \text{ pm} = 4.0 \times 10^{-12}$

$$W = \int_{1.0 \times 10^{-12}}^{4.0 \times 10^{-12}} \frac{9.0 \times 10^9 (1.6 \times 10^{-19})^2}{x^2} \, dx$$

$$= 9.0 \times 10^9 (1.6 \times 10^{-19})^2 \left(-\frac{1}{x} \right) \bigg|_{1.0 \times 10^{-12}}^{4.0 \times 10^{-12}}$$

$$= -23.04 \times 10^{-29} [0.25 \times 10^{12} - 10^{12}]$$

$$= -23.04 \times 10^{-29} (-0.75 \times 10^{12})$$

$$= 1.7 \times 10^{-16} \text{ J}$$

7. $f(x) = \frac{9.0 \times 10^9 q_1 q_2}{x^2}$

$$= \frac{9.0 \times 10^9 \times 1.6 \times 10^{-19} \times 1.3 \times 10^{-18}}{x^2}$$

$$= \frac{1.87 \times 10^{-27}}{x^2}$$

$$W = \int_{2.0 \times 10^{-6}}^{1.0} (1.87 \times 10^{-27} x^{-2})\, dx$$

$$= -1.9 \times 10^{-27} x^{-1} \big|_{2.0 \times 10^{-6}}^{1.0}$$

$$= -1.9 \times 10^{-27} - \frac{-1.9 \times 10^{-27}}{2.0 \times 10^{-6}}$$

$$= -1.9 \times 10^{-27} + 0.94 \times 10^{-21}$$

$$= -1.9 \times 10^{-27} + 9.4 \times 10^{-22}$$

$$= 9.4 \times 10^{-22} \text{ N} \cdot \text{m}$$

8. $-0.8 \text{ nC} = -0.8 \times 10^{-9} \text{ C};$
$$= -8 \times 10^{-10} \text{ C}$$
$$-0.6 \text{ nC} = -0.6 \times 10^{-9} \text{ C}$$
$$= -6 \times 10^{-10} \text{ C}$$

$100 \text{ mm} = 0.1 \text{ m}; \ 200 \text{ mm} = 0.2 \text{ m}$

$$W = \int_{0.1}^{0.2} k \frac{q_1 q_2}{x^2}\, dx$$

$$= 9 \times 10^9 (-8 \times 10^{10})(-6 \times 10^{-10}) \left(\frac{-1}{x}\right)\Big|_{0.1}^{0.2}$$

$$W = -9 \times 10^9 (8 \times 10^{-10})(6 \times 10^{-10}) \left(-\frac{1}{0.2} + \frac{1}{0.1}\right)$$

$$W = -2.16 \times 10^{-8} \text{ J}$$

9. $1 \text{ cm} = 0.01 \text{ m}$

$$W = \int_{0.01}^{1} \frac{k}{x^2}\, dx = -k \frac{1}{x}\Big|_{0.01}^{1}$$

$$= -k \left(\frac{1}{1} - \frac{1}{0.01}\right) = 99k$$

$$W = 99k \text{ J}$$

10. $F = \dfrac{k}{x^2}$

$$W = k \int_{10}^{100} \frac{dx}{x^2} = k \int_{10}^{100} x^{-2}\, dx$$

$$= -kx^{-1} = \frac{k(-1)}{x}\Big|_{10}^{100}$$

$$= -k \left(\frac{1}{100} - \frac{1}{10}\right) = \frac{-k(1-10)}{100}$$

$$= \frac{9k}{100} = 0.09k \text{ ft} \cdot \text{lb}$$

11. Weighs 100 lb: $\dfrac{100 \text{ lb}}{200 \text{ ft}} = 0.5 \text{ lb/ft}$

$$f(x) = 0.5(200 - x)$$

$$W = \int_{0}^{200} 0.5(200 - x)\, dx = \int_{0}^{200} (100 - 0.5x)\, dx$$

$$= 100x - 0.25x^2 \big|_{0}^{200} = 10{,}000 \text{ ft} \cdot \text{lb}$$

12. $W = 10{,}000 + 50(200)$
$W = 20{,}000 \text{ ft} \cdot \text{lb}$

13. Let $x = $ length of rope wound up
Weight/length $= 6.00 \text{ N/m}$

$$F = 6.00(25 - x)$$

$$W = \int_{0}^{20} 6.00(25 - x)\, dx = 150x - 3.00x^2 \big|_{0}^{20}$$

$$= 3000 - 1200 = 1800 \text{ N} \cdot \text{m}$$

14. Let x be positive downward from the pulley, then

$$W = \int_{20}^{50} 12(50 - x)\, dx$$

$$= 12 \left(50x - \frac{x^2}{2}\right)\Big|_{20}^{50} + 12(20)(30)$$

$$W = 12 \left[50(50) - \frac{50^2}{2} - 50(20) + \frac{20^2}{2}\right] + 7200$$

$$W = 12{,}600 \text{ J}$$

15. $f(x) = 32.5 - (1.25 \times 10^{-3})x$

$$W = \int_{0}^{12000} [32.5 - (1.25 \times 10^{-3})x]\, dx$$

$$= 32.5x - 0.63 \times 10^{-3} x^2 \big|_{0}^{12000} = 300{,}000 - 0$$

$$= 3.00 \times 10^5 \text{ ft} \cdot \text{ton}$$

16. $F = 550 \text{ N}, \ \dfrac{7.50 \text{ N}}{100 \text{ m}} = 7.50 \times 10^{-2} \text{ N/m}$

$$F = kx, F = 7.50 \times 10^{-2} x$$

$$W = \int_{0}^{1000} (550 + 7.50 \times 10^{-2} x)\, dx$$

$$= 550x + 7.50 \times 10^{-2} \frac{x^2}{2}\Big|_{0}^{1000}$$

$$= 550\,000 + \frac{7.50 \times 10^{-2}}{2}(10^6)$$

$$= 550\,000 + 3.75 \times 10^4$$

$$= 550\,000 + 37\,500$$

$$= 5.88 \times 10^5 \text{ N} \cdot \text{m}$$

17. Weight of element $= dV(62.4)$
$w = \pi 3.00^2 (62.4)dx = 9.00\pi(62.4)dx$

$$w = \int_{0}^{10.0} 9.00\pi(62.4)(10.0 - x)\, dx$$

$$= 9.00(62.4)\pi \left(10.0x - \frac{x^2}{2}\right)\Big|_{0}^{1.0}$$

$$= 9.00\pi(62.4)(100 - 50)$$

$$= 8.82 \times 10^4 \text{ ft} \cdot \text{lb}$$

18. $w = \int_0^{\frac{10.0}{2}} \pi (3.00)^2 (62.4)(10.0 - y)\,dy$

$w = \pi (3.00)^2 (62.4) \left(10.0y - \frac{y^2}{2} \right) \Big|_0^{\frac{10.0}{2}}$

$w = 6.62 \times 10^4 \text{ ft} \cdot \text{lb}$

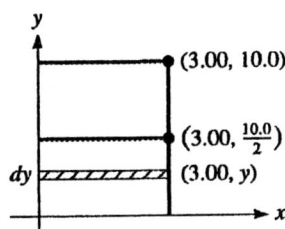

19. $w = \int_0^{4.00} 6270\pi (6.00h^2 - h^3)\,dh$ since each disk must be raised $6.00 - h$.

$w = 6270\pi \left(\frac{6.00}{3}h^3 - \frac{1}{4}h^4 \right) \Big|_0^{4.00}$

$w = 1.26 \times 10^6 \text{ N} \cdot \text{m} = 1.26 \text{ MJ}$

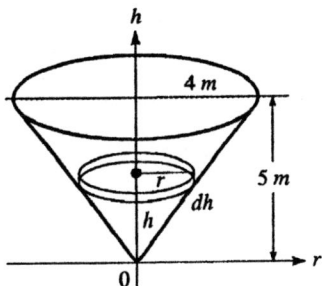

20. weight of an element
$= 60.0(\pi x^2)\,dy$
$= 60.0\pi (10.0^2 - y^2)\,dy$

$w = \int_0^{-10} 60.0\pi (10.0^2 - y^2) y\,dy$

$= 60.0\pi \left(\frac{10.0^2 y^2}{2} - \frac{y^4}{4} \right) \Big|_0^{-10}$

$w = 4.71 \times 10^5 \text{ ft} \cdot \text{lb}$

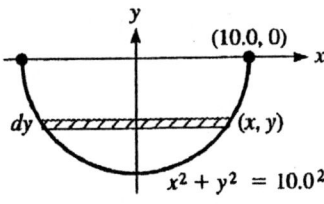

6.7   Force Due To Liquid Pressure

1. $F = 62.4 \int_0^{2.50} (12.0y)\,dy$

$= 62.4 \left(6.00y^2 \right) \Big|_0^{2.50}$

$= 62.4(37.5 - 0)$

$= 2340 \text{ lb}$

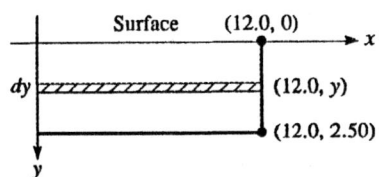

2. $F = 4 \int_0^2 9800y(2\,dy)$

$= 39{,}200y^2 \Big|_0^2$

$= 156{,}800 \text{ N}$

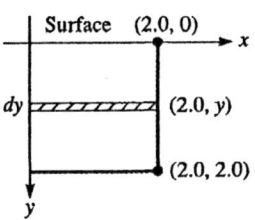

3. $F = \int_{2.00}^{8.00} 62.4(y)(15.0\,dy)$

$F = 62.4(15.0) \left(\frac{y^2}{2} \right) \Big|_{2.00}^{8.00}$

$F = 28{,}100 \text{ lb}$

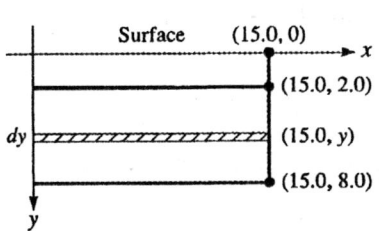

4. density $= 133 \times 10^3$ N/m^3 $\left(\dfrac{\text{m}^3}{100^3 \, \text{cm}^3}\right)$

$$= 133 \times 10^3 (100^{-3}) \text{N/cm}^3$$

$$F = 133 \times 10^3 (100^{-3}) \int_0^{6.00} (6.00 - y)(6.00 \, dy)$$

$$F = 133 \times 10^3 (100^{-3})(6.00) \left(\dfrac{y^2}{2}\right)\Big|_0^{6.00}$$

$$F = 14.4 \text{ N}$$

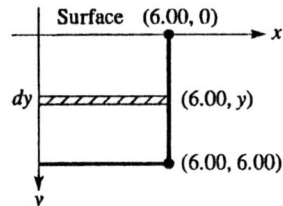

5. $F = 64.0 \displaystyle\int_{4.00}^{9.00} 10.0 y \, dy$

$$= 64.0(10.0)\dfrac{y^2}{2}\Big|_{4.00}^{9.00}$$

$$= 640(40.5 - 8.0) = 20,800 \text{ lb}$$

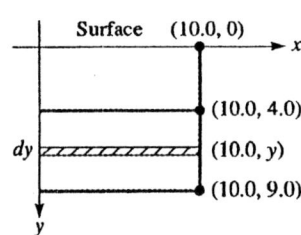

6. $F = 64.0 \displaystyle\int_{29.0}^{34.0} 10.0 y \, dy$

$$= 64.0(10.0) \left(\dfrac{y^2}{2}\right)\Big|_{29.0}^{34.0}$$

$$F = 101,000 \text{ lb}$$

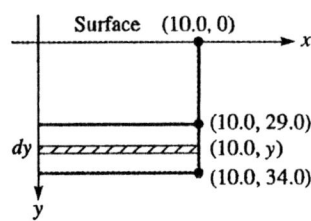

7. $F = \displaystyle\int_0^{8.00} 62.4 y (8.00 - y) \, dy$

$$= 62.4 \left(8.00\dfrac{y^2}{2} - \dfrac{y^3}{3}\right)\Big|_0^{8.00}$$

$$F = 5320 \text{ lb}$$

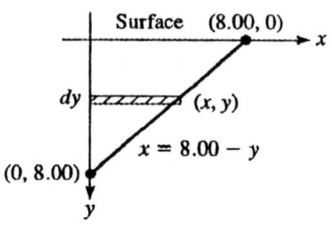

8. $F = \displaystyle\int_{3.00}^{8.00} 62.4 y (8.00 - y) \, dy$

$$= 62.4 \left(8.00\dfrac{y^2}{2} - \dfrac{y^3}{3}\right)\Big|_{3.00}^{8.00}$$

$$F = 3640 \text{ lb}$$

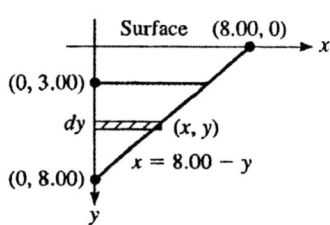

9. $F = 9800 \displaystyle\int_0^{1.0} (xy)(dy); \; x = 2.0y$

$$F = 9800 \int_0^{1.0} (2.0y)(y) \, dy$$

$$= 9800 \left[\dfrac{2.0}{3}y^3\right]\Big|_0^{1.0}$$

$$= 9800 \left(\dfrac{2.0}{3}\right)$$

$$= 6500 \text{ N}$$

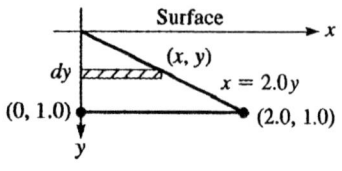

10. $F = 9800 \int_{3.0}^{4.0} y\,(x\,dy)$

$F = 9800 \int_{3.0}^{4.0} (2.0y - 6.0) y\, dy$

$= 9800 \int_{3.0}^{4.0} (2.0y^2 - 6.0y)\,dy$

$= 9800 \left(\dfrac{2.0}{3} y^3 - \dfrac{6.0 y^2}{2} \right) \Big|_{3.0}^{4.0}$

$= 35{,}900 \text{ N}$

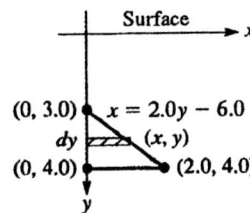

11. $F = \int_0^{4.00} 42.0(y)(2x\,dy)$

$F = \int_0^{4.00} 42.0(y) \left(2\dfrac{12.0 - y}{4.00} \right) dy$

$F = \dfrac{84.0}{4.00} \left(12.0\dfrac{y^2}{2} - \dfrac{y^3}{3} \right) \Big|_0^{4.00}$

$F = 1570 \text{ lb}$

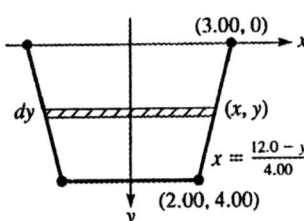

12. $F = \int_0^{4.00} 60.0y(2x)\,dy, \quad x^2 + y^2 = 4.00^2$

$F = -60.0 \int_0^{4.00} \sqrt{4.00^2 - y^2}\,(-2y\,dy)$

$F = -60.0 \left(\dfrac{2}{3} \right) (4.00^2 - y^2)^{3/2} \Big|_0^{4.00}$

$= 2560 \text{ lb}$

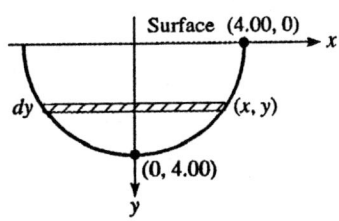

13. $F = (\text{density})(\text{height})(\text{area})$
$F_{\text{top}} = 1.00(9800)(2.00)^2 = 3.92 \times 10^4 \text{ N}$
$F_{\text{bottom}} = 1.00(9800)(2.00)^2 = 1.18 \times 10^5 \text{ N}$

The difference of the two forces is the buoyant force.

14. $F = \int_0^2 62.4(2 - y)(x\,dy)$

$= \int_0^2 62.4(2 - y)(2y - y^2)\,dy$

$F = \int_0^2 62.4(4y - 4y^2 + y^3)\,dy$

$F = 62.4 \left(2y^2 - 4\dfrac{y^3}{3} + \dfrac{y^4}{4} \right) \Big|_0^2$

$F = 83.2 \text{ lb}$

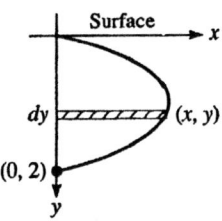

15. $y = x^2, y = 20$
$w = 62.4 \text{ lb/ft}^3$
$l = 2x, h = 20 - y$
$dh = -dy$
Top element: $h = 16, y = 4$

Bottom element: $h = 20, y = 0$

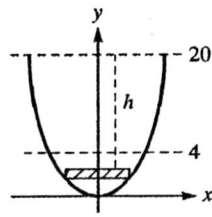

$F = 62.4 \int_4^0 (2x)(20 - y)(-dy)$

$= -62.4 \int_4^0 (2y^{1/2})(20 - y)\,dy$

$= -124.8 \int_4^0 (20y^{1/2} - y^{3/2})\,dy$

$= -124.8 \left(\dfrac{40}{3}y^{3/2} - \dfrac{2}{5}y^{5/2} \right) \Big|_4^0$

$= 124.8(93.87) = 11{,}700 \text{ lb}$

16. $\dfrac{x^2}{4} + \dfrac{y^2}{9} = 1$

$9x^2 + 4y^2 = 36$

$F = 50.0 \displaystyle\int_0^{3.0} y(2x\,dy) = 100.0 \int_0^{3.0} \sqrt{\dfrac{36 - 4y^2}{9}}\, y\,dy$

$ = \dfrac{-100.0}{8.3} \displaystyle\int_0^{3.0} (36 - 4y^2)^{1/2}(-8y\,dy)$

$ = \dfrac{-100.0}{24} \cdot \dfrac{2}{3}(36 - 4y^2)^{3/2}\Big|_0^{3.0}$

$ = \dfrac{-100.0}{36}(36 - 4y^2)^{3/2}\Big|_0^{3.0}$

$ = \dfrac{-100.0}{36}(0 - 216)$

$ = 6.00 \times 10^2 = 600\ \text{lb}$

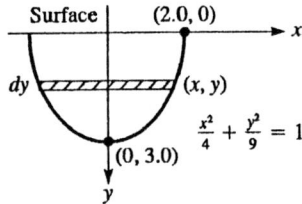

6.8   Other Applications

1. $s = \displaystyle\int_a^b \sqrt{1 + \left(\dfrac{dy}{dx}\right)^2}\,dx,\ a = 3, b = 8$

$y = \dfrac{2}{3}x^{3/2},\ \dfrac{dy}{dx} = x^{1/2}$

$s = \displaystyle\int_3^8 [1 + (x^{1/2})^2]^{1/2}\,dx = \int_3^8 [1 + x]^{1/2}\,dx$

$s = \dfrac{2}{3}[1 + x]^{3/2}\Big|_3^8 = \dfrac{2}{3}(27 - 8) = \dfrac{38}{3}$

2. $s = \displaystyle\int_a^b \sqrt{1 + \left(\dfrac{dy}{dx}\right)^2}\,dx,\ a = 1, b = 4$

$y = \dfrac{2}{3}(x - 1)^{3/2},\ \dfrac{dy}{dx} = (x - 1)^{1/2}$

$s = \displaystyle\int_1^4 \sqrt{1 + [(x-1)^{1/2}]^2}\,dx = \int_1^4 \sqrt{1 + x - 1}\,dx$

$s = \dfrac{2}{3}x^{3/2}\Big|_1^4 = \dfrac{2}{3}(4^{3/2} - 1^{3/2}) = \dfrac{14}{3}$

3. $s = \displaystyle\int_a^b \sqrt{1 + \left(\dfrac{dy}{dx}\right)^2}\,dx,\ a = 0, b = 3$

$y = \dfrac{2}{3}(x^2 + 1)^{3/2},\ \dfrac{dy}{dx} = 2x(x^2 + 1)^{1/2}$

$s = \displaystyle\int_0^3 [1 + (2x(x^2 + 1)^{1/2}]^{1/2}\,dx$

$s = \displaystyle\int_0^3 (1 + 4x^4 + 4x^2)^{1/2}\,dx = \int_0^3 [(1 + 2x^2)^2]^{1/2}\,dx$

$s = \displaystyle\int_0^3 (1 + 2x^2)\,dx = x + \dfrac{2}{3}x^3\Big|_0^3 = 3 + 18 = 21$

4. $s = \displaystyle\int_a^b \sqrt{1 + \left(\dfrac{dy}{dx}\right)^2}\,dx,\ a = 1, b = 3$

$y = \dfrac{1}{6}x^3 + \dfrac{1}{2x},\ \dfrac{dy}{dx} = \dfrac{1}{2}x^2 - \dfrac{1}{2x^2} = \dfrac{x^4 - 1}{2x^2}$

$s = \displaystyle\int_1^3 \sqrt{1 + \left(\dfrac{x^4 - 1}{2x^2}\right)^2}\,dx = \int_1^3 \sqrt{1 + \dfrac{(x^4 - 1)^2}{4x^4}}\,dx$

$s = \displaystyle\int_1^3 \sqrt{\dfrac{4x^4 + x^2 - 2x^4 + 1}{4x^4}}\,dx = \int_1^3 \sqrt{\dfrac{(x^4 + 1)^2}{4x^4}}\,dx$

$s = \displaystyle\int_1^3 \dfrac{x^4 + 1}{2x^2}\,dx = \int_1^3 \left(\dfrac{1}{2}x^2 + \dfrac{1}{2}x^{-2}\right)dx$

$s = \dfrac{1}{2}\dfrac{x^3}{3} - \dfrac{1}{2x}\Big|_1^3 = \dfrac{27}{6} - \dfrac{1}{6} - \dfrac{1}{6} + \dfrac{1}{2}$

$s = \dfrac{14}{3}$

5. $S = 2\pi \displaystyle\int_a^b y\sqrt{1 + \left(\dfrac{dy}{dx}\right)^2}\,dx$

$y = x,\ \dfrac{dy}{dx} = 1,\ a = 0, b = 2$

$S = 2\pi \displaystyle\int_0^2 x(1 + 1^2)^{1/2}\,dx = 2\pi\sqrt{2}\int_0^2 x\,dx$

$S = 2\pi\sqrt{2}\,\dfrac{x^2}{2}\Big|_0^2 = 4\pi\sqrt{2}$

6. $S = 2\pi \displaystyle\int_a^b y\sqrt{1 + \left(\dfrac{dy}{dx}\right)^2}\,dx$

$y = \dfrac{4}{3}x,\ \dfrac{dy}{dx} = \dfrac{4}{3},\ a = 0, b = 3$

$S = 2\pi \displaystyle\int_0^3 \dfrac{4}{3}x\sqrt{1 + \left(\dfrac{4}{3}\right)^2}\,dx = \dfrac{8\pi}{3}\dfrac{5}{3}\int_0^3 x\,dx$

$S = \dfrac{40\pi}{9}\dfrac{x^2}{2}\Big|_0^3 = \dfrac{20\pi}{9}(3^2) = 20\pi$

7. $S = 2\pi \int_a^b y \sqrt{1 + \left(\dfrac{dy}{dx}\right)^2}\, dx$

$y = 2\sqrt{x}, \dfrac{dy}{dx} = \dfrac{1}{x^{1/2}}, a = 0, b = 3$

$S = 2\pi \int_0^3 (2x^{1/2}) \left[1 + \left(\dfrac{1}{x^{1/2}}\right)^2\right]^{1/2} dx$

$S = 4\pi \int_0^3 x^{1/2} \left(1 + \dfrac{1}{x}\right)^{1/2} dx = 4\pi \int_0^3 (x+1)^{1/2} dx$

$S = 4\pi \left(\dfrac{2}{3}\right) (x+1)^{3/2} \Big|_0^3 = \dfrac{8\pi}{3}(8-1) = \dfrac{56\pi}{3}$

8. $S = 2\pi \int_a^b y \sqrt{1 + \left(\dfrac{dy}{dx}\right)^2}\, dx$

$y = x^3, \dfrac{dy}{dx} = 3x^2, a = 1, b = 3$

$S = 2\pi \int_1^3 x^3 \sqrt{1 + (3x^2)^2}\, dx = 2\pi \int_1^3 x^3 \sqrt{1 + 9x^4}\, dx$

$S = \dfrac{\pi}{18} \int_1^3 \sqrt{1 + 9x^4}(36x^3\, dx)$

$S = \dfrac{\pi}{18} \left(\dfrac{2}{3}\right)(1 + 9x^4)^{3/2} \Big|_1^3$

$\quad = \dfrac{\pi}{27}[(1 + 9(3)^4)^{3/2} - (1 + 9(1)^4)^{3/2}]$

$S = 2291$

9. $i = 4t - t^2$. Find i_{av} with respect to time for $t = 0$ to $t = 4$.

$i_{av} = \dfrac{\int_0^{4.0} i\, dt}{4.0 - 0} = \dfrac{\int_0^{4.0} (4t - t^2)\, dt}{4.0}$

$\quad = \dfrac{2t^2 - \frac{1}{3}t^3 \Big|_0^{4.0}}{4.0}$

$\quad = \dfrac{2(16) - \frac{1}{3}(64)}{4.0} = \dfrac{10.7}{4.0} = 2.7 \text{ A}$

10. $T = 0.001\,00t^4 - 0.280t^2 + 25.0, -12h \le t \le +12 \text{ h}$

$T_{AV} = \dfrac{\int_{-12}^{12} (0.001\,00t^4 - 0.280t^2 + 25.0)dt}{24}$

$\quad = \dfrac{1}{12}\left(\dfrac{0.001\,00t^5}{5} - \dfrac{0.280t^3}{3} + 25.0t\right)\Big|_0^{12}$

$\quad = \dfrac{1}{12}\left(\dfrac{0.001 \times 12^5}{5} - \dfrac{0.280}{3} \cdot 12^3 + 25(12)\right)$

$\quad = 15.7°\text{C}$

11. $e = 0.768s - 0.00004s^3, s_1 = 30.0 \text{ km/h}, s_2 = 90.0 \text{ km/h}$

$e_{av} = \int_{30.0}^{90.0} \dfrac{(0.768s - 0.00004s^3)}{60.0}\, ds$

$e = \dfrac{1}{60.0}\left(\dfrac{0.768s^2}{2} - 0.00001s^4\right)\Big|_{30.0}^{90.0}$

$\quad = \dfrac{1}{60.0}\left\{\dfrac{0.768}{2}(90.0)^2 - 0.00001(90.0)^4\right.$

$\qquad \left. - \left(\dfrac{0.768}{2}(30.0)^2 - 0.00001(30.0)^4\right)\right\}$

$\quad = \dfrac{1}{60.0}(2.454 \times 10^3 - 3.375 \times 10^2)$

$\quad = 35.3\%$

12. $V = \dfrac{4}{3}\pi x^3$

$V_{Avg} = \dfrac{4}{3}\pi \int_{-r}^{r} \dfrac{x^3\, dx}{2r} \quad \leftarrow\text{variable } x$
$\qquad\qquad\qquad\qquad \leftarrow\text{constant } r$

$\quad = \dfrac{4}{3}\dfrac{\pi}{r}\int_0^r x^3\, dx$

$\quad = \dfrac{4\pi}{3r}\dfrac{x^4}{4}\Big|_0^r$

$\quad = \dfrac{\pi x^4}{3r}\Big|_0^r$

$\quad = \dfrac{\pi r^4}{3r}$

$\quad = \dfrac{1}{3}\pi r^3$

The average volume is $\frac{1}{4}$ of the volume.

13. $\quad s = \int_a^b \sqrt{1 + \left(\dfrac{dy}{dx}\right)^2}\, dx;$

$\quad y = 0.4x^{3/2};$

$\quad \dfrac{dy}{dx} = 0.06x^{1/2}$

$\quad s = \int_0^{100} \sqrt{1 + (0.06x^{1/2})^2}\, dx$

$\quad\quad = \int_0^{100} \sqrt{(1 + 0.0036x)}\, dx$

$\quad\quad = \int_0^{100} (1 + 0.0036x)^{1/2}\, dx$

$\quad\quad = \dfrac{1}{0.0036} \int_0^{100} (1 + 0.0036x)^{1/2}(0.0036)(dx)$

$\quad\quad = \dfrac{1}{0.0036} \left[\dfrac{2}{3}(1 + 0.0036x)^{3/2}\right]\Big|_0^{100}$

$\quad\quad = \dfrac{1}{0.0036} \left[\dfrac{2}{3}(1.586) - \dfrac{2}{3}(1)^{3/2}\right]$

$\quad\quad = \dfrac{1}{0.0036} \left[\dfrac{2}{3}(0.586)\right]$

$\quad\quad = \dfrac{1}{0.0036}[0.391]$

$\quad\quad = 109 \text{ ft}$

14. $y = \dfrac{2}{3}(x^2 - 1)^{3/2}, x = 1.0 \text{ km to } x = 3.0 \text{ km}$

$\quad s = \int_a^b \sqrt{1 + \left(\dfrac{dy}{dx}\right)^2}\, dx$

$\quad \dfrac{dy}{dx} = (x^2 - 1)^{1/2}(2x)$

$\quad \left(\dfrac{dy}{dx}\right)^2 = (x^2 - 1)4x^2$

$\quad\quad\quad = 4x^4 - 4x^2$

$\quad s = \int_{1.0}^{3.0} \sqrt{1 + (x^2 - 1)4x^2}\, dx$

$\quad\quad = \int_{1.0}^{3.0} (2x^2 - 1)\, dx$

$\quad\quad = \left(\dfrac{2}{3}x^3 - x\right)\Big|_{1.0}^{3.0}$

$\quad\quad = (18.0 - 3.0) - (0.7 - 1.0)$

$\quad\quad = 15.3 \text{ km}$

15. $y = \dfrac{r}{h}x$

$\quad \dfrac{dy}{dx} = \dfrac{r}{h}; \left(\dfrac{dy}{dx}\right)^2 = \dfrac{r^2}{h^2}$

$\quad S = 2\pi \int_0^h \dfrac{r}{h}x \sqrt{1 + \dfrac{r^2}{h^2}}\, dx$

$\quad\quad = 2\pi \dfrac{r}{h}\dfrac{1}{h} \sqrt{h^2 + r^2} \int x\, dx$

$\quad\quad = \dfrac{2\pi r}{h^2} \sqrt{h^2 + r^2} \dfrac{x^2}{2}\Big|_0^h$

$\quad\quad = \dfrac{\pi r}{h^2} \sqrt{h^2 + r^2}\, x^2\Big|_0^h$

$\quad\quad = \dfrac{\pi r}{h^2} \sqrt{h^2 + r^2}\, h^2$

$\quad\quad = \pi r \sqrt{h^2 + r^2}$

16. $y = 0.2x^3$
$\quad x = 0 \text{ cm to } x = 2.0 \text{ cm}$

$\quad \dfrac{dy}{dx} = 0.6x^2; \left(\dfrac{dy}{dx}\right)^2 = 0.36x^4$

$\quad S = 2\pi \int_0^{2.0} 0.2x^3 \sqrt{1 + 0.36x^4}\, dx$

$\quad\quad = 0.4\pi \int_0^{2.0} (1 + 0.36x^4)^{1/2} x^3\, dx$

$\quad u = 1 + 0.36x^4; \, du = 1.44x^3 dx$

$\quad S = \dfrac{0.4\pi}{1.44} \int_0^{2.0} (1 + 0.36x^4)^{1/2} 1.44x^3\, dx$

$\quad\quad = \dfrac{0.4\pi}{1.44} \cdot \dfrac{2}{3}(1 + 0.36x^4)^{3/2}\Big|_0^{2.0}$

$\quad\quad = \dfrac{0.8\pi}{4.32}(6.76^{3/2} - 1)$

$\quad\quad = 9.6 \text{ cm}^3$

Chapter 6 Review Exercises

1. $\quad v = \int a\, dt; \, a = 32 \text{ ft/s}^2;$

$\quad\quad v = at + c, \text{ at } t = 0,$
$\quad\quad v = 0 \Rightarrow c = 0$
$\quad\quad v = at$
$\quad\quad\quad = 32t$
$\quad 140 = 32t$
$\quad\quad t = 4.4s$

2. $s = \int v \, dt = \int \sqrt{400 - 20t} \, dt$

$\qquad = -\dfrac{1}{20} \int \sqrt{400 - 20t} \, (-20 \, dt)$

$s = -\dfrac{1}{20} \left(\dfrac{2}{3} \right) (400 - 20t)^{3/2} + C$, at $t = 0, s = 0$

$0 = -\dfrac{1}{30} (400)^{3/2} + C, C = \dfrac{400^{3/2}}{30}$

$s = -\dfrac{1}{30} (400 - 20t)^{3/2} + \dfrac{400^{3/2}}{30}$

$v = 0 = \sqrt{400 - 20t}, t = 20$

$s = -\dfrac{1}{30} (400 - 20(20))^{3/2} + \dfrac{400^{3/2}}{30}$

$s = 267$ m

3. $\dfrac{d^2 s}{dt^2} = -32$

$\dfrac{ds}{dt} = -32t + v_0$, at $t = 0, \dfrac{ds}{dt} = 20, v_0 = 20$

$\dfrac{ds}{dt} = -32t + 20$

$s = -16t^2 + 20t + s_0$, at $t = 0, s = 200, s_0 = 200$

$s = -16t^2 + 20t + 200, t \geq 0$

$-16t^2 + 20t + 200 = 0$

$\qquad\qquad\qquad t = 4.2$ s

4. $\dfrac{d^2 s}{dt^2} = a$

$\dfrac{ds}{dt} = at + v_0$

$s = \dfrac{1}{2} at^2 + v_0 t + s_0$

$s = \dfrac{1}{2} (0.020)(30)^2 + 5.0(30) + 0$

$s = 160$ ft

5. $i = 0.25(2\sqrt{t} - t); q = \int i \, dt$

$q = \displaystyle\int_0^2 0.25(2\sqrt{t} - t) \, dt$

$\quad = \displaystyle\int_0^2 0.25(2t^{1/2} - t) \, dt$

$\quad = \displaystyle\int_0^2 (0.50 t^{1/2} - 0.25t) \, dt$

$\quad = \left(\dfrac{1}{3} t^{3/2} - \dfrac{1}{8} t^2 \right) \Big|_0^2$

$\quad = \dfrac{\sqrt{8}}{3} - \dfrac{1}{2} = 0.44$ C

6. $i = \sqrt{1 + 4t} = \dfrac{dq}{dt}$

$q = \int \sqrt{1 + 4t} \, dt = \dfrac{1}{6} (1 + 4t)^{3/2} + q_0$

$q = q_0 = 0$ when $t = 0$ determines the charge passing a point in 2 seconds.

$0 = \dfrac{1}{6} (1 + 4(0))^{3/2} + q_0$ from which $q_0 = -\dfrac{1}{6}$

$q = \dfrac{1}{6} (1 + 4t)^{3/2} - \dfrac{1}{6}$ and in $2.0s = 3.0s - 1.0s$

$q = \dfrac{1}{6} (1 + 4(2))^{3/2} - \dfrac{1}{6}$

$q = 4.33$ C

7. $V_C = \dfrac{1}{5.5 \times 10^{-9}} \int 12 \times 10^{-3} \, dt$

$V_C = \dfrac{12 \times 10^{-3}}{5.5 \times 10^{-9}} t + V_0$, at $t = 0, V_C = 0$ so $V_0 = 0$

$V_C = \dfrac{12 \times 10^{-3}}{5.5 \times 10^{-9}} t \Big|_{2.5 \times 10^{-6}}$

$V_C = 55$ V

8. $V_C = \dfrac{1}{C} \int t(t^2 + 1)^{-1/2} \, dt$

$\quad = \dfrac{1}{2C} \int (t^2 + 1)^{-1/2} (2t \, dt)$

$V_C = \dfrac{1}{2C} (2)(t^2 + 1)^{1/2} + V_0$, at $t = 0, V_C = 0$,

$0 = \dfrac{1}{C} + V_0, V_0 = -\dfrac{1}{C}$

$V_C = \dfrac{1}{C} (t^2 + 1)^{1/2} - \dfrac{1}{C}$, at $t = 8.00 \times 10^{-3}$,

$V_C = 2.50$

$2.50 = \dfrac{1}{C} ((8.00 \times 10^{-3})^2 + 1)^{1/2} - \dfrac{1}{C}$ from which

$C = 12.8 \ \mu$F

9. $dy = \left(20 + \dfrac{1}{40} x^2 \right) dx;$

$y = \int \left(20 + \dfrac{1}{40} x^2 \right) dx = 20x + \dfrac{1}{120} x^3 + C$

$y = 0$ when $x = 0$

$20(0) + \dfrac{1}{120} (0)^3 + C = 0; C = 0$

$y = 20x + \dfrac{1}{120} x^3$

10. $\dfrac{dR}{dt} = -2.5(0.05t + 1)^{-1.5}$

$R = -2.5\left(\dfrac{1}{0.05}\right) \int (0.05t + 1)^{-1.5}(0.05\,dt)$

$R = \dfrac{-2.5}{0.05}\left(\dfrac{1}{-0.5}\right)(0.05t + 1)^{-0.5} + R_0$

at $t = 0, R = 100$

$100 = \dfrac{-2.5}{0.05}\left(\dfrac{1}{-0.5}\right)(0.05(0) + 1)^{-0.5} + R_0$

$R_0 = 0$

$R = \dfrac{-2.5}{0.05}\left(\dfrac{1}{-0.5}\right)(0.05t + 1)^{-0.5}\Big|_{t=100}$

$R = 40.8\%$

11. $A = \displaystyle\int_0^1 \sqrt{1 - x}\,dx = -\dfrac{2}{3}(1 - x)^{3/2}\Big|_0^1$

$A = -\dfrac{2}{3}[(1 - 1)^{3/2} - (1 - 0)^{3/2}]$

$A = \dfrac{2}{3}$

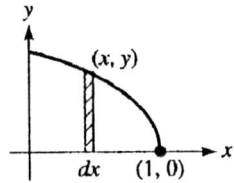

12. $A = \displaystyle\int_0^3 (3x^2 - x^3)\,dx$

$A = x^3 - \dfrac{x^4}{4}\Big|_0^3$

$A = \dfrac{27}{4}$

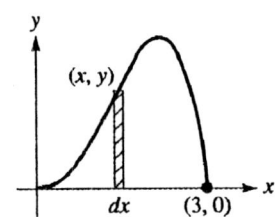

13. $y^2 = 2x, y = x - 4$

$x_1 = \dfrac{1}{2}y^2; \; x_2 = y + 4;$

$A = \displaystyle\int_{-2}^4 (x_2 - x_1)\,dy$

$= \displaystyle\int_{-2}^4 \left(y + 4 - \dfrac{1}{2}y^2\right)\,dy$

$= -\dfrac{1}{6}y^3 + \dfrac{1}{2}y^2 + 4y\Big|_{-2}^4$

$= -\dfrac{1}{6}(64) + \dfrac{1}{2}(16) + 4(4) + \dfrac{1}{6}(-8) - \dfrac{1}{2}(4) - 4(-2)$

$= 18$

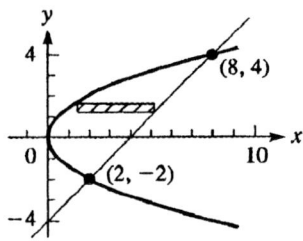

14. $A = \displaystyle\int_1^2 \dfrac{1}{(2x + 1)^2}\,dx$

$= \dfrac{1}{2}\displaystyle\int_1^2 (2x + 1)^{-2}(2\,dx)$

$A = \dfrac{-1}{2}(2x + 1)^{-1}\Big|_1^2$

$= \dfrac{-1}{2}(2(2) + 1)^{-1} + \dfrac{1}{2}(2(1) + 1)^{-1}$

$A = \dfrac{1}{15}$

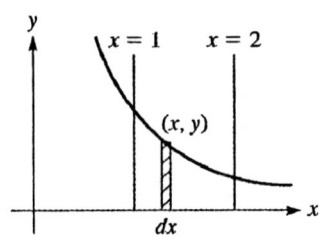

15. $y = x^2$ and $y = x^3 - 2x^2$ intersect at $(0,0)$ and $(3,9)$

$$A = \int_0^3 [x^2 - (x^3 - 2x^2)]dx$$

$$= \int_0^3 [x^2 - x^3 + 2x^2]dx$$

$$A = \int_0^3 [3x^2 - x^3]dx$$

$$= x^3 - \frac{x^4}{4}\Big|_0^3$$

$$= 3^3 - \frac{3^4}{4}$$

$$= \frac{27}{4}$$

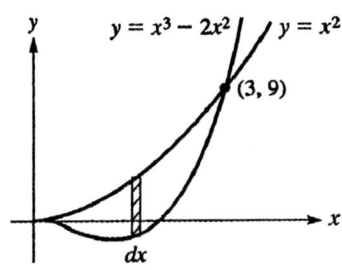

16. $y = x^2 + 2$ and $y = 3x^2$ intersects at $(-1,3)$ and $(1,3)$

$$A = \int_{-1}^1 [(x^2 + 2) - 3x^2]dx$$

$$= \int_{-1}^1 (-2x^2 + 2)dx$$

$$A = -2\frac{x^3}{3} + 2x\Big|_{-1}^1$$

$$= -2\frac{1^3}{3} + 2(1) + 2\frac{(-1)^3}{3} - 2(-1)$$

$$A = \frac{8}{3}$$

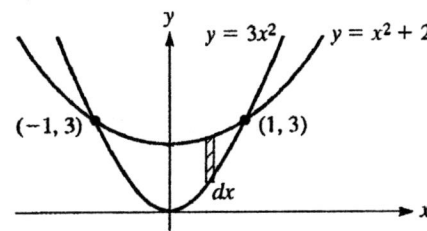

17.

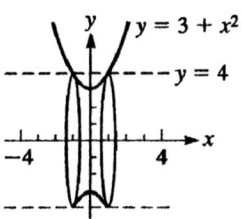

$$V = \pi \int_{-1}^1 4^2 dx - \pi \int_{-1}^1 y^2 dx = \pi \int_{-1}^1 (4^2 - y^2)\, dx$$

$$= \pi \int_{-1}^1 [16 - (3 + x^2)^2]\, dx$$

$$= \pi \int_{-1}^1 (16 - 9 - 6x^2 - x^4)\, dx$$

$$= \pi \int_{-1}^1 (7 - 6x^2 - x^4)\, dx$$

$$= \pi \left(7x - 2x^3 - \frac{1}{5}x^5\right)\Big|_{-1}^1$$

$$= \pi \left(7 - 2 - \frac{1}{5}\right) - \pi \left(-7 + 2 + \frac{1}{5}\right)$$

$$= \frac{24\pi}{5} + \frac{24\pi}{5} = \frac{48\pi}{5}$$

18. $y = 8x - x^4$ has $(0,0)$ and $(2,0)$ as intercepts. Using disks,

$$V = \int_0^2 \pi(8x - x^4)^2 dx = \int_0^2 \pi(64x^2 - 16x^5 + x^8)dx$$

$$V = \pi \left[64\frac{x^3}{3} - 16\frac{x^6}{6} + \frac{x^9}{9}\right]\Big|_0^2$$

$$= \pi \left[64\frac{2^3}{3} - 16\frac{2^6}{6} + \frac{2^9}{9}\right]$$

$$V = \frac{512\pi}{9}$$

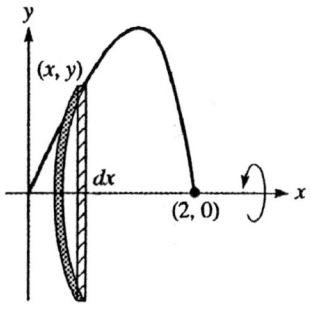

19. $y = x^3 - 4x^2$ has $(0,0)$ and $(4,0)$ as intercepts. Using shells,

$$V = \int_0^4 2\pi(-y\,dy) = \int_0^4 -2\pi x(x^3 - 4x^2)dx$$

$$V = -2\pi\int_0^4 (x^4 - 4x^3)dx = -2\pi\left(\frac{x^5}{5} - x^4\right)\Big|_0^4$$

$$V = -2\pi\left(\frac{4^5}{5} - 4^4\right) = \frac{512\pi}{5}$$

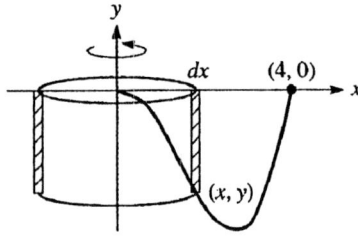

20. $y = x$ and $y = 3x - x^2$ intersect at $(0,0)$ and $(2,2)$. Using shells,

$$V = \int_0^2 2\pi x[(3x - x^2) - x]dx$$

$$= 2\pi\int_0^2 x(2x - x^2)dx$$

$$= 2\pi\int_0^2 (2x^2 - x^3)dx$$

$$V = 2\pi\left[2\frac{x^3}{3} - \frac{x^4}{4}\right]\Big|_0^2$$

$$= 2\pi\left[2\frac{2^3}{3} - \frac{2^4}{4}\right]$$

$$V = \frac{8\pi}{3}$$

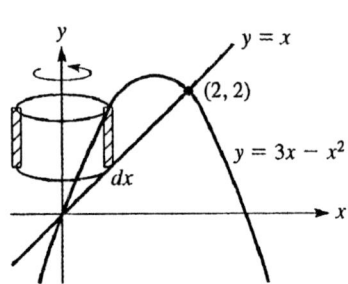

21.
$$\frac{x^2}{a^2} - \frac{y^2}{b^2} = 1$$

$$b^2x^2 + a^2y^2 = a^2b^2$$

$$y = \sqrt{\frac{a^2b^2 - b^2x^2}{a^2}}$$

$$V = \pi\int_{-a}^a y^2\,dx$$

$$= \pi\int_{-a}^a \left(\frac{a^2b^2 - b^2x^2}{a^2}\right)dx$$

$$= \pi\int_{-a}^a \left(b^2 - \frac{b^2}{a^2}x^2\right)dx$$

$$= \pi\left(b^2x - \frac{b^2}{3a^2}x^3\right)\Big|_{-a}^a$$

$$= \pi\left[\left(ab^2 - \frac{ab^2}{3}\right) - \left(-ab^2 + \frac{ab^2}{3}\right)\right]$$

$$= \pi\left(2ab^2 - \frac{2ab^2}{3}\right)$$

$$= \pi\left(\frac{4ab^2}{3}\right)$$

$$= \frac{4\pi ab^2}{3}$$

$$= \frac{4}{3}\pi ab^2$$

22. $V_{\text{hole}} = V_{\text{cylinder}} + 2(V_{\text{spherical cap}})$
Using disks for the spherical cap,

$$V_{\text{hole}} = \pi(1.00)^2(2\sqrt{15.0}) + 2\int_{\sqrt{15.0}}^{4.00} \pi x^2\,dy$$

$$V_{\text{hole}} = 2.00\pi\sqrt{15.0} + 2\int_{\sqrt{15.0}}^{4.00} \pi(16.0 - y^2)dy$$

$$V_{\text{hole}} = 2.00\pi\sqrt{15.0} + 2\pi\left(16.0y - \frac{y^3}{3}\right)\Big|_{\sqrt{15.0}}^{4.00}$$

$$V_{\text{hole}} = 2.00\pi\sqrt{15.0} + 2\pi\cdot\left(16.0(4.00) - \frac{(4.00)^3}{3}\right.$$
$$\left. -16.0\sqrt{15.0} + \frac{\sqrt{15.0}^3}{3}\right)$$

$$V_{\text{hole}} = 2.00\pi\sqrt{15.0} + 2\pi$$
$$\left(64.0 - \frac{64.0}{3} - 16.0\sqrt{15.0} + \frac{15.0\sqrt{15.0}}{3}\right)$$

$$V_{\text{hole}} = 2.00\pi\sqrt{15.0} + 2\pi$$

$$\left(\frac{2(64.0)}{3} - 16.0\sqrt{15.0} - 5.00\sqrt{15.0}\right)$$

$$V_{\text{hole}} = 2.00\pi\left(\sqrt{15.0} + \frac{2(64.0)}{3} - 11.0\sqrt{15.0}\right)$$

$$V_{\text{hole}} = 2.00\pi\left(\frac{2(64.0)}{3} - 10.0\sqrt{15.0}\right)$$

$$V_{\text{hole}} = \frac{4.00\pi}{3}(64.0 - 15.0\sqrt{15.0})$$

$$V_{\text{hole}} = 24.7\ \text{cm}^3$$

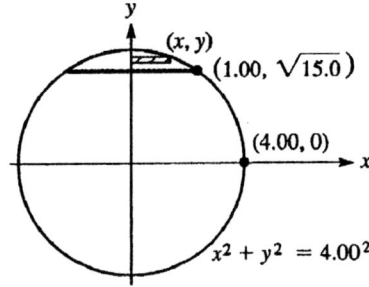

23. $y^2 = x^3$ and $y = 2x$ intersect at $(0,0)$ and $(4,8)$

$$\overline{x} = \frac{\displaystyle\int_0^4 x(2x - x^{3/2})\,dx}{\displaystyle\int_0^4 (2x - x^{3/2})\,dx}$$

$$= \frac{\displaystyle\int_0^4 (2x^2 - x^{5/2})\,dx}{\displaystyle\int_0^4 (2x - x^{3/2})\,dx}$$

$$\overline{x} = \frac{\frac{2}{3}x^3 - \frac{2}{7}x^{7/2}\Big|_0^4}{x^2 - \frac{2}{5}x^{5/2}\Big|_0^4}$$

$$= \frac{\frac{128}{3} - \frac{256}{7}}{16 - \frac{64}{5}}$$

$$\overline{x} = \frac{128}{21}\left(\frac{5}{16}\right) = \frac{40}{21}$$

$$\overline{y} = \frac{\displaystyle\int_0^8 y(y^2 - \tfrac{1}{2}y)\,dy}{\frac{16}{5}} = \frac{\displaystyle\int_0^8 (y^{5/3} - \tfrac{1}{2}y^2)\,dy}{\frac{16}{5}}$$

$$\overline{y} = \frac{\frac{3}{8}y^{8/3} - \frac{1}{6}y^3\Big|_0^8}{\frac{16}{5}} = \frac{96 - \frac{256}{3}}{\frac{16}{5}}$$

$$\overline{y} = \frac{32}{3}\left(\frac{5}{16}\right) = \frac{10}{3}$$

Centroid $\left(\frac{40}{21}, \frac{10}{3}\right)$

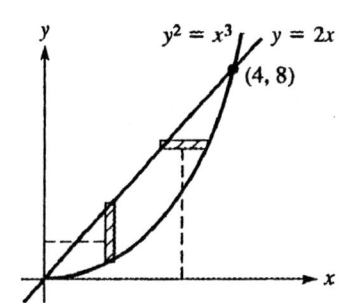

24. $y = 2x - 4, x = 1, y = 0$

$$\overline{x} = \frac{\displaystyle\int_1^2 x(0 - (2x - 4))\,dx}{\displaystyle\int_1^2 (0 - (2x - 4)\,dx}$$

$$\overline{x} = \frac{\displaystyle\int_1^2 (-2x^2 + 4)\,dx}{\displaystyle\int_1^2 (-2x + 4)\,dx}$$

$$\overline{x} = \frac{-2\frac{x^3}{3} + 2x^2\Big|_1^2}{-x^2 + 4x\Big|_1^2}$$

$$= \frac{-2\frac{2^3}{3} + 2(2)^2 + 2\frac{1^3}{3} - 2(1)^2}{-(2)^2 + 4(2) + 1^2 - 4(1)} = \frac{\frac{4}{3}}{1}$$

$$\overline{x} = \frac{4}{3}$$

$$\overline{y} = \frac{\displaystyle\int_{-2}^0 y\left(\frac{y+4}{2} - 1\right)dy}{1} = \int_{-2}^0 \left(\frac{1}{2}y^2 + y\right)dy$$

$$= \frac{1}{6}y^3 + \frac{1}{2}y^2\Big|_{-2}^0$$

$$= -\frac{1}{6}(-2)^3 - \frac{1}{2}(-2)^2$$

$$= -\frac{2}{3}$$

centroid $\left(\frac{4}{3}, -\frac{2}{3}\right)$

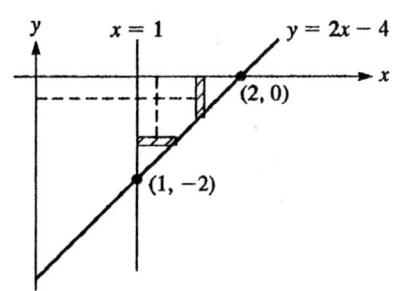

25. $y = \sqrt{x}, x = 1, x = 4, y = 0$

Volume symmetric to x-axis: $\bar{y} = 0$

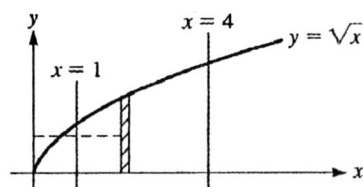

$$\bar{x} = \frac{\displaystyle\int_1^4 x y^2 \, dx}{\displaystyle\int_1^4 y^2 \, dx} = \frac{\displaystyle\int_1^4 x(x) \, dx}{\displaystyle\int_1^4 x \, dx}$$

$$= \frac{\displaystyle\int_1^4 x^2 \, dx}{\displaystyle\int_1^4 x \, dx}$$

$$\bar{x} = \frac{\left.\dfrac{x^3}{3}\right|_1^4}{\left.\dfrac{x^2}{2}\right|_1^4} = \frac{\dfrac{4^3}{3} - \dfrac{1^3}{3}}{\dfrac{4^2}{2} - \dfrac{1^2}{2}} = \frac{\dfrac{63}{3}}{\dfrac{15}{2}} = \frac{14}{5}$$

centroid $\left(\frac{14}{5}, 0\right)$

26. $y x^4 = 1, y = 1, y = 4$

Volume symmetric to y-axis: $\bar{x} = 0$

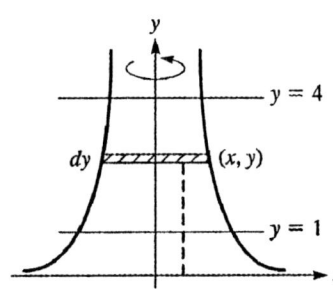

$$\bar{y} = \frac{\displaystyle\int_1^4 y x^2 \, dy}{\displaystyle\int_1^4 x^2 \, dy} = \frac{\displaystyle\int_1^4 y \frac{1}{\sqrt{y}} \, dy}{\displaystyle\int_1^4 \frac{1}{\sqrt{y}} \, dy}$$

$$\bar{y} = \frac{\displaystyle\int_1^4 y^{1/2} \, dy}{\displaystyle\int_1^4 y^{-1/2} \, dy} = \frac{\left.\frac{2}{3} y^{3/2}\right|_1^4}{\left.2 y^{1/2}\right|_1^4}$$

$$\bar{y} = \frac{1}{3} \frac{4^{3/2} - 1^{3/2}}{4^{1/2} - 1^{1/2}} = \frac{7}{3}$$

centroid $\left(0, \frac{7}{3}\right)$

27. $y = 3x - x^2$ and $y = x$ intersect

when $3x - x^2 = x$
$$x^2 - 2x = 0$$
$$x(x - 2) = 0$$
$$x = 2, x = 0$$

points of intersection: $(0,0), (2,0)$

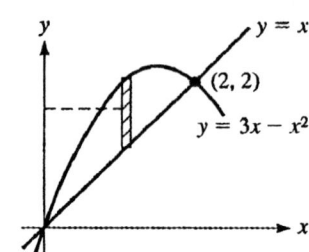

$$I_y = k \int_0^2 x^2[(3x - x^2) - x] \, dx$$

$$I_y = k \int_0^2 (2x^3 - x^4) \, dx = \left.\frac{1}{2} x^4 - \frac{1}{5} x^5\right|_0^2$$

$$I_y = k\left(8 - \frac{32}{5}\right) = \frac{8}{5} k$$

28. $I_y = k \int_0^2 x^2 (y_2 - y_1) \, dx = k \int_0^2 x^2 (8 - x^3 - 0) \, dx$

$$I_y = k \int_0^2 (8x^2 - x^5) \, dx$$

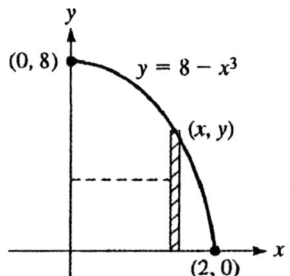

$$I_y = k\left[8\frac{x^3}{3} - \frac{x^6}{6}\right]\Big|_0^2 \quad I_y = k\left[8\frac{2^3}{3} - \frac{2^6}{6}\right] = \frac{32k}{3}$$

$$m = k \int_0^2 (8 - x^3) \, dx = k\left[8x - \frac{x^4}{4}\right]\Big|_0^2$$

$$m = k\left[8(2) - \frac{2^4}{4}\right] = 12k$$

$$R_y^2 = \frac{I_y}{m} = \frac{\frac{32k}{3}}{12k} = \frac{8}{9}$$

$$R_y = \frac{2\sqrt{2}}{3}$$

29. $I_x = 2\pi k \displaystyle\int_c^d (x_2 - x_1)y^3\, dy$

$I_x = 2\pi(0.0114) \displaystyle\int_0^{3.00(20.0)^{0.10}}$

$\left(20.0 - \left(\dfrac{y}{3.00}\right)^{10}\right) y^3\, dy$

$I_x = 68.7 \text{ g}\cdot\text{mm}^2$

30. $I_x = 2\pi k \displaystyle\int_{0.25}^{1.00} y^3 \left(\dfrac{1}{y} - 1.00\right) dy$

$I_x = 2\pi k \displaystyle\int_{0.25}^{1.00} (y^2 - y^3)\, dy$

$I_x = 2\pi k \left[\dfrac{y^3}{3} - \dfrac{y^4}{4} \right]\Big|_{0.25}^{1.00}$

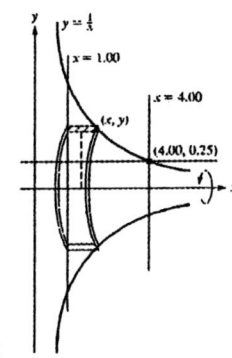

$I_x = 2\pi k \left[\dfrac{1.00^3}{3} - \dfrac{1.00^4}{4} - \dfrac{0.25^3}{3} + \dfrac{0.25^4}{4} \right]$

$m = 2\pi k \displaystyle\int_{0.25}^{1.00} y \left(\dfrac{1}{y} - 1.00\right) dy$

$= 2\pi k \left[y - \dfrac{y^2}{2} \right]\Big|_{0.25}^{1.00}$

$m = 2\pi k \left[1.00 - \dfrac{1.00^2}{2} - 0.25 + \dfrac{0.25^2}{2} \right]$

$R_x^2 = \dfrac{I_x}{m}$

$= \dfrac{2\pi k \left[\dfrac{1.00^3}{3} - \dfrac{1.00^4}{4} - \dfrac{0.25^3}{3} + \dfrac{0.25^4}{4} \right]}{2\pi k \left[1.00 - \dfrac{1.00^2}{2} - 0.25 + \dfrac{0.25^2}{2} \right]}$

$R_x = 0.53 \text{ in.}$

31. rope has mass per unit length $= \dfrac{1}{10}$ lb/ft

$W = W_{\text{bucket}} + W_{\text{rope}}$

$W = 80(100) + \displaystyle\int_0^{100} \dfrac{1}{10}(100 - x)\, dx$

$W = 8000 + \left(10x - \dfrac{x^2}{20}\right)\Big|_0^{100}$

$= 8500 \text{ ft}\cdot\text{lb}$

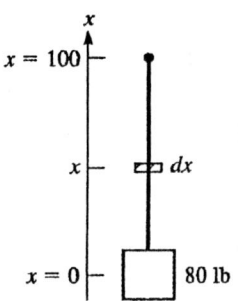

32. $F = \dfrac{10^{11}}{x^2};$

$W = \displaystyle\int_{3960}^{5960} \dfrac{10^{11}}{x^2}\, dx$

$= -\dfrac{10^{11}}{x}\Big|_{3960}^{5960}$

$W = 10^{11}\left(-\dfrac{1}{5960} + \dfrac{1}{3960}\right) = 8.47 \times 10^6 \text{ mi}\cdot\text{lb}$

33. $y_2 = 3x^2 - x^3;\ y_1 = 0;$

$\bar{x} = \dfrac{\displaystyle\int_0^3 x(y_2^2 - y_1)\, dx}{\displaystyle\int_0^3 x(y_2^2 - y_1)\, dx}$

$= \dfrac{\displaystyle\int_0^3 x(3x^2 - x^3)\, dx}{\displaystyle\int_0^3 (3x^2 - x^3)\, dx}$

$= \dfrac{\displaystyle\int_0^3 (3x^3 - x^4)\, dx}{\displaystyle\int_0^3 (3x^2 - x^3)\, dx}$

$= \dfrac{\left(\dfrac{3}{4}x^4 - \dfrac{1}{5}x^5\right)\Big|_0^3}{\left(x^3 - \dfrac{1}{4}x^4\right)\Big|_0^3}$

$= \dfrac{\dfrac{243}{4} - \dfrac{243}{5}}{27 - \dfrac{81}{4}} = \dfrac{12.15}{6.75} = 1.8 \text{ m}$

34. Using disks,

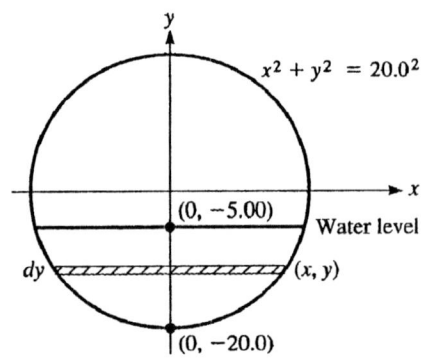

$$V = \int_{-20.0}^{-5.00} \pi x^2 \, dy$$

$$V = \pi \int_{-20.0}^{-5.00} (20.0^2 - y^2) \, dy$$

$$V = \pi \left[20.0^2 y - \frac{y^3}{3} \right]\Bigg|_{-20.0}^{-5.00}$$

$$V = \left[20.0^2(-5.000) - \frac{(-5.00)^3}{3} \right.$$

$$\left. - 20.0^2(-20.0) + \frac{(-20.0)^3}{3} \right]$$

$$V = 10{,}600 \text{ m}^3$$

35. Using disks and $\dfrac{x^2}{1.5^2} + \dfrac{y^2}{10^2} = 1$ as equation of ellipse

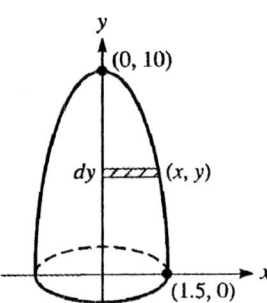

$$V = \int_0^{10} \pi x^2 \, dy$$

$$V = \pi \int_0^{10} 1.5^2 \left(1 - \frac{y^2}{10^2} \right) dy$$

$$V = \pi(1.5)^2 \left(y - \frac{1}{3}\frac{y^3}{10^2} \right)\Bigg|_0^{10}$$

$$V = \pi(1.5)^2 \left(10 - \frac{1}{3}\frac{10^3}{10^2} \right)$$

$$V = 47 \text{ m}^3$$

36. $x^2 = 4py$

$8^2 = 4p(-10)$

$4p = -6.4$

$x^2 = -6.4y$

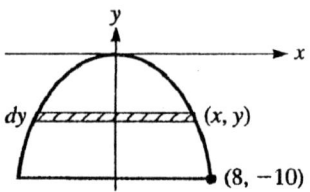

$$A = \int_{-10}^0 2x \, dy = 2\int_{-10}^0 \sqrt{-6.4y} \, dy$$

$$A = 2\sqrt{6.4} \int_{-10}^0 (-y)^{1/2} \, dy$$

$$A = -2\sqrt{6.4}\left(\frac{2}{3}\right)(-y^{3/2})\Bigg|_{-10}^0$$

$$A = 0 + \frac{4}{3}\sqrt{6.4}(10)^{3/2}$$

$$A = 107 \text{ ft}^2$$

37. The circumference of the bottom, $c = 2\pi r = 9\pi$, equates to l, the length of the vertical surface area.

$$F = 68.0 \int_0^{3.25} (9\pi h) \, dh$$

$$= 68.0 \left[4.50\pi h^2 \right]\Big|_0^{3.25}$$

$$= 68.0[4.50\pi(3.25)^2 - 0]$$

$$= 10{,}200 \text{ lb}$$

38. $$F = 62.4 \int_0^{4.00} yx \, dy$$

$$F = 62.4 \int_0^{4.00} y(6.00)\left(1 - \frac{y}{4.00}\right) dy$$

$$F = 62.4(6.00)\left[\frac{y^2}{2} - \frac{1}{3}\frac{y^3}{4.00}\right]\Bigg|_0^{4.00}$$

$$F = 998 \text{ lb}$$

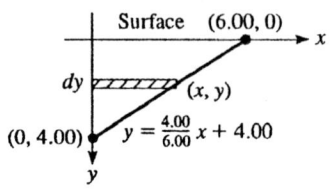

39. $R = \dfrac{k}{r^2}$, $R = 0.30 \; \Omega$ for $r = 2.00$ mm

$0.30 = \dfrac{k}{2.00^2}$, $k = (0.30)(2.00)^2 \; \Omega \cdot \text{mm}^2$

$R_{av} = \dfrac{\displaystyle\int_{2.0}^{2.1} \frac{k}{r^2}\, dr}{2.1 - 2.0} = \dfrac{-\frac{k}{r}\Big|_{2.0}^{2.1}}{0.1}$

$R_{av} = \dfrac{-(0.30)(2.00)^2 \left(\frac{1}{2.1} - \frac{1}{2.0} \right)}{0.1}$

$R_{av} = 0.29 \; \Omega$

40. $I_y = \displaystyle\int_{-\frac{L}{2}}^{\frac{L}{2}} x^2\, dm, \; dm = \dfrac{m}{L}\, dx$

$I_y = \displaystyle\int_{-\frac{L}{2}}^{\frac{L}{2}} x^2 \cdot \dfrac{m}{L}\, dx$

$I_y = \dfrac{m}{L} \left(\dfrac{x^3}{3} \right) \Big|_{-\frac{L}{2}}^{\frac{L}{2}}$

$ = \dfrac{m}{3L} \left(\left(\dfrac{L}{2} \right)^3 - \left(\dfrac{-L}{2} \right)^3 \right)$

$I_y = \dfrac{1}{12} m L^2$

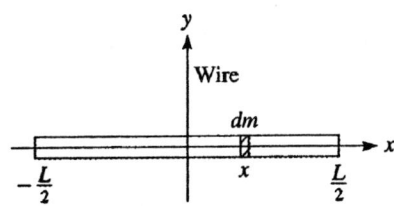

41. The formula is $V = 2 \int_0^a (2 \cdot \sqrt{a^2 - x^2})^2 \, dx$ where x is the distance from the center of the circular top to the cross section.

DIFFERENTIATION OF THE TRIGONOMETRIC AND INVERSE TRIGONOMETRIC FUNCTIONS

7.1 The Trigonometric Functions

1. $15°12' = 15° + 12' \cdot \dfrac{1°}{60'} = 15.2°$

2. $301°16' = 301\dfrac{16°}{60} = 301.27°$

3. $315.8° = 315° + \dfrac{8°}{10}\left(\dfrac{60'}{1°}\right) = 315°48'$

4. $24.92° = 24° + \dfrac{92°}{100}\left(\dfrac{60'}{1°}\right) = 24°55'$

5. $15° = \dfrac{\pi}{180°}(15) = \dfrac{\pi}{12}$

$150° = \dfrac{\pi}{180°}(150) = \dfrac{5\pi}{6}$

6. $12° = \dfrac{\pi}{180}(12) = \dfrac{12\pi}{180} = \dfrac{\pi}{15}$

$225° = \dfrac{\pi}{180}(225) = \dfrac{225\pi}{180} = \dfrac{5\pi}{4}$

7. $75° = \dfrac{\pi}{180}(75) = \dfrac{75\pi}{180} = \dfrac{5\pi}{12}$

$330° = \dfrac{\pi}{180}(330) = \dfrac{330\pi}{180} = \dfrac{11\pi}{6}$

8. $36° = \dfrac{\pi}{180}(36) = \dfrac{36\pi}{180} = \dfrac{\pi}{5}$

$315° = \dfrac{\pi}{180}(315) = \dfrac{315\pi}{180} = \dfrac{7\pi}{4}$

9. $\dfrac{2\pi}{5} = \dfrac{180°}{\pi}\left(\dfrac{2\pi}{2}\right) = 72°$

$\dfrac{3\pi}{2} = \dfrac{180°}{\pi}\left(\dfrac{3\pi}{2}\right) = 270°$

10. $\dfrac{3\pi}{10} = \dfrac{180°}{\pi}\left(\dfrac{3\pi}{10}\right) = \dfrac{540\pi°}{10\pi} = 54°$

$\dfrac{5\pi}{6} = \dfrac{180°}{\pi}\left(\dfrac{5\pi}{6}\right) = \dfrac{900\pi°}{6\pi} = 150°$

11. $\dfrac{\pi}{18} = \dfrac{180°}{\pi}\left(\dfrac{\pi}{18}\right) = \dfrac{180\pi°}{18\pi} = 10°$

$\dfrac{7\pi}{4} = \dfrac{180°}{\pi}\left(\dfrac{7\pi}{4}\right) = \dfrac{1260\pi°}{4\pi} = 315°$

12. $\dfrac{7\pi}{15} = \dfrac{180°}{\pi}\left(\dfrac{7\pi}{15}\right) = \dfrac{1260\pi°}{15\pi} = 84°$

$\dfrac{4\pi}{3} = \dfrac{180°}{\pi}\left(\dfrac{4\pi}{3}\right) = \dfrac{720\pi°}{3\pi} = 240°$

13. $23.0° = \dfrac{\pi}{180°}(23.0) = 0.401$

14. $178.5° = \dfrac{\pi}{180}(178.5) = 3.115$

15. $0.750 = \dfrac{180°}{\pi}(0.750) = 43.0°$

16. $16.4 = \dfrac{180°}{\pi}(16.4) = 939.7°$

17. $\cos 214° = -\cos(214° - 180°)$
$\qquad = -\cos 34° = -0.8290$

18. $\sin 286° = \sin(286° - 360°)$
$\qquad = \sin(-74°) = -0.9613$

19. $\csc 137° = \csc(180° - 137°)$
$\qquad = \csc 43° = \dfrac{1}{\sin 43°}$
$\qquad = 1.4663$

20. $\sec 476° = \sec(476° - 360°)$
$\qquad = \sec 116° = \dfrac{1}{\cos 116°}$
$\qquad = -2.281$

21. $\sin\dfrac{\pi}{4} = \sin\left[\left(\dfrac{\pi}{4}\right)\left(\dfrac{180}{\pi}\right)\right] = \sin 45°$
$\qquad = 0.7071$

22. $\cos\dfrac{\pi}{6} = \cos\left[\left(\dfrac{\pi}{6}\right)\left(\dfrac{180}{\pi}\right)\right] = \cos 30°$
$\qquad = 0.8660$

23. $\tan\dfrac{5\pi}{12} = \tan\left[\left(\dfrac{5\pi}{12}\right)\left(\dfrac{180}{\pi}\right)\right] = \tan 75°$
$\qquad = 3.732$

24. $\sin\dfrac{7\pi}{18} = \sin\left[\left(\dfrac{7\pi}{18}\right)\left(\dfrac{180}{\pi}\right)\right] = \sin 70°$
$\qquad = 0.9397$

25. $\cot \dfrac{5\pi}{6} = \cot\left[\dfrac{5\pi}{6}\left(\dfrac{180°}{\pi}\right)\right]$

$\qquad = \cot 150°$

$\qquad = -\cot(180° - 150°)$

$\qquad = -\cot 30° = -\dfrac{1}{\tan 30°}$

$\qquad = -1.732$

26. $\tan \dfrac{4\pi}{3} = \tan\left[\left(\dfrac{4\pi}{3}\right)\left(\dfrac{180}{\pi}\right)\right] = \tan 240°$

$\qquad = 1.732$

27. $\cos(4.596) = -\cos(4.596 - \pi)$

$\qquad = -\cos(1.4544)$

$\qquad = -0.1161$

28. $\cot(3.27) = \cot(3.27 + \pi)$

$\qquad = \cot(6.4116)$

$\qquad = 7.745$

29. $\sin\theta = 0.3090, \theta = 0.3141$

$\sin\theta$ is positive, θ is in QI and QII

In QI, $\theta = 0.3141$

In QII $\theta = \pi - 0.3141 = 2.827$

30. $\cos\theta = -0.9135, \theta = 0.4190$

$\cos\theta$ is negative, θ is in Quad II and
Quad III.

Quad II $= \pi - 0.4190 = 2.723$

Quad III $= \pi + 0.4190 = 3.561$

31. $\tan\theta = -0.2126, \theta = 0.2095$

$\tan\theta$ is negative, θ is in Quad II and Quad IV.

Quad II $= \pi - 0.2095 = 2.932$

Quad IV $= 2\pi - 0.2095 = 6.074$

32. $\sin\theta = -0.0436, \theta = 0.0436$

$\sin\theta$ is negative, θ is in Quad III and Quad IV.

Quad III $= \pi + 0.0436 = 3.185$

Quad IV $= 2\pi - 0.0436 = 6.240$

33. $\cos\theta = 0.6742, \theta = 0.8309$

$\cos\theta$ is positive, θ is in QI and QIV.

In QI, $\theta = 0.8309$

In QIV, $\theta = 2\pi - 0.8309 = 5.452$

34. $\cot\theta = 1.860, \theta = 0.493$

$\cot\theta$ is positive, θ is in Quad I and Quad III.

Quad I $= 0.4933$

Quad III $= \pi + 0.493 = 3.635$

35. $\sec\theta = -1.307, \theta = 0.700$

$\sec\theta$ is negative, θ is in Quad II and Quad III.

Quad II $= \pi - 0.700 = 2.442$

Quad III $= \pi + 0.700 = 3.841$

36. $\csc\theta = 3.940, \theta = 0.2566$

$\csc\theta$ is positive, θ is in Quad I and Quad II.

Quad I $= 0.2566$

Quad II $= \pi - 0.2566 = 2.885$

37. $y = 3\sin x$ has amplitude 3 and x-intercepts
$0, \pi, 2\pi$.

x	0	$\frac{\pi}{2}$	π	$\frac{3\pi}{2}$	2π
y	0	3	0	-3	0

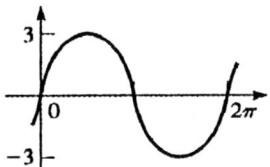

38. $y = 2\cos 2x$

$a = 2, b = 2, c = 0$

amplitude: $|2| = 2$

period: $\dfrac{2\pi}{2} = \pi$

displacement: $\dfrac{-0}{2} = 0$

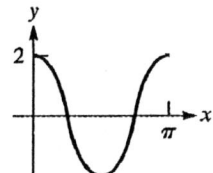

39. $y = -4\sin 3x$

$a = -4, b = 3, c = 0$

amplitude: $|-4| = 4$

period: $\dfrac{2\pi}{3}$

displacement: $-\dfrac{0}{3} = 0$

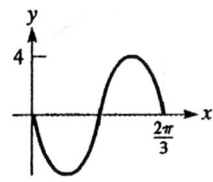

40. $y = 6\cos\pi x$

$a = 6, b = \pi, c = 0$

amplitude: $|6| = 6$

period: $\dfrac{2\pi}{\pi} = 2$

displacement: $\dfrac{-0}{\pi} = 0$

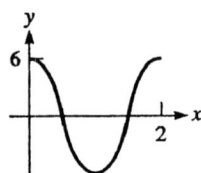

41. $y = \sin\left(3x - \dfrac{\pi}{2}\right)$

$a = 2, b = 3, c = -\dfrac{\pi}{2}$

amplitude: $|2| = 2$

period: $\dfrac{2\pi}{3}$

displacement: $-\dfrac{\frac{\pi}{2}}{3} = \dfrac{\pi}{6}$

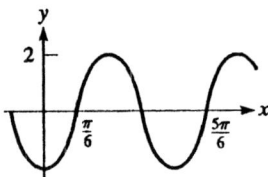

42. $y = 4\cos\left(\dfrac{x}{2} + \dfrac{\pi}{8}\right)$

$a = 4, b = \dfrac{1}{2}, c = \dfrac{\pi}{8}$

amplitude: $|4| = 4$

period: $\dfrac{2}{\frac{1}{2}} = 4\pi$

displacement: $\dfrac{-\frac{\pi}{8}}{\frac{1}{2}} = -\dfrac{\pi}{4}$

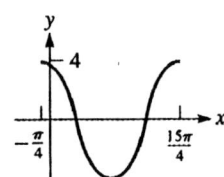

43. $y = \cos\left(\pi x - \dfrac{\pi}{2}\right)$

$a = 1, b = \pi, c = -\dfrac{\pi}{2}$

amplitude: $|1| = 1$

period: $\dfrac{2\pi}{\pi} = 2$

displacement: $-\dfrac{-\frac{\pi}{2}}{\pi} = \dfrac{1}{2}$

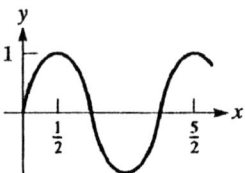

44. $y = \sin(\pi x + 1)$

$a = 1, b = \pi, c = 1$

amplitude: $|1| = 1$

period: $\dfrac{2\pi}{\pi} = 2$

displacement: $\dfrac{-1}{\pi}$

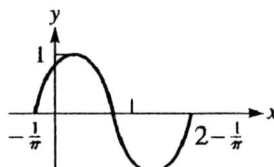

45. $y = 2.00\sin 2\pi\left(\dfrac{t}{0.100} - \dfrac{5.00}{20.0}\right); a = 2.00,$

$b = \dfrac{2\pi}{0.100}, c = \dfrac{-5.00(2\pi)}{20.0}$

Amplitude $= |a| = 2.00,$

period $= \dfrac{2\pi}{b} = 0.100$

displacement $= -\dfrac{c}{b} = 0.025$

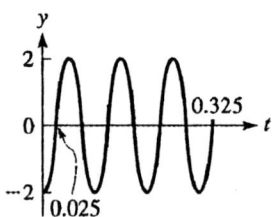

46. $i = 3.8 \cos 2\pi \, (t + 0.20)$; $a = 3.8, b = 2\pi, c = 0.4\pi$

Amplitude is $|a| = 3.8$; period is $\dfrac{2\pi}{b} = \dfrac{2\pi}{2\pi} = 1$;

displacement is $-\dfrac{c}{b} = -\dfrac{0.4\pi}{2\pi} = -0.2$.

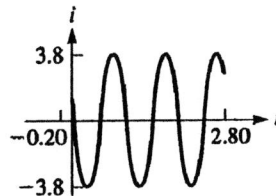

47. $y = 4500 \cos(0.025t - 0.25)$; $a = 4500$, $b = 0.025$,

$c = -0.25$

Amplitude is $|a| = 4500$;

period is $\dfrac{2\pi}{b} = \dfrac{2\pi}{0.025} = 80\pi$;

displacement is $-\dfrac{c}{b} = -\left(\dfrac{-0.25}{0.025}\right) = 10$.

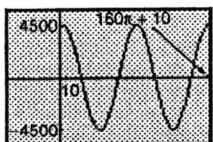

48. $I = A\sin(\omega t + \theta) = 5\sin(2 \times 10^5 t + 0.4)$

Amplitude is 5;

period is $\dfrac{2\pi}{2 \times 10^5} = 3.14 \times 10^{-5}$;

displacement is $\dfrac{-0.4}{2 \times 10^5} = -2 \times 10^{-6}$.

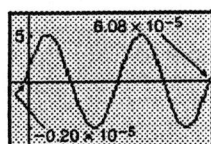

49. $y = a\sin(bx + c)$

Amplitude is 5,

period $= \dfrac{2\pi}{b} = 16, b = \dfrac{\pi}{8}$

displacement $= -\dfrac{c}{b} = -1, c = \dfrac{\pi}{8}$.

$y = 5\sin\left(\dfrac{\pi}{8}x + \dfrac{\pi}{8}\right)$

50. $y = a\cos(bx + c)$. Amplitude is 5.

Period is $\dfrac{2\pi}{b} = 16$; $16b = 2\pi$; $b = \dfrac{\pi}{8}$.

Displacement is $-\dfrac{c}{b} = 3$; $c = -\dfrac{3\pi}{8}$.

$y = 5\cos\left(\dfrac{\pi}{8}x - \dfrac{3\pi}{8}\right)$

51. $y = a\cos(bx + c)$. Amplitude is $|-0.8| = 0.8$.

Period is $\dfrac{2\pi}{b} = \pi$; $b = 2$.

Displacement is $-\dfrac{c}{b} = 0$; $c = 0$.

$y = -0.8\cos 2x$

52. $y = a\sin(bx + c)$. Amplitude is 0.8.

Period is $\dfrac{2\pi}{b} = \pi$; $b = 2$.

Displacement is $-\dfrac{c}{b} = \dfrac{\pi}{4}$; $c = -\dfrac{2\pi}{4} = -\dfrac{\pi}{2}$.

$y = 0.8\sin\left(2x - \dfrac{\pi}{2}\right)$

7.2   Basic Trigonometric Relations

1. $\dfrac{\cot\theta}{\cos\theta} = \cot\theta\dfrac{1}{\cos\theta}$

$\qquad = \dfrac{\cos\theta}{\sin\theta}\dfrac{1}{\cos\theta}$

$\qquad = \dfrac{1}{\sin\theta} = \csc\theta$

2. $\dfrac{\sin x}{\tan x} = \sin x\dfrac{1}{\tan x} = \sin x\dfrac{\cos x}{\sin x} = \cos x$

3. $\tan y \csc y = \dfrac{\tan y}{\sin y} = \tan y\dfrac{1}{\sin y}$

$\qquad = \dfrac{\sin y}{\cos y}\dfrac{1}{\sin y}$

$\qquad = \dfrac{1}{\cos y} = \sec y$

4. $\sin x \cot x = \dfrac{\sin x}{1}\dfrac{\cos x}{\sin x}$

$\qquad = \cos x$

5. $\cos^2 x - \sin^2 x = 1 - \sin^2 x - \sin^2 x = 1 - 2\sin^2 x$

6. $\cos^2\theta(1 + \tan^2\theta) = \cos^2\theta\left(1 + \dfrac{\sin^2\theta}{\cos^2\theta}\right)$

$\qquad\qquad = \cos^2\theta + \sin^2\theta = 1$

7. $\sin x \tan x + \cos x = \dfrac{\sin x}{1} \dfrac{\sin x}{\cos x} + \dfrac{\cos x}{1}$

$$= \frac{\sin^2 x}{\cos x} + \frac{\cos x}{1}$$

$$= \frac{\sin^2 x + \cos^2 x}{\cos x}$$

$$= \frac{1}{\cos x} = \sec x$$

8. $\tan x + \cot x = \dfrac{\sin x}{\cos x} + \dfrac{\cos x}{\sin x}$

$$= \frac{\sin^2 x + \cos^2 x}{\cos x \sin x}$$

$$= \frac{1}{\cos x \sin x}$$

$$= \sec x \csc x$$

9. $\tan x (\cot x + \tan x) = \tan x \cot x + \tan^2 x$

$$= \tan x \left(\frac{1}{\tan x} \right) + \tan^2 x$$

$$= 1 + \tan^2 x = \sec^2 x$$

10. $\tan^2 y \sec^2 y - \tan^4 y = \tan^2 y (\sec^2 y - \tan^2 y)$

$$= \tan^2 y (1) = \tan^2 y$$

11. $\dfrac{\sec \theta}{\cos \theta} - \dfrac{\tan \theta}{\cot \theta}$

$$= \frac{1}{\cos \theta} \left(\frac{1}{\cos \theta} \right) - \frac{\sin \theta}{\cos \theta} \left(\frac{\sin \theta}{\cos \theta} \right)$$

$$= \frac{1}{\cos^2 \theta} - \frac{\sin^2 \theta}{\cos^2 \theta}$$

$$= \frac{\cos^2 \theta}{\cos^2 \theta}$$

$$= 1$$

12. $\dfrac{\csc \theta}{\sin \theta} - \dfrac{\cot \theta}{\tan \theta} = \dfrac{1}{\sin \theta} \left(\dfrac{1}{\sin \theta} \right) - \dfrac{\cos \theta}{\sin \theta} \left(\dfrac{\cos \theta}{\sin \theta} \right)$

$$= \frac{1 - \cos^2 \theta}{\sin^2 \theta} = \frac{\sin^2 \theta}{\sin^2 \theta}$$

$$= 1$$

13. $\sin(x + y) \sin(x - y)$

$= (\sin x \cos y + \cos x \sin y)(\sin x \cos y - \cos x \sin y)$

$= \sin^2 x \cos^2 y - \cos^2 x \sin^2 y$

$= \sin^2 x (1 - \sin^2 y) - (1 - \sin^2 x)(\sin^2 y)$

$= \sin^2 x - \sin^2 x \sin^2 y - \sin^2 y + \sin^2 x \sin^2 y$

$= \sin^2 x - \sin^2 y$

14. $\cos(x + y) \cos(x - y)$

$= (\cos x \cos y - \sin x \sin y)(\cos x \cos y + \sin x \sin y)$

$= \cos^2 x \cos^2 y - \sin^2 x \sin^2 y$

$= \cos^2 x (1 - \sin^2 y) - (1 - \cos^2 x) \sin^2 y$

$= \cos^2 x - \cos^2 x \sin^2 y - \sin^2 y + \cos^2 x \sin^2 y$

$= \cos^2 x - \sin^2 y$

15. $\cos(x - y) + \sin(x + y)$

$= \cos x \cos y + \sin x \sin y + \sin x \cos y + \cos x \sin y$

$= \cos x (\cos y + \sin y) + \sin x (\cos y + \sin y)$

$= (\cos x + \sin x)(\cos y + \sin y)$

16. $2 \sin x + \sin 2x$

$= 2 \sin x + 2 \sin x \cos x$

$= 2 \sin x (1 + \cos x)$

$$= \frac{2 \sin x (1 + \cos x)(1 - \cos x)}{(1 - \cos x)}$$

$$= \frac{2 \sin x (1 - \cos^2 x)}{1 - \cos x}$$

$$= \frac{2 \sin x \sin^2 x}{1 - \cos x}$$

$$= \frac{2 \sin^3 x}{1 - \cos x}$$

17. $\cos^4 x - \sin^4 x = \cos 2x$

$(\cos^2 x + \sin^2 x)(\cos^2 x - \sin^2 x)$

$$= 1(\cos^2 x - \sin^2 x) = \cos 2x$$

18. $2 \csc 2x \tan x = \dfrac{2}{\sin 2x} \dfrac{\sin x}{\cos x}$

$$= \frac{2 \sin x}{2 \sin x \cos x \cos x}$$

$$= \frac{1}{\cos^2 x}$$

$$= \sec^2 x$$

19. $\dfrac{\sin 3x}{\sin x} + \dfrac{\cos 3x}{\cos}$

$$= \frac{\sin 3x \cos x + \cos 3x \sin x}{\sin x \cos x}$$

$$= \frac{\sin(3x + x)}{\frac{1}{2}(2 \sin x \cos x)}$$

$$= \frac{2 \sin 4x}{\sin 2x}$$

20. $\dfrac{\sin 3x}{\sin x} - \dfrac{\cos 3x}{\cos x} = \dfrac{\sin 3x \cos x - \cos 3x \sin x}{\sin x \cos x}$

$$= \frac{\sin(3x - x)}{\frac{1}{2} \sin 2x} = \frac{\sin 2x}{\frac{1}{2} \sin 2x} = 2$$

21. $\dfrac{1-\cos\alpha}{2\sin\dfrac{\alpha}{2}} = \dfrac{1-\cos\alpha}{2\sqrt{\dfrac{1-\cos\alpha}{2}}}\sqrt{\dfrac{\dfrac{1-\cos\alpha}{2}}{\dfrac{1-\cos\alpha}{2}}}$

$$= \dfrac{(1-\cos\alpha)\sqrt{\dfrac{1-\cos\alpha}{2}}}{2\left(\dfrac{1-\cos\alpha}{2}\right)}$$

$$= \sqrt{\dfrac{1-\cos\alpha}{2}} = \sin\dfrac{\alpha}{2}$$

22. $\tan\dfrac{\alpha}{2} = \dfrac{\sin\frac{\alpha}{2}}{\cos\frac{\alpha}{2}} = \dfrac{\pm\sqrt{\frac{1-\cos\alpha}{2}}}{\pm\sqrt{\frac{1+\cos\alpha}{2}}}$

$$= \pm\sqrt{\dfrac{1-\cos\alpha}{1+\cos\alpha}}$$

23. $2\sin^2\dfrac{x}{2} + \cos x = 2\left(\sqrt{\dfrac{1-\cos x}{2}}\right)^2 + \cos x$

$$= 2\left(\dfrac{1-\cos x}{2}\right) + \cos x$$

$$= 1 - \cos x + \cos x = 1$$

24. $2\cos^2\dfrac{\theta}{2}\sec\theta = 2\left(\sqrt{\dfrac{1+\cos\theta}{2}}\right)^2\sec\theta$

$$= 2\left(\dfrac{1+\cos\theta}{2}\right)\sec\theta$$

$$= (1+\cos\theta)\sec\theta$$

$$= (1+\cos\theta)\dfrac{1}{\cos\theta}$$

$$= \dfrac{1}{\cos\theta} + 1 = \sec\theta + 1$$

25. For $\theta = 90°$,

$$\cos\theta = \cos 90° = 0$$
$$= \cos A\cos B\cos C + \sin A\sin B$$

$$\cos A\cos B\cos C = -\sin A\sin B$$

$$\cos C = \dfrac{-\sin A\sin B}{\cos A\cos B}$$

$$\cos C = -\tan A\tan B$$

26. $(r - R\cos\theta)^2 (R\sin\theta)^2$
$$= r^2 - 2rR\cos\theta + R^2\cos^2\theta + R^2\sin^2\theta$$
$$= r^2 - 2rR\cos\theta + R^2(\cos^2\theta + \sin^2\theta)$$
$$= r^2 - 2rR\cos\theta + R^2$$

27. $i_0\sin(\omega t + \alpha) = i_0[\sin\omega t\cos\alpha + \sin\alpha\cos\omega t]$
$$= i_0\cos\alpha\sin\omega t + i_0\sin\alpha\cos\omega t$$
$$= i_1\sin\omega t + i_2\cos\omega t$$

28. $w = T\cos\alpha - F\sin\theta$

$$= T\cos\alpha - \left(T\dfrac{\sin\alpha}{\cos\theta}\right)\sin\theta$$

$$= \dfrac{T\cos\alpha\cos\theta - T\sin\alpha\sin\theta}{\cos\theta}$$

$$= \dfrac{T(\cos\alpha\cos\theta - \sin\alpha\sin\theta)}{\cos\theta}$$

$$= \dfrac{T\cos(\alpha+\theta)}{\cos\theta}$$

29. $p = vi\sin\omega t\sin\left(\omega t - \dfrac{\pi}{2}\right)$

$$= vi\sin\omega t\left[\sin\omega t\cos\dfrac{\pi}{2} - \cos\omega t\sin\dfrac{\pi}{2}\right]$$

$$= vi\sin\omega t[\sin\omega t(0) - \cos\omega t(1)]$$

$$= -vi\sin\omega t\cos\omega t = -\dfrac{1}{2}vi(2\sin\omega t\cos\omega t)$$

$$= -\dfrac{1}{2}vi\sin 2\omega t$$

30. $s = a\cos^2\theta + b\sin^2\theta - 2t\sin\theta\cos\theta$

$$= a\left(\dfrac{1}{2}(1+\cos 2\theta)\right) + b\left(\dfrac{1}{2}(1-\cos 2\theta)\right)$$

$$\quad - t(2\sin\theta\cos\theta)$$

$$= \dfrac{a}{2}[1 + \cos^2\theta - \sin^2\theta] + \dfrac{b}{2}[1 - \cos^2\theta + \sin^2\theta]$$

$$\quad - t\sin 2\theta$$

$$= \dfrac{a}{2} + \dfrac{a}{2}\cos^2\theta - \dfrac{a}{2}\sin^2\theta + \dfrac{b}{2} - \dfrac{b}{2}\cos^2\theta$$

$$\quad + \dfrac{b}{2}\sin^2\theta - t\sin 2\theta$$

$$= \dfrac{a}{2} + \dfrac{b}{2} + \cos^2\theta\left(\dfrac{a}{2} - \dfrac{b}{2}\right) - \sin^2\theta\left(\dfrac{a}{2} - \dfrac{b}{2}\right)$$

$$\quad - t\sin 2\theta$$

$$= \dfrac{1}{2}(a+b) + \left(\dfrac{a}{2} - \dfrac{b}{2}\right)(\cos^2\theta - \sin^2\theta) - t\sin 2\theta$$

$$= \dfrac{1}{2}(a+b) + \dfrac{1}{2}(a-b)\cos 2\theta - t\sin 2\theta$$

31. $\sin^2\omega t = \sin^2\left[\left(\dfrac{1}{2}\right)(2\omega t)\right] = \left(\sqrt{\dfrac{1-\cos 2\omega t}{2}}\right)^2$

$$= \dfrac{1-\cos 2\omega t}{2}$$

32. $2E^2 - 2E^2\cos(\pi - \theta)$
$$= 2E^2 - 2E^2(\cos\pi\cos\theta + \sin\pi\sin\theta)$$
$$= 2E^2 - 2E^2((-1)\cos\theta + (0)\sin\theta)$$
$$= 2E^2 + 2E^2\cos\theta$$
$$= 2E^2(1 + \cos\theta)$$
$$= 2E^2\left(2\cos^2\frac{\theta}{2}\right)$$
$$= 4E^2\cos^2\frac{\theta}{2}$$

33.

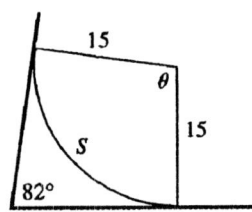

$$82.0° + 2\cdot 90° + \theta = 360°$$
$$\theta = 98.0°$$
$$s = r\theta$$
$$s = 15.0 \cdot 98.0° \cdot \frac{\pi}{180°}$$
$$s = 25.7 \text{ ft}$$

34. $r = \dfrac{d}{2} = \dfrac{2.38}{2} = 1.19 \text{ ft}$

$\theta = \dfrac{s}{r} = \dfrac{18.5}{1.19} = 15.5 \text{ rad}$

35. $s = r\theta;\ \theta = \dfrac{s}{r} = \dfrac{7.535}{8.25} = 0.9133$

$$A = \frac{1}{2}r^2\theta$$

$$A_1 = \frac{1}{2}(8.250 + 3.755)^2(0.9133) = 65.81 \text{ m}^2$$

$$A_2 = \frac{1}{2}(8.250)^2(0.9133) = 31.08 \text{ m}^2$$

$$A = A_1 - A_2 = 65.81 - 31.08 = 34.73 \text{ m}^2$$

36. $x = \sqrt{2.70^2 - 2.13^2} = 1.66,$

$$\theta\cos^{-1}\left(\frac{2.13}{2.70}\right) = 37.9° \times \frac{\pi}{180°} = 0.6618 \text{ rad}$$

$$A_{\text{sector}} = \frac{1}{2}(2.70)^2(2 \times 0.6618) = 4.82$$

$$A_{\text{triangle}} = \frac{1}{2}(2 \times 1.66)(2.13) = 3.53$$

$$A_{\text{segment}} = 4.82 - 3.53 = 1.29$$
$$A_{\text{illuminated}} = 1.29 \times 2 = 2.58 \text{ ft}^2$$

7.3 Derivatives of The Sine and Cosine Functions

1. $y = \sin(x + 2);$
$$\frac{dy}{dx} = \cos(x + 2)\frac{d(x + 2)}{dx} = \cos(x + 2)$$

2. $y = 3\sin 4x;\ \dfrac{dy}{dx} = 3\cos 4x(4) = 12\cos 4x$

3. $y = 2\sin(2x^3 - 1)$
$$\frac{dy}{dx} = 2\cos(2x^3 - 1)(6x^2) = 12x^2\cos(2x^3 - 1)$$

4. $y = 5\sin(2 - 3t);$
$$\frac{dy}{dt} = 5\cos(2 - 3t)(-3) = -15\cos(2 - 3t)$$

5. $y = 6\cos\left(\dfrac{1}{2}x\right);$
$$\frac{dy}{dx} = -6\left[\sin\left(\frac{1}{2}x\right)\right]\left[\frac{1}{2}\right] = -3\sin\left(\frac{1}{2}x\right)$$

6. $y = \cos(1 - x);$
$$\frac{dy}{dx} = -\sin(1 - x)(-1) = \sin(1 - x)$$

7. $y = 2\cos(3x - 1)$
$$\frac{dy}{dx} = 2\left[-\sin(3x - 1)(3)\right] = -6\sin(3x - 1)$$

8. $y = 4\cos(6x^2 + 5)$
$$\frac{dy}{dx} = -4\sin(6x^2 + 5)(12x) = -48x\sin(6x^2 + 5)$$

9. $r = \sin^2(3\pi\theta)$
$$\frac{dr}{d\theta} = 2\sin(3\pi\theta)\cos(3\pi\theta)(3\pi)$$
$$\frac{dr}{d\theta} = 6\pi\sin(3\pi\theta)\cos(3\pi\theta) = 3\pi\sin 6\pi\theta$$

10. $y = 3\sin^3(2x^4 + 1);$
$$\frac{dy}{dx} = 3.3\sin^2(2x^4 + 1)\cos(2x^4 + 1)(8x^3)$$
$$\frac{dy}{dx} = 72x^3\sin^2(2x^4 + 1)\cos(2x^4 + 1)$$

11. $y = 3\cos^3(5x + 2);$
$$\frac{dy}{dx} = 3.3\cos^2(5x + 2)[-\sin(5x + 2)(5)]$$
$$\frac{dy}{dx} = -45\cos^2(5x + 2)\sin(5x + 2)$$

12. $y = 4\cos^2\sqrt{x}$;

$$\frac{dy}{dx} = 4(2)\cos\sqrt{x}\left(-\sin\sqrt{x}\cdot\frac{1}{2}x^{-1/2}\right)$$

$$= -\frac{4\cos\sqrt{x}\sin\sqrt{x}}{\sqrt{x}}$$

13.　$y = x\sin 3x$;

$$\frac{dy}{dx} = x(3\cos 3x) + (\sin 3x)(1)$$

$$\frac{dy}{dx} = \sin 3x + 3x\cos 3x$$

14. $y = x^2\sin 2x$;

$$\frac{dy}{dx} = x^2(2\cos 2x) + \sin 2x(2x)$$

$$\frac{dy}{dx} = 2x^2\cos 2x + 2x\sin 2x$$

15. $y = 3x^3\cos 5x$;

$$\frac{dy}{dx} = 3[x^3(-5\sin 5x) + \cos 5x(3x^2)]$$

$$\frac{dy}{dx} = 9x^2\cos 5x - 15x^3\sin 5x$$

16. $y = 0.5\theta\cos\left(2\theta + \frac{\pi}{4}\right)$

$$\frac{dy}{d\theta} = 0.5\theta\left(-\sin\left(2\theta + \frac{\pi}{4}\right)2\right)$$

$$+\cos\left(2\theta + \frac{\pi}{4}\right)(0.5)$$

$$\frac{dy}{d\theta} = -\theta\sin\left(2\theta + \frac{\pi}{4}\right) + 0.5\cos\left(2\theta + \frac{\pi}{4}\right)$$

17.　$y = \sin x^2\cos 2x$;

$$\frac{dy}{dx} = \sin x^2(-2\sin 2x) + \cos 2x(2x\cos x^2)$$

$$\frac{dy}{dx} = 2x\cos x^2\cos 2x - 2\sin x^2\sin 2x$$

18. $y = 6\sin x\cos 4x$;

$$\frac{dy}{dx} = 6[\sin x(-4\sin 4x) + \cos 4x(\cos x)]$$

$$\frac{dy}{dx} = 6(\cos x\cos 4x - 4\sin x\sin 4x)$$

19.　$y = \sqrt{1 + \sin 4x} = (1 + \sin 4x)^{1/2}$

$$\frac{dy}{dx} = \frac{1}{2}(1 + \sin 4x)^{-1/2}(4\cos 4x)$$

$$\frac{dy}{dx} = \frac{2\cos 4x}{\sqrt{1 + \sin 4x}}$$

20.　$y = (x - \cos^2 x)^4$;

$$\frac{dy}{dx} = 4(x - \cos^2 x)^3[1 - 2\cos x\cdot(-\sin x)]$$

$$\frac{dy}{dx} = 4(x - \cos^2 x)^3(1 + 2\sin x\cos x)$$

21.　$r = \dfrac{\sin\left(3t - \frac{\pi}{3}\right)}{2t}$

$$\frac{dr}{dt} = \frac{2t\cos\left(3t - \frac{\pi}{3}\right)(3) - \sin\left(3t - \frac{\pi}{3}\right)(2)}{(2t)^2}$$

$$\frac{dr}{dt} = \frac{3t\cos\left(3t - \frac{\pi}{3}\right) - \sin\left(3t - \frac{\pi}{3}\right)}{2t^2}$$

22.　$y = \dfrac{2x + 3}{\sin 4x}$

$$\frac{dy}{dx} = \frac{\sin 4x(2) - (2x + 3)4\cos 4x}{\sin^2 4x}$$

$$\frac{dy}{dx} = \frac{2\sin 4x - 4(2x + 3)\cos 4x}{\sin^2 4x}$$

$$\frac{dy}{dx} = \frac{2\sin 4x - 8x\cos 4x - 12\cos 4x}{\sin^2 4x}$$

23.　$y = \dfrac{2\cos x^2}{3x - 1}$

$$\frac{dy}{dx} = \frac{(3x - 1)(2)(-\sin x^2)(2x) - 2\cos x^2(3)}{(3x - 1)^2}$$

$$\frac{dy}{dx} = \frac{-4x(3x - 1)\sin x^2 - 6\cos x^2}{(3x - 1)^2}$$

$$\frac{dy}{dx} = \frac{4x(1 - 3x)\sin x^2 - 6\cos x^2}{(3x - 1)^2}$$

24.　$y = \dfrac{\cos^2 3x}{(1 + 2\sin^2 2x)}$;

$$\frac{dy}{dx} = \frac{(1 + 2\sin^2 2x)(2\cos 3x)(-3\sin 3x) - (\cos^2 3x)(4\sin 2x)(2\cos 2x)}{(1 + 2\sin^2 2x)^2}$$

$$\frac{dy}{dx} = \frac{-2\cos 3x[3\sin 3x(1 + 2\sin^2 2x) + 4\cos 3x\sin 2x\cos 2x]}{(1 + 2\sin^2 2x)^2}$$

25. $y = 2\sin^2 3x \cos 2x$

$$\frac{dy}{dx} = (2\sin^2 3x)(-\sin 2x)(2) + (\cos 2x)(2)(2\sin 3x)(\cos 3x)(3)$$

$$\frac{dy}{dx} = -4\sin^2 3x \sin 2x + 12\cos 2x \sin 3x \cos 3x$$

$$\frac{dy}{dx} = 4\sin 3x(3\cos 3x \cos 2x - \sin 3x \sin 2x)$$

26. $y = \cos^3 4x \sin^2 2x$;

$$\frac{dy}{dx} = (\cos^3 4x)(2\sin 2x \cos 2x)(2) + (\sin^2 2x)(3\cos^2 4x)(-\sin 4x)(4)$$

$$\frac{dy}{dx} = 4\sin 2x \cos 2x \cos^3 4x - 12\sin 4x \sin^2 2x \cos^2 4x$$

27. $s = \sin(\sin 2t)$

$$\frac{ds}{dt} = \cos(\sin 2t)\cos(2t)(2)$$

$$\frac{ds}{dt} = 2\cos 2t \cos(\sin 2t)$$

28. $z = 0.2\cos(\sin 3\phi)$

$$\frac{dz}{d\phi} = 0.2[-\sin(\sin 3\phi)\cos 3\phi(3)]$$

$$\frac{dz}{d\phi} = -0.6\sin(\sin 3\phi)\cos 3\phi$$

29. $y = \sin^3 x - \cos 2x$;

$$\frac{dy}{dx} = 3\sin^2 x(\cos x) - (-\sin 2x)(2); \quad \frac{dy}{dx} = 3\sin^2 x \cos x + 2\sin 2x$$

30. $y = x\sin x + \cos x$; $\dfrac{dy}{dx} = x\cos x + \sin x - \sin x = x\cos x$

31. $p = \dfrac{1}{\sin s} + \dfrac{1}{\cos s}$

$$\frac{dp}{ds} = \frac{\sin s(0) - \cos s}{\sin^2 s} + \frac{\cos s(0) - (-\sin s)}{\cos^2 s}$$

$$\frac{dp}{ds} = \frac{-\cos s}{\sin^2 s} + \frac{\sin s}{\cos^2 s}$$

32. $y = 2x\sin x + 2\cos x - x^2 \cos x$

$$\frac{dy}{dx} = 2x\cos x + \sin x(2) - 2\sin x$$
$$\quad - [x^2(-\sin x) + \cos x(2x)]$$

$$\frac{dy}{dx} = 2x\cos x + 2\sin x - 2\sin x + x^2 \sin x - 2x\cos x$$

$$\frac{dy}{dx} = x^2 \sin x$$

33. (a) Set mode to radian. Set range to $x_{\min} = -4$, $x_{\max} = 4$, $y_{\min} = -2$, $y_{\max} = 2$, and enter $y_1 = \sin x \div x$.

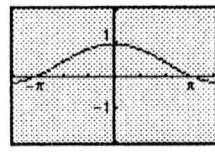

(b) Check the values on p. 779, using a calculator. When $x = 0.5, y = 0.95885108$; when $x = 0.1$, $y = 0.99833417$; when $x = 0.05, y = 0.99958339$; when $x = 0.01$, $y = 0.99998333$; when $x = 0.001$, $y = 0.99999983$.

34. $\displaystyle\lim_{\theta \to 0} \frac{\tan\theta}{\theta} = \lim_{\theta \to 0} \frac{\dfrac{\sin\theta}{\cos\theta}}{\theta} = \lim_{\theta \to 0}\left(\frac{\sin\theta}{\theta} \cdot \frac{1}{\cos\theta}\right) = \left(\lim_{\theta \to 0}\frac{\sin\theta}{\theta}\right)\left(\lim_{\theta \to 0}\frac{1}{\cos\theta}\right) = (1)(1) = 1$

35. (a) $\cos 1.0000 = 0.5403023$

Represents $\dfrac{d}{dx}(\sin x)$ at $x = 1$ (the derivative, slope of T.L. to sine curve at $x = 1$).

(b) $(\sin 1.0001 - \sin 1.0000)/0.0001 = 0.5402602$

Represents $\dfrac{f(x + \Delta x) - f(x)}{\Delta x}$ = slope of secant line through the two points on sine curve

where $x = 1.000$ and $x = 1.0001$.

36. (a) $-\sin 1.0000 = -0.8414710$

Value of derivative of $\cos x$ at $x = 1.0000$

(b) $(\cos 1.0001 - \cos 1.0000)/0.0001 = -0.8414980$

Value of slope of secant line through curve of $\cos x$, for $x = 1.0000$ and $x = 1.001$.

37. $y = \sin x$

$m_{\tan}\big|_{x=0} = 1.0$	$m_{\tan}\big|_{x=\pi/4} = 0.7$	$m_{\tan}\big|_{x=\pi/2} = 0.0$
$m_{\tan}\big|_{x=3\pi/4} = -0.7$	$m_{\tan}\big|_{x=\pi} = -1.0$	$m_{\tan}\big|_{x=5\pi/4} = -0.7$
$m_{\tan}\big|_{x=3\pi/2} = 0.0$	$m_{\tan}\big|_{x=7\pi/4} = 0.7$	$m_{\tan}\big|_{x=2\pi} = 1.0$

Plot points: $(0, 1.0), \left(\frac{\pi}{4}, 0.7\right), \left(\frac{\pi}{2}, 0.0\right), \left(\frac{3\pi}{4}, -0.7\right), (\pi, -1.0), \left(\frac{5\pi}{4}, -0.7\right), \left(\frac{3\pi}{2}, 0.0\right), \left(\frac{7\pi}{4}, 0.7\right), (2\pi, 1.0)$.

Resulting curve is $y = \cos x$.

38. $y = \cos x$

$m_{\tan}\big|_{x=0} = 0$	$m_{\tan}\big|_{x=\pi/4} = -0.7$	$m_{\tan}\big|_{x=\pi/2} = -1$
$m_{\tan}\big|_{x=3\pi/4} = -0.7$	$m_{\tan}\big|_{x=\pi} = 0$	$m_{\tan}\big|_{x=5\pi/4} = 0.7$
$m_{\tan}\big|_{x=3\pi/2} = 1$	$m_{\tan}\big|_{x=7\pi/4} = 0.7$	$m_{\tan}\big|_{x=2\pi} = 0$

Plot points: $(0, 0.0), \left(\frac{\pi}{4}, -0.7\right), \left(\frac{\pi}{2}, -1.0\right), \left(\frac{3\pi}{4}, -0.7\right), (\pi, 0.0), \left(\frac{5\pi}{4}, 0.7\right), \left(\frac{3\pi}{2}, 1.0\right), \left(\frac{7\pi}{4}, 0.7\right), (2\pi, 0.0)$.

Resulting curve is $y = -\sin x$.

39. $\sin(xy)\cos 2y = x^2$

$$\cos(xy)\left(x\frac{dy}{dx} + y\right) + (-\sin 2y)\left(2\frac{dy}{dx}\right) = 2x$$

$$x\frac{dy}{dx}\cos(xy) + y\cos(xy) - 2\frac{dy}{dx}\sin 2y = 2x$$

$$x\frac{dy}{dx}\cos(xy) - 2\frac{dy}{dx}\sin 2y = 2x - y\cos(xy)$$

$$\frac{dy}{dx}(x\cos(xy) - 2\sin 2y) = 2x - y\cos(xy)$$

$$\frac{dy}{dx} = \frac{2x - y\cos xy}{x\cos xy - 2\sin 2y}$$

40. $x \cos 2y + \sin x \cos y = 1$

$$x\left(-\sin 2y\, 2\frac{dy}{dx}\right) + \cos 2y + \sin x\left(-\sin y\frac{dy}{dx} + \cos y \cos x\right) = 0$$

$$-2x\sin 2y\frac{dy}{dx} - \sin x \sin y\frac{dy}{dx} = -\cos y \cos x - \cos 2y$$

$$\frac{dy}{dx}(2x\sin 2y + \sin x \sin y) = \cos x \cos y + \cos 2y$$

$$\frac{dy}{dx} = \frac{\cos 2y + \cos x \cos y}{2x\sin 2y + \sin x \sin y}$$

41. $\dfrac{d}{dx}\sin x = \cos x;$

$$\frac{d^2}{dx^2}\sin x = -\sin x$$

$$\frac{d^3}{dx^3}\sin x = -\cos x;$$

$$\frac{d^4}{dx^4}\sin x = \sin x$$

42. $y = \cos 2x$

$$\frac{dy}{dx} = -\sin 2x(2) = -2\sin 2x$$

$$\frac{d^2y}{dx^2} = -2\cos 2x(2)$$

$$= -4\cos 2x = -4y$$

43. $y = 3\sin 2x;$

$$\frac{dy}{dx} = 3(\cos 2x)(2x) = 6\cos 2x;$$

$$dy = (6\cos 2x)dx = \left(6\cos\frac{\pi}{4}\right)(0.02)$$

$$= \frac{6\sqrt{2}}{2}(0.02) = 0.085$$

44. $f(x) = \sin(\cos x)$

$$f\left(\frac{\pi}{2}\right) = \sin\left(\cos\frac{\pi}{2}\right) = \sin 0 = 0$$

$$f'(x) = \cos(\cos x)(-\sin x)$$

$$f'\left(\frac{\pi}{2}\right) = -1, L(x) - 0 = -1\left(x - \frac{\pi}{2}\right)$$

$$L(x) = -x + \frac{\pi}{2}$$

45. $y = x\cos 2x;\ x = 1.20$

$$\frac{dy}{dx} = x(-2\sin 2x) + \cos 2x$$

$$m_{\mathrm{TL}} = \cos 2x - 2x\sin x$$

$$m_{\mathrm{TL}}|_{x=1.20} = -2.36$$

```
nDeriv(Xcos(2X),
X,1.2)
     -2.358502793
```

46. $y = \dfrac{2\sin 3x}{x};\ x = 0.15$

$$\frac{dy}{dx} = 2\left(\frac{x3\cos 3x - \sin 3x}{x^2}\right);$$

$$m_{\mathrm{TL}} = \frac{6x\cos 3x - 2\sin 3x}{x^2}$$

$$m_{\mathrm{TL}}|_{x=0.15} = -2.646$$

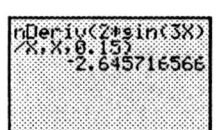

```
nDeriv(2sin(3X)
/X,X,0.15)
     -2.645716566
```

47. $V = 3.0\sin 188t \cos 188t$

$V = 1.5\sin 376t$

$$\frac{dV}{dt} = 1.5(376)\cos 376t|_{2.0\times 10^{-3}}$$

$$\frac{dV}{dt} = 410\ \mathrm{V/s}$$

48. $y = 2.0 + 2.0\cos(0.53x + 0.40), 0 \le x \le 5.0$

$$\frac{dy}{dx} = -2.0(0.53)\sin(0.53x + 0.40)|_{2.5}$$

$$\frac{dy}{dx} = -1.0 = m_{TL}$$

from which the angle with the horizontal is $45°$.

49. $y = 1.85 \sin 36\pi t$;

$$v = \frac{dy}{dx} = (1.85 \cos 36\pi t)(36\pi)$$

$$v\big|_{t=0.0250} = [1.85 \cos(36\pi \cdot 0.025)][36\pi] = -199 \text{ cm/s}$$

50. $i = 0.10c \cos\left(120\pi t + \frac{\pi}{6}\right)$;

$2.0 \text{ mH} - L; \ L = 2.0 \times 10^{-3} \text{ H}$

$$V_L = L\frac{di}{dt}$$

$$\frac{di}{dt} = -0.10 \sin\left(120\pi t + \frac{\pi}{6}\right)(120\pi)$$

$$\frac{di}{dt} = -12.0\pi \sin\left(120\pi t + \frac{\pi}{6}\right)$$

$$V_L = 2.0 \times 10^{-3}\left[-12.0\pi \sin\left(120\pi t + \frac{\pi}{6}\right)\right]$$

$$V_L = -0.024\pi \sin\left(120\pi t + \frac{\pi}{6}\right)$$

51. $r = \dfrac{100}{1 - \cos\theta} = 100(1 - \cos\theta)^{-1}$

$$\frac{dr}{d\theta} = -100(1 - \cos\theta)^{-2}(\sin\theta)$$

$$\frac{dr}{d\theta} = \frac{-100 \sin\theta}{(1 - \cos\theta)^2}$$

$$\theta = 120° = \frac{2\pi}{3}$$

$$\frac{dr}{d\theta}\bigg|_{\theta=120°} = \frac{-100 \sin\frac{2\pi}{3}}{\left(1 - \cos\frac{2\pi}{3}\right)^2} = -38.5 \text{ km}$$

52. $N = \dfrac{L\sin\theta}{d\sqrt{n^2 - \sin^2\theta}} = \dfrac{L\sin\theta}{d(n^2 - \sin^2\theta)^{1/2}}$;

$$\frac{dN}{d\theta} = \frac{d\sqrt{n^2 - \sin^2\theta}(L\cos\theta) - L\sin\theta \cdot d \cdot \frac{1}{2}(n^2 - \sin^2\theta)^{-1/2}(-2\sin\theta\cos\theta)}{d^2(n^2 - \sin^2\theta)}$$

$$\frac{dN}{d\theta} = \frac{dL\cos\theta\sqrt{n^2 - \sin^2\theta} + \frac{dL\sin^2\theta\cos\theta}{\sqrt{n^2-\sin^2\theta}}}{d^2(n^2 - \sin^2\theta)}; \quad \frac{dN}{d\theta} = \frac{\frac{dL\cos(n^2-\sin^2\theta)+dL\sin^2\theta\cos\theta}{\sqrt{n^2-\sin^2\theta}}}{d^2(n^2 - \sin^2\theta)}$$

$$\frac{dN}{d\theta} = \frac{n^2 dL\cos\theta - dL\sin^2\theta\cos\theta + dL\sin^2\theta\cos\theta}{d^2(n^2 - \sin^2\theta)\sqrt{n^2 - \sin^2\theta}}; \quad \frac{dN}{d\theta} = \frac{Ln^2\cos\theta}{d(n^2 - \sin^2\theta)^{3/2}}$$

7.4   Derivatives Of The Other Trigonometric Functions

1. $y = \tan 5x$;

$$\frac{dy}{dx} = \sec^2 5x\frac{d\,5x}{dx} = 5\sec^2 5x$$

2. $y = 3\tan(3x + 2)$

$$\frac{dy}{dx} = 3\sec^2(3x + 2)(3) = 9\sec^2(3x + 2)$$

3. $y = 5\cot(0.25\pi - \theta)$

$$\frac{dy}{d\theta} = -5\csc^2(0.25\pi - \theta) \cdot (-1)$$

$$= 5\csc^2(0.25\pi - \theta)$$

4. $y = 3\cot 6x$

$$\frac{dy}{dx} = 3[-\csc^2 6x \cdot (6)] = -18\csc^2 6x$$

5. $y = 3\sec 2x$;

$$\frac{dy}{dx} = 3\sec 2x\tan 2x\frac{d\,2x}{dx} = 6\sec 2x\tan 2x$$

6. $y = \sec\sqrt{1 - x}$

$$\frac{dy}{dx} = \sec\sqrt{1 - x}\tan\sqrt{1 - x} \cdot \frac{1}{2}(1 - x)^{-1/2}(-1)$$

$$\frac{dy}{dx} = -\frac{\sec\sqrt{1 - x}\tan\sqrt{1 - x}}{2\sqrt{1 - x}}$$

7. $y = -3 \csc \sqrt{2x+3}$

$\dfrac{dy}{dx} = -3\left[-\csc\sqrt{2x+3}\cot\sqrt{2x+3}\cdot\dfrac{1}{2}(2x+3)^{-1/2}(2)\right]$

$\dfrac{dy}{dx} = \dfrac{3\csc\sqrt{2x+3}\cot\sqrt{2x+3}}{\sqrt{2x+3}}$

8. $h = 0.5\csc(1-2\pi t)$

$\dfrac{dh}{dt} = -0.5\csc(1-2\pi t)\cot(1-2\pi t)\cdot(-2\pi)$

$\dfrac{dh}{dt} = \pi\csc(1-2\pi t)\cot(1-2\pi t)$

9. $y = 5\tan^2 3x$

$\dfrac{dy}{dx} = 5(2\tan 3x)(\sec^2 3x)\dfrac{d3x}{dx}$

$\qquad = 30\tan 3x\sec^2 3x$

10. $y = 2\tan^2(x^2)$

$\dfrac{dy}{dx} = 2[2\tan(x^2)\sec^2(x^2)(2x)]$

$\dfrac{dy}{dx} = 8x\tan(x^2)\sec^2(x^2)$

11. $y = 2\cot^4\dfrac{1}{2}x$

$\dfrac{dy}{dx} = 2(4)\cot^3\dfrac{1}{2}x\left[-\csc^2\dfrac{1}{2}x\left(\dfrac{1}{2}\right)\right]$

$\qquad = -4\cot^3\dfrac{1}{2}x\csc^2\dfrac{1}{2}x$

12. $y = \cot^2(1-x^2)$;

$\dfrac{dy}{dx} = 2\cot(1-x^2)[-\csc^2(1-x^2)(-2x)]$

$\dfrac{dy}{dx} = 4x\cot(1-x^2)\csc^2(1-x^2)$

13. $y = \sqrt{\sec 4x}$;

$\dfrac{dy}{dx} = \dfrac{1}{2}(\sec 4x^{-1/2})\sec 4x\tan 4x\dfrac{d4x}{dx}$

$\dfrac{dy}{dx} = \dfrac{2\sec 4x\tan 4x}{\sqrt{\sec 4x}} = 2\tan 4x\sqrt{\sec 4x}$

14. $y = 0.8\sec^3 5u$

$\dfrac{dy}{du} = 2.4\sec^2(5u)\sec(5u)\tan(5u)\cdot 5$

$\dfrac{dy}{du} = 12\sec^3(5u)\tan(5u)$

15. $y = 3\csc^4 7x$

$\dfrac{dy}{dx} = 3\cdot 4\csc^3 7x[-\csc 7x\cot 7x(7)]$

$\qquad = -84\csc^4 7x\cot 7x$

16. $y = \csc^2(2x^2)$

$\dfrac{dy}{dx} = 2\csc(2x^2)[-\csc(2x^2)\cot(2x^2)4x]$

$\qquad = -8x\csc^2(2x^2)\cot(2x^2)$

17. $r = t^2\tan(0.5t)$

$\dfrac{dr}{dt} = t^2\cdot\sec^2(0.5t)\cdot 0.5 + \tan(0.5t)\cdot 2t$

$\dfrac{dr}{dt} = \dfrac{t^2}{2}\sec^2(0.5t) + 2t\tan(0.5t)$

18. $y = 3x\sec 4x$;

$\dfrac{dy}{dx} = 3x\sec 4x\tan 4x(4) + \sec 4x(3)$

$\dfrac{dy}{dx} = 3\sec 4x(4x\tan 4x + 1)$

19. $y = 4\cos x\csc x^2$

$\dfrac{dy}{dx} = 4[\cos x(-\csc x^2\cot x^2\cdot 2x) + \csc x^2(-\sin x)]$

$\dfrac{dy}{dx} = -4\csc x^2(2x\cos x\cot x^2 + \sin x)$

20. $y = \dfrac{1}{2}\sin 2x\sec x$

$\dfrac{dy}{dx} = \dfrac{1}{2}(\sin 2x\sec x\tan x + \sec x2\cos 2x)$

$\dfrac{dy}{dx} = \dfrac{1}{2}\sec x(\sin 2x\tan x + 2\cos 2x)$

$\dfrac{dy}{dx} = \dfrac{1}{2}\sec x(2\sin x\cos x\tan x + 2\cos 2x)$

$\dfrac{dy}{dx} = \sin x\tan x + \sec x\cos 2x$

21. $y = \dfrac{\csc}{x}$;

$\dfrac{dy}{dx} = \dfrac{x\left(\frac{d\,\csc x}{dx}\right) - \csc x\left(\frac{dx}{dx}\right)}{x^2}$

$\dfrac{dy}{dx} = \dfrac{x(-\csc x\cot x) - \csc x}{x^2}$

$\qquad = -\dfrac{\csc x(x\cot x + 1)}{x^2}$

22. $u = \dfrac{\cot(0.25z)}{2z}$

$\dfrac{du}{dz} = \dfrac{2z(-\csc^2(0.25z)(0.25)) - \cot(0.25z)(2)}{(2z)^2}$

$\dfrac{du}{dz} = \dfrac{-z\csc^2(0.25z) - 4\cot(0.25z)}{8z^2}$

23. $y = \dfrac{2\cos 4x}{1 + \cot 3x}$

$\dfrac{dy}{dx} = \dfrac{(1+\cot 3x)[-2\sin 4x(4)] - 2\cos 4x(-\csc^2 3x)(3)}{(1 + \cot 3x)^2}$

$\dfrac{dy}{dx} = \dfrac{-8\sin 4x(1 + \cot 3x) + 6\cos 4x\csc^2 3x}{(1 + \cot 3x)^2}$

$\dfrac{dy}{dx} = \dfrac{2(-4\sin 4x - 4\sin 4x\cot 3x + 3\cos 4x\csc^2 3x)}{(1 + \cot 3x)^2}$

24. $y = \dfrac{\tan^2 3x}{2 + \sin x^2}$

$\dfrac{dy}{dx} = \dfrac{(2+\sin x^2)2\tan 3x\sec^2 3x(3) - \tan^2 3x\cos x^2(2x)}{(2 + \sin x^2)^2}$

$\dfrac{dy}{dx} = \dfrac{2\tan 3x(6\sec^2 3x + 3\sin x^2\sec^2 3x - x\tan 3x\cos x^2)}{(2 + \sin x^2)^2}$

25. $y = \dfrac{1}{3}\tan 3x - \tan x;$

$\dfrac{dy}{dx} = 3\left(\dfrac{1}{3}\right)\tan^2 x\dfrac{d(\tan x)}{dx} - \sec^2 x$

$\dfrac{dy}{dx} = \tan^2 x\sec^2 x - \sec^2 x$

$\qquad = \sec^2 x(\tan^2 x - 1)$

26. $y = \csc 2x - 2\cot 2x$

$\dfrac{dy}{dx} = -\csc 2x\cot 2x(2) - 2[-\csc^2 2x(2)]$

$\dfrac{dy}{dx} = -2\csc 2x(\cot 2x - 2\csc 2x)$

27. $r = \tan(\sin 2\theta)$

$\dfrac{dr}{d\theta} = \sec^2(\sin 2\theta)\cos(2\theta)(2)$

$\qquad = 2\cos 2\theta\sec^2(\sin 2\theta)$

28. $y = x\tan x + \sec^2 2x$

$\dfrac{dy}{dx} = x\sec^2 x + \tan x + 2\sec 2x\sec 2x\tan 2x(2)$

$\dfrac{dy}{dx} = x\sec^2 x + \tan x + 4\sec^2 2x\tan 2x$

29. $y = \sqrt{2x + \tan 4x} = (2x + \tan 4x)^{1/2}$

$\dfrac{dy}{dx} = \dfrac{1}{2}(2x + \tan 4x)^{-1/2}(2 + 4\sec^2 4x)$

$\dfrac{dy}{dx} = \dfrac{1 + 2\sec^2 4x}{\sqrt{2x + \tan 4x}}$

30. $y = (1 - \csc^2 3x)^3$

$\dfrac{dy}{dx} = 3(1 - \csc^2 3x)^2(-2\csc 3x)(-\csc 3x\cot 3x \cdot 3)$

$\dfrac{dy}{dx} = 18(1 - \csc^2 3x)^2\csc^2 3x\cot 3x$

31. $x\sec y - 2y = \sin 2x$

$x\sec y\tan y\dfrac{dy}{dx} + \sec y - 2\dfrac{dy}{dx} = 2\cos 2x$

$x\sec y\tan y\dfrac{dy}{dx} - 2\dfrac{dy}{dx} = 2\cos 2x - \sec y$

$\dfrac{dy}{dx}(x\sec y\tan y - 2) = 2\cos 2x - \sec y$

$\dfrac{dy}{dx} = \dfrac{2\cos 2x - \sec y}{x\sec y\tan y - 2}$

32. $3\cot(x + y) = \cos y^2$

$-3\csc^2(x + y)\left(1 + \dfrac{dy}{dx}\right) = -\sin y^2\left(2y\dfrac{dy}{dx}\right)$

$-3\csc^2(x + y) - 3\csc^2(x + y)\dfrac{dy}{dx} = -2y\sin y^2\dfrac{dy}{dx}$

$2y\sin y^2\dfrac{dy}{dx} - 3\csc^2(x + y)\dfrac{dy}{dx} = 3\csc^2(x + y)$

$\dfrac{dy}{dx}[2y\sin y^2 - 3\csc^2(x + y)] = 3\csc^2(x + y)$

$\dfrac{dy}{dx} = \dfrac{3\csc^2(x + y)}{2y\sin y^2 - 3\csc^2(x + y)}$

33. $y = 4\tan^2 3x$

$\dfrac{dy}{dx} = 4(2\tan 3x)(\sec^2 3x)(3) = 24\tan 3x\sec^2 3x;$

$dy = 24\tan 3x\sec^2 3x\,dx$

34. $y = 2.5\sec^3 2t$

$\dfrac{dy}{dt} = 7.5\sec^2 2t\sec 2t\tan 2t(2)$

$dy = 15\sec^3 2t\tan 2t\,dt$

35. $y = \tan 4x\sec 4x$

$\dfrac{dy}{dx} = \tan 4x \cdot 4\sec 4x\tan 4x + \sec 4x \cdot 4\sec^2 4x$

$dy = 4\sec 4x(\tan^2 4x + \sec^2 4x)\,dx$

36. $y = 2x \cot 3x$

$\frac{dy}{dx} = 2[x(-\csc^2 3x \cdot 3) + \cot 3x]$

$dy = (2 \cot 3x - 6x \csc^2 3x) dx$

37. (a) $\sec^2 1.000 = \frac{1}{\cos^2 1.0000} = 3.4255188.$

This is the slope of a tangent line to the curve $f(x) = \tan x$ at $x = 1.0000$. It is the value of $f'(x) = \sec^2 x$ at $x = 1.0000$ since $\frac{d(\tan x)}{dx} = \sec^2 x.$

(b) $\frac{\tan 1.0001 - \tan 1.0000}{0.0001} = 3.4260524.$

This is the slope of a secant line through the curve $f(x) = \tan x$ at $x = 1.000$, where $\Delta x = 0.0001$.

$\lim\limits_{\Delta x \to 0} \frac{\tan(x + \Delta x) - \tan x}{\Delta x} = \frac{d \tan x}{dx} = \sec^2 x$

For $\Delta x = 0.0001$, the slope of the tangent line is approximately equal to the slope of the secant line. (Final digits may vary.)

38. (a) $\sec 1.0000 \tan 1.0000 = 2.8824747$; value of derivative of $\sec x$ at $x = 1.0000$.

(b) $(\sec 1.0001 - \sec 1.0000)/0.0001 = 2.88302$; value of slope of secant line for function $y = \sec x$ passing through curve for $x = 1.0000$ and $x = 1.0001$.

39. $1 + \tan^2 x = \sec^2 x$

$0 + 2 \tan x \sec^2 x = 2 \sec x \sec x \tan x$

$2 \tan x \sec^2 x = 2 \sec^2 x \tan x$

$2 \tan x \sec^2 x = 2 \tan x \sec^2 x$; Q.E.D.

40. $1 + \cot^2 x = \csc^2 x$

$0 + 2 \cot x(-\csc^2 x) = 2 \csc x(-\csc x \cot x)$

$-2 \cot x \csc^2 x = -2 \cot x \csc^2 x$; Q.E.D.

41. $y = 2 \cot 3x$; $x = \frac{\pi}{12}$;

$\frac{dy}{dx} = 2(-\csc^2 3x)(3) = -6 \csc^2 3x$;

$\frac{dy}{dx}\Big|_{x = \pi/12} = -6 \csc^2 \frac{\pi}{4} = -6(\sqrt{2})^2 = -12$

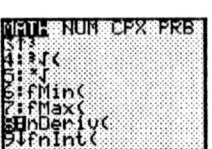

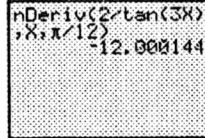

42. $y = \csc \sqrt{2x + 1}$; $x = 0.45$

$\frac{dy}{dx} = -\csc \sqrt{2x + 1} \cot \sqrt{2x + 1} \cdot \frac{1}{2}(2x + 1)^{-1/2}(2)$

$\frac{dy}{dx} = m_{TL} = \frac{-\csc \sqrt{2x + 1} \cot \sqrt{2x + 1}}{\sqrt{2x + 1}}$

$m_{TL}\big|_{x = 0.45} = -0.144$

$m_{NL} = 6.945$

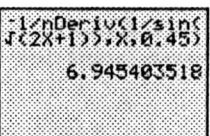

43. $y = 2 \tan x - \sec x$; satisfies $\frac{dy}{dx} = \frac{2 - \sin x}{\cos^2 x}$

$\frac{dy}{dx} = 2 \sec^2 x - \sec x \tan x = \frac{2}{\cos^2 x} - \frac{1}{\cos x} \cdot \frac{\sin x}{\cos x}$

$\frac{dy}{dx} = \frac{2}{\cos^2 x} - \frac{\sin x}{\cos^2 x} = \frac{2 - \sin x}{\cos^2 x}$; Q.E.D.

44. $y = \cos^3 x \tan x$;

satisfies $\cos x \frac{dy}{dx} + 3y \sin x - \cos^2 x = 0$.

$\frac{dy}{dx} = \cos^3 x \sec^2 x + \tan x \cdot 3 \cos^2 x(-\sin x)$

$\frac{dy}{dx} = \cos^3 x \sec^2 x - 3 \cos^2 x \sin x \tan x$

$\frac{dy}{dx} = \cos^3 x \frac{1}{\cos^2 x} - 3 \cos^2 x \sin x \frac{\sin x}{\cos x}$

$\frac{dy}{dx} = \cos x - 3 \cos x \sin^2 x = \cos x(1 - 3 \sin^2 x)$

Given: $\cos x \frac{dy}{dx} + 3y \sin x - \cos^2 x = 0$

$\cos x(\cos x - 3 \cos x \sin^2 x) + 3(\cos^3 x \tan x) \sin x - \cos^2 x = 0$

$\cos^2 x - 3 \cos^2 x \sin^2 x + 3 \cos^3 x \frac{\sin x}{\cos x} \sin x - \cos^2 x = 0$

$\cos^2 x - 3 \cos^2 x \sin^2 x + 3 \cos^2 x \sin^2 x - \cos^2 x = 0$

Four terms cancel to zero. Q.E.D.

45. $y = 2t^{1.5} - \tan 0.1t$; $v = \frac{dy}{dt} = 3t^{0.5} - 0.1 \sec^2 0.1t$;

$v\big|_{t = 15} = 3(15)^{0.5} - 0.1 \sec^2[0.1(15)] = -8.4$ cm/s

46. $q = t \sec \sqrt{0.2t^2 + 1}$; $t = 0.80$ s

$i = \dfrac{dq}{dt} = t \sec \sqrt{0.2t^2 + 1} \tan \sqrt{0.2t^2 + 1}$

$\qquad \times \dfrac{1}{2}(0.2t^2 + 1)^{-1/2}(0.4t) + \sec \sqrt{0.2t^2 + 1}$

$i = \dfrac{dq}{dt} = \sec \sqrt{0.2t^2 + 1} \left(\dfrac{0.2t^2 \tan \sqrt{0.2t^2 + 1}}{\sqrt{0.2t^2 + 1}} + 1 \right)$

$i|_{t=0.8} = \sec \sqrt{1.128} \left(\dfrac{0.128 \tan \sqrt{1.128}}{\sqrt{1.128}} + 1 \right)$

$\qquad = 2.5$ A

47. $\theta = \dfrac{3t}{(2t + 10)}$

$h = 1000 \tan \dfrac{3t}{2t + 10}$; $t = 5.0$ s

$\dfrac{dh}{dt} = 1000 \left[\sec^2 \left(\dfrac{3t}{2t + 10} \right) \right] \left(\dfrac{(2t + 10)3 - 3t(2)}{(2t + 10)^2} \right)$

$\dfrac{dh}{dt} \bigg|_{t=5.0} = \left(\sec^2 \dfrac{15}{20} \right) \left(\dfrac{30\,000}{400} \right)$

$\qquad = 140$ ft/s

48. $d = 189.00 \csc \dfrac{1}{2}\theta$

$\theta_1 = 98.20°, \theta_2 = 98.45°$;
$\Delta\theta = 0.25° = 0.004\ 36$ rad

$\dfrac{dd}{d\theta} = 189.00 \left(-\csc \dfrac{1}{2}\theta \cot \dfrac{1}{2}\theta \right) \dfrac{1}{2}$

$dd = \dfrac{-189.00}{2} \csc \dfrac{1}{2}\theta \cot \dfrac{1}{2}\theta \, d\theta$

$dd|_{\theta=98.20°} = \dfrac{-189.00}{2} \csc 49.1° \cot 49.1°(0.004\ 36)$

$\qquad = -0.47$ m

7.5   The Inverse Trigonometric Functions

1. y is an angle whose tangent is x.

2. y is an angle whose secant is x.

3. y is an angle whose cotangent is $3x$.

4. y is an angle whose cosecant is $4x$.

5. y is twice the angle whose sine is x.

6. y is three times the angle whose tangent is x.

7. y is five times the angle whose cosine is $2x - 1$.

8. y is four times the angle whose sine is $3x + 2$.

9. $\cos^{-1} 0.5 = \dfrac{\pi}{3}$ since $\cos \dfrac{\pi}{3} = 0.5$ and

$\qquad 0 \le \dfrac{\pi}{3} \le \pi$

10. $\sin^{-1} 1 = \dfrac{\pi}{2}$ since $\sin \dfrac{\pi}{2} = 1$ and $-\dfrac{\pi}{2} \le \dfrac{\pi}{2} \le \dfrac{\pi}{2}$

11. $\tan^{-1} 1 = \dfrac{\pi}{4}$ since $\tan \dfrac{\pi}{4} = 1$ and $-\dfrac{\pi}{2} < \dfrac{\pi}{4} < \dfrac{\pi}{2}$

12. $\cos^{-1} 0 = \dfrac{\pi}{2}$ since $\cos \dfrac{\pi}{2} = 0$ and $0 \le \dfrac{\pi}{2} \le \pi$

13. $\tan^{-1}(-\sqrt{3}) = -\dfrac{\pi}{3}$ since $\tan \left(-\dfrac{\pi}{3} \right) = -\sqrt{3}$ and

$\qquad -\dfrac{\pi}{2} < -\dfrac{\pi}{3} < \dfrac{\pi}{2}$

14. $\sin^{-1}(-0.5) = -\dfrac{\pi}{6}$ since $\sin \left(-\dfrac{\pi}{6} \right) = -0.5$ and

$\qquad -\dfrac{\pi}{2} \le -\dfrac{\pi}{6} \le \dfrac{\pi}{2}$

15. $\sec^{-1} 2 = \dfrac{\pi}{3}$ since $\sec \dfrac{\pi}{3} = 2$ and $0 \le \dfrac{\pi}{3} \le \pi$

16. $\cot^{-1} \sqrt{3} = \dfrac{\pi}{6}$ since $\cot \dfrac{\pi}{6} = \sqrt{3}$ and $0 < \dfrac{\pi}{6} < \pi$

17. $\sin^{-1} \left(-\dfrac{\sqrt{2}}{2} \right) = -\dfrac{\pi}{4}$ since $\sin \left(-\dfrac{\pi}{4} \right) = -\dfrac{\sqrt{2}}{2}$
and

$\qquad -\dfrac{\pi}{2} \le -\dfrac{\pi}{4} \le \dfrac{\pi}{2}$

18. $\cos^{-1} \left(-\dfrac{\sqrt{3}}{2} \right) = \dfrac{5\pi}{6}$ since $\cos \dfrac{5\pi}{6} = -\dfrac{\sqrt{3}}{2}$ and

$\qquad 0 \le \dfrac{5\pi}{6} \le \pi$

19. $\csc^{-1} \sqrt{2} = \dfrac{\pi}{4}$ since $\dfrac{\pi}{4} = \sqrt{2}$ and $-\dfrac{\pi}{2} \le \dfrac{\pi}{4} \le \dfrac{\pi}{2}$

20. $\cot^{-1}(-\sqrt{3}) = \dfrac{5\pi}{6}$ since $\cot \dfrac{5\pi}{6} = -\sqrt{3}$ and

$\qquad 0 < \dfrac{5\pi}{6} < \pi$

21. $\tan(\sin^{-1} 0) = \tan 0 = 0$

22. $\csc(\tan^{-1} 1) = \csc \dfrac{\pi}{4} = \sqrt{2}$

23. $\sin(\tan^{-1} \sqrt{3}) = \sin \dfrac{\pi}{3} = \dfrac{1}{2}\sqrt{3}$

24. $\tan \left[\sin^{-1} \left(\dfrac{\sqrt{2}}{2} \right) \right] = \tan \dfrac{\pi}{4} = 1$

25. $\cos[\tan^{-1}(-1)] = \cos \left(-\dfrac{\pi}{4} \right) = \dfrac{1}{2}\sqrt{2}$

26. $\sec[\cos^{-1}(-0.5)] = \sec\dfrac{2\pi}{3} = -2$

27. $\cos(2\sin^{-1}1) = \cos\left(2\dfrac{\pi}{2}\right) = \cos\pi = -1$

28. $\sin(2\tan^{-1}2) = \sin(2 \times 1.107\,15) = 0.8000$

29. $\tan^{-1}x = \sin^{-1}\dfrac{2}{5}$

$x = \tan\left(\sin^{-1}\dfrac{2}{5}\right)$

$x = \dfrac{2}{\sqrt{21}}$

30. $\cot^{-1}x = \cos^{-1}\dfrac{1}{3}$

$\cot(\cot^{-1}x) = \cot\left(\cos^{-1}\dfrac{1}{3}\right)$

$x = \dfrac{1}{\sqrt{8}} = \dfrac{1}{2\sqrt{2}} \cdot \dfrac{\sqrt{2}}{\sqrt{2}}$

$= \dfrac{\sqrt{2}}{4}$

31. $\sec^{-1}(x) = \sin^{-1}\left(-\dfrac{1}{2}\right)$

$\sec(\sec^{-1}(x)) = \sec\left(\sin^{-1}-\dfrac{1}{2}\right) = \dfrac{2}{\sqrt{3}}$

$x = \dfrac{2}{\sqrt{3}}$

32. $\sin^{-1}x = -\tan^{-1}(-1)$

$\sin(\sin^{-1}x) = \sin(-\tan^{-1}(-1))$

$x = \sin\left(-\left(-\dfrac{\pi}{4}\right)\right) = \sin\dfrac{\pi}{4}$

$x = \dfrac{\sqrt{2}}{2}$

33. $\tan^{-1}(-3.7321) = -1.3090$

34. $\cos^{-1}(-0.6561) = 2.286$

35. $\sin^{-1}(-0.8326) = -0.9838$

36. $\tan^{-1}(0.2846) = 0.2773$

37. $\cos^{-1}0.1291 = 1.4413$

38. $\sin^{-1}(0.2119) = 0.2135$

39. $\tan^{-1}(8.2614) = 1.4503$

40. $\sin^{-1}(-0.8881) = -1.093$

41. $\tan[\cos^{-1}(-0.6281)] = \tan 2.250 = -1.2389$

42. $\cos[\tan^{-1}(-1.2256)] = \cos(-0.88642) = 0.63219$

43. $\sin[\tan^{-1}(-0.2297)] = \sin(-0.22578) = -0.2239$

44. $\tan[\sin^{-1}(-0.3019)] = \tan(-0.30669) = -0.3167$

45. $y = \sin 3x;\ 3x = \sin^{-1}y;\ x = \dfrac{1}{3}\sin^{-1}y$

46. $y = \cos(x - \pi)$
$\cos^{-1}y = x - \pi$
$x = \pi + \cos^{-1}y$

47. $y = \tan^{-1}\left(\dfrac{x}{4}\right)$

$\tan y = \dfrac{x}{4}$

$x = 4\tan y$

48. $y = 2\sin^{-1}\left(\dfrac{x}{6}\right)$

$\dfrac{1}{2}y = \sin^{-1}\left(\dfrac{x}{6}\right)$

$\sin\dfrac{1}{2}y = \dfrac{x}{6}$

$x = 6\sin\dfrac{1}{2}y$

49. $y = 1 + 3\sec 3x;\ \sec 3x = \dfrac{y-1}{3}$

$3x = \sec^{-1}\left(\dfrac{y-1}{3}\right);\ x = \dfrac{1}{3}\sec^{-1}\left(\dfrac{y-1}{3}\right)$

50. $4y = 5 - 2\csc 8x$
$4y - 5 = -2\csc 8x$
$\dfrac{4y-5}{-2} = \csc 8x$
$\dfrac{5-4y}{2} = \csc 8x$
$\csc^{-1}\dfrac{5-4y}{2} = 8x$
$x = \dfrac{1}{8}\csc^{-1}\dfrac{5-4y}{2}$

51. $1 - y = \cos^{-1}(1 - x)$
$\cos(1 - y) = 1 - x$
$x = 1 - \cos(1 - y)$

52. $2y = \cot^{-1}3x - 5$
$2y + 5 = \cot^{-1}3x$
$\cot(2y + 5) = 3x$
$x = \dfrac{1}{3}\cot(2y + 5)$

53. $\tan(\sin^{-1}x) = \tan\theta = \dfrac{x}{\sqrt{1-x^2}}$

In a triangle, θ is set up such that its sine is x. This gives an opposite side x, hypotenuse 1, and adjacent side $\sqrt{1-x^2}$.

54. $\sin(\csc^{-1}x) = \sin\theta = \dfrac{\sqrt{1-x^2}}{1} = \sqrt{1-x^2}$

In a triangle, θ is set up such that its cosine is x. This gives an adjacent side x, hypotenuse 1, and opposite side $\sqrt{1-x^2}$.

55. $\cos(\sec^{-1}x) = \cos\theta = \dfrac{1}{x}$

In a triangle, θ is set up such that its secant is x. This gives an adjacent side 1, hypotenuse x, and opposite side $\sqrt{x^2-1}$.

56. $\cot(\cot^{-1}x) = \cot\theta = \dfrac{x}{1} = x$

Meaning of $\cot^{-1}x$ is the angle whose cotangent is x.

57. $\sec(\csc^{-1}3x) = \sec\theta = \dfrac{3x}{\sqrt{9x^2-1}}$

In a triangle, θ is set up such that is cosecant is $3x$. This gives an opposite side 1, hypotenuse $3x$, and adjacent side $\sqrt{9x^2-1}$.

58. $\tan(\sin^{-1}2x) = \tan\theta = \dfrac{2x}{\sqrt{1-4x^2}}$

In triangle, θ is set up such that its sine is $2x$. This gives an opposite side of $2x$, hypotenuse 1, and adjacent side $\sqrt{1-4x^2}$.

59. $\sin(2\sin^{-1}x) = \sin 2\theta = 2\sin\theta\cos\theta$

$$= 2\left(\frac{x}{1}\right)\left(\frac{\sqrt{1-x^2}}{1}\right) = 2x\sqrt{1-x^2}$$

In a triangle, θ is set up such that its sine is $2x$. This gives an opposite side of $2x$, hypotenuse 1, and adjacent side $\sqrt{1-4x^2}$.

60. $\cos(2\tan^{-1}x) = \cos 2\theta = 2\cos^2\theta - 1$

$$= 2\left(\frac{1}{\sqrt{1+x^2}}\right)^2 - 1$$

$$= 2\left(\frac{1}{1+x^2}\right) - 1$$

$$= \frac{2-(1+x^2)}{1+x^2}$$

$$= \frac{1-x^2}{1+x^2}$$

In a triangle, θ is set up such that its tangent is x. This gives an opposite side x, adjacent side 1, and hypotenuse $\sqrt{1+x^2}$.

61. $y = A\cos 2(\omega t + \phi)$

$$\frac{y}{A} = \cos 2(\omega t + \phi)$$

$$\cos^{-1}\frac{y}{A} = 2(\omega t + \phi) = 2\omega t + 2\phi$$

$$\cos^{-1}\frac{y}{A} - 2\phi = 2\omega t$$

$$\frac{\cos^{-1}\dfrac{y}{A} - 2\phi}{2\omega} = t$$

$$t = \frac{1}{2\omega}\cos^{-1}\frac{y}{A} - \frac{\phi}{\omega}$$

62. $\mu\omega\cos\theta = \omega\sin\theta$

$$\frac{\mu\omega}{\omega} = \frac{\sin\theta}{\cos\theta}$$

$$\mu = \tan\theta$$

$$\theta = \tan^{-1}\mu$$

63. $i = I_m[\sin(\omega t + \alpha)\cos\phi + \cos(\omega t + \alpha)\sin\phi]$

$$\frac{i}{I_m} = \sin(\omega t + \alpha + \phi)$$

$$\sin^{-1}\left(\frac{i}{I_m}\right) = \omega t + \alpha + \phi$$

$$\omega t = \sin^{-1}\left(\frac{i}{I_m}\right) - \alpha - \phi$$

$$t = \frac{1}{\omega}\left(\sin^{-1}\left(\frac{i}{I_m}\right) - \alpha - \phi\right)$$

64. $t = \dfrac{1}{2\pi f}\cos^{-1}\dfrac{d}{A}$

$$2\pi ft = \cos^{-1}\frac{d}{A}$$

$$\cos 2\pi ft = \frac{d}{A}$$

$$A\cos 2\pi ft = d$$

65. Let $\alpha = \sin^{-1}\dfrac{3}{5}$ and $\beta = \sin^{-1}\dfrac{5}{13}$; $\sin\alpha = \dfrac{3}{5}$

$$\cos\alpha = \sqrt{1 - \frac{9}{25}} = \sqrt{\frac{16}{25}} = \frac{4}{5}; \ \sin\beta = \frac{5}{13}$$

$$\cos\beta = \sqrt{1 - \frac{25}{169}} = \sqrt{\frac{144}{169}} = \frac{12}{13}$$

$$\sin^{-1}\frac{3}{5} + \sin^{-1}\frac{5}{13} = \alpha + \beta$$

$$\sin(\alpha+\beta) = \sin\alpha\cos\beta + \cos\alpha\sin\beta$$

$$= \frac{3}{5}\left(\frac{12}{13}\right) + \frac{4}{5}\left(\frac{5}{13}\right)$$

$$= \frac{36}{65} + \frac{20}{65} = \frac{56}{65}$$

66. $\tan^{-1}\dfrac{1}{3} + \tan^{-1}\dfrac{1}{2} = \dfrac{\pi}{4}$

$$\sin\left(\tan^{-1}\frac{1}{3} + \tan^{-1}\frac{1}{2}\right)$$

$$= \sin\left(\tan^{-1}\frac{1}{3}\right)\cos\left(\tan^{-1}\frac{1}{2}\right)$$

$$+ \cos\left(\tan^{-1}\frac{1}{3}\right)\sin\left(\tan^{-1}\frac{1}{2}\right)$$

$$\frac{1}{\sqrt{10}}\times\frac{2}{\sqrt{5}} + \frac{3}{\sqrt{10}}\times\frac{1}{\sqrt{5}} = \frac{2}{\sqrt{50}} + \frac{3}{\sqrt{50}}$$

$$= \frac{5}{\sqrt{50}} = \frac{5}{5\sqrt{2}}$$

$$= \frac{1}{\sqrt{2}} = \sin\frac{\pi}{4}$$

67. $\sin^{-1}0.5 + \cos^{-1}0.5 = \dfrac{\pi}{6} + \dfrac{\pi}{3} = \dfrac{\pi+2\pi}{6} = \dfrac{3\pi}{6} = \dfrac{\pi}{2}$

68. $\tan^{-1}\sqrt{3} + \cos^{-1}\sqrt{3} = \dfrac{\pi}{3} + \dfrac{\pi}{6} = \dfrac{\pi+2\pi}{6}$

$$= \frac{3\pi}{6} = \frac{\pi}{2}$$

69. Since $\sin A = \dfrac{a}{c}$, $A = \sin^{-1}\left(\dfrac{a}{c}\right)$

70. Using law of sines: $\dfrac{a}{\sin A} = \dfrac{b}{\sin B}$

$$b\sin A = a\sin B$$

$$\sin A = \frac{a\sin B}{b}$$

$$A = \sin^{-1}\left(\frac{a\sin B}{b}\right)$$

71. Let y = height to top of pedestal

$$\tan\alpha = \frac{151+y}{d}$$

$$\tan\beta = \frac{y}{d}$$

$$y = d\tan\beta$$

$$\tan\alpha = \frac{151 + d\tan\beta}{d}$$

$$\alpha = \tan^{-1}\left(\frac{151 + d\tan\beta}{d}\right) = \tan^{-1}\left(\frac{151}{d} + \tan\beta\right)$$

72. In triangle, $\sin\theta = \dfrac{5}{13}$, $\theta = \sin^{-1}\dfrac{5}{13}$

For big circle, $S = R\theta$

$$S = 8\sin^{-1}\frac{5}{13}$$

$$2S = 16\sin^{-1}\frac{5}{13}$$

$$\frac{1}{2}\text{ diameter } \pi R = 8\pi$$

For small circle, $\dfrac{1}{2}$ diameter $\pi R = 3\pi$

$$S_1 = R\theta$$

$$S_1 = 3\sin^{-1}\frac{5}{13}$$

$$2S_1 = 6\sin^{-1}\frac{5}{13}$$

$$L = 12 + 12 + 8\pi + 3\pi + 16\sin^{-1}\frac{5}{13} - 6\sin^{-1}\frac{5}{13}$$

$$L = 24 + 11\pi + 10\sin^{-1}\frac{5}{13}$$

7.6 Derivatives Of the Inverse Trigonometric Functions

1. $y = \sin^{-1}(x^2)$;

$$\frac{dy}{dx} = \frac{1}{\sqrt{1 - (x^2)^2}}\ \frac{dx^2}{dx} = \frac{2x}{\sqrt{1 - x^4}}$$

2. $y = \sin^{-1}(1 - x^2)$

$$\frac{dy}{dx} = \frac{1}{\sqrt{1 - (1 - x^2)^2}}(-2x) = \frac{-2x}{\sqrt{x^2(2 - x^2)}}$$

$$= \frac{-2}{\sqrt{2 - x^2}}$$

3. $y = 2\sin^{-1} 3x^3$

$$\frac{dy}{dx} = 2\frac{1}{\sqrt{1 - 9x^6}}(9x^2) = \frac{18x^2}{\sqrt{1 - 9x^6}}$$

4. $y = \sin^{-1}\sqrt{1 - 2x}$

$$\frac{dy}{dx} = \frac{1}{\sqrt{1 - (1 - 2x)}}\left[\frac{1}{2}(1 - 2x)^{-1/2}(-2)\right]$$

$$\frac{dy}{dx} = \frac{-1}{\sqrt{2x}\sqrt{1 - 2x}} = \frac{-1}{\sqrt{2x - 4x^2}}$$

5. $y = 3.6\cos^{-1} 0.5s$

$$\frac{dy}{ds} = 3.6\frac{-1}{\sqrt{1 - (-.5s)^2}}(0.5)$$

$$\frac{dy}{ds} = \frac{-1.8}{\sqrt{1 - \frac{s^2}{4}}}$$

$$\frac{dy}{ds} = \frac{-1.8}{\sqrt{1 - 0.25s^2}}$$

6. $\theta = 0.2\cos^{-1} 5t$

$$\frac{d\theta}{dt} = -0.2\frac{1}{\sqrt{1 - (5t)^2}}(5) = \frac{-1}{\sqrt{1 - 25t^2}}$$

7. $y = 2\cos^{-1}\sqrt{2 - x}$

$$\frac{dy}{dx} = 2 \cdot \frac{-1}{\sqrt{1 - (2 - x)}}\left[\frac{1}{2}(2 - x)^{-1/2}(-1)\right]$$

$$\frac{dy}{dx} = \frac{1}{\sqrt{x - 1}\sqrt{2 - x}} = \frac{1}{\sqrt{(x - 1)(2 - x)}}$$

8. $y = 3\cos^{-1}(x^2 + 0.5)$

$$\frac{dy}{dx} = 3 \cdot \frac{1}{\sqrt{1 - (x^2 + 0.5)^2}}(2x)$$

$$\frac{dy}{dx} = \frac{-6x}{\sqrt{1 - (x^2 + 0.5)^2}}$$

9. $y = \tan^{-1}\sqrt{x} = \tan^{-1} x^{1/2};$

$$\frac{dy}{dx} = \frac{1}{1 + (\sqrt{x})^2}\frac{dx^{1/2}}{dx}$$

$$\frac{dy}{dx} = \frac{1}{1 + x}\left(\frac{1}{2}x^{-1/2}\right)$$

$$= \frac{1}{2\sqrt{x}(1 + x)}$$

10. $y = \tan^{-1}(1 - x)$

$$\frac{dy}{dx} = \frac{1}{1 + (1 - x)^2}(-1) = \frac{-1}{x^2 - 2x + 2}$$

11. $= 6\tan^{-1}\left(\frac{1}{x}\right)$

$$\frac{dy}{dx} = 6\frac{1}{1 + \frac{1}{x^2}}\left(-\frac{1}{x^2}\right) = \frac{-\frac{6}{x^2}}{(x^2 + 1)/x^2} = \frac{-6}{x^2 + 1}$$

12. $y = 4\tan^{-1}(3x^4)$

$$\frac{dy}{dx} = 4 \cdot \frac{1}{1 + 9x^8}(12x^3) = \frac{48x^3}{1 + 9x^8}$$

13. $y = 5x\sin^{-1} x;$

$$\frac{dy}{dx} = 5x\left(\frac{1}{\sqrt{1 - x^2}}\right)(1) + (\sin^{-1})(5)$$

$$\frac{dy}{dx} = \frac{5x}{\sqrt{1 - x^2}} + 5\sin^{-1} x$$

14. $y = x^2\cos^{-1} x$

$$\frac{dy}{dx} = x^2\frac{-1}{\sqrt{1 - x^2}} + \cos^{-1} x(2x)$$

$$\frac{dy}{dx} = \frac{-x^2}{\sqrt{1 - x^2}} + 2x\cos^{-1} x$$

15. $y = 0.4u\tan^{-1} 2u$

$$\frac{dy}{du} = 0.4u \cdot \frac{2}{1 + 4u^2} + 0.4\tan^{-1} 2u$$

$$\frac{dy}{du} = \frac{0.8u}{1 + 4u^2} + 0.4\tan^{-1} 2u$$

16. $y = (x^2 + 1)\sin^{-1} 4x$

$$\frac{dy}{dx} = (x^2 + 1)\frac{1}{\sqrt{1 - 16x^2}}(4) + \sin^{-1} 4x(2x)$$

$$\frac{dy}{dx} = \frac{4(x^2 + 1)}{\sqrt{1 - 16x^2}} + 2x\sin^{-1} 4x$$

17. $y = \dfrac{3x - 1}{\sin^{-1} 2x}$

$$\frac{dy}{dx} = \frac{(\sin^{-1} 2x)(3) - (3x - 1)\frac{1}{\sqrt{1 - 4x^2}}(2)}{(\sin^{-1} 2x)^2}$$

$$\frac{dy}{dx} = \left(3\sin^{-1} 2x - \frac{6x - 2}{\sqrt{1 - 4x^2}}\right)\frac{1}{(\sin^{-1} 2x)^2}$$

$$\frac{dy}{dx} = \frac{3\sqrt{1 - 4x^2}\sin^{-1} 2x - 6x + 2}{\sqrt{1 - 4x^2}(\sin^{-1} 2x)^2}$$

18. $\theta = \dfrac{\tan^{-1} 2r}{\pi r}$

$\dfrac{d\theta}{dr} = \dfrac{\pi r \frac{2}{1+4r^2} - \pi(\tan^{-1}(2r))}{(\pi r)^2}$

$\dfrac{d\theta}{dr} = \dfrac{2r - (1+4r^2)\tan^{-1}(2r)}{\pi r^2(1+4r^2)}$

19. $y = \dfrac{\sin^{-1} 2x}{\cos^{-1} 2x}$

$\dfrac{dy}{dx} = \dfrac{\cos^{-1} 2x \frac{1}{\sqrt{1-4x^2}}(2) - \sin^{-1} 2x \frac{1}{\sqrt{1-4x^2}}(2)}{(\cos^{-1} 2x)^2}$

$\dfrac{dy}{dx} = \dfrac{2}{\sqrt{1-4x^2}}\left[\dfrac{\cos^{-1} 2x + \sin^{-1} 2x}{(\cos^{-1} 2x)^2}\right]$

$\dfrac{dy}{dx} = \dfrac{2(\cos^{-1} 2x + \sin^{-1} 2x)}{\sqrt{1-4x^2}(\cos^{-1} 2x)^2}$

20. $y = \dfrac{x^2+1}{\tan^{-1} x}$

$\dfrac{dy}{dx} = \dfrac{\tan^{-1} x(2x) - (x^2+1)\frac{1}{(1+x^2)}(1)}{(\tan^{-1} x)^2}$

$\dfrac{dy}{dx} = \dfrac{2x \tan^{-1} x - 1}{(\tan^{-1} x)^2}$

21. $y = 2(\cos^{-1} 4x)^3$

$\dfrac{dy}{dx} = 6(\cos^{-1} 4x)^2 \dfrac{d(\cos^{-1} 4x)}{dx}$

$\dfrac{dy}{dx} = 6(\cos^{-1} 4x)^2 \left(-\dfrac{1}{\sqrt{1-(4x)^2}}\right)(4)$

$\dfrac{dy}{dx} = -\dfrac{24(\cos^{-1} 4x)^2}{\sqrt{1-16x^2}}$

22. $r = 0.5(\sin^{-1}(3t))^4$

$\dfrac{dr}{dt} = 2(\sin^{-1}(3t))^3 \dfrac{1}{\sqrt{1-9t^2}}(3)$

$\dfrac{dr}{dt} = \dfrac{6(\sin^{-1}(3t))^3}{\sqrt{1-9t^2}}$

23. $y = [\sin^{-1}(4x+1)]^2$

$\dfrac{dy}{dx} = 2[\sin^{-1}(4x+1)]^1 \dfrac{1}{\sqrt{1-(4x+1)^2}}(4)$

$\dfrac{dy}{dx} = \dfrac{8\sin^{-1}(4x+1)}{\sqrt{1-16x^2-8x-1}} = \dfrac{8\sin^{-1}(4x+1)}{\sqrt{4(-4x^2-2x)}}$

$\dfrac{dy}{dx} = \dfrac{4\sin^{-1}(4x+1)}{\sqrt{-4x^2-2x}}$

24. $y = \sqrt{\sin^{-1}(x-1)}$

$\dfrac{dy}{dx} = \dfrac{1}{2}[\sin^{-1}(x-1)]^{-1/2} \dfrac{1}{\sqrt{1-(x-1)^2}}(1)$

$\dfrac{dy}{dx} = \dfrac{1}{2\sqrt{\sin^{-1}(x-1)}\sqrt{1-x^2+2x-1}}$

$\dfrac{dy}{dx} = \dfrac{1}{2\sqrt{(2x-x^2)\sin^{-1}(x-1)}}$

25. $y = \tan^{-1}\left(\dfrac{1-t}{1+t}\right)$

$\dfrac{dy}{dt} = \dfrac{1}{1+\left(\frac{1-t}{1+t}\right)^2} \cdot \dfrac{(1+t)(-1) - (1-t)(1)}{(1+t)^2}$

$\dfrac{dy}{dt} = \dfrac{-1-t-1+t}{(1+t)^2 + (1-t)^2}$

$\dfrac{dy}{dt} = \dfrac{-2}{1+2t+t^2+1-2t+t^2}$

$\dfrac{dy}{dt} = \dfrac{-2}{2+2t^2}; \ \dfrac{dy}{dt} = \dfrac{-1}{1+t^2}$

26. $y = \dfrac{1}{\cos^{-1} 2x} = (\cos^{-1} 2x)^{-1}$

$\dfrac{dy}{dx} = -1(\cos^{-1} 2x)^{-2}\dfrac{-1}{\sqrt{1-4x^2}}(2)$

$\dfrac{dy}{dx} = \dfrac{2}{\sqrt{1-4x^2}(\cos^{-1} 2x)^2}$

27. $y = \dfrac{1}{1+4x^2} - \tan^{-1} 2x$

$\qquad = (1+4x^2)^{-1} - \tan^{-1} 2x$

$\dfrac{dy}{dx} = -(1+4x^2)^{-2}(8x) - \dfrac{1}{1+4x^2}(2)$

$\dfrac{dy}{dx} = \dfrac{-8x}{(1+4x^2)^2} - \dfrac{2}{1+4x^2} = \dfrac{-8x - 2(1+4x^2)}{(1+4x^2)^2}$

$\dfrac{dy}{dx} = \dfrac{-8x-2-8x^2}{(1+4x^2)^2} = \dfrac{-2(1+4x+4x^2)}{(1+4x^2)^2}$

$\qquad = \dfrac{-2(1+2x)^2}{(1+4x^2)^2}$

28. $y = \sin^{-1} x - \sqrt{1-x^2}$

$\dfrac{dy}{dx} = \dfrac{1}{\sqrt{1-x^2}} - \dfrac{1}{2}(1-x^2)^{-1/2}(-2x)$

$\dfrac{dy}{dx} = \dfrac{1}{\sqrt{1-x^2}} + \dfrac{x}{\sqrt{1-x^2}} = \dfrac{1+x}{\sqrt{1-x^2}}$

29. $y = 3(4 - \cos^{-1} 2x)^3$

$$\frac{dy}{dx} = 9(4 - \cos^{-1} 2x)^2 \left(\frac{1}{\sqrt{1 - (2x)^2}} \right)(2)$$

$$= \frac{18(4 - \cos^{-1} 2x)^2}{\sqrt{1 - 4x^2}}$$

30. $\sin^{-1}(x + y) + y = x^2$

$$\frac{1}{\sqrt{1 - (x+y)^2}} \left(1 + \frac{dy}{dx} \right) + \frac{dy}{dx} = 2x$$

$$\frac{1}{\sqrt{1 - (x+y)^2}} + \frac{1}{\sqrt{1 - (x+y)^2}} \frac{dy}{dx} + \frac{dy}{dx} = 2x$$

$$\frac{dy}{dx} \left[\frac{1}{\sqrt{1 - (x+y)^2}} + 1 \right] = 2x - \frac{1}{\sqrt{1 - (x+y)^2}}$$

$$\frac{dy}{dx} \left[\frac{1 + \sqrt{1 - (x+y)^2}}{\sqrt{1 - (x+y)^2}} \right] = \frac{2x\sqrt{1 - (x+y)^2} - 1}{\sqrt{1 - (x+y)^2}}$$

$$\frac{dy}{dx} = \frac{2x\sqrt{1 - (x+y)^2} - 1}{1 + \sqrt{1 - (x+y)^2}}$$

31. $2\tan^{-1} xy + x = 3$

$$2\frac{1}{1 + x^2 y^2} \left(x \frac{dy}{dx} + y \right) + 1 = 0$$

$$\frac{2x}{1 + x^2 y^2} \frac{dy}{dx} + \frac{2y}{1 + x^2 y^2} = -1$$

$$\frac{2x}{1 + x^2 y^2} \frac{dy}{dx} = -1 - \frac{2y}{1 + x^2 y^2}$$

$$\frac{2x}{1 + x^2 y^2} \frac{dy}{dx} = \frac{-1 - x^2 y^2 - 2y}{1 + x^2 y^2}$$

$$2x \frac{dy}{dx} = -1 - x^2 y^2 - 2y; \quad \frac{dy}{dx} = \frac{-(x^2 y^2 + 2y + 1)}{2x}$$

32. $y = \sqrt{1 - \sin^{-1} 4x}$

$$\frac{dy}{dx} = \frac{1}{2}(1 - \sin^{-1} 4x)^{-1/2} \left[0 - \frac{1}{\sqrt{1 - 16x^2}}(4) \right]$$

$$\frac{dy}{dx} = \frac{-2}{\sqrt{1 - 16x^2}\sqrt{1 - \sin^{-1} 4x}}$$

$$\frac{dy}{dx} = \frac{-2}{\sqrt{(1 - 16x^2)(1 - \sin^{-1} 4x)}}$$

33. (a) $\dfrac{1}{\sqrt{1 - 0.5^2}} = 1.1547005;$

value of derivative

(b) $\dfrac{\sin^{-1} 0.5001 - \sin^{-1} 0.5000}{0.0001} = 1.1547390;$

slope of secant line

34. (a) $\dfrac{1}{(1 + 0.5)^2} = 0.8$

the derivative of $\tan^{-1} x$ at $x = 0.5$.

(b) $\dfrac{\tan^{-1} 0.5001 - \tan^{-1} 0.5000}{0.0001} = 0.79996801$

the slope of the secant line passing through point where $x_1 = 0.5000$ and $x_2 = 0.5001$

35. $y = (\sin^{-1} x)^3$

$$\frac{dy}{dx} = 3(\sin^{-1} x)^2 \frac{1}{\sqrt{1 - x^2}}$$

$$dy = \frac{3(\sin^{-1} x)^2 dx}{\sqrt{1 - x^2}}$$

36. $f(x) = 2x \cos^{-1} x$

$$f'(x) = 2x \cdot \frac{-1}{\sqrt{1 - x^2}} + 2\cos^{-1} x$$

$$f(0) = 0, f'(0) = \pi$$
$$L(x) = \pi x$$

37. $y = \dfrac{x}{\tan^{-1} x};$

$$m_{\tan} = \frac{dy}{dx} = \frac{\tan^{-1} x(1) - x\left[\frac{1}{1 + x^2}\right]}{(\tan^{-1} x)^2} = \frac{\tan^{-1} x - \frac{x}{1 + x^2}}{(\tan^{-1} x)^2}$$

When $x = 0.80, m_{\tan} = \dfrac{\tan^{-1}(0.8) - \frac{0.8}{1 + 0.8^2}}{(\tan^{-1} 0.8)^2}$

$$= 0.41$$

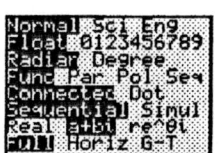 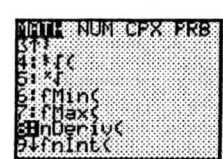

38. $y = \sin^{-1}(x^2 + 1)$

The domain of $\sin^{-1} u$ is $-1 \le u \le 1$.
Here $x^2 + 1$ is always equal to or greater than 1.
$\sin^{-1}(x^2 + 1)$ is defined for $x = 0$ only.

39. $y_1 = \sin^{-1} x, y_2 = \dfrac{1}{\sqrt{(1 - x^2)}}$

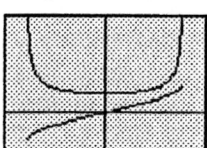

40. $y_1 = \tan^{-1} x, y_2 = \dfrac{1}{(1+x^2)}$

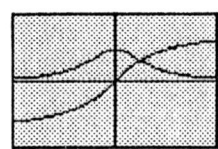

41. $y = \tan^{-1} 2x;\quad \dfrac{dy}{dx} = \dfrac{1}{1+(2x)^2}(2) = \dfrac{2}{1+4x^2};$

$\dfrac{d^2y}{dx^2} = \dfrac{(1+4x^2)(0) - 2(8x)}{(1+4x^2)^2} = \dfrac{-16x}{(1+4x^2)^2}$

42. $\dfrac{d}{dx}(\cot^{-1} u) = -\dfrac{1}{1+u^2}\dfrac{du}{dx}?$

Change $\cot^{-1}$ into $\tan^{-1}$ by definition.

Let $\cot^{-1} u = y$; then $u = \cot y$; $\tan y = \dfrac{1}{u}$

Therefore, $\tan^{-1} \dfrac{1}{u} = y = \cot^{-1} u.$

$\dfrac{d}{dx}\left(\dfrac{1}{u}\right) = \dfrac{d}{dx}(u)^{-1} = -1u^{-2}\dfrac{du}{dx} = \dfrac{-1}{u^2}\dfrac{du}{dx}$

$\dfrac{dy}{dx} = \dfrac{d}{dx}\cot^{-1} u = \dfrac{d}{dx}\tan^{-1}\dfrac{1}{u}$

$\dfrac{d}{dx}\tan^{-1}\dfrac{1}{u} = \dfrac{1}{\left(1+\frac{1}{u^2}\right)}\left(-\dfrac{1}{u^2}\dfrac{du}{dx}\right)$

$\qquad = \dfrac{-1}{\left(\frac{u^2+1}{u^2}\right)u^2}\dfrac{du}{dx} = \dfrac{-1}{1+u^2}\dfrac{du}{dx}$

43. $\dfrac{d}{dx}(\sec^{-1} u) = \dfrac{1}{\sqrt{u^2(u^2-1)}}\dfrac{du}{dx}$

Let $\sec^{-1} u = y$, therefore $u = \sec y$, therefore $\cos y = \dfrac{1}{u}$

Therefore, $\cos^{-1}\dfrac{1}{u} = y = \sec^{-1} u.$

$\dfrac{dy}{dx} = \dfrac{-1}{\sqrt{1-\frac{1}{u^2}}}\left(\dfrac{-1}{u^2}\right)\dfrac{du}{dx} = \dfrac{1}{\frac{\sqrt{u^2-1}}{u}}\left(\dfrac{1}{u^2}\right)\dfrac{du}{dx}$

$\dfrac{dy}{dx} = \dfrac{1}{\sqrt{u^2-1}(u)}\dfrac{du}{dx};\ u = \sqrt{u^2}$

$\dfrac{dy}{dx} = \dfrac{1}{\sqrt{u^2(u^2-1)}}\dfrac{du}{dx}$

44. $\dfrac{d}{dx}(\csc^{-1} u) = -\dfrac{1}{\sqrt{u^2(u^2-1)}}\dfrac{du}{dx}$

Let $\csc^{-1} u = y$, therefore $u = \csc y$, $\dfrac{1}{u} = \sin y$

Therefore, $\sin^{-1}\dfrac{1}{u} = y = \csc^{-1} u.$

$\dfrac{dy}{dx} = \dfrac{1}{\sqrt{1-\frac{1}{u^2}}}\left(\dfrac{-1}{u^2}\right)\dfrac{du}{dx} = \dfrac{-1}{\frac{\sqrt{u^2-1}}{u}}\left(\dfrac{1}{u^2}\right)\dfrac{du}{dx}$

$\dfrac{dy}{dx} = \dfrac{-1}{\sqrt{u^2-1}(u)}\dfrac{du}{dx} = \dfrac{-1}{\sqrt{u^2(u^2-1)}}\dfrac{du}{dx}$

45. $t = \dfrac{1}{\omega}\sin^{-1}\dfrac{A-E}{mE}$

$= \dfrac{1}{\omega}\sin^{-1}\left(\dfrac{A-E}{E}\right)\left(\dfrac{1}{m}\right)$

$= \dfrac{1}{\omega}\sin^{-1}\left(\dfrac{A-E}{E}\right)m^{-1}$

$u = \left(\dfrac{A-E}{E}\right)m^{-1};\ \dfrac{du}{dm} = -\left(\dfrac{A-E}{E}\right)m^{-2}$

$\dfrac{dt}{dm} = \dfrac{1}{\omega\sqrt{1-\left(\frac{-A+E}{E}\right)^2 m^{-2}}}\left(\dfrac{-A+E}{Em^2}\right)$

$= \dfrac{E-A}{\omega Em^2\sqrt{1-\frac{(A-E)^2}{E^2m^2}}}$

$= \dfrac{E-A}{\omega Em^2\sqrt{\frac{E^2m^2-(A-E)^2}{E^2m^2}}}$

$\dfrac{dt}{dm} = \dfrac{E-A}{\omega m\sqrt{E^2m^2-(A-E)^2}}$

46. $\alpha = \cos^{-1}\dfrac{2f-r}{r}$

$\dfrac{d\alpha}{dr} = \dfrac{-1}{\sqrt{1-\left(\frac{2f-r}{r}\right)^2}}\left(\dfrac{d}{dr}\right)\left(\dfrac{2f-r}{r}\right)$

Simplify under radical and the derivative:

$1-\left(\dfrac{2f-r}{r}\right)^2 = 1 - \dfrac{(4f^2-4fr+r^2)}{r^2}$

$= \dfrac{r^2-4f^2+4fr-r^2}{r^2}$

$= \dfrac{4fr-4f^2}{r^2}$

$\dfrac{d}{dr}\dfrac{2f-r}{e} = \dfrac{r(-1)-(2f-r)(1)}{r^2}$

$= \dfrac{-r-2f+r}{r^2} = \dfrac{-2f}{r^2}$

Therefore,

$\dfrac{d\alpha}{dr} = \dfrac{-1}{\sqrt{\frac{4(fr-f^2)}{r^2}}}\left(-\dfrac{2f}{r^2}\right)$

$\dfrac{d\alpha}{dr} = \dfrac{2f}{\frac{2}{r}\sqrt{fr-f^2}(r^2)} = \dfrac{f}{r\sqrt{fr-f^2}} = \dfrac{f}{r\sqrt{f(r-f)}}$

47. $\tan\theta = \dfrac{h}{x}$

$\theta = \tan^{-1}\dfrac{h}{x}$

$\dfrac{d}{dx}\dfrac{h}{x} = \dfrac{d}{dx}hx^{-1} = -hx^{-2} = -\dfrac{h}{x^2}$

$\dfrac{d\theta}{dx} = \dfrac{1}{1+\frac{h^2}{x^2}} \cdot -\dfrac{h}{x^2}$

$\dfrac{d\theta}{dx} = \dfrac{-h}{\frac{x^2+h^2}{x^2}(x^2)}$

$\dfrac{d\theta}{dx} = \dfrac{-h}{h^2+x^2}$

48. $a^2 = b^2 + c^2 - 2bc\cos A$ (Law of Cosines)

$x^2 = 5^2 + 8^2 - 2(5)(8)\cos A$

Therefore,

$\cos A = \dfrac{5^2 + 8^2 - x^2}{80}$

$\cos A = \dfrac{89 - x^2}{80}$

$A = \cos^{-1}\dfrac{89 - x^2}{80}$

$\dfrac{89 - x^2}{80} = \dfrac{89}{80} - \dfrac{x^2}{80}; \dfrac{d}{dx}\dfrac{89 - x^2}{80} = -\dfrac{x}{40}$

$\dfrac{dA}{dx} = \dfrac{-1}{\sqrt{1 - \left(\frac{89-x^2}{80}\right)^2}}\left(-\dfrac{x}{40}\right)$

$\dfrac{dA}{dx}\Big|_{x=6} = \dfrac{-1}{\sqrt{1 - \left(\frac{89-36}{80}\right)^2}}\dfrac{6}{40} = (1.335)\left(\dfrac{6}{40}\right)$

$= 0.20 \text{ rad/cm}$

7.7   Applications

1. Points of intersection occur when $\sin x = \cos x$.

$y_1 = \sin x; y_2 = \cos x; \dfrac{dy_1}{dx} = \cos x; \dfrac{dy_2}{dx} = -\sin x$

At points of intersection, $\dfrac{dy_1}{dx} = -\dfrac{dy_2}{dx}$.

2. $y = \tan x$

$\dfrac{dy}{dx} = (\sec x)^2 \geq 1$ for all x for which y is defined.

3. $y = \tan^{-1} x$

$\dfrac{dy}{dx} = \dfrac{1}{1+x^2} > 0$ for all x.

4. $y = \sin x + \cos x = f(x); (0 \leq x \leq 2\pi)$

$f'(x) = \cos x - \sin x; f''(x) = -\sin x - \cos x$

(1) $f'(x) = 0, \sin x = \cos x; \tan x = 1$ for

$x = \dfrac{\pi}{4}, \dfrac{5\pi}{4}$

$f''\left(\dfrac{\pi}{4}\right) < 0$ concave down, therefore max. $\left(\dfrac{\pi}{4}, 1.414\right)$

$f''\left(\dfrac{5\pi}{4}\right) > 0$ concave up, therefore min. $\left(\dfrac{5\pi}{4}, -1.414\right)$

(2) $f''(x) = 0, \sin x = -\cos x; \tan x = -1$ for

$x = \dfrac{3\pi}{4}, \dfrac{7\pi}{4}$

Therefore, infl. pt. $\left(\dfrac{3\pi}{4}, 0\right)$ and $\left(\dfrac{7\pi}{4}, 0\right)$

(3) $f(0) = 1$, y-intercept $= 1$

(4) $f(2\pi) = 1$

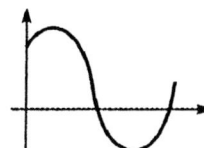

5. $y = x - \tan x; \left(-\dfrac{\pi}{2} < x < \dfrac{\pi}{2}\right); \dfrac{dy}{dx} = 1 - \sec^2 x;$

$\dfrac{d^2y}{dx^2} = 0 - 2\sec x(\sec x \tan x) = -2\sec^2 x \tan x.$

(1) $\dfrac{dy}{dx} < 0$ for $\dfrac{-\pi}{2} < x < 0; \dfrac{dy}{dx} = 0$ for $x = 0;$ $\dfrac{dy}{dx} < 0$ for $0 < x < \dfrac{\pi}{2}$. The function decreases from $-\dfrac{\pi}{2}$ to $\dfrac{\pi}{2}$ $(x \neq 0)$ and has one critical point at $x = 0$.

(2) Since $0 - \tan = 0$, there is an intercept at $(0,0)$.

(3) $\dfrac{d^2y}{dx^2} = 0$ when $\sec^2 x = 0$ or $\tan x = 0$. Since $|\sec x| \geq 1$ for all x, $\sec^2 x \geq 1$ for all x, and $\sec^2 x = 0$ has no solution. Between $-\dfrac{\pi}{2}$ and $\dfrac{\pi}{2}$, $\tan x = 0$ when $x = 0$.

(4) $-2\sec^2 x \tan x > 0$ when $\tan x < 0$. $\tan x < 0$ for $-\dfrac{\pi}{2} < x < 0$. The graph is concave up in this region. $-2\sec^2 x \tan x < 0$ when $\tan x > 0$ or x is between 0 and $\dfrac{\pi}{2}$. The graph is concave down in this region. Since there is a change of sign in the second derivative, the only critical point is an inflection point, and there are no maximum or minimum points.

(5) As $x \to -\frac{\pi}{2}$ from the right, $x - \tan x \to$ $-\frac{\pi}{2} - (-\infty)$ or $+\infty$; $x = -\frac{\pi}{2}$ is an asymptote.

Summarizing, the function decreases, intersects the x-axis at 0, is concave up for $-\frac{\pi}{2} < x < 0$, concave down for $0 < x < \frac{\pi}{2}$, has point of inflection at $x = 0$, and asymptotes at $x = -\frac{\pi}{2}$ and $\frac{\pi}{2}$.

Dec., $x > 0, x < 0$

Infl. $(0, 0)$

Asym., $x = \frac{\pi}{2}, x = -\frac{\pi}{2}$

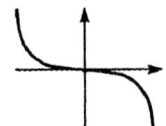

6. $y = f(x) = 2 \sin x + \sin 2x \, (0 \le x \le 2\pi)$

$f'(x) = 2 \cos x + 2 \cos 2x = 0$

$2 \cos^2 x + \cos x - 1 = 0$

$(2 \cos x - 1)(\cos x + 1) = 0$

$\cos x = \frac{1}{2} \qquad\qquad \cos x = -1$

$x = \frac{\pi}{3}, \frac{5\pi}{3} \qquad\qquad x = \pi$

$f''(x) = -2 \sin x - 4 \sin 2x$

$\sin x + 4 \sin x \cos x = 0$

$\sin x (1 + 4 \cos x) = 0$

$\sin x = 0 \qquad\qquad \cos x = \frac{-1}{4}$

$x = \pi \qquad\qquad x = 1.82, 4.46$

$f''\left(\frac{\pi}{3}\right) = -5.2 < 0, \left(\frac{\pi}{3}, \frac{3\sqrt{3}}{2}\right)$ max

$f''\left(\frac{5\pi}{3}\right) = 5.2 > 0, \left(\frac{5\pi}{3}, \frac{-3\sqrt{3}}{2}\right)$ min

$f''(x)$ changes sign for $x = \pi, 1.82, 4.46$.

$(\pi, 0), (1.82, 1.46), (4.46, -1.45)$

are inflection points.

$f(x) = 0$ for $x = 0, \pi, 2\pi$

7. $y = \sin 2x$ at $x = \frac{5\pi}{8}$

$\frac{dy}{dx} = 2 \cos 2x = m_{\text{TL}}$

$m_{\text{TL}}|_{x=5\pi/8} = -1.414$
$= -\sqrt{2};$

$f\left(\frac{5\pi}{8}\right) = -0.707$

$= -\frac{\sqrt{2}}{2}$

Equation of T.L.: $y + \frac{\sqrt{2}}{2} = -\sqrt{2}\left(x - \frac{5\pi}{8}\right)$

$8y + 4\sqrt{2} = -8\sqrt{2}x + 5\sqrt{2}\pi$
$8\sqrt{2}x + 8y + 4\sqrt{2} - 5\pi\sqrt{2} = 0$

8. $y = \tan^{-1}\left(\frac{x}{2}\right)$ at $x = 3$

$\frac{dy}{dx} = \frac{1}{1 + \frac{x^2}{4}}\left(\frac{1}{2}\right) = \frac{2}{4 + x^2}$

$= m_{\text{TL}}|_{x=3} = \frac{2}{13}$

$m_{\text{NL}} = -\frac{13}{2};$

$f(3) = \tan^{-1}\left(\frac{3}{2}\right) = 0.9828$

Equation of N.L.: $y - 0.9828 = -\frac{13}{2}(x - 3)$

$2y - 1.9656 = -13x + 39; \; 13x + 2y - 40.97 = 0$

9. $x^2 - 4 \sin x = 0$

By a rough sketch of the graph using $f(0) = 0$, $f(1) = -2.37$, and $f(2) = 0.36$, the root is between 1 and 2 and is possibly closer to 2. Let $x_1 = 1.7$:

n	x_n	$f(x_n)$	$f'(x_n)$	$x_n - \frac{f(x_n)}{f'(x_n)}$
1	1.7	-1.0766592	3.915378	1.9749822
2	1.9749822	0.2228632	5.5230459	1.9346307
3	1.9346307	0.0046391	5.2927023	1.9337542
4	1.9337542	0.0000023	5.2876722	1.9337538

$x_4 = 1.9337538$

10. $\tan x = 2x$

Smallest positive root lies between $\dfrac{\pi}{4}$ and $\dfrac{3\pi}{8}$.

Let $x_1 = \dfrac{5\pi}{16} = 0.98$

$f(x) = \tan x - 2x$

$f'(x) = \sec^2 x - 2$

n	x_n	$f(x_n)$	$f'(x_n)$	$x_n - \frac{f(x_n)}{f'(x_n)}$
1	0.98	-0.4690417	1.2229566	1.3635310
2	1.3635310	2.0283843	21.6142692	1.2696863
3	1.2696863	0.6806907	9.3688080	1.1970313
4	1.1970313	0.1556509	5.5010389	1.1687365
5	1.1687365	0.0142322	4.5305174	1.1655951
6	1.1655951	0.0001504	4.4351489	1.1655612
7	1.1655612	θ		

Therefore, root is 1.1655612.

11. $y = 6\cos x - 8\sin x$; maximum value occurs where $f'(x) = 0$.

$f'(x) = -6\sin x - 8\cos x = 0$

$\sin x = -\dfrac{8}{6}\cos x$

$\tan x = -\dfrac{8}{6} = -\dfrac{4}{3}$

$\alpha = 0.927$ (a $3, 4, 5$ triangle)

$Q_1 = 2.214, Q_2 = 5.356$

$f''(x) = -6\cos x + 8\sin$

$f''(2.214) > 0$, min; $f''(5.356) < 0$, max.

Maximum occurs where $x = 5.356$ rad.

$f(5.356) = 6\left(\dfrac{3}{5}\right) - 8\left(-\dfrac{4}{5}\right) = \dfrac{18}{5} + \dfrac{32}{5} = \dfrac{50}{5} = 10$

Maximum value of $y = 10$ units.

12. $y = \cos 2x + 2\sin x$; minimum value?

$f'(x) = -2\sin 2x + 2\cos x$;

$f''(x) = -4\cos 2x - 2\sin x$

$f'(x) = 0$; $-2\sin 2x + 2\cos x = 0$

$-2(2\sin x\cos x) + 2\cos x = 0$

$-4\sin x\cos x + 2\cos x = 0$

$-2\cos x(2\sin x - 1) = 0$

$\cos x = 0$; $x = \dfrac{\pi}{2}, \dfrac{3\pi}{2}$

$\sin x = \dfrac{1}{2}$; $x = \dfrac{\pi}{6}, \dfrac{5\pi}{6}$

$f''\left(\dfrac{\pi}{2}\right) > 0$, min.; $f''\left(\dfrac{3\pi}{2}\right) > 0$, min.

$f''\left(\dfrac{\pi}{6}\right) < 0$, max.; $f''\left(\dfrac{5\pi}{6}\right) < 0$, max.

$f\left(\dfrac{\pi}{6}\right) = 1.5$, $f\left(\dfrac{5\pi}{6}\right) = 1.5$

$f\left(\dfrac{\pi}{2}\right) = 1$, $f\left(\dfrac{3\pi}{2}\right) = -3$

Therefore, the minimum value of y is -3.

13. $y = 0.50\sin 2t + 0.30\cos t$;

$v = \dfrac{dy}{dt} = 1.00\cos 2t - 0.30\sin t$

$v\big|_{t=0.40s} = 1.00\cos 0.80 - 0.30\sin 0.40$
$= 0.58$ ft/s

$a = \dfrac{d^2y}{dt^2} = -2.00\sin 2t - 0.30\cos t$

$a\big|_{t=0.40s} = -200\sin 0.80 - 0.30\cos 0.40$
$= -1.7$ ft/s^2

14. $h = 12.1 + 2.0\sin\left[\dfrac{\pi}{6}(x - 2.7)\right]$

$\dfrac{dh}{dx} = 2.0\cos\left[\dfrac{\pi}{6}(x - 2.7)\right] \cdot \dfrac{\pi}{6} = 0$

$x = 5.7, 11.7$

$5.7 \rightarrow$ June 21st is longest day with 14.1 hours

$11.7 \rightarrow$ Dec 22nd is shortest day with 10.1 hours.

15. $T_x = 46.6\cos\theta$

$\dfrac{dT_x}{dt} = 46.6\left(-\sin\theta\dfrac{d\theta}{dt}\right)$

$\dfrac{dT_x}{dt} = -46.6\sin\theta\dfrac{d\theta}{dt}$

$\dfrac{d\theta}{dt} = 0.36°$/s; $\theta = 14.2°$; $\dfrac{d\theta}{dt} = \dfrac{0.36°\pi}{180°}$ r/s

$\dfrac{dT_x}{dt} = -46.6(\sin 14.2°)\dfrac{0.36\pi}{180}$

$\dfrac{dT_x}{dt} = -0.072$ lb/s

16. $P_a = P\sec\theta$; $P = 12$ W; $\dfrac{d\theta}{dt} = 0.050$ rad/min;

$\theta = 40.0°$

$\dfrac{dPa}{dt} = P\sec\theta\tan\theta\dfrac{d\theta}{dt}$

$\dfrac{dPa}{dt}\bigg|_{\theta=40.0°} = 12\sec 40.0\tan 40.0(0.050)$

$\dfrac{dPa}{dt} = 0.657$ W/min

17. $x = 2.625 \cos 12\pi t;$

$$\frac{dx}{dt} = -2.625(12\pi) \sin 12\pi t = -31.50\pi \sin 12\pi t;$$

$$y = 2.625 \sin 12\pi t;$$

$$\frac{dy}{dt} = 2.625(12\pi) \cos 12\pi t = 31.50\pi \cos 12\pi t$$

$$\left.\frac{dx}{dy}\right|_{t=1.250} = -31.50\pi \sin 47.12 = 0$$

$$\left.\frac{dy}{dt}\right|_{t=1.250} = -31.50\pi \cos 47.12$$
$$= 31.50\pi(-1.000) = -98.96 \text{ in/s}$$

$$v = \sqrt{(0)^2 + (-98.96)^2} = 98.96 \text{ in/s};$$
$$\theta = 270° \, (v_x = 0, v_y < 0)$$

18. $x = 2\cos 3t; \quad v_x = -6\sin 3t; \quad v_x|_{t=4.1} = 1.579$
$y = \cos 2t; \quad v_y = -2\sin 2t; \quad v_y|_{t=4.1} = -1.881$
$v = \sqrt{(1.579)^2 + (-1.881)^2}$
$v = 2.5 \text{ cm/s}$

$$\alpha = \tan^{-1} \frac{1.881}{1.579} = 49.988°$$

$$\theta = 310.01°$$

Therefore, $v = 2.5$ cm/s at $\theta = 310°$

19. $x = 2.625 \cos 12\pi t; \quad v_x = 2.625(-12\pi \sin 12\pi t)$
$v_x = -31.5\pi \sin 12\pi t$
$a_x = -31.5\pi(12\pi \cos 12\pi t) = -387\pi^2 \cos 12\pi t$
$a_x|_{t=1.250} = 3730.71$
$y = 2.625 \sin 12\pi t; \quad v_y = 2.625(12\pi \cos 12\pi t)$
$v_y = 31.5\pi \cos 12\pi t$
$a_y = 31.5\pi(-12\pi \sin 12\pi t) = -378\pi^2 \sin 12\pi t$
$a_y = 0$

Therefore, $a = 3731$ in/s^2 at $\theta = 0°$

20. $v_x = -6\sin 3t; \quad a_x = -18\cos 3t; \quad a_x|_{t=4.1} = -17.365$
$v_y = -2\sin 2t; \quad a_y = -4\cos 2t; \quad a_y = 1.357$
$a = 17.418$

$$\alpha = \tan^{-1} \frac{a_y}{a_x} = 0.078 = -4°$$

Therefore $a = 17.4$ cm/s^2 at $\theta = 176°$

21. $$s = 16t^2$$

$$\tan \theta = \frac{200 - s}{100} = \frac{200 - 16t^2}{100}$$
$$= 2 - 0.16t^2$$
$$\theta = \tan^{-1}(2 - 0.16t^2)$$
$$\frac{d\theta}{dt} = \frac{1}{1 + (2 - 0.16t^2)^2}(-0.32t)$$
$$= \frac{-0.32t}{5 - 0.64(t)^2 + 0.0256(t)^4}$$
$$\left.\frac{d\theta}{dt}\right|_{t=1.0} = \frac{-0.32(1.0)}{5 - 0.64(1.0)^2 + 0.0256(1.0)^4}$$
$$= -0.073 \text{ rad/s}$$

22. $\tan \theta = \dfrac{450}{x}; \quad \dfrac{d\theta}{dt} = -0.215$ rad/s

$x = 450 \cot \theta$
$x = 50.0$ ft; $\theta = 83.66°$

$$\frac{dx}{dt} = 450\left(-\csc^2 \theta \frac{d\theta}{dt}\right)$$

$$\left.\frac{dx}{dt}\right|_{x=50.0} = -450 \csc^2 83.66°(-0.215) = 97.9 \text{ ft/s}$$

23. $\cos \theta = \dfrac{225}{x}; \quad \dfrac{d\theta}{dt} = 1.5°/s$

$x = 225 \sec \theta; \quad x = 315, \theta = 44.415°$

$$\frac{dx}{dt} = 225 \sec \theta \tan \theta \frac{d\theta}{dt}$$

$$\frac{dx}{dt} = 225 \sec 44.4° \tan 44.4° \left(\frac{1.5\pi}{180}\right)$$

$$\frac{dx}{dt} = 8.08 \text{ ft/s}$$

24. $\tan \theta = \dfrac{x}{50}; \quad \dfrac{dx}{dt} = 12.0$ ft/s; $t = 10.0$ s; $x = 120$ ft

$$\theta = \tan^{-1} \frac{x}{50}$$

$$\frac{d\theta}{dt} = \frac{1}{1 + \dfrac{x^2}{2500}} \cdot \frac{1}{50} \cdot \frac{dx}{dt}$$

$$\frac{d\theta}{dt} = \frac{50}{2500 + x^2} \cdot \frac{dx}{dt}$$

$$\frac{d\theta}{dt} = \frac{50}{2500 + 120^2}(12)$$

$$\frac{d\theta}{dt} = 0.036 \text{ rad/s}$$

25.
$$\mu = \tan\theta; \ \theta = 20° = \frac{\pi}{9} = 0.349 \text{ rad}$$

$$d\theta = 1° = \frac{\pi}{180} = 0.0175 \text{ rad};$$

$$\frac{d\mu}{d\theta} = \sec^2\theta$$

$$d\mu = \sec^2\theta \, d\theta$$

$$d\mu|_{\theta=20°} = 0.349 = (\sec^2 0.349)(0.0175)$$
$$= (1.064)^2(0.0175) = 0.020$$

26. $p(t) = 0.0307\cos^2 120\pi t$
$p(0.0010) = 0.0265$
$p'(t) = 0.0614\cos 120\pi t(-\sin 120\pi t)(120\pi)$
$p'(0.0010) = -7.923$
$p - 0.0265 = -7.923(t - 0.0010)$

The linearization is $p(t) = -7.923t + 0.0344$.

27. $a^2 = b^2 + c^2 - 2bc\cos\theta; \ \theta = 38.38°; \ d\theta = 0.15°$
$a^2 = 75.37^2 + 82.04^2 - 2(75.37)(82.04)\cos\theta$
$a = 52.12$

$$2a\frac{da}{d\theta} = 0 + 0 + 2(75.37)(82.04)\sin\theta$$

$$da = \frac{2(75.37)(82.04)\sin\theta \, d\theta}{2(a)}$$

$$da|_{d\theta=0.15°} = \frac{2(75.37)(82.04)\sin 38.38(0.15\pi/180)}{2(52.12)}$$

$$= 0.19 \text{ m}$$

28. $2b = $ side of square; $a = x\sin\theta; \ x = b\sec\theta$
$y = 2b - 2a = 2b - 2x\sin\theta = 2b - 2b\tan\theta$
$L = 4x + y = 4b\sec\theta + 2b - 2b\tan\theta$

$$\frac{dL}{d\theta} = 4b\sec\theta\tan\theta - 2b\sec^2\theta$$

$$0 = 2b\sec\theta(2\tan\theta - \sec\theta), \sec\theta \neq 0$$

$$2\tan\theta = \sec\theta; \ \sin\theta = \frac{1}{2}; \ \theta = 30°$$

29. $s = kwd^2; \ d^2 = 256 - w^2; \ \cos\theta = \dfrac{w}{16.0};$

$w = 16.0\cos\theta;$

$S = k(16.0\cos\theta)[256 - (16.0\cos\theta)^2];$

$S = 4100k[\cos\theta - \cos^3\theta];$

$$\frac{dS}{d\theta} = 4100k[-\sin\theta - 3\cos^2\theta(-\sin\theta)];$$

$$\frac{dS}{d\theta} = 4100k(-\sin\theta + 3\sin\theta\cos^2\theta)$$

Maximum occurs when $\dfrac{dS}{d\theta} = 0$.

$4100k(-\sin\theta + 3\sin\theta\cos^2\theta) = 0;$

$-\sin\theta(1 - 3\cos^2\theta) = 0 - \sin\theta = 0;$

$\sin\theta = 0; \ \theta = 0, \theta = \pi; \ 1 - 3\cos^2\theta = 0;$

$$\cos^2\theta = \frac{1}{3}; \ \cos\theta = \sqrt{\frac{1}{3}}$$

$$w = 16.0\sqrt{\frac{1}{3}} = 9.24 \text{ in}$$

$$d = \sqrt{16.0^2 - 16.0^2\left(\frac{1}{3}\right)} = 13.1 \text{ in}$$

30. $2a + 2b = 60, a + b = 30, h = b\cot\theta$

$$A = \frac{1}{2}(2b)h = b(b\cot\theta) = b^2\cot\theta$$

$$a = b\csc\theta; \ b\csc\theta + b = 30; \ b = \frac{30}{1 + \csc\theta}$$

$$A = \left(\frac{30}{1 + \csc\theta}\right)^2\cot\theta = \frac{900\cot\theta}{(1 + \csc\theta)^2}$$

$$\frac{dA}{d\theta}$$

$$= \frac{900[(1+\csc\theta)^2(-\csc^2\theta) - \cot\theta(2)(1+\csc\theta)(-\csc\theta)\cot\theta]}{(1 + \csc\theta)^4}$$

$$= \frac{900[(1 + \csc\theta)(-\csc^2\theta) + 2\cot^2\theta\csc\theta]}{(1 + \csc\theta)^3}$$

Set $\dfrac{dA}{d\theta} = 0$; $(1 + \csc\theta)(-\csc\theta) + 2\cot^2\theta = 0$

$-\csc\theta - \csc^2\theta + 2(\csc^2\theta - 1) = 0$
$\csc^2\theta - \csc\theta - 2 = 0$
$(\csc\theta - 2)(\csc\theta + 1) = 0$
$\csc\theta = 2, \csc\theta \neq -1$

$\sin\theta = \dfrac{1}{2}, \theta = 30°$, vertex angle $= 60°$

31. $y = 6.0\csc\theta + 4.0\sec\theta$

$$\frac{dy}{d\theta} = 6.0(-\csc\theta\cot\theta) + 4(\sec\theta\tan\theta)$$

$$\frac{dy}{d\theta} = -6.0\csc\theta\cot\theta + 4\sec\theta\tan\theta$$

$$\frac{dy}{d\theta} = -6.0\frac{1}{\sin\theta}\cdot\frac{\cos\theta}{\sin\theta} + 4\frac{1}{\cos\theta}\cdot\frac{\sin\theta}{\cos\theta}$$

$$\frac{dy}{d\theta} = \frac{-6.0\cos^3\theta + 4\sin^3\theta}{\sin^2\theta\cos^2\theta}$$

$$-6.0\cos^3\theta + 4\sin^3\theta = 0; \quad 4\sin^3\theta = 6\cos^3\theta$$

$$\sin^3\theta = \frac{6}{4}\cos^3\theta; \quad \sin\theta = \sqrt[3]{\frac{3}{2}}\cos\theta$$

$$\tan\theta = \sqrt[3]{\frac{3}{2}}; \quad \theta = \tan^{-1}\sqrt[3]{\frac{3}{2}} = 0.853 \text{ rad}$$

$$y_{\theta=0.853} = 14 \text{ ft}$$

32. $\theta = \tan^{-1}\dfrac{8.0+25.0}{x} - \tan^{-1}\dfrac{25.0}{x}$

$$\frac{d\theta}{dx} = \frac{1}{1+\dfrac{33.0^2}{x^2}}\left(-\frac{33.0}{x^2}\right) - \frac{1}{1+\dfrac{25.0^2}{x^2}}\left(-\frac{25.0}{x^2}\right)$$

$$0 = \frac{-33.0}{x^2+33.0^2} + \frac{25.0}{x^2+25.0^2};$$

$$\frac{33.0}{x^2+33.0^2} = \frac{25.0}{x^2+25.0^2}$$

$$33.0x^2 + 33.0(25.0)^2 = 25.0x^2 + 25.0(33.0)^2$$
$$8.0x^2 = 25.0(33.0)(33.0-25.0)$$
$$x^2 = 25.0(33.0); \quad x = 28.7 \text{ ft}$$

Chapter 7 Review Exercises

1. $y = 3\cos(4x-1);$

$$\frac{dy}{dx} = [-3\sin(4x-1)][4] = -12\sin(4x-1)$$

2. $y = 4\sec(1-x^3)$

$$\frac{dy}{dx} = 4\sec(1-x^3)\tan(1-x^3)(-3x^2)$$

$$\frac{dy}{dx} = -12x^2\sec(1-x^3)\tan(1-x^3)$$

3. $u = 0.2\tan\sqrt{3-2v}$

$$\frac{du}{dv} = 0.2\sec^2\sqrt{3-2v}\,\frac{1}{2\sqrt{3-2v}}(-2)$$

$$\frac{du}{dv} = \frac{-0.2\sec^2\sqrt{3-2v}}{\sqrt{3-2v}}$$

4. $y = 5\sin(1-6x)$

$$\frac{dy}{dx} = 5\cos(1-6x)(-6)$$

$$\frac{dy}{dx} = -30\cos(1-6x)$$

5. $y = \csc^2(3x+2);$

$$\frac{dy}{dx} = 2\csc(3x+2)[-\csc(3x+2)\cot(3x+2)](3)$$
$$= -6\csc^2(3x+2)\cot(3x+2)$$

6. $r = \cot^2 5\pi\theta$

$$\frac{dr}{d\theta} = 2\cot 5\pi\theta(-\csc^2 5\pi\theta)(5\pi)$$

$$\frac{dr}{d\theta} = -10\pi\cot 5\pi\theta\csc^2 5\pi\theta$$

7. $y = 3\cos^4 x^2$

$$\frac{dy}{dx} = 3(4\cos^3 x^2)(-\sin x^2)(2x)$$

$$\frac{dy}{dx} = -24x\cos^3 x^2\sin x^2$$

8. $y = 2\sin^3\sqrt{x}$

$$\frac{dy}{dx} = 2(3)\sin^2\sqrt{x}\cos\sqrt{x}\,\frac{1}{2\sqrt{x}}$$

$$\frac{dy}{dx} = \frac{3\sin^2\sqrt{x}\cos\sqrt{x}}{\sqrt{x}}$$

9. $y = 3\tan^{-1}\left(\dfrac{x}{3}\right);$

$$\frac{dy}{dx} = 3\left[\frac{1}{1+\left(\frac{x}{3}\right)^2}\right]\frac{1}{3} = \frac{1}{1+\left(\frac{x}{3}\right)^2}$$

$$= \frac{1}{1+\frac{x^2}{9}} = \frac{9}{9+x^2}$$

10. $y = 0.4\cos^{-1}(2\pi t+1)$

$$\frac{dy}{dt} = \frac{-0.40}{\sqrt{1-(2\pi t+1)^2}}(2\pi)$$

$$= \frac{-0.4(2\pi)}{\sqrt{1-4\pi^2 t^2-4\pi t-1}}$$

$$\frac{dy}{dt} = \frac{-0.4(2\pi)}{\sqrt{4\pi(-t)(\pi t+1)}}$$

$$= \frac{-0.4(2\pi)}{2\sqrt{\pi}\sqrt{-t(\pi t+1)}}\frac{\sqrt{\pi}}{\sqrt{\pi}}$$

$$\frac{dy}{dt} = \frac{-0.4\sqrt{\pi}}{\sqrt{-t(\pi t+1)}}$$

11. $y = \sin^{-1}(\cos x)$

$$\frac{dy}{dx} = \frac{1}{\sqrt{1-\cos^2 x}}(-\sin x)$$

$$= \frac{-\sin x}{\sqrt{\sin^2 x}}$$

$$\frac{dy}{dx} = \frac{-\sin x}{|\sin x|}$$

$$\frac{dy}{dx} = \pm 1, x \neq k\pi$$

12. $y = \sin(\tan^{-1} x)$

$$\frac{dy}{dx} = \cos(\tan^{-1} x)\frac{1}{1 + x^2}$$

$$\frac{dy}{dx} = \frac{\cos(\tan^{-1} x)}{1 + x^2}$$

13. $y = \sqrt{\csc 4x + \cot 4x} = (\csc 4x + \cot 4x)^{1/2}$

$$\frac{dy}{dx} = \frac{1}{2}(\csc 4x + \cot 4x)^{-1/2}(-4\csc 4x \cot 4x - 4\csc^2 4x)$$

$$= \frac{1}{2}(\csc 4x + \cot 4x)^{-1/2}(-4\csc 4x)(\csc 4x + \cot 4x)$$

$$= -2\csc 4x(\csc 4x + \cot 4x)^{1/2}$$

$$= (-2\csc 4x)\sqrt{\csc 4x + \cot 4x}$$

14. $y = \cos^2(\tan x)$

$$\frac{dy}{dx} = 2\cos(\tan x)(-\sin(\tan x))\sec^2 x$$

$$\frac{dy}{dx} = -2\cos(\tan x)\sin(\tan x)\sec^2 x$$

15. $y = (1 + \sin 2x)^4$

$$\frac{dy}{dx} = 4(1 + \sin 2x)^3 \cos 2x(2)$$

$$\frac{dy}{dx} = 8\cos 2x(1 + \sin 2x)^3$$

16. $y = \sqrt{\dfrac{1 + \cos 2x}{2}} = \pm\cos x$

$$\frac{dy}{dx} = \pm\sin x$$

17. $y = \dfrac{x^2}{\tan^{-1} 2x}$

$$\frac{dy}{dx} = \frac{\tan^{-1} 2x(2x) - x^2\frac{1}{1+4x^2}(2)}{(\tan^{-1} 2x)^2}$$

$$\frac{dy}{dx} = \frac{2x\left(\tan^{-1} 2x - \frac{x}{1+4x^2}\right)}{(\tan^{-1} 2x)^2} \cdot \frac{1 + 4x^2}{1 + 4x^2}$$

$$\frac{dy}{dx} = \frac{2x(1 + 4x^2)\tan^{-1} 2x - 2x^2}{(1 + 4x^2)(\tan^{-1} 2x)^2}$$

18. $y = \dfrac{\sin^{-1} x}{4x}$

$$\frac{dy}{dx} = \frac{4x\frac{1}{\sqrt{1-x^2}} - \sin^{-1} x(4)}{16x^2}$$

$$\frac{dy}{dx} = \frac{x - \sqrt{1 - x^2}\sin^{-1} x}{4x^2\sqrt{1 - x^2}}$$

19.

$$y^2 \sin 2x + \tan x = 0$$

$$y^2 \cos 2x(2) + \sin 2x(2yy') + \sec^2 x = 0$$

$$2y^2 \cos 2x + 2y \sin 2x y' + \sec^2 x = 0$$

$$y' = -\frac{2y^2 \cos 2x + \sec^2 x}{2y \sin 2x}$$

20. $y = x(\sin^{-1} x)^2 + 2\sqrt{1 - x^2}\sin^{-1} x - 2x$

$$\frac{dy}{dx} = x(2\sin^{-1} x)\frac{1}{\sqrt{1 - x^2}} + (\sin^{-1} x)^2$$

$$+ 2\sqrt{1 - x^2}\frac{1}{\sqrt{1 - x^2}} + \sin^{-1} x\left(2\frac{1}{2\sqrt{1 - x^2}}\right)($$

$$\frac{dy}{dx} = (\sin^{-1} x)^2$$

21. $y = x\cos^{-1} x - \sqrt{1 - x^2}$

$$\frac{dy}{dx} = x\left(\frac{-1}{\sqrt{1 - x^2}}\right) + \cos^{-1} x - \frac{-2x}{2\sqrt{1 - x^2}}$$

$$\frac{dy}{dx} = \cos^{-1} x$$

22. $x + \sec^2(xy) = 1$

$$1 + 2\sec(xy)\sec(xy)\tan(xy)(xy' + y) = 0$$

$$2x\sec^2(xy)\tan(xy)y' + 2\sec^2(xy)\tan(xy)y = -1$$

$$y' = \frac{-1 - 2\sec^2(xy)\tan(xy)y}{2x\sec^2(xy)\tan(xy)}$$

23. $x\sin 2y = y\cos 2x$

$$x(\cos 2y)(2y') + (\sin 2y)(1) = y(-\sin 2x)(2) + (\cos 2x)y'$$

$$(2x\cos 2y - \cos 2x)y' = -2y\sin 2x - \sin 2y$$

$$y' = \frac{2y\sin 2x + \sin 2y}{\cos 2x - 2x\cos 2y}$$

24. $\tan^{-1}\dfrac{y}{x} = x^2 y$

$$\frac{1}{1 + \frac{y^2}{x^2}}\frac{xy' - y(1)}{x^2} = x^2 y' + y(2x)$$

$$\frac{xy' - y}{x^2 + y^2} = x^2 y' + 2xy$$

$$xy' - y = x^4 y' + 2x^3 y + x^2 y^2 y' + 2xy^3$$

$$(x - x^4 - x^2 y^2)y' = y + 2x^3 y + 2xy^3$$

$$y' = \frac{y + 2x^3 y + 2xy^3}{x - x^2 y^2 - x^4}$$

25. $y = x - \cos x$; $\dfrac{dy}{dx} = 1 + \sin x$; $\dfrac{d^2y}{dx^2} = \cos x$ Infl.:

$$\left(\frac{1}{2}\pi, \frac{1}{2}\pi\right), \left(\frac{3}{2}\pi, \frac{3}{2}\pi\right)$$

(a) $x = 0, y = 0 - \cos 0 = 0 - 1 = -1$; $(0, -1)$ is an intercept. $x - \cos x = 0$ when $x = \cos x$; $x = 0.74$ (see Table 3); $(0.74, 0)$ is an intercept.

(b) y is defined for all x; no asymptotes.

(c) Critical points occur at $1 + \sin x = 0$,

$\sin x = -1$, $x = -\dfrac{\pi}{2}, \dfrac{3\pi}{2}, \dfrac{7\pi}{2}$, etc.

(d) Inflections occur at $\cos x = 0$, $x = -\dfrac{\pi}{2}, \dfrac{\pi}{2}, \dfrac{3\pi}{2}$,

$\dfrac{5\pi}{2}$, etc., since the second derivative undergoes a change of sign at each of these points.

(e) All critical points are inflections; no maximum or minimum points.

(f) Checking concavity at $x = 0$, $-\cos 0 = -1$; the graph is concave up at $(0, -1)$ and on each side of this point up to the inflection points at $\left(-\frac{\pi}{2}, -\frac{\pi}{2}\right)$ and $\left(\frac{\pi}{2}, \frac{\pi}{2}\right)$. It will switch concavity again at each subsequent inflection point.

x	$-\frac{3\pi}{2}$	$-\pi$	$-\frac{\pi}{2}$	0	0.7	$\frac{\pi}{2}$	π	$\frac{3\pi}{2}$
y	-4.7	-4.1	-1.6	-1	0	1.6	4.1	4.7

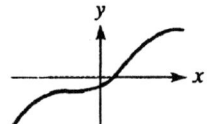

26. $y = \sin 4x + \cos 2x$ has period π.

$\dfrac{dy}{dx} = 4\cos 4x - 2\sin 2x$

$\dfrac{dy}{dx} = -16\sin 4x - 4\cos 2x$

(a) y is defined for all x; no asymptotes

(b) at $x = 0, y = 1$; y-intercept is $(0, 1)$

(c) $y = \sin 4x + \cos 2x = 2\sin 2x \cos 2x + \cos 2x$

Set $y = \cos 2x (2\sin 2x + 1) = 0$ for y-intercepts.

$\cos 2x = 0$

$\qquad 2x = \dfrac{\pi}{2} + k\pi$

$\qquad x = \dfrac{\pi}{4} + \dfrac{k\pi}{2}$

$2\sin 2x + 1 = 0$

$\qquad \sin 2x = -\dfrac{1}{2}$

$\qquad\quad 2x = \dfrac{7\pi}{6} + 2k\pi, 2x = \dfrac{11\pi}{6} + 2k\pi$

$\qquad\quad x = \dfrac{7\pi}{12} + k\pi, x = \dfrac{11\pi}{12} + k\pi$

For $0 \le x \le \pi$, one period, the intercepts are $\left(\frac{\pi}{4}, 0\right), \left(\frac{3\pi}{4}, 0\right), \left(\frac{7\pi}{12}, 0\right), \left(\frac{11\pi}{12}, 0\right)$.

(d) $\dfrac{dy}{dx} = 4\cos 4x - 2\sin 2x = 4(1 - 2\sin^2 2x) - 2\sin 2x$

Set $\dfrac{dy}{dx} = -8\sin^2 2x - 2\sin 2x + 4 = 0$ for critical numbers.

$4\sin^2 2x + \sin 2x - 2 = 0$

$\sin 2x = \dfrac{-1 + \sqrt{33}}{8}, \sin 2x = \dfrac{-1 - \sqrt{33}}{8}$

$2x = \sin^{-1}\left(\dfrac{-1 + \sqrt{33}}{8}\right) + 2k\pi,$

$2x = \sin^{-1}\left(\dfrac{-1 - \sqrt{33}}{8}\right) + 2k\pi$

$x = \dfrac{1}{2}\sin^{-1}\left(\dfrac{-1 + \sqrt{33}}{8}\right) + k\pi,$

$x = \dfrac{1}{2}\sin^{-1}\left(\dfrac{-1 - \sqrt{33}}{8}\right) + k\pi$

or $2x = \pi - \sin^{-1}\left(\dfrac{-1 + \sqrt{33}}{8}\right) + 2k\pi,$

or $2x = -\pi - \sin^{-1}\left(\dfrac{-1 - \sqrt{33}}{8}\right) + 2k\pi$

$x = \dfrac{\pi}{2} - \dfrac{1}{2}\sin^{-1}\left(\dfrac{-1 + \sqrt{33}}{8}\right) + k\pi,$

$x = \dfrac{-\pi}{2} - \dfrac{1}{2}\sin^{-1}\left(\dfrac{-1 - \sqrt{33}}{8}\right) + k\pi$

For $0 \le x \le \pi$, the first two critical numbers are found using $k = 0$ in

$x = \dfrac{1}{2}\sin^{-1}\left(\dfrac{-1 + \sqrt{33}}{8}\right) + k\pi\Big|_{k=0} = 0.317$ and $k = 0$ in

$x = \dfrac{\pi}{2} - \dfrac{1}{2}\sin^{-1}\left(\dfrac{-1 - \sqrt{33}}{8}\right) + k\pi\Big|_{k=0} = 1.25$

For $0 \le x \le \pi$, the last two critical numbers are found using $k = 1$ in

$x = \dfrac{1}{2}\sin^{-1}\left(\dfrac{-1 - \sqrt{33}}{8}\right) + k\pi\Big|_{k=1} = 2.64$ and $k = 1$ in

$x = \dfrac{-\pi}{2} - \dfrac{1}{2}\sin^{-1}\left(\dfrac{-1 - \sqrt{33}}{8}\right) + k\pi\Big|_{k=0} = 2.07$

(e) Evaluate $\dfrac{d^2y}{dx^2}$ at critical numbers

$\dfrac{d^2y}{dx^2} = -16\sin 4x - 4\cos 2x\big|_{0.317} = -18.5 < 0, (0.317, 1.76)$ is a local maximum

$\dfrac{d^2y}{dx^2} = -16\sin 4x - 4\cos 2x\big|_{1.25} = 18.5 > 0, (1.25, -1.76)$ is a local minimum

$\dfrac{d^2y}{dx^2} = -16\sin 4x - 4\cos 2x\big|_{2.07} = -12.4 < 0, (2.07, 0.367)$ is a local maximum

$\dfrac{d^2y}{dx^2} = -16\sin 4x - 4\cos 2x\big|_{2.64} = 12.4 > 0, (2.64, -0.369)$ is a local minimum

(f) Set $\dfrac{d^2y}{dx^2} = -16\sin 4x - 4\cos 2x = 0$ for inflection points.

$$-16\sin 4x - 4\cos 2x = 0$$
$$-16(2\sin 2x\cos 2x) - 4\cos 2x = 0$$
$$8\sin 2x\cos 2x + \cos 2x = 0$$

$\cos 2x = 0$, $8\sin 2x + 1 = 0$

$x = \dfrac{\pi}{4} + \dfrac{k\pi}{2}$ (previously) , $8\sin 2x + 1 = 0$

$\sin 2x = -\dfrac{1}{8}$

$k = 0, x = \dfrac{\pi}{4}$, $2x = \sin^{-1}\left(-\dfrac{1}{8}\right) + 2k\pi$

$k = 1, x = \dfrac{3\pi}{4}$ $x = \dfrac{1}{2}\sin^{-1}\left(-\dfrac{1}{8}\right) + k\pi$

for $k = 1$ $x = 3.08$

also, $2x = \dfrac{\pi}{2} - \sin^{-1}\left(-\dfrac{1}{8}\right) + 2k\pi$

$x = \dfrac{\pi}{2} - \dfrac{1}{2}\sin^{-1}\left(-\dfrac{1}{8}\right) + 2k\pi$

for $k = 0$, $x = 1.63$

inflection points: $\left(\dfrac{\pi}{4}, 0\right), (1.63, -0.758), \left(\dfrac{3\pi}{4}, 0\right), (3.08, 0.749)$

(g)

x	0	0.317	$\frac{\pi}{4}$	1.25	1.63	2.07	$\frac{3\pi}{4}$	2.64	3.08	π	$\frac{7\pi}{12}$	$\frac{11\pi}{12}$
y	1	1.11	0	−1.76	−0.758	0.367	0	−0.369	0.749	1	0	0

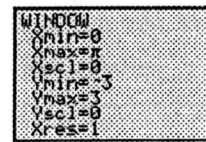  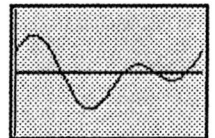

27. $y = 4\cos^2(x^2)$; slope $= \dfrac{dy}{dx} = 2[4\cos(x^2)][-\sin(x^2)](2x) = -16x\cos x^2 \sin x^2$

$\left.\dfrac{dy}{dx}\right|_{x=1} = -16\cos(1^2)\sin(1^2) = -16(0.5403)(0.8415) = -7.27$

$f(1) = 4\cos^2(1^2) = 4(0.5403)^2 = 1.168$

$y = -7.27x + b;\ 1.168 = -7.27(1) + b,\ b = 8.44;\ y = -7.27x + 8.44;\ 7.27x + y - 8.44 = 0$

28. $y = \tan^{-1}x,\ x = 1,\ y = \dfrac{\pi}{4}$

$\dfrac{dy}{dx} = \left.\dfrac{1}{1+x^2}\right|_{x=1} = \dfrac{1}{1+1^2} = \dfrac{1}{2} = m_{TL}$

$m_{NL} = -2,\ y - y_1 = m(x - x_1)$

$y - \dfrac{\pi}{4} = -2(x - 1)$

$2x + y - 2 - \dfrac{\pi}{4} = 0$

29. $\sin^2 x + \cos^2 x = 1$

$\dfrac{d(\sin^2 x + \cos^2 x)}{dx} = \dfrac{d(1)}{dx}$

$\dfrac{d\sin^2 x}{dx} + \dfrac{d\cos^2 x}{dx} = 0$

$2\sin x \cos x + 2\cos x(-\sin x) = 0$

$2\sin x \cos x - 2\cos x \sin x = 0;\ 0 = 0$

30. $\dfrac{d(\sin(x+1))}{dx} = \cos(x+1)$

$\dfrac{d}{dx}(\sin x \cos 1 + \cos x \sin 1) = \cos x \cos 1 - \sin x \sin 1 = \cos(x+1)$

31. $\sin x = 1 - x$

$f(x) = 1 - x - \sin x$

$f'(x) = -1 - \cos x$

$f(0) = 1, f(1) = -0.84$

Let $x_1 = 0.5$

n	x_n	$f(x_n)$	$f'(x_n)$	$x_n - \frac{f(x_n)}{f'(x_n)}$
1	0.5	0.0205745	-1.8775826	0.5109580
2	0.5109580	0.0000290	-1.8722765	0.5109734
3	0.5109734	1.76×10^{-10}	-1.8722689	0.5109734
4	0.5109734			

$x_4 = x_3 = 0.5109734$

32. $x^2 = \tan^{-1}x$

$f(x) = x^2 - \tan^{-1}x$

$f'(x) = 2x - \dfrac{1}{1+x^2}$

$f(0) = 0, x = 0$ is one solution

$f(0.5) = -0.213647609$

$f(1) = 0.2146018366.$ Let $x_1 = 0.75$

n	x_n	$f(x_n)$	$f'(x_n)$	$x_n - \frac{f(x_n)}{f'(x_n)}$
1	0.75	-0.0810011088	0.8599999399	0.8441873424
2	0.8441873424	0.0115424464	1.104484921	0.8337368182
3	0.8337368182	$1.407308373 \times 10^{-4}$	1.07754381	0.8336062148
4	0.8336062148	2.199693×10^{-8}	1.077206809	0.8336061944
5	0.8336061944	0	1.077206756	0.8336061944
6	0.8336061944			

$x_5 = x_6 = 0.8336061944$

33. $T = 17.2 + 5.2 \cos\left[\dfrac{\pi}{6}(x - 0.50)\right]$

$\dfrac{dT}{dt} = \dfrac{dT}{dx}\dfrac{dx}{dt} = -5.2 \sin\left[\dfrac{\pi}{6}(x - 0.50)\right]\left(\dfrac{\pi}{6}\right)\dfrac{dx}{dt}$

$\dfrac{dT}{dt} = -5.2 \sin\left[\dfrac{\pi}{6}(2 - 0.50)\right]\left(\dfrac{\pi}{6}\right)(0.033)$

$\dfrac{dT}{dt} = -0.064°\ \text{C/day}$

34. $W = 10\cos 2t$

$P = \dfrac{dW}{dt} = 10(-\sin 2t)(2)$

$P = -20\sin 2t$

35. $I = kE_0^2 \cos^2 \dfrac{1}{2}\theta$

$\dfrac{dI}{d\theta} = 2kE_0^2 \cos^2 \dfrac{1}{2}\theta\left(-\sin \dfrac{1}{2}\theta\right)\left(\dfrac{1}{2}\right)$

$\dfrac{dI}{d\theta} = -kE_0^2 \cos \dfrac{1}{2}\theta \sin \dfrac{1}{2}\theta$

36. $\theta = \sin^{-1}\left(\dfrac{Ff}{R}\right)$

$\dfrac{d\theta}{dF} = \dfrac{1}{\sqrt{1 - \dfrac{F^2 f^2}{R^2}}}\left(\dfrac{f}{R}\right)$

$\quad = \dfrac{1}{\sqrt{\dfrac{R^2 - F^2 f^2}{R^2}}}\left(\dfrac{f}{R}\right)$

$\dfrac{d\theta}{dF} = \dfrac{f}{\sqrt{R^2 - F^2 f^2}}$

37. $R = \dfrac{v_0^2}{g}\sin 2\theta,\ 0^0 \le \theta \le 90°$

$\dfrac{dR}{d\theta} = v_0^2 \cos 2\theta(2) = 0 \text{ for maximum}$

$\cos 2\theta = 0$

$2\theta = 90°$

$\theta = 45°$

38. $\omega = \sqrt{\dfrac{g}{l\cos\theta}}$

$d\omega = \dfrac{1}{2\sqrt{\dfrac{g}{l\cos\theta}}}\left(\dfrac{g}{l}\right)\sec\theta\cos\theta\,d\theta$

$d\omega = \dfrac{1}{2}\sqrt{\dfrac{g}{l\cos\theta}}\tan\theta\,d\theta$

$d\omega = \dfrac{1}{2}\sqrt{\dfrac{9.800}{0.6375\cos 32.5°}}$

$\qquad \cdot \tan 32.5°\,(32.75° - 32.5°)\left(\dfrac{\pi}{180}\right)$

$d\omega = 0.005934\ \text{rad/s}$

39. $x = \sin 2\pi t,\ y = 2\cos^2 \pi t,\ t = \dfrac{13}{12}$

$v_x = 2\pi \cos 2\pi t,$

$v_y = (4\pi \cos \pi t)(-\sin \pi t) = -2\pi \sin 2\pi t$

For $t = \dfrac{13}{12},\ v_x = 2\pi\cos\dfrac{13\pi}{6}$

$\qquad\qquad = 2\pi\left(\dfrac{\sqrt{3}}{2}\right) = \pi\sqrt{3}$

$\qquad v_y = -2\pi\sin\dfrac{13\pi}{6}$

$\qquad\quad = -2\pi\left(\dfrac{1}{2}\right) = -\pi$

$v = \sqrt{(\pi\sqrt{3})^2 + (-\pi)^2} = 2\pi$

$\tan\theta = \dfrac{-\pi}{\pi\sqrt{3}} = -\dfrac{1}{\sqrt{3}},\ (v_x > 0, v_y < 0)$

$\theta = 330°$

40. $v_x = 2\pi\cos 2\pi t,\qquad v_y = -2\pi\sin 2\pi t$

$a_x = -4\pi^2\sin 2\pi t,\quad a_y = -4\pi t^2\cos 2\pi t$

For $t = \dfrac{13}{12}$,

$a_x = -4\pi^2\sin\dfrac{13\pi}{6} = -4\pi^2\left(\dfrac{1}{2}\right) = -2\pi^2$

$a_y = -4\pi^2\cos\dfrac{13\pi}{6} = -4\pi^2\left(\dfrac{\sqrt{3}}{2}\right) = -2\pi^2\sqrt{3}$

$a = \sqrt{(-2\pi^2)^2 + (-2\pi^2\sqrt{3})^2} = 4\pi^2$

$\tan\theta = \dfrac{-2\pi^2\sqrt{3}}{-2\pi^2} = \sqrt{3},\ (a_x < 0, a_y < 0)$

$\theta = 240°$

41. Let $x = $ distance football has moved along sideline.

$$92 \tan \theta = x$$

$$92 \sec^2 \theta \frac{d\theta}{dt} = \frac{dx}{dt}$$

$$\frac{d\theta}{dt} = \frac{\frac{dx}{dt} \cos^2 \theta}{92} = \frac{56 \cos^2 15°}{92}$$

$$\frac{d\theta}{dt} = 0.568 \text{ rad/s}$$

42. $P = \dfrac{(0.20)(50)}{0.20 \sin \theta + \cos \theta}, 0° < \theta < 90°$

$$\frac{dP}{d\theta} = \frac{(0.20 \sin \theta + \cos \theta)(0) - (0.20)(50)(0.20 \cos \theta - \sin \theta)}{(0.20 \sin \theta + \cos \theta)^2}$$

$$\frac{dP}{d\theta} = 0 \text{ for minimum.}$$

$$-(0.20)(50)(0.20 \cos \theta - \sin \theta) = 0$$

$$\tan \theta = 0.20$$

$$\theta = 11.3°$$

43.
$$\tan \theta = \frac{6800}{x}$$

$$\sec^2 \theta \frac{d\theta}{dt} = \frac{-6800}{x^2} \frac{dx}{dt}$$

$$\frac{d\theta}{dt} = \frac{-6800}{x^2} \cos^2 \theta \frac{dx}{dt}$$

For $\theta = 13°$, $\dfrac{d\theta}{dt} = \dfrac{-6800}{\left(\frac{6800}{\tan 13°}\right)^2}(\cos^2 13°)(880)$

$$\frac{d\theta}{dt} = -0.0065 \text{ rad/s}$$

44. $x = r(\theta - \sin \theta); y = r(1 - \cos \theta); r = 5.500$ cm;

$\dfrac{d\theta}{dt} = 0.12$ rad/s; $\theta = 35°$

horizontal component of velocity,

$$v_x = \frac{dx}{dt} = \frac{d[5.500(\theta - \sin \theta]}{dt}$$

$$= 5.500\left(\frac{d\theta}{dt} - \cos \theta \frac{d\theta}{dt}\right)$$

$$= 5.500(0.12 - 0.12 \cos 35°)$$

$$= 0.119 \text{ cm/s}$$

vertical component of velocity,

$$v_y = \frac{dy}{dt} = \frac{d[5.500(1 - \cos \theta)]}{dt}$$

$$= 5.500 \sin \theta \frac{d\theta}{dt}$$

$$= (5.500)(0.12) \sin 35° = 0.379$$

$$v = \sqrt{0.119^2 + 0.379^2} = 0.40 \text{ cm/s.}$$

$$\theta = \tan^{-1} \frac{0.379}{0.119} = 72.5°$$

45. $I = \dfrac{k \cos \theta}{r^2}, \dfrac{10}{r} = \sin \theta, \tan \theta = \dfrac{10}{h}$

$$I = \frac{k \cos \theta}{\left(\frac{10}{\sin \theta}\right)^2} = \frac{k}{100} \cos \theta \sin^2 \theta$$

$$\frac{dI}{d\theta} = \frac{k}{100}\left[\cos \theta(2 \sin \theta \cos \theta) + \sin^2 \theta(-\sin \theta)\right]$$

$$\frac{dI}{d\theta} = \frac{k}{100} \sin \theta(2 \cos^2 \theta - \sin^2 \theta)$$

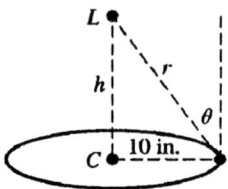

$$\sin \theta(2 \cos^2 \theta - \sin^2 \theta) = 0$$

$$\sin \theta = 0$$

$$\theta = 0, L \text{ infinitely for }, I = 0$$

$$2 \cos^2 \theta - \sin^2 \theta = 0$$

$$\tan^2 \theta = 2$$

$$\tan \theta = \sqrt{2}$$

$$\theta = 54.7°$$

$$h = \frac{10}{\tan \theta} = \frac{10}{\sqrt{2}} = 7.07 \text{ in.}$$

46.
$$L = 2x + y$$
$$x^2 = 3.00^2 + (10.0 - y)^2$$
$$x = \sqrt{3.00^2 + (10.0 - y)^2}$$
$$L = 2\sqrt{3.00^2 + (10.0 - y)^2} + y$$

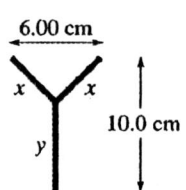

$$\frac{dL}{dy} = \frac{2(10.0 - y)(-1)}{\sqrt{3.00^2 + (10.0 - y)^2}} + 1 = 0$$

$$-2(10.0 - y) + \sqrt{3.00^2 + (10.0 - y)^2} = 0$$

$$4(10.0 - y)^2 = 3.00^2 + (10.0 - y)^2$$

$$3(10.0 - y)^2 = 3.00^2$$

$$\sqrt{3}(10.0 - y) = \sqrt{3.00^2}$$

$$10.0 - y = \frac{\sqrt{3.00^2}}{\sqrt{3}}$$

$$y = 10.0 - \frac{\sqrt{3.00^2}}{\sqrt{3}}$$

$$y = 8.27 \text{ cm}$$

Let $\theta =$ half angle,

$$\tan \theta = \frac{3.00}{10 - y} = \frac{3.00}{\frac{\sqrt{3.00^2}}{\sqrt{3}}}$$

$$\theta = 120°$$

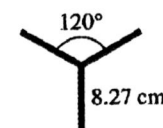

120°

8.27 cm

47.

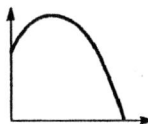

$$\cos \theta = \frac{y}{4}; \ y = 4 \cos \theta$$

$$\sin \theta = \frac{y}{4}; \ x = 4 \sin \theta$$

$$A = (4 + x)y = 4y + xy = 16 \cos \theta + 16 \sin \theta \cos \theta = 16 \cos \theta (1 + \sin \theta)$$
$$\frac{dA}{d\theta} = -16 \sin \theta + 16 \sin \theta (- \sin \theta) + 16 \cos \theta (\cos \theta) = -16 \sin \theta - 16 \sin^2 \theta + 16 \cos^2 \theta$$
$$= 16(\cos^2 \theta - \sin^2 \theta - \sin \theta) = 16(1 - 2 \sin^2 \theta - \sin \theta)$$

(1) Not valid for negative θ or A. Domain and range are positive real numbers.

(2) A-intercept at $\theta = 0$, $A = 16$; To find θ-intercept, $A = 0$

$$16 \cos \theta (1 + \sin \theta) = 0$$
$$16 \cos \theta = 0 \qquad 1 + \sin \theta = 0$$
$$\theta = \frac{\pi}{2} \qquad \sin \theta = -1, \ \theta = -\frac{\pi}{2} \quad (\text{not in domain})$$

(3) Critical value is

$$16(1 - 2 \sin^2 \theta - \sin \theta) = 0$$
$$1 - 2 \sin^2 \theta - \sin \theta = 0$$
$$-1 + 2 \sin^2 \theta + \sin \theta = 0$$
$$(2 \sin \theta - 1)(\sin \theta + 1) = 0$$
$$2 \sin \theta = 1 \qquad \sin \theta = -1$$
$$\sin \theta = \frac{1}{2} \qquad \theta = -\frac{\pi}{2} \quad (\text{not in domain})$$
$$\theta = \frac{\pi}{6}$$

(4) $\dfrac{d^2 A}{d\theta^2} = 16(-4\sin\theta\cos\theta - \cos\theta) = -16\cos\theta(4\sin\theta + 1) = -16(4\sin\theta + 1)\big|_{\theta=\pi/6}$

$\qquad = -16[4(0.5) + 1] = -16(3) = -48$

Curve is concave down at $\frac{\pi}{6}$ and $\theta = \frac{\pi}{6}$ is a maximum.

48.

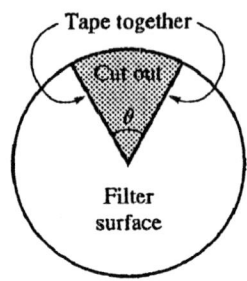

Tape together

Cut out

θ

Filter surface

when taped,

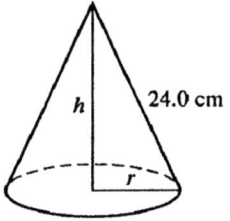

h 24.0 cm

r

Bottom circumference $= 2\pi r = 24.0(2\pi - \theta)$

$$r = \frac{12.0(2\pi - \theta)}{\pi}$$

$h = \sqrt{24.0^2 - r^2}$

$h = \sqrt{576\left[\dfrac{12.0(2\pi - \theta)}{\pi}\right]^2} = \dfrac{12.0}{\pi}\sqrt{4\pi\theta - 2}$

$V = \dfrac{1}{3}\pi r^2 h = \dfrac{1}{3}\pi\left(\dfrac{12.0^2}{\pi^2}\right)(2\pi - \theta)^2\left(\dfrac{12.0}{\pi}\right)\sqrt{4\pi\theta - \theta^2}$

$V = \dfrac{576}{\pi^2}(2\pi - \theta)^2(4\pi\theta - \theta)^{1/2}$

$\dfrac{dV}{d\theta} = \dfrac{576}{\pi^2}\left[(2\pi - \theta)^2\left(\dfrac{1}{2}\right)(4\pi\theta - \theta^2)^{-1/2}(4\pi - 2\theta) + (4\pi\theta - \theta^2)^{1/2}(2)(2\pi - \theta)(-1)\right]$

$\dfrac{dV}{d\theta} = \dfrac{576}{\pi^2}\left[\dfrac{(2\pi - \theta)^3}{(4\pi\theta - \theta^2)^{1/2}} - 2(2\pi - \theta)(4\pi\theta - \theta^2)^{1/2}\right]$

$\dfrac{dV}{d\theta} = \dfrac{576(2\pi - \theta)[(2\pi - \theta)^2 - 2(4\pi - \theta^2)(4\pi\theta - \theta^2)]}{\pi^2\pi^2(4\pi\theta - \theta^2)^{1/2}}$

$\dfrac{dV}{d\theta} = \dfrac{576(2\pi - \theta)(3\theta^2 - 12\pi\theta + 4\pi^2)}{\pi^2(4\pi\theta - \theta^2)^{1/2}} = 0$

$2\pi - \theta = 0$

$\qquad \theta = 2\pi$ gives $V = 0$

$3\theta^2 - 12\pi\theta + 4\pi^2 = 0$

$\theta = \dfrac{12\pi \pm \sqrt{144\pi^2 - 48\pi^2}}{6} = 2\pi \pm \dfrac{2}{3}\pi\sqrt{6}, 0 < \theta < 2\pi$

$\theta = 2\pi - \dfrac{2}{3}\pi\sqrt{6} = 1.153$

$V = \dfrac{576}{\pi^2}(2\pi - 1.153)^2[4\pi(1.153) - (1.153)^2]^{1/2}$

$V = 5570 \text{ cm}^3$

49.

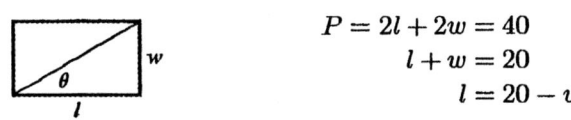

$$P = 2l + 2w = 40$$
$$l + w = 20$$
$$l = 20 - w$$

(a) Using algebra, $A = lw = (20 - w)w = 20w - w^2$

$$\frac{dA}{dw} = 20 - 2w = 0 \Rightarrow w = 10 \text{ and } l = 10$$

The area is maximum when the rectangle is a square.

(b) Using trigonometry $\tan\theta = \dfrac{w}{l} = \dfrac{w}{20 - w}$ from which $w = \dfrac{20\tan\theta}{1 + \tan\theta}$, then

$$A = \frac{20^2 \tan\theta}{1 + \tan\theta} - \frac{20^2 \tan^2\theta}{(1 + \tan\theta)^2} \quad \text{and taking} \quad \frac{dA}{d\theta} = 0$$

gives a maximum for $\theta = 45°$; again, a square.

DIFFERENTIATION OF THE EXPONENTIAL AND LOGARITHMIC FUNCTIONS

8.1 Exponential and Logarithmic Functions

1. $4^4 = 256$ has base 4, exponent 4 and number 256.

$$\log_4 256 = 4$$

2. $3^{-2} = \dfrac{1}{9}$ has base 3, exponent -2 and number $\dfrac{1}{9}$.

$$\log_3 \left(\dfrac{1}{9} \right) = -2$$

3. $\log_8 16 = \dfrac{4}{3}$ has base 8, exponent $\dfrac{4}{3}$ and number 16.

$$16 = 8^{4/3}$$

4. $\log_7 \left(\dfrac{1}{49} \right) = -2$ has base 7, exponent -2 and number $\dfrac{1}{49}$.

$$\dfrac{1}{49} = 7^{-2}$$

5. $8^{1/3} = 2$ has base 8, exponent $\dfrac{1}{3}$, and number 2.

$$\log_8 2 = \dfrac{1}{3}$$

6. $(81)^{3/4} = 27$ has base 81, exponent $\dfrac{3}{4}$ and number 27.

$$\log_{81} 27 = \dfrac{3}{4}$$

7. $\log_{0.5} 16 = -4$ has base 0.5, exponent -4 and number 16.

$$16 = (0.5)^{-4}$$

8. $\log_{243} 3 = \dfrac{1}{5}$ has base 243, exponent $\dfrac{1}{5}$ and number 3.

$$3 = 243^{1/5}$$

9. $(12)^0 = 1$ has base 12, exponent 0 and number 1.

$$\log_{12} 1 = 0$$

10. $\left(\dfrac{1}{2} \right)^{-2} = 4$ has base $\dfrac{1}{2}$, exponent -2 and number 4.

$$\log_{1/2} 4 = -2$$

11. $\log_8 \left(\dfrac{1}{512} \right) = -3$ has base 8, exponent -3, and number $\dfrac{1}{512}$.

$$\dfrac{1}{512} = 8^{-3}$$

12. $\log_{1/32} \left(\dfrac{1}{8} \right) = \dfrac{3}{5}$ has base $\dfrac{1}{32}$, exponent $\dfrac{3}{5}$ and number $\dfrac{1}{8}$.

$$\left(\dfrac{1}{32} \right)^{3/5} = \dfrac{1}{8}$$

13. $\log_7 y = 3$ has base 7, exponent 3, and number 3.

$$7^3 = y$$
$$y = 343$$

14. $\log_b 4 = \dfrac{2}{3}$ has base b, exponent $\dfrac{2}{3}$ and number 4.

$$b^{2/3} = 4, b = 8$$

15. $\log_{10} 10^{0.2} = x$ has base 10, exponent x, and number $10^{0.2}$.

$$10^x = 10^{0.2}$$
$$x = 0.2$$

16. $\log_8 y = -\dfrac{2}{3}$ has base 8, exponent $-\dfrac{2}{3}$ and number y

$$8^{-2/3} = y$$
$$y = \dfrac{1}{4}$$

17. $y = 3^x$

x	y
-3	0.037
-2	0.111
-1	0.333
0	1.000
1	3.000
2	9.000
3	27.000

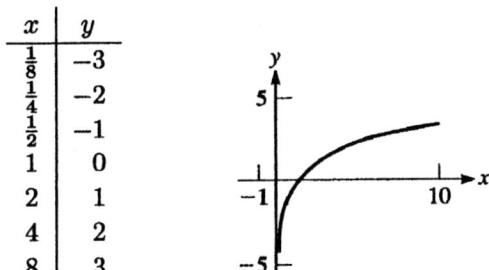

18. $y = \left(\dfrac{1}{3}\right)^x$

x	y
-3	27.000
-2	9.000
-1	3.000
0	1.000
1	0.333
2	0.111
3	0.037

19. $y = \log_2 x$

x	y
$\frac{1}{8}$	-3
$\frac{1}{4}$	-2
$\frac{1}{2}$	-1
1	0
2	1
4	2
8	3

20. $y = 2\log_4 x = \log_4 x^2 \Leftrightarrow 4^y = x^2$
$$2^{2y} = x^2$$
$$2y = \log_2 x^2$$
$$2y = 2\log_2 x$$
$$y = \log_2 x$$

The graph of $y = 2\log_4 x$ is the same as the graph of $y = \log_2 x$ for which, see exercise 19.

21. $\log_5 9 - \log_5 3 = \log_5 \dfrac{9}{3}$
$$= \log_5 3$$

22. $\log_2 3 + \log_2 x = \log_2 3x$

23. $\log_6 x^2 + \log_6 \sqrt{x} = \log_6 (x^2\sqrt{x})$
$$= \log_6 x^{5/2}$$

24. $\log_7 3^4 - \log_7 9 = \log_7 \dfrac{3^4}{9} = \log_7 \dfrac{3^4}{3^2}$
$$= \log_7 3^2$$

25. $2\log_b 2 + \log_b y = \log_b 2^2 + \log_b y$
$$= \log_b 4 + \log_b y$$
$$= \log_b 4y$$

26. $3\log_b 4 - \log_b x = \log_b 3^4 - \log_b x$
$$= \log_b 81 - \log_b x$$
$$= \log_b \frac{81}{x}$$

27. $2\log_2 x + \log_2 3 - \log_2 5 = \log_2 x^2 + \log_2 \dfrac{3}{5}$
$$= \log_2 \frac{3x^2}{5}$$

28. $\log_4 7 - 2\log_4 t + \log_4 a = \log_4 7 - \log_4 t^2 + \log_4 a$
$$= \log_4 \frac{7}{t^2} + \log_4 a$$
$$= \log_4 \frac{7a}{t^2}$$

29. $\ln 3 = \dfrac{\log 3}{\log e} = 1.099$

30. $\ln 0.9 = \dfrac{\log 0.9}{\log e} = -0.1054$

31. $\log_7 42 = \dfrac{\log 42}{\log 7} = \dfrac{1.6232}{0.8451} = 1.921$

32. $\log_{12} 122 = \dfrac{\log 122}{\log 12} = \dfrac{2.0864}{1.0792} = 1.933$

33. $m_1 - m_2 = 2.5\log_{10}\left(\dfrac{b_2}{b_1}\right)$ or $\dfrac{m_1 - m_2}{2.5} = \log_{10}\left(\dfrac{b_2}{b_1}\right)$
has base 10, exponent $\dfrac{m_1 - m_2}{2.5}$ and number $\dfrac{b_2}{b_1}$.

$10^{(m_1-m_2)/2.5} = \dfrac{b_2}{b_1}$, $b_2 = b_1(10)^{0.4(m_1-m_2)}$

34. $v = u\log_e\left(\dfrac{w_0}{w}\right)$ or $\dfrac{v}{u} = \log_e\left(\dfrac{w_0}{w}\right)$ has base e,
exponent $\dfrac{v}{u}$, and number $\dfrac{w_0}{w}$.

$e^{v/u} = \dfrac{w_0}{w}$; $w = \dfrac{w_0}{e^{v/u}} = w_0 e^{-v/u}$

35. $\log_e\left(\dfrac{N}{N_0}\right) = -kt \Rightarrow e^{-kt} = \dfrac{N}{N_0}$, $N = N_0 e^{-kt}$

36. $q = q_0(1 - e^{-at})$ or $\dfrac{q}{q_0} = (1 - e^{-at})$ or

$\dfrac{q}{q_0} - 1 = -e^{-at}$

$1 - \dfrac{q}{q_0} = e^{-at}$ has base e, exponent $-at$ and number $1 - \dfrac{q}{q_0}$.

$-at = \log_e\left(1 - \dfrac{q}{q_0}\right)$; $t = -\dfrac{1}{a}\log_e\left(1 - \dfrac{q}{q_0}\right)$

37. $p = 10(1.2^{-t} + 1), 0 \le t \le 10$

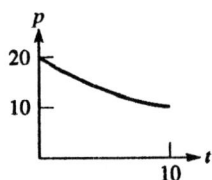

38. $i = 16(1 - e^{-250t})$

i	t
0	0
3.5	0.001
11.4	0.005
14.7	0.01
16.0	0.05

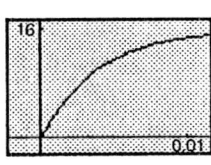

39. $t = N + \log_2 N$ where $n > 0$ and $t > 0$.

N	t
1	1
2	3
4	6
8	11

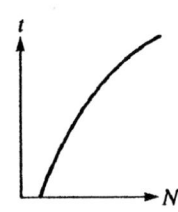

40. $t = 2350(\log_e 100 - \log_e N)$

i	N
10822	1
3258	25
1629	50
676	75
0	100

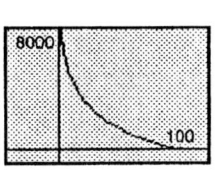

41. $x = \dfrac{1}{k} \ln(kv_0 t + 1); x = 150$ m,

$k = 6.80 \times 10^{-3}$ m

$v_0 = 12.0$ m/s

$kx = \ln(kv_0 t + 1)$

$e^{kx} = kv_0 t + 1$

$kv_0 t = e^{kx} - 1$

$$t = \frac{e^{kx} - 1}{kv_0} = \frac{e^{(6.80 \times 10^{-3})(150)} - 1}{(6.80 \times 10^{-3})(12.0)}$$

$$t = \frac{e^{1.02} - 1}{0.0816} = \frac{1.773}{0.0816} = 21.7 \text{ s}$$

42. $i = 0.6(1 - e^{-10t})$

$$\frac{i}{0.6} = 1 - e^{-10t}$$

$$e^{-10t} = 1 - \frac{i}{0.6}$$

$$-10t = \ln\left(1 - \frac{i}{0.6}\right)$$

$$t = -\frac{1}{10} \ln\left(1 - \frac{i}{0.6}\right)$$

$$t = -0.1 \ln\left(1 - \frac{i}{0.6}\right)$$

43. $N = (6.700)(3400)$
$\log N = \log[(6.700)(3400)]$
$\log N = \log 6.700 + \log 3400$
$\log N = 0.8261 + 3.5315 = 4.3576$
$N = 10^{4.3576} = 2.3 \times 10^4$

44. $N = (6.700)(3400)$
$\ln N = \ln[(6.700)(3400)]$
$\ln N = \ln 6.7000 + \ln 3400$
$\ln N = 1.9021 + 8.1315 = 10.0336$
$N = e^{10.0336} = 2.3 \times 10^4$

The final product required raising e to the appropriate power.

45.

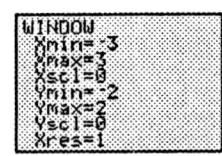 

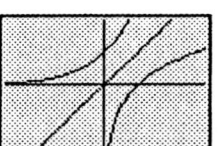

46.

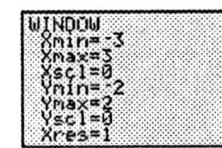 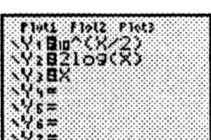

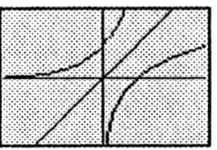

47. $\cosh^2 u - \sinh^2 u$

$$= \frac{1}{4}(e^u + e^{-u})^2 - \frac{1}{4}(e^u - e^{-u})^2$$

$$= \frac{1}{4}(e^{2u} + 2e^0 + e^{-2u}) - \frac{1}{4}(e^{2u} - 2e^0 + e^{-2u})$$

$$= \frac{1}{4}[0 + 2e^0 + 2e^0 + 0] = \frac{1}{4}(4e^0) = \frac{1}{4}(4) = 1$$

48. $\sinh u = \frac{1}{2}(e^u - e^{-u})$

$$\sinh(0.25) = \frac{1}{2}(e^{0.25} - e^{-0.25})$$

$$\cosh u = \frac{1}{2}(e^u + e^{-u})$$

$$\cosh(0.25) = \frac{1}{2}(e^{0.25} + e^{-0.25})$$

$$\cosh^2(0.25) - \sinh^2(0.25)$$

$$= \left[\frac{1}{2}(e^{0.25} + e^{-0.25})\right]^2 - \left[\frac{1}{2}(e^{0.25} - e^{-0.25})\right]^2$$

$$= \frac{1}{4}(e^{0.50} + 2e^0 + e^{-0.50}) - \frac{1}{4}(e^{0.50} - 2e^0 + e^{-0.50})$$

$$= \frac{1}{4}e^{0.50} + \frac{1}{2} + \frac{1}{4}^{-0.50} - \frac{1}{4}e^{0.50} + \frac{1}{2} - \frac{1}{4}e^{-0.50}$$

$$= 1$$

8.2   Derivative of the Logarithmic Function

1. $y = \log x^2; \quad u = x^2; \quad \frac{du}{dx} = 2x$

$$\frac{dy}{dx} = \frac{1}{x^2}(\log e)(2x) = \frac{2\log e}{x}$$

2. $y = \log_2 6x; \quad \frac{dy}{dx} = \frac{1}{6x}\log_2 e(1) = \frac{1}{x}\log_2 e$

3. $y = 2\log_5(3x + 1)$

$$\frac{dy}{dx} = 2\frac{1}{3x+1}\log_5 e(3) = \frac{6}{3x+1}\log_5 e$$

4. $y = 3\log_7(x^2 + 1)$

$$\frac{dy}{dx} = 3\frac{1}{x^2+1}\log_7 e(2x) = \frac{6x\log_7 e}{x^2+1}$$

5. $y = 0.2\ln(1 - 3x); \quad \frac{dy}{dx} = \frac{0.2}{1-3x}(-3) = \frac{-0.6}{1-3x}$

6. $y = 2\ln(3x^2 - 1); \quad \frac{dy}{dx} = 2\frac{1}{3x^2-1}(6x) = \frac{12x}{3x^2-1}$

7. $y = 2\ln\tan 2x$

$$\frac{dy}{dx} = 2\frac{1}{\tan 2x}\sec^2 2x(2) = \frac{4\sec^2 2x}{\tan 2x} = \frac{4\sec^2 2x}{\dfrac{\sec 2x}{\csc 2x}}$$

$$\frac{dy}{dx} = 4\sec 2x \csc 2x$$

8. $y = \ln\sin^2 x$

$$\frac{dy}{dx} = \frac{1}{\sin^2 x}2\sin x\cos x = \frac{2\cos x}{\sin x} = 2\cot x$$

9. $y = \ln\sqrt{x} = \ln(x^{1/2}) = \frac{1}{2}\ln x;$

$$\frac{dy}{dx} = \frac{1}{2}\left(\frac{1}{x}\right) = \frac{1}{2x}$$

10. $y = 5\ln\sqrt{4x - 3}$

$$\frac{dy}{dx} = 5\frac{1}{\sqrt{4x-3}}(4x - 3)^{-1/2}(4) = \frac{10}{4x-3}$$

11. $y = \ln(x^2 + 2x)^3$

$$\frac{dy}{dx} = \frac{1}{(x^2+2x)^3}3(x^2+2x)^2(2x+2) = \frac{6(x+1)}{x^2+2x}$$

12. $s = [\ln(2t^3 - t)]^2$

$$\frac{ds}{dt} = 2[\ln(2t^3 - t)]\frac{1}{2t^3 - t}\cdot(6t^2 - 1)$$

$$= \frac{2(6t^2 - 1)\ln(2t^3 - t)}{t(2t^2 - 1)}$$

13. $y = 3[t + \ln t^2]^2$

$$\frac{dy}{dt} = 6[t + \ln t^2]\left[1 + \frac{1}{t^2}\cdot 2t\right]$$

$$\frac{dy}{dt} = 6[t + \ln t^2]\left[1 + \frac{2}{t}\right] = \frac{6(t+2)(t + \ln t^2)}{t}$$

14. $y = x^2\ln 2x$

$$\frac{dy}{dx} = x^2\frac{1}{2x}(2) + \ln 2x(2x) = x + 2x\ln 2x$$

15. $y = \frac{3x}{\ln(2x + 1)}$

$$\frac{dy}{dx} = \frac{\ln(2x+1)(3) - (3x)\dfrac{1}{2x+1}(2)}{[\ln(2x+1)]^2}$$

$$\frac{dy}{dx} = \frac{3\ln(2x+1) - \dfrac{6x}{2x+1}}{[\ln(2x+1)]^2}$$

$$\frac{dy}{dx} = \frac{3(2x+1)\ln(2x+1) - 6x}{(2x+1)[\ln(2x+1)]^2}$$

16. $y = \dfrac{8\ln x}{x}$

$$\frac{dy}{dx} = \frac{x\left(\dfrac{8}{x}\right) - 8\ln x(1)}{x^2} = \frac{8 - 8\ln x}{x^2}$$

$$= \frac{8(1 - \ln x)}{x^2}$$

17. $y = \ln(\ln x)$; let $u = \ln x$

$$\frac{dy}{dx} = \frac{1}{\ln x}\left(\frac{d(\ln x)}{dx}\right)$$

$$= \frac{1}{\ln x}\left(\frac{1}{x}\right) = \frac{1}{x\ln x}$$

18. $r = 0.5\ln[\cos(\pi\theta^2)]$

$$\frac{dr}{d\theta} = 0.5 \cdot \frac{1}{\cos(\pi\theta^2)} \cdot (-\sin(\pi\theta^2)) \cdot 2\pi\theta$$

$$\frac{dr}{d\theta} = -\pi\theta\tan(\pi\theta^2)$$

19. $y = \ln\dfrac{2x}{1+x} = \ln 2x - \ln(1+x)$

$$\frac{dy}{dx} = \frac{1}{2x}(2) - \frac{1}{1+x} = \frac{1+x-x}{x(1+x)}$$

$$= \frac{1}{x(1+x)} = \frac{1}{x^2+x}$$

20. $y = \ln(x\sqrt{x+1})$

$$\frac{dy}{dx} = \frac{1}{(x\sqrt{x+1})}\left[x\frac{1}{2}(x+1)^{-1/2} + \sqrt{x+1}\right]$$

$$\frac{dy}{dx} = \frac{1}{x\sqrt{x+1}}\left[\frac{x}{2\sqrt{x+1}} + \sqrt{x+1}\right]$$

$$\frac{dy}{dx} = \frac{1}{x\sqrt{x+1}}\left[\frac{x+2(x+1)}{2\sqrt{x+1}}\right] = \frac{3x+2}{2x(x+1)}$$

21. $y = \sin(\ln x)$

$$\frac{dy}{dx} = \cos(\ln x)\frac{d(\ln x)}{dx} = \cos(\ln x)\left(\frac{1}{x}\right)$$

$$= \frac{\cos(\ln x)}{x}$$

22. $y = \tan^{-1}(\ln 2x)$

$$\frac{dy}{dx} = \frac{1}{1 + \ln^2 2x}\left(\frac{1}{2x}\right)(2) = \frac{1}{x(1 + \ln^2 2x)}$$

23. $u = 3v\ln^2 2v$

$$\frac{du}{dv} = 3v \cdot 2\ln 2v \cdot \frac{1}{2v} \cdot 2 + 3\ln^2 2v$$

$$\frac{du}{dv} = 6\ln 2v + 3\ln^2 2v$$

24. $h = 0.1s\ln^4 s$

$$\frac{dh}{ds} = 0.1s \cdot 4\ln^3 s \cdot \frac{1}{s} + 0.1\ln^4 s$$

$$\frac{dh}{ds} = 0.4\ln^3 s + 0.1\ln^4 s$$

25. $y = \ln(x\tan x)$

$$\frac{dy}{dx} = \frac{1}{x\tan x}[x(\sec^2 x) + \tan x]$$

$$= \frac{x\sec^2 x + \tan x}{x\tan x}$$

26. $y = \ln(x + \sqrt{x^2 - 1})$

$$\frac{dy}{dx} = \frac{1}{x + \sqrt{x^2 - 1}}\left[1 + \frac{1}{2}(x^2 - 1)^{-1/2}(2x)\right]$$

$$\frac{dy}{dx} = \frac{1}{x + \sqrt{x^2 - 1}}\left[1 + \frac{x}{\sqrt{x^2 - 1}}\right]$$

$$\frac{dy}{dx} = \frac{1}{x + \sqrt{x^2 - 1}}\left[\frac{\sqrt{x^2 - 1} + x}{\sqrt{x^2 - 1}}\right]$$

$$= \frac{1}{\sqrt{x^2 - 1}}$$

27. $y = \ln\dfrac{x^2}{x + 2}$

$$= \ln x^2 - \ln(x + 2)$$

$$\frac{dy}{dx} = \frac{1}{x^2}(2x) - \frac{1}{x + 2}$$

$$= \frac{2}{x} - \frac{1}{x + 2}$$

$$= \frac{2x + 4 - x}{x(x + 2)}$$

$$\frac{dy}{dx} = \frac{x + 4}{x(x + 2)}$$

28. $y = \sqrt{x + \ln 3x}$

$$\frac{dy}{dx} = \frac{1}{2}(x + \ln 3x)^{-1/2}\left(1 + \frac{1}{x}\right)$$

$$= \frac{\dfrac{x + 1}{2x}}{\sqrt{x + \ln 3x}}$$

$$\frac{dy}{dx} = \frac{x + 1}{2x\sqrt{x + \ln 3x}}$$

29. $y = \sqrt{x^2+1} - \ln \dfrac{1+\sqrt{x^2+1}}{x}$

$= (x^2+1)^{1/2} - \ln(1+\sqrt{x^2+1}) + \ln x$

$\dfrac{dy}{dx} = \dfrac{1}{2}(x^2+1)^{-1/2}(2x) - \dfrac{\frac{1}{2}(x^2+1)^{-1/2}(2x)}{1+\sqrt{x^2+1}} + \dfrac{1}{x}$

$= \dfrac{x}{(x^2+1)^{1/2}} - \dfrac{x}{(x^2+1)^{1/2}(1+\sqrt{x^2+1})} + \dfrac{1}{x}$

$= \dfrac{x^2(1+\sqrt{x^2+1}) - x^2 + (x^2+1)^{1/2}(1+\sqrt{x^2+1})}{x(x^2+1)^{1/2}(1+\sqrt{x^2+1})}$

$= \dfrac{x^2\sqrt{x^2+1} + \sqrt{x^2+1} + (x^2+1)}{x(x^2+1)^{1/2}(1+\sqrt{x^2+1})}$

$= \dfrac{(x^2+1)\sqrt{x^2+1} + (x^2+1)}{x(x^2+1)^{1/2}(1+\sqrt{x^2+1})}$

$= \dfrac{(x^2+1)(\sqrt{x^2+1}+1)}{x(x^2+1)^{1/2}(1+\sqrt{x^2+1})}$

$= \dfrac{\sqrt{x^2+1}}{x}$

30. $3\ln xy + \sin y = x^2$

$3\dfrac{1}{xy}\left(x\dfrac{dy}{dx}+y\right) + \cos y\dfrac{dy}{dx} = 2x$

$\dfrac{3}{y}\dfrac{dy}{dx} + \dfrac{3}{x} + \cos y\dfrac{dy}{dx} = 2x$

$\dfrac{3}{y}\dfrac{dy}{dx} + \cos y\dfrac{dy}{dx} = 2x - \dfrac{3}{x}$

$\dfrac{dy}{dx}\left(\dfrac{3}{y}+\cos y\right) = \dfrac{2x^2-3}{x}$

$\dfrac{dy}{dx} = \dfrac{2x^2-3}{x} \div \dfrac{3+y\cos y}{y}$

$\dfrac{dy}{dx} = \dfrac{y(2x^2-3)}{x(3+y\cos y)} = \dfrac{2x^2y-3y}{3x+xy\cos y}$

31. $y = x - \ln^2(x+y)$

$\dfrac{dy}{dx} = 1 - 2\ln(x+y)\dfrac{1}{x+y}\left(1+\dfrac{dy}{dx}\right)$

$\dfrac{dy}{dx} = 1 - \dfrac{2\ln(x+y)}{x+y} - \dfrac{2\ln(x+y)}{x+y}\dfrac{dy}{dx}$

$\dfrac{dy}{dx} + \dfrac{2\ln(x+y)}{x+1} = 1 - \dfrac{2\ln(x+y)}{x+1}$

$= \dfrac{x+y-2\ln(x+y)}{x+y}$

$\dfrac{dy}{dx}\left[\dfrac{x+y+2\ln(x+y)}{x+y}\right] = \dfrac{x+y-2\ln(x+y)}{x+y}$

$\dfrac{dy}{dx} = \dfrac{x+y-2\ln(x+y)}{x+y+2\ln(x+y)}$

32. $y = \ln(x+\ln x)$

$\dfrac{dy}{dx} = \dfrac{1}{x+\ln x}\left(1+\dfrac{1}{x}\right) = \dfrac{1}{x+\ln x}\left(\dfrac{x+1}{x}\right)$

$= \dfrac{x+1}{x(x+\ln x)}$

33. $\dfrac{\ln 2.0001 - \ln 2.0000}{0.0001} = 0.4999875$

Slope of secant line through $(2.0000, \ln 2.0000)$ and $(2.0001, \ln 2.0001)$

$0.5 = \dfrac{d\ln x}{dx}$ for $x = 2$ and is slope of tangent line through $(2.0000, \ln 2.0000)$

34. $\dfrac{\ln 0.5001 - \ln 0.5000}{0.0001} = 1.99980006;$

slope of secant line

$\dfrac{d}{dx}\ln x = \dfrac{1}{x}\Big|_{x=0.5} = 2;$ the value of the derivative at $x = 0.5$

35. $y = (1+x)^{(1/x)}$

(a)   (b) See table.

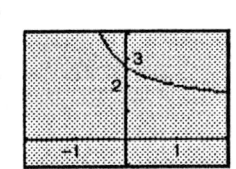

36. $y = x^2\ln x;$

$\dfrac{dy}{dx} = x^2\dfrac{1}{x} + \ln x(2x) = x + 2x\ln x$

$\dfrac{d^2y}{dx^2} = 1 + \left[2x\dfrac{1}{x} + \ln x(2)\right] = 1 + 2 + 2\ln x$

$= 3 + 2\ln$

37. $y = \sin^{-1}2x + \sqrt{1-4x^2};\ x = 0.250;$

$\dfrac{dy}{dx} = \dfrac{2}{\sqrt{1-(2x)^2}} + \dfrac{1}{2}(1-4x^2)^{-1/2}(-8x)$

$= \dfrac{2-4x}{\sqrt{1-4x^2}}$

$\dfrac{dy}{dx}\Big|_{x=0.250} = \dfrac{2-4(0.250)}{\sqrt{1-4(0.250)^2}} = 1.15$

38. $y = \ln\sqrt{\dfrac{2x+1}{3x+1}} \quad \ln\sqrt{2x+1} - \ln\sqrt{3x+1}$

$y = \dfrac{1}{2}\ln(2x+1) - \dfrac{1}{2}\ln(3x+1)$

$\dfrac{dy}{dx} = \left(\dfrac{1}{2}\right)\dfrac{1}{2x+1}(2) - \left(\dfrac{1}{2}\right)\dfrac{1}{3x+1}(3)$

$\dfrac{dy}{dx} = \dfrac{1}{2x+1} - \dfrac{\frac{3}{2}}{3x+1}; \quad \left.\dfrac{dy}{dx}\right|_{x=2.75} = -0.00832$

39. $f(x) = 2\ln(\tan x), \ f\left(\dfrac{\pi}{4}\right) = 0$

$f'(x) = 2\cot x + 2\tan x, \ f'\left(\dfrac{\pi}{4}\right) = 4$

$L(x) = 4\left(x - \dfrac{\pi}{4}\right) = 4x - \pi$

40. $y = 6\log_x 2$; therefore $\dfrac{y}{6} = \log_x 2$; $x^{y/6} = 2$

$\dfrac{y}{6}\ln x = \ln 2; \ y = \dfrac{6\ln 2}{\ln x} = 6\ln 2(\ln x)^{-1}$

$dy = 6\ln 2(-1)(\ln x)^{-2}\left(\dfrac{1}{x}\right)dx = \dfrac{-6\ln 2}{x\ln^2 x}dx$

41. $y = \tan^{-1} 2x + \ln(4x^2 + 1)$;

$m_{\tan} = \dfrac{dy}{dx} = \dfrac{1}{4x^2+1}(2) + \dfrac{1}{4x^2+1}(8x)$

$= \dfrac{2}{1+4x^2} + \dfrac{8x}{4x^2+1} = \dfrac{2+8x}{1+4x^2}$

When $x = 0.625$,

$m_{\tan} = \dfrac{2+8(0.625)}{1+4(0.625)^2} = \dfrac{7}{2.5625} = 2.73$

42. $y = x\ln 2x$ at $x = 2$; $\dfrac{dy}{dx} = x\dfrac{1}{2x}(2) + \ln 2x$

$m_{\text{TL}} = 1 + \ln 2x; \ m_{\text{TL}}|_{x=2} = 2.386$

43. $y = x^x$; $\ln y = x\ln x$

$\dfrac{1}{y}\dfrac{dy}{dx} = x\dfrac{1}{x} + \ln x = 1 + \ln x$

$\dfrac{dy}{dx} = (1 + \ln x)y = x^x(1 + \ln x)$

Eq. (3-15); $\dfrac{d}{dx}u^n$, n is a constant. In x^x both the base and the exponent are variable.

44. $y = (\sin x)^x$; $\ln y = x\ln\sin x$

$\dfrac{1}{y}\dfrac{dy}{dx} = x\dfrac{1}{\sin x}\cos x + \ln\sin x = x\cot x + \ln\sin x$

$\dfrac{dy}{dx} = (\sin x)^x(x\cot x + \ln\sin x)$

Cannot use Eq. (3-15) because exponent of $(\sin x)^x$ is variable.

45. $b = 10\log\left(\dfrac{I}{I_0}\right) = 10\log(I_0^{-1}I); \ u = (I_0^{-1}I);$

$\dfrac{db}{dt} = 10\left(\dfrac{I_0}{I}\right)(\log e)\left(\dfrac{I}{I_0}\right)\dfrac{dI}{dt} = \dfrac{10}{I}\log e\dfrac{dI}{dt}$

46. $t = kN\ln N$

$\dfrac{dt}{dN} = k\left(N\dfrac{1}{N} + \ln N\right) = k(1 + \ln N)$

47. $t = 5\ln\dfrac{16}{16 - 0.1v}; \ v = 100 \text{ ft/s}$

$t = 5[\ln 16 - \ln(16 - 0.1v)]$

$\dfrac{dt}{dv} = 5\left[0 - \dfrac{1}{16 - 0.1v}(-0.1)\right]$

$\dfrac{dt}{dv} = \dfrac{0.5}{16 - 0.1v}; \ \left.\dfrac{dt}{dv}\right|_{v=100} = 0.083 \text{ s}^2/\text{ft}$

48. $V = k\ln\dfrac{\sqrt{a^2+x^2}+a}{\sqrt{a^2+x^2}-a}$

$V = k[\ln(\sqrt{a^2+x^2}+a) - \ln(\sqrt{a^2+x^2}-a)]$

$\dfrac{dV}{dx} = \left[k\dfrac{1}{\sqrt{a^2+x^2}+a}\dfrac{1}{2}(a^2+x^2)^{-1/2}(2x)\right.$

$\left. - \dfrac{1}{\sqrt{a^2+x^2}-a}\dfrac{1}{2}(a^2+x^2)^{-1/2}(2x)\right]$

$= \dfrac{kx}{\sqrt{a^2+x^2}(\sqrt{a^2+x^2}+a)}$

$- \dfrac{kx}{\sqrt{a^2+x^2}(\sqrt{a^2+x^2}-a)}$

$= \dfrac{kx[a^2+x^2-a\sqrt{a^2+x^2}-(a^2+x^2+a\sqrt{a^2+x^2})]}{(a^2+x^2)(a^2+x^2-a^2)}$

$= \dfrac{kx[a^2+x^2-a\sqrt{a^2+x^2}-a^2-x^2-a\sqrt{a^2+x^2}}{(a^2+x^2)(x^2)}$

$\dfrac{dV}{dx} = \dfrac{kx(-2a\sqrt{a^2+x^2})}{x^2(a^2+x^2)} = \dfrac{-2ka}{x\sqrt{a^2+x^2}}$

$E = -\dfrac{dV}{dx} = \dfrac{2ka}{x\sqrt{a^2+x^2}}$

8.3   Derivative of the Exponential Function

1. $y = 3^{2x}$;

$$\frac{dy}{dx} = 3^{2x} \ln 3 \frac{d\,2x}{dx} = (2 \ln 3)3^{2x}$$

2. $y = 3^{1-x}$; $\dfrac{dy}{dx} = 3^{1-x} \ln 3(-1) = -(\ln 3)3^{1-x}$

3. $y = 4^{6x}$; $\dfrac{dy}{dx} = 4^{6x} \ln 4(6) = 6(\ln 4)4^{6x}$

4. $y = 10^{x^2}$; $\dfrac{dy}{dx} = 10^{x^2} \ln 10(2x) = (2x \ln 10)10^{x^2}$

5. $y = e^{\sqrt{x}}$;

$$\frac{dy}{dx} = e^{\sqrt{x}}\frac{1}{2}x^{-1/2} = \frac{e^{\sqrt{x}}}{2\sqrt{x}}$$

6. $r = 0.3e^{\theta^2}$; $\dfrac{dr}{d\theta} = 0.3e^{\theta^2}(2\theta) = 0.6\theta e^{\theta^2}$

7. $y = 4e^t(e^{2t} - e^t) = 4e^{3t} - 4e^{2t}$

$$\frac{dy}{dt} = 12e^{3t} - 8e^{2t} = 4e^{2t}(3e^t - 2)$$

8. $y = 0.2 \ln(e^{5x} + 1)$

$$\frac{dy}{dx} = 0.2 \cdot \frac{1}{e^{5x}+1} \cdot 5e^{5x} = \frac{e^{5x}}{e^{5x}+1}$$

9. $y = xe^{-x}$

$$\frac{dy}{dx} = x(e^{-x})(-1) + (1)(e^{-x})$$
$$= e^{-x} - xe^{-x} = e^{-x}(1-x)$$

10. $y = 5x^2 e^{2x}$

$$\frac{dy}{dx} = 5[x^2 e^{2x} \cdot 2 + e^{2x}(2x)]$$
$$\frac{dy}{dx} = 10x^2 e^{2x} + 10xe^{2x} = 10xe^{2x}(x+1)$$

11. $y = xe^{\sin x}$

$$\frac{dy}{dx} = xe^{\sin x} \cdot \cos x + e^{\sin x}(1)$$
$$\frac{dy}{dx} = e^{\sin x}(x \cos x + 1)$$

12. $y = 4e^x \sin \dfrac{1}{2}x$

$$\frac{dy}{dx} = 4\left[e^x \frac{1}{2}\cos\frac{1}{2}x + \sin\frac{1}{2}xe^x\right]$$
$$\frac{dy}{dx} = 2e^x \left(\cos\frac{1}{2}x + 2\sin\frac{1}{2}x\right)$$

13. $r = \dfrac{2(e^{2s} - e^{-2s})}{e^{2s}}$

$$r = 2(1 - e^{-4s})$$
$$\frac{dr}{ds} = 2(4e^{-4s})$$
$$\frac{dr}{ds} = 8e^{-4s}$$

14. $u = \dfrac{e^{0.5v}}{2v}$

$$\frac{du}{dv} = \frac{2v \cdot e^{0.5v} \cdot 0.5 - e^{0.5v} \cdot 2}{4v^2}$$
$$\frac{du}{dv} = \frac{e^{0.5v}(v-2)}{4v^2}$$

15. $y = e^{-3x} \sin 4x$

$$\frac{dy}{dx} = e^{-3x}4\cos 4x + \sin 4xe^{-3x}(-3)$$
$$\frac{dy}{dx} = e^{-3x}(4\cos 4x - 3\sin 4x)$$

16. $y = (\cos 2x)(e^{x^2-1})$

$$\frac{dy}{dx} = \cos 2x(e^{x^2-1} \cdot 2x) + e^{x^2-1}(-2\sin 2x)$$
$$\frac{dy}{dx} = 2e^{x^2-1}(x\cos 2x - \sin 2x)$$

17. $y = \dfrac{2e^{3x}}{4x+3}$;

$$\frac{dy}{dx} = \frac{(4x+3)(2e^{3x})(3) - (2e^{3x})(4)}{(4x+3)^2}$$
$$\frac{dy}{dx} = \frac{(12x+9)(2e^{3x}) - 8e^{3x}}{(4x+3)^2}$$
$$= \frac{2e^{3x}(12x+5)}{(4x+3)^2}$$

18. $y = \dfrac{7 \ln 2x}{e^{2x}+2}$

$$\frac{dy}{dx} = \frac{(e^{2x}+2)7 \cdot \frac{1}{2x}2 - 7\ln 2x(e^{2x}(2))}{(e^{2x}+2)^2}$$
$$\frac{dy}{dx} = \frac{\frac{7(e^{2x}+2)}{x} - 14\ln 2x(e^{2x})}{(e^{2x}+2)^2}$$
$$\frac{dy}{dx} = \frac{7(e^{2x}+2) - 14x\ln 2x(e^{2x})}{x(e^{2x}+2)^2}$$
$$\frac{dy}{dx} = \frac{7(e^{2x}+2 - 2xe^{2x}\ln 2x)}{x(e^{2x}+2)^2}$$

19. $y = \ln(e^{x^2}+4)$; $\dfrac{dy}{dx} = \dfrac{1}{e^{x^2}+4}e^{x^2}(2x) = \dfrac{2xe^{x^2}}{e^{x^2}+4}$

20. $p = (3e^{2n} + e^2)^3$

$$\frac{dp}{dn} = 3(3e^{2n} + e^2)^2 \cdot 6e^{2n}$$

$$= 18e^{2n}(3e^{2n} + e^2)^2$$

$$\frac{dp}{dn} = 162e^{6n} + 108e^{4n+2} + 18e^{2n+4}$$

21. $y = (2e^{2x})^3 \sin x^2$
       $= 8e^{6x} \sin x^2$

$$\frac{dy}{dx} = 8e^{6x}(\cos x^2)(2x) + \sin x^2 (8e^{6x})(6)$$

$$= 16e^{6x}(x \cos x^2 + 3 \sin x^2)$$

22. $y = (e^{3/x} \cos x)^2$

$$\frac{dy}{dx} = 2(e^{3/x} \cos x)\left[e^{3/x}(-\sin x) + \cos x e^{3/x}\left(\frac{-3}{x^2}\right)\right]$$

$$\frac{dy}{dx} = -2e^{3/x}e^{3/x}\left(\sin x + \frac{3 \cos x}{x^2}\right)$$

$$\frac{dy}{dx} = \frac{-2e^{6/x}(x^2 \sin x + 3 \cos x)}{x^2}$$

23. $u = 4\sqrt{\ln 2t + e^{2t}}$

$$\frac{du}{dt} = 4\left(\frac{1}{2}\right)(\ln 2t + e^{2t})^{-1/2}\left(\frac{1}{2t}(2) + 2e^{2t}\right)$$

$$\frac{du}{dt} = \frac{2(1 + 2te^{2t})}{t\sqrt{\ln 2t + e^{2t}}}$$

24. $y = (2e^{x^2} + x^2)^3$

$$\frac{dy}{dx} = 3(2e^{x^2} + x^2)^2(2e^{x^2} \cdot 2x + 2x)$$

$$\frac{dy}{dx} = 6x(2e^{x^2} + x^2)^2(2e^{x^2} + 1)$$

25. $y = xe^{xy} + \sin y$

$$\frac{dy}{dx} = x(e^{xy})\left(x\frac{dy}{dx} + y\right) + (1)e^{xy} + \cos y \frac{dy}{dx}$$

$$= x(e^{xy})\left(x\frac{dy}{dx}\right) + x(e^{xy})(y) + e^{xy} + \cos y \frac{dy}{dx}$$

$$\frac{dy}{dx} - x(e^{xy})\left(x\frac{dy}{dx}\right) - \cos y \frac{dy}{dx} = x(e^{xy})y + e^{xy}$$

$$\frac{dy}{dx}(1 - x(e^{xy})(x) - \cos y) = x(e^{xy})y + e^{xy}$$

$$\frac{dy}{dx} = \frac{xy(e^{xy}) + e^{xy}}{1 - x^2e^{xy} - \cos y} = \frac{e^{xy}(xy + 1)}{1 - x^2e^{xy} - \cos y}$$

26. $y = 4e^{-2/x} \ln y + 1$

$$\frac{dy}{dx} = 4\left[e^{-2/x}\frac{1}{y}\frac{dy}{dx} + \ln y e^{-2/x}\left(\frac{2}{x^2}\right)\right]$$

$$\frac{dy}{dx} = \frac{4e^{-2/x}}{y}\frac{dy}{dx} + \frac{8 \ln y e^{-2/x}}{x^2}$$

$$\frac{dy}{dx} - \frac{4e^{-2/x}}{y}\frac{dy}{dx} = \frac{8 \ln y e^{-2/x}}{x^2}$$

$$\frac{dy}{dx}\left(1 - \frac{4e^{-2/x}}{y}\right) = \frac{8 \ln y e^{-2/x}}{x^2}$$

$$\frac{dy}{dx} = \frac{8y \ln y e^{-2/x}}{x^2} \div \frac{y - 4e^{-2/x}}{y} = \frac{8 \ln y e^{-2/x}y}{x^2(y - 4e^{-2/x})}$$

$$\frac{dy}{dx} = \frac{8y \ln y e^{-2/x}}{x^2(y - 4e^{-2/x})} = \frac{8y \ln y e^{-2/x}}{x^2 y - 4x^2 e^{-2/x}}$$

27. $y = 3e^{2x} \ln x$

$$\frac{dy}{dx} = 3\left(e^{2x}\frac{1}{x} + \ln x \cdot 2e^{2x}\right)$$

$$\frac{dy}{dx} = \frac{3e^{2x}}{x} + 6e^{2x} \ln x$$

28. $r = 0.4e^{2\theta} \ln(\cos \theta)$

$$\frac{dr}{d\theta} = 0.4e^{2\theta} \cdot \frac{1}{\cos \theta} \cdot (-\sin \theta)$$

$$+ \ln(\cos \theta) \cdot (0.8e^{2\theta})$$

$$\frac{dr}{d\theta} = e^{2\theta}(-0.4 \tan \theta + 0.8 \ln(\cos \theta))$$

29. $y = \ln \sin 2e^{6x}$

$$\frac{dy}{dx} = \frac{1}{\sin 2e^{6x}}(\cos 2e^{6x})(2e^{6x})(6)$$

$$\frac{dy}{dx} = \frac{12e^{6x} \cos 2e^{6x}}{\sin 2e^{6x}} = 12e^{6x} \cot 2e^{6x}$$

30. $y = 6 \tan e^{x+1}$

$$\frac{dy}{dx} = 6 \sec^2 e^{x+1} \cdot e^{x+1} = 6e^{x+1} \sec^2 e^{x+1}$$

31. $y = 2 \sin^{-1} e^{2x}$

$$\frac{dy}{dx} = 2\frac{1}{\sqrt{1 - e^{4x}}} \cdot 2e^{2x} = \frac{4e^{2x}}{\sqrt{1 - e^{4x}}}$$

32. $y = \tan^{-1} e^{3x}$

$$\frac{dy}{dx} = \frac{1}{1 + e^{6x}} \cdot 3e^{3x} = \frac{3e^{3x}}{1 + e^{6x}}$$

33. (a) $e = e^x = 2.7182818$ when $x = 1.0000$. This is the slope of a tangent line to the curve $f(x) = e^x$ when $x = 1.0000$. It is the value of $f'(x) = e^x$, since $\frac{d\,e^x}{dx} = e^x$.

(b) $\dfrac{e^{1.0001} - e^{1.0000}}{0.0001} = 2.7184178$ This is

the slope of a secant line through the curve $f(x) = e^x$ at $x = 1.0000$, where $\Delta x = 0.0001$.

$$\lim_{\Delta x \to 0} \frac{e^{(x+\Delta x)} - e^x}{\Delta x} = \frac{d\,e^x}{dx} = e^x$$

For $\Delta x = 0.0001$, the slope of the tangent line is approximately equal to the slope of the secant line.

34. (a) $e^2 = 7.3890561$; value of $\dfrac{d}{dx}e^x$ at $x = 2$

(b) $\dfrac{e^{2.0001} - e^{2.0000}}{0.0001} = 7.3894256$; slope of secant

line for $y = e^x$ through $x_1 = 2$ and $x_2 = 2.0001$.

35. $y = e^{-2x}\cos 2x$; $x = 0.625$

$$\frac{dy}{dx} = e^{-2x}(-2\sin 2x) + \cos 2x(-2e^{-2x})$$

$$\frac{dy}{dx} = m_{\text{TL}} = -2e^{-2x}(\sin 2x + \cos 2x)$$

$$m_{\text{TL}}\big|_{x=0.625} = -0.724$$

36. $y = \dfrac{e^{-x}}{1 + \ln 4x}$; $x = 1.842$

$$\frac{dy}{dx} = \frac{(1 + \ln 4x)(-e^{-x}) - e^{-x}\left(\dfrac{1}{x}\right)}{(1 + \ln 4x)^2}$$

$$\frac{dy}{dx} = m_{\text{TL}} = \frac{-e^{-x}\left(1 + \ln 4x + \dfrac{1}{x}\right)}{(1 + \ln 4x)^2}$$

$$m_{\text{TL}}\big|_{x=1.842} = -0.06246$$

37. $y = \dfrac{2e^{4x}}{(x+2)}$;

$$\frac{dy}{dx} = \frac{(x+2)8e^{4x} - 2e^{4x}}{(x+2)^2}$$

$$\frac{dy}{dx} = 2e^{4x}\frac{[4(x+2) - 1]}{(x+2)^2} = \frac{2e^{4x}(4x+7)}{(x+2)^2}$$

$$dy = \frac{2e^{4x}(4x+7)}{(x+2)^2}\,dx$$

38. $f(x) = \dfrac{6e^{4x}}{2x+3}$

$f(0) = 2$

$$f'(x) = \frac{12(4x+5)e^{4x}}{(2x+3)^2}$$

$$f'(0) = \frac{20}{3}$$

$$L(x) = \frac{20}{3}(x - 0) + 2 = \frac{20x}{3} + 2$$

39.   **40.**

41. $y = xe^{-x}$; $\dfrac{dy}{dx} = x(e^{-x})(-1) + (e^{-x})(1)$

$$= -xe^{-x} + e^{-x}$$

Substituting, $\dfrac{dy}{dx} + y = (-xe^{-x} + e^{-x}) + (xe^{-x})$

$$= e^{-x}$$

42. $y = e^{-x}\sin x$ (1) $\dfrac{d^2y}{dx^2} + 2\dfrac{dy}{dx} + 2y = 0$ (2)

$$\frac{dy}{dx} = e^{-x}\cos x + \sin x(-e^{-x}) = e^{-x}(\cos x - \sin x)$$

$$\frac{d^2y}{dx^2} = e^{-x}(-\sin x - \cos x) + (\cos x - \sin x)(-e^{-x})$$

$$\frac{d^2y}{dx^2} = e^{-x}(-\sin x - \cos x - \cos x + \sin x)$$

$$= e^{-x}(-2\cos x)$$

$$\frac{d^2y}{dx^2} = -2e^{-x}\cos x$$

Put into (2).

$$-2e^{-x}\cos x + 2e^{-x}(\cos x - \sin x) + 2e^{-x}\sin x = 0$$
$$-2e^{-x}\cos x + 2e^{-x}\cos x - 2e^{-x}\sin x + 2e^{-x}\sin x = 0$$

43. $y = \dfrac{e^{2x} - 1}{e^{2x} + 1}$

$\dfrac{dy}{dx} = 1 - y^2$

$\dfrac{dy}{dx} = \dfrac{(e^{2x} + 1)2e^{2x} - (e^{2x} - 1)2e^{2x}}{(e^{2x} + 1)^2}$

$\dfrac{dy}{dx} = \dfrac{2e^{2x}(e^{2x} + 1 - e^{2x} + 1)}{(e^{2x} + 1)^2} = \dfrac{4e^{2x}}{(e^{2x} + 1)^2}$

$1 - y^2 = 1 - \dfrac{(e^{2x} - 1)^2}{(e^{2x} + 1)^2}$

$ = \dfrac{(e^{2x} + 1)^2 - (e^{2x} - 1)^2}{(e^{2x} + 1)^2}$

$1 - y^2 = \dfrac{e^{4x} + 2e^{2x} + 1 - e^{4x} + 2e^{2x} - 1}{(e^{2x} + 1)^2}$

$ = \dfrac{4e^{2x}}{(e^{2x} + 1)^2}$

Q.E.D

44. $e^x + e^y = e^{x+y}$ (1)

$e^x + e^y \dfrac{dy}{dx} = e^{x+y}\left(1 + \dfrac{dy}{dx}\right) = e^{x+y} + e^{x+y}\dfrac{dy}{dx}$

$e^y \dfrac{dy}{dx} - e^{x+y}\dfrac{dy}{dx} = e^{x+y} - e^x$; substituting (1) into this.

$e^y \dfrac{dy}{dx} - (e^x + e^y)\dfrac{dy}{dx} = e^x + e^y - e^x$

$\dfrac{dy}{dx}(e^y - e^x - e^y) = e^y$; $\dfrac{dy}{dx} = \dfrac{-e^y}{-e^x} = -e^{y-x}$

45. $R = e^{-0.002t}; t = 100 \text{ h}: R(0 \le R \le 1)$

$\dfrac{dR}{dt} = e^{-0.002t}(-0.002)$

$\dfrac{dR}{dt} = -0.002e^{-0.002t}$; $\left.\dfrac{dR}{dt}\right|_{t=100} = -0.00164/\text{h}$

46. $T = 8.0(3.0 - 5.0e^{-0.50t})$

$\left.\dfrac{dT}{dt} = 20e^{-t/2}\right|_{t=6.0} = 1.0°\text{C/min}$

47. $i = 4.42e^{-66.7t}\sin(226t)$

$\dfrac{di}{dt} = 4.42e^{-66.7t}\cos(226t)\cdot 226 - 294.8e^{-66.7t}\sin(226t)$

$\dfrac{di}{dt} = e^{-66.7t}(999\cos(226t) - 295\sin(226t))$

48. $P = \dfrac{10}{1 + 0.65e^{-0.060t}}$

$\dfrac{dP}{dt} = \dfrac{(1 + 0.65e^{-0.060t})(0) - 10(-0.039e^{-0.060t})}{(1 + 0.65e^{-0.060t})^2}$

$\dfrac{dP}{dt} = \dfrac{0.39e^{-0.060t}}{(1 + 0.65e^{-0.060t})^2}$

49. $\dfrac{d}{dx}\sinh u = \dfrac{d}{dx}\dfrac{1}{2}(e^u - e^{-u}) = \dfrac{1}{2}\left(e^u\dfrac{du}{dx} - e^{-u}\cdot -\dfrac{du}{dx}\right)$

$\dfrac{d}{dx}\sinh u = \dfrac{1}{2}e^u\dfrac{du}{dx} + \dfrac{1}{2}e^{-u}\dfrac{du}{dx} = \left[\dfrac{1}{2}(e^u + e^{-u})\right]\dfrac{du}{dx}$

$\dfrac{d}{dx}\sinh u = \cosh u\dfrac{du}{dx}$ Q.E.D.

50. $\dfrac{d}{dx}\cosh u = \dfrac{d}{dx}\dfrac{1}{2}(e^u + e^{-u}) = \dfrac{1}{2}\left(e^u\dfrac{du}{dx} + e^{-u}\cdot -\dfrac{du}{dx}\right)$

$\dfrac{d}{dx}\cosh u = \dfrac{1}{2}e^u\dfrac{du}{dx} - \dfrac{1}{2}e^{-u}\dfrac{du}{dx} = \left[\dfrac{1}{2}(e^u - e^{-u})\right]\dfrac{du}{dx}$

$\dfrac{d}{dx}\cosh u = \sinh u\dfrac{du}{dx}$ Q.E.D.

51. $y = x\sinh 2x$

$\dfrac{dy}{dx} = x\cosh 2x(2) + \sinh 2x(1)$

$\dfrac{dy}{dx} = 2x\cosh 2x + \sinh 2x$

52. $y = \sinh x$

$\dfrac{d\sinh x}{dx} = \cosh x$

$\dfrac{d^2\sinh x}{dx^2} = \dfrac{d}{dx}(\cosh x) = \sinh x$

$y = \cosh x$

$\dfrac{d\cosh x}{dx} = \sinh x$

$\dfrac{d^2\cosh x}{dx^2} = \dfrac{d\sinh x}{dx} = \cosh x$

8.4 Applications

1. $y = \ln\cos x$; $\dfrac{dy}{dx} = \dfrac{1}{\cos x}(-\sin x) = \dfrac{-\sin x}{\cos x}$

$\phantom{y = \ln\cos x;\ \dfrac{dy}{dx}} = -\tan x$;

$\dfrac{d^2y}{dy^2} = -\sec^2 x$

(1) Since $\ln(1) = 0$ and $\cos 0 = 1$, there is an intercept at $(0, 0)$.

(2) Since ln functions are not defined for negatives, y is undefined for $\cos x < 0$. The function is defined for x between $-\frac{\pi}{2}$ and $\frac{\pi}{2}$, etc.

(3) As $\cos x$ approaches 0, $\ln\cos x$ approaches negative infinity.

$\cos x = 0$ when $x = -\frac{\pi}{2}$, $\frac{\pi}{2}$, and their odd multiples, so $x = -\frac{\pi}{2}$, $x = \frac{\pi}{2}$, $x = \frac{3\pi}{2}$, etc., are asymptotes.

(4) Critical points exist where $-\tan x = 0$; i.e., where $x = 0, 2\pi, 4\pi$, etc.

(5) $-\sec^2 x$ is negative at all critical points so the graph is concave down, and all critical points are maximum points.

(6) Maximum points are $(0,0), (2\pi, 0), (4\pi, 0)$, etc.

Summary:
Int. $(0,0)$, max. $(0,0)$, not defined for $\cos x < 0$, asym. $x = -\frac{1}{2}\pi, \frac{1}{2}\pi, \ldots$

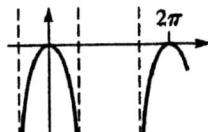

2. $y = \dfrac{2\ln x}{x}$

(1) x-int $= (1,0)$; y-int, none

(2) no symmetry

(3) As $x \to +\infty$, $y \to 0$ as horizontal asymptote

(4) $x = 0$ is vertical asymptote

(5) domain is $x > 0$

(6) $\dfrac{dy}{dx} = \dfrac{2 - 2\ln x}{x^2} = 0$ when $\ln x = 1, x = e$

$\dfrac{d^2 y}{dx^2} = \dfrac{4\ln x - 6}{x^3}\bigg|_{x=e} < 0 \Rightarrow (e, \frac{2}{e})$ is a maximum

$\dfrac{d^2 y}{dx^2} = \dfrac{4\ln x - 6}{x^3} = 0$ when $x = e^{3/2} \Rightarrow \left(e^{3/2}, \dfrac{3}{e^{3/2}}\right)$

is an inflection point

(7) range: $-\infty < y < \dfrac{2}{e}$

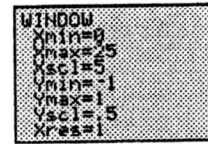 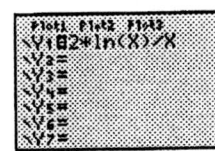

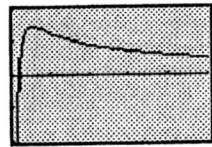

3. $y = xe^{-x} = \dfrac{x}{e^x}$

$\dfrac{dy}{dx} = x(-e^{-x}) + e^{-x} = e^{-x}(1 - x)$

$\dfrac{d^2 y}{dx^2} = e^{-x}(-1) + (1-x)(-e^{-x})$
$= e^{-x} - e^{-x} + xe^{-x} = e^{-x}(x - 2)$

(1) Intercepts: $x = 0, y = 0$ (origin)

(2) Symmetry: none

(3) As $x \to +\infty, y \to 0$ positively. Horizontal asymptote $y = 0$. As $x \to -\infty, y \to -\infty$.

(4) Vertical asymptote: none

(5) Domain: all x, range to be determined.

(6) $\dfrac{dy}{dx} = 0$; $e^{-x}(1 - x) = 0$; $x = 1$; $f''(1) < 0$,

max. $\left(1, \dfrac{1}{e}\right)$

$\dfrac{d^2 y}{dx^2} = 0$; $e^{-x}(x - 2) = 0$; $x = 2$; infl. $\left(2, \dfrac{2}{e^2}\right)$

Therefore, range: $-\infty < y \le \dfrac{1}{e}$

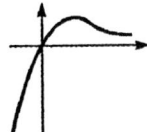

4. $y = \dfrac{e^x}{x}$; $\dfrac{dy}{dx} = \dfrac{xe^x - e^x}{x^2} = \dfrac{e^x(x-1)}{x^2}$

$\dfrac{d^2 y}{dx^2} = \dfrac{x^2[e^x + (x-1)e^x] - e^x(x-1)2x}{x^4}$

$\dfrac{d^2 y}{dx^2} = \dfrac{xe^x(x^2 + 2x - 2)}{x^4}$

(1) Intercepts: $x \ne 0, y \ne 0$ **(2)** No symmetry.

(3) As $x \to +\infty, y \to +\infty$

As $x \to -\infty, y \to 0$ negatively; horizontal asymptote
$y = 0$.

(4) Vertical asymptote at $x = 0$.

(5) Domain: all x except $x = 0$; range: to be determined.

(6) $\dfrac{dy}{dx} = 0$; $x = 1$; $f''(1) > 0$, min. $(1, e)$

$\dfrac{d^2y}{dx^2} = 0$; $x \neq 0$; $e^x \neq 0$; $x^2 + 2x - 2$, no real roots, no inflection points.

Therefore, range: $y < 0, 1 \leq y < \infty$

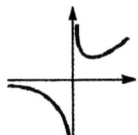

5. $y = \ln \dfrac{1}{x^2+1} = -\ln(x^2+1)$;

$\dfrac{dy}{dx} = -\dfrac{1}{x^2+1}(2x) = \dfrac{-2x}{x^2+1}$;

$\dfrac{d^2y}{dx^2} = \dfrac{(x^2+1)(-2) - (-2x)(2x)}{(x^2+1)^2} = \dfrac{2x^2-2}{(x^2+1)^2}$

(1) Since $\frac{1}{x^2+1} > 0$ for all numbers, $\ln\frac{1}{x^2+1}$ is defined for all numbers; there are no asymptotes.

(2) When $x = 0$, $y = \ln\frac{1}{0+1} = \ln 1 = 0$; $(0,0)$ is an intercept.

(3) Critical points; $\frac{dy}{dx} = \frac{-2x}{x^2+1} = 0$ when $-2x = 0$; $(0,0)$ is a critical point.

(4) $\frac{d^2y}{dx^2} = \frac{2x^2-2}{(x^2+1)^2} = 0$ when $2x^2 - 2 = 0$; $x^2 = 1, x = -1, x = 1$.

Inflections occur at $x = -1$.

$y = \ln\left(\dfrac{1}{(-1)^2+1}\right) = \ln\dfrac{1}{2} = \ln(2^{-1}) = -\ln 2$

and $x = 1$, $y = \ln\left(\dfrac{1}{1^2+1}\right) = -\ln 2$

(5) Since $\frac{d^2y}{dx^2}$ is negative at the critical point $(0,0)$, it is a maximum. The graph is concave down between the points $(-1, -\ln 2)$ and $(1, -\ln 2)$. It is concave up for $x < -1$ and $x > 1$. Since the second derivative goes through a change of sign at $x = -1$ and again at $x = 1$, $(-1, -\ln 2)$ and $(1, -\ln 2)$ are inflection points.

Summary:
Int. $(0,0)$, max. $(0,0)$, infl. $(-1, -\ln 2), (1, -\ln 2)$

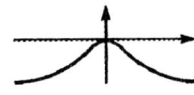

6. $y = \ln\dfrac{e}{x} = \ln e - \ln x = 1 - \ln x$; $\dfrac{dy}{dx} = -\dfrac{1}{x}$;

$\dfrac{d^2y}{dx^2} = \dfrac{1}{x^2}$

(1) Intercepts: $x \neq 0$; $y = 0$ when $x = e$, $(e, 0)$ is x intercept.

(2) Symmetry: none

(3) As $x \to +\infty$, y decreases slowly, remains negative.

(4) Vertical asymptote $x = 0$.

(5) Domain: all $x > 0$; range: to be determined

(6) $\dfrac{dy}{dx} = 0$; $-\dfrac{1}{x} \neq 0$; $\dfrac{d^2y}{dx^2} \neq 0$

$\dfrac{dy}{dx} < 0$ for all x; decreasing $x > 0$

$\dfrac{d^2y}{dx^2} > 0$ for all x; concave up $x > 0$.

Therefore, range: $-\infty < y < \infty$

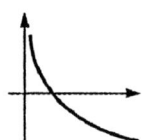

7. $y = 4e^{-x^2}$; $\dfrac{dy}{dx} = 4e^{-x^2}(-2x) = \dfrac{-8x}{e^{x^2}}$

$\dfrac{d^2y}{dx^2} = \dfrac{e^{x^2}(-8) + 8x(2xe^{x^2})}{e^{2x^2}} = 8e^{x^2}(2x^2 - 1)$

(1) Intercepts: $x = 0, y = 4, (0, 4)$ intercept.

(2) Symmetry: yes, with respect to y-axis.

(3) As $x \to \pm\infty, y \to 0$ positively; x-axis is a horizontal asymptote.

(4) No vertical asymptote.

(5) Domain: all x; range: to be determined.

(6) $\dfrac{dy}{dx} = 0$; $\dfrac{-8x}{e^{x^2}} = 0$; $f''(0) < 0$, max. $(0, 4)$

$\dfrac{d^2y}{dx^2} = 0$; $2x^2 - 1 = 0$; $x = \pm\sqrt{\dfrac{1}{2}} = \pm\dfrac{\sqrt{2}}{2}$;

$\left(-\dfrac{\sqrt{2}}{2}, \dfrac{4}{\sqrt{e}}\right), \left(\dfrac{\sqrt{2}}{2}, \dfrac{4}{\sqrt{e}}\right)$ are inflection points.

Range: $0 < y \leq 4$

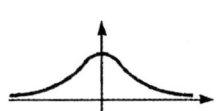

8. $y = x - e^x$; $\dfrac{dy}{dx} = 1 - e^x$; $\dfrac{d^2y}{dx^2} = -e^x$

    (1) Intercepts: $x = 0, y = -1, (0, -1)$

    (2) Symmetry: none.

    (3) As $x \to -\infty, y \to -\infty$

    As $x \to -\infty, y \to -\infty$; asymptote at $y = x$

    (4) Vertical asymptote: none.

    (5) Domain: all x; range: to be determined.

    $\dfrac{dy}{dx} = 0$; $e^x = 1$; $x = 0$; $f''(0) < 0$, max. $(0, -1)$

    $\dfrac{d^2y}{dx^2} =< 0$ for all x, therefore concave down for all x.

    Therefore, range: $-\infty < y \le -1$

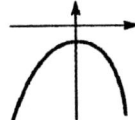

9. $y = \ln x - x$; $\dfrac{dy}{dx} = \dfrac{1}{x} - 1$; $\dfrac{d^2y}{dx^2} = -x^{-2} = -\dfrac{1}{x^2}$

    (1) $x \not< 0$ since ln is undefined for those values. $y \ne 0$ since $\ln x \ne x$ for any number.

    (2) There is an asymptote at $x = 0$.

    (3) Critical points occur at $\dfrac{1}{x} - 1 = 0$ or $x = 1$, $y = -1$.

    (4) Since $\dfrac{1}{x^2} \ne 0$ for any x, there is no point of inflection.

    (5) Since $-\dfrac{1}{x^2} < 0$, for all x, the graph is concave down, and the critical point $(1, -1)$ is a maximum.

    Max. $(1, -1)$, asymptote $x = 0$

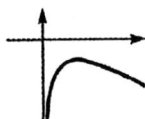

10. $y = e^{-x} \sin x$

    $\dfrac{dy}{dx} = e^{-x} \cos x + \sin x(e^{-x}) = e^{-x}(\cos x - \sin x)$

    $\dfrac{d^2y}{dx^2} = e^{-x}(-\sin x - \cos x) + (\cos x - \sin x) - e^{-x}$

    $\phantom{\dfrac{d^2y}{dx^2}} = e^{-x}(-\sin x - \cos x - \cos x + \sin x)$

    $\phantom{\dfrac{d^2y}{dx^2}} = -2e^{-x} \cos x$

    (1) Intercepts: $(0, 0), (\pi, 0)$, etc.

    (2) No symmetry.

    (3) $x \to +\infty, e^{-x} \to 0$; therefore, $y \to 0$ from above and below.

    (4) No vertical asymptotes.

    (5) Domain: all x; range: to be determined.

    (6) $\dfrac{dy}{dx} = 0$; $\sin x = \cos x$; $\tan x = 1$; $x = \dfrac{\pi}{4}, \dfrac{5\pi}{4}, \dfrac{9\pi}{4}$, etc..

    $f''\left(\dfrac{\pi}{4}\right) < 0$, max.; $f''\left(\dfrac{5\pi}{4}\right) > 0$, min.'

    $\left(\dfrac{\pi}{4}, 0.322\right)$ max.; $\left(\dfrac{5\pi}{4}, -0.014\right)$ min.

    Range: $-0.014 \le y \le 0.322$ over $(0 \le x \le 2\pi)$

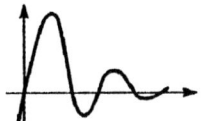

11. $y = \dfrac{1}{2}(e^x - e^{-x})$; $\dfrac{dy}{dx} = \dfrac{1}{2}(e^x + e^{-x})$

    $\dfrac{d^2y}{dx^2} = \dfrac{1}{2}(e^x - e^{-x})$

    (1) Intercepts: $x = 0, y = 0, (0, 0)$

    (2) No symmetry.

    (3) As $x \to +\infty, y \to +\infty, x \to -\infty, y \to -\infty$

    (4) No vertical asymptote.

    (5) Domain: all x; range: all y.

    (6) $\dfrac{dy}{dx} = 0$; $e^x = -e^{-x} = -\dfrac{1}{e^x}$; $(e^x)^2 = -1$;

    $e^x = \sqrt{-1}$. $\dfrac{dy}{dx} > 0$, inc for all x.

    (imaginary), no max,. no min.

    $\dfrac{d^2y}{dx^2} = 0, e^x = e^{-x} = \dfrac{1}{e^x}$; $(e^x)^2 = 1$; $e^x = \pm 1$

    $e^x \ne -1$; $e^x = 1$; $x = 0$, therefore infl. at $(0, 0)$

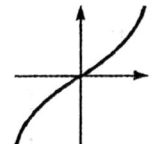

12. $y = \frac{1}{2}(e^x + e^{-x})$; $\frac{dy}{dx} = \frac{1}{2}(e^x + e^{-x})$

$\frac{d^2y}{dx^2} = \frac{1}{2}(e^x + e^{-x})$

(1) Intercepts: $x = 0, y = 1, (0,1)$

(2) Symmetry: yes, with respect to y-axis, $f(x) = f(-x)$

(3) As $x \to +\infty, y \to +\infty$; $x \to -\infty, y \to +\infty$

(4) No vertical asymptote.

(5) Domain: all x; range: to be determined.

(6) $\frac{dy}{dx} = 0$; $e^x = e^{-x} = \frac{1}{e^x}$; $(e^x)^2 = 1$; $e^x = \pm 1$;

$e^x \neq -1$; $e^x = 1$; $x = 0$; $f''(0) > 0$; min, $(0,1)$

$\frac{d^2y}{dx^2} = 0$; $e^x \neq -\frac{1}{e^x}$, no infl. points, therefore range: $1 \leq y$

$\frac{d^2y}{dx^2}$ always > 0, therefore $f(x)$ concave up for

all x.

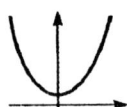

13. $y = x^2 \ln x$;

$\frac{dy}{dx} = x^2\left(\frac{1}{x}\right) + (\ln 2)(2x) = x + 2x \ln x$;

$\left.\frac{dy}{dx}\right|_{x=1} = 1 + 2\ln 1 = 1 + 2(0) = 1$

Slope is 1, $x = 1, y = 0$; using slope intercept form of the equation and substituting gives $0 = 1(1) + b$ or $b = 1$. The equation is $y = (1)x - 1$ or $y = x - 1$.

14. $y = \tan^{-1} 2x$; $x = 1$; $f(1) = 1.107$

$\frac{dy}{dx} = \frac{1}{1 + 4x^2}(2) = \frac{2}{1 + 4x^2}$; $\left.m_{TL}\right|_{x=1} = \frac{2}{5}$

Equation of T.L.: $y - 1.107 = \frac{2}{5}(x - 1)$

$5y - 5.536 = 2x - 2$; $2x - 5y + 3.54 = 0$

15. $y = 2\sin\frac{1}{2}x$; $x = \frac{3\pi}{2}$; $f\left(\frac{3\pi}{2}\right) = 1.414 = \sqrt{2}$

$\frac{dy}{dx} = 2\cos\frac{1}{2}x\left(\frac{1}{2}\right) = \cos\frac{1}{2}x$;

$\left.m_{TL}\right|_{3\pi/2} = -0.707 = -\frac{1}{\sqrt{2}}$

Therefore, $m_{NL} = \sqrt{2}$

Equation of N.L.:

$y - \sqrt{2} = \sqrt{2}\left(x - \frac{3\pi}{2}\right)$

$\qquad = \sqrt{2}x - \frac{3\sqrt{2}}{2}\pi$

$2y - 2\sqrt{2} = 2\sqrt{2}x - 3\sqrt{2}\pi$;
$2\sqrt{2}x - 2y + 2\sqrt{2} - 3\pi\sqrt{2} = 0$

16. $y = \frac{e^{2x}}{x}$;

$\frac{dy}{dx} = \frac{x\,2e^{2x} - e^{2x}}{x^2}$

$\qquad = \frac{e^{2x}(2x - 1)}{x^2}$

$\qquad = \left.m_{TL}\right|_{x=1} = e^2$

$m_{NL} = -\frac{1}{e^2}$; $f(1) = e^2$

$y - e^2 = -\frac{1}{e^2}(x - 1)$; $e^2 y - e^4 = -x + 1$

$x + e^2 y - 1 - e^4 = 0$

17. $f(x) = x^2 - 2 + \ln x$; $f'(x) = 2x + \frac{1}{x}$;

$f(1) = 1^2 - 2 + \ln 1 = -1$;

$f(2) = 2^2 - 4 + \ln 2 = 0.69$

Therefore we choose $x_1 = 1.5$

n	x_n	$f(x_n)$	$f'(x_n)$	$x_n - \frac{f(x_n)}{f'(x_n)}$
1	1.5	0.6554651	3.6666667	1.3212368
2	1.3212368	0.0242349	3.3993402	1.3141075
3	1.3141075	0.0000362	3.3891878	1.3140968

Therefore, the root is 1.3140968, which is correct to the number of decimal places shown.

18. $e^{-2x} - \tan^{-1} x = 0$; $e^{-2x} = \tan^{-1} x$

Let $y_1 = e^{-2x}$ and $y_2 = \tan^{-1} x$. Let $x_1 = 0.6$ from sketch.

$f(x) = e^{-2x} - \tan^{-1} x$

$$f'(x) = -2e^{-2x} - \frac{1}{1 + x^2}$$

x	y_1	y_2
0	1	0
1	0.135	$\frac{\pi}{4} = 0.785$

n	x_n	$f(x_n)$	$f'(x_n)$	$x_n - \frac{f(x_n)}{f'(x_n)}$
1	0.6	-0.239225288	-1.337682541	0.421164379
2	0.4211644	0.032088983	-1.710756554	0.439921567
3	0.4399216	0.000406821	-1.667546018	0.440165531
4	0.4401655	ϕ		

Therefore, root = 0.4401655.

19. $P = 100e^{-0.005t}$; $t = 100$ days;

$$\frac{dP}{dt} = 100e^{-0.005t}(-0.005)$$

$$\frac{dP}{dt} = -0.5e^{-0.005t}\Big|_{t=100} = -0.303 \text{ W/day}$$

20. $V_L = L\dfrac{di}{dt} = L\left[120\pi e^{-5.0t}\cos 120\pi t - 5.0e^{-5.0t}\sin 120\pi t\right]$

$V_L = 0.50\left[120\pi e^{-5.0(1.0\times 10^{-3})}\cos 120\pi(1\times 10^{-3}) - 5.0e^{-5.0(1\times 10^{-3})}\sin 120\pi(1\times 10^{-3})\right]\times 10^{-3}$

$V_L = 0.17 \text{ V}$

21. $\ln p = \dfrac{a}{T} + b\ln T + c$; $p = e^{(a/T + b\ln T + c)}$

$$\frac{dp}{dT} = e^{(a/T + b\ln T + c)}\left(-aT^{-2} + \frac{b}{T}\right)$$

$$= p\left(\frac{-a + bT}{T^2}\right)$$

$$= e^{(a/T + b\ln T + c)}\left(\frac{-a}{T^2} + \frac{b}{T}\right)$$

$$= \frac{p(-a + bT)}{T^2}$$

22. $q = CE(1 - e^{-t/RC})$; $R\dfrac{dq}{dt} + \dfrac{q}{C} = E$

$$\frac{dq}{dt} = CE(0 - e^{-t/RC})\left(-\frac{1}{RC}\right) = \left(\frac{CE}{RC}\right)e^{-t/RC}$$

$$R\left(\frac{CE}{RC}\right)e^{-t/RC} + \frac{CE}{C}(1 - e^{-t/RC}) = E$$

$$Ee^{-t/RC} + E - Ee^{-t/RC} = E;\ E = E$$

Therefore, the equation for q satisfies the differential equation given.

23. $F = ka$; $x = ae^{kt} + be^{-kt}$

$$\frac{dx}{dt} = ae^{kt}\cdot k + be^{-kt}(-k)$$

$v = ake^{kt} - bke^{-kt} = k(ae^{kt} - be^{-kt})$

$$\frac{d^2x}{dt^2} = k[ae^{kt}\cdot k - be^{-kt}(-k)]$$

$a = k(ake^{kt} + bke^{-kt}) = k^2(ae^{kt} + be^{-kt}) = k^2 x$

$F = ka = k(k^2 a) = k^3 a$; therefore, $F \propto a$ given

$x = ae^{kt} + be^{-kt}$

24. $R = \dfrac{[1 + (dy/dx)^2]^{3/2}}{d^2y/dx^2}$

$y = \ln \sec x$; $-1.5 \le x \le 1.5$ dm, $x = 0.85$ dm

$\dfrac{dy}{dx} = \dfrac{1}{\sec x}\sec x \tan x = \tan x$; $\dfrac{d^2y}{dx^2} = \sec^2 x$

$R = \dfrac{[1 + (\tan x)^2]^{3/2}}{\sec^2 x} = \dfrac{(\sec^2 x)^{3/2}}{\sec^2 x} = \dfrac{\sec^3 x}{\sec^2 x}$

$= \sec x$

$R|_{x=0.85} = \sec 0.85 = 1.515$ dm $= 1.5$ dm

25. $y = \ln \sec x;\ -1.5 \le x \le 1.5;\ u = \sec x;$

$\dfrac{dy}{dx} = \dfrac{1}{\sec x} \cdot \sec x \tan x = \tan x = 0$ at $x = 0;$
$x = 0$ is a critical value; also, multiples of $2\pi.$
$\dfrac{d^2 y}{dx^2} = \sec^2 x;\ \sec^2(0) = 1$ so the curve is concave up and there is a minimum point at $x = 0,$
$y = \ln \sec 0 = \ln 1 = 0$ recurring at multiples of
$x = 2\pi;\ (2\pi, 0), (4\pi, 0), \ldots (0,0)$ is an intercept.

Asymptotes occur where $\dfrac{dy}{dx} = \tan x$ is undefined.

These values are odd multiples of $\dfrac{\pi}{2};\ -\dfrac{\pi}{2}, \dfrac{\pi}{2},$
$\dfrac{3\pi}{2} \cdots$

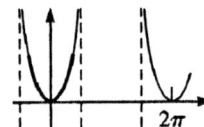

26. $i_m = AT^2 e^{k/t};$

$\dfrac{di_m}{dt} = A\left[T^2 e^{k/t} \dfrac{(-k\frac{dT}{dt})}{T^2} + e^{k/t} 2T \dfrac{dT}{dt} \right]$

$\dfrac{di_m}{dt} = A\left(2T e^{k/t} \dfrac{dT}{dt} - k e^{k/t} \dfrac{dT}{dt} \right)$

$di_m = A e^{k/t}(2T - k)\, dT$

27. $y = 6.0 e^{-0.020x} \sin(0.20x),\ 0 \le x \le 60$

$\dfrac{dy}{dx} = e^{-0.020x}\left[\dfrac{6\cos(0.2x)}{5} - \dfrac{3\sin(0.2x)}{25} \right] = 0$

$\tan 0.2x = 10$
$\quad 0.2x = \tan^{-1} 10 + k\pi$
$\qquad x = 5\tan^{-1} 10 + 5k\pi$

$k = 0 \quad x_1 = 7.355638372, y_1 = 5.153476505$
$k = 1 \quad x_2 = 23.06360164, y_2 = -3.764113107$
$k = 2 \quad x_3 = 38.77156491, y_3 = 2.749318343$
$k = 3 \quad x_4 = 54.47952818, y_4 = -2.008109516$

$(x_1, y_1) = 117.6°\ \text{W}\ 50.2°\text{N, maximum}$
$(x_2, y_2) = 101.9°\ \text{W}\ 41.2°\ \text{N, minimum}$
$(x_3, y_3) = 86.2°\ \text{W}\ 47.7°\ \text{N, maximum}$
$(x_4, y_4) = 70.5°\ \text{W}\ 43.0°\ \text{N, minimum}$

28. $R = 3e^{-0.004t} - 2e^{-0.006t} = 0.8$
$3e^{-0.004t} = 0.8 + 2e^{-0.006t}$
$f(t) = 3e^{-0.004t} - 2e^{-0.006t} - 0.8$

Using a graphing calculator, estimate $f(t) = 0$ for
$t_1 = 160$ h

$f'(t) = 3e^{-0.004t}(-0.004) - 2e^{-0.006t}(-0.006)$

$f'(t) = -0.012 e^{-0.004t} + 0.012 e^{-0.006t}$

n	t_n	$f(t_n)$	$f'(t_n)$	$t_n - \frac{f(t_n)}{f'(t_n)}$
1	160	0.0160915	-0.001732794	169.286
2	169.286	0.00009		

Therefore, $t = 169$ h

29. $y = e^{-0.5t}(0.4\cos 6t - 0.2\sin 6t);$

$v = \dfrac{dy}{dt} = e^{-0.5t}(-2.4\sin 6t - 1.2\cos 6t)$
$\qquad\qquad + (0.4\cos 6t - 0.2\sin 6t) \times (e^{-0.5t})(-0.5)$

$\dfrac{dy}{dt} = -2.4 e^{-0.5t}\sin 6t - 1.2 e^{-0.5t}\cos 6t$
$\qquad\quad + 0.1 e^{-0.5t}\sin 6t - 0.2 e^{-0.5t}\cos 6t$

$\dfrac{dy}{dt} = -2.3 e^{-0.5t}\sin 6t - 1.4 e^{-0.5t}\cos 6t;$

$\dfrac{dy}{dt} = -e^{-0.5t}(1.4\cos 6t + 2.3\sin 6t)$

$\left.\dfrac{dy}{dt}\right|_{t=-026} = -e^{-0.5(0.26)}[1.4\cos 6(0.26)$
$\qquad\qquad\qquad + 2.3\sin 6(0.26)]$
$\qquad\qquad = -e^{-0.13}[0.011 + 2.3(1.00)]$
$\qquad\qquad = -2.03\ \text{cm/s}$

30. $y = \dfrac{10t}{e^{0.4t} + 1};\ \dfrac{dy}{dx} = ?;\ t = 8.0$ min

$\dfrac{dy}{dt} = \dfrac{(e^{0.4t} + 1)10 - 10t(0.4e^{0.4t})}{(e^{0.4t} + 1)^2}$

$\dfrac{dy}{dt} = vy = \dfrac{10(e^{0.4t} + 1 - 0.4t e^{0.4t})}{(e^{0.4t} + 1)^2}$

$vy|_{t=8.0} = -0.81\ \text{km/min}$

31. $s = kx^2 \ln \dfrac{1}{x} = k[x^2(\ln 1 - \ln x)] = -kx^2 \ln x$

$\dfrac{ds}{dx} = -k\left(x^2\dfrac{1}{x} + \ln x\, 2x \right) = -k(x + 2x\ln x)$

$\dfrac{ds}{dx} = -kx(1 + 2\ln x) = 0$

For max., min.:

$x = 0;\ \ln x = -\dfrac{1}{2};\ x = e^{-1/2} = \dfrac{1}{\sqrt{e}} = 0.607$

32. $y = e^{-x};$ largest rectangle has max. area

$A = xy = xe^{-x}$

$\dfrac{dA}{dx} = x(-e^{-x}) + e^{-x} = e^{-x}(1 - x) = 0$

$e^{-x} \ne 0;\ 1 - x = 0;\ x = 1$

$x = 1, y = \dfrac{1}{e},$ therefore area $= \dfrac{1}{e}$ units^2

Chapter 8 Review Exercises

1. $y = 3\ln(x^2 + 1)$

$$\frac{dy}{dx} = \frac{3}{x^2 + 1}(2x) = \frac{6x}{x^2 + 1}$$

2. $y = \ln(1 - 5x)^3 = 3\ln(1 - 5x)$

$$\frac{dy}{dx} = \frac{3}{1 - 5x}(-5) = \frac{-15}{1 - 5x}$$

3. $y = (e^{x-3})^2;\ \dfrac{dy}{dx} = 2(e^{x-3})(e^{x-3})(1) = 2e^{2(x-3)}$

4. $y = e^{\sqrt{1-x}}$

$$\frac{dy}{dx} = e^{\sqrt{1-x}}\frac{1}{2\sqrt{1-x}}(-1) = \frac{-e^{\sqrt{1-x}}}{2\sqrt{1-x}}$$

5. $r = 2\ln(\csc t^2)$

$$\frac{dr}{dt} = 2\frac{1}{\csc t^2}(-\csc t^2 \cot t^2 (2t))$$

$$= -4t\cot t^2$$

6. $y = \ln(3 + \sin x^2)$

$$\frac{dy}{dx} = \frac{1}{3 + \sin x^2}(\cos x^2)(2x)$$

$$= \frac{2x\cos x^2}{3 + \sin x^2}$$

7. $y = [\ln(3 + \sin x)]^2$

$$\frac{dy}{dx} = 2[\ln(3 + \sin x)]\frac{1}{(3 + \sin x)}(\cos x)$$

$$\frac{dy}{dx} = \frac{2\cos x\ln(3 + \sin x)}{3 + \sin x}$$

8. $y = \ln(3 + \sin x)^2$

$$= 2\ln(3 + \sin x)$$

$$\frac{dy}{dx} = \frac{2\cos x}{3 + \sin x}$$

9. $R = 6e^{\sin 2\theta}$

$$\frac{dR}{d\theta} = 6e^{\sin 2\theta}(\cos 2\theta)(2)$$

$$= 12e^{\sin 2\theta}\cos 2\theta$$

10. $y = 3e^{\sec 3x}$

$$\frac{dy}{dx} = 3e^{\sec 3x}\sec 3x\tan 3x(3)$$

$$= 9e^{\sec 3x}\sec 3x\tan 3x$$

11. $y = x^3 e^x$

$$\frac{dy}{dx} = x^3 e^x + 3x^2 e^x$$

$$= x^2(x + 3)e^x$$

12. $y = 2xe^{4x}$

$$\frac{dy}{dx} = 2xe^{4x}(4) + 2e^{4x}$$

$$= 2(4x + 1)e^{4x}$$

13. $y = \ln(x - e^{-x})^2$

$$= 2\ln(x - e^{-x})$$

$$\frac{dy}{dx} = \frac{2}{x - e^{-x}}(1 - e^{-x}(-1))$$

$$= \frac{2(1 + e^{-x})}{x - e^{-x}}$$

14. $p = \ln\sqrt{\sin 2\phi}$

$$= \frac{1}{2}\ln(\sin 2\phi)$$

$$\frac{dp}{d\phi} = \frac{1}{2\sin 2\phi}(\cos 2\phi)(2)$$

$$= \cot 2\phi$$

15. $y = \dfrac{\cos^2 x}{e^{3x} + 1}$

$$\frac{dy}{dx} = \frac{(e^{3x} + 1)[2\cos x(-\sin x)] - (\cos^2 x)(e^{3x})(3)}{(e^{3x} + 1)^2}$$

$$= \frac{(e^{3x} + 1)[-2\sin x\cos x] - 3e^{3x}\cos^2 x}{(e^{3x} + 1)^2}$$

$$= \frac{-\cos x[(e^{3x} + 1)(2\sin x) + 3e^{3x}\cos x]}{(e^{3x} + 1)^2}$$

$$= \frac{-\cos x[2e^{3x}\sin x + 2\sin x + 3e^{3x}\cos x]}{(e^{3x} + 1)^2}$$

$$= \frac{-\cos x(2e^{3x}\sin x + 3e^{3x}\cos x + 2\sin x)}{(e^{3x} + 1)^2}$$

16. $y = \dfrac{\ln\sqrt{3x + 1}}{3x + 1}$

$$\frac{dy}{dx} = \frac{(3x + 1)\frac{1}{\sqrt{3x+1}}\frac{1}{2\sqrt{3x+1}}(3) - \ln\sqrt{3x + 1}(3)}{(3x + 1)^2}$$

$$\frac{dy}{dx} = \frac{\frac{3}{2} - 3\ln(3x + 1)^{1/2}}{(3x + 1)^2}\left(\frac{2}{2}\right)$$

$$= \frac{3 - 6\ln(3x + 1)^{1/2}}{2(3x + 1)^2}$$

$$\frac{dy}{dx} = \frac{3(1 - \ln(3x + 1))}{2(3x + 1)^2}$$

17. $y = x^2 \ln x$

$$\frac{dy}{dx} = x^2 \left(\frac{1}{x}\right) + 2x \ln x$$

$$= x + 2x \ln x$$

$$\frac{dy}{dx} = x(1 + 2 \ln x)$$

18. $s = (2t + 1) \ln(t^3 + 3)$

$$\frac{ds}{dt} = (2t + 1)\frac{1}{t^3 + 3}(3t^2) + 2\ln(t^3 + 3)$$

$$\frac{ds}{dt} = \frac{3t^2(2t + 1)}{t^3 + 3} + 2\ln(t^3 + 3)$$

19. $y = \dfrac{e^{2x}}{x^2 + 1}$

$$\frac{dy}{dx} = \frac{(x^2 + 1)2e^{2x} - e^{2x}(2x)}{(x^2 + 1)^2}$$

$$= \frac{2e^{2x}(x^2 + 1 - x)}{(x^2 + 1)^2}$$

$$\frac{dy}{dx} = \frac{2e^{2x}(x^2 - x + 1)}{(x^2 + 1)^2}$$

20. $y = \dfrac{\ln x}{2x + 3}$

$$\frac{dy}{dx} = \frac{(2x + 3)\left(\frac{1}{x}\right) - 2\ln x}{(2x + 3)^2}$$

$$= \frac{2 + \frac{3}{x} - \ln x^2}{(2x + 3)^2}$$

$$\frac{dy}{dx} = \frac{2x + 3 - x \ln x^2}{x(2x + 3)^2}$$

21. $v = 3e^{-\theta} \sec 2\theta$

$$\frac{dv}{d\theta} = 3e^{-\theta} \sec 2\theta \tan 2\theta(2) + 3\sec 2\theta(-e^{-\theta})$$

$$\frac{dv}{d\theta} = 3e^{-\theta} \sec 2\theta(2 \tan 2\theta - 1)$$

22. $y = e^{3x} \ln x$

$$\frac{dy}{dx} = e^{3x} \left(\frac{1}{x}\right) + \ln x(3e^{3x})$$

$$\frac{dy}{dx} = e^{3x} \left(\frac{1}{x} + \ln x\right)$$

23. $x^2 \ln x = y + x$

$$x^2 \left(\frac{1}{y}\right)\frac{dy}{dx} + 2x \ln y = \frac{dy}{dx} + 1$$

$$\left(\frac{x^2}{y} - 1\right)\frac{dy}{dx} = 1 - 2x \ln y$$

$$\frac{dy}{dx} = \frac{1 - 2x \ln y}{\frac{x^2}{y} - 1}$$

$$\frac{dy}{dx} = \frac{y - 2xy \ln y}{x^2 - y}$$

$$= \frac{y(1 - 2x \ln y)}{x^2 - y}$$

24. $y = x^2 (e^{\cos^2 x})^2$

$$\frac{dy}{dx} = x^2 \left[2e^{\cos^2 x}(e^{\cos^2 x})(2 \cos x)(-\sin x)\right]$$

$$+ (e^{\cos^2 x})^2(2x)$$

$$\frac{dy}{dx} = 2x(e^{\cos^2 x})^2(1 - 2x \sin x \cos x)$$

25. $y = \ln \cos x, x = \dfrac{\pi}{6}$

$y = \ln \cos \dfrac{\pi}{6} = \ln \dfrac{\sqrt{3}}{2}, \left(\dfrac{\pi}{6}, \ln \dfrac{\sqrt{3}}{2}\right)$ is point of

tangency

$$\frac{dy}{dx} = \frac{1}{\cos x}(-\sin x)$$

$$= -\tan x\big|_{\frac{\pi}{6}} = -\tan \frac{\pi}{6}$$

$$= -\frac{1}{\sqrt{3}} = M_{TL}$$

$$y - y_1 = m(x - x_1)$$

$$y - \ln \frac{\sqrt{3}}{2} = -\frac{1}{\sqrt{3}}\left(x - \frac{\pi}{6}\right)$$

$$y - \ln \frac{\sqrt{3}}{2} = -\frac{1}{\sqrt{3}}x + \frac{\pi}{6\sqrt{3}}$$

$$\frac{1}{\sqrt{3}}x + y - \ln \frac{\sqrt{3}}{2} - \frac{\pi}{6\sqrt{3}} = 0$$

$$x + \sqrt{3}y + \sqrt{3}\left(-\ln \frac{\sqrt{3}}{2} - \frac{\pi}{6\sqrt{3}}\right) = 0$$

$$\sqrt{3}x + 3.00y + 3.00\left(-\ln \frac{\sqrt{3}}{2} - \frac{\pi}{6\sqrt{3}}\right) = 0$$

$$1.73x + 3.00y - 0.48 = 0$$

26. $y = e^{x^2}, x = \dfrac{1}{2}$

$y = e^{(1/2)^2} = e^{1/4}$

$\left(\dfrac{1}{2}, e^{1/4}\right)$ is point of normal line

$\dfrac{dy}{dx} = 2xe^{x^2}\bigg|_{1/2} = 2\left(\dfrac{1}{2}\right)e^{(1/2)^2} = e^{1/4} = m_{TL}$

$m_{NL} = -e^{-1/4}$

$y - e^{1/4} = -e^{-1/4}\left(x - \dfrac{1}{2}\right)$

$y - e^{1/4} = -e^{-1/4}x + \dfrac{1}{2}e^{-1/4}$

$e^{-1/4}x + y - \dfrac{1}{2}e^{-1/4} - e^{1/4} = 0$

$x + e^{1/4}y - \dfrac{1}{2} - e^{1/2} = 0$

$2x + 2e^{1/4}y - 1 - 2e^{1/2} = 0$

$2x + 2.57y - 4.30 = 0$

27. $y = \ln(1 + x)$

Intercepts: $(0,0)$
For $x = 0, y = \ln 1 = 0$

Symmetry: None

Behavior as x becomes large:
As $x \to +\infty, \ln(1 + x) \to +\infty, y \to +\infty$
$\ln(x + 1)$ not defined for $x \le -1$

Vertical asymptotes:
As $x \to -1, \ln(x + 1) \to -\infty, y \to -\infty$
$x = -1$ is an asymptote

Derivatives: $y' = \dfrac{1}{1 + x}$

$x > -1, y' > 0, y$ inc.

$y'' = \dfrac{1}{(1 + x)^2}$

$x > -1, y'' < 0$, conc. down

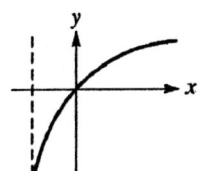

28. $y = x(\ln x)^2$

Intercepts: $(1,0)$
For $x = 1, y = 1(\ln 1)^2 = 0$

Symmetry: none

Behavior as x becomes large:
As $x \to +\infty, x(\ln x)^2 \to +\infty, y \to +\infty$
$\ln x$ not defined for $x \le 0$

Derivatives:

$\dfrac{dy}{dx} = (\ln x)^2 + 2\ln x$

$\dfrac{d^2y}{dx^2} = \dfrac{2\ln x}{x} + \dfrac{2}{x}$

Critical Numbers:

$\dfrac{dy}{dx} = \ln x(\ln x + 2) = 0$

$\ln x = 0 \qquad \ln x + 2 = 0$
$\quad x = 1 \qquad\qquad x = e^{-2}$

Inflection Points:

$\dfrac{d^2y}{dx^2} = \dfrac{2\ln x}{x} + \dfrac{2}{x} = 0$

$\ln x = -1$
$x = e^{-1}$, possible inflection point

Maximum, Minimum:

$\dfrac{d^2y}{dx^2} = \dfrac{2\ln x}{x} + \dfrac{2}{x}\bigg|_{x=1} = \dfrac{2\ln 1}{1} + \dfrac{2}{1} = 2 > 0$

$(1,0)$ minimum

$\dfrac{d^2y}{dx^2} = \dfrac{2\ln x}{x} + \dfrac{2}{x}\bigg|_{x=e^{-2}} = \dfrac{2\ln e^{-2}}{e^{-2}} + \dfrac{2}{e^{-2}}$

$= \dfrac{-4 + 2}{e^{-2}} < 0$

$(e^{-2}, 4e^{-2})$ maximum

Inflection Points:

$\dfrac{d^2y}{dx^2} > 0$ for $x > e^{-1}$

$\dfrac{d^2y}{dx^2} < 0$ for $x < e^{-1}$

(e^{-1}, e^{-1}) is an inflection point.

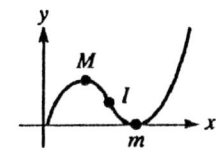

29. $e^x - x^2 = 0$
$f(x) = e^x - x^2$
$f'(x) = e^x - 2x$
$f(0) = 1, f(-1) = -0.63$
Let $x_1 = -0.7$

n	x_n	$f(x_n)$	$f'(x_n)$	$x_n - \frac{f(x_n)}{f'(x_n)}$
1	-0.7	0.0065853	1.8965853	-0.7034722
2	-0.7034722	-0.0000091	1.9018084	-0.7034674
3	-0.7034674			

$$x_3 = x_2 = -0.70347 \text{ (to 5 decimal places)}$$

30. $f(x) = x - 5(1 - e^{-x})$, by inspection $x = 0$ is a solution
$f'(x) = 1 - 5e^{-x}$
$f(4) = -0.908$
$f(5) = 0.034$

Let $x_1 = 4.5$

n	x_n	$f(x_n)$	$f'(x_n)$	$x_n - \frac{f(x_n)}{f'(x_n)}$
1	4.5	-0.4444550173	0.9444550081	4.970594166
2	4.970594166	0.0052892859	0.9653048749	4.965114772
3	4.965114772	5.2176×10^{-7}	0.9651142448	4.965114232
4	4.965114232	0	0.965114226	4.965114232
5	4.965114232			

$$x_5 = x_4 = 4.965114232$$

31. $y = e^{-x^2}$
$A = xy = xe^{-x^2}$

$$\frac{dA}{dx} = xe^{-x^2}(-2x) + e^{-x^2}(1) = e^{-x^2}(1 - 2x^2)$$

$$\frac{dA}{dx} = 0, 1 - 2x^2 = 0, x = \sqrt{\frac{1}{2}} = \frac{1}{2}\sqrt{2}$$

$$A = \frac{1}{2}\sqrt{2}e^{-1/2} = 0.429$$

32. $x = e^{-t}\sin 2t, y = e^{-t}\cos 2t$
$v_x = e^{-t}(2\cos 2t - \sin 2t)$
$v_y = e^{-t}(-2\sin 2t - \cos 2t)$
$a_x = e^{-t}(-4\sin 2t - 2\cos 2t - 2t\cos 2t + \sin 2t) = e^{-t}(-3\sin 2t - 4\cos 2t)$
$a_y = e^{-t}(-4\cos 2t + 2\sin 2t + 2\sin 2t + \cos 2t) = e^{-t}(4\sin 2t - 3\cos 2t)$

For $t = \frac{\pi}{4}$:

$$a_x = e^{-\frac{\pi}{4}}[-3(1) - 4(0)] = -3e^{-\pi/4} = -1.368$$

$$a_y = e^{-\frac{\pi}{4}}[4(1) - 3(0)] = 4e^{-\pi/4} = 1.824$$

$$a = \sqrt{(-1.368)^2 + (1.824)^2} = 2.28$$

$$\tan\theta = \frac{1.824}{-1.368} = -1.333, \theta = 126.9°$$
$$(a_x < 0, a_y > 0)$$

33. $y = 150(1 - e^{-0.05t})$

$$v = \frac{dy}{dt} = 150(0.05)e^{-0.05t}\Big|_{t=10.0}$$

$$= 4.55 \text{ mi/min}$$

34. $V = 1000e^{0.06t}$

$$\frac{dV}{dt} = 1000(0.06)e^{0.06t}\Big|_{t=2}$$

$$= \$67.65/\text{year}$$

35. For $N = 8, n = xN \log_x N$ becomes $n = 8x \log_x 8$ with $1 < x < 10$. For graph, use $y = 8x\dfrac{\ln 8}{\ln x}$.

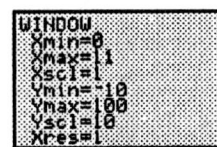 

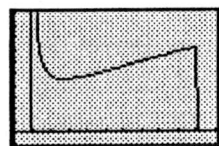

The VL's may be eliminated by using the Dot mode rather than Connected.

 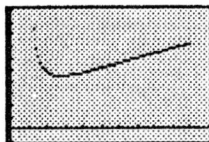

36. $V = -k \ln\left(1 + \dfrac{L}{x}\right)$

$$\frac{dV}{dx} = \frac{-k}{1 + \frac{L}{x}}\left(-\frac{L}{x^2}\right)$$

$$\frac{dV}{dx} = \frac{kL}{x(x+L)}$$

37. $p = 0.05 \ln(2 + 24t - t^2)$

$$dp = \frac{0.05(24 - 2t)}{2 + 24t - t^2} dt\Big|_{\substack{t = 10 \\ dt = 0.5}}$$

$$= \frac{0.05(24 - 2(10))}{2 + 24(10) - 10^2}(0.5)$$

$$dp = 0.0007 \text{ ppm}$$

38. $T = 80 + 120(0.5)^{0.2t};\ u = 0.2t,\ \dfrac{du}{dt} = 0.2$

$$T = 80 + 120(0.5)^u$$

$$\frac{dT}{dt} = 120(0.5)^{0.2t}(\ln(0.5))(0.2)$$

$$\frac{dT}{dt}\Big|_{t=5.00} = 120(0.5)^{0.2(5.00)}(-0.693)(0.2)$$

$$= 60(-0.693)(0.2)$$
$$= -8.32°\text{F/min}$$
$$L(t) = -8.32(t - 5.00) + 140$$

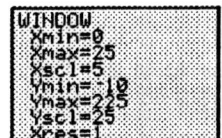 

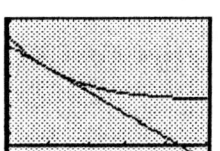

39. $q = e^{-0.1t}(0.2 \sin 120\pi t + 0.8 \cos 120\pi t)$

$$i = \frac{dq}{dt} = e^{-0.1t}(0.2(120\pi) \cos 120\pi t$$

$$- 0.8(120\pi) \sin 120\pi t)$$
$$+ (0.2 \sin 120\pi t + 0.8 \cos 120\pi t)$$
$$\cdot (-0.1)e^{-0.1t}$$
$$i = e^{-0.1t}(75.3 \cos 120\pi t - 301.6 \sin 120\pi t)$$

40. $n = 160 - 140e^{-0.30t}$

$$\frac{dn}{dt} = -140(-0.30)e^{-0.30t}\Big|_{t=10}$$

$$= 2.09 \text{ in millions}$$

The annual rate of increase in 2010 is 2,100,000/year

41. $A = xy = 3xe^{-0.5x^2}$ (Working with $\frac{1}{2}$ the actual are

$$\frac{dA}{dx} = 3x\, d(e^{-0.5x^2}) + e^{-0.5x^2}\left(\frac{d\, 3x}{dx}\right)$$

$$= 3x(e^{-0.5x^2})(-1.0x) + e^{-0.5x^2}(3)$$
$$= -3x^2 e^{-0.5x^2} + 3e^{-0.5x^2}$$

The maximum value will occur when $\frac{dA}{dx} = 0$

$$-3x^2 e^{-0.5x^2} + 3e^{-0.5x^2} = 0$$
$$(e^{-0.5x^2})(-3x^2 + 3) = 0$$
$$-3x^2 + 3 = 0; \ x^2 = 1, x = 1.00$$
$$e^{-0.5x^2} = 0 \text{ has no real solution}$$
$$y = 3e^{-0.5(1)^2}$$
$$= 3e^{-0.5} = 1.82$$
$$W = 2x = 2.00 \text{ m wide};$$
$$H = 1.82 \text{ m high}$$

42. $\quad i = i_0 e^{-Rt/L}$

$$\frac{di}{dt} = i_0 \left(-\frac{R}{L}\right) e^{-Rt/L} = -\frac{R}{L} i$$

43. $\quad y = \dfrac{H}{w} \cosh \dfrac{wx}{H}$

$$\frac{dy}{dx} = \frac{H}{w} \frac{w}{H} \sinh \frac{wx}{H} = \sinh \frac{wx}{H}$$

$$\frac{d^2 y}{dx^2} = \frac{w}{H} \cosh \frac{wx}{H}$$

$$\frac{w}{H} \sqrt{1 + \left(\frac{dy}{dx}\right)^2} = \frac{w}{H} \sqrt{1 + \sinh^2 \frac{wx}{H}}$$

$$= \frac{w}{H} \sqrt{\cosh^2 \frac{wx}{H}}$$

$$= \frac{w}{H} \cosh \frac{wx}{H}$$

$$= \frac{d^2 y}{dx^2}$$

44. $\quad y = 3.0te^{-0.20t}$

$$\frac{dy}{dt} = 3.0t(-0.20)e^{-0.20t} + 3.0e^{-0.20t}$$
$$= 0 \text{ for maximum}$$
$$3.0t(-0.20) + 3.0 = 0$$
$$t = \frac{3.0}{3.0(0.20)}$$
$$t = \frac{1}{0.20}$$

$$y_{\max} = 3.0 \left(\frac{1}{0.20}\right) e^{-0.20(1/0.20)}$$

$$y_{\max} = 5.5 \text{ in.}$$

45. $\quad V = 1000(0.95)^t, t \geq 0$
$$\ln V = \ln[1000(0.95)^t]$$
$$= \ln 1000 + t \ln 0.95$$

$$\frac{1}{V} \frac{dV}{dt} = \ln 0.95$$

$$\frac{dV}{dt} = V \ln(0.95)$$
$$= 1000(0.95)^t (\ln 0.95)$$

The change in the purchasing power during the nth year may be approximated by evaluating the derivative at $t = n$.

Consider the following table.

t (years)	V	change in V for n^{th} year	
0	1000		
1	950	-50.00	1st
2	902.50	-47.50	2nd
3	857.38	-45.12	3rd
4	814.51	-42.87	4th
5	773.78	-40.73	5th

$$\left.\frac{dV}{dt}\right|_{t=5} = 1000(0.95)^5 (\ln 0.95)$$

$$= -39.69 \text{ as compared to } -40.73$$

INTEGRATION BY STANDARD FORMS

9.1 The General Power Formula

1. $u = \sin x;\ du = \cos x\, dx$

$$\int \sin^4 x \cos x\, dx = \frac{1}{5}\sin^5 x + C$$

2. $u = \cos x;\ n = 5;\ du = -\sin x\, dx$

$$\int \cos^5 x(-\sin x\, dx) = \frac{1}{6}\cos^6 x + C$$

3. $u = \cos x;\ n = \frac{1}{2};\ du = -\sin x\, dx$

$$0.4\int \sqrt{\cos x}\,\sin x\, dx = 0.4\int (\cos x)^{1/2}(-\sin x\, dx)$$

$$= -\frac{0.8}{3}(\cos x)^{3/2} + C$$

4. $u = \sin x;\ n = \frac{1}{3};\ du = \cos x\, dx$

$$\int 8\sin^{1/3} x \cos x\, dx = 8\int \sin^{1/3} x \cos x\, dx$$

$$= 8(\sin x)^{4/3}\cdot\frac{3}{4} + C$$

$$= 6\sin^{4/3} x + C$$

5. $\displaystyle\int 4\tan^2 x \sec^2 x\, dx = 4\int \tan^2 x \sec^2 x\, dx$

$u = \tan x;\ du = \sec^2 x\, dx$

$$4\int \tan^2 x \sec^2 x\, dx = 4\left(\frac{1}{3}\tan^3 x + c\right)$$

$$= \frac{4}{3}\tan^3 x + c$$

6. $u = \sec x;\ n = 3;\ du = \sec x \tan x\, dx$

$$\int \sec^3 x(\sec x \tan x)\, dx = \frac{1}{4}\sec^4 x + C$$

7. Let $u = \cos 2x;\ n = 1;\ du = -2\sin 2x\, dx$

$$\int_0^{\pi/8} \cos 2x \sin 2x\, dx$$

$$= -\frac{1}{2}\int_0^{\pi/8} (\cos 2x)^{-1}(-2\sin 2x\, dx)$$

$$= -\frac{1}{2}\frac{(\cos 2x)^2}{2}\bigg|_0^{\pi/8}$$

$$= -\frac{1}{4}\cos^2 2x\bigg|_0^{\pi/8}$$

$$= -\frac{1}{4}\left(\cos^2\frac{\pi}{4} - \cos^2 0\right)$$

$$= -\frac{1}{4}\left(\frac{1}{2} - 1\right) = \frac{1}{8}$$

8. $u = \cot x;\ n = \frac{1}{2};\ du = -\csc^2 x\, dx$

$$\int_{\pi/6}^{\pi/4} 3\sqrt{\cot x}\,\csc^2 x\, dx$$

$$= -3\int_{\pi/6}^{\pi/4} (\cot x)^{1/2}(-\csc^2 x)\, dx$$

$$= -3(\cot x)^{3/2}\cdot\frac{2}{3}\bigg|_{\pi/6}^{\pi/4} = -2(\cot x)^{3/2}\bigg|_{\pi/6}^{\pi/4}$$

$$-2[1^{3/2} - (\sqrt{3})^{3/2}] = 2[3^{3/4} - 1] = 2.559$$

9. $u = \sin^{-1} x;\ du = \dfrac{1}{\sqrt{1-x^2}}\, dx = \dfrac{dx}{\sqrt{1-x^2}}$

$$\int (\sin^{-1} x)^3\left(\frac{dx}{\sqrt{1-x^2}}\right) = \frac{1}{4}(\sin^{-1} x)^4 + C$$

10. $u = \cos^{-1} 2t;\ du = \dfrac{-2\, dt}{\sqrt{1-4t^2}};\ n = 4$

$$20\int \frac{(\cos^{-1} 2t)^4 dt}{\sqrt{1-4t^2}} = 20\int (\cos^{-1} 2t)^4\frac{dt}{\sqrt{1-4t^2}}$$

$$= -10\int (\cos^{-1} 2t)^4\frac{-2\, dt}{\sqrt{1-4t^2}}$$

$$= -10\frac{(\cos^{-1} 2t)^5}{5} + C$$

$$= -2(\cos^{-1} 2t)^5 + C$$

11. $u = \tan^{-1} 5x;\; du = \dfrac{1}{1+25x^2}(5\,dx);\; n = 1$

$$\int \frac{5\tan^{-1}5x}{1+25x^2}dx = \int (\tan^{-1}5x)^1 \frac{5\,dx}{1+25x^2}$$
$$= \frac{1}{2}(\tan^{-1}5x)^2 + C$$

12. $u = \sin^{-1} 4x;\; du = \dfrac{1}{\sqrt{1-16x^2}}(4\,dx);\; n = 1$

$$\int \frac{\sin^{-1}4x\,dx}{\sqrt{1-16x^2}} = \frac{1}{4}\int (\sin^{-1}4x)^1 \frac{4\,dx}{\sqrt{1-16x^2}}$$
$$= \frac{1}{4}\frac{(\sin^{-1}4x)^2}{2} + C$$
$$= \frac{1}{8}(\sin^{-1}4x)^2 + C$$

13. $u = \ln(x+1);$

$$du = \frac{1}{x+1}(1)dx = \frac{dx}{x+1}$$

$$\int [\ln(x+1)^2 \frac{dx}{x+1} = \frac{1}{3}[\ln(x+1)]^3 + C$$

14. Let $x = 3 + 2\ln u$

$$dx = 2 \cdot \frac{du}{u}$$

$$\int 0.8(3+2\ln u)^3 \frac{du}{u} = \int 0.8x^3 \cdot \frac{dx}{2}$$
$$= 0.4\frac{x^4}{4} + C$$
$$= 0.1(3+2\ln u)^4 + C$$

15. $u = \ln(2x+3);\; du = \dfrac{1}{2x+3}(2\,dx);\; n = 1$

$$\frac{1}{2}\int_0^{1/2} [\ln(2x+3)]\frac{2\,dx}{2x+3} = \frac{1}{2}\frac{[\ln(2x+3)]^2}{2}\Big|_0^{1/2}$$
$$= \frac{1}{4}\ln^2(2x+3)\Big|_0^{1/2}$$
$$= \frac{1}{4}(\ln^2 4 - \ln^2 3)$$
$$= 0.179$$

16. $u = 1 - 2\ln x;\; du = \dfrac{2}{x}dx;\; n = 1$

$$\int_1^e \frac{(1-2\ln x)dx}{x} = \int_1^e (1-2\ln x)^1 \frac{dx}{x}$$
$$= \frac{1}{2}\int_1^e (1-2\ln x)^1 \frac{2\,dx}{x}$$
$$= \frac{1}{2}\frac{(1-2\ln x)^2}{2}\Big|_1^e$$
$$= \frac{1}{4}(1-2\ln x)^2\Big|_1^e$$
$$= \frac{1}{4}[(1-2)^2 - (1-0)^2]$$
$$= \frac{1}{4}(0)$$
$$= 0$$

17. $u = 4 + e^x;\; du = e^x dx;$

$$\int (4+e^x)^3 e^x\,dx = \frac{1}{4}(4+e^x)^4 + C$$

18. $u = 1 - e^{-x};\; n = \dfrac{1}{2};\; du = e^{-x}dx$

$$\int 2\sqrt{1-e^{-x}}(e^{-x}dx) = 2\int (1-e^{-x})^{1/2}e^{-x}dx$$
$$= 2(1-e^{-x})^{3/2}\cdot\frac{2}{3} + C$$
$$= \frac{4}{3}(1-e^{-x})^{3/2} + C$$

19. $\displaystyle\int \frac{\sqrt{(1+e^{-r})(1-e^{-r})}}{e^{2r}}dr$

$$= \frac{1}{2}\int (1-e^{-2r})^{1/2}(2e^{-2r}dr)$$
$$= \frac{1}{3}(1-e^{-2r})^{3/2} + C$$

20. $u = 1 + 3e^{-2x};\; du = -6e^{-2x}dx;\; n = 4$

$$\int \frac{(1+3e^{-2x})^4 dx}{e^{2x}} = -\frac{1}{6}\int (1+3e^{-2x})^4 - 6\frac{dx}{x}$$
$$= -\frac{1}{6}\frac{(1+3e^{-2x})^5}{5} + C$$
$$= -\frac{1}{30}(1+3e^{-2x})^5 + C$$

21. $u = 1 + \sec^2 x$;

$$\frac{du}{dx} = 2\sec x \frac{d(\sec x)}{dx} = 2\sec x \sec x \tan x;$$

$$du = 2\sec^2 x \tan x \, dx$$

$$\int (1 + \sec^2 x)^4 (\sec^2 x \tan x \, dx)$$

$$= \frac{1}{2} \int (1 + \sec^2 x)^4 2\sec^2 x \tan x \, dx$$

$$= \frac{1}{2} \times \frac{1}{5}(1 + \sec^2 x)^5 + C$$

$$= \frac{1}{10}(1 + \sec^2 x)^5 + C$$

22. $u = e^x + e^{-x}$; $du = (e^x - e^{-x})dx$; $n = \dfrac{1}{4}$

$$\int (e^x + e^{-x})^{1/4}(e^x - e^{-x})dx = \frac{4}{5}(e^x + e^{-x})^{5/4} + C$$

23. $u = \tan x$; $du = \sec^2 x \, dx$; $n = 1$

$$\int_0^{\pi/6} \frac{\tan x}{\cos^2 x}dx = \int_0^{\pi/6} (\tan x)^1 \sec^2 x \, dx$$

$$= \frac{1}{2}(\tan x)^2 \Big|_0^{\pi/6}$$

$$= \frac{1}{2}\left[\left(\frac{1}{\sqrt{3}}\right)^2 - 0\right]$$

$$= \frac{1}{6}$$

24. $u = 1 + \cos\theta$; $du = -\sin\theta \, d\theta$; $n = -\dfrac{1}{2}$

$$\int_{\pi/3}^{\pi/2} \frac{\sin\theta}{\sqrt{1 + \cos\theta}}d\theta$$

$$= -\int_{\pi/3}^{\pi/2} (1 + \cos\theta)^{-1/2} - \sin\theta \, d\theta$$

$$= -2(1 + \cos\theta)^{1/2} \Big|_{\pi/3}^{\pi/2}$$

$$= -2\left[(1 + 0)^{1/2} - \left(1 + \frac{1}{2}\right)^{1/2}\right]$$

$$= 0.449$$

25. $A = \displaystyle\int_0^2 \frac{1 + \tan^{-1} 2x}{1 + 4x^2}dx$; $u = \tan^{-1} 2x$;

$$du = \frac{1}{1 + (2x)^2} \times 2$$

$$A = \frac{1}{2}\int_0^2 1 + u \, du = \frac{1}{2}\left(u + \frac{1}{2}u^2\right)\Big|_0^2$$

$$= \frac{1}{2}\left[\tan^{-1} 2x + \frac{1}{2}(\tan^{-1} 2x)^2\right]\Big|_0^2$$

$$= \left[\frac{1}{2}\tan^{-1} 4 + \frac{1}{2}(\tan^{-1} 2x)^2\right.$$

$$\left. - \tan^{-1} 0 - \frac{1}{2}(\tan^{-1} 0)^2\right]$$

$$= \frac{1}{2}[1.326 + 0.879 - 0 - 0] = 1.102$$

26. $y = \dfrac{\ln(4x + 1)}{4x + 1}$ and $x = 2$

$$u = \ln(4x + 1); \quad du = \frac{1}{4x + 1}4 \, dx; \quad n = 1$$

$$A = \int_0^2 [\ln(4x + 1)]^1 \frac{dx}{4x + 1}$$

$$A = \frac{1}{4}\int_0^2 [\ln(4x + 1)]^1 \frac{dx}{4x + 1}$$

$$= \frac{\frac{1}{4}[\ln(4x + 1)]^2}{2}$$

$$A = \frac{1}{8}[\ln(4x + 1)]^2 \Big|_0^2$$

$$= \frac{1}{8}(4.828 - 0) = 0.6035$$

27. $\dfrac{dy}{dx} = m = \dfrac{(\ln x)^2}{x}$; passes through $(1, 2)$

$$dy = \frac{(\ln x)^2}{x}dx; \quad y = \int (\ln x)^2 \frac{dx}{x}; \quad u = \ln x;$$

$$du = \frac{dx}{x}; \quad n = 2$$

$$y = \frac{1}{3}(\ln x)^3 + C; \quad x = 1, y = 2$$

$$2 = \frac{1}{3}(\ln 1)^3 + C; \text{ therefore, } C = 2$$

Therefore, $y = \dfrac{1}{3}(\ln x)^3 + 2$

28. $\dfrac{dy}{dx} = (1 + \tan 2x)^2 \sec^2 2x$; passes through $(2,1)$

$u = 1 + \tan 2x$; $du = 2\sec^2 2x\, dx$; $n = 2$

$dy = (1 + \tan 2x)^2 \sec^2 2x\, dx$

$y = \dfrac{1}{2}\displaystyle\int (1 + \tan 2x)^2 2\sec^2 2x\, dx$

$y = \dfrac{1}{2}\dfrac{(1 + \tan 2x)^3}{3} + C = \dfrac{1}{6}(1 + \tan 2x)^3 + C$

$x = 2;\ y = 1$

$1 = \dfrac{1}{6}(1 + \tan 4)^3 + C$; therefore, $C = -0.6745$

$y = \dfrac{1}{6}(1 + \tan 2x)^3 - 0.6745$

29. $P = mnv^2 \displaystyle\int_0^{\pi/2} \sin\theta \cos^2\theta\, d\theta$; $n = 2$;

$\mu = \cos\theta$; $du = -\sin\theta\, d\theta$

$P = mnv^2 \displaystyle\int_0^{\pi/2} \cos^2\theta(-\sin\theta\, d\theta)$

$= -mnv^2 \left[\dfrac{\cos^3\theta}{3}\right]\Big|_0^{\pi/2}$

$= -mnv^2 \left[\dfrac{1}{3}\left(\cos^3\dfrac{\pi}{2} - \cos^3 0\right)\right]$

$= -mnv^2 \left[\dfrac{1}{3}(0 - 1)\right]$

$= -mnv^2 \left(-\dfrac{1}{3}\right) = \dfrac{1}{3}mnv^2$

30. $E = 2\pi I \displaystyle\int_0^{\pi/2} (\cos\theta)\sin\theta\, d\theta$; $n = 1$; $\mu = \cos\theta$

$du = -\sin\theta\, d\theta$

$E = -2\pi I \displaystyle\int_0^{\pi/2} (\cos\theta)^1 (-\sin\theta\, d\theta)$

$E = -2\pi I \dfrac{(\cos\theta)^2}{2} = -\pi I \cos^2\theta \Big|_0^{\pi/2}$

$= -\pi I(0 - 1) = \pi I$

31. $i = 3(1 - e^{-t})^2 e^{-t}$, $\dfrac{dq}{dt} = 3(1 - e^{-t})^2 e^{-t}$;

$t = 0,\ q = 0$

$dq = 3(1 - e^{-t})^2 e^{-t} dt$; $q = 3\displaystyle\int (1 - e^{-t})^2 e^{-t} dt$

$n = 2$; $\mu = 1 - e^{-t}$; $du = e^{-t} dt$

$q = \dfrac{3(1 - e^{-t})^3}{3} = (1 - e^{-t})^3 + C$; $t = 0, q = 0$

$0 = (1 - 1)^3 + C$; therefore, $C = 0$

Therefore, $q = (1 - e^{-t})^3$

32. $v = [\ln^2(t^3 + 1)]\dfrac{t^2}{t^3 + 1} = \dfrac{ds}{dt}$; $s = ?, t = 10$ s

$n = 2$; $\mu = \ln(t^3 + 1)$; $du = \dfrac{1}{t^3 + 1}3t^2 dt$

$s = \dfrac{1}{3}\displaystyle\int [\ln(t^3 + 1)]^2 \dfrac{3t^2}{t^3 + 1}$

$s = \dfrac{\frac{1}{3}[\ln(t^3 + 1)]^3}{3} + C$

$= \dfrac{1}{9}\ln^3(t^3 + 1) + C$

$t = 0, s = 0$; $t = 10, s = ?$

$0 = \dfrac{1}{9}\ln^3(1) + C$; therefore, $C = 0$

$s = \dfrac{1}{9}\ln^3(t^3 + 1)\Big|_{t=10}$

$= \dfrac{1}{9}\ln^3(1001) = 36.640$

$= 36.6$ km

9.2   The Basic Logarithmic Form

1. $u = 1 + 4x$; $du = 4\, dx$. Introduce a factor of 4.

$\displaystyle\int \dfrac{dx}{1 + 4x} = \dfrac{1}{4}\displaystyle\int \dfrac{4\, dx}{1 + 4x} = \dfrac{1}{4}\ln|1 + 4x| + C$

2. $\displaystyle\int \dfrac{dx}{1 - 4x}$; $u = 1 - 4x$; $du = -4\, dx$

$-\dfrac{1}{4}\displaystyle\int \dfrac{-4\, dx}{1 - 4x} = -\dfrac{1}{4}\ln|1 - 4x| + C$

3. $\displaystyle\int \dfrac{2x\, dx}{4 - 3x^2}$; $u = 4 - 3x^2$; $du = -6x\, dx$

$-\dfrac{1}{3}\displaystyle\int \dfrac{-6x\, dx}{4 - 3x^2} = -\dfrac{1}{3}\ln|4 - 3x^2| + C$

4. $\displaystyle\int \dfrac{4\sqrt{u}\,du}{1 + u\sqrt{u}}$

Let $x = 1 + u\sqrt{u} = 1 + u^{3/2}$

$dx = \dfrac{3}{2}u^{1/2}du = \dfrac{3}{2}\sqrt{u}\,du$

$\displaystyle\int \dfrac{4\sqrt{u}\,du}{1 + u\sqrt{u}} = 4\displaystyle\int \dfrac{\frac{2}{3}dx}{x} = \dfrac{8}{3}\ln|x| + C$

$= \dfrac{8}{3}\ln|1 + u\sqrt{u}| + C$

5. $u = 8 - 3x$; $du = -3\,dx$

$$\int_0^2 \frac{dx}{8-3x} = -\frac{1}{3}\int_0^2 \frac{-3\,dx}{8-3x}$$

$$= -\frac{1}{3}\ln|8-3x|\Big|_0^2$$

$$= -\frac{1}{3}\ln 2 + \frac{1}{3}\ln 8$$

$$= 0.462$$

6. $\displaystyle\int_{-1}^3 \frac{2x^3\,dx}{x^4+1}$; $u = x^4 + 1$; $du = 4x^3\,dx$

$$\frac{1}{2}\int \frac{4x^3\,dx}{x^4+1} = \frac{1}{2}\ln(x^4+1)\Big|_{-1}^3 = \frac{1}{2}(\ln 82 - \ln 2)$$

$$= \frac{1}{2}\ln 41 = 1.857$$

7. $\displaystyle 0.4\int \frac{\csc^2 2\theta\,d\theta}{\cot 2\theta}$; $u = \cot 2\theta$; $du = -2\csc^2 2\theta\,d\theta$

$$\frac{0.4}{2}\int \frac{2\csc^2 2\theta\,d\theta}{\cot 2\theta} = -0.2\ln|\cot 2\theta| + C$$

8. $\displaystyle\int \frac{7\sin x}{\cos x}$; $u = \cos x$; $du = -\sin x\,dx$

$$-7\int \frac{-\sin x\,dx}{\cos x} = -7\ln|\cos x| + C$$

9. $u = 1 + \sin x$; $du = \cos x\,dx$

$$\int_0^{\pi/2} \frac{\cos x\,dx}{1+\sin x} = \ln|\sin x|\Big|_0^{\pi/2}$$

$$= \ln\left|1 + \sin\frac{\pi}{2}\right| - \ln|1 + \sin 0|$$

$$= \ln|2| - \ln|1| = 0.693 - 0$$

$$= 0.693$$

10. $u = 4 + \tan x$; $du = \sec^2 x\,dx$

$$\int_0^{\pi/4} \frac{\sec^2 x\,dx}{4+\tan x} = \ln|4 + \tan x|\Big|_0^{\pi/4}$$

$$= \ln 5 - \ln 4 = \ln\frac{5}{4}$$

$$= 0.223$$

11. $u = 1 - e^{-x}$; $du = e^{-x}\,dx$

$$\int \frac{e^{-x}\,dx}{1-e^{-x}} = \ln|1-e^{-x}| + C$$

12. $u = 1 - e^{3x}$; $du = -3e^{3x}\,dx$

$$\int \frac{5e^{3x}}{1-e^{3x}}\,dx = -\frac{5}{3}\int \frac{-3e^{3x}\,dx}{1-e^{3x}}$$

$$= -\frac{5}{3}\ln|1-e^{3x}| + C$$

13. $u = x + e^x$; $du = (1 + e^x)\,dx$

$$\int \frac{1+e^x}{x+e^x}\,dx = \ln|x + e^x| + C$$

14. $\displaystyle\int \frac{3e^t\,dt}{\sqrt{e^{2t}+4e^t+4}} = \int \frac{1}{\sqrt{(e^t+2)^2}}3e^t\,dt$

$$= \int \frac{3e^t\,dt}{e^t+2}$$

$$= 3\ln(e^t+2) + C$$

15. $u = 1 + 4\sec x$; $du = 4\sec x\tan x\,dx$

$$\int \frac{\sec x\tan x\,dx}{1+4\sec x} = \frac{1}{4}\int \frac{4\sec x\tan x\,dx}{1+4\sec x}$$

$$= \frac{1}{4}\ln|1 + 4\sec x| + C$$

16. $u = 1 - \cos^2 x$

$du = -2\cos x(-\sin x)dx = 2\sin x\cos x\,dx$

$$\int \frac{\sin 2x}{1-\cos^2 x}\,dx = \int \frac{2\sin x\cos x\,dx}{1-\cos^2 x}$$

$$= \ln|1 - \cos^2 x| + C$$

17. $u = 4x + 2x^2$; $du = (4 + 4x)\,dx$

$$\int_1^3 \frac{1+x}{4x+2x^2}\,dx = \frac{1}{4}\int_1^3 \frac{4+4x}{4x+2x^2}\,dx$$

$$= \frac{1}{4}\ln|4x + 2x^2|\Big|_1^3$$

$$= \frac{1}{4}\ln 30 - \frac{1}{4}\ln 6$$

$$= 0.402$$

18. $\displaystyle\int_1^2 \frac{4x+6x^2}{x^2+x^3}\,dx$; $u = x^2 + x^3$; $du = (2x+3x^2)dx$

$$2\int_1^2 \frac{(2x+3x^2)\,dx}{x^2+x^3} = 2\ln|x^2+x^3|\Big|_1^2$$

$$= 2(\ln 12 - \ln 2)$$

$$= 2\ln 6 = 3.58$$

19. $u = \ln r$; $du = \dfrac{dr}{r}$

$$0.5\int \frac{dr}{r\ln r} = 0.5\int \frac{\frac{dr}{r}}{\ln r} = 0.5\ln|\ln r| + C$$

20. $u = 1 + 2\ln x$; $du = \dfrac{2\,dx}{x}$

$$\int \frac{dx}{x(1+2\ln x)} = \frac{1}{2}\int \frac{2\frac{dx}{x}}{1+2\ln x}$$

$$= \frac{1}{2}\ln|1 + 2\ln x| + C$$

21. $u = 2x + \tan x$; $du = (2 + \sec^2 x)\,dx$

$$\int \frac{2 + \sec^2 x}{2x + \tan x}\,dx = \ln|2x + \tan x| + C$$

22. $u = x^2 + \sin 2x$

$du = (2x + 2\cos 2x)dx = 2(x + \cos 2x)\,dx$

$$\int \frac{x + \cos 2x}{x^2 + \sin 2x}\,dx = \frac{1}{2}\int \frac{2(x + \cos 2x)}{x^2 + \sin 2x}\,dx$$

$$= \frac{1}{2}\ln|x^2 + \sin 2x| + C$$

23. $n = -\frac{1}{2}$; $u = 1 - 2x$; $du = -2\,dx$

$$\int \frac{2\,dx}{\sqrt{1 - 2x}} = -\int (1 - 2x)^{-1/2} = 2\,dx$$

$$= -(1 - 2x)^{1/2} \cdot 2 + C$$
$$= -2\sqrt{1 - 2x} + C$$

24. $n = -2$; $du = 2x\,dx$

$$\int \frac{4x\,dx}{(1 + x^2)^2} = 4\int (1 + x^2)^{-2} x\,dx$$

$$\frac{4}{2}\int (1 + x^2)^{-2} 2x\,dx = 2\frac{(1 + x^2)^{-1}}{-1} + C$$

$$= \frac{-2}{1 + x^2} + C$$

25.

$$\int \frac{x + 2}{x^2}\,dx = \int \frac{1}{x}\,dx + \int \frac{2}{x^2}\,dx$$

$$= \int \frac{1}{x}\,dx + \int (2x^{-2})\,dx$$

$$= \ln|x| - 2x^{-1} + C$$

$$= \ln|x| - \frac{2}{x} + C$$

26.

$$\int \frac{3v^2 - 2v}{v^2}\,dv = \int \left(3 - \frac{2}{v}\right)dv$$

$$= 3v - 2\ln|v| + C$$

27. $u = 4 + \tan 3x$; $du = 3\sec^2 3x\,dx$

$$\int_0^{\pi/12} \frac{\sec^2 3x}{4 + \tan 3x}\,dx = \frac{1}{3}\int_0^{\pi/12} \frac{3\sec^2 3x\,dx}{(4 + \tan 3x)}$$

$$= \frac{1}{3}\ln|4 + \tan 3x|\Big|_0^{\pi/12}$$

$$= \frac{1}{3}(\ln 5 - \ln 4) = \frac{1}{3}\ln\frac{5}{4}$$

$$= 0.0744$$

28. $u = x^3 + 3x$; $du = (3x^2 + 3)dx = 3(x^2 + 1)dx$

$$\int_1^2 \frac{x^2 + 1}{x^3 + 3x}\,dx = \frac{1}{3}\int_1^2 \frac{3(x^2 + 1)dx}{x^3 + 3x}$$

$$= \frac{1}{3}\ln|x^3 + 3x|\Big|_1^2$$

$$= \frac{1}{3}(\ln 14 - \ln 4)$$

$$= \frac{1}{3}\ln\frac{7}{2} = 0.418$$

29. $y = \frac{1}{x + 1}$

$$A = \int_0^2 \frac{1}{x + 1}\,dx = \ln(x + 1)\Big|_0^2$$

$$A = \ln 3 - \ln 1 = 1.10 - 0 = 1.10$$

30. $xy = 9$; $x = 1, x = 2, y = 0$

Therefore, $y = \frac{9}{x}$

$$A_{1,2} = \int_1^2 \frac{9}{x}\,dx = 9\int_1^2 \frac{dx}{x}$$

$$A = 9\ln x\Big|_1^2$$

$$= 9(\ln 2 - \ln 1) = 9\ln 2$$

$$A = 6.238$$

31. $y = \frac{1}{x^2 + 1}$; $x = 0, x = 1, y = 0$, rotated about y-axis;

shells

$r = x$; $h = y$

$$dV = 2\pi rh\,dx = 2\pi xy\,dx = 2\pi x\frac{1}{x^2 + 1}\,dx$$

$u = x^2 + 1$; $du = 2x\,dx$

$$V = 2\pi\int_0^1 \frac{x\,dx}{x^2 + 1}$$

$$= \pi\int_0^1 \frac{(2x\,dx)}{(x^2 + 1)}$$

$$V = \pi\ln|x^2 + 1|\Big|_0^1$$

$$= \pi(\ln 2 - \ln 1)$$
$$= \pi\ln 2 = 2.18$$

32. $y = \dfrac{2}{\sqrt{3x+1}}$; $x = 0, x = 3.5, y = 0$, rotated about

x-axis

$dV = \pi r^2\,dx$; $r = y$

$u = 3x + 1$; $du = 3\,dx$

$V = \pi \displaystyle\int_0^{3.5} \dfrac{4}{3x+1}\,dx$

$V = \dfrac{4\pi}{3} \displaystyle\int_0^{3.5} \dfrac{3\,dx}{3x+1} = \dfrac{4\pi}{3} \ln|3x+1|\Big|_0^{3.5}$

$V = \dfrac{4\pi}{3}(\ln 11.5 - \ln 1) = 10.2$

33. $m = \dfrac{dy}{dx} = \dfrac{\sin x}{3 + \cos x}$; $y = \displaystyle\int \dfrac{1}{3+\cos x} \times \sin x\,dx$

$y = -\displaystyle\int \dfrac{1}{3+\cos x}(-\sin x)\,dx$;

let $u = 3 + \cos x$; $du = -\sin x\,dx$

$y = -\displaystyle\int \dfrac{1}{u}\,du = -\ln|u| + C$

$\quad = -\ln(3 + \cos x) + C$

$2 = -\ln\left(3 + \cos\dfrac{\pi}{3}\right) + C$; substitute values of x

and y

$2 = -\ln(3 + 0.5) + C$; $C = 2 + \ln 3.5$

$y = -\ln(3 + \cos x) + \ln 3.5 + 2$; substituting for C

$y = \ln\dfrac{3.5}{3 + \cos x} + 2$

34. $xy = 4$; $x = 1$ to $x = 2$

$y_{av} = \dfrac{\displaystyle\int_1^2 \dfrac{4}{x}\,dx}{2 - 1} = 4\displaystyle\int_1^2 \dfrac{dx}{x} = 4\ln x\Big|_1^2$

$\quad = 4(\ln 2 - \ln 1) = 4\ln 2 = 2.7726$

35. $t = \dfrac{1}{k}\displaystyle\int \dfrac{dP}{P} = \dfrac{1}{k}\ln P + C$; $\dfrac{dP}{dt} = kP$

$t = 0, P = 249$ M; $t = 10, P = 275$ M;

$t = 30, P = ?$

(a) $0 = \dfrac{1}{k}\ln 249 + C$; therefore, $C = -\dfrac{1}{k}\ln 249$

(b) $10 = \dfrac{1}{k}\ln 275 - \dfrac{1}{k}\ln 249 = \dfrac{1}{k}\ln\dfrac{275}{249}$

Therefore, $k = \dfrac{1}{10}\ln\dfrac{275}{249}$

Therefore, $t = \dfrac{1}{\frac{1}{10}\ln\frac{275}{249}}\ln P - \dfrac{1}{\frac{1}{10}\ln\frac{275}{249}}\ln 249$

$t = \dfrac{10}{\ln\frac{275}{249}}(\ln P - \ln 249)$

$\ln P\big|_{t=30} = t\left(\dfrac{\ln\frac{275}{249}}{10}\right) + \ln 249 = 5.815$

$P = e^{5.815} = 335.4$ M $= 335$ million people

36. $\ln T = -\displaystyle\int \dfrac{dr}{r-1}$; $T = 273.16$; $r = 1.3361$

$r > 1$ for all T; $u = r - 1$; $du = dr$

$\ln T = -\ln|r-1| + C$; $\ln 273.16 + \ln 1.3361 - 1$

$= C$

Therefore, $C = \ln|273.16 \times 0.3361| = 4.520$

$\ln T = -\ln|r-1| + \ln|91.809|$; $T = \dfrac{91.809}{r-1}$

37. $t = L\displaystyle\int \dfrac{di}{E - iR}$; $u = E - iR$; $du = -R\,di$

$t = -\dfrac{L}{R}\displaystyle\int \dfrac{-R\,di}{E - iR} = \dfrac{-L}{R}\ln|E - iR| + C$;

$t = 0$ for $i = 0$

$0 = -\dfrac{L}{R}\ln E + C$; $C = \dfrac{L}{R}\ln|E|$

$t = \dfrac{L}{R}(-\ln|E - iR| + \ln|E|)$

$t = \dfrac{L}{R}\ln\dfrac{E}{E - iR}$; $\dfrac{R}{L}t = \ln\dfrac{E}{E - iR}$; $e^{Rt/L} = \dfrac{E}{E - iR}$

$i = \dfrac{E}{R} - \dfrac{E}{R}e^{-Rt/L}$;

$i = \dfrac{E}{R}(1 - e^{-Rt/L})$

38. $t = \displaystyle\int \dfrac{dv}{20 - v}$; $t = 0, v = 0$

$t = -\displaystyle\int \dfrac{(-dv)}{(20 - v)} = -\ln|20 - v| + C$

$0 = -\ln|20| + C$; therefore, $C = \ln 20$

$t = -\ln|20 - v| + \ln 20 = \ln\dfrac{20}{20 - v}$

$\dfrac{20}{20 - v} = e^t$; therefore, $20 - v = \dfrac{20}{e^t}$

Therefore, $v - 20 = -\dfrac{20}{e^t}$; $v = 20 - \dfrac{20}{e^t}$

Therefore, $v = 20(1 - e^{-t})$

39. $y = \dfrac{50}{x^2 + 20}; \; x = 3.00; \; A = 6.61 \text{ m}^2$

$u = x^2 + 20; \; du = 2x \, dx$

$$\bar{x} = \dfrac{\displaystyle\int_0^3 x(y)dx}{6.61}$$

$$\bar{x} = \dfrac{1}{6.61}\int_0^3 x\dfrac{50}{x^2 + 20}dx$$

$$\bar{x} = \dfrac{25}{6.61}\int_0^3 \dfrac{2x\,dx}{x^2 + 20} = \dfrac{25}{6.61}\ln\left|x^2 + 20\right|\Big|_0^3$$

$$\bar{x} = \dfrac{25}{6.61}(\ln 29 - \ln 20) = 1.41 \text{ m}$$

40. $p = 3\displaystyle\int \dfrac{\sin \pi t}{2 + \cos \pi t}dt; \; u = 2 + \cos \pi t; \; du = -\pi \sin \pi t \, dt$

$$p = -\dfrac{3}{\pi}\int \dfrac{-\pi \sin \pi t \, dt}{2 + \cos \pi t} = -\dfrac{3}{\pi}\ln(2 + \cos \pi t) + C$$

9.3 The Exponential Form

1. $u = 7x; \; du = 7\,dx; \; \displaystyle\int e^{7x}(7\,dx) = e^{7x} + C$

2. $u = x^4; \; du = 4x^3 dx; \; \displaystyle\int e^{x^4}(4x^3 dx) = e^{x^4} + C$

3. $u = 2x + 5; \; du = 2\,dx$

$$\int e^{2x+5}dx = \dfrac{1}{2}\int e^{2x+5}(2\,dx) = \dfrac{1}{2}e^{2x+5} + C$$

4. $u = -4x; \; du = -4\,dx$

$$\int 2e^{-4x}dx = -\dfrac{1}{2}\int e^{-4x}(-4\,dx) = -\dfrac{1}{2}e^{-4x} + C$$

5. $\displaystyle\int_{-2}^2 6e^{s/2}ds = \int_{-2}^2 12 \cdot e^{s/2} \cdot \dfrac{1}{2}\,ds = 12 \cdot e^{s/2}\Big|_{-2}^2$

$$= 12(e^{2/2} - e^{-2/2}) = 12\left(e - \dfrac{1}{e}\right)$$

$$= 28.2$$

6. $u = 4x; \; du = 4\,dx$

$$\int_1^2 3e^{4x}dx = 3\int_1^2 e^{4x}dx = \dfrac{3}{4}\int_1^2 e^{4x}(4\,dx)$$

$$= \dfrac{3}{4}e^{4x}\Big|_1^2 = \dfrac{3}{4}[e^8 - e^4]$$

$$= 2190$$

7. $y = x^3; \; du = 3x^2 dx$

$$\int 6x^2 e^{x^3} dx = 6\int e^{x^3}(x^2 dx) = \dfrac{6}{3}\int e^{x^3}(3x^2 dx)$$

$$= 2e^{x^3} + C$$

8. $u = -x^2; \; du = -2x\,dx$

$$\int xe^{-x^2}dx = -\dfrac{1}{2}\int e^{-x^2}(-2x\,dx) = -\dfrac{1}{2}e^{-x^2} + C$$

9. $u = \sqrt{x} = x^{1/2}; \; du = \dfrac{1}{2}x^{-1/2}dx = \dfrac{dx}{2\sqrt{x}};$

$$\int_1^4 \dfrac{e^{\sqrt{x}}}{\sqrt{x}}dx = 2\int_1^4 e^{\sqrt{x}}\dfrac{dx}{2\sqrt{x}} = 2e^{\sqrt{x}}\Big|_1^4$$

$$= 2e^2 - 2e = 9.34$$

10. $\displaystyle\int_0^1 4(\ln e^u)e^{-2u^2}du = 4\int_0^1 u \cdot \ln e \cdot e^{-2u^2}du$

$$= -\int_0^1 e^{-2u^2}(-4u\,du)$$

$$= -e^{-2u^2}\Big|_0^1 = -e^{-2\cdot 1^2} + e^{-2\cdot 0^2}$$

$$= 1 - \dfrac{1}{e^2} \approx 0.8647$$

11. $u = 2\sec\theta; \; du = 2\sec\theta\tan\theta\,d\theta$

$$\int 4(\sec\theta\tan\theta)e^{2\sec\theta}d\theta = \dfrac{4}{2}\int e^{2\sec\theta}2\sec\theta\tan\theta\,d\theta$$

$$= 2e^{2\sec\theta} + C$$

12. $u = \tan x; \; du = \sec^2 x\,dx$

$$\int (\sec^2 x)e^{\tan x}dx = \int e^{\tan x}(\sec^2 x\,dx)$$

$$= e^{\tan x} + C$$

13. $\displaystyle\int \sqrt{e^{2y} + e^{3y}}\,dy = \int \sqrt{e^{2y}(1 + e^y)}\,dy$

$$= \int \sqrt{1 + e^y} \cdot e^y dy$$

$$= \dfrac{2}{3}(1 + e^y)^{3/2} + C$$

14. $\displaystyle\int (e^x - e^{-x})^2 dx$

$$= \int (e^{2x} - 2e^x e^{-x} + e^{-2x})dx$$

$$= \int e^{2x}dx - 2\int dx + \int e^{-2x}dx$$

$$u = 2x; \; du = 2\,dx; \; u = -2x; \; du = -2\,dx$$

$$\dfrac{1}{2}\int e^{2x}(2\,dx) - 2\int dx - \dfrac{1}{2}\int e^{-2x}(-2\,dx)$$

$$= \dfrac{1}{2}e^{2x} - 2x - \dfrac{1}{2}e^{-2x} + C$$

15. $u = -2x;\ du = -2\,dx$

$$\int_1^3 3e^{2x}(e^{-2x}-1)dx = 3\int (e^0 - e^{2x})dx$$

$$= 3\int dx - \frac{3}{2}\int e^{2x}(2\,dx)$$

$$= 3x - \frac{3}{2}e^{2x}\Big|_1^3$$

$$= \left[9 - \frac{3}{2}e^6 - \left(3 - \frac{3}{2}e^2\right)\right]$$

$$= 9 - 3 - \frac{3}{2}e^6 + \frac{3}{2}e^2$$

$$= 6 - \frac{3}{2}(e^6 - e^2) = -588.06$$

16. $u = 2x+1;\ du = 2\,dx$

$$\int_0^{0.5} \frac{3e^{3x+1}}{e^x}\,dx = 3\int_0^{0.5} e^{2x+1}dx$$

$$= \frac{3}{2}\int_0^{0.5} e^{2x+1}(2\,dx)$$

$$= \frac{3}{2}e^{2x+1}\Big|_0^{0.5} = \frac{3}{2}(e^2 - e^1)$$

$$= 7.01$$

17. $\displaystyle\int \frac{2\,dx}{\sqrt{x}e^{\sqrt{x}}} = 2\int \frac{dx}{x^{1/2}e^{x^{1/2}}} = 2\int e^{-x^{1/2}}x^{-1/2}dx$

$u = -x^{1/2};\ du = -\dfrac{1}{2}x^{-1/2}dx$

$$2\int e^{-x^{1/2}}x^{-1/2}dx = 2(-2)\int e^{-x^{1/2}}\left(-\frac{1}{2}x^{-1/2}\right)$$

$$= -4e^{-x^{1/2}} + C = -\frac{4}{e^{x^{1/2}}} + C$$

$$= -\frac{4}{e^{\sqrt{x}}} + C$$

18. $u = -\sin x;\ du = -\cos x\,dx$

$$\int \frac{4\,dx}{\sec x e^{\sin x}} = 4\int e^{-\sin x}\cos x\,dx$$

$$= -4\int e^{-\sin x}(-\cos x\,dx)$$

$$= -4e^{-\sin x} + C$$

$$= \frac{-4}{e^{\sin x}} + C$$

19. $u = \tan^{-1}x;\ du = \dfrac{1}{1+x^2}dx$

$$\int \frac{e^{\tan^{-1}x}}{x^2+1}\,dx = \int e^{\tan^{-1}x}\frac{dx}{x^2+1} = e^{\tan^{-1}x} + C$$

20. $u = \sin^{-1}2x;\ du = \dfrac{1}{\sqrt{1-4x^2}}2\,dx$

$$\int \frac{e^{\sin^{-1}2x}}{\sqrt{1-4x^2}}\,dx = \frac{1}{2}\int e^{\sin^{-1}2x}\frac{2\,dx}{\sqrt{1-4x^2}}$$

$$= \frac{1}{2}e^{\sin^{-1}2x} + C$$

21. $u = \cos 3x\,dx;\ du = -\sin 3x\,(3\,dx) = \dfrac{-3}{\csc 3x}\,dx$

$$\int \frac{e^{\cos 3x}dx}{\csc 3x} = -\frac{1}{3}\int e^{\cos 3x}[-3\sin 3x(3\,dx)]$$

$$= -\frac{1}{3}e^{\cos 3x} + C$$

22. $\displaystyle\int 2e^{r^2+\ln r}\,dr = \int 2e^{r^2}\cdot e^{\ln r}\,dr$

$$= \int 2e^{r^2}\cdot r\,dr$$

$$= \int e^{r^2}\cdot 2r\,dr$$

$$= e^{r^2} + C$$

23. $u = \cos^2 x;$
$du = 2\cos x(-\sin x)dx = -2\sin x\cos x\,dx$
$\quad = -2\sin 2x\,dx$

$$\int_0^\pi (\sin 2x)e^{\cos^2 x}dx = -\frac{1}{2}\int_0^\pi e^{\cos^2 x}(-2\sin 2x\,dx)$$

$$= -\frac{1}{2}e^{\cos^2 x}\Big|_0^\pi$$

$$= -\frac{1}{2}[(\cos\pi)^2 - (\cos 0)^2]$$

$$= -\frac{1}{2}[(-1)^2 - 1^2] = 0$$

24. $\quad u = \dfrac{e^t}{1+e^t}$

$$du = \frac{(1+e^t)e^t - e^t(e^t)}{(1+e^t)^2}\,dt$$

$$du = \frac{1}{(1+e^t)}\frac{e^t}{(1+e^t)}\,dt$$

$$du = u\frac{dt}{(1+e^t)}$$

$$\frac{du}{u} = \frac{dt}{(1+e^t)}$$

$$\int_0^2 \frac{dt}{(1+e^t)} = \int_{\frac{1}{2}}^{\frac{e^2}{1+e^2}} \frac{du}{u}$$

$$\int_0^2 \frac{dt}{(1+e^t)} = \ln u \, \Big|_{\frac{1}{2}}^{\frac{e^2}{1+e^2}}$$

$$\int_0^2 \frac{dt}{(1+e^t)} = \ln \frac{e^2}{1+e^2} - \ln \frac{1}{2}$$

$$\int_0^2 \frac{dt}{(1+e^t)} \approx 0.5662$$

25. $A = \int_0^2 3e^x \, dx = 3e^x \big|_0^2 = 3e^2 - 3e^0 = 19.2$

26. $x = a; \; y = 0, x = b; \; y = e^x$

$$A_{a:b} = \int_a^b e^x \, dx$$

$$A = e^x \big|_a^b = e^b - e^a$$

The area is the difference between e^b and e^a, the ordinates.

27. $y = e^{x^2}, x = 1$
$y = 0, x = 2$

Rotated about y-axis
Generates a shell of volume dV
$dV = 2\pi r h \, dx; \; r = x; \; h = y$

$$V = 2\pi \int_1^2 xy \, dx = 2\pi \int_1^2 xe^{x^2} \, dx$$

$u = x^2; \; du = 2x \, dx$

$$V = \pi \int_1^2 e^{x^2}(2x \, dx) = \pi \, e^{x^2} \Big|_1^2 = \pi(e^4 - e^1)$$

$V = 163 \text{ units}^3$

28. $\dfrac{dy}{dx} = \sqrt{e^{x+3}}; \; (1,0)$

$dy = (e^{x+3})^{1/2} dx$

$y = \displaystyle\int e^{(1/2)x+(3/2)} dx; \; u = \frac{1}{2}x + \frac{3}{2}; \; du = \frac{1}{2}dx$

$y = 2 \displaystyle\int e^{(1/2)x+(3/2)} \frac{1}{2} dx = 2e^{(1/2)x+(3/2)} + C$

$y = 2\sqrt{e^{x+3}} + C; \; (1,0)$
$0 = 2\sqrt{e^4} + C; \text{ therefore, } C = -2e^2$

Therefore, $y = 2\sqrt{e^{x+3}} - 2e^2 = 2(\sqrt{e^{x+3}} - e^2)$

29. $y_{av} = \dfrac{\displaystyle\int_0^4 e^{2x} dx}{4-0} = \dfrac{\dfrac{1}{2}\displaystyle\int e^{2x}(2\,dx)}{4}$

$$= \dfrac{\dfrac{1}{2}e^{2x}\Big|_0^4}{4} = \dfrac{1}{8}e^{2x}\Big|_0^4$$

$$= \frac{1}{8}(e^8 - 1) = 372$$

30. $y = e^{x^3}; \; x = 1, \text{ axes}$

$$I_y = k \int_0^1 x^2 y \, dx$$

$$I_y = \frac{k}{3} \int_0^1 e^{x^3} 3x^2 \, dx$$

$u = x^3; \; du = 3x^2 \, dx$

$$I_y = \frac{k}{3} e^{x^3} \Big|_0^1 = \frac{k}{3}(e^1 - e^0) = 0.573k$$

Let $k = 1; \; I_y = 0.573$

31. $\displaystyle\int b^u du = \frac{b^u}{\ln b} + C; \; (b > 0; \; b \neq 1)$

Eq. 27-15: $\dfrac{d}{dx}b^u = b^u \ln b \dfrac{du}{dx}$

Therefore, $d(b^u) = b^u \ln b (du)$
Therefore, $(\ln b)b^u du = db^u$

$$b^u du = \frac{1}{\ln b} db^u; \; \int b^u du = \frac{1}{\ln b} \int db^u$$

$$\int b^u du = \frac{1}{\ln b} b^u + C; \; \text{Q.E.D.}$$

32. $y = 2^x; \; x = 3$

$$A_{0,3} = \int_0^3 2^x dx = \frac{2x}{\ln 2}\Big|_0^3 = \frac{2^3}{\ln 2} - \frac{2^0}{\ln 2}$$

$$A = \frac{1}{\ln 2}(8-1) = \frac{7}{\ln 2} = 10.10$$

33. $qe^{t/RC} = \dfrac{E}{R} \displaystyle\int e^{t/RC} dt; \; u = \dfrac{t}{RC}; \; du = \dfrac{1}{RC} dt$

$qe^{t/RC} = RC \cdot \dfrac{E}{R} \displaystyle\int e^{t/RC}\left(\dfrac{1}{RC}\right) dt$

$qe^{t/RC} EC(e^{t/RC}) + C_1$, where C_1 is the constant of integration.

$q = 0$ for $t = 0; \; 0 = EC + C_1; \; C_1 = -EC;$

$qe^{t/RC} = EC(r^{t/RC}) - EC$

$$q = EC - \frac{EC}{e^{t/RC}}; \; q = EC(1 - e^{-t/RC})$$

34. $E = a \int_0^{I_0} e^{-Tx} dx$; $u = -Tx$; $du = -T\,dx$

$$E = \frac{-a}{T} \int_0^{I_0} e^{-Tx}(-T\,dx) = \frac{-a}{T} e^{-Tx} \Big|_0^{I_0}$$

$$E = \frac{-a}{T}(e^{-TI_0} - e^0) = \frac{a}{T}(1 - e^{-I_0 T})$$

35. $v = 2e^{-2t} + 3e^{-5t}$; overdamped motion.

$$\frac{ds}{dt} = 2e^{-2t} + 3e^{-5t};\ s = 1.6, t = 0$$

$$ds = (2e^{-2t}+3e^{-5t})dt;\ s = 2\int e^{-2t}dt + 3\int e^{-5t}dt$$

$$u = -2t, du = -2\,dt;\ u = -5t, du = -5\,dt$$

$$s = -1\int e^{-2t}(-2\,dt) - \frac{3}{5}\int e^{-5t}(-5\,dt)$$

$$s = -e^{-2t} - \frac{3}{5}e^{-5t} + C;\ s = -1.6;\ t = 0$$

$$-1.6 = -e^0 - \frac{3}{5}e^0 + C = -1 - \frac{3}{5} + C;\ \text{therefore,}$$
$C = 0$

Therefore, $s = -e^{-2t} - 0.6e^{-5t}$

36. $F = 6\int e^{\sin \pi t}\cos \pi t\,dt$; $F = 0, t = 1.5$

$$u = \sin \pi t;\ du = \pi \cos \pi t\,dt$$

$$F = \frac{6}{\pi}\int e^{\sin \pi t}(\pi \cos \pi t\,dt) = \frac{6}{\pi}e^{\sin \pi t} + C$$

$$0 = \frac{6}{\pi}e^{\sin(3\pi/2)} + C;\ \text{therefore,}\ C = \frac{6}{\pi}e^{-1}$$

$$F = \frac{6}{\pi}e^{\sin \pi t} - \frac{6}{\pi}e^{-1} = \frac{6}{\pi}(e^{\sin \pi t} - e^{-1})$$

9.4   Basic Trigonometric Forms

1. $u = 2x$; $du = 2\,dx$

$$\int \cos 2x\,dx = \frac{1}{2}\int \cos 2x\,(dx)$$

$$= \frac{1}{2}\sin 2x + C$$

2. $u = 2 - x$; $du = -dx$

$$\int 4\sin(2-x)dx = -4\int \sin(2-x)(-dx)$$

$$= -4[-\cos(2-x)]$$
$$= 4\cos(2-x) + C$$

3. $u = 3\theta$; $du = 3\,d\theta$

$$\int 0.3\sec^2 3\theta\,d\theta = 0.3\int \sec^2 3\theta(3\,d\theta)$$

$$= 0.1\tan 3\theta + c$$

4. $u = 2x$; $du = 2\,dx$

$$\int \csc 2x \cot 2x\,dx = \frac{1}{2}\int \csc 2x \cot 2x(2\,dx)$$

$$= \frac{1}{2}(-\csc 2x) + C$$

$$= -\frac{1}{2}\csc 2x + C$$

5. $u = \frac{1}{2}x$; $du = \frac{1}{2}\,dx$

$$\int \sec \frac{1}{2}x \tan \frac{1}{2}x\,dx = 2\int \sec \frac{1}{2}x \tan \frac{1}{2}x \left(\frac{1}{2}dx\right)$$

$$= 2\sec \frac{1}{2}x + C$$

6. $u = e^x$; $du = e^x dx$

$$\int e^x \csc^2(e^x)dx = \int \csc^2(e^x)(e^x dx)$$

$$= -\cot e^x + C$$

7. $u = x^3$; $du = 3x^2 dx$

$$\int_{0.5}^1 x^2 \cot x^3\,dx = \frac{1}{3}\int_{0.5}^1 \cot x^3(3x^2 dx)$$

$$= \frac{1}{3}\ln|\sin x^3|\Big|_{0.5}^1$$

$$= \frac{1}{3}\left(\ln|\sin 1| - \ln\left|\sin \frac{1}{8}\right|\right)$$

$$= 0.6365$$

8. $u = \frac{1}{2}t$; $du = \frac{1}{2}\,dt$

$$\int_0^1 6\sin \frac{1}{2}t \sec \frac{1}{2}t\,dt = \int_0^{1/2} 12\tan u\,du$$

$$= -12\ln|\cos u|\big|_0^{1/2}$$

$$= -12\ln\left|\cos \frac{1}{2}\right| + 12\ln|\cos 0|$$

$$= 1.567$$

9. $\int 3\phi\sec^2 \phi^2 \cos \phi^2 d\phi$

$$= \frac{3}{2}\int \sec \phi^2 \cdot 2\phi\,d\phi$$

$$= \frac{3}{2}\ln|\sec \phi^2 + \tan \phi^2| + C$$

10. $u = 3x$; $du = 3\,dx$

$$\int 2\csc 3x\,dx = \frac{2}{3}\int \csc 3x(3dx)$$

$$= \frac{2}{3}\ln|\csc 3x - \cot 3x| + C$$

11. $u = \dfrac{1}{x} = x^{-1}$;

$$du = -1x^{-2}dx = -\frac{dx}{x^2}$$

$$\int \frac{\sin\left(\dfrac{1}{x}\right)}{x^2}dx = -\int \sin\left(\frac{1}{x}\right)\left(-\frac{dx}{x^2}\right)$$

$$= -\left[-\cos\left(\frac{1}{x}\right)\right] + C$$

$$= \cos\left(\frac{1}{x}\right) + C$$

12. $u = 4x$; $du = 4\,dx$

$$\int \frac{3\,dx}{\sin 4x} = 3\int \csc 4x\,dx$$

$$= \frac{3}{4}\int \csc 4x(4\,dx)$$

$$= \frac{3}{4}\ln|\csc 4x - \cot 4x| + C$$

13. $u = 2x$; $du = 2\,dx$

$$\int_0^{\pi/6} = \frac{dx}{\cos^2 2x} = \frac{1}{2}\int_0^{\pi/6}\sec^2 2x(2\,dx)$$

$$= \frac{1}{2}\tan 2x\Big|_0^{\pi/6} = \frac{1}{2}\left(\tan\frac{\pi}{3} - \tan 0\right)$$

$$= \frac{1}{2}(\sqrt{3} - 0)$$

$$= \frac{1}{2}\sqrt{3}$$

14. $u = e^s$; $du = e^s\,ds$

$$\int_0^1 \frac{2e^s\,ds}{\sec e^s} = 2\int_0^1 \cos e^s\,(e^s\,ds)$$

$$= 2\sin e^s\big|_0^1$$

$$= 2(\sin e^1 - \sin e^0)$$

$$= 2\sin e - 2\sin 1$$

$$= -0.861$$

15. $u = 5x$; $du = 5\,dx$

$$\int \frac{\sec 5x}{\cot 5x}dx = \int \sec 5x\tan 5x\,dx$$

$$= \frac{1}{5}\int \sec 5x\tan 5x(5\,dx)$$

$$= \frac{1}{5}\sec 5x + C$$

16. $\displaystyle\int \frac{\sin 2x}{\cos^2 x}dx = \int \frac{2\sin x\cos x}{\cos^2 x}dx$

$$= 2\int \tan x\,dx$$

$$= 2(-\ln|\cos x|) + C$$
$$= -2\ln|\cos x| + C$$

17. $\displaystyle\int \sqrt{\tan^2 2x + 1}\,dx = \int \sqrt{\sec^2 2x}\,dx = \int \sec 2x\,dx$

$u = 2x$; $du = 2\,dx$

$$\int \sec 2x\,dx = \frac{1}{2}\int \sec 2x(2\,dx)$$

$$= \frac{1}{2}\ln|\sec 2x + \tan 2x| + C$$

18. $\displaystyle\int 5(\tan u)\ln(\cos u)\,du$

let $x = \ln(\cos u)$

$$dx = \frac{1}{\cos u}\cdot(-\sin u)du$$

$$dx = -\tan u\,du$$

$$\int -5x\,dx = \frac{-5x^2}{2} + C$$

$$\int 5(\tan u)\ln(\cos u)\,du = \frac{-5(\ln(\cos u))^2}{2} + C$$

19. $\displaystyle\int \frac{1 + \sin 2x}{\tan 2x}dx$

$$= \int \frac{dx}{\tan 2x} + \int \frac{\sin 2x}{\tan 2x}dx$$

$$= \int \cot 2x\,dx + \int \cos 2x\,dx$$

$u = 2x$; $du = 2\,dx$

$$= \frac{1}{2}\int \cot 2x(2\,dx) + \frac{1}{2}\int \cos 2x(2\,dx)$$

$$= \frac{1}{2}\ln|\sin 2x| + \frac{1}{2}\sin 2x + C$$

$$= \frac{1}{2}(\ln|\sin 2x| + \sin 2x) + C$$

20. $\displaystyle\int \frac{1 - \cot^2 x}{\cos^2 x} = \int \sec^2 x\,dx - \int \csc^2 x\,dx$

$$= \tan x + \cot x + C$$

21. $\displaystyle\int \frac{1-\sin x}{1+\cos x}\,dx$

$\displaystyle = \int \frac{1-\sin x}{1+\cos x} \times \frac{1-\cos x}{1-\cos x}\,dx$

$\displaystyle = \int \frac{1-\sin x - \cos x + \sin x \cos x}{1-\cos^2 x}\,dx$

$\displaystyle = \int \frac{1-\sin x - \cos x + \sin x \cos x}{\sin^2 x}\,dx$

$\displaystyle = \int \left(\frac{1}{\sin^2 x} - \frac{\sin x}{\sin^2 x} - \frac{\cos x}{\sin^2 x} + \frac{\sin x \cos x}{\sin^2 x} \right) dx$

$\displaystyle = \int (\csc^2 x - \csc x - \cot x \csc x + \cot x)\,dx$

$= -\cot x - \ln|\csc x - \cot x| - (-\csc x)$
$\quad + \ln|\sin x| + C$

$= \csc x - \cot x - \ln|\csc x - \cot x|$
$\quad + \ln|\sin x| + C$

22. $u = x + \tan x;\ du = (1+\sec^2 x)dx$

$\displaystyle\int \frac{1+\sec^2 x}{x+\tan x}\,dx = \int \frac{(1+\sec^2 x)dx}{(x+\tan x)}$

$\displaystyle\qquad\qquad\qquad = \ln|x+\tan x| + C$

23. $\displaystyle \sin 3x\left(\frac{1}{\sin 3x} + \frac{1}{\cos 3x} \right) = 1 + \tan 3x;$

$u = 3x;\ du = 3\,dx$

$\displaystyle\int_0^{\pi/9} \sin 3x(\csc 3x + \sec 3x)dx$

$\displaystyle = \int_0^{\pi/9} (1+\tan 3x)dx$

$\displaystyle = \int_0^{\pi/9} dx + \int_0^{\pi/9} \tan 3x\,dx$

$\displaystyle = \int_0^{\pi/9} dx + \frac{1}{3}\int_0^{\pi/9} \tan 3x(3\,dx)$

$\displaystyle = \left(x - \frac{1}{3}\ln|\cos 3x| \right)\Big|_0^{\pi/9}$

$\displaystyle = \frac{\pi}{9} - \frac{1}{3}\ln\left|\cos\frac{\pi}{3}\right| - \left(0 - \frac{1}{3}\ln|\cos 0| \right)$

$\displaystyle = \frac{\pi}{9} - \frac{1}{3}\ln\left(\frac{1}{2} \right)$

$\displaystyle = \frac{\pi}{9} + \frac{1}{3}\ln 2$

$= 0.580$

24. $\displaystyle\int_{\pi/4}^{\pi/3} (1+\sec x)^2\,dx$

$\displaystyle = \int (1+2\sec x + \sec^2 x)dx$

$\displaystyle = \int dx + 2\int \sec x\,dx + \int \sec^2 x\,dx$

$= x + \ln|\sec x + \tan x| + \tan x\big|_{\pi/4}^{\pi/3}$

$\displaystyle = \frac{\pi}{3} + 2\ln|2+\sqrt{3}| + \sqrt{3}$

$\displaystyle \quad - \left(\frac{\pi}{4} + 2\ln|\sqrt{2}+1| + 1 \right)$

$= 1.865$

25. $\displaystyle A = \int_0^{\pi/4} y\,dx = \int_0^{\pi/4} 2\tan x\,dx$

$= 2(-\ln)|\cos x|\big|_0^{\pi/4}$

$\displaystyle = (-2)\left[\ln\left|\cos\frac{\pi}{4}\right| - 2\ln|\cos 0| \right]$

$= -2(\ln 0.7071 - \ln 1) = 0.693$

26. $y = \sin x;\ x = 0;\ x = \pi$

$\displaystyle A_{0;\pi} = \int_0^{\pi} \sin x\,dx = -\cos x\Big|_0^{\pi}$

$A = -\cos\pi - (-\cos 0)$
$\quad = -(-1) - (-1) = 1+1 = 2$

27. $y = \sec x;\ x = 0;\ x = \dfrac{\pi}{3}$

$y = 0$; rotated about x-axis, disks.
$dV = \pi r^2 dx;\ r = y$

$\displaystyle V = \int_0^{\pi/3} \pi y^2 dx = \pi \int_0^{\pi/3} \sec^2 x\,dx$

$\displaystyle V = \pi \tan x\Big|_0^{\pi/3} = \pi\sqrt{3} = 5.44$

28. $y = \cos x^2;\ x = 0;\ x = 1$

$y = 0$; rotated about y-axis, shells.
$dV = 2\pi rh\,dx;\ r = x, h = y$

$\displaystyle V = 2\pi\int_0^1 xy\,dx = 2\pi\int_0^1 x\cos x^2 dx$

$u = x^2;\ du = 2x\,dx$

$\displaystyle V = \frac{2\pi}{2}\int_0^1 \cos x^2(2x\,dx)$

$\displaystyle V = \pi\sin x^2\Big|_0^1$

$= \pi(\sin 1 - \sin 0)$

$= 2.644$

29. $\omega = -0.25\sin 2.5t$

$$\theta = \int -0.25\sin 2.5t\, dt;$$

$$u = 2.5t;\ du = 2.5\, dt$$

$$\theta = -0.10\int \sin 2.5t (2.5\, dt)$$

$$= -0.10(-\cos 2.5t) + C$$
$$\theta = 0.10\cos 2.5t + C$$
$$0.10 = 0.10\cos 0 + C;\ C = 0;$$
$$\theta = 0.10\cos 2.5t$$

30. $i = 110\cos 377t;\ 500\mu F;\ V_0 = 0;\ t = t_0 = 0$

$$V_C = \frac{1}{c}\int i\, dt;\ u = 377t;\ du = 377\, dt$$

$$V_C = \frac{110}{500\times 10^{-6}}\int \cos 377\, dt$$

$$= \frac{110}{377(500\times 10^{-6})}\int \cos 377t(377t\, dt)$$

$$= \frac{110}{377(500\times 10^{-6})}\sin 377t + C$$

$$V_C = 584\sin 377t + C;\ t = 0;\ V = 0;\ \text{therefore},\ C = 0$$
$$V_C = 584\sin 377t$$
$\sin 377t$ and $\cos 377t$ are 90° out of phase.

31. $y = \tan x^2;\ y = 0, x = 1$

$$\bar{x} = \frac{\int_0^1 xy\, dx}{\int_0^1 y\, dx} = \frac{\int_0^1 (\tan x^2)x\, dx}{\int_0^1 \tan x^2\, dx}$$

$$u = x^2;\ du = 2x\, dx$$

$$\bar{x} = \frac{\frac{1}{2}\int_0^1 \tan x^2(2x\, dx)}{0.3984} = \frac{-\frac{1}{2}\ln\left|\cos x^2\right|\Big|_0^1}{0.3984}$$

$$\bar{x} = \frac{-\frac{1}{2}(\ln\cos 1 - \ln\cos 0)}{0.3984} = \frac{0.3078}{0.3984}$$

$$= 0.7726\ \text{m}$$

32. $F = \dfrac{2 + \tan x}{\cos x};\ W = 0, x = 0;\ W = Fd$

$$W = \int \frac{2 + \tan x}{\cos x}\, dx = \int 2\sec x\, dx + \int \frac{\tan x}{\cos x}\, dx$$

$$W = 2\int \sec x\, dx + \int (\cos x)^{-2}\sin x\, dx$$

$$u = \cos x;\ du = -\sin x\, dx$$

$$W = 2\int \sec x\, dx - \int (\cos x)^{-2}(-\sin x\, dx)$$

$$W = 2\ln|\sec x + \tan x| - \frac{(\cos x)^{-1}}{-1} + C$$

$$W = 2\ln|\sec x + \tan x| + \sec x + C;\ x = 0, W = 0$$
$$0 = 2\ln|\sec 0 + \tan 0| + \sec 0 + C = 2\ln|1 + 0| + 1 + C;$$
therefore, $C = -1$
$$W = \sec x + 2\ln|\sec x + \tan x| - 1$$

9.5 Other Trigonometric Forms

1. $u = \sin x;\ du = \cos x\, dx;\ \displaystyle\int \sin^2 x\cos x\, dx = \frac{1}{3}\sin^3 x + C$

2. $n = 5;\ u = \cos x;\ du = -\sin x\, dx$

$$\int \sin x\cos^5 x\, dx = -\int (\cos x)^5(-\sin x\, dx) = -\frac{(\cos x)^6}{6} + C = -\frac{1}{6}\cos^6 x + C$$

3. $u = 2x;\ du = 2\, dx;\ u = \cos 2x;\ du = -2\sin 2x\, dx$

$$\int \sin^3 2x\, dx = \int \sin^2 2x\sin 2x\, dx = \int (1 - \cos^2 2x)\sin 2x\, dx = \int \sin 2x\, dx - \int \cos^2 2x\sin 2x\, dx$$

$$= \frac{1}{2}\int \sin 2x(2\, dx) + \frac{1}{2}\int \cos^2 2x(-2\sin 2x\, dx) = -\frac{1}{2}\cos 2x + \frac{1}{2}\frac{\cos^3 2x}{3} + C$$

$$= -\frac{1}{2}\cos 2x + \frac{1}{6}\cos^3 2x + C$$

4. $n = 2$; $u = \sin x$; $du = \cos x\, dx$

$$\int 3\cos^3 x\, dx = 3\int \cos^2 x \cos x\, dx = 3\int (1 - \sin^2 x)\cos x\, dx = 3\int \cos x\, dx - 3\int \sin^2 x \cos x\, dx$$

$$= 3\sin x - 3\frac{\sin^3 x}{3} + C = 3\sin x - \sin^3 x + C$$

5. $\displaystyle\int 4(\cos^4 \theta - \sin^4 \theta)\, d\theta = \int 4(\cos^2 \theta + \sin^2 \theta)(\cos^2 \theta - \sin^2 \theta)\, d\theta = \int 4(1)(1 - 2\sin^2 \theta)\, d\theta$

$$= \int 4\left(1 - 2\frac{1 - \cos 2\theta}{2}\right) d\theta = \int 4(1 - 1 + \cos 2\theta)\, d\theta = \int \frac{4}{2}\cos 2\theta(2\, d\theta)$$

$$= 2\sin(2\theta) + C$$

6. $u = \cos x$; $du = -\sin x\, dx$

$$\int \sin^3 x \cos^6 x\, dx = \int \sin x \sin^2 x \cos^6 x\, dx = \int \sin x(1 - \cos^2 x)\cos^6 x\, dx = \int \cos^6 x \sin x\, dx - \int \cos^8 x \sin x\, dx$$

$$= -\int \cos^6 x(-\sin x\, dx) + \int \cos^8 x(-\sin x\, dx) = -\frac{1}{7}\cos^7 x + \frac{1}{9}\cos^9 x + C$$

7. $u = \cos x$; $du = -\sin x\, dx$

$$\int_0^{\pi/4} 5\sin^5 x\, dx = 5\int_0^{\pi/4} \sin x \sin^4 x\, dx = 5\int_0^{\pi/4} \sin x(1 - \cos^2 x)^2\, dx = 5\int_0^{\pi/4} \sin x(1 - 2\cos^2 x + \cos^4 x)\, dx$$

$$= 5\left(\int \sin x\, dx - 2\int \cos^2 x \sin x\, dx + \int \cos^4 x \sin x\, dx\right) = 5(-\cos x) + 10\frac{\cos^3 x}{3} - 5\frac{\cos^5 x}{5}\Bigg|_0^{\pi/4}$$

$$= 5\left(-\frac{1}{\sqrt{2}}\right) + \frac{10}{3}\left(\frac{1}{\sqrt{2}}\right)^3 - \left(\frac{1}{\sqrt{2}}\right)^5 - \left[5(-1) + \frac{10}{3}(1) - 1\right]$$

$$= -\frac{5}{\sqrt{2}} + \frac{10}{3}\frac{1}{2\sqrt{2}} - \frac{1}{4\sqrt{2}} - \left(-5 + \frac{10}{3} - 1\right) = \frac{-60 + 10(2) - 3}{12\sqrt{2}} - \left[\frac{-15 + 10 - 3}{3}\right]$$

$$= \frac{-43 + 32\sqrt{2}}{12\sqrt{2}} = \frac{64 - 43\sqrt{2}}{24} = 0.1329$$

8. $\displaystyle\int_{\pi/3}^{\pi/2} 10\sin t(1 - \cos 2t)^2\, dt = 40\int_{\pi/3}^{\pi/2}(1 - \cos^2 t)^2(\sin t\, dt) = -40\int_{\pi/3}^{\pi/2}(1 - 2\cos^2 t + \cos^4 t)(-\sin t\, dt)$

$$= -40\left(\cos t - \frac{2\cos^3 t}{3} + \frac{\cos^5 t}{5}\right)\Bigg|_{\pi/3}^{\pi/2} = \frac{203}{12}$$

9. $\displaystyle\int \sin^2 x\, dx = \int \left[\frac{1}{2}(1 - \cos 2x)\right] dx = \frac{1}{2}\int dx - \frac{1}{2}\int \cos 2x\, dx = \frac{1}{2}\int dx - \frac{1}{4}\int \cos 2x(2\, dx)$

$$= \frac{1}{2}x - \frac{1}{4}\sin 2x + C$$

10. $u = 4x$; $du = 4\, dx$

$$2\cos^2 x = 1 + \cos 2x; \quad \cos^2 x = \frac{1}{2}(1 + \cos 2x)$$

$$\int \cos^2 2x\, dx = \frac{1}{2}\int (1 + \cos 4x)dx = \frac{1}{2}\int dx + \frac{1}{2}\int \cos 4x\, dx = \frac{1}{2}\int dx + \frac{1}{2}\frac{1}{4}\int \cos 4x(4\, dx)$$

$$= \frac{1}{2}x + \frac{1}{8}\sin 4x + C$$

11. $\displaystyle\int 2(1+\cos 3\phi)^2 d\phi = \int 2(1+2\cos 3\phi + \cos^2 3\phi)d\phi = \int (2+4\cos 3\phi + 2\cos^2 3\phi)d\phi$

$$= 2\phi + \frac{4\sin 3\phi}{3} + \int (1+\cos 6\phi)d\phi = 2\phi + \frac{4\sin 3\phi}{3} + \phi + \frac{1}{6}\int \cos 6\phi(6\,d\phi)$$

$$= 3\phi + \frac{4\sin 3\phi}{3} + \frac{\sin 6\phi}{6} + C = \frac{1}{3}(9\phi + 4\sin 3\phi + \sin 3\phi\cos 3\phi) + C$$

12. $u = 8x;\ du = 8\,dx;\ 2\sin^2 x = 1 - \cos 2x;\ \sin^2 x = \dfrac{1}{2}(1-\cos 2x)$

$$\int_0^1 \sin^2 4x\,dx = \frac{1}{2}\int_0^1 (1-\cos 8x)dx = \frac{1}{2}\int dx - \frac{1}{2}\cdot\frac{1}{8}\int \cos 8x(8\,dx) = \frac{1}{2}x - \frac{1}{16}\sin 8x\Big|_0^1$$

$$= \frac{1}{2} - \frac{1}{16}\sin 8 = 0.4382$$

13. $\displaystyle\int \tan x^3\,dx = \int \tan^2 x\tan x\,dx = \int (\sec^2 x - 1)\tan x\,dx = \int \tan x\sec^2 x\,dx - \int \tan x\,dx$

$$= \frac{1}{2}\tan^2 x - (-\ln|\cos x|) + C = \frac{1}{2}\tan^2 x + \ln|\cos x| + C$$

14. $u = \cot y;\ du = -\csc^2 y\,dy$

$$\int \frac{6\cot^2 y}{\tan y}dy = \int 6\cot^3 y\,dy = 6\int \cot y\cot^2 y\,dy = 6\int \cot y(\csc^2 y - 1)dy$$

$$= 6\int \csc^2 y\cot y\,dy - 6\int \cot y\,dy = 6\int (\cot y)\csc^2 y\,dy - 6\int \cot y\,dy$$

$$= -6\int (\cot y)^1(-\csc^2 y\,dy) - 6\int \cot y\,dy = -6\frac{(\cot y)^2}{2} - 6\ln|\sin y| + C$$

$$= -3\cot^2 y - 6\ln|\sin y| + C$$

15. $u = \tan x;\ du = \sec^2 x\,dx$

$$\int_0^{\pi/4} \tan x\sec^4 x\,dx = \int_0^{\pi/4} \tan x\sec^2 x(1+\tan^2 x)dx = \int_0^{\pi/4} (\tan x)^1\sec x\,dx + \int_0^{\pi/4} \tan^3 x\sec^2 x\,dx$$

$$= \frac{1}{2}(\tan x)^2 + \frac{1}{4}(\tan x)^4\Big|_0^{\pi/4} = \frac{1}{2}(1)^2 + \frac{1}{4}(1)^4 = \frac{1}{2} + \frac{1}{4} = \frac{3}{4}$$

16. $n = 3;\ u = \csc 4x;\ du = -4\csc 4x\cot 4x\,dx$

$$\int \cot 4x\csc^4 4x\,dx = \int \csc^3 4x\csc 4x\cot 4x\,dx = -\frac{1}{4}\int \csc^3 4x(-4\csc 4x\cot 4x\,dx) = -\frac{1}{4}\frac{\csc^4 4x}{4}$$

$$= -\frac{1}{16}\csc^4 4x + C$$

17. $\displaystyle\int \tan^4 2x\,dx = \int (\tan^2 2x)(\tan^2 2x)\,dx = \int (\tan^2 2x)(\sec^2 2x - 1)\,dx = \int (\tan^2 2x\sec^2 2x - \tan^2 2x)\,dx$

$$= \int \tan^2 2x\sec^2 2x\,dx - \int \tan^2 2x\,dx = \int \tan^2 2x\sec^2 2x\,dx - \int (\sec^2 2x - 1)dx$$

$$= \frac{1}{2}\int (\tan^2 2x\sec^2 2x)(2\,dx) - \int \sec^2 2x\,dx + \int 1\,dx$$

$$= \frac{1}{2}\int (\tan^2 2x\sec^2 2x)(2\,dx) - \frac{1}{2}\int \sec^2 2x(2\,dx) + \int 1\,dx$$

$$= \frac{1}{2}\times\frac{1}{3}\tan^3 2x - \frac{1}{2}\tan 2x + x + C = \frac{1}{6}\tan^3 2x - \frac{1}{2}\tan 2x + x + C$$

18. $u = \cot x;\; du = -\csc^2 x\, dx$

$$\int 4\cot^4 x\, dx = 4\int \cot^4 x\, dx = 4\int \cot^2 x(\csc^2 x - 1)dx = 4\int(\cot^2 x\csc^2 x\, dx - 4\int \cot^2 x\, dx$$

$$= 4\int(\cot x)^2\csc^2 x\, dx - 4\int(\csc^2 x - 1)dx = -4\int(\cot x)^2 - \csc^2 x\, dx - 4\int \csc^2 x\, dx + 4\int dx$$

$$= -4\frac{(\cot x)^3}{3} + 4\cot x + 4x + C = -\frac{4}{3}\cot^3 x + 4\cot x + 4x + C$$

19. $\displaystyle\int 0.5\sin s\sin 2s\, ds = \int 0.5\sin s\cdot 2\sin s\cos s\, ds = \int \sin^2 s\cos s\, ds = \frac{\sin^3 s}{3} + C$

20. $u = \tan x;\; du = \sec^2 x\, dx$

$$\int \sqrt{\tan x}\sec^4 x\, dx = \int(\tan x)^{1/2}\sec^2 x(1 + \tan^2 x)dx = \int(\tan x)^{1/2}\sec^2 x\, dx + \int(\tan x)^{5/2}\sec^2 x\, dx$$

$$= \frac{2}{3}(\tan x)^{3/2} + \frac{2}{7}(\tan x)^{7/2} + C = 2(\tan x)^{1/2}\left(\frac{1}{3}\tan x + \frac{1}{7}\tan^3 x\right) + C$$

21. $\displaystyle\int(\sin x + \cos x)^2 dx = \int(\sin^2 x + \cos^2 x)\, dx + \int 2\sin x\cos x\, dx = \int 1\, dx + \int \sin 2x\, dx$

$$= \int 1\, dx + \frac{1}{2}\int \sin 2x\,(2\, dx) = x - \frac{1}{2}\cos 2x + C$$

22. $\displaystyle\int(\tan 2x + \cot 2x)^2 dx = \int(\tan^2 2x + 2\tan 2x\cot 2x + \cot^2 2x)dx = \int \tan^2 2x\, dx + 2\int dx + \int \cot^2 2x\, dx$

$$= \int(\sec^2 2x - 1)dx + 2\int dx + \int(\csc^2 2x - 1)dx$$

$$= \int \sec^2 2x\, dx - \int dx + 2\int dx + \int \csc^2 2x\, dx - \int dx$$

$$= \frac{1}{2}\int \sec^2 2x(2\, dx) + \frac{1}{2}\int \csc^2 2x(2\, dx) = \frac{1}{2}\tan 2x - \frac{1}{2}\cot 2x + C$$

$$= \frac{1}{2}(\tan 2x - \cot 2x) + C$$

23. $\displaystyle\int \frac{1 - \cot x}{\sin^4 x}dx$

$$= \int(1 - \cot x)\csc^4 x\, dx = \int(1 - \cot x)\csc^2 x(1 + \cot^2 x)dx = \int(1 - \cot x + \cot^2 x - \cot^3 x\csc^2 x\, dx)$$

$$= \int \csc^2 x\, dx - \int \cot x\csc^2 x\, dx + \int \cot^2 x\csc^2 x\, dx - \int \cot^3 x\csc^2 x\, dx$$

$$= \int \csc^2 x\, dx + \int \cot x(-\csc^2 x\, dx) - \int \cot^2 x(-\csc^2 x\, dx) + \int \cot^3 x(-\csc^2 x\, dx)$$

$$= -\cot x + \frac{\cot^2 x}{2} - \frac{\cot^3 x}{3} + \frac{\cot^4 x}{4} + C = \frac{1}{4}\cot^4 x - \frac{1}{3}\cot^3 x + \frac{1}{2}\cot^2 x - \cot x + C$$

24. $\displaystyle\int \frac{(\sin u + \sin^2 u)^2}{\sec u}du = \int(\sin^2 u + 2\sin^3 u + \sin^4 u)\cos u\, du = \frac{\sin^3 u}{3} + \frac{\sin^4 u}{2} + \frac{\sin^5 u}{5} + C$

25. $\displaystyle\int_{\pi/6}^{\pi/4} \cot^5 x\,dx = \int_{\pi/6}^{\pi/4} \cot^3 x(\csc^2 x - 1)\,dx = \int_{\pi/6}^{\pi/4} \cot^3 x \csc^2 x\,dx - \int_{\pi/6}^{\pi/4} \cot^3 x\,dx$

$$= -\int_{\pi/6}^{\pi/4} \cot^3 x(-\csc^2 x\,dx) - \int_{\pi/6}^{\pi/4} \cot x(\csc^2 x - 1)\,dx$$

$$= -\frac{1}{4}\cot^4 x \Big|_{\pi/6}^{\pi/4} - \int_{\pi/6}^{\pi/4} \cot x \csc^2 x\,dx + \int_{\pi/6}^{\pi/4} \cot x\,dx$$

$$= -\frac{1}{4}\cot^4 x + \frac{1}{2}\cot^2 x + \ln|\sin x|\Big|_{\pi/6}^{\pi/4}$$

$$= -\frac{1}{4}\cot^4 \frac{\pi}{4} + \frac{1}{2}\cot^2 \frac{\pi}{4} + \ln\left|\sin\frac{\pi}{4}\right| - \left(-\frac{1}{4}\cot^4 \frac{\pi}{6} + \frac{1}{2}\cot^2 \frac{\pi}{6} + \ln\left|\sin\frac{\pi}{6}\right|\right)$$

$$= -\frac{1}{4} + \frac{1}{2} + \ln\frac{1}{2}\sqrt{2} + \frac{1}{4}(\sqrt{3})^4 - \frac{1}{2}(\sqrt{3})^2 + \ln\frac{1}{2} = 1.347$$

26. $\displaystyle\int_{\pi/6}^{\pi/3} \frac{2\,dx}{1 + \sin x} = 2\int \frac{(1 - \sin x)}{(1 - \sin^2 x)}\,dx = 2\int \frac{(1 - \sin x)\,dx}{\cos^2 x} = 2\int \sec^2 x\,dx - 2\int (\cos x)^{-2}\sin x\,dx$

$$= 2\tan x + 2\frac{(\cos x)^{-1}}{-1}\Big|_{\pi/6}^{\pi/3} = 2(\tan x - \sec x)\Big|_{\pi/6}^{\pi/3} = 2(\sqrt{3} - 2) - 2\left(\frac{1}{\sqrt{3}} - \frac{2}{\sqrt{3}}\right) = 0.6188$$

27. $\displaystyle\int \sec^6 x\,dx = \int \sec^4 x(1 + \tan^2 x)\,dx = \int \sec^4 x\,dx + \int \sec^4 x \tan^2 x\,dx$

$$= \int \sec^2 x(1 + \tan^2 x)\,dx + \int \sec^2 x(1 + \tan^2 x)x\,dx$$

$$= \int \sec^2 x\,dx + \int \tan^2 x \sec^2 x\,dx + \int \tan^2 x \sec^2 x\,dx + \int \tan^4 x \sec^2 x\,dx$$

$$= \tan x + \frac{1}{3}\tan^3 x + \frac{1}{3}\tan^3 x + \frac{1}{5}\tan^5 + C = \frac{1}{5}\tan^5 x + \frac{2}{3}\tan^3 x + \tan x + C$$

28. $\displaystyle\int \tan^7 x\,dx$

$$= \int \tan^5 x(\sec^2 x - 1)\,dx = \int \tan^5 x \sec^2 x\,dx - \int \tan^5 x\,dx = \int \tan^5 x \sec^2 x\,dx - \int \tan^3 x(\sec^2 x - 1)\,dx$$

$$= \int \tan^5 x \sec^2 x\,dx - \int \tan^3 x \sec^2 x\,dx + \int \tan^3 x\,dx$$

$$= \int \tan^5 x \sec^2 x\,dx - \int \tan^3 x \sec^2 x\,dx + \int \tan x(\sec^2 x - 1)\,dx$$

$$= \int \tan^5 x \sec^2 x\,dx - \int \tan^3 x \sec^2 x\,dx + \int \tan x \sec^2 x\,dx - \int \tan x\,dx$$

$$= \frac{1}{6}\tan^6 x - \frac{1}{4}\tan^4 x + \frac{1}{2}\tan^2 x + \ln|\cos x| + C$$

29. Rotate about x-axis, disks

$$V = \pi\int_0^\pi y^2\,dx = \pi\int_0^\pi \sin^2 x\,dx = \pi\int_0^\pi \frac{1}{2}(1 - \cos 2x)\,dx = \frac{\pi}{2}\int_0^\pi dx - \frac{\pi}{2}\int_0^\pi \cos 2x\,dx$$

$$= \frac{\pi}{2}x - \frac{\pi}{2} \times \frac{1}{2}\sin 2x\Big|_0^\pi = \frac{\pi^2}{2} - \frac{\pi}{4}\sin 2\pi - 0 + \frac{\pi}{4}\sin 0 = \frac{1}{2}\pi^2 = 4.935$$

30. $y = \tan^3(x^2)$; $y = 0, x = \dfrac{\pi}{4}$

Rotated about y-axis, shells.

$dV = 2\pi rh\, dx$; $r = x, h = y$

$$V = 2\pi \int_0^{\pi/4} xy\, dx = 2\pi \int_0^{\pi/4} x\tan^3(x^2)dx = 2\pi \int_0^{\pi/4} \tan^3(x^2)(x\, dx) = 2\pi \int_0^{\pi/4} (\sec^2(x^2) - 1)\tan(x^2)x\, dx$$

$$= 2\pi \left[\int_0^{\pi/4} \tan(x^2)\sec^2(x^2)x\, dx - \int_0^{\pi/4} \tan(x^2)x\, dx \right] = 2\pi \left[\frac{1}{2}\int_0^{\pi/4} \tan(x^2)\sec^2(x^2)2x\, dx - \frac{1}{2}\int_0^{\pi/4} \tan(x^2)2x\, dx \right]$$

$$= 2\pi \left[\frac{1}{2}\cdot\frac{1}{2}\tan^2(x^2) + \frac{1}{2}\ln\left|\cos(x^2)\right| \right]\Bigg|_0^{\pi/4} = 2\pi \left[\frac{1}{4}\tan^2(x^2) + \frac{1}{2}\ln\left|\cos(x^2)\right| \right]\Bigg|_0^{\pi/4}$$

$V = 2\pi[0.12573 - 0.10185] = 0.150$ units^3

31. $y = \sin x$; $y = \cos x, x = 0$; Quad I

$\sin x = \cos x$; $\tan x = 1$; $x = \dfrac{\pi}{4}$

$$A_{0,\pi4} = \int_0^{\pi/4} (y_1 - y_2)dx = \int_0^{\pi/4} (\cos x - \sin x)dx; \quad A = \int_0^{\pi/4} \cos x\, dx - \int_0^{\pi/4} \sin x\, dx = \sin x + \cos x\Bigg|_0^{\pi/4} = 0.414$$

32. $y = \ln\cos x$; $x = 0$; $x = \dfrac{\pi}{3}$; $\dfrac{dy}{dx} = \dfrac{1}{\cos x} - \sin x = -\tan x$

$$s = \int \sqrt{1 + \left(\frac{dy}{dx}\right)^2}\, dx = \int_0^{\pi/3} \sqrt{1 + \tan^2 x}\, dx = \int_0^{\pi/3} \sec x\, dx = \ln\left|\sec x + \tan x\right|\Bigg|_0^{\pi/3}$$

$$= \ln\left|\sec\frac{\pi}{3} + \tan\frac{\pi}{3}\right| - \ln|1|$$

$s = \ln\left|2 + \sqrt{3}\right| = 1.317$

33. $\displaystyle\int \sin x\cos x\, dx$; $u = \sin x$; $du = \cos x\, dx$

$$\int u\, du = \frac{1}{2}u^2 + C = \frac{1}{2}\sin^2 x + C_1$$

Let $u = \cos x$; $du = -\sin x\, dx$

$$-\int \cos x(-\sin x)\, dx = -\frac{1}{2}\cos^2 x + C_2$$

$$\frac{1}{2}\sin^2 x + C_1 = \frac{1}{2}(1 - \cos^2 x + C_1) = \frac{1}{2} - \frac{1}{2}\cos^2 x + C_1$$

$$-\frac{1}{2}\cos^2 x + C_2 = \frac{1}{2} - \frac{1}{2}\cos^2 x + C_1$$

$$C_2 = C_1 + \frac{1}{2}$$

34. $\displaystyle\int \sec^2 x\tan x\, dx$

(a) $\displaystyle\int (\sec x)^1 \sec x\tan x\, dx = \frac{1}{2}\sec^2 x + C_1$ **(b)** $\displaystyle\int (\tan x)^1 \sec^2 x\, dx = \frac{1}{2}\tan^2 x + C_2$

Identity: $1 + \tan^2 x = \sec^2 x$

Substitute in (a): $\dfrac{1}{2}\sec^2 x + C_1 = \dfrac{1}{2}(1 + \tan^2 x) + C_1 = \dfrac{1}{2} + \dfrac{1}{2}\tan^2 x + C_1 = \dfrac{1}{2}\tan^2 x + C_3$

$C_2 = C_3 = \dfrac{1}{2} + C_1$

35. $\displaystyle\int_0^\pi \sin^3\theta\, d\theta = \int_0^\pi \sin\theta(1-\cos^2\theta)d\theta = \int_0^\pi \sin\theta\, d\theta - \int_0^\pi \cos^2\theta\sin\theta\, d\theta = \int_0^\pi \sin\theta\, d\theta + \int_0^\pi \cos^2\theta(-\sin\theta\, d\theta)$

$$= -\cos\theta + \frac{1}{3}\cos^3\theta\Big|_0^\pi = -\cos\pi + \frac{1}{3}\cos^3\pi - \left(-\cos 0 + \frac{1}{3}\cos^3 0\right)$$

$$= -(-1) + \frac{1}{3}(-1) - \left(-1 + \frac{1}{3}\right) = 1 - \frac{1}{3} + 1 - \frac{1}{3} = \frac{4}{3}$$

36. $\displaystyle\int \frac{\sin^2\theta}{\cos^2\theta}d\theta = \int \frac{1-\cos^2\theta}{\cos^2\theta}d\theta = \int(\sec^2\theta\, d\theta - 1)d\theta = \int \sec^2\theta\, d\theta - \int d\theta = \tan\theta - \theta + C$

37. $\displaystyle V_{rms} = \sqrt{\frac{1}{1/60.0}\int_0^{1/60.0}(340\sin 120\pi t)^2 dt} = \sqrt{60}\sqrt{\int_0^{1/60.0}340^2\frac{1-\cos 240\pi t}{2}dt}$

$$= \sqrt{60}\sqrt{\frac{340^2}{2}\left(t - \frac{1}{240\pi}\sin 240\pi t\right)\Big|_0^{1/60.0}} = 240 \text{ V}$$

38. $\displaystyle y_{rms} = \sqrt{\frac{1}{T}\int_0^T y^2 dx}; \quad i = i_0\sin\omega t; \quad T = \frac{2\pi}{\omega}$

$$i_{rms} = \sqrt{\frac{1}{\frac{2\pi}{\omega}}\int_0^{2\pi/\omega}i_0^2\sin^2\omega t\, dt} \quad \cos 2x = 1 - 2\sin^2 x; \quad \sin^2 x = \frac{1}{2}(1-\cos 2x)$$

$$i_{rms} = \sqrt{\frac{wi_0^2}{2\pi}\int_0^{2\pi/\omega}\frac{1}{2}(1-\cos 2\omega t)dt} = \sqrt{\frac{wi_0^2}{4\pi}\left(t - \frac{1}{2\omega}\sin 2\omega t\right)\Big|_0^{2\pi/\omega}} = \sqrt{\frac{wi_0^2}{4\pi}\left(\frac{2\pi}{\omega}\right)} = \sqrt{\frac{i_0^2}{2}} = \frac{i_0}{\sqrt{2}}$$

39. $\displaystyle I = A\int_{-a/2}^{a/2}\cos^2[b\pi(c-x)]dx$

$$\cos 2x = 2\cos^2 x - 1; \quad \cos^2 x = \frac{1}{2}(\cos 2x + 1)$$

$$I = A\int_{-a/2}^{a/2}\frac{1}{2}[\cos 2b\pi(c-x) + 1]dx; \quad I = \frac{A}{2}\int_{-a/2}^{a/2}\cos 2b\pi(c-x)dx + \frac{A}{2}\int_{-a/2}^{a/2}dx$$

$$u = 2b\pi(c-x); \quad du = 2b\pi(-1)dx = -2b\, dx$$

$$I = A\cdot\frac{-1}{2b\pi}\int_{-a/2}^{a/2}\cos 2b\pi(c-x)(-2b\pi)dx + \frac{A}{2}\int_{-a/2}^{a/2}dx = \frac{-A}{4b\pi}\sin 2b\pi(c-x) + \frac{A}{2}x\Big|_{-a/2}^{a/2}$$

$$= \frac{-A}{4b\pi}\sin 2b\pi\left(c - \frac{a}{2}\right) + \frac{A}{2}\cdot\frac{a}{2} - \left[-\frac{A}{4b\pi}\sin 2b\pi\left(c + \frac{a}{2}\right) - \frac{A}{2}\cdot\frac{a}{2}\right]$$

$$= \frac{-A}{4b\pi}\sin 2b\pi\left(c - \frac{a}{2}\right) + \frac{Aa}{4} + \frac{A}{4b\pi}\sin 2b\pi\left(c + \frac{a}{2}\right) + \frac{Aa}{4}$$

$$I = \frac{Aa}{2} + \frac{A}{4b\pi}\left[\sin 2b\pi\left(c + \frac{a}{2}\right) - \sin 2b\pi\left(c - \frac{a}{2}\right)\right]$$

$$\sin(2b\pi c + b\pi a) - \sin(2b\pi c - b\pi a)$$
$$\sin(2b\pi c + b\pi a) = \sin 2b\pi c\cos b\pi a + \cos 2b\pi c\sin b\pi a$$
$$\sin(2b\pi c - b\pi a) = \sin 2b\pi c\cos b\pi a - \cos 2b\pi c\sin b\pi a$$
$$\sin(2b\pi c + b\pi a) - \sin(2b\pi c - b\pi a) = 2\cos 2b\pi c\sin b\pi a$$

Therefore, $\displaystyle I = \frac{Aa}{2} + \frac{A}{4b\pi}(2\cos 2b\pi c\sin b\pi a)$

$$I = \frac{Aa}{2} + \frac{A}{2b\pi}\sin ab\pi\cos 2bc\pi$$

40. $L = k \int_0^{2\pi} (a \sin \theta + b \sin^2 \theta - b \sin^3 \theta) d\theta = k \left[a \int \sin \theta \, d\theta + b \int \frac{1}{2}(1 - \cos 2\theta) d\theta - b \int \sin \theta (1 - \cos^2 \theta) d\theta \right]$

$= k \left[a \int \sin \theta \, d\theta + \frac{b}{2} \int d\theta - \frac{b}{2} \int \cos 2\theta \, d\theta - b \int \sin \theta \, d\theta + b \int \cos^2 \theta \sin \theta \, d\theta \right]$

$= k \left[a \int \sin \theta \, d\theta + \frac{b}{2} \int d\theta - \frac{b}{4} \int \cos 2\theta (2 \, d\theta) - b \int \sin \theta \, d\theta - b \int \cos^2 \theta (- \sin \theta \, d\theta) \right]$

$= k \left[-a \cos \theta + \frac{b}{2}\theta - \frac{b}{4} \sin 2\theta + b \cos \theta - \frac{b}{3} \cos^3 \theta \right] \Big|_0^{2\pi}$

$= k \left[-a \cos 2\pi + \frac{b}{2}(2\pi) - \frac{b}{4} \sin 4\pi + b \cos 2\pi - \frac{b}{3} \cos^3 2\pi - (-a \cos 0 + \frac{b}{2} 0 - \frac{b}{4} \sin 0 + b \cos 0 - \frac{b}{3} \cos^3 0) \right]$

$= k \left(-a + b\pi - 0 + b - \frac{b}{3} + a - 0 + 0 - b + \frac{b}{3} \right)$

$L = k(b\pi) = kb\pi$

9.6   Inverse Trigonometric Forms

1. $a = 2; u = x;$

$\int \frac{dx}{\sqrt{4 - x^2}} = \int \frac{dx}{\sqrt{2^2 - x^2}} = \sin^{-1} \frac{x}{2} + C$

2. $a = 7; u = x; du = dx;$

$\int \frac{dx}{\sqrt{49 - x^2}} = \sin^{-1} \frac{x}{7} + C$

3. $a = 8; u = x; du = dx;$ $\int \frac{dx}{64 + x^2} = \frac{1}{8} \tan^{-1} \frac{x}{8} + C$

4. $\int \frac{6p^2 \, dp}{4 + p^6} = \int \frac{2 \cdot 3p^2 \, dp}{4 + (p^3)^2},$ let $x = p^3, dx = 3p^2 \, dp$

$= \int \frac{2 \, dx}{4 + x^2} = 2 \cdot \frac{1}{2} \tan^{-1} \frac{x}{2} + C$

$= \tan^{-1} \frac{p^3}{2} + C$

5. $a = 1; u = 4x; du = 4 \, dx$

$\int \frac{dx}{\sqrt{1 - (4x)^2}} = \frac{1}{4} \int \frac{4 \, dx}{\sqrt{1 - (4x)^2}}$

$= \frac{1}{4} \sin^{-1} \frac{4x}{1} + C$

$= \frac{1}{4} \sin^{-1} 4x + C$

6. $a = 3; u = 2x; du = 2 \, dx;$ $\int_0^1 \frac{2 \, dx}{\sqrt{9 - 4x^2}} = \sin^{-1} \frac{2x}{3} \Big|_0^1 = \sin^{-1} \frac{2}{3} = 0.7297$

7. $\int_0^2 \frac{3e^{-t} dt}{1 + 9e^{-2t}}$

$= -\int_0^2 \frac{-3e^{-t} dt}{1 + (3e^{-t})^2}$ which has form $\int \frac{dx}{1 + x^2}$

$= -\tan^{-1}(3e^{-t}) \Big|_0^2$

$= -\tan^{-1} \frac{3}{e^2} + \tan^{-1} 3$

$= 0.8634$

8. $a = 7; u = 2x; du = 2dx$

$\int_1^3 \frac{dx}{49 + 4x^2} = \frac{1}{2} \int_1^3 \frac{2 \, dx}{49 + 4x^2}$

$= \frac{1}{2} \cdot \frac{1}{7} \tan^{-1} \frac{2x}{7} = \frac{1}{14} \tan^{-1} \frac{2x}{7} \Big|_1^3$

$= \frac{1}{14} \left(\tan^{-1} \frac{6}{7} - \tan^{-1} \frac{2}{7} \right)$

$= 0.0307$

9. $\dfrac{2}{\sqrt{5}} \displaystyle\int_0^{0.4} \dfrac{\sqrt{5}\,dx}{\sqrt{4-5x^2}} = \dfrac{2}{\sqrt{5}} \left(\sin^{-1} \dfrac{\sqrt{5}x}{2} \right) \Big|_0^{0.4}$

$\qquad\qquad\qquad = \dfrac{2\sqrt{5}}{5} [\sin^{-1}(0.2\sqrt{5}) - \sin^{-1} 0]$

$\qquad\qquad\qquad = 0.415$

10. $a = 1;\ u = \sqrt{x};\ du = \dfrac{dx}{2\sqrt{x}}$

$\quad \displaystyle\int \dfrac{dx}{2\sqrt{x}\sqrt{1-x}} = \int \dfrac{\dfrac{dx}{2\sqrt{x}}}{\sqrt{1-(\sqrt{x})^2}} = \sin^{-1}\sqrt{x} + C$

11. $u = 9x^2 + 16;\ du = 18x\,dx$

$\quad \displaystyle\int \dfrac{8x\,dx}{9x^2 + 16} = \dfrac{8}{18} \int \dfrac{18x\,dx}{9x^2 + 16}$

$\qquad\qquad\qquad = \dfrac{4}{9} \ln |9x^2 + 16| + C$

12. $\displaystyle\int \dfrac{4y\,dy}{\sqrt{25-16y^2}} = \int (25 - 16y^2)^{-1/2}(4y\,dy)$

$\qquad\qquad\qquad = -\dfrac{1}{8} \int (25 - 16y^2)^{-1/2}(-32y\,dy)$

$\qquad\qquad\qquad = -\dfrac{1}{8} \cdot 2(25 - 16y^2)^{1/2} + C$

$\qquad\qquad\qquad = -\dfrac{\sqrt{25-16y^2}}{4} + C$

13. $\displaystyle\int_1^e \dfrac{3\,du}{u[1 + (\ln u)^2]} = 3 \int_1^e \dfrac{1}{1 + (\ln u)^2} \cdot \dfrac{du}{u} =$

$\qquad\qquad\qquad = 3 \cdot \tan^{-1}(\ln u)\big|_1^e$

$\qquad\qquad\qquad = 3[\tan^{-1}(\ln e) - \tan^{-1}(\ln 1)]$

$\qquad\qquad\qquad = 3 \cdot [\tan^{-1} 1 - \tan^{-1} 0]$

$\qquad\qquad\qquad = 3 \cdot \left[\dfrac{\pi}{4} - 0 \right] = \dfrac{3\pi}{4} = 2.356$

14. $a = 1;\ u = x^2;\ du = 2x\,dx$

$\quad \displaystyle\int_0^1 \dfrac{4x\,dx}{1 + x^4} = \dfrac{4}{2} \int_0^1 \dfrac{2x\,dx}{1 + x^4}$

$\qquad\qquad\qquad = 2 \cdot \dfrac{1}{1} \tan^{-1} x^2$

$\qquad\qquad\qquad = 2 \tan^{-1} x^2 \big|_0^1 = 2(\tan^{-1} 1)$

$\qquad\qquad\qquad = 1.5708$

15. $a = 1;\ u = e^x;\ du = e^x dx$

$\quad \displaystyle\int \dfrac{e^x\,dx}{\sqrt{1 - e^{2x}}} = \sin^{-1} e^x + C$

16. $a = 1;\ u = \tan x;\ du = \sec^2 x\,dx$

$\quad \displaystyle\int \dfrac{\sec^2 x\,dx}{\sqrt{1 - \tan^2 x}} = \sin^{-1}(\tan x) + C$

17. $a = 1;\ u = x + 1;\ du = dx$

$\quad \displaystyle\int \dfrac{dx}{x^2 + 2x + 2} = \int \dfrac{dx}{(x^2 + 2x + 1) + 1}$

$\qquad\qquad\qquad = \int \dfrac{dx}{(x+1)^2 + 1^2}$

$\qquad\qquad\qquad = \dfrac{1}{1} \tan^{-1} \dfrac{(x+1)}{1} + C$

$\qquad\qquad\qquad = \tan^{-1}(x+1) + C$

18. $a = 1;\ u = x + 4;\ du = dx$

$\quad \displaystyle\int \dfrac{2\,dx}{x^2 + 8x + 17} = \int \dfrac{2\,dx}{(x+4)^2 + 1}$

$\quad 2 \displaystyle\int \dfrac{dx}{(x+4)^2 + 1} = 2 \cdot \dfrac{1}{1} \tan^{-1}(x+4) + C$

$\qquad\qquad\qquad = 2 \tan^{-1}(x+4) + C$

19. $a = 2;\ u = x + 2;\ du = dx$

$\quad \displaystyle\int \dfrac{4\,dx}{\sqrt{-4x - x^2}} = \int \dfrac{4\,dx}{\sqrt{4 - (x+2)^2}}$

$\qquad\qquad\qquad = 4 \displaystyle\int \dfrac{dx}{\sqrt{4 - (x+2)^2}}$

$\qquad\qquad\qquad = 4 \sin^{-1} \left(\dfrac{x+2}{2} \right) + C$

20. $a = 1;\ u = s - 1;\ du = ds$

$\quad \displaystyle\int \dfrac{0.3\,ds}{\sqrt{2s - s^2}} = \int \dfrac{0.3\,ds}{\sqrt{1 - (s-1)^2}}$

$\qquad\qquad\qquad = 0.3 \sin^{-1}(s - 1) + C$

21. $a = 1;\ u = \sin 2\theta;\ du = \cos 2\theta(2\,d\theta)$

$\quad \displaystyle\int_{\pi/6}^{\pi/2} \dfrac{2\cos 2\theta}{1 + \sin^2 2\theta}\,d\theta = \int_{\pi/6}^{\pi/2} \dfrac{\cos 2\theta\,(2d\theta)}{1 + \sin^2 2\theta}$

$\qquad\qquad\qquad = \tan^{-1} \dfrac{\sin 2\theta}{1} \Big|_{\pi/6}^{\pi/2}$

$\qquad\qquad\qquad = (\tan^{-1} \sin \pi - \tan^{-1} \sin \dfrac{\pi}{3})$

$\qquad\qquad\qquad = -0.714$

22. $a = 1;\ u = x + 2;\ du = dx$

$\quad \displaystyle\int_{-4}^0 \dfrac{dx}{x^2 + 4x + 5} = \int_{-4}^0 \dfrac{dx}{(x+2)^2 + 1}$

$\qquad\qquad\qquad = \dfrac{1}{1} \tan^{-1}(x+2) \Big|_{-4}^0$

$\qquad\qquad\qquad = \tan^{-1} 2 - \tan^{-1}(-2)$

$\qquad\qquad\qquad = 2.214$

23. $a = 2;\ u = x;\ du = dx;\ n = -\dfrac{1}{2};\ u = 4 - x^2;$

$du = -2x\,dx$

$$\int \frac{2-x}{\sqrt{4-x^2}}\,dx = \int \frac{2\,dx}{\sqrt{4-x^2}} - \int \frac{x\,dx}{\sqrt{4-x^2}}$$

$$= 2\int \frac{dx}{\sqrt{4-x^2}} + \frac{1}{2}\int \frac{(-2x\,dx)}{(4-x^2)^{1/2}}$$

$$= 2\sin^{-1}\frac{x}{2} + \frac{1}{2}(4-x^2)^{1/2}\cdot 2 + C$$

$$= 2\sin^{-1}\frac{x}{2} + \sqrt{4-x^2} + C$$

24. $a = 1;\ u = 2x;\ du = 2\,dx;\ u = 1 + 4x^2;\ du = 8x\,dx$

$$\int \frac{3-2x}{1+4x^2}\,dx = \int \frac{3\,dx}{1+4x^2} - \int \frac{2x\,dx}{1+4x^2}$$

$$= \frac{3}{2}\int \frac{2\,dx}{1+4x^2} - \frac{1}{4}\int \frac{8x\,dx}{1+4x^2}$$

$$= \frac{3}{2}\tan^{-1}2x - \frac{1}{4}\ln\left|1+4x^2\right| + C$$

25. (a) Inverse tangent, $\int \frac{du}{a^2+u^2}$ where $u = 3x$,
$du = 3\,dx,\ a = 2$; numerator cannot fit
du of denominator.
Positive $9x^2$ leads to inverse tangent form.

(b) $\displaystyle\int \frac{2\,dx}{4+9x} = 2\int \frac{dx}{4+9x};\ u = 4 + 9x;\ du = 9\,dx$

Therefore, the integral is logarithmic.

(c) $\displaystyle\int \frac{2x\,dx}{\sqrt{4+9x^2}} = 2\int (4+9x^2)^{-1/2}(x\,dx);$

$u = 4 + 9x^2;\ du = 18x\,dx$

Therefore, the form of the integral is general
power.

26. (a) Logarithmic, $\dfrac{du}{u}$ where $u = 4 - 9x^2$;

$du = -18x\,dx.$

Negative $9x^2$ shows it is not inverse tangent
form.

(b) General power; $\displaystyle\int u^n\,du;\ n = -\frac{1}{2};\ u = 4 - 9x;$

$du = -9\,dx$

(c) Logarithmic; $u = 4 + 9x^2;\ du = 18x\,dx$
$2x\,dx$ becomes $18x\,dx$, multiply by 9 and $\frac{1}{9}$.

27. (a) General power, $\displaystyle\int u^{-1/2}\,du$ where $u = 4 - 9x^2$.

$du = -18x\,dx$; numerator can fit du of
denominator.
Square root becomes $-1/2$ power.
Does not fit inverse sine form.

(b) Inverse sine; $a = 2;\ u = 3x;\ du = 3\,dx$

(c) Logarithmic; $u = 4 - 9x;\ du = -9\,dx$

28. (a) Inverse tangent $\displaystyle\int \frac{du}{a^2+u^2}$ where $u = 3x$,

$du = 3\,dx,\ a = 2$; numerator cannot fit
du of denominator.
Positive $9x^2$ leads to inverse tangent form.

(b) General power; $\displaystyle\int u^n\,du;\ n = -\frac{1}{2}$

$u = 9x^2 - 4;\ du = 18x\,dx$

(c) Logarithmic; $u = 9x^2 - 4;\ du = 18x\,dx$

29. $y = \dfrac{1}{1+x^2};\ A = \displaystyle\int_0^2 \frac{1}{1+x^2}\,dx;$

$a\ = 1;\ u = x;\ du = dx$

$$A = \frac{1}{1}\tan^{-1}\frac{x}{1}\bigg|_0^2 = \tan^{-1}2 - \tan^{-1}0 = 1.11$$

30. $y\sqrt{4-x^2} = 1;\ x = 0, x = 1, y = 0$

$$y = \frac{1}{\sqrt{4-x^2}}$$

$$A_{0,1} = \int_0^1 \frac{dx}{\sqrt{4-x^2}}$$

$a = 2;\ u = x;\ du = dx$

$$A = \sin^{-1}\frac{x}{2}\bigg|_0^1 = \sin^{-1}\frac{1}{2} = \frac{\pi}{6}$$

31. $a = d;\ u = x;\ du = dx$

$$kd\int \frac{dx}{d^2+x^2} = kd\cdot\frac{1}{d}\tan^{-1}\frac{x}{d} + C$$

$$= k\tan^{-1}\left(\frac{x}{d}\right) + C$$

32. $y = \dfrac{24}{\sqrt{16 + x^2}}$; $x = 0$; $x = 3$; $y = 0$; rotated about

x-axis; disks.

$dV = \pi r^2 dx$; $r = y$

$V = \pi \displaystyle\int_0^3 y^2 dx = \pi \int_0^3 \dfrac{2y^2}{16 + x^2} dx$

$a = 4$; $u = x$; $du = dx$

$V = 24^2 \pi \dfrac{1}{4} \tan^{-1} \dfrac{x}{4}\Big|_0^3 = 144\pi \tan^{-1} \dfrac{x}{4}\Big|_0^3$

$\qquad = 144\pi \tan^{-1} \dfrac{3}{4}$

$V = 291$ m^3

33. $\displaystyle\int \dfrac{dx}{\sqrt{A^2 - x^2}} = \int \sqrt{\dfrac{k}{m}} dt$; $\sin^{-1} \dfrac{x}{A} = \sqrt{\dfrac{k}{m}} t + C$

Solve for C by letting $x = x_0$ and $t = 0$.

$\sin^{-1} \dfrac{x_0}{A} = \sqrt{\dfrac{k}{m}}(0) + C$; $C = \sin^{-1} \dfrac{x_0}{A}$;

therefore, $\sin^{-1} \dfrac{x}{A} = \sqrt{\dfrac{k}{m}} t + \sin^{-1} \dfrac{x_0}{A}$

34. $v = 2t - \dfrac{12}{2 + t^2}$; $s = 0, t = 0$

$\dfrac{ds}{dt} = 2t - \dfrac{12}{2 + t^2}$; $s = \displaystyle\int \left(2t - \dfrac{12}{2 + t^2} \right) dt$

$s = 2 \displaystyle\int t\, dt - 12 \int \dfrac{dt}{2 + t^2}$;

$a = \sqrt{2}$; $u = t$; $du = dt$

$s = t^2 - 12 \dfrac{1}{\sqrt{2}} \tan^{-1} \dfrac{t}{\sqrt{2}} + C$

$\quad = t^2 - 6\sqrt{2} \tan^{-1} \left(\dfrac{\sqrt{2}}{2} t \right) + C$

$0 = 0 - 6\sqrt{2} \tan^{-1} 0 + C$; therefore, $C = 0$

$s = t^2 - 8.49 \tan^{-1} 0.707t$

35. $y = \dfrac{1}{1 + x^6}$; x-axis, $x = 1$; $x = 2$, WRT y-axis

$I_y = k \displaystyle\int_1^2 x^2 y\, dx = k \int_1^2 x^2 \dfrac{1}{1 + x^6} dx = k \int_1^2 \dfrac{x^2 dx}{1 + x^6}$

$a = 1$; $u = x^3$; $du = 3x^2 dx$

$I_y = \dfrac{k}{3} \displaystyle\int_1^2 \dfrac{3x^2 dx}{1 + x^6} = \dfrac{k}{3} \cdot \dfrac{1}{1} \tan^{-1} x^3 \Big|_1^2$

$\quad = \dfrac{k}{3} \tan^{-1} x^3 \Big|_1^2$

$I_y = \dfrac{k}{3}(\tan^{-1} 8 - \tan^{-1} 1) = 0.22k$

36. $y = \sqrt{1 - x^2}$; $x = 0$; $x = 1$; $y^2 = 1 - x^2$; $y^2 + x^2 = 1$

A circle, center at origin with $r = 1$.

$y = (1 - x^2)^{1/2}$;

$\dfrac{dy}{dx} = \dfrac{1}{2}(1 - x^2)^{-1/2}(-2x)$

$\qquad = \dfrac{-x}{\sqrt{1 - x^2}}$

$s = \displaystyle\int_0^1 \sqrt{1 + \dfrac{x^2}{1 - x^2}}\, dx$

$\quad = \displaystyle\int_0^1 \sqrt{\dfrac{1 - x^2 + x^2}{1 - x^2}}\, dx$

$\quad = \displaystyle\int_0^1 \dfrac{dx}{\sqrt{1 - x^2}}$

$s = \sin^{-1} x \Big|_0^1$

$s = \sin^{-1} 1 = \dfrac{\pi}{2} = $ quarter circle circumference

Circumference of circle $= 2\pi r$

Chapter 9 Review Exercises

1. $u = -2x$, $du = -2\, dx$

$\displaystyle\int e^{-2x} dx = -\dfrac{1}{2} \int e^{-2x}(-2\, dx) = -\dfrac{1}{2} e^{-2x} + C$

2. $\displaystyle\int e^{\cos 2x} \sin x \cos x\, dx = \int e^u \left(-\dfrac{1}{4} du \right)$

$\qquad\qquad\qquad\qquad = -\dfrac{1}{4} e^u + C$

$\qquad\qquad\qquad\qquad = -\dfrac{1}{4} e^{\cos 2x} + C$

$u = \cos 2x$

$du = -2 \sin 2x\, dx$

$du = -2(2 \sin x \cos x\, dx)$

$-\dfrac{1}{4} du = \sin x \cos x\, dx$

3. $\displaystyle\int \dfrac{dx}{x(\ln 2x)^2} = \int (\ln 2x)^{-2} \left(\dfrac{dx}{x} \right)$

$\qquad\qquad\quad = \dfrac{(\ln 2x)^{-1}}{-1} + C$

$\qquad\qquad\quad = -\dfrac{1}{\ln 2x} + C$

4. $\displaystyle\int_1^8 y^{1/3}\sqrt{y^{4/3}+9}\,dy$

$u = y^{4/3}+9$

$du = \dfrac{4}{3}y^{1/3}\,dy$

$\displaystyle\int_{10}^{25} u^{1/2}\left(\dfrac{3}{4}du\right) = \dfrac{3}{4}\left(\dfrac{2}{3}u^{3/2}\right)\Big|_{10}^{25}$

$\qquad = \dfrac{1}{2}\left[25^{3/2}-10^{3/2}\right]$

$\qquad = \dfrac{125}{2} - 5\sqrt{10}$

5. $\displaystyle\int_0^{\pi/2}\dfrac{4\cos\theta\,d\theta}{1+\sin\theta}$

$u = 1+\sin\theta$

$du = \cos\theta\,d\theta$

$\displaystyle\int_1^2 \dfrac{4\,du}{u} = 4\ln u\,\Big|_1^2$

$\qquad = 4[\ln 2 - \ln 1]$

$\qquad = 4\ln 2$

$\qquad = 2.773$

6. $\displaystyle\int\dfrac{\sec^2 x\,dx}{2+\tan x}$

$u = 2+\tan x$

$du = \sec^2 x\,dx$

$\displaystyle\int\dfrac{\sec^2 x\,dx}{2+\tan x} = \int\dfrac{du}{u}$

$\qquad = \ln|u| + C$

$\qquad = \ln|2+\tan x| + C$

7. $\displaystyle\int\dfrac{2\,dx}{25+49x^2} = \dfrac{2}{7}\int\dfrac{7\,dx}{25+49x^2}$

$\qquad = \dfrac{2}{7}\left(\dfrac{1}{5}\right)\tan^{-1}\dfrac{7x}{5} + C$

$\qquad = \dfrac{2}{35}\tan^{-1}\dfrac{7x}{5} + C$

8. $\displaystyle\int\dfrac{dx}{\sqrt{1-4x^2}}$

$u = 2x,\ u^2 = 4x^2$

$du = 2\,dx$

$\displaystyle\int\dfrac{dx}{\sqrt{1-4x^2}} = \int\dfrac{\frac{1}{2}du}{\sqrt{1-u^2}} = \dfrac{1}{2}\sin^{-1}\dfrac{u}{1} + C = \dfrac{1}{2}\sin^{-1}2x + C$

9. $\displaystyle\int_0^{\pi/2}\cos^3 2x\,dx = \int_0^{\pi/2}\cos^2 2x\cos 2x\,dx = \int_0^{\pi/2}(1-\sin^2 2x)\cos 2x\,dx$

$\qquad = \displaystyle\int_0^{\pi/2}\cos 2x\,dx - \int_0^{\pi/2}\sin^2 2x\cos 2x\,dx$

$\qquad = \dfrac{1}{2}\displaystyle\int_0^{\pi/2}\cos 2x(2\,dx) - \dfrac{1}{2}\int_0^{\pi/2}\sin^2 2x\cos 2x(2\,dx) = \dfrac{1}{2}\left[\sin 2x - \dfrac{1}{3}\sin^3 2x\right]\Big|_0^{\pi/2}$

$\qquad = \dfrac{1}{2}\left[\left(\sin\pi - \dfrac{1}{3}\sin^3\pi\right) - \left(\sin 0 - \dfrac{1}{3}\sin^3 0\right)\right] = \dfrac{1}{2}(0) = 0$

10. $\displaystyle\int_0^{\pi/8}\sec^3 2x\tan 2x\,dx = \dfrac{1}{2}\int_0^{\pi/8}\sec^2 2x(\sec 2x\tan 2x(2\,dx)) = \dfrac{1}{2}\dfrac{\sec^3 2x}{3}\Big|_0^{\pi/8}$

$\qquad = \dfrac{1}{6}\left[\sec^3\left(2\left(\dfrac{\pi}{8}\right)\right) - \sec^3(2(0))\right] = \dfrac{1}{6}\left[\sec^3\dfrac{\pi}{4} - \sec^3 0\right]$

$\qquad = \dfrac{1}{6}[2^{3/2}-1] = \dfrac{1}{6}[\sqrt{8}-1] = \dfrac{1}{6}[2\sqrt{2}-1]$

11. $\displaystyle\int_0^2 \frac{x\,dx}{4+x^2} = \frac{1}{2}\int_0^2 \frac{2x\,dx}{4+x^2} = \frac{1}{2}\ln\left|4+x^2\right|\Big|_0^2 = \frac{1}{2}\ln 8 - \frac{1}{2}\ln 4 = \frac{1}{2}\ln 2 = 0.3466$

12. $\displaystyle\int_1^e \frac{\ln v^2\,dv}{\ln e^v} = \int_1^e \frac{2\ln v\,dv}{v\ln e} = 2\int_1^e \ln v\left(\frac{dv}{v}\right) = 2\left.\frac{(\ln v)^2}{2}\right|_1^e = (\ln e)^2 - (\ln 1)^2 = 1$

13. $\displaystyle\int (\sin t + \cos t)^2 \cdot \sin t\,dt = \int (\sin^2 t + 2\sin t\cos t + \cos^2 t)\cdot \sin t\,dt = \int (1 + 2\sin t\cos t)\cdot \sin t\,dt$

$\displaystyle = \int (\sin t + 2\sin^2 t\cos t)dt = \int \sin t\,dt + 2\int \sin^2 t(\cos t\,dt) = -\cos t + \frac{2\sin^3 t}{3} + C$

14. $\displaystyle\int \frac{\sin^3 x\,dx}{\sqrt{\cos x}} = \int \sin^2 x(\cos x)^{-1/2}(\sin x\,dx) = \int (1-\cos^2 x)(\cos x)^{-1/2}(\sin x\,dx)$

$\displaystyle = -\int ((\cos x)^{-1/2} - (\cos x)^{3/2})(-\sin x\,dx) = -\frac{(\cos x)^{1/2}}{\frac{1}{2}} + \frac{(\cos x)^{5/2}}{\frac{5}{2}} + C$

$\displaystyle = -2\sqrt{\cos x} + \frac{2}{5}(\cos x)^{5/2} + C = 2\sqrt{\cos x}\left(\frac{1}{5}\cos^2 x - 1\right) + C$

15. $\displaystyle\int \frac{e^x\,dx}{1+e^{2x}} = \int \frac{e^x\,dx}{1+(e^x)^2}, u = e^x, du = e^x\,dx$

$\displaystyle = \int \frac{du}{1+u^2} = \tan^{-1}u + C = \tan^{-1}e^x + C$

16. $\displaystyle\int \sec^4 3x\tan 3x\,dx = \int \sec^3 3x\sec 3x\tan 3x\,dx, u = \sec 3x, du = 3\sec 3x\tan 3x\,dx$

$\displaystyle\int \sec^4 3x\tan 3x\,dx = \int u^3\left(\frac{1}{3}du\right) = \frac{u^4}{12} + C = \frac{\sec^4 3x}{12} + C$

17. $\displaystyle\int \sec^4 3x\,dx = \int \sec^2 3x\sec^2 3x\,dx = \int (1+\tan^2 3x)\sec^2 3x\,dx$

$\displaystyle = \frac{1}{3}\int \sec^2 3x(3\,dx) + \frac{1}{3}\int \tan^2 3x\sec^2 3x(3\,dx)$

$\displaystyle = \frac{1}{3}\tan 3x + \frac{1}{3}\frac{\tan^3 3x}{3} + C = \frac{1}{9}\tan^3 3x + \frac{1}{3}\tan 3x + C$

18. $\displaystyle\int \frac{(1-\cos^2\theta)d\theta}{1+\cos 2\theta} = \int \frac{(1-\cos^2\theta)d\theta}{1+2\cos^2\theta - 1} = \int \frac{(1-\cos^2\theta)d\theta}{2\cos^2\theta} = \int \frac{1}{2}\sec^2\theta\,d\theta - \int \frac{1}{2}d\theta$

$\displaystyle = \frac{1}{2}\tan\theta - \frac{\theta}{2} + C$

19. $\displaystyle\int_0^{0.5} \frac{2e^{2x}-3e^x}{e^{2x}}\,dx = \int_0^{0.5} \frac{e^{2x}(2-3e^{-x})dx}{e^{2x}} = \int_0^{0.5} (2-3e^{-x})dx = \int_0^{0.5} 2\,dx + 3\int_0^{0.5} e^{-x}(-dx)$

$\displaystyle = 2x\Big|_0^{0.5} + 3e^{-x}\Big|_0^{0.5} = 2(0.5) + 3[e^{-(0.5)} - e^{-(0)}] = 1 + \frac{3}{\sqrt{e}} - 3$

$\displaystyle = \frac{3}{\sqrt{e}} - 2 = -0.1804$

20. $\displaystyle\int \frac{4-e^{\sqrt{x}}}{\sqrt{x}e^{\sqrt{x}}}\,dx = 4\int \frac{e^{-\sqrt{x}}}{\sqrt{x}}\,dx - \int \frac{dx}{\sqrt{x}} = 4(-2)\int e^{-\sqrt{x}}\left(\frac{dx}{-2\sqrt{x}}\right) - \int \frac{dx}{\sqrt{x}} = -8e^{-\sqrt{x}} - 2\sqrt{x} + C$

21. $\displaystyle\int \frac{3x\,dx}{4+x^4} = 3\int \frac{x\,dx}{4+x^4} = 3\int \frac{1}{2^2+(x^2)^2}\,x\,dx = \frac{3}{2}\int \frac{1}{2^2+(x^2)^2}\,2x\,dx$

$\displaystyle\qquad\qquad = \frac{3}{2}\left(\frac{1}{2}\tan^{-1}\frac{x^2}{2} + C_1\right) = \frac{3}{4}\tan^{-1}\frac{x^2}{2} + C \text{ where } C = \frac{3}{2}C_1.$

22. $\displaystyle\int_1^3 \frac{2\,dx}{\sqrt{x}(1+x)} = \int_1^{\sqrt{3}} \frac{4\,du}{1+u^2} = 4\tan^{-1}u\Big|_1^{\sqrt{3}} = 4(\tan^{-1}\sqrt{3} - \tan^{-1}1) = 1.047$

$\qquad$ let $u = \sqrt{x},\, du = \dfrac{1}{2\sqrt{x}}\,dx,\, 4\,du = \dfrac{2\,dx}{\sqrt{x}}$

$\qquad u^2 = x$

23. $\displaystyle\int_1^3 \frac{x\,dx}{2(3+x^2)} = \frac{1}{4}\int_1^3 \frac{2x\,dx}{3+x^2} = \frac{1}{4}\ln(3+x^2)\Big|_1^3 = \frac{1}{4}[\ln(3+3^2) - \ln(3+1^2)]$

$\displaystyle\qquad\qquad = \frac{1}{4}[\ln 12 - \ln 4] = \frac{1}{4}\ln\frac{12}{4} = \frac{1}{4}\ln 3$

24. $\displaystyle\int \frac{e^x\,dx}{\sqrt{9-e^{2x}}} = \int \frac{e^x\,dx}{\sqrt{9-(e^x)^2}} = \sin^{-1}\frac{e^x}{3} + C$

25. $\displaystyle\int \frac{2+4e^{2x}}{x+e^{2x}}\,dx = \int \frac{2(1+2e^{2x})\,dx}{x+e^{2x}},\, u = x+e^{2x},\, du = (1+2e^{2x})\,dx$

$\displaystyle\qquad\int \frac{2+4e^{2x}}{x+e^{2x}}\,dx = 2\int \frac{du}{u} = 2\ln|u| + C$

$\displaystyle\qquad\int \frac{2+4e^{2x}}{x+e^{2x}}\,dx = 2\ln|x+e^{2x}| + C$

26. $\displaystyle\int \frac{\tan^2 x\,dx}{\tan x - x} = \int \frac{du}{u} = \ln|u| + C = \ln|\tan x - x| + C$

$\qquad u = \tan x - x,\, du = (\sec^2 x - 1)\,dx = \tan^2 x\,dx$

27. $\displaystyle\int \frac{dx}{\sqrt{5-(4x+x^2)}} = \int \frac{dx}{\sqrt{5+4-(4+4x+x^2)}} = \int \frac{dx}{\sqrt{3^2-(x+2)^2}} = \sin^{-1}\frac{x+2}{3} + C$

28. $\displaystyle\int_{1/2}^{e/2} \frac{(4+\ln 2u)^3}{u}\,du = \int_4^5 x^3\,dx = \frac{x^4}{4}\Big|_4^5 = \frac{5^4}{4} - \frac{4^4}{4} = \frac{369}{4}$

$\qquad x = 4 + \ln 2u,\, dx = \dfrac{du}{u}$

29. $\displaystyle\int_0^{\pi/6} 3\sin^2 3\theta\,d\theta = \int_0^{\pi/6} 3\cdot\frac{(1-\cos 6\theta)}{2}\,d\theta = \int_0^{\pi/6} \frac{3}{2}\,d\theta - \frac{1}{4}\int_0^{\pi/6} \cos 6\theta(6\,d\theta)$

$\displaystyle\qquad\qquad = \frac{3}{2}\theta\Big|_0^{\pi/6} - \frac{1}{4}\sin 6\theta\Big|_0^{\pi/6} = \frac{3}{2}\left[\frac{\pi}{6} - 0\right] - \frac{1}{4}[\sin\pi - \sin 0] = \frac{\pi}{4} = 0.7845$

30. $\displaystyle\int \sin^4 x\, dx = \int (\sin^2 x)^2 dx$

$\displaystyle\int \sin^4 x\, dx = \int \left(\frac{1-\cos 2x}{2}\right)^2 dx$

$\displaystyle\int \sin^4 x\, dx = \frac{1}{4}\int (1 - 2\cos 2x + \cos^2 2x)\, dx$

$\displaystyle\int \sin^4 x\, dx = \frac{1}{4}\int \left[1 - 2\cos 2x + \frac{1}{2}(1 + \cos 4x)\right] dx$

$\displaystyle\int \sin^4 x\, dx = \frac{1}{4}\int \left(\frac{3}{2} - 2\cos 2x + \frac{1}{2}\cos 4x\right) dx$

$\displaystyle\int \sin^4 x\, dx = \frac{1}{4}\left[\frac{3}{2}x - \sin 2x + \frac{1}{8}\sin 4x\right] + C$

31. $\displaystyle\int e^{2x}\cos e^{2x}\, dx = \frac{1}{2}\int \cos e^{2x}(2e^{2x}\, dx) = \frac{1}{2}\sin e^{2x} + C$

32. $\displaystyle\int \frac{3\, dx}{x^2 + 6x + 10} = \int \frac{3\, dx}{x^2 + 6x + 9 + 1} = \int \frac{3\, dx}{(x+3)^2 + 1} = 3\tan^{-1}(x+3) + C$

33. $\displaystyle\int_1^e 3\cos(\ln x)\frac{dx}{x} = 3\sin(\ln x)\Big|_1^e = 3\sin(\ln e) - 3\sin(\ln 1) = 3\sin(1) - 3\sin(0)$

$$= 3\sin 1 - 3(0) = 3\sin 1 = 2.524$$

34. $\displaystyle\int_1^3 \frac{2\, dx}{x^2 - 2x + 5} = \int_1^3 \frac{2\, dx}{x^2 - 2x + 1 + 4} = \int_1^3 \frac{2\, dx}{(x-1)^2 + 2^2} = 2\left(\frac{1}{2}\right)\tan^{-1}\frac{x-1}{2}\Big|_1^3$

$$= \tan^{-1}\frac{3-1}{2} - \tan\frac{1-1}{2} = \tan^{-1}1 - \tan^{-1}0 = \frac{\pi}{4}$$

35.

$$
\begin{array}{r}
x - 2 \\
x + 2\overline{)x^2 - 1} \\
\underline{x^2 + 2x} \\
-2x - 1 \\
\underline{-2x - 4} \\
3
\end{array}
$$

$\displaystyle\int \frac{x^2 - 1}{x + 2}\, dx = \int \left(x - 2 + \frac{3}{x+2}\right) dx = \frac{1}{2}x^2 - 2x + 3\ln|x + 2| + C$

36. $\displaystyle\int \frac{\log_x 2\, dx}{x\ln x} = \int \frac{\frac{\ln 2}{\ln x}}{x\ln x}\, dx = \ln 2 \int (\ln x)^{-2}\frac{dx}{x} = -\frac{\ln 2}{\ln x} + C$

37. Use the general power formula. Let $u = e^x + 1$, $du = e^x dx$, $n = 2$.

$\displaystyle\int e^x(e^x + 1)^2 dx = \int (e^x + 1)^2 e^x dx = \frac{(e^x + 1)^3}{3} + C_1 = \frac{e^{3x} + 3e^{2x} + 3e^x + 1}{3} + C_1$

The second method used is to multiply the factors before integrating:

$\displaystyle\int e^x(e^x + 1)^2 dx = \int e^x(e^{2x} + 2e^x + 1)dx = \int (e^{3x} + 2e^{2x} + e^x)dx = \int e^{3x}dx + \int 2e^{2x}dx + \int e^x dx$

$$= \frac{1}{3}\int e^{3x}(3\, dx) + 2\left(\frac{1}{2}\right)\int e^{2x}(2\, dx) + \int e^x dx = \frac{1}{3}e^{3x} + e^{2x} + e^x + C_2; \ C_2 = C_1 + \frac{1}{3}$$

38. $\int \frac{1}{x}(1+\ln x)dx$

(1) Let $u = 1 + \ln x, du = \frac{dx}{x}$

$$\int \frac{1}{x}(1+\ln x)dx = \int u\,du = \frac{1}{2}u^2 + C_1 = \frac{1}{2}(1+\ln x)^2 + C$$

$$= \frac{1}{2}(1+2\ln x+(\ln x)^2) + C_1 = \frac{1}{2} + \ln x + \frac{1}{2}(\ln x)^2 + C_1$$

$$\int \frac{1}{x}(1+\ln x)dx = \ln x + \frac{1}{2}(\ln x)^2 + C_1 + \frac{1}{2}$$

(2) $\int \frac{1}{x}(1+\ln x)dx = \int \frac{1}{x}dx + \int \ln x \frac{dx}{x} = \ln x + \frac{1}{2}(\ln x)^2 + C_2$

(1) and **(2)** give same result with $C_2 = C_1 + \frac{1}{2}$

39. $\int \frac{1}{1+\sin x}dx = \int \frac{1}{1+\sin x}\frac{(1-\sin x)}{(1-\sin x)}dx = \int \frac{1-\sin x}{1-\sin^2 x}dx = \int \frac{1-\sin x}{\cos^2 x}dx$

$$= \int \frac{1}{\cos^2 x}dx - \int \frac{1}{\cos x}\frac{\sin x}{\cos x}dx = \int \sec^2 x\,dx - \int \sec x \tan x\,dx$$

$$= \tan x - \sec x + C$$

40. $\frac{dy}{dx} = e^x(2-e^x)^2$

$$y = -\int (2-e^x)^2(-e^x dx)$$

$$y = -\frac{(2-e^x)^3}{3} + C$$

$$4 = -\frac{(2-e^0)^3}{3} + C$$

$$4 = -\frac{1}{3} + C, C = 4 + \frac{1}{3} = 4\frac{1}{3} = \frac{13}{3}$$

$$y = -\frac{(2-e^x)^3}{3} + \frac{13}{3}$$

41. $\int \sec^4 x\,dx = \frac{\sec^2 \tan x}{3} + \frac{2}{3}\int \sec^2 x\,dx$ Formula 37 in table of integrals.

$$y = \frac{\sec^2 x \tan x}{3} + \frac{2}{3}\tan x + C = \frac{1}{3}(1+\tan^2 x)(\tan x) + \frac{2}{3}\tan x + C$$

$$= \frac{1}{3}\tan x + \frac{1}{3}\tan^3 x + \frac{2}{3}\tan x + C = \frac{1}{3}\tan^3 x + \frac{2}{3}\tan x + C$$

$$0 = \frac{1}{3}\tan^3(0) + \frac{2}{3}\tan 0 + C$$

$$0 = 0 + 0 + C; C = 0; y = \frac{1}{3}\tan^3 x + \tan x$$

42. $A = \int_0^{1.5} y\,dx = \int_0^{1.5} 4e^{2x}\,dx = 2\int_0^{1.5} e^{2x}(2\,dx) = 2e^{2x}\Big|_0^{1.5} = 2(e^3 - e^0) = 38.17$

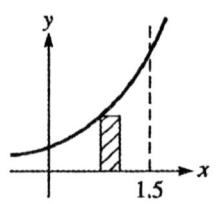

43. $y\ln(\sin x)$

$$\frac{dy}{dx} = \frac{1}{\sin x}\cos x$$

$$L = \int_{\pi/3}^{2\pi/3} \sqrt{1 + \left(\frac{\cos x}{\sin x}\right)^2}\,dx = \int_{\pi/3}^{2\pi/3} \sqrt{\frac{\sin^2 x + \cos^2 x}{\sin^2 x}}\,dx = \int_{\pi/3}^{2\pi/3} \frac{1}{\sin x}\,dx = \int_{\pi/3}^{2\pi/3} \csc x\,dx$$

$$= \ln|\csc x - \cot x|\Big|_{\pi/3}^{2\pi/3} = \ln\left|\csc\frac{2\pi}{3} - \cot\frac{2\pi}{3}\right| - \ln\left|\csc\frac{\pi}{3} - \cot\frac{\pi}{3}\right|$$

$$= \ln\left|\frac{2}{\sqrt{3}} - \left(\frac{-1}{\sqrt{3}}\right)\right| - \ln\left|\frac{2}{\sqrt{3}} - \frac{1}{\sqrt{3}}\right| = \ln\left|\frac{3}{\sqrt{3}}\right| - \ln\left|\frac{1}{\sqrt{3}}\right| = \ln\left(\frac{\frac{3}{\sqrt{3}}}{\frac{1}{\sqrt{3}}}\right) = \ln 3 = 1.10$$

44. $\Delta S = \int (c_v/T)dT = \int \frac{a + bT + cT^2}{T}\,dT = \int \frac{a}{T}\,dT + \int b\,dT + \int cT\,dT$

$$= a\ln T + bT + \frac{1}{2}cT^2 + C$$

45. $$\int \frac{dv}{32 - 0.5v} = \int dt$$

$$-\frac{1}{0.5}\ln|32 - 0.5v| = t + C$$

$$-\frac{1}{0.5}\ln|32 - 0.5(0)| = 0 + C$$

$$C = -\frac{1}{0.5}\ln|32|$$

$$-\frac{1}{0.5}\ln|32 - 0.5v| = t - \frac{1}{0.5}\ln(32)$$

$$\frac{1}{0.5}\ln 32 - \frac{1}{0.5}\ln|32 - 0.5v| = t$$

$$\ln\frac{32}{|32 - 0.5v|} = 0.5t$$

$$\frac{32}{|32 - 0.5v|} = e^{0.5t}$$

$$32e^{-0.5t} = |32 - 0.5v|$$

$$32 - 0.5v = \pm 32e^{-0.5t}$$

$$0.5v = 32(1 \pm e^{-0.5t})$$

$$v = 64(1 \pm e^{-0.5t}) \text{ where the negative must be chosen to satisfy } v = 0 \text{ at } t = 0$$

$$v = 64(1 - e^{-0.5t})$$

46. $P_{av} = \dfrac{4}{\pi} \displaystyle\int_0^{\pi/4} 20\cos 2t (3\sin 2t)\,dt = \dfrac{4}{\pi} \displaystyle\int_0^{\pi/4} 30(2\sin 2t \cos 2t)\,dt = \dfrac{1}{\pi} \displaystyle\int_0^{\pi/4} 30(\sin 4t)4\,dt)$

$= \dfrac{30}{\pi}(-\cos 4t)\Big|_0^{\pi/4} = \dfrac{30}{\pi}\left(-\cos\left(4\left(\dfrac{\pi}{4}\right)\right) + \cos(4(0))\right) = \dfrac{30}{\pi}(-(-1)+1) = \dfrac{60}{\pi} = 19.1 \text{ units}$

47.
$$y_{rms} = \sqrt{\dfrac{1}{T}\int i^2\,dt} = \sqrt{\dfrac{1}{T}\int (2\sin t)^2\,dt}$$

$$\int_0^{2\pi} (2\sin t)^2\,dt = \int_0^{2\pi} 4\sin^2 t\,dt = 4\int_0^{2\pi} \sin^2 t\,dt$$

$$= 4\left(\dfrac{t}{2} - \dfrac{1}{2}\sin t \cos t\right)\Big|_0^{2\pi} \qquad \text{Formula 29 in table of integrals.}$$

$$= 2t - 2\sin t \cos t\Big|_0^{2\pi} = 4\pi; \; T = 2\pi$$

$$y_{rms} = \sqrt{\dfrac{1}{2\pi}(4\pi)} = \sqrt{2}$$

48. $n = A\displaystyle\int_0^{\pi} e^{a\cos\theta}\sin\theta\,d\theta$

$n = \dfrac{-A}{a}\displaystyle\int_0^{\pi} e^{a\cos\theta}(-a\sin\theta\,d\theta)$

$n = \dfrac{-A}{a}e^{a\cos\theta}\Big|_0^{\pi}$

$n = \dfrac{-A}{a}\left(e^{a\cos\pi} - e^{a\cos 0}\right)$

$n = \dfrac{-A}{a}(e^{-a} - e^a)$

$n = \dfrac{A}{a}(e^a - e^{-a})$

49. $\displaystyle\int_0^{\pi} (1 + k\cos\theta)\cos\theta\sin\theta\,d\theta$

$= \displaystyle\int_0^{\pi} \sin\theta\cos\theta\,d\theta - \displaystyle\int_0^{\pi} k\cos^2\theta(-\sin\theta\,d\theta)$

$= \dfrac{1}{2}\sin^2\theta\Big|_0^{\pi} - \dfrac{k}{3}\cos^3\theta\Big|_0^{\pi}$

$= \dfrac{1}{2}(\sin^2\pi - \sin^2 0) - \dfrac{k}{3}(\cos^3\pi - \cos^3 0)$

$= -\dfrac{k}{3}(-1-1) = \dfrac{2k}{3}$

50. $v = \dfrac{ds}{dt} = 4\cos 3t; \; t = 0, s = 0$

(a) $s = \displaystyle\int_0^{0.5} 4\cos 3t\,dt$

$s = \dfrac{4}{3}\displaystyle\int_0^{0.5} \cos 3t(3\,dt)$

$s = \dfrac{4}{3}\sin 3t\Big|_0^{0.5}$

$s = \dfrac{4}{3}(\sin(3(0.5)) - \sin(3(0)))$

$s = 1.3$

(b) $s = \dfrac{4}{3}\sin 3t + C$

$0 = \dfrac{4}{3}\sin(3(0)) + C, C = 0$

$s = \dfrac{4}{3}\sin 3t\Big|_{t=0.5}$

$s = \dfrac{4}{3}\sin(3(0.5))$

$s = \dfrac{4}{3}, \text{ the results are the same.}$

51. $V = \pi \int_{2.00}^{4.00} y^2 \, dx = \pi \int_{2.00}^{4.00} e^{-0.2x} \, dx$; and

$$u = -0.2x \quad du = -0.2 \, dx = -\frac{\pi}{0.2} \int_{2.00}^{4.00} e^{-0.2x}(-0.2) \, dx$$

$$= \frac{-\pi}{0.2} e^{-0.2x} \Big|_{2.00}^{4.00} = -\frac{\pi}{2} [e^{-0.8} - e^{-0.4}] = 3.47 \text{ cm}^3$$

52. $\bar{x} = \dfrac{\int_a^b x(y_1 - y_2) \, dx}{\int_a^b (y_1 - y_2) \, dx}$

$$\bar{x} = \frac{\int_0^1 x \left(\frac{1}{x^2+1} - 0\right) dx}{\int_0^1 \left(\frac{1}{x^2+1} - 0\right) dx}$$

$$\bar{x} = \frac{\frac{1}{2} \int_0^1 \frac{2x \, dx}{x^2+1}}{\int_0^1 \frac{1}{x^2+1} \, dx}$$

$$\bar{x} = \frac{\frac{1}{2} \ln(x^2 + 1)\big|_0^1}{\tan^{-1} x \, \big|_0^1}$$

$$\bar{x} = \frac{\frac{1}{2}(\ln(1^2 + 1) - \ln(0^2 + 1))}{\tan^{-1} 1 - \tan^{-1} 0}$$

$$\bar{x} = \frac{\frac{1}{2} \ln 2}{\frac{\pi}{4}} = \frac{2}{\pi} \ln 2$$

$$\bar{x} = 0.441 \text{ in.}$$

53. (a) $\displaystyle\int_0^{2.8} \frac{4}{1 + e^x} \, dx = \int_0^{2.8} \frac{-4e^{-x}}{e^{-x} + 1} \cdot (-e^{-x} \, dx)$

$$= -4 \ln(e^{-x} + 1)\big|_0^{2.8}$$

$$= 2.536$$

(b)
$$\begin{array}{r} 4 \\ 1 + e^x \overline{)4 } \\ \underline{4 + 4e^x} \\ -4e^x \end{array} \quad\Rightarrow\quad \frac{4}{1 + e^x} = 4 + \frac{-4e^x}{1 + e^x}$$

$$\int_0^{2.8} \frac{4}{1 + e^x} \, dx = \int_0^{2.8} 4 \, dx - 4 \int_0^{2.8} \frac{e^x}{1 + e^x} \, dx$$

The second integral has the form $\displaystyle\int \frac{du}{u}$ with $u = 1 + e^x$.

$$\int_0^{2.8} \frac{4}{1 + e^x} \, dx = 4x \Big|_0^{2.8} - 4 \ln(1 + e^x)\big|_0^{2.8}$$

$$\int_0^{2.8} \frac{4}{1 + e^x} \, dx = 2.536$$

In (a) the integrand was multiplied by $\dfrac{e^{-x}}{e^{-x}}$.

In (b) long division led to two standard forms.

METHODS OF INTEGRATION

10.1 Integration By Parts

1. $u = \theta$; $du = d\theta$; $dv = \cos\theta\, d\theta$;

$$v = \int \cos\theta\, d\theta = \sin\theta$$

$$\int (\theta)(\cos\theta\, d\theta) = \theta(\sin\theta) - \int \sin\theta\, d\theta$$
$$= \theta(\sin\theta) - (-\cos\theta) + C$$
$$= \cos\theta + \theta\sin\theta + C$$

2. $\int x\sin 2x\, dx$; $u = x$; $du = dx$; $dv = \sin 2x\, dx$

$$v = \frac{1}{2}\int \sin 2x(2\, dx) = -\frac{1}{2}\cos 2x$$

$$\int x\sin 2x\, dx = -\frac{1}{2}x\cos 2x + \frac{1}{2}\int \cos 2x\, dx$$
$$= -\frac{1}{2}x\cos 2x + \frac{1}{4}\int \cos 2x(2\, dx)$$
$$= -\frac{1}{2}x\cos 2x + \frac{1}{4}\sin 2x + C$$

3. $\int xe^{2x} dx$; $u = x$; $du = dx$; $dv = e^{2x} dx$

$$v = \frac{1}{2}\int e^{2x}(2\, dx) = \frac{1}{2}e^{2x}$$

$$\int xe^{2x} dx = \frac{1}{2}xe^{2x} - \frac{1}{2}\int e^{2x} dx$$
$$= \frac{1}{2}xe^{2x} - \frac{1}{4}\int e^{2x}(2\, dx)$$
$$= \frac{1}{2}xe^{2x} - \frac{1}{4}e^{2x} + C$$

4. $\int 3xe^x dx = 3\int xe^x dx$; $u = x$; $du = dx$;

$dv = e^x dx$; $v = e^x$

$$3\int xe^x dx = 3\left(xe^x - \int e^x dx\right)$$
$$= 3(xe^x - e^x) + C$$
$$= 3e^x(x-1) + C$$

5. $u = x$; $du = dx$; $dv = \sec^2 x\, dx$;

$$v = \int \sec^2 x\, dx = \tan x$$

$$\int (x)(\sec^2 x\, dx) = x\tan x - \int \tan x\, dx$$
$$= x\tan x - (-\ln|\cos x| + C)$$
$$= x\tan x + \ln|\cos x| + C$$

6. $\int_0^{\pi/4} x\sec x\tan x\, dx$; $u = x$; $du = dx$;

$dv = \sec x\tan x dx$; $v = \sec x$

$$\int_0^{\pi/4} x\sec x\tan x\, dx$$

$$= x\sec x - \int \sec x\, dx$$

$$= x\sec x - \ln|\sec x + \tan x|\Big|_0^{\pi/4}$$

$$= \frac{\pi}{4}\sec\frac{\pi}{4} - \ln\left|\sec\frac{\pi}{4} + \tan\frac{\pi}{4}\right| - 0$$

$$= \frac{\pi}{4}\sqrt{2} - \ln\left|\sqrt{2} + 1\right| = 0.229$$

7. $\int 2\tan^{-1} x\, dx$; $u = \tan^{-1} x$; $du = \dfrac{1}{1+x^2}dx$;

$dv = dx$;

$v = x$

$$2\int \tan^{-1} x\, dx$$

$$= 2\left(x\tan^{-1} x - \int \frac{x\, dx}{1+x^2}\right)$$

$u = x^2$; $du = 2x\, dx$

$$= 2\left(x\tan^{-1} x - \frac{1}{2}\int \frac{2x\, dx}{1+x^2}\right)$$

$$= 2x\tan^{-1} x - \ln|1+x^2| + C$$

OR

$$= 2x\tan^{-1} x - 2\ln\sqrt{1+x^2} + C$$

8. $\displaystyle\int \ln s\, ds$; $u = \ln s$; $du = \dfrac{ds}{s}$; $dv = ds$; $v = s$

$$\int \ln s\, ds = s\ln s - \int \frac{s\, ds}{s} = s\ln s - \int ds$$

$$= s\ln s - s + C = s(\ln s - 1) + C$$

9. $\displaystyle\int_{-3}^{0} \frac{4t\, dt}{\sqrt{1-t}} = 4\int \frac{t\, dt}{\sqrt{1-t}}$; $u = t$; $du = dt$

$$dv = \frac{1}{\sqrt{1-t}}\, dt;\quad v = \int (1-t)^{-1/2}dt$$

$$v = -\int (1-x)^{-1/2}(-dt) = -(1-t)^{1/2}(2)$$

$$= -2(1-t)^{1/2}$$

$$4\int (t)(1-t)^{-1/2}dt$$

$$= 4t[-2(1-t)^{1/2}] - 4\int -2(1-t)^{1/2}dt$$

$$= -8t(1-t)^{1/2} + 8\int (1-x)^{1/2}dt$$

$$= -8t(1-t)^{1/2} - 8\int (1-t)^{1/2}(-dt)$$

$$= -8t(1-t)^{1/2} - 8(1-t)^{3/2}\left(\frac{2}{3}\right)\Big|_{-3}^{0}$$

$$= -\frac{32}{3}$$

10. $\displaystyle\int x\sqrt{x+1}\,dx$; $u = x$; $du = dx$; $dv = (x+1)^{1/2}dx$;

$$v = \frac{2}{3}(x+1)^{3/2}$$

$$\int x\sqrt{x+1}\,dx$$

$$= \frac{2}{3}x(x+1)^{3/2} - \frac{2}{3}\int (x+1)^{3/2}dx$$

$$= \frac{2}{3}x(x+1)^{3/2} - \frac{2}{3}\cdot\frac{2}{5}(x+1)^{5/2} + C$$

$$= \frac{2}{3}(x+1)^{3/2}\left[x - \frac{2}{5}(x+1)\right] + C$$

$$= \frac{2}{3}(x+1)^{3/2}\left[\frac{5x - 2x - 2}{5}\right]$$

$$= \frac{2}{15}(x+1)^{3/2}(3x-2) + C$$

11. $\displaystyle\int x\ln x\, dx$; $u = \ln x$; $du = \dfrac{dx}{x}$; $dv = x\, dx$; $v = \dfrac{x^2}{2}$

$$\int x\ln x\, dx = \frac{1}{2}x^2\ln x - \frac{1}{2}\int x^2\frac{dx}{x}$$

$$= \frac{1}{2}x^2\ln x - \frac{1}{2}\int x\, dx$$

$$= \frac{1}{2}x^2\ln x - \frac{1}{2}\cdot\frac{x^2}{2} + C$$

$$= \frac{1}{2}x^2\ln x - \frac{1}{4}x^2 + C$$

12. $\displaystyle\int x^2\ln 4x\, dx$; $u = \ln 4x$; $du = \dfrac{1}{4x}4\, dx = \dfrac{dx}{x}$;

$$dv = x^2 dx;\quad v = \frac{1}{3}x^3$$

$$\int x^2\ln 4x\, dx = \frac{1}{3}x^3\ln 4x - \frac{1}{3}\int x^3\frac{dx}{x}$$

$$= \frac{1}{3}x^3\ln 4x - \frac{1}{3}\int x^2 dx$$

$$= \frac{1}{3}x^3\ln 4x - \frac{1}{9}x^3 + C$$

13. $\displaystyle\int 2\phi^2\sin\phi\cos\phi\, d\phi = \int \phi^2\sin(2\phi)\, d\phi$

let $\quad u = \phi^2 \qquad dv = \sin(2\phi)\, d\phi$
$\qquad du = 2\phi\, d\phi$

$$v = \int \sin(2\phi)\, d\phi;$$

$$= \frac{1}{2}\sin(2\phi)\, 2\, d\phi$$

$$v = -\frac{1}{2}\cos(2\phi)$$

$$\int 2\phi^2\sin\phi\cos\phi\, d\phi$$

$$= \int \phi^2\sin(2\phi)\, d\phi$$

$$= -\frac{1}{2}\cdot\phi^2\cos(2\phi) + \int \phi\cos(2\phi)\, d\phi$$

let $\quad u = \phi \qquad dv = \cos(2\phi)\, d\phi$
$\qquad du = d\phi$

$$v = \int \cos(2\phi)\, d\phi$$

$$= \frac{1}{2}\sin(2\phi)$$

$$\int 2\phi^2\sin\phi\cos\phi\, d\phi$$

$$= -\frac{1}{2}\phi^2\cos(2\phi) + \frac{\phi}{2}\sin(2\phi) - \frac{1}{2}\int \sin(2\phi)\, d\phi$$

$$= -\frac{\phi^2}{2}\cos(2\phi) + \frac{\phi}{2}\sin(2\phi) + \frac{1}{4}\cos(2\phi) + C$$

$$= \frac{\phi}{2}\sin(2\phi) - \frac{1}{4}(2\phi^2 - 1)\cos(2\phi) + C$$

14. $\displaystyle\int r^2 e^{2r}\,dr;\ u=r^2;\ du=2r\,dr;\ dv=e^{2r}\,dr;$

$$v=\frac{1}{2}e^{2r}$$

$$\int r^2 e^{2r}\,dr=\frac{1}{2}r^2 e^{2r}-\frac{1}{2}\int 2re^{2r}\,dr$$

$$=\frac{1}{2}r^2 e^{2r}-\int re^{2r}\,dr$$

$$u=r;\ du=dr;\ dv=e^{2r}\,dr;\ v=\frac{1}{2}e^{2r}$$

$$\int r^2 e^{2r}\,dr=\frac{1}{2}r^2 e^{2r}-\left(\frac{1}{2}re^{2r}-\frac{1}{2}\int e^{2r}\,dr\right)$$

$$\int_0^1 r^2 e^{2r}\,dr=\frac{1}{2}r^2 e^{2r}-\frac{1}{2}re^{2r}+\frac{1}{4}e^{2r}\Big|_0^1$$

$$=1.597$$

15. $\displaystyle\int_0^{\pi/2} e^x\cos x\,dx;\ u=e^x;\ du=e^x dx;\ dv=\cos x\,dx;$

$$v=\sin x$$

$$\int_0^{\pi/2} e^x\cos x\,dx=e^x\sin x-\int \sin x e^x dx$$

$$u=e^x;\ du=e^x dx;\ dv=\sin x\,dx;\ v=-\cos x$$

$$\int_0^{\pi/2} e^x\cos x\,dx=e^x\sin x-\left(-e^x\cos x+\int e^x\cos x\,dx\right)$$

$$\int_0^{\pi/2} e^x\cos x\,dx=e^x\sin x+e^x\cos x-\int e^x\cos x\,dx$$

$$2\int e^x\cos x\,dx=e^x\sin x+e^x\cos x$$

$$\int_0^{\pi/2} e^x\cos x\,dx=\frac{1}{2}e^x(\sin x+\cos x)\Big|_0^{\pi/2}$$

$$=\frac{1}{2}e^{\pi/2}(1+0)-\frac{1}{2}e^0(0+1)$$

$$=\frac{1}{2}e^{\pi/2}-\frac{1}{2}$$

$$=\frac{1}{2}(e^{\pi/2}-1)=1.91$$

16. $\displaystyle\int e^{-x}\sin 2x\,dx;\ u=e^{-x};\ du=-e^{-x}dx;$

$$dv=\sin 2x\,dx;\ v=-\frac{1}{2}\cos 2x$$

$$\int e^{-x}\sin 2x\,dx=-\frac{1}{2}e^{-x}\cos 2x-\frac{1}{2}\int e^{-x}\cos 2x\,dx$$

$$u=e^{-x};\ du=-e^{-x}dx;\ dv=\cos 2x\,dx;$$

$$v=\frac{1}{2}\sin x$$

$$\int e^{-x}\sin 2x\,dx$$

$$=-\frac{1}{2}e^{-x}\cos 2x-\frac{1}{2}\left(\frac{1}{2}e^{-x}\sin 2x+\frac{1}{2}\int e^{-x}\sin 2x\,dx\right)$$

$$=-\frac{1}{2}e^{-x}\cos 2x-\frac{1}{4}e^{-x}\sin 2x-\frac{1}{4}\int e^{-x}\sin 2x\,dx$$

$$\frac{5}{4}\int e^{-x}\sin 2x\,dx=-\frac{1}{2}e^{-x}\cos 2x-\frac{1}{4}e^{-x}\sin 2x$$

$$=-\frac{1}{2}e^{-x}\left(\cos 2x+\frac{1}{2}\sin 2x\right)$$

$$\int e^{-x}\sin 2x\,dx=\frac{4}{5}\left(-\frac{1}{2}e^{-x}\right)\frac{(2\cos 2x+\sin 2x)}{2}$$

$$=-\frac{1}{5}e^{-x}(\sin 2x+2\cos 2x)+C$$

17. $\displaystyle A=\int_0^2 xe^{-x}dx;\ u=x;\ du=dx;\ dv=e^{-x}dx;$

$$v=\int e^{-x}dx=-e^{-x}$$

$$A=-xe^{-x}\Big|_0^2=\int_0^2 -e^{-x}dx=-xe^{-x}-e^{-x}\Big|_0^2$$

$$=-2e^{-2}-e^{-2}-(0-1)$$

$$=1-\frac{3}{e^2}$$

$$=0.594$$

18. $\displaystyle y=\frac{2(\ln x)}{x^2};\ y=0,x=3$

$$A_{1,3}=\int_1^3 \frac{2\ln x}{x^2}\,dx$$

$$2\int \ln x\frac{dx}{x^2};\ u=\ln x;\ du=\frac{dx}{x}$$

$$dv=\frac{dx}{x^2}=x^{-2}dx;\ v=-\frac{1}{x}$$

$$2\int \ln\frac{dx}{x^2}=2\left(-\frac{\ln x}{x}+\int \frac{1}{x}\frac{dx}{x}\right)$$

$$=2\left(\frac{-\ln x}{x}+\int \frac{dx}{x^2}\right)$$

$$=\frac{-2\ln x}{x}-\frac{2}{x}=-\frac{2}{x}(\ln x+1)\Big|_1^3$$

$$=\frac{-2}{3}(\ln 3+1)+2(\ln 1+1)$$

$$=0.6009$$

19. $y = \tan^2 x;\ y = 0, x = 0.5$; rotated about y-axis; shells.

$$dV = 2\pi xy\,dx$$

$$V = 2\pi \int_0^{0.5} x \tan^2 x\,dx = 2\pi \int_0^{0.5} x(\sec^2 x - 1)dx$$

$$u = x;\ du = dx;\ dv = (\sec^2 x - 1)dx;\ v = \tan x - x$$

$$V = 2\pi \int x \tan^2 x\,dx = x(\tan x - x) - \int (\tan x - x)dx$$

$$= 2\pi\left(x \tan x - x^2 - \int \tan x\,dx + \int x\,dx\right)$$

$$= 2\pi\left(x \tan x - x^2 + \ln|\cos x| + \frac{1}{2}x^2\right)$$

$$= 2\pi\left(x \tan x + \ln\left|-\frac{1}{2}x^2\right|\right)\Big|_0^{0.5} = 0.1104$$

20. Exercise 3: $\displaystyle\int xe^{2x}\,dx$

First choice:

$$u = x \qquad dv = e^{2x}\,dx$$

$$du = dx \qquad v = \frac{1}{2}e^{2x}\,dx$$

$$\int xe^{2x}\,dx = \frac{x}{2}e^{2x} - \int \frac{1}{2}e^{2x}\,dx = \frac{x}{2}e^{2x} - \frac{1}{4}e^{2x} + C$$

Second choice:

$$u = e^{2x} \qquad dv = x\,dx$$

$$du = 2e^{2x}\,dx \qquad v = \frac{x^2}{2}$$

$$\int xe^{2x}\,dx = \frac{x^2}{2}e^{2x} - \int \frac{x^2}{2}(2e^{2x}\,dx) \text{ and since}$$

$\displaystyle\int \frac{x^2}{2}(2e^{2x}\,dx)$ cannot be integrated, the first choice is necessary.

Exercise 15: $\displaystyle\int e^x \cos x\,dx$

First choice:

$$u = e^x \qquad dv = \cos x\,dx$$
$$du = e^x\,dx \qquad v = \sin x$$

$$\int e^x \cos x\,dx = e^x \sin x - \int \sin x\, e^x\,dx$$

$$u = e^x \qquad dv = \sin x\,dx$$
$$du = e^x\,dx \qquad v = -\cos x$$

$$\int e^x \cos x\,dx = e^x \sin x - \left[-e^x \cos x + \int e^x \cos x\,dx\right]$$

$$\int e^x \cos x\,dx = e^x \sin x + e^x \cos x - \int e^x \cos x\,dx$$

$$2\int e^x \cos x\,dx = e^x(\sin x + \cos x)$$

$$\int e^x \cos x\,dx = \frac{1}{2}e^x(\sin x + \cos x)$$

Second choice:

$$u = \cos x \qquad dv = e^x\,dx$$
$$du = -\sin x\,dx \qquad v = e^x$$

$$\int e^x \cos x\,dx = e^x \cos x + \int e^x \sin x\,dx$$

$$u = \sin x \qquad dv = e^x\,dx$$
$$du = \cos x\,dx \qquad v = e^x$$

$$\int e^x \cos x\,dx = e^x \cos x + \left[e^x \sin x - \int e^x \cos x\,dx\right]$$

$$2\int e^x \cos x\,dx = e^x(\cos x + \sin x)$$

$$\int e^x \cos x\,dx = \frac{1}{2}e^x(\sin x + \cos x).$$

Either choice will work.

21. $\displaystyle \bar{x} = \frac{\displaystyle\int_0^{\pi/2} x(\cos x)\,dx}{\displaystyle\int_0^{\pi/2} \cos x\,dx}$

Let $u = x;\ du = dx;\ dv = \cos x;\ v = \sin x$

$$\bar{x} = \frac{x \sin x\big|_0^{\pi/2} - \displaystyle\int_0^{\pi/2} \sin x\,dx}{\sin x\big|_0^{\pi/2}}$$

$$= \frac{x \sin x\big|_0^{\pi/2} - (-\cos x)\big|_0^{\pi/2}}{1}$$

$$= x \sin x + \cos x\big|_0^{\pi/2} = \frac{\pi}{2} - 1 = 0.571$$

22. $y = e^x;\ x = 1$; rotated about y-axis

$$I_y = 2\pi k \int_0^1 yx^3\,dx = 2\pi k \int_0^1 x^3 e^x\,dx$$

Successive applications of parts method, where $u = x^n;\ dv = e^x\,dx$.

$$I_y = 2\pi k \int_0^1 x^3 e^x\,dx$$

$$= 2\pi k(x^3 e^x - 3x^2 e^x + 6xe^x - 6e^x)\big|_0^1$$

$$= 2\pi k e^x(x^3 - 3x^2 + 6x - 6)\big|_0^1 = 3.54k$$

23. $p = \sqrt{\sin^{-1} x}$; $x = 0, x = 1$; $T = 1$

$$p_{rms} = \sqrt{\frac{1}{T} \int_0^T y^2 dx} = \sqrt{\frac{1}{1} \int_0^1 \sin^{-1} x \, dx}$$

$$\int_0^1 \sin^{-1} x \, dx; \quad u = \sin^{-1} x; \quad du = \frac{1}{\sqrt{1-x^2}} dx;$$

$$dv = dx; \quad v = x$$

$$\int_0^1 \sin^{-1} x \, dx = x \sin^{-1} x - \int \frac{x \, dx}{\sqrt{1-x^2}}$$

$$= x \sin^{-1} x + \frac{1}{2} \int (1-x^2)^{-1/2}(-2x \, dx)$$

$$n = -\frac{1}{2}; \quad u = 1 - x^2; \quad du = -2x \, dx$$

$$\int_0^1 \sin^{-1} dx = x \sin^{-1} x + (1-x^2)^{1/2}\Big|_0^1$$

$$= 1 \sin^{-1} 1 + 0 - (0+1)$$
$$= \sin^{-1} 1 - 1$$
$$p_{rms} = \sqrt{\sin^{-1} 1 - 1} = 0.756$$

24. $\dfrac{dy}{dx} = k x^3 \sqrt{1+x^2}$

$$dy = x^2 \sqrt{1+x^2} (kx \, dx)$$

$$y = \int x^2 \sqrt{1+x^2} (kx \, dx)$$

$$u = x^2 \qquad dv = \sqrt{1+x^2}(kx \, dx)$$

$$du = 2x \, dx \qquad v = \frac{k}{2} \int (1+x^2)^{1/2}(2x \, dx)$$

$$v = \frac{k}{2}(1+x^2)^{3/2}$$

$$y = \frac{kx^2}{3}(1+x^2)^{3/2} - \int \frac{k}{3}(1+x^2)^{3/2}(2x \, dx)$$

$$y = \frac{kx^2}{3}(1+x^2)^{3/2} - \frac{k}{3}\left(\frac{2}{5}\right)(1+x^2)^{5/2} + C$$

$$y = \frac{k}{3}(1+x^2)^{3/2}\left(x^2 - \frac{2(1+x^2)}{5}\right) + C$$

$$y = \frac{k}{3}(1+x^2)^{3/2}\left(\frac{5x^2 - 2 - 2x^2}{5}\right) + C$$

$$y = \frac{k}{3}(1+x^2)^{3/2}\left(\frac{3x^2 - 2}{5}\right) + C$$

$$y = \frac{k}{15}(1+x^2)^{3/2}(3x^2 - 2) + C$$

$$x = 0, y = 0$$

$$0 = \frac{k}{15}(1+0^2)^{3/2}(3(0^2) - 2) + C$$

$$0 = \frac{k}{15}(1)(-2) + C, C = \frac{2k}{15}$$

$$y = \frac{k}{15}(1+x^2)^{3/2}(3x^2 - 2) + \frac{2k}{15}$$

$$y = \frac{k}{15}((1+x^2)^{3/2}(3x^2 - 2) + 2)$$

25. $v = \dfrac{ds}{dt} = \dfrac{t^3}{\sqrt{t^2+1}}$; $s = \displaystyle\int \dfrac{t^3 dt}{\sqrt{t^2+1}}$

Let $u = t^2$; $du = 2t \, dt$; $dv = \dfrac{t \, dt}{(t^2+1)^{1/2}}$

$$v = \frac{1}{2}\int \frac{2t \, dt}{(t^2+1)^{1/2}} = \frac{1}{2}(2)(t^2+1)^{1/2}$$

$$= (t^2+1)^{1/2}$$

$$s = t^2(t^2+1)^{1/2} - \int (t^2+1)^{1/2}(2t \, dt)$$

$$= t^2(t^2+1)^{1/2} - \frac{2}{3}(t^2+1)^{3/2} + C$$

$$s = 0 \text{ for } t = 0; \quad 0 = -\frac{2}{3} + C; \quad C = \frac{2}{3}$$

$$s = \frac{1}{3}[3t^2(t^2+1)^{1/2} - 2(t^2+1)^{3/2} + 2]$$

$$= \frac{1}{3}[(t^2 - 2)(t^2+1)^{1/2} + 2]$$

26. $y = \ln x$; $y = 0$; $x = 9.5$; rotated about x-axis disks.

$$dV = \pi y^2 dx$$

$$V = \pi \int_1^{9.5} (\ln x)^2 dx; \quad u = \ln^1 x; \quad du = \frac{2\ln x}{x} dx;$$

$$dv = dx; \quad v = x$$

$$\pi \int_1^{9.5} (\ln x)^2 dx = \pi \left(x \ln^2 x - 2 \int \ln x \, dx\right)$$

$$u = \ln x; \quad du = \frac{dx}{x}; \quad dv = dx; \quad v = x$$

$$\pi \int \ln^2 x \, dx = \pi \left[x \ln^2 x - 2\left(x \ln x - \int dx\right)\right]$$

$$\pi \int \ln^2 x \, dx = \pi \left[x \ln^2 x - 2x \ln x + 2x\right]\Big|_1^{9.5}$$

$$V = \pi[9.5 \ln^2 9.5 - 2(9.5)\ln 9.5 + 2(9.5)] = \pi(2)$$
$$= 70.3 \text{ m}^3$$

27. $i = e^{-2t}\cos t = \dfrac{dy}{dt}$; $t = 0, q = q_0 = 0$

$q = \displaystyle\int e^{-2t}\cos t\, dt$; $u = e^{-2t}$; $du = -2e^{-2t}dt$;

$dv = \cos x\, dt$; $v = \sin t$

$q = \displaystyle\int e^{-2t}\cos t\, dt = e^{-2t}\sin t + 2\int e^{-2t}\sin t\, dt$;

$u = e^{-2t}$; $du = -2e^{-2t}dt$; $dv = \sin t\, dt$;

$v = -\cos t$

$\displaystyle\int e^{-2t}\cos t\, dt$

$= e^{-2t}\sin t + 2\left(-e^{-2t}\cos t - \displaystyle\int 2e^{-2t}\cos t\, dt\right)$

$\displaystyle\int e^{-2t}\cos t\, dt$

$= e^{-2t}\sin t - 2e^{-2t}\cos t - 4\displaystyle\int e^{-2t}\cos t\, dt$

$5\displaystyle\int e^{-2t}\cos t\, dt = e^{-2t}(\sin t - 2\cos t)$

$\displaystyle\int e^{-2t}\cos t\, dt = \frac{1}{5}e^{-2t}(\sin t - 2\cos t) + C$;

$t = 0$; $q = 0$; $q = \frac{1}{5}e^{-2t}(\sin t - 2\cos t) + C$

$0 = \frac{1}{5}(1)(\sin 0 - 2\cos 0) + C$;

therefore, $C = \dfrac{2}{5}$

$q = \frac{1}{5}e^{-2t}(\sin t - 2\cos t) + \frac{2}{5}$

$= \frac{1}{5}[e^{-2t}(\sin t - 2\cos t) + 2]$

28. $\bar{x} = \displaystyle\lim_{b\to\infty}\left[0.1\int_0^b x^3 e^{-x^2/8}dx\right]$; $x \to 3.2$ as $b \to \infty$

$\displaystyle\int_0^b x^3 e^{-x^2/8}dx$; $u = x^2$; $du = 2x\, dx$; $dv = e^{-x^2/8}x\, dx$

$v = -4\displaystyle\int e^{-x^2/8}\left(-\frac{1}{4}x\, dx\right) = -4e^{-x^2/8}$

$\displaystyle\int_0^b x^2(e^{-x^2/8}x\, dx)$

$= -4x^2 e^{-x^2/8} + 8\displaystyle\int xe^{-x^2/8}dx$

$= -4x^2 e^{-x^2/8} - 32\displaystyle\int e^{-(1/8)x^2} - \frac{1}{4}x\, dx$

$= -4x^2 e^{-x^2/8} - 32e^{-(1/8)x^2}\Big|_0^b$

$= -4b^2 e^{-b^2/8} - 32e^{-b^2/8} + 32$

$\bar{x} = \displaystyle\lim_{b\to\infty}[0.1(32 - 4b^2 e^{-b^2/8} - 32e^{-b^2/8})]$

$= \displaystyle\lim_{b\to\infty}(3.2 - 0.4b^2 e^{-b^2/8} - 3.2e^{-b^2/8})$

$= \displaystyle\lim_{b\to\infty}[3.2 - 0.4e^{-b^2/8}(b^2 + 8)]$

$= \displaystyle\lim_{b\to\infty}3.2 - \lim_{b\to\infty}\frac{0.4(b^2 + 8)}{e^{b^2/8}}$

$= 3.2 - 0$

$= 3.2$ nm

b	$\bar{x}$
10	3.1998
10^2	3.2
10^3	3.2
10^4	3.2

10.2   Integration By Substitution

1. $\displaystyle\int x\sqrt{x+1}\,dx$

$u = \sqrt{x+1}$
$u^2 = x + 1$
$x = u^2 - 1$
$dx = 2u\, du$

$\displaystyle\int x\sqrt{x+1}\,dx = \int (u^2 - 1)u(2u\, du)$

$= 2\displaystyle\int (u^4 - u^2)du$

$= 2\left(\dfrac{u^5}{5} - \dfrac{u^3}{3} + C\right)$

$= \dfrac{2}{5}(x+1)^{5/2} - \dfrac{2}{3}(x+1)^{3/2} + C$

$= 2(x+1)^{3/2}\left(\dfrac{1}{5}(x+1) - \dfrac{1}{3}\right) + C$

$= 2(x+1)^{3/2}\left(\dfrac{1}{5}x + \dfrac{1}{5} - \dfrac{1}{3}\right) + C$

$= 2(x+1)^{3/2}\left(\dfrac{1}{5}x - \dfrac{2}{15}\right) + C$

$= 2(x+1)^{3/2}\left(\dfrac{3x - 2}{15}\right) + C$

$= \dfrac{2}{15}(x+1)^{3/2}(3x - 2) + C$

2. $\int x\sqrt{x+3}\,dx$

$$u = \sqrt{x+3}$$
$$u^2 = x + 3$$
$$2u\,du = dx$$
$$x = u^2 - 3$$

$$\int x\sqrt{x+3}\,dx = \int (u^2 - 3)u(2u\,du)$$
$$= 2\int (u^4 - 3u^2)\,du$$
$$= 2\left(\frac{u^5}{5} - u^3\right) + C$$
$$= \frac{2}{5}(u^5 - 5u^3) + C$$
$$= \frac{2}{5}((x+3)^{5/2} - 5(x+3)^{3/2}) + C$$
$$= \frac{2}{5}(x+3)^{3/2}(x+3-5) + C$$
$$= \frac{2}{5}(x+3)^{3/2}(x-2) + C$$

3. $\int x\sqrt{2x+1}\,dx$

$$u = \sqrt{2x+1}$$
$$u^2 = 2x + 1$$
$$2u\,du = 2\,dx, u\,du = dx$$
$$x = \frac{1}{2}(u^2 - 1)$$

$$\int x\sqrt{2x+1}\,dx = \int \frac{1}{2}(u^2-1)(u)(u\,du)$$
$$= \frac{1}{2}\int (u^4 - u^2)\,du$$
$$= \frac{1}{2}\left(\frac{u^5}{5} - \frac{u^3}{3}\right) + C$$
$$= \frac{1}{10}u^5 - \frac{1}{6}u^3 + C$$
$$= \frac{1}{10}(2x+1)^{5/2} - \frac{1}{6}(2x+1)^{3/2} + C$$
$$= (2x+1)^{3/2}\left(\frac{1}{10}(2x+1) - \frac{1}{6}\right) + C$$
$$= (2x+1)^{3/2}\left(\frac{x}{5} + \frac{1}{10} - \frac{1}{6}\right) + C$$
$$= (2x+1)^{3/2}\left(\frac{x}{5} - \frac{1}{15}\right) + C$$
$$= \frac{1}{15}(2x+1)^{3/2}(3x-1) + C$$

4. $\int x\sqrt{3-x}\,dx$

$$u = \sqrt{3-x}$$
$$u^2 = 3 - x, x = 3 - u^2$$
$$2u\,du = -dx$$

$$\int x\sqrt{3-x}\,dx = \int (3-u^2)(u)(-2u\,du)$$
$$= \int (2u^4 - 6u^2)\,du$$
$$= \frac{2}{5}u^5 - 2u^3 + C$$
$$= \frac{2}{5}(3-x)^{5/2} - 2(3-x)^{3/2} + C$$
$$= 2(3-x)^{3/2}\left(\frac{1}{5}(3-x) - 1\right) + C$$
$$= 2(3-x)^{3/2}\left(\frac{3}{5} - \frac{1}{5}x - 1\right) + C$$
$$= 2(3-x)^{3/2}\left(-\frac{1}{5}x - \frac{2}{5}\right) + C$$
$$= -\frac{2}{5}(3-x)^{3/2}(x+2) + C$$

5. $\int \frac{x\,dx}{\sqrt{x+3}}$

$$u = \sqrt{x+3}$$
$$u^2 = x + 3, x = u^2 - 3$$
$$2u\,du = dx$$

$$\int \frac{x\,dx}{\sqrt{x+3}} = \int \frac{(u^2-3)(2u\,du)}{u}$$
$$= 2\int (u^2 - 3)\,du$$
$$= 2\frac{u^3}{3} - 6u + C$$
$$= \frac{2}{3}(x+3)^{3/2} - 6(x+3)^{1/2} + C$$
$$= (x+3)^{1/2}\left(\frac{2}{3}(x+3) - 6\right) + C$$
$$= (x+3)^{1/2}\left(\frac{2}{3}x + 2 - 6\right) + C$$
$$= (x+3)^{1/2}\left(\frac{2}{3}x - 4\right) + C$$
$$= \frac{2}{3}(x+3)^{1/2}(x-6) + C$$

6. $\int \dfrac{x\,dx}{\sqrt{2x+7}}$

$$u = \sqrt{2x+7}$$

$$u^2 = 2x+7, x = \frac{1}{2}(u^2-7)$$

$$2u\,du = 2\,dx, u\,du = dx$$

$$\int \frac{x\,dx}{\sqrt{2x+7}} = \int \frac{\frac{1}{2}(u^2-7)(u\,du)}{u} = \frac{1}{2}\int (u^2-7)du = \frac{1}{2}\left(\frac{u^3}{3} - 7u\right) + C$$

$$= \frac{1}{2}\left(\frac{1}{3}(2x+7)^{3/2} - 7(2x+7)^{1/2}\right) + C = \frac{1}{2}(2x+7)^{1/2}\left(\frac{1}{3}(2x+7) - 7\right) + C$$

$$= \frac{1}{2}(2x+7)^{1/2}\left(\frac{2}{3}x + \frac{7}{3} - 7\right) + C = \frac{1}{2}(2x+7)^{1/2}\left(\frac{2}{3}x - \frac{14}{3}\right) + C$$

$$= \frac{1}{3}(2x+7)^{1/2}(x-7) + C$$

7. $\int \dfrac{x^2\,dx}{\sqrt{x-2}}$

$$u = \sqrt{x-2}$$
$$u^2 = x-2, x = u^2+2$$
$$2u\,du = dx$$

$$\int \frac{x^2\,dx}{\sqrt{x-2}} = \int \frac{(u^2+2)^2(2u\,du)}{u} = 2\int (u^4 + 4u^2 + 4)du = 2\left(\frac{u^5}{5} + \frac{4}{3}u^3 + 4u\right) + C$$

$$= 2\left(\frac{1}{5}(x-2)^{5/2} + \frac{4}{3}(x-2)^{3/2} + 4(x-2)^{1/2}\right) + C = 2(x-2)^{1/2}\left(\frac{1}{5}(x-2)^2 + \frac{4}{3}(x-2) + 4\right) + C$$

$$= 2(x-2)^{1/2}\left(\frac{1}{5}(x^2 - 4x + 4) + \frac{4}{3}x - \frac{8}{3} + 4\right) + C = 2(x-2)^{1/2}\left(\frac{1}{5}x^2 - \frac{4}{5}x + \frac{4}{5} + \frac{4}{3}x + \frac{4}{3}\right) + C$$

$$= 2(x-2)^{1/2}\left(\frac{1}{5}x^2 + \frac{8}{15}x + \frac{32}{15}\right) + C\frac{2}{15}(x-2)^{1/2}(3x^2 + 8x + 32) + C$$

8. $\int_3^{11} \dfrac{3x\,dx}{\sqrt{2x+3}}$

$$u = \sqrt{2x+3}, u^2 = 2x+3, x = \frac{1}{2}(u^2-3)$$

$$2u\,du = 2\,dx$$
$$u\,du = dx$$

$$\int_3^{11} \frac{3x\,dx}{\sqrt{2x+3}} = \int_3^5 \frac{\frac{3}{2}(u^2-3)(u\,du)}{u} = \frac{3}{2}\left(\frac{u^3}{3} - 3u\right)\Big|_3^5 = \frac{3}{2}\left(\frac{5^3}{3} - 3(5) - \frac{3^3}{3} + 3(3)\right) = 40$$

9. $\displaystyle\int_0^3 x^2\sqrt{1+x}\,dx$

$u = \sqrt{1+x},\ u^2 = 1+x,\ x = u^2 - 1$
$\qquad 2u\,du = dx$

$\displaystyle\int_0^3 x^2\sqrt{1+x}\,dx = \int_1^2 (u^2 - 1)^2(u)(2u\,du) = \int_1^2 (u^4 - 2u^2 + 1)(2u^2)du = \int_1^2 (2u^6 - 4u^4 + 2u^2)du$

$\displaystyle = \frac{2}{7}u^7 - \frac{4}{5}u^5 + \frac{2}{3}u^3 \Big|_1^2 = \frac{2}{7}(2)^7 - \frac{4}{5}(2)^5 + \frac{2}{3}(2)^3 - \frac{2}{7} + \frac{4}{5} - \frac{2}{3} = \frac{1696}{105}$

10. $\displaystyle\int \frac{3x^2\,dx}{\sqrt{1-x}}$

$u = \sqrt{1-x},\ u^2 = 1-x,\ x = 1 - u^2$
$\qquad 2u\,du = -dx$

$\displaystyle\int \frac{3x^2\,dx}{\sqrt{1-x}} = \int \frac{3(1-u^2)^2(-2u\,du)}{u} = -6\int (1 - 2u^2 + u^4)du = -6\left(u - \frac{2}{3}u^3 + \frac{1}{5}u^5\right) + C$

$\displaystyle = -6\left((1-x)^{1/2} - \frac{2}{3}(1-x)^{3/2} + \frac{1}{5}(1-x)^{5/2}\right) + C = -6(1-x)^{1/2}\left(1 - \frac{2}{3}(1-x) + \frac{1}{5}(1-x)^2\right) + C$

$\displaystyle = -6(1-x)^{1/2}\left(1 - \frac{2}{3} + \frac{2}{3}x + \frac{1}{5}(1 - 2x + x^2)\right) + C = -6(1-x)^{1/2}\left(\frac{1}{5}x^2 + \frac{4}{15}x + \frac{8}{15}\right) + C$

$\displaystyle = -\frac{2}{5}(1-x)^{1/2}(3x^2 + 4x + 8) + C$

11. $\displaystyle\int x\sqrt[3]{x-1}\,dx$

$u = \sqrt[3]{x-1},\ u^3 = x - 1,\ x = u^3 + 1$
$\qquad 3u^2\,du = dx$

$\displaystyle\int x\sqrt[3]{x-1}\,dx = \int (u^3 + 1)(u)(3u^2\,du) = 3\int (u^6 + u^3)du = \frac{3}{7}u^7 + \frac{3}{4}u^4 + C = \frac{3}{7}(x-1)^{7/3} + \frac{3}{4}(x-1)^{4/3} + C$

$\displaystyle = 3(x-1)^{4/3}\left(\frac{1}{7}(x-1) + \frac{1}{4}\right) + C = 3(x-1)^{4/3}\left(\frac{1}{7}x - \frac{1}{7} + \frac{1}{4}\right) + C = 3(x-1)^{4/3}\left(\frac{1}{7}x + \frac{3}{28}\right) + C$

$\displaystyle = \frac{3}{28}(x-1)^{4/3}(4x + 3) + C$

12. $\displaystyle\int x\sqrt[3]{8-x}\,dx$

$u = \sqrt[3]{8-x},\ u^3 = 8 - x,\ x = 8 - u^3$
$\qquad 3u^2\,du = -dx$

$\displaystyle\int x\sqrt[3]{8-x}\,dx = \int (8 - u^3)(u)(-3u^2\,du) = 3\int (u^3 - 8)u^3\,du = 3\int (u^6 - 8u^3)du = \left(\frac{1}{7}u^7 - 2u^4\right) + C$

$\displaystyle = 3\left(\frac{1}{7}(8-x)^{7/3} - 2(8-x)^{4/3}\right) + C = 3(8-x)^{4/3}\left(\frac{1}{7}(8-x) - 2\right) + C = 3(8-x)^{4/3}\left(\frac{8}{7} - \frac{1}{7}x - 2\right)$

$\displaystyle = 3(8-x)^{4/3}\left(-\frac{1}{7}x - \frac{6}{7}\right) + C = -\frac{3}{7}(8-x)^{4/3}(x + 6) + C$

13. $\displaystyle\int \frac{x\,dx}{\sqrt[4]{2x+3}}$

$$u = \sqrt[4]{2x+3},\, u^4 = 2x+3,\, x = \frac{1}{2}(u^4-3)$$

$$4u^3\,du = 2\,dx$$
$$2u^3\,du = dx$$

$$\int \frac{x\,dx}{\sqrt[4]{2x+3}} = \int \frac{\frac{1}{2}(u^4-3)(2u^3\,du)}{u} = \int (u^6 - 3u^2)\,du = \frac{1}{7}u^7 - u^3 + C = \frac{1}{7}(2x+3)^{7/4} - (2x+3)^{3/4} + C$$

$$= (2x+3)^{3/4}\left(\frac{1}{7}(2x+3) - 1\right) + C = (2x+3)^{3/4}\left(\frac{2}{7}x + \frac{3}{7} - 1\right) + C = (2x+3)^{3/4}\left(\frac{2}{7}x - \frac{4}{7}\right) + C$$

$$= \frac{2}{7}(2x+3)^{3/4}(x-2) + C$$

14. $\displaystyle\int x(x-4)^{2/3}\,dx$

$$u = (x-4)^{1/3},\, u^3 = x-4,\, x = u^3 + 4$$
$$3u^2\,du = dx$$

$$\int x(x-4)^{2/3}\,d = \int (u^3+4)(u^2)(3u^2\,du) = 3\int (u^7 + 4u^4)\,du = 3\left(\frac{1}{8}u^8 + \frac{4}{5}u^5\right) + C$$

$$= 3\left(\frac{1}{8}(x-4)^{8/3} + \frac{4}{5}(x-4)^{5/3}\right) + C = 3(x-4)^{5/3}\left(\frac{1}{8}(x-4) + \frac{4}{5}\right) + C$$

$$= 3(x-4)^{5/3}\left(\frac{1}{8}x - \frac{1}{2} + \frac{4}{5}\right) + C = 3(x-4)^{5/3}\left(\frac{1}{8}x + \frac{3}{10}\right) + C$$

$$= \frac{3}{40}(x-4)^{5/3}(5x + 12) + C$$

15. $\displaystyle\int_0^7 x(x+1)^{2/3}\,dx$

$$u = (x+1)^{1/3},\, u^3 = x+1,\, x = u^3 - 1$$
$$3u^2\,du = dx$$

$$\int_0^7 x(x+1)^{2/3}\,dx = \int_1^2 (u^3-1)(u^2)(3u^2\,du) = 3\int_1^2 (u^7 - u^4)\,du = 3\left(\frac{1}{8}u^8 - \frac{1}{5}u^5\right)\Big|_1^2$$

$$= 3\left(\frac{1}{8}(2)^8 - \frac{1}{5}(2)^5 - \frac{1}{8}(1)^8 + \frac{1}{5}(1)^5\right) = \frac{3081}{40}$$

16. $\displaystyle\int_0^2 \frac{x^2\,dx}{(4x+1)^{5/2}}$

$$u = (4x+1)^{1/2},\, u^2 = 4x+1,\quad x = \frac{1}{4}(u^2-1)$$

$$2u\,du = 4\,dx \qquad x^2 = \frac{1}{16}(u^2-1)^2$$
$$u\,du = 2\,dx$$

$$\int_0^2 \frac{x^2\,dx}{(4x+1)^{5/2}} = \int_1^3 \frac{\frac{1}{16}(u^2-1)^2\left(\frac{1}{2}u\,du\right)}{u^5} = \frac{1}{32}\int_1^3 \frac{(u^4 - 2u^2 + 1)u\,du}{u^5} = \frac{1}{32}\int_1^3 (1 - 2u^{-2} + u^{-4})\,du$$

$$= \frac{1}{32}\left(u + 2u^{-1} - \frac{1}{3}u^{-3}\right)\Big|_1^3 = \frac{1}{32}\left(3 + 2(3)^{-1} - \frac{1}{3}(3)^{-3} - 1 - 2(1)^{-1} + \frac{1}{3}(1)^{-3}\right) = \frac{5}{162}$$

17. $A = \int_1^2 y\,dx = \int_1^2 x^3\sqrt{x-1}\,dx;\ u = \sqrt{x-1}, x = u^2 + 1, dx = 2u\,du$

When $x = 1, u = 0$, and when $x = 2, u = 1$.

$A = \int_0^1 (u^2+1)^3(u)(2u\,du) = 2\int_0^1 (u^8 + 3u^6 + 3u^4 + u^2)\,du = 2\left(\frac{1}{9}u^9 + \frac{3}{7} + \frac{3}{5}u^5 + \frac{1}{3}u^3\right)\Big|_0^1$

$= 2\left(\frac{1}{9} + \frac{3}{7} + \frac{3}{5} + \frac{1}{3}\right) - 2(0) = \frac{928}{315} = 2.946$

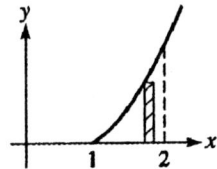

18. $A = \int_0^8 y\,dx = \int_0^8 x\sqrt[3]{x+8}\,dx$

$u = \sqrt[3]{x+8}, u^3 = x+8, x = u^3 - 8$
$3u^2\,du = dx$

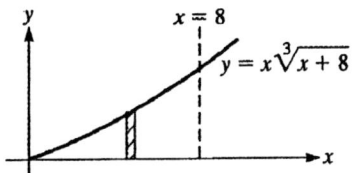

$A = \int_2^{\sqrt[3]{16}} (u^3 - 8)(u)(3u^2\,du) = 3\int_2^{\sqrt[3]{16}} (u^6 - 8u^3)\,du = 3\left(\frac{1}{7}u^7 - 2u^4\right)\Big|_2^{\sqrt[3]{16}}$

$A = 3\left(\frac{1}{7}(16)^{7/3} - 2(16)^{4/3} - \frac{1}{7}(2)^7 + 2(2)^4\right) = 3\left(\frac{1}{7}(2^4)^{7/3} - 2(2^4)^{4/3} - \frac{1}{7}(2^7) + 2^5\right)$

$A = 3\left(\frac{1}{7}(2^{28/3}) - 2^{19/3} - \frac{1}{7}(2^7) + 2^5\right) = 3(2^5)\left(\frac{1}{7}(2)^{13/3} - 2^{4/3} - \frac{1}{7}(2^2) + 1\right)$

$A = \frac{3(2^5)}{7}(2^{13/3} - 7(2^{4/3}) - 4 + 7) = \frac{96}{7}(2^{4/3}(8-7) - 4 + 7) = \frac{96}{7}(2\sqrt[3]{2} + 3) = 75.7$

19. $V = 2\pi\int_1^9 xy\,dx = 2\pi\int_1^9 x^2\sqrt[3]{x-1}\,dx;\ u = \sqrt[3]{x-1}, x = u^3 + 1, dx = 3u^2\,du$

When $x = 1, u = 0$, and when $x = 9,, u = 2$.

$V = 2\pi\int_0^2 (u^3+1)^2(u)(3u^2\,du) = 6\pi\int_0^2 (u^9 + 2u^6 + u^3)\,du = 6\pi\left(\frac{1}{10}u^{10} + \frac{2}{7}u^7 + \frac{1}{4}u^4\right)\Big|_0^2$

$= 6\pi\left[\frac{1}{10}(1024) + \frac{2}{7}(128) + \frac{1}{4}(16)\right] - 6\pi(0) = 6\pi\left(\frac{512}{5} + \frac{256}{7} + 4\right) = \frac{30024\pi}{35} = 2695$

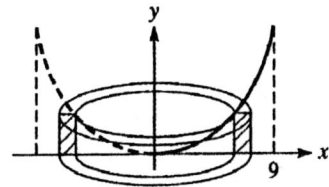

20. $V = \displaystyle\int_1^9 \pi y^2 dx$

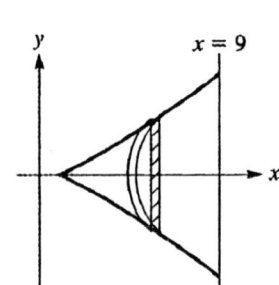

$V = \pi \displaystyle\int_1^9 x^2 (x-1)^{2/3} dx$

$u = (x-1)^{2/3}, u^{3/2} = x - 1$

$\dfrac{3}{2} u^{1/2} du = dx$

$x = u^{3/2} + 1$
$x^2 = (u^{3/2} + 1)^2 = u^3 + 2u^{3/2} + 1$

$V = \pi \displaystyle\int_1^4 (u^3 + 2u^{3/2} + 1)(u)\left(\dfrac{3}{2} u^{1/2} du\right) = \dfrac{3\pi}{2} \int_1^4 (u^{9/2} + 2u^3 + u^{3/2}) du$

$V = \dfrac{3\pi}{2} \left[\dfrac{2}{11} u^{1/2} + \dfrac{1}{2} u^4 + \dfrac{2}{5} u^{5/2}\right]\Bigg|_0^4 = \dfrac{42{,}336\pi}{55} = 2418$

21. $W = \displaystyle\int_0^{2.50} 4s\sqrt{4s+3}\,ds$; $u = \sqrt{4s+3}, s = \dfrac{1}{4}(u^2 - 3), ds = \dfrac{1}{2} u\,du$; when $s = 0, u = \sqrt{3}$, and when $s = 2.50$,

$u = \sqrt{13.0}$.

$W = \displaystyle\int_0^{2.50} 4s\sqrt{4s+3}\,ds = \int_{\sqrt{3}}^{\sqrt{13}} 4\left(\dfrac{1}{4}\right)(u^2 - 3)(u)\left(\dfrac{1}{2} u\,du\right) = \dfrac{1}{2}\int_{\sqrt{3}}^{\sqrt{13}} (u^4 - 3u^2) du = \dfrac{1}{10} u^5 - \dfrac{1}{2} u^3 \Bigg|_{\sqrt{3}}^{\sqrt{13}}$

$= \dfrac{1}{10}(\sqrt{13})^5 - \dfrac{1}{2}(\sqrt{13})^3 - \left[\dfrac{1}{10}(\sqrt{3})^5 - \dfrac{1}{2}(\sqrt{3})^3\right] = 38.5 \text{ ft} \cdot \text{lb}$

22. $i = \dfrac{dq}{dt} = t^3\sqrt{6t+1}$

$q = \displaystyle\int t^3\sqrt{6t+1}\,dt$

$u = \sqrt{6t+1}, u^2 = 6t + 1, \quad t = \dfrac{1}{6}(u^2 - 1)$

$2u\,du = 6\,dt \qquad t^3 = \dfrac{1}{216}(u^2 - 1)^3$

$\dfrac{1}{3} u\,du = dt \qquad t^3 = \dfrac{1}{216}(u^6 - 3u^4 + 3u^2 - 1)$

$q = \displaystyle\int \dfrac{1}{216}(u^6 - 3u^4 + 3u^2 - 1)(u)\left(\dfrac{1}{3} u\,du\right) = \dfrac{1}{648}\int (u^8 - 3u^6 + 3u^4 - u^2) du$

$q = \dfrac{1}{648}\left(\dfrac{1}{9} u^9 - \dfrac{3}{7} u^7 + \dfrac{3}{5} u^5 - \dfrac{1}{3} u^3\right) + C$

$q = \dfrac{1}{648}\left(\dfrac{1}{9}(6t+1)^{9/2} - \dfrac{3}{7}(6t+1)^{7/2} + \dfrac{3}{5}(6t+1)^{5/2} - \dfrac{1}{3}(6t+1)^{3/2}\right) + C$

$q = \dfrac{1}{648}(6t+1)^{3/2}\left(\dfrac{1}{9}(6t+1)^3 - \dfrac{3}{7}(6t+1)^2 + \dfrac{3}{5}(6t+1) - \dfrac{1}{3}\right) + C$

$q = \dfrac{1}{648}(6t+1)^{3/2}\left(\dfrac{8}{315}(945t^3 - 135t^2 + 18t - 2)\right) + C$

$q = \dfrac{(6t+1)^{3/2}}{25{,}515}(945t^3 - 135t^2 + 18t - 2) + C$

at $t = 0, q = 0$ from which $C = \dfrac{2}{25,515}$

$q = \dfrac{(6t+1)^{3/2}(945t^3 - 135t^2 + 18t - 2) + 2}{25,515}\Bigg|_{t=2.50}$

$q = 35.0 \times 10^{-15}\,\text{C}$

23. $\displaystyle\int x\sqrt{1-x}\,dx;\ v = 1 - x, x = 1 - v, dx = -dv$

$\displaystyle\int x\sqrt{1-x}\,dx = \int (1-v)(v^{1/2})(-dv) = -\int (v^{1/2} - v^{3/2})dv = \int (v^{3/2} - v^{1/2})dv = \frac{2}{5}v^{5/2} - \frac{2}{3}v^{3/2} + C$

$\qquad = \frac{2}{5}(1-x)^{5/2} - \frac{2}{3}(1-x)^{3/2} + C = 2(1-x)^{3/2}\left[\frac{1}{5}(1-x) - \frac{1}{3}\right] + C = \frac{-2}{15}(1-x)^{3/2}(2 + 3x) + C$

24. $\displaystyle\int \frac{2x\,dx}{(x+3)^{2/3}}$

$v = x + 3, x = v - 3$
$dv = dx$

$\displaystyle\int \frac{2x\,dx}{(x+3)^{2/3}} = \int \frac{2(v-3)dx}{v^{2/3}} = 2\int (v^{1/3} - 3v^{-2/3})dv = 2\left(\frac{3}{4}v^{4/3} - 3\left(\frac{3}{1}\right)v^{1/3}\right) + C = 2\left(\frac{3}{4}(x+3)^{4/3} - 9(x+3)\right)$

$\qquad = 2(x+3)^{1/3}\left(\frac{3}{4}(x+3) - 9\right) + C = 2(x+3)^{1/3}\left(\frac{3}{4}x + \frac{9}{4} - 9\right) + C = 2(x+3)^{1/3}\left(\frac{3}{4}x - \frac{27}{4}\right) + C$

$\qquad = 2\left(\frac{3}{4}\right)(x+3)^{1/3}(x-9) + C = \frac{3}{2}(x+3)^{1/3}(x-9) + C$

10.3   Integration By Trigonometric Substitution

1. Let $x = \sin\theta$; $dx = \cos\theta\,d\theta$

$\displaystyle\int \frac{\sqrt{1-x^2}}{x^2}\,dx = \int \frac{\sqrt{1 - \sin^2\theta}}{\sin^2\theta}\cos\theta\,d\theta = \int \frac{\cos^2\theta}{\sin^2\theta}\,d\theta = \int \cot^2\theta\,d\theta = \int (\csc^2\theta - 1)\,d\theta$

$\qquad = \int \csc^2\theta\,d\theta - \int d\theta = -\cot\theta - \theta + C = \frac{-\sqrt{1-x^2}}{x} - \sin^{-1}x + C$

2. $\displaystyle\int \frac{dt}{(t^2+9)^{3/2}};\ a = 3$

Let $t = 3\tan\theta$; therefore, $\tan\theta = \dfrac{t}{3}$; $dt = 3\sec^2\theta\,d\theta$

$\displaystyle\int \frac{3\sec^2\theta\,d\theta}{(9\tan^2\theta + 9)^{3/2}} = \int \frac{3\sec^2\theta\,d\theta}{[9(\tan^2\theta + 1)]^{3/2}} = \int \frac{3\sec^2\theta\,d\theta}{3^3\sec^3\theta};\ \int \frac{1}{9}\frac{d\theta}{9\sec\theta} = \frac{1}{9}\int \cos\theta\,d\theta = \frac{1}{9}\sin\theta + C$

$\displaystyle\int_0^4 \frac{dt}{(t^2+9)^{3/2}} = \frac{1}{9}\frac{t}{9\sqrt{t^2+9}}\Bigg|_0^4 = \frac{4}{45}$

3. $\displaystyle\int \frac{2\,dx}{\sqrt{x^2-4}}; \; a=2$

Let $x = 2\sec\theta$; therefore, $\sec\theta = \dfrac{x}{2}$

$dx = 2\sec\theta\tan\theta\,d\theta$

$$2\int \frac{2\sec\theta\tan\theta\,d\theta}{\sqrt{4\sec^2\theta-4}} = 2\int \frac{2\sec\theta\tan\theta\,d\theta}{2\tan\theta} = \int 2\sec\theta\,d\theta = 2\ln|\sec\theta+\tan\theta|+C_1 = 2\ln\left|\frac{x}{2}+\frac{\sqrt{x^2-4}}{2}\right|+C_1$$

$$= 2\ln\left|\frac{x+\sqrt{x^2-4}}{2}\right|+C_1 = 2\ln\left|x+\sqrt{x^2-4}\right|+C$$

4. $\displaystyle\int \frac{\sqrt{x^2-25}}{x}\,dx; \; a=5$

Let $x = 5\sec\theta$; $\sec\theta = \dfrac{x}{5}$

$dx = 5\sec\theta\tan\theta\,d\theta$

$$\int \frac{\sqrt{25\sec^2\theta-25}\,(5\sec\theta\tan\theta\,d\theta)}{5\sec\theta} = 5\int \tan\theta\tan\theta\,d\theta = 5\int \tan^2\theta\,d\theta = 5\int(\sec^2\theta-1)d\theta$$

$$= 5\int \sec^2\theta\,d\theta - 5\int d\theta = 5\tan\theta - 5\theta + C$$

$$= 5\frac{\sqrt{x^2-25}}{5} - 5\sec^{-1}\frac{x}{5} + C = \sqrt{x^2-25} - 5\sec^{-1}\left(\frac{x}{5}\right)+C$$

5. Let $z = 3\tan\theta$; $dz = 3\sec^2\theta\,d\theta$

$$\int \frac{6\,dz}{z^2\sqrt{z^2+9}} = 6\int \frac{3\sec^2\theta\,d\theta}{9\tan^2\theta\sqrt{9\tan^2\theta+9}} = 6\int \frac{3\sec^2\theta\,d\theta}{27\tan^2\theta\sqrt{\tan^2\theta+1}} = \frac{6}{9}\int \frac{\sec\theta\,d\theta}{\tan^2\theta} = \frac{6}{9}\int \frac{\cos\theta\,d\theta}{\sin^2\theta}$$

$$= \frac{6}{9}\int \csc\theta\cot\theta\,d\theta = -\frac{6}{9}\csc\theta + C = \frac{-6}{9\sin\theta}+C$$

$\tan\theta = \dfrac{z}{3}$; $\sin\theta = \dfrac{z}{\sqrt{9+z^2}}$

$$\frac{-6}{9\sin\theta}+C = \frac{-6}{\dfrac{9z}{\sqrt{9+z^2}}}+C = -\frac{2\sqrt{z^2+9}}{3z}+C$$

6. $\displaystyle\int \frac{3\,dx}{x\sqrt{4-x^2}}; \; a=2$

$x = 2\sin\theta$; $\sin\theta = \dfrac{x}{2}$; $dx = 2\cos\theta\,d\theta$

$$3\int \frac{2\cos\theta\,d\theta}{2\sin\theta\sqrt{4-4\sin^2\theta}} = 3\int \frac{\cos\theta\,d\theta}{\sin\theta\cdot 2\cdot\cos\theta} = \frac{3}{2}\int \csc\theta\,d\theta = \frac{3}{2}\ln|\csc\theta-\cot\theta|+C$$

$$= \frac{3}{2}\ln\left|\frac{2}{x}-\frac{\sqrt{4-x^2}}{x}\right|+C = \frac{3}{2}\ln\left|\frac{2-\sqrt{4-x^2}}{x}\right|+C$$

7. $\displaystyle\int \frac{4\,dx}{(4-x^2)^{3/2}}$; $a=2$; $x=2\sin\theta$; $\sin\theta=\dfrac{x}{2}$; $dx=2\cos\theta\,d\theta$

$$4\int\frac{2\cos\theta\,d\theta}{(4-4\sin^2\theta)^{3/2}}=8\int\frac{\cos\theta\,d\theta}{2^3\cos^3\theta}=\int\frac{d\theta}{\cos^2\theta}=\int\sec^2\theta\,d\theta=\tan\theta+C=\frac{x}{\sqrt{4-x^2}}+C$$

8. $\displaystyle\int\frac{6p^3\,dp}{\sqrt{9+p^2}}$; $a=3$; $p=3\tan\theta$; $\tan\theta=\dfrac{p}{3}$; $dp=3\sec^2\theta\,d\theta$

$$6\int\frac{27\tan^3\theta(3\sec^2\theta)d\theta}{\sqrt{9+9\tan^2\theta}}=18\int\frac{27\tan^3\theta\sec^2\theta\,d\theta}{3\sec\theta}=162\int\tan^3\theta\sec\theta\,d\theta=162\int\tan^2\theta\tan\theta\sec\theta\,d\theta$$

$$=162\int\sec^2(\sec\theta\tan\theta\,d\theta)-162\int\sec\theta\tan\theta\,d\theta=162\frac{\sec^3\theta}{3}-162\sec\theta+C$$

$$=54\left(\frac{\sqrt{9+p^2}}{3}\right)^3-162\left(\frac{\sqrt{9+p^2}}{3}\right)+C=\frac{54}{3^3}(9+p^2)\sqrt{9+p^2}-54\sqrt{9+p^2}+C$$

$$=2\sqrt{9+p^2}(p^2-18)+C$$

9. $\displaystyle\int_0^{0.5}\frac{x^3\,dx}{\sqrt{1-x^2}}$, $x=\sin\theta$; $dx=\cos\theta\,d\theta$

$$\int\frac{\sin^3\theta\cos\theta\,d\theta}{\sqrt{1-\sin^2\theta}}=\int\sin^3\theta\,d\theta=\int\sin\theta\sin^2\theta\,d\theta=\int\sin\theta(1-\cos^2\theta)\,d\theta=\int\sin\theta\,d\theta-\int\cos^2\theta\sin\theta\,d\theta$$

$$=-\cos\theta+\frac{\cos^3\theta}{3}$$

$$\cos\theta=\sqrt{1-x^2};\ -\sqrt{1-x^2}+\frac{1}{3}(\sqrt{1-x^2})^3\Big|_0^{0.5}=-\sqrt{1-0.5^2}+\frac{1}{3}(\sqrt{1-0.5^2})^3+\sqrt{1}-\frac{1}{3}\sqrt{1}=0.017$$

10. $\displaystyle\int_4^5\frac{\sqrt{x^2-16}}{x^2}\,dx$; $a=4$; $x=4\sec\theta$; $\sec\theta=\dfrac{x}{4}$; $dx=4\sec\theta\tan\theta\,d\theta$

$$\int\frac{\sqrt{16\sec^2\theta-16}}{16\sec^2\theta}4\sec\theta\tan\theta\,d\theta=\int\frac{4\tan\theta 4\sec\theta\tan\theta\,d\theta}{16\sec^2\theta}=\int\frac{\tan^2\theta\,d\theta}{\sec\theta}=\int\frac{(\sec^2\theta-1)d\theta}{\sec\theta}$$

$$=\int\sec\theta\,d\theta-\int\cos\theta\,d\theta=\ln|\sec\theta+\tan\theta|-\sin\theta$$

$$=\ln\left|\frac{x}{4}+\frac{\sqrt{x^2-16}}{4}\right|-\frac{\sqrt{x^2-16}}{x}$$

$$\ln\left|\frac{x+\sqrt{x^2-16}}{4}\right|-\frac{\sqrt{x^2-16}}{x}\Big|_4^5=\ln\left|\frac{5+3}{4}\right|-\frac{3}{5}-\left(\ln\left|\frac{4+0}{4}\right|-0\right)=\ln 2-\frac{3}{5}=0.0931$$

11. $\displaystyle\int\frac{5\,dx}{\sqrt{x^2+2x+2}}=\int\frac{5\,dx}{\sqrt{(x+1)^2+1}}$

$a=1$; $x+1=\tan\theta$; $dx=\sec^2\theta\,d\theta$

$$5\int\frac{\sec^2\theta\,d\theta}{\sqrt{\tan^2\theta+1}}=5\int\frac{\sec^2\theta\,d\theta}{\sec\theta}=5\int\sec\theta\,d\theta=5\ln|\sec\theta+\tan\theta|=5\ln\left|\sqrt{x^2+2x+2}+x+1\right|+C$$

12. $\displaystyle\int\frac{dx}{\sqrt{x^2+2x}}=\int\frac{dx}{\sqrt{(x^2+1)^2-1}}$

$a=1$; $u=x+1$; $x+1=\sec\theta$; $dx=\sec\theta\tan\theta\,d\theta$

$$=\int\frac{\sec\theta\tan\theta\,d\theta}{\sqrt{\sec^2\theta-1}}=\int\frac{\sec\theta\tan\theta\,d\theta}{\tan\theta}=\int\sec\theta\,d\theta=\ln|\sec\theta+\tan\theta|+C=\ln\left|x+1+\sqrt{x^2+2x}\right|+C$$

13. $\int \dfrac{dy}{y\sqrt{4y^2-9}}$; $2y = 3\sec\theta$; $y = \dfrac{3}{2}\sec\theta$; $dy = \dfrac{3}{2}\sec\theta\tan\theta\,d\theta$

$$\int \dfrac{\frac{3}{2}\sec\theta\tan\theta\,d\theta}{\frac{3}{2}\sec\theta\sqrt{4\left(\frac{3}{2}\sec\theta\right)^2-9}} = \int \dfrac{\tan\theta\,d\theta}{\sqrt{9\sec^2\theta-9}} = \int \dfrac{\tan\theta\,d\theta}{3\sqrt{\sec^2\theta-1}} = \int \dfrac{\tan\theta\,d\theta}{3\tan\theta} = \dfrac{1}{3}\int d\theta$$

$$= \dfrac{1}{3}\theta + C = \dfrac{1}{3}\sec^{-1}\dfrac{2}{3}y + C$$

$$\int_{2.5}^{3} \dfrac{dy}{y\sqrt{4y^2-9}} = \dfrac{1}{3}\sec^{-1}\left(\dfrac{2y}{3}\right)\Big|_{2.5}^{3} = \dfrac{1}{3}\cos^{-1}\left(\dfrac{3}{2y}\right)\Big|_{2.5}^{3} = 0.03997$$

14. $\int \sqrt{16-x^2}\,dx$; $a = 4$; $x = 4\sin\theta$; $dx = 4\cos\theta\,d\theta$

$$\int \sqrt{16-16\sin^2\theta}\cdot 4\cos\theta\,d\theta; \quad \cos 2x = 2\cos^2 x - 1$$

$$\int 4\cos\theta\cdot 4\cos\theta\,d\theta = 16\int\cos^2\theta\,d\theta$$

$$16\int\dfrac{1}{2}(\cos 2\theta+1)\,d\theta = 8\left(\int\cos 2\theta\,d\theta + \int d\theta\right) = 8\left(\dfrac{1}{2}\sin 2\theta + \theta\right) = 8(\sin\theta\cos\theta+\theta)$$

$$= 8\left(\dfrac{x}{4}\cdot\dfrac{\sqrt{16-x^2}}{4} + \sin^{-1}\dfrac{x}{4}\right) = 8\sin^{-1}\left(\dfrac{x}{4}\right) + \dfrac{1}{2}x\sqrt{16-x^2} + C$$

15. $\int \dfrac{2\,dx}{\sqrt{e^{2x}-1}}$; $a = 1$; $e^x = \sec\theta$; $\theta = \sec^{-1}e^x$; $e^x\dfrac{dx}{d\theta} = \sec\theta\tan\theta$

$$dx = \dfrac{\sec\theta\tan\theta\,d\theta}{e^x} = \dfrac{\sec\theta\tan\theta\,d\theta}{\sec\theta} = \tan\theta\,d\theta = 2\int\dfrac{\tan\theta\,d\theta}{\sqrt{\sec^2\theta-1}} = 2\int\dfrac{\tan\theta\,d\theta}{\tan\theta} = 2\theta + C = 2\sec^{-1}e^x + C$$

16. $\int \dfrac{12\sec^2 u\,du}{(4-\tan^2 u)^{3.2}}$; $a = 2$; $\tan u = 2\sin\theta$; $\sin\theta = \dfrac{\tan u}{2}$; $\sec^2 u\,du = 2\cos\theta\,d\theta$

$$12\int\dfrac{2\cos\theta\,d\theta}{(4-4\sin^2\theta)^{3/2}} = 12\int\dfrac{2\cos\theta\,d\theta}{8\cos^3\theta} = 3\int\sec^2\theta\,d\theta = 3\tan\theta + C = 3\dfrac{\tan u}{\sqrt{4-\tan^2 u}} + C$$

17. $A = 4\int_0^1 y\,dx = 4\int_0^1 \sqrt{1-x^2}\,dx$

Let $x = \sin\theta$, $dx = \cos\theta\,d\theta$

$$\int \sqrt{1-x^2}\,dx = \int\sqrt{1-\sin^2\theta}\cos\theta\,d\theta = \int\cos^2\theta\,d\theta = \dfrac{1}{2}\int(1+\cos 2\theta)\,d\theta$$

$$= \dfrac{1}{2}\theta + \dfrac{1}{4}\sin 2\theta + C = \dfrac{1}{2}\theta + \dfrac{1}{2}\sin\theta\cos\theta + C$$

$$= \dfrac{1}{2}\sin^{-1}x + \dfrac{1}{2}x\sqrt{1-x^2} + C$$

$$A = 4\int_0^1\sqrt{1-x^2}\,dx = 4\left(\dfrac{1}{2}\sin^{-1}x + \dfrac{1}{2}x\sqrt{1-x^2}\right)\Big|_0^1 = 2\sin^{-1}1 + 2(1)\sqrt{0} - [2\sin^{-1}0 + 2(0)]$$

$$= 2\sin^{-1}1 = 2\left(\dfrac{\pi}{2}\right) = \pi$$

18. $A_{\sqrt{2},\sqrt{5}} = \displaystyle\int_{\sqrt{2}}^{\sqrt{5}} \frac{dx}{x^2\sqrt{x^2-1}}$; $x = \sec\theta$; $dx = \sec\theta\tan\theta\,d\theta$

$$\int \frac{\sec\theta\tan\theta\,d\theta}{\sec^2\theta\sqrt{\sec^2\theta-1}} = \int \frac{\tan\theta\,d\theta}{\sec\theta\tan\theta} = \int \cos\theta\,d\theta = \sin\theta = \left.\frac{\sqrt{x^2-1}}{x}\right|_{\sqrt{2}}^{\sqrt{5}} = \frac{\sqrt{4}}{\sqrt{5}} - \frac{\sqrt{1}}{\sqrt{2}}$$

$$= \frac{2\sqrt{2}-\sqrt{5}}{\sqrt{10}} = 0.187$$

19. (a) general power rule

$$\int \frac{x}{\sqrt{x^2+4}}\,dx = \frac{1}{2}\int (x^2+4)^{-1/2}(2x\,dx) = \frac{1}{2}\frac{(x^2+4)^{-1/2+1}}{-\frac{1}{2}+1} + C = \sqrt{x^2+4} + C$$

(b) substitution $u = (x^2+4)^{1/2}, u^2 = x^2+4$

$$2u\,du = 2x\,dx$$
$$u\,du = x\,dx$$

$$\int \frac{x\,dx}{\sqrt{x^2+4}} = \int \frac{u\,du}{u} = \int du = u + C = \sqrt{x^2+4} + C$$

The general poer rule would appear to be a bit simpler.

20. $x^2 + y^2 = a^2$

$$I_y = 2\pi k \int_0^a 2yx^3\,dx = 4\pi k \int_0^a \sqrt{a^2-x^2}\,x^3\,dx = 4\pi k \int_0^a \sqrt{a^2-a^2\sin^2\theta}\,a^3\sin^3\theta\,a\cos\theta\,d\theta$$

$$= 4\pi k \int a\cos\theta\,a^3\sin^3\theta\,a\cos\theta\,d\theta = 4\pi k a^5 \int \sin^3\theta\cos^2\theta\,d\theta$$

$x = a\sin\theta$; $dx = a\cos\theta\,d\theta$

By Example 1, p. 298, Sec. 9-5:

$$I_y = 4\pi k a^4 \int \sin^3\theta\cos^2\theta\,d\theta = 4\pi k a^4 \left(-\frac{1}{3}\cos^3\theta + \frac{1}{5}\cos^5\theta\right) = 4\pi k a^4 \left[-\frac{1}{3}\frac{(a^2-x^2)^{3/2}}{a^3} + \frac{1}{5}\frac{(a^2-x^2)^{5/2}}{a^5}\right]\Bigg|_0^a$$

$$I_y = \frac{8}{15}\pi k a^5$$

$$m = 4\pi k \int_0^a xy\,dx = 4\pi k \int_0^a x\sqrt{a^2-x^2}\,dx = -\frac{4\pi k}{2}\int_0^a (a^2-x^2)^{1/2}(-2x\,dx) = \left.-2\pi k\frac{2}{3}(a^2-x^2)^{3/2}\right|_0^a$$

$$= \left.-\frac{4}{3}\pi k(a^2-x^2)^{3/2}\right|_0^a = -\left[-\frac{4}{3}\pi k(a^2)^{3/2}\right] = \frac{4}{3}\pi k a^3$$

$$I_y = \frac{8}{15}\pi k a^5 = \frac{4}{3}\pi k a^3\left(\frac{2}{5}a^2\right) = \frac{2}{5}ma^2$$

21. $V = 2\pi \displaystyle\int_4^5 \frac{x(\sqrt{x^2 - 16})}{x^2}\, dx$, disks

(The limits of integration are $x = 4$ since $x = 4$ when $y = 0$; and $x = 5$.)
Let $x = 4\sec\theta$; $dx = 4\sec\theta\tan\theta\, d\theta$

$$V = 2\pi \int_4^5 \frac{\sqrt{x^2 - 16}}{x}\, dx$$

$$2\pi \int \frac{\sqrt{x^2 - 16}}{x}\, dx = 2\pi \int \frac{\sqrt{16\sec^2\theta - 16}}{4\sec\theta}(4\sec\theta\tan\theta\, d\theta) = 2\pi \int \sqrt{16\sec^2\theta - 16}\tan\theta\, d\theta$$

$$= 2\pi \int 4\sqrt{\sec^2\theta - 1}\tan\theta\, d\theta = 8\pi \int (\tan\theta)\tan\theta\, d\theta = 8\pi \int \tan^2\theta\, d\theta$$

$$= 8\pi \int (\sec^2\theta - 1)\, d\theta = 8\pi(\tan\theta - \theta)$$

Since $x = 4\sec\theta$; $\sec = \dfrac{x}{4}$; and $\tan\theta = \dfrac{\sqrt{x^2 - 16}}{4}$; and

$$2\pi \int \frac{\sqrt{x^2 - 16}}{x}\, dx = 8\pi(\tan\theta - \theta) = 8\pi\left(\frac{\sqrt{x^2 - 16}}{4} - \sec^{-1}\frac{x}{4}\right)$$

$$V = 2\pi \int_4^5 \frac{x\sqrt{x^2 - 16}}{x^2}\, dx = 8\pi\left(\frac{\sqrt{x^2 - 16}}{4} - \sec^{-1}\frac{x}{4}\right)\Bigg|_4^5 = 8\pi\left(\frac{3}{4} - \sec^{-1}\frac{5}{4} - 0\right) = 2.68$$

22. $1.00x^2 + 9.00y^2 = 9.00$

$$9.00y^2 = 9.00 - 1.00x^2$$

$$y^2 = 1.00 - \frac{1.00}{9.00}x^2$$

$$y^2 = \frac{9.00 - 1.00x^2}{9.00}$$

$$y = \frac{\sqrt{9.00 - 1.00x^2}}{3.00}$$

$$A = 4 \int_{0.00}^{3.00} \frac{\sqrt{9.00 - 1.00x^2}}{3.00}\, dx$$

$$x = 3.00\sin\theta, \quad dx = 3.00\cos\theta\, d\theta$$
$$x = 0 = 3.00\sin\theta, \quad \theta = 0$$
$$x = 3.00 = 3.00\sin\theta, \quad \theta = \frac{\pi}{2}$$

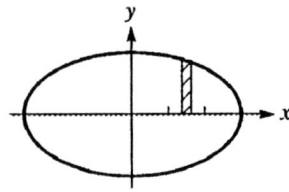

$$A = \frac{4}{3.00} \int_0^{\pi/2} \sqrt{9.00 - 1.00(9.00 \sin^2 \theta)}(3.00 \cos \theta \, d\theta)$$

$$A = \frac{4}{3.00} \int_0^{\pi/2} 3.00 \sqrt{1 - \sin^2 \theta}(3.00 \cos \theta \, d\theta)$$

$$A = 12.0 \int_0^{\pi/2} \cos^2 \theta \, d\theta$$

$$A = 12.0 \int_0^{\pi/2} \frac{1}{2}(1 + \cos 2\theta) d\theta$$

$$A = 6.00 \left(\theta + \frac{1}{2} \sin 2\theta \right) \Big|_0^{\pi/2}$$

$$A = 6.00 \left(\frac{\pi}{2} + \frac{1}{2} \sin \left(2 \left(\frac{\pi}{2} \right) \right) - 0 - \frac{1}{2} \sin(2(0)) \right)$$

$$A = 3.00\pi$$
$$A = 9.42 \text{ m}^2$$

23. $V = kQ \int_{-a}^{a} \frac{dx}{\sqrt{b^2 + x^2}}$

$x = b \tan \theta; \ dx = b \sec^2 \theta \, d\theta$

$$V = kQ \int \frac{b \sec^2 \theta \, d\theta}{\sqrt{b^2 + b^2 \tan^2 \theta}} = kQ \int \frac{b \sec^2 \theta \, d\theta}{b \sec \theta} = kQ \int \sec \theta \, d\theta = kQ \ln |\sec \theta + \tan \theta|$$

$$= kQ \ln \left| \frac{\sqrt{x^2 + b^2}}{b} + \frac{x}{b} \right| \Big|_{-a}^{a} = kQ \ln \left| \frac{\sqrt{x^2 + b^2} + x}{x} \right| \Big|_{-a}^{a}$$

$$= kQ \left(\ln \left| \frac{\sqrt{x^2 + b^2} + a}{a} \right| - \ln \left| \frac{\sqrt{x^2 + b^2} - a}{-a} \right| \right)$$

$$= kQ(\ln |\sqrt{a^2 + b^2} + a| - \ln |a| - \ln |\sqrt{a^2 + b^2}| + \ln |a|)$$

$$V = kQ \left(\ln \left| \frac{\sqrt{a^2 + b^2} + a}{\sqrt{a^2 + b^2} - a} \right| \right)$$

24. $y = \sqrt{x^2 - 4}; \ y = 0, x = 2.5$, rotated about y-axis; shells.

$dV = 2\pi xy \, dx$

$$V = 2\pi \int_2^{2.5} (x)x^2 \sqrt{x^2 - 4} dx = 2\pi \int_2^{2.5} x^3 \sqrt{x^2 - 4} dx$$

$x = 2 \sec \theta; \ dx = 2 \sec \theta \tan \theta \, d\theta$

$$V = 2\pi \int 8 \sec^3 \theta \sqrt{4 \sec^2 \theta - 4} \cdot 2 \sec \theta \tan \theta \, d\theta = 64\pi \int \sec^3 \tan \theta \sec \theta \tan \theta \, d\theta = 64\pi \int \sec^4 \theta \tan^2 \theta \, d\theta$$

$$= 64\pi \int \sec^2 \theta (1 + \tan^2 \theta) \tan^2 \theta \, d\theta = 64\pi \left(\int \tan^2 \theta \sec^2 \theta \, d\theta + \int \tan^4 \theta \sec^2 \theta \, d\theta \right)$$

$$= 64\pi \left[\frac{1}{3} \tan^3 \theta + \frac{1}{5} \tan^5 \theta \right] = 64\pi \left[\frac{1}{3} \frac{(x^2 - 4)^{3/2}}{8} + \frac{1}{5} \frac{(x^2 - 4)^{5/2}}{32} \right] \Big|_2^{2.5}$$

$$= 64\pi \left[\frac{1}{24} (x^2 - 4)^{3/2} + \frac{1}{160} (x^2 - 4)^{5/2} \right] \Big|_2^{2.5} = 64\pi \left[\frac{1}{24} (2.25)^{3/2} + \frac{1}{160} (2.25)^{5/2} - 0 \right]$$

$$V = 37.8 \text{ cm}^3$$

10.4   Integration by Partial Fractions: Nonrepeated Linear Factors

1. $\dfrac{x+3}{(x+1)(x+2)} = \dfrac{A}{x+1} + \dfrac{B}{x+2}$

$\qquad\qquad = \dfrac{A(x+2) + B(x+1)}{(x+1)(x+2)}$

$\qquad x+3 = A(x+2) + B(x+1)$

$\quad x = -2, \quad 1 = -B, \quad B = -1$

$\quad x = -1, \quad 2 = A$

$\displaystyle\int \frac{x+3}{(x+1)(x+2)} = \int \frac{2}{x+1}\,dx + \int \frac{-1}{x+2}\,dx$

$\qquad\qquad = 2\ln|x+1| - \ln|x+2| + C$

$\qquad\qquad = \ln(x+1)^2 - \ln|x+2|$

$\qquad\qquad = \ln\dfrac{(x+1)^2}{|x+2|} + C$

2. $\displaystyle\int \frac{x+2}{x(x+1)}\,dx = \int \frac{2}{x}\,dx - \int \frac{dx}{x+1}$

$\qquad\qquad = 2\ln|x| - \ln|x+1| + C$

$\qquad\qquad = \ln\dfrac{x^2}{|x+1|} + C$

3. $\displaystyle\int \frac{dx}{x^2-4} = \int \frac{dx}{(x+2)(x-2)}$

$\qquad\qquad = \int \frac{-\frac{1}{4}}{x+2}\,dx + \int \frac{\frac{1}{4}}{x-2}\,dx$

$\qquad\qquad = -\frac{1}{4}\ln|x+2| + \frac{1}{4}\ln|x-2| + C$

$\qquad\qquad = \frac{1}{4}\ln\left|\dfrac{x-2}{x+2}\right| + C$

4. $\displaystyle\int \frac{p-9}{2p^2-3p+1}\,dp$

$\quad = \displaystyle\int \frac{p-9}{(2p-1)(p-1)}\,dp$

$\quad = \frac{1}{2}\displaystyle\int \frac{17}{2p-1}\cdot 2\,dp - \int \frac{8}{p-1}\,dp$

$\quad = \frac{17}{2}\ln|2p-1| - 8\ln|p-1| + C$

5.

$$x^2 + 3x \,\overline{)\,x^2 + 0x + 3\,}$$
$$\underline{x^2 + 3x}$$
$$-3x + 3$$

$\dfrac{x^2+3}{x^2+3x} = 1 + \dfrac{-3x+3}{x(x+3)}$

$\dfrac{-3x+3}{x(x+3)} = \dfrac{A}{x} + \dfrac{B}{x+3} = \dfrac{A(x+3)+Bx}{x(x+3)}$

$\qquad -3x+3 = A(x+3) + Bx$

$\quad x = 0, 3 = 3A, = A = 1$

$x = -3, 12 = -3B = B = -4$

$\displaystyle\int \frac{x^2+3}{x^2+3x} = \int dx + \int \frac{dx}{x} + \int \frac{-4}{x+3}\,dx$

$\qquad\qquad = x + \ln|x| - 4\ln|x+3| + C$

$\qquad\qquad = x + \ln\dfrac{|x|}{(x+3)^4} + C$

6. $\displaystyle\int \frac{x^3}{x^2+3x+2}\,dx$

$\quad = \displaystyle\int (x-3)\,dx + \int \frac{8}{x+2}\,dx - \int \frac{dx}{x+1}$

$\quad = \dfrac{x^2}{2} - 3x + 8\ln|x+2| - \ln|x+1| + C$

$\quad = \dfrac{x^2}{2} - 3x + \ln\dfrac{(x+2)^8}{|x+1|} + C$

7. $\displaystyle\int_0^1 \frac{2t+4}{3t^2+5t+2}\,dt$

$\quad = \displaystyle\int_0^1 \frac{8}{3t+2}\,dt - \int_0^1 \frac{2}{t+1}\,dt$

$\quad = \dfrac{8\ln(3t+2)}{3}\Big|_0^1 - 2\ln(t+1)\Big|_0^1$

$\quad = \dfrac{8\ln 5}{3} - \dfrac{8\ln 2}{3} - 2\ln 2 + 2\ln 1$

$\quad = 1.057$

8. $\displaystyle\int_1^3 \frac{x-1}{4x^2+x}\,dx$

$\quad = \displaystyle\int_1^3 \frac{x-1}{x(4x+1)}\,dx = \int_1^3 \frac{5\,dx}{4x+1} - \int_1^3 \frac{dx}{x}$

$\quad = \dfrac{5\ln(4x+1)}{4}\Big|_1^3 - \ln x \Big|_1^3$

$\quad = \dfrac{5\ln 13}{4} - \dfrac{5\ln 5}{4} - \ln 3 + \ln 1 = 0.09578$

9. $\dfrac{4x^2 - 10}{x(x+1)(x-5)}$

$$= \frac{A}{x} + \frac{B}{x+1} + \frac{C}{x-5}$$

$$= \frac{A(x+1)(x-5) + B \cdot x(x-5) + Cx(x+1)}{x(x+1)(x-5)}$$

$4x^2 - 10 = A(x+1)(x-5) + Bx(x-5) + Cx(x+1)$

$x = 0, -10 = -5A, \ A = 2$

$x = 5, \quad 90 = 30C, \ C = 3$

$4x^2 - 10 = 2(x+1)(x-5) + Bx(x-5) + 3x(x+1)$

$x = 1, -6 = -16 - 4B + 6, \ B = -1$

$$\int \frac{4x^2 - 10}{x(x+1)(x-5)} \, dx$$

$$= \int \frac{2}{x} \, dx - \int \frac{dx}{x+1} + \int \frac{3}{x-5} \, dx$$

$$= 2\ln|x| - \ln|x+1| + 3\ln|x-5| + C$$

$$= \ln \frac{x^2 \, |x-5|^3}{|x+1|} + C$$

10. $\displaystyle \int \frac{4x^2 + 21x + 6}{(x+2)(x-3)(x+4)} \, dx$

$$= \int \frac{2}{x+2} \, dx + \int \frac{3}{x-3} \, dx - \int \frac{1}{x+4} \, dx$$

$$= 2\ln|x+2| + 3\ln|x-3| - \ln|x+4| + C$$

$$= \ln \left| \frac{(x+2)^2 (x-3)^3}{x+4} \right| + C$$

11. $\displaystyle \int \frac{6x^2 - 2x - 1}{4x^3 - x} \, dx$

$$= \int \frac{6x^2 - 2x - 1}{x(4x^2 - 1)} \, dx = \int \frac{6x^2 - 2x - 1}{x(2x+1)(2x-1)} \, dx$$

$$= \int \frac{dx}{x} + \int \frac{\frac{3}{2}}{2x+1} \, dx - \frac{\frac{1}{2}}{2x-1} \, dx$$

$$= \ln|x| + \frac{3\ln|2x+1|}{4} - \frac{\ln|2x-1|}{4} + C$$

$$= \frac{4\ln|x|}{4} + \frac{3\ln|2x+1|}{4} - \frac{\ln|2x-1|}{4} + C$$

$$= \frac{1}{4} \ln \left| \frac{x^4 (2x+1)^3}{2x-1} \right| + C$$

12. $\displaystyle \int_2^3 \frac{dR}{R^3 - R}$

$$= \int_2^3 \frac{dR}{R(R-1)(R+1)}$$

$$= \int_2^3 \frac{\frac{1}{2}}{R-1} \, dR + \int_2^3 \frac{\frac{1}{2}}{R+1} \, dR - \int_2^3 \frac{dR}{R}$$

$$= \frac{1}{2} \ln|R-1| \Big|_2^3 + \frac{1}{2} \ln|R+1| \Big|_2^3 - \ln|R| \Big|_2^3$$

$$= \frac{1}{2} \ln 2 - \frac{1}{2} \ln 1 + \frac{1}{2} \ln 4 - \frac{1}{2} \ln 3 - \ln 3 + \ln 2$$

$$= \frac{1}{2} \ln 4 - \frac{3}{2} \ln 3 + \frac{3}{2} \ln 2 = \frac{\ln \frac{32}{27}}{2}$$

$$= 0.08495$$

13. $\dfrac{x^3 + 7x^2 + 9x + 2}{x(x^2 + 3x + 2)} = \dfrac{x^3 + 7x^2 + 9x + 2}{x^3 + 3x^2 + 2x}$

$$
\begin{array}{r}
1 \hspace{3.5cm} \\
x^3 + 3x^2 + 2x \overline{)x^3 + 7x^2 + 9x + 2} \\
\underline{x^3 + 3x^2 + 2x} \hspace{0.5cm} \\
4x^2 + 7x + 2
\end{array}
$$

$$\frac{x^3 + 7x^2 + 9x + 2}{x(x^2 + 3x + 2)} = 1 + \frac{4x^2 + 7x + 2}{x(x^2 + 3x + 2)}$$

$$= 1 + \frac{4x^2 + 7x + 2}{x(x+1)(x+2)}$$

$\dfrac{4x^2 + 7x + 2}{x(x+1)(x+2)}$

$$= \frac{A}{x} + \frac{B}{x+1} + \frac{C}{x+2}$$

$$= \frac{A(x+1)(x+2) + Bx(x+2) + Cx(x+1)}{x(x+1)(x+2)}$$

$4x^2 + 7x + 2$

$= A(x+1)(x+2) + Bx(x+2) + Cx(x+1)$

$x = 0, \qquad 2 = 2A, \ A = 1$

$x = -1, \quad -1 = -B, \ B = 1$

$x = -2, \qquad 4 = 2C, \ C = 2$

$$\int_1^2 \frac{x^3 + 7x^2 + 9x + 2}{x(x^2 + 3x + 2)} \, dx$$

$$= \int_1^2 dx + \int_1^2 \frac{dx}{x} + \int_1^2 \frac{dx}{x+1} + \int_1^2 \frac{2}{x+2} \, dx$$

$$= x + \ln|x| + \ln|x+1| + 2\ln|x+2| \Big|_1^2$$

$$= 2.674$$

14. $\displaystyle\int \frac{2x^3 + x - 1}{x^3 + x^2 - 4x - 4}\, dx$

$\displaystyle = \int \frac{\frac{17}{12}}{x - 2}\, dx - \int \frac{\frac{19}{4}}{x + 2}\, dx + \int \frac{\frac{4}{3}}{x + 1}\, dx$

$\displaystyle\qquad + \int 2\, dx$

$\displaystyle = \frac{17}{12} \ln |x - 2| - \frac{19}{4} \ln |x + 2| + \frac{4}{3} \ln |x + 1|$

$\displaystyle\qquad + 2x + C$

15. $\displaystyle\int \frac{dV}{(V^2 - 4)(V^2 - 9)}$

$\displaystyle = \int \frac{dV}{(V - 2)(V + 2)(V + 3)(V - 3)}$

$\displaystyle = \int \frac{\frac{1}{30}}{V - 3}\, dV - \int \frac{\frac{1}{30}}{V + 3}\, dV - \int \frac{\frac{1}{20}}{V - 2}\, dV + \int \frac{\frac{1}{20}}{V + 2}\, dV$

$\displaystyle = \frac{1}{30} \ln |V - 3| - \frac{1}{30} \ln |V + 3|$

$\displaystyle\qquad - \frac{1}{20} \ln |V - 2| + \frac{1}{20} \ln |V + 2| + C$

$\displaystyle = \frac{2}{60} \ln |V - 3| - \frac{2}{60} \ln |V + 3| - \frac{3}{60} \ln |V - 2|$

$\displaystyle\qquad + \frac{3}{60} \ln |V + 2| + C$

$\displaystyle = \frac{1}{60} \ln \left| \frac{(V + 2)^3 (V - 3)^2}{(V - 2)^3 (V + 3)^2} \right| + C$

16. $\displaystyle\int \frac{5x^3 - 2x^2 - 15x + 24}{x^4 - 2x^3 - 11x^2 + 12x}\, dx$

$\displaystyle = \int \frac{5x^3 - 2x^2 - 15x + 24}{x(x - 1)(x + 3)(x - 4)}\, dx$

$\displaystyle = \int \frac{3}{x - 4}\, dx + \int \frac{dx}{x + 3} + \int \frac{2\, dx}{x} - \int \frac{dx}{x - 1}$

$\displaystyle = 3 \ln |x - 4| + \ln |x + 3| + 2 \ln |x| - \ln |x - 1| + C$

$\displaystyle = \ln \left| \frac{(x - 4)^3 (x + 3) x^2}{x - 1} \right| + C$

17. $\displaystyle A = \int_2^4 \frac{x - 16}{x^2 - 5x - 14}\, dx$

$\displaystyle = \int_2^4 \frac{-1}{x - 7}\, dx + \int_2^4 \frac{2}{x + 2}\, dx$

$\displaystyle = 1.322$

18. $\displaystyle A = \int_1^3 \frac{dx}{x^3 + 3x^2 + 2x} = \int_1^3 \frac{dx}{x(x + 1)(x + 2)}$

$\displaystyle = \int_1^3 \frac{\frac{1}{2}}{x + 2}\, dx + \int_1^3 \frac{\frac{1}{2}}{x}\, dx - \int_1^3 \frac{dx}{x + 1}$

$\displaystyle = \frac{1}{2} \ln(x + 2) \Big|_1^3 + \frac{1}{2} \ln x \Big|_1^3 - \ln(x + 1) \Big|_1^3$

$\displaystyle = \frac{1}{2} \ln 5 - \frac{1}{2} \ln 3 + \frac{1}{2} \ln 3 - \frac{1}{2} \ln 1 - \ln 4 + \ln 2$

$\displaystyle = \frac{1}{2} \ln 5 - \ln 4 + \ln 2 = \ln \frac{2\sqrt{5}}{4} = 0.1116$

19. $\displaystyle V = \int_1^3 \frac{2\pi x\, dx}{(x^3 + 3x^2 + 2x)}$

$\displaystyle = \int_1^3 \frac{2\pi\, dx}{x + 1} - \int_1^3 \frac{2\pi\, dx}{x + 2}$

$\displaystyle V = 2\pi \ln(x + 1) \big|_1^3 - 2\pi \ln(x + 2) \big|_1^3$

$\displaystyle = 2\pi [\ln 4 - \ln 2 - \ln 5 + \ln 3]$

$\displaystyle V = 2\pi \ln \frac{4 \cdot 3}{2 \cdot 5}$

$\displaystyle = 2\pi \ln \frac{6}{5}$

$\displaystyle = 1.146$

20. $\displaystyle \bar{x} = \frac{\displaystyle\int_2^4 x\, \frac{1}{x^2 - 1}\, dx}{\displaystyle\int_2^4 \frac{1}{x^2 - 1}\, dx}$

$\displaystyle \bar{x} = \frac{\displaystyle\frac{1}{2} \int_2^4 \frac{1}{x^2 - 1}\, (2x\, dx)}{\displaystyle\frac{1}{2} \ln \frac{x - 1}{x + 1} \Big|_2^4}$

$\displaystyle \bar{x} = \frac{\displaystyle\frac{1}{2} \ln(x^2 - 1) \Big|_2^4}{\displaystyle\ln \sqrt{\frac{x - 1}{x + 1}} \Big|_2^4}$

$\displaystyle \bar{x} = \frac{\displaystyle\ln \sqrt{x^2 - 1} \Big|_2^4}{\displaystyle\ln \sqrt{\frac{3}{5}} - \ln \sqrt{\frac{1}{3}}}$

$\displaystyle \bar{x} = \frac{\ln \sqrt{15} - \ln \sqrt{3}}{\displaystyle\ln \sqrt{\frac{3}{5}} - \ln \sqrt{\frac{1}{3}}}$

$\displaystyle \bar{x} = 2.738$

21. $\dfrac{3x+5}{x^2+5x} = \dfrac{3x+5}{x(x+5)} = \dfrac{A}{x} + \dfrac{B}{x+5}$

$\qquad\quad = \dfrac{A(x+5)+Bx}{x(x+5)}$

$\qquad 3x+5 = A(x+5)+Bx$

$x=0, \qquad 5=5A, \quad A=1$

$x=-5, \ -10=-5B, \ B=2$

$y = \displaystyle\int \dfrac{3x+5}{x^2+5x}\,dx = \int \dfrac{dx}{x} + \int \dfrac{2}{x+5}\,dx$

$0 = \ln|x| + 2\ln|x+5| + C = \ln|1| + 2\ln|6| + C, \ C=-2\ln 6$

$y = \ln|x| + 2\ln|x+5| - 2\ln 6 = \ln|x| + \ln|x+5|^2 - \ln 36$

$y = \ln \dfrac{|x|\,(x+5)^2}{36}$

22. $q = \displaystyle\int_0^1 \dfrac{(4t+3)\,dt}{2t^2+3t+1}$

$q = \ln(2t^2+3t+1)\,\Big|_0^1$

$q = \ln(2(1)^2+3(1)+1) - \ln(2(2)^2+3(0)+1)$

$q = 1.792 \ \mathrm{C}$

23. $w = \displaystyle\int_0^{0.5} \dfrac{4x\,dx}{x^2+3x+2} = \int_0^{0.5} \dfrac{8\,dx}{x+2} - \int_0^{0.5} \dfrac{4\,dx}{x+1}$

$w = 8\ln(x+2)\,\Big|_0^{0.5} - 4\ln(x+1)\,\Big|_0^{0.5}$

$w = 8\ln(2.5) - 8\ln(2) - 4\ln(1.5) + \ln 1$

$w = 0.1633 \ \mathrm{N\cdot m}$

24. $t = \displaystyle\int \dfrac{dx}{(4-x)(2-x)} = \dfrac{1}{2}\ln\left|\dfrac{x-4}{x-2}\right| + C$

$0 = \dfrac{1}{2}\ln\left|\dfrac{0-4}{0-2}\right| + C \Rightarrow C = -\dfrac{1}{2}\ln 2$

$t = \dfrac{1}{2}\ln\left|\dfrac{x-4}{x-2}\right| - \dfrac{1}{2}\ln 2 = \dfrac{1}{2}\ln\left|\dfrac{x-4}{2(x-2)}\right|$

10.5   Integration by Partial Fractions: Other Cases

1. $\dfrac{x-8}{x^3-4x^2+4x} = \dfrac{x-8}{x(x^2-4x+4)} = \dfrac{x-8}{x(x-2)(x-2)} = \dfrac{A}{x} + \dfrac{B}{x-2} + \dfrac{C}{(x-2)^2}$

$\qquad x-8 = A(x-2)^2 + Bx(x-2) + Cx$

$x=0, \ -8=4A, \ A=-2$

$x=2, \ -6=2C, \ C=-3$

$\qquad x-8 = -2(x-2)^2 + Bx(x-2) - 3x$

$x=1, \ -7=-2-B-3, \ B=2$

$\displaystyle\int \dfrac{x-8}{x^3-4x^2+4x} = \int \dfrac{-2}{x}\,dx + \int \dfrac{2}{x-2}\,dx - \int \dfrac{3}{(x-2)^2}\,dx = -2\ln|x| + 2\ln|x-2| + \dfrac{3}{x-2} + C$

$\qquad\qquad = 2\ln\left|\dfrac{x-2}{x}\right| + \dfrac{3}{x-2} + C$

2. $\displaystyle\int \dfrac{dT}{T^3-T^2} = \int \dfrac{dT}{T^2(T-1)} = \int \dfrac{dT}{T-1} - \int \dfrac{dT}{T^2} - \int \dfrac{dT}{T} = \ln|T-1| + \dfrac{1}{T} - \ln|T| + C$

$\qquad\qquad = \ln\left|\dfrac{T-1}{T}\right| + \dfrac{1}{T} + C$

3. $\displaystyle\int \dfrac{2\,dx}{x^2(x^2-1)} = \int \dfrac{2\,dx}{x^2(x-1)(x+1)} = \int \dfrac{dx}{x-1} - \int \dfrac{dx}{x+1} - \int \dfrac{2\,dx}{x^2} = \ln|x-1| - \ln|x+1| + \dfrac{2}{x} + C$

$\qquad\qquad = \ln\left|\dfrac{x-1}{x+1}\right| + \dfrac{2}{x} + C$

4. $\displaystyle \int_1^3 \frac{3x^3 + 8x^2 + 10x + 2}{x(x+1)^3}\,dx = \int_1^3 \frac{3\,dx}{(x+1)^3} + \int_1^3 \frac{dx}{x+1} + \int_1^3 \frac{2}{x}\,dx = \frac{-3}{2(x+1)^2} + \ln|x+1| + 2\ln|x|\,\Big|_1^3$

$$= 3.172$$

5. $\displaystyle \frac{2s}{(s-3)^3} = \frac{A}{s-3} + \frac{B}{(s-3)^2} + \frac{C}{(s-3)^3} = \frac{A(s-3)^2 + B(s-3) + C}{(s-3)^3}$

$$2s = A(s-3)^2 + B(s-3) + C$$

$$s = 3, \quad 6 = C$$

$$2s = A(s-3)^2 + B(s-3) + 6$$

$$\left. \begin{array}{ll} s = 1, & 2 = 4A - 2B + 6 \\ s = 2, & 4 = A - B + 6 \end{array} \right\} A = 0,\ B = 2$$

$$\int_1^2 \frac{2s}{(s-3)^3}\,ds = \int_1^2 \frac{2}{(s-3)^3}\,ds + \int_1^2 \frac{6}{(s-3)^3}\,ds = 2 \cdot \frac{(s-3)^{-2+1}}{-2+1} + 6 \cdot \frac{(s-3)^{-3+1}}{-3+1}\,\Big|_1^2$$

$$= \frac{-2}{(s-3)} - \frac{3}{(s-3)^2}\,\Big|_1^2 = -\frac{5}{4}$$

6. $\displaystyle \int \frac{x\,dx}{(x+2)^4} = \int \frac{dx}{(x+2)^3} - \int \frac{2\,dx}{(x+2)^4} = \frac{-1}{2(x+2)^2} + \frac{2}{3(x+2)^3} + C = -\frac{3x+2}{6(x+2)^3} + C$

7. $\displaystyle \int \frac{x^3 - 2x^2 - 7x + 28}{(x+1)^2(x-3)^2}\,dx = \int \frac{dx}{(x-3)^2} + \int \frac{2\,dx}{(x+1)^2} + \int \frac{dx}{x+1} = \frac{-1}{x-3} - \frac{2}{x+1} + \ln|x+1| + C$

8. $\displaystyle \int \frac{4\,dx}{(x+1)^2(x-1)^2}$

$$= \int \frac{dx}{(x-1)^2} - \int \frac{dx}{x-1} + \int \frac{dx}{(x+1)^2} + \int \frac{dx}{x+1} = \frac{-1}{x-1} - \ln|x-1| - \frac{1}{x+1} + \ln|x+1| + C$$

$$= \ln\left|\frac{x+1}{x-1}\right| - \frac{2x}{(x+1)(x-1)} + C$$

9. $\displaystyle \frac{x^2 + x + 5}{(x+1)(x^2+4)} = \frac{A}{x+1} + \frac{Bx+C}{x^2+4} = \frac{A(x^2+4) + (Bx+C)(x+1)}{(x+1)(x^2+4)}$

$$x^2 + x + 5 = A(x^2 + 4) + (Bx + C)(x + 1)$$

$$\left. \begin{array}{lr} x = 0, & 5 = 4A + \qquad C \\ x = 1, & 7 = 5A + 2B + 2C \\ x = 2, & 11 = 8A + 6B + 3C \end{array} \right\} A = 1,\ B = 0,\ C = 1$$

$$\int_0^2 \frac{x^2 + x + 5}{(x+1)(x^2+4)}\,dx = \int_0^2 \frac{dx}{x+1} + \int_1^2 \frac{1}{x^2+4}\,dx = \ln|x+1| + \frac{\tan^{-1}\frac{x}{2}}{2}\,\Big|_0^2 = 1.491$$

10. $\displaystyle \int \frac{v^2 + v - 1}{(v^2+1)(v-2)}\,dv = \int \frac{dv}{v^2+1} + \int \frac{dv}{v-2} = \tan^{-1}v + \ln|v-2| + C$

11. $\displaystyle \int \frac{5x^2 + 8x + 16}{x^2(x^2 + 4x + 8)}\,dx = \int \frac{3\,dx}{x^2 + 4x + 8} + \int \frac{2\,dx}{x^2} = \frac{3}{2}\tan^{-1}\left(\frac{x+2}{2}\right) - \frac{2}{x} + C$

12. $\displaystyle \int \frac{2x^2 + x + 3}{(x^2+2)(x-1)}\,dx = \int \frac{dx}{x^2+2} + \int \frac{2\,dx}{x-1} = \frac{\sqrt{2}}{2}\tan^{-1}\frac{\sqrt{2}x}{2} + 2\ln|x-1| + C$

13. $\dfrac{10x^3 + 40x^2 + 22x + 7}{(4x^2 + 1)(x^2 + 6x + 10)} = \dfrac{Ax + B}{4x^2 + 1} + \dfrac{Cx + D}{x^2 + 6x + 10}$

$10x^3 + 40x^2 + 22x + 7 = (Ax + B)(x^2 + 6x + 10) + (Cx + D)(4x^2 + 1)$

$10x^3 + 40x^2 + 22x + 7 = Ax^3 + 6Ax^2 + 10Ax + Bx^2 + 6Bx + 10B + 4Cx^3 + Cx + 4Dx^2 + D$

$10x^3 + 40x^2 + 22x + 7 = (A + 4C)x^3 + (6A + B + 4D)x^2 + (10A + 6B + C)x + 10B + D$

$$
\left.
\begin{array}{ll}
(1) & A + 4C = 10 \\
(2) & 6A + B + 4D = 40 \\
(3) & 10A + 6B + C = 22 \\
(4) & 10B + D = 7
\end{array}
\right\} A = 2, B = 0, C = 2, D = 7
$$

$$\int \frac{10x^3 + 40x^2 + 22x + 7}{(4x^2 + 1)(x^2 + 6x + 10)}\, dx = \int \frac{2x}{4x^2 + 1}\, dx + \int \frac{2x + 7}{x^2 + 6x + 10}\, dx$$

$$\int \frac{2x}{4x^2 + 1}\, dx = \frac{1}{4} \int \frac{8x}{4x^2 + 1}\, dx = \frac{\ln(4x^2 + 1)}{4}$$

$$\int \frac{2x + 7}{x^2 + 6x + 10}\, dx = \int \frac{2x + 7}{x^2 + 6x + 9 + 1}\, dx = \int \frac{2x + 7}{(x + 3)^2 + 1}\, dx \quad \text{let} \quad \begin{array}{l} u = x + 3,\ x = u - 3 \\ du = dx \end{array}$$

$$\int \frac{2x + 7}{x^2 + 6x + 10}\, dx = \int \frac{2(u - 3) + 7}{u^2 + 1}\, du$$

$$\int \frac{2x + 7}{x^2 + 6x + 10}\, dx = \int \frac{2u - 6 + 7}{u^2 + 1}\, du$$

$$\int \frac{2x + 7}{x^2 + 6x + 10}\, dx = \int \frac{2u}{u^2 + 1}\, du + \int \frac{1}{u^2 + 1}\, du$$

$$\int \frac{2x + 7}{x^2 + 6x + 10}\, dx = \ln(u^2 + 1) + \tan^{-1} u + C$$

$$\int \frac{2x + 7}{x^2 + 6x + 10}\, dx = \ln(x^2 + 6x + 10) + \tan^{-1}(x + 3) + C$$

$$\int \frac{10x^3 + 40x^2 + 22x + 7}{(4x^2 + 1)(x^2 + 6x + 10)}\, dx = \frac{\ln(4x^2 + 1)}{4} + \ln(x^2 + 6x + 10) + \tan^{-1}(x + 3) + C$$

14. $\displaystyle\int_3^4 \frac{5x^3 - 4x}{x^4 - 16}\, dx = \int_3^4 \frac{x(5x^2 - 4)\, dx}{(x + 2)(x - 2)(x^2 + 4)} = \int_3^4 \frac{3x\, dx}{x^2 + 4} + \int_3^4 \frac{dx}{x - 2} + \int_3^4 \frac{dx}{x + 2}$

$$= \frac{3}{2} \ln(x^2 + 4) \Big|_3^4 + \ln(x - 2) \Big|_3^4 + \ln(x + 2) \Big|_3^4$$

$$= \frac{3}{2} \ln(20) - \frac{3}{2} \ln(13) + \ln(2) - \ln 1 + \ln 6 - \ln 5 = \frac{3}{2} \ln \frac{20}{13} + \ln \frac{12}{5} = 1.522$$

15. $\displaystyle\int \frac{2r^3}{(r^2 + 1)^2}\, dr = \int \frac{2r\, dr}{r^2 + 1} - \int \frac{2r\, dr}{(r^2 + 1)^2} = \ln(r^2 + 1) + \frac{1}{r^2 + 1} + C$

16. $\displaystyle\int \frac{-x^3 + x^2 + x + 3}{(x + 1)(x^2 + 1)^2}\, dx = \int \frac{2\, dx}{(x^2 + 1)^2} - \int \frac{x\, dx}{x^2 + 1} + \int \frac{dx}{x + 1} = \tan^{-1} x + \frac{x}{x^2 + 1} - \frac{1}{2} \ln(x^2 + 1) + \ln|x + 1| + C$

17.
$$A = -\int_1^3 \frac{x-3}{x^3+x^2}\, dx = -\int_1^3 \frac{x-3}{x^2(x+1)}\, dx$$

$$\frac{x-3}{x^2(x+1)} = \frac{Ax+B}{x^2} + \frac{C}{x+1} = \frac{(Ax+B)(x+1)+Cx^2}{x^2(x+1)} = \frac{Ax^2+Ax+Bx+B+Cx^2}{x^2(x+1)}$$

$$x - 3 = (A+C)x^2 + (A+B)x + B$$

(1) $A + C = 0,\ C = -4$
(2) $A + B = 1,\ A = 4$
(3) $B = -3$

$$\frac{x-3}{x^2(x+1)} = \frac{4x-3}{x^2} - \frac{4}{x+1} = \frac{4}{x} - \frac{3}{x^2} - \frac{4}{x+1}$$

$$\int_1^3 \frac{x-3}{x^3+x^2}\, dx = \int_1^3 \frac{4}{x}\, dx - 3\int_1^3 \frac{dx}{x^2} - 4\int_1^3 \frac{dx}{x+1} = 4\ln|x|\Big|_1^3 + \frac{3}{x}\Big|_1^3 - 4\ln(x+1)\Big|_1^3$$

$$= 4(\ln 3 - \ln 1) + \frac{3}{3} - \frac{3}{1} - 4\ln(4) + 4\ln 2 = \ln 3^4 + 1 - 3 - \ln 4^4 + \ln 2^4$$

$$= -2 + \ln 81 - \ln 256 + \ln 16 = -2 + \ln \frac{81 \cdot 16}{256} = -2 + \ln \frac{81}{16} = -0.3781$$

$$A = 2 - \ln \frac{81}{16} = 0.3781$$

18. $A = \int_0^2 \frac{3x^2+2x+9}{(x^2+9)(x+1)}\, dx = \int_0^2 \frac{2x\, dx}{x^2+9} + \int_0^2 \frac{dx}{x+1} = \ln(x^2+9)\Big|_0^2 + \ln(x+1)\Big|_0^2$

$$= \ln 13 - \ln 9 + \ln 3 - \ln 1 = \ln \frac{13 \cdot 3}{9} = \ln \frac{13}{3} = 1.466$$

19. $V = \int_0^2 2\pi x \frac{4}{x^4+6x^2+5}\, dx = 8\pi \int_0^2 \frac{x}{(x^2+5)(x^2+1)}\, dx$

$$V = \pi \int_0^2 \left[\frac{2x}{(x^2+1)} - \frac{2x}{(x^2+5)} \right] dx = \pi \left[\ln(x^2+1) - \ln(x^2+5) \right]\Big|_0^2$$

$$V = \pi \left[\ln \frac{x^2+1}{x^2+5} \right]\Big|_0^2 = \pi \left[\ln \frac{5}{9} - \ln \frac{1}{5} \right] = \pi \ln \frac{25}{9} = 3.210$$

20. $V = \int_0^3 \pi \left(\frac{x}{(x+3)^2} \right)^2 dx = \pi \int_0^3 \frac{x^2}{(x+3)^4}\, dx = \pi \int_0^3 \left[\frac{9}{(x+3)^4} - \frac{6}{(x+3)^3} + \frac{1}{(x+3)^2} \right] dx$

$$V = \pi \left[\frac{-3}{(x+3)^3} + \frac{3}{(x+3)^2} - \frac{1}{(x+3)} \right]\Big|_0^3 = \pi \left[\frac{-3}{6^3} + \frac{3}{6^2} - \frac{1}{6} + \frac{3}{3^3} - \frac{3}{3^2} + \frac{1}{3} \right]$$

$$V = \frac{\pi}{72} = 0.0436$$

21. $\dfrac{t^2+14t+27}{(2t+1)(t+5)^2} = \dfrac{A}{2t+1} + \dfrac{B}{t+5} + \dfrac{C}{(t+5)^2}$

$$\frac{t^2+14t+27}{(2t+1)(t+5)^2} = \frac{A(t+5)^2 + B(t+5)(2t+1) + C(2t+1)}{(2t+1)(t+5)^2}$$

$$t^2 + 14t + 27 = At^2 + 10At + 25A + 2Bt^2 + 11Bt + 5B + 2Ct + C$$

$$t^2 + 14t + 27 = (A+2B)t^2 + (10A+11B+2C)t + 25A + 5B + C$$

(1) $A + 2B = 1$
(2) $10A + 11B + 2C = 14$ $\Big\}$ $A = 1,\ B = 0,\ C = 2$
(3) $25A + 5B + C = 27$

$$\frac{ds}{dt} = \frac{t^2 + 14t + 27}{(2t + 1)(t + 5)^2}$$

$$= \frac{1}{2t + 1} + \frac{2}{(t + 5)^2}$$

$$s = \frac{1}{2}\int_0^{2.00} \frac{2}{2t + 1}\, dt + \int_0^{2.00} \frac{2}{(t + 5)^2}\, dt;$$

$$s = \frac{1}{2}\ln|2t + 1|\Big|_0^{2.00} - 2 \cdot \frac{1}{(t + 5)}\Big|_0^{2.00}$$

$$s = \frac{1}{2}\ln 5.00 - \frac{1}{2}\ln 1 - \frac{2}{7.00} + \frac{2}{5.00}$$

$$= 0.919 \text{ m}$$

22. $q = \displaystyle\int_0^{0.250} \frac{0.0010(7t^2 + 16t + 48)}{(t + 4)(t^2 + 16)}\, dt$

$$q = 0.0010 \int_0^{0.250} \left[\frac{4t}{t^2 + 16} + \frac{3}{t + 4}\right] dt$$

$$q = 0.0010\left[2\ln(t^2 + 16) + 3\ln(t + 4)\right]\Big|_0^{0.250}$$

$$q = 0.0010[2\ln(0.250^2 + 16) \\ + 3\ln(0.250 + 4) - 2\ln 16 - 3\ln 4]$$

$$q = 190\mu A$$

23. $\bar{x} = \dfrac{\displaystyle\int_1^2 x\dfrac{4}{x^3 + x}\, dx}{\displaystyle\int_1^2 \dfrac{4}{x^3 + x}\, dx}$

$$\bar{x} = \frac{\displaystyle\int_1^2 \frac{4x}{x(x^2 + 1)}\, dx}{\displaystyle\int_1^2 \frac{4}{x(x^2 + 1)}\, dx}$$

$$= \frac{\displaystyle\int_1^2 \frac{4}{x^2 + 1}\, dx}{\displaystyle\int_1^2 \frac{4}{x(x^2 + 1)}\, dx}$$

$$\bar{x} = \frac{4\tan^{-1}x\Big|_1^2}{\displaystyle\int_1^2 \left[\frac{4}{x} - \frac{4x}{x^2 + 1}\right] dx}$$

$$= \frac{4(\tan^{-1}2 - \tan^{-1}1)}{[4\ln x - 2\ln(x^2 + 1)]\Big|_1^2}$$

$$\bar{x} = \frac{4\left(\tan^{-1}2 - \dfrac{\pi}{4}\right)}{4\ln 2 - 2\ln 5 - 4\ln 1 + 2\ln 2}$$

$$\bar{x} = 1.369$$

24. $y = \displaystyle\int \frac{29x^2 + 36}{4x^4 + 9x^2}\, dx$

$$y = \int \frac{29x^2 + 36}{x^2(4x^2 + 9)}\, dx$$

$$y = \int \left[\frac{13}{(2x)^2 + 3^2} + \frac{4}{x^2}\right]$$

$$y = \frac{13}{6}\tan^{-1}\frac{2x}{3} - \frac{4}{x} + C$$

$$5 = \frac{13}{6}\tan^{-1}\frac{2(1)}{3} - \frac{4}{1} + C \Rightarrow$$

$$C = 9 - \frac{13}{6}\tan^{-1}\frac{2}{3}$$

$$y = \frac{13}{6}\tan^{-1}\frac{2x}{3} - \frac{4}{x} + 9 - \frac{13}{6}\tan^{-1}\frac{2}{3}$$

10.6   Integration by Use of Tables

1. Formula #1; $u = x$; $a = 2$; $b = 5$; $du = dx$

$$\int \frac{3x\, dx}{2 + 5x} = 3\int \frac{3x\, dx}{2 + 5x}$$

$$= 3\left\{\frac{1}{25}[(2 + 5x) - 2\ln|2 + 5x|\right\} + C$$

$$= \frac{3}{25}[2 + 5x - 2\ln|2 + 5x|] + C$$

2. $\displaystyle\int \frac{4x\, dx}{(1 + x)^2}$; Formula #3; $a = 1$; $b = 1$; $u = x$; $du = dx$

$$4\int \frac{x\, dx}{(1 + x)^2} = 4\left[\frac{1}{1}\left(\frac{1}{1 + x} + \ln(1 + x)\right)\right] + C$$

$$= 4\left[\frac{1}{1 + x} + \ln(1 + x)\right] + C$$

3. $\displaystyle\int_2^7 4x\sqrt{2 + x}\, dx$; Formula #5; $a = 2$; $b = 1$; $u = x$;

$$du = dx$$

$$4\int_2^7 x\sqrt{2 + x}\, dx = 4\left[\frac{-2(4 - 3x)(2 + x)^{3/2}}{15}\right]\Big|_2^7$$

$$= 4\left[\frac{-2(-17)27}{15} + \frac{2(-2)8}{15}\right]$$

$$= \frac{4}{15}(918 - 32) = \frac{3544}{15}$$

$$= 236.3$$

4. $\int \dfrac{dx}{x^2 - 4}$; Formula #9; $u = x$; $du = dx$; $a = 2$

$$\int \frac{dx}{x^2 - 4} = \frac{1}{4} \ln \frac{x - 2}{x + 2} + C$$

5. Formula #24; $u = y$, $a = 2$

$$\int \frac{dy}{(y^2 + 4)^{3/2}} = \frac{y}{4\sqrt{y^2 + 4}} + C$$

6. $\int_0^{\pi/3} \sin^3 x\, dx$; Formula #30; $u = x$; $du = dx$

$$\int_0^{\pi/3} \sin^3 x\, dx$$

$$= -\cos x + \frac{1}{3}\cos^3 x \Big|_0^{\pi/3}$$

$$= -\cos \frac{\pi}{3} + \frac{1}{3}\left(\cos \frac{\pi}{3}\right)^3 - \left(-\cos 0 + \frac{1}{3}\cos^3 0\right)$$

$$= -\frac{1}{2} + \frac{1}{3}\left(\frac{1}{2}\right)^3 - \left(-1 + \frac{1}{3}\right)$$

$$= -\frac{1}{2} + \frac{1}{24} + \frac{2}{3} = \frac{5}{24}$$

7. $\int \sin 2x \sin 3x\, dx$; Formula #39; $a = 2$; $b = 3$; $u = x$;

$du = dx$; $\sin(-x) = -\sin x$

$$\int \sin 2x \sin 3x\, dx = \frac{\sin(-x)}{-2} - \frac{\sin((5x))}{10} + C$$

$$= \frac{-\sin x}{-2} - \frac{\sin 5x}{10} + C$$

$$= \frac{1}{2}\sin x - \frac{1}{10}\sin 5x + C$$

8. $\int 6\sin^{-1} 3x\, dx$; $u = 3x$; $du = 3\,dx$; Formula #51

$$\frac{6}{3}\int \sin^{-1} 3x (3\, dx)$$

$$= 2(3x\sin^{-1} 3x + \sqrt{1 - 9x^2}) + C$$
$$= 6x\sin^{-1} 3x + 2\sqrt{1 - 9x^2} + C$$

9. Formula #17; $u = 2x$; $du = 2\,dx$; $a = 3$

$$\int \frac{\sqrt{4x^2 - 9}}{x}\, dx = \int \frac{\sqrt{(2x)^2 - 3^2}}{2x}\, dx$$

$$= \sqrt{4x^2 - 9} - 3\sec^{-1}\left(\frac{2x}{3}\right) + C$$

10. $\int \dfrac{(9x^2 + 16)^{3/2}}{x}\, dx$; $u = 3x$; $du = 3\,dx$; $a = 4$;
Formula #21

$$\int \frac{(9x^2 + 16)^{3/2}\, 3x\, dx}{3x} = \frac{1}{3}(9x^2 + 16)^{3/2} + 16\sqrt{9x^2 + 16}$$

$$-64\ln\left(\frac{4 + \sqrt{9x^2 + 16}}{3x}\right) + C$$

11. $\int \cos^5 4x\, dx$; $u = 4x$; $du = 4\,dx$; $n = 5$;

Formula #34

$$\frac{1}{4}\int \cos^5 4x (4\, dx)$$

$$= \frac{1}{4}\left(\frac{1}{5}\cos^4 4x \sin 4x\, dx + \frac{4}{5}\int \cos^3 4x\, 4\, dx\right);$$

Formula #33

$$= \frac{1}{4}\left[\frac{1}{5}\cos^4 4x \sin 4x + \frac{4}{5}\left(\sin 4x - \frac{1}{3}\sin^3 4x\right)\right] + C$$

$$= \frac{1}{20}\cos^4 4x \sin 4x + \frac{1}{5}\sin 4x - \frac{1}{15}\sin^3 4x + C$$

12. $\int 0.2\tan^2(2\phi)\, d\phi = \frac{1}{10}\int \tan^2(2\phi)(2\, d\phi)$

Formula #35, $n = 2$, $u = 2\phi$, $du = 2\, d\phi$

$$\int 0.2\tan^2(2\phi)\, d\phi$$

$$= \frac{1}{10}\left[\frac{\tan 2\phi}{1} - \int \tan^0(2\phi)(2\, d\phi)\right]$$

$$= \frac{1}{10}\tan 2\phi - \frac{1}{5}\phi + C$$

13. Formula #52; $u = r^2$; $du = 2r\, dr$

$$6\int \tan^{-1} r^2 (r\, dr)$$

$$= 3\int \tan^{-1} r^2 (2r\, dr)$$
$$= 3\left[r^2\tan^{-1} r^2 - \frac{1}{2}\ln(1 + r^4)\right] + C$$
$$= 3r^2\tan^{-1} r^2 - \frac{3}{2}\ln(1 + r^4) + C$$

14. $\int 5xe^{4x}\, dx$; $a = 4$; $u = x$; $du = dx$; Formula #44

$$5\int xe^{4x}\, dx = 5\left[\frac{e^{4x}(4x - 1)}{16}\right] + C$$

$$= \frac{5}{16}e^{4x}(4x - 1) + C$$

15. $\int_1^2 (4-x^2)^{3/2} dx$; $a=2$; $u=x$; $du=dx$;

Formula #20

$\int_1^2 (4-x^2)^{3/2} dx$

$= \frac{x}{4}(4-x^2)^{3/2} + \frac{12x}{8}\sqrt{4-x^2} + \frac{48}{8}\sin^{-1}\frac{x}{2}\Big|_1^2$

$= \frac{x}{4}(4-x^2)^{3/2} + \frac{3}{2}x\sqrt{4-x^2} + 6\sin^{-1}\left(\frac{x}{2}\right)\Big|_1^2$

$= 6\sin^{-1}1 - \left(\frac{1}{4}\cdot 3^{3/2} + \frac{3}{2}\cdot 3^{1/2} + 6\sin^{-1}\frac{1}{2}\right)$

$= 6\left(\frac{\pi}{2}\right) - \left[\frac{9}{4}\sqrt{3} + 6\left(\frac{\pi}{6}\right)\right] = 3\pi - \frac{9}{4}\sqrt{3} - \pi$

$= 2\pi - \frac{9}{4}\sqrt{3} = \frac{1}{4}(8\pi - 9\sqrt{3}) = 2.386$

16. $\int \frac{3\,dx}{9-16x^2} = -3\int \frac{dx}{16x^2-9}$; $a=3$; $u=4x$; $du=4\,dx$

Formula #9

$-\frac{3}{4}\int \frac{4\,dx}{16x^2-9} = -\frac{3}{4}\left[\frac{1}{6}\ln\frac{4x-3}{4x+3}\right] + C$

$= -\frac{1}{8}\ln\frac{4x-3}{4x+3} + C$

17. Formula #11; $u=2x$; $du=2\,dx$; $a=1$

$\int \frac{dx}{x\sqrt{4x^2+1}} = \int \frac{2\,dx}{2x\sqrt{(2x)^2+1^2}}$

$= -\ln\left(\frac{1+\sqrt{4x^2+1}}{2x}\right) + C$

18. $\int \frac{\sqrt{4+x^2}}{x}dx = \int \frac{\sqrt{x^2+4}\,dx}{x}$; $u=x$; $du=dx$; $a=2$

Formula #16

$\int \frac{\sqrt{x^2+4}\,dx}{x} = \sqrt{4+x^2} - 2\ln\left(\frac{2+\sqrt{x^2+4}}{x}\right) + C$

19. $\int \frac{8\,dx}{x\sqrt{1-4x^2}}$; $a=1$; $u=2x$; $du=2\,dx$;

Formula #13

$8\int \frac{2\,dx}{2x\sqrt{1-4x^2}} = 8\left[-\frac{1}{1}\ln\left(\frac{1+\sqrt{1-4x^2}}{2x}\right)\right]$

$= -8\ln\left(\frac{1+\sqrt{1-4x^2}}{2x}\right) + C$

20. $\int \frac{dx}{x(1+4x)^2}$; $u=x$; $du=dx$; $a=1$; $b=4$;

Formula #4

$\int \frac{dx}{x(1+4x)^2} = \frac{1}{1(1+4x)} - \frac{1}{1}\ln\frac{1+4x}{x} + C$

$= \frac{1}{1+4x} - \ln\left(\frac{1+4x}{x}\right) + C$

21. Formula #40; $a=1$; $u=x$; $du=dx$; $b=5$

$\int_0^{\pi/12} \sin\theta\cos 5\theta\,d\theta = -\frac{\cos(-4\theta)}{2(-4)} - \frac{\cos 6\theta}{12}$

$= \frac{1}{8}\cos 4\theta - \frac{1}{12}\cos 6\theta\Big|_0^{\pi/12}$

$= 0.0208$

22. $\int_0^2 x^2 e^{3x} dx$; $u=x$; $du=dx$; $a=3$; Formula #45

$\int_0^2 x^2 e^{3x} dx = \frac{e^{3x}}{27}(9x^2 - 6x + 2)\Big|_0^2$

$= \frac{e^6}{27}(36 - 12 + 2) - \frac{e^0}{27}(2)$

$= \frac{1}{27}(26e^6 - 2) = 388$

23. $\int x^5 \cos x^3\,dx$; $u=x^3$; $du=3x^2dx$; Formula #48

$\int x^3 \cos x^3 x^2\,dx = \frac{1}{3}\int x^3 \cos x^3 (3x^2 dx)$

$= \frac{1}{3}x^3 \cos x^3 (3x^2 dx)$

$= \frac{1}{3}(\cos x^3 + x^3 \sin x^3) + C$

24. $\int 5\sin^3 t \cos^2 t\,dt$; $u=t$; $du=dt$; $m=3$;

$n=2$;

Formula #43

$5\int \sin^3 t \cos^2 t\,dt$

$= 5\left(-\frac{\sin^2 t \cos^3 t}{5} + \frac{2}{5}\int \sin t \cos^2 t\,dt\right)$

$= -\sin^2 t \cos^3 t - 2\int \cos^2 t(-\sin t\,dt)$

$= -\sin^2 t \cos^3 t - 2\cdot\frac{1}{3}\cos^3 t + C$

$= -\frac{1}{3}\cos^3 t(3\sin^2 t + 2) + C$

25. let $u = x^2$, $du = 2x\, dx$
$$u^2 = x^4$$

$$\int \frac{2x\, dx}{(1-x^4)^{3/2}} = \int \frac{du}{(1-u^2)^{3/2}}$$

Formula #25: $a = 1$

$$\int \frac{2x\, dx}{(1-x^4)^{3/2}} = \frac{u}{\sqrt{1-u^2}} + C;$$
$$\int \frac{2x\, dx}{(1-x^4)^{3/2}} = \frac{x^2}{\sqrt{1-x^4}} + C$$

26. $\displaystyle\int \frac{dx}{x(1-4x)}$; $u = x$; $du = dx$; $b = -4$; $a = 1$;

Formula #2

$$\int \frac{dx}{x(1-4x)} = -\frac{1}{1} \ln \frac{1-4x}{x} + C$$
$$= -\ln\left(\frac{1-4x}{x}\right) + C$$

27. $\displaystyle\int_1^3 \frac{\sqrt{3+5x^2}\, dx}{x}$; $a^2 = 3$; $a = \sqrt{3}$; $u = \sqrt{5}x$;

$du = \sqrt{5}\, dx$;

Formula #16

$$\int_1^3 \frac{\sqrt{3+5x^2}}{\sqrt{5}x} \sqrt{5}\, dx$$

$$= \sqrt{5x^2+3} - \sqrt{3} \ln\left(\frac{\sqrt{3}+\sqrt{5x^2+3}}{\sqrt{5}x}\right)\Bigg|_1^3$$

$$= \sqrt{48} - \sqrt{3} \ln\left(\frac{\sqrt{3}+\sqrt{48}}{3\sqrt{5}}\right)$$

$$\quad - \left[\sqrt{8} - \sqrt{3} \ln\left(\frac{\sqrt{3}+\sqrt{8}}{\sqrt{5}}\right)\right]$$

$$= 4.892$$

28. $\displaystyle\int_0^1 \frac{\sqrt{9-4x^2}\, dx}{x}$; $a = 3$; $u = 2x$; $du = 2\, dx$;

Formula #18

$$\int_0^1 \frac{\sqrt{9-4x^2} \cdot 2\, dx}{2x}$$

$$= \sqrt{9-4x^2} - 3 \ln\left(\frac{3+\sqrt{9-4x^2}}{2x}\right)\Bigg|_0^1$$

$$= \sqrt{5} - 3 \ln\left(\frac{3+\sqrt{5}}{2}\right) - \left[3 - 3 \ln\left(\frac{3+3}{0}\right)\right]$$

Division by zero, undefined.

29. Formula #46; $u = x^2$; $du = 2x\, dx$; $n = 1$

$$\int x^3 \ln x^2\, dx = \frac{1}{2} \int x^2 \ln x^2 (2x\, dx)$$

$$= \frac{1}{2}\left[(x^2)^2 \left(\frac{\ln x^2}{2} - \frac{1}{4}\right)\right]$$

$$= \frac{1}{2}\left[\frac{x^4}{2}\left(\ln x^2 - \frac{1}{2}\right)\right]$$

$$= \frac{1}{4}x^4\left(\ln x^2 - \frac{1}{2}\right) + C$$

30. Let $x = u^2$ and $dx = 2u\, du$, then

$$\int \frac{1.2u\, du}{u^2\sqrt{u^4-9}} = \int \frac{0.6\, dx}{x\sqrt{x^2-9}} \text{ which}$$

is Formula #12 with $a = 3$

$$\int \frac{1.2u\, du}{u^2\sqrt{u^4-9}} = \frac{0.6}{3} \sec^{-1}\frac{x}{3} + C$$

$$\int \frac{1.2u\, du}{u^2\sqrt{u^4-9}} = \frac{1}{5} \sec^{-1}\frac{u^2}{3} + C$$

31. $\displaystyle\int \frac{9x^2\, dx}{(x^6-1)^{3/2}}$; $a = 1$; $u = x^3$; $du = 3x^2\, dx$; Formula #24

$$3 \int \frac{3x^2\, dx}{(x^6-1)^{3/2}} = 3\left(-\frac{x^3}{\sqrt{x^6-1}}\right) + C$$

$$= \frac{-3x^3}{\sqrt{x^6-1}} + C$$

32. $\displaystyle\int x^7 \sqrt{x^4+4}\, dx = \int x^4 \sqrt{x^4+4} \cdot x^3\, dx$

$$= \int x^4 \sqrt{4+x^4} \cdot x^3\, dx$$

$u = x^4$; $du = 4x^3\, dx$; $a = 4$; $b = 1$; Formula #5

$$\frac{1}{4} \int x^4 \sqrt{4-x^4} \cdot 4x^3\, dx$$

$$= \frac{1}{4}\left[-\frac{2(8-3x^4)(4+x^4)^{3/2}}{15}\right] + C$$

$$= -\frac{1}{30}(8-3x^4)(4+x^4)^{3/2} + C$$

33. From Exercise 17 of Section 26-6,

$$s = \int_a^b \sqrt{1 + \left(\frac{dy}{dx}\right)^2} \, dx; \quad y = x^2; \quad \frac{dy}{dx} = 2x;$$

$$s = \int_0^1 \sqrt{1 + (2x)^2} \, dx)$$

$$= \frac{1}{2} \int_0^1 \sqrt{(2x)^2 + 1} \,(2dx)$$

Formula #14; $u = 2x$; $du = 2\,dx$

$$s = \frac{1}{2} \left[\frac{2x}{2} \sqrt{4x^2 + 1} + \frac{1}{2} \ln(2x + \sqrt{4x^2 + 1}) \right]\Big|_0^1$$

$$= \frac{1}{2} \left[\left(1\sqrt{5} + \frac{1}{2} \ln(2 + \sqrt{5}) \right) - \frac{1}{2} \ln 1 \right]$$

$$= \frac{1}{4} [2\sqrt{5} + \ln(2 + \sqrt{5})] = 1.479$$

34. $y = 3 \ln x$; $x = e$; x-axis; rotated about y-axis

$$I_y = 2\pi k \int_1^e y x^3 \, dx$$

$$I_y = 2\pi k \int_1^e 3 \ln x x^3 \, dx = 6\pi k \int_1^e x^3 \ln x \, dx$$

$u = x$; $du = dx$; $n = 3$; Formula #46

$$6\pi k \int_1^e x^3 \ln x \, dx$$

$$= 6\pi k x^4 \left(\frac{\ln x}{4} - \frac{1}{16} \right)\Big|_1^e$$

$$= 6\pi k \left[e^4 \left(\frac{1}{4} - \frac{1}{16} \right) - 1 \left(0 - \frac{1}{16} \right) \right]$$

$$= 6\pi k \left(\frac{3e^4}{16} + \frac{1}{16} \right) = \frac{3k\pi}{8} (3e^4 + 1)$$

35. $A = 4 \int_0^a b\sqrt{1 - \frac{x^2}{a^2}} \, dx = \frac{4b}{a} \int_0^a \sqrt{a^2 - x^2} \, dx$

$$= \frac{4b}{a} \left(\frac{x}{2} \sqrt{a^2 - x^2} + \frac{a^2}{2} \sin^{-1} \frac{x}{a} \right)\Big|_0^a$$

$$= \pi a b$$

(Formula #15)

36. $i = \dfrac{dq}{dt} = \dfrac{d(CV)}{dt} = \tan^{-1} 2t$

$$CV = \int_0^{5.00 \times 10^{-6}} \tan^{-1} 2t \, dt$$

$$V = \frac{1}{5.0 \times 10^{-6}} \left(\frac{1}{2}\right) \int_0^{5.00 \times 10^{-6}} \tan^{-1}(2t)(2\,dt),$$

Formula #52

$$V = \frac{1}{5.0 \times 10^{-6}} \left[2t \tan^{-1}(2t) - \frac{1}{2} \ln(1 + 4t^2) \right]\Big|_0^{5.00 \times 10}$$

$$V = \frac{1}{5.0 \times 10^{-6}} \left[t \tan^{-1} 2t - \frac{1}{4} \ln(1 + 4t^2) \right]\Big|_0^{5.00 \times 10^{-6}}$$

$$V = 5.0 \times 10^{-6} \text{ V}$$

37. $F = w \int_0^3 lh \, dh = w \int_0^3 x(3 - y) \, dy$

$$= w \int_0^3 \frac{3 - y}{\sqrt{1 + y}} \, dy$$

(Formula #6)

$$\int \frac{3 - y}{\sqrt{1 + y}} \, dy = 3 \int \frac{dy}{\sqrt{1 + y}} - \int \frac{y \, dy}{\sqrt{1 + y}}$$

$$= 3\frac{(1 + y)^{1/2}}{\frac{1}{2}} - \left[\frac{-2(2 - y)\sqrt{1 + y}}{3(1)^2} \right] + C$$

$$F = w \int_0^3 \frac{3 - y}{\sqrt{1 + y}} \, dy$$

$$= w \left[6(1 + y)^{1/2} + \frac{2}{3}(2 - y)(1 + y)^{1/2} \right]\Big|_0^3$$

$$= w \left[6(2) + \frac{2}{3}(-1)(2) - 6(1) - \frac{2}{3}(2)(1) \right]$$

$$F = w \left(12 - \frac{4}{3} - 6 - \frac{4}{3} \right) = \frac{10w}{3}$$

$$= \frac{10(62.4)}{3} = 208 \text{ lb}$$

38. $t = 560 \int_3^6 \dfrac{dx}{x(x + 4)} = 560 \int_3^6 \dfrac{dx}{x(4 + x)}$

$u = x$; $du = dx$; $b = 1$; $a = 4$; Formula #2

$$560 \int_3^6 \frac{dx}{x(4 + x)} = 560 \left(-\frac{1}{4} \ln \frac{4 + x}{x} \right)\Big|_3^6$$

$$= 560 \left(-\frac{1}{4} \ln \frac{10}{6} + \frac{1}{4} \ln \frac{7}{3} \right)$$

$$= 47.1 \text{ min}$$

39. $V = \displaystyle\int_0^{80} 2\pi x \cdot 20\cos(0.0196x)\,dx$

$\qquad = \dfrac{40\pi}{(0.0196)^2} \displaystyle\int_0^{80} (0.0196x)\cos(0.0196x)(0.0196x\,dx)$

$V = \dfrac{40\pi}{(0.0196)^2}\left(\cos(0.0196x) + 0.0196x\sin(0.0196x)\right)\Big|_0^{80}$

$V = 187,000\text{ m}^3$

40. $F = kqQ\displaystyle\int \dfrac{b\,dx}{(b^2 + x^2)^{3/2}} = kqQ\displaystyle\int \dfrac{b\,dx}{(x^2 + b^2)^{3/2}}$

$\qquad u = x;\ du = dx;\ a = b;$ Formula #24

$kqQ\displaystyle\int \dfrac{b\,dx}{(x^2+b^2)^{3/2}} = kqQ\left(\dfrac{bx}{b^2\sqrt{x^2+b^2}}\right) + C$

$\qquad\qquad\qquad\qquad = \dfrac{kqQx}{b\sqrt{x^2+b^2}} + C$

10.7   Improper Integrals

1. $\displaystyle\int_1^{+\infty} \dfrac{dx}{(x+2)^2} = \lim_{b\to+\infty}\int_1^b \dfrac{dx}{(x+2)^2}$

$\qquad\qquad = \lim_{b\to+\infty}\int_1^b (x+2)^{-2}\,dx$

$\qquad\qquad = \lim_{b\to+\infty} -\dfrac{1}{x+2}\Big|_1^b$

$\qquad\qquad = \lim_{b\to+\infty} -\dfrac{1}{b+2} - \left(-\dfrac{1}{3}\right)$

$\qquad\qquad = \dfrac{1}{3}$

2. $\displaystyle\int_2^{+\infty} \dfrac{dx}{3x+4} = \lim_{b\to+\infty}\dfrac{1}{3}\int_2^b \dfrac{3\,dx}{3x+4}$

$\qquad\qquad = \lim_{b\to+\infty}\dfrac{1}{3}\left[\ln(3x+4)\right]\Big|_2^b$

$\qquad\qquad = \dfrac{1}{3}\lim_{b\to+\infty}\left[\ln(3b+4) - \ln 10\right]$

$\qquad\qquad = +\infty\text{ (divergent)}$

3. $\displaystyle\int_{-\infty}^0 \dfrac{dx}{\sqrt{1-3x}}$

$\qquad = \lim_{a\to-\infty}\int_a^0 \dfrac{dx}{\sqrt{1-3x}}$

$\qquad = \lim_{a\to-\infty} -\dfrac{1}{3}\int_a^0 (1-3x)^{-1/2}(-3\,dx)$

$\qquad = \lim_{a\to-\infty} -\dfrac{1}{3}\dfrac{(1-3x)^{1/2}}{1/2}\Big|_a^0$

$\qquad = \lim_{a\to-\infty} -\dfrac{2}{3}(1-3x)^{1/2}\Big|_a^0$

$\qquad = \lim_{a\to-\infty} -\dfrac{2}{3}(1-3a)^{1/2} - \left(-\dfrac{2}{3}\right)[1-3(0)]$

$\qquad = -\infty\text{ (Divergent)}$

4. $\displaystyle\int_0^{+\infty} \dfrac{dx}{(1+x)^3} = \lim_{b\to+\infty}\int_0^b \dfrac{dx}{(1+x)^3}$

$\qquad\qquad = \lim_{b\to+\infty}\left[\dfrac{-1}{2(1+x)^2}\right]\Big|_0^b$

$\qquad\qquad = \lim_{b\to+\infty}\left[\dfrac{-2}{(1+b)^2} + \dfrac{1}{2}\right]$

$\qquad\qquad = \dfrac{1}{2}$

5. $\displaystyle\int_1^{+\infty} \dfrac{x\,dx}{1+x^2} = \lim_{b\to+\infty}\int_1^b \dfrac{x\,dx}{1+x^2}$

$\qquad\qquad = \lim_{b\to+\infty}\dfrac{1}{2}\dfrac{2x\,dx}{1+x^2}$

$\qquad\qquad = \lim_{b\to+\infty}\ln|1+x^2|\,\Big|_1^b$

$\qquad\qquad = \lim_{b\to+\infty}\left(\ln|1+b^2| - \ln|1+1|\right)$

$\qquad\qquad = +\infty\text{ (Divergent)}$

6. $\displaystyle\int_3^{+\infty} \dfrac{x\,dx}{\sqrt{x^2-4}} = \lim_{b\to+\infty}\dfrac{1}{2}\int_3^b \dfrac{2x\,dx}{\sqrt{x^2-4}}$

$\qquad\qquad = \dfrac{1}{2}\lim_{b\to+\infty} 2\sqrt{x^2-4}\,\Big|_3^b$

$\qquad\qquad = \lim_{b\to+\infty}\left[\sqrt{b^2-4} - \sqrt{5}\right]$

$\qquad\qquad = +\infty\text{ (divergent)}$

7. $\displaystyle\int_0^{+\infty} xe^{-x^2}\,dx = \lim_{b\to+\infty} \int_0^b xe^{-x^2}\,dx$

$$= \lim_{b\to+\infty} -\frac{1}{2}\int_0^b e^{-x^2}(-2x\,dx)$$

$$= \lim_{b\to+\infty} -\frac{1}{2}e^{-x^2}\Big|_0^b$$

$$= \lim_{b\to+\infty} -\frac{1}{2}(e^{-b^2} - e^0)$$

$$= -\frac{1}{2}(0 - 1)$$

$$= \frac{1}{2}$$

8. $\displaystyle\int_{-\infty}^{-1} \frac{-x\,dx}{(1+x^2)^2}$

$$= \lim_{b\to-\infty} \int_b^{-1} \frac{1}{2}\frac{-2x\,dx}{(1+x^2)^2}$$

$$= \frac{-1}{2}\lim_{b\to-\infty}\left[\frac{-1}{(1+x^2)}\right]\Big|_b^{-1}$$

$$= \frac{-1}{2}\lim_{b\to-\infty}\left[\frac{-1}{(1+(-1)^2)} + \frac{1}{(1+b^2)}\right]$$

$$= -\frac{1}{2}\left(-\frac{1}{2}\right)$$

$$= \frac{1}{4}$$

9. $\displaystyle\int_{-\infty}^0 \frac{2\,dx}{(1-x)^3} = \lim_{a\to-\infty}\int_a^0 \frac{2\,dx}{(1-x)^3}$

$$= \lim_{a\to-\infty} -2\int_a^0 (1-x)^{-3}(-dx)$$

$$= \lim_{a\to-\infty} -2\frac{(1-x)^{-2}}{-2}\Big|_a^0$$

$$= \lim_{a\to-\infty} (1-x)^{-2}\Big|_a^0$$

$$= \lim_{a\to-\infty}\left[\frac{1}{1^2} - \frac{1}{(1-a)^2}\right]$$

$$= 1 - 0$$

$$= 1$$

10. $\displaystyle\int_0^{+\infty} \frac{x\,dx}{\sqrt{x+1}}$

$$u = \sqrt{x+1}, u^2 = x+1, x = u^2 - 1$$
$$2u\,du = dx$$

$$\int \frac{x\,dx}{\sqrt{x+1}} = \int \frac{(u^2-1)(2u\,du)}{u}$$

$$= \int 2(u^2 - 1)\,du$$

$$= \frac{2}{3}u^3 - 2u$$

$$= \frac{2}{3}(x+1)^{3/2} - 2(x+1)^{1/2}$$

$$= 2(x+1)^{1/2}\left(\frac{1}{3}(x+1) - 1\right)$$

$$= 2(x+1)^{1/2}\left(\frac{1}{3}x + \frac{1}{3} - 1\right)$$

$$= 2(x+1)^{1/2}\left(\frac{1}{3}x - \frac{2}{3}\right)$$

$$= \frac{2}{3}(x+1)^{1/2}(x - 2)$$

$$\int_0^{+\infty} \frac{x\,dx}{\sqrt{x+1}} = \lim_{b\to+\infty}\int_0^b \frac{x\,dx}{\sqrt{x+1}}$$

$$= \lim_{b\to+\infty}\left[\frac{2}{3}(x+1)^{1/2}(x-2)\right]\Big|_0^b$$

$$= \frac{2}{3}\lim_{b\to+\infty}[(b+1)^{1/2}(b-2) - (1)(-2)]$$

$$= +\infty \text{ (Divergent)}$$

11. $\displaystyle\int_{-\infty}^{+\infty} e^{-x}\,dx$

$$= \lim_{a\to-\infty}\int_a^0 e^{-x}\,dx + \lim_{b\to+\infty}\int_0^b e^{-x}\,dx$$

$$= \lim_{a\to-\infty} -e^{-x}\Big|_a^0 + \lim_{b\to+\infty} -e^{-x}\Big|_0^b$$

$$= \lim_{a\to-\infty}[-e^0 - (-e^{-a})] + \lim_{b\to+\infty}[-e^{-b} - (-e^0)]$$

$$= +\infty \text{ (since } \lim_{a\to-\infty} e^{-a} = +\infty)$$

(Divergent)

12. $\displaystyle\int_{-\infty}^{+\infty} \frac{dx}{x^2+4} = \lim_{a\to-\infty}\int_a^0 \frac{dx}{x^2+4} + \lim_{b\to+\infty}\int_0^b \frac{dx}{x^2+4}$

$\displaystyle = \lim_{a\to-\infty}\frac{1}{2}\tan^{-1}\frac{x}{2}\Big|_a^0 + \lim_{b\to+\infty}\frac{1}{2}\tan^{-1}\frac{x}{2}\Big|_0^b$

$\displaystyle = \lim_{a\to-\infty}\frac{1}{2}\left[\tan^{-1}0 - \tan^{-1}\frac{a}{2}\right] + \lim_{b\to+\infty}\frac{1}{2}\left[\tan^{-1}\frac{b}{2} - \tan^{-1}0\right]$

$\displaystyle = -\lim_{a\to-\infty}\frac{1}{2}\tan^{-1}\frac{a}{2} + \lim_{b\to+\infty}\frac{1}{2}\tan^{-1}\frac{b}{2}$

$\displaystyle = -\frac{1}{2}\left(-\frac{\pi}{2}\right) + \frac{1}{2}\left(\frac{\pi}{2}\right)$

$\displaystyle = \frac{\pi}{2}$

13. $\displaystyle\int_0^9 \frac{dx}{\sqrt{9-x}} = \lim_{h\to0}\int_0^{9-h}\frac{dx}{\sqrt{9-x}} = \lim_{h\to0}-\int_0^{9-h}(9-x)^{-1/2}(-dx) = \lim_{h\to0}-\frac{(9-x)^{1/2}}{1/2}\Big|_0^{9-h}$

$\displaystyle = \lim_{h\to0}-2(9-x)^{1/2}\Big|_0^{9-h} = \lim_{h\to0}\left[-2(h)^{1/2} - (-2)(9)^{/2}\right] = 0 + 6 = 6$

14. $\displaystyle\int_0^1 \frac{dx}{x} = \lim_{b\to0+}\int_b^1 \frac{dx}{x} = \lim_{b\to0+}\ln|x|\Big|_b^1 = \lim_{b\to0+}[\ln 1 - \ln b] = \lim_{b\to0+}[-\ln b]$

$= +\infty$ (Divergent)

15. $\displaystyle\int_{\pi/4}^{\pi/3}\tan 2x\,dx = \lim_{h\to0}\int_{\pi/4+h}^{\pi/3}\tan 2x\,dx = \lim_{h\to0}\frac{1}{2}\int_{\pi/4+h}^{\pi/3}\tan 2x(2\,dx)$

$\displaystyle = \lim_{h\to0}-\frac{1}{2}\ln|\cos 2x|\Big|_{\pi/4+h}^{\pi/3} = \lim_{h\to0}-\frac{1}{2}\left[\ln\cos\frac{2\pi}{3} - \ln\cos\left(\frac{\pi}{2}+2h\right)\right];$

$\ln\cos\dfrac{2\pi}{3}$ undefined since $\cos\dfrac{2\pi}{3} < 0$, same for $\ln\cos\left(\dfrac{\pi}{2}+2h\right)$; (Divergent)

16. $\displaystyle\int_0^3 \frac{dx}{\sqrt{9-x^2}} = \lim_{b\to3-}\int_0^b \frac{dx}{\sqrt{9-x^2}} = \lim_{b\to3-}\sin^{-1}\frac{x}{3}\Big|_0^b = \lim_{b\to3-}\left[\sin^{-1}\frac{b}{3} - \sin^{-1}\frac{0}{3}\right]$

$\displaystyle = \lim_{b\to3-}\left[\sin^{-1}\frac{b}{3}\right] = \frac{\pi}{2}$

17. $\displaystyle\int_0^3 \frac{dx}{(x-2)^2} = \lim_{h\to0}\int_0^{2-h}\frac{dx}{(x-2)^2} + \lim_{h'\to0}\int_{2+h'}^3\frac{dx}{(x-2)^2} = \lim_{h\to0}\left[-\frac{1}{x-2}\right]\Big|_0^{2-h} + \lim_{h'\to0}\left[-\frac{1}{x-2}\right]\Big|_{2+h'}^3$

$\displaystyle = \lim_{h\to0}\left(\frac{1}{h} - \frac{1}{2}\right) + \lim_{h'\to0}\left(-1 + \frac{1}{h'}\right) = +\infty$ (Neither limit exists) (Divergent)

18. $\displaystyle\int_{-1}^1 \frac{dx}{x^{1/3}} = \lim_{b\to0-}\int_{-1}^b\frac{dx}{x^{1/3}} + \lim_{a\to0+}\int_a^1\frac{dx}{x^{1/3}} = \lim_{b\to0-}\frac{3}{4}x^{4/3}\Big|_{-1}^b + \lim_{a\to0+}\frac{3}{4}x^{4/3}\Big|_a^1$

$\displaystyle = \lim_{b\to0-}\left[\frac{3}{4}b^{4/3} - \frac{3}{4}(1)\right] + \lim_{a\to0+}\left[\frac{3}{4}(1) - \frac{3}{4}a^{4/3}\right] = -\frac{3}{4} + \frac{3}{4} = 0$

19. $\displaystyle\int_0^1 \frac{2\,dx}{1-x^2} = \lim_{h\to 0}\int_0^{1-h}\frac{2\,dx}{1-x^2} = \lim_{h\to 0}\int_0^{1-h}\left(\frac{1}{1+x}+\frac{1}{1-x}\right)dx$

$$= \lim_{h\to 0}\ \ln|1+x| - \ln|1-x|\ \Big|_0^{1-h} = \lim_{h\to 0}\ \ln\left|\frac{1+x}{1-x}\right|\Big|_0^{1-h}$$

$$= \lim_{h\to 0}\left[\ln\frac{2-h}{h} - \ln 1\right] = +\infty \text{ (Divergent)}$$

20. $\displaystyle\int_2^3 \frac{dx}{\sqrt{x-2}} = \lim_{b\to 2^+}\int_b^3\frac{dx}{\sqrt{x-2}} = \lim_{b\to 2^+}\ [2\sqrt{x-2}]\big|_b^3 = \lim_{b\to 2^+}\ [2\sqrt{3-2}-2\sqrt{b-2}]$

$$= 2 - 2\lim_{b\to 2^+}\ \sqrt{b-2} = 2$$

21. $\displaystyle\int_1^2 \frac{dx}{\sqrt{x^2-1}} = \lim_{h\to 0}\int_{1+h}^2\frac{dx}{\sqrt{x^2-1}} = \lim_{h\to 0}\ (\ln(x+\sqrt{x^2-1})\big|_{1+h}^2\quad\text{(Trig. sub)}$

$$= \lim_{h\to 0}[\ln(2+\sqrt 3) - \ln(1+h+\sqrt{2h+h^2})] = \ln(2+\sqrt 3)$$

22. $\displaystyle\int_1^3 \frac{3\,dx}{3x-x^2} = \int_1^3\frac{3\,dx}{x(3-x)} = \int_1^3\frac{1}{x}\,dx - \int_1^3\frac{dx}{x-3} = \ln|x|\ \Big|_1^3 - \lim_{b\to 3^-}\int_1^b\frac{dx}{x-3}$

$$= \ln 3 - \ln 1 - \lim_{b\to 3^-}\ \ln|x-3|\ \Big|_1^b = \ln 3 - \lim_{b\to 3^-}\ [\ln|b-3| - \ln|1-3|]$$

$$= \ln 3 - \lim_{b\to 3^-}\ \ln|b-3| - \ln 2 = \ln\frac{3}{2} - (-\infty) = +\infty \text{ (Divergent)}$$

23. $\displaystyle\int_1^3 \frac{dx}{(x-2)^{1/3}} = \lim_{h\to 0}\int_1^{2-h}\frac{dx}{(x-2)^{1/3}} + \lim_{h'\to 0}\int_{2+h'}^3\frac{dx}{(x-2)^{1/3}}$

$$= \lim_{h\to 0}\frac{3}{2}(x-2)^{2/3}\Big|_1^{2-h} + \lim_{h'\to 0}\frac{3}{2}(x-2)^{2/3}\Big|_{2+h'}^3$$

$$= \lim_{h\to 0}\left[\frac{3}{2}(-h)^{2/3} - \frac{3}{2}(-1)^{2/3}\right] + \lim_{h'\to 0}\left[\frac{3}{2}(1)^{2/3} - \frac{3}{2}(h)^{2/3}\right]$$

$$= -\frac{3}{2} + \frac{3}{2} = 0$$

24. $\displaystyle\int_{-\pi/2}^{\pi/2} \sec x\,dx = \lim_{b\to -\pi/2^+}\int_b^0 \sec x\,dx + \lim_{a\to \pi/2^-}\int_0^a \sec x\,dx$

$$= \lim_{b\to -\pi/2^+}\ \ln|\sec x + \tan x|\ \Big|_b^0 + \lim_{a\to \pi/2^-}\ \ln|\sec x + \tan x|\ \Big|_0^a$$

$$= \lim_{b\to -\pi/2^+}[\ln|\sec 0 + \tan 0| - \ln|\sec b + \tan b|] + \lim_{a\to \pi/2^-}[\ln|\sec a + \tan a| - \ln|\sec 0 + \tan 0|]$$

$$= \lim_{b\to -\pi/2^+}[\ln|1+0| - \ln|\sec b + \tan b|] + \lim_{a\to \pi/2^-}[\ln|\sec a + \tan a| - \ln|1+0|]$$

$$= \lim_{b\to -\pi/2^+}[-\ln|\sec b + \tan b|] + \lim_{a\to \pi/2^-}[\ln|\sec a + \tan a|]$$

$$= \lim_{b\to -\pi/2^+}[-\ln|\sec b + \tan b|] = +\infty \text{ (Divergent)}$$

25. (a) $A = \displaystyle\int_1^\infty y\,dx = \int_1^\infty \frac{dx}{x} = \lim_{b\to+\infty} \int_1^b \frac{dx}{x} = \lim_{b\to+\infty} \ln x \Big|_1^b = \lim_{b\to+\infty} (\ln b - \ln 1) = +\infty$ (Divergent)

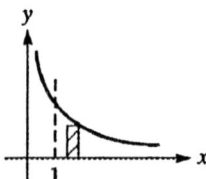

(b) $V = \pi \displaystyle\int_1^\infty y^2 dx = \pi \int_1^\infty \frac{dx}{x^2} = \lim_{b\to+\infty} \pi \int_1^b \frac{dx}{x^2} = \lim_{b\to+\infty} -\frac{\pi}{x}\Big|_1^b = \lim_{b\to+\infty} \left(-\frac{\pi}{b} + \frac{\pi}{1}\right) = \pi$

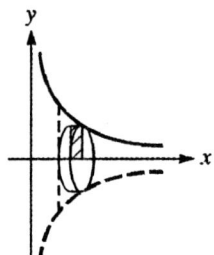

26. $A = \displaystyle\int_4^{+\infty} \frac{1}{\sqrt{x}}\,dx = \lim_{b\to+\infty} \int_4^b \frac{1}{\sqrt{x}}\,dx = \lim_{b\to+\infty} 2\sqrt{x}\Big|_4^b = \lim_{b\to+\infty} [2\sqrt{b} - 2\sqrt{4}] = +\infty$

Area (to right) is undefined.

27. $A = \displaystyle\int_1^2 y\,dx = \int_1^2 \frac{x\,dx}{\sqrt{x^2-1}} = \lim_{h\to 0} \int_{1+h}^2 \frac{x\,dx}{\sqrt{x^2-1}} = \lim_{h\to 0} (x^2-1)^{1/2}\Big|_{1+h}^2 = \lim_{h\to 0} (\sqrt{3} - \sqrt{2h+h^2}) = \sqrt{3}$

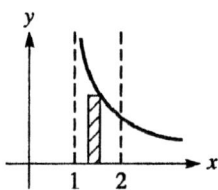

28. $V = \displaystyle\int \pi y^2 dx$

$V = \displaystyle\int \pi \frac{dx}{(x-2)^2}$

$V = \lim_{b\to 2^+} \displaystyle\int_b^5 \pi \frac{dx}{(x-2)^2} + \lim_{a\to+\infty} \int_5^a \pi \frac{dx}{(x-2)^2}$

$V = \pi \lim_{b\to 2^+} \left[\frac{1}{2-x}\right]\Big|_b^5 + \pi \lim_{a\to+\infty} \left[\frac{1}{2-x}\right]\Big|_5^a$

$V = \pi \lim_{b\to 2^+} \left[\frac{1}{2-5} - \frac{1}{2-b}\right] + \pi \lim_{a\to+\infty} \left[\frac{1}{2-a} - \frac{1}{2-5}\right]$

$V = \pi(+\infty) + \pi\left(\frac{1}{3}\right) = +\infty$ (Volume is undefined)

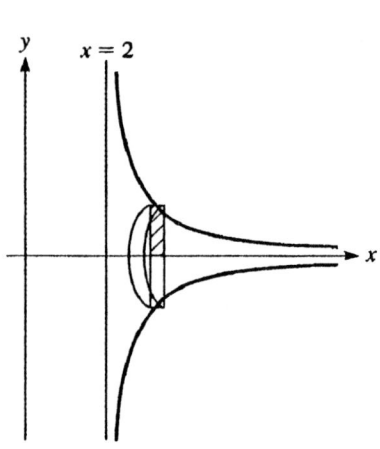

29. $W = \displaystyle\int_0^\infty F\,dx = \int_0^\infty \frac{10^{11}\,dx}{(x+3960)^2} = \lim_{b\to+\infty} \int_0^b \frac{10^{11}\,dx}{(x+3960)^2} = \lim_{b\to+\infty} -\frac{10^{11}}{(x+3960)^2}\Big|_0^b$

$\qquad = \displaystyle\lim_{b\to+\infty}\left[\frac{-10^{11}}{b+3960} - \left(\frac{-10^{11}}{3960}\right)\right] = \frac{10^{11}}{3960} = 2.53\times 10^7 \text{ mi}\cdot\text{lb}$

30. $\displaystyle\int \frac{dx}{(a^2+x^2)^{3/2}}$

$x = a\tan\theta$
$dx = a\sec^2\theta\,d\theta$

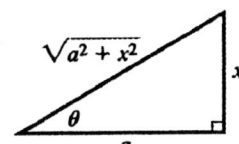

$\displaystyle\int \frac{dx}{(a^2+x^2)^{3/2}} = \int \frac{a\sec^2\theta\,d\theta}{(a^2+a^2\tan^2)^{3/2}} = \int \frac{a\sec^2\theta\,d\theta}{(a^2\sec^2\theta)^{3/2}} = \int \frac{\cos\theta}{a^2}\,d\theta = \frac{\sin\theta}{a^2} = \frac{x}{a^2\sqrt{a^2+x^2}}$

$\displaystyle\int_{-\infty}^{+\infty} \frac{dx}{(a^2+x^2)^{3/2}} = \lim_{t\to-\infty}\int_t^0 \frac{dx}{(a^2+x^2)^{3/2}} + \lim_{b\to+\infty}\int_0^b \frac{dx}{(a^2+x^2)^{3/2}}$

$\qquad = \displaystyle\lim_{t\to-\infty} \frac{x}{a^2\sqrt{a^2+x^2}}\Big|_t^0 + \lim_{b\to+\infty} \frac{x}{a^2\sqrt{a^2+x^2}}\Big|_0^b$

$\qquad = \displaystyle\lim_{t\to-\infty}\left[\frac{0}{a^2\sqrt{a^2+0^2}} - \frac{t}{a^2\sqrt{a^2+t^2}}\right]$

$\qquad\quad + \displaystyle\lim_{b\to+\infty}\left[\frac{b}{a^2\sqrt{a^2+b^2}} - \frac{0}{a^2\sqrt{a^2+0^2}}\right]$

$\qquad = \dfrac{1}{a^2} + \dfrac{1}{a^2} = \dfrac{2}{a^2}$

$E = kiam\displaystyle\int_{-\infty}^{+\infty} \frac{dx}{(a^2+x^2)^{3/2}} = kiam\left(\frac{2}{a^2}\right)$

$E = \dfrac{2\,kim}{a}$

31. $\displaystyle\int_0^{+\infty} \cos x\,dx = \lim_{b\to+\infty}\int_0^b \cos x\,dx = \lim_{b\to+\infty} \sin x\Big|_0^b = \lim_{b\to+\infty}(\sin b - \sin 0) = \lim_{b\to+\infty}\sin b$

(As $b\to+\infty$, $\sin b$ takes on all values from -1 to $+1$. Therefore, since the limit does not exist, the integral is divergent.)

32. (a) $\displaystyle\lim_{b\to+\infty}\int_{-b}^b x\,dx = \lim_{b\to+\infty}\frac{x^2}{2}\Big|_{-b}^b = \lim_{b\to+\infty}\left[\frac{b^2}{2} - \frac{(-b)^2}{2}\right] = \lim_{b\to+\infty}(0) = 0$

(b) $\displaystyle\int_{-\infty}^{+\infty} x\,dx = \lim_{t\to-\infty}\int_t^0 x\,dx + \lim_{b\to+\infty}\int_0^b x\,dx = \lim_{t\to-\infty}\frac{x^2}{2}\Big|_t^0 + \lim_{b\to+\infty}\frac{x^2}{2}\Big|_0^b$

$\qquad = \displaystyle\lim_{t\to-\infty}\frac{-t^2}{2} + \lim_{b\to+\infty}\frac{t^2}{2} = -\infty + \infty \text{ (divergent)}$

Chapter 10 Review Exercises

1. $\displaystyle\int x\csc^2 2x\,dx;\ u=x,\ du=dx,\ dv=\csc^2 2x\,dx,\ v=\frac{1}{2}\int\csc^2 2x(2\,dx)=-\frac{1}{2}\cot 2x$

$$\int x\csc^2 2x\,dx = x\left(-\frac{1}{2}\cot 2x\right)-\int\left(-\frac{1}{2}\cos 2x\right)dx = -\frac{1}{2}x\cot 2x+\frac{1}{4}\int\cot 2x(2\,dx)$$

$$=\frac{1}{2}x\cot 2x+\frac{1}{4}\ln|\sin 2x|+C$$

2. $\displaystyle\int x\tan^{-1}x\,dx$

$$u=\tan^{-1}x \qquad dv=x\,dx$$

$$du=\frac{1}{1+x^2}\,dx \qquad v=\frac{1}{2}x^2$$

$$\int x\tan^{-1}x\,dx=\frac{1}{2}x^2\tan^{-1}x-\frac{1}{2}\int\frac{x^2}{1+x^2}\,dx=\frac{1}{2}x^2\tan^{-1}x-\frac{1}{2}\left[\int\left(1-\frac{1}{1+x^2}\right)dx\right]$$

$$=\frac{1}{2}x^2\tan^{-1}x-\frac{1}{2}[x-\tan^{-1}x]+C=\frac{1}{2}(x^2+1)\tan^{-1}x-\frac{1}{2}x+C$$

3. $\displaystyle\int x\sqrt{x-4}\,dx;\ u=\sqrt{x-4},\ x=u^2+4,\ dx=2u\,du$

$$\int x\sqrt{x-4}\,dx=\int(u^2+4)(u)(2u\,du)=2\int(u^4+4u^2)du=\frac{2}{5}u^5+\frac{8}{3}u^3+C$$

$$=\frac{2}{5}(x-4)^{5/2}+\frac{8}{3}(x-4)^{3/2}+C=2(x-4)^{3/2}\left[\frac{1}{5}(x-4)+\frac{4}{3}\right]+C$$

$$=\frac{2}{15}(3x+8)(x-4)^{3/2}+C$$

4. $\displaystyle\int\frac{2x\,dx}{\sqrt{2x+5}}$

$$u=\sqrt{2x+5},\ u^2=2x+5,\ x=\frac{1}{2}(u^2-5)$$

$$2u\,du=2\,dx$$

$$u\,du=dx$$

$$\int\frac{2x\,dx}{\sqrt{2x+5}}=\int\frac{2\left(\frac{1}{2}\right)(u^2-5)u\,du}{u}=\int(u^2-5)du=\frac{u^3}{3}-5u+C$$

$$=\frac{1}{3}(2x+5)^{3/2}-5(2x+5)^{1/2}+C=(2x+5)^{1/2}\left(\frac{1}{3}(2x+5)-5\right)+C$$

$$=\sqrt{2x+5}\left(\frac{2}{3}x+\frac{5}{3}-5\right)+C=\sqrt{2x+5}\left(\frac{2}{3}x-\frac{10}{3}\right)+C$$

$$=\frac{2}{3}\sqrt{2x+5}(2x+5)+C$$

5. $\int \dfrac{dx}{\sqrt{4x^2-9}}$; $Let\, x = \dfrac{3}{2}\sec\theta,\, dx = \dfrac{3}{2}\sec\theta\tan\theta\, d\theta$

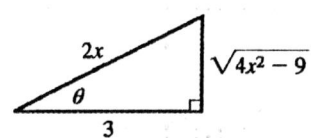

$$\int \frac{dx}{\sqrt{4x^2-9}} = \int \frac{\left(\frac{3}{2}\right)\sec\theta\tan\theta\, d\theta}{\sqrt{9\sec^2\theta-9}} = \frac{1}{2}\int \frac{\sec\theta\tan\theta\, d\theta}{\tan\theta} = \frac{1}{2}\int \sec\theta\, d\theta$$

$$= \frac{1}{2}\ln|\sec\theta+\tan\theta| + C_1 = \frac{1}{2}\ln\left|\frac{2x}{3}+\frac{\sqrt{4x^2-9}}{3}\right| + C_1$$

$$= \frac{1}{2}\ln\left|2x+\sqrt{4x^2-9}\right| + C;\ \left(C = C_1 - \frac{1}{2}\ln 3\right)$$

6. $\int \dfrac{x^2\, dx}{\sqrt{9-x^2}}$ $\qquad x = 3\sin\theta, x^2 = 9\sin^2\theta$
$\qquad\qquad\qquad\qquad dx = 3\cos\theta\, d\theta$

$$\int \frac{x^2\, dx}{\sqrt{9-x^2}} = \int \frac{9\sin^2\theta(3\cos\theta\, d\theta)}{\sqrt{9-9\sin^2\theta}} = \int \frac{27\sin^2\theta\cos\theta\, d\theta}{\sqrt{9(1-\sin^2\theta)}} = \int \frac{9\sin^2\theta\cos\theta\, d\theta}{\sqrt{\cos^2\theta}}$$

$$= 9\int \sin^2\theta\, d\theta = 9\int \frac{1}{2}(1-\cos 2\theta)d\theta = \frac{9}{2}\int d\theta - \frac{9}{4}\int \cos 2\theta(2\, d\theta)$$

$$= \frac{9}{2}\theta - \frac{9}{4}\sin 2\theta + C = \frac{9}{2}\theta - \frac{9}{4}(2\sin\theta\cos\theta) + C$$

$$= \frac{9}{2}\sin^{-1}\frac{x}{3} - \frac{9}{2}\frac{x}{3}\frac{\sqrt{9-x^2}}{3} + C = \frac{9}{2}\sin^{-1}\frac{x}{3} - \frac{1}{2}x\sqrt{9-x^2} + C$$

7. $\int \dfrac{x+25}{x^2-25}\, dx$

$$\frac{x+25}{x^2-25} = \frac{A}{x+5} + \frac{B}{x-5}$$

$x+25 = A(x-5) + B(x+5),\quad x=-5: 20 = -10A, A = -2$
$\qquad\qquad\qquad\qquad\qquad\qquad x=5: 30 = 10B, B = 3$

$$\int \frac{x+25}{x^2-25}\, dx = -\int \frac{2\, dx}{x+5} + \int \frac{3\, dx}{x-5} = -2\ln|x+5| + 3\ln|x-5| + C = \ln\left|\frac{(x-5)^3}{(x+5)^2}\right| + C$$

8. $\int \dfrac{2x^2+6x+1}{2x^3-x^2-x}\, dx = \int \dfrac{2x^2+6x+1}{x(x-1)(2x+1)}\, dx$

$$\frac{2x^2+6x+1}{x(x-1)(2x+1)} = \frac{A}{x} + \frac{B}{x-1} + \frac{C}{2x+1} = \frac{A(x-1)(2x+1) + Bx(2x+1) + Cx(x-1)}{x(x-1)(2x+1)}$$

$2x^2+6x+1 = A(x-1)(2x+1) + Bx(2x+1) + Cx(x-1)$
$x = 0, 1 = A(-1)(1), A = -1$
$x = 1, 9 = B(1)(3), B = 3$

$$x = -\frac{1}{2}, -\frac{3}{2} = C\left(-\frac{1}{2}\right)\left(-\frac{3}{2}\right), C = -2$$

$$\int \frac{2x^2+6x+1}{2x^3-x^2-x}\, dx = \int \frac{-1}{x}\, dx + \int \frac{3}{x-1}\, dx + \int \frac{2}{2x+1}\, dx = -\ln|x| + 3\ln|x-1| - \ln|2x+1| + C$$

$$= \ln|x-1|^3 - \ln|x| - \ln|2x+1| + C = \ln\left|\frac{(x-1)^3}{x(2x+1)}\right| + C$$

9. $\int \dfrac{x^2-2x+3}{(x-1)^3}\,dx$; Let $u=x-1, x=u+1, dx=du$

$$\int \frac{x^2-2x+3}{(x-1)^3}\,dx = \int \frac{(u+1)^2-2(u+1)+3}{u^3}\,du = \int \frac{u^2+2}{u^3}\,du = \int \frac{du}{u}+2\int \frac{du}{u^3}$$

$$= \ln|u| - \frac{1}{u^2} + C = \ln|x-1| - \frac{1}{(x-1)^2} + C$$

10. $\int \dfrac{2x^2+3x+18}{x^3+9x}\,dx = \int \dfrac{2x^2+3x+18}{x(x^2+9)}$

$$\frac{2x^2+3x+18}{x(x^2+9)} = \frac{A}{x}+\frac{Bx+C}{x^2+9} = \frac{A(x^2+9)+x(Bx+C)}{x(x^2+9)}$$

$$2x^2+3x+18 = A(x^2+9)+x(Bx+C)$$

$x=0, 18=A(9), A=2$
$x=1, 23=2(1^2+9)+B+C, B+C=3$
$x=-1, 17=2((-1)^2+9)+-1(B(-1)+C),$
$\qquad 17=20+B-C, B-C=-3$

$$\begin{array}{r} B+C=\ \ 3 \\ B-C=-3 \\ \hline 2B\quad\ \ =\ \ 0 \\ B=\ \ 0\ , C=3 \end{array}$$

$$\int \frac{2x^2+3x+18}{x^3+9x} = \int \frac{2}{x}\,dx + \int \frac{3}{x^2+9}\,dx = 2\ln|x|+3\left(\frac{1}{3}\tan^{-1}\frac{x}{3}\right)+C = 2\ln|x|+\tan^{-1}\frac{x}{3}+C$$

11. $\displaystyle\int_1^{+\infty}\frac{dx}{2x+7} = \lim_{b\to+\infty}\int_1^b \frac{dx}{2x+7} = \lim_{b\to+\infty}\frac{1}{2}\ln|2x+7|\Big|_1^b = \lim_{b\to+\infty}\frac{1}{2}|2b+7|-\frac{1}{2}\ln 9$
$$= +\infty \text{ (Divergent)}$$

12. $\displaystyle\int_{-1}^1\frac{dx}{\sqrt{x+1}} = \lim_{t\to-1^+}\int_t^1\frac{dx}{\sqrt{x+1}} = \lim_{t\to-1^+}2\sqrt{x+1}\Big|_t^1 = \lim_{t\to-1^+}[2\sqrt{1+1}-2\sqrt{t+1}]$
$$= 2\sqrt{2}$$

13. $\int \dfrac{x\,dx}{(x+3)^{2/3}}$; $u=(x+3)^{1/3}, x=u^3-3, dx=3u^2du$

$$\int \frac{x\,dx}{(x+3)^{2/3}} = \int \frac{(u^3-3)(3u^2du)}{u^2} = 3\int(u^3-3)du = \frac{3}{4}u^4-9u+C = \frac{3}{4}(x+3)^{4/3}-9(x+3)^{1/3}+C$$

$$= 3(x+3)^{1/3}\left[\frac{1}{4}(x+3)-3\right]+C = \frac{3}{4}(x+3)^{1/3}(x-9)+C$$

14. $\displaystyle\int x\tan^2 x\,dx$

$$u = x, dv = \int \tan^2 x\,dx$$

$$du = dx, dv = \int (\sec^2 x - 1)dx$$

$$dv = \int \sec^2 x\,dx - \int dx$$

$$v = \tan x - x$$

$$\int x\tan^2 x\,dx = x(\tan x - x) - \int(\tan x - x)dx = x\tan x - x^2 - \left(-\ln|\cos x| - \frac{x^2}{2}\right) + C$$

$$= x\tan x - x^2 + \ln|\cos x| + \frac{x^2}{2} + C = x\tan x + \ln|\cos x| - \frac{x^2}{2} + C$$

15. $\displaystyle\int \frac{2\,dx}{(4x^2+1)^{3/2}}$; Let $x = \frac{1}{2}\tan\theta$, $dx = \frac{1}{2}\sec^2\theta\,d\theta$

$$\int \frac{2\,dx}{(4x^2+1)^{3/2}} = \int \frac{2\left(\frac{1}{2}\right)\sec^2\theta\,d\theta}{(\tan^2\theta+1)^{3/2}} = \int \frac{\sec^2\theta\,d\theta}{\sec^3\theta} = \int \cos\theta\,d\theta = \sin\theta + C = \frac{2x}{\sqrt{4x^2+1}} + C$$

16. $\displaystyle\frac{3x^2 - 6x - 2}{x^2(3x+1)} = \frac{Ax+B}{x^2} + \frac{C}{3x+1} = \frac{(Ax+B)(3x+1) + Cx^2}{x^2(3x+1)}$

$$3x^2 - 6x - 2 = 3Ax^2 + Ax + 3Bx + B + Cx^2$$
$$3x^2 - 6x - 2 = (3A+C)x^2 + (A+3B)x + B$$

(1) $\quad 3A + C = 3$, $3(0) + C = 3$, $C = 3$
(2) $\quad A + 3B = -6$; $A + 3(-2) = -6$, $A = 0$
(3) $\qquad\quad B = -2$

$$\int \frac{3x^2 - 6x - 2}{x^2(3x+1)}\,dx = \int \frac{-2}{x^2}\,dx + \int \frac{3}{3x+1}\,dx = \frac{2}{x} + \ln|3x+1| + C$$

17. $\displaystyle\int \frac{x^2+3}{x^4+3x^2+2}\,dx$; $x^4 + 3x^2 + 2 = (x^2+1)(x^2+2)$; $\displaystyle\frac{x^2+3}{x^4+3x^2+2} = \frac{Ax+B}{x^2+1} + \frac{Cx+D}{x^2+2}$

$$x^2 + 3 = (Ax+B)(x^2+2) + (Cx+D)(x^2+1)$$

$$
\begin{aligned}
&\text{Coeff. } x^3: &&0 = A + C &&\Big\} A = 0, C = 0\\
&\text{Coeff. } x: &&0 = 2A + C\\
&\text{Coeff. } x^2: &&1 = B + D &&\Big\} B = 2, D = -1\\
&\text{Const.:} &&3 = 2B + D
\end{aligned}
$$

$$\int \frac{x^2+3}{x^4+3x^2+2}\,dx = \int \frac{2\,dx}{x^2+1} - \int \frac{dx}{x^2+1} = 2\tan^{-1}x - \frac{1}{\sqrt{2}}\tan^{-1}\frac{x}{\sqrt{2}} + C = 2\tan^{-1}x - \frac{1}{2}\sqrt{2}\tan^{-1}\frac{1}{2}x\sqrt{2} + C$$

18. $\displaystyle\int x^3\sqrt{4-x^2}\,dx$

$$u = \sqrt{4-x^2}, u^2 = 4 - x^2, x^2 = 4 - u^2$$
$$2u\,du = -2x\,dx$$
$$-u\,du = x\,dx$$

$$\int x^3\sqrt{4-x^2}\,d = \int x^2\sqrt{4-x^2}(x\,dx) = \int (4-u^2)(u)(-u\,du) = \int (u^4 - 4u^2)\,du$$

$$= \frac{1}{5}u^5 - \frac{4}{3}u^3 + C = \frac{1}{5}(4-x^2)^{5/2} - \frac{4}{3}(4-x^2)^{3/2} + C$$

$$= (4-x^2)^{3/2}\left(\frac{1}{5}(4-x^2) - \frac{4}{3}\right) + C = (4-x^2)^{3/2}\left(-\frac{8}{15} - \frac{1}{5}x^2\right) + C$$

$$= -\frac{1}{15}(3x^2 + 8)(4-x^2)^{3/2} + C$$

19. $\displaystyle\int \ln(x+2)\,dx; \ u = \ln(x+2), du = \frac{dx}{x+2}, dv = dx, v = \int dx = x$

$$\int \ln(x+2)\,dx = x\ln(x+2) - \int x\left(\frac{dx}{x+2}\right) = x\ln(x+2) - \int\left(1 - \frac{2}{x+2}\right)dx$$

$$= x\ln(x+2) - x + 2\ln(x+2) + C = (x+2)\ln(x+2) - x + C$$

$$\int_0^2 \ln(x+2)\,dx = (x+2)\ln(x+2) - x \,\Big|_0^2 = 4\ln 4 - 2 - (2\ln 2 - 0)$$

$$= 8\ln 2 - 2\ln 2 - 2 = 6\ln 2 - 2 = 2.159$$

20. $\displaystyle\int_{-2}^1 \frac{x^2}{\sqrt{2-x}}\,dx$

$$u = \sqrt{2-x}, u^2 = 2 - x, x = 2 - u^2$$
$$2u\,du = -dx$$
$$-2u\,du = dx$$

$$\int_{-2}^1 \frac{x^2\,dx}{\sqrt{2-x}} = \int_2^1 \frac{(2-u^2)^2(-2u\,du)}{u} = \int_1^2 2(2-u^2)\,du = 2\int_1^2 (4 - 4u^2 + u^4)\,du$$

$$= 2\left[4u - \frac{4}{3}u^3 + \frac{1}{5}u^5\right]\Big|_1^2 = 2\left[4(2) - \frac{4}{3}(2)^3 + \frac{1}{5}(2)^5 - 4(1) + \frac{4}{3}(1)^3 - \frac{1}{5}(1)^5\right]$$

$$= \frac{26}{15}$$

21. $\displaystyle\int \frac{6(2-x^2)}{(x^2-1)(x^2-4)}\,dx$

$$(x^2-1)(x^2-4) = (x+1)(x-1)(x+2)(x-2)$$

$$\frac{12 - 6x^2}{(x^2-1)(x^2-4)} = \frac{A}{x+1} + \frac{B}{x-1} + \frac{C}{x+2} + \frac{D}{x-2}$$

$$12 - 6x^2 = A(x-1)(x+2)(x-2) + B(x+1)(x+2)(x-2)$$
$$+ C(x+1)(x-1)(x-2) + D(x+1)(x-1)(x+2)$$

$$x = -2: \ -12 = -12C, C = 1$$
$$x = -1: \ 6 = 6A, A = 1$$
$$x = 1: \ 6 = -6B, B = -1$$
$$x = 2: \ -12 = 12D, D = -1$$

$$\int \frac{6(2-x^2)}{(x^2-1)(x^2-4)} \, dx = \int \frac{dx}{x+1} - \int \frac{dx}{x-1} + \int \frac{dx}{x+2} - \int \frac{dx}{x-2}$$

$$= \ln|x+1| - \ln|x-1| + \ln|x+2| - \ln|x-2| + C$$

$$= \ln\left|\frac{(x+1)(x+2)}{(x-1)(x-2)}\right| + C$$

22. $\displaystyle\int \frac{-x^3+9x^2-24x+36}{x^2(x-3)^2} \, dx$

$$\frac{-x^3+9x^2-24x+36}{x^2(x-3)^2} = \frac{Ax+B}{x^2} + \frac{C}{x-3} + \frac{D}{(x-3)^2} = \frac{(Ax+B)(x-3)^2 + Cx^2(x-3) + Dx^2}{x^2(x-3)^2}$$

$$-x^3+9x^2-24x+36 = (Ax+B)(x-3)^2 + Cx^2(x-3) + Dx^2$$

$$x = 0, \quad 36 = 9B, B = 4$$
$$x = 3 \quad 18 = 9D, D = 2$$
$$x = -1, \quad 20 = (A+4)(4) + C(-2) + 2$$
$$4A - 2C = 2 \quad \text{(a)}$$
$$x = -1, 70 = (-A+4)(16) + C(-4) + 2$$
$$-4A - C = 1 \quad \text{(b)}$$

Adding (a) and (b) $-3C = 3, C = -1$
         (a) $4A - 2(-2) = 2, A = 0$

$$\int \frac{-x^3+9x^2-24x+36}{x^2(x-3)^2} \, dx = \int \frac{4}{x^2} \, dx - \int \frac{dx}{x-3} + \int \frac{2\,dx}{(x-3)^2} = -\frac{4}{x} - \ln|x-3| - \frac{2}{x-3} + C$$

$$= -\left(\frac{4}{x} + \ln|x+3| + \frac{2}{x-3}\right) + C$$

23. $\displaystyle\int_0^8 \frac{x\,dx}{(8-x)^{1/3}} = \lim_{h\to 0} \int_0^8 \frac{x\,dx}{(8-x)^{1/3}}$

$$(8-x)^{1/3} = u, x = 8 - u^3, dx = -3u^2 du$$

$$\int \frac{x\,dx}{(8-x)^{1/3}} = \int \frac{(8-u^3)(-3u^2 du)}{u} = 3\int (u^4 - 8u)du = \frac{3}{5}u^5 - 12u^2 + C$$

$$= \frac{3}{5}(8-x)^{5/3} - 12(8-x)^{2/3} + C = 3(8-x)^{2/3}\left[\frac{1}{5}(8-x) - 4\right] + C$$

$$= -\frac{3}{5}(8-x)^{2/3}(x+12) + C$$

$$\lim_{h\to 0}\int_0^{8-h} \frac{x\,dx}{(8-x)^{1/3}} = \lim_{h\to 0} -\frac{3}{5}(8-x)^{2/3}(x+12)\Big|_0^{8-h}$$

$$= \lim_{h\to 0} -\frac{3}{5}h^{2/3}(20-h) - \left[-\frac{3}{5}(8)^{2/3}(12)\right]$$

$$= 0 + \frac{144}{5} = \frac{144}{5}$$

24. $\displaystyle\int_1^{+\infty} \frac{dx}{x^2(x^2+1)}$

$$\frac{1}{x^2(x^2+1)} = \frac{Ax+B}{x^2} + \frac{Cx+D}{x^2+1} = \frac{(Ax+B)(x^2+1)+(Cx+D)x^2}{x^2(x^2+1)}$$

$$1 = (Ax+B)(x^2+1)+(Cx+D)x^2$$
$$1 = Ax^3 + Ax + Bx^2 + B + Cx^3 + Dx^2$$
$$1 = (A+C)x^3 + (B+D)x^2 + Ax + B$$

$$A+C=0, 0+C=0, C=0$$
$$B+D=0, 1+D=0, D=-1$$
$$A=0$$
$$B=1$$

$$\int_1^{+\infty} \frac{dx}{x^2(x^2+1)} = \int_1^{+\infty} \frac{1}{x^2}\,dx - \int_1^{+\infty} \frac{1}{x^2+1}\,dx = \lim_{a\to+\infty} \int_1^a \frac{1}{x^2}\,dx - \lim_{b\to+\infty} \int_1^b \frac{1}{x^2+1}\,dx$$

$$= \lim_{a\to+\infty} -\frac{1}{x}\Big|_1^a - \lim_{b\to+\infty} \tan^{-1}x\Big|_1^b$$

$$= \lim_{a\to+\infty}\left[-\frac{1}{a}+\frac{1}{1}\right] - \lim_{b\to+\infty}[\tan^{-1}b - \tan^{-1}1] = 0+1-\frac{\pi}{2}+\frac{\pi}{4} = 1-\frac{\pi}{4}$$

25. $A = \displaystyle\int_2^6 y\,dx = \int_2^6 \frac{dx}{\sqrt{x-2}} = \lim_{h\to 0} \int_{2+h}^6 \frac{dx}{\sqrt{x-2}} = \lim_{h\to 0} 2(x-2)^{1/2}\Big|_{2+h}^6$

$$= \lim_{h\to 0} 2(2) - 2(h)^{1/2} = 4$$

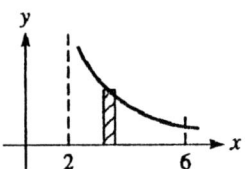

26. $\quad A = \displaystyle\int_0^4 \frac{x}{(1+x)^2}\,dx$

$$\frac{x}{(1+x)^2} = \frac{A}{1+x} + \frac{B}{(1+x)^2}$$

$$x = A(1+x) + B$$
$$x = A + Ax + B$$
$$x = -1, -1 = A - A + B, B = -1$$

$$x = 0, A+B = 0, A-1 = 0, A = 1$$

$$A = \int_0^4 \frac{x}{(1+x)^2}\,dx = \int_0^4 \frac{dx}{1+x} + \int_0^4 \frac{-dx}{(1+x)^2}$$

$$A = \ln|1+x|\ \Big|_0^4 + \frac{1}{1+x}\Big|_0^4$$

$$A = \ln|1+4| - \ln|1+0| + \frac{1}{1+4} - \frac{1}{1+0}$$

$$A = \ln 5 - \frac{4}{5}$$

$$A = 0.809$$

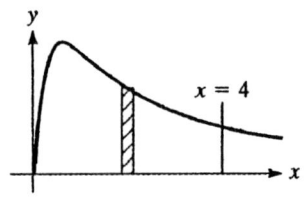

27. $A = 2\displaystyle\int_3^5 y\,dx = 2\int_3^5 \sqrt{25-x^2}\,dx$; Let $x = 5\sin\theta$, $dx = 5\cos\theta\,d\theta$

$$\int (25-x^2)^{1/2}dx = \int (25-25\sin^2\theta)^{1/2}(5\cos\theta\,d\theta) = 25\int \cos^2\theta\,d\theta$$

$$= \frac{25}{2}\int (1+\cos 2\theta)d\theta = \frac{25}{2}\theta + \frac{25}{4}\sin 2\theta + C$$

$$= \frac{25}{4}(2\theta + 2\sin\theta\cos\theta) + C = \frac{25}{4}\left(\sin^{-1}\frac{x}{5} + \frac{x}{5}\frac{\sqrt{25-x^2}}{5}\right) + C$$

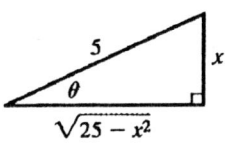

$$A = 2\left[\frac{25}{2}\left(\sin^{-1}\frac{x}{5} + \frac{x\sqrt{25-x^2}}{25}\right)\right]\Bigg|_3^5$$

$$= 25\sin^{-1}\frac{x}{5} + x\sqrt{25-x^2}\Bigg|_3^5 = 25\sin^{-1}1 + 0 - \left[25\sin^{-1}\frac{3}{5} + 3(4)\right]$$

$$= 25\left(\frac{\pi}{2}\right) - 25(0.6435) - 12 = 11.18$$

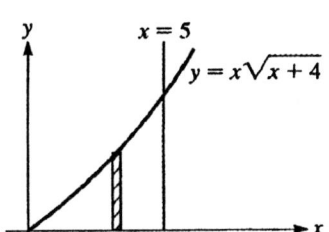

28. $y = x\sqrt{x+4}, y = 0, x = 5$

$$A = \int_0^5 y\,dx = \int_0^5 x\sqrt{x+4}\,dx$$

$$u = \sqrt{x+4}, u^2 = x+4, x = u^2 - 4$$

$$2u\,du = dx$$

$$A = \int_2^3 (u^2 - 4)(u)(2u\,du)$$

$$A = 2\int_2^3 (u^4 - 4u^2)du$$

$$A = 2\left[\frac{1}{5}u^5 - \frac{4}{3}u^3\right]\Bigg|_2^3$$

$$A = 2\left[\frac{1}{5}(3)^5 - \frac{4}{3}(3)^3 - \frac{1}{5}(2)^5 + \frac{4}{3}(2)^3\right]$$

$$A = \frac{506}{15}$$

29. $V = 2\pi\displaystyle\int_0^2 xy\,dx = 2\pi\int_0^2 x^2 e^x\,dx$; $u = x^2, du = 2x\,dx, dv = e^x\,dx, v = e^x$

$$\int x^2 e^x\,dx = x^2 e^x - \int e^x(2x\,dx) = x^2 e^x - 2\int xe^x\,dx$$

$$u = x, du = dx, dv = e^x\,dx, v = e^x; \int xe^x\,dx = xe^x - \int e^x\,dx = xe^x - e^x - \frac{C}{2}$$

$$\int x^2 e^x\,dx = x^2 e^x - 2\left(xe^x - e^x - \frac{C}{2}\right) = e^x(x^2 - 2x + 2) + C$$

$$V = 2\pi\left[e^x(x^2 - 2x + 2)\right]\Big|_0^2 = 2\pi\left[e^2(4 - 4 + 2) - e^0(0 - 0 + 2)\right]$$

$$= 4\pi(e^2 - 1) = 80.29$$

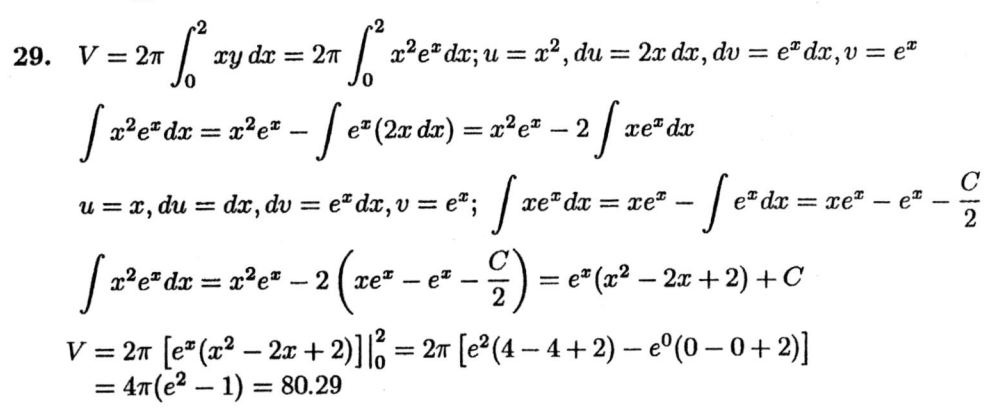

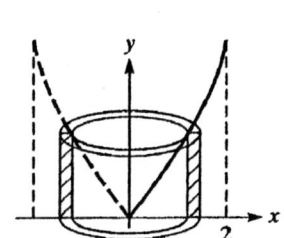

30. $V = \displaystyle\int_0^3 2\pi xy\,dx$

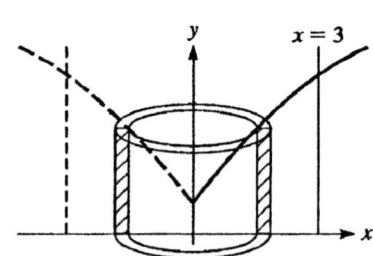

$$V = \int_0^3 2\pi x(x + \sqrt{x+1})dy$$

$$u = \sqrt{x+1}, u^2 = x+1, x = u^2 + 1$$
$$2u\,du = dx$$

$$V = \int_1^2 2\pi(u^2 - 1)(u^2 - 1 + u)(2u\,du)$$

$$V = 4\pi \int_1^2 (u^5 + u^4 - 2u^3 - u^2 + u)du$$

$$V = 4\pi \left[\frac{1}{6}u^6 + \frac{1}{5}u^5 - \frac{1}{2}u^4 - \frac{1}{3}u^3 + \frac{1}{2}u^2\right]\Big|_1^2$$

$$V = 4\pi \left[\frac{1}{6}(2)^6 + \frac{1}{5}(2)^5 - \frac{1}{2}(2)^4 - \frac{1}{3}(2)^3 + \frac{1}{2}(2)^2 - \frac{1}{6} - \frac{1}{5} + \frac{1}{2} + \frac{1}{3} - \frac{1}{2}\right]$$

$$= \frac{502\pi}{15}$$

31. $W = \displaystyle\int_0^{0.500} F\,dx = \int_0^{0.500} \frac{1000x\,dx}{(2x+1)^2} = 1000 \int_0^{0.500} \frac{x\,dx}{(2x+1)^2}; \; u = 2x+1, x = \frac{1}{2}(u-1), dx = \frac{1}{2}\,du$

$$\int \frac{x\,dx}{(2x+1)^2} = \int \frac{\left(\frac{1}{2}\right)(u-1)\left(\frac{1}{2}\right)(du)}{u^2} = \frac{1}{4}\int \left(\frac{1}{u} - \frac{1}{u^2}\right)du = \frac{1}{4}\left(\ln|u| + \frac{1}{u}\right) + C$$

When $x = 0, u = 1$, and when $x = 0.500, u = 2.00$.

$$W = (1000)\frac{1}{4}\left(\ln u + \frac{1}{u}\right)\Big|_1^{2.00} = 250\left[\ln 2.00 + \frac{1}{2.00} - (\ln 1 + 1)\right] = 48.3 \text{ N·cm}$$

32. $\displaystyle\int_0^\infty Ri^2\,dt = \int_0^\infty R(Ie^{-Rt/L})^2 dt = I^2 R \lim_{b\to\infty} \int_0^b e^{-2Rt/L}\,dt$

$$= I^2 R\left(-\frac{L}{2R}\right)\lim_{b\to\infty}\int_0^b e^{-2Rt/L}\left(\frac{-2R}{L}\,dt\right) = \frac{-I^2 L}{2}\lim_{b\to\infty}\left[e^{-2Rt/L}\right]\Big|_0^b$$

$$= -\frac{1}{2}I^2 L \lim_{b\to\infty}\left[e^{-2Rb/L} - e^0\right] = -\frac{1}{2}I^2 L \lim_{b\to\infty}[0 - 1] = \frac{1}{2}I^2 L$$

33. $A = \displaystyle\int_1^2 \ln x\,dx = x\ln x - x\big|_1^2 = 2\ln 2 - 2 - (\ln 1 - 1) = 2\ln 2 - 1 = 0.3863$

$$\bar{x} = \frac{\int_1^2 x\ln x\,dx}{A} = \frac{\frac{1}{2}x^2\left(\ln x - \frac{1}{2}\right)\big|_1^2}{A} = \frac{2\left(\ln 2 - \frac{1}{2}\right) - \frac{1}{2}\left(\ln 1 - \frac{1}{2}\right)}{A}$$

$$= \frac{2\ln 2 - \frac{3}{4}}{A} = \frac{0.6363}{0.3863} = 1.65 \text{ Integrals found from Formula 46.}$$

$$\bar{y} = \frac{\int_0^{\ln 2}(2-x)y\,dy}{A} = \frac{\int_0^{\ln 2} 2y\,dy - \int_0^{\ln 2} ye^y\,dy}{A} = \frac{y^2 - e^y(y-1)\big|_0^{\ln 2}}{A}$$

$$= \frac{\ln^2 2 - 2(\ln 2 - 1) - [0 - e^0(-1)]}{A} = \frac{0.09416}{0.3863} = 0.244$$

34. From $r = 1.5x^{2/3}$, the equation of the curve in QIV is

$$f(x) = -1.5^{-1.5}x^{3/2}$$

$$f'(x) = -\frac{3}{2}(1.5)^{-1.5}x^{1/2}$$

$$(f'(x))^2 = \frac{9}{4}(1.5)^{-3}x$$

$$S = \int_0^{3/2\sqrt[3]{256}} 2\pi x\sqrt{1 + \frac{9}{4}(1.5)^{-3}x}\, dx$$

$$S = \int_0^{3/2\sqrt[3]{256}} 2\pi x\sqrt{1 + \frac{2}{3}x}\, dx$$

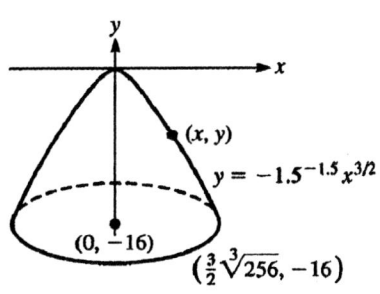

$$u = \sqrt{1 + \frac{2}{3}x},\, u^2 = 1 + \frac{2}{3}x,\, x = \frac{3}{2}(u^2 - 1)$$

$$2u\,du = \frac{2}{3}\,dx$$

$$3u\,du = dx$$

$$S = \int_1^{\sqrt{1+\sqrt[3]{256}}} 2\pi\left(\frac{3}{2}(u^2 - 1)\right)(u)(3u\,du)$$

$$S = \int_1^{\sqrt{1+\sqrt[3]{256}}} (u^4 - u^2)du = 9\pi\left(\frac{1}{5}u^5 - \frac{1}{3}u^3\right)\Bigg|_1^{\sqrt{1+\sqrt[3]{256}}}$$

$$S = 9\pi\left(\frac{1}{5}(\sqrt{1+\sqrt[3]{256}})^5 - \frac{1}{3}(\sqrt{1+\sqrt[3]{256}})^3\right) - 9\pi\left(\frac{1}{5} - \frac{1}{3}\right)$$

$$S = 644 \text{ ft}^2$$

35. $\dfrac{y^2}{1.25^2} + \dfrac{x^2}{0.56^2} = 1 \qquad A = 2\int_0^{0.56} y\,dx(1 - 3.189x^2)^{1/2}dx$

$y = 1.25\sqrt{1 - 3.189x^2}$

$$= 2.50\int_0^{0.56}(1 - 3.189x^2)^{1/2}dx$$

See Fig. 10-14

$$= \frac{2.50}{1.786}\int_0^{0.56}(1 - 3.189x^2)^{1/2}(1.786\,dx)$$

$$= 1.400\left[\frac{1.786x}{2}\sqrt{1 - 3.189x^2} + \frac{1}{2}\sin^{-1}1.786x\right]\Bigg|_0^{0.56} \quad \text{(Formula 15)}$$

$$= 1.400(0.500\sqrt{0} + \frac{1}{2}\sin^{-1}1.00 - 0 - 0)$$

$$= 1.400\left(\frac{1}{2}\right)\left(\frac{\pi}{2}\right) = 1.10 \text{ m}^2$$

36.   Assume $k = 1$.

$$I_y = \int_a^b x^2(y_1 - y_2)dx$$

$$I_y = \int_0^3 x^2(\sqrt{x+1} - 0)dx$$

$$u = x + 1, u^2 = x + 1, \quad x = u^2 - 1$$
$$2u\,du = dx, \quad x^2 = u^4 - 2u^2 + 1$$

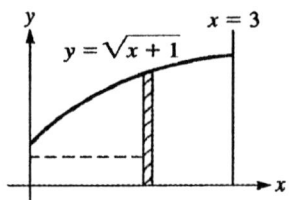

$$I_y = \int_1^2 (u^4 - 2u^2 + 1)(u)(2u\,du)$$

$$I_y = \int_1^2 (2u^6 - 4u^4 + 2u^2)du$$

$$I_y = \frac{2}{7}u^7 - \frac{4}{5}u^5 + \frac{2}{3}u^3 \Big|_1^2$$

$$I_y = \frac{2}{7}(2)^7 - \frac{4}{5}(2)^5 + \frac{2}{3}(2)^3 - \frac{2}{7} + \frac{4}{5} - \frac{2}{3}$$

$$I_y = \frac{1696}{105}$$

$$m = \int_a^b y\,dx = \int_0^3 \sqrt{x+1}\,dx$$

$$m = \frac{2}{3}(x+1)^{3/2}\Big|_0^3 = \frac{2}{3}(3+1)^{3/2} - \frac{2}{3}$$

$$m = \frac{14}{3}$$

$$R_y = \sqrt{\frac{I_y}{m}} = \sqrt{\frac{\frac{1696}{105}}{\frac{14}{3}}}$$

$$R_y = 1.86$$

37. $A = \int_a^b y\,dx = \int_0^1 \frac{x}{4 - x^2}\,dx$

   (a) $\int_0^1 \frac{x}{4 - x^2}\,dx$

$$u = 4 - x^2 \quad \text{when } x = 0, u = 4$$
$$du = -2x\,dx \qquad x = 1, u = 3$$

$$-\frac{1}{2}\,du = x\,dx$$

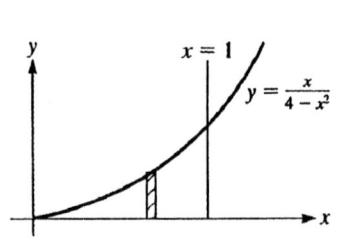

$$\int_0^1 \frac{x}{4 - x^2}\,dx = \int_4^3 \frac{-\frac{1}{2}\,du}{u} = -\frac{1}{2}\ln|u|\Big|_4^3 = -\frac{1}{2}\ln 3 + \frac{1}{2}\ln 4 = \ln 4^{1/2} - \ln 3^{1/2}$$

$$= \ln 2 - \ln \sqrt{3} = \ln \frac{2}{\sqrt{3}}$$

(b) $x = 2\sin\theta,$ $\qquad x^2 = 4\sin^2\theta,$ when $x = 0, \theta = 0$

$\qquad\qquad dx = 2\cos\theta\, d\theta$ $\qquad\qquad\qquad\qquad\qquad\qquad x = 1, \theta = \dfrac{\pi}{6}$

$$\int_0^1 \frac{x\,dx}{4 - x^2} = \int_0^{\pi/6} \frac{2\sin\theta(2\cos\theta\, d\theta)}{4 - 4\sin^2\theta} = \int_0^{\pi/6} \frac{4\sin\theta\cos\theta\, d\theta}{4(1 - \sin^2\theta)} = \int_0^{\pi/6} \frac{\sin\theta\cos\theta}{\cos^2\theta}\, d\theta$$

$$= -\int_0^{\pi/6} \frac{1}{\cos\theta}(-\sin\theta\, d\theta) = -\ln|\cos\theta|\big|_0^{\pi/6} = -\ln\left|\cos\frac{\pi}{6}\right| + \ln|\cos 0|$$

$$= -\ln\left|\frac{\sqrt{3}}{2}\right| + \ln 1 = -\ln\sqrt{3} + \ln 2 = \ln\frac{2}{\sqrt{3}} \text{ as in (a)}$$

INTRODUCTION TO PARTIAL DERIVATIVES AND DOUBLE INTEGRALS

11.1 Functions Of Two Variables

1. From geometry: $V = \pi r^2 h$

2. From geometry: $V = \dfrac{1}{3}\pi r^2 h$

3. From geometry: $A = \dfrac{1}{2}bh$

4. From Pythagorean theorem: $d = \sqrt{l^2 + w^2}$

5. From geometry:

$$A = 2\pi rh + 2\pi r^2; \; V = \pi r^2 h, h = \frac{V}{\pi r^2}$$

$$A = 2\pi r\left(\frac{V}{\pi r^2}\right) + 2\pi r^2 = \frac{2V}{r} + 2\pi r^2$$

6. From Law of Cosines: See Fig. 11-3

$$R^2 = F_1^2 + F_2^2 - 2F_1 F_2 \cos 150^\circ$$
$$= F_1^2 + F_2^2 - 2F_1 F_2(-0.8660)$$
$$R = \sqrt{F_1^2 + F_2^2 + 1.732F_1 F_2}$$

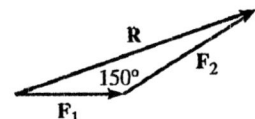

7. Let R = radius of cycliner

$$r^2 = R^2 + \left(\frac{h}{2}\right)^2 = R^2 + \frac{h^2}{4},$$

$$R^2 = r^2 - \frac{h^2}{4} = \frac{4r^2 - h^2}{4}$$

$$V = \pi R^2 h = \pi\frac{4r^2 - h^2}{4}h$$

$$V = \frac{1}{4}\pi h(4r^2 - h^2)$$

8. $T = F + 100h$

9. $f(x,y) = 2x - 6y$
   $f(0,-4) = 2(0) - 6(-4) = 24$

10. $F(x,y) = x^2 - 5y + y^2$
   $F(2,-2) = 2^2 - 5(-2) + (-2)^2$
   $F(2,-2) = 18$

11. $g(r,s) = r - 2rs - r^2 s$
   $g(-2,1) = -2 - (-2)(1) - (-2)^2(1)$
   $\qquad\quad = -2 + 4 - 4 = -2$

12. $f(r,\theta) = 2r(r\tan\theta - \sin 2\theta)$

$$f\left(3, \frac{\pi}{4}\right) = 2(3)\left[3\tan\frac{\pi}{4} - \sin\left(2\left(\frac{\pi}{4}\right)\right)\right]$$

$$f\left(3, \frac{\pi}{4}\right) = 6[3(1) - 1]$$

$$f\left(3, \frac{\pi}{4}\right) = 12$$

13. $Y(y,t) = \dfrac{2 - 3y}{t - y} + 2y^2 t$

$$Y(y,2) = \frac{2 - 3y}{t - y} + 2y^2(2)$$

$$= 2 - 3y + 4y^2$$

14. $f(r,t) = r^3 - 3r^2 t + 3rt + 3rt^2 - t^3$
   $f(3,t) = (3)^3 - 3(3)^2 t + 3(3)t^2 - t^3$
   $f(3,t) = 27 - 27t + 9t^2 - t^3$

15. $X(x,t) = -6xt + xt^2 - t^3$
   $X(x,-t) = -6x(-t) + x(-t)^2 - (-t)^3$
   $\qquad\quad = 6xt + xt^2 + t^3$

16. $g(y,z) = 2yz^2 - 6y^2 z - y^2 z^2$
   $g(y,2y) = 2y(2y)^2 - 6y^2(2y) - y^2(2y)^2$
   $g(y,2y) = 2y(4y^2) - 6y^2(2y) - y^2(4y^2)$
   $g(y,2y) = 8y^3 - 12y^3 - 4y^4$
   $g(y,2y) = -4y^4 - 4y^3$

17. $H(p,q) = p - \dfrac{p - 2q^2 - 5q}{p + q}$

$H(p, q + k)$

$$= p - \frac{p - 2(q+k)^2 - 5(q+k)}{p + q + k}$$

$$= \frac{p(p+q+k) - p + 2(q^2 + 2kq + k^2) + 5(q+k)}{p + q + k}$$

$$= \frac{p^2 + pq + pk - p + 2q^2 + 4kq + 2k^2 + 5q + 5k}{p + q + k}$$

18.
$$g(x,z) = z\tan^{-1}(x^2 + xz)$$
$$g(-x,z)) = z\tan^{-1}((-x)^2 + (-x)z)$$
$$g(-x,z) = z\tan^{-1}(x^2 - xz)$$

19. $f(x,y) = x^2 - 2xy - 4x$
$$f(x+h, y+k) - f(x,y)$$
$$= (x+h)^2 - 2(x+h)(y+k) - 4(x+h)$$
$$\quad - (x^2 - 2xy - 4x)$$
$$= x^2 + 2hx + h^2 - 2xy - 2kx - 2hy - 2hk$$
$$\quad - 4x - 4h - x^2 + 2xy + 4x$$
$$= 2hx - 2kx - 2hy + h^2 - 2hk - 4h$$

20. $g(y,z) = 4yz - z^3 + 4y$
$$g(y+1, z+2) - g(y,z)$$
$$= 4(y+1)(z+2) - (z+2)^3 + 4(y+1)$$
$$\quad - 4yz + z^3 - 4y$$
$$= 4(yz + 2y + z + 2) - (z^3 + 6z^2 + 12z + 8)$$
$$\quad + 4y + 4 - 4yz + z^3 - 4y$$
$$= 4yz + 8y + 4z + 8 - z^3 - 6z^2 - 12z - 8 + 4$$
$$\quad - 4yz + z^3$$
$$= 8y - 8z + 4 - 6z^2$$
$$= 8y - 6z^2 - 8z + 4$$

21. $f(x,y) = xy + x^2 - y^2$
$$f(x,x) - f(x,0)$$
$$\quad = x(x) + x^2 - x^2 - [x(0) + x^2 - 0^2]$$
$$\quad = x^2 + x^2 - x^2 - x^2$$
$$\quad = 0$$

22. $f(x,y) = 4x^2 - xy - 2y$
$$f(x,x^2) = 4x^2 - x(x^2) - 2(x^2)$$
$$f(x,x^2) = 4x^2 - x^3 - 2x^2$$
$$f(x,x^2) = 2x^2 - x^3$$
$$f(x,1) = 4x^2 - x(1) - 2(1)$$
$$\quad = 4x^2 - x - 2$$
$$f(x,x^2) - f(x,1) = 2x^2 - x^3 - 4x^2 + x + 2$$
$$f(x,x^2) - f(x,1) = -x^3 - 2x^2 + x + 2$$

23. $g(y,z) = 3y^3 - y^2z + 5z^2$
$$g(3z^2, z) - g(z,z)$$
$$= 3(3z^2)^3 - (3z^2)^2(z) + 5z^2 - [3z^3 - (z)^2z + 5z^2]$$
$$= 81z^6 - 9z^5 + 5z^2 - 3z^3 + z^3 - 5z^2$$
$$= 81z^6 - 9z^5 - 2z^3$$

24. $X(x,t) = 2x - \dfrac{t^2 - 2x^2}{x}$
$$X(2t,t) = 2(2t) - \frac{t^2 - 2(2t)^2}{2t} = 4t - \frac{t^2 - 8t^2}{2t}$$
$$X(2t,t) = 4t + \frac{7}{2}t = \frac{15}{2}t$$
$$X(2t^2, t) = 2(2t^2) - \frac{t^2 - 2(2t^2)^2}{2t^2}$$
$$= 4t^2 - \frac{t^2 - 8t^4}{2t^2}$$
$$X(2t^2, t) = 4t^2 - \frac{1}{2} + 4t^2 = 8t^2 - \frac{1}{2}$$
$$X(2t,t) - X(2t^2, t) = -8t^2 + \frac{15}{2}t + \frac{1}{2}$$

25. $f(x,y) = \frac{\sqrt{y}}{2x}$; considering $\sqrt{y}$, $y \geq 0$ for real values of $f(x,y)$; considering $2x, x \neq 0$ to avoid division by zero. Thus, $y \geq 0$ and $x \neq 0$.

26. $f(x,y) = \dfrac{x^2 - 4y^2}{x^2 + 9}$

There are no values of x and y which are not permissable.

27. $f(x,y) = \sqrt{x^2 + y^2 - x^2y - y^3}$
$$= \sqrt{x^2(1-y) + y^2(1-y)}$$
$$= \sqrt{(x^2 + y^2)(1-y)}$$

$x^2 + y^2 \geq 0$ for all real x and y; $1 - y \geq 0$, or $y \leq 1$ for real values of $f(x,y)$.

28. $f(x,y) = \dfrac{1}{xy - y}$
$$xy - y \neq 0$$
$$y(x-1) \neq 0$$
$$y \neq 0, x \neq 1$$

29. $v = iR$; $i = 3$ A, $R = 6 \, \Omega$
$$v = 3(6) = 18 \text{ V}$$

30. $a = \dfrac{v^2}{R} = \dfrac{6^2}{4} = 9 \text{ ft/s}^2$

31. $p = \dfrac{nRT}{V}$; $n = 3$ mol, $R = 8.31$ J/mol $\cdot$ K
$$\qquad\qquad T = 300 \text{ K}, V = 50 \text{ m}^3$$
$$p = \frac{3(8.31)(300)}{50} = 150 \text{ Pa}$$

32. $P = i^2R = 4^2(2.4) = 38.4$ W

33. For a, b, T with the same sign: circle if $a = b$, ellipse if $a \neq b$.
For a and b with different signs: hyperbola.

34. $p_i = \dfrac{F}{A}$

$p_f = \dfrac{2F}{\frac{2}{3}A}$

$\dfrac{p_i}{p_f} = \dfrac{\frac{F}{A}}{\frac{2F}{\frac{2}{3}A}} = \dfrac{F}{A}\dfrac{\frac{2}{3}A}{2F} = \dfrac{1}{3}$

35. $i = \dfrac{E}{R+0.25} = \dfrac{150}{1.20+0.25} = 1.03 \text{ A}$

$i = \dfrac{E}{R+0.25} = \dfrac{1.60}{1.05+0.25} = 1.23 \text{ A}$

36. $\dfrac{1}{p} + \dfrac{1}{q} = \dfrac{1}{f}$; $q = \dfrac{pf}{p-f}$; $p = 20$ cm, $f = 5$ cm

$q = \dfrac{20(5)}{20-5} = 6.67 \text{ cm}$

37. $p = 2\ell + 2w$; $\ell = \dfrac{p-2w}{2}$

$A = \ell w = \dfrac{p-2w}{2}w = \dfrac{pw - 2w^2}{2}$

$p = 250$ cm, $w = 55$ cm

$A = \dfrac{250(55) - 2(55)^2}{2} = 3850 \text{ cm}^2$

38. $V = \pi r^2 h + \dfrac{4}{3}\pi r^3$

$V = \pi(3.75)^2(12.5) + \dfrac{4}{3}\pi(3.75)^3$

$V = 773 \text{ ft}^3$

39. $L = \dfrac{kr^4}{\ell^2}$; $L = 20$ ton, $\ell = 20$ ft, $r = 0.5$ ft

$20 = \dfrac{k(0.5)^4}{20^2}$; $k = 1.28 \times 10^5 \text{ ton/ft}^2$

$L = \dfrac{(1.28 \times 10^5)r^4}{\ell^2}$

40. $f = \dfrac{k}{\sqrt{LC}}$

$10 = \dfrac{k}{\sqrt{4(64 \times 10^{-6})}}$, $k = 0.16$

$f = \dfrac{0.16}{\sqrt{LC}}$

11.2   Curves and Surfaces In Three Dimensions

1. $z = x^2 + y^2$
$z = x^2 + y^2 = 1$, circle, $r = 1$
$z = x^2 + y^2 = 4$, circle, $r = 2$
$z = x^2 + y^2 = 9$, circle, $r = 4$

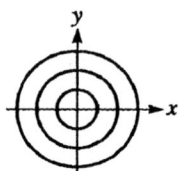

2. $z = x - 2y$
$z = -2, \ -2 = x - 2y$

$\qquad y = \dfrac{1}{2}x + 1$, line, $m = \dfrac{1}{2}, b = 1$

$z = 2,$
$\quad 2 = x - 2y$

$\qquad y = \dfrac{1}{2}x - 1$, line, $m = \dfrac{1}{2}, b = -1$

$z = 5, 5 = x - 2y$

$\qquad y = \dfrac{1}{2}x - \dfrac{5}{2}$, line, $m = \dfrac{1}{2}, b = -\dfrac{5}{2}$

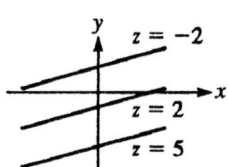

3. $z = y - x^2$
$z = 0$; $y = x^2$, parabola, $V(0,0)$
$z = 2$; $y - 2 = x^2$, parabola, $V(0,2)$
$z = 4$; $y - 4 = x^2$, parabola, $V(0,4)$

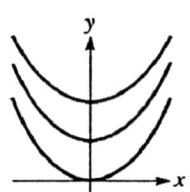

4. $z = x^2 - 4y^2$

$z = -4, \ -4 = x^2 - 4y^2$

$y^2 - \dfrac{x^2}{4} = 1$, hyperbola, $V(0 \pm 1)$

$z = 1, 1 = x^2 - 4y^2$, hyperbola, $V(\pm 1, 0)$

$z = 4, 4 = x^2 - 4y^2$

$\dfrac{x^2}{4} - y^2 = 1$, hyperbola, $V(\pm 2, 0)$

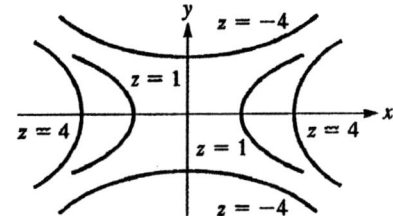

5. $x + y + 2z - 4 = 0$; plane

Intercepts: $(4, 0, 0), (0, 4, 0), (0, 0, 2)$

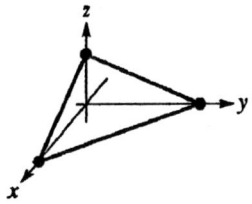

6. $2x - y - z + 6 = 0$; plane

Intercepts: $(-3, 0, 0), (0, 6, 0), (0, 0, -6)$

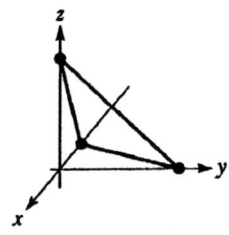

7. $4x - 2y + z - 8 = 0$; plane

Intercepts: $(2, 0, 0), (0, -4, 0), (0, 0, 8)$

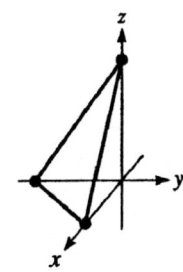

8. $3x + 3y - 2z - 6 = 0$; plane

Intercepts: $(2, 0, 0), (0, 2, 0), (0, 0, -3)$

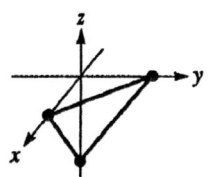

9. $z = y - 2x - 2$; plane

Intercepts: $(-1, 0, 0), (0, 2, 0), (0, 0, -2)$

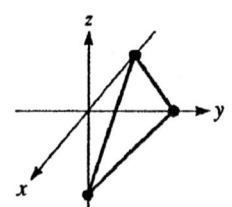

10. $z = x - 4y$; plane

Intercepts: $(0, 0, 0)$

Traces:

xy $-$ plane: $y = \dfrac{1}{4}x$, line, $m = \dfrac{1}{4}, b = 0$

xz $-$ plane: $z = x$, line, $m = 1, b = 0$

yz $-$ plane: $z = -4y$, line, $m = -4, b = 0$

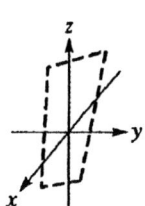

11. $x + 2y = 4$; plane

Intercepts: $(4, 0, 0), (0, 2, 0)$, no z-int.
plane parallel to z-axis

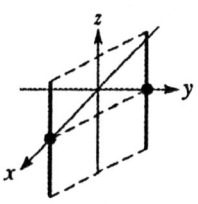

12. $2x - 3z = 6$; plane

Intercepts: $(0, 0, -2), (3, 0, 0)$
no y-int.
plane parallel to y-axis

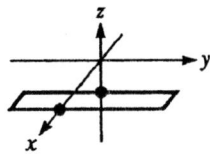

13. $x^2 + y^2 + z^2 = 4$

Intercepts: $(\pm 2, 0, 0), (0, \pm 2, 0), (0, 0, \pm 2)$

Traces:
yz-plane: $y^2 + z^2 = 4$, circle, $r = 2$
xz-plane: $x^2 + z^2 = 4$, circle, $r = 2$
xy-plane: $x^2 + y^2 = 4$, circle, $r = 2$

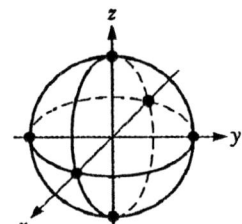

14. $2x^2 + 2y^2 + z^2 = 8$

Intercepts: $(\pm 2, 0, 0), (0, \pm 2, 0), (0, 0, \pm 2\sqrt{2})$

Traces:

xy-plane: $x^2 + y^2 = 4$, circle, $r = 2$

yz-plane: $\dfrac{y^2}{4} + \dfrac{z^2}{8} = 1$; ellipse: $a = 2, b = 2\sqrt{2}$

xz-plane: $\dfrac{x^2}{4} + \dfrac{z^2}{8} = 1$; ellipse: $a = 2, b = 2\sqrt{2}$

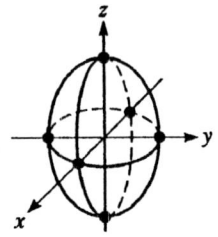

15. $z = 4 - x^2 - y^2$

Intercepts: $(\pm 2, 0, 0), (0, \pm 2, 0), (0, 0, 4)$

Traces:

yz-plane: $z = 4 - y^2$: parabola, $V(0, 0, 4)$
xz-plane: $z = 4 - x^2$: parabola, $V(0, 0, 4)$
xy-plane: $x^2 + y^2 = 4$: circle, $r = 2$

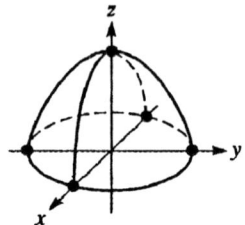

16. $z = x^2 + y^2$

Intercept: $(0, 0, 0)$

Traces:

xy-plane: single point $(0, 0, 0)$
yz-plane: parabola, $z = y^2, V(0, 0, 0)$
xz-plane: parabola, $z = x^2, V(0, 0, 0)$

Section:

For $z = 4, x^2 + y^2 = 4$, circle, $r = 2$

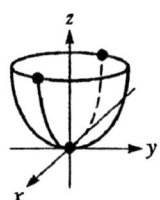

17. $z = 2x^2 + y^2 + 2$

Intercepts: No x-int., no y-int., $(0, 0, 2)$

Traces:

yz-plane:: $z = y^2 + 2$; parabola, $V(0, 0, 2)$
xz-plane: $z = 2x^2 + 2$, parabola, $V(0, 0, 2)$
xy-plane: No trace, $(2x^2 + y^2 + 2 \neq 0)$

Section: For $z = 4, 2x^2 + y^2 = 2$, ellipse

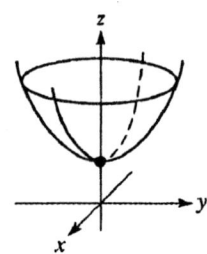

18. $x^2 + y^2 - 4z^2 = 4$

Intercepts: $(\pm 2, 0, 0), (0, \pm 2, 0)$, No z-intercept

Traces:

xy-plane: $x^2 + y^2 = 4$, circle, $r = 2$

xz-plane: $\dfrac{x^2}{4} - \dfrac{z^2}{1} = 1$, hyperbola, $V(\pm 2, 0, 0)$

yz-plane: $\dfrac{y^2}{4} - z^2 = 1$, hyperbola, $V(0, \pm 2, 0)$

Section:

For $z = \pm 1, x^2 + y^2 = 8$, circle, $r = 2\sqrt{2}$

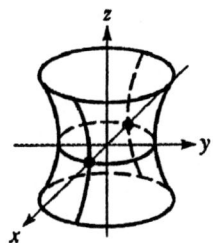

19. $x^2 - y^2 - x^2 = 9$

Intercepts: $(\pm 3, 0, 0)$, no y-int., no z int.

Traces:

yz-plane: No trace, $(-y^2 - x^2 \neq 9)$
xz-plane: $x^2 - z^2 = 9$, hyperbola, $a = 3, b = 3$
xy-plane: $x^2 - y^2 = 9$, hyperbola, $a = 3, b = 3$

Section:

For $x = \pm 5, y^2 + z^2 = 16$, circles, $r = 4$

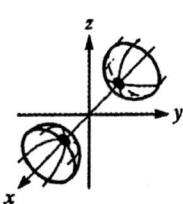

20. $z^2 = 9x^2 + 4y^2$

Intercepts: $(0, 0, 0)$

Traces:

xy-plane, single point $(0, 0, 0)$
yz-plane, lines, $z = \pm 2y$
xz-plane, lines, $z = \pm 3x$

Section: For $z = \pm 6$,

$9x^2 + 4y^2 = 36$

$\dfrac{x^2}{4} + \dfrac{y^2}{9} = 1$, ellipse, $a = 2, b = 3$

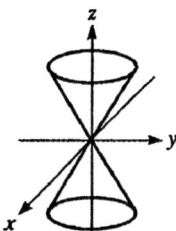

21. $x^2 + y^2 = 16$

Intercepts: $(\pm 4, 0, 0), (0, \pm 4, 0)$, no z-int.

Traces and Sections:

Since z is not present in the equation, the trace and sections are circles $x^2 + y^2 = 16$, with $r = 4$, for all z. This is a cylindrical surface.

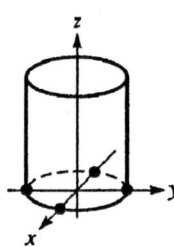

22. $4z = x^2$

Intercepts: $(0, 0, 0)$

Traces and Sections:

Since y is not present in the equation, the traces and sections are parabolas $4z = x^2$ with $V(0, y, 0)$ for all y. This is a cylindrical surface.

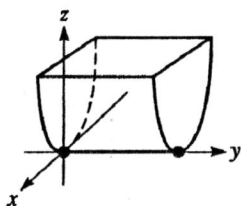

23. $y^2 + 9z^2 = 9$

Intercepts: No x-int., $(0, \pm 3, 0), (0, 0, \pm 1)$

<u>Traces and Sections</u>:

Since x is not present in the equation, the trace and sections are ellipses $y^2 + 9z^2 = 9$, with $a = 3$ and $b = 1$, for all x. This is cylindrical surface.

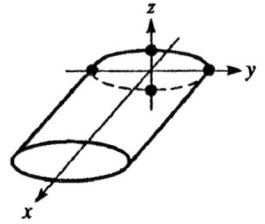

24. $xy = 2$

Intercepts: none

<u>Traces and Sections</u>:

Since z is not present in the equation the trace and sections are $y = \frac{2}{x}$ for all z. This is a cylindrical surface.

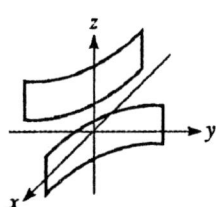

25. $t = 4x - y^2$
$t = -4;\ y^2 = 4(x + 1)$, parabola, $V(-1, 0)$
$t = 0;\ y^2 = 4x$, parabola, $V(0, 0)$
$t = 8;\ y^2 = 4(x - 2)$, parabola, $V(2, 0)$

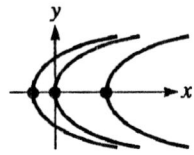

26. $V = y^2 - x^2$
$V = -9,\ -9 = y^2 - x^2, x^2 - y^2 = 9$

$$\frac{x^2}{9} - \frac{y^2}{9} = 1,\ \text{hyperbola, } V(\pm 3, 0)$$

$V = 0,\quad 0 = y^2 - x^2, y^2 = x^2$
$\qquad y = \pm x$, lines with $m = \pm 1, b = 0$
$V = 9,\quad 9 = y^2 - x^2,$

$$\frac{y^2}{9} - \frac{x^2}{9} = 1,\ \text{hyperbola, } V(0, \pm 3)$$

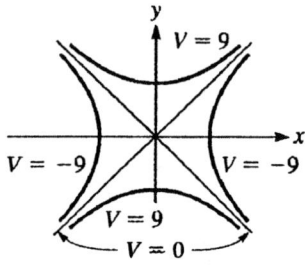

27. $2x^2 + 2y^2 + 3z^2 = 6$

Intercepts: $(\pm\sqrt{3}, 0, 0), (0, \pm\sqrt{3}, 0), (0, 0, \pm\sqrt{2})$

<u>Traces</u>:

yz-plane: $2y^2 + 3z^2 = 6$,
ellipse, $a = \sqrt{3}, b = \sqrt{2}$
xz-plane: $2x^2 + 3z^2 = 6$,
ellipse, $a = \sqrt{3}, b = \sqrt{2}$
xy-plane: $2x^2 + 2y^2 = 6$,
circle, $x^2 + y^2 = 3, r = \sqrt{3}$

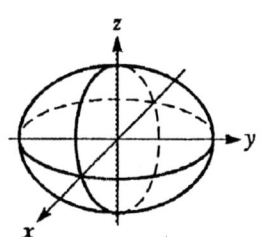

28. $z(2x^2 + y^2 + 100) = 1500$

Intercepts: No x-int., no y-int., $(0,0,15)$

Traces:

yz-plane: $z = \dfrac{1500}{y^2 + 100}$

xz-plane: $z = \dfrac{1500}{2x^2 + 100}$

xy-plane: No trace ($z \neq 0$)

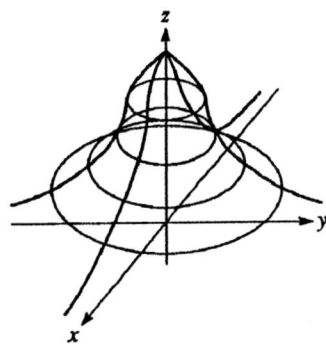

29. $p = \dfrac{T}{2V}$

Sections:

for constant p: $T = $ constant $\cdot V$, a line
for constant V: $p = $ constant $\cdot T$, a line
for constant T:

$p = \dfrac{\text{constant}}{V}$, one branch of a rotated hyperbola.

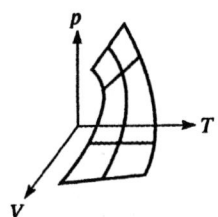

30. $x + 2y + 3z - 6 = 0$, plane

Intercepts: $(6,0,0), (0,3,0), (0,0,2)$
$2x + y + z - 4 = 0$, plane

Intercepts: $(2,0,0), (0,4,0), (0,0,4)$
The intersection of these two planes is the required line.

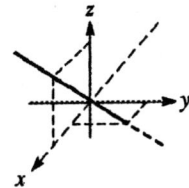

31. $x^2 + y^2 - 2y = 0$

Since z does not appear in the equation, all the traces and sections are circles,

$$x^2 + y^2 - 2y + 1 = 1$$
$$x^2 + (y-1)^2 = 1$$

with center $(0,1,z)$ for z and $r = 1$. This is a cylindrical surface.

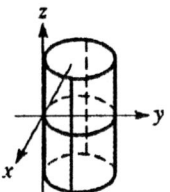

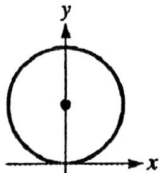

32. $x^2 + z^2 - 2z + 1 = 1$

Since y does not appear in the equation, the trace and sections are circles,

$$x^2 + (z-1)^2 = 1,$$

with center $(0,y,1)$ and $r = 1$ for all y. This is a cylindrical surface.

$z = 4 - x^2 - y^2 = 4 - (x^2 + y^2).$

z has maximum value 4.

Intercepts: $(0,0,4), (0,\pm 2,0), (\pm 2,0,0)$

Sections parallel to xz-plane and yz-plane are parabolas opening downward. Sections parallel to xy-plane are circles.

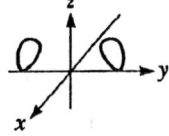

11.3 Partial Derivatives

1. $z = 5x + 4x^2 y$

$$\frac{\partial z}{\partial x} = 5 + 4(2x)y = 5 + 8xy$$

$$\frac{\partial z}{\partial y} = 0 + 4x^2(1) = 4x^2$$

2. $z = 3x^2 y^3 - 3x + 4y$

$$\frac{\partial z}{\partial x} = 6xy^3 - 3$$

$$\frac{\partial z}{\partial y} = 9x^2 y^2 + 4$$

3. $z = \dfrac{x^2}{y} - 2xy$

$$\frac{\partial z}{\partial x} = \frac{2x}{y} - 2(1)y = \frac{2x}{y} - 2y$$

$$\frac{\partial z}{\partial y} = \frac{x^2}{y^2} - 2x(1) = -\frac{x^2}{y^2} - 2x$$

4. $f(x,y) = \dfrac{x+y}{x-y}$

$$\frac{\partial f}{\partial x} = \frac{(x-y)(1) - (x+y)(1)}{(x-y)^2}$$

$$= \frac{x - y - x - y}{(x-y)^2}$$

$$\frac{\partial f}{\partial x} = \frac{-2y}{(x-y)^2}$$

$$\frac{\partial f}{\partial y} = \frac{(x-y)(1) - (x+y)(-1)}{(x-y)^2}$$

$$= \frac{x - y + x + y}{(x-y)^2}$$

$$\frac{\partial f}{\partial x} = \frac{2x}{(x-y)^2}$$

5. $f(x,y) = xe^{2y}$

$$\frac{\partial f}{\partial x} = (1)e^{2y} = e^{2y}$$

$$\frac{\partial f}{\partial y} = xe^{2y}(2) = 2xe^{2y}$$

6. $z = 3y \cos 2x$

$$\frac{\partial z}{\partial x} = 3y(-\sin 2x)(2)$$

$$\frac{\partial z}{\partial x} = -6y \sin 2x$$

$$\frac{\partial z}{\partial y} = 3\cos 2x$$

7. $f(x,y) = \dfrac{2 + \cos x}{1 - \sec 3y}$

$$\frac{\partial f}{\partial x} = \frac{-\sin x}{1 - \sec 3y}$$

$$\frac{\partial f}{\partial y} = -(2 + \cos x)(1 - \sec 3y)^{-2}$$

$$\cdot (-\sec 3y \tan 3y)(3)$$

$$= \frac{3(2 + \cos x)\sec 3y \tan 3y}{(1 - \sec 3y)^2}$$

8. $f(x,y) = \dfrac{\tan^{-1} y}{2 + x^2}$

$$\frac{\partial f}{\partial x} = \frac{(2 + x^2)(0) - \tan^{-1} y(2x)}{(2 + x^2)^2}$$

$$\frac{\partial f}{\partial x} = -\frac{2x \tan^{-1} y}{(2 + x^2)^2}$$

$$\frac{\partial f}{\partial y} = \frac{1}{2 + x^2}\left(\frac{1}{1 + y^2}\right)$$

$$\frac{\partial f}{\partial y} = \frac{1}{(2 + x^2)(1 + y^2)}$$

9. $\phi = r\sqrt{1 + 2rs}$

$$\frac{\partial \phi}{\partial r} = r\left(\frac{1}{2}\right)(1 + 2rs)^{-1/2}(2s)$$

$$+ (1 + 2rs)^{1/2}$$

$$= \frac{rs}{(1 + 2rs)^{1/2}} + (1 + 2rs)^{1/2}$$

$$= \frac{1 + 3rs}{\sqrt{1 + 2rs}}$$

$$\frac{\partial \phi}{\partial s} = r\left(\frac{1}{2}\right)(1 + 2rs)^{-1/2}(2r)$$

$$= \frac{r^2}{\sqrt{1 + 2rs}}$$

10. $w = uv^2\sqrt{1 - u}$

$$\frac{\partial w}{\partial u} = uv^2\left(\frac{-1}{2\sqrt{1-u}}\right) + v^2\sqrt{1 - u}$$

$$= \frac{-uv^2 + 2v^2(1 - u)}{2\sqrt{1-u}}$$

$$\frac{\partial w}{\partial u} = \frac{-uv^2 + 2v^2 - 2uv^2}{2\sqrt{1-u}}$$

$$= \frac{2v^2 - 3uv^2}{2\sqrt{1-u}}$$

$$\frac{\partial w}{\partial u} = \frac{v^2(2 - 3u)}{\sqrt{1-u}}$$

$$\frac{\partial w}{\partial v} = 2uv\sqrt{1 - u}$$

11. $z = (x^2 + xy^3)^4$

$$\frac{\partial z}{\partial x} = 4(x^2 + xy^3)^3(2x + y^3)$$
$$= 4(2x + y^3)(x^2 + xy^3)^3$$
$$\frac{\partial z}{\partial y} = 4(x^2 + xy^3)^3[x(3y^2)]$$
$$= 12xy^2(x^2 + xy^3)^3$$

12. $f(x, y) = (2xy - x^2)^5$

$$\frac{\partial f}{\partial x} = 5(2xy - x^2)^4(2y - 2x)$$
$$\frac{\partial f}{\partial x} = 10(2xy - x^2)^4(y - x)$$
$$\frac{\partial f}{\partial y} = 5(2xy - x^2)^4(2x)$$
$$\frac{\partial f}{\partial y} = 10x(2xy - x^2)^4$$

13. $z = \sin xy$

$$\frac{\partial z}{\partial x} = (\cos xy)(y)$$
$$= y \cos xy$$
$$\frac{\partial z}{\partial y} = (\cos xy)(x)$$
$$= x \cos xy$$

14. $y = \tan^{-1} \dfrac{x}{t}$

$$\frac{\partial y}{\partial x} = \frac{1}{1 + \left(\frac{x}{t}\right)^2}\left(\frac{1}{t}\right)$$
$$= \frac{1}{1 + \frac{x^2}{t^2}}\left(\frac{1}{t}\right)$$
$$\frac{\partial y}{\partial x} = \frac{1}{t + \frac{x^2}{t}}$$
$$\frac{\partial y}{\partial x} = \frac{t}{t^2 + x^2}$$
$$\frac{\partial y}{\partial t} = \frac{1}{1 + \left(\frac{x}{t}\right)^2}\left(\frac{-x}{t^2}\right)$$
$$= \frac{-x}{\left(1 + \frac{x^2}{t^2}\right)t^2}$$
$$\frac{\partial y}{\partial t} = -\frac{x}{t^2 + x^2}$$

15. $y = \ln(r^2 + s)$

$$\frac{\partial y}{\partial r} = \frac{2r}{r^2 + s}$$
$$\frac{\partial y}{\partial s} = \frac{1}{r^2 + s}$$

16. $u = e^{x+2y}$

$$\frac{\partial u}{\partial x} = e^{x+2y}$$
$$\frac{\partial u}{\partial y} = 2e^{x+2y}$$

17. $f(x, y) = \dfrac{2\sin^3 2x}{1 - 3y}$

$$\frac{\partial f}{\partial x} = \frac{2(3)(\sin^2 2x)(\cos 2x)(2)}{1 - 3y}$$
$$= \frac{12 \sin^2 2x \cos 2x}{1 - 3y}$$
$$\frac{\partial f}{\partial y} = -(2\sin^3 2x)(1 - 3y)^{-2}(-3)$$
$$= \frac{6 \sin^3 2x}{(1 - 3y)^2}$$

18. $f(x, y) = \dfrac{3x + \ln y}{x^2 + y^2}$

$$\frac{\partial f}{\partial x} = \frac{(x^2 + y^2)(3) - (3x + \ln y)(2x)}{(x^2 + y^2)^2}$$
$$= \frac{3x^2 + 3y^2 - 6x^2 - 2x \ln y}{(x^2 + y^2)^2}$$
$$\frac{\partial f}{\partial x} = \frac{3y^2 - 3x^2 - 2x \ln y}{(x^2 + y^2)^2}$$
$$\frac{\partial f}{\partial y} = \frac{(x^2 + y^2)\left(\frac{1}{y}\right) - (3x + \ln y)(2y)}{(x^2 + y^2)^2}$$
$$= \frac{x^2 + y^2 - 6xy^2 - 2y^2 \ln y}{y(x^2 + y^2)}$$

19. $z = \dfrac{\sin^{-1} xy}{3 + x^2}$

$$\frac{\partial z}{\partial x} = \frac{(3 + x^2)\frac{1}{[1-(xy)^2]^{1/2}}(y) - (\sin^{-1} xy)(2x)}{(3 + x^2)^2}$$
$$= \frac{3y + x^2 y - 2x\sqrt{1 - x^2 y^2}\, \sin^{-1} xy}{(3 + x^2)^2\sqrt{1 - x^2 y^2}}$$
$$\frac{\partial z}{\partial y} = \frac{1}{3 + x^2}\left(\frac{1}{\sqrt{1 - x^2 y^2}}\right)(x)$$
$$= \frac{x}{(3 + x^2)\sqrt{1 - x^2 y^2}}$$

20. $z = \dfrac{\sqrt{1 - \tan xy}}{xy + y^2}$

$$\frac{\partial z}{\partial x} = \frac{(xy + y^2)\left(\frac{1}{2\sqrt{1 - \tan xy}}\right)(-y \sec^2 xy) - \sqrt{1 - \tan xy}(y)}{(xy + y^2)^2}$$

$$\frac{\partial z}{\partial x} = \frac{(xy + y^2)(-y \sec^2 xy) - 2y(1 - \tan xy)}{2\sqrt{1 - \tan xy}(y^2)(x + y^2)}$$

$$\frac{\partial z}{\partial x} = \frac{-[y(x + y)\sec^2 xy + 2(1 - \tan xy)]}{2y\sqrt{1 - \tan xy}(x + y)^2}$$

$$\frac{\partial z}{\partial y} = \frac{(xy + y^2)\left(\frac{1}{2\sqrt{1 - \tan xy}}\right)(-x \sec^2 xy) - \sqrt{1 - \tan xy}(x + 2y)}{(xy + y^2)^2}$$

$$\frac{\partial z}{\partial y} = \frac{-[xy(x + y)\sec^2 xy + 2(x + 2y)(1 - \tan xy)]}{2y^2(x + y)^2\sqrt{1 - \tan xy}}$$

21. $z = \sin x + \cos xy - \cos y$

$$\frac{\partial z}{\partial x} = \cos x - (\sin xy)(y)$$
$$= \cos x - y \sin xy$$

$$\frac{\partial z}{\partial y} = -(\sin xy)(x) + \sin y$$
$$= -x \sin xy + \sin y$$

22. $t = 2re^{rs^2} - \tan(2r + s)$

$$\frac{\partial t}{\partial r} = 2r(s^2 e^{rs^2}) + 2e^{rs^2} - \sec^2(2r + s)(2)$$

$$\frac{\partial t}{\partial r} = 2(rs^2 + 1)e^{rs^2} - 2\sec^2(2r + s)$$

$$\frac{\partial t}{\partial s} = 2r(2rs)e^{rs^2} - \sec^2(2r + s)(1)$$

$$\frac{\partial t}{\partial s} = 4r^2 s e^{rs^2} - \sec^2(2r + s)$$

23. $f(x, y) = e^x \cos xy + e^{-2x} \tan y$

$$\frac{\partial f}{\partial x} = e^x[(-\sin xy)(y)] + (\cos xy)e^x + e^{-2x}\tan y(-2) = e^x(\cos xy - y \sin xy) - 2e^{-2x}r \tan y$$

$$\frac{\partial f}{\partial y} = e^x[-(\sin xy)(x)] + e^{-2x}\sec^2 y = -xe^x \sin xy + e^{-2x}\sec^2 y$$

24. $u = \ln \dfrac{y^2}{x - y} + e^{-x}(\sin y - \cos 2y)$

$$\frac{\partial u}{\partial x} = \frac{1}{\frac{y^2}{x-y}}\frac{(x - y)(0) - y^2(1)}{(x - y)^2} + e^{-x}(0) + (-e^{-x})(\sin y - \cos 2y)$$

$$\frac{\partial u}{\partial x} = -\frac{1}{x - y} - e^{-x}(\sin y - \cos 2y)$$

$$\frac{\partial u}{\partial y} = \frac{1}{\frac{y^2}{x-y}}\frac{(x - y)(2y) - y^2(-1)}{(x - y)^2} + e^{-x}(\cos y + 2\sin 2y)$$

$$\frac{\partial u}{\partial y} = \frac{2xy - 2y^2 + y^2}{y^2(x - y)} + e^{-x}(\cos y + 2\sin 2y)$$

$$\frac{\partial u}{\partial y} = \frac{y(2x - y)}{y^2(x - y)} + e^{-x}(\cos y + 2\sin 2y)$$

$$\frac{\partial u}{\partial y} = \frac{2x - y}{y(x - y)} + e^{-x}(\cos y + 2\sin 2y)$$

25. $z = 3xy - x^2$

$$\frac{\partial z}{\partial x} = 3y - 2x$$

$$\left.\frac{\partial z}{\partial y}\right|_{(1,-2,-7)} = 3(-2) - 2(1) = -8$$

26. $z = x^2 \cos 4y$

$$\frac{\partial z}{\partial y} = -4x^2 \sin 4y$$

$$\left.\frac{\partial z}{\partial y}\right|_{(2,\frac{\pi}{2},4)} = -4(2)^2 \sin\left(4\left(\frac{\pi}{2}\right)\right) = 0$$

27. $z = x\sqrt{x^2 - y^2}$

$$\frac{\partial z}{\partial x} = x\frac{1}{2\sqrt{x^2 - y^2}}(2x) + \sqrt{x^2 - y^2}$$

$$\left.\frac{\partial z}{\partial y}\right|_{(5,3,20)} = \frac{(5)^2}{\sqrt{5^2 - 3^2}} + \sqrt{5^2 - 3^2} = \frac{41}{4}$$

28. $z = e^y \ln xy$

$$\frac{\partial z}{\partial y} = e^y\left(\frac{1}{xy}\right)(x) + e^y \ln xy$$

$$= e^y\left(\frac{1}{y} + \ln xy\right)$$

$$\left.\frac{\partial z}{\partial y}\right|_{(e,1,e)} = e\left(\frac{1}{1} + \ln e\right) = 2e$$

29. $z = 2xy^3 - 3x^2y$

$$\frac{\partial z}{\partial x} = 2y^3 - 6xy$$

$$\frac{\partial z}{\partial y} = 6xy^2 - 3x^2$$

$$\frac{\partial^2 z}{\partial x^2} = -6y, \frac{\partial^2 z}{\partial x^2} = 12xy$$

$$\frac{\partial^2 z}{\partial x \partial y} = \frac{\partial^2 z}{\partial x \partial y} = 6y^2 - 6x$$

30. $f(x,y) = y\ln(x + 2y)$

$$\frac{\partial f}{\partial x} = \frac{y}{x + 2y}, \frac{\partial f}{\partial y} = \frac{2y}{x + 2y} + \ln(x + 2y)$$

$$\frac{\partial^2 f}{\partial x^2} = \frac{-y}{(x + 2y)^2}$$

$$\frac{\partial^2 f}{\partial y^2} = \frac{(x + 2y)(2) - 2y(2)}{(x + 2y)^2} + \frac{2}{x + 2y} = \frac{4x + 4y}{(x + 2y)^2}$$

$$\frac{\partial^2 f}{\partial y \partial x} = \frac{(x + 2y) - y(2)}{(x + 2y)^2} = \frac{x}{(x + 2y)^2}$$

$$\frac{\partial^2 f}{\partial x \partial y} = \frac{-2y}{(x + 2y)^2} + \frac{1}{x + 2y} = \frac{-2y + x + 2y}{(x + 2y)^2} = \frac{x}{(x + 2y)^2}$$

31. $z = \dfrac{x}{y} + e^x \sin y$

$$\frac{\partial z}{\partial x} = \frac{1}{y} + e^x \sin y, \frac{\partial z}{\partial y} = \frac{-x}{y^2} + e^x \cos y$$

$$\frac{\partial^2 z}{\partial x^2} = e^x \sin y, \frac{\partial^2 z}{\partial y^2} = \frac{2x}{y^3} - e^x \sin y$$

$$\frac{\partial^2 z}{\partial x2y} = \frac{\partial^2 z}{\partial y2x} = -\frac{1}{y^2} + e^x \cos y$$

32. $f(x,y) = \dfrac{2 + \cos y}{1 + x^2}$

$$\frac{\partial f}{\partial x} = \frac{(1 + x^2)(0) - (2 + \cos y)(2x)}{(1 + x^2)^2} = \frac{-2x(2 + \cos y)}{(1 + x^2)^2}$$

$$\frac{\partial^2 f}{\partial x^2} = \frac{(1 + x^2)(-2(2 + \cos y)) + 2x(2 + \cos y)(2(1 + x^2))(2x)}{(1 + x^2)^4}$$

$$\frac{\partial^2 f}{\partial x^2} = \frac{(2 + \cos y)[-2(1 + x^2) + 8x^2]}{(1 + x^2)^3} = \frac{2(3x^2 - 1)(2 + \cos y)}{(1 + x^2)^3}$$

$$\frac{\partial f}{\partial y} = \frac{-\sin y}{1 + x^2}$$

$$\frac{\partial^2 f}{\partial y^2} = \frac{-\cos y}{1 + x^2}$$

$$\frac{\partial^2 f}{\partial x \partial y} = \frac{\partial^2 f}{\partial y \partial x} = \frac{(1 + x^2)(0) + \sin y(2x)}{(1 + x^2)^2}$$

$$\frac{\partial^2 f}{\partial x \partial y} = \frac{\partial^2 f}{\partial y \partial x} = \frac{2x \sin y}{(1 + x^2)^2}$$

33. $z = 9 - x^2 - y^2$

$$\frac{\partial z}{\partial y} = -2y$$

$$\left.\frac{\partial z}{\partial y}\right|_{(1,2,4)} = -4$$

$$\left.\frac{\partial z}{\partial y}\right|_{(2,2,1)} = -4$$

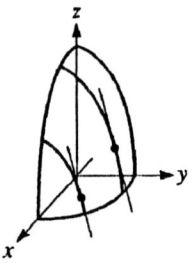

34. $r^2 = (r - h)^2 + \left(\frac{w}{2}\right)^2$

$$r^2 = r^2 - 2rh + h^2 + \frac{w^2}{4}$$

$$w^2 = 4(2rh - h^2)$$
$$w = 2\sqrt{2rh - h^2}$$

$$\frac{\partial w}{\partial r} = 2\left[\frac{2h}{2\sqrt{2rh - h^2}}\right] = \frac{2h}{\sqrt{2rh - h^2}}$$

$$\frac{\partial w}{\partial h} = 2\left[\frac{2r - 2h}{2\sqrt{2rh - h^2}}\right] = \frac{2(r - h)}{\sqrt{2rh - h^2}}$$

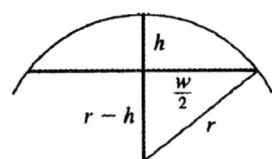

35. $\dfrac{1}{R_T} = \dfrac{1}{R_1} + \dfrac{1}{R_2}$

$$R_T = \frac{R_1 R_2}{R_1 + R_2}$$

$$\frac{\partial R_T}{\partial R_1} = \frac{(R_1 + R_2)(R_2) - R_1 R_2(1)}{(R_1 + R_2)^2}$$

$$= \frac{R_2^2}{(R_1 + R_2)^2}$$

36. $z = 4x^2 - 8$

$$\frac{\partial z}{\partial y} = 0.$$

Any line tangent to the surface and parallel to the yz-plane is in the surface for all values of x.

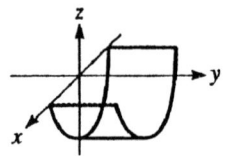

37. $V = \pi r^2 h + \dfrac{1}{2}\left(\dfrac{4}{3}\pi r^3\right)$

$$\frac{\partial V}{\partial R} = 2\pi r h + 2\pi r^2\big|_{r=2.65, h=4.20}$$

$$\frac{\partial V}{\partial R} = 2\pi(2.65)(4.20) + 2\pi(2.65)^2$$

$$\frac{\partial V}{\partial R} = 114 \text{ cm}^2$$

38. $a = \dfrac{M - m}{M + m}g$

$$\frac{\partial a}{\partial M} = \frac{(M + m) - (M - m)}{(M + m)^2}g$$

$$= \frac{2mg}{(M + m)^2}$$

$$\frac{\partial a}{\partial m} = \frac{(M + m)(-1) - (M - m)(1)}{(M + m)^2}g$$

$$= \frac{-2Mg}{(M + m)^2}$$

$$M\left[\frac{2mg}{(M + m)^2}\right] + m\left[\frac{-2Mg}{(M + m)^2}\right] = 0; \ 0 = 0$$

39.

$$\frac{\partial d}{\partial x} = \frac{1}{2}(x^2 + y^2)^{-1/2}(2x) = \frac{x}{\sqrt{x^2 + y^2}}$$

For $x = 6.50$ ft, $y = 4.75$ ft

$$\frac{\partial d}{\partial x} = \frac{6.50}{\sqrt{6.50^2 + 4.75^2}} = 0.807$$

40. $f_o = f_s \dfrac{v + v_o}{v - v_s}; f_s = f_o \dfrac{v - v_s}{v + v_0}$

$$\frac{\partial f_o}{\partial v_s} = f_s \left[\frac{(v - v_s)(0) - (v + v_o)(-1)}{(v - v_s)^2} \right]$$

$$\frac{\partial f_o}{\partial v_s} = f_s \frac{v + v_o}{(v - v_s)^2}$$

$$f_s \frac{\partial f_o}{\partial v_s} = f_s^2 \frac{v + v_o}{(v - v_s)^2}$$

$$= f_0^2 \frac{(v - v_s)^2}{(v + v_o)^2} \frac{(v + v_o)}{(v - v_s)^2}$$

$$f_s \frac{\partial f_o}{\partial v_s} = \frac{f_0^2}{v + v_o} \qquad (1)$$

$$\frac{\partial f_0}{\partial v_0} = f_s \left[\frac{(v - v_s)(1) - (v + v_o)(0)}{(v - v_s)^2} \right]$$

$$= \frac{f_s}{v - v_s}$$

$$f_0 \frac{\partial f_0}{\partial v_0} = f_0 \frac{f_s}{v - v_s}$$

$$= f_0 \frac{f_0 \frac{v - v_o}{v - v_s}}{v - v_s}$$

$$= \frac{f_o^2}{v + v_o}$$

$$f_0 \frac{\partial f_0}{\partial v_0} = \frac{f_o^2}{v + v_o} \qquad (2)$$

Comparing (1) and (2) shows $f_s \dfrac{\partial f_0}{\partial v_s} = f_0 \dfrac{\partial f_0}{\partial v_0}$

41. $i_b = 50(e_b + 5e_c)^{1.5}$

$$\frac{\partial i_b}{\partial e_c} = 50(1.5)(e_b + 5e_c)^{0.5}(5)$$
$$= 375(e_b + 5e_c)^{0.5}$$

For $e_b = 200$ V and $e_c = -20$ V

$$\frac{\partial i_b}{\partial e_c} = 375(200 - 100)^{0.5}$$
$$= 3750 \ \mu\text{A/V} = 3.75 \ 10^{-3} \ 1/\Omega$$

42. $i_b = 50(e_b + 5e_c)^{1.5}$

$$e_b = \left(\frac{i_b}{50} \right)^{2/3} - 5e_c$$

$$\mu = -\frac{\partial e_b}{\partial e_c} = -(-5) = 5$$

43. $\alpha = \dfrac{1}{L} \dfrac{\partial L}{\partial T}, L = L_o + k_1 F + k_2 T + k_3 FT^2$

$$\alpha = \frac{1}{L_o + k_1 F + k_2 T + k_3 FT^2} (k_2 + 2k_3 FT)$$

$$\alpha = \frac{k_2 + 2k_3 FT}{L_o + k_1 F + k_2 T + k_3 FT^2}$$

44. $f = \dfrac{k\sqrt{T}}{L}, 30 = \dfrac{k\sqrt{65}}{60}, k = \dfrac{1800}{\sqrt{65}}$

$$\frac{\partial f}{\partial T} = \frac{k}{L} \frac{1}{2\sqrt{T}} = \frac{1800}{\sqrt{65}L} \frac{1}{2\sqrt{T}}$$

$$= \frac{900}{\sqrt{65}L\sqrt{T}} \Bigg|_{L=60, T=65}$$

$$\frac{\partial f}{\partial T} = \frac{900}{\sqrt{65}(60)\sqrt{65}}$$

$$\frac{\partial f}{\partial T} = 0.231 \ \text{Hz/N}$$

45. $pV = nRT, p = \dfrac{nRT}{V}$

$$\frac{\partial p}{\partial V} = -nRTV^{-2} = -\frac{nRT}{V^2}$$

46. $pV = nRT, V = \dfrac{nRT}{p}, T = \dfrac{pV}{nR}, p = \dfrac{nRT}{V}$

$$\left(\frac{\partial V}{\partial T} \right) \left(\frac{\partial T}{\partial p} \right) \left(\frac{\partial p}{\partial V} \right) = \left(\frac{nR}{p} \right) \left(\frac{V}{nR} \right) \left(\frac{-nRT}{V^2} \right)$$

$$= -\left(\frac{nR}{p} \right) \left(\frac{V}{nR} \right) \left(\frac{pV}{V^2} \right)$$

$$= -1$$

47. $u(x,t) = 5e^{-t}\sin 4x$

$$\frac{\partial u}{\partial x} = 5e^{-t}(4\cos 4x) = 20e^{-t}\cos 4x$$

$$\frac{\partial^2 u}{\partial x^2} = 20e^{-t}(-4\sin 4x) = -80e^{-t}\sin 4x$$

$$\frac{\partial u}{\partial t} = -5e^{-t}\sin 4x$$

$$\frac{\partial u}{\partial t} = k\frac{\partial^2 u}{\partial x^2}, k = \frac{1}{16}$$

$$-5e^{-t}\sin 4x = \frac{1}{16}(-80e^{-t}\sin 4x)$$

$$-5e^{-t}\sin 4x = -5e^{-t}\sin 4x$$

48. $y(x, t) = 2 \sin 2x \cos 4t$

$$\frac{\partial y}{\partial t} = 2 \sin 2x (-4 \sin 4t)$$
$$= -8 \sin 2x \sin 4t$$

$$\frac{\partial^2 y}{\partial t^2} = -8 \sin 2x (4 \cos 4t)$$
$$= -32 \sin 2x \cos 4t$$

$$\frac{\partial^2 y}{\partial t^2} = -32 \sin 2x \cos 4t \qquad (1)$$

$$\frac{\partial y}{\partial x} = 4 \cos 2x \cos 4t$$

$$\frac{\partial^2 y}{\partial x^2} = -8 \sin 2x \cos 4t,$$

$$4\frac{\partial^2 y}{\partial x^2} = -32 \sin 2x \cos 4t \qquad (2)$$

$$\frac{\partial^2 y}{\partial t^2} = a^2 \frac{\partial^2 y}{\partial x^2}, \text{ with } a = 2$$

$$\frac{\partial^2 y}{\partial t^2} = 4\frac{\partial^2 y}{\partial x^2} \qquad (3)$$

comparing (1) and (2) verfies (3).

11.4   Certain Applications of Partial Derivatives

1. $z = 2x^2 - y^2 + 3x$

$$\frac{\partial z}{\partial x} = 4x + 3, \frac{\partial z}{\partial y} = -2y$$
$$dz = (4x + 3)dx - 2y \, dy$$

2. $z = xy - 3yx^2 + y^3$

$$\frac{\partial z}{\partial x} = y - 6yx, \frac{\partial z}{\partial y}$$
$$= x - 3x^2 + 3y^2$$
$$dz = (y - 6xy)dx + (x - 3x^2 + 3y^2)dy$$

3. $z = xe^y - y^2$

$$\frac{\partial z}{\partial x} = e^y, \frac{\partial z}{\partial y}$$
$$= xe^y - 2y$$
$$dz = e^y dx + (xe^y - 2y)dy$$

4. $z = x^3 - y \ln x$

$$\frac{\partial z}{\partial x} = 3x^2 - \frac{y}{x}, \frac{\partial z}{\partial y}$$
$$= -\ln x$$
$$dz = \left(3x^2 - \frac{y}{x}\right)dx - \ln x \, dy$$

5. $z = x(y - 2x^2)^5$

$$\frac{\partial z}{\partial x} = 5x(y - 2x^2)^4(-4x) + (y - 2x^2)^5$$
$$= (y - 22x^2)(y - 2x^2)^4$$
$$\frac{\partial z}{\partial y} = 5x(y - 2x^2)^4$$
$$dz = (y - 22x^2)(y - 2x^2)^4 dx$$
$$+ 5x(y - 2x^2)^4 dy$$

6. $z = y^2\sqrt{1 - 2xy}$

$$\frac{\partial z}{\partial x} = \frac{y^2}{2\sqrt{1 - 2xy}}(-2y)$$
$$= \frac{-y^3}{\sqrt{1 - 2xy}}$$
$$\frac{\partial z}{\partial y} = y^2 \frac{1}{2\sqrt{1 - 2xy}}(-2x) + 2y\sqrt{1 - 2xy}$$
$$dz = \frac{-y^3}{\sqrt{1 - 2xy}}dx + \left(\frac{-xy^2}{\sqrt{1 - 2xy}} + 2y\sqrt{1 - 2xy}\right)dy$$
$$dz = \frac{-y^3 dx}{\sqrt{1 - 2xy}} + \frac{-xy^2 + 2y(1 - 2xy)}{\sqrt{1 - 2xy}}dy$$
$$dz = \frac{-y^3 dx + (2y - 5xy^2)dy}{\sqrt{1 - 2xy}}$$

7. $z = \sin xy - y \cos x$

$$\frac{\partial z}{\partial x} = y \cos xy + y \sin x, \frac{\partial z}{\partial y} = x \cos xy - \cos x$$
$$dz = (y \cos xy + y \sin x)dx + (x \cos xy - \cos x)dy$$

8. $z = x \tan 2y - x^3$

$$\frac{\partial z}{\partial x} = \tan 2y - 3x^2, \frac{\partial z}{\partial y} = 2x \sec^2 2y$$
$$dz = (\tan 2y - 3x^2)dx + 2x \sec^2 2y \, dy$$

9. $z = \dfrac{x - 3y^2}{1 - \sin y}$

$\dfrac{\partial z}{\partial x} = \dfrac{1}{1 - \sin y}, \dfrac{\partial z}{\partial y} = \dfrac{(1 - \sin y)(-6y) - (x - 3y^2)(-\cos y)}{(1 - \sin y)^2} = \dfrac{6y \sin y - 6y + x \cos y - 3y^2 \cos y}{(1 - \sin y)^2}$

$dz = \dfrac{dx}{1 - \sin y} + \dfrac{(6y - 6y + x \cos y - 3y^2 \cos y)dy}{(1 - \sin y)^2}$

10. $z = \dfrac{\sec x^2}{x - y}$

$\dfrac{\partial z}{\partial x} = \dfrac{(x - y) \sec x^2 \tan x^2 (2x) - \sec x^2}{(x - y)^2}$

$\dfrac{\partial z}{\partial x} = \dfrac{\sec x^2}{(x - y)^2}(2x(x - y) \tan x^2 - 1)$

$\dfrac{\partial z}{\partial y} = \dfrac{(x - y)(0) - \sec x^2(-1)}{(x - y)^2} = \dfrac{\sec x^2}{(x - y)^2}$

$dz = \dfrac{\sec x^2}{(x - y)^2}\left[(2x(x - y) \tan x^2 - 1)dx + dy]\right]$

11. $z = \tan^{-1} \dfrac{y}{x^2}$

$\dfrac{\partial z}{\partial x} = \dfrac{y}{1 + (y/x^2)^2}\left(\dfrac{-2y}{x^3}\right) = \dfrac{-2xy^2}{x^4 + y^2}$

$\dfrac{\partial z}{\partial y} = y\left[\dfrac{1}{1 + (y/x^2)^2}\right]\left(\dfrac{1}{x^2}\right) + \tan^{-1}\dfrac{y}{x^2}$

$\qquad = \dfrac{yx^2}{x^4 + y^2} + \tan^{-1}\dfrac{y}{x^2}$

$dz = \dfrac{-2xy^2 dx}{x^4 + y^2} + \left(\dfrac{yx^2}{x^4 + y^2} + \tan^{-1}\dfrac{y}{x^2}\right)dy$

12. $z = e^{2xy} \ln(x^2 + y^2)$

$\dfrac{\partial z}{\partial x} = e^{2xy}\dfrac{2x}{x^2 + y^2} + 2ye^{2xy} \ln(x^2 + y^2)$

$\dfrac{\partial z}{\partial x} = 2e^{2xy}\left[y \ln(x^2 + y^2) + \dfrac{x}{x^2 + y^2}\right]$

$\dfrac{\partial z}{\partial y} = e^{2xy}\dfrac{2y}{x^2 + y^2} + 2xe^{2xy} \ln(x^2 + y^2)$

$\dfrac{\partial z}{\partial y} = 2e^{2xy}\left[x \ln(x^2 + y^2) + \dfrac{y}{x^2 + y^2}\right]$

$dz = 2e^{2xy}\left[y \ln(x^2 + y^2) + \dfrac{y}{x^2 + y^2}\right]dx$

$\qquad + 2e^{2xy}\left[x \ln(x^2 + y^2) + \dfrac{y}{x^2 + y^2}\right]dy$

13. $z = x^2 + y^2 - 2x - 6y + 10$

$\dfrac{\partial z}{\partial x} = 2x - 2, \dfrac{\partial z}{\partial y} = 2y - 6$

$2x - 2 = 0, x = 1; 2y - 6 = 0, y = 3$

$\dfrac{\partial^2 z}{\partial x^2} = 2, \dfrac{\partial^2 z}{\partial y^2} = 2, \dfrac{\partial^2 z}{\partial x \partial y} = 0$

$D = 2(2) - 0^2 = 4$

$D > 0, \dfrac{\partial^2 z}{\partial x^2} > 0, Min.(1, 3, 0)$

14. $z = x^2 + xy + y^2 + 3x - 3y + 1$

$\dfrac{\partial z}{\partial x} = 2x + y + 3, \dfrac{\partial z}{\partial y} = x + 2y - 3$

$\begin{array}{ll} 2x + y = -3 & 2x + y = -3 \\ x + 2y = 3 & \underline{-2x - 4y = -6} \\ & -3y = -9 \\ & y = 3 \\ & x + 2(3) = 3 \\ & x = -3 \end{array}$

$\dfrac{\partial^2 z}{\partial x^2} = 2, \dfrac{\partial^2 z}{\partial y^2} = 2, \dfrac{\partial^2 z}{\partial y \partial x} = 1$

$D = 2(2) - (1)^2 = 3 > 0$

$Min.(-3, 3, -8)$

15. $z = x^2 + xy + y^2 - 3y + 2$

$\dfrac{\partial z}{\partial x} = 2x + y, \dfrac{\partial z}{\partial y} = x + 2y - 3$

$\begin{array}{lll} 2x + y = 0 & 4x + 2y = 0 & x = -1, y = 2 \\ x + 2y = 3 & \underline{x + 2y = 3} & \\ & 3y = -3 & \end{array}$

$\dfrac{\partial^2 z}{\partial x^2} = 2, \dfrac{\partial^2 z}{\partial y^2} = 2, \dfrac{\partial^2 z}{\partial x \partial y} = 1$

$D = 2(2) - 1^2 = 3$

$D > 0, \dfrac{\partial^2 z}{\partial x^2} > 0, Min.(-1, 2, -1)$

16. $z = x^4 - 4x + y^2$

$$\frac{\partial z}{\partial x} = 4x^3 - 4, \frac{\partial z}{\partial y} = 2y$$

$$4x^3 - 4 = 0 \qquad\qquad 2y = 0$$
$$x = 1 \qquad\qquad\qquad y = 0$$

$$\frac{\partial^2 z}{\partial x^2} = 12x^2 \qquad\qquad \frac{\partial^2 z}{\partial y^2} = 2$$

$$\frac{\partial^2 z}{\partial y \partial y} = 0$$

$$D = 12(1)^2(2) - 0^2 = 24 > 0$$

$$\frac{\partial^2 z}{\partial x^2} = 12(1)^2 = 12 > 0$$
Min. $(1, 0, -3)$

17. $z = 2y - xy + x^3$

$$\frac{\partial z}{\partial x} = -y + 3x^2, \frac{\partial z}{\partial y} = 2 - x$$

$$-y + 3x^2 = 0, y = 3x^2$$
$$2 - x = 0, x = 2, y = 12$$

$$\frac{\partial^2 z}{\partial x^2} = 6x, \frac{\partial^2 z}{\partial y^2} = 0, \frac{\partial^2 z}{\partial x \partial y} = -1$$

$$D = 12(0) - (-1)^2 = -1$$
$D < 0$, no. max. or min.

18. $z = xy + \dfrac{1}{x} + \dfrac{8}{y}$

$$\frac{\partial z}{\partial x} = y - \frac{1}{x^2}, \frac{\partial z}{\partial y} = x - \frac{8}{y^2}$$

$$y - \frac{1}{x^2} = 0 \qquad\qquad x - \frac{8}{y^2} = 0$$

$$x^2 y = 1 \qquad\qquad xy^2 = 8$$

$$y = \frac{1}{x^2} \qquad\qquad x\left(\frac{1}{x^4}\right) = 8$$

$$y^2 = \frac{1}{x^4} \qquad\qquad \frac{1}{x^3} = 8$$

$$x = \frac{1}{2}$$

$$y = 4$$

$$\frac{\partial^2 z}{\partial x^2} = \frac{2}{x^3}, \frac{\partial^2 z}{\partial y^2} = \frac{16}{y^3}, \frac{\partial^2 z}{\partial x \partial y} = 1$$

$$D = \frac{2}{\frac{1}{8}} \frac{16}{4^3} - (1)^2 = 3 > 0$$

$$\frac{\partial^2 z}{\partial x^2} = \frac{2}{\frac{1}{8}} = 16 > 0$$

Min. $\left(\dfrac{1}{2}, 4, 6\right)$

19. $z = x^2 + 4 - 4x \cos y$

$$\frac{\partial z}{\partial x} = 2x - 4\cos y, \frac{\partial z}{\partial y} = 4x \sin y$$

$$2x - 4\cos y = 0, \cos y = \frac{x}{2}$$
$$4x \sin y = 0, x = 0, \sin y = 0$$
$$\sin y = 0, y = 0, \pi$$

$$x = 0, \cos y = 0, y = \frac{\pi}{2}, \frac{3\pi}{2}$$

$$\frac{\partial^2 z}{\partial x^2} = 2, \frac{\partial^2 z}{\partial y^2} = 4x \cos y, \frac{\partial^2 z}{\partial x \partial y} = 4 \sin y$$

$$D = 8x \cos y - 16 \sin^2 y$$

For $y = 0, x = 2$: $D = 16$, Min. $(2, 0, 0)$
For $y = \pi, x = -2$: $D = 16$, Min. $(-2, \pi, 0)$

For $y = \dfrac{\pi}{2}, x = 0$: $D = -16$, No max. or min.

For $y = \dfrac{3\pi}{2}, x = 0$: $D = -16$, No max. or min.

20. $z = x - y \ln x$

$$\frac{\partial z}{\partial x} = 1 - \frac{y}{x}, \frac{\partial z}{\partial y} = -\ln x$$

$$1 - \frac{y}{x} = 0 \qquad\qquad -\ln x = 0$$

$$1 - \frac{y}{1} = 0 \qquad\qquad x = 1$$

$$y = 1$$

$$\frac{\partial^2 z}{\partial x^2} = \frac{y}{x^2}, \frac{\partial^2 z}{\partial y^2} = 0, \frac{\partial^2 z}{\partial x \partial y} = -\frac{1}{x}$$

$$D = \frac{1}{1^2}(0) - \left(\frac{-1}{1}\right)^2 = -1 < 0.$$

Neither max or min.

21. $a = r\omega^2$
$$da = \omega^2 dr + 2r\omega \, d\omega$$
$$r = 10.0 \text{ cm}, dr = 0.3 \text{ cm}$$
$$\omega = 400 \text{ rad/min}, d\omega = 6 \text{ rad/min}$$

$$da = 400^2(0.3) + 2(10.0)(400)(6)$$
$$= 9.60 \times 10^4 \text{ rad/min}^2$$

22. $i = \dfrac{V}{R}$

$$di = \frac{\partial i}{\partial V} dV + \frac{\partial i}{\partial R} dR$$

$$di = \frac{1}{R} dV - \frac{V}{R^2} dR$$

$$di = \frac{1}{20.0}(225 - 220) - \frac{220}{(20.0)^2}(21.0 - 20.0)$$

$$di = -0.30 \text{ A}$$

23. $n = \dfrac{\sin i}{\sin r}$

$$dn = \frac{\cos i\, di}{\sin r} - \frac{\sin i \cos r\, dr}{\sin^2 r}$$

$i = 45°, r = 30°, di = dr = -1° = -0.01745$

$$dn = \left(\frac{\cos 45°}{\sin 30°} - \frac{\sin 45° \cos 30°}{\sin^2 30°}\right)(-0.01745)$$

$$= 0.0181$$

$$n = \frac{\sin 45°}{\sin 30°} = 1.414; \quad \frac{dn}{n} = 0.013 = 1.3\%$$

24. $i = 10.0(2 - e^{-0.100Rt}) = 10.0 - 10.0e^{-0.100Rt}$

$$di = \frac{\partial i}{\partial R}\, dR + \frac{\partial i}{\partial t}\, dt$$

$di = 1.00te^{-0.100Rt}dR + 1.00Re^{-0.100Rt}dt$

$di = 1.00e^{-0.100Rt}(t\, dR + R\, dt)$

$di = 1.00e^{-0.100(2.00)(10.0)}(10.0(2.04 - 2.00)$

$\qquad\qquad + 2.00(11.0 - 10.0))$

$di = 0.325$ A

25.

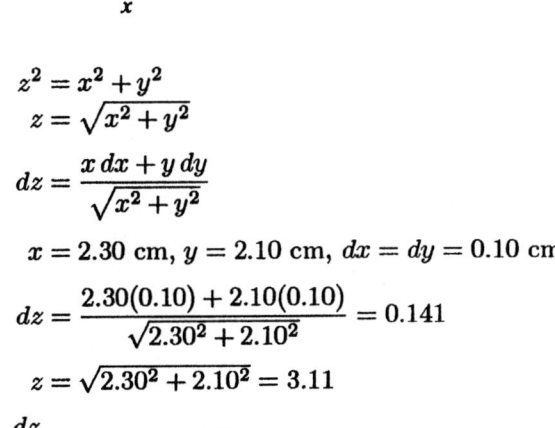

$z^2 = x^2 + y^2$

$z = \sqrt{x^2 + y^2}$

$$dz = \frac{x\, dx + y\, dy}{\sqrt{x^2 + y^2}}$$

$x = 2.30$ cm, $y = 2.10$ cm, $dx = dy = 0.10$ cm

$$dz = \frac{2.30(0.10) + 2.10(0.10)}{\sqrt{2.30^2 + 2.10^2}} = 0.141$$

$z = \sqrt{2.30^2 + 2.10^2} = 3.11$

$$\frac{dz}{z} = 0.045 = 4.5\%$$

26. $V = \pi r^2 h$

$$dV = \frac{\partial V}{\partial r}\, dr + \frac{\partial V}{\partial h}\, dh$$

$dV = 2\pi rh\, dr + \pi r^2 dh$

$dV = 2\pi(3.00)(5.00)(3.10 - 3.00)$

$\qquad\quad + \pi(3.00)^2(5.20 - 5.00)$

$dV = 15.1$ ft^3

27. $\ell + w + h = 3$

$V = \ell wh = (3 - w - h)wh = 3wh - w^2h - wh^2$

$$\frac{\partial V}{\partial w} = 3h - 2wh - h^2 = 0$$

$$3 - 2w - h = 0$$

$$2w + h = 3 \qquad (1)$$

$$\frac{\partial V}{\partial h} = 3w - w^2 - 2wh = 0, \quad 3 - w - 2h = 0$$

$$w + 2h = 3 \qquad (2)$$

Solving (1) and (2) gives $w = 1, h = 1$. From $\ell + w + h = 3, \ell = 1$.

$$\frac{\partial^2 V}{\partial w^2} = -2h = -2(2) = -4 < 0$$

$$\frac{\partial^2 V}{\partial h^2} = -2w = -2(1) = -2$$

$$\frac{\partial^2 V}{\partial h\partial w} = 3 - 2w - 2h = 3 - 2(1) - 2(1) = -1$$

$$D = \frac{\partial^2 V}{\partial w^2}\frac{\partial^2 V}{\partial h^2} - \left(\frac{\partial^2 V}{\partial w\partial h}\right)^2$$

$D = (-2(1))(-2(1)) - (-1)^2 = 4 - 1 = 3 > 0$

Max. $\ell = 1$ m, $w = 1$ m, $h = 1$ m

28. $S = 2\ell w + 2\ell h + 2wh = 64.0$

$$\ell w + \ell h + wh = 32.0$$

$$\ell(w + h) = 32.0 - wh$$

$$\ell = \frac{32.0 - wh}{w + h}$$

$$V = \ell wh = \frac{(32 - wh)wh}{w + h} = \frac{32wh - w^2h^2}{w + h}$$

$$\frac{\partial V}{\partial w} = \frac{(w + h)(32h - 2wh^2) - (32wh - w^2h^2)(1)}{(w + h)^2}$$

$$\frac{\partial V}{\partial w} = \frac{32wh + 32h^2 - 2w^2h^2 - 2wh^3 - 32wh + w^2h^2}{(w + h)^2}$$

$$\frac{\partial V}{\partial w} = \frac{32h^2 - w^2h^2 - 2wh^3}{(w + h)^2}$$

$$= \frac{h^2(32 - w^2 - 2wh)}{(w + h)^2} = 0$$

$$w^2 + 2wh - 32 = 0 \qquad (1)$$

$$\frac{\partial V}{\partial h} = \frac{(w + h)(32w - 2w^2h) - (32wh - w^2h^2)(1)}{(w + h)^2}$$

$$\frac{\partial V}{\partial h} = \frac{32w^2 + 32wh - 2w^3h - 2w^2h^2 - 32wh + w^2h^2}{(w + h)^2}$$

$$\frac{\partial V}{\partial h} = \frac{32w^2 - 2w^3h - w^2h^2}{(w + h)^2}$$

$$= \frac{w^2(32 - 2wh + h^2)}{(w + h)^2} = 0$$

$$h^2 + 2wh - 32 = 0$$

$$h = \frac{-2w + \sqrt{4w^2 - 4(-32)}}{2} = \frac{-2w + \sqrt{4(w^2 + 32)}}{2}$$

$h = -w + \sqrt{w^2 + 32}$ from which (1) becomes

$$w^2 + 2w(-w + \sqrt{w^2 + 32}) - 32 = 0$$
$$w^2 - 2w^2 + 2w\sqrt{w^2 + 32} - 32 = 0$$
$$w^2 - 2w\sqrt{w^2 + 32} + 32 = 0.$$

Solve with Newton's Method
$$w = 3.27$$

$$h = -w + \sqrt{w^2 + 32}\Big|_{w=3.27} = 3.27$$

$$\ell = \frac{32 - wh}{w + h}\Big|_{w=h=3.27} = 3.27$$

$\ell = w = h = 327$, a cube

$$\frac{\partial^2 V}{\partial w^2} = \frac{-2h^2(h^2 + 32)}{(w + h)^3}\Big|_{w=h=3.27} = -3.26 < 0$$

$$\frac{\partial^2 V}{\partial h^2} = \frac{-w^2(w^2 + 32)}{(w + h)^2}\Big|_{w=h=3.27} = -3.26$$

$$\frac{\partial^2 V}{\partial w \partial h} = \frac{2wh(32 - w^2 - h^2 - 3wh)}{(w + h)^3}\Big|_{w=h=3.27} = -1.64$$

$$D = \frac{\partial^2 V}{\partial w^2}\frac{\partial^2 V}{\partial x^2} - \left(\frac{\partial^2 V}{\partial h \partial w}\right)^2$$

$$D = (-3.26)(-3.26) - (1.64) = 12.3 > 0$$

Max. for $\ell = w = h = 3.27$ ft

29. $xyz = 32.0$ m³

$$z = \frac{32.0}{xy}$$

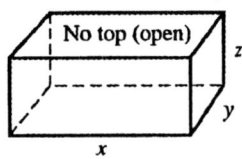

Least material means surface area is minimum

$$A = xy + 2yz + 2xz$$

$$A = xy + 2y\frac{32.0}{xy} + 2x\left(\frac{32.0}{xy}\right)$$

$$A = xy + \frac{64.0}{x} + \frac{64.0}{y}$$

$$\frac{\partial A}{\partial x} = y - \frac{64.0}{x^2}, \frac{\partial A}{\partial y} = x - \frac{64.0}{y^2}$$

$$y = \frac{64.0}{x^2}, x = \frac{64.0}{y^2} = \frac{64.0}{\left(\frac{64.0}{x^2}\right)^2} = \frac{x^4}{64.0}$$

$$x^4 - 64.0x = 0$$
$$x(x^3 - 64.0) = 0$$

$x = 0, x = 4.00$. For $x = 4.00, y = \dfrac{64.0}{4.00^2} = 4.00$

$$\frac{\partial^2 A}{\partial x^2} = \frac{128}{x^3}, \frac{\partial^2 A}{\partial y^2} = \frac{128}{y^3}, \frac{\partial^2 A}{\partial x \partial y} = 1$$

$$D = \frac{\partial^2 A}{\partial x^2}\frac{\partial^2 A}{\partial y^2} - \left(\frac{\partial^2 A}{\partial x \partial y}\right)^2 = 2(2) - 1^2 = 3 > 0$$

$$\frac{\partial^2 A}{\partial x^2} = 2 > 0$$

Min. for $x = 4.00$ m, $y = 4.00$ m,

$$z = \frac{32.0}{4.00(4.00)} = 2.00 \text{ m}$$

30. The distance from any point (x, y, z) to the origin is

$$d = \sqrt{x^2 + y^2 + x^2}.$$

If the point (x, y, z) is on the plane $2x + 3y - z = 12$, then $z = 2x + 3y - 12$ and
$d = \sqrt{x^2 + y^2 + (2x + 3y - 12)^2}$ from which
$d = \sqrt{5x^2 + 12xy + +10y^2 - 48x - 72y + 144}$.

Let $D = d^2 = 5x^2 + 12xy + 10y^2 - 48x - 72y + 144$.

Minimizing D will also minimize d.

$$\frac{\partial D}{\partial x} = 10x + 12y - 48, \qquad \frac{\partial D}{\partial y} = 12x + 20y - 72$$

$$\frac{\partial^2 D}{\partial x^2} = 10 \qquad\qquad \frac{\partial^2 D}{\partial y^2} = 20$$

(1) $10x + 12y = 48$ and (2) $12x + 20y = 72$
have $x = \frac{12}{7}, y = \frac{18}{7}$ as a solution. From

$$z = 12 - 2x - 3y$$

$$z = -\frac{6}{7}$$

$$\frac{\partial^2 D}{\partial y 2x} = 12 = \frac{\partial^2 D}{\partial x 2y}$$

$$\frac{\partial^2 D}{\partial x^2}\frac{\partial^2 D}{\partial y^2} - \left(\frac{\partial D}{\partial x \partial y}\right)^2 = 10(20) - 12^2 = 56 > 0$$

$$\frac{\partial^2 D}{\partial x^2} = 10 > 0$$

$\left(\frac{12}{7}, \frac{18}{7}, -\frac{6}{7}\right)$ is the point on the plane
$2x + 3y - z = 12$ that is closest to the origin.

31. $P = x^4 + y^2 - 4x + 20$

$$\frac{\partial P}{\partial x} = 4x^3 - 4, \frac{\partial P}{\partial y} = 2y$$

$$4x^3 - 4 = 0, x = 1; 2y = 0, y = 0$$

$$\frac{\partial^2 P}{\partial x^2} = 12x^2, \frac{\partial^2 P}{\partial y^2} = 2, \frac{\partial^2 P}{\partial x \partial y}$$

For $x = 1, y = 0$

$$D = 12(2) - 0^2 = 24, \frac{\partial^2 P}{\partial x^2} = 12$$

Min. at $(1,0)$; $P = 1^4 + 0^2 - 4(1) + 20$
$$= 17 \text{ Pa}$$

32. $T = x^2 + 2y^2 - x$

$$\frac{\partial T}{\partial x} = 2x - 1 = 0 \quad , \quad \frac{\partial T}{\partial y} = 4y = 0$$

$$x = \frac{1}{2} \qquad\qquad y = 0$$

$$\frac{\partial^2 T}{\partial x^2} = 2 > 0 \qquad \frac{\partial^2 T}{\partial y^2} = 4$$

$$\frac{\partial^2 T}{\partial y \partial x} = 0$$

$$D = \frac{\partial^2 T}{\partial x^2}\frac{\partial^2 T}{\partial y^2} - \left(\frac{\partial^2 T}{\partial x \partial y}\right)^2 = 2(4) - 0 = 8 > 0$$

$$T_{\text{coldest}} = \left(\frac{1}{2}\right)^2 + 2(0)^2 - \frac{1}{2} = -\frac{1}{4}$$

$$T_{\text{coldest}} = -0.25 \text{ at } (0.5, 0.00)$$

33. $T = 2xy - 5x^2 - 2y^2 + 4x + 4y - 4$

$$\frac{\partial T}{\partial x} = 2y - 10x + 4 = 0 \quad , \quad \frac{\partial T}{\partial y} = 2x - 4y + 4 = 0$$

$$(1) \quad 10x - 2y = 4 \qquad (2) \quad 2x - 4y = -4$$

$\left(\frac{2}{3}, \frac{4}{3}\right)$ is the solution to (1) and (2).

$$\frac{\partial^2 T}{\partial x^2} = -10 < 0, \frac{\partial^2 T}{\partial y^2} = -4$$

$$\frac{\partial^2 T}{\partial x \partial y} = -4 < 0$$

$$D = \frac{\partial^2 T}{\partial x^2}\frac{\partial^2 T}{\partial y^2} - \left(\frac{\partial^2 T}{\partial x \partial y}\right)$$
$$= (-10)(-4) - (-4)^2$$
$$= 24 > 0$$

$$T_{\text{warmest}} = 2xy - 5x^2 - 2y^2 + 4x + 4y - 4\big|_{\left(\frac{2}{3}, \frac{4}{3}\right)} = 0$$

$$T_{\text{warmest}} = 0°\text{C at } (0.67, 1.33)$$

34. The cross-section is a trapezoid with area

$$A = \frac{1}{2}(b_1 + b_2)h$$

$$A = \frac{1}{2}(9 - 2x + 9 - 2x + 2x \cos\theta)x \sin\theta$$

$$A = \frac{1}{2}(18x - 4x^2 + 2x^2 \cos\theta)\sin\theta$$

$$\frac{\partial A}{\partial x} = 2x \sin\theta \cos\theta + (9 - 4x)\sin\theta = 0$$

from which $x = \dfrac{9}{4 - 2\cos\theta}$ (1)

$$\frac{\partial A}{\partial x} = 2x^2 \cos^2\theta + x(9 - 2x)\cos\theta - x^2 = 0$$
$$2x \cos^2\theta + (9 - 2x)\cos\theta - x = 0,$$

substituting x from (1) gives

$$2\left(\frac{9}{4 - 2\cos\theta}\right)\cos^2\theta + \left(9 - \frac{18}{4 - 2\cos\theta}\right)\cos\theta$$

$$- \frac{9}{4 - 2\cos\theta} = 0$$

which may be solved using Newton's method to obtain

$$\theta = \frac{\pi}{3} = 60°, x = \frac{9}{4 - 2\cos\frac{\pi}{3}} = 3$$

$$\frac{\partial^2 A}{\partial x^2} = 2\sin\theta\cos\theta - 4\sin\theta\bigg|_{\theta = \frac{\pi}{3}} = -2.60 < 0$$

$$\frac{\partial^2 A}{\partial\theta^2} = x(2x - 9)\sin\theta - 4x^2\sin\theta\cos\theta\bigg|_{x = 3, \theta = \frac{\pi}{3}} = -23.4$$

$$\frac{\partial^2 A}{\partial x\partial\theta} = 4x\cos^2\theta + (9 - 4x)\cos\theta - 2x\bigg|_{x = 3, \theta = \frac{\pi}{3}} = -4.5$$

$$D = \frac{\partial^2 A}{\partial x^2}\frac{\partial^2 A}{\partial\theta^2} - \left(\frac{\partial^2 A}{\partial x\partial\theta}\right)^2$$

$$D = (-2.6)(-23.4) - (-4.5)^2 = 40.6 > 0$$

The area is a maximum for $x = 3$ in. and an angle of $120°(\theta = 60°)$ between the bottom and the edges.

35. $z = x^2 - xy + y^2$

$$\frac{\partial z}{\partial x} = 2x - y, \frac{\partial z}{\partial y} = -x + 2y$$

$$x = 1 + t^2, y = 1 - t^2$$

$$\frac{dx}{dt} = 2t, \frac{dy}{dt} = -2t$$

$$\frac{dz}{dt} = (2x - y)(2t) + (-x + 2y)(-2t)$$
$$= 4xt - 2yt + 2xt - 4yt$$
$$= 6xt - 6yt$$

36. $z = \ln(x + y^2), x = \sqrt{1+t}, y = 1 + \sqrt{t}$

$$\frac{\partial z}{\partial x} = \frac{1}{x + y^2}$$

$$\frac{dx}{dt} = \frac{1}{2\sqrt{1+t}}$$

$$\frac{\partial z}{\partial y} = \frac{2y}{x + y^2}$$

$$\frac{dy}{dt} = \frac{1}{2\sqrt{t}}$$

$$\frac{dz}{dt} = \frac{\partial z}{\partial x}\frac{dx}{dt} + \frac{\partial z}{\partial y}\frac{dy}{dt}$$

$$\frac{dz}{dt} = \frac{1}{(x + y^2)}\Big)\frac{1}{2\sqrt{1+t}} + \frac{2y}{x + y^2}\frac{1}{2\sqrt{t}}$$

$$\frac{dz}{dt} = \frac{2y\sqrt{t+1} + \sqrt{t}}{2\sqrt{t}\sqrt{t+1}(x + y^2)}$$

37. $z = x^2 + y^2, x = \cos t, y = \sin t$

$$\frac{\partial z}{\partial x} = 2x, \frac{\partial z}{\partial y} = 2y, \frac{dx}{dt} = -\sin t, \frac{dy}{dt} = \cos t$$

$$\frac{dz}{dt} = \frac{\partial z}{\partial x}\frac{dx}{dt} + \frac{\partial z}{\partial y}\frac{dy}{dt} = 2x(-\sin t) + 2y \cos t$$

$$\frac{dz}{dt} = -2x\sin t + 2y\cos t$$

when $t = \pi, x = \cos \pi = -1, y = \sin \pi = 0$

$$\frac{dz}{dt} = -2x\sin t + 2y\cos t\Big|_{x=-1, t=\pi, y=0}$$

$$\frac{dz}{dt} = -2(-1)\sin \pi + 2(0)\cos \pi = 0$$

38. $z = \sin(xy), x = t^2 - 2t, y = t^3 - 5t$

$$\frac{\partial z}{\partial x} = y\cos(xy), \frac{\partial z}{\partial y} = x\sin(xy),$$

$$\frac{dx}{dt} = 2t - 2, \frac{dy}{dt} = 3t^2 - 5$$

when $t = 2, x = 2^2 - 2(2) = 0, y = 2^3 - 5(2) = -2$

$$\frac{dz}{dt} = \frac{\partial z}{\partial x}\frac{dx}{dt} + \frac{\partial z}{\partial y}\frac{dy}{dt}$$

$$\frac{dz}{dt} = y\cos(xy)(2t-2) + x\sin(xy)(3t^2 - 5)\Big|_{x=0, y=-2, t=2}$$

$$\frac{dz}{dt} = (-2)\cos(0)(2(2) - 2) + 0\sin(0)(3((2)^2 - 5)$$

$$\frac{dz}{dt} = -4$$

39. $I = \dfrac{V}{R}$

$$\frac{dI}{dt} = \frac{\partial I}{\partial V}\frac{dV}{dt} + \frac{\partial I}{\partial R}\frac{dR}{dt}$$

$$\frac{dI}{dt} = \frac{1}{R}\frac{dV}{dt} - \frac{V}{R^2}\frac{dR}{dt} = \frac{1}{R}\frac{dV}{dt} - \frac{IR}{R^2}\frac{dR}{dt}$$

$$\frac{dI}{dt} = \frac{1}{R}\frac{dV}{dt} - \frac{I}{R}\frac{dR}{dt}$$

$$\frac{dI}{dt} = \frac{1}{600}(-0.010) - \frac{0.040}{600}(0.50)$$

$$\frac{dI}{dt} = -5.0 \times 10^{-5} \text{ A/s}$$

40. $V = \pi r^2 h$

$$\frac{dV}{dt} = \frac{\partial V}{2r}\frac{dr}{dt} + \frac{\partial V}{\partial h}\frac{dh}{dt}$$

$$\frac{dV}{dt} = 2\pi r\frac{dr}{dt} + \pi r^2\frac{dh}{dt}$$

$$\frac{dV}{dt} = 2\pi(6.00)(3.00)(0.050) + \pi(6.00)^2(0.20)$$

$$\frac{dV}{dt} = 28 \text{ ft}^3/\text{s}$$

11.5   Double Integrals

1.
$$\int_2^4 \int_0^1 xy^2 dx\, dy = \int_2^4 y^2 \left(\frac{1}{2}x^2\right)\Big|_0^1 dy$$
$$= \int_2^4 y^2 \left(\frac{1}{2} - 0\right) dy$$
$$= \frac{1}{2}\int_2^4 y^2 dy = \frac{1}{6}y^3\Big|_2^4$$
$$= \frac{1}{6}(64 - 8) = \frac{28}{3}$$

2.
$$\int_0^2 \int_0^1 \frac{y}{(xy + 1)^2} dx\, dy = \int_0^2 \frac{-1}{(xy + 1)^2}\Big|_0^1 dy$$
$$= \int_0^2 \left[\frac{-1}{y + 1} + \frac{1}{1}\right] dy$$
$$= -\ln(y + 1) + y\Big|_0^2$$
$$= -\ln(3) + 2 + \ln(1) - 0$$
$$= 2 - \ln 3 = 0.9014$$

3. $\int_0^1 \int_0^x 2y\,dy\,dx = \int_0^1 y^2\Big|_0^x dx$

$= \int_0^1 (x^2 - 0)dx$

$= \int_0^1 x^2 dx$

$= \frac{1}{3}x^3\Big|_0^1$

$= \frac{1}{3} - 0 = \frac{1}{3}$

$= \frac{1}{3}$

4. $\int_0^2 \int_1^x 2x\,dy\,dx = \int_0^2 2xy\Big|_1^x dx$

$= \int_0^2 [2x(x) - 2x(1)]dx$

$= \int_0^2 [2x^2 - 2x]dx$

$= \frac{2}{3}x^3 - x^2\Big|_0^2$

$= \frac{2}{3}(2)^3 - 2^2$

$= \frac{4}{3}$

5. $\int_1^2 \int_0^{y^2} xy^2\,dx\,dy = \int_1^2 y^2\left(\frac{1}{2}x^2\right)\Big|_0^{y^2} dy$

$= \int_1^2 y^2\left(\frac{1}{2}y^4\right)dy$

$= \frac{1}{2}\int_1^2 y^6 dy$

$= \frac{1}{14}y^7\Big|_1^2$

$= \frac{127}{14}$

6. $\int_0^4 \int_0^{\sqrt{y}} (x-y)dx\,dy = \int_0^4 \left[\frac{x^2}{2} - xy\right]\Big|_1^{\sqrt{y}} dy$

$= \int_0^4 \left[\frac{y}{2} - y^{3/2} - \frac{1}{2} + y\right]dy$

$= \int_0^4 \left[\frac{3y}{2} - y^{3/2} - \frac{1}{2}\right]dy$

$= \frac{3y^2}{4} - \frac{2}{5}y^{5/2} - \frac{y}{2}\Big|_0^4$

$= \frac{3(4)^2}{4} - \frac{2}{5}4^{5/2} - \frac{4}{2} = -\frac{14}{5}$

7. $\int_0^1 \int_0^{\sqrt{1-x^2}} y\,dy\,dx = \int_0^1 \frac{1}{2}y^2\Big|_0^{\sqrt{1-x^2}} dx$

$= \int_0^1 \frac{1}{2}(1 - x^2)dx$

$= \frac{1}{2}x - \frac{1}{6}x^3\Big|_0^1$

$= \frac{1}{2} - \frac{1}{6}$

$= \frac{1}{3}$

8. $\int_4^9 \int_0^x \sqrt{x - y}\,dy\,dx$

$= \int_4^9 \frac{-2(x-y)^{3/2}}{3}\Big|_0^x dx$

$= \int_4^9 \left[\frac{-2(x-x)^{3/2}}{3} + \frac{2(x-0)^{3/2}}{3}\right]dx$

$= \int_4^9 \frac{2x^{3/2}}{3}dx$

$= \frac{2}{3}\left(\frac{5}{2}x^{5/2}\right)\Big|_4^9$

$= \frac{4}{15}[9^{5/2} - 4^{5/2}]$

$= \frac{844}{15}$

9. $\int_0^{\pi/6} \int_{\pi/3}^y \sin x\,dx\,dy$

$= \int_0^{\pi/6} (-\cos x)\Big|_{\pi/3}^y dy$

$= -\int_0^{\pi/6} \left(\cos y - \cos\frac{\pi}{3}\right)dy$

$= -\int_0^{\pi/6} \left(\cos y - \frac{1}{2}\right)dy$

$= -\sin y + \frac{1}{2}y\Big|_0^{\pi/6}$

$= -\sin\frac{\pi}{6} + \frac{\pi}{12}$

$= \frac{\pi}{12} - \frac{1}{2}$

$= \frac{\pi - 6}{12}$

10. $\displaystyle\int_0^{\sqrt3}\int_{x^2/3}^1 (4-x^2-y^2)dy\,dx = \int_0^{\sqrt3}\left(4y-x^2y-\frac{y^3}{3}\right)\Big|_{x^2/3}^1 dx$

$$= \int_0^{\sqrt3}\left[4(1)-x^2(1)-\frac13(1)^3-\frac{4x^2}{3}+x^2\left(\frac{x^2}{3}\right)+\frac13\left(\frac{x^2}{3}\right)^3\right]dx$$

$$= \int_0^{\sqrt3}\left[4-x^2-\frac13-\frac{4x^2}{3}+\frac{x^4}{3}+\frac{x^6}{81}\right]dx = \int_0^{\sqrt3}\left[\frac{11}{3}-\frac{7x^2}{3}+\frac{x^4}{3}+\frac{x^6}{81}\right]dx$$

$$= \frac{11x}{3}-\frac{7x^3}{9}+\frac{x^5}{15}+\frac{x^7}{567}\Big|_0^{\sqrt3} = \frac{11\sqrt3}{3}-\frac{7(3)\sqrt3}{9}+\frac{9\sqrt3}{15}+\frac{27\sqrt3}{567}=\frac{208\sqrt3}{105}$$

11. $\displaystyle\int_1^e\int_0^y \frac1x dx\,dy = \int_1^e \ln x\Big|_0^y dy = \int_1^e \ln y\,dy = y(\ln y-1)\Big|_1^e$ (Parts or Formula 46)

$$= e(\ln e-1)-(\ln1-1)=e(0)+1=1$$

12. $\displaystyle\int_{-1}^1\int_1^{e^x}\frac1{xy}dy\,dx = \int_{-1}^1\frac1x \ln y\Big|_1^{e^x}dx = \int_{-1}^1\frac1x\ln y\Big|_1^{e^x}dx = \int_{-1}^1\frac1x[\ln e^x-\ln1]dx = \int_{-1}^1\frac1x[x\ln e-0]dx$

$$= \int_{-1}^1 dx = x\Big|_{-1}^1 = (1-(-1))=2$$

13. $\displaystyle\int_1^2\int_0^x yx^3 e^{xy^2}dy\,dx = \frac12\int_1^2\int_0^x x^2(2xye^{xy^2}dy)dx = \frac12\int_1^2 x^2(e^{xy^2})\Big|_0^x dx = \frac12\int_1^2 x^2(e^{x^3}-1)dx$

$$= \frac16\int_1^2 3x^2 e^{x^3}dx - \frac12\int_1^2 x^2 dx = \frac16 e^{x^3}-\frac16 x^3\Big|_1^2 = \frac16[e^8-8-(e-1)]=495.2$$

14. $\displaystyle\int_0^{\pi/6}\int_0^1 y\sin x\,dy\,dx = \int_0^{\pi/6}\frac12 y^2\sin x\Big|_0^1 dx = \int_0^{\pi/6}\sin x\left(\frac12(1)^2-\frac12(0)^2\right)dx = \int_0^{\pi/6}\frac12\sin x\,dx$

$$= -\frac12\cos x\Big|_0^{\pi/6} = -\frac12\left[\cos\frac\pi6-\cos0\right] = -\frac12\left[\frac{\sqrt3}{2}-1\right]=\frac12-\frac{\sqrt3}{4}=0.0670$$

15. $\displaystyle\int_0^{\ln3}\int_0^x e^{2x+3y}dy\,dx = \frac13\int_0^{\ln3}e^{2x+3y}\Big|_0^x dx = \frac13\int_0^{\ln3}(e^{5x}-e^{2x})dx = \frac1{15}e^{5x}-\frac16 e^{2x}\Big|_0^{\ln3}$

$$= \frac1{15}(e^{5\ln3}-1)-\frac16(e^{2\ln3}-1) = \frac1{15}(3^5-1)-\frac16(3^2-1) = \frac{242}{15}-\frac43=\frac{74}{5}$$

16. $\displaystyle\int_0^{1/2}\int_y^{y^2}\frac1{\sqrt{y^2-x^2}}dx\,dy = \int_0^{1/2}\sin^{-1}\left(\frac xy\right)\Big|_y^{y^2}dy = \int_0^{1/2}\left[\sin^{-1}\frac{y^2}{y}-\sin^{-1}\frac{y}{y}\right]dy = \int_0^{1/2}[\sin^{-1}y-\sin^{-1}1]dy$

$$= \int_0^{1/2}\left[\sin^{-1}y-\frac\pi2\right]dy = y\sin^{-1}y+\sqrt{1-y^2}-\frac\pi2 y\Big|_0^{1/2}$$

$$= \frac12\sin^{-1}\frac12+\sqrt{1-\left(\tfrac12\right)^2}-\frac\pi2\left(\frac12\right)-0\sin^{-1}0-\sqrt{1-0^2}+\frac\pi2(0)$$

$$= \frac\pi{12}+\frac{\sqrt3}{2}-\frac\pi4-1 = -\frac\pi6+\frac{\sqrt3}{2}-1=-0.658$$

17. $V = \int_0^4 \int_0^{4-x} z\,dy\,dx = \int_0^4 \int_0^{4-x} (4-x-y)dy\,dx = \int_0^4 \left(4y - xy - \frac{1}{2}y^2\right)\bigg|_0^{4-x} dx$

$$= \int_0^4 \left[4(4-x) - x(4-x) - \frac{1}{2}(4-x)^2\right] dx = \int_0^4 \left(8 - 4x + \frac{1}{2}x^2\right) dx$$

$$= 8x - 2x^2 + \frac{1}{6}x^3 \bigg|_0^4 = 32 - 32 + \frac{64}{6} = \frac{32}{3}$$

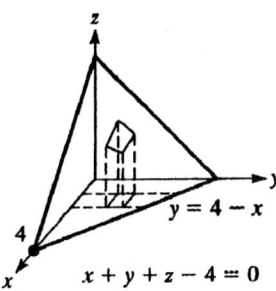

$y = 4 - x$

$x + y + z - 4 = 0$

18.

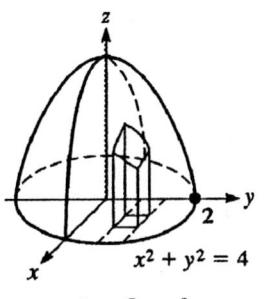

$z = y^2$

$y = 2$

$x = 3$

$V \int_0^2 \int_0^3 y^2\,dy\,dx = \int_0^2 \frac{y^3}{3}\bigg|_0^3 dx = \int_0^2 9\,dx = 9x\bigg|_0^2 = 18$

19. $V = 4\int_0^2 \int_0^{\sqrt{4-y^2}} z\,dx\,dy = 4\int_0^2 \int_0^{\sqrt{4-y^2}} (4 - x^2 - y^2)dx\,dy = \int_0^2 \left(16x - \frac{4}{3}x^3 - 4y^2 x\right)\bigg|_0^{\sqrt{4-y^2}} dy$

$$= \int_0^2 \left(16\sqrt{4-y^2} - \frac{4}{3}(4-y^2)^{3/2} - 4y^2\sqrt{4-y^2}\right)dy = \frac{1}{3}\int_0^2 (32 - 8y^2)\sqrt{4-y^2}\,dy$$

$$= \frac{8}{3}\int_0^2 (4-y^2)^{3/2}dy = \frac{8}{3}\left[\frac{y}{4}(4-y^2)^{3/2} + \frac{3(4)y}{8}(4-y^2)^{1/2} + \frac{3(16)}{8}\sin^{-1}\frac{y}{2}\right]\bigg|_0^2 \quad \text{(Formula 20)}$$

$$= \frac{8}{3}[0 + 0 + 6\sin^{-1}1 - 0 - 0 - 0] = \frac{8}{3}(6)\left(\frac{\pi}{2}\right) = 8\pi$$

Total volume is 4 times
volume in first octant.

$x^2 + y^2 = 4$

$z = 4 - x^2 - y^2$

20.

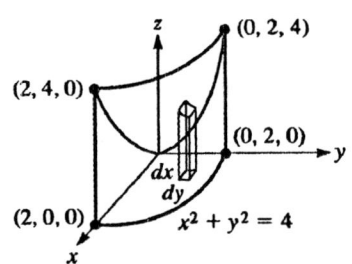

The parabolid of revolution, $z = x^2 + y^2$, is inside and behind the cylinder $x^2 + y^2 = 4$.

$$V = 4 \int_0^2 \int_0^{\sqrt{4-y^2}} (x^2 + y^2)\,dx\,dy \text{ where the 4 is}$$

introduced because the total volume is 4 times the first octant volume.

$$V = 4 \int_0^2 \left(\frac{x^3}{3} + xy^2 \right) \Big|_0^{\sqrt{4-y^2}} dy$$

$$V = 4 \int_0^2 \left(\frac{(4-y^2)^{3/2}}{3} + y^2\sqrt{4-y^2} \right) dy$$

$$V = \frac{4}{3} \int_0^2 (4-y^2)^{3/2}\,dy + 4 \int_0^2 y^2\sqrt{4-y^2}\,dy$$

The first integral is

$$\frac{4}{3} \int_0^2 (4-y^2)^{3/2}\,dy = \frac{4}{3} \int_0^{\pi/2} (4 - 4\sin^2\theta)^{3/2}(2\cos\theta\,d\theta)$$

where $\quad y = 2\sin\theta$
$\qquad\quad dy = 2\cos\theta\,d\theta$

$$\frac{4}{3} \int_0^2 (4-y^2)^{3/2}\,dy = \frac{8}{3} \int_0^{\pi/2} 4^{3/2}(1-\sin^2\theta)^{3/2}\cos\theta\,d\theta = \frac{64}{3} \int_0^{\pi/2} (\cos^2\theta)^{3/2}\cos\theta\,d\theta = \frac{64}{3} \int_0^{\pi/2} \cos^4\theta\,d\theta$$

$$= \frac{64}{3} \int_0^{\pi/2} \cos^2\theta \cos^2\theta\,d\theta = \frac{64}{3} \int_0^{\pi/2} \frac{(1+\cos 2\theta)}{2} \frac{(1+\cos 2\theta)}{2}\,d\theta$$

$$= \frac{64}{3} \int_0^{\pi/2} (1 + 2\cos 2\theta + \cos^2 2\theta)\,d\theta = \frac{16}{3} \left[\theta\Big|_0^{\pi/2} + \int_0^{\pi/2} \cos 2\theta(2\,d\theta) + \int_0^{\pi/2} \cos^2 2\theta\,d\theta \right]$$

$$= \frac{16}{3} \left[\frac{\pi}{2} + \sin 2\theta\Big|_0^{\pi/2} + \int_0^{\pi/2} \frac{1+\cos 4\theta}{2}\,d\theta \right] = \frac{16}{3} \left[\frac{\pi}{2} \sin\pi - \sin 0 + \frac{\theta}{2}\Big|_0^{\pi/2} + \frac{1}{8} \int_0^{\pi/2} \cos 4\theta(4\,d\theta) \right]$$

$$= \frac{16}{3} \left[\frac{\pi}{2} + \frac{\pi}{4} + \frac{1}{8}\sin 4\theta\Big|_0^{\pi/2} \right] = \frac{16}{3} \left[\frac{3\pi}{4} + \frac{1}{8}\sin 2\pi - \frac{1}{8}\sin 0 \right] = 4\pi$$

$V = 4\pi + 4 \int_0^2 y^2\sqrt{4-y^2}\,dy.$ Evaluate the second integral

$$4 \int_0^2 y^2\sqrt{4-y^2}\,dy = 4 \int_0^{\pi/2} 4\sin^2\theta\sqrt{4 - 4\sin^2\theta}(2\cos\theta)\,d\theta$$

where $\quad y = 2\sin\theta, y^2 = 4\sin^2\theta$
$\qquad\quad dy = 2\cos\theta\,d\theta$

$$4 \int_0^2 y^2 \sqrt{4-y^2}\, dy = 64 \int_0^{\pi/2} \sin^2\theta \cos^2\theta\, d\theta = 64 \int_0^{\pi/2} \sin^2\theta(1-\sin^2\theta)d\theta$$

$$= 64 \int_0^{\pi/2} \frac{(1-\cos 2\theta)}{2}\left(1 - \frac{(1-\cos 2\theta)}{2}\right) d\theta = 16 \int_0^{\pi/2}(1-\cos^2 2\theta)d\theta$$

$$= 16 \int_0^{\pi/2} d\theta - 16 \int_0^{\pi/2} \cos^2 2\theta\, d\theta = 8\pi - 16 \int_0^{\pi/2} \frac{1+\cos 4\theta}{2}\, d\theta$$

$$= 8\pi - \int_0^{\pi/2} 8\, d\theta - 2 \int_0^{\pi/2} \cos 4\theta(4\, d\theta) = 8\pi - 8\left(\frac{\pi}{2}\right) - 2[\sin 2\pi - \sin 0]$$

$$= 4\pi$$

$V = 4\pi + 4\pi$

$V = 8\pi$

21. $V = \displaystyle\int_0^2 \int_x^2 z\, dy\, dx \int_0^2 \int_x^2 (2 + x^2 + y^2)dy\, dx = \int_0^2 \left(2y + x^2 y + \frac{1}{3}y^3\right)\Big|_x^2 dx$

$$= \int_0^2 \left(4 + 2x^2 + \frac{8}{3} - 2x - x^3 - \frac{1}{3}x^3\right) dx = \int_0^2 \left(\frac{20}{3} - 2x + 2x^2 - \frac{4}{3}x^3\right) dx$$

$$= \frac{20}{3}x - x^2 + \frac{2}{3}x^3 - \frac{1}{3}x^4\Big|_0^2 = \frac{40}{3} - 4 + \frac{16}{3} - \frac{16}{3} = \frac{28}{3}$$

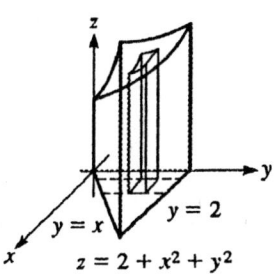

22.

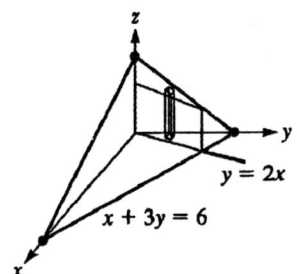

From $y = 2x$ and $x + 3y = 6$, $x + 3(2x) = 6$

$$7x = 6$$

$$x = \frac{6}{7}$$

$$y = \frac{12}{7}$$

$x + 3y = 6, y = 2x$ intersect at $\left(\dfrac{6}{7}, \dfrac{12}{7}\right)$

$$V = \int_0^{6/7} \int_{2x}^{\frac{6-x}{3}} \frac{6-x-3y}{2} \, dy \, dx$$

$$V = \int_0^{6/7} \left[\frac{12y - 3y^2 - 2xy}{4} \right]\Big|_{2x}^{\frac{6-x}{3}} dx$$

$$V = \int_0^{6/7} \left[\frac{12\left(\frac{6-x}{3}\right) - 3\left(\frac{6-x}{3}\right)^2 - 2x\left(\frac{6-x}{3}\right)}{4} - \frac{12(2x) - 3(2x)^2 - 2x(2x)}{4} \right] dx$$

$$V = \int_0^{6/7} \left[\frac{4(6-x) - \frac{1}{3}(36 - 12x + x^2) - \frac{2}{3}(6x - x^2)}{4} - \frac{24x - 12x^2 - 4x^2}{4} \right] dx$$

$$V = \int_0^{6/7} \frac{24 - 4x - 12 + 4x - \frac{x^2}{3} - 4x + \frac{2}{3}x^2 - 24x + 16x^2}{4} \, dx$$

$$V = \int_0^{6/7} \frac{12 - 28x + \frac{49}{3}x^2}{4} \, dx$$

$$V = 3x - \frac{7x^2}{2} + \frac{49x^3}{9} \Big|_0^{6/7}$$

$$V = \frac{6}{7}$$

23. $$V = \int_0^3 \int_0^{\sqrt{9-x^2}} z \, dy \, dx = \int_0^3 \int_0^{\sqrt{9-x^2}} (x+y) \, dy \, dx$$

$$= \int_0^3 \left(xy + \frac{1}{2}y^2 \right)\Big|_0^{\sqrt{9-x^2}} = \int_0^3 \left[x\sqrt{9-x^2} + \frac{1}{2}(9 - x^2) \right] dx$$

$$= -\frac{1}{3}(9 - x^2)^{3/2} + \frac{9}{2}x - \frac{1}{6}x^3 \Big|_0^3$$

$$= 0 + \frac{27}{2} - \frac{27}{6} + \frac{1}{3}(27) - 0 + 0 = \frac{27}{2} - \frac{9}{2} + 9 = 18$$

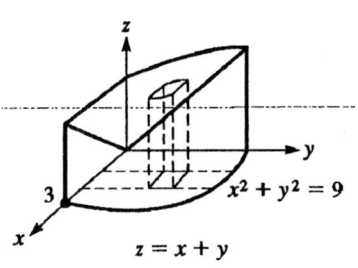

24.

Solve $y = 8x^2$ and $x = y^2$ to obtain $\left(\frac{1}{4}, \frac{1}{2}, 0 \right)$.

Integrating over y first

$$V = \int_0^{1/4} \int_{8x^2}^{\sqrt{x}} (x^2 + 1) \, dy \, dx$$

$$V = \int_0^{1/4} (x^2 + 1)y \Big|_{8x^2}^{\sqrt{x}} dx$$

$$V = \int_0^{1/4} [(x^2 + 1)\sqrt{x} - (x^2 + 1)8x^2] \, dx$$

$$V = \int_0^{1/4} (x^{5/2} + x^{1/2} - 8x^4 - 8x^2) \, dx$$

$$V = \frac{2}{7}x^{7/2} + \frac{2}{3}x^{3/2} - \frac{8}{5}x^5 - \frac{8}{3}x^3 \Big|_0^{1/4}$$

$$V = \frac{2}{7}\left(\frac{1}{4}\right)^{7/2} + \frac{2}{3}\left(\frac{1}{4}\right)^{3/2} - \frac{8}{5}\left(\frac{1}{4}\right)^5 - \frac{8}{3}\left(\frac{1}{4}\right)^3$$

$$V = \frac{569}{13,440}$$

Integrating over x first

$$V = \int_0^{1/2} \int_{y^2}^{\sqrt{y/8}} (x^2 + 1)dx\, dy$$

$$V = \int_0^{1/2} \left(\frac{x^3}{3} + x\right)\bigg|_{y^2}^{\sqrt{y/8}} dy$$

$$V = \int_0^{1/2} \left(\frac{1}{3}\frac{y^{3/2}}{8^{3/2}} + \frac{y^{1/2}}{\sqrt{8}} - \frac{y^6}{3} - y^2\right) dy$$

$$V = \frac{2y^{5/2}}{3(8^{3/2})(5)} + \frac{2y^{3/2}}{3\sqrt{8}} - \frac{y^7}{21} - \frac{y^3}{3}\bigg|_0^{1/2}$$

$$V = \frac{2\left(\frac{1}{2}\right)^{5/2}}{3(8^{3/2})(5)} + \frac{2\left(\frac{1}{2}\right)^{3/2}}{3\sqrt{8}} - \frac{\left(\frac{1}{2}\right)^7}{21} - \frac{\left(\frac{1}{2}\right)^3}{3}$$

$$V = \frac{569}{13,440}$$

25. $V = \int_0^5 \int_0^{12} z\, dx\, dy = \int_0^5 \int_0^{12} (10 - 2y)dx\, dy = \int_0^5 (10 - 2y)x\bigg|_0^{12} dy = 12\int_0^5 (10 - 2y)dy$

$= 12(10y - y^2)\bigg|_0^5 = 12(50 - 25) = 300 \text{ cm}^2$

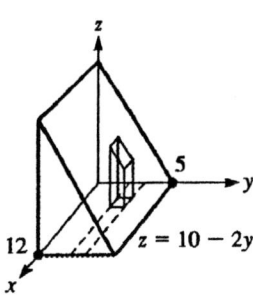

26.

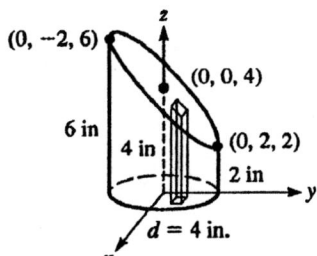

Let the points $(0, -2, 6), (0, 0, 4), (0, 2, 2)$ determine
the plane with equation

$x + ay + bz = c$,   then

$\begin{aligned} -2a + 6b &= c &\Rightarrow& -2a + 6b = 4b \Rightarrow & -2a &= -2b \\ 4b &= c & & & a &= b \\ 2a + 2b &= c &\Rightarrow& 2a + 2b = 4b \\ & & & 2a = 2b \end{aligned}$

let $a = b = 1$, then $c = 4$ and an equation of the plane is
$x + y + z = 4$ or $z = 4 - x - y$.

The equation of the circle in the xy-plane is $x^2 + y^2 = 4$ from which $y = \pm\sqrt{4 - x^2}$.

$$V = \int_{-2}^{2} \int_{-\sqrt{4-x^2}}^{\sqrt{4-x^2}} (4 - x - y)\, dy\, dx$$

$$V = \int_{-2}^{2} \left(4y - xy - \frac{y^2}{2}\right)\Bigg|_{-\sqrt{4-x^2}}^{\sqrt{4-x^2}}\, dx$$

$$V = \int_{-2}^{2} \left[4\sqrt{4-x^2} - x\sqrt{4-x^2} - \frac{4-x^2}{2} - \left(4(-\sqrt{4-x^2}) - x\left(-\sqrt{4-x^2}\right) - \frac{4-x^2}{2}\right)\right] dx$$

$$V = \int_{-2}^{2} \left[4\sqrt{4-x^2} - x\sqrt{4-x^2} - \frac{4-x^2}{2} + 4\sqrt{4-x^2} - x\sqrt{4-x^2} + \frac{4-x^2}{2}\right] dx$$

$$V = \int_{-2}^{2} (8\sqrt{4-x^2} - 2x\sqrt{4-x^2})\, dx$$

$$V = \int_{-2}^{2} 8\sqrt{4-x^2}\, dx + \int_{-2}^{2} \sqrt{4-x^2}\, dx$$

For the first integral, let $x = 2\sin\theta,\ dx = 2\cos\theta\, d\theta$
$$x^2 = 4\sin^2\theta$$

$$\int_{-2}^{2} 8\sqrt{4-x^2}\, dx = \int_{-\pi/2}^{\pi/2} 8\sqrt{4 - 4\sin^2\theta}(2\cos\theta\, d\theta) = 16\int_{-\pi/2}^{\pi/2} 2\cos\theta(2\cos\theta)\, d\theta = 64\int_{-\pi/2}^{\pi/2} \cos^2\theta\, d\theta$$

$$= 64\int_{-\pi/2}^{\pi/2} \frac{1 + \cos 2\theta}{2}\, d\theta = 32\left[\theta\Big|_{-\pi/2}^{\pi/2} + \frac{1}{2}\sin 2\theta\Big|_{-\pi/2}^{\pi/2}\right]$$

$$= 32\left[\frac{\pi}{2} - \left(-\frac{\pi}{2}\right) + \frac{1}{2}\sin\pi - \frac{1}{2}\sin(-\pi)\right] = 16\pi$$

For the second integral

$$\int_{-2}^{2} \sqrt{4-x^2}(-2x\, dx) = \frac{2}{3}(4-x^2)^{3/2}\Big|_{-2}^{2} = \frac{2}{3}(4 - (2)^2)^{3/2} - \frac{2}{3}(4 - (-2)^2)^{3/2} = 0$$

$$V = 16\pi = 50.3 \text{ in}^3$$

27. $z = 4 - x - 2y$ from $y = x^2$ to $y = 1$, from $x = 0$ to $x = \dfrac{1}{2}$

Intercepts of plane: $(4,0,0), (0,2,0), (0,0,4)$

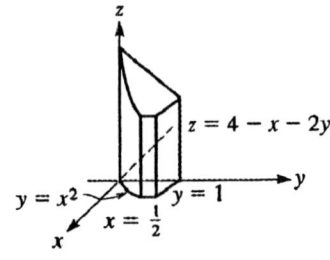

28. The region in the xy-plane is from $x = 0$ to the line $x + y = 2$, $(x = 2 - y)$ and from $y = 1$ to $y = 2$ on y. The volume is above the region and below the surface given by $z = \sqrt{1 + x^2 + y^2}$. The volume is bounded by the planes $x = 0$, $x + y = 2$, and $y = 1$, which is indicated by the dashed lines in the figure.

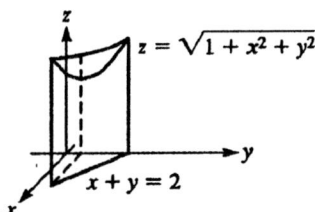

11.6   Centroids and Moments of Inertia By Double Integration

1. $A = \displaystyle\int_0^2 \int_0^{2-y} dx\, dy = \int_0^2 x\Big|_0^{2-y} dy = \int_0^2 (2-y)dy = 2y - \frac{1}{2}y^2\Big|_0^2 = 2$

$A\bar{x} = \displaystyle\int_0^2 \int_0^{2-y} x\, dx\, dy = \int_0^2 \frac{1}{2}x^2\Big|_0^{2-y} dy = \frac{1}{2}\int_0^2 (4 - 4y + y^2)dy = \frac{1}{2}\left(4y - 2y^2 + \frac{1}{3}y^3\right)\Big|_0^2$

$\qquad = \frac{1}{2}\left(8 - 8 + \frac{8}{3}\right) = \frac{8}{6} = \frac{4}{3}; \ \bar{x} = \frac{4}{3}\left(\frac{1}{2}\right) = \frac{2}{3}$

$A\bar{y} = \displaystyle\int_0^2 \int_0^{2-y} y\, dx\, dy = \int_0^2 x\Big|_0^{2-y} y\, dy = \int_0^2 (2y - y^2)dy = y^2 - \frac{1}{3}y^3\Big|_0^2 = 4 - \frac{8}{3} = \frac{4}{3}; \ \bar{y} = \frac{4}{3}\left(\frac{1}{2}\right) = \frac{2}{3}$

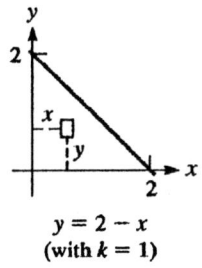

$y = 2 - x$
(with $k = 1$)

2. $y = x^2, y = 4, x = 0$

$A = \displaystyle\int_0^2 \int_{x^2}^4 dy\, dx = \int_0^2 y\Big|_{x^2}^4 dx = \int_0^2 (4 - x^2)dx = 4x - \frac{x^3}{3}\Big|_0^2 = 4(2) - \frac{2^3}{3} = \frac{16}{3}$

$A\bar{x} = \displaystyle\int_0^2 \int_{x^2}^4 x\, dy\, dx = \int_0^2 xy\Big|_{x^2}^4 dx = \int_0^2 (4x - x^3)dx = 2x^2 - \frac{x^4}{4}\Big|_0^2 = 2(2)^2 - \frac{2^4}{4} = 4$

$\bar{x} = \frac{4}{A} = \frac{3}{4}$

$A\bar{y} = \displaystyle\int_0^2 \int_{x^2}^4 y\, dy\, dx = \int_0^2 \frac{y^2}{2}\Big|_{x^2}^4 dx = \int_0^2 \left(\frac{4^2}{2} - \frac{x^4}{2}\right) dx = 8x - \frac{x^5}{10}\Big|_0^2 = 8(2) - \frac{2^5}{10} = \frac{64}{5}$

$\bar{y} = \frac{\frac{64}{5}}{A} = \frac{\frac{64}{5}}{\frac{16}{3}} = \frac{12}{5}$

$$(\bar{x}, \bar{y}) = \left(\frac{3}{4}, \frac{12}{5}\right)$$

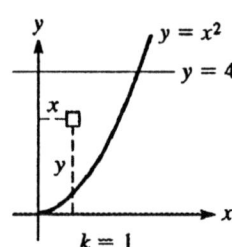

3. $A = \int_0^3 \int_0^{9-x^2} dy\, dx = \int_0^3 y\Big|_0^{9-x^2} dx = \int_0^3 (9-x^2)dx = 9x - \frac{1}{3}x^3 \Big|_0^3 = 27 - 9 = 18$

$A\bar{x} = \int_0^3 \int_0^{9-x^2} x\, dy\, dx = \int_0^3 y\Big|_0^{9-x^2} x\, dx = \int_0^3 (9x - x^3)dx = \frac{9}{2}x^2 - \frac{1}{4}x^4 \Big|_0^3 = \frac{81}{2} - \frac{81}{4} = \frac{81}{4}$

$\bar{x} = \frac{81}{4}\left(\frac{1}{18}\right) = \frac{9}{8}$

$A\bar{y} = \int_0^3 \int_0^{9-x^2} y\, dy\, dx = \int_0^3 \frac{1}{2}y^2 \Big|_0^{9-x^2} dx = \frac{1}{2}\int_0^3 (81 - 18x^2 + x^4)dx = \frac{1}{2}\left(81x - 6x^3 + \frac{1}{5}x^5\right)\Big|_0^3$

$= \frac{1}{2}\left(243 - 162 + \frac{243}{5}\right) = \frac{324}{5}; \quad \bar{y} = \frac{324}{5}\left(\frac{1}{18}\right) = \frac{18}{5}$

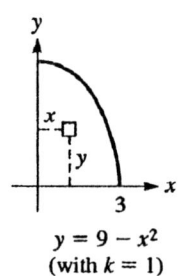

$y = 9 - x^2$
(with $k = 1$)

4. $y = 2x, x = 1, x = 2$

$A = \int_1^2 \int_0^{2x} dy\, dx = \int_1^2 y\Big|_0^{2x} dx = \int_1^2 2x\, dx = x^2\Big|_1^2 = 2^2 - 1^2 = 3$

$A\bar{x} = \int_1^2 \int_0^{2x} x\, dy\, dx = \int_1^2 xy\Big|_0^{2x} dx = \int_1^2 2x^2 dx; \quad 3\bar{x} = \frac{2x^3}{3}\Big|_1^2 = \frac{2}{3}(2^3 - 1^3) = \frac{14}{3}; \quad x = \frac{14}{9}$

$A\bar{y} = \int_1^2 \int_0^{2x} y\, dy\, dx = \int_1^2 \frac{y^2}{2}\Big|_0^{2x} dx = \int_1^2 2x^2 dx; \quad 3\bar{y} = \frac{2x^3}{3}\Big|_1^2 = \frac{2}{3}(2^3 - 1^3) = \frac{14}{3}; \quad \bar{y} = \frac{14}{9}$

$(\bar{x}, \bar{y}) = \left(\frac{14}{9}, \frac{14}{9}\right)$

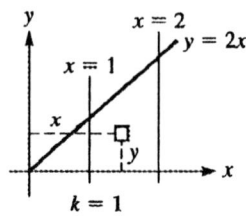

5. See Figure for #1

$$I_y = \int_0^2 \int_0^{2-x} x^2 dy\, dx = \int_0^2 y\Big|_0^{2-x} x^2 dx = \int_0^2 (2x^2 - x^3) = \frac{2}{3}x^3 - \frac{1}{4}x^4\Big|_0^2 = \frac{16}{3} - 4 = \frac{4}{3}$$

$$I_x = \int_0^2 \int_0^{2-x} y^2 dy\, dx = \int_0^2 x\Big|_0^{2-y} y^2 dx = \int_0^2 (2y^2 - y^3)dy = \frac{2}{3}y^3 - \frac{1}{4}y^4\Big|_0^2 = \frac{16}{3} - 4 = \frac{4}{3}$$

6. See Figure for #2

$$I_y = \int_0^2 \int_{x^2}^4 x^2 dy\, dx = \int_0^2 x^2 y\Big|_{x^2}^4 dx = \int_0^{2:} (4x^2 - x)dx$$

$$I_x = \frac{4x^3}{3} - \frac{x^5}{5}\Big|_0^2 = \frac{4(2)^3}{3} - \frac{2^5}{5} = \frac{64}{15} = \int_0^2 \int_{x^2}^4 y^2 dy\, dx = \int_0^2 \frac{y^3}{3}\Big|_{x^3}^4 dx$$

$$I_x = \int_0^2 \left(\frac{64}{3} - \frac{x^6}{3}\right) dx$$

$$I_x = \frac{64x}{3} - \frac{x^7}{21}\Big|_0^2 = \frac{64(2)}{3} - \frac{2^7}{21} = \frac{256}{7}$$

7. See Figure for #3.

$$A = \int_0^3 \int_0^{9-x^2} dy\, dx = 18 \text{ (see #3)}$$

$$I_y = \int_0^3 \int_0^{9-x^2} x^2 dy\, dx = \int_0^3 y\Big|_0^{9-x^2} x^2 dx = \int_0^3 (9x^2 - x^4)dx = 3x^3 - \frac{1}{5}x^5\Big|_0^3$$

$$I_y = 81 - \frac{243}{5} = \frac{162}{5}$$

$$R_y^2 = \frac{162}{5}\left(\frac{1}{18}\right) = \frac{9}{5}, R_y = \sqrt{\frac{9}{5}} = 1.34$$

8. See Figure for #4.

$A = 3$, see #4.

$$I_x = \int_1^2 \int_0^{2x} y^2 dy\, dx = \int_1^2 \frac{y^3}{3}\Big|_0^{2x} dx = \int_1^2 \frac{8x^3}{3} dx = \frac{2x^4}{3}\Big|_1^2 = \frac{2(2)^4}{3} - \frac{2(1)^4}{3}$$

$$= \frac{32}{3} - \frac{2}{3} = \frac{30}{3} = 10$$

$$R_x^2 = \frac{I_x}{A} = \frac{10}{3}$$

$$R_x = \sqrt{\frac{10}{3}} = 1.83$$

9. $A = \displaystyle\int_0^1 \int_{1-x^2}^1 dy\, dx = \int_0^1 y\Big|_{1-x^2}^1 dx = \int_0^1 (1 - 1 + x^2)dx = \int_0^1 x^2 dx = \frac{1}{3}x^3\Big|_0^1 = \frac{1}{3}$

$A\bar{x} = \displaystyle\int_0^1 \int_{1-x^2}^1 x\, dy\, dx = \int_0^1 y\Big|_{1-x^2}^1 x\, dx = \int_0^1 x^3 dx = \frac{1}{4}x^4\Big|_0^1 = \frac{1}{4}; \; \bar{x} = \frac{1}{4}\left(\frac{3}{1}\right) = \frac{3}{4}$

$A\bar{y} = \displaystyle\int_0^1 \int_{1-x^2}^1 y\, dy\, dx = \int_0^1 \frac{1}{2}y^2\Big|_{1-x^2}^1 dx = \frac{1}{2}\int_0^1 [1 - (1-x^2)^2]dx = \frac{1}{2}\int_0^1 (2x^2 - x^4)dx$

$= \frac{1}{3}x^3 - \frac{1}{10}x^5\Big|_0^1 = \frac{1}{3} - \frac{1}{10} = \frac{7}{30}; \; \bar{y} = \frac{7}{30}\left(\frac{3}{1}\right) = \frac{7}{10}$

$I_y = \displaystyle\int_0^1 \int_{1-x^2}^1 x^2\, dy\, dx = \int_0^1 y\Big|_{1-x^2}^1 x^2 dx = \int_0^1 x^4 dx = \frac{1}{5}x^5\Big|_0^1 = \frac{1}{5}; \; R_y^2 = \frac{1}{5}\left(\frac{3}{1}\right) = \frac{3}{5}; \; R_y = 0.775$

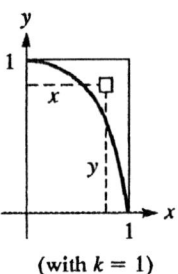

(with $k = 1$)

10. $A = \displaystyle\int_0^3 \int_{3-x}^{3+2x-x^2} dy\, dx = \int_0^3 y\Big|_{3-x}^{3+2x-x^2} dx = \int_0^3 (3 + 2x - x^2 - 3 + x)dx$

$A = \displaystyle\int_0^3 (3x - x^2)dx = \frac{3x^2}{2} - \frac{x^3}{3}\Big|_0^3 = \frac{3(3)^2}{2} - \frac{3^3}{3} = \frac{9}{2}$

$I_y = \displaystyle\int_0^3 \int_{3-x}^{3+2x-x^2} x^2\, dy\, dx = \int_0^3 x^2 y\Big|_{3-x}^{3+2x-x^2} = \int_0^3 (3x^3 - x^4)dx$

$I_y = \dfrac{3x^4}{4} - \dfrac{x^5}{5}\Big|_0^3 = \dfrac{3(3)^4}{4} - \dfrac{3^5}{5} = \dfrac{243}{20}$

$AR_y^2 = I_y$

$R_y = \sqrt{\dfrac{I_y}{A}} = \sqrt{\dfrac{\frac{243}{20}}{\frac{9}{2}}} = 1.64$

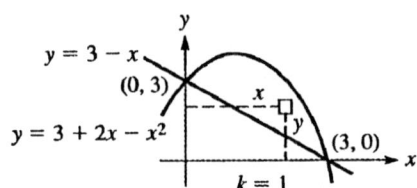

11. $A = \displaystyle\int_0^a \int_0^{b-\frac{b}{a}x} dy\, dx = \int_0^a \int_0^{b-\frac{b}{a}x} dx = \int_0^a \left(b - \frac{b}{a}x\right) dx = bx - \frac{b}{2a}x^2 \Big|_0^2 = \frac{1}{2}ab$

$A\bar{x} = \displaystyle\int_0^a \int_0^{b-\frac{b}{a}x} x\, dy\, dx = \int_0^a y \Big|_0^{b-\frac{b}{a}x} x\, dx = \int_0^a \left(bx - \frac{b}{a}x^2\right) dx = \frac{1}{2}bx^2 - \frac{b}{3a}x^3 \Big|_0^a = \frac{a^2 b}{6}$

$A\bar{y} = \displaystyle\int_0^a \int_0^{b-\frac{b}{a}x} y\, dy\, dx = \frac{1}{2}\int_0^a y^2 \Big|_0^{b-\frac{b}{a}x} dx = \frac{1}{2}\int_0^a \left(b^2 - \frac{2b^2}{a}x + \frac{b^2}{a^2}x^2\right) dx$

$\qquad = \dfrac{1}{2}\left(b^2 x - \dfrac{b^2}{a}x^2 + \dfrac{b^2}{3a^2}x^3\right)\Big|_0^a = \dfrac{b^2 a}{6}$

$\bar{x} = \dfrac{b^2 a}{6}\left(\dfrac{2}{ab}\right) = \dfrac{a}{3}; \quad \bar{y} = \dfrac{ab^2}{6}\left(\dfrac{2}{ab}\right) = \dfrac{b}{3}$

<!-- figure: triangle with y = b - (b/a)x, vertices at (0,b) and (a,0), centroid marked with x̄, ȳ -->

$y = b - \dfrac{b}{a}x$

(with $k = 1$)

12. $A = \dfrac{1}{4}\pi a^2$ (quarter circle)

$A\bar{x} = \displaystyle\int_0^a \int_0^{\sqrt{a^2-x^2}} x\, dy\, dx = \int_0^a xy \Big|_0^{\sqrt{a^2-x^2}} dx = \int_0^a x\sqrt{a^2 - x^2}\, dx$

$A\bar{x} = -\dfrac{1}{2}\displaystyle\int_0^a \sqrt{a^2 - x^2}(-2x\, dx)$

$A\bar{x} = -\dfrac{1}{2}\left(\dfrac{2}{3}\right)(a^2 - x^2)^{3/2}\Big|_0^a = -\dfrac{1}{3}(a^2 - a^2)^{3/2} + \dfrac{1}{3}(a^2 - 0^2)^{3/2} = \dfrac{1}{3}a^3$

$\bar{x} = \dfrac{\frac{1}{3}a^3}{\frac{\pi a^2}{4}} = \dfrac{4a}{3\pi}$

$A\bar{y} = \displaystyle\int_0^a \int_0^{\sqrt{a^2-x^2}} y\, dy\, dx = \int_0^a \dfrac{y^2}{2}\Big|_0^{\sqrt{a^2-x^2}} dx = \int_0^a \dfrac{a^2 - x^2}{2}\, dx = \dfrac{a^2 x}{2} - \dfrac{x^3}{6}\Big|_0^a$

$A\bar{y} = \dfrac{a^3}{2} - \dfrac{a^3}{3} = \dfrac{a^3}{3}$

$\bar{y} = \dfrac{\frac{a^3}{3}}{\frac{\pi a^2}{4}} = \dfrac{4a}{3\pi}$

$(\bar{x}, \bar{y}) = \left(\dfrac{4a}{3\pi}, \dfrac{4a}{3\pi}\right)$

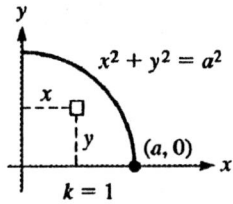

13. $A = \int_0^a \int_0^b dy\,dx = \int_0^a y\Big|_0^b dx = \int_0^a b\,dx = bx\Big|_0^a = ab;\ m = A$ (with $k = 1$)

$I_x = \int_0^a \int_0^b y^2\,dy\,dx = \int_0^a \frac{1}{3}y^3\Big|_0^b dx = \int_0^a \frac{1}{3}b^3\,dx = \frac{1}{3}b^3 x\Big|_0^a = \frac{1}{3}ab^3 = \frac{1}{3}(ab)b^2 = \frac{1}{3}mb^2$

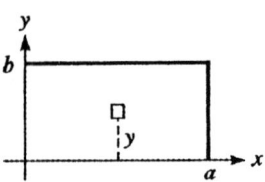

14. $k = \dfrac{m}{A} = \dfrac{m}{\frac{1}{2}a^2}$

$I_x = k \int_0^a \int_0^x y^2\,dy\,dx$

$I_x = \dfrac{2m}{a^2} \int_0^a \dfrac{y^3}{3}\Big|_0^x dx$

$I_x = \dfrac{2m}{a^2} \int_0^a \dfrac{x^3}{3}\,dx$

$I_x = \dfrac{2m}{a^2} \dfrac{x^4}{12}\Big|_0^a = \dfrac{2m}{a^2}\dfrac{a^4}{12}$

$I_x = \dfrac{ma^2}{6}$

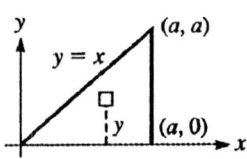

15. Solution for A and I_x same as in #13, with $a = b = 4$.

$A = 16;\ I_x = \dfrac{1}{3}(4)(64) = \dfrac{256}{3};\ R_x^2 = \dfrac{256}{3}\left(\dfrac{1}{16}\right) = \dfrac{1}{16};\ R_x = \sqrt{\dfrac{16}{3}} = \dfrac{4}{3}\sqrt{3} = 2.31$

16. $A = bh = 2\sqrt{2}$

$I_x = \int_0^{\sqrt{2}} \int_y^{y+2} y^2\,dx\,dy = \int_0^{\sqrt{2}} y^2 x\Big|_y^{y+2} dy = \int_0^{\sqrt{2}} [y^2(y+2) - y^2(y)]dy$

$I_x = \int_0^{\sqrt{2}} 2y^2\,dy = \dfrac{2y^3}{3}\Big|_0^{\sqrt{2}} = \dfrac{4\sqrt{2}}{3}$

$A R_x^2 = I_x$

$2\sqrt{2}R_x^2 = \dfrac{4\sqrt{2}}{3}$

$R_x = \sqrt{\dfrac{2}{3}} = 0.816$

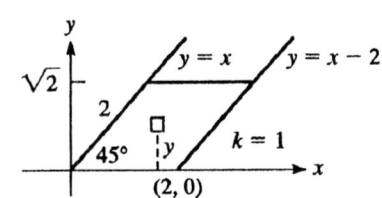

17. $A = \int_0^2 \int_y^{6-y} dx\, dy = \int_0^2 x \Big|_y^{6-y} dy = \int_0^2 (6 - 2y) dy = 6y - y^2 \Big|_0^2 = 8$

$A\bar{x} = \int_0^2 \int_y^{6-y} x\, dx\, dy = \int_0^2 \frac{1}{2}x^2 \Big|_y^{6-y} dy = \frac{1}{2} \int_0^2 (36 - 12y) dy = 18y - 3y^2 \Big|_0^2 = 36 - 12 = 24$

$A\bar{y} = \int_0^2 \int_y^{6-y} y\, dx\, dy = \int_0^2 x \Big|_y^{6-y} y\, dy = \int_0^2 (6y - 2y^2) dy = 3y^2 - \frac{2}{3}y^2 \Big|_0^2 = \frac{20}{3}$

$\bar{x} = \frac{24}{8} = 3, \; \bar{y} = \frac{20}{3}\left(\frac{1}{8}\right) = \frac{5}{6}$

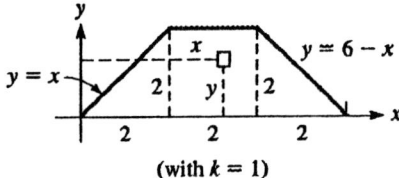

(with $k = 1$)

18. By symmetry, $\bar{x} = 0$
First determine the equation of the parabola from $y = ax^2 + bx + c$.

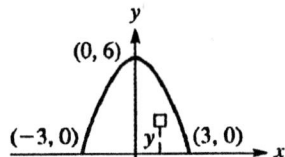

$(0, 6), \; 6 = a(0)^2 + b(0) + c$

$c = 6$

$y = ax^2 + bx + 6$

$(-3, 0), 0 = a(-3)^2 + b(-3) + 6, \qquad 9a - 3b = -6$

$(3, 0), 0 = a(3)^2 + b(3) + 6, \qquad \underline{\;9a + 3b = -6\;}$

$\qquad\qquad\qquad\qquad\qquad\qquad 18a \qquad = -12$

$\qquad\qquad\qquad\qquad\qquad\qquad a = -\frac{2}{3}$

$\qquad\qquad\qquad\qquad 9\left(-\frac{2}{3}\right) + 3b = -6$

$\qquad\qquad\qquad\qquad\qquad\qquad b = 0$

$y = -\frac{2}{3}x^2 + 6$ is the equaion of the parabola.

$A = \int_{-3}^3 \int_0^{-\frac{2}{3}x^2 + 6} dy\, dx = \int_{-3}^3 \left(-\frac{2}{3}x^2 + 6\right) dx = -\frac{2}{9}x^3 + 6x \Big|_{-3}^3$

$A = -\frac{2}{9}(3)^3 + 6(3) + \frac{2}{9}(-3)^3 - 6(-3)$

$A = 24$

$$A\bar{y} = \int_{-3}^{3}\int_{0}^{-\frac{2}{3}x^2+6} y\, dy\, dx = \int_{-3}^{3} \frac{y^2}{2}\Big|_{0}^{-\frac{2}{3}x^2+6} dx$$

$$A\bar{y} = \frac{1}{2}\int_{-3}^{3}\left(\frac{4}{9}x^4 - 8x^2 + 36\right)dx$$

$$A\bar{y} = \frac{1}{2}\left[\frac{4}{45}x^5 - \frac{8}{3}x^3 + 36x\right]\Big|_{-3}^{3}$$

$$24\bar{y} = \frac{1}{2}\left[\frac{4}{45}(3)^5 - \frac{8}{3}(3)^3 + 36(3) - \frac{4}{45}(-3)^5 + \frac{8}{3}(-3)^3 - 36(-3)\right]$$

$$24\bar{y} = \frac{288}{5}$$

$$\bar{y} = 2.4,\ (\bar{x},\bar{y}) = (0, 2.4)$$

The centroid is on the center line 3.6 ft from top.

19. $I_x = \displaystyle\iint y^2 dy\, dx$

$$I_x = \int_{0}^{\frac{a\sqrt{2}}{2}}\int_{-x}^{x} y^2 dy\, dx + \int_{\frac{a\sqrt{2}}{2}}^{a\sqrt{2}}\int_{x-a\sqrt{2}}^{-x+a\sqrt{2}} y^2 dy\, dx$$

$$I_x = \int_{0}^{\frac{a\sqrt{2}}{2}} \frac{y^3}{3}\Big|_{-x}^{x} dx + \int_{\frac{a\sqrt{2}}{2}}^{a\sqrt{2}} \frac{y^3}{3}\Big|_{x-a\sqrt{2}}^{-x+a\sqrt{2}} dx$$

$$I_x = \int_{0}^{\frac{a\sqrt{2}}{2}}\left(\frac{x^3}{3} - \frac{(-x)^3}{3}\right)dx + \int_{\frac{a\sqrt{2}}{2}}^{a\sqrt{2}}\left(\frac{(-x+a\sqrt{2})^3}{3} - \frac{(x-a\sqrt{2})^3}{3}\right)dx$$

$$I_x = \int_{0}^{\frac{a\sqrt{2}}{2}} \frac{2x^3}{3}dx - \int_{\frac{a\sqrt{2}}{2}}^{a\sqrt{2}} \frac{2(\sqrt{2}a - x)^3}{3}(-dx)$$

$$I_x = \frac{x^4}{6}\Big|_{0}^{\frac{a\sqrt{2}}{2}} - \frac{(\sqrt{2}a - x)^4}{6}\Big|_{a\sqrt{2}/2}^{a\sqrt{2}}$$

$$I_x = \frac{\left(\frac{a\sqrt{2}}{2}\right)^4}{6} - \frac{(\sqrt{2}a - a\sqrt{2})^4}{6} + \frac{\left(\sqrt{2}a - \frac{a\sqrt{2}}{2}\right)^4}{6}$$

$$I_x = \frac{\left(\frac{a\sqrt{2}}{2}\right)^4}{6} + \frac{\left(\frac{a\sqrt{2}}{2}\right)^4}{6} = \frac{\left(\frac{a\sqrt{2}}{2}\right)^4}{3} = \frac{\frac{a^4(4)}{16}}{3}$$

$$I_x = \frac{a^4}{12}$$

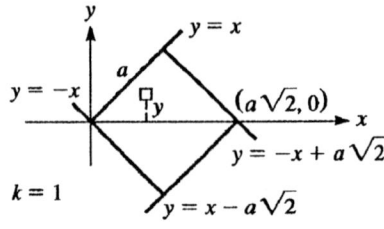

20. $A = \lim\limits_{a \to -\infty} \displaystyle\int_a^0 \int_0^{e^x} dy\, dx$

$A = \lim\limits_{a \to -\infty} \displaystyle\int_a^0 y \Big|_0^{e^x} dx$

$A = \lim\limits_{a \to -\infty} \displaystyle\int_a^0 e^x dx = \lim\limits_{a \to -\infty} e^x \Big|_a^0$

$A = \lim\limits_{a \to -\infty} [e^0 - e^a] = 1$

$A\bar{x} = \lim\limits_{a \to -\infty} \displaystyle\int_a^0 \int_0^{e^x} x\, dy\, dx$

$\quad = \lim\limits_{a \to -\infty} \displaystyle\int_a^0 xy \Big|_0^{e^x} dx$

$\bar{x} = \lim\limits_{a \to -\infty} \displaystyle\int_a^0 x\, e^x dx$ (use parts for $\int xe^x$)

$\bar{x} = \lim\limits_{a \to -\infty} \left[e^x (x - 1) \Big|_a^0 \right]$

$\bar{x} = \lim\limits_{a \to -\infty} [e^0(0 - 1) - e^a(a - 1)]$

$\bar{x} = -1$

$\bar{y} = \lim\limits_{a \to -\infty} \displaystyle\int_a^0 \int_0^{e^x} y\, dy\, dx$

$\bar{y} = \lim\limits_{a \to -\infty} \displaystyle\int_a^0 \frac{y^2}{2} \Big|_0^{e^x} dx$

$\bar{y} = \lim\limits_{a \to -\infty} \displaystyle\int_a^0 \frac{1}{2}(e^x)^2 dx = \lim\limits_{a \to -\infty} \displaystyle\int_a^0 \frac{1}{2} e^{2x} dx$

$\bar{y} = \lim\limits_{a \to -\infty} \frac{1}{4} \displaystyle\int_a^0 e^{2x}(2\,dx)$

$\bar{y} = \lim\limits_{a \to -\infty} \frac{1}{4} e^{2x} \Big|_a^0$

$\bar{y} = \lim\limits_{a \to -\infty} \frac{1}{4}[e^0 - e^{2a}]$

$\bar{y} = \frac{1}{4}$

$(\bar{x}, \bar{y}) = \left(-1, \dfrac{1}{4}\right)$

Chapter 11 Review Exercises

1. $z = 5x^3 y^2 - 2xy^4$

$\dfrac{\partial z}{\partial x} = 5y^2(3x^2) - 2y^4(1)$

$\quad = 15x^2 y^2 - 2y^4$

$\dfrac{\partial z}{\partial y} = 5x^3(2y) - 2x(4y^3)$

$\quad = 10x^3 y - 8xy^3$

2. $z = 2x\sqrt{y} - x^2 y$

$\dfrac{\partial z}{\partial x} = 2\sqrt{y} - 2xy;$

$\dfrac{\partial z}{\partial y} = \dfrac{2x}{2\sqrt{y}} - x^2$

$\quad = \dfrac{x}{\sqrt{y}} - x^2$

3. $z = \sqrt{x^2 - 3y^2}$

$\dfrac{\partial z}{\partial x} = \dfrac{1}{2}(x^2 - 3y^2)^{-1/2}(2x)$

$\quad = \dfrac{x}{\sqrt{x^2 - 3y^2}}$

$\dfrac{\partial z}{\partial y} = \dfrac{1}{2}(x^2 - 3y^2)^{-1/2}(-6y)$

$\quad = \dfrac{-3y}{\sqrt{x^2 - 3y^2}}$

4. $u = \dfrac{r}{(r - 3s)^2}$

$\dfrac{\partial u}{\partial r} = \dfrac{(3 - 3s)^2(1) - r(2(r - 3s)(1))}{(r - 3s)^4}$

$\quad = \dfrac{r - 3s - 2r}{(r - 3s)^3}$

$\dfrac{\partial u}{\partial r} = \dfrac{-(r + 3s)}{(r - 3s)^3}$

$\dfrac{\partial u}{\partial s} = \dfrac{(r - 3s)^3(0) - r(2(r - 3s)(-3))}{(r - 3s)^4}$

$\dfrac{\partial u}{\partial s} = \dfrac{6r(r - 3s)}{(r - 3s)^4}$

$\quad = \dfrac{6r}{(r - 3s)^3}$

5. $z = \dfrac{2x - 3y}{x^2y + 1}$

$\dfrac{\partial z}{\partial x} = \dfrac{(x^2y + 1)(2) - (2x - 3y)(2xy)}{(x^2y + 1)^2}$

$= \dfrac{2 - 2x^2y + 6xy^2}{(x^2y + 1)^2}$

$\dfrac{\partial z}{\partial y} = \dfrac{(x^2y + 1)(-3) - (2x - 3y)(2x^2)}{(x^2y + 1)^2}$

$= \dfrac{-(3 + 2x^3)}{(x^2y + 1)^2}$

6. $z = x(y^2 + xy + 2)^4$

$\dfrac{\partial z}{\partial x} = x(4(y^2 + xy + 2)^3 y) + (y^2 + xy + 2)^4$

$\dfrac{\partial z}{\partial x} = 4xy(y^2 + xy + 2)^3 + (y^2 + xy + 2)^4$

$\dfrac{\partial z}{\partial x} = (y^2 + xy + 2)^3(4xy + y^2 + xy + 2)$

$\dfrac{\partial z}{\partial x} = (y^2 + xy + 2)^3(y^2 + 5xy + 2)$

$\dfrac{\partial z}{\partial y} = 4x(y^2 + xy + 2)^3(2y + x)$

7. $u = y\ln\sin(x^2 + 2y)$

$\dfrac{\partial u}{\partial x} = \dfrac{y\cos(x^2 + 2y)}{\sin(x^2 + 2y)}(2x)$

$= 2xy\cot(x^2 + 2y)$

$\dfrac{\partial u}{\partial y} = y\left[\dfrac{\cos(x^2 + 2y)}{\sin(x^2 + 2y)}\right](2) + \ln\sin(x^2 + 2y)$

$= 2y\cot(x^2 + 2y) + \ln\sin(x^2 + 2y)$

8. $q = p\ln(r + 1) - \dfrac{rp}{r + 1}$

$\dfrac{\partial q}{\partial p} = \ln(r + 1) - \dfrac{r}{r + 1}$

$\dfrac{\partial q}{\partial r} = \dfrac{p}{r + 1} - \dfrac{(r + 1)(p) - rp(1)}{(r + 1)^2}$

$= \dfrac{p}{r + 1} - \dfrac{rp + p - rp}{(r + 1)^2}$

$\dfrac{\partial q}{\partial r} = \dfrac{p}{r + 1} - \dfrac{p}{(r + 1)^2}$

$= \dfrac{p(r + 1) - p}{(r + 1)^2}$

$\dfrac{\partial q}{\partial r} = \dfrac{pr + p - p}{(r + 1)^2}$

$\dfrac{\partial q}{\partial r} = \dfrac{pr}{(r + 1)^2}$

9. $z = \sin^{-1}\sqrt{x + y}$

$\dfrac{\partial z}{\partial x} = \dfrac{1}{\sqrt{1 - (x + y)}}\left(\dfrac{1}{2}\right)(x + y)^{-1/2}(1)$

$= \dfrac{1}{2\sqrt{(x + y)(1 - x - y)}}$

$\dfrac{\partial z}{\partial y} = \dfrac{1}{\sqrt{1 - (x + y)}}\left(\dfrac{1}{2}\right)(x + y)^{-1/2}(1)$

$= \dfrac{1}{2\sqrt{(x + y)(1 - x - y)}}$

10. $z = ye^{xy}\sin(2x - y)$

$\dfrac{\partial z}{\partial x} = y[e^{xy}\cos(2x - y)(2) + ye^{xy}\sin(2x - y)]$

$\dfrac{\partial z}{\partial x} = ye^{xy}[2\cos(2x - y) + y\sin(2x - y)]$

$\dfrac{\partial z}{\partial y} = e^{xy}\sin(2x - y) + yxe^{xy}\sin(2x - y)$
$\qquad + ye^{xy}\cos(2x - y)(-1)$

$\dfrac{\partial z}{\partial y} = e^{xy}[(1 + xy)\sin(2x - y) - y\cos(2x - y)]$

11. $z = 3x^2y - y^3 + 2xy$

$\dfrac{\partial z}{\partial x} = 6xy + 2y, \dfrac{\partial z}{\partial y} = 3x^2 - 3y^2 + 2x$

$\dfrac{\partial^2 z}{\partial x^2} = 6y, \dfrac{\partial^2 z}{\partial y^2} = -6y,$

$\dfrac{\partial^2 z}{\partial y\partial x} = \dfrac{\partial^2 z}{\partial x\partial y} = 6x + 2$

12. $z = x\sqrt{2y + 1} + y^2(x - 2)^3$

$\dfrac{\partial z}{\partial x} = \sqrt{2y + 1} + 3y^2(x - 2)^2$

$\dfrac{\partial^2 z}{\partial x^2} = 6y^2(x - 2)$

$\dfrac{\partial z}{\partial y} = \dfrac{x(2)}{2\sqrt{2y + 1}} + 2y(x - 2)^3$

$\dfrac{\partial z}{\partial y} = \dfrac{x}{\sqrt{2y + 1}} + 2y(x - 2)^3$

$\dfrac{\partial^2 z}{\partial y^2} = -\dfrac{x}{2}(2y + 1)^{-3/2}(2) + 2(x - 2)^3$

$\dfrac{\partial^2 z}{\partial y^2} = \dfrac{-x}{(2y + 1)^{3/2}} + 2(x - 2)^3$

$\dfrac{\partial^2 z}{\partial y\partial x} = \dfrac{\partial^2 z}{\partial x\partial y} = \dfrac{(2)}{2\sqrt{2y + 1}} + 6y(x - 2)^2$

$\dfrac{\partial^2 z}{\partial y\partial x} = \dfrac{\partial^2 z}{\partial x\partial y} = \dfrac{1}{\sqrt{2y + 1}} + 6y(x - 2)^2$

13. $\int_0^2 \int_1^2 (3y + 2xy)dx\,dy = \int_0^2 (3xy + x^2y)\Big|_1^2 dy = \int_0^2 (6y + 4y - 3y - y)dy = \int_0^2 6y\,dy = 3y^2\Big|_0^2 = 12$

14. $\int_2^7 \int_0^1 x\sqrt{2 + x^2y}dx\,dy = \int_2^7 \int_0^1 \frac{1}{2y}\sqrt{2 + x^2y}(2xy\,dx)dy = \int_2^7 \frac{1}{2y}\left(\frac{2}{3}\right)(2 + x^2y)^{3/2}\Big|_0^1 dy$

$$= \int_2^7 \frac{1}{3y}[(2 + y)^{3/2} - 2^{3/2}]dy = \int_2^7 \frac{(2 + y)^{3/2}dy}{3y} - \int_2^7 \frac{2^{3/2}}{3}\frac{1}{y}dy$$

for the first integral,   let $u = (2 + y)^{1/2}$

$$u^2 = 2 + y, y = u^2 - 2$$

$$2u\,du = dy$$

$$= \int_2^3 \frac{u^3(2u\,du)}{3(u^2 - 2)} - \frac{2^{3/2}}{3}\ln y\Big|_2^7 = \frac{2}{3}\int_2^3 \frac{u^4\,du}{u^2 - 2} - \frac{2^{3/2}}{3}\ln\frac{7}{2},$$

$$u^2 - 2\overline{)u^4} \quad \begin{array}{c} u^2 + 2 \\ \hline \end{array}$$
$$\underline{u^4 - 2u^2}$$
$$2u^2$$
$$\underline{2u^2 - 4}$$
$$4$$

$$= \frac{2}{3}\int_2^3 \frac{4\,du}{u^2 - 2} + \frac{2}{3}\int_2^3 u^2\,du + \frac{2}{3}\int_2^3 2\,du - \frac{2^{3/2}}{3}\ln\frac{7}{2}$$

$$= \frac{8}{3}\int_2^3 \frac{du}{(u + \sqrt{2})(u - \sqrt{2})} + \frac{2}{9}u^3\Big|_2^3 + \frac{4}{3}u\Big|_2^3 - \frac{2^{2/3}}{3}\cdot\ln\frac{7}{2}$$

$$\frac{1}{(u + \sqrt{2})(u - \sqrt{2})} = \frac{A}{u - \sqrt{2}} + \frac{B}{u + \sqrt{2}}$$

$$1 = A(u + \sqrt{2})B(u - \sqrt{2})$$

$$u = -\sqrt{2}, 1 = 0 - 2\sqrt{2}B, B = \frac{-1}{2\sqrt{2}}$$

$$u = \sqrt{2}, 1 = A(2\sqrt{2}), A = \frac{1}{2\sqrt{2}}$$

$$= \frac{8}{3}\int_2^3 \frac{\frac{1}{2\sqrt{2}}}{u - \sqrt{2}}du - \frac{8}{3}\int_2^3 \frac{\frac{1}{2\sqrt{2}}}{u + \sqrt{2}}du + \frac{2}{9}(3^3 - 2^3) + \frac{4}{3}(3 - 2) - \frac{2^{3/2}}{3}\ln\frac{7}{2}$$

$$= \frac{4}{3\sqrt{2}}\int_2^3 \frac{du}{u - \sqrt{2}} - \frac{4}{3\sqrt{2}}\int_2^3 \frac{du}{u + \sqrt{2}} + \frac{38}{9} + \frac{4}{3} - \frac{2^{3/2}}{3}\ln\frac{7}{2}$$

$$= \frac{4}{3\sqrt{2}}\left[\ln(u - \sqrt{2})\Big|_2^3 - \ln(u + \sqrt{2})\Big|_2^3\right] + \frac{50}{9} - \frac{2^{3/2}}{3}\ln\frac{7}{2}$$

$$= \frac{4}{3\sqrt{2}}\left[\ln\frac{3 - \sqrt{2}}{2 - \sqrt{2}} - \ln\frac{3 + \sqrt{2}}{2 + \sqrt{2}}\right] + \frac{50}{9} - \frac{2^{3/2}}{3}\ln\frac{7}{2} = \frac{50}{9} + \frac{4}{3\sqrt{2}}\ln\frac{\frac{3 - \sqrt{2}}{2 - \sqrt{2}}}{\frac{3 + \sqrt{2}}{2 + \sqrt{2}}} - \frac{2\sqrt{2}}{3}\ln\frac{7}{2}$$

$$= \frac{50}{9} + \frac{4}{3\sqrt{2}}\ln\left[\frac{(3 - \sqrt{2})(2 + \sqrt{2})}{(2 - \sqrt{2})(3 + \sqrt{2})}\right] - \frac{2\sqrt{2}}{3}\ln\frac{7}{2} = \frac{50}{9} + \frac{2\sqrt{2}}{3}\ln\frac{6 + \sqrt{2} - 2}{6 - \sqrt{2} - 2} - \frac{2\sqrt{2}}{3}\ln\frac{7}{2}$$

$$= \frac{50}{9} + \frac{2\sqrt{2}}{3}\left[\ln\frac{4 + \sqrt{2}}{4 - \sqrt{2}} - \ln\frac{7}{2}\right] = \frac{50}{9} + \frac{2\sqrt{2}}{3}\ln\frac{(4 + \sqrt{2})(2)}{(4 - \sqrt{2})(7)} = \frac{50}{9} + \frac{2\sqrt{2}}{3}\ln\frac{8 + 2\sqrt{2}}{28 - 7\sqrt{2}} = 5.071173245$$

$$\int_2^7 \int_0^1 x\sqrt{2 + x^2y}dx\,dy = 5.071$$

15. $\displaystyle\int_0^3\int_1^x (x+2y)\,dx\,dy = \int_0^3 (xy+y^2)\Big|_1^x\,dx = \int_0^3 (x^2+x^2-x-1)\,dx = \int_0^3 (2x^2-x-1)\,dx$

$$= \frac{2}{3}x^3 - \frac{1}{2}x^2 - x\Big|_0^3 = \frac{2}{3}(27) - \frac{1}{2}(9) - 3 - 0 = \frac{21}{2}$$

16. $\displaystyle\int_1^2\int_0^{\pi/4} r\sec^2\theta\,d\theta\,dr = \int_1^2 r\tan\theta\Big|_0^{\pi/4}\,dr = \int_1^2 r\left(\tan\frac{\pi}{4} - \tan\theta\right)dr = \int_1^2 r\,dr = \frac{r^2}{2}\Big|_1^2$

$$= \frac{2^2}{2} - \frac{1^2}{2} = \frac{3}{2}$$

17. $\displaystyle\int_0^1\int_0^{2x} x^2 e^{xy}\,dy\,dx = \int_0^1\int_0^{2x} x(e^{xy}x\,dy)dx = \int_0^1 xe^{xy}\Big|_0^{2x}\,dx = \int_0^1 x(e^{2x^2}-1)\,dx$

$$= \frac{1}{4}e^{2x^2} - \frac{1}{2}x^2\Big|_0^1 = \frac{1}{4}e^2 - \frac{1}{2} - \frac{1}{4} = \frac{1}{4}(e^2-3) = 1.097$$

18. $\displaystyle\int_{\pi/4}^{\pi/2}\int_1^{\sqrt{\cos\theta}} r\sin\theta\,dr\,d\theta = \int_{\pi/4}^{\pi/2} \sin\theta\frac{r^2}{2}\Big|_1^{\sqrt{\cos\theta}}\,d\theta = \int_{\pi/4}^{\pi/2}\sin\theta\left(\frac{\sqrt{\cos\theta}^2}{2} - \frac{1^2}{2}\right)d\theta$

$$= \int_{\pi/4}^{\pi/2}\frac{1}{2}(\sin\theta\cos\theta - \sin\theta)d\theta = \left(\frac{1}{4}\sin^2 + \frac{1}{2}\cos\theta\right)\Big|_{\pi/4}^{\pi/2}$$

$$= \frac{1}{4}\sin^2\frac{\pi}{2} + \frac{1}{2}\cos\frac{\pi}{2} - \frac{1}{4}\sin^2\frac{\pi}{4} - \frac{1}{2}\cos\frac{\pi}{4}$$

$$= \frac{1}{4} + \frac{1}{2}(0) - \frac{1}{4}\left(\frac{\sqrt{2}}{2}\right)^2 - \frac{1}{2}\left(\frac{\sqrt{2}}{2}\right) = \frac{1}{8} - \frac{\sqrt{2}}{4} = -0.23$$

19. $\displaystyle\int_1^e\int_1^x \frac{\ln y}{xy}\,dy\,dx = \int_1^e\int_1^x \frac{1}{x}\left[\ln y\left(\frac{dy}{y}\right)\right]dx = \int_1^e \frac{1}{x}\left(\frac{\ln^2 y}{2}\right)\Big|_1^x\,dx = \int_1^e \frac{\ln^2 x}{2x}\,dx$

$$= \frac{1}{6}\ln^3 x\Big|_1^e = \frac{1}{6}\ln^3 e - 0 = \frac{1}{6}$$

20. $\displaystyle\int_1^3\int_1^x \frac{2}{x^2+y^2}\,dy\,dx = \int_1^3 \frac{2}{x}\tan^{-1}\frac{y}{x}\Big|_0^x\,dx = \int_1^3 \frac{2}{x}[\tan^{-1}1 - \tan^{-1}0]dx = \int_1^3 \frac{2}{x}\frac{\pi}{4}\,dx = \frac{\pi}{2}\ln x\Big|_1^3$

$$= \frac{\pi}{2}[\ln 3 - \ln 1] = \frac{\pi}{2}\ln 3 = 1.73$$

21. $z = \sqrt{x^2+4y^2}$

<u>Intercepts:</u> $(0,0,0)$

<u>Traces:</u>

$z \geq 0$ for all x and y
(defined by positive square root)
In yz-plane: $z = \pm 2y$
In xz-plane: $z = \pm x$
In xy-plane: $(0,0,0)$

<u>Section:</u> For $z = 2$

$x^2 + 4y^2 = 4$ (ellipse)

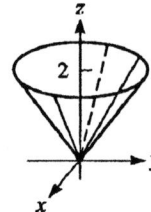

22. $z = \sqrt{x^2 + 4y^2}$

Parallel to yz-plane means x is constant.

$$\frac{\partial z}{\partial y} = \frac{8y}{2\sqrt{x^2 + 4y^2}}\bigg|_{(2,1,2\sqrt{2})}$$

$$= \frac{4(1)}{\sqrt{2^2 + 4(1)^2}}$$

$$= \frac{4}{2\sqrt{2}}$$

$$z - 2\sqrt{2} = \frac{4}{2\sqrt{2}}(y - 1)$$

$$2\sqrt{2}z - 8 = 4y - 4$$

$$2y - \sqrt{2}z + 2 = 0$$

23. $z = \sqrt{x^2 + 4y^2}$

$$dz = \frac{1}{2}(x^2 + 4y^2)^{-1/2}(2x)dx$$

$$+ \frac{1}{2}(x^2 + 4y^2)^{-1/2}(8y)dy$$

$$= \frac{x\,dx + 4y\,dy}{\sqrt{x^2 + 4y^2}}$$

For $x = 2.00, dx = 0.04, y = 1.00, dy = 0.06$

$$dz = \frac{2.00(0.04) + 4(1.00)(0.06)}{\sqrt{2.00^2 + 4(1.00)^2}} = 0.113$$

24. $z = \sqrt{x^2 + 4y^2}$

$$dz = \frac{\partial z}{\partial x}\,dx + \frac{\partial y}{\partial y}\,dy = \frac{x\,dx}{\sqrt{x^2 + 4y^2}} + \frac{4y\,dy}{\sqrt{x^2 + 4y^2}}$$

$$dz = \frac{x\,dx + 4y\,dy}{\sqrt{x^2 + 4y^2}}$$

25. $z = e^{x+y}$

Parallel to xz-plane means y is constant.

$$\frac{\partial z}{\partial x} = e^{x+y}, \frac{\partial z}{\partial x}\bigg|_{(1,1,e^2)} = e^2$$

$$z - e^2 = e^2(x - 1)$$

$$z = e^2 x$$

26. $z = e^{x+y}$

Intercepts: $(0, 0, 1)$

Traces:

$z > 0$ for all x and y since z is defined by an exponential.

In xy-plane: $z = e^x$

In yz-plane: $z = e^y$

Section: $z = 2 = e^{x+y}$

$\ln 2 = x + y$ (a line)

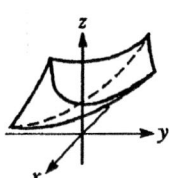

27. $V = \displaystyle\int_0^1 \int_0^x z\,dy\,dx$

$$= \int_0^1 \int_0^x e^{x+y}\,dy\,dx$$

$$= \int_0^1 e^{x+y}\bigg|_0^x dx$$

$$= \int_0^1 (e^{2x} - e^x)\,dx$$

$$= \frac{1}{2}e^{2x} - e^x \bigg|_0^1$$

$$= \frac{1}{2}e^2 - e - \frac{1}{2} + 1$$

$$= 1.48$$

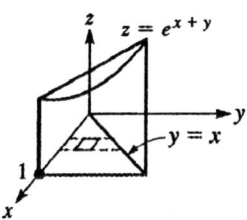

28. $z = e^{x+2y}$

$$dz = \frac{\partial z}{\partial x}\,dx + \frac{\partial z}{\partial y}\,dy = e^{x+y}dx + e^{x+y}dy$$

$$dz = e^{1.00+1.00}(1.01 - 1.00)$$

$$+ e^{1.00+1.00}(1.02 - 1.00)$$

$$dz = 0.222$$

29. $v = \dfrac{rE}{r + R}$

$$\frac{\partial v}{\partial r} = \frac{(r + R)E - rE(1)}{(r + R)^2} = \frac{ER}{(r + R)^2}$$

$$\frac{\partial v}{\partial R} = -rE(r + R)^{-2}(1) = \frac{-rE}{(r + R)^2}$$

30. $V = a + \dfrac{bT}{p} - \dfrac{c}{T^2}$

$\beta = \dfrac{1}{V}\dfrac{\partial V}{\partial T} = \dfrac{1}{V}\left(\dfrac{b}{p} + \dfrac{2c}{T^3}\right) = \dfrac{1}{V}\dfrac{bT^3 + 2cp}{pT^3}$

$\beta = \dfrac{1}{a + \frac{bT}{p} - \frac{c}{T^2}}\dfrac{bT^3 + 2cp}{pT^3}$

$\beta = \dfrac{bT^3 + 2cp}{apT^3 + bT^4 - pcT} = \dfrac{bT^3 + 2cp}{T(apT^2 + bT^3 - pc)}$

31. $i_c = i_e(1 - e^{-2v_c})$

$\alpha = \dfrac{\partial i_c}{\partial i_e} = 1 - e^{-2v_c}$

For $v_c = 2$ V, $\alpha = 1 - e^{-4} = 0.982$

32. $F = \dfrac{-0.0100T}{L^2}, \dfrac{\partial F}{\partial L} = \dfrac{0.0100(2T)}{L^3}$

$Y = \dfrac{L}{A}\dfrac{\partial F}{\partial L} = \dfrac{1.10}{1.00 \times 10^{-6}} = \dfrac{0.0100(2)(300)}{(1.10)^3}$

$Y = 4.96 \times 10^6$ Pa

33. $T = 2\pi\sqrt{\dfrac{\ell}{g}}$

$\dfrac{\partial T}{\partial \ell} = 2\pi\sqrt{\dfrac{1}{g}}\left(\dfrac{1}{2}\ell^{-1/2}\right) = \dfrac{\pi}{\sqrt{g\ell}}$

$\dfrac{T}{2\ell} = \dfrac{2\pi\sqrt{\ell/g}}{2\ell} = \dfrac{\pi}{\sqrt{g\ell}}$

$\dfrac{\partial T}{\partial \ell} = \dfrac{T}{2\ell}$

34. $V = \dfrac{1}{3}\pi r^2 h$

$\dfrac{\partial V}{\partial r} = \dfrac{2}{3}\pi r h\dfrac{r}{r} = 2\left(\dfrac{1}{3}\pi r^2 h\right)\dfrac{1}{r} = \dfrac{2V}{r}$

35. $q = \dfrac{pf}{p - f}$

$dq = \dfrac{(p - f)f - pf(1)}{(p - f)^2}dp + \dfrac{(p - f)p - pf(-1)}{(p - f)^2}$

$= \dfrac{-f^2 dp + p^2 df}{(p - f)^2}$

For $f = 20$ cm, $df = 1$ cm, $p = 100$ cm, $dp = 5$ cm

$dq = \dfrac{-(20^2)(5) + (100^2)(1)}{(100 - 20)^2} = 1.25$ cm

36. $Z = \sqrt{R^2 + X^2}$

$dZ = \dfrac{\partial Z}{\partial R}dR + \dfrac{\partial Z}{\partial X}dX$

$dZ = \dfrac{R}{\sqrt{R^2 + X^2}}dR + \dfrac{X}{\sqrt{R^2 + X^2}}dX$

$dZ = \dfrac{6}{\sqrt{6^2 + 8^2}}(6(0.02)) + \dfrac{8}{\sqrt{6^2 + 8^2}}(8(0.02))$

$dZ = 0.2$

The maximum possible error in the impedance is 2%.

37. $T = 2xy - y^2 - 2x^2 + 3y + 2$

$\dfrac{\partial T}{\partial x} = 2y - 4x, \dfrac{\partial T}{\partial y} = 2x - 2y + 3$

$2y - 4x = 0, 2x - 2y + 3 = 0$

$y = 2x, 2x - 2(2x) + 3 = 0$

$x = \dfrac{3}{2}, y = 3$

$\dfrac{\partial^2 T}{\partial x^2} = -4, \dfrac{\partial^2 T}{\partial y^2} = -2$

$\dfrac{\partial^2 T}{\partial y \partial x} = 2; D = -4(-2) - 2^2 = 4$

$D > 0, \dfrac{\partial^2 T}{\partial x^2} < 0$, Max. at $\left(\dfrac{3}{2}, 3\right)$

38. $V = xyz = 1000, z = \dfrac{1000}{xy}$

$S = 2xy + 2xz + 2yz = 2xy + \dfrac{2000}{y} + \dfrac{2000}{x}$

$\dfrac{\partial S}{\partial x} = 2y - \dfrac{2000}{x^2} = 0; \qquad \dfrac{\partial S}{\partial y} = 2x - \dfrac{2000}{y^2} = 0$

$y = \dfrac{1000}{x^2} \qquad x = \dfrac{1000}{y^2}$

$x = \dfrac{1000}{\left(\frac{1000}{x^2}\right)} = \dfrac{x^4}{1000}$

$x^4 - 1000x = 0$

$x(x^3 - 1000) = 0$

$x^3 - 1000 = 0$

$x^3 = 1000$

$x = 10$

$y = \dfrac{1000}{10^2} = 10$

$z = \dfrac{1000}{10(10)} = 10$

$$\frac{\partial^2 S}{\partial x^2} = \frac{4000}{x^3}; \frac{\partial^2 S}{\partial y^2} = \frac{4000}{y^3}$$

$$\frac{\partial S}{\partial y \partial x} = \frac{\partial S}{\partial x \partial y} = 2$$

$$D = \frac{\partial^2 S}{\partial x^2} \frac{\partial^2 S}{\partial y^2} - \left(\frac{\partial S}{\partial x \partial y}\right)^2 = \frac{4000}{10^3} \frac{4000}{10^3} - 2^3 = 8 > 0$$

$$\frac{\partial^2 S}{\partial x^2} = \frac{4000}{10^3} = 4 > 0$$

Minimum for $x = y = z = 2$ in., a cube.

39. $\displaystyle V = \int_0^2 \int_0^{4-x^2} z \, dy \, dx = \int_0^2 \int_0^{4-x^2} (6 - x - y) dy \, dx = \int_0^2 (6y - xy - \frac{1}{2}y^2)\Big|_0^{4-x^2} dx$

$\displaystyle = \int_0^2 \left[6(4 - x^2) - x(4 - x^2) - \frac{1}{2}(4 - x^2)^2\right] dx = \int_0^2 \left(16 - 4x - 2x^2 + x^3 - \frac{1}{2}x^4\right) dx$

$\displaystyle = 16x - 2x^2 - \frac{2}{3}x^3 + \frac{1}{4}x^4 - \frac{1}{10}x^5 \Big|_0^2 = 32 - 8 - \frac{16}{3} + 4 - \frac{32}{10} - 0 = 28 - \frac{16}{3} - \frac{16}{5} = \frac{292}{15}$

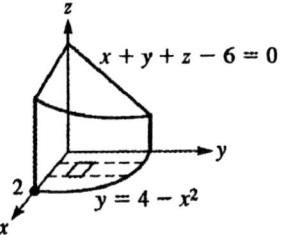

40. $\displaystyle V = \int_0^1 \int_0^1 (4 - x) dx \, dy = \int_0^1 \left(4x - \frac{x^2}{2}\right)\Big|_0^1 dy = \int_0^1 \left(4 - \frac{1}{2}\right) dy = \frac{7}{2}y\Big|_0^1 = \frac{7}{2}$

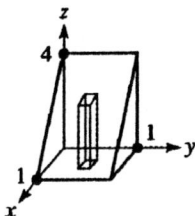

41. $\displaystyle V = \int_0^4 \int_0^{\sqrt{16-x^2}} z \, dy \, dx = \int_0^4 \int_0^{\sqrt{16-x^2}} (8 - x) dy \, dx = \int_0^4 (8 - x)y \Big|_0^{\sqrt{16-x^2}} dx = \int_0^4 (8 - x)\sqrt{16 - x^2} dx$

$\displaystyle = 8\int_0^4 \sqrt{16 - x^2} dx - \int_0^4 x\sqrt{16 - x^2} dx = 8\left[\frac{x}{2}\sqrt{16 - x^2} + \frac{16}{2}\sin^{-1}\frac{x}{4}\right] + \frac{1}{3}(16 - x^2)^{3/2}\Big|_0^4$

$\displaystyle = 8[2(0) + 8\sin^{-1}1] + \frac{1}{3}(0) - 8[0 + 8\sin^{-1}0] - \frac{1}{3}(16)^{3/2} = 64\sin^{-1}1 - \frac{64}{3} = 64\left(\frac{\pi}{2}\right) - \frac{64}{3} = 32\left(\pi - \frac{2}{3}\right) = 79.2$

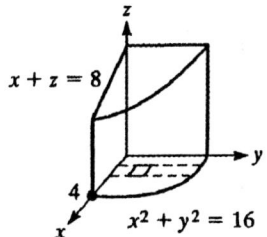

42. $V = \int_0^2 \int_0^{2-x} (4 - y^2) dy\, dx = \int_0^2 \left(4y - \frac{y^3}{3}\right)\Big|_0^{2-x} dx = \int_0^2 \left(4(2-x) - \frac{(2-x)^3}{3}\right) dx$

$V = \int_0^2 \left(\frac{16}{3} - 2x^2 + \frac{x^3}{8}\right) dx = \frac{16}{3}x - \frac{2}{3}x^3 + \frac{x^4}{12}\Big|_0^2 = \frac{16}{3}(2) - \frac{2}{3}(2)^3 + \frac{2^4}{12} = \frac{20}{3}$

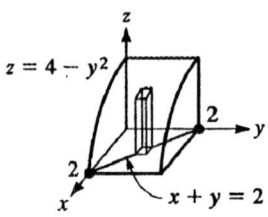

43. $A = \int_0^2 \int_0^{4-y} dx\, dy = \int_0^2 x\Big|_0^{4-y} dy = \int_0^2 (4-y) dy = 4y - \frac{1}{2}y^2\Big|_0^2 = 6$

$A\overline{x} = \int_0^2 \int_0^{4-y} x\, dx\, dy = \frac{1}{2}\int_0^2 x^2\Big|_0^{4-y} dy = \frac{1}{2}\int_0^2 (16 - 8y + y^2) dy = 8y - 2y^2 + \frac{1}{6}y^3\Big|_0^2 = 16 - 8 + \frac{8}{6} = \frac{28}{3}$

$\overline{x} = \frac{28}{3}\left(\frac{1}{6}\right) = \frac{14}{9}$

$A\overline{y} = \int_0^2 \int_0^{4-y} y\, dy\, dx = \int_0^2 yx\Big|_0^{4-y} dy = \int_0^2 (4y - y^2) dy = 2y^2 - \frac{1}{3}y^3\Big|_0^2 = 8 - \frac{8}{3} = \frac{16}{3}; \ y = \frac{16}{3}\left(\frac{1}{6}\right) = \frac{8}{9}$

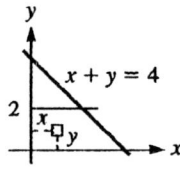

44. $A = \int_0^3 \int_x^{4x-x^2} dy\, dx = \int_0^3 y\Big|_x^{4x-x^2} dx = \int_0^3 (4x - x^2 - x) dx = \int_0^3 (3x - x^2) dx = \frac{3}{2}x^2 - \frac{x^3}{3}\Big|_0^3 = \frac{9}{2}$

$A\overline{x} = \int_0^3 \int_0^{4x-x^2} x\, dy\, dx = \int_0^3 xy\Big|_x^{4x-x^2} dx$

$\frac{9}{2}\overline{x} = \int_0^3 [x(4x - x^2) - x^2] dx = \int_0^3 (3x^2 - x^3) dx$

$\frac{9}{2}\overline{x} = x^3 - \frac{x^4}{4}\Big|_0^3 = 3^3 - \frac{3^4}{4} = \frac{27}{4}$

$\overline{x} = \frac{3}{2}$

$$A\bar{y} = \int_0^3 \int_x^{4x-x^2} y\, dy\, dx = \int_0^3 \frac{y^2}{2}\Big|_x^{4x-x^2} dx$$

$$\frac{9}{2}\bar{y} = \int_0^3 \frac{1}{2}[(4x-x^2)^2 - x^2]dx$$

$$9\bar{y} = \int_0^3 (15x^2 - 8x^3 + x^4)dx$$

$$= 5x^3 - 2x^4 + \frac{x^5}{5}\Big|_0^3$$

$$9\bar{y} = 5(3)^3 - 2(3)^4 + \frac{3^5}{5} = \frac{108}{5}$$

$$\bar{y} = \frac{12}{5}, (\bar{x}, \bar{y}) = \left(\frac{3}{2}, \frac{12}{5}\right)$$

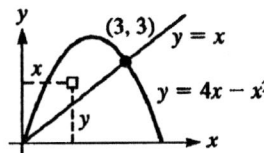

45. See Figure for #43

$$I_y = \int_0^2 \int_0^{4-y} x^2 dx\, dy = \frac{1}{3}\int_0^2 x^3\Big|_0^{4-y} dy$$

$$= \frac{1}{3}\int_0^2 (4-y)^3 dy = -\frac{1}{12}(4-y)^4\Big|_0^2$$

$$= -\frac{1}{12}(2^4 - 4^4) = 20$$

$$R_y^2 = \frac{20}{6} = \frac{10}{3},$$

$$R_y = 1.83$$

46. See Figure for #44

$$I_x = \int_0^3 \int_x^{4x-x^2} y^2 dy\, dx$$

$$I_x = \int_0^3 \frac{y^3}{3}\Big|_x^{4x-x^2} dx$$

$$= \int_0^3 \frac{1}{3}((4x-x^2)^3 - x^3)dx$$

$$I_x = \int_0^3 \left(-\frac{x^6}{3} + 4x^5 - 16x^4 + 21x^3)dx\right)$$

$$I_x = -\frac{x^7}{21} + \frac{2x^6}{3} - \frac{16x^5}{5} + \frac{21x^4}{4}\Big|_0^3$$

$$I_x = -\frac{3^7}{21} + \frac{2(3^6)}{3} - \frac{16(3^5)}{5} + \frac{21(3^4)}{4} = \frac{4131}{140}$$

$$AR_x^2 = I_x$$

$$R_x^2 = \frac{I_x}{A}$$

$$R_x^2 = \frac{\frac{4131}{140}}{\frac{9}{2}}$$

$$R_x^2 = \frac{459}{70}$$

$$R_x = 2.56$$

47. $P = \dfrac{V^2}{R}$

$$\frac{dP}{dt} = \frac{2V}{R}\frac{dV}{dt} - \frac{V^2}{R^2}\frac{dR}{dt}$$

For $V = 60$ V, $\dfrac{dV}{dt} = 1$ V/min, $R = 20\ \Omega$,

$$\frac{dR}{dt} = 2\ \Omega/\text{min}$$

$$\frac{dP}{dt} = \frac{2(60)}{20}(1) - \frac{60^2}{20^2}(2) = -12\ \text{W/min}$$

48. $E = \dfrac{1}{2}mv^2$

$$\frac{dE}{dt} = \frac{\partial E}{\partial m}\frac{dm}{dt} + \frac{\partial E}{\partial v}\frac{dv}{dt}$$

$$= \frac{1}{2}v^2\frac{dm}{dt} + mv\frac{dv}{dt}$$

$$= \frac{1}{2}(100)^2(0.0100) + (5.00)(100)(5.00)$$

$$= 2550\ \text{J/s}$$

49. From the drawing

$$\int_0^4 \int_0^{\sqrt{y}} f(x,y)dy\, dx = \int_0^2 \int_{x^2}^4 f(x,y)dy\, dx$$

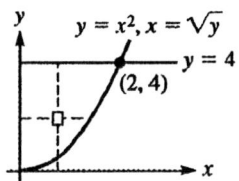

POLAR AND CYLINDRICAL COORDINATES

12.1 Polar Coordinates

1. $\left(3, \frac{\pi}{6}\right)$; $r = 3$, $\theta = \frac{\pi}{6}$

2. $(2, \pi)$; $r = 2$, $\theta = \pi$

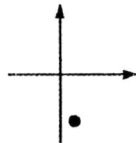

3. $\left(\frac{5}{2}, -\frac{2\pi}{5}\right)$; $r = \frac{5}{2}$, $\theta = -\frac{2\pi}{3}$

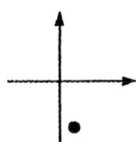

4. $\left(5 - \frac{\pi}{3}\right)$; $r = 5$, $\theta = -\frac{\pi}{3}$

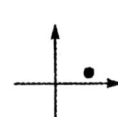

5. $\left(-2, \frac{7\pi}{6}\right)$; negative r is reversed in direction from positive r.

6. $\left(-5, \frac{\pi}{4}\right)$

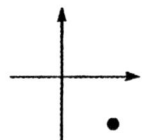

7. $\left(-3, -\frac{5\pi}{4}\right)$

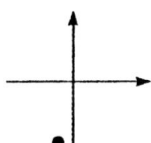

8. $\left(-4, -\frac{5\pi}{3}\right)$

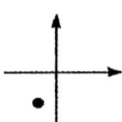

9. $\left(0.5, -\frac{8\pi}{3}\right)$

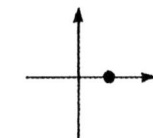

10. $(2.2, -6\pi)$

11. $(2,2)$; $\dfrac{\pi}{180°} = \dfrac{2}{\theta}$; $\theta = \dfrac{360}{\pi} = 114.6°$

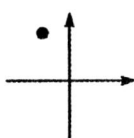

12. $(-1,-1)$; $\dfrac{\pi}{180°} = \dfrac{-1}{\theta}$; $\theta = \dfrac{-180}{\pi} = -57.3°$

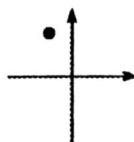

13. $(\sqrt{3},1)$ is (x,y), quadrant I

$\tan\theta = \dfrac{y}{x}$

$\theta = \tan^{-1}\dfrac{y}{x} = \tan^{-1}\dfrac{1}{\sqrt{3}} = \tan^{-1}\dfrac{\sqrt{3}}{3}$;

$\theta = 30° = \dfrac{\pi}{6}$

$r = \sqrt{x^2 + y^2} = \sqrt{(\sqrt{3})^2 + 1^2}$
$= \sqrt{3+1} = \sqrt{4} = 2$

(r,θ) is $\left(2, \dfrac{\pi}{6}\right)$

14. $(-1,-1)$ is (x,y), quadrant III
$\alpha = 45°$, therefore $\theta = 180° + 45° = 225°$

$\alpha = \dfrac{\pi}{4}$, therefore $\theta = \pi + \dfrac{\pi}{4} = \dfrac{5\pi}{4}$

$r = \sqrt{1^2 + 1^2} = \sqrt{2}$

Therefore, $(-1,-1)$ is $\left(\sqrt{2}, \dfrac{5\pi}{4}\right)$

15. $\left(-\dfrac{\sqrt{3}}{2}, -\dfrac{1}{2}\right)$, quadrant III

$\alpha = 30°$
Therefore, $\theta = 180° + 30° = 210°$

$\theta = \pi + \dfrac{\pi}{6} = \dfrac{7\pi}{6}$

$r = \sqrt{\left(\dfrac{\sqrt{3}}{2}\right)^2 + \left(\dfrac{1}{2}\right)^2} = \sqrt{\dfrac{3}{4} + \dfrac{1}{4}} = 1$

Therefore, $\left(-\dfrac{\sqrt{3}}{2}, -\dfrac{1}{2}\right)$ is $\left(1, \dfrac{7\pi}{6}\right)$

16. $(-5,4)$, quadrant II

$\alpha = \arctan\dfrac{4}{5} = 38.7° = 0.675$ rad
$\theta = 180° - 38.7° = 141.3° = 2.47$
$r = \sqrt{4^2 + 5^2} = \sqrt{41} = 6.40$
Therefore, $(-5,4)$ is $(6.40, 2.47)$

17. (r,θ) is $\left(8, \dfrac{4\pi}{3}\right)$, quadrant III

$x = r\cos\theta = 8\cos\dfrac{4\pi}{3} = 8\left(-\dfrac{1}{2}\right) = -4$

$y = r\sin\theta = 8\left(-\dfrac{\sqrt{3}}{2}\right) = -4\sqrt{3}$

(x,y) is $\left(-4, -4\sqrt{3}\right)$

18. $(-4,-\pi)$ is $(4,0)$
x-axis

19. $(3.0, -0.40)$, quadrant IV
$0.40 = 22.9°$
$x = 3.0\cos 0.4 = 2.76$
$y = -3.0\sin 0.4 = -1.17$
Therefore, $(3.0 - 0.40)$ is $(2.76, -1.17)$

20. $(-1.0, 1.0)$, quadrant III
$x = -1\cos 1.0 = -0.540$
$y = --1\sin 1.0 = -0.841$
Therefore, $(-1.0, 1.0)$ is $(-0.54, -0.84)$

21. $x = 3$

$r\cos\theta = x = 3$; $r = \dfrac{3}{\cos\theta} = 3\sec\theta$

22. $y = x$

$r\sin\theta = r\cos\theta$
$\tan\theta = 1$

$\theta = \dfrac{\pi}{4}$

23. $x + 2y = 3$
$r\cos\theta + 2r\sin\theta = 3$

$r = \dfrac{3}{\cos\theta + 2\sin\theta}$

24. $x^2 + y^2 = 0.81$; $r = 0.9$; circle, $C(0,0)$
$x = 0.9\cos\theta$; $y = 0.9\sin\theta$
$(r\cos\theta)^2 + (r\sin\theta)^2 = 0.81$;
$r^2\cos^2\theta + r^2\sin^2\theta = 0.81$
$r^2(\cos^2\theta + \sin^2\theta) = 0.81$;
$r^2(1) = 0.81$; $r = 0.9$

25. $x^2 + (y-2)^2 = 4$
$$x^2 + y^2 - 4y + 4 = 4$$
$$r^2 - 4 \cdot r \sin\theta = 0$$
$$r = 4\sin\theta$$

26. $x^2 - y^2 = 0.01$
$$(r\cos\theta)^2 - (r\sin\theta)^2 = 0.01;$$
$$r^2\cos^2\theta - r^2\sin^2\theta = 0.01$$
$$r^2(\cos^2\theta - \sin^2) = 0.01$$
$$r^2 = \frac{0.01}{\cos^2\theta - \sin^2\theta}$$
$$= \frac{0.01}{\cos 2\theta} = 0.01\sec 2\theta$$

27. $x^2 + 4y^2 = 4$
$$(r\cos\theta)^2 + 4(r\sin\theta)^2 = 4;\ r^2\cos^2\theta + 4r^2\sin^2\theta = 4$$
$$r^2(\cos^2\theta + 4\sin^2\theta) = 4;$$
$$r^2 = \frac{4}{\cos^2\theta + 4\sin^2\theta}$$
$$= \frac{4}{1 - \sin^2\theta + 4\sin^2\theta}$$
$$= \frac{4}{1 + 3\sin^2\theta}$$

28. $y^2 = 4x$
$$r^2\sin^2\theta = 4r\cos\theta$$
$$r^2\sin^2\theta - 4r\cos\theta = 0$$
$$r(r\sin^2\theta - 4\cos\theta) = 0$$
$$r = 0, r = \frac{4\cos\theta}{\sin^2\theta}$$

29. $r = \sin\theta;\ r^2 = r\sin\theta;\ r^2 = x^2 + y^2$
$x^2 + y^2 = r^2 = r\sin\theta = y;\ x^2 + y^2 - y = 0$,
circle

30. $r = 4\cos\theta;\ x = r\cos\theta$
$$\sqrt{x^2 + y^2} = 4\frac{x}{r} = \frac{4x}{\sqrt{x^2 + y^2}}$$
Therefore, $x^2 + y^2 = 4x$

31. $r\cos\theta = 4$; therefore $x = 4$, straight line

32. $r\sin\theta = -2;\ y = -2$, straight line

33. $r = \dfrac{2}{\cos\theta - 3\sin\theta}$
$$r\cos\theta - 3r\sin\theta = 2$$
$$x - 3y = 2, \text{ line}$$

34. $r = \csc\theta e^{r\cos\theta} = \dfrac{1}{\sin\theta} \cdot e^{r\cos\theta}$
$$r\sin\theta = e^{r\cos\theta}$$
$$y = e^x, \text{ natural exponential}$$

35. $r = 4\cos\theta + 2\sin\theta = 4 \cdot \dfrac{x}{r} + 2 \cdot \dfrac{y}{r}$
$$r^2 = 4x + 2y$$
$$x^2 + y^2 = 4x + 2y$$
$$x^2 + y^2 - 4x - 2y = 0, \text{ circle}$$

36. $r\sin\left(\theta + \dfrac{\pi}{6}\right) = 3$
$$r\left(\sin\theta\cos\frac{\pi}{6} + \sin\frac{\pi}{6}\cos\theta\right) = 3$$
$$\frac{\sqrt{3}}{2}r\sin\theta + \frac{1}{2}r\cos\theta = 3$$
$$\sqrt{3}y + x = 6, \text{ line}$$

37. $r = 2(1 + \cos\theta);\ x = r\cos\theta;\ \dfrac{x}{r} = \cos\theta$
$$r^2 = x^2 + y^2;\ r = \sqrt{x^2 + y^2}$$
$$r = 2(1+\cos\theta) = 2\left(1+\frac{x}{r}\right) = 2 + \frac{2x}{r};\ r^2 = 2r + 2x$$
Multiply through by r.
$$x^2 + y^2 = 2\sqrt{x^2 + y^2} + 2x;\ x^2 + y^2 - 2x = 2\sqrt{x^2 + y^2}$$
$$(x^2 + y^2 - 2x)^2 = 4(x^2 + y^2)$$
$$x^4 + y^4 - 4x^3 + 2x^2y^2 - 4xy^2 + 4x^2 = 4x^2 + 4y^2$$
$$x^4 + y^4 - 4x^3 + 2x^2y^2 - 4xy^2 + 4x^2 - 4x^2 - 4y^2 = 0$$
$$x^4 + y^4 - 4x^3 + 2x^2y^2 - 4xy^2 - 4y^2 = 0$$

38. $r = 1 - \sin\theta;\ y = r\sin\theta$
$$\sqrt{x^2 + y^2} = 1 - \frac{y}{r} = 1 - \frac{y}{\sqrt{x^2 + y^2}} = \frac{\sqrt{x^2 + y^2} - y}{\sqrt{x^2 + y^2}}$$
Therefore,
$$x^2 + y^2 = \sqrt{x^2 + y^2} - y;\ x^2 + y^2 + y = \sqrt{x^2 + y^2}$$
Square both sides.
$$x^4 + 2x^2y^2 + 2x^2y + y^4 + 2y^3 + y^2 = x^2 + y^2$$
$$x^4 + 2x^2y^2 + 2x^2y + y^4 + 2y^3 - x^2 = 0$$

39. $r^2 = \sin 2\theta = 2\sin\theta\cos\theta;\ x = r\cos\theta;\ y = r\sin\theta$
$$x^2 + y^2 = 2\left(\frac{y}{r}\right)\left(\frac{x}{r}\right) = \frac{2xy}{r^2} = \frac{2xy}{x^2 + y^2}$$
Therefore, $(x^2 + y^2)^2 = 2xy$

40. $r^2 = 16\cos 2\theta = 16(1 - 2\sin^2\theta)$
$$x^2 + y^2 = 16\left[1 - 2\left(\frac{y}{r}\right)^2\right] = 16 - 32\frac{y^2}{r^2}$$
$$x^2 + y^2 = \frac{16r^2 - 32y^2}{r^2} = \frac{16(x^2 + y^2) - 32y^2}{x^2 + y^2}$$
$$(x^2 + y^2)^2 = 16x^2 + 16y^2 - 32y^2$$
$$(x^2 + y^2)^2 = 16x^2 - 16y^2 = 16(x^2 - y^2)$$

41. $B_x = \dfrac{-ky}{x^2 + y^2} = -\dfrac{-ky}{r^2}$

$\qquad = \dfrac{-kr \sin \theta}{r^2} = -\dfrac{k \sin \theta}{r}$

$\quad B_y = \dfrac{kx}{x^2 + y^2} = \dfrac{kx}{r^2}$

$\qquad = \dfrac{kr \cos \theta}{r^2} = \dfrac{k \cos \theta}{r}$

42. $x^2 + \dfrac{y^2}{k^2} = 1;\ (r \cos \theta)^2 + \dfrac{(r \sin \theta)^2}{k^2} = 1$

$\quad k^2 r^2 \cos^2 \theta + r^2 \sin^2 \theta = k^2;$
$\quad r^2 (k^2 \cos^2 \theta + \sin^2 \theta) = k^2$

$\quad r^2 = \dfrac{k^2}{k^2 \cos^2 \theta + \sin^2 \theta}$

43. $r = a \sin \theta + b \cos \theta;\ a > 0, b > 0$

$\quad r = a \dfrac{y}{r} + b \dfrac{x}{r}$

$\quad r^2 = ay + bx$
$\quad x^2 + y^2 - ay - bx = 0$ which represents a circle.

44. $r = \dfrac{4800}{1 + 0.14 \cos \theta};\ \sqrt{x^2 + y^2} = \dfrac{4800}{1 + 0.14 \left(\dfrac{x}{r} \right)}$

$\quad \sqrt{x^2 + y^2} = \dfrac{4800}{\left(\dfrac{r + 0.14x}{r} \right)} = \dfrac{4800r}{r + 0.14x}$

$\quad \sqrt{x^2 + y^2} = \dfrac{4800 \sqrt{x^2 + y^2}}{\sqrt{x^2 + y^2} + 0.14x}$

$\quad 1 = \dfrac{4800}{\left(\sqrt{x^2 + y^2} + 0.14x \right)}$

$\quad \sqrt{x^2 + y^2} + 0.14x = 4800;$
$\quad \sqrt{x^2 + y^2} = 4800 - 0.14x$

Square both sides.

$\quad x^2 + y^2 = 23.04 \times 10^6 - 1344x + 0.0196x^2$
$\quad 0.9804x^2 + y^2 + 1344x - 23.04 \times 10^6 = 0$
$\quad 0.98x^2 + y^2 + 1340x - 2.3 \times 10^7 = 0$

12.2   Curves in Polar Coordinates

1. $r = 4$ for all θ. Graph is a circle with radius 4.

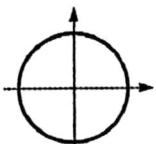

2. $r = 2$ for all θ; circle

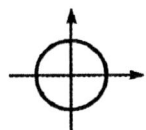

3. $\theta = \dfrac{3\pi}{4}$ for all r; straight line

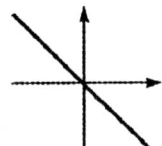

4. $\theta = -1.5;\ \dfrac{\pi}{180} = \dfrac{-1.5}{x};\ x = -85.9°$ for all r;

straight line

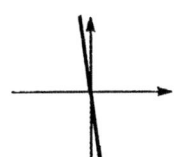

5. $r = 4 \sec \theta = \dfrac{4}{\cos \theta}$

θ	r
0	4
$\dfrac{\pi}{6}$	4.6
$\dfrac{\pi}{4}$	5.7
$\dfrac{\pi}{3}$	8
$\dfrac{\pi}{2}$	*
$\dfrac{2\pi}{3}$	-8
$\dfrac{3\pi}{4}$	-5.7
$\dfrac{5\pi}{6}$	4.6
π	-4
$\dfrac{5\pi}{4}$	-5.7
$\dfrac{3\pi}{2}$	*
$\dfrac{7\pi}{4}$	5.7
2π	4

*denotes undefined

6. $r = 4 \csc \theta = \dfrac{4}{\sin \theta}$; straight line

θ	r
0	undefined
$\frac{\pi}{6}$	8
$\frac{\pi}{4}$	5.66
$\frac{\pi}{3}$	4.62
$\frac{\pi}{2}$	4
$\frac{2\pi}{3}$	4.62
$\frac{3\pi}{4}$	5.66
$\frac{5\pi}{6}$	8
π	undefined
$\frac{5\pi}{4}$	-5.66
$\frac{3\pi}{2}$	-4
$\frac{7\pi}{4}$	-5.66
2π	undefined

7. $r = 2 \sin \theta$; circle centered at $\left(1, \dfrac{\pi}{2}\right)$

θ	0	$\frac{\pi}{4}$	$\frac{\pi}{2}$	$\frac{3\pi}{4}$	π	$\frac{5\pi}{4}$	$\frac{3\pi}{2}$	$\frac{7\pi}{4}$	2π	$\frac{\pi}{6}$	$\frac{\pi}{3}$
r	0	1.41	2	1.41	0	-1.41	-2	-1.41	0	1	1.73

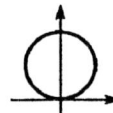

8. $r = 3 \cos \theta$; circle centered at $(1.5, 0)$

θ	0	$\frac{\pi}{6}$	$\frac{\pi}{4}$	$\frac{\pi}{3}$	$\frac{\pi}{2}$	$\frac{3\pi}{4}$	π	$\frac{5\pi}{4}$	$\frac{3\pi}{2}$	$\frac{7\pi}{4}$	2π
r	3	2.60	2.12	1.5	0	-2.12	-3	-2.12	0	2.12	3

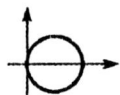

9. $r = 1 - \cos \theta$; cardioid

θ	r
0	0
$\frac{\pi}{4}$	0.3
$\frac{\pi}{2}$	1
$\frac{3\pi}{4}$	1.7
π	2
$\frac{5\pi}{4}$	1.7
$\frac{3\pi}{2}$	1
$\frac{7\pi}{4}$	0.3

10. $r = \sin\theta - 1$; cardoid

θ	r
0	-1
$\frac{\pi}{6}$	-0.5
$\frac{\pi}{4}$	-0.29
$\frac{\pi}{3}$	-0.13
$\frac{\pi}{2}$	0
$\frac{2\pi}{3}$	-0.13
$\frac{3\pi}{4}$	-0.29
$\frac{5\pi}{6}$	-0.5
π	-1
$\frac{7\pi}{6}$	-1.5
$\frac{5\pi}{4}$	-1.71
$\frac{4\pi}{3}$	-1.87
$\frac{3\pi}{2}$	-2
$\frac{5\pi}{3}$	-1.87
$\frac{7\pi}{4}$	-1.71
$\frac{11\pi}{6}$	-1.5
2π	-1

12. $r = 2 + 3\sin\theta$

θ	r	θ	r
0	2	$\frac{13\pi}{12}$	1.22
$\frac{\pi}{12}$	2.78	$\frac{7\pi}{6}$	0.5
$\frac{\pi}{6}$	3.5	$\frac{5\pi}{4}$	-0.12
$\frac{\pi}{4}$	4.12	$\frac{4\pi}{3}$	-0.60
$\frac{\pi}{3}$	4.60	$\frac{17\pi}{12}$	-0.90
$\frac{5\pi}{12}$	4.90	$\frac{3\pi}{2}$	-1
$\frac{\pi}{2}$	5	$\frac{19\pi}{12}$	-0.90
$\frac{7\pi}{12}$	4.90	$\frac{5\pi}{3}$	-0.60
$\frac{2\pi}{3}$	4.60	$\frac{7\pi}{4}$	-0.12
$\frac{3\pi}{4}$	4.12	$\frac{11\pi}{6}$	0.5
$\frac{5\pi}{6}$	3.5	$\frac{23\pi}{12}$	1.22
$\frac{11\pi}{12}$	2.78	2π	2
π	2		

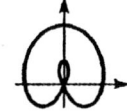

11. $r = 2 - \cos\theta$; limaçon

θ	r
0	1
$\frac{\pi}{6}$	1.13
$\frac{\pi}{4}$	1.29
$\frac{\pi}{3}$	1.5
$\frac{\pi}{2}$	2
$\frac{2\pi}{3}$	2.5
$\frac{3\pi}{4}$	2.71
$\frac{5\pi}{6}$	2.87
π	3
$\frac{7\pi}{6}$	2.87
$\frac{5\pi}{4}$	2.71
$\frac{4\pi}{3}$	2.5
$\frac{3\pi}{2}$	2
$\frac{5\pi}{3}$	1.5
$\frac{7\pi}{4}$	1.29
$\frac{11\pi}{6}$	1.13
2π	1

13. $r = 4\sin 2\theta$; rose (4 petals)

θ	r
0	0
$\frac{\pi}{8}$	2.8
$\frac{\pi}{4}$	4
$\frac{3\pi}{8}$	-2.8
$\frac{\pi}{2}$	0
$\frac{5\pi}{8}$	2.8
$\frac{3\pi}{4}$	-4
$\frac{7\pi}{8}$	-2.8
π	0
$\frac{9\pi}{8}$	2.8
$\frac{5\pi}{4}$	4
$\frac{11\pi}{8}$	2.8
$\frac{3\pi}{2}$	0
$\frac{13\pi}{8}$	-2.8
$\frac{7\pi}{4}$	-4
2π	-2.8

14. $r = 2\sin 3\theta$; rose

θ	r	θ	r
0	0	$\frac{13\pi}{12}$	-1.41
$\frac{\pi}{12}$	1.41	$\frac{7\pi}{6}$	-2
$\frac{\pi}{6}$	2	$\frac{5\pi}{4}$	-1.41
$\frac{\pi}{4}$	1.41	$\frac{4\pi}{3}$	0
$\frac{\pi}{3}$	0	$\frac{17\pi}{12}$	1.41
$\frac{5\pi}{12}$	-1.41	$\frac{3\pi}{2}$	2
$\frac{\pi}{2}$	-2	$\frac{19\pi}{12}$	1.41
$\frac{7\pi}{12}$	-1.41	$\frac{5\pi}{3}$	0
$\frac{2\pi}{3}$	0	$\frac{7\pi}{4}$	-1.41
$\frac{3\pi}{4}$	1.41	$\frac{11\pi}{6}$	-2
$\frac{5\pi}{6}$	2	$\frac{23\pi}{12}$	-1.41
$\frac{11\pi}{12}$	1.41	2π	0
π	0		

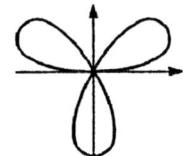

15. $r^2 = 4\sin 2\theta$; $r = \pm\sqrt{4\sin 2\theta}$; leminiscate

θ	r	θ	r
0	0	$\frac{13\pi}{12}$	1.41
$\frac{\pi}{12}$	1.41	$\frac{7\pi}{6}$	1.86
$\frac{\pi}{6}$	1.86	$\frac{5\pi}{4}$	2
$\frac{\pi}{4}$	2	$\frac{4\pi}{3}$	1.86
$\frac{\pi}{3}$	1.86	$\frac{17\pi}{12}$	1.41
$\frac{5\pi}{12}$	1.41	$\frac{3\pi}{2}$	0
$\frac{\pi}{2}$	0	$\frac{19\pi}{12}$	$\sim$
$\frac{7\pi}{12}$	$\sim$	$\frac{5\pi}{3}$	$\sim$
$\frac{2\pi}{3}$	$\sim$	$\frac{7\pi}{4}$	$\sim$
$\frac{3\pi}{4}$	$\sim$	$\frac{11\pi}{6}$	$\sim$
$\frac{5\pi}{6}$	$\sim$	$\frac{23\pi}{12}$	$\sim$
$\frac{11\pi}{12}$	$\sim$	2π	0
π	0		

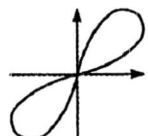

16. $r^2 = 2\sin\theta$; $r = \pm\sqrt{2\sin\theta}$

θ	r	θ	r
0	0	$\frac{7\pi}{8}$	±0.87
$\frac{\pi}{8}$	±0.87	π	0
$\frac{\pi}{4}$	±1.19	$\sim$	$\sim$
$\frac{3\pi}{8}$	±1.36	2π	0
$\frac{\pi}{2}$	±1.41		
$\frac{5\pi}{8}$	±1.36		
$\frac{3\pi}{4}$	±1.19		

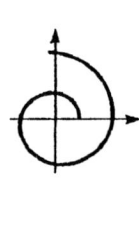

17. $r = 2^\theta$; spiral

θ	r
0	1
$\frac{\pi}{4}$	1.7
$\frac{\pi}{2}$	3.0
$\frac{3\pi}{4}$	5.1
π	8.8
$\frac{5\pi}{4}$	15.2
$\frac{3\pi}{2}$	26.2
$\frac{7\pi}{4}$	45.2
2π	77.9

18. $r = 1.5^{-\theta}$; spiral

θ	r	θ	r
0	1	$\frac{13\pi}{12}$	0.25
$\frac{\pi}{12}$	0.90	$\frac{7\pi}{6}$	0.23
$\frac{\pi}{6}$	0.81	$\frac{5\pi}{4}$	0.20
$\frac{\pi}{4}$	0.73	$\frac{4\pi}{3}$	0.18
$\frac{\pi}{3}$	0.65	$\frac{17\pi}{12}$	0.16
$\frac{5\pi}{12}$	0.59	$\frac{3\pi}{2}$	0.15
$\frac{\pi}{2}$	0.53	$\frac{19\pi}{12}$	0.13
$\frac{7\pi}{12}$	0.48	$\frac{5\pi}{3}$	0.12
$\frac{2\pi}{3}$	0.43	$\frac{7\pi}{4}$	0.11
$\frac{3\pi}{4}$	0.38	$\frac{11\pi}{6}$	0.10
$\frac{5\pi}{6}$	0.35	$\frac{23\pi}{12}$	0.09
$\frac{11\pi}{12}$	0.311	2π	0.08
π	0.28		

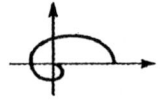

19. $r = 4|\sin 3\theta|$

θ	r
$\pm\frac{\pi}{12}$	2.82
$\pm\frac{\pi}{6}$	4
$\pm\frac{\pi}{4}$	2.82
$\pm\frac{\pi}{3}$	0
$\pm\frac{5\pi}{12}$	2.82
$\pm\frac{\pi}{2}$	4
$\pm\frac{7\pi}{12}$	2.82
$\pm\frac{2\pi}{3}$	0
$\pm\frac{3\pi}{4}$	2.82
$\pm\frac{5\pi}{6}$	4
$\pm\frac{11\pi}{12}$	2.82
$\pm\pi$	0

20. $r = 2\sin\theta\tan\theta$

θ	r
0	0
$\pm\frac{\pi}{6}$	0.58
$\pm\frac{\pi}{4}$	1.41
$\pm\frac{\pi}{3}$	3
$\pm\frac{\pi}{2}$	*

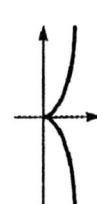

* = undefined

21. $r = \dfrac{1}{2 - \cos\theta}$; ellipse

θ	r
0	1
$\frac{\pi}{4}$	0.77
$\frac{\pi}{2}$	0.50
$\frac{3\pi}{4}$	0.37
π	0.33
$\frac{5\pi}{4}$	0.37
$\frac{3\pi}{2}$	0.50
$\frac{7\pi}{4}$	0.77
2π	1

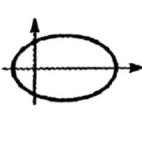

22. $r = \dfrac{1}{1 - \cos\theta}$; parabola

θ	r	θ	r
0	$\sim$	$\frac{7\pi}{6}$	0.54
$\frac{\pi}{6}$	7.46	$\frac{5\pi}{4}$	0.59
$\frac{\pi}{4}$	3.41	$\frac{4\pi}{3}$	0.69
$\frac{\pi}{3}$	2	$\frac{3\pi}{2}$	1
$\frac{\pi}{2}$	1	$\frac{5\pi}{3}$	2
$\frac{2\pi}{3}$	0.67	$\frac{7\pi}{4}$	3.41
$\frac{3\pi}{4}$	0.59	$\frac{11\pi}{6}$	7.46
$\frac{5\pi}{6}$	0.54	2π	$\sim$
2π	0.5		

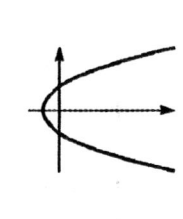

23. $r = \dfrac{6}{1 - 2\cos\theta}$; hyperbola

θ	r	θ	r
0	-6	$\frac{7\pi}{6}$	2.20
$\frac{\pi}{6}$	-8.20	$\frac{5\pi}{4}$	2.49
$\frac{\pi}{4}$	-14.49	$\frac{4\pi}{3}$	3
$\frac{\pi}{3}$	$\sim$	$\frac{3\pi}{2}$	6
$\frac{\pi}{2}$	6	$\frac{5\pi}{3}$	$\sim$
$\frac{2\pi}{3}$	3	$\frac{7\pi}{4}$	-14.49
$\frac{3\pi}{4}$	2.49	$\frac{11\pi}{6}$	-8.20
$\frac{5\pi}{6}$	2.20	2π	-6
π	2		

24. $r = \dfrac{6}{3 - 2\sin\theta}$; ellipse

θ	r	θ	r
0	2	$\frac{7\pi}{6}$	1.5
$\frac{\pi}{6}$	3	$\frac{5\pi}{4}$	1.36
$\frac{\pi}{4}$	3.78	$\frac{4\pi}{3}$	1.27
$\frac{\pi}{3}$	4.73	$\frac{3\pi}{2}$	1.2
$\frac{5\pi}{12}$	5.62		
$\frac{\pi}{2}$	6	$\frac{5\pi}{3}$	1.27
$\frac{7\pi}{12}$	5.62		
$\frac{2\pi}{3}$	4.73	$\frac{7\pi}{4}$	1.36
$\frac{3\pi}{4}$	3.78	$\frac{11\pi}{6}$	1.5
$\frac{5\pi}{6}$	3	2π	2
π	2		

25. $r = 4\cos\frac{1}{2}\theta$

θ	r	θ	r
0	4.0	$\frac{13\pi}{6}$	-3.9
$\frac{\pi}{6}$	3.9	$\frac{9\pi}{4}$	-3.7
$\frac{\pi}{4}$	3.7	$\frac{7\pi}{3}$	-3.5
$\frac{\pi}{3}$	3.5	$\frac{5\pi}{2}$	-2.8
$\frac{\pi}{2}$	2.8	$\frac{8\pi}{3}$	-2.0
$\frac{2\pi}{3}$	2.0	$\frac{11\pi}{4}$	-1.5
$\frac{3\pi}{4}$	1.5	$\frac{17\pi}{6}$	-1.0
$\frac{5\pi}{6}$	1.0	3π	0
π	0	$\frac{19\pi}{6}$	1.0
$\frac{7\pi}{6}$	-1.0	$\frac{13\pi}{4}$	1.5
$\frac{5\pi}{4}$	-1.5	$\frac{10\pi}{3}$	2.0
$\frac{4\pi}{3}$	-2.0	$\frac{7\pi}{2}$	2.8
$\frac{3\pi}{2}$	-2.8	$\frac{11\pi}{3}$	3.5
$\frac{5\pi}{3}$	-3.5	$\frac{15\pi}{4}$	3.7
$\frac{7\pi}{4}$	-3.7	$\frac{23\pi}{6}$	3.9
$\frac{11\pi}{6}$	-3.9	4π	4.0
2π	-4.0		

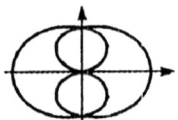

26. $r = 2 + \cos 3\theta$

θ	r	θ	r
0	3	$\frac{13\pi}{12}$	1.29
$\frac{\pi}{12}$	2.71	$\frac{7\pi}{6}$	2
$\frac{\pi}{6}$	2	$\frac{5\pi}{4}$	2.71
$\frac{\pi}{4}$	1.29	$\frac{4\pi}{3}$	3
$\frac{\pi}{3}$	1	$\frac{17\pi}{12}$	2.71
$\frac{5\pi}{12}$	1.29	$\frac{3\pi}{2}$	2
$\frac{\pi}{2}$	2	$\frac{19\pi}{12}$	1.29
$\frac{7\pi}{12}$	2.71	$\frac{5\pi}{3}$	1
$\frac{2\pi}{3}$	3	$\frac{7\pi}{4}$	1.29
$\frac{3\pi}{4}$	2.71	$\frac{11\pi}{6}$	2
$\frac{5\pi}{6}$	2	$\frac{23\pi}{12}$	2.71
$\frac{11\pi}{12}$	1.29	2π	3
π	1		

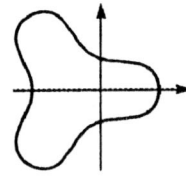

27. $r = 2\left(1 - \sin\left(\theta - \frac{\pi}{4}\right)\right)$

θ	r
0	3.41
$\frac{\pi}{6}$	2.52
$\frac{\pi}{3}$	1.48
$\frac{\pi}{2}$	0.59
$\frac{2\pi}{3}$	0.068
$\frac{5\pi}{6}$	0.068
π	0.59
$\frac{7\pi}{6}$	1.48
$\frac{4\pi}{3}$	2.52
$\frac{3\pi}{2}$	3.41
$\frac{5\pi}{3}$	3.93
$\frac{11\pi}{6}$	3.93
2π	3.41

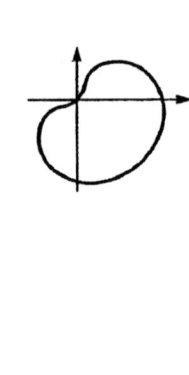

28. $r = 4\tan\theta$

θ	r	θ	r
0	0	$\frac{13\pi}{12}$	1.07
$\frac{\pi}{12}$	1.07	$\frac{7\pi}{6}$	2.31
$\frac{\pi}{6}$	2.31	$\frac{5\pi}{4}$	4
$\frac{\pi}{4}$	4	$\frac{4\pi}{3}$	6.93
$\frac{\pi}{3}$	6.9	$\frac{17\pi}{12}$	14.93
$\frac{5\pi}{12}$	14.93	$\frac{3\pi}{2}$	$\sim$
$\frac{\pi}{2}$	$\sim$	$\frac{19\pi}{12}$	-14.93
$\frac{7\pi}{12}$	-14.93	$\frac{5\pi}{3}$	-6.93
$\frac{2\pi}{3}$	-6.93	$\frac{7\pi}{4}$	-4
$\frac{3\pi}{4}$	-4	$\frac{11\pi}{6}$	-2.31
$\frac{5\pi}{6}$	-2.31	$\frac{23\pi}{12}$	-1.07
$\frac{11\pi}{12}$	-1.07	2π	0
π	0		

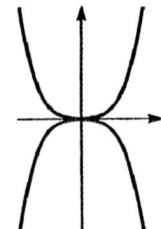

29. $r = \theta$ $(-20 \le \theta \le 20)$

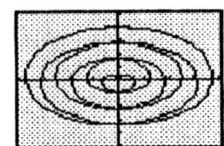

30. $r = 0.5^{\sin\theta}$

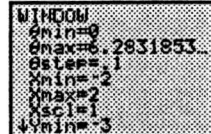
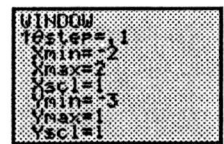

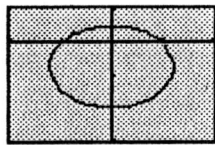

31. $r = 2\sec\theta + 1$

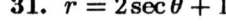
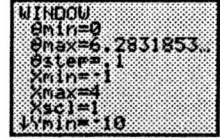
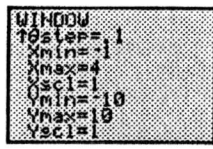

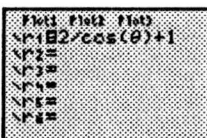
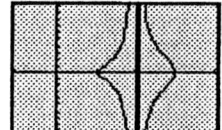

32. $r = 2\cos(\cos 2\theta)$

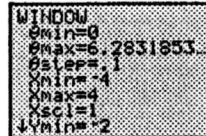
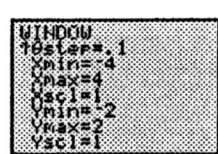

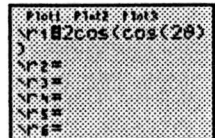
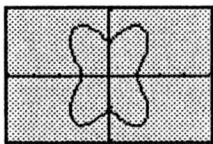

33. $r = 4.0 - \sin\theta$

θ	r
0	4.0
$\frac{\pi}{4}$	3.3
$\frac{\pi}{2}$	3.0
$\frac{3\pi}{4}$	3.3
π	4.0
$\frac{5\pi}{4}$	4.7
$\frac{3\pi}{2}$	5.0
$\frac{7\pi}{4}$	4.7

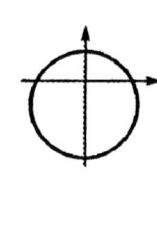

34. $r = \dfrac{70\sin\theta}{(1-\cos\theta)^2}$

$\dfrac{\pi}{4} \le \theta \le \pi$

"Relative pursuit curve"

θ	r
$\frac{\pi}{4}$	577
$\frac{\pi}{3}$	242
$\frac{5\pi}{12}$	123
$\frac{\pi}{2}$	70
$\frac{7\pi}{12}$	43
$\frac{2\pi}{3}$	27
$\frac{3\pi}{4}$	17
$\frac{5\pi}{6}$	10
$\frac{11\pi}{12}$	5
π	0

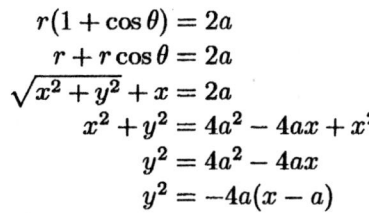

35. $r = a\sec^2\dfrac{\theta}{2} = \dfrac{a}{\cos^2\frac{\theta}{2}}$

$= \dfrac{a}{\frac{1+\cos\theta}{2}} = \dfrac{2a}{1+\cos\theta}$

$r(1+\cos\theta) = 2a$

$r + r\cos\theta = 2a$

$\sqrt{x^2+y^2} + x = 2a$

$x^2 + y^2 = 4a^2 - 4ax + x^2$

$y^2 = 4a^2 - 4ax$

$y^2 = -4a(x-a)$

which is a horizontal parabola with vertex at $(a, 0)$.
For $a = 1$

$$y^2 = -4(x-1)$$

a horizontal parabola opening left with vertex at $(1, 0)$.

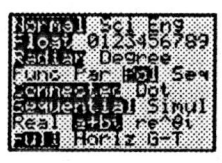
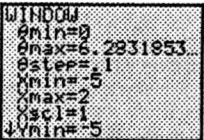

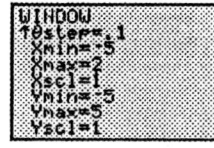
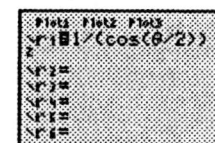

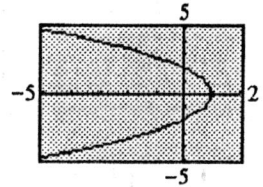

36.

θ	r_1	r_2
0	1	0
$\frac{\pi}{12}$	0.99	0.13
$\frac{\pi}{6}$	0.97	0.26
$\frac{\pi}{4}$	0.92	0.38
$\frac{\pi}{3}$	0.87	0.5
$\frac{5\pi}{12}$	0.79	0.61
$\frac{\pi}{2}$	0.71	0.71
$\frac{7\pi}{12}$	0.61	0.79
$\frac{2\pi}{3}$	0.5	0.87
$\frac{3\pi}{4}$	0.38	0.92
$\frac{5\pi}{6}$	0.26	0.97
$\frac{11\pi}{12}$	0.13	0.99
π	0	1
$\frac{13\pi}{12}$	0.13	0.99
$\frac{7\pi}{6}$	0.26	0.97
$\frac{5\pi}{4}$	0.38	0.92
$\frac{4\pi}{3}$	0.5	0.87
$\frac{17\pi}{12}$	0.61	0.79
$\frac{3\pi}{2}$	0.71	0.71
$\frac{19\pi}{12}$	0.79	0.61
$\frac{5\pi}{3}$	0.87	0.5
$\frac{7\pi}{4}$	0.92	0.38
$\frac{11\pi}{6}$	0.97	0.26
$\frac{23\pi}{12}$	0.99	0.13
2π	1	0

$$4(x^2 + y^2)^3 - 4(x^2 + y^2)^2 + y^2 = 0$$
$$4r^6 - 4r^4 + r^2 \sin^2 \theta = 0$$
$$r^2(4r^4 - 4r^2 + \sin^2 \theta) = 0$$
$$4r^4 - 4r^2 = -\sin^2 \theta$$

Complete the square.

$$r^4 - r^2 + \frac{1}{4} = -\frac{\sin^2 \theta}{4} + \frac{1}{4}$$

$$\left(r^2 - \frac{1}{2}\right)^2 = \frac{1 - \sin^2 \theta}{4} = \frac{\cos^2 \theta}{4}$$

$$r^2 - \frac{1}{2} = \pm \frac{\cos \theta}{2}$$

$$r^2 = \frac{1}{2} \pm \frac{\cos \theta}{2} = \frac{1 \pm \cos \theta}{2}$$

Therefore, $r_1 = \pm \sqrt{\dfrac{1 + \cos \theta}{2}}$;

$$r_2 = \pm \sqrt{\frac{1 - \cos \theta}{2}}$$

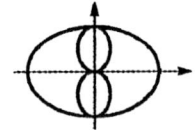

12.3 Applications of Differentiation and Integration in Polar Coordinates

1. $r = 3t,\ \theta = t^3,\ t = 1$

$$v_r = \frac{dr}{dt}, v_\theta = r\frac{d\theta}{dt}$$

$$v_r = 3, v_\theta = (3t)(3t^2) = 9t^3$$

For $t = 1, v_r = 3, v_\theta = 9$

$$v = \sqrt{3^2 + 9^2} = 9.49$$

2. $r = 4t,\ \theta = \sqrt{t+1},\ t = 3$

$$v_r = \frac{dr}{dt} = 4, v_\theta = r\frac{d\theta}{dt} = \frac{4t}{2\sqrt{t+1}}$$

For $t = 3, v_r = 4, v_\theta = \frac{4t}{2\sqrt{3+1}} = \frac{4(3)}{4} = 3$

$$v = \sqrt{v_r^2 + v_\theta^2} = \sqrt{4^2 + (3)^2} = 5$$

3. $r = e^{0.1t}, \theta = \cos 2t, t = 2$

$$v_r = \frac{dr}{dt}, v_\theta = r\frac{d\theta}{dt}$$

$$v_r = 0.1e^{0.1t}, v_\theta = e^{0.1t}(-2\sin 2t)$$

For $t = 2$:

$$v_r = 0.1e^{0.2} = 0.122$$
$$v_\theta = -2e^{0.2}\sin 4 = 1.849$$
$$v = \sqrt{0.122^2 + 1.849^2} = 1.85$$

4. $r = \sin\frac{\pi t}{6}, \theta = \frac{1}{2t+1}, t = 2$

$$\frac{dr}{dt} = \frac{\pi}{6}\cos\frac{\pi t}{6}; \frac{d\theta}{dt} = \frac{-2}{(2t+1)^2}$$

$$v_r = \frac{\pi}{6}\cos\frac{\pi t}{6}; v_\theta = r\frac{d\theta}{dt} = -2\sin\frac{\pi t}{6}\frac{1}{(2t+1)^2}$$

For $t = 2$

$$v_r = \frac{\pi}{6}\cos\frac{\pi(2)}{6}$$

$$v_\theta = -2\sin\frac{\pi(2)}{6}\frac{1}{(2(2)+1)^2}$$

$$v_r = \frac{\pi}{12}, v_\theta = \frac{-\sqrt{3}}{25}$$

$$v = \sqrt{v_r^2 + v_\theta^2} = \sqrt{\left(\frac{\pi}{12}\right)^2 + \left(\frac{-\sqrt{3}}{25}\right)^2}$$

$$v = 0.271$$

5. $r = 2, \theta = \frac{\pi}{3}, \omega = 2.00$ rad/s

$$\frac{dr}{dt} = 0, \frac{d\theta}{dt} = 2$$

$$v_r = 0, v_\theta = 2.00(2) = 4.00 \text{ ft/s}$$
$$v = \sqrt{0^2 + 4^2} = 4.00 \text{ ft/s}$$

6. $r = 4\theta$

$$v_r = \frac{dr}{dt} = \frac{dr}{d\theta}\frac{d\theta}{dt} = 4(2.00)$$

$$v_\theta = r\frac{d\theta}{dt} = 4\theta(2.00)$$

For $\theta = 2, r = 4(2)$

$$v_r = 4(2.00)$$
$$v_\theta = 4(2)(2.00)$$
$$v_r = \sqrt{v_r^2 + v_\theta^2}$$
$$v = \sqrt{(4(2.00))^2 + (4(2)(2.00))^2}$$
$$= 17.9 \text{ ft/s}$$

7. $r = \sin\theta$

$$v_r = \frac{dr}{dt} = \frac{dr}{d\theta}\frac{d\theta}{dt}, v_\theta = r\frac{d\theta}{dt}$$

$$v_r = \frac{dr}{dt} = \cos\theta(2.00), v_\theta = \sin\theta(2.00)$$

For $\theta = \frac{\pi}{6}$

$$v = \sqrt{v_r^2 + v_\theta^2}$$

$$= \sqrt{\left(\cos\frac{\pi}{6}(2.00)\right)^2 + \left(\sin\frac{\pi}{6}(2.00)\right)^2}$$

$$v = 2.00 \text{ ft/s}$$

8. $r = \cos 2\theta$

$$v_r = \frac{dr}{dt} = \frac{dr}{d\theta}\frac{d\theta}{dt} = 2\sin 2\theta(2.00)$$

$$v_\theta = r\frac{d\theta}{dt} = \cos 2\theta(2.00)$$

For $\theta = \frac{\pi}{4}$

$$v = \sqrt{v_r^2 + v_\theta^2}$$

$$= \sqrt{\left(2\sin\left(2\left(\frac{\pi}{4}\right)\right)(2.00)\right)^2 + \left(\cos\left(2\left(\frac{\pi}{4}\right)\right)(2.00)\right)^2}$$

$$v = 4.00 \text{ ft/s}$$

9. $r = 4 - \cos\theta$

$$v_r = \frac{dr}{dt} = \frac{dr}{d\theta}\frac{d\theta}{dt} = \sin\theta(2.00)$$

$$v_\theta = r\frac{d\theta}{dt} = (4 - \cos\theta)(2.00)$$

For $\theta = \frac{\pi}{2}$

$$v = \sqrt{v_r^2 + v_\theta^2}$$

$$= \sqrt{\left(\sin\frac{\pi}{2}(2.00)\right)^2 + \left(\left(4 - \cos\frac{\pi}{2}\right)(2.00)\right)^2}$$

$$v = 8.25 \text{ ft/s}$$

10. $r = 5 + 2\sin 2\theta$

$$v_r = \frac{dr}{dt} = \frac{dr}{d\theta}\frac{d\theta}{dt} = 4\cos(2\theta)(2.00)$$

$$v_\theta = r\frac{d\theta}{dt} = (5 + 2\sin 2\theta)(2.00)$$

For $\theta = \frac{\pi}{6}$

$$v = \sqrt{v_r^2 + v_\theta^2}$$

$$v = \sqrt{\left(4\cos\frac{\pi}{3}(2.00)\right)^2 + \left(\left(5 + 2\sin\frac{\pi}{3}\right)(2.00)\right)^2}$$

$$v = 14.1 \text{ ft/s}$$

11. $r = \dfrac{1}{2 - \sin\theta}, \theta = \dfrac{\pi}{6}, \omega = 2.00$ rad/s

$$\frac{dr}{dt} = \frac{\cos\theta}{(2 - \sin\theta)^2}\frac{d\theta}{dt}$$

$$v_r = \frac{\cos\left(\frac{\pi}{6}\right)}{\left[2 - \sin\left(\frac{\pi}{6}\right)\right]^2}(2.00) = 0.770 \text{ ft/s}$$

$$v_\theta = \frac{1}{2 - \sin\left(\frac{\pi}{6}\right)}(2.00) = 1.33 \text{ ft/s}$$

$$v = \sqrt{0.770^2 + 1.33^2} = 1.54 \text{ ft/s}$$

12. $r = 1 + \cos^2\theta$

$$v_r = \frac{dr}{dt} = \frac{dr}{d\theta}\frac{d\theta}{dt} = 2\cos\theta\sin\theta(2.00) = \sin(2\theta)(2.00)$$

$$v_\theta = r\frac{d\theta}{dt} = (1 + \cos^2\theta)(2.00)$$

For $\theta = \dfrac{\pi}{4}$

$$v = \sqrt{v_r^2 + v_\theta^2}$$

$$v = \sqrt{\left(\sin\left(\frac{\pi}{2}\right)(2.00)\right)^2 + \left(\left(1 + \cos^2\frac{\pi}{4}\right)(2.00)\right)^2}$$

$$v = 3.61 \text{ ft/s}$$

13. $\theta = \dfrac{1}{6}\pi, \theta = \dfrac{2}{3}\pi, r = 2$

$$A = \frac{1}{2}\int_{\pi/6}^{2\pi/3} 2^2 d\theta$$

$$= 2\theta\Big|_{\pi/6}^{2\pi/6}$$

$$= \frac{4\pi}{3} - \frac{\pi}{3}$$

$$\pi$$

14. $A = \dfrac{1}{2}\int_0^{4\pi/3} 3^2 d\theta$ Wrong

$$A = \frac{9}{2}\theta\Big|_0^{4\pi/3}$$

$$A = \frac{9}{2}\left(\frac{4\pi}{3}\right)$$

$$A = 6\pi$$

15. $\theta = 0, \theta = \dfrac{1}{4}\pi, r = \sec\theta$

$$A = \frac{1}{2}\int_0^{\pi/4} \sec^2\theta\, d\theta$$

$$= \frac{1}{2}\tan\theta\Big|_0^{\pi/4}$$

$$= \frac{1}{2}\tan\frac{\pi}{4} = \frac{1}{2}(1) = \frac{1}{2}$$

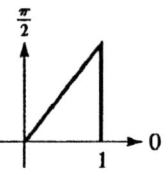

16. $\theta = \dfrac{\pi}{6}, \theta = \dfrac{\pi}{2}, r\sin\theta = 2$

$$A = \frac{1}{2}\int_{\pi/6}^{\pi/2} \frac{4}{\sin^2\theta}\, d\theta$$

$$A = 2\int_{\pi/6}^{\pi/2} \csc^2\theta\, d\theta$$

$$A = 2(-\cot\theta)\Big|_{\pi/6}^{\pi/2}$$

$$= -2\left(\cot\frac{\pi}{2} - \cot\frac{\pi}{6}\right)$$

$$A = -2\left(\frac{\cos\frac{\pi}{2}}{\sin\frac{\pi}{2}} - \frac{\cos\frac{\pi}{6}}{\sin\frac{\pi}{6}}\right)$$

$$= 2\frac{\frac{\sqrt{3}}{2}}{\frac{1}{2}}$$

$$A = 2\sqrt{3}$$

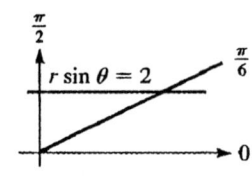

17. $r = 2\cos\theta$

$$A = 2\left[\frac{1}{2}\int_0^{\pi/2}(2\cos\theta)^2 d\theta\right]$$

$$= 4\int_0^{\pi/2}\cos^2\theta\, d\theta$$

$$= 2\int_0^{\pi/2}(1+\cos 2\theta)d\theta$$

$$= 2\theta + \sin 2\theta\,\Big|_0^{\pi/2}$$

$$= \pi + \sin\pi - 0$$

$$= \pi$$

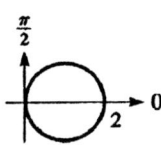

19. $\theta = 0, \theta = \frac{1}{2}\pi, r = e^{2\theta}$

$$A = \frac{1}{2}\int_0^{\pi/2}(e^{2\theta})^2 d\theta$$

$$= \frac{1}{2}\int_0^{\pi/2}e^{4\theta}d\theta$$

$$= \frac{1}{8}e^{4\theta}\,\Big|_0^{\pi/2}$$

$$= \frac{1}{8}(e^{2\pi} - 1)$$

$$= 66.8$$

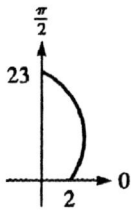

18. $\theta = 0, \theta = \frac{\pi}{3}, r = 4\sin\theta$

$$A = \frac{1}{2}\int_0^{\pi/3}16\sin^2\theta\, d\theta$$

$$A = 8\int_0^{\pi/3}\frac{1-\cos 2\theta}{2}d\theta$$

$$A = 4\int_0^{\pi/3}d\theta - 2\int_0^{\pi/3}\cos(2\theta)(2\,d\theta)$$

$$A = 4\theta\,\Big|_0^{\pi/3} - 2\sin(2\theta)\,\Big|_0^{\pi/3}$$

$$A = \frac{4\pi}{3} - 2\left(\sin\frac{2\pi}{3} - \sin\theta\right)$$

$$A = \frac{4\pi}{3} - 2\frac{\sqrt{3}}{2}$$

$$= \frac{4\pi}{3} - \sqrt{3}$$

$$= \frac{1}{3}(4\pi - 3\sqrt{3})$$

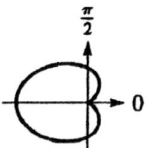

20. $r = 1 - \cos\theta$

$$A = \frac{1}{2}\int_0^{2\pi}(1-\cos\theta)^2 d\theta$$

$$A = \frac{1}{2}\int_0^{2\pi}(1 - 2\cos\theta + \cos^2\theta)d\theta$$

$$A = \frac{1}{2}\int_0^{2\pi}d\theta - \int_0^{2\pi}\cos\theta\, d\theta + \frac{1}{2}\int_0^{2\pi}\cos^2\theta\, d\theta$$

$$A = \frac{1}{2}\theta\,\Big|_0^{2\pi} - \sin\theta\,\Big|_0^{2\pi} + \frac{1}{2}\int_0^{2\pi}\frac{1+\cos 2\theta}{2}d\theta$$

$$A = \pi + \frac{1}{4}\int_0^{2\pi}d\theta + \frac{1}{8}\int_0^{2\pi}\cos(2\theta)(2\,d\theta)$$

$$A = \pi + \frac{\pi}{2} + \frac{1}{8}\sin 2\theta\,\Big|_0^{2\pi}$$

$$A = \frac{3\pi}{2}$$

21.

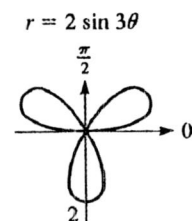

$r = 2 \sin 3\theta$

$$A = 3\left[\frac{1}{2}\int_0^{\pi/3}(2\sin 3\theta)^2 d\theta\right]$$

$$= 6\int_0^{\pi/3}\sin^2 3\theta\, d\theta$$

$$= 3\int_0^{\pi/3}(1 - \sin 6\theta)d\theta$$

$$= 3\theta + \frac{1}{2}\cos 6\theta\Big|_0^{\pi/3}$$

$$= \pi + \frac{1}{2}\cos 2\pi - 0 - \frac{1}{2}\cos 0$$

$$= \pi$$

22.

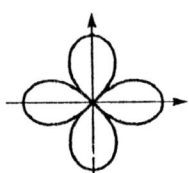

$r = 3\cos 2\theta$

$$A = \frac{1}{2}\int_0^{2\pi}9\cos^2 2\theta\, d\theta$$

$$A = \frac{9}{2}\int_0^{2\pi}\frac{1 + \cos 4\theta}{2}d\theta$$

$$A = \frac{9}{4}\int_0^{2\pi}d\theta + \frac{9}{16}\int_0^{2\pi}\cos(4\theta)(4\, d\theta)$$

$$A = \frac{9}{4}(2\pi) + \frac{9}{16}\sin(4\theta)\Big|_0^{2\pi}$$

$$A = \frac{9\pi}{2}$$

23.

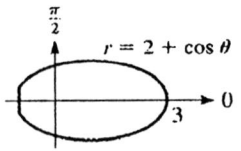

$$A = 2\left[\frac{1}{2}\int_0^{\pi}(2 + \cos\theta)^2 d\theta\right]$$

$$= \int_0^{\pi}(4 + 4\cos\theta + \cos^2\theta)d\theta$$

$$= \int_0^{\pi}\left[4 + 4\cos\theta + \frac{1}{2}(1 + \cos 2\theta)\right]d\theta$$

$$= \frac{1}{2}\int_0^{\pi}(9 + 8\cos\theta + \cos 2\theta)d\theta$$

$$= \frac{1}{2}\left(9\theta + 8\sin\theta + \frac{1}{2}\sin 2\theta\right)\Big|_0^{\pi}$$

$$= \frac{1}{2}\left(9\pi + 8\sin\pi + \frac{1}{2}\sin 2\pi - 0\right)$$

$$= \frac{9}{2}\pi$$

24.

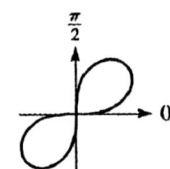

$$A = 2\left[\frac{1}{2}\int_0^{\pi/2}4\sin 2\theta\, d\theta\right]$$

$$A = 2\int_0^{\pi/2}\sin(2\theta)d\theta$$

$$A = 2[-\cos(2\theta)]\Big|_0^{\pi/2}$$

$$A = 2[-\cos\pi + \cos 0]$$

$$A = 4$$

25. $r = 1 + \sin\theta, r = 1$

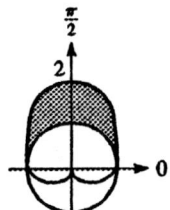

$$A = \frac{1}{2}\int_0^{\pi}[(1+\sin\theta)^2 - 1^2]d\theta$$

$$= \frac{1}{2}\int_0^{\pi}(2\sin\theta + \sin^2\theta)d\theta$$

$$= \frac{1}{2}\int_0^{\pi}\left[2\sin\theta + \frac{1}{2}(1 - \cos 2\theta)\right]d\theta$$

$$= \frac{1}{4}\int_0^{\pi}(1 + 4\sin\theta - \cos 2\theta)d\theta$$

$$= \frac{1}{4}\left(\theta - 4\cos\theta - \frac{1}{2}\sin 2\theta\right)\Big|_0^{\pi}$$

$$= \frac{1}{4}\left(\pi - 4\cos\pi - \frac{1}{2}\sin 2\pi\right)$$

$$\quad - \frac{1}{4}\left(0 - 4\cos 0 - \frac{1}{2}\sin 0\right)$$

$$= \frac{1}{4}(\pi + 4 + 4) = \frac{1}{4}(8 + \pi) = 2.79$$

26.

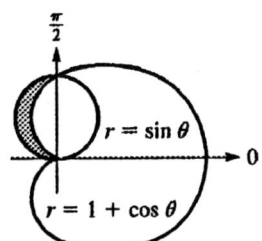

$r = \sin\theta$

$r = 1 + \cos\theta$

$$A = \frac{1}{2}\int_{\pi/2}^{\pi}(\sin^2\theta - (1+\cos\theta)^2)d\theta$$

$$A = \frac{1}{2}\int_{\pi/2}^{\pi}(\sin^2\theta - 1 - 2\cos\theta - \cos^2\theta)d\theta$$

$$A = \frac{1}{2}\int_{\pi/2}^{\pi}(1 - \cos^2\theta - 1 - 2\cos\theta - \cos^2\theta)d\theta$$

$$A = \frac{1}{2}\int_{\pi/2}^{\pi}(-2\cos\theta)d\theta - \frac{1}{2}\int_{\pi/2}^{\pi}2\cos^2\theta\, d\theta$$

$$A = -\sin\theta\Big|_{\pi/2}^{\pi} - \int_{\pi/2}^{\pi}\frac{1+\cos 2\theta}{2}d\theta$$

$$A = -\sin\pi + \sin\frac{\pi}{2} - \int_{\pi/2}^{\pi}\frac{1}{2}d\theta - \frac{1}{4}\int_{\pi/2}^{\pi}\cos(2\theta)(2\,d\theta)$$

$$A = 1 - \frac{1}{2}\theta\Big|_{\pi/2}^{\pi} - \frac{1}{4}(2\theta)\Big|_{\pi/2}^{\pi}$$

$$A = 1 - \frac{\pi}{2} + \frac{\pi}{4} - \frac{1}{4}(\sin 2\pi - \sin\pi)$$

$$A = 1 - \frac{\pi}{4}$$

27.

$r = 2\sin\theta, r = 2\cos\theta$

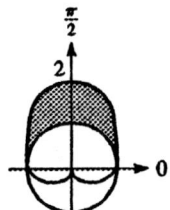

$$A = 2\left[\frac{1}{2}\int_0^{\pi/4}(2\sin\theta)^2 d\theta\right] = 4\int_0^{\pi/4}\sin^2\theta\, d\theta$$

$$= 2\int_0^{\pi/4}(1 - \cos 2\theta)d\theta = 2\theta - \sin 2\theta\Big|_0^{\pi/4}$$

$$= \frac{\pi}{2} - \sin\frac{\pi}{2} - 0 = \frac{\pi}{2} - 1 = 0.571$$

28.

$r = 2 + \cos\theta$ $r = 5\cos\theta$

$$A = 2\left(\frac{1}{2}\right)\int_0^{\pi/3}(25\cos^2\theta - (2+\cos\theta)^2)d\theta$$

$$A = \int_0^{\pi/3}(25\cos^2\theta - 4 - 4\cos\theta - \cos^2\theta)d\theta$$

$$A = \int_0^{\pi/3}(24\cos^2\theta - 4\cos\theta - 4)d\theta$$

$$A = \int_0^{\pi/3}(12(1 + \cos 2\theta) - 4\cos\theta - 4)d\theta$$

$$A = \int_0^{\pi/3}8\,d\theta - \int_0^{\pi/3}4\cos\theta\, d\theta + 6\int_0^{\pi/3}\cos(2\theta)(2\,d\theta)$$

$$A = 8\theta\Big|_0^{\pi/3} - 4\sin\theta\Big|_0^{\pi/3} + 6\sin 2\theta\Big|_0^{\pi/3}$$

$$A = \frac{8\pi}{3} - 4\left(\sin\frac{\pi}{3} - \sin 0\right) + 6\left(\sin\frac{2\pi}{3} - \sin 0\right)$$

$$A = \frac{8\pi}{3} - 4\frac{\sqrt{3}}{2} + 6\frac{\sqrt{3}}{2} = \frac{8\pi}{3} + \sqrt{3}$$

$$A = \frac{1}{3}(8\pi + 3\sqrt{3})$$

29. $r = 2.00 \sin \theta$

$$v_r = \frac{dr}{dt} = \frac{dr}{dt}\frac{d\theta}{dt} = 2.00 \cos \theta \frac{d\theta}{dt}$$

$$v_\theta = r\frac{d\theta}{dt} = 2.00 \sin \theta \frac{d\theta}{dt}$$

$$v = 10.0 = \sqrt{\left(2.00 \cos \theta \frac{d\theta}{dt}\right)^2 + \left(2.00 \sin \theta \frac{d\theta}{dt}\right)^2}$$

$$10.0 = \sqrt{2.00^2(\cos^2 \theta + \sin^2 \theta)\left(\frac{d\theta}{dt}\right)^2}$$

$$10.0 = 2.00 \frac{d\theta}{dt}, \frac{d\theta}{dt} = 5.00$$

For $\theta = 1.05$

$$v_r = 2.00 \cos(1.05)(5.00) = 4.98 \text{ ft/s}$$
$$v_\theta = 2.00 \sin(1.05)(5.00) = 8.67 \text{ ft/s}$$

30. $r = \dfrac{4.00}{1 + \cos \theta}$

$$v_r = \frac{dr}{dt} = \frac{dr}{d\theta}\frac{d\theta}{dt} = \frac{4.00 \sin \theta}{(1+\cos \theta)^2}\frac{d\theta}{dt}$$

$$v_\theta = r\frac{d\theta}{dt} = \frac{4.00}{1+\cos \theta}\frac{d\theta}{dt}$$

$$v = 4.00 = \sqrt{\left(\frac{4.00 \sin \theta}{(1+\cos \theta)^2}\frac{d\theta}{dt}\right)^2 + \left(\frac{4.00}{1+\cos \theta}\frac{d\theta}{dt}\right)^2}$$

$$v = 4.00 = \sqrt{\left(\frac{4.00 \sin \theta}{(1+\cos \theta)^2}\frac{d\theta}{dt}\right)^2 + \left(\frac{4.00}{1+\cos \theta}\frac{d\theta}{dt}\right)^2}$$

$$4.00 = \sqrt{\left(\frac{4.00^2 \sin^2 \theta}{(1+\cos \theta)^4} + \frac{4.00^2}{(1+\cos \theta)2}\right)\left(\frac{d\theta}{dt}\right)^2}$$

$$4.00 = 4.00\sqrt{\frac{\sin^2 \theta + (1+\cos \theta)^2}{(1+\cos \theta)^4}}\frac{d\theta}{dt}$$

$$\frac{d\theta}{dt} = \sqrt{\frac{(1+\cos \theta)^4}{\sin^2 \theta + 1 + 2\cos \theta + \cos^2 \theta}}$$

$$\frac{d\theta}{dt} = \sqrt{\frac{(1+\cos \theta)^4}{2(1+\cos \theta)}} = \sqrt{\frac{(1+\cos \theta)^3}{2}}$$

For $\theta = 1.57$

$$v_r = \frac{4.00 \sin 1.57}{(1+\cos 1.57)^2}\sqrt{\frac{(1+\cos(1.57))^3}{2}}$$

$$= 2.83 \text{ m/s}$$

$$v_\theta = \frac{4.00}{1+\cos 1.57}\sqrt{\frac{(1+\cos(1.57))^3}{2}}$$

$$= 2.83 \text{ m/s}$$

31.

$$r = 2.00\,(3 - \cos\theta)$$

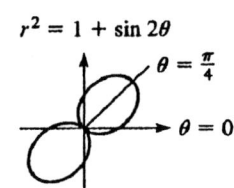

$$A = 2\left(\frac{1}{2}\right)\int_0^\pi (2.00(3 - \cos \theta))^2 d\theta$$

$$A = \int_0^\pi (36.0 - 24.0 \cos \theta + 4.00 \cos^2 \theta)\,d\theta$$

$$A = \int_0^\pi [36.0 - 24.0 \cos \theta + 2.00(1 + \cos 2\theta)]d\theta$$

$$A = \int_0^\pi (38.0 - 24.0 \cos \theta + 4.00 \cos 2\theta)\,d\theta$$

$$A = 38.0\theta - 24.0 \sin \theta + 2.00 \sin 2\theta \Big|_0^\pi$$

$$A = 38.0\pi - 24.0 \sin \pi + 2.00 \sin 2\pi - 0 = 38.0\pi$$
$$A = 119 \text{ cm}^2$$

32.

$$r^2 = 1 + \sin 2\theta$$

$$\theta = \frac{\pi}{4}$$

$$\theta = 0$$

$$(x^2 + y^2)^2 - (x^2 + y^2) = 2xy$$

$$(r^2)^2 - r^2 = 2(r \cos \theta)(r \sin \theta)$$
$$r^4 - r^2 = 2r^2 \sin \theta \cos \theta$$
$$r^2 = 1 + \sin 2\theta$$

$$A = \frac{1}{2}\int_0^\pi (1 + \sin 2\theta)\,d\theta$$

$$= \frac{1}{2}\int_0^\pi d\theta + \frac{1}{4}\int_0^\pi \sin(2\theta)(2\,d\theta)$$

$$A = \frac{1}{2}\theta\Big|_0^{\pi/4} + \frac{1}{4}(-\cos 2\theta)\Big|_0^{\pi/4}$$

$$= \frac{\pi}{8} + \frac{1}{4}\left(-\cos\frac{\pi}{2} + \cos 0\right)$$

$$A = \frac{\pi}{8} + \frac{1}{4}$$

$$= \frac{\pi + 2}{8}$$

33. $x = r \cos \theta, y = r \sin \theta$

$$\frac{dx}{dr} = \cos \theta \frac{dr}{d\theta} - r \sin \theta$$

$$\frac{dy}{dr} = \sin \theta \frac{dr}{d\theta} + r \cos \theta$$

$$\frac{dy}{dx} = \frac{dy/dr}{dx/dr}$$

$$= \frac{r' \sin \theta + r \cos \theta}{r' \cos \theta - r \sin \theta}$$

$$= \frac{r' \tan \theta + r}{r' - r \tan \theta}$$

(divide by $\cos \theta$), $r' = dr/d\theta$

34. $r = \cos \theta, \dfrac{dr}{d\theta} = -\sin \theta$

$$\frac{dy}{dx} = \frac{r' \tan \theta + r}{r' - r \tan \theta} = \frac{-\sin \theta \tan \theta + \cos \theta}{-\sin \theta - \cos \theta \tan \theta}$$

$$\frac{dy}{dx} = \frac{-\sin \frac{\pi}{6} \tan \frac{\pi}{6} + \cos \frac{\pi}{6}}{-\sin \frac{\pi}{6} - \cos \frac{\pi}{6} \tan \frac{\pi}{6}}$$

$$\frac{dy}{dx} = \frac{\sqrt{3}}{3}$$

35. $r = \cos^2(\theta/2)$;

$$r' = \frac{dr}{d\theta} = 2 \left(\cos \frac{\theta}{2} \left(-\sin \frac{\theta}{2} \right) \right) \left(\frac{1}{2} \right) = -\sin \frac{\theta}{2} \cos \frac{\theta}{2}$$

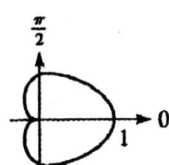

$$s = 2 \int_0^\pi \sqrt{\cos^4 \frac{\theta}{2} + \left(-\sin \frac{\theta}{2} \cos \frac{\theta}{2} \right)^2} \, d\theta$$

$$= 2 \int_0^\pi \sqrt{\left(\cos^2 \frac{\theta}{2} \right) \left(\cos^2 \frac{\theta}{2} + \sin^2 \frac{\theta}{2} \right)} \, d\theta$$

$$= 2 \int_0^\pi \cos \frac{\theta}{2} \, d\theta = 4 \sin \frac{\theta}{2} \Big|_0^\pi = 4 \sin \frac{\pi}{2} = 4$$

36. $r = \sin \theta + \cos \theta, r' = \cos \theta - \sin \theta$

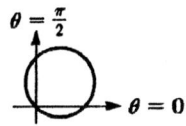

$$s = \int_0^\pi \sqrt{r^2 + (r')^2} \, d\theta$$

$$= \int_0^\pi \sqrt{(\sin \theta + \cos \theta)^2 + (\cos \theta - \sin \theta)^2} \, d\theta$$

$$= \int_0^\pi \sqrt{\sin^2 \theta + 2\sin \theta \cos \theta + \cos^2 \theta + \cos^2 \theta - 2\sin \theta \cos \theta + \sin^2 \theta} \, d\theta$$

$$= \int_0^\pi \sqrt{2} \, d\theta = \sqrt{2} \theta \Big|_0^\pi = \pi\sqrt{2}$$

12.4 Cylindrical Coordinates

1. (a) $\left(2, \dfrac{\pi}{2}, 4 \right)$

$$x = r \cos \theta = 2 \cos \frac{\pi}{2} = 0$$

$$y = r \sin \theta = 2 \sin \frac{\pi}{2} = 2$$

$$z = z = 4$$

$$(0, 2, 4)$$

(b) $\left(3, -\dfrac{\pi}{4}, 5 \right)$

$$x = r \cos \theta = 3 \cos \left(-\frac{\pi}{4} \right) = \frac{3\sqrt{2}}{2}$$

$$y = r \sin \theta = 3 \sin \left(-\frac{\pi}{4} \right) = \frac{-3\sqrt{2}}{2}$$

$$z = z = 5$$

$$\left(\frac{3\sqrt{2}}{2}, \frac{-3\sqrt{2}}{2}, 5 \right)$$

2. (a) $\left(4, \dfrac{\pi}{3}, 2 \right)$

$$x = r \cos \theta = 4 \cos \frac{\pi}{3} = 4 \left(\frac{1}{2} \right) = 2$$

$$y = r \sin \theta = 4 \sin \frac{\pi}{3} = 4 \left(\frac{\sqrt{3}}{2} \right) = 2\sqrt{3}$$

$$z = z = 2$$

$$(2, 2\sqrt{3}, 2)$$

(b) $\left(3, \dfrac{5\pi}{6}, 2\right)$

$$x = r\cos\theta = 3\cos\dfrac{5\pi}{6} = 3\left(\dfrac{-\sqrt{3}}{2}\right) = \dfrac{-3\sqrt{3}}{2}$$

$$y = r\sin\theta = 3\sin\dfrac{5\pi}{6} = 3\left(\dfrac{1}{2}\right) = \dfrac{3}{2}$$

$$z = z = 2$$

$$\left(\dfrac{-3\sqrt{3}}{2}, \dfrac{3}{2}, 2\right)$$

3. (a) $(6, 8, 5)$

$$\tan\theta = \dfrac{y}{x} = \dfrac{8}{6} = \dfrac{4}{3}, \theta = \tan^{-1}\dfrac{4}{3}$$

$$r^2 = x^2 + y^2 = 6^2 + 8^2 = 100, r = 10$$

$$z = z = 5$$

$$\left(10, \tan^{-1}\dfrac{4}{3}, 5\right)$$

(b) $(4, 4, -3)$

$$\tan\theta = \dfrac{y}{x} = \dfrac{4}{4} = 1, \theta = \tan^{-1}1 = \dfrac{\pi}{4}$$

$$r^2 = x^2 + y^2 = 4^2 + 4^2 = 32, r = \sqrt{32} = 4\sqrt{2}$$

$$z = z = -3$$

$$\left(4\sqrt{2}, \dfrac{\pi}{4}, -3\right)$$

4. (a) $(8, 15, -6)$

$$\tan\theta = \dfrac{y}{x} = \dfrac{15}{8}, \theta = \tan^{-1}\dfrac{15}{8}$$

$$r^2 = x^2 + y^2 = 8^2 + 15^2 = 289, r = 17$$

$$z = z = -6$$

$$\left(17, \tan^{-1}\dfrac{15}{8}, -6\right)$$

(b) $(\sqrt{3}, -2, 1)$

$$\tan\theta = \dfrac{y}{x} = \dfrac{-2}{\sqrt{3}}, \theta = \tan^{-1}\dfrac{-2}{\sqrt{3}}$$

$$r^2 = x^2 + y^2 = \sqrt{3}^2 + (-2)^2 = 7, r = \sqrt{7}$$

$$z = z = 1$$

$$\left(\sqrt{7}, \tan^{-1}\dfrac{-2}{\sqrt{3}}, 1\right)$$

5. (a) $r = 2$ describes a cylinder with z-axis as axis and radius 2.

(b) $\theta = 2$ describes a plane with $\theta = 2$ for all r and z.

6. (a) $\theta = -2$ describes a plane with $\theta = -2$ for all r and z.

(b) $z = -2$ describes a plane parallel to xy-plane and 2 units below the pole.

7. (a) $x = 2 = r\cos\theta, r = 2\sec\theta$ which describes a plane 2 units in front of yz-plane

(b) $z = 2$ describes a plane 2 units above xy-plane.

8. (a) $y = x$

$$r\sin\theta = r\cos\theta$$

$$\tan\theta = 1$$

$\theta = \dfrac{\pi}{4}$ describes the plane $\theta = \dfrac{\pi}{4}$ for all r and z

(b) $x^2 + y^2 = 4$

$$r^2 = 4$$

$r = 2$ describes a cylinder with z-axis as axis and radius 2.

9. $x^2 + y^2 = 16$

$$r^2 = 16$$

$$r = 4$$

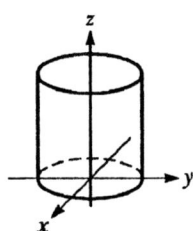

10. $x^2 + y^2 = 4z$

$$r^2 = 4z$$

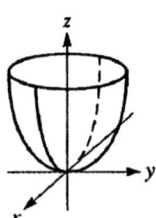

11. $x^2 + y^2 + 4z^2 = 4$

$$r^2 + 4z^2 = 4$$

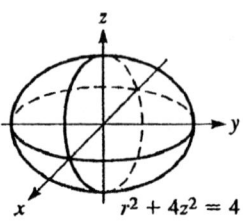

$r^2 + 4z^2 = 4$

12. $x = 2 = r\cos\theta$
  $r\cos\theta = 2$
    $r = 2\sec\theta$

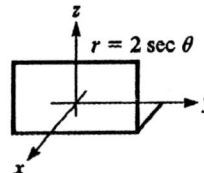

13. $9z = 4x^2 + 4y^2 = 4(x^2 + y^2)$
  $9z = 4r^2$

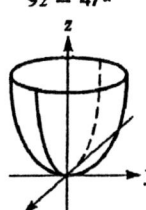

14. $z^2 + 9 = x^2 + y^2 = r^2$
  $r^2 = z^2 + 9$

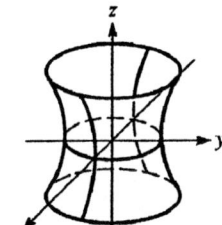

15. $r^2 = 4z$
  $x^2 + y^2 = 4z$

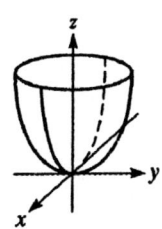

16. $r = 4\cos\theta$
  $r^2 = 4r\cos\theta$
  $x^2 + y^2 = 4x$

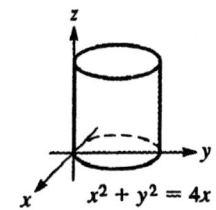

17. $r^2 + 4z^2 = 16$
  $x^2 + y^2 + 4z^2 = 16$

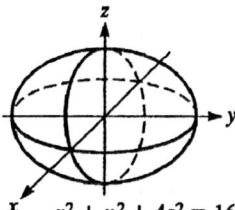

18. $r = z$
  $r^2 = z^2$
  $x^2 + y^2 = z^2$

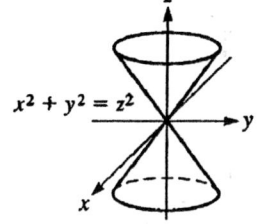

19. $r\sin\theta = 3$
  $y = 3$

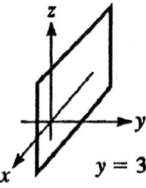

20. $r = \dfrac{1}{\sin\theta - \cos\theta}$
  $r\sin\theta - r\cos\theta = 1$
  $y - x = 1$
  $y = x + 1$

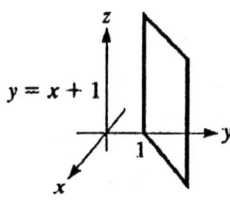

21.

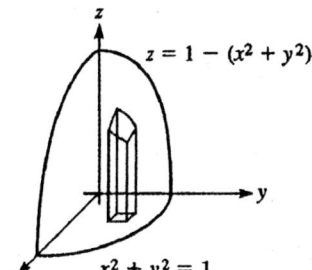

$z = 1 - r^2 = 1 - (x^2 + y^2)$

$$V = 4 \int_0^1 \int_0^{\sqrt{1-x^2}} (1 - (x^2 + y^2)) dy\, dx$$

$$= 4 \int_0^1 \int_0^{\sqrt{1-x^2}} (1 - x^2 - y^2) dy\, dx$$

$$= 4 \int_0^1 \left[(1 - x^2)y - \frac{y^3}{3} \right] \Big|_0^{\sqrt{1-x^2}} dx$$

$$= 4 \int_0^1 \left[(1 - x^2)\sqrt{1-x^2} - \frac{1}{3}(\sqrt{1-x^2})^3 \right] dx$$

$$= 4 \int_0^1 \left[(1 - x^2)^{3/2} - \frac{1}{3}(1 - x^2)^{3/2} \right] dx$$

$$= \frac{8}{3} \int_0^1 (1 - x^2)^{3/2} dx$$

let $x = \sin\theta$,
$$(1 - x^2)^{3/2} = (1 - \sin^2\theta)^{3/2}$$
$$= (\cos^2\theta)^{3/2}$$
$$= \cos^3\theta$$
$$dx = \cos\theta\, d\theta$$

$$V = \frac{8}{3} \int_0^{\pi/2} \cos^3\theta \cos\theta\, d\theta$$

$$= \frac{8}{3} \int_0^{\pi/2} (\cos^2\theta)^2 d\theta$$

$$= \frac{8}{3} \int_0^{\pi/2} \left(\frac{1 + \cos 2\theta}{2} \right)^2 d\theta$$

$$= \frac{2}{3} \int_0^{\pi/2} (1 + 2\cos 2\theta + \cos^2 2\theta) d\theta$$

$$= \frac{2}{3}\theta \Big|_0^{\pi/2} + \frac{2}{3} \int_0^{\pi/2} \cos(2\theta)(2\, d\theta)$$
$$\qquad + \frac{2}{3} \int_0^{\pi/2} \frac{1 + \cos 4\theta}{2} d\theta$$

$$= \frac{2}{3} \left(\frac{\pi}{2} - 0 \right) + \frac{2}{3} \sin(2\theta) \Big|_0^{\pi/2} + \frac{1}{3} \int_0^{\pi/2} d\theta$$
$$\qquad + \frac{1}{12} \int_0^{\pi/2} \cos(4\theta)(4\, d\theta)$$

$$= \frac{\pi}{3} + \frac{2}{3}(\sin\pi - \sin 0) + \frac{1}{3}\theta \Big|_0^{\pi/2} + \frac{1}{12}(4\theta) \Big|_0^{\pi/2}$$

$$= \frac{\pi}{3} + \frac{\pi}{6} + \frac{1}{12}(\sin 2\pi - \sin 0)$$

$$V = \frac{\pi}{2}$$

22. From exercise #18, $z = r$ is the top cone since the surface $z = 0$ is a lower boundary. From $r = 2\cos\theta$

$$r^2 = 2r\cos\theta$$
$$x^2 + y^2 = 2x$$
$$x^2 - 2x + 1 + y^2 = 1$$
$$(x - 1)^2 + y^2 = 1 \quad \text{is a circle with center } (1,0)$$
$$\text{and radius } 1.$$

The volume above the xy-plane ($z = 0$), inside the cylinder $r = 2\cos\theta$, and below the cone $z = r$ is

$$V = 2 \int_0^2 \int_0^{\sqrt{2x-x^2}} \sqrt{x^2 + y^2}\, dy\, dx$$

and since $\sqrt{x^2 + y^2} = z$, $dy\, dx = r\, dr\, dt$

$$= 2 \int_0^{\pi/2} \int_0^{2\cos\theta} z(r\, dr\, d\theta)$$

$$= 2 \int_0^{\pi/2} \int_0^{2\cos\theta} r(r\, dr\, d\theta)$$

$$= 2 \int_0^{\pi/2} \int_0^{2\cos\theta} r^2\, dr\, d\theta$$

$$= \frac{2}{3} \int_0^{\pi/2} r^3 \Big|_0^{2\cos\theta} d\theta$$

$$= \frac{2}{3} \int_0^{\pi/2} 8\cos^3\theta\, d\theta$$

$$= \frac{8}{3} \int_0^{\pi/2} \cos^2\theta \cos\theta\, d\theta$$

$$= \frac{16}{3} \int_0^{\pi/2} (1 - \sin^2\theta)\cos\theta\, d\theta$$

$$= \frac{16}{3} \left[\sin\theta \Big|_0^{\pi/2} - \frac{1}{3}\sin^{-3}\theta \Big|_0^{\pi/2} \right]$$

$$= \frac{16}{3} \left[\frac{\pi}{2} - \sin 0 - \frac{1}{3}\sin^3\frac{\pi}{2} + \frac{1}{3}\sin 0 \right]$$

$$= \frac{16}{3} \left[1 - \frac{1}{3} \right] = \frac{16}{3} \left(\frac{2}{3} \right) = \frac{32}{9}$$

23. $V = 4 \displaystyle\int_0^4 \int_0^{\sqrt{16-x^2}} \sqrt{25 - x^2 - y^2}\, dy\, dx$

$V = 4 \displaystyle\int_0^{\pi/2} \int_0^4 \sqrt{25 - r^2}\, (r\, dr\, d\theta)$

$V = \dfrac{4}{-2} \displaystyle\int_0^{\pi/2} \int_0^4 \sqrt{25 - r^2}(-2r\, dr)\, d\theta$

$V = -2 \displaystyle\int_0^{\pi/2} \dfrac{2}{3}(25 - r^2)^{3/2} \Big|_0^4 \, d\theta$

$V = \dfrac{-4}{3} \displaystyle\int_0^{\pi/2} [(25 - 4^2)^{3/2} - 25^{3/2}]\, d\theta$

$V = \dfrac{-4}{3} \displaystyle\int_0^{\pi/2} (27 - 125)\, d\theta = -\dfrac{4}{3} \int_0^{\pi/2} (-98)\, d\theta$

$V = \dfrac{392}{3} \displaystyle\int_0^{\pi/2} d\theta = \dfrac{392}{3} \theta \Big|_0^{\pi/2} = \dfrac{392}{3}\left(\dfrac{\pi}{2}\right) = \dfrac{196\pi}{3}$

$V = 205 \text{ m}^3$

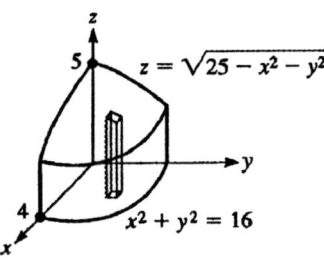

24. Write $\displaystyle\int_0^{\pi/2} \int_0^{2\cos\theta} r^3\, dr\, d\theta$ as

$\displaystyle\int_0^{\pi/2} \int_0^{2\cos\theta} r^3 (r\, dr\, d\theta)$

Which represents the volume bounded by the surfaces $z = 0$, $z = r^2$, and $r = 2\cos\theta$ in the first octant.

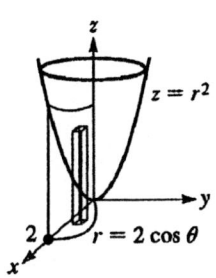

Chapter 12 Review Exercises

1. $y = 2x$

$r\sin\theta = 2r\cos\theta$

$\dfrac{\sin\theta}{\cos\theta} = 2$

$\tan\theta = 2$

$\theta = \tan^{-1} 2 = 1.11$

2. $2xy = 2$

$2(r\cos\theta)(r\sin\theta) = 1$

$r^2(2\sin\theta\cos\theta) = 1$

$r^2\sin 2\theta = 1$

3. $x^2 - y^2 = 16$

$r^2\cos^2\theta - r^2\sin^2\theta = 16$

$r^2(\cos^2\theta - \sin^2\theta) = 16$

$r^2\cos 2\theta = 16$

4. $x^2 + y^2 = 7 - 6y$

$r^2 = 7 - 6r\sin\theta$

$r^2 + 6r\sin\theta - 7 = 0$

5. $r = 2\sin 2\theta = 4\sin\theta\cos\theta$

$r^3 = 4(r\sin\theta)(r\cos\theta)$

$(x^2 + y^2)^{3/2} = 4yx$

$(x^2 + y^2)^3 = 16x^2y^2$

6. $r^2 = \sin\theta$

$r^2 = \dfrac{y}{r}$

$r^3 = y$

$(x^2 + y^2)^3 = y$

7. $r = \dfrac{4}{2 - \cos\theta}$

$2r - r\cos\theta = 4$

$2\sqrt{x^2 + y^2} - x = 4$

$2\sqrt{x^2 + y^2} = x + 4$

$4(x^2 + y^2) = x^2 + 8x + 16$

$3x^2 + 4y^2 - 8x - 16 = 0$

8. $r = \dfrac{2}{1 - \sin\theta}$

$r(1 - \sin\theta) = 2$

$r - r\sin\theta = 2$

$\sqrt{x^2 + y^2} - y = 2$

$\sqrt{x^2 + y^2} = y + 2$

$x^2 + y^2 = y^2 + 4y + 4$

$x^2 - 4y - 4 = 0$

9. $r = 4(1 + \sin\theta)$

θ	0	$\dfrac{\pi}{4}$	$\dfrac{\pi}{2}$	$\dfrac{3\pi}{4}$	π
r	4	6.8	8	6.8	4

θ	$\dfrac{5\pi}{4}$	$\dfrac{3\pi}{2}$	$\dfrac{7\pi}{4}$	2π
r	1.2	0	1.2	4

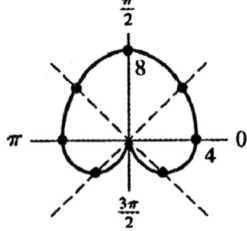

10. $r = 1 + 2\cos\theta$

θ	r
0	3
$\dfrac{\pi}{4}$	2.4
$\dfrac{\pi}{2}$	1
$\dfrac{3\pi}{4}$	−0.4
π	−1
$\dfrac{5\pi}{4}$	−0.4
$\dfrac{3\pi}{4}$	1
$\dfrac{7\pi}{4}$	2.4
2π	3

11. $r = 4\cos 3\theta$

θ	0	$\dfrac{\pi}{12}$	$\dfrac{\pi}{6}$	$\dfrac{\pi}{4}$	$\dfrac{\pi}{3}$	$\dfrac{5\pi}{12}$	$\dfrac{\pi}{2}$	$\dfrac{7\pi}{12}$	$\dfrac{2\pi}{3}$
r	4	2.8	0	−2.8	−4	−2.8	0	2.8	4

Values repeat using multiples of $\pi/12$.

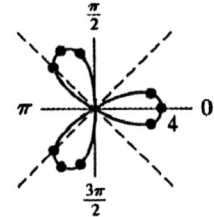

12. $r = -3\sin\theta$

θ	r
0	0
$\dfrac{\pi}{4}$	−2.1
$\dfrac{\pi}{2}$	−3
$\dfrac{3\pi}{4}$	−2.1
π	0
$\dfrac{\pi}{3}$	−2.6
$\dfrac{2\pi}{3}$	−2.6

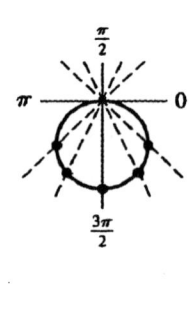

13. $r = \cot\theta$

θ	0	$\dfrac{\pi}{6}$	$\dfrac{\pi}{3}$	$\dfrac{\pi}{2}$	$\dfrac{2\pi}{3}$	$\dfrac{5\pi}{6}$	π
r	undef.	1.7	0.6	0	−0.6	−1.7	undef.

θ	$\dfrac{7\pi}{6}$	$\dfrac{2\pi}{3}$	$\dfrac{4\pi}{3}$	$\dfrac{3\pi}{2}$	$\dfrac{5\pi}{3}$	$\dfrac{11\pi}{6}$	2π
r	1.7	4	0.6	0	−1.7	−0.6	undef.

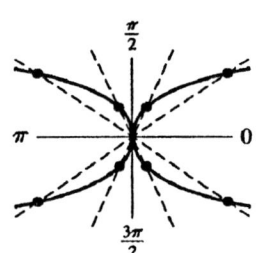

14. $r = \dfrac{1}{2(\sin\theta - 1)}$

θ	r
0	−0.5
$\dfrac{\pi}{4}$	−1.7
$\dfrac{3\pi}{4}$	−1.7
π	−0.5
$\dfrac{5\pi}{4}$	−0.3
$\dfrac{3\pi}{2}$	−0.3
$\dfrac{7\pi}{4}$	−0.3

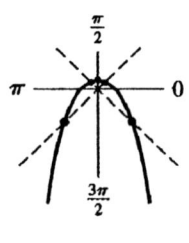

15. $r = 2\sin\left(\dfrac{\theta}{2}\right)$

θ	0	$\dfrac{\pi}{2}$	π	$\dfrac{3\pi}{2}$	2π	$\dfrac{5\pi}{2}$	3π	$\dfrac{7\pi}{2}$	4π
r	0	1.4	2	1.4	0	-1.4	-2	-1.4	0

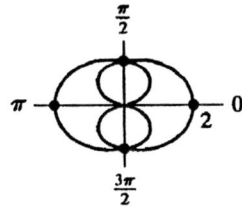

16. $r = \theta$

θ	r
0	0
$\dfrac{\pi}{4}$	0.8
$\dfrac{\pi}{2}$	1.6
$\dfrac{3\pi}{4}$	2.4
π	3.1
$\dfrac{5\pi}{4}$	3.9
$\dfrac{3\pi}{2}$	4.7
$\dfrac{7\pi}{4}$	5.5
2π	6.3
$\dfrac{9\pi}{4}$	7.1
$\dfrac{5\pi}{2}$	7.9

θ	r
0	0
$-\dfrac{\pi}{4}$	-0.8
$-\dfrac{\pi}{2}$	-1.6
$-\dfrac{3\pi}{4}$	-2.4
$-\dfrac{5\pi}{4}$	-3.9
$-\dfrac{3\pi}{2}$	-4.7
$-\dfrac{7\pi}{4}$	-5.5
-2π	-6.3
$-\dfrac{9\pi}{4}$	-7.1
$\dfrac{5\pi}{2}$	-7.9

17. $r = 2 + \cos 3\theta$

θ	0	$\dfrac{\pi}{12}$	$\dfrac{\pi}{6}$	$\dfrac{\pi}{4}$	$\dfrac{\pi}{3}$	$\dfrac{5\pi}{12}$	$\dfrac{\pi}{2}$	$\dfrac{7\pi}{12}$	$\dfrac{2\pi}{3}$
r	3	2.7	2	1.3	1	1.3	2	2.7	3

Values repeat using multiplies of $\pi/12$.

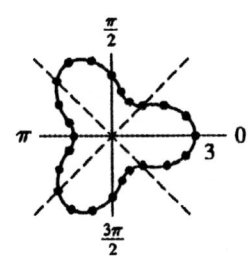

18. $r = 3 - \cos 2\theta$

θ	r
0	2.0
$\dfrac{\pi}{6}$	2.5
$\dfrac{\pi}{4}$	3.0
$\dfrac{\pi}{2}$	4.0
$\dfrac{3\pi}{4}$	4.0
$\dfrac{5\pi}{6}$	2.5
π	2.0
$\dfrac{7\pi}{6}$	2.5
$\dfrac{5\pi}{4}$	3.0
$\dfrac{3\pi}{2}$	4.0
$\dfrac{7\pi}{4}$	3.0
$\dfrac{11\pi}{6}$	2.5
2π	2.0

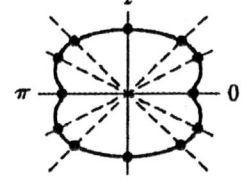

19. $r^2 = \theta$

θ	0	$\dfrac{\pi}{4}$	$\dfrac{\pi}{2}$	$\dfrac{3\pi}{4}$	π
r	0	±0.9	±1.3	±1.5	±1.8

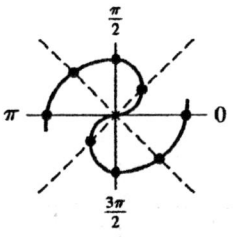

20. $r^2 = 4\cos\theta$

θ	r
0	± 2
$\dfrac{\pi}{6}$	± 1.9
$\dfrac{\pi}{4}$	± 1.7
$\dfrac{\pi}{3}$	± 1.4
$\dfrac{\pi}{2}$	0
$\dfrac{5\pi}{3}$	± 1.4
$\dfrac{7\pi}{4}$	± 1.7
$\dfrac{11\pi}{6}$	± 1.9
2π	2.0

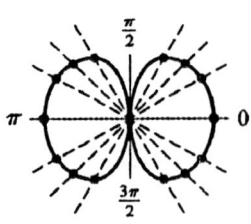

21. $r = 2 + \csc\theta$

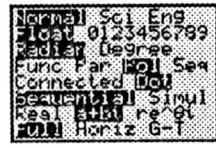 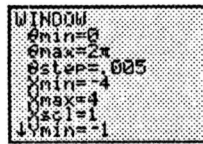 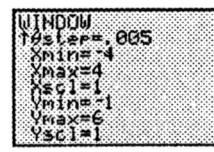 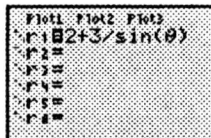

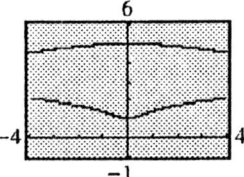

22. $r = 2 - 3^{\sin\theta}$

 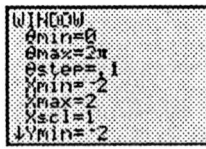 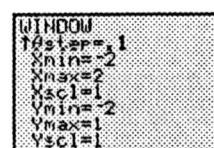 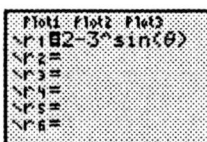

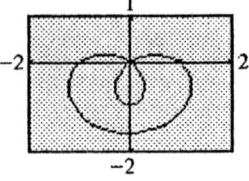

23. $r = \dfrac{\sin\theta}{1+\cos\theta}$

 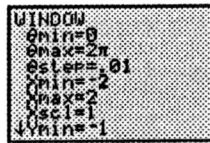 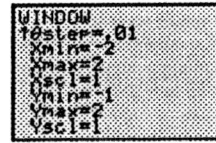 

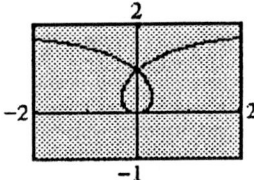

24. $r = 1 + 3\sin(\cos(3\theta))$

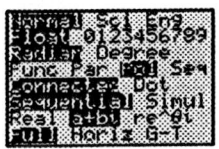 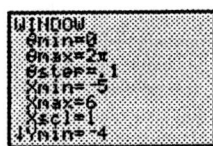 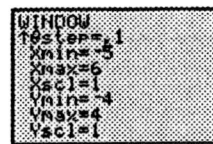 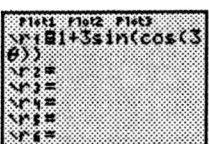

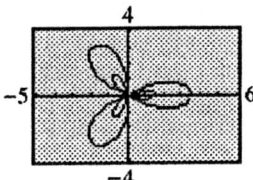

25. $r = 4\cos\theta, \theta = 0, \theta = \dfrac{\pi}{4}$

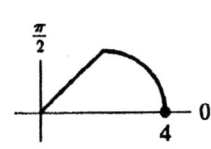

$$A = \frac{1}{2}\int_0^{\pi/4} (4\cos\theta)^2 d\theta = 8\int_0^{\pi/4}\cos^2\theta\, d\theta$$

$$= 4\int_0^{\pi/4}(1+\cos 2\theta)d\theta = 4\theta + 2\sin 2\theta \Big|_0^{\pi/4}$$

$$= \pi + 2\sin\frac{\pi}{2} - 0 = \pi + 2$$

26. $\quad r = 2\csc\theta = \dfrac{2}{\sin\theta}$

$r\sin\theta = 2$

$y = 2$

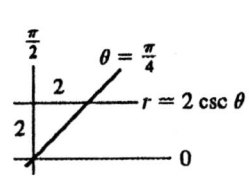

$$A = \frac{1}{2}\int_{\pi/4}^{\pi/2}\left(\frac{2}{\sin\theta}\right)d\theta$$

$$A = -\frac{2}{\tan\theta}\Big|_{\pi/4}^{\pi/2} = -\frac{2}{\tan\frac{\pi}{2}} + \frac{2}{\tan\frac{\pi}{2}}$$

$$A = -\frac{2}{\infty} + \frac{2}{1} = -0 + 2$$

$$A = 2$$

27. $r = 3 - \sin\theta$

$$A = \frac{1}{2}\int_0^{2\pi}(3-\sin\theta)^2\,d\theta = \frac{1}{2}\int_0^{2\pi}(9 - 6\sin\theta + \sin^2\theta)\,d\theta$$

$$= \frac{1}{2}\int_0^{2\pi}\left[9 - 6\sin\theta + \frac{1}{2}(1-\cos 2\theta)\right]d\theta = \frac{1}{4}\int_0^{2\pi}(19 - 12\sin\theta - \cos 2\theta)\,d\theta$$

$$= \frac{1}{4}\left(19\theta + 12\cos\theta - \frac{1}{2}\sin 2\theta\right)\Bigg|_0^{2\pi} = \frac{1}{4}\left(38\pi + 12\cos 2\pi - \frac{1}{2}\sin 4\pi\right)$$

$$-\frac{1}{4}\left(0 + 12\cos 0 - \frac{1}{2}\sin 0\right)$$

$$= \frac{1}{4}(38\pi + 12 - 12) = \frac{19\pi}{2}$$

28. $r = \sqrt{\sin\theta}$

$$A = \int_0^\pi \frac{1}{2}r^2\,d\theta = \int_0^\pi \frac{1}{2}\sin\theta\,d\theta = \frac{1}{2}(-\cos\theta)\Big|_0^\pi$$

$$A = \frac{1}{2}(-\cos\pi + \cos 0) = \frac{1}{2}(-(-1)+1) = 1$$

29. $r = 6, r\cos\theta = 5$

$$A = 2\left[\frac{1}{2}\int_0^{\cos^{-1}5/6}[6^2 - (5\sec\theta)^2]\,d\theta\right] = \int_0^{\cos^{-1}5/6}(36 - 25\sec^2\theta)\,d\theta$$

$$= 36\theta - 25\tan\theta\Big|_0^{\cos^{-1}5/6} = 36\left(\cos^{-1}\frac{5}{6}\right) - 25\tan\left(\cos^{-1}\frac{5}{6}\right)$$

$$= 4.502 \qquad \left(\cos^{-1}\frac{5}{6} = 0.5856855\right)$$

30. $\displaystyle A = 2\int_0^{\pi/3}\frac{1}{2}r^2\,d\theta = \int_0^{\pi/3}(1 - 2\cos\theta)^2\,d\theta = \int_0^{\pi/3}(1 - 4\cos\theta + 4\cos^2\theta)\,d\theta$

$$A = \int_0^{\pi/3}d\theta - 4\int_0^{\pi/3}\cos\theta\,d\theta + 4\int_0^{\pi/3}\cos^2\theta\,d\theta$$

$$A = \theta\Big|_0^{\pi/3} - 4\sin\Big|_0^{\pi/3} + 4\int_0^{\pi/3}\frac{1+\cos 2\theta}{2}\,d\theta$$

$$A = \frac{\pi}{3} - 4\left(\sin\frac{\pi}{3} - \sin 0\right) + 2\int_0^{\pi/3}d\theta + \int_0^{\pi/3}\cos(2\theta)(2\,d\theta)$$

$$A = \frac{\pi}{3} - 4\left(\frac{\sqrt{3}}{2}\right) + 2\theta\Big|_0^{\pi/3} + \sin 2\theta\Big|_0^{\pi/3}$$

$$A = \frac{\pi}{3} - 2\sqrt{3} + \frac{2\pi}{3} + \sin\frac{2\pi}{3} - \sin 0 = \pi - 2\sqrt{3} + \frac{\sqrt{3}}{2} = \pi - \frac{3\sqrt{3}}{2}$$

31. $r = \theta, \theta = \dfrac{\pi}{2}, \theta \geq 0$

$$A = \frac{1}{2}\int_0^{\pi/2} \theta^2 d\theta = \frac{1}{6}\theta^3 \Big|_0^{\pi/2} = \frac{\pi^3}{48}$$

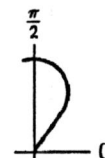

32. $A = \displaystyle\int_0^{2\pi} \frac{1}{2}r^2 d\theta = \int_0^{2\pi} \frac{1}{2}(e^\theta)^2 d\theta = \frac{1}{2}\int_0^{2\pi} e^{2\theta} d\theta = \frac{1}{4}\int_0^{2\pi} e^{2\theta}(2\,d\theta)$

$A = \dfrac{1}{4}e^{2\theta}\Big|_0^{2\pi} = \dfrac{1}{4}(e^{2(2\pi)} - e^{2(0)}) = \dfrac{1}{4}(e^{4\pi} - e^0) = \dfrac{1}{4}(e^{4\pi} - 1)$

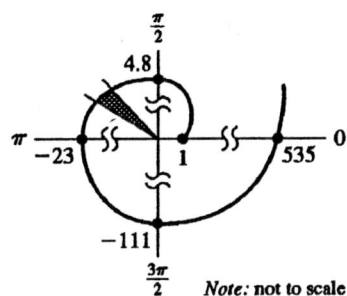

Note: not to scale

33. (a) $(5, 12, 3)$

$$r = \sqrt{x^2 + y^2} = \sqrt{5^2 + 12^2} = \sqrt{25 + 144} = \sqrt{169} = 13$$

$$\theta = \tan^{-1}\frac{y}{x} = \tan^{-1}\frac{12}{5} = \tan^{-1} 2.4$$

$(13, \tan^{-1} 2.4, 3)$

(b) $(3, -3, 2)$

$$r^2 = \sqrt{x^2 + y^2} = \sqrt{(3)^2 + (-3)^2} = \sqrt{9 + 9} = \sqrt{18} = \sqrt{9(2)} = 3\sqrt{2}$$

$$\theta = \tan^{-1}\frac{-3}{3} = \tan^{-1}(-1)$$

$(3\sqrt{2}, \tan^{-1}(-1), 2)$

34. (a) $\left(8, \dfrac{\pi}{6}, 3\right)$

$$x = r\cos\theta = 8\cos\frac{\pi}{6} = 8\left(\frac{\sqrt{3}}{2}\right) = 4\sqrt{3}$$

$$y = r\sin\theta = 8\sin\frac{\pi}{6} = 8\left(\frac{1}{2}\right) = 4$$

$(4\sqrt{3}, 4, 3)$

(b) $\left(5, \dfrac{\pi}{6}, 4\right)$

$$x = r\cos\theta = 5\cos\left(-\frac{\pi}{2}\right) = 5(0) = 0$$

$$y = r\sin\theta = 5\sin\left(-\frac{\pi}{2}\right) = 5(-1) = -5$$

$(0, -5, 4)$

35.

 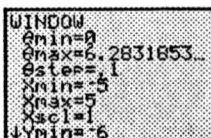 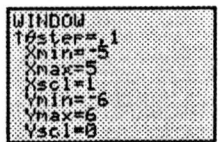 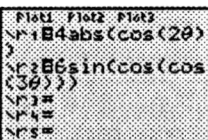

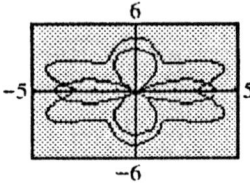

From the graphs the number of points of intersection is 4.

36. The solid described by $1 \le x \le 2 - r^2$ or $1 \le z \le 2 - (x^2 + y^2)$ is the solid below the surface $z = 2 - r^2$, a paraboloid, and above the plane $z = 1$.

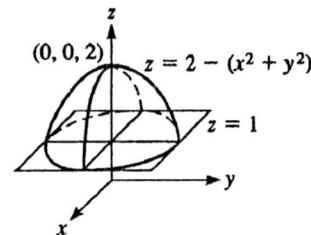

37. $r = t^3 - 6t, \theta = \dfrac{t}{t+1}, t = 1$

$$v_r = \frac{dr}{dt} = 3t^2 - 6$$

$$v_\theta = r\frac{d\theta}{dt} = (t^3 - 6t)\left[\frac{1}{(t+1)^2}\right]$$

For $t = 1$: $v_r = 3(1) - 6 = -3$

$$v_\theta = (1 - 6)\left(\frac{1}{4}\right) = -\frac{5}{4}$$

$$v = \sqrt{(-3)^2\left(-\frac{5}{4}\right)^2} = \frac{13}{4}$$

38. From exercise 32 on page 51, $a = 3.7$ and $c = 0.9$

$$a^2 = b^2 + c^2$$
$$3.7^2 = b^2 + 0.9^2$$
$$b = 3.6$$

The equation in rectangular coordinates is

$$\frac{x^2}{3.7^2} + \frac{y^2}{3.6^2} = 1 \text{ and in polar coordinates}$$

$$\frac{r^2 \cos^2 \theta}{3.7^2} + \frac{r^2 \sin^2 \theta}{3.6^2} = 1$$

39. $V = \left(\dfrac{a}{r^2} - br\right)\cos\theta$

$$E_r = -\frac{\partial V}{\partial r} = -\left(-\frac{2a}{r^2} - b\right)\cos\theta$$

$$E_r = \left(\frac{2a}{r^2} + b\right)\cos\theta$$

$$E_\theta = -\frac{1}{r}\frac{\partial V}{\partial\theta} = -\frac{1}{r}\left(\frac{a}{r^2} - br\right)(-\sin\theta)$$

$$E_\theta = \left(\frac{a}{r^3} - b\right)\sin\theta$$

40.

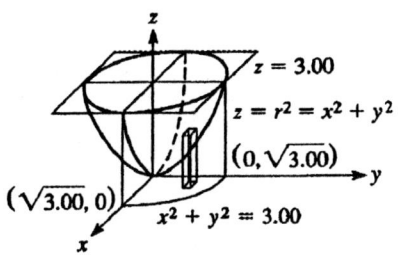

$$V = \pi(\sqrt{3.00})^2(3.00) - 4\int_0^{\sqrt{3.00}} \int_0^{\sqrt{3.00-y^2}} (x^2 + y^2)\,dx\,dy$$

$$\int_0^{\sqrt{3.00}} \left(\frac{x^3}{3} + y^2 x\right)\bigg|_0^{\sqrt{3.00-y^2}} dy$$

$$= \int_0^{\sqrt{3.00}} \left[\frac{(3.00 - y^2)^{3/2}}{3} + y^2\sqrt{3.00-y^2}\right] dy = \int_0^{\sqrt{3.00}} \sqrt{3.00-y^2}\left[\frac{(3.00-y^2)}{3} + y^2\right] dy$$

$$= \int_0^{\sqrt{3.00}} \sqrt{3.00-y^2}\left(\frac{3.00 - y^2 + 3y^2}{3}\right) dy = \int_0^{\sqrt{3.00}} \sqrt{3.00-y^2}\left(\frac{3.00 + 2y^2}{3}\right) dy$$

$$= \int_0^{\sqrt{3.00}} \sqrt{3.00-y^2}(1.00\,dy) + \frac{2}{3}\int_0^{\sqrt{3.00}} \sqrt{3.00-y^2}(y^2\,dy)$$

#15 may be used for the first integral

$$= \left(\frac{y}{2}\sqrt{3.00-y^2} + \frac{3.00}{2}\sin^{-1}\frac{y}{\sqrt{3.00}}\right)\bigg|_0^{\sqrt{3.00}} + \frac{2}{3}\int_0^{\sqrt{3.00}} \sqrt{3.00-y^2}(y^2\,dy)$$

$$= \frac{\sqrt{3.00}}{2}\sqrt{3.00-3.00} + \frac{3.00}{2}\sin^{-1}\frac{\sqrt{3.00}}{\sqrt{3.00}} + \frac{2}{3}\int_0^{\sqrt{3.00}} \sqrt{3.00-y^2}(y^2\,dy)$$

$$= \frac{3.00\pi}{4} + \frac{2}{3}\int_0^{\sqrt{3.00}} \sqrt{3.00-y^2}(y^2\,dy)$$

let $y = \sqrt{3.00}\sin\theta, y^2 = 3.00\sin^2\theta$
$\quad dy = \sqrt{3.00}\cos\theta\,d\theta$

$$= \frac{3.00\pi}{4} + \frac{2}{3}\int_0^{\pi/2} \sqrt{3.00(1-\sin^2\theta)}(3.00\sin^2\theta)\sqrt{3.00}\cos\theta\,d\theta = \frac{3.00\pi}{4} + \frac{2}{3}\int_0^{\pi/2} 9.00\sqrt{\cos^2\theta}\sin^2\theta\cos\theta\,d\theta$$

$$= \frac{3.00\pi}{4} + 6.00\int_0^{\pi/2} \sin^2\theta\cos^2\theta\,d\theta = \frac{3.00\pi}{4} + 6.00\int_0^{\pi/2} \frac{(1-\cos 2\theta)}{2}\frac{(1+\cos 2\theta)}{2}\,d\theta$$

$$= \frac{3.00\pi}{4} + 1.50\int_0^{\pi/2} (1-\cos^2 2\theta)d\theta = \frac{3.00\pi}{4} + 1.50\int_0^{\pi/2} d\theta - 1.50\int_0^{\pi/2} \frac{1+\cos 4\theta}{2}\,d\theta$$

$$= \frac{3.00\pi}{4} + 1.50\theta\bigg|_0^{\pi/2} - 0.750\int_0^{\pi/2} d\theta - \frac{1.50}{8}\int_0^{\pi/2} \cos(4\theta)(4\,d\theta) = \frac{3.00\pi}{4} + \frac{1.50\pi}{2} - \frac{0.750\pi}{2} - \frac{1.50}{8}\sin(4\theta)\bigg|_0^{\pi/2}$$

$$= \frac{9.00\pi}{4}$$

$$V = 9.00\pi - 4\frac{9.00\pi}{8}$$

$$V = 14.1 \text{ cm}^3$$

41. $r = 115 + 10\cos\theta$

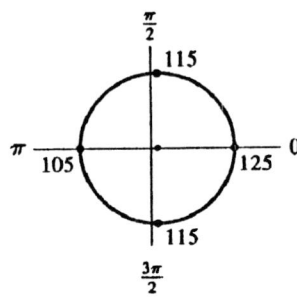

42. See Figure 12-33, $r = 10\cos\theta$

$40° = 0.6981, 90° = 1.5708$

$$A = \frac{1}{2}\int_{0.6981}^{1.5708}(10\cos\theta)^2 d\theta = 50\int_{0.6981}^{1.5708}\cos^2\theta\, d\theta = 25\int_{0.6981}^{1.5708}(1+\cos 2\theta)d\theta$$

$$= 25\theta + \frac{25}{2}\sin 2\theta\Big|_{0.6981}^{1.5708} = 25(1.5708 - 0.6981) + 12.5(\sin 3.1416 - \sin 1.3962)$$

$$= 9.507 \text{ mi}^2$$

43. See Figure 12-34.

$$A = (8.0)(6.0) + \int_0^\pi \frac{1}{2}r^2 d\theta$$

$$\int_0^\pi \frac{1}{2}r^2 d\theta = \frac{1}{2}\int_0^\pi (4.0 + \sin\theta)^2 d\theta = \frac{1}{2}\int_0^\pi (4.0^2 + 8.0\sin\theta + \sin^2\theta)d\theta$$

$$= \frac{1}{2}\int_0^\pi 4.0^2 d\theta + 4.0\int_0^\pi \sin\theta\, d\theta + \frac{1}{2}\int_0^\pi \frac{1-\cos 2\theta}{2}\, d\theta$$

$$= 8.0\theta\Big|_0^\pi + 4.0(-\cos\theta)\Big|_0^\pi + \frac{1}{4}\int_0^\pi d\theta - \frac{1}{8}\int_0^\pi \cos(2\theta)(2\, d\theta)$$

$$= 8.0\pi + 4.0(-\cos\pi + \cos 0) + \frac{1}{4}\theta\Big|_0^\pi - \frac{1}{8}\sin(2\theta)\Big|_0^\pi$$

$$= 8.0\pi + 4.0(2) + \frac{1}{4}\pi - \frac{1}{8}(\sin 2\pi - \sin 0)$$

$$A = (8.0)(6.0) + 8.0\pi + 8.0 + \frac{\pi}{4}$$

$$A = 81.9 \text{ ft}^2$$

44.

 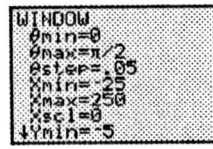 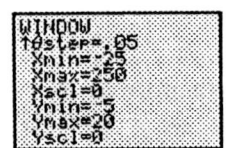 

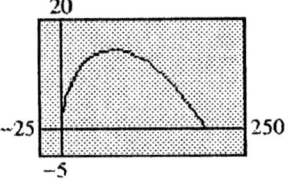

45. $\dfrac{1}{r} = a + b\cos\theta, 1 = ar + br\cos\theta$

$1 = a\sqrt{x^2 + y^2} + bx, (1 - bx)^2 = (a\sqrt{x^2 + y^2})^2$

$1 - 2bx + b^2x^2 = a^2x^2 + a^2y^2$

$(a^2 - b^2)x^2 + a^2y^2 + 2bx - 1 = 0$

$a = b$, parabola (no x^2-term)

$a^2 > b^2$, ellipse (x^2 and y^2 terms positive, but different values)

$a^2 > b^2$, hyperbola (x^2 and y^2 terms of different signs)

EXPANSION OF FUNCTIONS IN SERIES

13.1 Infinite Series

1. $a_n = n^2; n = 1, 2, 3 \ldots$
$a_1 = (1)^1 = 1; a_2 = (2)^2 = 4;$
$a_3 = (3)^2 = 9; a_4 = (4)^2 = 16$

2. $a_n = \dfrac{2}{3^n}; n = 1, 2, 3, 4 \ldots$

$a_1 = \dfrac{2}{3^1} = \dfrac{2}{3}; a_2 = \dfrac{2}{3^2} = \dfrac{2}{9};$

$a_3 = \dfrac{2}{3^3} = \dfrac{2}{27}; a_4 = \dfrac{2}{3^4} = \dfrac{2}{81}$

3. $a_n = \dfrac{1}{n+2}; n = 0, 1, 2, 3, \ldots$

$a_0 = \dfrac{1}{0+2} = \dfrac{1}{2} \qquad a_2 = \dfrac{1}{2+2} = \dfrac{1}{4}$

$a_1 = \dfrac{1}{1+2} = \dfrac{1}{3} \qquad a_3 = \dfrac{1}{3+2} = \dfrac{1}{5}$

4. $a_n = \dfrac{n^2+1}{2n+1}; n = 0, 1, 2, 3, \ldots$

$a_0 = \dfrac{0+1}{0+1} = 1 \qquad a_2 = \dfrac{2^2+1}{4+1} = \dfrac{5}{5} = 1$

$a_1 = \dfrac{1+1}{2+1} = \dfrac{2}{3} \qquad a_3 = \dfrac{3^2+1}{6+1} = \dfrac{10}{7}$

5. $a_n = \left(-\dfrac{2}{5}\right)^n; n = 0, 1, 2, 3, \ldots$

(a) $a_1 = \left(-\dfrac{2}{5}\right)^1 = -\dfrac{2}{5}$

$a_2 = \left(-\dfrac{2}{5}\right)^2 = \dfrac{4}{25}$

$a_3 = \left(-\dfrac{2}{5}\right)^3 = -\dfrac{8}{125}$

$a_4 = \left(-\dfrac{2}{5}\right)^4 = \dfrac{16}{625}$

(b) $S = -\dfrac{2}{5} + \dfrac{4}{25} - \dfrac{8}{125} + \dfrac{16}{625} - \cdots$

6. $a_n = \dfrac{1}{n} + \dfrac{1}{n+1}; n = 0, 1, 2, 3, 4, \ldots$

$a_1 = \dfrac{1}{1} + \dfrac{1}{2} = \dfrac{3}{2}$

$a_2 = \dfrac{1}{2} + \dfrac{1}{3} = \dfrac{5}{6}$

$a_3 = \dfrac{1}{3} + \dfrac{1}{4} = \dfrac{7}{12}$

$a_4 = \dfrac{1}{4} + \dfrac{1}{5} = \dfrac{9}{20}$

(a) $\dfrac{3}{2}, \dfrac{5}{6}, \dfrac{7}{12}, \dfrac{9}{20},$

(b) $\dfrac{3}{2} + \dfrac{5}{6} + \dfrac{7}{12} + \dfrac{9}{20} + \cdots$

7. $a_n = \cos\dfrac{n\pi}{2}, n = 0, 1, 2, 3, \ldots$

$a_0 = \cos\dfrac{0\cdot\pi}{2} = 1$

$a_1 = \cos\dfrac{1\cdot\pi}{2} = 0$

$a_2 = \cos\dfrac{2\cdot\pi}{2} = -1$

$a_3 = \cos\dfrac{3\cdot\pi}{2} = 0$

(a) $1, 0, -1, 0$

(b) $1 + 0 - 1 + 0 - \cdots$

8. $a_n = \dfrac{1}{n(n+1)}; n = 2, 3, 4, 5, \ldots$

$a_2 = \dfrac{1}{2(3)} = \dfrac{1}{6}$

$a_3 = \dfrac{1}{3(4)} = \dfrac{1}{12}$

$a_4 = \dfrac{1}{4(5)} = \dfrac{1}{20}$

$a_5 = \dfrac{1}{5(6)} = \dfrac{1}{30}$

(a) $\dfrac{1}{6}, \dfrac{1}{12}, \dfrac{1}{20}, \dfrac{1}{30}$

(b) $\dfrac{1}{6} + \dfrac{1}{12} + \dfrac{1}{20} + \dfrac{1}{30} + \cdots$

9. $\dfrac{1}{2} + \dfrac{1}{3} + \dfrac{1}{4} + \dfrac{1}{5} + \cdots$

$n = 1, a_1 = \dfrac{1}{2} = \dfrac{1}{1+1};$

$n = 2, a_2 = \dfrac{1}{3} = \dfrac{1}{2+1}$

$a_n = \dfrac{1}{n+1}$

10. $\dfrac{1}{2} + \dfrac{1}{4} + \dfrac{1}{8} + \dfrac{1}{16} + \cdots$

$n = 1, a_1 = \dfrac{1}{2} = \dfrac{1}{2^n} = \dfrac{1}{2^1};$

$n = 2, a_2 = \dfrac{1}{4} = \dfrac{1}{2^n} = \dfrac{1}{2^2}$

$a_n = \dfrac{1}{2^n}$

11. $\dfrac{1}{2 \times 3} + \dfrac{1}{3 \times 4} + \dfrac{1}{4 \times 5} + \dfrac{1}{5 \times 6} + \cdots$

$n = 1, a_2 = \dfrac{1}{(1+1)(1+2)} = \dfrac{1}{2 \times 3}$

$n = 2, a_3 = \dfrac{1}{(2+1)(2+2)} = \dfrac{1}{3 \times 4}$

$a_n = \dfrac{1}{(n+1)(n+2)}$

12. $-\dfrac{2}{3} + \dfrac{4}{9} - \dfrac{8}{27} + \dfrac{16}{81} - \cdots$

$n = 1, \left(-\dfrac{2}{3}\right)^1 = -\dfrac{2}{3}$

$n = 2, \left(-\dfrac{2}{3}\right)^2 = \dfrac{4}{9}; \ a_n = \left(-\dfrac{2}{3}\right)^n$

13. $1 + \dfrac{1}{8} + \dfrac{1}{27} + \dfrac{1}{64} + \dfrac{1}{15} + \cdots$

$S_1 = 1; \ S_2 = 1 + \dfrac{1}{8} = 1.125;$

$S_3 = 1 + \dfrac{1}{8} + \dfrac{1}{27} = 1.1620370;$

$S_4 = 1 + \dfrac{1}{8} + \dfrac{1}{27} + \dfrac{1}{64} = 1.776620;$

$S_5 = 1 + \dfrac{1}{8} + \dfrac{1}{27} + \dfrac{1}{64} + \dfrac{1}{125} = 1.1856620$

Values appear to approach 1.2. Convergent to approx. 1.2.

14. $1 + 2 + 5 + 10 + 17 + \cdots$

$S_0 = 1; \ S_1 = 1 + 2 = 3;$
$S_2 = 1 + 2 + 5 = 8;$
$S_3 = 1 + 2 + 5 + 10 = 18;$
$S_4 = 1 + 2 + 5 + 10 + 17 = 35$
Divergent

15. $1 + \dfrac{1}{2} + \dfrac{2}{3} + \dfrac{3}{4} + \dfrac{4}{5} + \cdots$

$S_0 = 1; \ S_1 = 1 + \dfrac{1}{2} = \dfrac{3}{2} = 1.5$

$S_2 = 1 + \dfrac{1}{2} + \dfrac{2}{3} = \dfrac{13}{6} = 2.1666667$

$S_3 = 1 + \dfrac{1}{2} + \dfrac{2}{3} + \dfrac{3}{4} = \dfrac{35}{12} = 2.9166667$

$S_4 = 1 + \dfrac{1}{2} + \dfrac{2}{3} + \dfrac{3}{4} + \dfrac{4}{5} = \dfrac{223}{60} = 3.7166667$

Divergent

16. $\dfrac{1}{3} - \dfrac{1}{9} + \dfrac{1}{27} - \dfrac{1}{81} + \dfrac{1}{243} - \cdots$

$S_0 = \dfrac{1}{3} = 0.3333333$

$S_1 = \dfrac{1}{3} - \dfrac{1}{9} = \dfrac{2}{9} = 0.2222222$

$S_2 = \dfrac{1}{3} - \dfrac{1}{9} + \dfrac{1}{27} = \dfrac{7}{27} = 0.2592593$

$S_3 = \dfrac{1}{3} - \dfrac{1}{9} + \dfrac{1}{27} - \dfrac{1}{81} = \dfrac{20}{81} = 0.2469136$

$S_4 = \dfrac{1}{3} - \dfrac{1}{9} + \dfrac{1}{27} - \dfrac{1}{81} + \dfrac{1}{243}$

$\quad = \dfrac{61}{243} = 0.2510288$

Convergent to 0.25

17. $\displaystyle\sum_{n=0}^{\infty} \sqrt{n} = \sqrt{0} + \sqrt{1} + \sqrt{2} + \sqrt{3} + \cdots$

$S_0 = 0$
$S_1 = 1$
$S_2 = 2.4142$
$S_3 = 4.1463$
$S_4 = 6.1463$

Series appears divergent

18. $\sum_{n=1}^{\infty} \dfrac{2}{n(n+1)}$

The first five terms are:

$a_1 = \dfrac{2}{1(2)} = 1$; $a_2 = \dfrac{2}{2(3)} = \dfrac{1}{3}$;

$a_3 = \dfrac{2}{3(4)} = \dfrac{1}{6}$; $a_4 = \dfrac{2}{4(5)} = \dfrac{1}{10}$;

$a_5 = \dfrac{2}{5(6)} = \dfrac{1}{15}$

First five partial sums:

$S_0 = 1$

$S_1 = 1 + \dfrac{1}{3} = \dfrac{4}{3} = 1.3333333$

$S_2 = 1 + \dfrac{1}{3} + \dfrac{1}{6} = \dfrac{9}{6} = \dfrac{3}{2} = 1.5$

$S_3 = 1 + \dfrac{1}{3} + \dfrac{1}{6} + \dfrac{1}{10} = \dfrac{16}{10} = \dfrac{8}{5} = 1.6$

$S_4 = 1 + \dfrac{1}{3} + \dfrac{1}{6} + \dfrac{1}{10} + \dfrac{1}{15} = \dfrac{25}{15} = 1.6666667$

Convergent 1.7

19. $\sum_{n=1}^{\infty} \dfrac{2n+1}{n^2(n+1)^2}$

First five terms:

$a_1 = \dfrac{3}{4}$; $a_2 = \dfrac{5}{36}$; $a_3 = \dfrac{7}{144}$; $a_4 = \dfrac{9}{400}$; $a_5 = \dfrac{11}{900}$

First five partial sums:

$S_1 = 0.75$;

$S_2 = \dfrac{3}{4} + \dfrac{5}{36} = 0.8888889$

$S_3 = \dfrac{3}{4} + \dfrac{5}{36} + \dfrac{7}{144} = 0.9375000$

$S_4 = \dfrac{3}{4} + \dfrac{5}{36} + \dfrac{7}{144} + \dfrac{9}{400} = 0.9600000$

$S_5 = \dfrac{3}{4} + \dfrac{5}{36} + \dfrac{7}{144} + \dfrac{9}{400} + \dfrac{11}{900} = 0.9722222$

Convergent, converging to 1 (approx. sum)

20. $\sum_{n=1}^{\infty} \dfrac{n}{2n+1}$

$a_1 = \dfrac{1}{3}$; $a_2 = \dfrac{2}{5}$; $a_3 = \dfrac{3}{7}$; $a_4 = \dfrac{4}{9}$; $a_5 = \dfrac{5}{11}$

$S_1 = \dfrac{1}{3} = 0.3333333$

$S_2 = \dfrac{1}{3} + \dfrac{2}{5} = 0.7333333$

$S_3 = \dfrac{1}{3} + \dfrac{2}{5} + \dfrac{3}{7} = 1.1619048$

$S_4 = \dfrac{1}{3} + \dfrac{2}{5} + \dfrac{3}{7} + \dfrac{4}{9} = 1.6063492$

$S_5 = \dfrac{1}{3} + \dfrac{2}{5} + \dfrac{3}{7} + \dfrac{4}{9} + \dfrac{5}{11} = 2.0608947$

Divergent

21. $1 + 2 + 4 + \cdots + 2^n + \cdots$; $n = 0, 1, 2, 3, \cdots$

$S_0 = 1$
$S_1 = 3$
$S_2 = 7$
$S_3 = 15$
$\vdots$
$S_n = 2^{n+1} - 1$

$\lim_{n \to \infty} S_n = \lim_{n \to \infty} (2^{n+1} - 1) = \infty$, divergent

22. $1 + \dfrac{1}{2} + \dfrac{1}{4} + \cdots + \dfrac{1}{2^n} + \cdots$

$S = \lim_{n \to \infty} S_n = \lim_{n \to \infty} \dfrac{1}{2^n} = 0$; convergent.

$a = 1, r = \dfrac{1}{2}$

$S = \dfrac{a}{1-r} = \dfrac{1}{1 - \frac{1}{2}} = 2$

23. $1 - \dfrac{1}{3} + \dfrac{1}{9} - \cdots + \left(-\dfrac{1}{3}\right)^n + \cdots$

$S = \lim_{n \to \infty} S_n = \lim_{n \to \infty} \left(-\dfrac{1}{3}\right)^n = \lim_{n \to \infty} \dfrac{(-1)^n}{3^n} = 0$

Convergent. $a = 1, n = -\dfrac{1}{3}$

$S = \dfrac{a}{1-r} = \dfrac{1}{1 - \left(-\frac{1}{3}\right)} = \dfrac{1}{\frac{4}{3}} = \dfrac{3}{4}$

24. $1 - \dfrac{3}{2} + \dfrac{9}{4} - \cdots + \left(-\dfrac{3}{2}\right)^n + \cdots$

$S = \lim_{n \to \infty} S_n = \lim_{n \to \infty} \dfrac{(-1)^n}{\left(\frac{2}{3}\right)^n} \to \pm\infty$; divergent

25. $10 + 9 + 8.1 + 7.29 + 6.561 + \cdots + 10(0.9)^n + \cdots$; $n = 0, 1, 2, 3, \ldots$ $S_n = 100 - 100(0.9)^{n+1}$; $n = 0, 1, 2, 3, \ldots$

$\lim_{n \to \infty} S_n = \lim_{n \to \infty} 100 - 100(0.9)^{n+1} = 100$, convergent

26. $4 + 1 + \dfrac{1}{4} + \dfrac{1}{16} + \dfrac{1}{64} + \cdots$

$a_n = \dfrac{4}{4^n}$; $n = 0, 1, 2, 3, \ldots$

$S = \lim_{n \to \infty} \left(\dfrac{4}{4^n}\right) = 0$; therefore, convergent.

$a = 4, r = \dfrac{1}{4}$; $S = \dfrac{4}{1 - \frac{1}{4}} = \dfrac{4}{\frac{3}{4}} = \dfrac{16}{3}$

27. $512 - 64 + 8 - 1 + \dfrac{1}{8} - \cdots$

$a_n = \dfrac{512}{(-8)^n}; \ n = 0, 1, 2, 3, \ldots$

$S = \displaystyle\lim_{n \to \infty} \dfrac{512}{(-8)^n} = 0; \ \text{convergent.}$

$a = 512, r = -\dfrac{1}{8}; \ S = \dfrac{512}{1 + \frac{1}{8}} = \dfrac{512(8)}{9} = \dfrac{4096}{9}.$

28. $16 + 12 + 9 + \dfrac{27}{4} + \dfrac{81}{6} + \cdots; \ n = 0, 1, 2, 3, \ldots$

$a_n = \dfrac{16}{\left(\frac{4}{3}\right)^n}; \ \displaystyle\lim_{n \to \infty} \dfrac{16}{\left(\frac{4}{3}\right)^n} = 0; \ \text{convergent.}$

$a = 16, r = \dfrac{3}{4}; \ S = \dfrac{16}{1 - \frac{3}{4}} = \dfrac{16}{\frac{1}{4}} = 64$

29. Successive square roots of 2 are approximately:
1.414214, 1.189207, 1.090508, 1.044274,
1.021897, 1.010889, 1.005430, 1.002711,
1.001355, 1.000677, 1.000339, 1.000169,
1.000085, 1.000042, 1.000021, 1.000011,
1.000005, 1.000003, 1.000001, 1.000000

(a) It appears that the value of $\displaystyle\lim_{n \to \infty} 2^{(1/2)n}$ is 1.

(b) The infinite series will diverge since each successive term will increase the sum by approximately 1.

30. $\sqrt{0.01}; \ (0.01)^{1/n}, \ n = 2, 4, 8, \ldots; \ (0.01)^{1/2n},$
$n = 1, 2, 3, 4, \ldots$

Successive square roots of (0.01) are:

0.1, 0.316228, 0.562341, 0.749894, 0.865964,
0.930572, 0.964662, 0.982172, 0.991046, 0.995513,
0.997754, 0.998876, 0.999438, 0.999719, 0.999859,
0.999930, 0.999965, 0.999982, 0.999991, 0.999996

(a) It appears the value is approaching 1.

(b) Diverges as each term adds on to previous term by approx. value of 1.

Successive square roots of 100 are approximately:

10.000000, 3.162278, 1.778279, 1.333521, 1.154782,
1.074608, 1.036633, 1.018152, 1.009035, 1.004507,
1.002251, 1.001125, 1.000562, 1.000281, 1.000141,
1.000070, 1.000035, 1.000018, 1.000009, 1.000004

(a) It appears that the value of $\displaystyle\lim_{n \to \infty} 100^{1/n}$ is 1.

(b) The successive square roots of any real number approach 1 in value.

31. $S_n = \dfrac{a_1(1 - r^n)}{(1 - r)}; \ r \neq 1; \ \text{geometric series}$

Series: $\dfrac{1}{2} + \dfrac{1}{4} + \dfrac{1}{8} + \cdots; \ a_n = \dfrac{1}{2^n}, a = \dfrac{1}{2}, r = \dfrac{1}{2}$

$f(x) = \dfrac{a_1(1 - r^x)}{(1 - r)}; \ f(x) = \dfrac{\frac{1}{2}(1 - r^x)}{\left(1 - \frac{1}{2}\right)} = (1 - r^x)$

x	$f(x)$
0	0
1	$\frac{1}{2}$
2	$\frac{3}{4}$
3	$\frac{7}{8}$
4	$\frac{15}{16}$
5	$\frac{31}{32}$

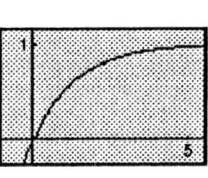

The infinite series approaches 1.

32. $\displaystyle\sum_{n=0}^{\infty} \dfrac{1}{5^n} = \dfrac{1}{5^0} + \dfrac{1}{5^1} + \dfrac{1}{5^2} + \cdots + \dfrac{1}{5^n} + \cdots$

$= 1 + \dfrac{1}{5} + \dfrac{1}{25} + \cdots; \ a = 1, r = \dfrac{1}{5}$

$f(x) = \dfrac{a_1(1 - r^{x+1})}{(1 - r)}; \ f(x) = \dfrac{1\left[1 - \left(\frac{1}{5}\right)^{x+1}\right]}{\left(1 - \frac{1}{5}\right)}$

$f(x) = \dfrac{\left(1 - \frac{1}{5^{x+1}}\right)}{\frac{4}{5}}; \ f(x) = \dfrac{5}{4}\left(1 - \dfrac{1}{5^{x+1}}\right)$

x	$f(x)$
0	1
1	$\frac{6}{5}$
2	$\frac{31}{25}$
3	$\frac{156}{125}$
4	$\frac{781}{625}$
5	$\frac{3906}{3125}$

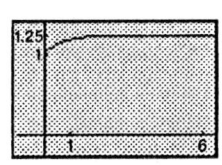

33. $a_1 = 105, r = 1.05;$

$$S = \frac{105(1 - 1.05^n)}{1 - 1.05} = 2100(1.05^n - 1) \to \infty,$$

diverges

On a graphing calculator, let $y_1 = 2100(1.05^x - 1)$, $x_{min} = 0$, $x_{max} = 10$, $y_{min} = 0$, $y_{max} = 1500$.

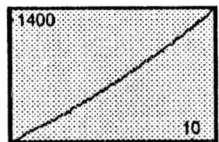

34. Balmer series: $\dfrac{1}{\lambda} = 1.097 \times 10^{-2}\left(\dfrac{1}{2^2} - \dfrac{1}{n^2}\right)$

$n = 3, 4, 5\ldots;\ \lambda = \dfrac{4n^2 \times 10^2}{1.097(n^2 - 4)}$

Line one $(n = 3)$; $\lambda = \dfrac{3600}{1.097(5)} = 656$ nm

Line two $(n = 4)$; $\lambda = \dfrac{6400}{1.097(12)} = 486$ nm

Line three $(n = 5)$; $\lambda = \dfrac{10000}{1.097(21)} = 434$ nm

$$\lim_{n\to\infty} \frac{4n^2 \times 10^2}{1.097(n^2 - 4)} = \lim_{n\to\infty} \frac{4 \times 10^2}{1.097\left(1 - \frac{4}{n^2}\right)}$$

$$= \frac{400}{1.097} = 365 \text{ nm}$$

35. $\displaystyle\sum_{n=0}^{\infty} x^n = 1 + x + x^2 + \cdots + x^n + \cdots$

For $|x| < 1, a_1 = 1, r = x,$

$$S = \frac{1}{1 - x}$$

$$\sum_{n=0}^{\infty} x^n = \frac{1}{1 - x}$$

36. $\displaystyle\sum_{n=0}^{\infty} (-1)^n x^n = 1 - x + x^2 - \cdots + x^n + \cdots$

For $|x| < 1, a_1 = 1, r = -x,$

$$S = \frac{1}{1 - (-x)} = \frac{1}{1 + x}$$

$$\sum_{n=0}^{\infty} (-1)^n x^n = \frac{1}{1 + x}$$

13.2 Maclaurin Series

1. $\quad f(x) = e^x \qquad\qquad f(0) = e^0 = 1$
$\quad\ f'(x) = e^x \qquad\qquad f'(0) = e^0 = 1$
$\quad\ f''(x) = e^x \qquad\qquad f''(0) = e^0 = 1$

$$f(x) = 1 + (1)x + \frac{(1)x^2}{2!} + \cdots$$

$$= 1 + x + \frac{1}{2}x^2 + \cdots$$

2. $\quad f(x) = \sin x \qquad\qquad f(0) = 0$
$\quad\ f'(x) = \cos x \qquad\qquad f'(0) = 1$
$\quad\ f''(x) = -\sin x \qquad\quad f''(0) = 0$
$\quad\ f'''(x) = -\cos x \qquad\quad f'''(0) = -1$
$\quad\ f^{iv}(x) = \sin x \qquad\quad f^{iv}(0) = 0$
$\quad\ f^v(x) = \cos x \qquad\quad f^v(0) = 1$

Therefore,

$$f(x) = \sin x = f'(0)x + f'''(0)\frac{x^3}{3!} + f^v(0)\frac{x^5}{5!} + \cdots$$

$$\sin x = 1x - 1\frac{x^3}{6} + 1\frac{x^5}{120} + \cdots$$

$$\sin x = x - \frac{1}{6}x^3 + \frac{1}{120}x^5 - \cdots$$

3. $\quad f(x) = \cos x \qquad\qquad f(0) = 1$
$\quad\ f'(x) = -\sin x \qquad\quad f'(0) = 0$
$\quad\ f''(x) = -\cos x \qquad\quad f''(0) = -1$
$\quad\ f'''(x) = \sin x \qquad\qquad f'''(0) = 0$
$\quad\ f^{iv}(x) = \cos x \qquad\quad f^{iv}(0) = 1$

$$f(x) = \cos x = f(0) + f''(0)\frac{x^2}{2!} + f^{iv}(0)\frac{x^4}{4!} - \cdots$$

$$\cos x = 1 - 1\frac{x^2}{2} + 1\frac{x^4}{24} - \cdots$$

$$\cos x = 1 - \frac{1}{2}x^2 + \frac{1}{24}x^4 - \cdots$$

4. $\quad f(x) = \ln(1 + x) \qquad\qquad f(0) = 0$

$$f'(x) = \frac{1}{(1 + x)} \qquad\qquad f'(0) = 1$$

$$f''(x) = -\frac{1}{(1 + x)^2} \qquad\qquad f''(0) = -1$$

$$f'''(x) = \frac{2}{(1 + x)^3} \qquad\qquad f'''(0) = 2$$

$$\ln(1 + x) = f'(0)x + f''(0)\frac{x^2}{2!} + f'''(0)\frac{x^3}{3!} + \cdots$$

$$= 1x - 1\frac{x^2}{2} + 2\frac{x^3}{6} + \cdots$$

$$= x - \frac{1}{2}x^2 + \frac{1}{3}x^3 - \cdots$$

5. $f(x) = (1+x)^{1/2}$ $f(0) = 1$

$f'(x) = \dfrac{1}{2}(1+x)^{-1/2}$ $f'(0) = \dfrac{1}{2}$

$f''(x) = -\dfrac{1}{4}(1+x)^{-3/2}$ $f''(0) = -\dfrac{1}{4}$

$f(x) = 1 + \dfrac{1}{2}x - \dfrac{1}{4}\dfrac{x^2}{2!} + \cdots$

$\qquad = 1 + \dfrac{1}{2}x - \dfrac{1}{8}x^2 + \cdots$

6. $f(x) = \dfrac{1}{(1-x)^{1/3}}$ $f(0) = 1$

$f'(x) = \dfrac{1}{3(1-x)^{4/3}}$ $f'(0) = \dfrac{1}{3}$

$f''(x) = \dfrac{1}{9(1-x)^{7/3}}$ $f''(0) = -\dfrac{4}{9}$

$\dfrac{1}{(1-x)^{1/3}} = 1 + \dfrac{1}{3}x + \dfrac{4}{9}\dfrac{x^2}{2} + \cdots$

$\qquad\qquad = 1 + \dfrac{1}{3}x + \dfrac{2}{9}x^2 + \cdots$

7. $f(x) = e^{-2x}$ $f(0) = 1$
$f'(x) = -2e^{-2x}$ $f'(0) = -2$
$f''(x) = 4e^{-2x}$ $f''(0) = 4$

$e^{-2x} = 1 - 2x + 4\dfrac{x^2}{2} - \cdots = 1 - 2x + 2x^2 - \cdots$

8. $f(x) = \dfrac{1}{2}(e^x + e^{-x})$ $f(0) = 1$

$f'(x) = \dfrac{1}{2}(e^x - e^{-x})$ $f'(0) = 0$

$f''(x) = \dfrac{1}{2}(e^x + e^{-x})$ $f''(0) = 1$

$f'''(x) = \dfrac{1}{2}(e^x - e^{-x})$ $f'''(0) = 0$

$f^{iv}(x) = \dfrac{1}{2}(e^x + e^{-x})$ $f^{iv}(0) = 1$

$\dfrac{1}{2}(e^x + e^{-x}) = f(0) + f''(0)\dfrac{x^2}{2!} + f^{iv}(0)\dfrac{x^4}{4!} + \cdots$

$\qquad\qquad = 1 + 1 \cdot \dfrac{x^2}{2} + 1 \cdot \dfrac{x^4}{24} + \cdots$

$\qquad\qquad = 1 + \dfrac{1}{2}x^2 + \dfrac{1}{24}x^4 + \cdots$

9. $f(x) = \cos 4\pi x$
$f'(x) = -4\pi \sin 4\pi x$
$f''(x) = -16\pi^2 \cos 4\pi x$
$f'''(x) = 64\pi^3 \sin 4\pi x$
$f^{iv}(x) = 256\pi^4 \cos 4\pi x$
$f(0) = 1$
$f'(0) = 0$
$f''(0) = -16\pi^2$
$f'''(0) = 0$
$f^{iv}(0) = 256\pi^4$

$f(x) = 1 + 0 \cdot x - \dfrac{16\pi^2}{2!} \cdot x^2 + 0 \cdot \dfrac{x^3}{3!}$

$\qquad + \dfrac{256\pi^4}{4!}x^4 + \cdots$

$\qquad = f(x) = 1 - 8\pi^2 x^2 + \dfrac{32\pi^4 x^4}{3} - \cdots$

10. $f(x) = e^x \sin x$
$f'(x) = e^x(\cos x + \sin x)$
$f''(x) = 2e^x \cos x$
$f'''(x) = 2e^x(\cos x - \sin x)$
$f(0) = 0$
$f'(0) = 1$
$f''(0) = 2$
$f'''(0) = 2$

$e^x \sin x = f(0) + 1x + \dfrac{2x^2}{2!} + \dfrac{2x^3}{3!} + \cdots$

$\qquad\qquad = x + x^2 + \dfrac{1}{3}x^3 + \cdots$

11. $f(x) = \dfrac{1}{(1-x)}$ $f(0) = 1$

$f'(x) = \dfrac{1}{(1-x)^2}$ $f'(0) = 1$

$f''(x) = \dfrac{2}{(1-x)^3}$ $f''(0) = 2$

$\dfrac{1}{(1-x)} = 1 + x + \dfrac{2x^2}{2} + \cdots = 1 + x + x^2 + \cdots$

12. $f(x) = \dfrac{1}{(1+x)^2}$ $f(0) = 1$

$f'(x) = \dfrac{-2}{(1+x)^3}$ $f'(0) = -2$

$f''(x) = \dfrac{6}{(1+x)^4}$ $f''(0) = 6$

$\dfrac{1}{(1+x)^2} = 1 - 2x + \dfrac{6x^2}{2} - \cdots$

$\qquad\qquad = 1 - 2x + 3x^2 - \cdots$

13. $f(x) = \ln(1 - 2x)$

$$f'(x) = \frac{1}{1 - 2x}(-2) = \frac{-2}{1 - 2x}$$

$$f''(x) = \frac{0 - (-2)(-2)}{(1 - 2x)^2} = \frac{-4}{(1 - 2x)^2}$$

$$f'''(x) = \frac{0 - (-4)2(1 - 2x)(-2)}{(1 - 2x)^4} = \frac{-16(1 - 2x)}{(1 - 2x)^4}$$

$$f(0) = \ln 1 = 0$$

$$f'(0) = -2$$

$$f''(0) = -4$$

$$f'''(0) = -16$$

$$\ln(1 - 2x) = 0 + (-2)x + (-4)\frac{x^2}{2!} + (-16)\frac{x^3}{3!} + \cdots$$

$$= -2x - 2x^2 - \frac{8}{3}x^3 - \cdots$$

14. $f(x) = (1 + x)^{3/2}$ $\qquad$ $f(0) = 1$

$$f'(x) = \frac{3}{2}(1 + x)^{1/2} \qquad f'(0) = \frac{3}{2}$$

$$f''(x) = \frac{3}{4(1 + x)^{1/2}} \qquad f''(0) = \frac{3}{4}$$

$$(1 + x)^{3/2} = 1 + \frac{3}{2}x + \frac{3}{4}\frac{x^2}{2} - \cdots$$

$$= 1 + \frac{3}{2}x + \frac{3}{8}x^2 - \cdots$$

15. $f(x) = \cos^2 x$ $\qquad$ $f(0) = 1$
$f'(x) = -2\sin x \cos x \qquad f'(0) = 0$
$f''(x) = 2 - 4\cos^2 x \qquad f''(0) = -2$
$f'''(x) = 8\sin x \cos x \qquad f'''(0) = 0$
$f^{iv}(x) = 16\cos^2 x - 8 \qquad f^{iv}(0) = 8$

$$\cos^2 x = 1 - 2\frac{x^2}{2!} + 8\frac{x^4}{4!} - \cdots$$

$$= 1 - x^2 + \frac{1}{3}x^4 - \cdots$$

16. $f(x) = \ln(1 + 4x)$ $\qquad$ $f(0) = 0$

$$f'(x) = \frac{4}{(1 + 4x)} \qquad f'(0) = 4$$

$$f''(x) = \frac{-16}{(1 + 4x)^2} \qquad f''(0) = -16$$

$$f'''(x) = \frac{128}{(1 + 4x)^3} \qquad f'''(0) = 128$$

$$\ln(1 + 4x) = 4x - 16\frac{x^2}{2!} + 128\frac{x^3}{3!} - \cdots$$

$$= 4x - 8x^2 + \frac{64}{3}x^3 - \cdots$$

17. $f(x) = \tan^{-1} x$

$$f'(x) = \frac{1}{1 + x^2} = (1 + x^2)^{-1}$$

$$f''(x) = -(1 + x^2)^{-2}2x = -2x(1 + x^2)^{-2}$$
$$f'''(x) = -2x[-2(1 + x^2)^{-3}(2x)] + (1 + x^2)^{-2}(-2)$$
$$f(0) = 0$$

$$f'(0) = 1$$

$$f''(0) = 0$$
$$f'''(0) = -2$$

$$f(x) = 0 + 1x + \frac{0x^2}{2!} - \frac{2x^3}{3!} + \cdots = x - \frac{1}{3}x^3 + \cdots$$

18. $f(x) = \cos x^2$
$f'(x) = -2x\sin x^2$
$f''(x) = -2(2x^2\cos x^2 + \sin x^2)$
$f'''(x) = 8x^3\sin x^2 - 12x\cos x^2$
$f^{iv}(x) = 16x^4\cos x^2 + 48x^2\sin x^2 - 12\cos x^2$
$f(0) = 1$
$f'(0) = 0$
$f''(0) = 0$
$f'''(0) = 0$
$f^{iv}(0) = -12$

$$\cos x^2 = 1 + \frac{(-12x^4)}{4!} = 1 - \frac{12x^4}{24} + \cdots$$

$$= 1 - \frac{1}{2}x^4 + \cdots$$

19. $f(x) = \tan x$
$f'(x) = \sec^2 x$
$f''(x) = 2\sec x \sec x \tan x$
$\qquad = 2\sec^2 x \tan x$
$f'''(x) = 2\sec^4 x + 2\sec^2 x \tan^2 x$
$f^{iv}(x) = 8\sec^4 \tan x + 4\sec^4 x \tan x + 4\sec^2 \tan^3 x$
$f(0) = 0$
$f'(0) = 1$
$f''(0) = 0$
$f'''(0) = 2$
$f^{iv}(0) = 0$

$$\tan x = x + 2\frac{x^3}{3!} + \cdots = x + \frac{1}{3}x^3 + \cdots$$

20. $f(x) = \sec x$ $f(0) = 1$
$f'(x) = \sec x \tan x$ $f'(0) = 0$
$f''(x) = \sec^3 x + \sec x \tan^2 x$ $f''(0) = 1$

$$\sec x = 1 + \frac{x^2}{2} + \cdots$$

21. $f(x) = \ln \cos x$

$$f'(x) = -\frac{1}{\cos x} \sin x = -\tan x$$

$f''(x) = -\sec^2 x$
$f'''(x) = -2 \sec x \sec x \tan x = -2 \sec^2 x \tan x$
$f^{iv}(x) = -2 \sec^2 x \sec^2 x - 2 \tan x (2 \sec x \sec x \tan x)$
$f(0) = \ln 1 = 0$

$f'(0) = 0$

$f''(0) = -1$
$f'''(0) = 0$
$f^{iv}(0) = -2 - 0 = -2$

$$f(x) = 0 + 0x - \frac{1x^2}{2!} + \frac{0x^3}{3!} - \frac{2x^4}{4!} + \cdots$$

$$= -\frac{1}{2}x^2 - \frac{1}{12}x^4 - \cdots$$

22. $f(x) = xe^{\sin x}$
$f'(x) = e^{\sin x}(x \cos x + 1)$
$f''(x) = e^{\sin x}[\cos x - x \sin x + \cos x(x \cos x + 1)]$
$f(0) = 0$
$f'(0) = 1$
$f''(0) = 2$

$$xe^{\sin x} = x + \frac{2x^2}{2} + \cdots = x + x^2 + \cdots$$

23. $f(x) = \sin^2 x$ $f(0) = 0$
$f'(x) = 2 \sin x \cos x = \sin 2x$ $f'(0) = 0$
$f''(x) = 2 \cos 2x$ $f''(0) = 2$
$f'''(x) = -4 \sin 2x$ $f'''(0) = 0$
$f^{iv}(x) = -8 \cos 2x$ $f^{iv}(0) = -8$

$$\sin^2 x = \frac{2x^2}{2!} - \frac{8x^4}{4!} + \cdots = x^2 - \frac{1}{3}x^4 + \cdots$$

24. $f(x) = e^{-x^2}$ $f(0) = 1$
$f'(x) = -2xe^{-x^2}$ $f'(0) = 0$
$f''(x) = 2e^{-x^2}(2x^2 - 1)$ $f''(0) = -2$

$$e^{-x^2} = 1 - \frac{2x^2}{2!} + \cdots = 1 - x^2 + \cdots$$

25. (a) It is not possible to find a Maclaurin's expansion for $f(x) = \csc x$ since the formula is not defined when $x = 0$.

(b) $f(x) = \ln x$ is not defined when $x = 0$.

26. (a) $f(x) = \sqrt{x}$ $f(0) = 0$

$$f'(x) = \frac{1}{2\sqrt{x}}$$ $f'(0)$ undefined

$$f''(x) = -\frac{1}{4x^{3/2}}$$ $f''(0)$ undefined

Maclaurin's expansion not possible; derivatives undefined for $x = 0$

(b) $f(x) = \sqrt{1+x}$ $f(0) = 1$

$$f'(x) = \frac{1}{2\sqrt{1+x}}$$ $f'(0) = \frac{1}{2}$

$$f''(x) = -\frac{1}{4(1+x)^{3/2}}$$ $f''(0) = -\frac{1}{4}$

Maclaurin's expansion is possible; $f(x)$ and its derivatives are defined at $x = 0$.

27. (a) $f(x) = e^x$ $f(0) = 1$
$f'(x) = e^x$ $f'(0) = 1$
$f''(x) = e^x$ $f''(0) = 1$

$$e^x = 1 + x + \frac{1}{2}x^2 + \cdots$$

(b) $f(x) = e^{x^2}$ $f(0) = 1$
$f'(x) = 2xe^{x^2}$ $f'(0) = 0$
$f''(x) = 2e^{x^2}(2x^2 + 1)$ $f''(0) = 2$
$f'''(x) = 4xe^{x^2}(2x^2 + 3)$ $f'''(0) = 0$
$f^{iv}(x) = 4e^{x^2}(4x^4 + 12x^2 + 3)$ $f^{iv}(0) = 12$

$$e^{x^2} = 1 + \frac{2x^2}{2} + \frac{12x^4}{4!} + \cdots = 1 + x^2 + \frac{1}{2}x^4 + \cdots$$

28. $f(x) = (1+x)^n$
$f'(x) = n(1+x)^{n-1}$
$f''(x) = n(n-1)(1+x)^{n-2}$
$f'''(x) = n(n-1)(n-2)(1+x)^{n-3}$
$f(0) = 1^n = 1$
$f'(0) = n$
$f''(0) = n(n-1)$
$f'''(0) = n(n-1)(n-2)$

$f(x) = (1+x)^n$

$$= 1 + nx + n(n-1)\frac{x^2}{2} + n(n-1)(n-2)\frac{x^3}{6} + \cdots$$

29. The Maclaurin's expansion of $f(x) = e^{3x}$ is

$$f(x) = 1 + 3x + \frac{9}{2}x^2 + \frac{9}{2}x^3 + \cdots. \text{ The linearization}$$

is $L(x) = 1 + 3x$, the first two terms of the expansion.

30.

$$f(x) = x^4 + 2x^2 \qquad f(0) = 0$$
$$f'(x) = 4x^3 + 4x \qquad f'(0) = 0$$
$$f''(x) = 12x^2 + 4 \qquad f''(0) = 4$$
$$f'''(x) = 24x \qquad f'''(0) = 0$$
$$f^{iv}(x) = 24 \qquad f^{iv}(0) = 24$$
$$f^{(n)}(x) = 0, n \geq 5$$

$$f(x) = \frac{4x^2}{2!} + \frac{24x^4}{4!} = 2x^2 + x^4$$

31. $0 \leq R \leq 1$; $R = e^{-0.001t}$

$$\frac{dR}{dt} = -0.001e^{-0.001t}; \quad \frac{d^2R}{dt^2} = 1 \times 10^{-6}e^{-0.001t}$$

$$f(0) = 1; \quad f'(0) = -0.001; \quad f''(0) = 1 \times 10^6$$

$$e^{-0.001t} = 1 - 0.001t + 1 \times 10^{-6}\frac{t^2}{2!} - \cdots$$

$$= 1 - 0.001t + 10 \times 10^{-7}\frac{t^2}{2} - \cdots$$

$$= 1 - 0.001t + (5 \times 10^{-7})t^2 - \cdots$$

32. $c^2 = a^2 + (a+b)^2 - 2a(a+b)\cos\dfrac{s}{a}$; Let $\dfrac{s}{a} = x$

$$f(x) = \cos x \qquad f(0) = 1$$
$$f'(x) = -\sin x \qquad f'(0) = 0$$
$$f''(x) = -\cos x \qquad f''(0) = -1$$

$$\cos x = 1 - \frac{x^2}{2} + \cdots$$

$$\cos\frac{s}{a} = 1 - \frac{s^2}{2a^2} + \cdots = \frac{2a^2 - s^2}{2a^2}$$

$$c^2 = a^2 + (a+b)^2 - 2a(a+b)\left(\frac{2a^2 - s^2}{2a^2}\right)$$

$$= a^2 + a^2 + 2ab + b^2 - (a+b)\left(\frac{2a^2 - s^2}{a}\right)$$

$$= 2a^2 + 2ab + b^2 - \frac{(2a^3 + 2a^2b - as^2 - bs^2)}{a}$$

$$= \frac{2a^3 + 2a^2b + ab^2 - 2a^3 - 2a^2b + as^2 + bs^2}{a}$$

$$= \frac{ab^2 + as^2 + bs^2}{a}$$

13.3   Certain Operations with Series

1. $f(x) = e^{3x}$; $e^x = 1 + x + \dfrac{x^2}{2!} + \dfrac{x^3}{3!} + \cdots$

$$f(x) = e^{3x} = 1 + 3x + \frac{(3x)^2}{2!} + \frac{(3x)^3}{3!} + \cdots$$

$$= 1 + 3x + \frac{9}{2}x^2 + \frac{9}{2}x^3 + \cdots$$

2. $f(x) = e^{-2x}$

$$f(x) = 1 + (-2x) + \frac{(-2x)^2}{2} + \frac{(-2x)^3}{6} + \cdots$$

$$f(x) = 1 - 2x + 2x^2 - \frac{4}{3}x^3 + \cdots$$

3. $f(x) = \sin\left(\dfrac{1}{2}x\right)$; $\sin x = x - \dfrac{x^3}{3!} + \dfrac{x^5}{5!} - \dfrac{x^7}{7!} + \cdots$

$$f(x) = \sin\left(\frac{1}{2}x\right)$$

$$= \frac{1}{2}x - \frac{\left(\frac{1}{2}x\right)^3}{6} + \frac{\left(\frac{1}{2}x\right)^5}{120} - \frac{\left(\frac{1}{2}x\right)^7}{7!} + \cdots$$

$$= \frac{1}{2}x - \frac{x^3}{2^3 3!} + \frac{x^5}{2^5 5!} - \frac{x^7}{2^7 7!} + \cdots$$

4. $f(x) = \sin x^4 = x^4 - \dfrac{(x^4)^3}{3!} + \dfrac{(x^4)^5}{5!} - \dfrac{(x^4)^7}{7!} + \cdots$

$$= x^4 - \frac{1}{6}x^{12} + \frac{1}{120}x^{20} - \frac{1}{5040}x^{28} + \cdots$$

5. $f(x) = 1 - \dfrac{x^2}{2!} + \dfrac{x^4}{4!} - \dfrac{x^6}{6!} + \cdots$

$$\cos 4x = f(4x) = 1 - \frac{(4x)^2}{2!} + \frac{(4x)^4}{4!} - \frac{4x)^6}{6!} + \cdots$$

$$= 1 - 8x^2 + \frac{32}{3}x^4 - \frac{256}{45}x^6 + \cdots$$

$$x\cos 4x = xf(4x)$$

$$= x\left(1 - 8x^2 + \frac{32}{3}x^4 - \frac{256}{45}x^6 + \cdots\right)$$

$$= x - 8x^3 + \frac{32}{3}x^5 - \frac{256}{45}x^7 + \cdots$$

6. $f(x) = \sqrt{1 - x^4} = [1 + (-x^4)]^{1/2}$

$$(1 + x)^n$$

$$= 1 + nx + \frac{n(n-1)}{2!}x^2 + \frac{n(n-1)(n-2)}{3!}x^3 + \cdots$$

$$[1 + (-x^4)]^{1/2}$$

$$= 1 + \frac{1}{2}(-x^4) + \frac{\frac{1}{2}\left(-\frac{1}{2}\right)}{2!}(-x^4)^2$$

$$+ \frac{\frac{1}{2}\left(-\frac{1}{2}\right)\left(-\frac{3}{2}\right)}{6}(-x^4)^3 + \cdots$$

$$= 1 - \frac{1}{2}x^4 - \frac{1}{8}x^8 - \frac{1}{16}x^{12} - \cdots$$

7. $f(x) = \ln(1+x^2)$; $\ln(1+x) = x - \dfrac{x^2}{2} + \dfrac{x^3}{3} - \dfrac{x^4}{4} + \cdots$

$$\ln(1+x^2) = x^2 - \frac{(x^2)^2}{2} + \frac{(x^2)^3}{3} - \frac{(x^2)^2}{4} + \cdots$$
$$= x^2 - \frac{1}{2}x^4 + \frac{1}{3}x^6 - \frac{1}{4}x^8 + \cdots$$

8. $f(x) = x^2 \cdot \ln(1-x)^2$
$$= 2x^2 \cdot \ln(1-x)$$
$$= 2x^2 \cdot \left(-x - \frac{x^2}{2} - \frac{x^3}{3} - \frac{x^4}{4} - \cdots \right)$$
$$= -2x^3 - x^4 - \frac{2x^5}{3} - \frac{x^6}{2} - \cdots$$

9. $\displaystyle\int_0^1 \sin x^2\, dx = \int_0^1 \left(x^2 - \frac{(x^2)^3}{3!} + \frac{(x^2)^5}{5!} \right) dx$
$$= -\int_0^1 \left(x^2 - \frac{x^6}{6} + \frac{x^{10}}{120} \right) dx$$
$$= \left(\frac{1}{3}x^3 - \frac{1}{42}x^7 + \frac{1}{1320}x^{11} \right)\Big|_0^1$$
$$= \frac{1}{3} - \frac{1}{42} + \frac{1}{1320} = 0.3103$$

10. $\displaystyle\int_0^{0.4} \sqrt[4]{1-2x^2}\, dx$
$$= \int_0^{0.4} (1-2x^2)^{1/4}\, dx$$
$$= \int_0^{0.4} [1 + (-2x^2)]^{1/4}\, dx$$
$$= \int_0^{0.4} \left[1 + \frac{1}{4}(-2x^2) + \frac{1}{4}\left(-\frac{3}{4}\right)\frac{(-2x^2)^2}{2} \right] dx$$
$$= \int_0^{0.4} \left(1 - \frac{1}{2}x^2 - \frac{3}{8}x^4 \right) dx$$
$$= \left(x - \frac{1}{6}x^3 - \frac{3}{40}x^5 \right)\Big|_0^{0.4}$$
$$= 0.4 - \frac{1}{6}(0.4)^3 - \frac{3}{40}(0.4)^5 = 0.3886$$

11. $\displaystyle\int_0^{0.2} \cos\sqrt{x}\, dx = \int_0^{0.2} \left(1 - \frac{(\sqrt{x})^2}{2} + \frac{(\sqrt{x})^4}{24} \right) dx$
$$= \int_0^{0.2} \left(1 - \frac{1}{2}x - \frac{1}{24}x^2 \right) dx$$
$$= \left(x - \frac{1}{4}x^2 + \frac{1}{72}x^3 \right)\Big|_0^{0.2}$$
$$= 0.2 - \frac{1}{4}(0.2)^2 + \frac{1}{72}(0.2)^3$$
$$= 0.1901$$

12. $\displaystyle\int_{0.1}^{0.2} \frac{\cos x - 1}{x}\, dx$
$$= \int_{0.1}^{0.2} \frac{\left(1 - \frac{x^2}{2} + \frac{x^4}{24} - 1 \right)}{x}\, dx$$
$$= \int_{0.1}^{0.2} \left(-\frac{1}{2}x^1 + \frac{1}{24}x^3 \right) dx$$
$$= \left(-\frac{1}{4}x^2 + \frac{1}{96}x^4 \right)\Big|_{0.1}^{0.2}$$
$$= -\frac{1}{4}(0.2)^2 + \frac{1}{96}(0.2)^4 + \frac{1}{4}(0.1)^2 - \frac{1}{96}(0.1)^4$$
$$= -0.00748$$

13. $f(x) = \dfrac{2}{1-x^2} = \dfrac{1}{1+x} + \dfrac{1}{1-x}$

$f(x) = 1 - x + x^2 - x^3 + \cdots + 1 + x + x^2 + x^3 + \cdots$
$f(x) = 1 - x + x^2 - x^3 + x^4 - x^5 + x^6 - x^7 + x^8 + \cdots$
$\qquad\quad + 1 + x + x^2 + x^3 + x^4 + x^5 + x^6 + x^7 + x^8 + \cdots$
$f(x) = 2(1 + x^2 + x^4 + x^6 + \cdots)$

14. $f(x) = \dfrac{1}{2}(e^x + e^{-x})$

$$e^x = 1 + x + \frac{x^2}{2} + \frac{x^3}{3!} + \frac{x^4}{4!} + \frac{x^5}{5!} + \frac{x^6}{6!}$$
$$+ \frac{x^7}{7!} + \cdots$$
$$e^{-x} = 1 - x + \frac{x^2}{2} - \frac{x^3}{3!} + \frac{x^4}{4!} + \frac{x^5}{5!} + \frac{x^6}{6!}$$
$$- \frac{x^7}{7!} + \cdots$$

$$f(x) = \frac{1}{2}(e^x - e^{-x})$$
$$= \frac{1}{2}\left(2x + 2\cdot\frac{x^3}{3!} + 2\cdot\frac{x^5}{5!} + 2\cdot\frac{x^7}{7!} \right)$$
$$\frac{1}{2}(e^x + e^{-x}) = x + \frac{x^3}{3!} + \frac{x^5}{5!} + \frac{x^7}{7!} + \cdots$$

15. $e^x \sin x = f(x)$

$$e^x = 1 + x + \frac{x^2}{2!} + \frac{x^3}{3!} + \cdots \qquad (1)$$
$$\sin x = x - \frac{x^3}{3!} + \frac{x^5}{5!} - \cdots \qquad (2)$$

Multiply (1) by x from (2): $x + x^2 + \dfrac{x^3}{2!} + \dfrac{x^4}{3!} + \cdots$

Multiply (1) by $-\dfrac{x^3}{3!}$ from (2): $-\dfrac{x^3}{3!} - \dfrac{x^4}{3!} - \cdots$

Combine:

$$e^x \sin x = x + x^2 + \frac{x^3}{2!} + \frac{x^4}{3!} - \frac{x^3}{3!} - \frac{x^4}{3!}$$

$$= x + x^2 + \frac{1}{2}x^3 - \frac{1}{6}x^3$$

$$= x + x^2 + \frac{2}{6}x^3 + \cdots$$

$$= x + x^2 + \frac{1}{3}x^3 + \cdots$$

16. $f(x) = \tan x = \dfrac{\sin x}{\cos x}$

$$\sin x = x - \frac{x^3}{3!} + \frac{x^5}{5!} - \frac{x^7}{7!} + \cdots$$

$$\cos x = 1 - \frac{x^2}{2!} + \frac{x^4}{4!} - \cdots$$

$$
\begin{array}{r}
x + \frac{1}{3}x^3 + \frac{4}{30}x^5 + \\[4pt]
\hline
1 - \frac{x^2}{2!} + \frac{x^4}{4!}\,\big)\ x - \frac{x^3}{3!} + \frac{x^5}{5!} - \frac{x^7}{7!} \\[4pt]
x - \frac{x^3}{2!} + \frac{x^5}{4!} \\[4pt]
\hline
\frac{x^3}{3} - \frac{4x^5}{120} - \frac{x^7}{7!} \\[4pt]
\frac{x^3}{3} - \frac{1}{6}x^5 + \frac{1}{82}x^7 \\[4pt]
\hline
\frac{4x^5}{30} -
\end{array}
$$

$$\tan x = x + \frac{1}{3}x^3 + \frac{2}{15}x^5 + \cdots$$

17. $\dfrac{d(\sin x)}{dx} = \dfrac{d}{dx}\left(x - \dfrac{x^3}{3!} + \dfrac{x^5}{5!} - \cdots \right)$

$$= 1 - \frac{3x^2}{3!} + \frac{5x^4}{4!} - \cdots$$

$$= 1 - \frac{x^2}{2} + \frac{x^4}{24} - \cdots = \cos x$$

18. $e^x = 1 + x + \dfrac{x^2}{2!} + \dfrac{x^3}{3!} + \dfrac{x^4}{4!} + \cdots$

$$\frac{d}{dx}e^x = 0 + 1 + \frac{2x}{2!} + \frac{3x^2}{6} + \frac{4x^3}{24} + \cdots$$

$$= 1 + x + \frac{x^2}{2} + \frac{x^3}{6} + \cdots$$

$$= 1 + x + \frac{x^2}{2!} + \frac{x^3}{3!} + \cdots = e^x$$

19. $\cos x = 1 - \dfrac{x^2}{2!} + \dfrac{x^4}{4!} - \dfrac{x^6}{6!} + \cdots$

$$\int \cos x\, dx$$

$$= \int dx - \frac{1}{2}\int x^2\, dx + \frac{1}{4!}\int x^4\, dx$$

$$\quad - \frac{1}{6!}\int x^6\, dx + \cdots$$

$$= x - \frac{1}{2}\frac{x^3}{3} + \frac{1}{4!}\frac{x^5}{5} - \frac{1}{6!}\frac{x^7}{7} + \cdots$$

$$= x - \frac{x^3}{3!} + \frac{x^5}{5!} - \frac{x^7}{7!} + \cdots = \sin x$$

20. $\dfrac{-1}{1-x} = -\left(\dfrac{1}{1-x}\right)$

[by Exercise #11, section 29-2]

$$-(1 + x + x^2 + \cdots) = -1 - x - x^2 - \cdots$$

$$\int -\frac{dx}{(1-x)} = -\int dx - \int x\, dx - \int x^2\, dx$$

$$= -x - \frac{x^2}{2} - \frac{x^3}{3}$$

$$\ln(1 + x) = x - \frac{x^2}{2} + \frac{x^3}{3} - \cdots$$

$$\ln(1 - x) = -x - \frac{(-x)^2}{2} + \frac{(-x)^3}{3} - \cdots$$

$$= -x - \frac{x^2}{2} - \frac{x^3}{3} - \cdots$$

Therefore, $\displaystyle\int -\frac{dx}{(1-x)} = \ln(1-x)$

21. $\displaystyle\int_0^1 e^x\, dx = e^x\Big|_0^1 = e - e^0 = e - 1$

$$= 2.7182818 - 1 = 1.7182818$$

$$
\begin{array}{ll}
f(x) = e^x & f(0) = e^0 = 1 \\
f'(x) = e^x & f'(0) = 1 \\
f''(x) = e^x & f''(0) = 1 \\
f'''(x) = e^x & f'''(0) = 1
\end{array}
$$

$$e^x = 1 + x + \frac{x^2}{2!} + \frac{x^3}{3!} + \cdots$$

$$\int_0^1 \left(1 + x + \frac{x^2}{2} + \frac{x^3}{6}\right) dx$$

$$= x + \frac{x^2}{2} + \frac{x^3}{6} + \frac{x^4}{24}\Big|_0^1$$

$$= 1 + \frac{1}{2} + \frac{1}{6} + \frac{1}{24}$$

$$= 1.7083333$$

22. $\lim\limits_{x \to 0} \dfrac{\sin x}{x} = \lim\limits_{x \to 0} \dfrac{x - \frac{x^3}{3!} + \frac{x^5}{5!} - \frac{x^7}{7!} + \cdots}{x}$

$= \lim\limits_{x \to 0} \left[1 - \dfrac{x^2}{3!} + \dfrac{x^4}{5!} - \dfrac{x^6}{7!} + \cdots \right]$

$= 1 - 0 + 0 - 0 + \cdots = 1$

Eq. (27-1); $\lim\limits_{x \to 0} \dfrac{\sin \theta}{\theta} = 1$; the same result.

23. $y = x^2 e^x$; $x = 0.2$, x-axis

$A_{0,0.2} = \displaystyle\int_0^{0.2} x^2 e^x \, dx$

$= \displaystyle\int_0^{0.2} x^2 \left(1 + x + \dfrac{x^2}{2} \right) dx$

$= \displaystyle\int_0^{0.2} \left(x^2 + x^3 + \dfrac{1}{2} x^4 \right) dx$

$= \dfrac{x^3}{3} + \dfrac{x^4}{4} + \dfrac{x^5}{10} \Big|_0^{0.2}$

$= \dfrac{1}{3}(0.2)^3 + \dfrac{1}{4}(0.2)^4 + \dfrac{1}{10}(0.2)^5$

$= 0.003099$

24. $y = e^{-x^2}$; $x = -1, x = 1, y = 0$

$e^x = 1 + x + \dfrac{x^2}{2!} + \cdots$

$e^{-x^2} = 1 + (-x^2) + \dfrac{(-x^2)^2}{2!} + \cdots$

$= 1 - x^2 + \dfrac{x^4}{2} + \cdots$

$A_{-1,1} = \displaystyle\int_{-1}^{1} e^{-x^2} dx = \int_{-1}^{1} \left(1 - x^2 + \dfrac{1}{2} x^4 \right) dx$

$= x - \dfrac{1}{3} x^3 + \dfrac{1}{10} x^5 \Big|_{-1}^{1}$

$= 1 - \dfrac{1}{3}(1) + \dfrac{1}{10}(1)$

$\qquad - \left[-1 + \dfrac{1}{3}(1) - \dfrac{1}{10}(1) \right]$

$= 1 - \dfrac{1}{3} + \dfrac{1}{10} + 1 - \dfrac{1}{3} + \dfrac{1}{10} = 1.53$

25. $\displaystyle\int_0^x \cos t^2 \, dt = \int_0^{0.2} \left[1 - \dfrac{(t^2)^2}{2!} \right] dt$

$= \displaystyle\int_0^{0.2} \left(1 - \dfrac{t^4}{2} \right) dt$

$\left(t - \dfrac{t^5}{10} \right) \Big|_0^{0.2} = 0.199968$

26. $V = \displaystyle\int_0^{80} 2\pi x \cdot 20 \cos(0.0196x) \, dx$

$= 40\pi \displaystyle\int_0^{80} x \left(1 - \dfrac{0.0196^2 x^2}{2!} + \dfrac{0.0196^4 x^4}{4!} - \cdots \right) dx$

$= 40\pi \displaystyle\int_0^{80} \left(x - \dfrac{0.0196^2 x^3}{2!} + \dfrac{0.0196^4 x^5}{4!} - \cdots \right) dx$

$= 40\pi \left(\dfrac{x}{2} - \dfrac{0.0196^2 x^4}{2! \cdot 4} + \dfrac{0.0196^4 x^6}{4! \cdot 6} - \cdots \right) \Big|_0^{80}$

$= 40\pi \left(\dfrac{80^2}{2} - \dfrac{0.0196^2 \cdot 80^4}{2! \cdot 4} + \dfrac{0.0196^4 \cdot 80^6}{4! \cdot 6} - \cdots \right)$

$= 189,000 \text{ m}^3$ as compared to $187,000 \text{ m}^3$ previously

27. $Z = R + j(X_L - X_C)$ is in rectangular form with $a = R$ and $b = (X_L - X_C)$, hence

$\tan \theta = \dfrac{b}{a} = \dfrac{X_L - X_C}{R}$

$\theta = \tan^{-1} \dfrac{X_L - X_C}{R}$

28. $q = ce^{-at} \sin 6at$

$ce^{-at} = c \left[1 + (-at) + \dfrac{(-at)^2}{2!} + \dfrac{(-at)^3}{3!} + \dfrac{(-at)^4}{4!} \right]$

$\sin 6at = 6at - \dfrac{(6at)^3}{3!} + \dfrac{(6at)^5}{5!}$

$ce^{-at} \sin 6t$

$= c \left[6at - 6a^2 t^2 + 6at \left(\dfrac{a^2 t^2}{2} \right) - 6at \left(\dfrac{a^3 t^3}{6} \right) \right.$

$\qquad \left. - 36a^3 t^3 + at(36a^3 t^3) \right]$

$= c(6at - 6a^2 t^2 + 3a^3 t^3 - a^4 t^4 - 36a^3 t^3 + 36a^4 t^4)$

$= c(6at - 6a^2 t^2 - 33a^3 t^3 + 35a^4 t^4)$

29. $y_1 = e^x$, $y_2 = 1$, $y_3 = 1 + x$, $y_4 = 1 + x + \dfrac{1}{2}x^2$; $x_{min} = -5$, $x_{max} = 5$, $y_{min} = -1$, $y_{max} = 3$

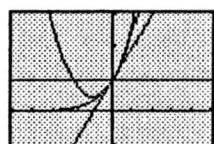

30. $y_1 = \cos x,$

$\quad y_1 = 1,$

$\quad y_3 = 1 - \dfrac{1}{2}x^2,$

$\quad y_4 = 1 - \dfrac{1}{2}x^2 + \dfrac{1}{24}x^4$

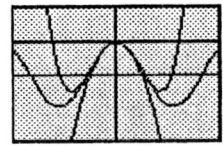

31. $y_1 = \ln(1+x),$

$\quad y_2 = x,$

$\quad y_3 = x - \dfrac{1}{2}x^2,$

$\quad y_4 = x - \dfrac{1}{2}x^2 + \dfrac{1}{3}x^3$

32. $y_1 = \sqrt{1+x},$

$\quad y_2 = 1,$

$\quad y_3 = 1 + \dfrac{1}{2}x,$

$\quad y_4 = 1 + \dfrac{1}{2}x - \dfrac{1}{8}x^2$

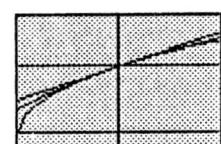

13.4 Computations by Use of Series Expansions

1. $e^x = 1 + x + \dfrac{x^2}{2!} + \cdots$

$\quad e^{0.2} = 1 + 0.2 + \dfrac{(0.2)^2}{2!} = 1.22$

$\quad$ (1.2214028 calculator)

2. $(1+x)^{-1} = 1 - x + x^2 - x^3 + x^4 - \cdots$

$\quad 1.01^{-1} = (1 + 0.01)^{-1}$

$\quad\quad\quad = 1 - 0.01 + 0.01^2 - \cdots$

$\quad\quad\quad = 0.9901$

$\quad 1.01^{-1} = 0.9900990099,$ calculator

3. $\sin 0.1,$ (2 terms); $\sin x = x - \dfrac{x^3}{3!}$

$\quad \sin 0.1 = 0.1 - \dfrac{(0.1)^3}{6} = 0.09983333$

$\quad$ (0.0998334 calculator)

4. $\cos 0.05,$ (2 terms); $\cos x = 1 - \dfrac{x^2}{2!}$

$\quad \cos 0.05 = 1 - \dfrac{(0.05)^2}{2} = 0.99875;$

$\quad$ (0.9987503 calculator)

5. $e^x = 1 + x + \dfrac{x}{2!} + \dfrac{x}{3!} + \dfrac{x}{4!} + \dfrac{x}{5!} + \dfrac{x}{6!} + \cdots$

$\quad e^1 = 1 + 1 + \dfrac{1}{2} + \dfrac{1}{6} + \dfrac{1}{24} + \dfrac{1}{120} + \dfrac{1}{720} + \cdots$

$\quad\quad = 2.7180556;$ (2.7182818 calculator)

6. $\dfrac{1}{\sqrt{e}} = e^{-1/2},$ (5 terms)

$\quad e^x = 1 + x + \dfrac{x^2}{2!} + \dfrac{x^3}{3!} + \dfrac{x^4}{4!}$

$\quad e^{-1/2} = 1 + \left(-\dfrac{1}{2}\right) + \dfrac{\left(-\frac{1}{2}\right)^2}{2!} + \dfrac{\left(-\frac{1}{2}\right)^3}{3!} + \dfrac{\left(-\frac{1}{2}\right)^4}{4!}$

$\quad\quad = 0.6068$

$\quad$ (0.6065307 calculator)

7. $\cos 3°,$ (2 terms); $3° = \dfrac{3}{180}\pi$ radians

$\quad \cos x = 1 - \dfrac{x^2}{2!}$

$\quad \cos 3° = 1 - \dfrac{\left(\frac{3\pi}{180}\right)^2}{2} = 0.9986292$

$\quad$ (0.9986295 calculator)

8. $\sin 4°,$ (2 terms); $\sin x = x - \dfrac{x^3}{3!}$

$\quad \sin 4° = \sin\left(\dfrac{4\pi}{180}\right) = \dfrac{4\pi}{180} - \dfrac{\left(\frac{4\pi}{180}\right)^3}{6}$

$\quad\quad = 0.0697565$

$\quad$ (0.0697565 calculator)

9. $\ln(1 + x) = x - \dfrac{x^2}{2} + \dfrac{x^3}{3} - \dfrac{x^4}{4} + \cdots$

$\ln(1 + 0.4)$

$= 0.4 - \dfrac{(0.4)^2}{2} + \dfrac{(0.4)^3}{3} - \dfrac{(0.4)^4}{4} + \cdots$

$= 0.3349333; \ (0.3364722 \text{ calculator})$

10. $\ln(0.95)$, (4 terms)

$\ln(1 + x) = x - \dfrac{x^2}{2} + \dfrac{x^3}{3} - \dfrac{x^4}{4}$

$\ln 0.95 = \ln(1 - 0.05)$; therefore, $x = -0.05$

$\ln 0.95 = -0.05 - \dfrac{(-0.05)^2}{2} + \dfrac{(-0.05)^3}{3} - \dfrac{(-0.05)^4}{4}$

$= -0.051293 (-0.0512933 \text{ calculator})$

11. $\sin 0.3625$, (3 terms): $\sin x = x - \dfrac{x^3}{3!} + \dfrac{x^5}{5!}$

$\sin 0.3625 = 0.3625 - \dfrac{(0.3625)^3}{6} + \dfrac{(0.3625)^5}{5!}$

$= 0.3546130$

$(0.3546129 \text{ calculator})$

12. $\cos 0.4072$, (3 terms): $\cos x = 1 - \dfrac{x^2}{2} + \dfrac{x^4}{4}$

$\cos 0.4072 = 1 - \dfrac{(0.4072)^2}{2} + \dfrac{(0.4072)^4}{4}$

$= 0.91824$

$(0.9182333 \text{ calculator})$

13. $\ln 0.9861$

$\ln(1 + x) = x - \dfrac{x^2}{2} + \dfrac{x^3}{3} - \dfrac{x^4}{4} + \cdots$

Let $x = -0.0139$

$\ln[1 + (-0.0139)]$

$= -0.0139 - \dfrac{(-0.0139)^2}{2} + \dfrac{(-0.0139)^3}{3} + \cdots$

$= -0.0139 - 0.0000966 - 0.0000009 + \cdots$

$= -0.0139975; \ (-0.0139975 \text{ calculator})$

14. $\ln 1.0534$, (3 terms); $\ln(1 + x) = x - \dfrac{x^2}{2} + \dfrac{x^3}{3}$

$\ln(1 + 0.0534) = 0.0534 - \dfrac{(0.0534)^2}{2} + \dfrac{(0.0534)^3}{3}$

$= 0.052025$

$(0.0520230 \text{ calculator})$

15. $(1 + x)^6 = 1 + 6x + 15x^2 + 20x^3 + 15x^4$
$\qquad\qquad + 6x^5 + x^6$

$(1.032)^6 = 1 + 6(0.032) + 15(0.032)^2$

$\qquad\quad = 1.20736$

$(1.032)^6 = 1.20803, \text{ calculator}$

16. $(1 - x)^n = 1 - nx + \dfrac{n(n - 1)}{2!} x^2 + \cdots$

$0.9982^8 = (1 - 0.0018)^8$

$= 1 - 8(0.0018) + \dfrac{8(7)}{2!}(0.0018)^2$

$= 0.98569072$

$0.9982^8 = 0.98569039, \text{ calculator}$

17. $\sqrt{1.1076} = 1.1076^{1/2} = (1 + 0.1076)^{1/2}$

$(1 + x)^n = 1 + nx + \dfrac{n(n - 1)x^2}{2!} + \cdots$

$x = 0.1076$ and $n = \frac{1}{2}$

$\sqrt{1.1076}$

$= 1 + \dfrac{1}{2}(0.1076) + \dfrac{\frac{1}{2}\left(-\frac{1}{2}\right)(0.1076)^2}{2} + \cdots$

$= 1 + 0.0538000 - 0.0014472 + \cdots$

$= 1.0523528$

18. $\sqrt{1 + x} = (1 + x)^{1/2}$

$(1 + x)^n = 1 + nx + \dfrac{n(n - 1)}{2!} x^2$

$(1 + x)^{1/2} = 1 + \dfrac{1}{2}x + \dfrac{\frac{1}{2}\left(-\frac{1}{2}\right)}{2!} x^2$

$= 1 + \dfrac{1}{2}x - \dfrac{1}{8}x^2$

$\sqrt{0.7915} = \sqrt{1 - 0.2085} = (1 - 0.2085)^{1/2}$

$(1 - 0.2085)^{1/2} = 1 + \dfrac{1}{2}(-0.2085) - \dfrac{1}{8}(-0.2085)^2$

$= 0.8903$

$(0.8896629 \text{ calculator})$

19. $\sqrt[3]{1 + x} = (1 + x)^{1/3} = 1 + \dfrac{1}{3}x + \dfrac{\frac{1}{3}\left(-\frac{2}{3}\right)x^2}{2!}$

$= 1 + \dfrac{1}{3}x - \dfrac{1}{9}x^2$

$\sqrt[3]{0.9628} = (1 - 0.0372)^{1/3}$

$= 1 + \dfrac{1}{3}(-0.0372) - \dfrac{1}{9}(-0.0372)^2$

$= 0.9874462$

$(0.9874430 \text{ calculator})$

20. $\sqrt[3]{1.1392} = (1 + 0.1392)^{1/3}$

$$= 1 + \frac{1}{3}(0.1392) - \frac{1}{9}(0.1392)^2$$

$$= 1.0442$$

(1.0443995 calculator)

21. From Exercise 3, $\sin(0.1) = 0.1 + \frac{0.1^3}{6} = 0.1001667$

The maximum possible error is the value of the first term omitted,

$$\frac{x^5}{5!} = \left|\frac{0.1^5}{120}\right| = 8.3 \times 10^{-8}$$

22. Max. error is 4th term $= |-0.01^3| = 0.000001$

23. $\cos 3° = 0.9986292$ to 2 terms.

Max. error is 3rd term $= \left|\frac{\left(\frac{3\pi}{180}\right)^4}{4!}\right| = 3.1 \times 10^{-7}$

24. $\ln(1.4) = 0.334\,933\,3$ to 4 terms.

Max. error is 5th term $= \left|\frac{(0.4)^5}{5}\right|$

$$= 0.002048$$

$$= 0.002$$

25. $\tan^{-1}\frac{1}{2} = \frac{1}{2} - \frac{1}{3}\left(\frac{1}{2}\right)^3 + \frac{1}{5}\left(\frac{1}{2}\right)^5 = 0.4646$

$\tan^{-1}\frac{1}{3} = \frac{1}{3} - \frac{1}{3}\left(\frac{1}{3}\right)^3 + \frac{1}{5}\left(\frac{1}{3}\right)^5 = 0.3218$

$\pi = 4(0.4646 + 0.3218) = 3.146$

26. $\frac{1}{4}\pi = \tan^{-1}\frac{1}{7} + 2\tan^{-1}\frac{1}{3}$

$\pi = 4\left(\tan^{-1}\frac{1}{7} + 2\tan^{-1}\frac{1}{3}\right) = 4\tan^{-1}\frac{1}{7} + 8\tan^{-1}\frac{1}{3}$

$\tan^{-1}\frac{1}{7} = \frac{1}{7} - \frac{1}{3}\left(\frac{1}{7}\right)^3 + \frac{1}{5}\left(\frac{1}{7}\right)^5 = 0.141897$

$\tan^{-1}\frac{1}{3} = \frac{1}{3} - \frac{1}{3}\left(\frac{1}{3}\right)^3 + \frac{1}{5}\left(\frac{1}{3}\right)^5 = 0.321811$

$4(0.141897) + 8(0.321811) = 3.142$

27. $t = \frac{\ln 1.1}{0.06} = \frac{\ln(1 + 0.1)}{0.06}$

$$= \frac{0.1 - \frac{(0.1)^2}{2} + \frac{(0.1)^3}{3} - \frac{(0.1)^4}{4}}{0.06}$$

$t = 1.59$ years

28. $T = 2\pi\sqrt{\frac{L}{g}}\left(1 + \frac{1}{4}\sin^2\frac{\theta}{2} + \frac{9}{24}\sin^4\frac{\theta}{2} + \cdots\right)$

(a) $T = 2\pi\sqrt{\frac{1.0000}{9.800}}(1) = 2.007$ s

(b) $T = 2\pi\sqrt{\frac{1.0000}{9.800}}\left[1 + \frac{1}{4}\left(\frac{\theta}{2}\right)^2\right]$

$\theta = 10.0°; \quad \frac{\theta}{2} = 5.0° = \frac{5.0\pi}{180}$

$T = 2\pi\sqrt{\frac{1.0000}{9.800}}\left[1 + \frac{1}{4}\left(\frac{5.0\pi}{180}\right)^2\right] = 2.011$ s

29. $f(t) = \frac{E}{R}(1 - e^{-Rt/L}); \quad e^x = 1 + x + \frac{x^2}{2} + \cdots$

$e^{-Rt/L} = 1 - \frac{Rt}{L} + \frac{R^2t^2}{2L^2} + \cdots$

$i = \frac{E}{R}\left[1 - \left(1 - \frac{Rt}{L} + \frac{R^2t^2}{2L^2}\right)\right] = \frac{E}{L}\left(t - \frac{Rt^2}{2L}\right)$

The approximation will be valid for small values of t.

30. $q = \frac{20p}{(p - 20)} = 20p(p - 20)^{-1}$

$(p - 20)^{-1} = (-20 + p)^{-1} = \left[-20\left(1 - \frac{p}{20}\right)\right]^{-1}$

$$= -\frac{1}{20}\left(1 - \frac{p}{20}\right)^{-1}$$

$20p(p - 20)^{-1}$

$= 20p\left[-\frac{1}{20}\left(1 - \frac{p}{20}\right)^{-1}\right] = -p\left(1 - \frac{p}{20}\right)^{-1}$

$= -p\left[1 + (-1)\left(-\frac{p}{20}\right) + \frac{(-1)(-2)}{2!}\left(-\frac{p}{20}\right)^2\right]$

$q = f(p) = -p - \frac{p^2}{20} - \frac{p^3}{400}$

$f(2) = -2 - \frac{4}{20} - \frac{8}{400} = -2.22$ cm

$q_{p=2} = \frac{20(2)}{(2 - 20)} = -2.22$ cm

31. $s = r\theta$

$\theta = \frac{s}{r} = \frac{15}{6400}$

$\cos\frac{15}{6400} = 1 - \frac{\left(\frac{15}{6400}\right)^2}{2!} = 0.999997$

$\cos\theta = \frac{6400}{6400 + x}$

$$x = \frac{6400}{\cos\theta} - 6400$$

$$x = \frac{6400}{0.999\,997} - 6400 = 0.017578 \text{ km}$$

$$x = 18 \text{ m}$$

32. $E = 100(1 - c^{-0.4})$; $c = 6.00$, error $= 0.50$

$(6 + x)^{-0.4}$

$$= \left[6\left(1 + \frac{x}{6}\right)\right]^{-0.4} = 6^{-0.4}\left(1 + \frac{x}{6}\right)^{-0.4}$$

$$= 6^{-0.4}\left[1 + (0.4)\left(\frac{x}{6}\right) + \frac{(-0.4)(-1.4)}{2}\left(\frac{x}{6}\right)^2\right]$$

$6.5^{-0.4}$

$$= 6^{-0.4}\left[1 - (0.4)\left(\frac{0.5}{6}\right) + \frac{(0.4)(1.4)}{2}\left(\frac{0.5}{6}\right)^2\right]$$

$$= 0.473$$

$$E_{c=6} = 100(1 - 6^{-0.4}) = 51.164\%$$
$$E_{c=6.5} = 100(1 - 0.473) = 52.70\%$$
$$\Delta E = 52.70 - 51.16 = 1.54\%$$

13.5   Taylor Series

1. $e^x = e\left[1 + (x - 1) + \frac{(x-1)^2}{2} + \frac{(x-1)^3}{6} + \cdots\right]$

Let $x = 1.2$;

$e^{1.2}$

$$= 2.7183\left[1 + (1.2 - 1) + \frac{(1.2-1)^2}{2} + \frac{(1.2-1)^3}{6} + \cdots\right]$$

$$= 2.7183(1.2227) = 3.32$$

2. $e^{0.7}$; $e^x = e\left[1 + (x-1) + \frac{(x-1)^2}{2} + \frac{(x-1)3}{6} + \cdots\right]$

$$e^{0.7} = e\left[1 + (0.7 - 1) + \frac{(0.7-1)^2}{2} + \frac{(0.7-1)^3}{6} + \cdots\right]$$

$$= 2.013; \ (2.0137527 \text{ calculator})$$

3. $\sqrt{4.2}$; $\sqrt{x} = 2 + \frac{(x-4)}{4} - \frac{(x-4)^2}{64} + \frac{(x-3)^3}{512}$

$$\sqrt{4.2} = 2 + \frac{(4.2-4)}{4} - \frac{(4.2-4)^2}{64} + \frac{(4.2-4)^3}{512}$$

$$= 2.049; \ (2.04939 \text{ calculator})$$

4. $\sqrt{3.5} = 2 + \frac{(3.5-4)}{4} - \frac{(3.5-4)^2}{64} + \frac{(3.5-4)^3}{512}$

$$= 1.87; \ (1.87083 \text{ calculator})$$

5. Let $a = 30° = \frac{\pi}{6}$; from Example 4,

$$\sin x = \frac{1}{2} + \frac{\sqrt{3}}{2}\left(x - \frac{\pi}{6}\right) - \frac{1}{4}\left(x - \frac{\pi}{6}\right)^2 - \cdots$$

$$31° = \frac{\pi}{6} + \frac{\pi}{180}$$

$\sin 31°$

$$= \frac{1}{2} + \frac{\sqrt{3}}{2}\left(\frac{\pi}{6} + \frac{\pi}{180} - \frac{\pi}{6}\right)$$

$$- \frac{1}{4}\left(\frac{\pi}{6} + \frac{\pi}{180} - \frac{\pi}{6}\right)^2$$

$$= \frac{1}{2} + \frac{\sqrt{3}}{2}\left(\frac{\pi}{180}\right) - \frac{1}{4}\left(\frac{\pi}{180}\right)^2$$

$$= 0.5150388403 \ (0.5150380749, \text{ calculator})$$

$$= 0.5150$$

6. $\sin x = \frac{1}{2} + \frac{\sqrt{3}}{2}\left(x - \frac{\pi}{6}\right) - \frac{1}{4}\left(x - \frac{\pi}{6}\right)^2$

$$x = 28° = \frac{28}{180}\pi; \ 28° = \frac{\pi}{6} - \frac{\pi}{90}$$

$$\sin 28° = \frac{1}{2} + \frac{\sqrt{3}}{2}\left(\frac{\pi}{6} - \frac{\pi}{90} - \frac{\pi}{6}\right)$$

$$- \frac{1}{4}\left(\frac{\pi}{6} - \frac{\pi}{90} - \frac{\pi}{6}\right)^2$$

$$= \frac{1}{2} + \frac{\sqrt{3}}{2}\left(-\frac{\pi}{90}\right) - \frac{1}{4}\left(-\frac{\pi}{90}\right)^2$$

$$= 0.4695$$

$(0.4694716 \text{ calculator})$

7. $\sin x = \frac{1}{2} + \frac{\sqrt{3}}{2}\left(x - \frac{\pi}{6}\right) - \frac{1}{4}\left(x - \frac{\pi}{6}\right)^2$

$\sin 29.53°$

$$= \frac{1}{2} + \frac{\sqrt{3}}{2}\left(\frac{29.53\pi}{180} - \frac{\pi}{6}\right)$$

$$- \frac{1}{4}\left(\frac{29.53\pi}{180} - \frac{\pi}{6}\right)^2$$

$$= 0.49288; \ (0.4928792 \text{ calculator})$$

8. $\sqrt{3.8527}$; $\sqrt{x} = 2 + \frac{(x-4)}{4} - \frac{(x-4)^2}{64} + \frac{(x-4)^3}{512}$

$\sqrt{3.8527}$

$$= 2 + \frac{(3.8527 - 4)}{4} - \frac{(3.8527 - 4)^2}{64}$$

$$+ \frac{(3.8527 - 4)^3}{512}$$

$$= 1.9628$$

9. $f(x) = e^{-x}$ $f(2) = e^{-2}$
    $f'(x) = -e^{-x}$ $f'(2) = -e^{-2}$
    $f''(x) = e^{-x}$ $f''(2) = e^{-2}$

$$f(x) = e^{-2} - e^{-2}(x-2) + \frac{e^{-2}(x-2)^2}{2!} + \cdots$$

$$= e^{-2}\left[1 - (x-2) + \frac{(x-2)^2}{2!} - \cdots\right]$$

10. $\cos x; \ a = \dfrac{\pi}{4}$

$f(x) = \cos x$ $f\left(\dfrac{\pi}{4}\right) = \dfrac{1}{\sqrt{2}} = \dfrac{\sqrt{2}}{2}$

$f'(x) = -\sin x$ $f'\left(\dfrac{\pi}{4}\right) = -\dfrac{\sqrt{2}}{2}$

$f''(x) = -\cos x$ $f''\left(\dfrac{\pi}{4}\right) = -\dfrac{\sqrt{2}}{2}$

$$\sin x = \frac{\sqrt{2}}{2} - \frac{\sqrt{2}}{2}\left(x - \frac{\pi}{4}\right) - \frac{\sqrt{2}}{2}\frac{\left(x - \frac{\pi}{4}\right)^2}{2!} + \cdots$$

$$= \frac{\sqrt{2}}{2}\left[1 - \left(x - \frac{\pi}{4}\right) - \frac{1}{2}\left(x - \frac{\pi}{4}\right)^2 + \cdots\right]$$

11. $\sin x; \ a = \dfrac{\pi}{3}$

$f(x) = \sin x$ $f\left(\dfrac{\pi}{3}\right) = \dfrac{\sqrt{3}}{2}$

$f'(x) = \cos x$ $f'\left(\dfrac{\pi}{3}\right) = \dfrac{1}{2}$

$f''(x) = -\sin x$ $f''\left(\dfrac{\pi}{3}\right) = -\dfrac{\sqrt{3}}{2}$

$$\sin x = \frac{\sqrt{3}}{2} + \frac{1}{2}\left(x - \frac{\pi}{3}\right) - \frac{\sqrt{3}}{2!}\left(x - \frac{\pi}{3}\right)^2 - \cdots$$

$$= \frac{1}{2}\left[\sqrt{3} + \left(x - \frac{\pi}{3}\right) - \frac{\sqrt{3}}{2!}\left(x - \frac{\pi}{3}\right)^2 - \cdots\right]$$

12. $\ln x; \ a = 3$

$f(x) = \ln x$ $f(3) = \ln 3$

$f'(x) = \dfrac{1}{x}$ $f'(3) = \dfrac{1}{3}$

$f''(x) = -\dfrac{1}{x^2}$ $f''(3) = -\dfrac{1}{9}$

$$\ln x = \ln 3 + \frac{1}{3}(x-3) - \frac{1}{9}\frac{(x-3)^2}{2!} + \cdots$$

$$= \ln 3 + \frac{1}{3}(x-3) - \frac{1}{18}(x-3)^2 + \cdots$$

13. $f(x) = x^{1/3}$

$$f'(x) = \frac{1}{3}x^{-2/3}$$

$$f''(x) = -\frac{2}{9}x^{-5/3}$$

$$f(8) = 8^{1/3} = 2$$

$$f'(8) = \frac{1}{3(8)^{2/3}} = \frac{1}{12}$$

$$f''(8) = \frac{-2}{9(8^{5/3})} = -\frac{1}{144}$$

$$f(x) = 2 + \frac{1}{12}(x-8) - \frac{1}{288}(x-8)^2 + \cdots$$

14. $\dfrac{1}{x}; \ a = 2$

$f(x) = \dfrac{1}{x}$ $f(2) = \dfrac{1}{2}$

$f'(x) = -\dfrac{1}{x^2}$ $f'(2) = -\dfrac{1}{4}$

$f''(x) = \dfrac{2}{x^3}$ $f''(2) = \dfrac{1}{4}$

$$\frac{1}{x} = \frac{1}{2} - \frac{1}{4}(x-2) + \frac{\frac{1}{4}(x-2)^2}{2!} - \cdots$$

$$= \frac{1}{2} - \frac{1}{4}(x-2) + \frac{1}{8}(x-2)^2 - \cdots$$

15. $\tan x; \ a = \dfrac{\pi}{4}$

$f(x) = \tan x$ $f\left(\dfrac{\pi}{4}\right) = 1$

$f'(x) = \sec^2 x$ $f'\left(\dfrac{\pi}{4}\right) = (\sqrt{2})^2 = 2$

$f''(x) = 2\sec x \sec x \tan x = 2\sec^2 x \tan x$
$f''(x) = 2(\sqrt{2})^2(1) = 4$

$$\tan x = 1 + 2\left(x - \frac{\pi}{4}\right) + \frac{4\left(x - \frac{\pi}{4}\right)^2}{2!} + \cdots$$

$$= 1 + 2\left(x - \frac{\pi}{4}\right) + 2\left(x - \frac{\pi}{4}\right)^2 + \cdots$$

16. $\ln \sin x; \ a = \dfrac{\pi}{2}$

$f(x) = \ln \sin x,$ $f\left(\dfrac{\pi}{2}\right) = 0$

$f'(x) = \dfrac{1}{\sin x}\cos x = \cot x,$ $f'\left(\dfrac{\pi}{2}\right) = 0$

$f''(x) = -\csc^2 x,$ $f''\left(\dfrac{\pi}{2}\right) = -1$

$f'''(x) = -2\csc x(-\csc \cot x) = 2\csc^2 x \cot x$

$$= 2\frac{\cos x}{\sin^3 x}$$

$$f'''\left(\frac{\pi}{2}\right) = 0$$

$$f^{iv}(x) = 2\left[\frac{\sin^3 x(-\sin x) - \cos x\, 3\sin^2 x \cos x}{\sin^6 x}\right]$$

$$= 2\left(\frac{-\sin^4 x - 3\sin^2 x \cos^2 x}{\sin^6 x}\right)$$

$$= -2\csc^2 x - 6\frac{\cos^2 x}{\sin^4 x}$$

$$f^{iv}\left(\frac{\pi}{2}\right) = -2$$

$$f^v(x)$$
$$= -4\csc x(-\csc x \, \cot \, x)$$

$$- 6\left[\frac{\sin^4 x\, 2\cos x(-\sin x) - \cos^2 x\, 4\sin^3 x \cos x}{\sin^8 x}\right]$$

$$= 4\csc^2 x \, \text{ctn}\, x + 12\frac{\cos x}{\sin^3 x} + 24\frac{\cos^3 x}{\sin^5 x}$$

$$f^v(x) = 16\frac{\cos x}{\sin^3 x} + 24\frac{\cos^3 x}{\sin^5 x}$$
$$f^v\left(\frac{\pi}{2}\right) = 0$$

$$f^{vi}(x)$$

$$= 16\left[\frac{\sin^3 x(-\sin x) - \cos x\, 3\sin^2 x \cos x}{\sin^6 x}\right]$$

$$+24\left[\frac{\sin^5 x\, 3\cos^2 x(-\sin x) - \cos^3 x\, 5\sin^4 x \cos x}{\sin^{10} x}\right]$$

$$f^{vi}(x) = 16\left(\frac{-\sin^4 x - 3\sin^2 x \cos^2 x}{\sin^6 x}\right)$$

$$+24\left(\frac{-3\sin^6 x \cos^2 x - 5\sin^4 x \cos^4 x}{\sin^{10} x}\right)$$

$$f^{vi}\left(\frac{\pi}{2}\right) = 16(-1) + 24(0) = -16$$

$$\ln \sin x$$

$$= -1\frac{\left(x-\frac{\pi}{2}\right)^2}{2!} - 2\frac{\left(x-\frac{\pi}{2}\right)^4}{4!} - 16\frac{\left(x-\frac{\pi}{2}\right)^6}{6!}$$

$$= -\frac{1}{2}\left(x-\frac{\pi}{2}\right)^2 - \frac{1}{12}\left(x-\frac{\pi}{2}\right)^4 - \frac{1}{45}\left(x-\frac{\pi}{2}\right)^6 - \cdots$$

17. $e^x = e^3\left[1 + (x-3) + \frac{(x-3)^2}{2}\right]$

$$e^\pi = e^3\left[1 + (\pi-3) + \frac{(\pi-3)^2}{2}\right]$$

$$e^\pi = 23.13084363 \text{ as compared with}$$

$$e^\pi = 23.14069263 \text{ on a calculator.}$$

18. $\quad \ln x = \ln 3 + \frac{1}{3}(x-3) - \frac{1}{18}(x-3)^2$

$$\ln 3.1 = \ln 3 + \frac{1}{3}(0.1) - \frac{1}{18}(0.1)^2$$

$$= 1.0986 + \frac{1}{3}(0.1)^2 = -\frac{1}{18}(0.1)^2 = 1.131$$

19. $\sqrt{9.3};\ a = 9$

$$f(x) = \sqrt{x} \qquad\qquad f(9) = 3$$

$$f'(x) = \frac{1}{2\sqrt{x}} \qquad\qquad f'(9) = \frac{1}{6}$$

$$f''(x) = -\frac{1}{4x^{3/2}} \qquad f''(9) = -\frac{1}{108}$$

$$\sqrt{x} = 3 + \frac{1}{6}(x-9) - \frac{1}{108}\frac{(x-9)^2}{2!}$$

$$\sqrt{9.3} = 3 + \frac{1}{6}(0.3) - \frac{1}{108}\frac{(0.3)^2}{2} = 3.0496$$

20. $f(x) = \frac{1}{x} = \frac{1}{2} - \frac{1}{4}(x-2) + \frac{1}{8}(x-2)^2 - \cdots$

$$\frac{1}{2.056} = \frac{1}{2} - \frac{1}{4}(2.056-2) + \frac{1}{8}(2.056-2)^2$$

$$= 0.486392$$

21. $\sqrt[3]{8.3};\ f(x) = x^{1/3}, f(8) = 2$

$$f'(x) = \frac{1}{3}x^{-2/3}, f'(8) = \frac{1}{12}$$

$$f''(x) = -\frac{2}{9}x^{-5/3}, f''(8) = -\frac{1}{144}$$

$$f(8.3) = f(8) + f'(8)(8.3-8) + \frac{f''(8)(8.3-8)^2}{2!}$$

$$\sqrt[3]{8.3} = 2 + \frac{1}{12}(8.3-8) - \frac{1}{288}(8.3-8)^2$$

$$= 2.0246875$$
$$\sqrt[3]{8.3} = 2.0247$$

22. $\tan 46°$; $a = 45° = \dfrac{\pi}{4}$

$$f(x) = \tan x = 1 + 2\left(x - \frac{\pi}{4}\right) + 2\left(x - \frac{\pi}{4}\right)^2$$

$$46° = 45° + 1° = \frac{\pi}{4} + \frac{\pi}{180}$$

$$\tan 46° = 1 + 2\left(\frac{\pi}{4} + \frac{\pi}{180} - \frac{\pi}{4}\right) + 2\left(\frac{\pi}{4} + \frac{\pi}{180} - \frac{\pi}{4}\right)^2 = 1 + 2\left(\frac{\pi}{180}\right) + 2\left(\frac{\pi}{180}\right)^2 = 1.036$$

23. $\sin x = \dfrac{1}{2}\left[\sqrt{3} + \left(x - \dfrac{\pi}{3}\right) - \dfrac{\sqrt{3}}{2}\left(x - \dfrac{\pi}{3}\right)^2\right]$; $a = \dfrac{\pi}{3}$

$$61° = 60° + 1° = \frac{\pi}{3} + \frac{\pi}{180}$$

$$\sin 61° = \frac{1}{2}\left[\sqrt{3} + \frac{\pi}{180} - \frac{\sqrt{3}}{2}\left(\frac{\pi}{180}\right)^2\right] = 0.87462$$

24. $\cos 42°$; $a = 45° = \dfrac{\pi}{4}$

$$42° = 45° - 3° = \frac{\pi}{4} - \frac{3\pi}{180} = \frac{\pi}{4} - \frac{\pi}{60}$$

$$\cos x = \frac{\sqrt{2}}{2}\left[1 - \left(x - \frac{\pi}{4}\right) - \frac{1}{2}\left(x - \frac{\pi}{4}\right)^2\right]$$

$$\cos 42° = \frac{\sqrt{2}}{2}\left[1 - \left(-\frac{\pi}{60}\right) - \frac{1}{2}\left(-\frac{\pi}{60}\right)^2\right] = 0.7432$$

25.
$$f(x) = c_0 + c_1(x - a) + c_2(x - a)^2 + c_3(x - a)^3 + c_4(x - a)^4 + c_5(x - a)^5 + \cdots + c_n(x - a)^n$$
$$f'(x) = c_1 + 2c_2(x - a) + 3c_3(x - a)^2 + 4c_4(x - a)^3 + 5c_5(x - a)^4 + \cdots + nc_n(x - a)^{n-1}$$
$$f''(x) = 2c_2 + 2 \times 3c_3(x - a) + 3 \times 4c_4(x - a)^2 + 4 \times 5c_5(x - a)^3 + \cdots + (n - 1)nc_n(x - a)^{n-2}$$
$$f'''(x) = 2 \times 3c_3 + 2 \times 3 \times 4c_4(x - a) + 3 \times 4 \times 5c_5(x - a)^2 + \cdots + (n - 2)(n - 1)nc_n(x - a)^{n-3}$$
$$f^{iv}(x) = 2 \times 3 \times 4c_4 + 2 \times 3 \times 4 \times 5c_5(x - a) + \cdots + (n - 3)(n - 2)(n - 1)nc_n(x - a)^{n-4}$$

Let $x = a$; $f(a) = c_0$; $f'(a) = c_1$; $f''(a) = 2c_2$; $c_2 = \dfrac{f''(a)}{2!}$

$$f'''(a) = 2 \times 3c_3; \quad c_3 = \frac{f'''(a)}{3!}$$

$$f^{iv}(a) = 2 \times 3 \times 4c_4; \quad c_4 = \frac{f^{iv}(a)}{4!}$$

$$f(x) = f(a) + f'(a)(x - a) + \frac{f''(a)(x - a)^2}{2!} + \frac{f'''(a)(x - a)^3}{3!} + \frac{f^{iv}(a)(x - a)^4}{4!} + \cdots$$

26. $f(x) = \ln x$, $L(x) = x - 1$

Taylor: $\ln x = \ln 1 + f'(1)(x - 1) + \dfrac{f''(1)(x - 2)^2}{2!}$

$$\ln x = (x - 1) - \frac{(x - 1)^2}{2}$$

$y_1 = f(x) = \ln x$

$y_2 = L(x) = x - 1$

$y_3 = (x - 1) - \dfrac{(x-1)^2}{2}$, Taylor

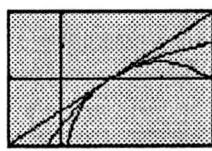

27. $\sin 31°$; $a = \dfrac{\pi}{6}$; $31° = 30° + 1° = \dfrac{\pi}{6} + \dfrac{\pi}{180}$

Maclaurin's: $\sin x = x - \dfrac{x^3}{3!} + \dfrac{x^5}{5!}$

$\sin 31° = \dfrac{31\pi}{180} - \dfrac{\left(\frac{31\pi}{180}\right)^3}{6} + \dfrac{\left(\frac{31\pi}{180}\right)^5}{5!}$

$= 0.5150408$

Taylor: $\sin x = \dfrac{1}{2} + \dfrac{\sqrt{3}}{2}\left(x - \dfrac{\pi}{6}\right) - \dfrac{1}{4}\left(x - \dfrac{\pi}{6}\right)^2$

$\sin 31° = \dfrac{1}{2} + \dfrac{\sqrt{3}}{2}\left(\dfrac{\pi}{180}\right) - \dfrac{1}{4}\left(\dfrac{\pi}{180}\right)^2$

$= 0.5150388$

Calculator: $\sin 31° = 0.5150381$

28. $\ln \dfrac{x+L}{x}$; $a = L$

$f(x) = \ln \dfrac{x+L}{x}$ $f(L) = \ln \dfrac{2L}{L} = \ln 2$

$f'(x) = \dfrac{-L}{x^2 + Lx}$ $f'(L) = -\dfrac{1}{2L}$

$f''(x) = \dfrac{L(2x+L)}{(x^2+Lx)^2}$ $f''(L) = \dfrac{3}{4L^2}$

$\ln \dfrac{(x+L)}{x} = \ln 2 - \dfrac{1}{2L}(x-L) + \dfrac{3}{4L^2}\dfrac{(x-L)^2}{2!}$

$= \ln 2 - \dfrac{1}{2L}(x-L) + \dfrac{3}{8L^2}(x-L)^2$

29. $f(x) = \sin x$; $x = 0$ to $x = 2$

(a) $y_1 = \sin x$

(b) $y_2 = \dfrac{\sqrt{3}}{2} + \dfrac{1}{2}\left(x - \dfrac{\pi}{3}\right)$

     $x_{\min} = 0$, $x_{\max} = 2$, $y_{\min} = 0$, $y_{\max} = 1.2$

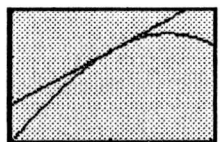

The series gives a good approximation near $x = \dfrac{\pi}{3}$ and deteriorates as x moves away from $x = \dfrac{\pi}{3}$.

30. $f(x) = \sqrt[3]{x}$; $x = 0$ to $x = 16$

(a) $y_1 = \sqrt[3]{x}$

(b) $y_2 = 2 + \dfrac{1}{12}(x - 8)$

Graph in part (b) will fit the graph of part (a) well for x close to $x = 8$.

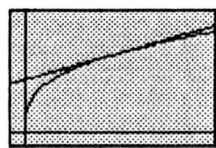

31. $f(x) = \dfrac{1}{x}$; $x = 0$ to $x = 4$

(a) $y_1 = \dfrac{1}{x}$

(b) $y_2 = \dfrac{1}{2} - \dfrac{1}{4}(x - 2)$

Graph of part (b) will fit the graph of part (a) well for values of x close to $x = 2$.

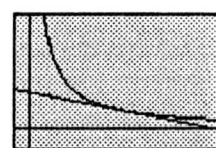

32. $f(x) = \tan x$; $x = 0$ to $x = 1.5$; $\dfrac{\pi}{4} = 0.785$

(a) $y_1 = \tan x$

(b) $y_2 = 1 + 2\left(x - \dfrac{\pi}{4}\right)$

Graph of part (b) will fit the graph of part (a) well for values of x close to $x = \dfrac{\pi}{4}$.

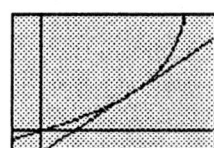

$$f(x) = \frac{1}{2} - \frac{2}{\pi}\sin x - \frac{2}{3\pi}\sin 3x - \cdots$$

13.6   Introduction to Fourier Series

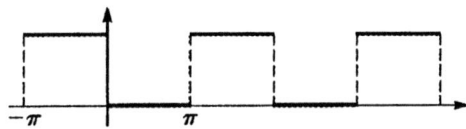

1. $a_0 = \dfrac{1}{2\pi}\displaystyle\int_{-\pi}^{0} 1\,dx + \dfrac{1}{2\pi}\int_{0}^{\pi} 0\,dx$

$\quad = \dfrac{1}{2\pi}x\Big|_{-\pi}^{0}$

$\quad = 0 - \dfrac{1}{2\pi}(-\pi)$

$\quad = \dfrac{1}{2}$

$a_1 = \dfrac{1}{\pi}\displaystyle\int_{-\pi}^{0}(1)\cos x\,dx + \dfrac{1}{\pi}\int_{0}^{\pi} 0\,dx$

$\quad = \dfrac{1}{\pi}\sin x\Big|_{-\pi}^{0} = 0$

$a_n = 0$ for all values of n since $\sin n\pi = 0$

$b_1 = \dfrac{1}{\pi}\displaystyle\int_{-\pi}^{0} 1\sin x\,dx + \int_{0}^{\pi} 0\,dx$

$\quad = \dfrac{1}{\pi}(-\cos x)\Big|_{-\pi}^{0}$

$\quad = \dfrac{1}{\pi}(-1-1)$

$\quad = -\dfrac{2}{\pi}$

$b_2 = \dfrac{1}{\pi}\displaystyle\int_{-\pi}^{0} 1\sin 2x\,dx + \int_{0}^{\pi} 0\,dx$

$\quad = \dfrac{1}{2\pi}\displaystyle\int_{-\pi}^{0}\sin 2x(2\,dx)$

$\quad = \dfrac{1}{2\pi}\displaystyle\int_{-\pi}^{0}\sin 2x(2\,dx)$

$\quad = \dfrac{1}{2\pi}(-\cos 2x)\Big|_{-\pi}^{0}$

$\quad = \dfrac{1}{2\pi}(-1+1) = 0$

$b_3 = \dfrac{1}{\pi}\displaystyle\int_{-\pi}^{0} 1\sin 3x\,dx + \int_{0}^{\pi} 0\,dx$

$\quad = \dfrac{1}{3\pi}\displaystyle\int_{-\pi}^{0}\sin 3x(3\,dx)$

$\quad = \dfrac{1}{3\pi}(-\cos 3x)\Big|_{-\pi}^{0}$

$\quad = \dfrac{1}{3\pi}(-1-1) = \dfrac{-2}{3\pi}$

$f(x) = \dfrac{1}{2} - \dfrac{2}{\pi}\sin x - \dfrac{2}{3\pi}\sin 3x - \cdots$

2. $f(x) = \begin{cases} 0 & -\pi \le x < 0 \\ 1 & 0 \le x < \pi \end{cases}$

$a_0 = \dfrac{1}{2\pi}\displaystyle\int_{-\pi}^{\pi} f(x)\,dx$

$\quad = \dfrac{1}{2\pi}\displaystyle\int_{-\pi}^{0} 0\,dx + \dfrac{1}{2\pi}\int_{0}^{\pi} dx$

$\quad = \dfrac{1}{2\pi}x\Big|_{0}^{\pi}$

$\quad = \dfrac{1}{2}$

$a_n = \dfrac{1}{\pi}\displaystyle\int_{-\pi}^{\pi} f(x)\cos nx$

$\quad = \dfrac{1}{\pi}\displaystyle\int_{-\pi}^{0} 0\cos nx\,dx + \dfrac{1}{\pi}\int_{0}^{\pi} 1\cdot\cos nx\,dx$

$a_n = \dfrac{\pi^2\cdot n^2\cdot 0}{3}$

$\quad = 0$

$b_n = \dfrac{1}{\pi}\displaystyle\int_{-\pi}^{\pi} f(x)\sin nx\,dx$

$\quad = \dfrac{1}{\pi}\displaystyle\int_{-\pi}^{0} 0\cdot\sin nx\,dx + \dfrac{1}{\pi}\int_{0}^{\pi} 1\cdot\sin nx\,dx$

$b_n = \dfrac{1-\cos n\pi}{n\pi}$

$b_1 = \dfrac{2}{\pi},$

$b_2 = 0,$

$b_3 = \dfrac{2}{3\pi}$

$f(x) = \dfrac{1}{2} + \dfrac{2}{\pi}\sin x + \dfrac{2}{3\pi}\sin(3x) + \cdots$

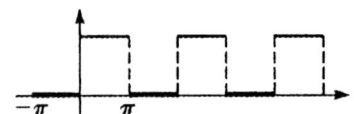

3. $f(x) = \begin{cases} 1 & -\pi \le x < 0 \\ 2 & 0 \le x < \pi \end{cases}$

$a_0 = \dfrac{1}{2\pi} \displaystyle\int_{-\pi}^{0} 1\, dx + \dfrac{1}{2\pi} \int_{0}^{\pi} 2\, dx$

$= \dfrac{x}{2\pi}\Big|_{-\pi}^{0} + \dfrac{2x}{2\pi}\Big|_{0}^{\pi}$

$= 0 + \dfrac{\pi}{2\pi} + \dfrac{2\pi}{2\pi} - 0 = \dfrac{1}{2} + 1 = \dfrac{3}{2}$

$a_1 = \dfrac{1}{\pi} \displaystyle\int_{-\pi}^{0} 1 \cos x\, dx + \dfrac{1}{\pi} \int_{0}^{\pi} 2 \cos x\, dx$

$= \dfrac{1}{\pi} \sin x \Big|_{-\pi}^{0} + \dfrac{2}{\pi} \sin x \Big|_{0}^{\pi}$

$= \dfrac{1}{\pi}(0-0) + \dfrac{2}{\pi}(0-0) = 0$

$a_n = 0$ since $\sin n\pi = 0$

$b_1 = \dfrac{1}{\pi} \displaystyle\int_{-\pi}^{0} 1 \sin x\, dx + \dfrac{1}{\pi} \int_{0}^{\pi} 2 \sin x\, dx$

$= -\dfrac{1}{\pi} \cos x \Big|_{-\pi}^{0} - \dfrac{2}{\pi} \cos x \Big|_{0}^{\pi}$

$= -\dfrac{1}{\pi}(1+1) - \dfrac{2}{\pi}(-1-1)$

$= -\dfrac{2}{\pi} + \dfrac{4}{\pi} = \dfrac{2}{\pi}$

$b_2 = \dfrac{1}{\pi} \displaystyle\int_{-\pi}^{0} 1 \sin 2x\, dx + \dfrac{1}{\pi} \int_{0}^{\pi} 2 \sin x\, dx$

$= -\dfrac{1}{2\pi} \cos 2x \Big|_{-\pi}^{0} - \dfrac{1}{\pi} \cos 2x \Big|_{0}^{\pi}$

$= -\dfrac{1}{2\pi}(1-1) - \dfrac{1}{\pi}(1-1) = 0$

$b_3 = \dfrac{1}{\pi} \displaystyle\int_{-\pi}^{0} \sin 3x\, dx + \dfrac{1}{\pi} \int_{0}^{\pi} 2 \sin 3x\, dx$

$= -\dfrac{1}{3\pi} \cos 3x \Big|_{-\pi}^{0} - \dfrac{2}{3\pi} \cos 3x \Big|_{0}^{\pi}$

$= -\dfrac{1}{3\pi}(1+1) - \dfrac{2}{3\pi}(-1-1) = \dfrac{2}{3\pi}$

Therefore, $b_n = 0$ for n even; $b_n = \dfrac{2}{n\pi}$ for n odd.

Therefore, $f(x) = \dfrac{3}{2} + \dfrac{2}{\pi} \sin x + \dfrac{2}{3\pi} \sin 3x + \cdots$

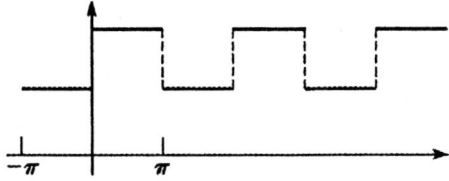

4. $f(x) = \begin{cases} 0 & -\pi \le x < 0, \frac{\pi}{2} < x < \pi \\ 1 & 0 \le x \le \frac{\pi}{2} \end{cases}$

$a_0 = \dfrac{1}{2\pi} \left(\displaystyle\int_{-\pi}^{0} 0\, dx + \int_{0}^{\pi/2} 1\, dx + \int_{\pi/2}^{\pi} 0\, dx \right)$

$= \dfrac{1}{2\pi} x \Big|_{0}^{\pi/2} = \dfrac{1}{2\pi} \dfrac{\pi}{2} = \dfrac{1}{4}$

$a_1 = \dfrac{1}{\pi} \displaystyle\int_{0}^{\pi/2} \cos x\, dx = \dfrac{1}{\pi} \sin x \Big|_{0}^{\pi/2} = \dfrac{1}{\pi}(1) = \dfrac{1}{\pi}$

$a_2 = \dfrac{1}{\pi} \displaystyle\int_{0}^{\pi/2} \cos 2x\, dx = \dfrac{1}{2\pi} \sin 2x \Big|_{0}^{\pi/2}$

$= \dfrac{1}{2\pi}(0) = 0$

$a_3 = \dfrac{1}{\pi} \displaystyle\int_{0}^{\pi/2} \cos 3x\, dx = \dfrac{1}{3\pi} \sin 3x \Big|_{0}^{\pi/2}$

$= \dfrac{1}{3\pi}(-1) = -\dfrac{1}{3\pi}$

$a_n = 0$ for n even; $a_n = \pm\dfrac{1}{n\pi}$ for n odd.

$b_1 = \dfrac{1}{\pi} \displaystyle\int_{0}^{\pi/2} \sin x\, dx = -\dfrac{1}{\pi} \cos x \Big|_{0}^{\pi/2}$

$= -\dfrac{1}{\pi}(0-1) = \dfrac{1}{\pi}$

$b_2 = \dfrac{1}{\pi} \displaystyle\int_{0}^{\pi/2} \sin 2x\, dx$

$= -\dfrac{1}{2\pi} \cos x \Big|_{0}^{\pi/2} - \dfrac{1}{2\pi}(-1-1)$

$= \dfrac{1}{\pi}$

$b_n = \dfrac{1}{\pi}$ for all n

$f(x) = \dfrac{1}{4} + \dfrac{1}{\pi} \cos x - \dfrac{1}{3\pi} \cos 3x + \dfrac{1}{5\pi} \cos 5x$

$\qquad + \dfrac{1}{\pi} \sin x + \dfrac{1}{\pi} \sin 2x + \dfrac{1}{\pi} \sin 3x + \cdots$

$= \dfrac{1}{4} + \dfrac{1}{\pi} \left(\cos x - \dfrac{1}{3} \cos 3x + \cdots \right.$

$\qquad \left. + \sin x + \sin 2x + \sin 3x + \cdots \right)$

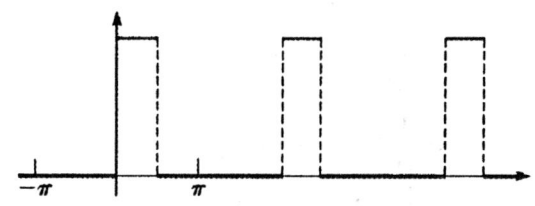

5. $a_0 = \dfrac{1}{2\pi}\displaystyle\int_{-\pi}^{0} 0\,dx + \dfrac{1}{2\pi}\displaystyle\int_{0}^{\pi} x\,dx$

$\qquad = \dfrac{1}{2\pi}\left(\dfrac{x^2}{2}\right)\Big|_{0}^{\pi} = \dfrac{\pi}{4}$; Eq. (29-25)

$a_1 = \dfrac{1}{\pi}\displaystyle\int_{-\pi}^{0} 0\cos x\,dx + \dfrac{1}{\pi}\displaystyle\int_{0}^{\pi} x\cos x\,dx$

$\qquad = \dfrac{1}{\pi}(\cos x + x\sin x)\Big|_{0}^{\pi} = \dfrac{1}{\pi}(-1-1) = -\dfrac{2}{\pi}$

$a_2 = \dfrac{1}{\pi}\displaystyle\int_{-\pi}^{0} 0\,dx + \dfrac{1}{\pi}\displaystyle\int_{0}^{\pi} x\cos 2x\,dx$

$\qquad = \dfrac{1}{4\pi}\displaystyle\int_{0}^{\pi} 2x\cos 2x\,(2\,dx)$

$\qquad = \dfrac{1}{4\pi}(\cos 2x + 2x\sin 2x)\Big|_{0}^{\pi}$

$\qquad = \dfrac{1}{4\pi}(\cos 2\pi - \cos 0) = 0$

$a_3 = \dfrac{1}{\pi}\displaystyle\int_{-\pi}^{0} 0\,dx + \dfrac{1}{\pi}\displaystyle\int_{0}^{\pi} x\cos 3x\,dx$

$\qquad = \dfrac{1}{9\pi}\displaystyle\int_{0}^{\pi} 3x\cos 3x\,(3\,dx)$

$\qquad = \dfrac{1}{9\pi}(\cos 3x + 3x\sin 3x)\Big|_{0}^{\pi}$

$\qquad = \dfrac{1}{9\pi}[(\cos 3\pi + 3\pi\sin 3\pi) - (\cos 0 + 0)]$

$\qquad = \dfrac{1}{9\pi}(-1-1) = -\dfrac{2}{9\pi}$

$b_1 = \dfrac{1}{\pi}\displaystyle\int_{-\pi}^{0} 0\,dx + \dfrac{1}{\pi}\displaystyle\int_{0}^{\pi} x\sin x\,dx$

$\qquad = \dfrac{1}{\pi}(\sin x - x\cos x)\Big|_{0}^{\pi}$

$\qquad = \dfrac{1}{\pi}[(\sin \pi - \pi\cos \pi) - (\sin 0 - 0)]$

$\qquad = \dfrac{1}{\pi}(-\pi)(-1) = 1$

$b_2 = \dfrac{1}{\pi}\displaystyle\int_{-\pi}^{0} 0\,dx + \dfrac{1}{\pi}\displaystyle\int_{0}^{\pi} x\sin 2x\,dx$

$\qquad = \dfrac{1}{4\pi}\displaystyle\int_{0}^{\pi} 2x\sin 2x\,(2\,dx)$

$\qquad = \dfrac{1}{4\pi}(\sin 2x - 2x\cos 2x)\Big|_{0}^{\pi}$

$\qquad = -\dfrac{1}{4\pi}[(\sin 2\pi - 2\pi\cos 2\pi) - (\sin 0 - 0)]$

$\qquad = \dfrac{1}{4\pi}(0 - 2\pi - 0) = -\dfrac{1}{2}$

$f(x) = \dfrac{\pi}{4} - \dfrac{2}{\pi}\cos x + 0\cos 2x - \dfrac{2}{9\pi}\cos 3x + \cdots$

$\qquad + \sin x - \dfrac{1}{2}\sin 2x + \cdots$

$\qquad = \dfrac{\pi}{4} - \dfrac{2}{\pi}\left(\cos x + \dfrac{1}{9}\cos 3x + \cdots\right)$

$\qquad + \left(\sin x - \dfrac{1}{2}\sin 2x + \cdots\right)$

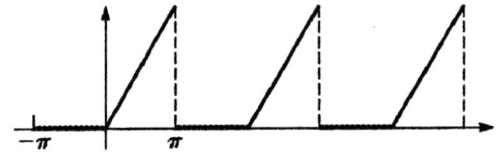

6. $f(x) = x;\ -\pi \le x < \pi$

$a_0 = \dfrac{1}{2\pi}\displaystyle\int_{-\pi}^{\pi} x\,dx = \dfrac{1}{4\pi}x^2\Big|_{-\pi}^{\pi} = 0$

$a_1 = \dfrac{1}{\pi}\displaystyle\int_{-\pi}^{\pi} x\cos x\,dx = \dfrac{1}{\pi}(\cos x + x\sin x)\Big|_{-\pi}^{\pi}$

$\qquad = \dfrac{1}{\pi}\{-1 + \pi(0) - [-1 - (-\pi^0)]\} = 0$

$a_2 = \dfrac{1}{\pi}\displaystyle\int_{-\pi}^{\pi} x\cos 2x\,dx$

$\qquad = \dfrac{1}{4\pi}\displaystyle\int_{-\pi}^{\pi} 2x\cos 2x\,(2\,dx)$

$\qquad = \dfrac{1}{4\pi}(\cos 2x + 2x\sin 2x)\Big|_{-\pi}^{\pi} = 0$

$a_n = 0$

$b_1 = \dfrac{1}{\pi}\displaystyle\int_{-\pi}^{\pi} x\sin x\,dx = \dfrac{1}{\pi}(\sin x - x\cos x)\Big|_{-\pi}^{\pi}$

$\qquad = \dfrac{1}{\pi}[0 - \pi(-1) - [0 + \pi(-1)]] = 2 = \dfrac{2}{(1)}$

$b_2 = \dfrac{1}{\pi}\displaystyle\int_{-\pi}^{\pi} x\sin 2x\,dx = \dfrac{1}{4\pi}\displaystyle\int_{-\pi}^{\pi} 2x\sin 2x\,(2\,dx)$

$\qquad = \dfrac{1}{4\pi}(\sin 2x - 2x\cos 2x)\Big|_{-\pi}^{\pi} = -1 = \dfrac{2}{(-2)}$

$b_3 = \dfrac{1}{\pi}\displaystyle\int_{-\pi}^{\pi} x\sin 3x\,dx = \dfrac{1}{9\pi}\displaystyle\int_{-\pi}^{\pi} 3x\sin 3x\,(3\,dx)$

$\qquad = \dfrac{1}{9\pi}(\sin 3x - 3x\cos 3x)\Big|_{-\pi}^{\pi} = \dfrac{2}{3} = \dfrac{2}{(3)}$

$b_n = \pm\dfrac{2}{n}$; (+ for n odd, $-$ for n even)

Therefore,

$$f(x) = 2\sin x - \frac{2}{2}\sin 2x + \frac{2}{3}\sin 3x - \cdots$$

$$= 2\left(\sin x - \frac{1}{2}\sin 2x + \frac{1}{3}\sin 3x - \cdots\right)$$

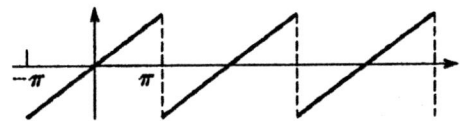

7. $f(x) = \begin{cases} -1 & -\pi \le x < 0 \\ 0 & 0 \le x < \frac{\pi}{2} \\ 1 & \frac{\pi}{2} \le x < \pi \end{cases}$

$a_0 = \frac{1}{2\pi}\int_{-\pi}^{0} -dx + \frac{1}{2\pi}\int_{\pi/2}^{\pi} dx$

$= \frac{1}{2\pi}x\Big|_{-\pi}^{0} + \frac{1}{2\pi}x\Big|_{\pi/2}^{\pi}$

$= -\frac{1}{2\pi}\left(x\Big|_{-\pi}^{0} - x\Big|_{\pi/2}^{\pi}\right)$

$= -\frac{1}{2\pi}\left[\pi - \left(\pi - \frac{\pi}{2}\right)\right] = -\frac{1}{4}$

$a_1 = \frac{1}{\pi}\int_{-\pi}^{0} -\cos x\, dx + \frac{1}{\pi}\int_{\pi/2}^{\pi} c\cos x\, dx$

$= -\frac{1}{\pi}\sin x\Big|_{-\pi}^{0} + \frac{1}{\pi}\sin x\Big|_{\pi/2}^{\pi}$

$= -\frac{1}{\pi}\left(\sin x\Big|_{-\pi}^{0} - \sin x\Big|_{\pi/2}^{\pi}\right) = -\frac{1}{\pi}$

$a_2 = \frac{1}{\pi}\int_{-\pi}^{0} -\cos 2x\, dx + \frac{1}{\pi}\int_{\pi/2}^{\pi} \cos 2x\, dx$

$= -\frac{1}{2\pi}\sin 2x\Big|_{-\pi}^{0} + \frac{1}{2\pi}\sin 2x\Big|_{\pi/2}^{\pi}$

$= -\frac{1}{2\pi}\left(\sin 2x\Big|_{-\pi}^{0} - \sin 2x\Big|_{\pi/2}^{\pi}\right) = 0$

$a_3 = \frac{1}{\pi}\int_{-\pi}^{0} -\cos 3x\, dx + \frac{1}{\pi}\int_{\pi/2}^{\pi} \cos 3x\, dx$

$= -\frac{1}{3\pi}\sin 3x\Big|_{-\pi}^{0} + \frac{1}{3\pi}\sin 3x\Big|_{\pi/2}^{\pi}$

$= -\frac{1}{3\pi}\left(\sin 3x\Big|_{-\pi}^{0} - \sin 3x\Big|_{\pi/2}^{\pi}\right) = \frac{1}{3\pi}$

Therefore, $a_n = \pm\frac{1}{n\pi}$ for n odd; $a_n = 0$ for n even.

$b_1 = \frac{1}{\pi}\int_{-\pi}^{0} -\sin x\, dx + \frac{1}{\pi}\int_{\pi/2}^{\pi} \sin x\, dx$

$= \frac{1}{\pi}\cos x\Big|_{-\pi}^{0} - \frac{1}{\pi}\cos x\Big|_{\pi/2}^{\pi}$

$= \frac{1}{\pi}\left(\cos x\Big|_{-\pi}^{0} - \cos x\Big|_{\pi/2}^{\pi}\right) = \frac{3}{\pi}$

$b_2 = \frac{1}{\pi}\int_{-\pi}^{0} -\sin 2x\, dx + \frac{1}{\pi}\int_{\pi/2}^{\pi} \sin 2x\, dx$

$= \frac{1}{2\pi}\cos 2x\Big|_{-\pi}^{0} - \frac{1}{2\pi}\cos 2x\Big|_{\pi/2}^{\pi}$

$= \frac{1}{2\pi}\left(\cos x\Big|_{-\pi}^{0} - \cos 2x\Big|_{\pi/2}^{\pi}\right) = -\frac{1}{\pi}$

$b_3 = \frac{1}{\pi}\int_{-\pi}^{0} -\sin 3x\, dx + \frac{1}{\pi}\int_{\pi/2}^{\pi} \sin 3x\, dx$

$= \frac{1}{3\pi}\cos 3x\Big|_{-\pi}^{0} - \frac{1}{3\pi}\cos 3x\Big|_{\pi/2}^{\pi}$

$= \frac{1}{3\pi}\left(\cos 3x\Big|_{-\pi}^{0} - \cos 3x\Big|_{\pi/2}^{\pi}\right) = -\frac{1}{\pi}$

$b_n = \pm\frac{1}{\pi}$ for $n > 1$

$f(x) = -\frac{1}{4} - \frac{1}{\pi}\cos x + \frac{1}{3\pi}\cos 3x - \cdots$

$\qquad + \frac{3}{\pi}\sin x - \frac{1}{\pi}\sin 2x + \frac{1}{\pi}\sin 3x - \cdots$

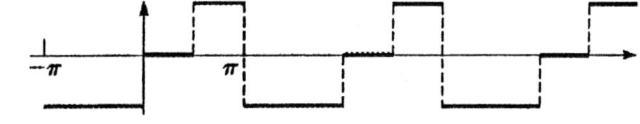

8. $f(x) = x^2;\ -\pi \le x < \pi$

$a_0 = \frac{1}{2\pi}\int_{-\pi}^{\pi} x^2\, dx = \frac{1}{2\pi}\frac{x^3}{3}\Big|_{-\pi}^{\pi}$

$= \frac{1}{6\pi}(\pi^3 + \pi^3) = \frac{\pi^2}{3}$

$u = x^2;\ du = 2x\, dx;\ dv = \cos x\, dx;\ v = \sin x$

$a_1 = \frac{1}{\pi}\int_{-\pi}^{\pi} x^2\cos x\, dx = \frac{1}{\pi}\int_{-\pi}^{\pi} x^2\cos x\, dx$

$= \frac{1}{\pi}\left(x^2\sin x - 2\int_{-\pi}^{\pi} x\sin x\, dx\right)$

$= \frac{1}{\pi}[x^2\sin x - 2(\sin x - x\cos x]$

$= \frac{1}{\pi}(x^2\sin x - 2\sin x + 2x\cos x)\Big|_{-\pi}^{\pi} = -4$

$u = x^2; \ du = 2x \, dx$

$dv = \cos 2x \, dx$

$v = \dfrac{1}{2} \sin 2x$

$a_2 = \dfrac{1}{\pi} \displaystyle\int_{-\pi}^{\pi} x^2 \cos 2x \, dx$

$\quad = \dfrac{1}{\pi} \left(\dfrac{1}{2} x^2 \sin 2x - \displaystyle\int x \sin 2x \, dx \right)$

$\quad = \dfrac{1}{\pi} \left(\dfrac{1}{2} x^2 \sin 2x - \dfrac{1}{4} \displaystyle\int 2x \sin 2x \, 2 \, dx \right)$

$\quad = \dfrac{1}{\pi} \left[\dfrac{1}{2} x^2 \sin 2x - \dfrac{1}{4} (\sin 2x - 2x \cos 2x) \right]$

$\quad = \dfrac{1}{\pi} \left(\dfrac{1}{2} x^2 \sin 2x - \dfrac{1}{4} \sin 2x + \dfrac{1}{2} x \cos 2x \right) \Big|_{-\pi}^{\pi}$

$\quad = 1$

$u = x^2; \ du = 2x \, dx$

$dv = \cos 3x \, dx$

$v = \dfrac{1}{3} \sin 3x$

$a_3 = \dfrac{1}{\pi} \displaystyle\int_{-\pi}^{\pi} x^2 \cos 3x \, dx$

$\quad = \dfrac{1}{\pi} \left(\dfrac{x^2}{3} \sin 3x - \dfrac{2}{3} \displaystyle\int_{-\pi}^{\pi} x \sin 3x \, dx \right)$

$\quad = \dfrac{1}{\pi} \left(\dfrac{x^2}{3} \sin 3x - \dfrac{2}{27} \displaystyle\int_{-\pi}^{\pi} 3x \sin 3x \, 3 \, dx \right)$

$\quad = \dfrac{1}{\pi} \left[\dfrac{x^2}{3} \sin 3x - \dfrac{2}{27} (\sin 3x - 3x \cos 3x) \right]$

$\quad = \dfrac{1}{\pi} \left(\dfrac{x^2}{3} \sin 3x - \dfrac{2}{27} \sin 3x + \dfrac{2}{9} x \cos 3x \right) \Big|_{-\pi}^{\pi}$

$\quad = -\dfrac{4}{9}$

$u = x^2; \ du = 2x \, dx$

$dv = \sin x \, dx; \ v = -\cos x$

$b_1 = \dfrac{1}{\pi} \displaystyle\int_{-\pi}^{\pi} x^2 \sin x \, dx$

$\quad = \dfrac{1}{\pi} \left(-x^2 \cos x + 2 \displaystyle\int x \cos x \, dx \right)$

$\quad = \dfrac{1}{\pi} [-x^2 \cos x + 2(\cos x + x \sin x)$

$\quad = \dfrac{1}{\pi} (-x^2 \cos x + 2 \cos x + 2x \sin x) \Big|_{-\pi}^{\pi}$

$\quad = 0$

$u = x^2; \ du = 2x; \ dv = \sin 2x \, dx;$

$v = -\dfrac{1}{2} \cos 2x$

$b_2 = \dfrac{1}{\pi} \displaystyle\int_{-\pi}^{\pi} x^2 \sin 2x \, dx$

$\quad = \dfrac{1}{\pi} \left(-\dfrac{x^2}{2} \cos 2x + \dfrac{1}{4} \displaystyle\int_{-\pi}^{\pi} 2x \cos 2x \, 2 \, dx \right)$

$\quad = \dfrac{1}{\pi} \left[-\dfrac{x^2}{2} \cos 2x + \dfrac{1}{4} (\cos 2x + 2x \sin 2x) \right]$

$\quad = \dfrac{1}{\pi} \left(-\dfrac{x^2}{2} \cos 2x + \dfrac{1}{4} \cos 2x + \dfrac{1}{2} x \sin 2x \right) \Big|_{-\pi}^{\pi}$

$\quad = 0$

$b_n = 0$ for all n.

$f(x) = \dfrac{\pi^2}{3} - 4 \cos x + \cos 2x - \dfrac{4}{9} \cos 3x$

$\qquad = \dfrac{\pi^2}{3} - 4 \left(\cos x - \dfrac{1}{4} \cos 2x + \dfrac{1}{9} \cos 3x - \cdots \right)$

9. $a_0 = \dfrac{1}{2\pi} \displaystyle\int_{-\pi}^{0} -x \, dx + \dfrac{1}{2\pi} \displaystyle\int_{0}^{\pi} x \, dx$

$\quad = -\dfrac{1}{2\pi} \left(\dfrac{x^2}{2} \right) \Big|_{0}^{\pi} + \dfrac{1}{2\pi} \left(\dfrac{x^2}{2} \right) \Big|_{0}^{\pi}$

$\quad = 0 - \left(-\dfrac{\pi}{4} \right) + \dfrac{\pi^2}{4\pi} - 0$

$\quad = \dfrac{\pi}{4} + \dfrac{\pi}{4}$

$\quad = \dfrac{\pi}{2}$

$a_1 = \dfrac{1}{\pi} \displaystyle\int_{-\pi}^{0} -x \cos x \, dx + \dfrac{1}{\pi} \displaystyle\int_{0}^{\pi} x \cos x \, dx$

$\quad = -\dfrac{1}{\pi} (\cos x + x \sin x) \Big|_{-\pi}^{0}$

$\qquad + \dfrac{1}{\pi} (\cos x + x \sin x) \Big|_{0}^{\pi}$

$\quad = -\dfrac{1}{\pi} [\cos 0 + 0 \sin 0 - \cos(-\pi) + \pi \sin(-\pi)]$

$\qquad + \dfrac{1}{\pi} (\cos \pi + \pi \sin \pi - \cos 0 - 0 \sin 0)$

$\quad = -\dfrac{4}{\pi}$

$$a_2 = \frac{1}{\pi}\int_{-\pi}^{0} -x\cos 2x\,dx + \frac{1}{\pi}\int_{0}^{\pi} x\cos 2x\,dx$$

$$= -\frac{1}{4\pi}\int_{-\pi}^{0} 2x\cos 2x(2\,dx)$$

$$+\frac{1}{4\pi}\int_{0}^{\pi} 2x\cos 2x\,(2\,dx)$$

$$= -\frac{1}{4\pi}\left.(\cos 2x + 2x\sin 2x)\right|_{-\pi}^{0}$$

$$+\frac{1}{4\pi}\left.(\cos 2x + 2x\sin 2x)\right|_{0}^{\pi} = 0$$

$$a_3 = \frac{1}{\pi}\int_{-\pi}^{0} -x\cos 3x\,dx + \frac{1}{\pi}\int_{0}^{\pi} x\cos 3x\,dx$$

$$= -\frac{1}{9\pi}\int_{-\pi}^{0} 3x\cos 3x(3\,dx)$$

$$+\frac{1}{9\pi}\int_{0}^{\pi} 3x\cos 3x(3\,dx)$$

$$= -\frac{1}{9\pi}\left.(\cos 3x + 3x\sin 3x)\right|_{-\pi}^{0}$$

$$+\frac{1}{9\pi}\left.(\cos 3x + 3x\sin 3x)\right|_{0}^{\pi}$$

$$= -\frac{4}{9\pi}$$

$$b_1 = \frac{1}{\pi}\int_{-\pi}^{0} -x\sin x\,dx + \frac{1}{\pi}\int_{0}^{\pi} x\sin x\,dx$$

$$= -\frac{1}{\pi}\left.(\sin x - x\cos x)\right|_{-\pi}^{0}$$

$$+\frac{1}{\pi}\left.(\sin x - x\cos x)\right|_{0}^{\pi}$$

$$= -1 + 1 = 0$$

$$b_2 = \frac{1}{\pi}\int_{-\pi}^{0} -x\sin 2x\,dx + \frac{1}{\pi}\int_{0}^{\pi} x\sin 2x\,dx$$

$$= -\frac{1}{4\pi}\int_{-\pi}^{0} 2x\sin 2x(2\,dx)$$

$$+\frac{1}{4\pi}\int_{0}^{\pi} 2x\sin 2x(2dx)$$

$$= -\frac{1}{4\pi}\left.(\sin 2x - 2x\cos 2x)\right|_{-\pi}^{0}$$

$$+\frac{1}{4\pi}\left.(\sin 2x - 2x\cos 2x)\right|_{0}^{\pi}$$

$$= \frac{1}{2} - \frac{1}{2} = 0$$

$b_n = 0$ for all values of n.

$$f(x) = \frac{\pi}{2} - \frac{4}{\pi}\cos x + 0\cos 2x - \frac{4}{9\pi}\cos 3x + \cdots$$

$$= \frac{\pi}{2} - \frac{4}{\pi}\cos x - \frac{4}{9\pi}\cos 3x - \cdots$$

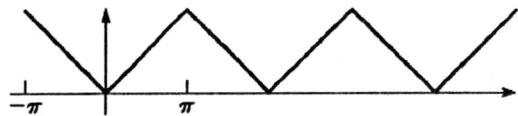

10. $f(x) = \begin{cases} 0 & -\pi \le x < 0 \\ x^2 & 0 \le x < \pi \end{cases}$

$$a_0 = \frac{1}{2\pi}\int_{0}^{\pi} x^2\,dx$$

$$= \left.\frac{1}{6\pi}x^3\right|_{0}^{\pi} = \frac{\pi^2}{6}$$

$$b_1 = \frac{1}{\pi}\int_{0}^{\pi} x^2\sin x\,dx$$

$$= \frac{1}{\pi}\left.(-x^2\cos x + 2\cos x + 2x\sin x)\right|_{0}^{\pi}$$

$$= \frac{\pi^2 - 4}{\pi}$$

$$a_1 = \frac{1}{\pi}\int_{0}^{\pi} x^2\cos x\,dx$$

$$= \frac{1}{\pi}\left.(x^2\sin x - 2\sin x + 2x\cos x)\right|_{0}^{\pi} = -2$$

$$a_2 = \frac{1}{\pi}\int_{0}^{\pi} x^2\cos 2x\,dx$$

$$= \frac{1}{\pi}\left.\left(\frac{1}{2}x^2\sin 2x - \frac{1}{4}\sin 2x + \frac{1}{2}x\cos 2x\right)\right|_{0}^{\pi}$$

$$= \frac{1}{2}$$

$$b_2 = \frac{1}{\pi}\int_{0}^{\pi} x^2\sin 2x\,dx$$

$$= \frac{1}{\pi}\left.\left(-\frac{x^2}{2}\cos 2x + \frac{1}{4}\cos 2x + \frac{1}{2}x\sin 2x\right)\right|_{0}^{\pi}$$

$$= -\frac{\pi}{2}$$

Therefore,

$$f(x) = \frac{\pi^2}{6} - 2\cos x + \frac{1}{2}\cos 2x - \cdots$$

$$+ \frac{\pi^2 - 4}{\pi}\sin x - \frac{\pi}{2}\sin 2x + \cdots$$

$$= \frac{\pi^2}{6} + 2\left(-\cos x + \frac{1}{4}\cos 2x - \cdots\right)$$

$$+ \frac{\pi^2 - 4}{\pi}\sin x - \frac{\pi}{2}\sin 2x + \cdots$$

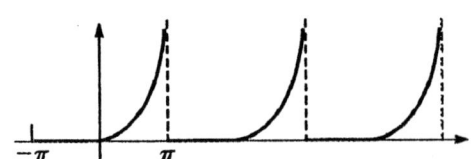

11. $f(x) = \begin{cases} -1 & -\pi \le x < 0 \\ 1 & 0 \le x < \pi \end{cases}$

$$f(x) = \frac{4}{\pi}\left(\sin x + \frac{1}{3}\sin 3x + \frac{1}{5}\sin 5x + \cdots\right)$$

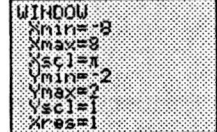 

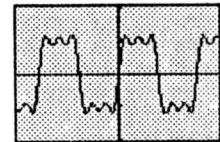

12. $f(x) = \dfrac{2+\pi}{4} - \dfrac{2}{\pi}\cos x - \dfrac{2}{9\pi}\cos 3x - \cdots$

$$+ \left(\frac{\pi-2}{\pi}\right)\sin x - \frac{1}{2}\sin 2x + \cdots$$

Let $y_1 = f(x)$ for graph

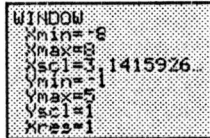 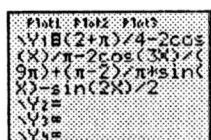

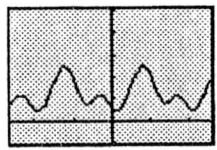

13. $y_1 = \dfrac{\pi}{4} - \left(\dfrac{2}{\pi}\right)(\cos x + 9^{-1}\cos 3x) + \sin x - 0.5\sin 2x$

$x_{\min} = -8, \; = x_{\max} = 8$

$y_{\min} = -1, \; = y_{\max} = 3$

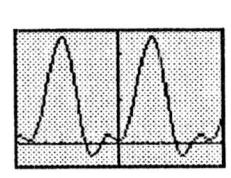

14. $f(x) = \begin{cases} -x & -\pi \le x < 0 \\ x & 0 \le x < \pi \end{cases}$

$$f(x) = \frac{\pi}{2} - \frac{4}{\pi}\cos x - \frac{4}{9\pi}\cos 3x - \cdots$$

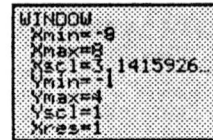 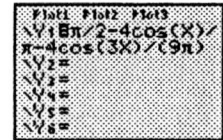

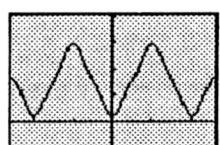

15. $a_n = \dfrac{1}{\pi}\displaystyle\int_{-\pi}^{0} -\sin t\cos nt\,dt + \dfrac{1}{\pi}\int_{0}^{\pi}\sin t\cos nt\,dt$

$$= \frac{1}{2\pi}\left[\frac{\cos(1-n)t}{1-n} + \frac{\cos(1+n)t}{1+n}\right]\Big|_{-\pi}^{0}$$

$$+ -\frac{1}{2\pi}\left[\frac{\cos(1-n)t}{1-n} + \frac{\cos(1+n)t}{1+n}\right]\Big|_{0}^{\pi}$$

$$= -\frac{1}{2\pi}\left(\frac{1}{1-n} + \frac{1}{1+n} - \frac{\cos(1-n)(-\pi)}{1-n}\right.$$

$$\left. - \frac{\cos(1+n)(-\pi)}{1+n}\right)$$

$$- \frac{1}{2\pi}\left(\frac{\cos(1-n)\pi}{1-n} + \frac{\cos(1+n)\pi}{1+n}\right.$$

$$\left. - \frac{1}{1-n} - \frac{1}{1+n}\right)$$

$$a_1 = \frac{1}{\pi}\int_{-\pi}^{0} -\sin t\cos t\,dt + \int_{0}^{\pi}\sin t\cos t\,dt$$

$$= -\frac{1}{2\pi}\sin^2 t\Big|_{-\pi}^{0} + \frac{1}{2\pi}\sin^2 t\Big|_{0}^{\pi} = 0$$

$$a_2 = \frac{1}{2\pi}\left[\frac{1}{-1} + \frac{1}{3} - \frac{(-1)}{-1} - \frac{(-1)}{3}\right]$$

$$- \frac{1}{2\pi}\left[\frac{-1}{-1} + \frac{(-1)}{3} - \frac{1}{-1} - \frac{1}{3}\right]$$

$$= \frac{1}{2\pi}\left(\frac{-4}{3}\right) - \frac{1}{2\pi}\left(\frac{4}{3}\right) = -\frac{4}{3\pi}$$

$$a_3 = \frac{1}{2\pi}\left(\frac{1}{-2} + \frac{1}{4} - \frac{1}{-2} - \frac{1}{4}\right)$$

$$- \frac{1}{2\pi}\left(\frac{1}{-2} + \frac{1}{4} - \frac{1}{-2} - \frac{1}{4}\right)$$

$$= \frac{1}{2\pi}(0) - \frac{1}{2\pi}(0) = 0$$

$$a_4 = \frac{1}{2\pi}\left[\frac{1}{-3} + \frac{1}{5} - \frac{(-1)}{-3} - \frac{(-1)}{5}\right]$$

$$- \frac{1}{2\pi}\left(\frac{-1}{-3} + \frac{-1}{5} - \frac{1}{-3} - \frac{1}{5}\right)$$

$$= \frac{1}{2\pi}\left(-\frac{2}{5}\right) - \frac{1}{2\pi}\left(\frac{2}{5}\right) = -\frac{4}{15\pi}$$

$$b_n = \frac{1}{\pi}\int_{-\pi}^{0} -\sin t \sin nt\, dt + \frac{1}{\pi}\int_{0}^{\pi}\sin t \sin nt$$

$$= -\frac{1}{2\pi}\left[\frac{\sin(1-n)t}{1-n} - \frac{\sin(1+n)t}{1+n}\right]\Big|_{-\pi}^{0}$$

$$+ \frac{1}{2\pi}\left[\frac{\sin(1-n)t}{1-n} - \frac{\sin(1+n)t}{1+n}\right]\Big|_{0}^{\pi}$$

$$= -\frac{1}{2\pi}\left[\frac{-\sin(1-n)(-\pi)}{1-n} - \frac{\sin(1+n)(-\pi)}{1+n}\right]$$

$$+ \frac{1}{2\pi}\left[\frac{\sin(1-n)\pi}{1-n} - \frac{\sin(1+n)\pi}{1+n}\right]$$

$$b_1 = \frac{1}{\pi}\int_{-\pi}^{0} -\sin^2 t\, dt + \frac{1}{\pi}\int_{0}^{\pi}\sin^2 t\, dt$$

$$= -\frac{1}{2\pi}(t - \sin t \cos t)\Big|_{-\pi}^{0} + \frac{1}{2\pi}(t - \sin t \cos t)\Big|_{0}^{\pi}$$

$$= -\frac{1}{2\pi}(\pi) + \frac{1}{2\pi}(\pi) = 0$$

$b_n = 0$ if $n > 1$, since each is evaluated in terms of the sine of a multiple of π.

$$f(t) = \frac{2}{\pi} - \frac{4}{3\pi}\cos 2t - \frac{4}{15\pi}\cos 4t - \cdots$$

16. $f(t) = 0, \ -\pi \le t < 0$

$$f(t) = 100t, \ 0 \le t < \frac{\pi}{2}$$

$$f(t) = 100(\pi - t), \ \frac{\pi}{2} \le t < \pi$$

$$a_0 = \frac{1}{2\pi}\int_{-\pi}^{0} 0 \cdot 0 dt + \frac{1}{2\pi}\int_{0}^{\pi/2} 100t\, dt$$

$$+ \frac{1}{2\pi}\int_{\pi/2}^{\pi} 100(\pi - t)dt$$

$$a_0 = \frac{25\pi}{2}$$

$$a_n = \frac{1}{\pi}\int_{0}^{\pi/2} 100t \cdot \cos nt\, dt$$

$$+ \frac{1}{\pi}\int_{\pi/2}^{\pi} 100(\pi - t)\cos nt\, dt$$

$$a_n = \frac{200\cos\frac{n\pi}{2} - 100\cos(n\pi) - 100}{n^2\pi}$$

$$a_1 = 0, a_2 = \frac{-100}{\pi}$$

$$b_n = \frac{1}{\pi}\int_{0}^{\pi/2} 100t \sin nt\, dt$$

$$+ \frac{1}{\pi}\int_{\pi/2}^{\pi} 100(\pi - t)\sin nt\, dt$$

$$b_n = \frac{200\sin\frac{n\pi}{2} - 100\sin n\pi}{n^2\pi}$$

$$b_1 = \frac{200}{\pi}, b_2 = 0, b_3 = \frac{-200}{9\pi}$$

$$f(t) = \frac{25\pi}{2} - \frac{100}{\pi}\cos(2t) + \frac{200}{\pi}\sin t - \frac{200}{9\pi}\sin(3t) + \cdots$$

13.7   More About Fourier Series

1. $f(x) = \begin{cases} 5 & -3 \le x < 0 \\ 0 & 0 \le x < 3 \end{cases}$

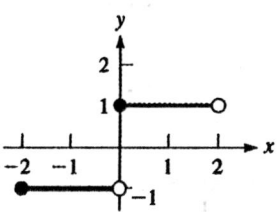

from the graph $f(x)$ is neither odd nor even.

2. $f(x) = \begin{cases} -1 & -2 \le x < 0 \\ 1 & 0 \le x < 2 \end{cases}$

from the graph $f(x) = -f(-x)$. $f(x)$ is odd.

3. $f(x) = \begin{cases} 2 & -1 \le x < 1 \\ 0 & -2 \le x < -1, 1 \le x < 2 \end{cases}$

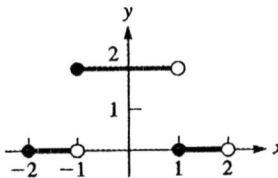

from the graph $f(x) = f(-x)$. $f(x)$ is even.

4. $f(x) = \begin{cases} 0 & -2 \le x < 0, 1 \le x < 2 \\ 1 & 0 \le x < 1 \end{cases}$

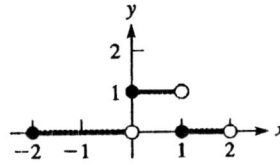

from the graph $f(x)$ is neither odd nor even.

5. $f(x) = |x| \quad -4 \le x < 4$

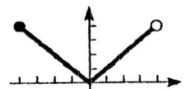

from the graph $f(x)$
is even.

6. $f(x) = \begin{cases} 0 & -1 \le x < 0 \\ e^x & 0 \le x < 1 \end{cases}$

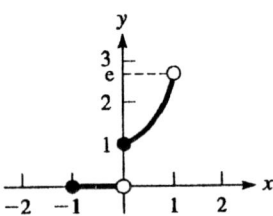

from the graph $f(x)$ is neither odd nor even.

7. $f(x) = -x \cos 3x \quad -3 \le x < 3$

$$f(-x) = -(-x) \cos(3(-x)) = x \cos(-3x)$$
$$= x \cos 3x = -f(x).$$

$f(x)$ is odd.

8. $f(x) = x \sin 2x \cos x \quad -4 \le x < 4$

$$f(-x) = -x \sin(2(-x)) \cos(-x)$$
$$= x \sin 2x \cos x = f(x).$$

$f(x)$ is even.

9. $f(x) \begin{cases} 2 & -\pi \le x < -\dfrac{\pi}{2}, \dfrac{\pi}{2} \le x < \pi \\ 3 & -\dfrac{\pi}{2} \le x < \dfrac{\pi}{2} \end{cases}$

From Example 2,

$$f(x) = \frac{1}{2} + \frac{2}{\pi}\left(\cos x - \frac{\cos 3x}{3} + \frac{\cos 5x}{5} - \cdots\right) + 2.$$

$$= \frac{5}{2} + \frac{2}{\pi}\left(\cos x - \frac{\cos 3x}{3} + \frac{\cos 5x}{5} - \cdots\right)$$

10. $f(x) = \begin{cases} -\frac{1}{2} & -\pi \le x < 0 \\ \frac{3}{2} & 0 \le x < \pi \end{cases}$

From Example 4

$$f(x) = \frac{4}{\pi}\left(\sin x + \frac{1}{3}\sin 3x + \frac{1}{5}\sin 5x - \cdots\right) + \frac{1}{2}$$

11. $f(x) = \begin{cases} -2 & -4 \le x < 0 \\ 0 & 0 \le x < 4 \end{cases}$

From Example 7

$$f(x) = 1 + \frac{4}{\pi}\left(\sin \frac{\pi x}{4} + \frac{\sin \frac{3\pi x}{4}}{3} + \cdots\right) - 2$$

$$f(x) = -1 + \frac{4}{\pi}\left(\sin \frac{\pi x}{4} + \frac{\sin \frac{3\pi x}{4}}{3} + \cdots\right)$$

12. $f(x) = \begin{cases} -\frac{1}{3} & -\pi \le x < -\frac{\pi}{2}, \frac{\pi}{2} \le x < \pi \\ \frac{2}{3} & -\frac{\pi}{2} \le x < \frac{\pi}{2} \end{cases}$

From Example 2

$$f(x) = \frac{1}{2} + \frac{2}{\pi}\left(\cos x - \frac{1}{3}\cos(3x) + \frac{1}{5}\cos(5x) - \cdots\right)$$

$$-\frac{1}{3}$$

$$f(x) = \frac{1}{6} + \frac{2}{\pi}\left(\cos x - \frac{1}{3}\cos(3x) + \frac{1}{5}\cos(5x) - \cdots\right)$$

13. $f(x) \begin{cases} 5 & -3 \le x < 0 \\ 0 & 0 \le x < 3 \end{cases}$

period $= 6 = 2L, L = 3$

$$a_0 = \frac{1}{2L}\int_{-L}^{L} f(x)\,dx$$

$$= \frac{1}{6}\int_{-3}^{0} 5\,dx + \frac{1}{6}\int_{0}^{3} 0 \cdot dx = \frac{5}{2}$$

$$a_n = \frac{1}{L}\int_{-L}^{L} f(x)\cos\frac{n\pi x}{L}\,dx$$

$$= \frac{1}{3}\int_{-3}^{0} 5\cos\frac{n\pi x}{3}\,dx + \frac{1}{3}\int_{0}^{3} 0\cdot\cos\frac{n\pi x}{3}\,dx$$

$$a_n = \frac{5\sin(n\pi)}{n\pi} = 0, n = 1,2,3\cdots$$

$$b_n = \frac{1}{L}\int_{-L}^{L} f(x)\sin\frac{n\pi x}{L}\,dx$$

$$= \frac{1}{3}\int_{-3}^{0} 5\sin\frac{n\pi x}{3}\,dx + \frac{1}{3}\int_{0}^{3} 0\cdot\sin\frac{2\pi x}{3}\,dx$$

$$b_n = \frac{5\cos(n\pi) - 5}{n\pi} = \frac{5}{\pi}\left(\frac{\cos(n\pi) - 1}{n}\right)$$

n	b_n
1	$\frac{5}{\pi}\cdot(-2) = \frac{-10}{\pi}$
2	0
3	$\frac{5}{\pi}\left(-\frac{2}{3}\right) = \frac{-10}{3\pi}$
4	0
5	$\frac{5}{\pi}\left(-\frac{2}{5}\right) = \frac{-10}{5\pi}$

$$f(x) = a_0 + a_1\cos\frac{\pi x}{L} + a_2\cos\frac{2\pi x}{L} + a_3\cos\frac{3\pi x}{L} + \cdots$$
$$+ b_1,\sin\frac{\pi x}{L} + b_2\frac{2\pi x}{L} + b_3\frac{3\pi x}{L} + \cdots$$

$$f(x) = \frac{5}{2}$$

$$-\frac{10}{\pi}\left(\sin\frac{\pi x}{3} + \frac{1}{3}\sin\frac{3\pi x}{3} + \frac{1}{5}\sin\frac{5\pi x}{3} + \cdots\right)$$

 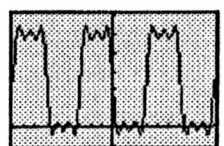

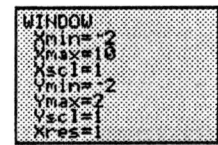

14. $f(x) = \begin{cases} -1 & -2 \le x < 0 \\ 1 & 0 \le x < 2 \end{cases}$, period $= 2L = 4 \Rightarrow$

$L = 2$

$$a_0 = \frac{1}{2\cdot 2}\int_{-2}^{2} f(x)\,dx$$

$$= \frac{1}{4}\int_{-2}^{0}(-1)\,dx + \frac{1}{4}\int_{0}^{2}(1)\,dx = 0$$

$$a_n = \frac{1}{2}\int_{-2}^{2} f(x)\cos\frac{n\pi x}{2}\,dx$$

$$= \frac{1}{4}\int_{-2}^{0} -\cos\frac{n\pi x}{2}\,dx + \frac{1}{4}\int_{0}^{2}\cos\frac{n\pi x}{2}\,dx$$

$$a_n = 0$$

$$b_n = \frac{1}{2}\int_{-2}^{2} f(x)\sin\frac{n\pi x}{2}\,dx$$

$$= \frac{1}{2}\int_{-2}^{0} -\sin\frac{n\pi x}{2}\,dx + \frac{1}{2}\int_{0}^{2}\sin\frac{n\pi x}{2}\,dx$$

$$b_n = \frac{2(1 - \cos n\pi)}{n\pi}$$

$$b_1 = \frac{4}{\pi}, b_2 = 0, b_3 = \frac{4}{3\pi}, b_4 = 0, b_5 = \frac{4}{5\pi}$$

$$f(x) = \frac{4}{\pi}\left(\sin\frac{\pi x}{2} + \frac{1}{3}\sin\frac{3\pi x}{2} + \frac{1}{5}\sin\frac{5\pi x}{2} + \cdots\right)$$

 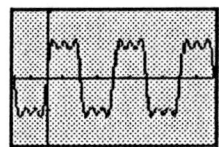

15. $f(x) = \begin{cases} 2 & -1 \leq x < 1 \\ 0 & -2 \leq x < -1, 1 \leq x < 2 \end{cases}$

period $= 2L = 4$
$$L = 2$$

$$a_0 = \frac{1}{2 \cdot 2} \int_{-2}^{2} f(x)\,dx = \frac{1}{4} \int_{-1}^{1} 2\,dx = 1$$

$$a_n = \frac{1}{2} \int_{-2}^{2} f(x) \cos \frac{n\pi x}{2}\,dx$$

$$= \frac{1}{2} \int_{-1}^{1} 2 \cos \frac{n\pi x}{2}\,dx$$

$$= \frac{4 \sin \frac{n\pi}{2}}{n\pi}$$

$$a_1 = \frac{4}{\pi}, a_2 = 0, a_3 = \frac{4}{\pi} \cdot \left(-\frac{1}{3} \right), a_4 = \frac{4}{\pi} \cdot \left(\frac{1}{5} \right)$$

$$b_n = \frac{1}{2} \int_{-2}^{2} f(x) \sin \frac{n\pi x}{2}\,dx$$

$$= \frac{1}{2} \int_{-1}^{1} 2 \sin \frac{n\pi x}{2}\,dx = 0$$

$$f(x) = 1 + \frac{4}{\pi} \left(\cos \frac{\pi x}{2} - \frac{1}{3} \cos \frac{3\pi x}{2} + \frac{1}{5} \cos \frac{5\pi x}{2} - \cdots \right)$$

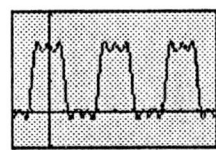

16. $f(x) = \begin{cases} 0 & -2 \leq x < 0, 1 \leq x < 2 \\ 1 & 0 \leq x < 1 \end{cases}$

period $= 2L = 4$
$$L = 2$$

$$a_0 = \frac{1}{2 \cdot 2} \int_{-2}^{2} f(x)\,dx = \frac{1}{4} \int_{0}^{1} 1\,dx = \frac{1}{4}$$

$$a_n = \frac{1}{2} \int_{-2}^{2} f(x) \cos \frac{n\pi x}{2}\,dx$$

$$= \frac{1}{2} \int_{0}^{1} \cos \frac{n\pi x}{2}\,dx = \frac{\sin \frac{n\pi}{2}}{n\pi}$$

$$a_1 = \frac{1}{\pi}, a_2 = 0, a_3 = \frac{-1}{3\pi}, a_4 = 0, a_5 = \frac{1}{5\pi}$$

$$b_n = \frac{1}{2} \int_{-2}^{2} f(x) \sin \frac{n\pi x}{2}\,dx$$

$$= \frac{1}{2} \int_{0}^{1} \sin \frac{n\pi x}{2}\,dx$$

$$= \frac{1 - \cos \frac{n\pi}{2}}{n\pi}$$

$$b_1 = \frac{1}{\pi},$$

$$b_2 = \frac{1}{\pi},$$

$$b_3 = \frac{1}{3\pi},$$

$$b_4 = 0,$$

$$b_5 = \frac{1}{5\pi}$$

$$f(x) = \frac{1}{4} + \frac{1}{\pi} \left(\cos \frac{\pi x}{2} - \frac{1}{3} \cos \frac{3\pi x}{2} \right.$$
$$+ \frac{1}{5} \cos \frac{5\pi x}{2} - \frac{1}{7} \cos \frac{7\pi x}{2} + \cdots \right)$$
$$+ \frac{1}{\pi} \left(\sin \frac{\pi x}{2} + \sin \pi x + \frac{1}{3} \sin \frac{3\pi x}{2} \right.$$
$$\left. + \frac{1}{5} \sin \frac{5\pi x}{2} + \frac{1}{7} \sin \frac{7\pi x}{2} + \cdots \right)$$

Let $y_1 = f(x)$ to graph

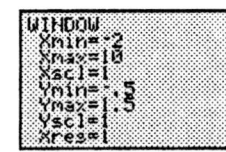

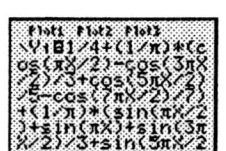

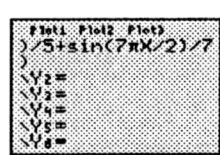

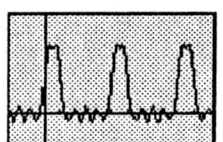

17. $f(x) = \begin{cases} -x & -4 \leq x < 0 \\ x & 0 \leq x < 4 \end{cases}$

$$a_0 = \frac{1}{8} \int_{-4}^{0} -x\,dx + \frac{1}{8} \int_{0}^{4} x\,dx$$

$$= -\frac{1}{16} x^2 \Big|_{-4}^{0} + \frac{1}{16} x^2 \Big|_{0}^{4} = 2$$

$$a_n = \frac{1}{4}\int_{-4}^{0} -x\cos\frac{n\pi x}{4}\,dx + \frac{1}{4}\int x\cos\frac{n\pi x}{4}\,dx$$

$$= -\frac{1}{4}\frac{16}{(n\pi)^2}\int_{-4}^{0}\frac{n\pi}{4}x\cos\frac{n\pi}{4}x\frac{n\pi}{4}\,dx$$

$$+ \frac{1}{4}\frac{16}{(n\pi)^2}\int_{0}^{4}\frac{n\pi}{4}x\cos\frac{n\pi}{4}x\frac{n\pi}{4}\,dx$$

$$= -\frac{4}{(n\pi)^2}\left(\cos\frac{n\pi x}{4} + \frac{n\pi x}{4}\sin\frac{n\pi x}{4}\right)\Big|_{-4}^{0}$$

$$+ \frac{4}{(n\pi)^2}\left(\cos\frac{n\pi x}{4} + \frac{n\pi x}{4}\sin\frac{n\pi x}{4}\right)\Big|_{0}^{4}$$

$$- \frac{4}{(n\pi)^2}(\cos 0 - [\cos(-n\pi) + n\pi\sin n\pi])$$

$$+ \frac{4}{(n\pi)^2}(\cos n\pi + n\pi\sin n\pi - [\cos 0])$$

$$= -\frac{4}{(n\pi)^2}(1 - \cos n\pi - n\pi\sin n\pi)$$

$$+ \frac{4}{(n\pi)^2}(\cos n\pi + n\pi\sin n\pi - 1)$$

$$= -\frac{4}{(n\pi)^2}(1 - \cos n\pi - n\pi\sin n\pi$$

$$- \cos n\pi - n\pi\sin n\pi + 1)$$

$$= -\frac{4}{(n\pi)^2}(2 - 2\cos n\pi - 2n\pi\sin n\pi)$$

$$a_1 = -\frac{16}{\pi^2}; \quad a_2 = 0; \quad a_3 = -\frac{16}{9\pi^2}$$

$$b_n = \frac{1}{4}\int_{-4}^{0}-x\sin\frac{n\pi x}{4}\,dx + \frac{1}{4}\int_{0}^{4}x\sin\frac{n\pi x}{4}\,dx$$

$$= -\frac{1}{4}\frac{16}{(n\pi)^2}\int_{-4}^{0}\frac{n\pi x}{4}\sin\frac{n\pi x}{4}\left(\frac{n\pi}{4}\,dx\right)$$

$$+ \frac{1}{4}\frac{16}{(n\pi)^2}\int_{0}^{4}\frac{n\pi x}{4}\sin\frac{n\pi x}{4}\cdot\frac{n\pi\,dx}{4}$$

$$= -\frac{4}{(n\pi)^2}\left(\sin\frac{n\pi x}{4} - \frac{n\pi x}{4}\cos\frac{n\pi x}{4}\right)\Big|_{-4}^{0}$$

$$+ \frac{4}{(n\pi)^2}\left(\sin\frac{n\pi x}{4} - \frac{n\pi x}{4}\cos\frac{n\pi x}{4}\right)\Big|_{0}^{4}$$

$$= -\frac{4}{(n\pi)^2}\{-[\sin(-n\pi) + n\pi\cos(-n\pi)]\}$$

$$+ \frac{4}{(n\pi)^2}(\sin n\pi - n\pi\cos n\pi)$$

$$= -\frac{4}{(n\pi)^2}(\sin n\pi - n\pi\cos n\pi)$$

$$+ \frac{4}{(n\pi)^2}(\sin n\pi - n\pi\cos n\pi) = 0, \text{ for all } n.$$

Therefore,

$$f(x) = 2 - \frac{16}{\pi^2}\cos\frac{\pi x}{4} - \frac{16}{9\pi^2}\cos\frac{3\pi x}{4}$$

$$= 2 - \frac{16}{\pi^2}\left(\cos\frac{\pi x}{4} + \frac{1}{9}\cos\frac{3\pi x}{4} + \cdots\right)$$

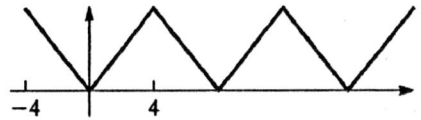

and comparing with calculator,

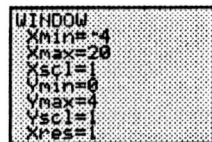

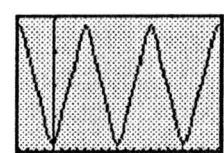

18. $f(x) = \begin{cases} 0 & -1 \le x < 0 \\ e^x & 0 \le x < 1 \end{cases}$

$$\text{period} = 2L = 2$$
$$L = 1$$

$$a_0 = \frac{1}{2.1}\int_{-1}^{1}f(x)\,dx = \frac{1}{2}\int_{0}^{1}e^x\,dx = \frac{e-1}{2}$$

$$a_n = \frac{1}{1}\int_{-1}^{1}f(x)\cos\frac{n\pi x}{1}$$

$$= \int_{0}^{1}e^x\cos(n\pi x)\,dx$$

$$a_n = \frac{e(\cos(n\pi) + n\pi\sin(n\pi)) - 1}{n^2\pi^2 + 1}$$

$$a_1 = \frac{-e-1}{\pi^2+1}, \, a_2 = \frac{e-1}{4\pi^2+1}, \, a_3 = \frac{-e-1}{9\pi^2+1},$$

$$b_n = \int_{0}^{1}e^x\sin(n\pi x)\,dx$$

$$= \frac{n\pi + e(\sin n\pi - n\pi\cos n\pi)}{\pi^2 n^2 + 1}$$

$$b_1 = \frac{\pi(1+e)}{\pi^2+1}, \, b_2 = \frac{2\pi(1-e)}{4\pi^2+1}, \, b_3 = \frac{3\pi(1+e)}{9\pi^2+1}$$

$$f(x) = \frac{e-1}{2} - \frac{e+1}{\pi^2+1}\cos \pi x + \frac{e-1}{4\pi^2+1}\cos 2\pi x$$

$$- \frac{e+1}{9\pi^2+1}\cos 3\pi x + \cdots$$

$$+ \frac{\pi(1+e)}{\pi^2+1}\sin \pi x + \frac{2\pi(1-e)}{4\pi^2+1}\sin 2\pi x$$

$$+ \frac{3\pi(1+e)}{9\pi^2+1}\sin 3\pi x + \cdots$$

To graph, let $y_1 = f(x)$

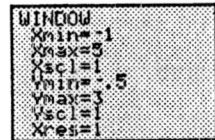 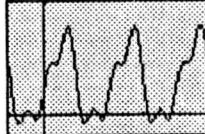

19. Expand $f(x) = 1$ in half-range since series for $0 \le x < 4$

$$b_n = \frac{2}{L}\int_0^L f(x)\sin\frac{n\pi x}{L}\,dx \quad (n = 1,2,3,\ldots)$$

$$b_n = \frac{2}{4}\int_0^4 1 \sin\frac{n\pi x}{4}\,dx$$

$$= \frac{1}{2}\frac{4}{n\pi}\int_0^4 \sin\frac{n\pi x}{4}\left(\frac{n\pi}{4}\right)dx$$

$$b_n = \frac{2}{n\pi}\left(-\cos\frac{n\pi x}{4}\right)\Big|_0^4$$

$$= \frac{2}{n\pi}\left(-\cos\frac{n\pi(4)}{4} + \cos 0\right)$$

$$b_n = \frac{2}{n\pi}(1 - \cos n\pi) = \begin{cases} 0 & \text{for } n \text{ even} \\ \dfrac{4}{n\pi} & \text{for } n \text{ odd} \end{cases}$$

$$b_1 = \frac{4}{\pi}, b_3 = \frac{4}{3\pi}, b_5 = \frac{4}{5\pi}$$

$$f(x) = \frac{4}{\pi}\sin\frac{\pi x}{4} + \frac{4}{3\pi}\sin\frac{3\pi x}{4} + \frac{4}{5\pi}\sin\frac{5\pi x}{4} + \cdots$$

$$f(x) = \frac{4}{\pi}\left(\sin\frac{\pi x}{4} + \frac{1}{3}\sin\frac{3\pi x}{4} + \frac{1}{5}\sin\frac{5\pi x}{4} + \cdots\right)$$

20. Expand $f(x) = \begin{cases} 1, & 0 \le x < 2 \\ 0, & 2 \le x < 4 \end{cases}$ into a half-range cosine series.

$$a_0 = \frac{1}{L}\int_0^L f(x)\,dx = \frac{1}{4}\int_0^2 1\,dx + \frac{1}{4}\int_0^4 0\,dx$$

$$= \frac{1}{4}x\Big|_0^2 = \frac{1}{2}$$

$$a_n = \frac{2}{L}\int_0^L f(x)\cos\frac{n\pi x}{L}\,dx, (n = 1,2,3\ldots)$$

$$a_n = \frac{2}{4}\int_0^2 1\cos\frac{n\pi x}{4}\,dx + \frac{2}{4}\int_0^4 0\cos\frac{n\pi x}{4}\,dx$$

$$a_n = \frac{1}{2}\frac{4}{n\pi}\int_0^2 \cos\frac{n\pi x}{4}\left(\frac{n\pi}{4}\right)dx$$

$$a_n = \frac{2}{n\pi}\sin\frac{n\pi x}{4}\Big|_0^4$$

$$= \frac{2}{n\pi}\left(\sin\frac{n\pi(2)}{4} - \sin\frac{n\pi(0)}{4}\right)$$

$$a_n = \frac{2}{n\pi}\sin\frac{n\pi}{2} = \begin{cases} \dfrac{2}{n\pi}, & n = 1,5,9,\ldots \\ -\dfrac{2}{n\pi}, & n = 3,7,11,\ldots \\ 0, & n \text{ even} \end{cases}$$

$$a_1 = \frac{2}{\pi}$$

$$a_2 = -\frac{2}{3\pi}$$

$$a_5 = \frac{2}{5\pi}$$

$$f(x) = \frac{1}{2} + \frac{2}{\pi}\left(\cos\frac{\pi x}{4} - \frac{1}{3}\cos\frac{3\pi x}{4} + \frac{1}{5}\cos\frac{5\pi x}{4} - \cdots\right)$$

21. Expand $f(x) = x^2$ in a half-range cosine series for $0 \le x < 2$.

$$a_0 = \frac{1}{L}\int_0^L f(x)\,dx = \frac{1}{2}\int_0^2 x^2\,dx$$

$$= \frac{1}{2}\frac{x^3}{3}\Big|_0^2 = \frac{1}{6}(2^3 - 0) = \frac{4}{3}$$

$$a_n = \frac{2}{L}\int_0^L f(x)\cos\frac{n\pi x}{L}\,dx, (n = 1,2,3,\ldots)$$

$$a_n = \frac{2}{2}\int_0^2 x^2\cos\frac{n\pi x}{2}\,dx$$

$$= \frac{2x}{\frac{n^2\pi^2}{4}}\cos\frac{n\pi x}{2} + \left(\frac{x^2}{\frac{n\pi}{2}} - \frac{2}{\frac{n^3\pi^3}{8}}\right)\sin\frac{n\pi x}{2}\Big|_0^2$$

$$a_n = \frac{8x}{n^2\pi 2}\cos\frac{n\pi x}{2} + \left(\frac{2x^2}{n\pi} - \frac{16}{n^3\pi^3}\right)\sin\frac{n\pi x}{2}\bigg|_0^2$$

$$a_n = \frac{16}{n^2\pi^2}\cos n\pi + \left(\frac{8}{n\pi} - \frac{16}{n^3\pi^3}\right)\sin n\pi$$

$$a_n = \frac{16}{n^2\pi^2}\cos n\pi$$

$$a_1 = \frac{16}{1^2\pi^2}\cos\pi = \frac{-16}{\pi^2}$$

$$a_2 = \frac{16}{2^2\pi^2}\cos 2\pi = \frac{4}{\pi^2}$$

$$a_3 = \frac{16}{3^2\pi^2}\cos 3\pi = \frac{-16}{9\pi^2}$$

$$f(x) = \frac{4}{3} - \frac{16}{\pi^2}\cos\frac{\pi x}{2} + \frac{4}{\pi^2}\cos\frac{2\pi x}{2}$$
$$- \frac{16}{9\pi^2}\cos\frac{3\pi x}{2} + \cdots$$

$$f(x) = \frac{4}{3} - \frac{16}{\pi^2}\left(\cos\frac{\pi x}{2} - \frac{1}{4}\cos\pi x + \frac{1}{9}\cos\frac{3\pi x}{2} - \cdots\right)$$

22. Expand $f(x) = x^2$ in a half-range sine series for $0 \le x < 2$.

$$b_n = \frac{2}{L}\int_0^L f(x)\sin\frac{n\pi x}{L}\,dx,\ (n = 1, 2, 3, \ldots)$$

$$b_n = \frac{2}{2}\int_0^2 x^2\sin\frac{n\pi x}{2}\,dx$$

$$b_n = \left[\frac{2x}{\frac{n^2\pi^2}{4}}\sin\frac{n\pi x}{2} + \left(\frac{2}{\frac{n^3\pi^3}{8}} - \frac{x^2}{\frac{n\pi}{2}}\right)\cos\frac{n\pi x}{2}\right]\bigg|_0^2$$

$$b_n = \left[\frac{8x}{n^2\pi^2}\sin\frac{n\pi x}{2} + \left(\frac{16}{n^3\pi^3} - \frac{2x^2}{n\pi}\right)\cos\frac{n\pi x}{2}\right]\bigg|_0^2$$

$$b_n = \frac{8(2)}{n^2\pi^2}\sin n\pi + \left(\frac{16}{n^3\pi^3} - \frac{8}{n\pi}\right)\cos n\pi - \frac{8(0)}{n^2\pi^2}\sin 0$$
$$- \left(\frac{16}{n^3\pi^3} - \frac{2(0)^2}{n\pi}\right)\cos 0$$

$$b_n = \left(\frac{16}{n^3\pi^3} - \frac{8}{n\pi}\right)\cos n\pi - \frac{16}{n^3\pi^3}$$

$$b_1 = \left(\frac{16}{1^3\pi^3} - \frac{8}{1\pi}\right)(-1) - \frac{16}{1^3\pi^3} = \frac{-32}{\pi 3} + \frac{8}{\pi}$$

$$b_2 = \left(\frac{16}{2^3\pi^3} - \frac{8}{2\pi}\right)(1) - \frac{16}{2^3\pi^3} = -\frac{8}{2\pi} = \frac{-4}{\pi}$$

$$b_3 = \left(\frac{16}{3^3\pi^3} - \frac{8}{3\pi}\right)(-1) - \frac{16}{3^3\pi^3} = \frac{-32}{27\pi^3} + \frac{8}{3\pi}$$

$$f(x) = \left(\frac{-32}{\pi^3} + \frac{8}{\pi}\right)\sin\frac{\pi}{2} - \frac{4}{\pi}\sin\pi$$
$$+ \left(\frac{-32}{27\pi^3} + \frac{8}{3\pi}\right)\sin\frac{3\pi x}{2} - \cdots$$

23. $f(x) = \begin{cases} 0 & -2 \le t < 0 \text{ and } 1 \le t \le 2 \\ 8 & 0 \le t < 1 \end{cases}$ $P = 4s$

$$a_0 = \frac{1}{4}\int_0^1 8\,dt = \frac{1}{4}(8t)\bigg|_0^1 = 2t\bigg|_0^1 = 2$$

$$a_n = \frac{1}{4}\int_0^1 8\cos\frac{n\pi t}{2}\,dt$$

$$= 4\cdot\frac{2}{n\pi}\int_0^1\cos\frac{n\pi}{2}t\left(\frac{n\pi}{2}\,dt\right)$$

$$= \frac{8}{n\pi}\sin\frac{n\pi t}{2}\bigg|_0^1$$

$$= \frac{8}{n\pi}\left(\sin\frac{n\pi}{2}\right)$$

Therefore, $a_1 = \frac{8}{\pi}\sin\frac{\pi}{2} = \frac{8}{\pi}$; $a_2 = \frac{8}{2\pi}\sin\pi = 0$;

$$a_3 = \frac{8}{3\pi}\sin\frac{3\pi}{2} = -\frac{8}{3\pi}$$

$$b_n = \frac{1}{2}\int_0^1 8\sin\frac{n\pi t}{2}\,dt$$

$$= 4\cdot\frac{2}{n\pi}\int_0^1\sin\frac{n\pi}{2}t\frac{n\pi}{2}\,dt$$

$$= \frac{8}{n\pi}\left(-\cos\frac{n\pi t}{2}\right)\bigg|_0^1$$

$$= -\frac{8}{n\pi}\left[\cos\frac{n\pi}{2} - \cos 0\right]$$

$$= -\frac{8}{n\pi}\left(\cos\frac{n\pi}{2} - 1\right)$$

$$b_1 = \frac{-8}{\pi}(-1) = \frac{8}{\pi};$$

$$b_2 = \frac{8}{2\pi}(\cos\pi - 1) = \frac{8}{\pi};$$

$$b_3 = \frac{8}{3\pi}\left(\cos\frac{3\pi}{2} - 1\right) = \frac{8}{3\pi}$$

$$f(t) = 2 + \frac{8}{\pi}\cos\frac{\pi t}{2} - \frac{8}{3\pi}\cos\frac{3\pi t}{2} + \cdots$$
$$+ \frac{8}{\pi}\sin\frac{\pi t}{2} + \frac{8}{\pi}\sin\pi t + \frac{8}{3\pi}\sin\frac{3\pi t}{2} + \cdots$$

$$= 2 + \frac{8}{\pi}\left(\cos\frac{\pi t}{2} - \frac{1}{3}\cos\frac{3\pi t}{2} + \cdots\right.$$
$$\left. + \sin\frac{\pi t}{2} + \sin\pi t + \frac{1}{3}\sin\frac{3\pi t}{2} + \cdots\right)$$

24. $i(t) = e^{-t}, -1 \le t \le 1,$ period $= 2L = 2, L = 1$

$$a_0 = \frac{1}{2 \cdot 1} \int_{-1}^{1} e^{-x} dx = \frac{e^2 - 1}{2e}$$

$$= \frac{1}{2} \cdot \frac{e^2 - 1}{e}$$

$$a_n = \frac{1}{1} \int_{-1}^{1} e^{-x} \cos \frac{n\pi x}{1} dx$$

$$= \frac{(e^2 - 1)\cos(n\pi) + \pi n(e^2 + 1)\sin(n\pi)}{e(n^2\pi^2 + 1)}$$

$$a_1 = \frac{-(e^2 - 1)}{e(\pi^2 + 1)},$$

$$a_2 = \frac{-(e^2 - 1)}{e(4\pi^2 + 1)},$$

$$a_3 = \frac{-(e^2 - 1)}{e(9\pi^2 + 1)}$$

$$b_n = \frac{1}{1} \int_{-1}^{1} e^{-x} \sin \frac{n\pi x}{1} dx$$

$$= \frac{n\pi(e^2 - 1)\cos(n\pi) - (e^2 + 1)\sin(n\pi)}{e(n^2\pi^2 + 1)}$$

$$b_1 = \frac{-\pi(e^2 - 1)}{e(\pi^2 + 1)},$$

$$b_2 = \frac{2\pi(e^2 - 1)}{e(4\pi^2 + 1)},$$

$$b_3 = \frac{-3\pi(e^2 - 1)}{e(9\pi^2 + 1)}$$

$$i(t) = \frac{e^2 - 1}{e}\left(\frac{1}{2} - \frac{1}{\pi^2 + 1}\cos \pi t + \frac{1}{4\pi^2 + 1}\cos 2\pi t\right.$$
$$- \frac{1}{9\pi^2 + 1}\cos 3\pi t + \cdots$$
$$- \frac{\pi}{\pi^2 + 1}\sin \pi t + \frac{2\pi}{4\pi^2 + 1}\sin 2\pi t$$
$$\left. - \frac{3\pi}{9\pi^2 + 1}\sin 3\pi t + \cdots\right)$$

Let $y_1 = e^{-x}$ and $y_2 = i(x)$.

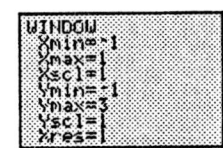 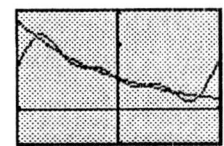

Chapter 13 Review Exercises

1. $f(x) = \dfrac{1}{1 + e^x} = (1 + e^x)^{-1}$

$$f'(x) = -(1 + e^x)^{-2}(e^x) = e^x(1 + e^x)^{-2}$$
$$f''(x) = -(1 + e^x)^{-2}e^x + e^x(2)(1 + e^x)^{-3}(e^x)$$
$$f'''(x) = -(1 + e^x)^{-2}e^x + e^x(2)(1 + e^x)^{-3}(e^x)$$
$$+ 2e^{2x}(-3)(1 + e^x)^{-4}e^x$$
$$+ (1 + e^x)^{-3}(2e^{2x})(2)$$

$$f(0) = \frac{1}{1 + 1} = \frac{1}{2}$$

$$f'(0) = -1(2^{-2}) = -\frac{1}{4}$$

$$f''(0) = -\frac{1}{4} + \frac{1}{4} = 0$$

$$f'''(0) = -\frac{1}{4} + \frac{1}{4} - \frac{3}{8} + \frac{1}{2} = \frac{1}{8}$$

$$f(x) = \frac{1}{2} - \frac{1}{4}x + \frac{0x^2}{2!} + \left(\frac{1}{8}\right)\frac{x^3}{3!} + \cdots$$

$$= \frac{1}{2} - \frac{1}{4}x + \frac{1}{48}x^3 - \cdots$$

2. $f(x) = e^{\cos x}$

$$f'(x) = e^{\cos x}(-\sin x) = -\sin x e^{\cos x}$$
$$f''(x) = \sin^2 e^{\cos x} - e^{\cos x}\cos x$$
$$= e^{\cos x}(\sin^2 x - \cos x)$$
$$f'''(x) = e^{\cos x}(2\sin x \cos x - \sin x)$$
$$+ (\sin^2 x - \cos x)e^{\cos x}(-\sin x)$$
$$f'''(x) = e^{\cos x}(2\sin x \cos x - \sin x$$
$$- \sin^3 x + \sin x \cos x)$$
$$f'''(x) = e^{\cos x}(3\sin x \cos x - \sin x - \sin^3 x)$$
$$f^{iv}(x) = e^{\cos x}(7\cos^2 x + \cos x$$
$$- 6\sin^2 x \cos x + \sin^4 x - 4)$$

$$f(0) = e^{\cos 0} = e$$
$$f'(0) = -\sin 0 e^{\cos 0} = 0$$
$$f''(0) = e^{\cos 0}(\sin^2 0 - \cos 0) = -e$$
$$f'''(0) = e^{\cos 0}(3\sin 0 \cos 0 - \sin 0 - \sin^3 0) = 0$$
$$f^{iv}(0) = e^{\cos 0}(7\cos^2 0 + \cos 0$$
$$- 6\sin^2 0 \cos 0 + \tan^4 0 - 4)$$
$$= 4e$$

$$f(x) = f(0) + f'(0)x + f''(0)\frac{x^2}{2!} + f'''(0)\frac{x^3}{3!}$$
$$+ f^{iv}(0)\frac{x^4}{4!} + \cdots$$

$$f(x) = e + 0x - e\frac{x^2}{2!} + 0\frac{x^3}{3!} + 4e\frac{x^4}{4!}$$

$$f(x) = e\left(1 - \frac{x^2}{2} + \frac{x^4}{6} - \cdots\right)$$

3. $F(x) = \sin x = x - \dfrac{x^3}{3!} + \dfrac{x^5}{5!} - \cdots$

$F(2x^2) = \sin 2x^2 = 2x^2 - \dfrac{(2x^2)^3}{3!} + \dfrac{(2x^2)^5}{5!} - \cdots$

$f(x) = \sin 2x^2 = 2x^2 - \dfrac{4}{3}x^6 + \dfrac{4}{15}x^{10} - \cdots$

4. $f(x) = \dfrac{1}{(1-x)^2}, f(0) = 1$

$f'(x) = \dfrac{2}{(1-x)^3}, f'(0) = 2$

$f''(0) = \dfrac{6}{(1-x)^4}, f''(0) = 6$

$f(x) = f(0) + f'(0)x + f''(0)\dfrac{x^2}{2!} + \cdots$

$f(x) = 1 + 2x + 6\dfrac{x^2}{2} + \cdots$

$f(x) = 1 + 2x + 3x^2 + \cdots$

5. $f(x) = (x+1)^{1/3} \qquad\qquad f(0) = 1$ $\qquad f(x) = 1 + \dfrac{1}{3}x - \dfrac{2x^2}{9(2)} + \cdots$

$f'(x) = \dfrac{1}{3}(x+1)^{-2/3} \qquad f'(0) = \dfrac{1}{3}$ $\qquad\qquad = 1 + \dfrac{1}{3}x - \dfrac{1}{9}x^2 + \cdots$

$f''(x) = -\dfrac{2}{9}(x+1)^{-5/3} \quad f''(0) = -\dfrac{2}{9}$

6. $f(x) = \dfrac{x^2}{1+x^2}, f(0) = 0$

$f'(x) = \dfrac{2x}{(1+x^2)^2}, f'(0) = 0$

$f''(x) = \dfrac{2(1-3x^2)}{(1+x^2)^3}, f''(0) = 2$

$f'''(x) = \dfrac{-24x(1-x^2)}{(1+x^2)^4}, f'''(0) = 0$

$f^{iv}(x) = \dfrac{-24(1 - 10x^2 + 5x^4)}{(1+x^2)^5}, f^{iv}(0) = -24$

$f^{(5)}(x) = \dfrac{240x(3x^4 - 10x^2 + 3)}{(1+x^2)^6}, f^{(5)}(0) = 0$

$f^{(6)}(x) = \dfrac{-720(7x^6 - 35x^4 + 21x^2 - 1)}{(1+x^2)^7}, f^{(6)}(0) = 720$

$f(x) = f(0) + f'(0)x + f''(0)\dfrac{x^2}{2!} + f'''(0)\dfrac{x^3}{3!} + f^{iv}(0)\dfrac{x^4}{4!}$

$\qquad\qquad f^{(5)}(0)\dfrac{x^5}{5!} + f^{(6)}\dfrac{x^6}{6!} + \cdots$

$f(x) = 2\dfrac{x^2}{2!} - 24\dfrac{x^4}{4!} + 720\dfrac{x^6}{6!} - \cdots$

$f(x) = x^2 - x^4 + x^6 - \cdots$

7. $f(x) = \sin^{-1}$ $f(0) = 0$

$f'(x) = \dfrac{1}{\sqrt{1-x^2}}$ $f'(0) = 1$

$f''(x) = \dfrac{x}{(1-x^2)^{3/2}}$ $f''(0) = 0$

$f'''(x) = \dfrac{(1-x^2)^{3/2} - x(3/2)(1-x^2)^{1/2}(-2x)}{(1-x^2)^3} = \dfrac{2x^2+1}{(1-x^2)^{5/2}}$ $f'''(0) = 1$

$f^{iv}(x) = \dfrac{(1-x^2)^{5/2}(4x) - (2x^2+1)(5/2)(1-x^2)^{3/2}(-2x)}{(1-x^2)^5} = \dfrac{6x^3+9x}{(1-x^2)^{7/2}}$ $f^{iv}(0) = 0$

$f^{v}(x) = \dfrac{(1-x^2)^{7/2}(18x^2+9) - (6x^3+9x)(7/2)(1-x^2)^{5/2}(-2x)}{(1-x^2)^7} = \dfrac{24x^4+72x^2+9}{(1-x^2)^{9/2}}$ $f^{v}(0) = 9$

$f(x) = x + \dfrac{x^3}{3!} + 9\left(\dfrac{x^5}{5!}\right) + \cdots = x + \dfrac{1}{6}x^3 + \dfrac{3}{40}x^5 + \cdots$

8. $f(x) = \dfrac{1}{1-\sin x}, f(0) = 1$

$f'(x) = \dfrac{\cos x}{(\sin x - 1)^2}, f'(0) = 1$

$f''(x) = \dfrac{-2\cos^2 x}{(\sin x - 1)^3} - \dfrac{\sin x}{(\sin x - 1)^2}, f''(0) = 2$

$f(x) = f(0) + f'(0)x + f''(0)\dfrac{x^2}{2!} = 1 + x + 2\dfrac{x^2}{2!} + \cdots$

$f(x) = 1 + x + x^2 + \cdots$

9. In $e^x = 1 + x + \dfrac{x^2}{2} + \cdots$ Let $x = -0.2$

$e^{-0.2} = 1 - 0.2 + \dfrac{(0.2)^2}{2} + \cdots = 0.82$

10. Using 13-10

$\ln(1.10) = \ln(1+0.01) = 0.10 - \dfrac{(0.10)^2}{2} + \dfrac{(0.10)^3}{3} - \cdots$

$\ln(1.10) = 0.0953$

11. See Exercise 5.

$\sqrt[3]{1+x} = 1 + \dfrac{1}{3}x - \dfrac{1}{9}x^2 + \cdots$

$\sqrt[3]{1+0.3} = 1 + \dfrac{1}{3}(0.3) - \dfrac{1}{9}(0.3)^2$

$\sqrt[3]{1.3} = 1.09$

12. $\sin x = x - \dfrac{x^3}{3!} + \dfrac{x^5}{5!} - \cdots$ $(13-8)$

$\sin 3.5° = \left(3.5°\left(\dfrac{\pi}{180°}\right)\right) = \dfrac{3.5\pi}{180} - \dfrac{\left(\frac{3.5\pi}{180}\right)^3}{3!} + \dfrac{\left(\frac{3.5\pi}{180}\right)^5}{5!} = 0.0610485395$

$\sin 3.5° = 0.0610485$

13. Taylor Series: $f(x) = f(a) + f'(a)(x-a) + \dfrac{f''(a)(x-a)^2}{2!} + \cdots$

Let $f(x) = \dfrac{1}{x}$

$f'(x) = -\dfrac{1}{x^2}$

$f''(x) = \dfrac{2}{x^3}$

and $a = 1$, then

$$f(x) = 1 - (x-1) + 2 \cdot \frac{(x-1)^2}{2!} + \cdots \text{ from which}$$

$$f(1.086) = \frac{1}{1.086} = 1 - (1.086 - 1) + (1.086 - 1)^2$$

$$1.086^{-1} = 0.9214$$

14. $(1+x)^n = 1 + nx + \dfrac{n(n-1)}{2!}x^2 + \cdots$ $(13-11)$

$$0.9839 = 1 - 0.0161$$
$$0.9839^{10} = (1 + (-0.0161))^{10}$$

$$= 1 + 10(-0.0161) + \frac{10(10-1)}{2!}(-0.0161)^2 = 0.85066445$$

$$= 0.8507$$

15. $\ln(1+x) = x - \dfrac{x^2}{2} + \dfrac{x^3}{3} - \cdots$

$$\ln[1 + (-0.1828)] = -0.1828 - \frac{(-0.1828)^2}{2} + \frac{(-0.1828)^3}{3} - \cdots$$

$$\ln 0.8172 = -0.2015$$

16. $\cos x = 1 - \dfrac{x^2}{2!} + \dfrac{x^4}{4!} - \cdots$ $(13-9)$

$$\cos 0.1376 = 1 - \frac{0.1376^2}{2!} + \frac{0.1376^4}{4!} = 0.9905481$$

$$\cos 0.1376 = 0.9905481$$

17. $f(x) = \tan x; \ a = \dfrac{\pi}{4}$ $\qquad\qquad\qquad f\left(\dfrac{\pi}{4}\right) = 1$

$f'(x) = \sec^2 x = 1 + \tan^2 x$ $\qquad\quad f'\left(\dfrac{\pi}{4}\right) = 2$

$f''(x) = 2\tan x \sec^2 x$ $\qquad\qquad\quad f''\left(\dfrac{\pi}{4}\right) = 4$

$$f(x) = 1 + 2\left(x - \frac{\pi}{4}\right) + \frac{4\left(x - \frac{\pi}{4}\right)^2}{2!} + \cdots$$

$$\tan 43.62° = \tan(45° - 1.38°) = \tan\left(\frac{\pi}{4} - \frac{1.38\pi}{180}\right) = 1 + 2\left(\frac{\pi}{4} - \frac{1.38\pi}{180} - \frac{\pi}{4}\right) + 2\left(\frac{\pi}{4} - \frac{1.38\pi}{180} - \frac{\pi}{4}\right)^2 + \cdots$$

$$= 1 + 2(-0.0240855) + 2(0.0005801) = 0.95299$$

18. $f(x) = \sqrt[4]{x}, a = 256$

$f(x) = \sqrt[4]{x}, f(a) = f(256) = \sqrt[4]{256} = 4$

$$f'(x) = \frac{1}{4\sqrt[4]{x^3}}, f'(256) = \frac{1}{4\sqrt[4]{256^3}}$$

$$f''(x) = \frac{-3}{16\sqrt[4]{x^7}}, f''(256) = \frac{-3}{16\sqrt[4]{256^7}}$$

$$\sqrt[4]{x} = 4 + \frac{1}{4\sqrt[4]{256^3}}(x - 256) - \frac{3}{16\sqrt[4]{256^7}}\frac{(x-256)^2}{2!}$$

$$\sqrt[4]{260} = 4 + \frac{(260-256)}{4(256)^{3/4}} - \frac{3(260-256)^2}{32(256)^{7/4}} = 4.015533447$$

$$\sqrt[4]{260} = 4.02$$

19. $f(x) = \sqrt{x}, a = 144$

$f(x) = \sqrt{x}, f(144) = 12$

$$f'(x) = \frac{1}{2}x^{-1/2}, f'(144) = \frac{1}{24}$$

$$f''(x) = -\frac{1}{4}x^{-3/2}, f''(144) = -\frac{1}{6912}$$

$$\sqrt{x} = 12 + \frac{1}{24}(x-144) - \frac{1}{6912}\frac{(x-144)^2}{2} + \cdots = 12 + \frac{x-144}{24} - \frac{(x-144)^2}{13,824}$$

$$\sqrt{148} = 12 + \frac{4}{24} - \frac{4^2}{13,824} = 12.1655$$

20. $f(x) = \cos x, a = \frac{\pi}{4}$

$$f(x) = \cos x, f(a) = f\left(\frac{\pi}{4}\right) = \cos\frac{\pi}{4} = \frac{\sqrt{2}}{2}$$

$$f'(x) = -\sin x, f'\left(\frac{\pi}{4}\right) = -\sin\frac{\pi}{4} = \frac{-\sqrt{2}}{2}$$

$$f''(x) = -\cos x, f''\left(\frac{\pi}{4}\right) = -\cos\frac{\pi}{4} = \frac{-\sqrt{2}}{2}$$

$$\cos x = \frac{\sqrt{2}}{2} - \frac{\sqrt{2}}{2}\left(x - \frac{\pi}{4}\right) - \frac{\sqrt{2}}{2}\frac{\left(x - \frac{\pi}{4}\right)^2}{2!}$$

$$\cos 47° = \cos\left(\frac{\pi}{4} + \frac{2\pi}{180}\right) = \cos\left(\frac{\pi}{4} + \frac{\pi}{90}\right)$$

$$= \frac{\sqrt{2}}{2} - \frac{\sqrt{2}}{2}\left(\frac{\pi}{4} + \frac{\pi}{90} - \frac{\pi}{4}\right) - \frac{\sqrt{2}}{4}\left(\frac{\pi}{4} + \frac{\pi}{90} - \frac{\pi}{4}\right)^2 = 0.6819933041 = 0.6820$$

21. $\displaystyle\int_{0.1}^{0.2}\frac{1 - \dfrac{x^2}{2} + \dfrac{x^4}{24} + \cdots}{\sqrt{x}}\,dx = \int_{0.1}^{0.2}\left(x^{-1/2} - \frac{x^{3/2}}{2} + \frac{x^{7/2}}{24} + \cdots\right)dx = \int_{0.1}^{0.2} x^{-1/2}dx - \frac{1}{2}\int_{0.1}^{0.2} x^{3/2}dx + \frac{1}{24}\int_{0.1}^{0.2} x^{7/2}d$

$$= 2x^{1/2} - \frac{1}{5}x^{5/2} + \frac{1}{108}x^{9/2} + \cdots\Big|_{0.1}^{0.2} = 0.259$$

22. $\displaystyle\int_{2}^{0.1} \sqrt[3]{1+x^2}\,dx$

$$f(x) = (1+x^2)^{1/3}, f(0) = 1$$

$$f'(x) = \frac{2x}{3(1+x^2)^{2/3}}, f'(0) = 0$$

$$f''(x) = \frac{-2(x^2-3)}{9(1+x^2)^{5/3}}, f''(0) = \frac{2}{3}$$

$$f'''(x) = \frac{8x(x^2-9)}{27(1+x^2)^{8/3}}, f'''(0) = 0$$

$$f^{iv}(x) = \frac{-8(7x^4 - 126x^2 + 27)}{81(1+x^2)^{11/3}}, f^{iv}(0) = \frac{-8(27)}{81} = -\frac{8}{3}$$

$$\sqrt[3]{1+x^2} = 1 + \frac{2}{3}\frac{x^2}{2!} - \frac{8}{3}\frac{x^4}{4} = 1 + \frac{x^2}{3} - \frac{x^4}{9}$$

$$\int_{0}^{0.1} \sqrt[3]{1+x^2}\,dx = \int_{0}^{0.1}\left(1 + \frac{x^2}{3} - \frac{x^4}{9}\right)dx = x + \frac{x^3}{9} - \frac{x^5}{45}\Big|_{0}^{0.1} = 0.1 + \frac{0.1^3}{9} - \frac{0.1^5}{45}$$

$$= 0.1001108899 = 0.1001109$$

23. $f(x) = \cos x, a = \dfrac{\pi}{3}$

$$f(x) = \cos x, f\left(\frac{\pi}{3}\right) = \frac{1}{2}$$

$$f'(x) = -\sin x, f'\left(\frac{\pi}{3}\right) = \frac{1}{2}\sqrt{3}$$

$$f''(x) = -\cos x, f''\left(\frac{\pi}{3}\right) = -\frac{1}{2}$$

$$f(x) = \frac{1}{2} + \frac{1}{2}\sqrt{3}\left(x - \frac{\pi}{3}\right) - \frac{1}{4}\left(x - \frac{\pi}{3}\right)^2 + \cdots$$

24. $f(x) = \ln(\cos x), a = \dfrac{\pi}{4}$

$$f(x) = \ln(\cos x), f(a) = f\left(\frac{\pi}{4}\right) = \ln\left(\cos\frac{\pi}{4}\right) = \ln\frac{1}{\sqrt{2}} = \ln\sqrt{\frac{1}{2}} = \frac{1}{2}\ln\frac{1}{2}$$

$$f'(x) = -\tan x, f'\left(\frac{\pi}{4}\right) = -1$$

$$f''(x) = -\tan^2 x - 1, f''\left(\frac{\pi}{4}\right) = -\tan^2\frac{\pi}{4} - 1 = -2$$

$$f'''(x) = -2\tan x - 2\tan^3 x$$

$$f'''\left(\frac{\pi}{4}\right) = -2\tan\frac{\pi}{4} - 2\tan^3\frac{\pi}{4} = -4$$

$$f(x) = f(a) + f'(a)(x-a) + f''(a)\frac{(x-a)^2}{2!} + \cdots$$

$$f(x) = \frac{1}{2}\ln\frac{1}{2} - \left(x - \frac{\pi}{4}\right) + (-2)\frac{\left(x - \frac{\pi}{2}\right)}{2!} + \cdots$$

$$f(x) = \frac{1}{2}\ln\frac{1}{2} - \left(x - \frac{\pi}{4}\right) - \left(x - \frac{\pi}{4}\right)^2 - \cdots$$

25. $f(x) = \begin{cases} 0, & -\pi \le x < 0 \\ x - 1, & 0 \le x < \pi \end{cases}$ is Example 2 of 13-6 shifted down 1 unit.

$$f(x) = -1 + \frac{2 + \pi}{4} - \frac{2}{\pi}\cos x - \frac{2}{9\pi}\cos 3x - \cdots + \left(\frac{\pi - 2}{\pi}\right)\sin x - \frac{1}{2}\sin 2x + \cdots$$

$$f(x) = -1 + \frac{2}{4} + \frac{\pi}{4} - \frac{2}{\pi}\left(\cos x + \frac{1}{9}\cos 3x + \cdots\right) + \left(\frac{\pi - 2}{\pi}\right)\sin x - \frac{1}{2}\sin 2x + \cdots$$

$$f(x) = \frac{\pi - 2}{4} - \frac{2}{\pi}\left(\cos x + \frac{1}{9}\cos 3x + \cdots\right) + \frac{\pi - 2}{\pi.}\sin x - \frac{1}{2}\sin 2x + \cdots$$

26. $f(x) = x^2 - 1, -1 \le x < 1$ is Example 8 in 13-7 shifted down 1 unit

$$f(x) = \frac{1}{3} - \frac{4}{\pi^2}\left(\cos \pi x - \frac{1}{4}\cos 2\pi x + \frac{1}{9}\cos 3\pi x - \ldots\right) - 1$$

$$f(x) = \frac{-2}{3} - \frac{4}{\pi^2}\left(\cos \pi x - \frac{1}{4}\cos 2\pi x + \frac{1}{9}\cos 3\pi x - \ldots\right)$$

27. $f(x) = \begin{cases} \pi - 1, & -4 \le x < 0 \\ \pi + 1, & 0 \le x < 4 \end{cases}$ is Example 7 in 13-7 shifted up $\pi - 1$ units

$$f(x) = \pi - 1 + 1 + \frac{4}{\pi}\sin\frac{\pi x}{4} + \frac{4}{3\pi}\sin\frac{3\pi x}{4} + \cdots$$

$$f(x) = \pi + \frac{4}{\pi}\left(\sin\frac{\pi x}{4} + \frac{1}{3}\sin\frac{3\pi x}{4} + \cdots\right)$$

28. $f(x) = \begin{cases} 1, & -\pi \le x < 0 \\ 1 + \sin x, & 0 \le x < \pi \end{cases}$ is Example 3 in 13-6 shifted up 1 unit

$$f(x) = 1 + \frac{1}{\pi} + \frac{1}{2}\sin x - \frac{2}{\pi}\left(\frac{1}{3}\cos 2x + \frac{1}{15}\cos 4x + \ldots\right)$$

29. $f(x) = \begin{cases} 0 & -\pi \le x < -\pi/2, \pi/2 < x < \pi \\ 1 & -\pi/2 \le x \le \pi/2 \end{cases}$

$$a_0 = \frac{1}{2\pi}\int_{-\pi}^{-\pi/2} 0\, dx + \frac{1}{2\pi}\int_{\pi/2}^{\pi/2} dx + \frac{1}{2\pi}\int_{-\pi/2}^{\pi/2} 0\, dx = \frac{x}{2\pi}\Big|_{-\pi/2}^{\pi/2} = \frac{1}{2\pi}\left(\frac{\pi}{2} + \frac{\pi}{2}\right) = \frac{1}{2}$$

$$a_n = \frac{1}{\pi}\int_{-\pi}^{\pi/2} 0\cos nx\, dx + \frac{1}{\pi}\int_{-\pi/2}^{\pi/2}\cos nx\, dx + \frac{1}{\pi}\int_{\pi/2}^{\pi/2} 0\cos nx\, dx = \frac{1}{n\pi}\sin nx\Big|_{-\pi/2}^{\pi/2}$$

$$= \frac{1}{n\pi}\left[\sin\frac{n\pi}{2} - \sin\frac{(-n\pi)}{2}\right] = \frac{2}{n\pi}\sin\frac{n\pi}{2}$$

$$a_1 = \frac{2}{\pi}\sin\frac{\pi}{2} = \frac{2}{\pi}, a_2 = \frac{2}{2\pi}\sin\pi = 0, a_3 = \frac{2}{3\pi}\sin\frac{3\pi}{2} = -\frac{2}{3\pi}$$

$$b_n = \frac{1}{\pi}\int_{-\pi}^{-\pi/2} 0\sin nx\, dx + \frac{1}{\pi}\int_{-\pi/2}^{\pi/2}\sin nx\, dx + \frac{1}{\pi}\int_{\pi/2}^{\pi/2} 0\sin nx\, dx = -\frac{1}{n\pi}\cos nx\Big|_{-\pi/2}^{\pi/2}$$

$$= -\frac{1}{n\pi}\left[\cos\frac{n\pi}{2} - \cos\frac{(-n\pi)}{2}\right] = 0; \text{ (for all } n \text{ since } \cos\theta = \cos(-\theta))$$

$$f(x) = \frac{1}{2} + \frac{2}{\pi}\left(\cos x - \frac{1}{3}\cos 3x + \cdots\right)$$

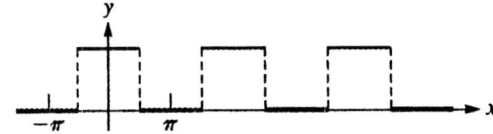

30. $f(x) = \begin{cases} -x, & -\pi \le x < 0 \\ 0, & 0 \le x < \pi \end{cases}$

$$a_0 = \frac{1}{2\pi} \int_{-\pi}^{0} -x\, dx + \int_{0}^{\pi} 0\, dx = \frac{1}{2\pi} \left. \frac{x^2}{2} \right|_{-\pi}^{0} = \frac{1}{4\pi}(0^2 - (-\pi)^2) = \frac{\pi}{4}$$

$$a_n = \frac{1}{\pi} \int_{-\pi}^{0} -x \cos nx\, dx = \frac{-1}{\pi} \int_{-\pi}^{0} x \cos nx\, dx$$

$$= -\frac{1}{\pi}\left(-\frac{\cos n\pi}{n^2} - \frac{\pi n\pi}{n} + \frac{1}{n^2}\right) = -\frac{1}{\pi}\left(\frac{\cos n\pi}{n^2} + \frac{1}{n^2}\right)$$

$$= -\frac{1}{\pi} \begin{cases} 0, & n \text{ even} \\ \dfrac{2}{n^2}, & n \text{ odd} \end{cases}$$

$$a_1 = -\frac{1}{\pi}\left(\frac{2}{1^2}\right) = \frac{-2}{\pi}$$

$$a_3 = -\frac{1}{\pi}\left(\frac{2}{3^2}\right) = \frac{-2}{9\pi}$$

$$b_n = \frac{1}{\pi} \int_{-\pi}^{0} -x \sin nx\, dx = -\frac{1}{\pi} \int_{-\pi}^{0} \sin nx\, dx$$

$$b_n = -\frac{1}{\pi}\left(\frac{\sin n\pi}{n^2} - \frac{\pi \cos n\pi}{n}\right) = \frac{-1}{\pi}\left(\frac{-\pi \cos n\pi}{n}\right) = \frac{\cos n\pi}{n}$$

$$b_1 = -1$$

$$b_2 = \frac{1}{2}$$

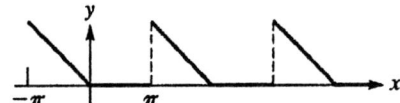

$$f(x) = \frac{\pi}{4} - \frac{2}{\pi}\cos x - \frac{2}{9\pi}\cos 3x \ldots - \sin x + \frac{1}{2}\sin 2x - \ldots$$

$$f(x) = \frac{\pi}{4} - \frac{2}{\pi}\left(\cos x + \frac{1}{9}\cos 3x + \cdots\right) - \left(\sin x - \frac{1}{2}\sin 2x + \ldots\right)$$

31. $f(x) = x \qquad -2 \le x < 2$, period $= 4$, $L = 2$; $a_0 = \frac{1}{4} \int_{-2}^{2} x\, dx = \frac{1}{8}x^2 \Big|_{-2}^{2} = \frac{1}{8}(4 - 4) = 0$

$$a_n = \frac{1}{2} \int_{-2}^{2} x \cos \frac{n\pi x}{2}\, dx = \frac{1}{2}\left(\frac{2}{n\pi}\right)^2 \left(\cos \frac{n\pi x}{2} + \frac{n\pi x}{2}\sin \frac{n\pi x}{2}\right)\Big|_{-2}^{2}$$

$$= \frac{2}{n^2\pi^2}[\cos n\pi + n\pi \sin n\pi - \cos(-n\pi) + n\pi \sin(-n\pi)] = 0 \text{ for all } n$$

$$b_n = \frac{1}{2} \int_{-2}^{2} x \sin \frac{n\pi x}{2}\, dx = \frac{1}{2}\left(\frac{2}{n\pi}\right)^2 \left(\sin \frac{n\pi x}{2} - \frac{n\pi x}{2}\cos \frac{n\pi x}{2}\right)\Big|_{-2}^{2}$$

$$= \frac{2}{n^2\pi^2}[\sin n\pi - n\pi \cos n\pi - \sin(-n\pi) - n\pi \cos(-n\pi)]$$

$$= \frac{2}{n^2\pi^2}(-n\pi \cos n\pi - n\pi \cos n\pi) = \frac{-4}{n\pi}\cos n\pi; \ b_1 = -\frac{4}{\pi}\cos \pi = \frac{4}{\pi},$$

$$b_2 = -\frac{4}{2\pi}\cos 2\pi = -\frac{2}{\pi}, b_3 = \frac{-4}{3\pi}\cos 3\pi = \frac{4}{3\pi}$$

$$f(x) = \frac{4}{\pi}\left(\sin\frac{\pi x}{2} - \frac{1}{2}\sin\pi x + \frac{1}{3}\sin\frac{3\pi x}{2} - \cdots\right)$$

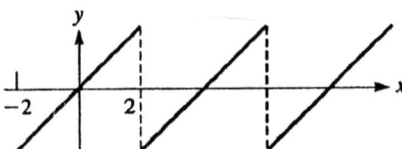

32. $f(x) = \begin{cases} -2, & -3 \le x < 0 \\ 2, & 0 \le x < 3 \end{cases}$

$$a_o = \frac{1}{2L}\int_{-L}^{L} f(x)\,dx = \frac{1}{6}\int_{-3}^{0} -2\,dx + \frac{1}{6}\int_{0}^{3} 2\,dx$$

$$a_o = -\frac{1}{3}x\Big|_{-3}^{0} + \frac{1}{3}x\Big|_{0}^{3} = -\frac{1}{3}(0-(-3)) + \frac{1}{3}(3-0) = -\frac{1}{3}(3) + \frac{1}{3}(3) = 0$$

$$a_o = 0$$

$$a_n = \frac{1}{L}\int_{-L}^{L} f(x)\cos\frac{n\pi x}{L}\,dx = \frac{1}{3}\int_{-3}^{0} -2\cos\frac{n\pi x}{3}\,dx + \frac{1}{3}\int_{0}^{3} 2\cos\frac{n\pi x}{3}\,dx$$

$$a_n = \frac{-2}{n\pi}\sin\frac{n\pi x}{3}\Big|_{-3}^{0} + \frac{2}{n\pi}\sin\frac{n\pi x}{3}\Big|_{0}^{3}$$

$$a_n = \frac{-2}{n\pi}\left[\sin 0 - \sin\frac{n\pi(-3)}{3}\right] + \frac{2}{n\pi}\left[\sin\frac{n\pi(3)}{3} - \sin 0\right] = 0$$

$$a_n = 0$$

$$b_n = \frac{1}{L}\int_{-L}^{L} f(x)\sin\frac{n\pi x}{L}\,dx = \frac{1}{3}\int_{-3}^{0} -2\sin\frac{n\pi x}{3}\,dx + \frac{1}{3}\int_{0}^{-3} 2\sin\frac{n\pi x}{3}\,dx$$

$$b_n = \frac{2}{n\pi}\cos\frac{n\pi x}{3}\Big|_{-3}^{0} - \frac{2}{n\pi}\cos\frac{n\pi x}{3}\Big|_{0}^{3}$$

$$b_n = \frac{2}{n\pi}\left[\cos 0 - \cos\frac{n\pi(-3)}{3}\right] - \frac{2}{n\pi}\left[\cos\frac{n\pi(3)}{3} - \cos 0\right]$$

$$b_n = \frac{2}{n\pi}[1 - \cos n\pi] - \frac{2}{n\pi}[\cos n\pi - 1]$$

$$b_n = \frac{2}{n\pi} - \frac{2}{n\pi}\cos n\pi - \frac{2}{n\pi}\cos n\pi + \frac{2}{n\pi}$$

$$b_n = \frac{4}{n\pi} - \frac{4}{n\pi}\cos n\pi$$

$$b_1 = \frac{4}{1\pi} - \frac{4}{1\pi}\cos(1\pi) = \frac{4}{\pi} - \frac{4}{\pi}(-1) = \frac{8}{\pi}$$

$$b_2 = \frac{4}{2\pi} - \frac{4}{2\pi}\cos(2\pi) = \frac{2}{\pi} - \frac{2}{\pi}(1) = 0$$

$$b_3 = \frac{4}{3\pi} - \frac{4}{3\pi}\cos(3\pi) = \frac{4}{3\pi} - \frac{4}{3\pi}(-1) = \frac{8}{3\pi}$$

$$f(x) = \frac{8}{\pi}\left(\sin\frac{\pi x}{3} + \frac{1}{3}\sin\pi x + \cdots\right)$$

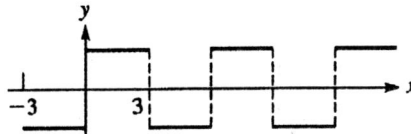

33. It is a geometric series for which $|r| < 1 \doteq 0.75$. Therefore the series converges.

$$S = \frac{64}{1 - 0.75} = 256$$

34. $\displaystyle\sum_{n=1}^{\infty}\frac{n}{3n+1} = \frac{1}{3(1)+1} + \frac{2}{3(2)+1} + \frac{3}{3(3)+1} + \frac{4}{3(4)+1} + \frac{5}{3(5)+1} + \cdots$

$$= \frac{1}{4} + \frac{2}{7} + \frac{3}{10} + \frac{4}{13} + \frac{5}{16} + \cdots$$

The first five partial sums are

$$\frac{1}{4} = 0.25, \frac{15}{28} = 0.5357, \frac{117}{140} = 0.8357, \frac{2081}{1820} = 1.1434, \frac{10{,}599}{7280} = 1.4559$$

divergent

35. $\sin x = x - \dfrac{x^3}{3!} + \cdots$

$$\sin(x+h) - \sin(x-h) = (x+h) - \frac{(x+h)^3}{3!} + \cdots - (x-h) + \frac{(x+h)^3}{3!} - \cdots$$

$$= x + h - \frac{x^3 + 3x^2h + 3xh^2 + h^3}{3!} + \cdots - x + h + \frac{x^3 - 3x^2h + 3xh^2 - h^3}{3!} - \cdots$$

$$= 2h - \frac{6x^2h}{3!} - \frac{2h^3}{3!} + \cdots = 2h\left(1 - \frac{x^2}{2} + \cdots\right) - \frac{2h^3}{3!} + \cdots$$

$$= 2h\cos x \quad\text{for small } h$$

36. $\sin x = x - \dfrac{x^3}{3!} + \dfrac{x^5}{5!} - \cdots$

$$x\sin x = x^2 - \frac{x^4}{3!} + \frac{x^6}{5!} - \cdots$$

$$x\cos x + \sin x = 2x - \frac{4x^3}{3!} + \frac{6x^5}{5!} - \cdots$$

$$= 2x - \frac{2}{3}x^3 + \frac{1}{20}x^5 - \cdots$$

37. $\ln(1+x)^4 = 4\ln(1+x) = 4\left(x - \dfrac{x^2}{2} + \dfrac{x^3}{3} - \dfrac{x^4}{4} + \cdots\right) = 4x - 2x^2 + \dfrac{4x^3}{3} - x^4 + \cdots$

38.
$$\sin x = x - \frac{x^3}{3!} + \frac{x^5}{5!} - \cdots, \cos x = 1 - \frac{x^2}{2!} + \frac{x^4}{4!} - \cdots$$

$$\sin 2x = (2x) - \frac{(2x)^3}{3!} + \frac{(2x)^5}{5!} - \cdots = 2x - \frac{4}{3}x^3 + \frac{4}{15}x^5 - \cdots$$

$$2\sin x \cos x = \left(2x - \frac{2x^3}{3!} + \frac{2x^5}{5!} - \cdots\right)\left(1 - \frac{x^2}{2!} + \frac{x^4}{4!} - \cdots\right)$$

$$2\sin x \cos x = 2x - \frac{2x^3}{2!} + \frac{2x^5}{4!} - \frac{2x^3}{3!} + \frac{2x^5}{3!2!} - \frac{2x^7}{3!4!} + \frac{2x^5}{5!} - \frac{2x^7}{5!2!} + \frac{2x^9}{5!4!}$$

$$2\sin x \cos x = 2x - \frac{4}{3}x^3 + \frac{4}{15}x^5 - \cdots = \sin 2x$$

39. $\cos^2 x = \frac{1}{2}(1 + \cos 2x) = \frac{1}{2}\left(1 + 1 - \frac{(2x)^2}{2!} + \frac{(2x)^4}{4!} - \frac{(2x)^6}{6!} + \cdots\right)$

$$= \frac{1}{2}\left(2 - \frac{4x^2}{2} + \frac{16x^4}{24} - \frac{64x^6}{720} + \cdots\right) = 1 - x^2 + \frac{1}{3}x^4 - \frac{2}{45}x^6 + \cdots$$

40. $\displaystyle\int_0^1 x \sin x \, dx = -x\cos x \Big|_0^1 - \int_0^1 -\cos x \, dx$

$$u = x \qquad\qquad dv = \sin x \, dx$$
$$du = dx \qquad\qquad v = -\cos x$$

$$\int_0^1 x \sin x \, dx = -\cos 1 + \sin x \Big|_0^1 = -\cos 1 + \sin 1$$

$$\int_0^1 x \sin x \, dx = \sin 1 - \cos 1$$

$$\int_0^1 x \sin x \, dx = \sin 1 - \cos 1 = 0.3011686789 = 0.3012$$

$$\int_0^1 x \sin x \, dx = \int_0^1 x\left(x - \frac{x^3}{3!} + \frac{x^5}{5!}\right) dx = \int_0^1 \left(x^2 - \frac{x^4}{6} + \frac{x^6}{120}\right) dx = \frac{x^3}{3} - \frac{x^5}{30} + \frac{x^7}{840}\Big|_0^1$$

$$= \frac{1}{3} - \frac{1}{30} + \frac{1}{840} = \frac{253}{840} = 0.3011904762 = 0.3012$$

41.
$$\begin{aligned}
f(x) &= \cos x & f(0) &= 1 \\
f'(x) &= -\sin x & f'(0) &= 0 \\
f''(x) &= -\cos x & f''(0) &= -1 \\
f'''(x) &= \sin x & f'''(0) &= 0 \\
f^{iv}(x) &= \cos x & f^{iv}(0) &= 1
\end{aligned}$$

$$f(x) = 1 + 0x - \frac{1}{2}x^2 + \frac{0x^3}{3!} + \frac{1x^4}{4!} = 1 - \frac{1}{2}x^2 + \frac{1}{24}x^4 + \cdots$$

$$\sec x = \frac{1}{\cos x} = \frac{1}{1 - \frac{1}{2}x^2 - \frac{1}{24}x^4} = 1 + \frac{1}{2}x^2 + \frac{5}{24}x^4 + \cdots$$

The long division process is shown below:

$$
\begin{array}{r}
1 + \dfrac{1}{2}x^2 + \dfrac{5}{24}x^4 \\[4pt]
\hline
1 - \dfrac{1}{2}x^2 + \dfrac{1}{24}x^4 \overline{)\, 1 + 0 \quad + 0 \quad + 0 \quad + 0} \\[4pt]
\dfrac{1 - \dfrac{1}{2}x^2 + \dfrac{1}{24}x^4}{} \\[4pt]
\dfrac{1}{2}x^2 - \dfrac{1}{24}x^4 \\[4pt]
\dfrac{\dfrac{1}{2}x^2 - \dfrac{1}{4}x^4}{} \\[4pt]
\dfrac{5}{24}x^4 + 0 \quad + 0 \\[4pt]
\dfrac{5}{24}x^4 - \dfrac{5}{48}x^2 + \dfrac{5}{576}x^4
\end{array}
$$

42.

$$\sin^2 x = \left(x - \frac{x^3}{3!} + \frac{x^5}{5!} - \cdots \right)\left(x - \frac{x^3}{3!} + \frac{x^5}{5!} - \cdots \right)$$

$$\sin^2 x = x^2 - \frac{x^4}{3!} + \frac{x^6}{5!} - \frac{x^4}{3!} + \frac{x^6}{3!3!} - \frac{x^8}{3!5!} + \frac{x^6}{5!} - \frac{x^8}{5!3!} + \frac{x^{10}}{5!5!}$$

$$\sin^2 x = x^2 - \frac{1}{3}x^4 + \frac{2}{45}x^6$$

$$\cos^2 x = \left(1 - \frac{x^2}{2!} + \frac{x^4}{4!} - \cdots \right)\left(1 - \frac{x^2}{2!} + \frac{x^4}{4!} - \cdots \right)$$

$$\cos^2 x = 1 - \frac{x^2}{2!} + \frac{x^4}{4!} - \frac{x^2}{2!} + \frac{x^4}{2!2!} - \frac{x^6}{2!4!} + \frac{x^4}{4!} - \frac{x^6}{2!4!} + \frac{x^8}{4!4!}$$

$$\cos^2 x = 1 - x^2 + \frac{1}{3}x^4$$

$$\sin^2 x + \cos^2 x = x^2 - \frac{1}{3}x^4 + 1 - x^2 + \frac{1}{3}x^4 = 1$$

43. $a_0 = \dfrac{1}{L}\displaystyle\int_0^L f(x)\,dx = \dfrac{1}{\pi}\int_0^\pi \sin x\,dx = \dfrac{1}{\pi}(-\cos x)\Big|_0^\pi$

$a_0 = \dfrac{1}{\pi}(-\cos \pi + \cos 0) = \dfrac{2}{\pi}$

$a_n = \dfrac{2}{L}\displaystyle\int_0^L f(x)\cos\frac{n\pi x}{L} = \dfrac{2}{\pi}\int_0^\pi \sin x \cos\frac{n\pi x}{\pi}\,dx = \dfrac{2}{\pi}\int_0^\pi \sin x \cos nx\,dx$

$a_1 = \dfrac{2}{\pi}\displaystyle\int_0^\pi \sin x \cos x\,dx = \dfrac{1}{\pi}\sin^2 x\Big|_0^\pi = \dfrac{1}{\pi}[\sin^2 \pi - \sin^2 0] = 0$

$a_2 = \dfrac{2}{\pi}\displaystyle\int_0^\pi \sin x \cos 2x\,dx = \dfrac{2}{\pi}\left[\dfrac{\cos x}{2} - \dfrac{\cos 3x}{6}\right]\Big|_0^\pi$

$a_2 = \dfrac{2}{\pi}\left[\dfrac{\cos \pi}{2} - \dfrac{\cos 3\pi}{6} - \dfrac{\cos 0}{2} + \dfrac{\cos 0}{6}\right] = \dfrac{2}{\pi}\left[-\dfrac{1}{2} + \dfrac{1}{6} - \dfrac{1}{2} + \dfrac{1}{6}\right] = \dfrac{-4}{3\pi}$

$$a_3 = \frac{2}{\pi} \int_0^\pi \sin x \cos 3x \, dx = \frac{2}{\pi} \left[\frac{\cos 2x}{4} - \frac{\cos 4x}{8} \right] \Big|_0^\pi$$

$$a_3 = \frac{2}{\pi} \left[\frac{\cos 2\pi}{4} - \frac{\cos 4\pi}{8} - \frac{\cos 0}{4} + \frac{\cos 0}{8} \right] = \frac{2}{\pi} \left[\frac{1}{4} - \frac{1}{8} - \frac{1}{4} + \frac{1}{8} \right] = 0$$

$$a_4 = \frac{2}{\pi} \int_0^\pi \sin x \cos 4x \, dx = \frac{2}{\pi} \left[\frac{\cos 3x}{6} - \frac{\cos 5x}{10} \right] \Big|_0^\pi$$

$$a_4 = \frac{2}{\pi} \left[\frac{\cos 3\pi}{6} - \frac{\cos 5\pi}{10} - \frac{\cos 0}{6} + \frac{\cos 0}{10} \right] = \frac{2}{\pi} \left[\frac{-2}{15} \right] = \frac{-4}{15\pi}$$

$$a_5 = \frac{2}{\pi} \int_0^\pi \sin x \cos 5x \, dx = \frac{2}{\pi} \left[\frac{\cos 4x}{8} - \frac{\cos 6x}{12} \right] \Big|_0^\pi$$

$$a_5 = \frac{2}{\pi} \left[\frac{\cos 4\pi}{8} - \frac{\cos 6\pi}{12} - \frac{\cos 0}{8} + \frac{\cos 0}{12} \right] = 0$$

$$f(x) = \frac{e^\pi + 1}{\pi} \sin x - \frac{4}{5\pi}(e^\pi - 1) \sin 2x + \frac{3}{5\pi}(e^\pi + 1) \sin 3x - \frac{8}{17\pi}(e^\pi - 1) \sin 4x + \dots$$

45. $\displaystyle A = \int_{0.1}^{0.2} \frac{x - \sin x}{x^2} \, dx = \int_{0.1}^{0.2} \frac{x - \left(x - \frac{x^3}{3!} + \frac{x^5}{5!} \right)}{x^2} \, dx$

$$= \int_{0.1}^{0.2} \frac{\frac{x^3}{3!} + \frac{x^5}{5!}}{x^2} \, dx = \int_{0.1}^{0.2} \left(\frac{x}{6} - \frac{x^3}{120} \right) dx$$

$$= \left(\frac{x^2}{12} - \frac{x^4}{480} \right) \Big|_{0.1}^{0.2} = \left(\frac{0.4}{12} - \frac{0.0016}{480} \right) - \left(\frac{0.01}{12} - \frac{0.0001}{480} \right)$$

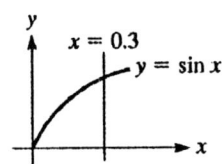

$$= 0.0025$$

46.

From 6-23

$$I_y = 2\pi k \int_a^b (y_2 - y_1) x^3 \, dx$$

$$I_y = 2\pi k \int_0^{0.3} (\sin x - 0) x^3 \, dx$$

$$I_y = 2\pi k \int_0^{0.3} x^3 \left(x - \frac{x^3}{3!} \right) dx$$

$$I_y = 2\pi k \int_0^{0.3} \left(x^4 - \frac{x^6}{6} \right) dx$$

$$I_y = 2\pi k \left(\frac{x^5}{5} - \frac{x^7}{42} \right) \Big|_0^{0.3}$$

$$I_y = 2\pi k \left(\frac{0.3^5}{5} - \frac{0.3^7}{42} \right) = 0.0030209106k$$

$$I_y = 0.00302k$$

47. $\tan^{-1} x = \displaystyle\int \frac{1}{1+x^2}\, dx = \int (1 - x^2 + x^4 - x^6 + \ldots)dx$

$\tan^{-1} x = x - \dfrac{x^3}{3} + \dfrac{x^5}{5} - \dfrac{x^7}{7} + \cdots$

48. $i = 0.5 \sin 0.5t - 0.2 \sin 0.4t$

$= 0.5 \left(0.5t - \dfrac{(0.5t)^3}{3!} + \dfrac{(0.5t)^5}{5!} \right) - 0.2 \left(0.4t - \dfrac{(0.4t)^3}{3!} + \dfrac{(0.4t)^5}{5!} \right)$

$= 0.17t - 0.0083t^3 - 0.00011t^5$

49. $N = N_0 e^{-\lambda t} \cdot N = N_0 \left[1 + (-\lambda t) + \dfrac{(-\lambda t)^2}{2!} + \dfrac{(-\lambda t)3}{3!} + \cdots \right] = N_0 \left[1 - \lambda t + \dfrac{\lambda^2 t^2}{2} - \dfrac{\lambda^3 t^3}{6} + \cdots \right]$

50.

$390 = 6400(2\theta)$

$\theta = 0.03046875 = 0.0305$

$\cos\theta = \dfrac{x}{6400} = \left(1 - \dfrac{\theta^2}{2} + \dfrac{\theta^4}{24} \right)$

$x = 6400 \left(1 - \dfrac{0.0305^2}{2} + \dfrac{0.0305^4}{24} \right)$

$x \doteq 6397$

error $= 6400 - 6397 = 3$ km

51. $f(x) = \ln \dfrac{1+x}{1-x}, f(0) = \ln \dfrac{1+0}{1-0} = \ln 1 = 0$

$f'(x) = -\dfrac{2}{x^2 - 1}, f'(0) = 2$

$f''(x) = \dfrac{4x}{(x+1)^2 (x-1)^2}, f''(0) = 0$

$f'''(x) = \dfrac{-4(3x^2 + 1)}{(x+1)^3 (x-1)^3}, f'''(0) = 4$

$f^{iv}(x) = \dfrac{48x(x^2 + 1)}{(x+1)^4 (x-1)^4}, f^{iv}(0) = 0$

$f^{(5)}(x) = \dfrac{-48(5x^4 + 10x^2 + 1)}{(x+1)^5 (x-1)5}, f^{(5)}(0) = 48$

$f^{(6)}(x) = \dfrac{480x(3x^4 + 10x^2 + 3)}{(x+1)^6 (x-1)^6}, f^{(6)}(0) = 0$

$f^{(7)}(x) = \dfrac{-1440(7x^6 + 35x^4 + 21x^2 + 1)}{(x+1)^7 (x-1)^7}, f^{(7)}(0) = 1440$

$V = \ln \dfrac{1+x}{1-x} = 2x + \dfrac{4x^3}{3!} + \dfrac{48x^5}{5!} + \dfrac{1440x^7}{7!}$

$V = \ln \dfrac{1+x}{1-x} = 2x + \dfrac{2}{3}x^3 + \dfrac{2}{5}x^5 + \dfrac{2}{7}x^7$

52. $\dfrac{m}{M} = kw \tan kw = kw \left(kw + \dfrac{(kw)^3}{3} + \dfrac{2(kw)^5}{15} + \cdots \right)$

$\dfrac{m}{M} = k^2 w^2 + \dfrac{k^4 w^4}{3} + \cdots$

53. $\dfrac{N_0}{1 - e^{-kt}} = N_0 \left(\dfrac{1}{1 - e^{-kt}} \right)$; Let $x = e^{-kt}$

$N_0 \left(\dfrac{N_0}{1 - e^{-kt}} \right) = N_0 \left(\dfrac{1}{1 - x} \right)$

The Maclaurin's expansion for $f(x) = \dfrac{1}{1 - x}$ is:

$$f(x) = \frac{1}{1 - x}; \qquad\qquad f(0) = 1$$

$$f'(x) = \frac{1}{(1 - x)^2}; \qquad\qquad f'(0) = 1$$

$$f''(x) = \frac{2}{(1 - x)^3}: \qquad\qquad f''(0) = 2$$

$$f(x) = 1 + x + \frac{2x^2}{2!} + \cdots = 1 + x + x^2 + \cdots$$

Substituting $e^{-k/t}$ for x; $f(x) = 1 + e^{-kt} + e^{-2kt} + \cdots$;

$$\therefore \frac{N_0}{1 - e^{-kt}} = N_0(1 + e^{-kt} + e^{-2kt} + \cdots)$$

54. $(x - R)^2 + (y - 0)^2 = R^2$
$x^2 - 2Rx + R^2 + y^2 = R^2$
$x^2 - 2Rx + y^2 = 0$, use quadratic formula

$$x = \frac{2R \pm \sqrt{4R^2 - 4y^2}}{2} = R \pm \sqrt{R^2 - y^2}$$

pick-since x must be 0 when y is 0.

$x = R - \sqrt{R^2 - y^2}$

Let $f(y) = (R^2 - y^2)^{1/2}, f(0) = R$

$$f'(y) = \frac{-y}{(R^2 - y^2)^{1/2}}, f'(0) = 0$$

$$f''(y) = \frac{-1}{(R^2 - y^2)^{1/2}} - \frac{y^2}{(R^2 - y^2)^{3/2}}, f''(0) = \frac{-1}{R}$$

$$f'''(y) = \frac{-3y}{(R^2 - y^2)^{3/2}} - \frac{3y^3}{(R^2 - y^2)^{5/2}}, f'''(0) = 0$$

$$f^{iv}(y) = \frac{-3}{(R^2 - y^2)^{3/2}} - \frac{18y^2}{(R^2 - y^2)^{5/2}} - \frac{15y^4}{(R^2 - y^2)^{7/2}}, f^{iv}(0) = -\frac{3}{R^3}$$

$$f^{(5)}(y) = \frac{-45y}{(R^2 - y^2)^{5/2}} - \frac{150y^3}{(R^2 - y^2)^{7/2}} - \frac{105y^5}{(R^2 - y^2)^{9/2}}, f^{(5)}(0) = 0$$

$$f^{(6)}(y) = \frac{-45y}{(R^2 - y^2)^{5/2}} - \frac{675y^2}{(R^2 - y^2)^{7/2}} - \frac{1575y^4}{(R^2 - y^2)^{9/2}} - \frac{945y^6}{(R^2 - y^2)^{11/2}}, f^{(6)}(0) = \frac{-45}{R^5}$$

$$x = R - \left(R - \frac{1}{R}\frac{y^2}{2!} - \frac{3}{R^3}\frac{y^4}{4!} - \frac{45}{R^5}\frac{y^6}{6!} \right)$$

$$x = \frac{1}{2R}y^2 + \frac{1}{8R^3}y^4 + \frac{1}{16R^5}y^6 + \cdots$$

55. $f(x) = \begin{cases} 0, & -\pi \le x < 0 \\ \sin t, & 0 < x < \pi/2 \\ 0, & \pi/2 < x < \pi \end{cases}$

$$a_0 = \frac{1}{2\pi} \int_{-\pi}^{\pi} f(x)\, dx = \frac{1}{2\pi} \int_{-\pi}^{0} 0\, dx + \frac{1}{2\pi} \int_{0}^{\pi/2} \sin t\, dt + \frac{1}{2\pi} \int_{\pi/2}^{\pi} 0\, dx$$

$$a_0 = \frac{-1}{2\pi} \cos t \Big|_{0}^{\pi/2} = \frac{1}{2\pi} \left[-\cos \frac{\pi}{2} + \cos 0 \right] = \frac{1}{2\pi}$$

$$a_n = \frac{1}{\pi} \int_{-\pi}^{\pi} f(t) \cos nt\, dt = \frac{1}{\pi} \int_{0}^{\pi/2} \sin t \cos nt\, dt$$

$$a_1 = \frac{1}{\pi} \left[\frac{\sin^2 t}{2} \right] \Big|_{0}^{\pi/2} = \frac{1}{2\pi} \left[\sin^2 \frac{\pi}{2} - \sin^2 0 \right] = \frac{1}{2\pi}$$

$$a_2 = \frac{1}{\pi} \left[\frac{\cos t}{2} - \frac{\cos 3t}{6} \right] \Big|_{0}^{\pi/2} = \frac{1}{\pi} \left[\frac{\cos \pi/2}{2} - \frac{\cos 3\pi/2}{6} - \frac{\cos 0}{2} + \frac{\cos(3(0))}{6} \right]$$

$$= \frac{1}{\pi} \left[-\frac{1}{2} + \frac{1}{6} \right] = -\frac{1}{3\pi}$$

$$b_n = \frac{1}{\pi} \int_{-\pi}^{\pi} f(x) \sin nt\, dt = \frac{1}{\pi} \int_{0}^{\pi/2} \sin t \sin nt\, dt$$

$$b_1 = \frac{1}{\pi} \int_{0}^{\pi/2} \sin^2 t\, dt = \frac{1}{\pi} \left[\frac{t}{2} - \frac{\sin t \cos t}{2} \right] \Big|_{0}^{\pi/2}$$

$$b_1 = \frac{1}{\pi} \left[\frac{\pi}{4} \right] = \frac{1}{4}$$

$$b_2 = \pi \int_{0}^{\pi/2} \sin t \sin 2t\, dt$$

$$b_2 = \frac{1}{\pi} \left[\frac{\sin t}{2} - \frac{\sin 3t}{6} \right] \Big|_{0}^{\pi/2} = \frac{1}{\pi} \left[\frac{\sin \pi/2}{2} - \frac{\sin 3\pi/2}{6} \right]$$

$$b_2 = \frac{1}{\pi} \left[\frac{1}{2} - \frac{-1}{6} \right] = \frac{2}{3\pi}$$

$$f(x) = \frac{1}{2\pi} + \frac{1}{\pi} \left(\frac{1}{2} \cos t - \frac{1}{3} \cos 2t + \cdots \right) + \frac{1}{4} \sin t + \frac{2}{3\pi} \sin 2t + \cdots$$

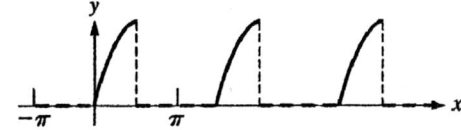

56. $F(t) = \begin{cases} \dfrac{t}{\pi}, & 0 \le t \le \pi \\[2mm] 0, & \pi < t < 2\pi \end{cases}$

$a_0 = \dfrac{1}{2\pi} \displaystyle\int_0^\pi \dfrac{t}{\pi}\, dt = \dfrac{1}{2\pi^2} \dfrac{t^2}{2}\bigg|_0^\pi = \dfrac{1}{4\pi^2}(\pi^2) = \dfrac{1}{4}$

$a_n = \dfrac{1}{\pi} \displaystyle\int_0^\pi \dfrac{t}{\pi} \cos nt\, dt = \dfrac{1}{\pi^2} \displaystyle\int_0^\pi t \cos nt\, dt$

$a_n = \dfrac{1}{\pi^2} \left[\dfrac{\cos n\pi}{n^2} + \dfrac{\pi n \sin n\pi}{n} - \dfrac{1}{n^2} \right]$

$a_n = \dfrac{1}{\pi^2} \left[\dfrac{\cos n\pi}{n^2} - \dfrac{1}{n^2} \right]$

$a_1 = \dfrac{1}{\pi^2} \left[\dfrac{\cos \pi}{n^2} - \dfrac{1}{1} \right] = \dfrac{-2}{\pi^2}$

$a_2 = \dfrac{1}{\pi^2} \left[\dfrac{\cos 2\pi}{2^2} - \dfrac{1}{2^2} \right] = 0$

$a_3 = \dfrac{1}{\pi^2} \left[\dfrac{\cos 3\pi}{3^2} - \dfrac{1}{3^2} \right] = \dfrac{-2}{9\pi^2}$

$b_n = \dfrac{1}{\pi} \displaystyle\int_0^\pi \dfrac{t}{\pi} \sin nt\, dt = \dfrac{1}{\pi^2} \displaystyle\int_0^\pi t \sin nt\, dt$

$b_n = \dfrac{1}{\pi^2} \left[\dfrac{\sin n\pi}{n^2} - \dfrac{\pi \cos n\pi}{n} \right]$

$b_n = \dfrac{-1}{\pi} \left[\dfrac{\cos n\pi}{n} \right]$

$b_1 = -\dfrac{1}{\pi}[\cos \pi] = \dfrac{1}{\pi}$

$b_2 = -\dfrac{1}{\pi} \left[\dfrac{\cos 2\pi}{2} \right] = \dfrac{-1}{2\pi}$

$F(t) = \dfrac{1}{4} - \dfrac{2}{\pi^2} \left(\cos t + \dfrac{1}{9} \cos 3t + \cdots \right) + \dfrac{1}{\pi} \left(\sin t - \dfrac{1}{2} \sin 2t + \cdots \right)$

57. Answers may vary. One way would be to use a Taylor Series and choosing an a-value from the known values which is close the angle for the particular trig function being computed.

FIRST—ORDER DIFFERENTIAL EQUATIONS

14.1 Solutions of Differential Equations

1. $y = e^{-x^2}$; $\dfrac{dy}{dx} = -2xe^{-x^2}$

Substitute y and $\dfrac{dy}{dx}$ into the differential equation.

$\dfrac{dy}{dx} + 2xy = 0$, $-2xe^{-x^2} + 2x(e^{-x^2}) = 0$; $0 = 0$

particular solution, order one, no c.

2. $y' \ln x - \dfrac{y}{x} = 0$; $y = c \ln x$; $y' = c\dfrac{1}{x} = \dfrac{c}{x}$

Substitute y and y' into the differential equation.

$\dfrac{c}{x} \ln - \dfrac{c \ln x}{x} = 0$; $\dfrac{c}{x} \ln x - \dfrac{c}{x} \ln x = 0$

$0 = 0$ identity. General solution; order one, c

3. $y'' + 3y' - 4y = 3e^x$; $y = c_1 e^x + c_2 e^{-4x} + \dfrac{3}{5} xe^x$

$y' = c_1 e^x - 4c_2 e^{-4x} + \dfrac{3}{5}(xe^x + e^x)$

$\quad = c_1 e^x - 4c_2 e^{-4x} + \dfrac{3}{5} xe^x + \dfrac{3}{5} e^x$

$y'' = c_1 e^x + 16c_2 e^{-4x} + \dfrac{3}{5} e^x + \dfrac{3}{5}(xe^x + e^x)$

$\quad = c_1 e^x + 16c_2 e^{-4x} + \dfrac{3}{5} e^x + \dfrac{3}{5} xe^x + \dfrac{3}{5} e^x$

$\quad = c_1 e^x + 16c_2 e^{-4x} + \dfrac{3}{5} xe^x + \dfrac{6}{5} e^x$

Substitute y, y', y'' into differential equation.

$c_1 e^x + 16c_2 e^{-4x} + \dfrac{3}{5} xe^x + \dfrac{6}{5} e^x$

$+ 3\left(c_1 e^x - 4c_2 e^{-4x} + \dfrac{3}{5} xe^x + \dfrac{3}{5} e^x \right)$

$- 4\left(c_1 e^x + c_2 e^{-4x} + \dfrac{3}{5} xe^x \right) = 3e^x$

$c_1 e^x + 16c_2 e^{-4x} + \dfrac{3}{5} xe^x + \dfrac{6}{5} e^x$

$+ 3c_1 e^x - 12c_2 e^{-4x} + \dfrac{9}{5} xe^x + \dfrac{9}{5} e^x$

$- 4c_1 e^x - 4c_2 e^{-4x} - \dfrac{12}{5} xe^x = 3e^x$

$\dfrac{15}{5} e^x = 3e^x$; $3e^x = 3e^x$ identity

General solution; order two, c_1 and c_2

4. $\dfrac{d^2 y}{dx^2} + 4y = 0$; $y = c_1 \sin 2x + 3\cos 2x$

$\dfrac{dy}{dx} = 2c_1 \cos 2x - 6 \sin 2x$;

$\dfrac{d^2 y}{dx^2} = -4c_1 \sin 2x - 12 \cos 2x$

Substitute y and y'' into differential equation

$-4c_1 \sin 2x - 12 \cos 2x + 4(c_1 \sin 2x + 3\cos 2x) = 0$
$-4c_1 \sin 2x - 12 \cos 2x + 4c_1 \sin 2x + 12 \cos 2x = 0$

$0 = 0$ identity; particular solution, order 2, $c_2 = 3$

5. $\dfrac{dy}{dx} - y = 1$; $y = e^x - 1$; $\dfrac{dy}{dx} = e^x$;

$y = 5e^x - 1$, $\dfrac{dy}{dx} = 5e^x$

Substitute y and y'.

$e^x - (e^x - 1) = 1$; $5e^x - 5(e^x - 1) = 1$;
$1 = 1$ identity for both.

6. $y = -\dfrac{1}{x^2}$; $\dfrac{dy}{dx} = \dfrac{2}{x^3}$; $y^2 = \dfrac{1}{x^4}$

$\dfrac{dy}{dx} = \dfrac{2}{x^3} = \dfrac{2x}{x^4} = 2x \cdot \dfrac{1}{x^4} = 2xy^2$

$y = \dfrac{-1}{x^2 + c}$; $\dfrac{dy}{dx} = \dfrac{2x}{(x^2 + c)^2}$; $y^2 = \dfrac{1}{(x^2 + c)^2}$

$\dfrac{dy}{dx} = \dfrac{2x}{(x^2 + c)^2} = 2x \cdot \dfrac{1}{(x^2 + c)^2} = 2xy^2$

7. $y = 3\cos 2x$; $y'' = -12 \cos 2x$
$y'' + 4y = -12 \cos 2x + 4(3 \cos 2x) = 0$
$y = c_1 \sin 2x + c_2 \cos 2x$;
$y'' = -4c_1 \sin 2x - 4c_2 \cos 2x$
$y'' + 4y = -4c_1 \sin 2x - 4c_2 \cos 2x$
$\qquad\qquad + 4(c_1 \sin 2x + c_2 \cos 2x)$
$\qquad = 0$

8. $y = 3e^{2x}$; $y' = 6e^{2x}$; $y'' = 12e^{2x}$
$y'' = 12e^{2x} = 2(6e^{2x}) = 2y'$
$y = 2e^{2x} - 5$; $y' = 4e^{2x}$; $y'' = 8e^{2x}$
$y'' = 8e^{2x} = 2(4e^{2x}) = 2y'$

9. $\dfrac{dy}{dx} = 2x$; $y = x^2 + 1$; $\dfrac{dy}{dx} = 2x$

Substitute. $2x = 2x$ identity

10. $y = 2 + x - x^3$

$\dfrac{dy}{dx} = 1 - 3x^2$

11. $\dfrac{dy}{dx} - 3 = 2x$; $y = x^2 + 3x$; $\dfrac{dy}{dx} = 2x + 3$

Substitute. $2x + 3 - 3 = 2x$; $2x = 2x$ identity

12. $xy' = 2y$; $y = cx^2$; $y' = 2cx$

Substitute, $xy' = x(2cx) = 2(cx^2) = 2y$

$2y = 2y$ identity

13. $y' + 2y = 2x$; $y = ce^{-2x} + x - \dfrac{1}{2}$;

$y' = -2ce^{-2x} + 1$

Substitute y and y'.

$-2ce^{-2x} + 1 + 2\left(ce^{-2x} + x - \dfrac{1}{2}\right) = 2x$

$-2ce^{-2x} + 1 + 2ce^{-2x} + 2x - 1 = 2x$

$2x = 2x$ identity

14. $y' - 3x^2 = 1$; $y = x^3 + x + c$; $y' = 3x^2 + 1$; substitute.
$3x^2 + 1 - 3x^2 = 1$; $1 = 1$ identity

15. $y'' + 9y = 4\cos x$; $2y = \cos x$; $y = \dfrac{1}{2}\cos x$;

$y' = -\dfrac{1}{2}\sin x$; $y'' = -\dfrac{1}{2}\cos x$

Substitute y and y''.

$-\dfrac{1}{2}\cos x + 9\left(\dfrac{1}{2}\cos x\right) = 4\cos x$

$-\dfrac{1}{2}\cos x + \dfrac{9}{2}\cos x = 4\cos x$

$\dfrac{8}{2}\cos x = 4\cos x$; $4\cos x = 4\cos x$ identity

16. $y'' - 4y' + 4y = e^{2x}$; $y = e^{2x}\left(c_1 + c_2 x + \dfrac{x^2}{2}\right)$

$y' = e^{2x}(c_2 + x) + \left(c_1 + c_2 x + \dfrac{x^2}{2}\right)2e^{2x}$

$= c_2 e^{2x} + xe^{2x} + 2c_1 e^{2x} + 2c_2 xe^{2x} + x^2 e^{2x}$
$y'' = 2c_2 e^{2x} + (x2e^{2x} + e^{2x}) + 4c_1 e^{2x}$
$\quad + 2c_2(2xe^{2x} + e^{2x})$
$\quad + (x^2 2e^{2x} + e^{2x}2x)$
$y'' = 2c_2 e^{2x} + 2xe^{2x} + e^{2x} + 4c_1 e^{2x} + 4c_2 xe^{2x}$
$\quad + 2c_2 e^{2x} + 2x^2 e^{2x} + 2xe^{2x}$
$y'' = 4c_2 e^{2x} + 4xe^{2x} + e^{2x} + 4c_1 e^{2x} + 4c_2 xe^{2x}$
$\quad + 2x^2 e^{2x}$

Substitute y, y', y'',

$4c_2 e^{2x} + 4xe^{2x} + e^{2x} + 4c_1 e^{2x} + 4c_2 xe^{2x}$
$\quad + 2x^2 e^{2x} - 4(c_2 e^{2x} + xe^{2x} + 2c_1 e^{2x}$
$\quad + 2c_2 xe^{2x} + x^2 e^{2x})$
$\quad + 4\left(c_1 e^{2x} + c_2 xe^{2x} + \dfrac{1}{2}x^2 e^{2x}\right)$
$\quad = e^{2x}$

$4c_2 e^{2x} + 4xe^{2x} + e^{2x} + 4c_1 e^{2x} + 4c_2 xe^{2x} + 2x^2 e^{2x}$
$\quad - 4c_2 e^{2x} - 4xe^{2x} - 8c_1 e^{2x} - 8c_2 xe^{2x} - 4x^2 e^{2x}$
$\quad + 4c_1 e^{2x} + 4c_2 xe^{2x} + 2x^2 e^{2x}$
$\quad = e^{2x}$
$e^{2x} = e^{2x}$ identity

17. $x^2 y' + y^2 = 0$; $xy = cx + cy$; $y = \dfrac{cx}{x - c}$

$y' = \dfrac{(x - c)(c) - cx(1)}{(x - c)^2} = \dfrac{-c^2}{(c - x)^2}$

Substituting,

$x^2 y' + y^2 = x^2\left[\dfrac{-c^2}{(x - c)^2}\right] + \left(\dfrac{cx}{x - c}\right)^2$

$= \dfrac{-c^2 x^2}{(x - c)^2} + \dfrac{c^2 x^2}{(x - c)^2} = 0$; $0 = 0$ identity

18. $xy' - 3y = x^2$; $y = cx^3 - x^2$; $y' = 3cx^2 - 2x$

Substitute y and y'.

$x(3cx^2 - 2x) - 3(cx^3 - x^2) = x^2$
$3cx^3 - 2x^2 - 3cx^3 + 3x^2 = x^2$; $x^2 = x^2$ identity

19. $x\dfrac{d^2y}{dx^2} + \dfrac{dy}{dx} = 0$; $y = c_1 \ln x + c_2$

$\dfrac{dy}{dx} = \dfrac{c_1}{x} = c_1 x^{-1}$; $\dfrac{d^2y}{dx^2} = -c_1 x^{-2} = -\dfrac{c_1}{x^2}$

Substitute $\dfrac{dy}{dx}$ and $\dfrac{d^2y}{dx^2}$.

$x\left(\dfrac{-c_1}{x^2}\right) + c_1 x^{-1} = 0$; $-\dfrac{c_1}{x} + \dfrac{c_1}{x} = 0$

$0 = 0$ identity

20. $y'' + 4y = 10e^x$

$y = c_1 \sin 2x + c_2 \cos 2x + 2e^x$

$y' = 2c_1 \cos 2x - 2c_2 \sin 2x + 2e^x$

$y'' = -4c_1 \sin 2x - 4c_2 \cos 2x + 2e^x$

Substitute y and y''.

$-4c_1 \sin 2x - 4c_2 \cos 2x + 2e^x$
$\quad + 4(c_1 \sin 2x + c_2 \cos 2x + 2e^x)$
$\quad = 10e^x$

$-4c_1 \sin 2x - 4c_2 \cos 2x + 2e^x + 4c_1 \sin 2x$
$\quad + 4c_2 \cos 2x + 8e^x$
$\quad = 10e^x$

$10e^x = 10e^x$ identity

21. $y' + y = 2\cos x$;

$y = \sin x + \cos x - e^{-x}$

$y' = \cos x - \sin x - e^{-x}(-1)$
$\quad = \cos x - \sin x + e^{-x}$

Substitute,

$y' + y = \cos x - \sin x + e^{-x} + \sin x + \cos x - e^{-x}$
$\quad = 2\cos x$

$2\cos x = 2\cos x$　identity

22. $(x + y) - xy' = 0$; $y = x \ln x - cx$

$y' = x\dfrac{1}{x} + \ln x - c = 1 + \ln x - c$;

substitute y and y'.

$x + x \ln x - cx - x(1 + \ln x - c) = 0$

$x + x \ln x - cx - x - x \ln x + cx = 0$

$0 = 0$ identity

23. $y'' + y' = 6 \sin 2x$

$y = e^{-x} - \dfrac{3}{5}\cos 2x - \dfrac{6}{5}\sin 2x$

$y' = -e^{-x} + \dfrac{6}{5}\sin 2x - \dfrac{12}{5}\cos 2x$

$y'' = e^{-x} + \dfrac{12}{5}\cos 2x + \dfrac{24}{5}\sin 2x$

Substitute y' and y''.

$e^{-x} + \dfrac{12}{5}\cos 2x + \dfrac{24}{5}\sin 2x - e^{-x}$

$\quad + \dfrac{6}{5}\sin 2x - \dfrac{12}{5}\cos 2x = 6 \sin 2x$

$\dfrac{30}{5}\sin 2x = 6 \sin 2x$; $6 \sin 2x = 6 \sin 2x$ identity

24. $xy'' + y' = 16x^3$

$y = x^4 + c_1 + c_2 \ln x$; $y' = 4x^3 + \dfrac{c_2}{x}$; $y'' = 12x^2 - \dfrac{c_2}{x^2}$

Substitute y' and y''.

$x\left(12x^2 - \dfrac{c_2}{x^2}\right) + 4x^3 + \dfrac{c_2}{x} = 16x^3$

$12x^3 - \dfrac{c_2}{x} + 4x^3 + \dfrac{c_2}{x} = 16x^3$; $16x^3 = 16x^3$ identity

25. $\cos x \dfrac{dy}{dx} + \sin x = 1 - y$; $y = \dfrac{x + c}{\sec x + \tan x}$

$\dfrac{dy}{dx} = \dfrac{(\sec x + \tan x)(1) - (x+c)(\sec x \tan x + \sec^2 x)}{(\sec x + \tan x)^2}$

$\quad = \dfrac{(\sec x + \tan x) - (x+c)(\sec x)(\tan x + \sec x)}{(\sec x + \tan x)^2}$

$\quad = \dfrac{(\sec x + \tan x)[1 - (x+c)(\sec x)]}{(\sec x + \tan x)(\sec x + \tan x)}$

$\quad = \dfrac{1 - (x+c)(\sec x)}{(\sec x + \tan x)}$

Substituting,

$\cos x \dfrac{dy}{dx} + \sin x$

$= \cos x\left[\dfrac{1 - (x+c)(\sec x)}{\sec x + \tan x}\right] + \sin x$

$= \dfrac{\cos x - (x + c)}{\sec x + \tan x} + \sin x$

$= \dfrac{\cos x - (x + c)}{\sec x + \tan x} + \dfrac{\sin x \sec x + \sin x \tan x}{\sec x + \tan x}$

$= \dfrac{\cos x - x - c + \dfrac{\sin x}{\cos x} + \dfrac{\sin^2 x}{\cos x}}{\sec x + \tan x}$

$= \dfrac{\dfrac{\cos^2 x}{\cos x} + \dfrac{\sin^2 x}{\cos x} + \tan x - x - c}{\sec x + \tan x}$

$= \dfrac{\dfrac{1}{\cos x} + \tan x - x - c}{\sec x + \tan x}$

$= \dfrac{\sec x + \tan x - x - c}{\sec x + \tan x}$

$= 1 - \dfrac{(x + c)}{\sec x + \tan x} = 1 - y$

26. $2xyy' + x^2 = y^2$

$x^2 + y^2 = cx; \ 2x + 2yy' = c; \ y' = \dfrac{c - 2x}{2y}$

Substitute y and y'.

$2xy\left(\dfrac{c - 2x}{2y}\right) + x^2 = y^2; \ x(c - 2x) + x^2 = y^2$

$cx - 2x^2 + x^2 = y^2; \ cx - x^2 = y^2; \ y^2 = y^2$ identity

27. $(y')^2 + xy' = y$

$y = cx + c^2; \ y' = c$

Substitute.

$(c)^2 + x(c) = y; \ c^2 + cx = y; \ y = y$ identity

28. $x^4(y')^2 - xy' = y$

$y = c^2 + \dfrac{c}{x}; \ y' = -\dfrac{c}{x^2}$

Substitute.

$x^4\left(-\dfrac{c}{x^2}\right)^2 - x\left(-\dfrac{c}{x^2}\right) = y; \ x^4\left(\dfrac{c^2}{x^4}\right) - \dfrac{cx}{x^2} = y$

$c^2 + \dfrac{c}{x} = y; \ y = y$ identity

14.2   Separation of Variables

1. $2x\,dx + dy = 0$; integrate. $x^2 + y = c$;
$y = c - x^2$

2. $y^2\,dy + x^3\,dx = 0$; integrate

$\dfrac{1}{3}y^3 + \dfrac{1}{4}x^4 = c_1$; multiply by 12; $c_1 \times 12 = c$

$4y^3 + 3x^4 = c$

3. $y^2\,dx + dy = 0$; divide by y^2;

$dx + \dfrac{dy}{y^2} = 0$; integrate

$x + \dfrac{y^{-1}}{-1} = c; \ x - \dfrac{1}{y} = c$

4. $y\,dx + x\,dy = 0$; divide by xy; $\dfrac{dx}{x} + \dfrac{dy}{y} = 0$;
integrate

$\ln x + \ln y = c; \ \ln(xy) = \ln c, \ xy = c$

5. $\dfrac{dV}{dP} = \dfrac{-V}{P^2}; \ \dfrac{-dV}{V} = \dfrac{dP}{P^2}; \ \dfrac{dP}{P^2} + \dfrac{dV}{V} = 0$

$\displaystyle\int P^{-2}\,dP + \int \dfrac{dV}{V} = 0; \ \dfrac{P^{-1}}{-1} + \ln V = c;$

$\ln V = \dfrac{1}{P} + c$

6. $\dfrac{2\,dy}{dx} = \dfrac{y(x+1)}{x}$; cross multiply

$\dfrac{2\,dy}{y} = \dfrac{(x+1)}{x}\,dx$; integrate

$2\ln y = x + \ln x + c$

7. $x^2 + (x^3 + 5)y' = 0; \ (x^3 + 5)\dfrac{dy}{dx} = -x^2$

$(x^3 + 5)dy = -x^2\,dx; \ dy = \dfrac{-x^2\,dx}{x^3 + 5}$; integrate

$y = -\dfrac{1}{3}\ln(x^3 + 5) + c$

$3y + \ln(x^3 + 5) = 3c_1; \ 3c_1$ is constant
$3y + \ln(x^3 + 5) = c$

8. $xyy' + \sqrt{1 + y^2} = 0; \ xy\dfrac{dy}{dx} = -\sqrt{1 + y^2}$

$\dfrac{x}{dx} = \dfrac{-\sqrt{1 + y^2}}{y\,dy}$; reciprocals

$\dfrac{dx}{x} = -\dfrac{y\,dy}{\sqrt{1 + y^2}}$; integrate

$\ln x = -\dfrac{1}{2}\dfrac{(1 + y^2)^{1/2}}{\frac{1}{2}} + c$

$\ln x = -\sqrt{1 + y^2} + c; \ \ln x + \sqrt{1 + y^2} = c$

9. $\qquad dy + \ln xy\,dx = (4x + \ln y)dx;$
$dy + (\ln x + \ln y)dx = 4x\,dx + \ln y\,dx$
$dy + \ln x\,dx + \ln y\,dy = 4x\,dx + \ln y\,dx$
$\qquad dy + \ln x\,dx = 4x\,dx$
Integrate: $y + x\ln x - x - 2x^2 = c$
$y = 2x^2 + x - x\ln x + c$

$\displaystyle\int dy + \int \ln x\,dx - \int 4x\,dx = 0;$

$y + x\ln x - x - 2x^2 = c$
$y = 2x^2 + x - x\ln x + c$

10. $r\sqrt{1 - \theta^2}\dfrac{dr}{d\theta} = \theta + 4; \ r\,dr = \dfrac{(\theta + 4)d\theta}{\sqrt{1 - \theta^2}}$

$r\,dr = \dfrac{\theta\,d\theta}{(1 - \theta^2)^{1/2}} + \dfrac{4\,d\theta}{(1 - \theta^2)^{1/2}}$; integrate

$\dfrac{1}{2}r^2 = -\dfrac{1}{2}\dfrac{(1 - \theta^2)^{1/2}}{\frac{1}{2}} + 4\sin^{-1}\dfrac{\theta}{1} + c$

$r^2 = -2\sqrt{1 - \theta^2} + 8\sin^{-1}\theta + c$

11. $e^{x^2} dy = x\sqrt{1-y}\,dx;\quad \dfrac{dy}{\sqrt{1-y}} = \dfrac{x\,dx}{e^{x^2}}$

$\dfrac{dy}{(1-y)^{1/2}} = e^{-x^2} x\,dx;$ integrate

$-\dfrac{(1-y)^{1/2}}{\frac{1}{2}} = -\dfrac{1}{2} e^{-x^2} + c$

$-2\sqrt{1-y} = -\dfrac{1}{2} e^{-x^2} + c;$ multiply by -2

$4\sqrt{1-y} = e^{-x^2} - 2c_1;\ -2c_1 = c$
$4\sqrt{1-y} = e^{-x^2} + c$

12. $\sqrt{1+4x^2}\,dy = y^3 x\,dx$

$\dfrac{dy}{y^3} = \dfrac{x\,dx}{\sqrt{1+4x^2}};$ integrate

$\dfrac{y^{-2}}{(-2)} = \dfrac{1}{8}\dfrac{(1+4x^2)^{1/2}}{\frac{1}{2}} + c$

$-\dfrac{1}{2y^2} = \dfrac{1}{4}\sqrt{1+4x^2} + c_1;\ \dfrac{1}{4}\sqrt{1+4x^2} + \dfrac{1}{2y^2} = c_1$

$\sqrt{1+4x^2} + \dfrac{2}{y^2} = 4c_1 = c$

13. $e^{x+y} dx + dy = 0$
$\quad\ e^x e^y dx + dy = 0$

$\quad\quad e^x dx + \dfrac{dy}{e^y} = 0$

$e^x dx + e^{-y} dy = 0$

Integrate:

$e^x - e^{-y} = c$

14. $e^{2x} dy + e^x dx = 0;\ dy + \dfrac{e^x dx}{e^{2x}} = 0;\ dy + e^{-x} dx = 0$

Integrate: $y - e^{-x} = c_1;\ y = e^{-x} + c$

15. $y' - y = 4;\ \dfrac{dy}{dx} = 4 + y;\ \dfrac{dy}{4+y} = dx;$ integrate

$\ln(4+y) = x + c$

16. $ds - s^2 dt = 9\,dt$

$ds = (9 + s^2)\,dt$

$\dfrac{ds}{9+s^2} = dt$

$\dfrac{1}{3}\tan^{-1}\dfrac{s}{3} = t + c$

17. $x\dfrac{dy}{dx} = y^2 + y^2 \ln x;$ divide by xy^2 and multiply by dx

$\dfrac{dy}{y^2} = \dfrac{dx}{x} + \dfrac{1}{x}\ln x\,dx;$

$\dfrac{dy}{y^2} = (1 + \ln x)\dfrac{dy}{dx};$ integrate

$-\dfrac{1}{y} = \dfrac{1}{2}(1 + \ln x)^2 + \dfrac{c}{2};$

$-2 = y(1 + \ln x)^2 + cy$

$y(1 + \ln x)^2 + cy + 2 = 0$

18. $(yx^2 + y)\dfrac{dy}{dx} = \tan^{-1} x;\ y(x^2+1)dy = \tan^{-1} x\,dx$

$y\,dy = \dfrac{\tan^{-1} x\,dx}{x^2 + 1};$ integrate

$\dfrac{y^2}{2} = \dfrac{(\tan^{-1} x)^2}{2} + c;\ y^2 = (\tan^{-1} x)^2 + 2c_1;$

$2c_1 = c$

$y^2 = (\tan^{-1} x)^2 + c$

19. $y\tan x\,dx + \cos^2 x\,dy = 0;\ \dfrac{\tan x\,dx}{\cos^2 x} + \dfrac{dy}{y} = 0$

$(\tan x)^1 \sec^2 x\,dx + \dfrac{dy}{y} = 0;$ integrate

$\dfrac{1}{2}\tan^2 x + \ln y = c_1;\ 2c_1 = c;\ \tan^2 x + 2\ln y = c$

20. $\sin x \sec y\,dx = dy;\ \sin x\,dx = \dfrac{dy}{\sec y};$

$\sin x\,dx = \cos y\,dy$

Integrate: $-\cos x = \sin y + c_1;\ -c_1 = c$
$\cos x + \sin y = c$

21. $y x^2 dx = y\,dx - x^2 dy$

Divide by y and by x^2; $dx = \dfrac{dx}{x^2} - \dfrac{dy}{y};$ integrate

$x + c = \dfrac{x^{-1}}{-1} - \ln y$

$x + \dfrac{1}{x} + \ln y = c;\ x^2 + 1 + x\ln y + cx = 0$

22. $e^{\cos\theta}\tan\theta\,d\theta + \sec\theta\,dy = 0$
$e^{\cos\theta}\sin\theta\,d\theta + dy = 0$
$-e^{\cos\theta} + y = c$

23. $y\sqrt{1-x^2}\,dy + 2\,dx = 0;\ y\,dy + \dfrac{2}{\sqrt{1-x^2}}\,dx = 0$

Integrate: $\dfrac{1}{2}y^2 + 2\sin^{-1} x = c_1;$

$y^2 + 4\sin^{-1} x = 2c_1;\ y^2 + 4\sin^{-1} x = c$

24. $(x^3 + x^2)dx + (x+1)y\,dy = 0$
$x^2(x+1)dx + (x+1)y\,dy = 0x^2\,dx + y\,dy = 0$

integrate: $\dfrac{1}{3}x^3 + \dfrac{1}{2}y^2 = c$; $2x^3 + 3y^2 = c$

25. $2\ln t\,dt + t\,di = 0$;

$2\ln t\dfrac{dt}{t} + di = 0$; integrate

$\dfrac{2(\ln t)^2}{2} + i = 0$; $(\ln t)^2 + i = c$;

$i = c - (\ln t)^2$

26. $2y(x^3 + 1)dy + 3x^2(y^2 - 1)dx = 0$

$\dfrac{2y\,dx}{y^2 - 1} + \dfrac{3x^2\,dx}{x^3 + 1} = 0$

Integrate: $\ln(y^2 - 1) + \ln(x^3 + 1) = c_1$
$\ln(y^2 - 1)(x^3 + 1) = c_1$; $(y^2 - 1)(x^3 + 1) = e^{c_1}$
$(y^2 - 1)(x^3 + 1) = c$

27. $y^2 e^x + (e^x + 1)\dfrac{dy}{dx} = 0$; $y^2 e^x = -(e^x + 1)\dfrac{dy}{dx}$

$\dfrac{e^x\,dx}{e^x + 1} = -\dfrac{dy}{y^2}$; integrate

$\ln(e^x + 1) = -\dfrac{y^{-1}}{-1} + c = \dfrac{1}{y} + c$; $\ln(e^x + 1) - \dfrac{1}{y} = c$

28. $y + 1 + \sec x(\sin x + 1)\dfrac{dy}{dx} = 0$;

$\sec x(\sin x + 1)\dfrac{dy}{dx} = -(y+1)$

$\dfrac{\sin x + 1}{\cos x\,dx} = -\dfrac{(y+1)}{dy}$; $\dfrac{\cos x\,dx}{(\sin x + 1)} = -\dfrac{dy}{y+1}$;

integrate
$\ln(\sin x + 1) = -\ln(y+1) + c_1$;
$\ln(\sin x + 1) + \ln(y+1) = c_1$
$(\sin x + 1)(y+1) = e^{c_1} = c$

29. $\dfrac{dy}{dx} + yx^2 = 0$; $\dfrac{dy}{y} + x^2\,dx = 0$

Integrate: $\ln y + \dfrac{x^3}{3} + c$

Substitute $x = 0, y = 1$; $\ln 1 = c$; $c = 0$
$3\ln y + x^3 = 0$

30. $\dfrac{dy}{dx} + 2y = 6$; $x = 0, y = 1$

$\dfrac{dy}{dx} = 6 - 2y$; $\dfrac{dy}{6 - 2y} = dx$; integrate

$-\dfrac{1}{2}\ln(6 - 2y) = x + c$

Substitute $x = 0, y = 1$; $-\dfrac{1}{2}\ln(4) = 0 + c$;

$c = -\dfrac{1}{2}\ln 4$

$-\dfrac{1}{2}\ln(6 - 2y) = x - \dfrac{1}{2}\ln 4$; $-\ln(6 - 2y) = 2x - \ln 4$

$\ln(6 - 2y) - \ln 4 = -2x$; $\ln\dfrac{6 - 2y}{4} = -2x$;

$\dfrac{6 - 2y}{4} = e^{-2x}$

$6 - 2y = 4e^{-2x}$; $2y = 6 - 4e^{-2x}$; $y = 3 - 2e^{-2x}$

31. $(xy^2 + x)\dfrac{dy}{dx} = \ln x$; $x = 1, y = 0$

$x(y^2 + 1)dy = \ln x\,dx$; $(y^2 + 1)dy = (\ln x)\dfrac{dx}{x}$;
integrate

$\dfrac{1}{3}y^3 + y = \dfrac{1}{2}(\ln x)^2 + c$; $\dfrac{1}{3}y^3 + y = \dfrac{1}{2}\ln^2 x + c$

Substitute $x = 1, y = 0$

$0 + 0 = \dfrac{1}{2}(\ln 1)^2 + c$; therefore, $c = 0$

$\dfrac{1}{3}y^3 + y = \dfrac{1}{2}\ln^2 x$

32. $\dfrac{ds}{dt} = \sec s$; $t = 0, s = 0$

$\dfrac{ds}{\sec s} = dt$; $\cos s\,ds = dt$; integrate: $\sin s = t + c$

Substitute $t = 0, s = 0$

$\sin 0 = 0 + c$; therefore, $c = 0, \sin s = t$

33. $\dfrac{dy}{dx} = (1 - y)\cos x$; $x = \dfrac{\pi}{6}$ when $y = 0$

$dy = (1 - y)\cos x\,dx$

$\dfrac{1}{1 - y}dy = \cos x\,dx$; integrate

$-\ln(1 - y) = \sin x + c$; $\sin x + \ln(1 - y) = c$

Substitute $x = \dfrac{\pi}{6}, y = 0$; $\sin\dfrac{\pi}{6} + \ln 1 = c$; $c = \dfrac{1}{2}$

$\sin x + \ln(1 - y) = \dfrac{1}{2}$; $2\ln(1 - y) = 1 - 2\sin x$

34. $x\,dy = y\ln y\,dx;\; x = 2, y = e$

$\dfrac{dx}{x} = \dfrac{dy}{y\ln y};$ integrate: $\ln x = \ln\ln y + c$

Substitute $x = 2, y = e$

$\ln 2 = \ln\ln e + c = \ln 1 + c;$ therefore, $c = \ln 2$
$\ln x = \ln(\ln y) + \ln 2 = \ln(2\ln y);\; x = 2\ln y$

35. $y^2 e^x\,dx + e^{-x}\,dy = y^2\,dx;\; x = 0, y = 2$

$y^2\,dx - y^2 e^x\,dx = e^{-x}\,dy;\; y^2(1 - e^x)\,dx = e^{-x}\,dy$

$\dfrac{(1 - e^x)}{e^{-x}}\,dx = \dfrac{dy}{y^2};\; e^x\,dx - e^{2x} = y^{-2}\,dy;$ integrate

$e^x - \dfrac{1}{2}e^{2x} = -\dfrac{1}{y} + c$

Substitute $x = 0, y = 2$

$e^0 - \dfrac{1}{2}e^0 = -\dfrac{1}{2} + c;$ therefore, $c = 1$

$e^x - \dfrac{1}{2}e^{2x} = -\dfrac{1}{y} + 1;\; 2e^x - e^{2x} = -\dfrac{2}{y} + 2$

$e^{2x} - \dfrac{2}{y} = 2(e^x - 1)$

36. $2y\cos y\,dy - \sin y\,dy = y\sin y\,dx;\; x = 0, y = \dfrac{\pi}{2}$

$\dfrac{2y\cos y\,dy}{y\sin y} - \dfrac{\sin y\,dy}{y\sin y} = dx$

$2\cot y\,dy - \dfrac{dy}{y} = dx;$ integrate

$2\ln|\sin y| - \ln y = x + c$

Substitute $x = 0, y = \pi/2$

$2\ln\left|\sin\dfrac{\pi}{2}\right| - \ln\dfrac{\pi}{2} = 0 + c$

$c = -\ln\dfrac{\pi}{2}$

$2\ln\sin y - \ln y = x - \ln\dfrac{\pi}{2}$

$x = \ln(\sin y)^2 - \ln y + \ln\dfrac{\pi}{2} = \ln\dfrac{\sin^2 y}{y}\left(\dfrac{\pi}{2}\right)$

$x = \ln\dfrac{\pi\sin^2 y}{2y}$

14.3   Integrating Combinations

1. $x\,dy + y\,dx + x\,dx = 0;\; d(xy) + x\,dx = 0$

$xy + \dfrac{x^2}{2} = c;\; 2xy + x^2 = c$

2. $(2y + x)\,dy + y\,dx = 0;\; 2y\,dy + (x\,dy + y\,dx) = 0$

$\dfrac{2y^2}{2} + \displaystyle\int d(xy) = 0$

$y^2 + xy = c$

3. $y\,dx - x\,dy + x^3\,dx = 2\,dx;$
$x\,dy - y\,dx - x^3\,dx = -2\,dx$

$\dfrac{(x\,dy - y\,dx)}{x^2} - x\,dx = -\dfrac{2\,dx}{x^2}$

$\dfrac{y}{x} - \dfrac{1}{2}x^2 = 2x^{-1} = \dfrac{2}{x} + c_1;\; y - \dfrac{1}{2}x^3 = 2 + c_1 x$

$2y - x^3 = 4 + 2c_1 x;\; x^3 - 2y = -2c_1 x - 4;\; -2c_1 = c$
$x^3 - 2y = cx - 4$

4. $x\,dy - y\,dx + y^2\,dx = 0;\; y\,dx - x\,dy - y^2\,dx = 0$

$\dfrac{y\,dx - x\,dy}{y^2} - dx = 0;\; d\left(\dfrac{x}{y}\right) - dx = 0$

$\dfrac{x}{y} - x = c_1;\; x - \dfrac{x}{y} = -c_1;\; -c = c;\; x - \dfrac{x}{y} = c$

5. $A^3\,dr + A^2 r\,dA + r\,dA - A\,dr = 0$

$A\,dr + r\,dA - \left(\dfrac{A\,dr - r\,dA}{A^2}\right) = 0$

$Ar - \dfrac{r}{A} = c;\; A^2 r - r = cA$

6. $\sec(xy)\,dx + (x\,dy + y\,dx) = 0;\; dx + \dfrac{x\,dy + y\,dx}{\sec(xy)} = 0$

$dx + \cos(xy)(x\,dy + y\,dx) = 0$
$x + \sin(xy) = c$

7. $x^3 y^4(x\,dy + y\,dx) = 3\,dy;$

$\displaystyle\int x^3 y^3(x\,dy + y\,dx) = \int \dfrac{3}{y}\,dy$

$(xy)^3(x\,dy + y\,dx) = 3\dfrac{dy}{y};\; \dfrac{(xy)^4}{4} = 3\ln y + c_1$

$(xy)^4 = 12\ln y + 4c_1;\; 4c_1 = c;\; (xy)^4 = 12\ln y + c$

8. $x\,dy + y\,dx + 4xy^3\,dy = 0;\; \dfrac{x\,dy + y\,dx}{xy} + 4y^2\,dy = 0$

$\dfrac{d(xy)}{xy} + 4y^2\,dy = 0;\; \ln xy + \dfrac{4y^3}{3} = c$

$3\ln xy + 4y^3 = 3c_1;\; 3c_1 = c;\; 3\ln xy + 4y^3 = c$

9. $\sqrt{x^2 + y^2}\,dx - 2y\,dy = 2x\,dx;$
$\quad (x^2 + y^2)\,dx = 2x\,dx + 2y\,dy$
$\quad\quad dx = (x^2 + y^2)^{-1/2}\,d(x^2 + y^2)$
$\quad\quad\quad x = 2(x^2 + y^2)^{1/2} + c$
$\quad 2\sqrt{x^2 + y^2} = x + c$

10. $R\,dR + (R^2 + T^2 + T)dT = 0$;

$R\,dR + (R^2 + T^2)dT + T\,dT = 0$

$R\,dR + T\,dT + (R^2 + T^2)dT = 0$

$$\frac{R\,dR + T\,dT}{(R^2 + T^2)} + dT = 0$$

$d(R^2 + T^2) = 2R\,dR + 2T\,dT$

$$\frac{1}{2}\frac{d(R^2 + T^2)}{R^2 + T^2} + dT = 0; \quad \frac{1}{2}\ln(R^2 + T^2) + T = c_1$$

$\ln(R^2 + T^2) + 2T = 2c_1; \quad 2c_1 = c$;

$\ln(R^2 + T^2) + 2T = c$

11. $\tan(x^2 + y^2)dy + x\,dx + y\,dy = 0$;

$$dy + \frac{x\,dx + y\,dy}{\tan(x^2 + y^2)} = 0$$

$d(x^2 + y^2) = 2x\,dx + 2y\,dy = 2(x\,dx + y\,dy)$

$dy + \cot(x^2 + y^2)(x\,dx + y\,dy) = 0$

$$dy + \frac{1}{2}\cot(x^2 + y^2)2\,d(x^2 + y^2)$$

$$y + \frac{1}{2}\ln\sin(x^2 + y^2) = c; \quad y = c - \frac{1}{2}\ln\sin(x^2 + y^2)$$

12. $(x^2 + y^3)^2 dy + 2x\,dx + 3y^2\,dy = 0$;

$$dy + \frac{2x\,dx + 3y^2\,dy}{(x^2 + y^3)^2} = 0$$

$dy + (x^2 + y^3)^{-2}(2x\,dx + 3y^2\,dy) = 0$

$dy + (x^2 + y^3)^{-2}d(x^2 + y^3) = 0$

$$y + \frac{(x^2 + y^3)^{-1}}{-1} = c; \quad y - \frac{1}{x^2 + y^3} = c$$

$y(x^2 + y^3) - 1 = c(x^2 + y^3)$;

$y(x^2 + y^3) = 1 + c(x^2 + y^3)$

13. $y\,dy - x\,dx + (y^2 - x^2)dx = 0$

$$\frac{y\,dy - x\,dx}{y^2 - x^2} + dx = 0$$

$$\frac{2y\,dy - 2x\,dx}{y^2 - x^2} + 2\,dx = 0$$

$$\ln(y^2 - x^2) + 2x = c$$

14. $e^{x+y}(dx+dy) + 4x\,dx = 0; \quad e^{x+y}d(x+y) + 4x\,dx = 0$

$e^{x+y} + 2x^2 = c$

15. $10x\,dy + 5y\,dx + 3y\,dy = 0$

$5(2x\,dy + y\,dx) + 3y\,dy = 0$; multiply by y

$5(2xy\,dy + y^2\,dx) + 3y^2\,dy = 0$;

$5d(xy^2) + 3y^2\,dy = 0$

$5xy^2 + y^3 = c$

16. $2(u\,dv + v\,du)\ln(uv) + 3u^3v\,du = 0$

$$2\left(\frac{u\,dv + v\,du}{uv}\right)\ln(uv) + u^2\,du = 0$$

$(\ln(uv))^2 + u^3 = c$

17. $2(x\,dy + y\,dx) + 3x^2\,dx = 0$

$$x\,dy + y\,dx + \frac{3}{2}x^2\,dx = 0$$

$$d(xy) - \frac{3}{2}x^2\,dx = 0$$

$$xy + \frac{1}{2}x^3 + c = 0$$

$$2xy + x^3 + c = 0$$

Substituting $x = 1$, $y = 2$; $4 + 1 + c = 0$;

$c = -5$

$2xy + x^3 = 5$

18. $t\,dt + s\,ds = 2(t^2 + s^2)dt$; $t = 1, s = 0$

$$\frac{t\,dt + s\,ds}{t^2 + s^2} = 2\,dt; \quad \frac{1}{2}\frac{2(t\,dt + s\,ds)}{t^2 + s^2} = 2\,dt$$

$$\frac{1}{2}\frac{d(t^2 + s^2)}{t^2 + s^2} = 2\,dt; \quad \frac{1}{2}\ln(t^2 + s^2) = 2t + c;$$

$s = 0, t = 1$

$\ln(t^2 + s^2) = 4t + c_1; \quad \ln(1) = 4 + c_1$;

therefore, $c_1 = -4$

$\ln(t^2 + s^2) = 4t - 4$

19. $y\,dx - x\,dy = y^3\,dx + y^2x\,dy$; $x = 2, y = 4$

$$\frac{y\,dx - x\,dy}{y^2} = y\,dx + x\,dy; \quad d\left(\frac{x}{y}\right) = d(xy)$$

$$\frac{x}{y} = xy + c; \quad x = 2, y = 4; \quad \frac{2}{4} = 2(4) + c;$$

$$c = -\frac{15}{2}$$

$$\frac{x}{y} = xy - \frac{15}{2}; \text{ multiply by } 2y$$

$$2x = 2xy^2 - 15y$$

20. $e^{x/y}(x\,dy - y\,dx) = y^4\,dy;\ x = 0, y = 2$

$$-e^{x/y}\left(\frac{y\,dx - x\,dy}{y^2}\right) = y^2\,dy;$$

$$-e^{x/y}d\left(\frac{x}{y}\right) = y^2\,dy$$

$$-e^{x/y} = \frac{1}{3}y^3 + c;\ x = 0, y = 2;$$

$$-e^0 = \frac{1}{3}(2^3) + c;$$

$$c = -\frac{11}{3}$$

$$-e^{x/y} = \frac{1}{3}y^3 - \frac{11}{3}$$

$$y^3 + 3e^{x/y} = 11$$

14.4 The Linear Differential Equation of the First Order

1. $dy + y\,dx = e^{-x}dx;\ P = 1,$

$Q = e^{-x}; e^{\int dx} = e^x$

$$ye^x = \int e^{-x}e^x dx + c$$

$$ye^x = \int dx + c;$$

$$ye^x = x + c$$
$$y = e^{-x}(x + c)$$

2. $dy + 3y\,dx = e^{-3x}dx;\ P = 3, Q = e^{-3x};$

$e^{\int 3dx} = e^{3x}$

$$ye^{3x} = \int e^{-3x}e^{3x}dx + c$$

$$ye^{3x} = \int e^0 dx + c = x + c;\ y = e^{-3x}(x + c)$$

3. $dy + 2y\,dx = e^{-4x}dx;\ P = 2, Q = e^{-4x};$

$e^{\int 2dx} = e^{2x}$

$$ye^{2x} = \int e^{-4x}e^{2x}dx$$

$$= -\frac{1}{2}\int e^{-2x}(-2\,dx)$$

$$= -\frac{1}{2}e^{-2x} + c$$

$$y = -\frac{1}{2}e^{-2x}e^{-2x} + ce^{-2x} = -\frac{1}{2}e^{-4x} + ce^{-2x}$$

4. $di + i\,dt = e^{-t}\cos t\,dt;\ P = 1, Q = e^{-t}\cos t;$

$e^{\int dt} = e^t$

$$ie^t = \int e^{-t}\cos t\,e^t dt + c = \int \cos t\,dt + c = \sin t + c$$

$$i = e^{-t}(\sin t + c)$$
$$ie^t = \sin t + c$$

5. $\dfrac{dy}{dx} - 2y = 4;\ dy - 2y\,dx = 4\,dx$

$P = -2, Q = 4;\ e^{\int -2dx} = e^{-2x}$

$$ye^{-2x} = \int 4e^{-2x}dx + c$$

$$ye^{-2x} = 4\left[-\frac{1}{2}\int e^{-2x}(-2\,dx)\right] + c$$

$$ye^{-2x} = -2e^{-2x} + c \quad \text{or} \quad y = -2 + ce^{2x}$$

6. $2\dfrac{dy}{dx} = 5 - 6y;\ 2\,dy = (5 - 6y)dx;\ dy = \frac{1}{2}(5 - 6y)dx$

$dy + 3y\,dx = \dfrac{5}{2}dx;\ P = 3, Q = \dfrac{5}{2};\ e^{\int 3\,dx} = e^{3x}$

$$ye^{3x} = \int \frac{5}{2}e^{3x}dx + c_1 = \frac{5}{2}\cdot\frac{1}{3}\int e^{3x}(3\,dx) + c_1$$

$$ye^{3x} = \frac{5}{6}e^{3x} + c_1;\ y = \frac{5}{6} + c_1 e^{-3x};\ 6c_1 = c$$

$$y = \frac{1}{6}(5 + 6c_1 e^{-3x}) = \frac{1}{6}(5 + ce^{-3x})$$

7. $x\,dy - y\,dx = 3x\,dx;\ dy - \left(\dfrac{y}{x}\right)dx = 3\,dx$

$P = -\dfrac{1}{x}, Q = 3;\ e^{-\int(dx/x)} = e^{-\ln x} = \dfrac{1}{x}$

$$dy + \left(-\frac{1}{x}\right)y\,dx = 3dx;\ y\left(\frac{1}{x}\right) = \int 3\frac{1}{x}dx + c$$

$$\frac{y}{x} = 3\ln x + c;\ y = x(3\ln x + c)$$

8. $x\,dy + 3y\,dx = dx;\ dy + \left(\dfrac{3}{x}\right)y\,dx = \left(\dfrac{1}{x}\right)dx$

$P = \dfrac{3}{x}, Q = \dfrac{1}{x};\ e^{\int(3\,dx/x)} = e^{3\ln x} = x^3$

$$yx^3 = \int\left(\frac{1}{x}\right)x^3 dx + c = \int x^2 dx + c = \frac{1}{3}x^3 + c_1$$

$$yx^3 - \frac{1}{3}x^3 = c_1;\ 3yx^3 - x^3 = 3c_1;\ 3c_1 = c$$

$$x^3(3y - 1) = c$$

9. $2x\,dy + y\,dx = 8x^3\,dx;\ dy + \dfrac{1}{2x}y\,dx = 4x^2\,dx$

$$P = \frac{1}{2x},\ Q = 4x^2$$

$$e^{\int (1/2)x\,dx} = e^{(1/2)\ln x} = e^{\ln x^{1/2}} = x^{1/2}$$

$$yx^{1/2} = \int 4x^2(x^{1/2})dx + c$$

$$= \int 4x^{5/2}\,dx + c = 4x^{7/2}\left(\frac{2}{7}\right) + c$$

$$= \frac{8}{7}x^{7/2} + c$$

$$y = \frac{8}{7}x^3 + \frac{c}{\sqrt{x}}$$

10. $3x\,dy - y\,dx = 9x\,dx;\ dy + \left(-\dfrac{1}{3x}\right)y\,dx = 3\,dx$

$$P = -\frac{1}{3x},\ Q = 3;$$

$$e^{(-1/3)\int (dx/x)} = e^{-(1/3)\ln x} = x^{-1/3}$$

$$yx^{-1/3} = \int 3x^{-1/3}\,dx + c_1;\ \frac{y}{x^{1/3}} = 3\frac{x^{2/3}}{\frac{2}{3}} + c_1$$

$$\frac{y}{x^{1/3}} = \frac{9}{2}x^{2/3} + c_1;\ y = \frac{9}{2}x^{2/3}x^{1/3} + c_1 x^{1/3}$$

$$2y = 9x + 2c_1 x^{1/3};\ 2c_1 = c;\ 2y = 9x + cx^{1/3}$$

11. $dr + r\cot\theta\,d\theta = d\theta;\ dr + \cot\theta\, r\,d\theta = d\theta$
$$P = \cot\theta,\ Q = 1,\ e^{\int \cot\theta\,d\theta} = e^{\ln\sin\theta} = \sin\theta$$

$$r\sin\theta = \int \sin\theta\,d\theta + c;\ r\sin\theta = -\cos\theta + c$$

$$r = -\frac{\cos\theta}{\sin\theta} + \frac{c}{\sin\theta} = -\cot\theta + c\csc\theta$$

12. $y' = x^2 y + 3x^2;\ \dfrac{dy}{dx} = x^2 y + 3x^2$

$$dy = x^2 y\,dx + 3x^2\,dx;\ dy - x^2 y\,dx = 3x^2\,dx$$
$$P = -x^2,\ Q = 3x^2;\ e^{-\int x^2\,dx} = e^{-(1/3)x^3}$$

$$ye^{-(1/3)x^3} = \int 3x^2 e^{-(1/3)x^3}\,dx$$

$$ye^{-x^3/3} = 3(-1)\int e^{-x^3/3}(-x^2)\,dx + c$$

$$= -3e^{-x^3/3} + c$$

$$y = -3 + ce^{x^3/3}$$

13. $$\sin x\frac{dy}{dx} = 1 - y\cos x$$

$$\sin x\,dy + y\cos x\,dx = dx$$

$$dy + y\frac{\cos x}{\sin x}\,dx = \frac{dx}{\sin x}$$

$$dy + \cot x\, y\,dx = \csc x\,dx$$

$$P = \cot x,\ Q = \csc x$$

$$e^{\int \cot x\,dx} = e^{\ln\sin x} = \sin x$$

$$y\sin x = \int \csc x\sin x\,dx$$

$$y\sin x = \int dx + c = x + c$$

$$y = \frac{x + c}{\sin x} = (x + c)\csc x$$

14. $\dfrac{dy}{dx} - \dfrac{y}{x} = \ln x;\ dy - \dfrac{1}{x}y\,dx = \ln x\,dx$

$$P = -\frac{1}{x},\ Q = \ln x;\ e^{-\int (dx/x)} = e^{-\ln x} = x^{-1}$$

$$yx^{-1} = \int \ln x\, x^{-1}\,dx + c;\ \frac{y}{x} = \int (\ln x)^1\frac{dx}{x} + c$$

$$\frac{y}{x} = \frac{1}{2}\ln^2 x + c;\ y = \frac{1}{2}x\ln^2 x + cx$$

$$y = \frac{1}{2}x(\ln^2 x + 2c);\ 2c = c_1, y = \frac{1}{2}x(\ln^2 x + c)$$

15. $y' + y = 3;\ \dfrac{dy}{dx} + y = 3;\ dy + y\,dx = 3\,dx$

$$P = 1, Q = 3;\ e^{\int dx} = e^x$$

$$ye^x = \int 3e^x\,dx + c = 3e^x + c;\ y = 3 + ce^{-x}$$

16. $y' + 2y = \sin x;\ \dfrac{dy}{dx} + 2y = \sin x;$

$$dy + 2y\,dx = \sin x\,dx$$

$$P = 2, Q = \sin x, e^{\int 2\,dx} = e^{2x}$$

$$ye^{2x} = \int e^{2x}\sin x\,dx = \frac{2}{5}e^{2x}\sin x - \frac{1}{5}e^{2x}\cos x + c$$

Integration by parts.

$$ye^{2x} = \frac{1}{5}e^{2x}(2\sin x - \cos x) + c;$$

$$y = \frac{1}{5}(2\sin x - \cos x) + ce^{-2x}$$

17. $ds = (te^{4t} + 4s)dt$; $ds - 4s\,dt = te^{4t}dt$;

$ds - 4s\,dt = te^{4t}dt$; $P = -4$; $Q = te^{4t}$;

$e^{\int -4dt} = e^{-4t}$

$se^{-4t} = \int te^{4t}e^{-4t}dt + c$; $se^{-4t} = \int t\,dt + c$;

$se^{-4t} = \dfrac{t^2}{2} + c$

$2s = e^{4t}(t^2 + c)$

18. $y' - 2y = 2e^{2x}$; $\dfrac{dy}{dx} - 2y = 2e^{2x}$;

$dy - 2y\,dx = 2e^{2x}dx$

$P = -2, Q = 2e^{2x}$; $e^{-\int 2\,dx} = e^{-2x}$

$ye^{-2x} = \int 2e^{2x}e^{-2x}dx + c = 2\int e^0 dx + c = 2x + c$

$y = e^{2x}(2x + c)$

19. $y' = x^3(1 - 4y)$; $\dfrac{dy}{dx} = x^3 - 4x^3y$;

$dy = x^3\,dx - 4x^3y\,dx$

$dy + 4x^3y\,dx = x^3\,dx$; $P = 4x^3, Q = x^3$;

$e^{4\int x^3 dx} = e^{x^4}$

$ye^{x^4} = \int x^3 e^{x^4}dx + c = \dfrac{1}{4}e^{x^4}4x^3\,dx + c$

$= \dfrac{1}{4}e^{x^4} + c$

$y = \dfrac{1}{4} + ce^{-x^4}$

20. $y' + y\tan x = -\sin x$; $\dfrac{dy}{dx} + \tan xy = -\sin x$

$dy + \tan xy\,dx = -\sin x\,dx$; $P = \tan x, Q = -\sin x$

$e^{\int \tan x\,dx} = e^{-\ln \cos x} = (\cos x)^{-1} = \sec x$

$y\sec x = \int -\sin x\sec x\,dx + c = -\int \tan x\,dx + c$

$y\sec x = \ln\cos x + c$

21. $x\dfrac{dy}{dx} = y + (x^2 - 1)^2$; $x\dfrac{dy}{dx} = y + x^4 - 2x^2 + 1$;

$dy = \dfrac{1}{x}y\,dx + \dfrac{1}{x}(x^4 - 2x^2 + 1)dx$;

$dy - \dfrac{1}{x}y\,dx = \left(x^3 - 2x + \dfrac{1}{x}\right)dx$;

$P = -\dfrac{1}{x}$; $Q = \left(x^3 - 2x + \dfrac{1}{x}\right)$;

$e^{\int(-1/x)dx} = e^{-\ln x} = e^{\ln x^{-1}} = \dfrac{1}{x}$

$\dfrac{y}{x} = \int\left(x^3 - 2x + \dfrac{1}{x}\right)\left(\dfrac{1}{x}\right)dx + c$

$\dfrac{y}{x} = \int(x^2 - 2 + x^{-2})dx + c = \dfrac{x^3}{3} - 2x - x^{-1} + c$

$y = \dfrac{x^4}{3} - 2x^2 - 1 + cx$; $3y = x^4 - 6x^2 - 3 + cx$

22. $dy = dt - \dfrac{y\,dt}{(1 + t^2)\tan^{-1}t}$

$dy + \dfrac{1}{(1 + t^2)\tan^{-1}t}\cdot y\,dt = dt$

$e^{\int \frac{1}{(1+t^2)\tan^{-1}t}dt} = \tan^{-1}t$

$\tan^{-1}t\,dy + \dfrac{y\,dt}{1 + t^2} = \tan^{-1}t\,dt$

$\int d(y\cdot\tan^{-1}t) = \int \tan^{-1}t\,dt$

$y\cdot\tan^{-1}t = t\cdot\tan^{-1}t - \dfrac{\ln(t^2 + 1)}{2} + c$

$y = t - \dfrac{\ln(t^2 + 1)}{2\tan^{-1}t} + \dfrac{c}{\tan^{-1}t}$

23. $x\,dy + (1 - 3x)y\,dx = 3x^2 e^{3x}dx$

$dy + \left(\dfrac{1 - 3x}{x}\right)y\,dx = 3xe^{3x}dx$

$P = \dfrac{1}{x} - 3$; $Q = 3xe^{3x}$

$e^{\int((1/x)-3)dx} = e^{\int(dx/x) - \int 3\,dx} = e^{\ln x - 3x}$

$= \dfrac{e^{\ln x}}{e^{3x}} = \dfrac{x}{e^{3x}}$

$y\dfrac{x}{e^{3x}} = \int 3xe^{3x}\dfrac{x}{e^{3x}}dx + c = 3\int x^2\,dx + c = x^3 + c$

$xy = e^{3x}(x^3 + c)$

24. $(1+x^2)dy + xy\,dx = x\,dx$

$$dy + \left(\frac{x}{1+x^2}\right)y\,dx = \frac{x}{1+x^2}dx$$

$$P = \frac{x}{1+x^2};\ Q = \frac{x}{1+x^2}$$

$$e^{(1/2)\int((2x\,dx)/(1+x^2))} = e^{(1/2)\ln(1+x^2)} = (1+x^2)^{1/2}$$
$$= \sqrt{1+x^2}$$

$$y\sqrt{1+x^2} = \int \frac{x}{1+x^2}(1+x^2)^{1/2}dx + c$$

$$y\sqrt{1+x^2} = \frac{1}{2}\int (1+x^2)^{-1/2}2x\,dx + c$$

$$= \frac{1}{2}\frac{(1+x^2)^{1/2}}{\frac{1}{2}} + c$$

$$y\sqrt{1+x^2} = \sqrt{1+x^2} + c$$
$$y\sqrt{1+x^2} - \sqrt{1+x^2} = c;\ \sqrt{1+x^2}(y-1) = c$$

25. $y' = 2(1-y)$; solve by separation of variables.

$$\frac{dy}{1-y} = 2\,dx;\ -\ln(1-y) = 2x - \ln c;$$

$$\ln\frac{c}{1-y} = 2x;$$

$$c = (1-y)e^{2x};\ 1-y = ce^{-2x};\ y = 1 - ce^{-2x}$$

Solve as a first order equation.

$$dy = 2\,dx - 2y\,dx;\ dy + 2y\,dx = 2\,dx$$
$$ye^{\int 2dx} = \int e^{\int 2dx}dx;\ ye^{2x} = \int 2e^{2x}dx$$
$$ye^{2x} = e^{2x} + c;\ y = 1 + ce^{-2x}$$

26. $x\,dy = (2x - y)dx;\ x\,dy = 2x\,dx - y\,dx$

(a) $x\,dy + y\,dx = 2x\,dx$; integrable combinations.

$$\int d(xy) = \int 2x\,dx;\ xy = x^2 + c$$

(b) $x\,dy + y\,dx = 2x\,dx$

$$dy = \left(\frac{1}{x}\right)y\,dx = 2\,dx;\ \text{linear differential eqn. of}$$
the first order.

$$P = \frac{1}{x}, Q = 2;\ e^{\int(dx/x)} = e^{\ln x} = x$$

$$yx = \int 2x\,dx + c;\ xy = x^2 + c$$

27. $\dfrac{dy}{dx} + 2y = e^{-x};\ x = 0, y = 1$

$$dy + 2y\,dx = e^{-x}dx;\ P = 2, Q = e^{-x};\ e^{\int 2\,dx} = e^{2x}$$

$$ye^{2x} = \int e^{-x}e^{2x}dx + c = \int e^x dx + c = e^x + c;$$

$$x = 0, y = 1$$
$$1(e^0) = e^0 + c;\ \text{therefore},\ c = 0$$
$$ye^{2x} = e^x;\ y = e^{-x};\ \text{particular solution}$$

28. $dq - 4q\,du = 2\,du;\ q = 2, u = 0$

$$qe^{-4u} = \int 2e^{-4u}du + c;\ P = -4, Q = 2;$$

$$e^{-4\int du} = e^{-4u}$$

$$qe^{-4u} = 2\left(-\frac{1}{4}\right)\int e^{-4u}(-4)\,du + c$$

$$= -\frac{1}{2}e^{-4u} + c$$

$$q = 2, u = 0$$

$$2(e^0) = -\frac{1}{2}(e^0) + c;\ \text{therefore},\ c = \frac{5}{2}$$

$$qe^{-4u} = -\frac{1}{2}e^{-4u} + \frac{5}{2};$$

$$q = -\frac{1}{2} + \frac{5}{2}e^{4u} = \frac{1}{2}(5e^{4u} - 1)$$

29. $\dfrac{dy}{dx} + 2y\cot x = 4\cos x;\ x = \dfrac{\pi}{2}, y = \dfrac{1}{3};$

$$dy + 2y\cot x\,dx = 4\cos x\,dx;\ P = 2\cot x;$$

$$Q = 4\cos x$$

$$e^{\int P dx} = e^{\int 2\cot x\,dx} = 2^{2\ln|\sin x|} = e^{\ln|\sin x|^2}$$

$$= |\sin x|^2$$

$$y(\sin x)^2 = \int 4\cos x(\sin x)^2 dx + c = \frac{4(\sin x)^3}{3} + c$$

$$y = \frac{4}{3}\sin x + c(\csc^2 x)$$

$$x = \frac{\pi}{2}\ \text{when}\ y = \frac{1}{3};\ \frac{1}{3} = \frac{4}{3}\sin\frac{\pi}{2} + c;\ c = -1$$

$$y = \frac{4}{3}\sin x - \csc^2 x$$

30. $y'\sqrt{x} + \dfrac{1}{2}y = e^{\sqrt{x}};\ x = 1, y = 3$

$$\frac{dy}{dx}\sqrt{x} + \frac{1}{2}y = e^{\sqrt{x}};\ \sqrt{x}\,dy + \frac{1}{2}y\,dx = e^{\sqrt{x}}dx$$

$$dy = \frac{1}{2\sqrt{x}}y\,dx = \frac{e^{\sqrt{x}}}{\sqrt{x}}dx$$

$$P = \frac{1}{2\sqrt{x}}, Q = \frac{e^{\sqrt{x}}}{\sqrt{x}};\ e^{(1/2)\int(dx/\sqrt{x})} = e^{\sqrt{x}}$$

$$ye^{\sqrt{x}} = \int \frac{e^{\sqrt{x}}}{\sqrt{x}} e^{\sqrt{x}} dx = \int e^{2\sqrt{x}} \frac{dx}{\sqrt{x}} + c$$

$x = 1, y = 3;\ ye^{\sqrt{x}} = e^{2\sqrt{x}} + c$
$3e^1 = e^2 + c$; therefore, $c = 3e - e^2$
$ye^{\sqrt{x}} = e^{2\sqrt{x}} + 3e - e^2;\ y = e^{\sqrt{x}} + (3e - e^2)e^{-\sqrt{x}}$

31. $\sin x \dfrac{dy}{dx} + y = \tan x;\ x = \dfrac{\pi}{4}, y = 0$

$\sin x\, dy + y\, dx = \tan x\, dx;\ dy + \dfrac{y\, dx}{\sin x} = \dfrac{\tan x\, dx}{\sin x}$

$dy + \csc x\, y\, dx = \sec x\, dx$
$P = \csc x, Q = \sec x;\ e^{\int \csc x\, dx} = e^{\ln(\csc x - \cot x)}$
$\quad = \csc x - \cot x$

$y(\csc x - \cot x)$

$= \displaystyle\int \sec x(\csc x - \cot x\, dx)$

$= \displaystyle\int \sec x \csc x\, dx - \int \sec x \cot x\, dx$

$= \displaystyle\int \sec x \csc x\, dx - \int \frac{1}{\cos x}\frac{\cos x}{\sin x}dx$

$= \displaystyle\int \sec x \csc x\, dx - \int \csc x\, dx$

$= \displaystyle\int \sec x \csc x\, dx - \ln(\csc x - \cot x) + c$

Simplify: $\sec x \csc x = \dfrac{1}{\cos x}\cdot\dfrac{1}{\sin x}$

$\sin x = \sqrt{\dfrac{1 - \cos 2x}{2}};\ \cos x = \sqrt{\dfrac{1 + \cos 2x}{2}}$

By half-angle formulas.

Therefore,

$\sec x \csc x = \dfrac{\sqrt{2}}{\sqrt{1 + \cos 2x}}\cdot\dfrac{\sqrt{2}}{\sqrt{1 - \cos 2x}}$

$\quad = \dfrac{2}{\sqrt{1 - \cos^2 2x}} = \dfrac{2}{\sin 2x} = 2\csc 2x$

Therefore,

$\displaystyle\int \sec x \csc x\, dx = \int \csc 2x(2\, dx)$

$\quad\quad\quad\quad\quad = \ln(\csc 2x - \cot 2x)$

$y(\csc x - \cot x)$

$\quad = \displaystyle\int \sec x \csc x\, dx - \ln(\csc x - \cot x) + c$

$y(\csc x - \cot x)$
$\quad = \ln(\csc 2x - \cot 2x) - \ln(\csc x - \cot x) + c$

$x = \dfrac{\pi}{4}, y = 0$

Therefore, $\ln(\csc 2x - \cot 2x)$ at $x = \dfrac{\pi}{4}$

$\ln\left(\csc \dfrac{\pi}{2} - \cot \dfrac{\pi}{2}\right) = \ln(1 - 0) = \ln 1 = 0$

$0(\sqrt{2} - 1) = \ln(1) - \ln(\sqrt{2} - 1) + c$

Therefore, $c = \ln(\sqrt{2} - 1)$

$y(\csc x - \cot x) = \ln(\csc 2x - \cot 2x)$
$\quad\quad\quad\quad\quad - \ln(\csc x - \cot x) + \ln(\sqrt{2} - 1)$

$y(\csc x - \cot x) = \ln\left[\dfrac{(\sqrt{2} - 1)(\csc 2x - \cot 2x)}{(\csc x - \cot x)}\right]$

32. $f(x)dy + 2yf'(x)dx = f(x)f'(x)dx;$
$f(x) = -1, y = 3$

$dy + \dfrac{2f'(x)}{f(x)}dx = f'(x)dx;$

$P = 2\dfrac{f'(x)}{f(x)}, Q = f'(x)$

$e^{2\int f'(x)(dx/f(x))} = e^{2\ln f(x)} = [f(x)]^2$

$[f(x)]^2 y = \displaystyle\int f'(x)[f(x)]^2\, dx = \int [f(x)]^2 f'(x)dx$

$[f(x)]^2 y = \dfrac{1}{3}[f(x)] + c$

$(-1)^2(3) = \dfrac{1}{3}(-1)^3 + c$; therefore, $c = \dfrac{10}{3}$

$[f(x)]^2 y = \dfrac{1}{3}[f(x)]^3 + \dfrac{10}{3}$

$y = \dfrac{1}{3}\dfrac{[f(x)]^3}{[f(x)]^2} + \dfrac{10}{3[f(x)]^2};\ 3y = f(x) + 10[f(x)]^{-2}$

14.5   Elementary Applications

1. $\dfrac{dy}{dx} = \dfrac{2x}{y};\ y\, dy = 2x\, dx;\ \dfrac{1}{2}y^2 = x^2 + c$

Substitute $x = 2, y = 3;\ \dfrac{1}{2}(9) = 4 + c;\ c = 0.5$

$\dfrac{1}{2}y^2 = x^2 + 0.5;\ y^2 = 2x^2 + 1;\ y = \pm\sqrt{2x^2 + 1}$

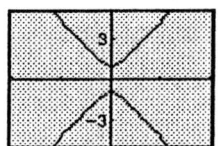

2. $\dfrac{dy}{dx} = -\dfrac{y}{(x+y)}$; passes through $(-1,3)$

$(x+y)dy = -y\,dx$

$x\,dy + y\,dy = -y\,dx$

$x\,dy + y\,dx + y\,dy = 0$

$d(xy) + y\,dy = 0$

$xy + \dfrac{1}{2}y^2 = c$

Substitute $x = -1, y = 3$

$(-1)(3) + \dfrac{1}{2}(3)^2 = c$

Therefore, $c = \dfrac{3}{2}$

$xy + \dfrac{1}{2}y^2 = \dfrac{3}{2}$

$2xy + y^2 = 3$

$y^2 + 2xy = 3$

$y^2 + (2x)y + x^2 = 3 + x^2$

$(y+x)^2 = 3 + x^2$

$y = -x \pm \sqrt{3 + x^2}$

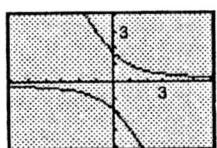

3. $\dfrac{dy}{dx} = y + x$; passes through $(0,1)$

$dy = (y+x)dx = y\,dx + x\,dx$

$dy - y\,dx = x\,dx$

$P = -1, Q = x, e^{-\int dx} = e^{-x}$

$ye^{-x} = \int xe^{-x}\,dx$

$u = x, du = dx; dv = e^{-x}dx, v = -e^{-x}$

$ye^{-x} = -xe^{-x} + \int e^{-x}\,dx = -xe^{-x} - e^{-x} + c$

$y = -x - 1 + ce^{x}; \ (0,1)$

$1 = -0 - 1 + ce^0; \ 2 = c$; therefore, $c = 2$

$y = -x - 1 + 2e^{x} = 2e^{x} - x - 1$

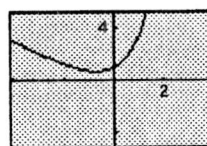

4. $\dfrac{dy}{dx} = -2y + e^{-x}$; passes through $(0, 2)$

$dy = -2y\,dx + e^{-x}dx; \ dy + 2y\,dx = e^{-x}dx$

$P = 2, Q = e^{-x}; e^{2\int dx} = e^{2x}$

$ye^{2x} = \int e^{-x}e^{2x}\,dx = \int e^{x}\,dx = e^{x} + c, (0,2)$

$2(e^0) = e^0 + c$; therefore, $c = 1$

$ye^{2x} = e^{x} + 1; \ y = e^{-x} + e^{-2x}$

5. See Example 2; $\dfrac{dy}{dx} = ce^{x}; \ y = ce^{x}; \ c = \dfrac{y}{e^{x}}$

Substitute for c in the equation for the derivative.

$\dfrac{dy}{dx} = \dfrac{y}{e^{x}}e^{x} = y; \ \dfrac{dy}{dx}\bigg|_{OT} = -\dfrac{1}{y}; \ y\,dy = -dx$

Integrating, $\dfrac{y^2}{2} = -x + \dfrac{c}{2}; \ y^2 = c - 2x$

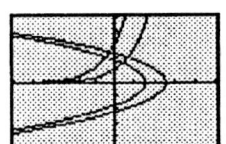

6. $y = cx^3; \ \dfrac{dy}{dx} = 3cx^2$; therefore, $c = \dfrac{y}{x^3}$

Substitute into $\dfrac{dy}{dx} = 3x^2\left(\dfrac{y}{x^3}\right) = \dfrac{3y}{x}$

$\dfrac{dy}{dx} = \dfrac{3y}{x}; \ \dfrac{dy}{dx}\bigg|_{OT} = -\dfrac{x}{3y}$

$3y\,dy = -x\,dx; \ \dfrac{3}{2}y^2 = -\dfrac{1}{2}x^2 + c$

$3y^2 = -x^2 + 2c; \ 2c = c_1; \ 3y^2 + x^2 = c$; ellipses

7. $y = c(\sec x + \tan x); \dfrac{dy}{dx} = c(\sec x \tan x + \sec^2 x)$

Therefore, $c = \dfrac{y}{\sec x + \tan x}$

Substitute into $\dfrac{dy}{dx}$.

$\dfrac{dy}{dx} = \dfrac{y}{\sec x + \tan x}(\sec x \tan x + \sec^2 x)$

$= y \sec x \dfrac{(\tan x + \sec x)}{(\sec x + \tan x)} = y \sec x$

$\left.\dfrac{dy}{dx}\right|_{OT} = -\dfrac{1}{y \sec x} = -\dfrac{\cos x}{y}$

$y\, dy = -\cos x\, dx; \dfrac{1}{2}y^2 = -\sin x + c_1$

$y^2 = -2\sin x + 2c_1; (2c_1 = c); y^2 = c - 2\sin x$

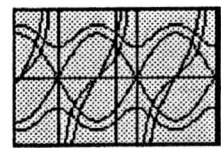

8. Family of circles, centered at origin.

$x^2 + y^2 = c$

$y = \sqrt{c - x^2}$

$\dfrac{dy}{dx} = \dfrac{1}{2}(c - x^2)^{-1/2}(-2x)$

$\dfrac{dy}{dx} = \dfrac{-x}{\sqrt{c - x^2}}$

$\left.\dfrac{dy}{dx}\right|_{OT} = \dfrac{\sqrt{c - x^2}}{x}$

Substitute:

$\dfrac{dy}{dx} = \dfrac{\sqrt{x^2 + y^2 - x^2}}{x} = \dfrac{y}{x}; \dfrac{dy}{y} = \dfrac{dx}{x};$

$\ln y = \ln x + \ln c$

$y = cx$

OR:

$x^2 + y^2 = c; 2x + 2y\dfrac{dy}{dx} = 0; \dfrac{dy}{dx} = -\dfrac{x}{y};$

$\left.\dfrac{dy}{dx}\right|_{OT} = \dfrac{y}{x}$

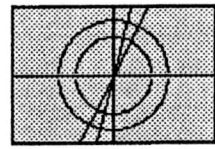

9. See Example 3; $N = N_0 e^{kt}$

Use the condition that half this isotope decays in 40.0 s.

$N = \dfrac{N_0}{2}$ when $t = 40.0$ s. $\dfrac{N_0}{2} = N_0 e^{40.0k};$

$0.5 = e^{40.0k}; 0.5^{1/40.0} = e^k; 0.5^{0.025} = e^k;$
$N = N_0(0.5)^{0.025}$

Evaluating when $t = 60.0$ s,

$N = N_0(0.5)^{0.025(60.0)} = N_0(0.5)^{1.50} = 0.354 N_0$

Therefore, 35.4% remains after 60.0 s.

10. Let N be amount at any time t N

Let N_0 be original amount 0 N_0

$N = N_0(0.90)^{t/246}$ 246 $0.10 N_0$ Decays

                                      246 $0.9 N_0$ Remains

For $N = \dfrac{1}{2}N_0$

$\dfrac{1}{2}N_0(0.9)^{t/246}; 0.5 = 0.9^{t/246}; \ln 0.5 = \dfrac{t}{246}\ln 0.9$

$t = 246\dfrac{\ln 0.5}{\ln 0.9} = 1620$ years

1620 years for half of original amount of radium 226 to disintegrate.

11. $N = N_0 e^{kt}$

$0.10 N_0 = N_0 e^{k(12.7)} \Rightarrow k = \dfrac{\ln 0.10}{12.7}$

$N = N_0 e^{\frac{\ln 0.10}{12.7}\cdot t}$ @ $t = t_{\text{half-life}}$, $N = \dfrac{N_0}{2}$

$\dfrac{N_0}{2} = N_0 e^{\frac{\ln 0.10}{12.7}}\cdot t_{\text{half-life}}$

$t_{\text{half-life}} = 3.82$ days

12. $\dfrac{dN}{dt} = r - kN$

$\dfrac{dN}{r - kN} = dt$

$-\dfrac{1}{k}\ln(r - kN) = t + c$

$N = 0$ for $t = 0$

$c = -\dfrac{1}{k}\ln r$

$-\dfrac{1}{k}\ln(r - kN) = t - \dfrac{1}{k}\ln r$

$\ln\dfrac{r - kN}{r} = -kt$

$r - kN = re^{-kt}$

$N = \dfrac{r}{k}(1 - e^{-kt})$

13. $r\dfrac{dS}{dr} = 2(a - S)$; $r\,dS = 2a\,dr - 2S\,dr$;

$dS + \dfrac{2}{r}S\,dx = 2a\dfrac{dr}{r}$; $y = S$, $x = r$, $P = \dfrac{2}{r}$,

$Q = \dfrac{2a}{r}$

$e^{\int P\,dx} = e^{\int (2/r)\,dr} = e^{2\ln x} = e^{\ln r^2} = r^2$

$Sr^2 = \int \left(2a\dfrac{dr}{r}\right) - 2a\ln r + c = \int 2ar\,dr$

$Sr^2 = ar^2 + c$; $S = a + \dfrac{c}{r^2}$

14. $v\dfrac{dv}{dr} = -\dfrac{GM}{r^2}$; $v = 0, r = r_0$

$\int v\,dv = -GM\int\dfrac{dr}{r^2} = -GM\int r^{-2}\,dr$

$\dfrac{1}{2}v^2 = -GM\dfrac{r^{-1}}{-1} + c = \dfrac{GM}{r} + c$; $v = 0, r = r_0$

$0 = \dfrac{GM}{r_0} + c$; therefore, $c = -\dfrac{GM}{r_0}$

$\dfrac{1}{2}v^2 = \dfrac{GM}{r} - \dfrac{GM}{r_0} = GM\left(\dfrac{r_0 - r}{r_0 r}\right)$

$v = \sqrt{2GM\left(\dfrac{r_0 - r}{r_0 r}\right)}$

15. $\dfrac{dM}{dt} = KM$

$\dfrac{dM}{M} = K\,dt$; $\ln M = Kt + c$; $M = 5250, t = 0$

$\ln 5250 = c$; $\ln M = Kt + \ln 5250$;

$M = 5460, t = 2.00$

$\ln 5460 = 2.00K + \ln 5250$; $K = \ln\dfrac{\dfrac{5460}{5250}}{2} = 0.0196$

$\ln M = 0.0196t + \ln 5250$; $\ln\dfrac{M}{5250} = 0.0196t$;

$\dfrac{M}{5250} = e^{0.0196t}$; $M = 5250e^{0.0196t}$

16. $\dfrac{dP}{dx} = e^{-x^2} - 2Px$; $P = 0, x = 0$

$dP = (e^{-x^2} - 2Px)\,dx$; first order differential equation.

$dP = e^{-x^2}\,dx - 2Px\,dx$; $dP + 2xP\,dx = e^{-x^2}\,dx$

$e^{2\int x\,dx} = e^{x^2}$

$Pe^{x^2} = \int e^{-x^2}e^{x^2}\,dx + c = \int dx + c = x + c$

$P = xe^{-x^2} + ce^{-x^2}$; $P = 0, x = 0$

$0 = 0 + c(1)$; therefore, $c = 0$

Therefore, $P = xe^{-x^2}$

17. $\dfrac{dT}{dt} = k(T - 80.0)$; $\dfrac{dT}{T - 80.0} = k\,dt$;

$\ln(T - 80.0) = kt + \ln c$; $T = 80.0 + ce^{kt}$;

$200 = 80.0 + c$; $T = 80.0 + 120e^{kt}$;

$140 = 80.0 + 120e^{5k}$; $e^{5k} = 0.5$; $e^k = (0.5)^{1/5}$;

$T = 80.0 + 120(0.5)^{t/5}$; $100 = 80.0 + 120(0.5)^{t/5}$;

$0.5^{t/5} = \dfrac{1}{6.0}$; $t = 5\dfrac{\ln\frac{1}{6.0}}{\ln\frac{1}{2}} = 13$ min

18. $\dfrac{dT}{dt} = k(T - T_M)$; T is temp. of object, T_M is temp of medium $= 20°$

$\dfrac{dT}{dt} = k(T - 20°)$; $\int\dfrac{dT}{T - 20} = \int k\,dt$

$\ln(T - 20) = kt + c$; $T = 100°\,\text{C}$ at $t = 0$

$\ln(100 - 20) = 0 + c$; therefore, $c = \ln 80$

Therefore, $\ln(T - 20) = kt + \ln 80$; $T = 50°\,\text{C}$ at $t = 10$

$\ln\dfrac{T - 20}{80} = kt$; $\dfrac{T - 20}{80} = e^{kt}$; $T = 20 + 80e^{kt}$

$50 = 20 + 80e^{10k}$; $80e^{10k} = 30$; $e^{10k} = \dfrac{3}{8}$

$e^k = \left(\dfrac{3}{8}\right)^{1/10} = (0.375)^{1/10}$

Therefore, $T = 20 + 80(0.375)^{t/10}$

19. $\dfrac{dP}{dt} = kP$; $\dfrac{dP}{P} = k\,dt$; $k =$ interest rate

$\ln P = kt + c$; $t = 0, P = P_0$, the original investment.

$\ln P_0 = c$; $\ln P = kt + \ln P_0$; $\ln\dfrac{P}{P_0} = kt$; $\dfrac{P}{P_0} = e^{kt}$

$P = P_0e^{0.04t}$; \$1000 invested for $t = 1$ year at $k = 4\%$

$P = 1000e^{0.04(1)} = \$1040.81$

20. $\dfrac{dI}{dy} = kI \Rightarrow I = I_0e^{ky}$

$0.5I_0 = I_0e^{k(15)} \Rightarrow k = \dfrac{\ln 0.5}{15}$

$I = I_0e^{\frac{\ln 0.5}{15}\cdot y}$

$0.15I_0 = I_0e^{\frac{\ln 0.5}{15}\cdot y}$

$y = 41.1$ ft

21. See Example 4; $i = \dfrac{E}{R}(1 - e^{-(R/L)t})$

$\lim_{i\to\infty}\left(\dfrac{E}{R} - \dfrac{E}{R}e^{-(R/L)t}\right) = \dfrac{E}{R} - 0 = \dfrac{E}{R}$

22. $L = 2.0$ H, $R = 30\ \Omega$, $i = 0.020$ A when
$t = 0, E = 0$

$$L\frac{di}{dt} + Ri + \frac{q}{c} = E;\ 2.0\frac{di}{dt} + 30i = E$$

$$2.0\,di + 30i\,dt = E\,dt;\ di = 15i\,dt = \frac{E}{2}dt$$

$$P = 15, Q = \frac{E}{2};\ e^{\int 15\,dt} = e^{15t}$$

$$ie^{15t} = \int \frac{E}{2}e^{15t}dt = \frac{E}{2}\frac{1}{15}e^{15t} + c$$

$$i = \frac{E}{30} + ce^{-15t};\ t = 0, E = 0\text{ V}, i = 0.020\text{ A}$$
$$0.020 = 0 + ce^0;\ \text{therefore, } c = 0.020;\ i = 0.020e^{-15t}$$

23. $V = E\sin\omega t;\ t = 0, i = 0$

$$L\frac{di}{dt} + Ri = E\sin\omega t;\ L\,di + Ri\,dt = E\sin\omega t\,dt$$

$$di + \frac{R}{L}i\,dt = \frac{E}{L}\sin\omega t\,dt$$

$$P = \frac{R}{L};\ Q = \frac{E}{L}\sin\omega t;\ e^{\int (R/L)dt} = e^{Rt/L}$$

$$ie^{Rt/L} = \int \frac{E}{L}\sin\omega t\,e^{Rt/L}dt = \frac{E}{L}\int e^{Rt/L}\sin\omega t\,dt$$

Integration by parts:

$$ie^{Rt/L} = \frac{E}{L}\frac{L^2\omega^2}{L^2\omega^2 + R^2}$$
$$\times \left(\frac{R}{L\omega^2}e^{Rt/L}\sin\omega t - \frac{1}{\omega}e^{Rt/L}\cos\omega t + c\right)$$

$$0 = \frac{EL\omega^2}{L^2\omega^2 + R^2}\left(0 - \frac{1}{\omega} + c\right) = -\frac{1}{\omega} + c$$

Therefore, $c = \dfrac{1}{\omega}$

$$ie^{Rt/L} = \frac{EL\omega^2}{R^2 + L^2\omega^2}$$
$$\times \left(\frac{R}{L\omega^2}e^{Rt/L}\sin\omega t - \frac{1}{\omega}e^{Rt/L}\cos\omega t + \frac{1}{\omega}\right)$$

$$ie^{Rt/L}$$
$$= \frac{E}{R^2 + L^2\omega^2}(Re^{Rt/L}\sin\omega t - L\omega e^{Rt/L}\cos\omega t + L\omega)$$

$$i = \frac{E}{R^2 + L^2\omega^2}(R\sin\omega t - L\omega\cos\omega t + L\omega e^{-Rt/L})$$

24. $R = 2.0\ \Omega;\ L = 100 - t;\ E = 4.0$ V; $0 \le t \le 100$ s

$$(100 - t)\frac{di}{dt} + 2.0i = 4.0;\ t = 0, i = 0$$

$$(100 - t)di + 2.0i\,dt = 4.0\,dt$$

$$di + \frac{2.0}{(100 - t)}i\,dt = \frac{4.0}{(100 - t)}dt$$

$$P = \frac{2.0}{100 - t}, Q = \frac{4.0}{100 - t}$$

$$e^{\int (2.0\,dt/(100-t))} = e^{-2\ln(100-t)} = (100 - t)^{-2}$$

$$i(100 - t)^{-2} = \int \frac{4.0}{100 - t}(100 - t)^{-2}dt$$

$$i(100 - t)^{-2} = 4.0\int (100 - t)^{-3}dt$$

$$i(100 - t)^{-2} = -4.0\int (100 - t)^{-3}(-dt)$$

$$i(100 - t)^{-2} = -4.0\frac{(100 - t)^{-2}}{-2} + c$$

$$= 2.0(100 - t)^{-2} + c$$
$$i = 2.0 + c(100 - t)^2;\ t = 0, i = 0$$
$$0 = 2.0 + c(100)^2;\ \text{therefore, } c = -2 \times 10^{-4}$$
$$\text{Therefore, } i = 2.0 - 2.0 \times 10^{-4}(100 - t)^2$$

25. $Ri + \dfrac{q}{C} = 0;\ i = \dfrac{dq}{dt};\ R\dfrac{dq}{dt} + \dfrac{q}{C} = 0;$

$$\frac{dq}{q} + \frac{1}{RC}dt = 0$$

Integrating, $\ln q = -\dfrac{1}{RC}t + c$

Let $q = q_0$ when $t = 0;\ \ln q_0 = c;$

$$\ln q = -\frac{1}{RC}t + \ln q_0;\ \ln q - \ln q_0 = -\frac{1}{RC}t;$$

$$\ln \frac{q}{q_0} = -\frac{1}{RC}t;\ \frac{q}{q_0} = e^{(-1/RC)t};\ q = q_0e^{-t/RC}$$

26. $C = 4.0\ \mu\text{F}, R = 450\ \Omega, E = 20.0$ mV
$t = 0, q = 20.0$ nC $= 20.0 \times 10^{-9}$ C

$$450i + \frac{q}{4.6 \times 10^{-6}} = 20.0 \times 10^{-3}$$

$$450\frac{dq}{dt} + \frac{10^6}{4.0}q = 20.0 \times 10^{-3}$$

$$450\frac{dq}{dt} + \frac{10^6}{4.0}q\,dt = 20.0 \times 10^{-3}dt$$

$$dq + \frac{10^6}{1800}q\,dt = \frac{2.0}{45} \times 10^{-3}dt$$

$$P = \frac{10^6}{1800}, Q = \frac{2.0 \times 10^{-3}}{45};$$

$$e^{\int (10^6/1800)dt} = e^{(10^6/1800)t}$$

$$qe^{(10^6/1800)t} = \int \frac{2.0 \times 10^{-3}}{45}e^{10^6t/1800}dt$$

$$qe^{10^6t/1800} = \frac{2.0 \times 10^{-3}}{45}\frac{1800}{10^6}\int e^{10^6t/1800}\frac{10^6}{1800}dt$$

$$qe^{10^6t/1800} = 80 \times 10^{-9}e^{10^6t/1800} + c$$
$$q = 80 \times 10^{-9} + ce^{-10^6t/1800}$$

$20.0 \times 10^{-9} = 80 \times 10^{-9} + c$; therefore,
$c = -60 \times 10^{-9}$ C
$q = 80 \times 10^{-9} - 60 \times 10^{-9}e^{-10^6 t/1800}$
$q = 80 \times 10^{-9} - 60 \times 10^{-9}(e^{-10^6(0.01)/1800})$
$\quad = 0.000\ 000\ 08$ C
$t = 0.010$ s, $q = 80 \times 10^{-9} = 80$ nC

27. 100 gal. of brine, 30 lb of salt when $t = 0$
5.0 gal of water enters tank and 5.0 gal of brine
leaves.

Let x be number of lb of salt in tank after time
t min. Then $\frac{x}{100}$ lb of salt in each gallon of brine
$\left(\frac{x}{100}\right.$ lb per gal$\left.\right)$. After time t, 5.0 dt gal of mixture
leave tank. Each of these gal of mixture contains
$\frac{x}{100}(5.0\ dt)$ lb salt.

Therefore, $-dx = \dfrac{5.0x}{100}dt$ where $-dx$ is amount of
salt leaving.

$\dfrac{dx}{x} = -\dfrac{dt}{20}$; $\ln x = -\dfrac{1}{20}t + c$; $t = 0, x = 30$ lb

$\ln 30 = 0 + c$; therefore, $c = \ln 30$

$\ln x = -\dfrac{1}{20}t + \ln 30$; $\ln \dfrac{x}{30} = -\dfrac{1}{20}t$; $\dfrac{x}{30} = e^{-t/20}$

$x = 30e^{-t/20}$ after 20 min
$x = 30e^{-1} = 11$ lb

28. Let x = number of pounds of salt after t min.

$dx = \left(1 - \dfrac{x}{100}\right)5.0\ dt$

$20\ dx = (100 - x)dt$

$\dfrac{20\ dx}{100 - x} = dt$

$-20\ln(100 - x) = t + c$
$x = 30$ lb for $t = 0$
$c = -20\ln 70$
$-20\ln(100 - x) = t - 20\ln 70$

$20\ln \dfrac{70}{100 - x} = t$

$\dfrac{70}{100 - x} = e^{t/20}$

$x = 100 - 70e^{-t/20}$
For $t = 20$ min, $x = 100 - 70e^{-1} = 74.2$ lb

29. $\dfrac{dv}{dt} = 32 - v$; $\dfrac{dv}{32 - v} = dt$; $\dfrac{-dv}{32 - v} = dt$

Integrating, $(-1)\ln(32 - v) = t - \ln c$;

$\ln \dfrac{32 - v}{c} = -t$;

$32 - v = ce^{-t}$; $-v = -32 + ce^{-t}$; $v = 32 - ce^{-t}$

Starting from rest means $v = 0$ when $t = 0$;
$0 = 32 - c$; $c = 32$; $v = 32 - 32e^{-t} = 32(1 - e^{-t})$

$\lim\limits_{t\to\infty} 32(1 - e^{-t}) = 32$

30. $a = \dfrac{dv}{dt} = -40\sqrt{v}$; bullet will stop when $v = 0$

$t = 0, v = 950$ ft/s

$\dfrac{dv}{\sqrt{v}} = -40dt$; $\displaystyle\int v^{-1/2}dv = -40t + c$;

$2v^{1/2} = -40t + c$

$2\sqrt{950} = c$; $2\sqrt{v} = -40t + 2\sqrt{950}$;
$\sqrt{v} = -20t + \sqrt{950}$
$v = (-20t + \sqrt{950})^2$; stops when $v = 0$
$(-20t + \sqrt{950}) = 0$; $20t = \sqrt{950}$; $t = 1.5$ s

31. Mass of boat = 10 slugs
Force driving boat = 20 lb;
water slows down by $2v$

$10\dfrac{dv}{dt} = 20 - 2v$

$dv = 2\ dt - \dfrac{1}{5}v\ dt$

$dv + \dfrac{1}{5}v\ dt = 2\ dt$

$P = \dfrac{1}{5}, Q = 2$

$e^{\int 1/5 dt} = e^{t/5}$

$ve^{t/5} = 2\displaystyle\int e^{t/5}dt$

$\quad = 10\displaystyle\int e^{t/5}\left(\dfrac{1}{5}dt\right)$

$ve^{t/5} = 10e^{t/5} + c$

at $t = 0, v = 8.0$ mi/h = 11.733 ft/s

$11.733e^{0/5} = 10e^{0/5} + c, c = 1.733$
$ve^{t/5} = 10e^{t/5} + 1.733$
at $t = 3.0$ min = 180 sec
$ve^{180/5} = 10e^{180/5} + 1.733$
$v = 10$ ft/s

32. $5.00\dfrac{dv}{dt} = 160 - 0.5v^2$

$10.0dv = (320 - v^2)dt$

$\dfrac{10.0dv}{v^2 - 320} = -dt$

Integrate using Formula 9 (p. A-19).

$\dfrac{10.0}{2\sqrt{320}}\ln\dfrac{v - \sqrt{320}}{v + \sqrt{320}} = -t + c$

$v = 196$ ft/s for $t = 0$

$c = 0.280\ln\dfrac{196 - 17.9}{196 + 17.9} = -0.0512$

$\ln\dfrac{v - 17.9}{v + 17.9} = -t - 0.0512$

$\dfrac{v - 17.9}{v + 17.9} = e^{-3.6t - 0.183} = 0.833e^{-3.6t}$

$v = 17.9\left(\dfrac{1 + 0.833e^{-3.6t}}{1 - 0.833e^{-3.6t}}\right)$

33. $\dfrac{dx}{dt} = 6t - 3t^2;\ dx = (6t - 3t^2);$

$x = \dfrac{6t^2}{2} - \dfrac{3t^3}{3} + c;\ x = 3t^2 - t^3 + c;$

$x = 0$ for $t = 0;\ 0 = c;\ x = 3t^2 - t^3;$

$y = 2(3t^2 - t^3) - (3t^2 - t^3)^2;$

$y = -t^6 + 6t^5 - 9t^4 - 2t^3 + 6t^2$

34. $y(x + 1) = 10;\ x = 4, t = 0, y = 2$

$(t + 1)dx = (x - 2)dt;\ \dfrac{dx}{x - 2} = \dfrac{dt}{t + 1}$

$\ln(x - 2) = \ln(t + 1) + c$

$\ln(2) = \ln 1 + c;$ therefore, $c = \ln 2$

$x + 1 = \dfrac{10}{y};\ x = \dfrac{10}{y} - 1 = \dfrac{10 - y}{y}$

$\ln(x - 2) = \ln(t + 1) + \ln 2;\ \dfrac{x - 2}{t + 1} = 2;$

$x - 2 = 2(t + 1)$

$x = 2 + 2t + 2 = 4 + 2t;$

$\dfrac{10 - y}{y} = 4 + 2t;$

$10 - y = (4 + 2t)y$

$10 - y = 4y + 2ty;$

$10 = 5y + 2ty = y(5 + 2t);$

$y = \dfrac{10}{5 + 2t}$

35. $\dfrac{dp}{dh} = kp, h = 0, p = 15$ lb/in^2

$h = 9800$ ft, $p = 10.0$ lb/in^2

$\displaystyle\int\dfrac{dp}{p} = \int k\,dh;\ \ln p = kh + c;\ \ln 15 = 0 + c$

Therefore $c = \ln 15$

$\ln p = kh + \ln 15;\ \ln p - \ln 15 = kh;$

$\ln\dfrac{p}{15} = kh;\ \dfrac{p}{15} = e^{kh}$

$p = 15e^{kh};\ 10.0 = 15e^{k9800};\ \dfrac{10.0}{15} = e^{k9800} = 0.667$

$e^k = (0.667)^{1/9800} = (0.667)^{10^{-4}};\ p = 15(0.667)^{10^{-4}h}$

36. Let $V =$ volume of water in tank

$a =$ radius of bottom of tank

$V = \pi a^2 h = kh$

$\dfrac{dV}{dt} = k\dfrac{dh}{dt} = 4.8Ah^{1/2}$

$h^{-1/2}dh = \dfrac{4.8A}{k}dt$

$2h^{-1/2} = \dfrac{4.8A}{k}t + c$

$h = 9.0$ ft when $t = 0, c = 2(9.0)^{1/2} = 6.0$

$2h^{1/2} = \dfrac{4.8A}{k}t + 6.0$

$h = 8.0$ m when $t = 16$ min

$2\sqrt{80} = \dfrac{4.8A}{k}(16) + 6.0$

$\dfrac{4.8A}{k} = -0.0214$

$2h^{1/2} = -0.0214t + 6.0$

$h^{1/2} = -0.0107t + 3.0$

For $h = 0$

$t = \dfrac{3.0}{0.0107} = 280$ min

37. $\dfrac{dv}{dt} = kv;\ \dfrac{dv}{v} = k\,dt;$ integrating, $\ln v = kt + \ln c$

Let $v = 16\,500$ when $t = 0;\ 16\,500 = c;$

$\ln v = kt + 16500;\ \ln v - \ln 16\,500 - kt;$

$\ln\dfrac{v}{16\,500} = kt;\ \dfrac{v}{16\,500} = e^{kt};\ v = 16\,500e^{kt}$

Let $v = 9850$ when $t = 3;\ 9850 = 16\,500e^{3k};$

$e^k = (0.597)^{1/3};\ v = 16\,500(0.597)^{t/3}$

After 11 years, $v = 16\,500(0.597)^{11/3} = \$2490$

38. Let x = grams of sugar remaining at time t

$$\frac{dx}{dt} = -kx; \quad \frac{dx}{x} = -k\,dt$$

$$\ln x = -kt + c$$

$$x = 525 \text{ g when } t = 0$$

$$c = \ln 525$$

$$x = 525e^{-kt}$$

$$x = 225 \text{ g when } t = 4.00 \text{ min}$$

$$225 = 525e^{-4.00k}$$

$$e^{-k} = \left(\frac{225}{525}\right)^{1/4.00} = 0.809$$

$$x = 525(0.809)^t$$

$$x = 150 \text{ g when } 375 \text{ g are dissolved}$$

$$150 = 525(0.809)^t$$

$$\ln\left(\frac{150}{525}\right) = t\ln 0.809$$

$$t = \frac{\ln(150/.525)}{\ln 0.809} = 5.91 \text{ min}$$

39. $\dfrac{dx}{dt} = 1 - 0.25x$

$$\frac{dx}{0.25x - 1} = -dt$$

$$-4\ln(0.25x - 1) = -t + c$$

$$x = 12 \text{ ft}^3 \text{ when } t = 0$$

$$c = 4\ln 2$$

$$4\ln(0.25x - 1) = -t + 4\ln 2$$

$$4\ln\frac{0.25x - 1}{2} = -t$$

$$0.25x - 1 = 2e^{-0.25t}$$

$$x = 4(1 + 2e^{-0.25t})$$

40. $x^2 + y^2 = cx$; therefore, $c = \dfrac{x^2 + y^2}{x}$

$$2x + 2y\frac{dy}{dx} = c; \quad \frac{dy}{dx} = \frac{c - 2x}{2y} = \frac{\dfrac{x^2 + y^2}{x} - 2x}{2y}$$

$$\frac{dy}{dx} = \frac{y^2 - x^2}{2xy}; \quad \text{therefore,} \quad \frac{dy}{dx}\bigg|_{OT} = \frac{2xy}{x^2 - y^2}$$

$$dy(x^2 - y^2) = 2xy\,dx; \quad x^2\,dy - y^2\,dy = 2xy\,dx$$

$$x^2\,dy - 2xy\,dx = y^2\,dy; \quad \frac{x^2\,dy - 2xy\,dx}{y^2} = dy$$

$$\frac{2xy\,dx - x^2\,dy}{y^2} = -dy; \quad d\left(\frac{x^2}{y}\right) = -dy$$

$$\frac{x^2}{y} = -y + c; \quad \text{therefore,}$$

$$x^2 = -y^2 + cy; \quad x^2 + y^2 = cy$$

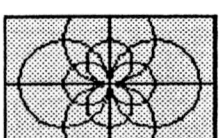

Chapter 14 Review Exercises

1. $4xy^3\,dx + (x^2 + 1)\,dy = 0$; divide by y^3 and $x^2 + 1$

$$\frac{4x}{x^2 + 1}\,dx + \frac{dy}{y^3} = 0; \text{ integrating,}$$

$$2\ln(x^2 + 1) - \frac{1}{2y^2} = c$$

2. $\dfrac{dy}{dx} = e^{x-y} = \dfrac{e^x}{e^y}$

$$\int e^y\,dy = \int e^x\,dx$$

$$e^y = e^x + c$$

3. $\sin 2x\,dx + y\sin x\,dy = \sin x\,dx$

$$\frac{\sin 2x}{\sin x}\,dx + y\,dy = dx$$

$$\frac{2\sin x\cos x}{\sin x}\,dx + y\,dy = dx$$

$$\int 2\cos x\,dx + \int y\,dy = \int dx$$

$$2\sin x + \frac{y^2}{2} = x + \frac{c}{2}$$

$$4\sin x + y = 2x + c$$

$$y^2 = 2x - 4\sin x + c$$

4. $x\,dy + y\,dx = y\,dy$

$$\int d(xy) = \int y\,dy$$

$$xy = \frac{y^2}{2} + \frac{c}{2}$$

$$2xy = y^2 + c$$

5. $(x+y)dx + (x+y^3)dy = 0;$

$x\,dx + y\,dx + x\,dy + y^3\,dy = 0;$

$x\,dx + d(xy) + y^3\,dy = 0$

Integrating, $\dfrac{1}{2}x^2 + xy + \dfrac{1}{4}y^4 = c_1;$

$2x^2 + 4xy + y^4 = c$

6. $\qquad y \ln x\, dx = x\, dx$

$\displaystyle\int \ln x \left(\frac{1}{x}\,dx\right) = \int \frac{1}{y}\,dy$

$\dfrac{\ln^2 x}{2} = \ln|y| + \dfrac{c}{2}$

$\ln^2 x = 2\ln|y| + c$

7. $\quad x\dfrac{dy}{dx} - 3y = x^2$

$dy - \dfrac{3}{x}y\,dx = x\,dx$

$P = -\dfrac{3}{x}, Q = x$

$e^{\int -3\,dx/x} = e^{-3\ln x} = x^{-3}$

$yx^{-3} = \displaystyle\int x\,x^{-3}dx = \int x^{-2}dx = -\dfrac{1}{x} + c$

$y = cx^3 - x^2$

8. $dy - 2y\,dx = (x-2)e^x\,dx$

$P = -2, Q = (x-2)e^x$

$ye^{\int -2\,dx} = \displaystyle\int (x-2)e^x e^{\int -2\,dx}dx + c$

$ye^{-2x} = \displaystyle\int (x-2)e^x e^{-2x}dx + c$

$ye^{-2x} = \displaystyle\int (x-2)e^{-x}dx + c$

$ye^{-2x} = \displaystyle\int xe^{-x}dx + 2\int e^{-x}(-dx) + c$

$ye^{-2x} = -(x+1)e^{-x} + 2e^{-x} + c$

$y = -(x+1)e^x + 2e^x + ce^{2x}$

$y = e^x(1-x) + ce^{2x}$

9. $dy = (2y+y^2)dx;\quad \dfrac{dy}{(2y+y^2)} = dx$

$-\dfrac{1}{2}\ln\left(\dfrac{2+y}{y}\right) = x + \ln c_1;$

$\ln\dfrac{2+y}{y} = -2x - \ln c_1^2 \left(\dfrac{2+y}{y}\right) = -2x;$

$\dfrac{2+y}{y} = \dfrac{e^{-2x}}{c_1^2};$

$y = c_1^2(y+2)e^{2x};\ y = c(y+2)e^{2x}$

10. $\qquad x^2 y\,dy = (1+x)\csc y\,dx$

$= \dfrac{1+x}{\sin y}\,dx$

$x^2 y \sin y\,dy = (1+x)dx$

$y \sin y\,dy = \dfrac{1+x}{x^2\,dx} = \left(\dfrac{1}{x^2} + \dfrac{1}{x}\right)dx$

$\displaystyle\int y\sin y\,dy = \int \dfrac{1}{x^2}\,dx + \int \dfrac{1}{x}\,dx$

$\sin y - y\cos y = -\dfrac{1}{x} + \ln|x| + c$

11. $y' + 4y = 2e^{-2x};\ \dfrac{dy}{dx} + 4y = 2e^{-2x};$

$dy + 4y\,dx = 2e^{-2x}\,dx$

$e^{\int 4\,dx} = e^{4x};\ e^{4x}dy + 4ye^{4x}dx = 2e^{-2x}e^{4x}dx;$

$d(ye^{4x}) = 2e^{2x}dx;\ \displaystyle\int d(ye^{4x}) = \int e^{2x}(2\,dx);$

$ye^{4x} = e^{2x} + c;\ y = e^{-2x} + ce^{-4x}$

12. $\qquad 2xy\,dx = (2y - \ln y)dy$

$\displaystyle\int 2x\,dx = \int \left(2 - \ln y\left(\dfrac{1}{y}\right)\right)dy$

$x^2 = 2y - \dfrac{\ln^2 y}{2} + \dfrac{c}{2}$

$2x^2 = 4y - \ln^2 y + c$

13. $\quad \sin x\dfrac{dy}{dx} + y\cos x + x = 0$

$dy + y\cot x\,dx = -x\csc x\,dx$

$P = \cot x, Q = -x\csc x\,dx$

$e^{\int \cot x\,dx} = e^{\ln \sin x}$

$y\sin x = -\displaystyle\int x\csc x \sin x\,dx$

$y\sin x = -\displaystyle\int x\,dx$

$y\sin x = -\dfrac{x^2}{2} + \dfrac{c}{2}$

$y = \dfrac{1}{2}(c - x^2)\csc x$

14. $\qquad y\,dy = (x^2 + y^2 - x)dx$

$y\,dy = (x^2 + y^2)dx - x\,dx$

$2(y\,dy + x\,dx) = 2(x^2 + y^2)dx$

$\dfrac{2(y\,dy + x\,dx)}{x^2 + y^2} = 2\,dx$

$\displaystyle\int \dfrac{d(x^2 + y^2)}{x^2 + y^2} = \int 2\,dx$

$\ln(x^2 + y^2) = 2x + c$

15. $(x^2 + 2x)dy = 2xy\,dx + 5y\,dx$

$(x^2 + 2x)dy = y(2x + 5)dx$

$$\frac{dy}{y} = \frac{2x + 5}{x^2 + 2x}dx; \quad \frac{2x + 5}{x^2 + 2x} = \frac{5}{2x} - \frac{1}{2(x + 2)}$$

$$\frac{dy}{y} = \frac{5\,dx}{2x} - \frac{dx}{2(x + 2)}$$

$$\ln y = \frac{5}{2}\ln x - \frac{1}{2}\ln(x + 2) + \ln c$$

$$y = \frac{cx^{5/2}}{(x + 2)^{1/2}}$$

16. $\dfrac{dy}{dx} = \dfrac{1}{1 + e^x} - y$

$$dy + y\,dx = \frac{1}{1 + e^x}dx$$

$$P = 1, Q = \frac{1}{1 + e^x}$$

$$ye^{\int dx} = \int \frac{1}{1 + e^x}e^{\int dx}dx + c$$

$$ye^x = \int \frac{1}{1 + e^x}e^x dx + c$$

$$ye^x = \ln(1 + e^x) + c$$

$$y = e^{-x}[\ln(1 + e^x) + c]$$

17. $x\,dy - y\,dx = x^3 y\,dx$

$$\frac{dy}{y} - \frac{dx}{x} = x^2 dx$$

$$\ln y - \ln x = \frac{1}{3}x^3 + \ln c$$

$$\ln\frac{y}{cx} = \frac{1}{3}x^3$$

$$y = cxe^{x^3/3}$$

18. $x\,dy + y\,dx = x^3 y^4 dy,$

note: $d\left(\dfrac{-1}{x^2 y^2}\right) = \dfrac{2xy^2 dx + 2x^2 y\,dy}{x^4 y^4}$

$$\left(\frac{xy}{xy}\right)\frac{x\,dy + y\,dx}{x^3 y^3} = y\,dy$$

$$\frac{2x^2 y\,dy + 2xy^2 dx}{x^4 y^4} = 2y\,dy$$

$$d\left(\frac{-1}{x^2 y^2}\right) = 2y\,dy$$

$$-\frac{1}{x^2 y^2} = y^2 + c$$

$$x^2 y^2(y^2 + c) + 1 = 0$$

19. $y\,dy = 2(x^2 + y^2)^2 dx - x\,dx$

$x\,dx + y\,dy = 2(x^2 + y^2)^2 dx$

$$\frac{2x\,dx + 2y\,dy}{(x^2 + y^2)^2} = 4\,dx$$

$$-\frac{1}{x^2 + y^2} = 4x + c$$

$$-1 = (x^2 + y^2)(4x + c)$$

$$(x^2 + y^2)(4x + c) + 1 = 0$$

20. $y' = 2 + xe^{3x}$

$$\frac{dy}{dx} = 2 + xe^{3x}$$

$$dy = (2 + xe^{3x})dx$$

$$y = 2x + \int xe^{3x}dx$$

$$y = 2x + \left[\frac{x}{3} - \frac{1}{9}\right]e^{3x} + c$$

$$y = 2x + \frac{1}{9}e^{3x}(3x - 1) + c$$

21. $3y' = 2y\cot x, \; x = \dfrac{\pi}{2}$ when $y = 2$

$$3\frac{dy}{y} = 2\cot x\,dx$$

$3\ln y = 2\ln(\sin x) + \ln c = \ln\sin^2 x + \ln c$
$\ln y^3 = \ln(c\sin^2 x)$

$$y^3 = c\sin^2 x, \; 2^3 = c\sin^3 \frac{\pi}{2}, \; c = 8$$

$$y^3 = 8\sin^2 x$$

22. $x\,dy - y\,dx = y^3 dy, \; x = 1$ when $y = 3$

$$\frac{x\,dy - y\,dx}{y^2} = y\,dy$$

$$-\frac{y\,dx - x\,dy}{y^2} = y\,dy$$

$$-d\left(\frac{x}{y}\right) = y\,dy$$

$$-\frac{x}{y} = \frac{y^2}{2} + c$$

$$-\frac{1}{3} = \frac{3^2}{2} + c$$

$$c = \frac{-29}{6}$$

$$-\frac{x}{y} = \frac{y^2}{2} - \frac{29}{6}$$

$$3y^3 - 29y + 6x = 0$$

23. $y' = 4x - 2y$

$dy + 2y\,dx = 4x\,dx$

$P = 2, Q = 4x, e^{\int 2\,dx} = e^{2x}$

$ye^{2x} = \int 4xe^{2x}dx = e^{2x}(2x - 1) + c$

(Formula 44)

$y = 2x - 1 + ce^{-2x}$

$x = 0$ when $y = -2; -2 = -1 + c; c = -1$

$y = 2x - 1 - e^{-2x}$

24. $xy^2\,dx + e^x\,dy = 0, x = 0$ when $y = 2$

$\int \dfrac{x\,dx}{e^x} + \int \dfrac{dy}{y^2} = 0$

$-(x+1)e^{-x} - \dfrac{1}{y} = -c$

$(x+1)e^{-x} + \dfrac{1}{y} = c$

$(0+1)e^{-0} + \dfrac{1}{2} = c$

$c = \dfrac{3}{2}$

$(x+1)e^{-x} + \dfrac{1}{y} = \dfrac{3}{2}$

$2y(x+1)e^{-x} + 2 = 3y$

$3y = 2ye^{-x} + 2xye^{-x} + 2$

25. $\dfrac{dy}{dx} = x - xy = x(1 - y)$

$\dfrac{dy}{1 - y} = x\,dx; \; -\ln(1 - y) = \dfrac{1}{2}x^2 + c$

$x = 2$ when $y = 0; \; -\ln 1 = 2 + c, c = -2$

$-\ln(1 - y) = \dfrac{1}{2}x^2 - 2$

$x^2 + 2\ln(1 - y) - 4 = 0$

26. $\dfrac{dy}{dx} = 2 - y, x = 0$, when $y = 1$

$\dfrac{dy}{y - 2} = -dx$

$\ln|y - 2| = -x + c$

$\ln|1 - 2| = -0 + c$

$\ln|-1| = \ln 1 = 0 = 0 + c, c = 0$

$\ln|y - 2| = -x$

$e^{\ln|y-2|} = e^{-x}$

$|y - 2| = e$

$|y - 2| = \begin{cases} y - 2, y - 2 \geq 0, y \geq 2 \\ -(y - 2), y - 2 < 0, y < 2 \end{cases}$

since $x = 0$ when $y = 1 < 2$, pick $-(y - 2)$

$-(y - 2) = e^{-x}$

$-y + 2 = e^{-x}$

$y = 2 - e^{-x}$

27. $xy' - 2y = x^3 \cos 4x$

$dy - \dfrac{2}{x}y\,dx = x^2 \cos 4x\,dx$

$P = -\dfrac{2}{x}, Q = x^2 \cos 4x$

$e^{\int -2\,dx/x} = e^{-2\ln x} = e^{\ln x^{-2}} = x^{-2}$

$\dfrac{y}{x^2} = \int x^{-2}x^2 \cos 4x\,dx = \int \cos 4x\,dx$

$= \dfrac{1}{4}\sin 4x + c; \; y = x^2\left(\dfrac{1}{4}\sin 4x + c\right)$

$x = \dfrac{\pi}{4}$ when $y = 0; \; 0 = \left(\dfrac{\pi}{4}\right)^2\left(\dfrac{1}{4}\sin \pi + c\right)$

$c = 0; \; y = \dfrac{1}{4}x^2 \sin 4x$

28. $2xy\,dy + dx = 2x^2 \sin x^2\,dx, x = 1$ when $y = 0$

$y\,dy + \dfrac{dx}{2x} = x \sin x^2\,dx$

$\dfrac{y^2}{2} + \dfrac{1}{2}\ln x = -\dfrac{\cos x^2}{2} + c$

$\dfrac{0^2}{2} + \dfrac{1}{2}\ln 1 = -\dfrac{\cos 1^2}{2} + c$

$c = \dfrac{\cos 1}{2}$

$\dfrac{y^2}{2} + \dfrac{1}{2}\ln x = -\cos x^2 + \cos 1$

$y^2 + \ln x = -\cos x^2 + \cos 1$

$y^2 = \cos 1 - \ln x - \cos x^2$

29. $\dfrac{dV}{dt} = kA$

$V = \dfrac{4}{3}\pi r^3, A = 4\pi r^2$

$\dfrac{dV}{dt} = 4\pi r^2 \dfrac{dr}{dt}$

$4\pi r^2 \dfrac{dr}{dt} = k(4\pi r^2)$

$dr = k\,dt; \; r = kt + c$

$r = r_0$ when $t = 0, c = r_0$

$r = r_0 + kt$

30. $56{,}200 = 262(T - 70) + 20{,}200 \frac{dT}{dt}$,

$T = 70°\text{F}$ when $t = 0$

$20{,}200 \frac{dT}{dt} + 262T - 18{,}340 = 56{,}600$

$20{,}200 \frac{dT}{dt} + 262T = 74{,}940$

$dT + 0.013T\, dt = 3.71\, dt$

$T e^{\int 0.013\, dt} = \int 3.71 e^{\int 0.013\, dt}\, dt + c$

$T e^{0.013\, t} = \int 3.71 e^{0.013\, t}\, dt + c$

$T e^{0.013\, t} = 285 e^{0.013t} + c$

$70 = 285 + c$

$c = -215$

$T e^{0.013\, t} = 285 e^{0.013t} - 215$

$T e^{0.013(24)} = 285 e^{0.013(24)} - 215$

$T = 128°\text{F}$

31. See Example 6, Section 14-5

$m = 1\text{ kg}, \ F = 4\text{ N}, \ k = 1$

$\frac{dV}{dt} = 4 - v; \quad \frac{dv}{4 - v} = dt$

$-\ln(4 - v) = t + c$

$v = 0$ when $t = 0, c = -\ln 4$

$-\ln(4 - v) = t - \ln 4$

$\ln \frac{4}{4 - v} = t$

$4 = (4 - v)e^t$

$v = 4 - 4e^{-t} = 4(1 - e^{-t})$

When $t = 4$ s,

$v = 4(1 - e^{-4}) = 3.93$ m/s

32. $192 - 2v = \frac{192}{32} \frac{dv}{dt} = 6 \frac{dv}{dt}$

$96 - v = 3 \frac{dv}{dt}$

$\frac{3\, dv}{96 - v} = dt$

$-3\ln(96 - v) = t + c$

$-3\ln(96) = 0 + c$

$-3\ln(96 - v) = t - 3\ln 96$

$\ln 96 - \ln(96 - v) = \frac{t}{3}$

$\ln \frac{96}{96 - v} = \frac{t}{3}$

$\frac{96}{96 - v} = e^{t/3}$

$96 e^{-t/3} = 96 - v$

$v = 96(1 - e^{-t/3})$

33. $N = N_0 e^{-kt}$

$\frac{N_0}{2} = N_0 e^{-kt(1.28 \times 10^9)}$

$e^{k(1.28 \times 10^9)} = 2$

$k(1.28 \times 10^9) = \ln 2$

$k = \frac{\ln 2}{1.28 \times 10^9}$

$N = N_0 e^{-\frac{\ln 2}{1.28 \times 10^9} \cdot t}$

$0.75 N_0 = N_0 e^{\frac{-\ln 2}{1.28 \times 10^9} \cdot t}$

$e^{\frac{\ln 2}{1.28 \times 10^9} \cdot t} = \frac{4}{3}$

$\frac{\ln 2}{1.28 \times 10^9} \cdot t = \ln \frac{4}{3}$

$t = \frac{\ln \frac{4}{3}}{\frac{\ln 2}{1.28 \times 10^9}}$

$t = 5.31 \times 10^8$ years.

34. $\frac{dp}{dV} = \frac{kp}{V}$

$\frac{dp}{p} = \frac{k\, dV}{V}$

$\ln p = k \ln V + \ln c$

$\qquad = \ln V^k + \ln c$

$\ln p = \ln(c V^k)$

$p = c V^k$

35. $\frac{dp}{dt} = kp$

$\frac{dp}{p} = k\, dt$

$\ln p = kt + c$

let $p = 5$ billion in 1987 $(t = 0)$

$\ln 5 = k(0) + c$

$\ln p = kt + \ln 5,$

let $p = 6$ billion in 1999 $(t = 12)$

$\ln 6 = k(12) + \ln 5$

$$k = \frac{\ln \frac{6}{5}}{12}$$

$\ln p = \dfrac{\ln \frac{6}{5}}{12} t + \ln 5$, in 2010, $t = 23$

$\ln p = \dfrac{\ln \frac{6}{5}}{12}(23) + \ln 5$

$p = 7.1$ billion

36.

$$V = \frac{4}{3}\pi r^3$$

$$V^{2/3} = \left(\frac{4}{3}\pi\right)^{2/3} r^2$$

$$r^2 = \frac{V^{2/3}}{\left(\frac{4}{3}\pi\right)^{2/3}}$$

$$\frac{dV}{dt} = KS = K4\pi r^2$$

$$\frac{dV}{dt} = K4\pi \frac{V^{2/3}}{\left(\frac{4}{3}\pi\right)^{2/3}}$$

$$\frac{dV}{dt} = kV^{2/3}$$

$$V^{-2/3}dV = k\,dt$$

$$3V^{1/3} = kt + c$$

$$V^{1/3} = \frac{kt + c}{3}$$

$$V = \left(\frac{kt + c}{3}\right)^3$$

37. See Example 2, Section 14-5

$$y = cx^5, c = \frac{y}{x^5}$$

$$y' = 5cx^4 = 5\left(\frac{y}{x^5}\right)x^4 = \frac{5y}{x}$$

$$y' \mid OT = -\frac{x}{5y}; \quad 5y\,dy = -x\,dx$$

$$\frac{5}{2}y^2 = -\frac{1}{2}x^2 + \frac{1}{2}c$$

$$5y^2 + x^2 = c$$

38. $m_{NL} = -\dfrac{1}{\frac{dy}{dx}} = \dfrac{y}{x}$

$$\frac{dy}{dx} = -\frac{x}{y}$$

$$2(y\,dy + x\,dx) = 0$$

$$d(x^2 + y^2) = 0$$

$$x^2 + y^2 = c$$

39. See Example 4, Section 14-5

$$Ri + \frac{1}{C}q = E$$

$$R\frac{dq}{dt} + \frac{1}{C}q = E; \; dq + \frac{1}{RC}q\,dt = \frac{E}{R}\,dt$$

$$P = \frac{1}{RC}, \; Q = \frac{E}{R}, \; e^{\int dt/RC} = e^{t/RC}$$

$$qe^{t/RC} = \int \frac{E}{R}e^{t/RC}dt = ECe^{t/RC} + c_1$$

$$q = c_1 e^{-t/RC} + EC$$

40. $L\dfrac{di}{dt} + Ri = E$

$$2\frac{di}{dt} + 40i = 20$$

$$\frac{di}{dt} + 20i = 10$$

$$di + 20i\,dt = 10\,dt$$

$$ie^{\int 20\,dt} = \int 10e^{\int 20\,dt} + c$$

$$ie^{20t} = 0.5e^{20t} + c$$

$$(0)e^{20(0)} = 0.5e^{20(0)} + c$$

$$c = -0.5$$

$$ie^{20t} = 0.5e^{20t} - 0.5 = 0.5(e^{20t} - 1)$$

$$i = 0.5(1 - e^{-20t})$$

41. Let V = volume in L of oxygen in container at time t. At $t = 0, V = 5.00$

$\dfrac{dV}{dr}$ = rate at which oxygen flows in—rate at which oxygen flows out

Let R = rate in L per unit time at which mixture flows in = rate in L per unit time at which mixture flows out

$$\frac{dV}{dt} = 0.2R - \frac{V}{5.00}R = R\left(0.2 - \frac{V}{5.00}\right)$$

$$\frac{dV}{0.2 - \frac{V}{5.00}} = R\,dt$$

$$\frac{5.00dV}{1.00 - V} = R\,dt$$

$-5.00\ln|1.00 - V| = Rt + c$ at $t = 0, V = 5.00$

$$-5.00\ln|1.00 - 5.00| = R(0) + c$$

$$c = -5.00\ln 4.00$$

$$-5.00\ln|1.00 - V| = Rt - 5.00\ln 4.00$$

$$\ln 4.00 - \ln|1.00 - V| = \frac{Rt}{5.00}$$

when 5.00 L of mixture have flowed in $Rt = 5.00$

$$\ln 4.00 - \ln |1.00 - V| = \frac{5.00}{5.00}$$

$$\ln \frac{4.00}{|1.00 - V|} = 1.00$$

$$\frac{4.00}{|1.00 - V|} = e^{1.00}$$

$$|1.00 - V| = \frac{4.00}{e^{1.00}}$$

$$1.00 - V = \pm \frac{4.00}{e^{1.00}}$$

$$V = 1.00 \pm \frac{4.00}{e^{1.00}}$$

pick $+$, since $-$ gives a negative volume

$$V = 2.47\text{L}$$

42. Acceleration is inversely proportional to square of radius

$$\frac{dv}{dt} = \frac{k}{r^2}$$

From chain rule, $\dfrac{dv}{dt} = \dfrac{dv}{dr}\dfrac{dr}{dt} = \dfrac{dr}{dt}\dfrac{dv}{dr} = v\dfrac{dv}{dr}$

$$v\frac{dv}{dr} = \frac{k}{r^2}$$

$$v\,dv = \frac{k}{r^2}\,dr$$

$$\frac{v^2}{2} = -\frac{k}{r} + c$$

when $r = R$, $\dfrac{dv}{dt} = -g = \dfrac{k}{R^2}$, $k = -gR^2$

$$\frac{v^2}{2} = \frac{gR^2}{r} + c$$

$v = v_0$ when $r = R$

$$\frac{v_0^2}{2} = \frac{gR^2}{R} + c$$

$$c = \frac{v_0^2}{2} - gR$$

$$\frac{v^2}{2} = \frac{gR^2}{r} + \frac{v_0^2}{2} - gR$$

$$v^2 = \frac{2gR^2}{r} + v_0^2 - 2gR,$$

as $r \to \infty$, $v_0^2 - 2gR \geq 0$ in order that $v^2 \geq 0$

$$v_0 = \sqrt{2gR}$$

43. See Fig. 14-9

$r = a\sin\theta, a = \dfrac{r}{\sin\theta}$

$\dfrac{dr}{d\theta} = a\cos\theta = \dfrac{r}{\sin\theta}\cos\theta = r\cot\theta; \; r\,\dfrac{d\theta}{dr} = \tan\theta$

$\phi_{OT} = \phi + \dfrac{\pi}{2}$ or $\tan\phi_{OT} = -\dfrac{1}{\tan\phi}$

$\dfrac{dr}{d\theta} = \dfrac{r}{\tan\phi}, \; \tan\phi = r\,\dfrac{d\theta}{dr}$

$r\,\dfrac{d\theta}{dr}\bigg|_{OT} = -\dfrac{1}{\tan\theta}$

$\tan\theta\,d\theta = -\dfrac{dr}{r}$

$-\ln\cos\theta = -\ln r + \ln c$

$r = c\cos\theta$

44. $M(x,y) = 3x^2\cos^2 y$

$\dfrac{\partial M}{\partial y} = 3x^2(2\cos y(-\sin y)) = -3x^2(2\sin y\cos y)$

$\dfrac{\partial M}{\partial y} = -3x^2\sin 2y$

$N(x,y) = -x^3\sin 2y$

$\dfrac{\partial N}{\partial x} = -3x^2\sin 2y = \dfrac{\partial M}{\partial y}$

45. $L\dfrac{di}{dt} + Ri + \dfrac{q}{C} = E$ for an inductor, resistor and voltage source is

$\boxed{1}$ $L\dfrac{di}{dt} + Ri = E$ and for a resistor, capacitor, and voltage source is

$\boxed{2}$ $Ri + \dfrac{q}{c} = E.$

$\boxed{1}$ may be written

$di + \dfrac{R}{L}i\,dt = E\,dt$ which may be solved as a linear differential equation of first order for $E = $ constant or $E = $ function of time.

$\boxed{2}$ may be written

$$R\dfrac{dq}{dt} + \dfrac{q}{c} = E$$

$$dq + \dfrac{R}{C}q\,dt = E\,dt$$

which may be solved as a linear differential equation of first order for $E = $ function of time or by separation for $E = $ constant.

HIGH–ORDER DIFFERENTIAL EQUATIONS

15.1 Higher-Order Homogeneous Equations

1. $D^2y - Dy - 6y = 0$; $m^2 - m - 6 = 0$;
$(m-3)(m+2) = 0$; $m_1 = 3$ and $m_2 = -2$;
$y = c_1 e^{3x} + c_2 e^{-2x}$

2. $\dfrac{d^2y}{dx^2} + \dfrac{dy}{dx} = 0$; $D^2y + D = 0$; $m^2 + m = 0$
$m(m+1) = 0$; $m_1 = 0, m_2 = -1$
$y = c_1 e^{0x} + c_2 e^{-x} = c_1 + c_2 e^{-x}$

3. $3\dfrac{d^2y}{dx^2} + 4\dfrac{dy}{dx} + y = 0$
$3D^2y + 4Dy + y = 0$;; $3m^2 + 4m + 1 = 0$
$(3m+1)(m+1) = 0$; $m_1 = -\dfrac{1}{3}, m_2 = -1$
$y = c_1 e^{-(1/3)x} + c_2 e^{-x}$

4. $\dfrac{d^2y}{dx^2} - 2\dfrac{dy}{dx} - 8y = 0$
$D^2y - 2Dy - ey = 0$; $m^2 - 2m - 8 = 0$
$(m-4)(m+2) = 0$; $m_1 = 4, m_2 = -2$
$y = c_1 e^{4x} + c_2 e^{-2x}$

5. $D^2y - 3Dy = 0$; $m^2 - m = 0$; $m_1 = 0$ and
$m_2 = 3$; $y = c_1 e^0 + c_2 e^{3x}$; $y = c_1 + c_2 e^{3x}$

6. $D^2y + 7Dy + 6y = 0$; $m^2 + 7m + 6 = 0$;
$(m+6)(m+1) = 0$
$m = -6, m = -1$; $y = c_1 e^{-6x} + c_2 e^{-x}$

7. $3D^2y + 12y = 20Dy$; $3D^2y - 20Dy + 12y = 0$
$3m^2 - 20m + 12 = 0$; $(3m-2)(m-6) = 0$
$m = \dfrac{2}{3}, m = 6$; $y = c_1 e^{(2/3)x} + c_2 e^{6x}$

8. $4D^2y + 12Dy - 7y = 0$; $4m^2 + 12m - 7 = 0$
$(2m-1)(2m+7) = 0$; $m = \dfrac{1}{2}, m = -\dfrac{7}{2}$
$y = c_1 e^{(1/2)x} + c_2 e^{-(7/2)x}$

9. $3D^2y + 8Dy - 3y = 0$; $3m^2 + 8m - 3 = 0$;
$(3m-1)(m+3) = 0$; $m_1 = \frac{1}{3}$ and $m_2 = -3$;
$y = c_1 e^{x/3} + c_2 e^{-3x}$

10. $8y'' + 6y' - 9y = 0$; $8D^2y + 6Dy - 9y = 0$;
$8m^2 + 6m - 9 = 0$
$(4m-3)(2m+3) = 0$; $m = \dfrac{3}{4}, m = -\dfrac{3}{2}$
$y = c_1 e^{(3/4)x} + c_2 e^{-(3/2)x}$

11. $3y'' + 2y' - y = 0$; $3D^2y + 2Dy - y = 0$;
$3m^2 + 2m - 1 = 0$
$(3m-1)(m+1) = 0$; $m = \dfrac{1}{3}, m = -1$
$y = c_1 e^{(1/3)x} + c_2 e^{-x}$

12. $2y'' - 7y' + 6y = 0$; $2D^2y - 7Dy + 6y = 0$;
$2m^2 - 7m + 6 = 0$
$(2m-3)(m-2) = 0$; $m = \dfrac{3}{2}, m = 2$
$y = c_1 e^{(3/2)x} + c_2 e^{2x}$

13. $2\dfrac{d^2y}{dx^2} - 4\dfrac{dy}{dx} + y = 0$; $2D^2y - 4Dy + y = 0$;
$2m^2 - 4m + 1 = 0$
Quadratic formula: $m = \dfrac{4 \pm \sqrt{16-8}}{4}$;
$m_1 = 1 + \dfrac{\sqrt{2}}{2}$,
$m_2 = 1 - \dfrac{\sqrt{2}}{2}$
$y = c_1 e^{(1+(\sqrt{2}/2))x} + c_2 e^{(1-(\sqrt{2}/2))x}$;
$y = c_1 e^x e^{(\sqrt{2}/2)x} + c_2 e^x e^{-(\sqrt{2}/2)x}$
$\quad = e^x (c_1 e^{x(\sqrt{2}/2)} + c_2 e^{-x(\sqrt{2}/2)})$

14. $\dfrac{d^2y}{dx^2} + \dfrac{dy}{dx} - 5y = 0$; $D^2y + D - 5 = 0$
$m^2 + m - 5 = 0$
Quadratic formula:
$m = \dfrac{-1 \pm \sqrt{1+20}}{2} = \dfrac{-1 \pm \sqrt{21}}{2}$
$m_1 = \dfrac{-1 + \sqrt{21}}{2}$,
$m_2 = \dfrac{-1 - \sqrt{21}}{2}$
$y = c_1 e^{(-(1/2)+(\sqrt{21}/2))x} + c_2 e^{(-(1/2)-(\sqrt{21}/2))x}$
$y = e^{-x/2} \left[c_1 e^{(\sqrt{21}/2)x} + c_2 e^{-(\sqrt{21}/2)x} \right]$

15. $4D^2y - 3Dy - 2y = 0$; $4m^2 - 3m - 2 = 0$

Quadratic formula:

$$m = \frac{3 \pm \sqrt{9 + 32}}{8} = \frac{3 \pm \sqrt{41}}{8}$$

$$m_1 = \frac{3}{8} + \frac{\sqrt{41}}{8},$$

$$m_2 = \frac{3}{8} - \frac{\sqrt{41}}{8}$$

$$y = c_1 e^{((3/8) + (\sqrt{41}/8))x} + c_2 e^{((3/8) - (\sqrt{41}/8))x}$$
$$y = e^{(3/8)x} \left[c_1 e^{(\sqrt{41}/8)x} + c_2 e^{-(\sqrt{41}/8)x} \right]$$

16. $2D^2y - 3Dy - y = 0$; $2m^2 - 3m - 1 = 0$

Quadratic formula:

$$\frac{3 \pm \sqrt{9 + 8}}{4} = \frac{3}{4} \pm \frac{\sqrt{17}}{4}$$

$$m_1 = \frac{3}{4} + \frac{\sqrt{17}}{4}x,$$

$$m_2 = \frac{3}{4} - \frac{\sqrt{17}}{4}x,$$

$$y = c_1 e^{((3/4) + (\sqrt{17}/4))x} + c_2 e^{((3/4) - (\sqrt{17}/4))x}$$
$$y = e^{(3/4)x} \left[c_1 e^{(\sqrt{17}/4)x} + c_2 e^{-(\sqrt{17}/4)x} \right]$$

17. $y'' = 3y' + y$; $D^2y - 3Dy - y = 0$; $m^2 - 3m - 1 = 0$

Quadratic formula:

$$m = \frac{3 \pm \sqrt{9 + 4}}{2}; m_1 = \frac{3}{2} + \frac{\sqrt{13}}{2}; m_2 = \frac{3}{2} - \frac{\sqrt{13}}{2}$$

$$y = c_1 e^{((3/2) + (\sqrt{13}/2))x} + c_2 e^{((3/2) - (\sqrt{13}/2))x};$$
$$y = e^{3x/2} \left(c_1 e^{x(\sqrt{13}/2)} + c_2 e^{-x(\sqrt{13}/2)} \right)$$

18. $5y'' - y' - 3y = 0$; $5D^2y - Dy - 3y = 0$:
$5m^2 - m - 3 = 0$

Quadratic formula:

$$m = \frac{1 \pm \sqrt{1 + 60}}{10}$$

$$m_1 = \frac{1}{10} + \frac{\sqrt{61}}{10}$$

$$m_2 = \frac{1}{10} - \frac{\sqrt{61}}{10}$$

$$y = c_1 e^{((1/10) + (\sqrt{61}/10))x} + c_2 e^{((1/10) - (\sqrt{61}/10))x}$$
$$y = e^{(1/10)x} \left[c_1 e^{(\sqrt{61}/10)x} + c_2 e^{-(\sqrt{61}/10)x} \right]$$

19. $y'' + y' - 8y = 0$; $D^2y + Dy - 8 = 0$
$m^2 + m - 8 = 0$

Quadratic formula:

$$m = \frac{-1 \pm \sqrt{1 + 32}}{2}$$

$$m_1 = \frac{-1 + \sqrt{33}}{2},$$

$$m_2 = \frac{-1 - \sqrt{33}}{2}$$

$$y = c_1 e^{(-(1/2) + (\sqrt{33}/2))x} + c_2 e^{(-(1/2) - (\sqrt{33}/2))x}$$
$$y = e^{-(1/2)x} \left[c_1 e^{(\sqrt{33}/2)x} + c_2 e^{-(\sqrt{33}/2)x} \right]$$

20. $8y'' - y' = 0$; $8D^2y - Dy - y = 0$;
$8m^2 - m - 1 = 0$

Quadratic formula:

$$m = \frac{1 \pm \sqrt{1 + 32}}{16}$$

$$m_1 = \frac{1 + \sqrt{33}}{16},$$

$$m_2 = \frac{1 - \sqrt{33}}{16}$$

$$y = c_1 e^{((1/16) + (\sqrt{33}/16))x} + c_2 e^{((1/16) - (\sqrt{33}/16))x}$$
$$y = e^{x/16} \left[c_1 e^{(\sqrt{33}/16)x} + c_2 e^{-(\sqrt{33}/16)x} \right]$$

21. $D^2y - 4Dy - 21y = 0$; $m^2 - 4m - 21 = 0$;
$(m - 7)(m + 3) = 0$; $m_1 = 7$ and $m_2 = -3$;
$y = c_1 e^{7x} + c_2 e^{-3x}$

The derivative of the general equation is:
$Dy = 7c_1 e^{7x} - 3c_2 e^{-3x}$
Substituting $Dy = 0$, $y = 2$ when $x = 0$:
$c_1 + c_2 = 2$ and $7c_1 - 3c_2 = 0$.

Solving this system: $c_1 = \frac{3}{5}$ and $c_2 = \frac{7}{5}$.

Therefore, $y = \frac{3}{5}e^{7x} + \frac{7}{5}e^{-3x}$; $y = \frac{1}{5}(3e^{7x} + 7e^{-3x})$

22. $4D^2y - Dy = 0$; $Dy = 2, y = 4$ when $x = 0$

$4m^2 - m = 0$; $m(4m - 1) = 0$; $m_1 = 0$; $m_2 = \frac{1}{4}$

$y = c_1 e^{0x} + c_2 e^{(1/4)x} = c_1 + c_2 e^{x/4}$

$Dy = \frac{c_2}{4}e^{x/4}$; therefore, $c_2 = 4e^{-x/4}Dy$

Substituting $Dy = 2, y = 4$ when $x = 0$;
$c_2 = 4(2)e^0 = 8$
$y = c_1 + c_2 e^{x/4} = c_1 + 8e^{x/4}$; $4 = c_1 + 8$;
therefore, $c_1 = -4$; $y = -4 + 8e^{x/4}$

23. $D^2y - Dy - 12y = 0$; $y = 0$ when $x = 0$; $y = 1$
when $x = 1$
$m^2 - m - 12 = 0$; $(m - 4)(m + 3) = 0$; $m_1 = 4$;
$m_2 = -3$
$y = c_1 e^{4x} + c_2 e^{-3x}$

Substituting given values: $0 = c_1 + c_2$;
therefore, $c_1 = -c_2$;
$1 = c_1 e^4 + c_2 e^{-3}$

$1 = -c_2 e^4 + c_2 e^{-3} = -c_2 e^4 + \dfrac{c_2}{e^3} = \dfrac{-c_2 e^7 + c_2}{e^3}$

$e^3 = c_1(1 - e^7)$; therefore,

$c_2 = \dfrac{e^3}{(1 - e^7)}$, $c_1 - \dfrac{e^3}{(1 - e^7)}$

$y = -\dfrac{e^3}{(1 - e^7)} e^{4x} + \dfrac{e^3}{(1 - e^7)} e^{-3x}$

$\quad = \dfrac{e^3}{e^7 - 1} e^{4x} - \dfrac{e^3}{e^7 - 1} e^{-3x}$

$y = \dfrac{e^3}{e^7 - 1} (e^{4x} - e^{-3x})$

24. $2D^2y + 5Dy = 0$; $y = 0, x = 0$; $y = 2, x = 1$

$2m^2 + 5m = 0$; $m(2m + 5) = 0$;

$m_1 = 0, m_2 = -\dfrac{5}{2}$

$y = c_1 e^{0x} + c_2 e^{-(5/2)x} = c_1 + c_2 e^{-(5/2)x}$

Substituting given values:

$0 = c_1 + c_2$; therefore, $c_1 = -c_2$
$2 = c_1 + c_2 e^{-5/2}$
$2 = -c_2 + c_2 e^{-5/2} = c_2(-1 + e^{-5/2})$

$c_2 = \dfrac{2}{(e^{-5/2} - 1)}, c_1 = \dfrac{2}{e^{-5/2} - 1}$

Therefore,

$y = -\dfrac{2}{(e^{-5/2} - 1)} + \dfrac{2}{(e^{-5/2} - 1)} e^{-(5/2)x}$

$y = \dfrac{2}{(e^{-5/2} - 1)} (-1 + e^{-(5/2)x})$

$y = \dfrac{2}{(e^{-5/2} - 1)} (e^{-(5/2)x} - 1)$

25. $D^3y - 2D^2y - 3Dy = 0$; $m^3 - 2m^2 - 3m = 0$;
$m(m + 1)(m - 3) = 0$; $m_1 = 0, m_2 = -1$, and
$m_3 = 3$; $y = c_1 e^0 + c_2 e^{-x} + c_3 e^{3x}$;
$y = c_1 + c_2 e^{-x} + c_3 e^{3x}$

26. $D^3y - 6D^2y + 11Dy - 6y = 0$
$m^3 - 6m^2 + 11m - 6 = 0$

Using methods of Chapter 15,
$f(m) = m^3 - 6m^2 + 11m - 6$

$f(1) = 0$; therefore, $m_1 = 1$
$f(2) = 0$; therefore, $m_2 = 2$;
$f(3) = 0$; therefore, $m_3 = 3$
$y = c_1 e^x + c_2 e^{2x} + c_3 e^{3x}$

27. $D^4y - 5D^2y + 4y = 0, m^4 - 5m^2 + 4 = 0$
$m = -2, -1, 1, 2$
$y = c_1 e^x + c_2 e^{-x} + c_3 e^{2x} + c_4 e^{-2x}$

28. $D^4y - D^3y - 9D^2y + 9Dy = 0$
$m^4 - m^3 - 9m^2 + 9m = 0$
$m = -3, 0, 1, 3$
$y = c_1 e^{-3x} + c_2 e^{0x} + c_3 e^x + c_4 e^{3x}$
$y = c_1 e^{-3x} + c_2 + c_3 e^x + c_4 e^{3x}$

15.2 Auxiliary Equation with Repeated or Complex Roots

1. $D^2y - 2Dy + y = 0$; $m^2 - 2m + 1 = 0$;
$(m - 1)^2 = 0$; $m = 1, 1$
$y = e^x(c_1 + c_2 x)$; $y = (c_1 + c_2 x)e^x$

2. $\dfrac{d^2y}{dx^2} - 6\dfrac{dy}{dx} + 9y = 0$; $D^2y - 6Dy + 9y = 0$
$m^2 - 6m + 9 = 0$; $(m - 3)^2 = 0$; $m = 3, 3$
Therefore, $y = e^{3x}(c_1 + c_2 x)$

3. $D^2y + 12Dy + 36y = 0$; $m^2 + 12m + 36 = 0$
$(m + 6)^2 = 0$; $m = -6, -6$
Therefore, $y = e^{-6x}(c_1 + c_2 x)$

4. $16D^2y + 8Dy + y = 0$; $16m^2 + 8m + 1 = 0$
$(4m + 1)^2 = 0$; $m = -\dfrac{1}{4}, -\dfrac{1}{4}$
$y = e^{-x/4}(c_1 + c_2 x)$

5. $D^2y + 9y = 0$; $m^2 + 9 = 0$; $m_1 = 3j$
and $m_2 = -3j$, $\alpha = 0, \beta = 3$
$y = e^{0x}(c_1 \sin 3x + c_2 \cos 3x)$;
$y = c_1 \sin 3x + c_2 \cos 3x$

6. $\dfrac{d^2y}{dx^2} + y = 0$; $D^2y + y = 0$
$m^2 + 1 = 0$; $m = \pm\sqrt{-1} = j$; $\alpha = 0, \beta = 1$
$y = e^0(c_1 \sin x + c_2 \cos x) = c_1 \sin x + c_2 \cos x$

7. $D^2y + Dy + 2y = 0$; $m^2 + m + 2 = 0$

Quadratic formula:

$$m = \frac{-1 \pm \sqrt{1-8}}{2} = \frac{-1 \pm \sqrt{-7}}{2}$$

$$m = -\frac{1}{2} \pm \frac{\sqrt{7}}{2}j; \; \alpha = -\frac{1}{2}, \beta = \frac{\sqrt{7}}{2}$$

$$y = e^{-x/2}\left(c_1 \sin \frac{\sqrt{7}}{2}x + c_2 \cos \frac{\sqrt{7}}{2}x\right)$$

8. $D^2y - 2Dy + 4y = 0$; $m^2 - 2m + 4 = 0$

$$m = \frac{2 \pm \sqrt{4-16}}{2} = \frac{2 \pm \sqrt{-12}}{2}$$

$$m = \frac{2 \pm 2\sqrt{-3}}{2} = 1 \pm \sqrt{3}j; \; \alpha = 1, \beta = \sqrt{3}$$

$$y = e^x(c_1 \sin \sqrt{3}x + c_2 \cos \sqrt{3}x)$$

9. $D^4y - y = 0$
$m^4 - 1 = 0$
$(m^2 - 1)(m^2 + 1) = 0$
$(m - 1)(m + 1)(m^2 + 1) = 0$
$m = \pm 1, \; m = \pm j$
$y = c_1 e^x + c_2 e^{-x} + c_3 \sin x + c_4 \cos(-x)$
$y = c_1 e^x + c_2 e^{-x} + c_3 \sin x + c_4 \cos x$

10. $4D^2y - 12Dy + 9y = 0$; $4m^2 - 12m + 9 = 0$

$$(2m - 3)^2 = 0; \; m = \frac{3}{2}, \frac{3}{2}$$

$$y = e^{3x/2}(c_1 + c_2 x)$$

11. $4D^2y + y = 0$; $4m^2 + 1 = 0$; $m = \sqrt{-\frac{1}{4}} = \pm \frac{1}{2}j$

$$\alpha = 0, \beta = \frac{1}{2}$$

$$y = e^0\left(c_1 \sin \frac{x}{2} + c_2 \cos \frac{x}{2}\right) = c_1 \sin \frac{x}{2} + c_2 \cos \frac{x}{2}$$

12. $9D^2y + 4y = 0$; $9m^2 + 4 = 0$; $m = \pm\sqrt{-\frac{4}{9}} = \pm\frac{2}{3}j$

$$\alpha = 0, \beta = \frac{2}{3}$$

$$y = e^0\left(c_1 \sin \frac{2}{3}x + c_2 \cos \frac{2}{3}x\right)$$

$$= c_1 \sin \frac{2}{3}x + c_2 \cos \frac{2}{3}x$$

13. $16D^2y - 24Dy + 9y = 0$; $16m^2 - 24m + 9 = 0$;

$$(4m - 3)^2 = 0; \; m = \frac{3}{4}, \frac{3}{4}$$

$$y = e^{3x/4}(c_1 + c_2 x)$$

14. $9y'' - 24y' + 16y = 0$; $9D^2y - 24Dy + 16y = 0$

$$9m^2 - 24m + 16 = 0; \; (3m - 4)^2 = 0; \; m = \frac{4}{3}, \frac{4}{3}$$

$$y = e^{(4/3)x}(c_1 + c_2 x)$$

15. $25y'' + 2y = 0$; $25D^2y + 2y = 0$; $25m^2 + 2 = 0$

$$m = \sqrt{-\frac{2}{25}} = \pm\frac{\sqrt{2}}{5}j; \; \alpha = 0, \beta = \frac{\sqrt{2}}{5}$$

$$y = e^0\left(c_1 \sin \frac{\sqrt{2}}{5}x + c_2 \cos \frac{\sqrt{2}}{5}x\right)$$

$$y = c_1 \sin \frac{\sqrt{2}}{5}x + c_2 \cos \frac{\sqrt{2}}{5}x$$

16. $y'' - 4y' = 0$; $D^2y - 4Dy + 5y = 0$; $m^2 - 4m + 5 = 0$

Quadratic formula:

$$m = \frac{4 \pm \sqrt{16-20}}{2} = \frac{4 \pm \sqrt{-4}}{2}$$

$$m = 2 \pm j; \; \alpha = 2, \beta = 1$$

$$y = e^{2x}(c_1 \sin x + c_2 \cos x)$$

17. $2D^2y + 5y = 4Dy$; $2D^2y + 5y - 4Dy = 0$;
$2m^2 - 4m + 5 = 0$

Quadratic formula:

$$m = \frac{4 \pm \sqrt{16-40}}{4} = \frac{4 \pm 2\sqrt{-6}}{4}$$

$$m_1 = 1 + \frac{\sqrt{6}}{2}j; \; m_2 = 1 - \frac{\sqrt{6}}{2}j; \; \alpha = 1, \beta = \frac{1}{2}\sqrt{6}$$

$$y = e^x\left(c_1 \cos \frac{1}{2}\sqrt{6}x + c_2 \sin \frac{1}{2}\sqrt{6}x\right)$$

18. $D^2y + 4Dy + 6y = 0$; $m^2 + 4m + 6 = 0$

$$m = \frac{-4 \pm \sqrt{16-24}}{2} = \frac{-4 \pm 2\sqrt{-2}}{2}$$

$$m = -2 \pm \sqrt{2}; \; \alpha = -2, \beta = \sqrt{2}$$
$$y = e^{-2x}(c_1 \sin \sqrt{2}x + c_2 \cos \sqrt{2}x)$$

19. $25y'' - 40y' + 16y = 0$; $25D^2y - 40 + 16y = 0$

$$25m^2 - 40m + 16 = 0; \; (5m - 4)^2 = 0; \; m = \frac{4}{5}, \frac{4}{5}$$

$$y = e^{4x/5}(c_1 + c_2 x)$$

20. $\quad 9y''' + 0.6y'' + 0.01y' = 0$
$9D^3y + 0.6D^2y + 0.01Dy = 0$
$9m^3 + 0.6m^2 + 0.01m = 0$
$m(9m^2 + 0.6m + 0.01) = 0$

$$m = 0, m = -\frac{1}{30} \text{ (double root)}$$

$$y = c_1 e^{0x} + e^{-x/30}(c_2 + c_3 x)$$
$$y = c_1 + e^{-x/30}(c_2 + c_3 x)$$

21. $2D^2y - 3Dy - y = 0$; $2m^2 - 3m - 1 = 0$

By the quadratic formula,

$$m = \frac{3 \pm \sqrt{9 + 8}}{4}$$

$$m_1 = \frac{3}{4} + \frac{\sqrt{17}}{4},$$

$$m_2 = \frac{3}{4} - \frac{\sqrt{17}}{4}$$

$$y = c_1 e^{((3/4)+(\sqrt{17}/4))x} + c_2 e^{((3/4)+(\sqrt{17}/4))x};$$
$$y = e^{(3/4)x}(c_1 e^{x(\sqrt{17}/4)} + c_2 e^{-x(\sqrt{17}/4)})$$

22. $D^2y - 5Dy - 4y = 0$; $m^2 - 5m - 4 = 0$

$$m = \frac{5 \pm \sqrt{25 + 16}}{2}$$

$$= \frac{5 \pm \sqrt{41}}{2}$$

$$m_1 = \frac{5 + \sqrt{41}}{2},$$

$$m_2 = \frac{5 - \sqrt{41}}{2}$$

$$y = c_1 e^{((5/2)+(\sqrt{41}/2))x} + c_2 e^{((5/2)-(\sqrt{41}/2))x}$$
$$y = e^{5x/2}(c_1 e^{\sqrt{41}x/2} + c_2 e^{-\sqrt{41}x/2})$$

23. $3D^2y + 12Dy - 2 = 0$; $3m^2 + 12m - 2 = 0$

$$m = \frac{-12 \pm \sqrt{144 + 24}}{6}$$

$$= \frac{-12 \pm \sqrt{168}}{6}$$

$$m_1 = -2 + \frac{1}{3}\sqrt{42},$$

$$m_2 = -2 - \frac{1}{3}\sqrt{42}$$

$$m = \frac{-6 \pm \sqrt{42}}{3}$$

$$y = c_1 e^{(-2+(1/3)\sqrt{42})x} + c_2 e^{(-2-(1/3)\sqrt{42})x}$$
$$y = c_1 e^{(-6+\sqrt{42})(x/3)} + c_2 e^{(-6-\sqrt{42})(x/3)}$$

24. $36D^2y - 25y = 0$; $36m^2 - 25 = 0$

$$m = \sqrt{\frac{25}{36}} = \pm\frac{5}{6};\ m_1 = \frac{5}{6},\ m_2 = -\frac{5}{6}$$

$$y = c_1 e^{(5/6)x} + c_2 e^{-(5/6)x}$$

25. $D^3y - 6D^2y + 12Dy - 8y = 0$

$$m^3 - 6m^2 + 12m - 8 = 0$$
$$(m - 2)(m^2 - 4m + 4) = 0$$
$$(m - 2)(m - 2)(m - 2) = 0$$
$$m = 2, 2, 2 \text{ repeated root}$$
$$y = e^{2x}(c_1 + c_2 x + c_3 x^2)$$

26. $D^4y - 2D^3y + 2D^2y - 2Dy + y = 0$

$$m^4 - 2m^3 + 2m^2 - 2m + 1 = 0$$
$$(m - 1)^2 \cdot (m^2 + 1) = 0$$
$$m = 1 \text{ (double root)},$$
$$m = 0 \pm j, \alpha = 0, \beta = 1$$

$$y = e^x(c_1 + c_2 x) + e^{0x}(c_3 \sin x + c_4 \cos x)$$
$$y = e^x(c_1 + c_2 x) + c_3 \sin x + c_4 \cos x$$

27. $D^4y + 2D^2y + y = 0$

$$m^4 + 2m^2 + 1 = 0$$
$$m = j \text{ (double root)},$$
$$m = -j \text{ (double root)}$$
$$y = (c_1 + c_2 x)\sin x + (c_3 + c_4 x)\cos x$$

28. $2D^4y - y = 0$

$$2m^4 - 1 = 0$$

$$m = \pm\frac{\sqrt[4]{8}}{2},$$

$$m = \pm\frac{\sqrt[4]{8}}{2}j$$

$$y = c_1 e^{(\sqrt[4]{8}/2)x} + c_2 e^{(-\sqrt[4]{8}/2)x} + c_3 \sin[(\sqrt[4]{8}/2)x]$$
$$+ c_4 \cos[(\sqrt[4]{8}/2)x]$$

29. $D^2y + 2Dy + 10y = 0$; $m^2 + 2m + 10 = 0$

By the quadratic formula, $m = \dfrac{-2 \pm \sqrt{4 - 40}}{2}$

$$m_1 = -1 + 3j;\ m_2 = -1 - 3j,\ \alpha = -1,\ \beta = 3;$$
$$y = e^{-x}(c_1 \sin 3x + c_2 \cos 3x)$$

Substituting $y = 0$ when $x = 0$;

$$0 = e^0(c_1 \sin 0 + c_2 \cos 0);\ c_2 = 0$$

Substituting $y = e^{-\pi/6}$, $x = \dfrac{\pi}{6}$;

$$e^{-\pi/6} = e^{-\pi/6}\left(c_1 \sin\frac{\pi}{2}\right);\ e^{-\pi/6} = e^{-\pi/6}c_1;$$
$$c_1 = 1$$
$$y = e^{-x}\sin 3x$$

30. $9D^2y + 16y = 0$; $Dy = 0$ and $y = 2$ when $x = \dfrac{\pi}{2}$

$$9m^2 + 16 = 0;\ m = \pm\sqrt{-\frac{16}{9}} = \frac{4}{3}j;\ \alpha = 0, \beta = \frac{4}{3}$$

$$y = e^0\left(c_1 \sin\frac{4}{3}x + c_2 \cos\frac{4}{3}x\right) = c_1 \sin\frac{4}{3}x + c_2 \cos\frac{4}{3}x$$

$$\frac{dy}{dx} = Dy = \frac{4}{3}c_1 \cos\frac{4}{3}x - \frac{4}{3}c_2 \sin\frac{4}{3}x$$

Substituting given values:

$$2 = c_1 \frac{\sqrt{3}}{2} - c_2 \frac{1}{2} \text{ and } 0 = -\frac{4}{3}c_1 \frac{1}{2} - \frac{4}{3}c_2 \frac{\sqrt{3}}{2}$$

Therefore, $c_1 = -\sqrt{3}, c_2 = -1$

$$y = \sqrt{3}\sin\frac{4}{3}x - \cos\frac{4}{3}x$$

31. $D^2y - 8Dy + 16y = 0$; $Dy = 2$ and $y = 4$ when
$x = 0$
$m^2 - 8m + 16 = 0$; $(m - 4)^2 = 0$; $m = 4, 4$
$y = e^{4x}(c_1 + c_2x)$
$Dy = e^{4x}(c_2) + (c_1 + c_2x)4e^{4x} = e^{4x}(c_2 + 4c_1 + 4c_2x)$

Substituting given values:

$4 = e^0(c_1 + 0)$; $c_1 = 4$
$2 = e^0[c_2 + 4(4) + 0]$; $c_2 = -14$

Therefore, $y = e^{4x}(4 - 14x)$

32. $D^4y + 3D^3y + 2D^2y = 0$; $y = 0, Dy = 4$,
$D^2y = -8, D^3y = 16$ when $x = 0$
$m^4 + 3m^3 + 2m^2 = 0$
$m^2(m + 1)(m + 2) = 0$
$m = 0$ (double root), $m = -1, m = -2$
$y = e^{0x}(c_1 + c_2x) + c_3e^{-x} + c_4e^{-2x}$
$y = c_1 + c_2x + c_3e^{-x} + c_4e^{-2x}$
$D^3y = -c_3e^{-x} - 8c_4e^{-2x}$
$D^2y = c_3e^{-x} + 4c_4e^{-2x}$
$Dy = c_2 - c_3e^{-x} - 2c_4e^{-2x}$
$\qquad y = 0$ when $x = 0 \Rightarrow c_1 \qquad + c_3 + c_4 = 0$
$\qquad Dy = 4$ when $x = 0 \Rightarrow \qquad c_2 - c_3 - 2c_4 = 4$
$\qquad D^2y = -8$ when $x = 0 \Rightarrow \qquad c_3 + 4c_4 = -8$
$\qquad D^3y = 16$ when $x = 0 \Rightarrow \qquad - c_3 - 8c_4 = 16$
$c_1 = 2, c_2 = 0, c_3 = 0, c_4 = -2$
$y = 2 - 2e^{-2x}$

33. $y = c_1e^{3x} + c_2e^{-3x}$
$(m - 3)(m + 3) = 0$
$m^2 - 9 = 0$; $(D^2 - 9)y = 0$

34. $y = c_1e^{3x} + c_2xe^{3x} = e^{3x}(c_1 + c_2x)$
$m = 3$, double root, $(m - 3)^2 = 0$
$y = m^2 - 6m + 9$, $(D^2 - 6D + 9)y = 0$

35. $y = c_1 \cos 3x + c_2 \sin 3x$
Complex roots: $\alpha = 0, \beta = 3j$ (imaginary)
$m = \pm 3j$; $m^2 = -9$; therefore, $m^2 + 9 = 0$
Therefore, $(D^2 + 9)y = 0$; $D^2y + 9y = 0$

36. $y = c_1e^{2x} \cos x + c_2e^{2x} \sin x$
Complex roots of form $\alpha \pm \beta j$; $\alpha = 2, \beta = 1$
$m_1 = 2 + j, m_2 = 2 - j$
$[m - (2 + j)][m - (2 - j)] = 0$
$m^2 - 2m + jm - 2m - jm + (2 + j)(2 - j) = 0$
$m_2 - 4m + 5 = 0$
$(D^2 - 4D + 5)y = 0$

15.3   Solutions of Nonhomogeneous Equations

1. $D^2y - Dy - 2y = 4$; $m^2 - m - 2 = 0$
$(m - 2)(m + 1) = 0$; $m_1 = 2, m_2 = -1$
$y_c = c_1e^{2x} + c_2e^{-x}$
$y_p = A$; $Dy_p = 0$; $D^2y_p = 0$

Substituting in diff. equation, $0 - 0 - 2A = 4$;
$A = -2$
Therefore, $y_p = -2$; $y = c_1e^{2x} + c_2e^{-x} - 2$

2. $D^2y - Dy - 6y = 4x$; $D^2y - Dy - 6y = 0$;
$m^2 - m - 6 = 0$
$(m - 3)(m + 2) = 0$; $m_1 = 3, m_2 = -2$
$y_c = c_1e^{3x} + c_2e^{-2x}$; $y_p = A + Bx$;
$Dy_p = B$; $D^2y_p = 0$

Substituting in diff. eq.:

$0 - B - 6(A + Bx) = 4x$; $-B - 6A - 6Bx = 4x$

$-(B + 6A) - 6Bx = 4x$; $-6B = 4$; $B = -\dfrac{2}{3}$

$-(B + 6A) = 0$; $6A = -B = \dfrac{2}{3}$; $A = \dfrac{2}{18} = \dfrac{1}{9}$

Therefore, $y = c_1e^{3x} + c_2e^{-2x} + \dfrac{1}{9} - \dfrac{2}{3}x$

3. $D^2y - y = 2 + x^2$; $m^2 - 1 = 0$; $m = \pm 1$,
$m_1 = 1, m_2 = -1$
$y_c = c_1e^x + c_2e^{-x}$; $y_p = A + Bx + Cx^2$;
$Dy_p = B + 2Cx$; $D^2y_p = 2C$; substituting in diff.
eq.:
$2C - (A + Bx + Cx^2) = 2 + x^2$;
$2C - A - Bx - Cx^2 = 2 + x^2$
$-C = 1$; $-B = 0$; $2C - A = 2$
$C = -1, B = 0, A = 2C - 2 = -4$
$y_p = -4 - x^2$
Therefore, $y = c_1e^x + c_2e^{-x} - 4 - x^2$

4. $D^2y + 4Dy + 3y = 2 + e^x$; $m^2 + 4m + 3 = 0$
$(m + 1)(m + 3) = 0$; $m_1 = -1, m_2 = -3$
$y_c = c_1e^{-x} + c_2e^{-3x}$
$y_p = A + Be^x$; $Dy_p = Be^x$; $D^2y_p = Be^x$

Substituting in diff. eq.:

$Be^x + 4Be^x + 3(A + be^x) = 2 + e^x$
$3A + 8Be^x = 2 + e^x$; $3A = 2$

$A = \dfrac{2}{3}, 8B = 1$; $B = \dfrac{1}{8}$; $y_p = \dfrac{2}{3} + \dfrac{1}{8}e^x$

$y = c_1e^{-x} + c_2e^{-3x} + \dfrac{1}{8}e^x + \dfrac{2}{3}$

5. $y'' - 3y' = 2e^x + xe^x$; $D^2y - 3Dy = 2e^x + xe^x$

$m^2 - 3m = 0$; $m(m-3) = 0$; $m_1 = 0, m_2 = 3$

$y_c = c_1e^0 + c_2e^{3x} = c_1 + c_2e^{3x}$; $y_p = Ae^x + Bxe^x$

$Dy_p = Ae^x + B(xe^x + e^x)$

$\quad = Ae^x + Bxe^x + Be^x$

$D^2y_p = Ae^x + Be^x + B(xe^x + e^x)$

$\quad = Ae^x + Be^x + Bxe^x + Be^x$

$D^2y_p = Ae^x + 2Be^x + Bxe^x$

Substituting in diff. equation:

$Ae^x + 2Be^x + Bxe^x - 3(Ae^x + Bxe^x + Be^x)$

$\quad = 2e^x + xe^x$

$Ae^x + 2Be^x + Bxe^x - 3Ae^x - 3Bxe^x - 3Be^x$

$\quad = 2e^x + xe^x$

$-2Ae^x - Be^x - 2Bxe^x$

$\quad = 2e^x + xe^x$

$e^x(-2A - B) + xe^x(-2B)$

$\quad 2e^x + xe^x$

$-2A - B = 2$; $-2A + \dfrac{1}{2} = 2$; $-2A = \dfrac{3}{2}$;

$A = -\dfrac{3}{4}$; $-2B = 1$; $B = -\dfrac{1}{2}$

$y = c_1 + c_2e^{3x} - \dfrac{3}{4}e^x - \dfrac{1}{2}xe^x$

6. $y'' + y' - 2y = 8 + 4x + 2xe^{2x}$

$D^2y + Dy - 2y$; $m^2 + m - 2 = 0$

$(m+2)(m-1) = 0$; $m_1 = 1, m_2 = -2$

$y_c = c_1e^{-2x} + c_2e^x + y_p$;

$y_p = A + Bx + Ce^{2x} + Exe^{2x}$

$Dy_p = B + 2Ce^{2x} + E(x2e^{2x} + e^{2x})$

$Dy_p = B + 2Ce^{2x} + 2Exe^{2x} + Ee^{2x}$

$D^2y_p = 4Ce^{2x} + 2Ee^{2x} + 2E(x2e^{2x} + e^{2x})$

$D^2y_p = 4Ce^{2x} + 2Ee^{2x} + 4Exe^{2x} + 2Ee^{2x}$

$D^2y_p = 4Ce^{2x} + 4Ee^{2x} + 4Exe^{2x}$

$\quad = e^{2x}(4C + 4E) + 4Exe^{2x}$

Substituting diff. eq."

$4Ce^{2x} + 4Ee^{2x} + 4Exe^{2x} + B + 2Ce^{2x} + 2Exe^{2x}$

$\quad + Ee^{2x} - 2 - 2(A + Bx + Ce^{2x} + Exe^{2x})$

$6Ce^{2x} + 5Ee^{2x} + 6Exe^{2x} + B - 2A - 2Bx - 2Ce^{2x}$

$\quad - 2Exe^{2x}$

$4Ce^{2x} + 4Exe^{2x} + 5Ee^{2x} + B - 2A - 2Bx$

$\quad = 8 + 4x + 2xe^{2x}$

$e^{2x}(4C + 5E) + xe^{2x}(4E) + (B - 2A) + x(-2B)$

$\quad = 8 + 4x + 2xe^{2x}$

$4C + 5E = 0$; $4C = -5E$;

$C = -\dfrac{5}{4}\left(\dfrac{1}{2}\right)$; $C = -\dfrac{5}{8}$

$4E = 2$; $E = \dfrac{1}{2}$

$B - 2A = 8$; $-2 - 2A = 8$; $-2A = 10$; $A = -5$

$-2B = 4$; $B = -2$

$y = c_1e^{-2x} + c_2e^x - 5 - 2x - \dfrac{5}{8}e^{2x} + \dfrac{1}{2}xe^{2x}$

7. $9D^2y - y = \sin x$; let $y_p = A\sin x + B\cos x$

$9m^2 - 1 = 0$; $m = -\dfrac{1}{3}, \dfrac{1}{3}$

$y_c = c_1e^{-(1/3)x} + c_2e^{(1/3)x}$; $y_p = A\sin x + B\cos x$

$Dy_p = A\cos x - B\cos x$; $D^2y_p = -A\sin x - B\cos x$

Substituting in diff. eq.:

$9(-A\sin x - B\cos x) - (A\sin x + B\cos x) = \sin x$

$-10\sin x - 10\cos x = \sin x$

$-10A = 1$; $A = -\dfrac{1}{10}$; $-10B = 0$; $B = 0$

$y_p = -\dfrac{1}{10}\sin x + 0\cos x = -\dfrac{1}{10}\sin x$

$y = c_1e^{(1/3)x} + c_2e^{-(1/3)x} - \dfrac{1}{10}\sin x$

8. $D^2y + 4y = \sin x + 4$; $m^2 + 4 = 0$; $m = \sqrt{-4} = 2j$

$\alpha = 0, \beta = 2$

$y_c = e^0(c_1\sin 2x + c_2\cos 2x) = c_1\sin 2x + c_2\cos 2x$

$y_p = A + B\sin x + c\cos x$; $dy_p = B\cos x - c\sin x$

$D^2y = -B\sin x - c\cos x$

Substituting in diff. eq.:

$-B\sin x - C\cos x + 4(A + B\sin x + C\cos x)$

$\quad = \sin x + 4$

$-B\sin x - C\cos x + 4A + 4B\sin x + 4C\cos x$

$\quad = \sin x + 4$

$3B\sin x + 3C\cos x + 4A = \sin x + 4$

$3B = 1$; $B = \dfrac{1}{3}$; $3C = 0$; $C = 0$; $4A = 4$; $A = 1$

$y = c_1\sin 2x + c_2\cos 2x + 1 + \dfrac{1}{3}\sin x$

9. $\dfrac{d^2y}{dx^2} - 2\dfrac{dy}{dx} + y = 2x + x^2 + \sin 3x$

$D^2y - 2Dy + y = 2x + x^2 + \sin 3x$

$m^2 - 2m + 1 = 0$; $(m-1)^2 = 0$; $m = 1, 1$

$y_c = e^x(c_1 + c_2x)$;

$y_p = A + Bx + Cx^2 + E\sin 3x + F\cos 3x$

$Dy_p = B + 3Cx + 3E\cos 3x - 3F\sin 3x$

$D^2y_p = 2c - 9E\sin 3x - 9F\cos 3x$

Substituting in diff. equation:

$2C - 9E \sin 3x - 9F \cos 3x$
$\quad - 2(B + 2Cx + 3E \cos 3x - 3F \sin 3x)$
$\quad + A + Bx + Cx^2 + E \sin 3x + F \cos 3x$
$\quad = 2x + x^2 + \sin 3x$

$2C - 9E \sin 3x - 9F \cos 3x - 2B - 4Cx - 6E \cos 3x$
$\quad + 6F \sin 3x + A + Bx + Cx^2 + E \sin 3x + F \cos 3x$
$\quad = 2x + x^2 + \sin 3x$

$2C - 2B + A - 8E \sin 3x + 6F \sin 3x - 8F \cos 3x$
$\quad - 6E \cos 3x - 4Cx + Bx + Cx^2$
$\quad = 2x + x^2 + \sin 3x$

$(2C - 2B + A) + \sin 3x(6F - 8E) + \cos 3x(-8F - 6E)$
$\quad + x(B - 4) + Cx^2$
$\quad = 2x + x^2 + \sin 3x$

$2C - 2B + A = 0;\ A = 10$

$6F - 8E = 1;\ -8F - 6E = 0;\ E = -\dfrac{2}{25};$

$F = \dfrac{3}{50};\ B - 4 = 2;\ B = 6;\ C = 1$

$y = e^x(c_1 + c_2x) + 10 + 6x + x^2$
$\quad - \dfrac{2}{25} \sin 3x + \dfrac{3}{50} \cos 3x$

10. $D^2y - y = e^{-x}$
$m^2 - 1 = 0, m = \pm 1$
$y_c = c_1e^x + c_2e^{-x}$
$y_p = Axe^{-x}$
$D^2y_p = Ae^{-x}(x - 2) = Axe^{-x} - 2Ae^{-x}$
$D^2y_p - y_p = Axe^{-x} - 2Ae^{-x} - Axe^{-x}$
$\qquad = e^{-x} \Rightarrow -2A = 1,$

$A = \dfrac{-1}{2}$

$y_p = \dfrac{-1}{2}xe^{-x}$

$y = c_1e^x + c_2e^{-x} - \dfrac{1}{2}xe^{-x}$

11. $D^2y + 4y = -12 \sin 2x$
$m^2 + 4 = 0,, m = 0 \pm 2j$
$y_c = c_1 \sin 2x + c_2 \cos 2x$
$y_p = Ax \sin 2x + Bx \cos 2x$
$D^2y_p = 4A \cos 2x - 4Bx \cos 2x - 4Ax \sin 2x - 4B \sin 2x$
$D^2y_p + 4y_p$
$\quad = 4A \cos 2x - 4Bx \cos 2x - 4A \sin 2x$
$\quad\quad - 4B \sin 2x + 4Ax \sin 2x + 4Bx \cos 2x$
$\quad = -12 \sin 2x$
$D^2y_p + 4y_p$
$\quad = 4A \cos 2x - 4B \sin 2x = -12 \sin 2x$
$4A = 0, A = 0;\ -4B = -12, B = 3$
$y = c_1 \sin 2x + c_2 \cos 2x + 3x \cos 2x$

12. $D^2y - 2Dy + y = 3 + e^x;\ m^2 - 2m + 1 = 0,$
$m = 1$ (double root)
$y_c = e^x(c_1 + c_2x);\ y_p = A + Bx^2e^x$
$D^2y_p = Bx^2e^x + 4Bxe^x + 2Be^x\ Dy_p$
$\qquad = Bx^2e^x + 2Bxe^x$
$D^2y_p - 2Dy_p + y_p$
$\qquad = Bx^2e^x + 4Bxe^x + 2Be^x - 2Bx^2e^x - 4Bxe^x$
$\qquad\quad + A + Bx^2e^x$
$D^2y_p - 2Dy_p + y_p$
$\qquad = A + 2Be^x = 3 + e^x \Rightarrow A = 3;\ 2B = 1, B = \dfrac{1}{2}$
$y = e^x(c_1 + c_2x) + 3 + \dfrac{1}{2}x^2e^x$

13. $D^2y - Dy - 30y = 10;\ m^2 - m - 30 = 0;$
$(m - 6)(m + 5) = 0;\ m_1 = -5;\ m_2 = 6$
$y_c = c_1e^{-5x} + c_2e^{6x};\ y_p = A;\ Dy_p = 0;$

$D^2y_p = 0;\ 0 = 0 - 30A = 10;\ A = -\dfrac{1}{3};$

$y_p = -\dfrac{1}{3};\ y = c_1e^{-5x} + c_2e^{6x} - \dfrac{1}{3}$

14. $2\dfrac{d^2y}{dx^2} + 11\dfrac{dy}{dx} - 6y = 8x;$ therefore, $y_p = A + Bx$

$2D^2y + 11Dy - 6y = 8x;\ 2m^2 + 11m - 6 = 0$

$m = \dfrac{-11 \pm \sqrt{121 + 48}}{4} = \dfrac{-11 \pm \sqrt{169}}{4}$

$m_1 = \dfrac{1}{2},\ m_2 = -6;\ y_c = c_1e^{-6x} + c_2e^{(1/2)x}$

$y_p = A + Bx;\ Dy_p = B;\ D^2y_p = 0$
$2(0) + 11B - 6(A + Bx) = 8x;\ 11B - 6A - 6Bx = 8x$

$11B - 6A = 0;\ A = -\dfrac{22}{9};\ -6B = 8;\ B = -\dfrac{4}{3}$

$y = c_1e^{-6x} + c_2e^{x/2} - \dfrac{22}{9} - \dfrac{4x}{3}$

15. $3\dfrac{d^2y}{dx^2} + 13\dfrac{dy}{dx} - 10y = 14e^{3x};\ y_p = Ae^{3x}$

$3D^2y + B\,Dy - 10y = 14e^{3x};\ 3m^2 + 13m - 10 = 0$

$(3m - 2)(m + 5) = 0;\ m_1 = -5,\ m_2 = \dfrac{2}{3}$

$y_c = c_1e^{-5x} + c_2e^{(2/3)x};\ y_p = Ae^{3x};\ Dy_p = 3Ae^{3x};$
$D^2y_p = 9Ae^{3x}$
$3(9Ae^{3x}) + 13(3Ae^{3x}) - 10(Ae^{3x}) = 14e^{3x}$
$27Ae^{3x} + 39Ae^{3x} - 10Ae^{3x} = 14e^{3x}$

$56Ae^{3x} = 14e^{3x};\ 56A = 14;\ A = \dfrac{14}{56} = \dfrac{1}{4}$

$y = c_1e^{-5x} + c_2e^{(2/3)x} + \dfrac{1}{4}e^{3x}$

16. $\dfrac{d^2y}{dx^2} + 4y = 2\sin 3x$; $y_p = A\sin 3x + B\cos 3x$

$D^2y + 4y = 2\sin 3x$; $m^2 + 4 = 0$; $m = \sqrt{-4} = 2j$;
$\alpha = 0, \beta = 2$
$y_c = c_1\sin 2x + c_2\cos 2x$; $y_p = A\sin 3x + B\cos 3x$
$Dy_p = 3A\cos 3x - 3B\sin 3x$;
$D^2y_p = -9A\sin 3x - 9B\cos 3x$
$-9A\sin 3x - 9B\cos 3x + 4(A\sin 3x + B\cos 3x)$
$\quad = 2\sin 3x$
$-9A\sin 3x - 9B\cos 3x + 4A\sin 3x + 4B\cos 3x$
$\quad = 2\sin 3x$
$\sin 3x(-9A + 4A) + \cos 3x(-9B + 4B) = 2\sin 3x$

$-5A = 2$; $A = -\dfrac{2}{5}$; $-5B = 0$; $B = 0$

$y = c_1\sin 2x + c_2\cos 2x - \dfrac{2}{5}\sin 3x$

17. $D^2y - 4y = \sin x + 2\cos x$; $m^2 - 4 = 0$;
$m_1 = 2$; $m_2 = -2$
$y_c = c_1e^{2x} + c_2e^{-2x}$
$y_p = A\sin x + B\cos x$; $Dy_p = A\cos x - B\sin x$
$D^2y_p = -A\sin x - B\cos x$
$-A\sin x - B\cos x - 4A\sin x - 4B\cos x$
$\quad = \sin x + 2\cos x$
$-5A\sin x - 5B\cos x$
$\quad = \sin x + 2\cos x$

$-5A = 1$; $A = -\dfrac{1}{5}$; $-5B = 2$; $B = -\dfrac{2}{5}$

$y_p = -\dfrac{1}{5}\sin x - \dfrac{2}{5}\cos x$

$y = c_1e^{2x} + c_2e^{-2x} - \dfrac{1}{5}\sin x - \dfrac{2}{5}\cos x$

18. $2D^2y + 5Dy - 3y = e^x + 4e^{2x}$; $y_p = Ae^x + Be^{2x}$
$2m^2 + 5m - 3 = 0$; $(2m - 1)(m + 3) = 0$;

$m_1 = \dfrac{1}{2}, m_2 = -3$

$y_c = c_1e^{x/2} + c_2e^{-3x}$; $y_p = Ae^x + Be^{2x}$
$Dy_p = Ae^x + 2Be^{2x}$; $D^2y_p = Ae^x + 4Be^{2x}$
$2(Ae^x + 4Be^{2x}) + 5(Ae^x + 2Be^{2x}) - 3(Ae^x + Be^{2x})$
$\quad = e^x + 4e^{2x}$
$2Ae^x + 8Be^{2x} + 5Ae^x + 10Be^{2x} - 3Ae^x - 3Be^{2x}$
$\quad = e^x + 4e^{2x}$
$4Ae^x + 15Be^{2x} = e^x + 4e^{2x}$

$4A = 1$; $A = \dfrac{1}{4}$; $15B = 4$; $B = \dfrac{4}{15}$

$y = c_1e^{x/2} + c_2e^{-3x} + \dfrac{1}{4}e^x + \dfrac{4}{15}e^{2x}$

19. $D^2y + y = 4 + \sin 2x$; $y_p = A + B\sin 2x + C\cos 2x$
$m^2 + 1 = 0$; $m = \sqrt{-1} = \pm j$; $\alpha = 0, \beta = 1$
$y_c = c_1\sin x + c_2\cos x$; $y_p = A + B\sin 2x + C\cos 2x$
$Dy_p = 2B\cos 2x - 2C\sin 2x$;
$D^2y_p = -4B\sin 2x - 4C\cos 2x$
$-4B\sin 2x - 4C\cos 2x + A + B\sin 2x + C\cos 2x$
$\quad = 4 + \sin 2x$
$\sin 2x(-4B + B) + \cos 2x(-4C + C) + A = 4 + \sin 2x$

$-3B = 1$; $B = -\dfrac{1}{3}$; $-3C = 0$; $C = 0$; $A = 4$

$y = c_1\sin x + c_2\cos x + 4 - \dfrac{1}{3}\sin 2x$

20. $D^2y - Dy + y = x + \sin x$;
$y_p = A + Bx + C\sin x + E\cos x$
$m^2 - m + 1 = 0$

$m = \dfrac{1 \pm \sqrt{1 - 4}}{2} = \dfrac{1 \pm \sqrt{-3}}{2} = \dfrac{1}{2} \pm \dfrac{\sqrt{3}}{2}j$

$\alpha = \dfrac{1}{2}$; $\beta = \dfrac{\sqrt{3}}{2}$

$y_c = e^{x/2}\left(c_1\sin\dfrac{\sqrt{3}}{2}x + c_2\cos\dfrac{\sqrt{3}}{2}x\right)$

$y_p = A + Bx + C\sin x + E\cos x$;
$Dy_p = B + C\cos x - E\sin x$
$D^2y_p = -C\sin x - E\cos x$
$-C\sin x - E\cos x - (B + C\cos x - E\sin x) + A + Bx$
$\quad + C\sin x + E\cos x = x + \sin x$
$-C\sin x - E\cos x - B - C\cos x + E\sin x + A + Bx$
$\quad + C\sin x + E\cos x = x + \sin x$
$\cos x(-C) + \sin x(E) + A - B + Bx = x + \sin x$
$-C = 0$; $C = 0$; $E = 1$; $A - B = 0$; $A = 1$; $B = 1$

$y = e^{x/2}\left(c_1\sin\dfrac{\sqrt{3}}{2}x + c_2\cos\dfrac{\sqrt{3}}{2}x\right) + 1 + x + \cos x$

21. $D^2y + 5Dy + 4y = xe^x + 4$; $m^2 + 5m + 4 = 0$;
$(m + 1)(m + 4) = 0$; $m_1 = -1, m_2 = -4$
$y_c = c_1e^{-x} + c_2e^{-4x}$; $y_p = Ae^x + Bxe^x + C$
$Dy_p = Ae^x + B(xe^x + e^x) = Ae^x + Bxe^x + Be^x$
$D^2y_p = Ae^x - B(xe^x + e^x) + Be^x$
$\quad = Ae^x + 2Be^x + Bxe^x$
$Ae^x + 2Be^x + Bxe^x + 5(Ae^x + Bxe^x + Be^x)$
$\quad + 4(Ae^x + Bxe^x + C) = xe^x + 4$
$(10A + 7B)e^x + 10Bxe^x + 4C = xe^x + 4$

$10A + 7B = 0$; $10B = 1$; $B = \dfrac{1}{10}$

$4C = 4$; $C = 1$; $10A + 7\left(\dfrac{1}{10}\right) = 0$

$A = -\dfrac{7}{100};\ y_p = -\dfrac{7}{100}e^x + \dfrac{1}{10}e^x + 1$

$y = c_1 e^{-x} + c_2 e^{-4x} - \dfrac{7}{100}e^x + \dfrac{1}{10}xe^x + 1$

22. $3D^2 y + Dy - 2y = 4 + 2x + e^x;\ y_p = A + Bx + Ce^x$
$3m^2 + m - 2 = 0;\ (3m - 2)(m + 1) = 0;$

$m_1 = \dfrac{2}{3},\ m_2 = -1$

$y_c = c_1 e^{(2/3)x} + c_2 e^{-x};\ y_p = A + Bx + Ce^x$
$Dy_p = B + Ce^x,\ D^2 y_p = Ce^x$
$3Ce^x + B + Ce^x - 2(A + Bx + Ce^x) = 4 + 2x + e^x$
$3Ce^x + B + Ce^x - 2A - 2Bx - 2Ce^x = 4 + 2x + e^x$
$2Ce^x + B - 2A - 2Bx = 4 + 2x + e^x$

$2C = 1;\ C = \dfrac{1}{2};\ B - 2A = 4;\ A = -\dfrac{5}{2}$

$-2B = 2;\ B = -1$

$y = c_1 e^{(2/3)x} + c_2 e^{-x} - \dfrac{5}{2} - x + \dfrac{1}{2}e^x$

23. $y''' - y' = \sin 2x;\ m^3 - m = 0,\ m = 0, m = \pm 1$
$y_c = c_1 + c_2 e^x + c_3 e^{-x}$
$y_p = A\cos 2x,\ y_p''' = 8A\sin 2x,\ y_p' = -2A\sin 2x$

$y_p''' - y_p' = 8A\sin 2x - (-2A\sin 2x)$

$\qquad = \sin 2x \Rightarrow A = \dfrac{1}{10}$

$y = c_1 + c_2 e^x + c_3 e^{-x} + \dfrac{1}{10}\cos 2x$

24. $D^4 y - y = x,\ m^4 - 1 = 0,\ m = \pm 1, \pm i$
$y_c = c_1 e^x + c_2 e^{-x} + e^{0 \cdot x}(c_3 \sin x + c_4 \cos x)$
$y_c = c_1 e^x + c_2 e^{-x} + c_3 \sin x + c_4 \cos x$
$y_p = A + Bx$
$D^4 y_p = 0,\ D^4 y_p - y_p = 0 - A - Bx$
$\qquad = x \Rightarrow B = -1, A = 0$
$y_p = -x$
$y = y_c + y_p = c_1 e^x + c_2 e^{-x} + c_3 \sin x + c_4 \cos x - x$

25. $D^2 y + y = \cos x$
$\qquad m^2 + 1 = 0$
$\qquad\qquad m = \pm i$

$y_c = c_1 \sin x + c_2 \cos x$

let $\quad y_p = x(A\sin x + B\cos x)$
$\quad Dy_p = x(A\cos x - B\sin x) + A\sin x + B\cos x$
$\quad D^2 y_p = x(-A\sin x - B\cos x) + A\cos x$
$\qquad\qquad - B\sin x + A\cos x - B\sin x$

$D^2 y_p + y_p = \cos x$

$-x(A\sin x + B\cos x) + 2A\cos x - 2B\sin x$
$\quad + x(A\sin x + B\cos x) = \cos x$

$2A\cos x - 2B\sin x = \cos x$
$2A = 1, B = 0$

$A = \dfrac{1}{2}$

$y_p = \dfrac{1}{2}x\sin x$

$y = y_c + y_p$

$y = c_1 \sin x + c_2 \cos x + \dfrac{1}{2}x\sin x$

26. $4y'' - 4y' + y = 4e^{x/2};\ 4m^2 - 4m + 1 = 0,$

$m = \dfrac{1}{2}$ (double root)

$y_c = e^{x/2}(c_1 + c_2 x)$

$y_p = Ax^2 e^{x/2},\ y_p'' = \dfrac{1}{4}Ae^{x/2}(x^2 + 8x + 8),$

$y_p' = \dfrac{1}{2}Axe^{x/2}(x + 4)$

$4y_p'' - 4y_p' + y_p$
$\quad = Ae^{x/2}(x^2 + 8x + 8) - 2Axe^{x/2}(x + 4) + Ax^2 e^{x/2}$
$\quad = 4e^{x/2}$

$8Ae^{x/2} - 4e^{x/2} \Rightarrow A = \dfrac{1}{2}$

$y = y_c + y_p = e^{x/2}(c_1 + c_2 x) + \dfrac{1}{2}x^2 e^{x/2}$

27. $D^2 y + 2Dy = 8x + e^{-2x};\ m^2 + 2m = 0;\ m = 0, -2$
$y_c = c_1 + c_2 e^{-2x}$
$y_p = Ax^2 + Bx + Cxe^{-2x};$
$y_p' = B + 2Ax + Ce^{-2x}(1 - 2x)$
$y_p'' = 2A + 4Ce^{-2x}(x - 1)$
$y_p'' + 2y_p' = 4Ax + 2A + 2B - 2Ce^{-2x} = 8x + e^{-2x}$
$4A = 8,\quad A + B = 0,\quad -2C = 1$

$A = 2\qquad\quad B = -2\qquad C = \dfrac{-1}{2}$

$y = c_1 + c_2 e^{-2x} + 2x^2 - 2x - \dfrac{1}{2}xe^{-2x}$

28. $D^3 y - Dy = 4e^{-x} + 3e^{2x};\ m^3 - m = 0, m = 0, \pm 1$
$y_c = c_1 + c_2 e^{-x} + c_3 e^x$
$y_p = Axe^{-x} + Be^{2x}$
$y_p' = 2Be^{2x} + Ae^{-x} - Axe^{-x}$
$y_p'' = 8Be^{2x} + 3Ae^{-x} - Axe^{-x}$
$y_p''' - y_p' = 6Be^{2x} + 2Ae^{-x} = 4e^{-x} + 3e^{2x}$
$2A = 4,\quad 6B = 3$

$A = 2\qquad B = \dfrac{1}{2}$

$y = c_1 + c_2 e^{-x} + c_3 e^x + 2xe^{-x} + \dfrac{1}{2}e^{2x}$

29. $D^2y - Dy - 6y = 5 - e^x$; $m^2 - m - 6 = 0$;
$(m-3)(m+2) = 0$; $m_1 = 3, m_2 = -2$
$y_c = c_1 e^{3x} + c_2 e^{-2x}$
$y_p = A + Be^x$; $Dy_p = Be^x$; $D^2 y_p = Be^x$
$Be^x - Be^x - 6(A + Be^x) = 5 - e^x$

$-6A - 6Be^x = 5 - e^x$; $-6A = 5$; $A = -\dfrac{5}{6}$;

$-6B = -1$; $B = \dfrac{1}{6}$

$y_p = -\dfrac{5}{6} + \dfrac{1}{6}e^x$; $y = c_1 e^{3x} + c_2 e^{-2x} + \dfrac{1}{6}e^x - \dfrac{5}{6}$

Substituting $x = 0$ when $y = 2$; $3c_1 + 3c_2 = 8$

$Dy = 3c_1 e^{3x} - 2c_2 e^{-2x} + \dfrac{1}{6}e^x$

Substituting, $Dy = 4$ when $x = 0$, $18c_1 - 12c_2 = 23$

Solving the two linear equations simultaneously,

$c_1 = \dfrac{11}{6}$, $c_2 = \dfrac{5}{6}$

$y = \dfrac{11}{6}e^{3x} + \dfrac{5}{6}e^{-2x} + \dfrac{1}{6}e^x - \dfrac{5}{6}$

$= \dfrac{1}{6}(11e^{3x} + 5e^{-2x} + e^x - 5)$

30. $3y'' - 10y' + 3y = xe^{-2x}$; @$x = 0, y' = \dfrac{-9}{35}$ and

$y = \dfrac{-13}{35}$

$3m^2 - 10m + 3 = 0, m = \dfrac{1}{3}, 3$

$y_c = c_1 e^{x/3} + c_2 e^{3x}$
$y_p = Axe^{-2x} + Be^{-2x}$
$y_p' = -e^{-2x}(2Ax - A + 2B)$
$y_p'' = 4e^{-2x}(Ax - A + B)$
$3y_p'' - 10y_p' + 3y_p = e^{-2x}(35Ax - 22A + 35B)$
$= xe^{-2x}$

$35A = 1$, $-22A + 35B = 0$

$A = \dfrac{1}{35}$, $B = \dfrac{22}{1225}$

$y_p = \dfrac{1}{35}xe^{-2x} + \dfrac{22}{1225}e^{-2x}$

$y = c_1 e^{x/3} + c_2 e^{3x} + \dfrac{1}{35}xe^{-2x} + \dfrac{22}{1225}e^{-2x}$

$\dfrac{-13}{35} = c_1 + c_2 + \dfrac{22}{1225}$

$y' = 3c_2 e^{3x} + \dfrac{1}{3}c_1 e^{x/3} - \dfrac{e^{-2x}(70x + 9)}{1225}$

$\dfrac{-9}{35} = \dfrac{c_1}{3} + 3c_2 - \dfrac{9}{1225}$

$c_1 = \dfrac{-135}{392}, c_2 = \dfrac{-9}{200}$

$y = \dfrac{-135}{392}e^{x/3} - \dfrac{9}{200}e^{3x} + \dfrac{1}{35}xe^{-2x} + \dfrac{12}{1225}e^{-2x}$

31. $y'' + y = x + \sin 2x$; $y_p = A + Bx + C2x + E\cos 2x$
$D^2y + y = x + \sin 2x$; $m^2 + 1 = 0$; $m = \pm j$;
$\alpha = 0, \beta = 1$
$y_c = c_1 \sin x + c_2 \cos x$
$Dy_p = B + 2C\cos 2x - 2E\sin 2x$
$D^2 y_p = -4C\sin 2x - 4E\cos 2x$
$-4C\sin 2x - 4E\cos 2x + A + Bx + C\sin 2x + E\cos 2x$
   $= x + \sin 2x$
$\sin 2x(-4C + C) + \cos 2x(-4E + E) + x(B) + A$
   $= x + \sin 2x$

$-3C = 1$; $C = -\dfrac{1}{3}$; $-3E = 0$; $E = 0$; $B = 1$;
$A = 0$;
$Dy = 1, y = 0, x = \pi$

$y = c_1 \sin x + c_2 \cos x + x - \dfrac{1}{3}\sin 2x$;

therefore, $c_2 = \pi$

$Dy = c_1 \cos x - c_2 \sin x + 1 - \dfrac{2}{3}\cos 2x$;

therefore, $c_1 = -\dfrac{2}{3}$

$y = -\dfrac{2}{3}\sin x + \pi \cos x + x - \dfrac{1}{3}\sin 2x$

32. $D^2y - 2Dy + y = xe^{2x} - e^{2x}$; $y_p = Ae^{2x} + Bxe^{2x}$
$m^2 - 2m + 1 = 0$; $(m-1)^2 = 0$; $m = 1, 1$
$y_c = e^x(c_1 + c_2x)$; $Dy_p = 2Ae^{2x} + 2Bxe^{2x} + Be^{2x}$
$D^2 y_p = 4Ae^{2x} + 2Be^{2x} + 4Bxe^{2x} + 2Be^{2x}$
$4Ae^{2x} + 4Be^{2x} + 4Bxe^{2x} - 4Ae^{2x} - 4Bxe^{2x}$
  $- 2Be^{2x} + Ae^{2x} + Bxe^{2x} = xe^{2x} - e^{2x}$
$e^{2x}(4A + 4B - 4A - 2B + A) + xe^{2x}(4B - 4B + B)$
  $= xe^{2x} - e^{2x}$
$A + 2B = -1$; $B = 1$; therefore $A = -3$
$Dy = 4$; $y = -2$; $x = 0$
$y = e^x(c_1 + c_2x) - 3e^{2x} + xe^{2x}$
$Dy = e^x(c_2) + (c_1 + c_2x)e^x - 6e^{2x} + 2xe^{2x} + e^{2x}$
From (1) we get $c_1 = 1$; from (2) we get $c_2 = 8$.
$y = e^x(1 + 8x) - 3e^{2x} + xe^{2x}$
$= e^x(1 + 8x) + e^{2x}(x - 3)$

15.4 Applications of Higher-Order Equations

1. $D^2\theta + \dfrac{g}{l}\theta = 0$; $g = 9.8$ m/s^2, $l = 0.1$ m;
   $D^2\theta + 9.8\theta = 0$; $m^2 + 9.8 = 0$

   $m_1 = \sqrt{9.8}j$, $m_2 = -\sqrt{9.8}j$; $\alpha = 0, \beta = \sqrt{9.8}$
   $x = c_1 \sin\sqrt{9.8}t + c_2 \cos\sqrt{9.8}t$

   Substituting $\theta = 0.1$ when $t = 0$;
   $0.1 = c_1\sin 0 + c_2\cos 0$; $0.1 = c_2$;
   $D_x = c_1\cos\sqrt{9.8}t - c_2\sin\sqrt{9.8}t$
   Substituting $D\theta = 0$ when $t = 0$;
   $0 = c_1\cos 0 - c_2\sin 0$; $c_1 = 0$;
   $\theta = 0.1\cos\sqrt{9.8}t = 0.1\cos 3.1t$

$\sqrt{9.8}t$	t	$\cos\sqrt{9.8}t$	$0.1\cos\sqrt{9.8}t$
0	0	1	0.1
$\frac{\pi}{2}$	0.50	0	0
π	1.00	-1	-0.1
$\frac{3\pi}{2}$	1.50	0	0
2π	2.00	1	0.1

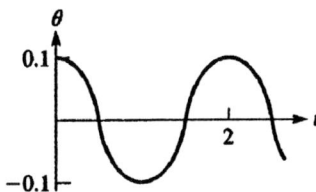

2. $D^2y + 8Dy + 3y = 0$; $y = 6.0, Dy = 0$ when $t = 0$
   $m^2 + 8m + 3 = 0$; $y = ?, t = 12$ s

   $m = \dfrac{-e \pm \sqrt{64 - 12}}{2} = \dfrac{-8 \pm \sqrt{52}}{2}$

   $m_1 = -0.394, m_2 = -7.803$;
   $y = c_1 e^{-0.394t} + c_2 e^{-7.803t}$
   $6.0 = c_1 + c_2$; $Dy = -0.394c_1 e^{-0.394t} - 7.803c_2 e^{-7.803t}$
   $0 = -0.394c_1 - 7.803c_2$; therefore,
   $c_1 = 6.319, c_2 = -0.319$
   $y = 6.319 e^{-0.394t} - 0.319 e^{-7.803t}$
   $y_{t=12} = 0.056$ in

3. $D^2y + bDy + 25y = 0$; $m^2 + bm + 25 = 0$

   $m = \dfrac{-b \pm \sqrt{b^2 - 4(25)}}{2}$
   $ = \dfrac{-b \pm \sqrt{b^2 - 100}}{2}$

   For critical damping: $b^2 - 100 = 0$; $b = 10$ $(b > 0)$

4. $F = kx$; $0.820(9.8) = k(0.250)$; $k = 32.144$ N/m
   $0.820D^2x = -32.144x$
   $0.820D^2x + 32.144x = 0$

   $0.820m^2 + 32.144 = 0, m = \pm 6.26j$

   $x = c_1\sin(6.26t) + c_2\cos(6.26t)$

   @ $t = 0, x = 0.150$
   $0.150 = c_1\sin 0 + c_2\cos 0 \Rightarrow c_2 = 0.150$

   $x = c_1\sin(6.26t) + 0.150\cos(6.26t)$

   $\dfrac{dx}{dt} = 6.26c_1\cos(6.26t) - 0.939\sin(6.26t)$

   @ $t = 0, \dfrac{dx}{dt} = 0$

   $0 = 6.26c_1\cos 0 - 0.939\sin 0 \Rightarrow c_1 = 0$

   $x = 0.150\cos(6.26t)$

5. To get spring constant: $F = kx$; $4.00 = k(0.125)$;
   $k = 32.0$ lb/ft
   To get mass of object:

   $F = ma$; $4.00 = m(32.0)$;
   $m = 0.125\dfrac{\text{lb} \cdot \text{s}^2}{\text{ft}}$ (a slug)

   Using Newton's Second Law:
   mass $\times$ accel. = restoring force

   $0.125\dfrac{d^2x}{dt^2} = -32.0x$; $D^2x + 256x = 0$;

   $m^2 + 256 = 0$;
   $m = \pm 16.0j$; $x = c_1\sin 16.0t + c_2\cos 16.0t$
   $D_x = 16.0c_1\cos 16.0t - 16.0c_2\sin 16.0t$

   Let $x = 0.250$ ft when $t = 0$;
   $0.250 = c_1\sin 0 + c_2\cos 0$;
   $c_2 = 0.250$
   Let $Dx = 0$ when $t = 0$,
   $0 = 16.0c_1\cos 0 - 16.0c_2\sin 0$;
   $c_1 = 0$; $x = 0.250\cos 16.0t$

6. $m = \dfrac{4.00 \text{ lb}}{32.0 \text{ ft/s}^2} = 0.125 \text{ slug}$

$F = kx, 4.00 = k(0.125)$

$k = 32.0 \text{ lb/ft}$

$mD^2x = -Dx - kx$

$0.125D^2x = -Dx - 32.0x$

$D^2x + 8.00Dx + 256x = 0$

$m^2 + 8.00m + 256 = 0$

$m = \dfrac{-8.00 \pm \sqrt{64.0 - 1024}}{2}$

$= -4.00 \pm 15.5j$

$x = e^{-4.00t}(c_1 \sin 15.5t + c_2 \cos 15.5t)$

$Dx = e^{-4.00t}(15.5c_1 \cos 15.5t - 15.5c_2 \sin 15.5t$
$\qquad - 4.00c_1 \sin 15.5t - 4.00c_2)$

$x = 0.250 \text{ ft}, Dx = 0, \text{ when } t = 0$

$0.250 = e^0(c_1 \sin 0 + c_2 \cos 0)$

$c_2 = 0.250$

$0 = e^0(15.5c_1 \cos 0 - 15.5c_2 \sin 0 - 4.00c_1 \sin 0$
$\qquad - 4.00c_2 \cos 0)$

$0 = 15.5c_1 - 4.00c_2$

$c_1 = \dfrac{4.00c_2}{15.5} = 0.065$

$x = e^{-4.00t}(0.065 \sin 15.5t + 0.250 \cos 15.5t)$

7. $m = \dfrac{4.00 \text{ lb}}{32.0 \text{ ft/s}^2} = 0.125 \text{ slug}$

$F = kx, 4.00 = k(0.125)$

$k = 32.0 \text{ lb/ft}$

$0.125D^2x = 4 \sin 2t - 32.0x$

$D^2x + 256x = 32.0 \sin 2t$

$m^2 + 256 = 0; \ m = \pm 16.0j$

$x_c = c_1 \sin 16.0t + c_2 \cos 16.0t$

$x_p = A \sin 2t + B \cos 2t$

$Dx_p = 2A \cos 2t - 2B \sin 2t$

$D^2x_p = -4A \sin 2t - 4B \cos 2t$

$-4A \sin 2t - 4B \cos 2t + 256(A \sin 2t + B \cos 2t)$
$\qquad = 32.0 \sin 2t$

$252A \sin 2t + 252B \cos 2t = 32.0 \sin 2t$

$A = \dfrac{32.0}{252} = 0.127, B = 0$

$x_p = 0.127 \sin 2.00t$

$x = c_1 \sin 16.0t + c_2 \cos 16.0t + 0.127 \sin 2.00t$

$Dx = 16.0c_1 \cos 16.0t - 16.0c_2 \sin 16.0 + 0.254 \cos 2.00t$

$x = 0.250 \text{ ft}, Dx = 0, \text{ when } t = 0$

$0.250 = c_1(0) + c_2 + 0.127(0)$

$c_2 = 0.250$

$0 = 16.0c_1 - 16.0c_2(0) + 0.254$

$c_1 = -0.016$

$x = 0.250 \cos 16.0t + 0.127 \sin 2.00t - 0.016 \sin 16.0t$

8. $m = \dfrac{4.00 \text{ lb}}{32.0 \text{ ft/s}^2} = 0.125 \text{ slug}$

$F = kx, 4.00 = k(0.125)$

$k = 32.0 \text{ lb/ft}$

$0.125D^2x = 4 \sin 2t - Dx - 32.0x$

$D^2x + 8.00Dx + 256x = 32.0 \sin 2t$

$x_c = e^{-4.00t}(c_1 \sin 15.5t + c_2 \cos 15.5t) \text{ (from \#6)}$

$x_p = A \sin 2t + B \cos 2t$

$Dx_p = 2A \cos 2t - 2B \sin 2t$

$D^2x_p = -4A \sin 2t - 4B \cos 2t$

$-4A \sin 2t - 4B \cos 2t + 8.00(2A \cos 2t - 2B \sin 2t)$
$\qquad + 256(A \sin 2t + B \cos 2t) = 32.0 \sin 2t$

$(252A - 16.0B) \sin 2t + (252B + 16.0A) \cos 2t$
$\qquad = 32.0 \sin 2t$

$252A - 16.0B = 32.0$

$16.0A + 252B = 0$

$A = 0.126, B = -0.008$

$x_p = 0.126 \sin 2.00t - 0.008 \cos 2.00t$

$x = e^{-4.00t}(c_2 \sin 15.5t + c_2 \cos 15.5t) + 0.126 \sin 2.00t$
$\qquad - 0.008 \cos 2.00t$

$Dx = e^{-4.00t}(15.5c_1 \cos 15.5t - 15.5c_2 \sin 15.5t$
$\qquad - 4.00c_1 \sin 15.5t - 4.00c_2 \cos 15.5t)$
$\qquad + 0.252 \cos 2.00t + 0.016 \sin 2.00t$

$x = 0.250 \text{ ft}, Dx = 0, \text{ when } t = 0$

$0.250 = c_2 - 0.008, c_2 = 0.258$

$0 = 15.5c_1 - 4.00c_2 + 0.252, c_1 = 0.050$

$x = e^{-4.00t}(0.050 \sin 15.5t + 0.258 \cos 15.5t)$
$\qquad + 0.126 \sin 2.00t - 0.008 \cos 2.00t$

9.

$$L\frac{d^2q}{dt^2} + R\frac{dq}{dt} + \frac{q}{C} = E$$

$$0.200\frac{d^2\theta}{dt^2} + 8.00\frac{dq}{dt} + 10^6 q = 0$$

$$0.200m^2 + 8.00m + 10^6 = 0$$

$$m = \frac{-8.00 \pm \sqrt{64.0 - 0.800 \times 10^6}}{0.400};$$

$m = -20.0 \pm 2240j; \ \alpha = -20.0, \ \beta = 2240$

$q = e^{-20.0t}(c_1 \sin 2240t + c_2 \cos 2240t); \ t = 0, q = 0$

Therefore, $c_2 = 0$

$\dfrac{dq}{dt} = e^{-20.0t}(2240c_1 \cos 2240t - 2240c_2 \sin 2240t)$
$\qquad + (c_1 \sin 2240t + c_2 \cos 2240t)(-20.0e^{-20.0t})$

$t = 0, i = 0.500; \text{ therefore, } c_1 = 2.24 \times 10^{-4};$

$q = e^{-20.0t}(2.24 \times 10^{-4}) \sin 2240t$

$q = 2.24 \times 10^{-4}e^{-20.0t} \sin 2240t$

10. $2 \times 10^{-3}\dfrac{d^2q}{dt^2} + \dfrac{q}{50 \times 10^{-9}} = 0$

$\dfrac{2}{10^3}\dfrac{d^2q}{dt^2} + \dfrac{10^9}{50}q = 0$

$\dfrac{2}{10^3}m^2 + \dfrac{10^9}{50} = 0$

$m^2 = -\dfrac{10^9}{50} \cdot \dfrac{10^3}{2}$

$\quad = -\dfrac{10^{+12}}{10^2}$

$\quad = -10^{10}$

$m = -10^5 j; \; \alpha = 0, \beta = 10^5$

$q = c_1 \sin 10^5 t + c_2 \cos 10^5 t; \; 10^5 = c_2$

$\dfrac{dq}{dt} = 10^5 c_1 \cos 10^5 t - 10^5 c_2 \sin 10^5 t$

$0 = 10^5 c_1 - 0;$ therefore, $c_1 = 0$

Therefore, $q = 10^5 \cos 10^5 t$

11. $L = 0.100H, R = 0, C = 100$

$\mu F = 10^{-4}, E = 100$ V

$0.100\dfrac{d^2q}{dt^2} + 0\dfrac{dq}{dt} + \dfrac{q}{10^{-4}} = 100$

$\dfrac{d^2q}{dt^2} + 10^5 q = 1000; \; m^2 + 10^5 = 0; \; m = \pm 316j$

$q_c = c_1 \sin 316t + c_2 \cos 316t; \; q_p = A; \;|$

$q_p' = 0; \; q_p'' = 0$

$0 + 0 + 10^5 A = 1000; \; A = 0.01$

$q = c_1 \sin 316t + c_2 \cos 316t + 0.01; \; q = 0$ when $t = 0; \; 0 = 0 + c_2(1) + 0.01$

$c_2 = -0.01; \dfrac{dq}{dt} = 316c_1 \cos 316t + 3.16 \sin 316t;$

$i = \dfrac{dq}{dt} = 0$ when $t = 0$

$0 = 316c_1 + 0; \; c_1 = 0;$

$y = 0 \sin 316t - 0.01 \cos 316t + 0.01$

$q = 0.01(1 - \cos 316t)$

12. From Ex. 13, $q = 0.01(1 - \cos 316t)$

$i = \dfrac{dq}{dt} = 0.01(0 + 316 \sin 316t)$

$\quad = 3.16 \sin 316t$

13. $0.500D^2q + 10.0Dq + \dfrac{q}{200 \times 10^6} = 120 \sin 120\pi t$

$\quad 1.00D^2q + 20.0Dq + 10^4 q = 240 \sin 120\pi t$

$\quad 1.00m^2 + 20.0m + 10^4 = 0$

$m = \dfrac{-20.0 \pm \sqrt{400 - 4 \times 10^4}}{200} = -10.0 \pm 99.5j;$

$q_c = e^{-10.0t}(c_1 \sin 99.5t + c_2 \cos 99.5t);$

$q_p = A \sin 120\pi t + B \cos 120\pi t$

$Dq_p = 120\pi A \cos 120\pi t - 120\pi B \sin 120\pi t$

$D^2q_p = -142\,000\,A \sin 120\pi t - 142\,000B \cos 120\pi t$

$\qquad - 142\,000\,A120\pi t - 142\,000\,B \cos 120\pi t$

$\qquad + 7540A \cos 120\pi t - 7540B \sin 120\pi t$

$\qquad + 10\,000A \sin 120\pi t + 10\,000B \cos 120\pi t$

$\qquad = 240 \sin 120\pi t$

$-132\,000A - 7540B = 240$

$-132\,000B + 7540A = 0$

$A = -1.81 \times 10^{-3}, B = -1.03 \times 10^{-4}$

$q = e^{-10.0t}(c_1 \sin 99.5t + c_2 \cos 99.5t)$

$\qquad -1.81 \times 10^{-3} \sin 120\pi t - 1.03 \times 10^{-4} \cos 120\pi t$

14. From Ex. 15:

$q_p = -1.81 \times 10^{-3} \sin 120\pi t - 1.03 \times 10^{-4} \cos 120\pi t$

$i_p = -1.81 \times 10^{-3}(120\pi) \cos 120\pi t$

$\qquad + 1.03 \times 10^4 (120\pi) \sin 120\pi t$

$\qquad = 0.039 \sin 120\pi t - 0.682 \cos 120\pi t$

15. $L = 8.00$ mH $= 8.00 \times 10^{-3}$ H; $R = 0$; $C = 0.50$ μF

$C = 5.00 \times 10^{-7}$ F

$E = 20.0e^{-200t}$ mV $= 2.00 \times 10^{-2}e^{-200t}$ V

$8.00 \times 10^{-3}\dfrac{d^2q}{dt^2} + 0\dfrac{dq}{dt} + \dfrac{q}{5.00 \times 10^{-7}}$

$\qquad = 2.00 \times 10^{-2}e^{-200t}$

$\dfrac{d^2q}{dt^2} + 2.50 \times 10^8 q = 2.50e^{-200t}$

$m^2 + 2.50 \times 10^8 = 0; \; m_1 = 1.58 \times 10^4 j$ and $m_2 = -1.58 \times 10^4 j$

$q_c = c_1 \sin 1.58 \times 10^4 t + c_2 \cos 1.58 \times 10^4 t$

$q_p = Ae^{-200t}; \; q_p' = -200Ae^{-200t};$

$q_p'' = 4.00 \times 10^4 Ae^{-200t}$

Substituting into the differential equation,

$4.00 \times 10^4 Ae^{-200t} + 2.50 \times 10^8 Ae^{-200t} = 2.50e^{-200t}$

$4.00 \times 10^4 A + 2.50 \times 10^8 A = 2.50; \; A = 10^{-8}$

$q = c_1 \sin 1.58 \times 10^4 t + c_2 \cos 1.58 \times 10^4 + 10^{-8}e^{-200t}$

$q = 0$ when $t = 0; \; 0 = c_1(0) + c_2(1) + 10^{-8};$

$c_2 = -10^{-8}$

$i = \dfrac{dq}{dt} = 1.58 \times 10^4 c_1 \cos 1.58 \times 10^4 t$

$\qquad - 1.58 \times 10^4 c_2 \sin 1.58 \times 10^4 t - 200 \times 10^{-8}e^{-200t}$

$i = 0$ when $t = 0;$

$0 = 1.58 \times 10^4 c_1(1) - 2.00 \times 10^{-6}(1)$

$c_1 = \dfrac{2.00 \times 10^{-6}}{1.58 \times 10^4}$

$$i = 1.58 \times 10^4 \times \frac{2 \times 10^{-6}}{1.58 \times 10^4} \cos 1.58 \times 10^4 t$$
$$- 1.58 \times 10^4 \times (-10^{-8}) \sin 1.58 \times 10^4 t$$
$$- 200 \times 10^{-6} e^{-200t}$$
$$= 2.00 \times 10^{-6} \cos 1.58 \times 10^4 t$$
$$+ 1.58 \times 10^{-4} \sin 1.58 \times 10^4 t$$
$$- 2.00 \times 10^{-6} e^{-200t}$$
$$= 10^{-6}[2.00 \cos(1.58 \times 10^4 t)$$
$$+ 158 \sin(1.58 \times 10^4 t) - 2.00 e^{-200t}]$$

16. $0.400 D^2 q + 60.0 Dq + \dfrac{q}{0.200 \times 10^{-6}} = 0.800 e^{-100t}$

$D^2 q + 150 Dq + 1.25 \times 10^7 q = 2.00 e^{-100t}$
$m^2 + 150 m + 1.25 \times 10^7 = 0$

$$m = \frac{-150 \pm \sqrt{150^2 - 5.00 \times 10^7}}{2}$$
$$= -75.0 \pm 3530 j$$

$q_c = e^{-75.0t}(c_1 \sin 3530t + c_2 \cos 3530t)$
$q_p = A e^{-100t}$
$Dq_p = -100 A e^{-100t}, D^2 q = 10^4 A e^{-100t}$
$10^4 A e^{-100t} + 150(-100 A e^{-100t}) + 1.25 \times 10^7 A e^{-100t}$
$\qquad = 2.00 e^{-100t}$
$(10^4 - 1.50 \times 10^4 + 1.25 \times 10^7)A = 2.00$
$A = 1.60 \times 10^{-7}$
$q = e^{-75.0t}(c_1 \sin 3530t + c_2 \cos 3530t)$
$\qquad + 1.60 \times 10^{-7} e^{-100t}$
$Dq = e^{-75.0t}(-75.0 c_1 \sin 3530t - 75.0 c_2 \cos 3530t$
$\qquad + 3530 c_1 \cos 3530t - 3530 \sin 3530t)$
$\qquad - 1.60 \times 10^{-5} e^{-100t}$
$q = 0, D_q = 5.00 \times 10^{-3}$ A for $t = 0$
$0 = c_2 + 1.60 \times 10^{-7}, c_2 = -1.60 \times 10^{-5}$
$5.00 \times 10^{-3} = -75.0 c_2 + 3530 c_1 - 1.60 \times 10^{-5}$;
$c_1 = 1.42 \times 10^{-6}$
$q = e^{-75.0t}(1.418 \times 10^{-6} \sin 3530t - 1.60 \times 10^{-7} \cos 3530t)$
$\qquad + 1.60 \times 10^{-7} e^{-100t}$
$i = e^{-75.0t}(-75.0 \times 1.42 \times 10^{-6} \sin 3530t$
$\qquad + 75.0 \times 1.60 \times 10^{-7} \cos 3530t$
$\qquad + 3530 \times 1.42 \times 10^{-6} \cos 3530t$
$\qquad + 3530 \times 1.60 \times 10^{-7} \sin 3530t)$
$\qquad - 1.60 \times 10^{-5} e^{-100t}$
$i = e^{-75.0t}(0.46 \times 10^{-3} \sin 3530t$
$\qquad + 5.00 \times 10^{-3} \cos 3530t) - 1.60 \times 10^{-5} e^{-100t}$
(e^{-100t} term is negligible for all t.)

17. $1.00 D^2 q + 5.00 Dq + \dfrac{q}{150 \times 10^{-6}} = 120 \sin 100t$

$\qquad 1.00 D^2 q + 5.00 Dq + 6670 q = 120 \sin 100t$

$q_p = A \sin 100t + B \cos 100t$
$Dq_p = 100 A \cos 100t - 100 B \sin 100t$
$D^2 q_p = -10^4 A \sin 100t - 10^4 B \cos 100t$

$-10^4 A \sin 100t - 10^4 B \cos 100t$
$\qquad + 5.00(100 A \cos 100t - 100 B \sin 100t)$
$\qquad + 6670(A \sin 100t + B \cos 100t)$
$\qquad = 120 \sin 100t$
$(-10^4 A - 500 B + 6670 A) \sin 100t$
$\qquad + (-10^4 B + 500 A + 6670 B) \cos 100t$
$\qquad = 120 \sin 100t$
$3330 A + 500 B = -120$
$500 A - 3330 B = 0$
$A = -0.0352$;
$B = 0.005\,28$;
$q_p = -0.0352 \sin 100t + 0.005\,28 \cos 100t$;
$i_p = -3.52 \cos 100t + 0.528 \sin 100t$

18. $2.00 D^2 q + 20.0 Dq + \dfrac{q}{20.0 \times 10^{-6}} = 200 \sin 10t$

$1.00 D^2 + 10.0 Dq + 25\,000 q = 100 \sin 10t$
$q_p = A \sin 10t + B \cos 10t$
$Dq_p = 10 A \cos 10t - 10 B \sin 10t$
$D^2 q_p = -100 A \sin 10t - 100 B \cos 10t$
$\qquad - 100 A \sin 10t - 100 B \cos 10t$
$\qquad + 10(10 A \cos 10t - 10 B \sin 10t)$
$\qquad + 25\,000(A \sin 10t + B \cos 10t)$
$\qquad = 100 \sin 10t$
$(-100 A - 100 B + 25\,000 A) \sin 10t$
$\qquad + (-100 B + 100 A + 25\,000 B) \cos 10t$
$\qquad = 100 \sin 10t$
$24\,900 A - 100 B = 100$
$100 A - 24\,900 B = 0$
$249 A - B = 1$
$A + 249 B = 0$
$A = 0.004\,02$; $B = -1.61 \times 10^{-5}$
$q_p = 0.004\,02 \sin 10t - 1.61 \times 10^{-5} \cos 10t$
$i_p = 4.02 \times 10^{-2} \cos 10t + 1.61 \times 10^{-4} \sin 10t$

19. $EI \dfrac{d^4 y}{dx^4} = w$

$y = c_1 + c_2 + c_3 x^2 + c_4 x^3 + \dfrac{w}{24EI} x^4$

@ $x = 0, y = 0 \Rightarrow c_1 = 0$

$y = c_2 x + c_3 x^2 + c_4 x^3 + \dfrac{w}{24EI} x^4$

$y' = c_2 + 2 c_3 x + 3 c_4 x^2 + \dfrac{4w}{24EI} x^3$

@ $x = 0, y' = 0 \Rightarrow c_2 = 0$

$y = c_3 x^2 + c_4 x^3 + \dfrac{w}{24EI} x^4$

$y'' = 2 c_3 + 6 c_4 + \dfrac{12w}{24EI} x^2$

@ $x = L, y'' = 0$

$$0 = 2c_3 + 6c_4 L + \frac{12w}{24EI}L^2$$

$$y''' = 6c_4 + \frac{24w}{24EI}x$$

@ $x = L, y''' = 0$

$$0 = 6c_4 + \frac{24w}{24EI} \cdot L \Rightarrow c_4 = \frac{-wL}{6EI}$$

$$0 = 2c_3 + 6 \cdot \frac{-wL}{6EI} \cdot L + \frac{12w}{24EI}L^2 \Rightarrow c_3 = \frac{wL^2}{4EI}$$

$$y = \frac{wL^2}{4EI}x^2 + \frac{-wL}{6EI}x^3 + \frac{w}{24EI}x^4$$

$$y = \frac{w}{24EI}(6L^2x^2 - 4Lx^3 + x^4)$$

20. $EI = \dfrac{d^4y}{dx^4} = kEIx \Rightarrow \dfrac{d^4y}{dx^4} = kx$

$y_c = c_1 + c_2 x + c_3 x^2 + c_4 x^3$

$y_p = Ax^5$

$y'_p = 5Ax^4$

$y''_p = 20Ax^3$

$y'''_p = 60Ax^2$

$y'''_p = 120Ax = kx \Rightarrow A = \dfrac{k}{120}$

$$y = c_1 + c_2 x + c_3 x^2 + c_4 x^3 + \frac{k}{120}x^5$$

@ $x = 0, y = 0 \Rightarrow c_1 = 0$

$\quad x = L, y = 0 \Rightarrow 0 = c_2 L + c_3 L^2 + c_4 L^3 + \dfrac{k}{120}L^5$

$$y' = c_2 + 2c_3 x + 3c_4 x^2 + \frac{k}{24}x^4$$

$$y'' = 2c_3 + 6c_4 x + \frac{k}{6}x^3$$

@ $x = 0, y'' = 0 \Rightarrow c_3 = 0$

@ $x = L, y'' = 0 \Rightarrow 0 = 6c_4 L + \dfrac{k}{6}L^3 \Rightarrow c_4 = \dfrac{-kL^2}{36}$

$$0 = c_2 L + \frac{-kL^2}{36}L^3 + \frac{k}{120}L^5 \Rightarrow c_2 = \frac{7kL^4}{360}$$

$$y = \frac{7kL^4}{360}x - \frac{kL^2}{36}x^3 + \frac{k}{120}x^5 \Bigg|_{\substack{k = 7.2 \times 10^{-4} \\ L = 10}}$$

$$y = 0.14x - 0.002x^3 + 6 \times 10^{-6}x^5$$

Chapter 15 Review Exercises

1. $2D^2 y + Dy = 0$. The auxiliary equation is

$$2m^2 + m = 0$$

$$m(2m + 1) = 0; \; m_1 = 0 \text{ and } m_2 = -\frac{1}{2}$$

$$y = c_1 e^0 + c_2 e^{-1/2x}; \; y = c_1 + c_2 e^{-x/2}$$

2. $2D_y^2 - 5Dy + 2y = 0$. The auxiliary equation is

$$2m^2 - 5m + 2 = 0$$
$$(m - 2)(2m - 1) = 0$$

$$m - 2 = 0 \quad \text{or} \quad 2m - 1 = 0$$
$$m = 2 \qquad\qquad m = \frac{1}{2}$$

$$y = c_1 e^{2x} + c_2 e^{1/2x}$$

3. $\quad y'' + 2y' + y = 0$
$\quad D^2 y + 2Dy + y = 0$
$\quad m^2 + 2m + 1 = 0$
$\quad (m + 1)^2 = 0, \; m = -1, -1$
$\quad y = (c_1 + c_2 x)e^{-x}$

4. $\quad y'' + 2y' + 2y = 0$
$\quad m^2 + 2m + 2 = 0$
$\quad m = -1 \pm j, \alpha = -1, \beta = 1$
$\quad y = e^{-x}(c_1 \sin x + c_2 \cos x)$

5. $D^2 y + 2Dy + 6y = 0$
$\quad m^2 + 2m + 6 = 0$
$\quad m = -1 \pm \sqrt{5}j, \; \alpha = -1, \beta = \sqrt{5}$
$\quad y = e^{-x}(c_1 \sin \sqrt{5}x + c_2 \cos \sqrt{5}x)$

6. $\quad 4D^2 y - 4Dy + y = 0$
$\quad 4m^2 - 4m + 1 = 0$

$\quad (2m - 1)(2m - 1) = 0, m = \dfrac{1}{2}$

$\quad y = e^{1/2x}(c_1 + c_2 x)$

7. $\dfrac{d^2 y}{dx^2} = 0.08y - 0.2\dfrac{dy}{dx}$

$\quad D^2 y + 0.2Dy - 0.08y = 0$
$\quad m^2 + 0.2m - 0.08 = 0$
$\quad (m - 0.2)(m + 0.4) = 0, m = 0.2, -0.4$
$\quad y = c_1 e^{0.2x} + c_2 e^{-0.4x}$

8. $\dfrac{d^2y}{dx^2} = 0.2\dfrac{dy}{dx} - 0.17y$

$\dfrac{d^2y}{dx^2} - 0.2\dfrac{dy}{dx} + 0.17y = 0$

$m^2 - 0.2m + 0.17 = 0$

$m = 0.1 \pm 0.4j,\ \alpha = 0.1, \beta = 0.4$

$y = e^{0.1x}(c_1 \sin 0.4x + c_2 \cos 0.4x)$

9. $2D^2y + Dy - 3y = 6;\ 2m^2 + m - 3 = 0;$

$(m - 1)(2m + 3) = 0;\ m_1 = 1,\ m_2 = -\dfrac{3}{2}$

$y_c = c_1 e^x + c_2 e^{-3x/2};\ y_p = A;\ y_p' = 0;\ y_p'' = 0$

Substituting into the differential equation,

$2(0) + 0 - 3A = 6;\ A = -2;\ y_p = -2.$

$y = c_1 e^x + c_2 e^{-3x/2} - 2$

10. $\dfrac{d^2y}{dx^2} + 6\dfrac{dy}{dx} + 9y = 3x$

$m^2 + 6m + 9 = 0$

$(m + 3)^2 = 0, m = -3, -3$

$y_c = (c_1 + c_2 x)e^{-3x}$

$y_p = A + Bx, y_p' = B, y_p'' = 0$

$0 + 6B + 9(A + Bx) = 3x$

$6B + 9A = 0, 9B = 3, B = \dfrac{1}{3}, A = -\dfrac{2}{9}$

$y = (c_1 + c_2 x)e^{-3x} - \dfrac{2}{9} + \dfrac{1}{3}x$

11. $y'' + y' - y = 2e^x$

$m^2 + m - 1 = 0, m = \dfrac{-1 \pm \sqrt{1 + 4}}{2} = -\dfrac{1}{2} \pm \dfrac{1}{2}\sqrt{5}$

$y = e^{-x/2}(c_1 e^{x\sqrt{5}/2} + c_2 e^{-x\sqrt{5}/2})$

$y_p = Ae^x, y_p' = Ae^x, y_p'' = Ae^x$

$Ae^x + Ae^x - Ae^x = 2e^x, A = 2$

$y = e^{-x/2}(c_1 e^{x\sqrt{5}/2} + c_2 e^{-x\sqrt{5}/2}) + 2e^x$

12. $4D^3y + 9Dy = xe^x$

$4m^3 + 9m = 0$

$m(4m^2 + 9) = 0$

$m = 0, m = 0 \pm \dfrac{3}{2}j, \alpha = 0, \beta = \dfrac{3}{2}$

$y_c = c_1 e^{0x} + e^{0x}\left(c_2 \sin \dfrac{3}{2}x + c_2 \cos \dfrac{3}{2}x\right)$

$y_c = c_1 + c_2 \sin \dfrac{3}{2}x + c_3 \cos \dfrac{3}{2}x$

$y_p = Axe^x + Be^x$

$y_p' = Axe^x + Ae^x + Be^x$

$y_p'' = Axe^x + 2Ae^x + Be^x$

$y_p''' = Axe^x + 3Ae^x + Be^x$

$4(Axe^x + 3Ae^x + Be^x) + 9(Axe^x + Ae^x + Be^x) = xe^x$

$4A + 9A = 1$

$A = \dfrac{1}{13}$

$12A + 4B + 9A + 9B = 0$

$21A + 13B = 0$

$21\left(\dfrac{1}{13}\right) + 13B = 0$

$21 + 169B = 0, B = \dfrac{-21}{169}$

$y_p = \dfrac{1}{13}xe^x - \dfrac{21}{169}e^x$

$y = c_1 + c_2 \sin \dfrac{3}{2}x + c_3 \cos \dfrac{3}{2}x + \dfrac{1}{13}xe^x - \dfrac{21}{169}e^x$

13. $9D^2y - 18Dy + 8y = 16 + 4x$

$D^2y - 2Dy + \dfrac{8}{9}y = \dfrac{16}{9} + \dfrac{4}{9}x$

$m^2 - 2m + \dfrac{8}{9} = 0;\ m = \dfrac{2 \pm \sqrt{4 - 4\left(\frac{8}{9}\right)}}{2};$

$m_1 = \dfrac{2}{3};\ m_2 = \dfrac{4}{3}$

$y_c = c_1 e^{2x/3} + c_2 e^{4x/3};\ y_p = A + Bx;$

$y_p' = B;\ y_p'' = 0$

Substituting into the differential equation,

$0 - 2B + \dfrac{8}{9}(A + Bx) = \dfrac{16}{9} + \dfrac{4}{9}x$

$\left(-2B + \dfrac{8}{9}A\right) + \dfrac{8}{9}Bx = \dfrac{16}{9} + \dfrac{4}{9}x;\ -2B + \dfrac{8}{9}A = \dfrac{16}{9}$

$\dfrac{8}{9}B = \dfrac{4}{9};\ B = \dfrac{1}{2};\ -2\left(\dfrac{1}{2}\right) + \dfrac{8}{9}A = \dfrac{16}{9};$

$A = \dfrac{25}{8};\ y_p = \dfrac{1}{2}x + \dfrac{25}{8}$

$y = c_1 e^{2x/3} + c_2 e^{4x/3} + \dfrac{1}{2}x + \dfrac{25}{8}$

14. $y'' + y = 4\cos 2x$

$m^2 + 1 = 0$

$m = 0 \pm j, \alpha = 0, \beta = 1$

$y_c = e^{0x}(c_1 \sin x + c_2 \cos x)$

$y_c = c_1 \sin x + c_2 \cos x$

$y_p = A \sin 2x + B \cos 2x$

$y_p' = 2A \cos 2x - 2B \sin 2x$

$y_p'' = -4B \cos 2x - 4A \sin 2x$

$-4B \cos 2x - 4A \sin 2x + A \sin 2x + B \cos 2x = 4\cos 2x$

$-4B + B = 4 \quad , \quad -4A + A = 0$

$-3B = 4 \qquad\qquad A = 0$

$B = -\dfrac{4}{3}$

$$y_p = -\frac{4}{3}\cos 2x$$

$$y = y_c + y_p$$

$$y = c_1 \sin x + c_2 \cos x - \frac{4}{3}\cos 2x$$

15. $D^3 y - D^2 y + 9Dy - 9y = \sin x$

$$m^3 - m^2 + 9m - 9 = 0$$
$$m^2(m-1) + 9(m-1) = 0$$
$$(m-1)(m^2 + 9) = 0$$
$$m = 1, m = 0 \pm 3j, \alpha = 0, \beta = 3$$
$$y_c = c_1 e^x + e^{0x}(c_2 \sin 3x + c_3 \cos 3x)$$
$$y_c = c_1 e^x + c_2 \sin 3x + c_3 \cos 3x$$
$$y_p = A \sin x + B \cos x$$
$$y_p' = A \cos x - B \sin x$$
$$y_p'' = -A \sin x - B \cos x$$
$$y_p''' = -A \cos x + B \sin x$$
$$-A\cos x + B\sin x - (-A\sin x - B\cos x) +$$
$$9(A\cos x - B\sin x) - 9(A\sin x + B\cos x)$$
$$= \sin x$$
$$(-A+B+9A-9B)\cos x + (B+A-9B-9A)\sin x$$
$$= \sin x$$
$$8A - 8B = 0$$
$$\underline{-8A - 8B = 1}$$
$$-16B = 1$$

$$B = -\frac{1}{16}$$

$$A - B = 0$$

$$A - \left(-\frac{1}{16}\right) = 0$$

$$A = -\frac{1}{16}$$

$$y_p = -\frac{1}{16}\sin x - \frac{1}{16}\cos x$$

$$y = c_1 e^x + c_2 \sin 3x + c_3 \cos 3x - \frac{1}{16}(\sin x + \cos x)$$

16. $y'' + y' = e^x + \cos 2x$

$$m^2 + m = 0$$
$$m(m+1) = 0$$
$$m = 0, m = -1$$
$$y_c = c_1 e^{0x} x + c_2 e^{-x} = c_1 x + c_2 e^{-x}$$
$$y_p = A e^x + B \sin 2x + C \cos 2x$$
$$y_p' = A e^x + 2B \cos 2x - 2C \sin 2x$$
$$y_p'' = A e^x - 4B \sin 2x - 4C \cos 2x$$
$$A e^x - 4B \sin 2x - 4C \cos 2x + A e^x + 2B \cos 2x$$
$$\quad - 2C \sin 2x = e^x + \cos 2x$$
$$2A e^x + (-4B - 2C)\sin 2x + (-4C + 2B)\cos 2x$$
$$= e^x + \cos 2x$$

$$2A = 1$$

$$A = \frac{1}{2}$$

$$-4B - 2C = 0$$

$$2B - 4C = 1, A = \frac{1}{10}, B = -\frac{1}{5}$$

$$y = c_1 x + c_2 e^{-x} + \frac{1}{10}\sin 2x - \frac{1}{5}\cos 2x + \frac{1}{2}e^x$$

17. $y'' - 7y' - 8y = 2e^{-x}$

$$D^2 y_c - 7D y_c - 8 y_c = 0$$
$$m^2 - 7m - 8 = 0$$
$$(m+1)(m-8) = 0$$
$$m = -1, m = 8$$
$$y_c = c_1 e^{-x} + c_2 e^{8x}$$

Let $y_p = Axe^{-x}, y_p' = Ae^{-x}(1-x), y_p'' = Ae^{-x}(x-2)$

$$Ae^{-x}(x-2) - 7Ae^{-x}(1-x) - 8Axe^{-x} = 2e^{-x}$$
$$A(x-2) - 7A(1-x) - 8Ax = 2$$
$$-9A = 2$$

$$A = \frac{-2}{9}$$

$$y_p = \frac{-2}{9}xe^{-x}$$

$$y = y_c + y_p$$

$$y = c_1 e^{-x} + c_2 e^{8x} + \frac{-2}{9}xe^{-x}$$

18. $3y'' - 6y' = 4 + xe^x$

$$3m^2 - 6m = 0$$
$$3m(m-2) = 0$$
$$m = 0, m = 2$$
$$y_c = c_1 e^{0x} + c_2 e^{2x} = c_1 + c_2 e^{2x}$$
$$y_p = Ax + Bxe^x$$
$$y_p' = A + Be^x + Bxe^x$$
$$y_p'' = 2Be^x + Bxe^x$$
$$3(2Be^x + Bxe^x) - 6(A + Be^x + Bxe^x) = 4 + xe^x$$
$$6Be^x + 3Bxe^x - 6A - 6Be^x - 6Bxe^x = 4 + xe^x$$
$$-6A = 4$$

$$A = -\frac{2}{3}$$

$$-3B = 1$$

$$B = -\frac{1}{3}$$

$$y_p = -\frac{2}{3}x - \frac{1}{3}xe^x$$

$$y = c_1 + c_2 e^{2x} - \frac{2}{3}x - \frac{1}{3}xe^x$$

19. $D^2 y + 25 y = 50 \cos 5x$

$\qquad m^2 + 25 = 0$

$\qquad\qquad m = 0 \pm 5j, \alpha = 0, \beta = 5$

$\qquad y_c = e^{0x}(c_1 \sin 5x + c_2 \cos 5x)$

$\qquad\quad = c_1 \sin 5x + c_2 \cos 5x$

$\qquad y_p = Ax \sin 5x + B \cos 5x$

$\qquad y_p' = 5Ax \cos 5x + A \sin 5x + B \cos 5x - Bx \sin 5x$

$\qquad y_p'' = 10A \cos 5x - 25Ax \sin 5x - 2B \sin 5x - Bx \cos 5x$

$\qquad 10A \cos 5x - 25Ax \sin 5x + 25Ax \sin 5x - 2B \sin 5x - Bx \cos 5x = 50 \cos 5x$

$\qquad 10A \cos 5x - 2B \sin 5x - Bx \cos 5x = 50 \cos 5x$

$\qquad A = 5, B = 0$

$\qquad y_p = 5x \sin 5x$

$\qquad y = c_1 \sin 5x + c_2 \cos 5x + 5x \cos 5x$

20. $D^2 y + 4y = 8x \sin 2x$

$\qquad m^2 + 4 = 0$

$\qquad\qquad m = 0 \pm 2j, \alpha = 0, \beta = 2$

$\qquad y_c = e^{0x}(c_1 \sin 2x + c_2 \cos 2x) = c_1 \sin 2x + c_2 \cos 2x$

$\qquad y_p = Ax \sin 2x + Bx^2 \cos 2x$

$\qquad y_p' = A \sin 2x + 2Ax \cos 2x + 2Bx \cos 2x - 2Bx^2 \sin 2x$

$\qquad y_p'' = 4A \cos 2x - 4Ax2x + 2B \cos 2x - 8Bx \sin 2x - 4Bx^2 \cos 2x$

$\qquad \cos 2x(4A + 2B) + x \sin 2x(-4A + 4A - 8B) + x^2 \cos 2x(-4B + 4B) = 8x \sin 2x$

$\qquad\quad 4A + 2B = 0$

$\qquad\qquad -8B = 8, B = -1$

$\qquad 4A + 2(-1) = 0$

$\qquad\qquad A = \dfrac{1}{2}$

$\qquad y_p = \dfrac{1}{2} x \sin 2x - x^2 \cos 2x$

$\qquad y = c_1 \sin 2x + c_2 \cos 2x + \dfrac{1}{2} x \sin 2x - x^2 \cos 2x$

21. $D^2 y + Dy + 4y = 0.$ $m^2 + m + 4 = 0;$ $m = \dfrac{-1 \pm \sqrt{-15}}{2}$

$\qquad m_1 = -\dfrac{1}{2} + \dfrac{\sqrt{15}}{2} j;$ $m_2 = -\dfrac{1}{2} - \dfrac{\sqrt{15}}{2} j;$ $\alpha = -\dfrac{1}{2}, \beta = \dfrac{\sqrt{15}}{2}$

$\qquad y = e^{-x/2} \left(c_1 \sin \dfrac{\sqrt{15}}{2} x + c_2 \cos \dfrac{\sqrt{15}}{2} x \right)$

$\qquad Dy = e^{-x/2} \left(\dfrac{\sqrt{15}}{2} c_1 \cos \dfrac{\sqrt{15}}{2} x - c_2 \sin \dfrac{\sqrt{15}}{2} x \right) + \left(c_1 \sin \dfrac{\sqrt{15}}{2} x + c_2 \cos \dfrac{\sqrt{15}}{2} x \right) \left(-\dfrac{1}{2} e^{-x} \right)$

Substituting $Dy = \sqrt{15}$, $y = 0$ when $x = 0$; $c_2 = 0$; $c_1 = 2$.

$\qquad y = 2 e^{-x/2} \sin \left(\dfrac{1}{2} \sqrt{15} x \right)$

22.
$$5y'' + 7y' - 6y = 0$$
$$5m^2 + 7m - 6 = 0$$
$$(m + 2)(5m - 3) = 0$$

$$m = -2, m = \frac{3}{5}$$

$$y = c_1 e^{-2x} + c_2 e^{3x/5}$$
$$= 2 \text{ when } x = 0$$
$$2 = c_1 e^0 + c_2 e^0, c_1 + c_2 = 2$$

$$y' = -2c_1 e^{-2x} + \frac{3}{5} c_2 e^{-3x/5}$$
$$= 10 \text{ when } x = 0$$

$$-2c_1 + \frac{3}{5} c_2 = 10$$

$$c_1 + c_2 = 2,$$

$$c_1 = \frac{-44}{13},$$

$$c_2 = \frac{70}{13}$$

$$y = \frac{-44}{13} e^{-2x} + \frac{70}{13} e^{3x/5}$$

23.
$$(D^2 + 4D + 4)y = 4\cos$$
$$m^2 + 4m + 4 = 0$$
$$(m + 2)^2 = 0, m = -2, -2$$
$$y_c = (c_1 + c_2 x)e^{-2x}$$
$$y_p = A\sin x + B\cos x$$
$$y_p' = A\cos x - B\sin x$$
$$y_p'' = -A\sin x - B\cos x$$
$$-A\sin x - B\cos x + 4(A\cos x - B\sin x)$$
$$\quad + 4(A\sin x + B\cos x) = 4\cos x$$
$$3A - 4B = 0$$

$$4A + 3B = 4, A = \frac{16}{25}, B = \frac{12}{25}$$

$$y = (c_1 + c_2 x)e^{-2x} + \frac{16}{25}\sin x + \frac{12}{25}\cos x$$

$$y = 0 \text{ when } x = 0;$$

$$0 = c_1 + \frac{16}{25}(0) + \frac{12}{25}, c_1 = -\frac{12}{25}$$

$$y = \left(-\frac{12}{25} + c_2 x\right)e^{-2x} + \frac{16}{25}\sin x + \frac{12}{25}\cos x$$

$$Dy = e^{-2x}\left(\frac{24}{25} - 2c_2 x + c_2\right) + \frac{16}{25}\cos x - \frac{12}{25}\sin x$$

$$Dy = 1 \text{ when } x = 0;$$

$$1 = \frac{24}{25} + c_2 + \frac{16}{25}, c_2 = -\frac{15}{25}$$

$$y = \frac{1}{25}[(-12 - 15x)e^{-2x} + 16\sin x + 12\cos x]$$
$$= \frac{1}{25}[16\sin x + 12\cos x - 3(4 + 5x)e^{-2x}]$$

24.
$$y'' - 2y' + y = e^x + x$$
$$m^2 - 2m + 1 = 0$$
$$(m - 1)(m - 1) = 0$$
$$m = 1$$
$$y_c = e^x(c_1 + c_2 x) = c_1 e^x + c_2 x e^x$$
$$y_p = Ax^2 e^x + Bx + C$$
$$y_p' = Ax^2 e^x + 2Axe^x + B$$
$$y_p'' = Ax^2 e^x + 4Axe^x + 2Ae^x$$
$$Ax^2 e^x + 4Axe^x + 2Ae^x - 2(Ax^2 e^x + 2Axe^x + B)$$
$$\quad + Ax^2 e^x + Bx + C$$
$$\quad = e^x + x$$
$$x^2 e^x(A - 2A + A) + xe^x(4A - 4A)$$
$$\quad + e^x(2A) + x(B) - 2B + C$$
$$\quad = e^x + x$$

$$2A = 1, A = \frac{1}{2}$$

$$B = 1$$
$$-2B + C = 0$$
$$-2(1) + C = 0, C = 2$$

$$y_p = \frac{1}{2}x^2 e^x + x + 2$$

$$y = c_1 e^x + c_2 x e^x + \frac{1}{2}x^2 e^x + x + 2 = 0$$

$$\text{when } x = 0, c_1 + 2 = 0, c_1 = -2$$

$$y = -2e^x + c_2 x e^x + \frac{1}{2}x^2 e^x + x + 2$$

$$y' = -2e^x + c_2(xe^x + e^x) + \frac{1}{2}(x^2 e^x + 2xe^x) + 1$$
$$\quad = 0$$

$$\text{when } x = 0$$

$$-2 + c_2 + 1 = 0$$
$$c_2 = 1$$

$$y = -2e^x + xe^x + \frac{1}{2}x^2 e^x + x + 2$$

$$y = e^x\left(\frac{1}{2}x^2 + x - 2\right) + x + 2$$

25. $F = kx$; $40 = 0.50k$; $k = 80$ N/m; $m = 4\,kg$; $4\dfrac{d^2x}{dt^2} = -16\dfrac{dx}{dt} - 80x$;

$4D^2x + 16Dx + 80 = 0$; $D^2x + 4Dx + 20 = 0$; $4m^2 + 16m + 80 = 0$; $m_1 = -2 + 4j$;

$m_2 = -2 - 4j$; $x = e^{-2t}(c_1 \sin 4t + c_2 \cos 4t)$

Let $x = 0.50$ when $t = 0$; $0.50 = c_2$

$Dx = e^{-2t}(4c_1 \cos 4t - 4c_2 \sin 4t) + (c_1 \sin 4t + c_2 \cos 4t)(-2)e^{-2t}$

Let $Dx = 0$ when $t = 0$; $0 = 4c_1 - 2c_2$;
$0 = 4c_1 - 1$; $c_1 = 0.25$

$x = e^{-2t}(0.25 \sin 4t + 0.5 \cos 4t) = 0.25e^{-2t}(\sin 4t + 2 \cos 4t)$ underdamped

26. $\qquad 4\dfrac{d^2x}{dt^2} = -80x$

$4D^2x + 80x = 0$
$4m^2 + 80 = 0$
$m^2 + 20 = 0$
$m = 0 \pm 2\sqrt{5}j, \alpha = 0, \beta = 2\sqrt{5}$
$x = e^{0x}(c_1 \sin 2\sqrt{5}t + c_2 \cos 2\sqrt{5}t)$
$x = c_1 \sin 2\sqrt{5}t + c_2 \cos 2\sqrt{5}t$
$= 0.50$ when $t = 0$
$c_2 = 0.50$
$x = c_1 \sin 2\sqrt{5}t + 0.50 \cos 2\sqrt{5}t$

$\dfrac{dx}{dt} = 2\sqrt{5}c_1 \cos 2\sqrt{5}t - 0.50(2\sqrt{5}) \sin 2\sqrt{5}t = 0$

when $t = 0$

$2\sqrt{5}c_1 = 0$
$c_1 = 0$
$x = 0.50 \cos 2\sqrt{5}t$

27. $m = 1.25$ kg, $x = 0.2.00$ m
$F = 1.25(9.80) = 12.25$ N
$12.25 = k(0.200), k = 61.25$ N/m
$1.25D^2x + 4Dx + 61.25x = 0$
$D^2x + 3.2Dx + 49x = 0$
$m^2 + 3.20m + 49 = 0$

$m = \dfrac{-3.20 \pm \sqrt{3.20^2 - 4(49)}}{2} = -1.60 \pm 6.81j$

$x = e^{-1.60t}(c_1 \sin 6.81t + c_2 \cos 6.81t)$
$x = 0$ when $t = 0, c_2 = 0$
$x = c_1 e^{-1.60t} \sin 6.81t$
$Dx = c_1 e^{-1.60t}(-1.60 \sin 6.81t + 6.81 \cos 6.81t)$
$Dx = 4.50$ m/s when $t = 0$
$4.50 = c_1(6.81), c_1 = 0.661$
$x = 0.661e^{-1.60t} \sin 6.81t$

28. $D^2\theta + 0.2D\theta + \dfrac{g}{l}\theta = 0$

$\qquad m^2 + 0.2m + \dfrac{9.8}{1.0} = 0$

$\qquad m = -0.10 \pm \dfrac{\sqrt{39}}{2}j, \alpha = -0.10, \beta = \dfrac{\sqrt{39}}{2} = 3.1; \theta = e^{-0.10t}(c_1 \sin 3.1t + c_2 \cos 3.1t) = 0.10$ when $t = 0$

$\qquad e^{-0.10(0)}(c_1 \sin 0 + c_2 \cos 0) = 0.10$

$\qquad\qquad\qquad\qquad\qquad c_2 = 0.10$

$\qquad \theta = e^{-0.10t}(c_1 \sin 3.1t + 0.10 \cos 3.1t)$

$\qquad D\theta = -0.10e^{-0.10t}(c_1 \sin 3.1t + 0.10 \cos 3.1t) + e^{-0.10t}(3.1c_1 \cos 3.1t - 0.10(3.1) \sin 3.1t) = 0$, when $t = 0$

$\qquad -0.10e^{-0.10(0)}(c_1 \sin 0 + 0.10) + e^{-0.10(0)}(3.1c_1 \cos 0 - 0.10(3.1) \sin 0) = 0$

$\qquad -0.01 + 3.1c_1 = 0, c_1 = 0.0032$

$\qquad \theta = e^{-0.10t}(0.0032 \sin 3.1t + 0.10 \cos 3.1t)$

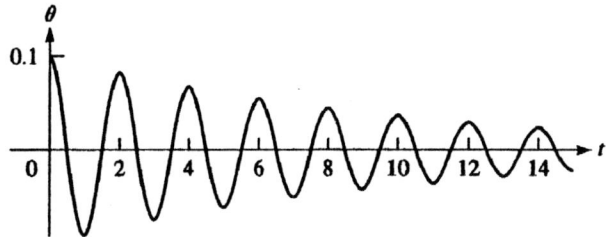

29. $D^2y + 0.2Dy + 4000y = 0$

$\qquad m^2 + 0.2m + 4000 = 0$

$\qquad m = -0.100 \pm 63.2j, \alpha = -0.100, \beta = 63.2$

$\qquad y = e^{-0.100t}(c_1 \sin 63.2t + c_2 \cos 63.2t) = 3.00$ when $t = 0$

$\qquad e^{-0.100(0)}(c_1 \sin 0 + c_2 \cos 0) = 3.00$

$\qquad\qquad\qquad\qquad\qquad c_2 = 3.00$

$\qquad y = e^{-0.100t}(c_1 \sin 63.2t + 3.00 \cos 63.2t)$

$\qquad \dfrac{dy}{dt} = -0.100e^{-0.100t}(c_1 \sin 63.2t + 3.00 \cos 63.2t) + e^{-0.100t}(63.2c_1 \sin 63.2t - 63.2(3.00) \sin 63.2t)$

$\qquad \dfrac{dy}{dt} = -0.300$ when $t = 0$

$\qquad -0.100e^0(c_1 \sin 0 + 3.00 \cos 0) + e^0(c_1 \cos 0 - 63.2(3.00) \sin 0) = -0.300$

$\qquad c_1 = 0$

$\qquad y = e^{-0.100t}(3.00 \cos 63.2t)$

30. $D^2x - 2Dx - 8x = 5 \sin 2t$

$\qquad m^2 - 2m - 8 = 0$

$\qquad (m + 2)(m - 4) = 0$

$\qquad m = -2, m = 4$

$\qquad x_c = c_1 e^{-2x} + c_2 e^{4x}$

$\qquad x_p = A \sin 2t + B \cos 2t$

$\qquad x_p' = 2A \cos 2t - 2B \sin 2t$

$\qquad x_p'' = -4A \sin 2t - 4B \cos 2t$

$\qquad -4A \sin 2t - 4B \cos 2t - 2(2A \cos 2t - 2B \sin 2t) - 8(A \sin 2t + B \cos 2t) = 5 \sin 2t$

$\qquad \sin 2t(-4A + 4B - 8A) + \cos 2t(-4B - 4A - 8B) = 5 \sin 2t$

$\qquad -12A + 4B = 5$

$\qquad -4A - 12B = 0, A = -\dfrac{3}{8}, B = \dfrac{1}{8}$

The steady-state solution is

$$x_p = -\frac{3}{8}\sin 2t + \frac{1}{8}\cos 2t$$

31. $LD^2q + RDq + \dfrac{q}{C} = E \quad \left(D = \dfrac{d}{dt} \right)$

$L = 0.5$ H, $R = 6\ \Omega$, $C = 20$ mF, $E = 24\sin 10t$
$0.5D^2q + 6Dq + 50q = 24\sin 10t,\ D^2q + 12Dq + 100q = 48\sin 10t$

$$m^2 + 12m + 100 = 0, m = \frac{-12 \pm \sqrt{144 - 400}}{2} = -6 \pm 8j$$

$q_c = e^{-6t}(c_1\sin 8t + c_2\cos 8t),\ q_p = A\sin 10t + B\cos 10t$
$Dq_p = 10A\cos 10t - 10B\sin 10t,\ D^2q_p = -100A\sin 10t - 100B\cos 10t$
$-100A\sin 10t - 100B\cos 10t + 12(10A\cos 10t - 10B\sin 10t)$
$\quad + 100(A\sin 10t + B\cos 10t) = 48\sin 10t$
$-120B = 48, B = -0.4;\ 120A = 0, A = 0$
$q = e^{-6t}(c_1\sin 8t + c_2\cos 8t) - 0.4\cos 10t$
$q = 0$ when $t = 0, 0 = c_2 - 0.4, c_2 = 0.4$
$q = e^{-6t}(c_1\sin 8t + 0.4\cos 8t) - 0.4\cos 10t$
$Dq = e^{-6t}(-6c_1\sin 8t - 2.4\cos 8t + 8c_1\cos 8t - 3.2\sin 8t) + 4\sin 10t$
$Dq = 0$ when $t = 0;\ 0 = -2.4 + 8c_1, c_1 = 0.3$
$\quad q = e^{-6t}(0.3\sin 8t + 0.4\cos 8t) - 0.4\cos 10t$

32. $L\dfrac{d^2q}{dt^2} + \dfrac{q}{C} = E$

$$5.00 \times 10^{-3}\frac{d^2q}{dt^2} + \frac{q}{10.0 \times 10^{-6}} = 0.200e^{-200t}$$

$$5.00 \times 10^{-3}\frac{d^2q}{dt^2} + 1.00 \times 10^5 q = 0.200e^{-200t}$$

$$5.00 \times 10^{-3}m^2 + 1.00 \times 10^5 = 0$$

$m = 0 \pm 4.47 \times 10^3 j, \alpha = 0, \beta = 4.47 \times 10^3$
$q_c = c_1\sin 4.47 \times 10^3 t + c_2\cos 4.47 \times 10^3 t$
$q_p = Ae^{-200t}$
$q_p' = -200Ae^{-200t}$
$q_p'' = 4.00 \times 10^4 Ae^{-200t}$
$5.00 \times 10^{-3}(4.00 \times 10^4 Ae^{-200t}) + 1.00 \times 10^5 Ae^{-200t} = 0.200e^{-200t}$
$\qquad\qquad (2.00 \times 10^2 A + 1.00 \times 10^5 A)e^{-200t} = 0.200e^{-200t}$
$\qquad\qquad\qquad\qquad\qquad\qquad 1.00 \times 10^5 A = 0.200$
$\qquad\qquad\qquad\qquad\qquad\qquad\qquad A = 2.00 \times 10^{-6}$

$q_p = 2.00 \times 10^{-6}e^{-200t}$
$q = c_1\sin 4.47 \times 10^3 t + c_2\cos 4.47 \times 10^3 t + 2.00 \times 10^{-6}e^{-200t}$

when $t = 0, q = 0$
$0 = c_1\sin 0 + c_2\cos 0 + 2.00 \times 10^{-6}$
$c_2 = -2.00 \times 10^{-6}$
$\quad q = c_1\sin 4.47 \times 10^3 t - 2.00 \times 10^{-6}\cos 4.47 \times 10^3 t + 2.00 \times 10^{-6}e^{-200t}$
$\quad q_1 = c_1(4.47 \times 10^3)\cos 4.47 \times 10^3 t + 8.94 \times 10^{-3}\sin 4.47 \times 10^3 t$
$\qquad\quad + 4.00 \times 10^{-4}e^{-200t}$

when $t = 0$, $q' = 4.00 \times 10^{-3}$

$4.00 \times 10^{-3} = c_1(4.47 \times 10^3) \cos 0 + 8.94 \times 10^{-3} \sin 0 + 4.00 \times 10^{-4}$

$\qquad c_1 = 8.05 \times 10^{-7}$

$q = 8.05 \times 10^{-7} \sin 4.47 \times 10^3 t - 2.00 \times 10^{-6} \cos 4.47 \times 10^3 t + 2.00 \times 10^{-6} e^{-200t}$

33.
$$L\frac{d^2q}{dt^2} + R\frac{dq}{dt} + \frac{q}{C} = E$$

$$4\frac{d^2q}{dt^2} + 20\frac{dq}{dt} + \frac{q}{100 \times 10^{-6}} = 100$$

$$4\frac{d^2q}{dt^2} + 20\frac{dq}{dt} + 10{,}000q = 100$$

$$\frac{d^2q}{dt^2} + 5\frac{dq}{dt} + 2500q = 25$$

$$m^2 + 5m + 2500 = 0$$

$m = -2.5 \pm 50j, \alpha = -2.5, \beta = 50$

$q_c = e^{-2.5t}(c_1 \sin 50t + c_2 \cos 50t)$

$q_p = A$

$q_p' = 0$

$q_p'' = 0$

$0 + 5(0) + 2500A = 25$

$\qquad A = 0.01$

$q_p = 0.01$; $q = e^{-2.5t}(c_1 \sin 50t + c_2 \cos 50t) + 0.01 = 10 \times 10^{-3}$ when $t = 0$

$10 \times 10^{-3} = c_1 \sin 0 + c_2 \cos 0 + 0.01$; $c_1 = 0$

$q = e^{-2.5t}c_2 \cos 50t + 0.01$

$i = \dfrac{dq}{dt} = (-2.5)e^{-2.5t}c_2 \cos 50t + e^{-2.5t}(-50c_2 \sin 50t) = 0$ when $t = 0$

$0 = (-2.5)e^0 c_2 \cos 0 + e^0(-50c_2 \sin 0)$

$0 = -2.5c_2$

$c_2 = 0$

$q = 0.01$

$i = \dfrac{dq}{dt} = 0$

34. $L\dfrac{d^2q}{dt^2} + \dfrac{q}{C} = E_0 \sin wt$

$\qquad Lm^2 + \dfrac{1}{C} = 0$

$\qquad m = 0 \pm \dfrac{1}{\sqrt{LC}}j, \alpha = 0, \beta = \dfrac{1}{\sqrt{LC}}$

$\qquad q = c_1 \sin \dfrac{t}{\sqrt{LC}} + c_2 \cos \dfrac{t}{\sqrt{LC}}$

$\qquad q_p = A \sin wt + B \cos wt$

$\qquad q_p' = Aw \cos wt - Bw \sin wt$

$\qquad q_p'' = -Aw^2 \sin wt - Bw^2 \cos wt$

$\qquad L(-Aw^2 \sin wt - Bw^2 \cos wt) + \dfrac{A \sin wt + B \cos wt}{C} = E_0 \sin wt$

$\qquad \sin wt \left(-LAw^2 + \dfrac{A}{C} \right) + \cos wt \left(-Bw^2 + \dfrac{B}{C} \right) = E_0 \sin wt$

$$A \left(\frac{1}{C} - Lw^2 \right) = E_0 \qquad , \quad -Bw^2 + \frac{B}{C} = 0$$

$$A = \frac{CE_0}{1 - CLw^2} \qquad\qquad B = 0$$

$$q = c_1 \sin \frac{t}{\sqrt{LC}} + c_2 \cos \frac{t}{\sqrt{LC}} + \frac{CE_0}{1 - CLw^2} \sin wt, \text{ at } t = 0, q = 0$$

$$0 = c_1 \sin 0 + c_2 \cos 0 + \frac{CE_0}{1 - CLw^2} \sin 0$$

$$c_2 = 0$$

$$q = c_1 \sin \frac{t}{\sqrt{LC}} + \frac{CE_0}{1 - CLw^2} \sin wt$$

$$q' = \frac{c_1}{\sqrt{LC}} \cos \frac{t}{\sqrt{LC}} + \frac{CE_0 w}{1 - CLw^2} \cos wt, \text{ at } t = 0, q_1 = i = 0$$

$$0 = \frac{c_1}{\sqrt{LC}} \cos 0 + \frac{CE_0 w}{1 - CLw^2} \cos 0$$

$$c_1 = \frac{-\sqrt{LC} E_0 w}{1 - w^2 LC}$$

$$q = \frac{-\sqrt{LC} E_0 w}{1 - w^2 LC} \sin \frac{t}{\sqrt{LC}} + \frac{CE_0}{1 - w^2 LC} \sin wt$$

$$q = \frac{CE_0}{1 - w^2 LC} \left(\sin wt - w\sqrt{LC} \sin \frac{t}{\sqrt{LC}} \right)$$

35. $EI \dfrac{d^2 y}{dx^2} = M, M = 2000x - 40x^2$

$$EID^2 y = 2000x - 40x^2$$

$$D^2 y = \frac{1}{EI} (2000x - 40x^2)$$

$$Dy = \frac{1}{EI} \left(1000x^2 - \frac{40}{3} x^3 \right) + c_1$$

$$y = \frac{1}{EI} \left(\frac{1000}{3} x^3 - \frac{10}{3} x^4 \right) + c_1 x + c_2$$

$$y = 0 \text{ for } x = 0 \text{ and } x = L$$

$$0 = \frac{1}{EI} (0 - 0) + 0 + c_2, c_2 = 0$$

$$0 = \frac{1}{EI} \left(\frac{1000}{3} L^3 - \frac{10}{3} L^4 \right) + c_1 L$$

$$c_1 = \frac{1}{EI} \left(\frac{10}{3} L^3 - \frac{1000}{3} L^2 \right)$$

$$y = \frac{1}{EI} \left(\frac{1000}{3} x^3 - \frac{10}{3} x^4 \right) + \frac{1}{EI} \left(\frac{1000}{3} L^3 - \frac{1000}{3} L^2 \right) x$$

$$= \frac{10}{3EI} (100x^3 - x^4 + L^3 x - 100L^2 x)$$

36. $\dfrac{1}{2}mr^2\dfrac{d^2\theta}{dt^2} = -k\theta$

$\dfrac{1}{2}mr^2\dfrac{d^2\theta}{dt^2} + k\theta = 0$

The auxiliary equation has roots.

$0 \pm \sqrt{\dfrac{2k}{mr^2}}\,j,\ \alpha = 0, \beta = \sqrt{\dfrac{2k}{mr^2}} = w$

$\theta = c_1 \sin wt + c_2 \cos wt$ at $t = 0, \theta = \theta_0$

$\theta_0 = c_1 \sin 0 + c_2 \cos 0$

$c_2 = \theta_0$

$\theta = c_1 \sin wt + \theta_0 \cos wt$

$\dfrac{d\theta}{dt} = c_1 w \cos wt - \theta_0 w \sin wt,$ at $t = 0, \dfrac{d\theta}{dt} = w_0$

$w_0 = c_1 w \cos 0 - \theta_0 w \sin 0$

$c_1 = \dfrac{w_0}{w}$

$\theta = \dfrac{w_0}{w} \sin wt + \theta_0 \cos wt$ where $w = \sqrt{\dfrac{2k}{mr^2}}$

37. In comparing

$mD^2x + 2.00D_x + kx = 0$ and

$LD^2q + RDq + \dfrac{1}{C}q = E\text{:}$

–position corresponds to charge

–mass corresponds to inductance

–electrical resistance corresponds to the constant of proportionality which must be multiplied times velocity to give the resisting force

–the reciprocal of the capacitance corresponds to the spring constant.

OTHER METHODS OF SOLVING DIFFERENTIAL EQUATIONS

16.1 Numerical Solutions

1. $\dfrac{dy}{dx} = x + 1$

x	y	$x+1$	dy	y(correct)
0.0	1.00	1.0	0.20	1.00
0.2	1.20	1.2	0.24	1.22
0.4	1.44	1.4	0.28	1.48
0.6	1.72	1.6	0.32	1.78
0.8	2.04	1.8	0.36	2.12
1.0	2.40	2.0	0.40	2.50

$$y = \frac{1}{2}x^2 + x + c$$

$$y = 1 \text{ when } x = 0$$

$$c = 1$$

$$y = \frac{1}{2}x^2 + x + 1$$

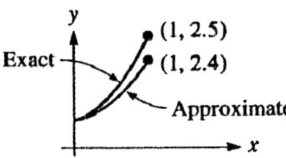

2. $\dfrac{dy}{dx} = \sqrt{2x+1} = (2x+1)^{1/2}$

x	y approximate	y exact
0.0	2.0000	2.0000
0.3	$2.000 + \sqrt{2(0)+1}(0.3) = 2.3000$	2.3413
0.6	$2.3000 + \sqrt{2(0.3)+1}(0.3) = 2.6795$	2.7544
0.9	$2.6795 + \sqrt{2(0.6)+1}(0.3) = 3.1244$	3.2284
1.2	$3.1244 + \sqrt{2(0.9)+1}(0.3) = 3.6264$	3.7564

$$y = \frac{1}{2}\int (2x+1)^{1/2}(2\,dx) = \frac{1}{2}\cdot\frac{2}{3}(2x+1)^{3/2} + c$$

$$y = 2 \text{ when } x = 0$$

$$2 = \frac{1}{3}(2(0)+1)^{3/2} + c$$

$$c = \frac{5}{3}$$

$$y = \frac{(2x+1)^{3/2}}{3} + \frac{5}{3}, \text{ exact}$$

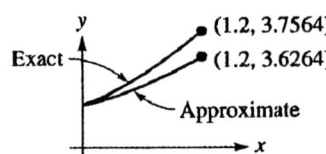

3. $\dfrac{dy}{dx} = y(0.4x + 1)$

x	y	$y(0.4x+1)$	dy	y(correct)
−0.2	2.0000	1.8400	0.1840	2.0000
−0.1	2.1840	2.0966	0.2097	2.1971
0.0	2.3937	2.3937	0.2394	2.4233
0.1	2.6330	2.7384	0.2738	2.6836
0.2	2.9069	3.1394	0.3139	2.9836
0.3	3.2208	3.6073	0.3607	3.3306

$$\frac{dy}{y} = (0.4x + 1)\,dx$$

$$\ln y = 0.2x^2 + x + \ln c$$

$y = 2$ when $x = -0.2$

$\ln 2 = 0.008 - 0.2 + \ln c,\, c = 2.4233$

$$y = 2.4233 e^{0.2x^2 + x}$$

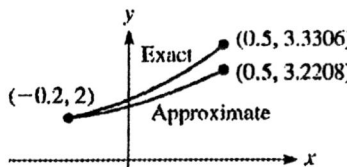

4.

x	y approximate	y exact
0.0	0	0
0.1	$(0 + e^0)(0.1) + 0 = 0.1000$	0.1105
0.2	$(0.1 + e^{0.1})(0.1) + 0.1 = 0.2205$	0.2443
0.3	$(0.2205 + e^{0.2})(0.1) + 0.2205 = 0.3647$	0.4050
0.4	$(0.3647 + e^{0.3})(0.1) + 0.3647 = 0.5362$	0.5967
0.5	$(0.5362 + e^{0.4})(0.1) + 0.5362 = 0.7390$	0.8244

$$\frac{dy}{dx} = y + e^x, \text{ linear}$$

$$y e^{\int -dx} = \int e^x \cdot e^{\int -dx}\,dx$$

$$y e^{-x} = x + c, (0,0)$$

$$0 \cdot e^{-0} = 0 + c$$

$$c = 0$$

$$y e^{-x} = x$$

$$y = x e^x, \text{ exact}$$

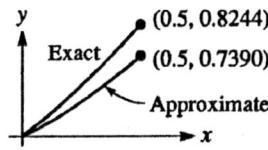

5.

x	y approximate	y exact
0.0	1	1
0.1	$1.00 + (0.0 + 1)(0.1) = 1.10$	0.105
0.2	$1.10 + (0.1 + 1)(0.1) = 1.21$	1.220
0.3	$1.21 + (0.2 + 1)(0.1) = 1.33$	1.345
0.4	$1.33 + (0.3 + 1)(0.1) = 1.46$	1.480
0.5	$1.46 + (0.4 + 1)(0.1) = 1.60$	1.625
0.6	$1.60 + (0.5 + 1)(0.1) = 1.75$	1.780
0.7	$1.75 + (0.6 + 1)(0.1) = 1.91$	1.945
0.8	$1.91 + (0.7 + 1)(0.1) = 2.08$	2.120
0.9	$2.08 + (0.8 + 1)(0.1) = 2.26$	2.305
1.0	$2.26 + (0.9 + 1)(0.1) = 2.45$	2.500

6.

x	y approximate	y exact
0.0	2	2
0.1	$2.0000 + \sqrt{2(0)+1}(0.1) = 2.1000$	2.1048
0.2	$2.1000 + \sqrt{2(0.1)+1}(0.1) = 2.2095$	2.2188
0.3	$2.2095 + \sqrt{2(0.2)+1}(0.1) = 2.3279$	2.3413
0.4	$2.3279 + \sqrt{2(0.3)+1}(0.1) = 2.4544$	2.4717
0.5	$2.4544 + \sqrt{2(0.4)+1}(0.1) = 2.5885$	2.6095
0.6	$2.5885 + \sqrt{2(0.5)+1}(0.1) = 2.7299$	2.7544
0.7	$2.7299 + \sqrt{2(0.6)+1}(0.1) = 2.8783$	2.9060
0.8	$2.8783 + \sqrt{2(0.7)+1}(0.1) = 3.0332$	3.0641
0.9	$3.0332 + \sqrt{2(0.8)+1}(0.1) = 3.1944$	3.2284
1.0	$3.1944 + \sqrt{2(0.9)+1}(0.1) = 3.3618$	3.3987
1.1	$3.3618 + \sqrt{2(1.0)+1}(0.1) = 3.5350$	3.5748
1.2	$3.5350 + \sqrt{2(1.1)+1}(0.1) = 3.7139$	3.7564

7.

x	y approximate	y exact
−0.20	2	2
−0.15	$2 + 2(0.4(-0.20)+1)(0.05) = 2.0920$	2.0952
−0.10	$2.0920 + 2.0920(0.4(-0.15)+1)(0.05) = 2.1903$	2.1971
−0.05	$2.1903 + 2.1903(0.4(-0.10)+1)(0.05) = 2.2955$	2.3063
0	$2.2955 + 2.2955(0.4(-0.05)+1)(0.05) = 2.4079$	2.4233
0.05	$2.4079 + 2.4079(0.4(0)+1)(0.05) = 2.5283$	2.5488
0.10	$2.5283 + 2.5283(0.4(0.05)+1)(0.05) = 2.6573$	2.6835
0.15	$2.6573 + 2.6573(0.4(0.10)+1)(0.05) = 2.7955$	2.8282
0.20	$2.7955 + 2.7955(0.4(0.15)+1)(0.05) = 2.9436$	2.9836
0.25	$2.9436 + 2.9436(0.4(0.20)+1)(0.05) = 3.1026$	3.1507
0.30	$3.1026 + 3.1026(0.4(0.25)+1)(0.05) = 3.2732$	3.3305

8.

x	y approximate	y exact
0	0	0
0.05	$(0 + e^{0})(0.05)(0.05) + 0 = 0.0500$	0.0526
0.10	$(0.0500 + e^{0.05})(0.05) + 0.0500 = 0.1051$	0.1105
0.15	$(0.1051 + e^{0.10})(0.05) + 0.1051 = 0.1656$	0.1743
0.20	$(0.1656 + e^{0.15})(0.05) + 0.1656 = 0.2319$	0.2443
0.25	$(0.2319 + e^{0.20})(0.05) + 0.2319 = 0.3046$	0.3210
0.30	$(0.3046 + e^{0.25})(0.05) + 0.3046 = 0.3840$	0.4050
0.35	$(0.3840 + e^{0.30})(0.05) + 0.3840 = 0.4707$	0.4967
0.40	$(0.4707 + e^{0.35})(0.05) + 0.4707 = 0.5652$	0.5967
0.45	$(0.5652 + e^{0.40})(0.05) + 0.5652 = 0.6681$	0.7057
0.50	$(0.6681 + e^{0.45})(0.05) + 0.6681 = 0.7799$	0.8244

9. $\dfrac{dy}{dx} = xy + 1, x = 0 \text{ to } x = 0.4, \Delta x = 0.1, (0,0)$

x	y	y to 4 places
0	0	0
0.1	0.1003339594	0.1003
0.2	0.20268804	0.2027
0.3	0.3091639819	0.3092
0.4	0.4220317248	0.4220

10. $\dfrac{dy}{dx} = x^2 + y^2$, $x = 0$ to $x = 0.4$, $\Delta x = 0.1$, $(0, 1)$

x	y	y to 4 places
0	1	1
0.1	1.111462856	1.1115
0.2	1.253015175	1.2530
0.3	1.439665975	1.4397
0.4	1.696097904	1.6961

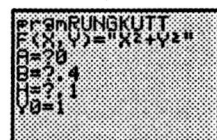

11. $\dfrac{dy}{dx} = e^y$, $x = 0$ to $x = 1$, $\Delta x = 0.2$, $(0, 0)$

x	y
0	0
0.2	0.2027
0.4	0.4232
0.6	0.6884
0.8	1.0588
1.0	1.7722

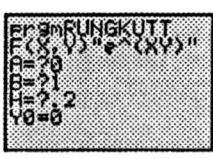

12. $\dfrac{dy}{dx} = \sqrt{1 + xy}$, $x = 0$ to $x = 0.2$, $\Delta x = 0.05$, $(0, 1)$

x	y
0	0
0.05	1.0506
0.10	1.1026
0.15	1.1560
0.20	1.2110

13. $\dfrac{dy}{dx} = \cos(x+y)$, $x = 0$ to $x = 0.5$, $\Delta x = 0.1$, $\left(0, \dfrac{\pi}{2}\right)$

x	y
0	$\frac{\pi}{2} = 1.5708$
0.1	1.5660
0.2	1.5521
0.3	1.5302
0.4	1.5011
0.5	1.4656

14. $\dfrac{dy}{dx} = y + \sin x$, $x = 0.5$ to $x = 1.0$, $\Delta x = 0.1$, $(0, 5), 0$

x	y
0.5	0
0.6	0.0549
0.7	0.1242
0.8	0.2089
0.9	1.3097
1.0	1.4278

15. $\dfrac{di}{dt} = +2i = \sin t$

t	i	$\sin t - 2i$	di
0.0	0.0000	0.0000	0.0000
0.1	0.0000	0.0998	0.0100
0.2	0.0100	0.1787	0.0179
0.3	0.0279	0.2398	0.0240
0.4	0.0518	0.2857	0.0286
0.5	0.0804	0.3186	0.0319

$i = 0.0804$ A for $t = 0.5$ s

$$di + 2i\,dt = \sin t\,dt; \quad e^{\int 2\,dt} = e^{2t}$$

$$ie^{2t} = \int e^{2t}\sin t\,dt = \frac{e^{2t}(2\sin t - \cos t)}{4 + 1} + c$$

$$i = \frac{1}{5}(2\sin t - \cos t) + ce^{-2t} \quad \text{(Formula 49)}$$

$$i = 0 \text{ for } t = 0, 0 = \frac{1}{5}(0 - 1) + c, c = \frac{1}{5}$$

$$i = \frac{1}{5}(2\sin t - \cos t + e^{-2t})$$

$$i = 0.0898 \text{ A for } t = 0.5 \text{ t}$$

16. $\dfrac{dT}{dt} = \sqrt[3]{1 + t^3}$, $t = 0$ to 5 min, $\Delta t = 1$, $(0, 0)$

t	T
0	0
1	1.0700
2	2.7170
3	5.2713
4	8.7989
5	13.3155

$T = 13.3°$ for $t = 5$ min

16.2   A Method of Successive Approximations

1. $y' = y^2, (0, 1)$

$dy = (1)dx$

$y_1 = x + c_1, 1 = 0 + c_1, c_1 = 1, y_1 = x + 1$

$dy = (x + 1)^2 dx$

$y_2 = \frac{1}{3}(x + 1)^3 + c_2, 1 = \frac{1}{3} + c_2 = \frac{2}{3}$

$y_2 = \frac{1}{3}(x + 1)^3 + \frac{2}{3} = \frac{1}{3}x^3 + x^2 + x + 1$

$y_2(0.1) = \frac{1}{3}(0.1)^3 + (0.1)^2 + 0.1 + 1 = 1.1103$

$\frac{dy}{y^2} = dx$

$-\frac{1}{y} = 0 + c, c = -1$

$-\frac{1}{y} = x - 1$

$y = \frac{1}{1 - x}$

$y(0.1) = 1.1111$

2. $y' = x - y$

$\frac{dy}{dx} = x - y = x - 1$

$y_1 = \frac{x^2}{2} - x + c$

$1 = \frac{0^2}{2} - 0 + c, c = 1$

$y_1 = \frac{x^2}{2} - x + 1$

$\frac{dy}{dx} = x - \frac{x^2}{2} + x - 1 = -\frac{x^2}{2} + 2x - 1$

$y_2 = -\frac{x^3}{6} + x^2 - x + c$

$1 = -\frac{0^3}{6} + 0^2 - 0 + c, c = 1$

$y_2 = -\frac{x^3}{6} + x^2 - x + 1$

$y^2\left(\frac{1}{2}\right) = -\frac{\left(\frac{1}{2}\right)^3}{6} + \left(\frac{1}{2}\right)^2 - \left(\frac{1}{2}\right) + 1 = 0.729$

$\frac{dy}{dx} = x - y$

$dy = x\,dx - y\,dx$

$dy + y\,dx = x\,dx$, linear

$ye^{\int dx} = \int xe^{\int dx}dx + c$

$ye^x = \int xe^x dx + c$

$ye^x = e^x(x - 1) + c$

$1e^0 = e^0(0 - 1) + c, c = 2$

$ye^x = e^x(x - 1) + 2$

$y = 2e^{-x} + x - 1$, exact

$y_{\text{exact}}\left(\frac{1}{2}\right) = 2e^{-1/2} + \frac{1}{2} - 1 = 0.713$

3. $y' = 2x(1 + y), (0, 0)$

$dy = 2x(1 + 0)dx = 2x\,dx$

$y_1 = x^2 + c_1, 0 = 0 + c_1, c_1 = 0, y_1 = x^2$

$dy = 2x(1 + x^2)dx = (2x + 2x^3)dx$

$y_2 = x^2 + \frac{1}{2}x^4 + c_2, 0 = 0 + 0 + c_2, c_2 = 0$

$y_2 = x^2 + \frac{1}{2}x^4$

$dy = 2x\left(1 + x^2 + \frac{1}{2}x^4\right)dx$

$\quad = (2x + 2x^3 + x^5)dx$

$y_3 = x^2 + \frac{1}{2}x^4 + \frac{1}{6}x^6 + c_3, c_3 = 0$

$y_3 = x^2 + \frac{1}{2}x^4 + \frac{1}{6}x^6;$

$y_3(1) = 1 + \frac{1}{2} + \frac{1}{6} = 1.67$

$\frac{dy}{1 + y} = 2x\,dx$

$\ln(1 + y) = x^2 + c$

$\ln(1 + 0) = 0 + c, c = 0$

$\ln(1 + y) = x^2, 1 + y = e^{x^2}$

$y = e^{x^2} - 1$

$y(1) = 1.72$

4. $y' = y + e^x$

$\frac{dy}{dx} = 0 + e^x$

$y_1 = e^x + c$

$0 = e^0 + c, c = -1$

$y_1 = e^x - 1$

$\frac{dy}{dx} = e^x - 1 + e^x = 2e^x - 1$

$y_2 = 2e^x - x + c$

$0 = 2e^0 - 0 + c, c = -2$

$y_2 = 2e^x - x - 2$

$$\frac{dy}{dx} = 2e^x - x - 2 + e^x$$

$$= 3e^x - x - 2$$

$$y_3 = 3e^x - \frac{x^2}{2} - 2x + c$$

$$0 = 3e^0 - \frac{0^2}{2} - 2(0) + c,$$

$$c = -3$$

$$y_3 = 3e^x - \frac{x^2}{2} - 2x - 3$$

$$y_3(0.2) = 3e^{0.2} - \frac{0.2^2}{2} - 2(0.2) - 3$$

$$y_3(0.2) = 0.244$$

$$\frac{dy}{dx} = y + e^x$$

$$dy = y\,dx + e^x dx$$

$$dy + (-y\,dx) = e^x dx, \text{ linear}$$

$$ye^{\int -dx} = \int e^x \cdot e^{\int -dx}\,dx + c$$

$$ye^{-x} = \int e^x \cdot e^{-x}dx + c$$

$$ye^{-x} = x + c$$

$$0 \cdot e^{-0} = 0 + c,$$

$$c = 0$$

$$y = xe^x, \text{ exact}$$

$$y_{\text{exact}}(0.2) = 0.2e^{0.2}$$

$$y_{\text{exact}}(0.2) = 0.244$$

5. $y' = x + y^2, (0,0)$

$$dy = x\,dx$$

$$y_1 = \frac{1}{2}x^2 + c_1,$$

$$c_1 = 0, y = \frac{1}{2}x^2$$

$$dy = \left[x + \left(\frac{1}{2}x^2\right)^2\right]dx$$

$$= \left(x + \frac{1}{4}x^4\right)dx$$

$$y_2 = \frac{1}{2}x^2 + \frac{1}{20}x^5 + c_2, c_2 = 0$$

$$dy = \left[x + \left(\frac{1}{2}x^2 + \frac{1}{20}x^5\right)^2\right]dx$$

$$= \left(x + \frac{1}{4}x^4 + \frac{1}{20}x^7 + \frac{1}{400}x^{10}\right)dx$$

$$y_3 = \frac{1}{2}x^2 + \frac{1}{20}x^5 + \frac{1}{160}x^8 + \frac{1}{4400}x^{11} + c_3$$

$$c_3 = 0$$

$$y_3 = \frac{1}{2}x^2 + \frac{1}{20}x^5 + \frac{1}{160}x^8 + \frac{1}{4400}x^{11}$$

6. $y' = 1 + xy^2, y_2, (0,0)$

$$y' = 1 + x(0)^2$$

$$y' = x + c$$

$$0 = 0 + c, c = 0$$

$$y' = x$$

$$\frac{dy}{dx} = 1 + x(x^2)$$

$$= 1 + x^3$$

$$y_2 = x + \frac{x^4}{4} + c$$

$$0 = 0 + \frac{0^4}{4} + c, c = 0$$

$$y_2 = x + \frac{x^4}{4}$$

7. $y' = y - \cos 2x, (0,1)$

$$dy = (1 - \cos 2x)dx$$

$$y_1 = x - \frac{1}{2}\sin 2x + c_1, c_1 = 1$$

$$y_1 = x - \frac{1}{2}\sin 2x + 1$$

$$dy = \left(x - \frac{1}{2}\sin 2x + 1 - \cos 2x\right)dx$$

$$y_2 = \frac{1}{2}x^2 + \frac{1}{4}\cos 2x + x - \frac{1}{2}\sin 2x + c_2$$

$$1 = 0 + \frac{1}{4} + 0 - 0 + c_2, c_2 = \frac{3}{4}$$

$$y_2 = \frac{3}{4} + x + \frac{1}{2}x^2 + \frac{1}{4}\cos 2x - \frac{1}{2}\sin 2x$$

8. $y' = x + y\cos x, y_2, (0,1)$

$$\frac{dy}{dx} = x + \cos x$$

$$y_1 = \frac{x^2}{2} + \sin x + c$$

$$1 = \frac{0^2}{2} + \sin 0 + c, c = 1$$

$$y_1 = \frac{x^2}{2} + \sin x + 1$$

$$\frac{dy}{dx} = x + \left(\frac{x^2}{2} + \sin x + 1\right)\cos x$$

$$\frac{dy}{dx} = x + \frac{x^2}{2}\cos x + \sin x \cos x + \cos x$$

$$y_2 = \frac{x^2}{2} + x\cos x + \frac{x^2}{2}\sin x$$

$$- \sin x + \frac{\sin^2 x}{2} + \sin x + c$$

$$1 = \frac{0^2}{2} + 0\cos 0 + \frac{0^2}{2}\sin 0$$

$$- \sin 0 + \frac{\sin^2 0}{2} + \sin 0 + c$$

$$c = 1$$

$$y_2 = \frac{x^2}{2} + x\cos x + \frac{x^2}{2}\sin x + \frac{\sin^2 x}{2} + 1$$

$$y_2 = \frac{1}{2}(x^2 + 2x\cos x + x^2\sin x + \sin^2 x + 2)$$

9. $e^x = 1 + x + \frac{1}{2}x^2 + \frac{1}{6}x^3 + \frac{1}{24}x^4 + \cdots$

$$2e^x - x - 1 = 2\left(1 + x + \frac{1}{2}x^2 + \frac{1}{6}x^3 + \frac{1}{24}x^4 + \cdots\right)$$

$$- x - 1$$

$$= 1 + x + x^2 + \frac{1}{3}x^3 + \frac{1}{12}x^4 + \cdots$$

From Example 1: $y = 1 + x + x^2 + \frac{1}{3}x^3 + \frac{1}{24}x^4$

10. From Exercise 3, $y = e^{x^2} - 1$

Maclaurin: $e^x = 1 + x + \frac{x^2}{2} + \frac{x^3}{6} + \cdots$

$$y = e^{x^2} - 1 = 1 + x^2 + \frac{(x^2)^2}{2} + \frac{(x^2)^3}{6} + \cdots - 1$$

$$y = x^2 + \frac{x^4}{2} + \frac{x^6}{6} + \cdots, \text{ exact}$$

$$y_3 = x^2 + \frac{1}{2}x^4 + \frac{x^6}{6}, \text{ approximate from Exercise 3.}$$

11. $\frac{dv}{dt} = 2 - v^3, v = 0$ when $t = 0$ (from rest)

$$dv = 2\,dt,$$
$$v_1 = 2t + c_1, c_1 = 0,$$
$$v_1 = 2t, dv = [2 - (2t)^3]dt$$
$$v_2 = 2t - 2t^4 + c_2, c_2 = 0,$$
$$v_2 = 2t - 2t^4;$$
$$v_2(0.3) = 2(0.3) - 2(0.3)^4 = 0.58 \text{ ft/s}$$

12. $\frac{di}{dt} = -2i + \sin t, (0,0)$

$$\frac{di}{dt} = \sin t$$

$$i_1 = -\cos t + c$$
$$0 = -\cos 0 + c, c = 1$$
$$i_1 = -\cos t + 1$$

$$\frac{di}{dt} = -2(-\cos t + 1) + \sin t = 2\cos t - 2 + \sin t$$

$$i_2 = 2\sin t - 2t - \cos t + c$$
$$0 = 2\sin 0 - 2(0) - \cos 0 + c$$
$$c = 1$$
$$i_2 = 2\sin t - 2t - \cos t + 1$$
$$i_2(0.5) = 2\sin 0.5 - 2(0.5) - \cos 0.5 + 1$$
$$i_2(0.5) = 0.0813 \text{ A approximate}$$

$$\frac{\text{exact} - \text{approximate}}{\text{exact}} = \frac{0.0898 - 0.0813}{0.0898} = 0.095$$

The approximate and exact values differ by 9.5%.

16.3　Laplace Transforms

1. $f(t) = 1; L(f) = L(t)$

$$F(s) = \int_0^\infty e^{-st}\,dt = -\frac{1}{s}\int_0^\infty e^{-st}(-s\,dt)$$

$$= \lim_{c\to\infty}\left[-\frac{1}{s}\int_0^c e^{-st}(-s\,dt)\right]$$

$$= \lim_{c\to\infty}\left[-\frac{1}{s}e^{-st}\right]\Big|_0^c$$

$$L(t) = \lim_{c\to\infty}\left[-\frac{1}{s}e^{-sc} + \frac{1}{s}\right] = 0 + \frac{1}{s} = \frac{1}{s}$$

2. $f(t) = e^{-at}; L(f) = L(t) = L(e^{-at})$

$$F(s) = \int_0^\infty e^{-st}e^{-at}\,dt = \int_0^\infty e^{(-s-a)t}\,dt$$

$$= \lim_{c\to\infty}\int_0^c e^{-(s+a)t}\,dt$$

$$= \lim_{c\to\infty}\frac{-1}{s+a}\int_0^c e^{-(s+a)t} - (s+a)dt$$

$$= \lim_{c\to\infty}\frac{-1}{s+a}e^{-(s+a)t}\Big|_0^c$$

$$= \lim_{c\to\infty}\frac{-1}{s+a}[e^{-(s+a)c} - e^{-(s+a)0}]$$

$$= \lim_{c\to\infty}\frac{-1}{s+a}[e^{-(s+a)c} - 1] = \lim_{c\to\infty}e^{-(s+a)c}$$

$$= 0$$

$$L(e^{-at}) = -\frac{1}{s+a}(0-1) = \frac{1}{s+a}$$

3. $f(t) = \sin at, L(f) = L(\sin at) = F(s)$

$$F(s) = L(\sin at) = \int_0^\infty e^{-st} \sin at\, dt$$

$$= \lim_{c \to \infty} \int_0^c e^{-st} \sin at\, dt$$

By Formula 49, Appendix D, $a = -s, b = a, u = t$

$$F(s) = \lim_{c \to \infty} \left[\frac{e^{-st}(-s \sin at - a \cos at)}{s^2 + a^2} \right] \Big|_0^c$$

$$= \lim_{c \to \infty} \left[\frac{e^{-sc}(-s \sin ac - a \cos ac)}{s^2 + a^2} - \left(\frac{-a}{s^2 + a^2} \right) \right]$$

$$= \lim_{c \to \infty} \frac{-e^{-sc}(s \sin ac + a \cos ac)}{s^2 + a^2} + \frac{a}{s^2 + a^2}$$

$$= 0 + \frac{a}{s^2 + a^2}$$

$$L(\sin at) = \frac{a}{s^2 + a^2}$$

4. $f(t) = te^{-at}; L(f) = L(te^{-at}) = F(s)$

$$F(s) = \int_0^\infty e^{-st} te^{-at} dt = \int_0^\infty te^{-(s+a)t} dt$$

By Formula 44, Appendix D, $u = t, a = -(s+a)$

$$F(s) = \lim_{c \to \infty} \int_0^c te^{-(s+a)t} dt$$

$$= \lim_{c \to \infty} \left\{ \frac{e^{-(s+a)t}[-(s+a)t - 1]}{(s+a)^2} \right\} \Big|_0^c$$

$$= \lim_{c \to \infty} \left\{ \frac{[-(s+a)c - 1]}{s^2 + a^2} e^{-(s+a)c} + \frac{1}{(s+a)^2} \right\}$$

$$\lim_{c \to \infty} -(s+a)c \to \infty \text{ at a slower rate than}$$

$$\lim_{c \to \infty} e^{-(s+a)c} \to 0$$

Therefore, $\displaystyle \lim_{c \to \infty} \frac{[-(s+a)c - 1]e^{-(s+a)c}}{s^2 + a^2} = 0$

$$F(s) = \frac{1}{(s+a)^2}; L(te^{-at}) = \frac{1}{(s+a)^2}$$

5. $f(t) = e^{3t}$; from transform (3) of the table,

$a = -3; L(3t) = \dfrac{1}{s - 3}$

6. $f(t) = 1 - \cos 2t$; from transform (7), $a = 2$

$$L(1 - \cos 2t) = \frac{4}{s(s^2 + 4)}$$

7. $f(t) = 5t^3 e^{-2t}$; from transform (12), $n - 1 = 3$, $n = 4; a = 2$

$$L(5t^3 e^{-2t}) = \frac{5(3!)}{(s+2)^4} = \frac{30}{(s+2)^4}$$

8. $f(t) = 2e^{-3t} \sin 4t$; from transform (19), $k = 2$, $a = 3, b = 4$

$$L(2e^{-3t} \sin 4t) = 2L(e^{-3t} \sin 4t) = 2 \left[\frac{4}{(s+3)^2 + 16} \right]$$

$$L(2e^{-3t} \sin 4t) = \frac{8}{(s+3)^2 + 16} = \frac{8}{s^2 + 6s + 25}$$

9. $f(t) = \cos 2t - \sin 2t; L(f) = L(\cos 2t) - L(\sin 2t)$

By transforms (5) and (6),

$$L(f) = \frac{s}{s^2 + 4} - \frac{2}{s^2 + 4}; L(f) = \frac{s - 2}{s^2 + 4}$$

10. $f(t) = 2t \sin 3t + e^{-3t} \cos t$

By transforms (16) and (20), $k = 2, a = 3, b = 1$

$$L(f) = F(s) = 2 \left[\frac{6s}{(s^2 + 9)^2} \right] + \frac{s + 3}{(s+3)^2 + 1}$$

$$F(s) = \frac{12s}{(s^2 + 9)^2} + \frac{(s + 3)}{s^2 + 6s + 10}$$

$$= \frac{12s(s^2 + 6s + 10) + (s + 3)(s^2 + 9)^2}{(s^2 + 9)^2(s^2 + 6s + 10)}$$

$$= \frac{12s^3 + 72s^2 + 120s + s^5 + 3s^4 + 18s^3 + 54s^2 + 815 + 243}{(s^2 + 9)^2(s^2 + 6s + 10)}$$

$$F(s) = \frac{s^5 + 3s^4 + 30s^3 + 126s^2 + 201s + 243}{(s^2 + 9)^2(s^2 + 6s + 10)}$$

11. $f(t) = 3t + 2t \cos 3t; L(f) = L(3) + 2Lt(\cos 3t)$

By transforms (1) and (18), $a = 3$

$$F(s) = \frac{3}{s} + 2 \left[\frac{s^2 - 9}{(s^2 + 9)^2} \right] = \frac{3}{s} + \frac{2(s^2 - 9)}{(s^2 + 9)^2}$$

12. $f(t) = r^3 - 3te^{-t}; L(f) = L(t^3) - 3L(te^{-t})$

By transforms (2), $n - 1 = 3(n - 1)! = 6; n = 4$; transform (11), $a = 1$

$$F(s) = \frac{6}{s^4} - 3 \left[\frac{1}{(s+1)^2} \right] = \frac{6}{s^4} - \frac{3}{(s+1)^2}$$

13. $y'' + y'; f(0) = 0; f'(0) = 0$

$L[f''(y) + f'(y)]$
$= L(f'') + L(f')$
$= s^2 L(f) - sf(0) - f'(0) + sL(f) - f(0)$
$= s^2 L(f) - s(0) - 0 + sL(f) - 0$
$= s^2 L(f) + sL(f)$

14. $y'' - 3y'; f(0) = 2, f'(0) = -1$

$L(y'' - 3y')$
$= L(y'') - 3L(y')$
$= s^2 L(y) - sf(0) - f'(0) - 3[sL(y) - f(0)]$
$= s^2 L(f) - 2s + 1 - 3[sL(f) - 2]$
$= s^2 L(f) - 2s + 1 - 3sL(f) + 6$
$= L(f)(s^2 - 3s) - 2s + 7$

15. $2y'' - y' + y; \ f(0) = 1, f'(0) = 0$

$L(2y'' - y' + y)$

$= 2L(f'') - L'(f') + L(f)$

$= 2[s^2 L(f) - sf(0) - f'(0)] - \{sL(f) - f(0)\} + L(f)$

$= 2s^2 L(f) - 2sf(0) - 2f'(0) - sL(f) + f(0) + L(f)$

$= L(f)(2s^2 - s + 1) - 2sf(0) + f(0) - 2f'(0)$

$= L(f)(2s^2 - s + 1) - 2s + 1 - 0$

$= L(f)(2s^2 - s + 1) - 2s + 1$

16. $y'' - 3y' + 2y; \ f(0) = -1, f'(0) = 2$

$L(y'' - 3y' + 2y)$

$= L(f'') - 3L(f') + 2L(f)$

$= s^2 L(f) - sf(0) - f'(0) - 3[sL(f) - f(0)] + 2L(f)$

$= s^2 L(f) + s - 2 - 3sL(f) + 3f(0) + 2L(f)$

$= s^2 L(f) + s - 22 - 3sL(f) - 3 + 2L(f)$

$= L(f)(s^2 - 3s + 2) + s - 5$

17. $L^{-1}(F) = L^{-1}\left(\dfrac{2}{s^3}\right) = 2L^{-1}\left(\dfrac{1}{s^3}\right);$

$L^{-1}(F) = \dfrac{2t^2}{2} = t^2; \ \text{transform (2)}$

18. $F(s) = \dfrac{3}{s^2 + 4}; \ L^{-1}\left[\dfrac{3}{2} \cdot \dfrac{2}{s^2 + 2^2}\right] = \dfrac{3}{2}\sin 2t;$

transform (6)

19. $F(s) = \dfrac{1}{2s + 6}$

$L^{-1}\left[\dfrac{1}{2(s + 3)}\right] = \dfrac{1}{2} \cdot \dfrac{1}{s + 3} = \dfrac{1}{2}e^{-3t};$

transform (3)

20. $F(s) = \dfrac{3}{s^4 + 4s^2}$

$L^{-1}\left[\dfrac{3}{8} \cdot \dfrac{8}{s^2(s^2 + 4)}\right] = \dfrac{3}{8}(2t - \sin 2t);$

transform (8)

21. $L^{-1}(F) = L^{-1}\dfrac{1}{(s + 1)^3} = L^{-1}\dfrac{1}{2}\left[\dfrac{2}{(s + 1)^3}\right]$

$= \dfrac{1}{2}t^2 e^{-t};$

transform (12)

22. $F(s) = \dfrac{s^2 - 1}{s^4 + 2s^2 + 1}$

$L^{-1}\left[\dfrac{s^2 - 1}{(s^2 + 1)^2}\right] = t\cos t; \ \text{transform 18}$

23. $F(s) = \dfrac{s + 2}{(s^2 + 9)^2} = \dfrac{s}{(s^2 + 9)^2} + \dfrac{2}{(s^2 + 9)^2}$

$L^{-1}[F(s)] = L^{-1}\left[\dfrac{s}{(s^2 + 9)^2}\right] + L^{-1}\left[\dfrac{2}{(s^2 + 9)^2}\right]$

$f(t) = L^{-1}\left[\dfrac{1}{6} \cdot \dfrac{6s}{(s^2 + 9)^2}\right] + L^{-1}\left[\dfrac{1}{27} \cdot \dfrac{54}{(s^2 + 9)^2}\right]$

$\qquad\qquad\quad \text{transform (16)} \qquad\qquad\quad \text{transform (15)}$

$f(t) = \dfrac{1}{6}t\sin 3t + \dfrac{1}{27}(\sin 3t - 3t\cos 3t)$

$= \dfrac{1}{6}t\sin 3t + \dfrac{1}{27}\sin 3t - \dfrac{3}{27}t\cos 3t$

$= \dfrac{9}{54}t\sin 3t + \dfrac{2}{54}\sin 3t - \dfrac{6}{54}t\cos 3t$

$f(t) = \dfrac{1}{54}(9t\sin 3t + 2\sin 3t - 6t\cos 3t)$

24. $F(s) = \dfrac{s + 3}{s^2 + 4s + 13}$

$= \dfrac{s + 3}{(s + 2)^2 + 9}$

$= \dfrac{s + 2}{(s + 2)^2 + 9} + \dfrac{1}{(s + 2)^2 + 9}$

$L^{-1}[F(s)]$

$= L^{-1}\left[\dfrac{s + 2}{(s + 2)^2 + 9}\right] + L^{-1}\left[\dfrac{1}{(s + 2)^2 + 9}\right]$

$\qquad\quad \text{transform (20)} \qquad\qquad\quad \text{transform (19)}$

$f(t) = L^{-1}\left[\dfrac{s + 2}{(s + 2)^2 + 3^2}\right] + L^{-1}\left[\dfrac{1}{3} \cdot \dfrac{3}{(s + 2)^2 + 3^2}\right]$

$f'(t) = e^{-2t}\cos 3t + \dfrac{1}{3}e^{-2t}\sin 3t$

$f(t) = e^{-2t}\left(\cos 3t + \dfrac{1}{3}\sin 3t\right)$

$f(t) = \dfrac{1}{3}e^{-2t}(3\cos 3t + \sin 3t)$

25. $F(s) = \dfrac{4s^2 - 8}{(s + 1)(s - 2)(s - 3)}$

$= \dfrac{-\frac{1}{3}}{s + 1} + \dfrac{-\frac{8}{3}}{s - 2} + \dfrac{7}{s - 3}$

$L^{-1}(F) = -\dfrac{1}{3}L^{-1}\left(\dfrac{1}{s + 1}\right) - \dfrac{8}{3}L^{-1}\left(\dfrac{1}{s - 2}\right)$

$\qquad\qquad + 7L^{-1}\dfrac{1}{s - 3}$

$f(t) = -\dfrac{1}{3}e^{-t} - \dfrac{8}{3}e^{2t} + 7e^{3t}$

26. $F(s) = \dfrac{3s+1}{(s-1)(s^2+1)}$

$f(t) = L^{-1}(F) = L^{-1}\left[\dfrac{3s+1}{(s-1)(s^2+1)}\right]$

$f(t) = L^{-1}\left(\dfrac{-2s}{s^2+1}\right)+L^{-1}\left(\dfrac{1}{s^2+1}\right)+L^{-1}\left(\dfrac{2}{s-1}\right)$

$f(t) = -2\cos t + \sin t + 2e^t$

27. $F(s) = \dfrac{2s+3}{s^2-2s+5}$

$= \dfrac{2s+3}{s^2-2s+1+4}$

$= \dfrac{2s+3}{(s-1)^2+2^2}$

$F(s) = 2\cdot\dfrac{s+\frac{3}{2}}{(s-1)^2+2^2}$

$= 2\cdot\dfrac{s-1+\frac{5}{2}}{(s-1)^2+2^2}$

$F(s) = 2\cdot\dfrac{s-1}{(s-1)^2+2^2} + \dfrac{5}{2}\cdot\dfrac{2}{(s-1)^2+2^2}$

$f(t) = L^{-1}(F(s)) = 2e^t\cos 2t + \dfrac{5}{2}e^t\sin 2t$

$f(t) = e^t\left(2\cos 2t + \dfrac{5}{2}\sin 2t\right)$

28. $F(s) = \dfrac{3s^4+3s^3+6s^2+s+1}{s^5+s^3}$

$= \dfrac{3s^4}{s^3(s^2+1)} + \dfrac{3s^3}{s^3(s^2+1)} + \dfrac{6s^2+s+1}{s^3(s^2+1)}$

$F(s) = \dfrac{3s}{s^2+1} + \dfrac{3}{s^2+1} - \dfrac{5s}{s^2+1} - \dfrac{1}{s^2+1}$

$\qquad + \dfrac{1}{s^3} + \dfrac{1}{s^2} + \dfrac{5}{s}$

$F(s) = \dfrac{-2s}{s^2+1} + \dfrac{2}{s^2+1} + \dfrac{1}{s^3} + \dfrac{1}{s^2} + \dfrac{5}{s}$

$f(t) = L^{-1}(s) = -2\cos t + 2\sin t + \dfrac{t^2}{2} + t + 5$

16.4   Solving Differential Equations by Laplace Transforms

1. $y' + y = 0;\ y(0) = 1;\ L(y') + L(y) = L(0);$
$L(y') + L(y) = 0;\ sL(y) - y(0) + L(y) = 0$
$sL(y) - 1 + L(y) = 0;\ (s+1)L(y) = 1;$

$L(y) = \dfrac{1}{s+1};\ a = -1,$ transforms (3); $y = e^{-t}$

2. $y' - 2y = 0;\ y(0) = 2;\ L(y') - L(2y) = L(0) = 0$
$L(y') - 2L(y) = 0;\ sL(y) - y(0) - 2L(y) = 0$
$sL(y) - 2L(y) - 2 = 0;\ L(y)(s-2) = 2$

$L(y) = \dfrac{2}{s-2} = 2\left(\dfrac{1}{s+(-2)}\right);\ a = -2,$ transform (3)

$y = L^{-1}\left(2\dfrac{1}{s-2}\right) = 2e^{2t}$

3. $2y' - 3y = 0;\ y(0) = -1;\ L(2y') - L(3y) = 0$
$2L(y') - 3L(y) = 0;\ 2[sL(y) - y(0)] - 3L(y) = 0$
$2[sL(y) + 1] - 3L(y) = 0;\ 2sL(y) + 2 - 3L(y) = 0$
$L(y)(2s - 3) = -2$

$L(y) = \dfrac{-2}{2s-3} = \dfrac{-1}{s-\frac{3}{2}} = -\left(\dfrac{1}{s-\frac{3}{2}}\right);\ a = -\dfrac{3}{2},$

transform (3)

$y = L^{-1}\left[(-1)\dfrac{1}{s-\frac{3}{2}}\right] = (-1)e^{(3/2)t} = -e^{3t/2}$

4. $y' + 2y = 1;\ y(0) = 0;\ L(y') + 2L = L(1)$

$sL(y) - f(0) + 2L(y) = \dfrac{1}{s};\ L(y)(s+2) = \dfrac{1}{s}$

$L(y) = \dfrac{1}{s(s+2)} = F(s)$

$y = L^{-1}\left[\dfrac{1}{s(s+2)}\right] = L^{-1}\left(\dfrac{1}{2}\cdot\dfrac{2}{s(s+2)}\right);\ a = 2,$

transform (4)

$y = \dfrac{1}{2}(1 - e^{-2t})$

5. $y' + 3y = e^{-3t};\ y(0) = 1;$
$L(y') + L(3y) = L(e^{-3t});$

$L(y') + 3L(y) = L(3^{-3t});$

$[sL(y-1] + 3L(y) = \dfrac{1}{s+3}$

$(s+3)L(y) = \dfrac{1}{s+3} + 1;$

$L(y) = \dfrac{1}{(s+3)^2} + \dfrac{1}{s+3}$

The inverse is found from transforms (11) and (3).

$y = te^{-3t} + e^{-3t} = (1+t)e^{-3t}$

6. $y' + 2y = te^{-2t}$; $y(0) = 0$

$L(y') + 2L(y) = L(te^{-2t})$; transform (11), $a = 2$

$$sL(y) - f(0) + 2L(y) = \frac{1}{(s+2)^2}$$

$$L(y)(s+2) = \frac{1}{(s+2)^2}; \quad L(y) = \frac{1}{(s+2)^3} = F(s)$$

$$y = L^{-1}[F(s)] = L^{-1}\left[\frac{1}{(s+2)^3}\right]$$

From transform (12), $a = 2, n = 3, n - 1 = 2$

$$y = L^{-1}\left[\frac{1}{2} \cdot \frac{2}{(s+2)^3}\right] = \frac{1}{2}t^2 e^{-2t}$$

7. $y'' + 4y = 0$; $y(0) = 0$; $y'(0) = 1$

$L(y'') + 4L(y) = 0$;

$s^2 L(y) - sy(0) - y'(0) + 4L(y) = 0$

$$L(y)(s^2 + 4) - 1 = 0; \quad L(y) = \frac{1}{s^2 + 4} = F(s)$$

$$y = L^{-1}\left(\frac{1}{s^2 + 4}\right); \text{ from transform (6), } a = 2$$

$$y = L^{-1}\left(\frac{1}{2} \cdot \frac{2}{s^2 + 4}\right) = \frac{1}{2}\sin 2t$$

8. $9y'' - 4y = 0, y(0) = 1, y'(0) = 0$

$9L(y'') - 4L(y) = 0$

$9[s^2 L(y) - sy(0) - y'(0)] - 4L(y) = 0$

$9s^2 L(y) - 9s - 4L(y) = 0$

$(9s^2 - 4)L(y) = 9s$

$$L(y) = \frac{9s}{9s^2 - 4} = \frac{9s}{(3s+2)(3s-2)}$$

$$= \frac{s}{\left(s + \frac{2}{3}\right)\left(s - \frac{2}{3}\right)}$$

$$L(y) = \frac{3}{4} \cdot \frac{s\left(\frac{2}{3} - \frac{-2}{3}\right)}{\left(s + \frac{2}{3}\right)\left(s - \frac{2}{3}\right)}$$

$$y = \frac{3}{4} \cdot \left(\frac{2}{3}e^{-(2/3)t} + \frac{2}{3}e^{(2/3)t}\right)$$

$$y = \frac{1}{2}\left(e^{-(2/3)t} + e^{(2/3)t}\right)$$

9. $y'' + 2y' = 0$; $y(0) = 0, y'(0) = 2$;

$L(y'') + L(2y') = 0$;

$L(y'') + 2L(y') = 0$

$[s^2 L(y) - 0 - 2] + 2sL(y) - 0 = 0$;

$(s^2 + 2s)L(y) = 2$;

$$L(y) = \frac{2}{s^2 + 2s} = \frac{2}{s(s+2)}$$

By transform (4), $a = 2$; $y = 1 - e^{-2t}$

10. $y'' + 2y' + y = 0$; $y(0) = 0$; $y'(0) = -2$

$L(y'') + 2L(y') = 0$

$s^2 L(y) - sy(0) - y'(0) + 2[sL(y) - y(0)] + L(y) = 0$

$s^2 L(y) - 0 + 2 + 2sL(y) - 0 + L(y) = 0$

$L(y)(s^2 + 2s + 1) = -2$

$$L(y) = \frac{-2}{(s+1)^2} = -2\frac{1}{(s+1)^2}; \text{ transform (11),}$$

$a = 1$

$$y = L^{-1}[F(s)] = L^{-1}\left[-2\frac{1}{(s+1)^2}\right] = -2te^{-t}$$

11. $y'' - 4y' + 5y = 0$; $y(0) = 1$; $y'(0) = 2$

$L(y'') - 4L(y') + 5L(y) = 0$

$s^2 L(y) - sy(0) - y'(0) - 4[sL(y) - y(0)] + 5L(y) = 0$

$L(y)(s^2 - 4s + 5) - s - 2 + 4(1) = 0$

$$L(y) = \frac{2 - s}{s^2 - 4s + 5} = -\frac{(s-2)}{(s-2)^2 + 1}; \text{ from}$$

transform (20), $a = -2, b = 1$

$$y = L^{-1}[F(s)] = L^{-1}\left[-\frac{(s-2)}{(s-2)^2 + 1}\right]$$

$$y = e^{2t}\cos t$$

12. $4y'' + 4y' + y = 0$; $y(0) = 1, y'(0) = 0$

$4L(y'') + 4L(y') + L(y) = 0$

$4[s^2 L(y) - sy(0) - y'(0)] + 4[sL(y) - y(0)] + L(y) = 0$

$4s^2 L(y) - 4sy(0) - 4y'(0) + 4sL(y) - 4y(0) + L(y) = 0$

$4s^2 L(y) - 4s - 0 + 4sL(y) - 4 + L(y) = 0$

$L(y)(4s^2 + 4s + 1) = 4 + 4s$

$$L(y) = \frac{4(1+s)}{4s^2 + 4s + 1} = \frac{s+1}{s^2 + s + \frac{1}{4}}$$

$$= \frac{s+1}{\left(s + \frac{1}{2}\right)^2} = F(s)$$

$$y = L^{-1}\left[\frac{s+1}{\left(s + \frac{1}{2}\right)^2}\right]$$

$$= L^{-1}\left[\frac{s+1}{\left(s + \frac{1}{2}\right)^2}\right] + L^{-1}\left[\frac{1}{\left(s + \frac{1}{2}\right)^2}\right]$$

transform (13), $a = \frac{1}{2}$ transform (11), $a = \frac{1}{2}$

$$y = e^{-(1/2)t}\left(1 - \frac{1}{2}t\right) + te^{-(1/2)t}$$

$$= e^{-(1/2)t} - \frac{1}{2}te^{-(1/2)t} + te^{-(1/2)t}$$

$$= e^{-(1/2)t} + \frac{1}{2}te^{-(1/2)t} = e^{-(1/2)t}\left(1 + \frac{1}{2}t\right)$$

$$= \frac{1}{2}e^{-(1/2)t}(2 + t)$$

13. $y'' + y = 1$; $y(0) = 1$; $y'(0) = 1$;
$L(y'') + L(y) = L(1)$;

$$s^2 L(y) - s - 1 + L(y) = \frac{1}{s}$$

$$(s^2 + 1)L(y) = \frac{1}{s} + s + 1;$$

$$L(y) = \frac{1}{s(s^2 + 1)} + \frac{s}{s^2 + 1} + \frac{1}{s^2 + 1}$$

By transforms (7), (5), and (6),
$y = 1 - \cos t + \cos t + \sin t$; $y = 1 + \sin t$

14. $y'' + 4y = 2t$; $y(0) = 0$; $y'(0) = 0$
$L(y'') + 4L(y) = 2L(t)$

$$s^2 L(y) - sy(0) - y'(0) + 4L(y) = \frac{2}{s^2}$$

$$L(y)(s^2 + 4) = \frac{2}{s^2}; \quad L(y) = \frac{2}{s^2(s^2 + 4)}; \text{ from}$$

transform (8), $a = 2$

$$L(y) = \frac{1}{4} \cdot \frac{8}{s^2(s^2 + 4)} = F(s)$$

$$y = L^{-1}\left[\frac{1}{4} \cdot \frac{8}{s^2(s^2 + 4)}\right] = \frac{1}{4}(2t - \sin 2t)$$

15. $y'' + 2y' + y = e^{-t}$; $y(0) = 1$; $y'(0) = 2$
$L(y'') + 2L(y') + L(y) = L(e^{-1})$; transform (3),
$a = 1$

$$s^2 L(y) - sy(0) - y'(0) + 2[sL(y) - y(0)] + L(y)$$
$$= \frac{1}{s + 1}$$

$$L(y)(s^2 + 2s + 1) - s - 2 - 2 = \frac{1}{s + 1}$$

$$L(y)[(s + 1)^2] = \frac{1}{s + 1} + s + 4$$

$$L(y) = \frac{1}{(s + 1)^3} + \frac{s + 4}{(s + 1)^2} = F(s)$$

From transform (12), $n = 3$; $n - 1 = 2$; $a = 1$;
from
transform (14), $a = 1$; $b = 4$

$$y = \frac{1}{2}(t^2 e^{-t}) + (3t + 1)e^{-t} = e^{-t}\left(\frac{1}{2}t^2 + 3t + 1\right)$$

16. $2y'' + 8y = 3\sin 2t$, $y(0) = y'(0) = 0$
$2L(y'') + 8L(y) = 3L(\sin 2t)$

$$2s^2 L(y) - 2sy(0) - 2y'(0) + 8L(y) = \frac{6}{s^2 + 4}$$

$$L(y) = \frac{3}{(s^2 + 4)^2} = \frac{3}{16}\frac{16}{(s^2 + 4)^2}$$

$$y = \frac{3}{16}(\sin 2t - 2t\cos 2t)$$

17. $y'' - 4y = 10e^{3t}$, $y(0) = 5$, $y'(0) = 0$;
$L(y'') - 4L(y) = 10 \cdot L(e^{3t})$

$$s^2 L(y) - sy(0) - y'(0) - 4L(y) = \frac{10}{s - 3};$$

$$s^2 L(y) - 5s - 0 - 4L(y) = \frac{10}{s - 3}$$

$$(s^2 - 4)L(y) = 5s + \frac{10}{s - 3}$$

$$L(y) = \frac{5s}{(s + 2)(s - 2)} + \frac{10}{(s + 2)(s - 2)(s - 3)}$$

$$L(y) = \frac{\frac{5}{2}}{s + 2} + \frac{\frac{5}{2}}{s - 2} + \frac{\frac{1}{2}}{s + 2} + \frac{-\frac{5}{2}}{s - 2} + \frac{2}{s - 3}$$

$$L(y) = \frac{3}{s + 2} + \frac{2}{s - 3}$$

$$y = 3e^{-2t} + 2e^{3t}$$

18. $y'' - 2y' + y = e^{2t}$, $y(0) = 1$, $y'(0) = 3$
$L(y'') - 2L(y') + L(y) = L(e^{2t})$
$s^2 L(y) - sy(0) - y'(0) - 2[sL(y) - y(0)] + L(y)$

$$= \frac{1}{s - 2}$$

$$s^2 L(y) - s - 3 - 2sL(y) + 2 + L(y) = \frac{1}{s - 2}$$

$$L(y)[s^2 - 2s + 1] = \frac{1}{s - 2} + s + 1 = \frac{s^2 - s - 1}{s - 2}$$

$$L(y) = \frac{s^2 - s - 1}{(s - 2)(s^2 - 2s + 1)} = \frac{s^2 - s - 1}{(s - 1)(s - 1)^2}$$

$$= \frac{1}{s - 2} + \frac{1}{(s - 1)^2}$$

$$y = e^{2t} + te^t$$

19. $y'' - y = 5\sin 2t$, $y(0) = 0$, $y'(0) = 1$
$L(y'') - L(y) = 5L(\sin 2t)$

$$s^2 L(y) - s \cdot y(0) - y'(0) - L(y) = 5 \cdot \frac{2}{s^2 + 4}$$

$$s^2 L(y) - 1 - L(y) = \frac{10}{s^2 + 4}$$

$$L(y)[s^2 - 1] = \frac{10}{s^2 + 4} + 1 = \frac{s^2 + 14}{s^2 + 4}$$

$$L(y) = \frac{s^2 + 14}{(s^2 + 4)(s - 1)(s + 1)}$$

$$= \frac{-2}{s^2 + 4} + \frac{\frac{3}{2}}{s - 1} - \frac{\frac{3}{2}}{s + 1}$$

$$y = -\sin 2t + \frac{3}{2}e^t - \frac{3}{2}e^{-t}$$

20. $y'' + y' - 2y = \sin 3t, y(0) = 0, y'(0) = 0$

$L(y'') + L(y') - 2L(y) = L(\sin 3t)$

$s^2 L(y) - s \cdot y(0) - y'(0) + sL(y) - y(0) - 2L(y) = \dfrac{3}{s^2 + 9}$

$s^2 L(y) + sL(y) - 2L(y) = \dfrac{3}{s^2 + 9}$

$L(y) = \dfrac{3}{(s^2 + 9)(s + 2)(s - 1)}$

$\qquad = \dfrac{-\frac{3}{130}s}{s^2 + 9} + \dfrac{-\frac{33}{130}}{s^2 + 9} + \dfrac{-\frac{1}{13}}{s + 2} + \dfrac{\frac{1}{10}}{s - 1}$

$y = -\dfrac{3}{130}\cos 3t - \dfrac{11}{130}\sin 3t - \dfrac{1}{13}e^{-2t} + \dfrac{1}{10}e^t$

21. $2v' = 6 - v$; since the object starts from rest,
$f(0) = 0, f'(0) = 0$

$2L(v') + L(v) = 6L(1); \quad 2sL(v) - 0 + L(v) = \dfrac{6}{s};$

$(2s + 1)L(v) = \dfrac{6}{s}$

$L(v) = \dfrac{6}{s(2s + 1)} = 6\left[\dfrac{\frac{1}{2}}{s\left(s + \frac{1}{2}\right)}\right]$

By transform (4), $v = 6(1 - e^{-t/2})$

22. $D^2\theta + 20\theta; \theta = 0; D\theta = 0.4$ rad/s; $t = 0$

$\theta'' + 20(\theta) = 0;$ let $\theta = y; y'' + 20y = 0$

$L(y'') + 20L(y) = 0$

$s^2 L(y) - sy(0) - y'(0) + 20L(y) = 0$

$s^2 L(y) - 0 - 0.4 + 20L(y) = 0$

$L(y)(s^2 + 20) = 0.4$

$L(y) = L(\theta) = \dfrac{0.4}{s^2 + 20};$ from transform (6),

$a = \sqrt{20}$

$F(s) = \dfrac{0.4}{\sqrt{20}} \cdot \dfrac{\sqrt{20}}{s^2 + 20}$

$f(t) = \dfrac{0.4}{\sqrt{20}}\sin\sqrt{20}t$

$\theta = 0.089\sin 4.5t$

23. $L\dfrac{d^2q}{dt^2} + R\dfrac{dq}{dt} + \dfrac{q}{i} = E$

$0(q'') + 50q' + \dfrac{q}{4 \times 10^{-6}} = 40$

$50q' + \dfrac{10^6}{4}q = 40$

$50L(q') + \dfrac{10^6}{4}L(q) = L(40)$

$50[sL(q) - q(0)] + \dfrac{10^6}{4}L(q) = \dfrac{40}{5}$

$L(q)\left(50s + \dfrac{10^6}{4}\right) = \dfrac{40}{s}$

$L(q) = \dfrac{40}{s\left(50s + \dfrac{10^6}{4}\right)}$

$L(q) = F(s) = \dfrac{40}{50s(s + 5000)}$

$\qquad = \dfrac{40}{50(5000)}\dfrac{5000}{s(s + 5000)}$

$\qquad = 0.00016 \cdot \dfrac{5000}{s(s + 5000)}$

$L^{-1}[F(s)] = f(t) = \dfrac{1.60}{10^4}(1 - e^{-5000t}) = 9;$ from

transform (4), $a = 5000$

$q = 1.60 \times 10^{-4}(1 - e^{-5000t})$

24. $L\dfrac{d^2q}{dt^2} + R\dfrac{dq}{dt} + \dfrac{q}{c} = E$

$2i' + 80i = 8$

$2L(i') + 80L(i) = L(8)$

$2[sL(i) - i(0)] + 80L(i) = \dfrac{8}{5}$

$2sL(i) - 2i(0) + 80L(i) = \dfrac{8}{5}$

$L(i)(2s + 80) = \dfrac{8}{5}$

$L(i) = \dfrac{8}{s(2s + 80)} = \dfrac{8}{2s(s + 40)} = \dfrac{4}{40} \cdot \dfrac{40}{s(s + 40)}$

From transform (4), $a = 40$

$i = L^{-1}\left[\dfrac{8}{s(2s + 80)}\right] = 0.1(1 - e^{-40t})$

$\qquad = 0.1(1 - e^{-40t})$

25. $10\dfrac{d^2q}{dt^2} + \dfrac{q}{4 \times 10^{-5}} = 100\sin 50t; q(0) = 0$ and

$q'(0) = 0$

$10L(q'') + 2.5 \times 10^4 L(q) = L(100\sin 50t)$

$10s^2 L(q) + 2.5 \times 10^4 L(q) = L(100\sin 50t)$

$(s^2 + 2.5 \times 10^3)L(q) = 10\left(\dfrac{50}{s^2 + 50}\right);$

$L(q) = \dfrac{500}{(s^2 + 50^2)^2}; L(q) = \dfrac{1}{500}\dfrac{2(50)^3}{(s^2 + 50^2)^2}$

By transforms (15),

$$q = \frac{1}{500}(\sin 50t - 50t \cos 50t)$$

$$i = \frac{dq}{dt}$$

$$= \frac{1}{500}\{50\cos 50t$$

$$- [50t(-50\sin 50t) + (\cos 50t)50]\}$$

$$= \frac{1}{500}(2500t \sin 50t) = 5t \sin 50t$$

26. $L\dfrac{d^2q}{dt^2} + R\dfrac{dq}{dt} + \dfrac{q}{c} = E$

$$20 \times 10^{-3}q'' + 40q' + \frac{10^6}{50}q = 100e^{-1000t}$$

$$20 \times 10^{-3}[s^2L(q) - sq(0) - q'(0)] + 40[sL(q) - q(0)]$$
$$+ \frac{10^6}{50}L(q) = \frac{100}{s+1000}$$

$$20 \times 10^{-3}s^2L(q) + 40sL(q) + \frac{10^6}{50}L(q) = \frac{100}{s+1000}$$

$$L(q)\left(20 \times 10^{-3}s^2 + 40s + \frac{10^6}{50}\right) = \frac{100}{s+1000}$$

$$L(q)\frac{20}{10^3}\left(s^2 + 2 \times 10^3 s + \frac{10^9}{1000}\right) = \frac{100}{s+1000}$$

$$L(q)\frac{20}{10^3}(s^2 + 200s + 10^6) = \frac{100}{s+1000}$$

$$L(q) = \frac{100\frac{10^3}{20}}{(s+1000)^3} = \frac{5000}{(s+1000)^3} = F(s)$$

$$q = L^{-1}\left[\frac{5000}{(s+1000)^3}\right]; \text{ transform (12), } n = 3,$$

$$a = 1000$$

$$q = \frac{5000}{2}L^{-1}\left[\frac{2}{(s+1000)^3}\right] = 2500t^2e^{-1000t}$$

27. $D^2y + 9y = 18\sin 3t; \ y = 0, Dy = 0, t = 0$

$$y'' + 9y = 18\sin 3t; \ L(y'') + 9L(y) = 18L(\sin 3t)$$

$$s^2L(y) - sy(0) - y'(0) + 9L(y) = 18\frac{3}{s^2+9} = \frac{54}{s^2+9}$$

$$L(y)(s^2 + 9) = \frac{54}{s^2+9}$$

$$L(y) = \frac{54}{(s^2+9)^2} = F(s); \text{ from transform (15),}$$

$$a = 3;$$

$$2a^3 = 54$$

$$y = L^{-1}\left[\frac{54}{(s^2+9)^2}\right] = L^{-1}\left[\frac{2(27)}{(s^2+9)^2}\right]$$

$$y = \sin 3t - 3t\cos 3t$$

28. Hooke's Law: $F = kx; \ 20.0 = k1;$ therefore,

$k = 20$

$$2D^2x + 12v + 19.6x = 0;$$
$$2D^2x + 12D'x + 19.6x = 0$$
$$2y'' + 12y' + 19.6y = 0;$$
$$2L(y'') + 12L(y') + 19.6L(y) = 0$$
$$2[s^2L(y) - sy(0) - y'(0)]$$
$$+ 12[sL(y) - y(0)] + 19.6L(y) = 0$$
$$2s^2L(y) + 12sL(y) + 19.6L(y) = s + 6$$

$$L(y) = \frac{s+6}{2s^2 + 12s + 20} = \frac{1}{2}\left(\frac{s+6}{s^2 + 6s + 10}\right)$$

$$= \frac{1}{2}\left[\frac{s+6}{(s+3)^2 + 1}\right]$$

$$F(s) = \frac{1}{2}\left[\frac{s+3}{(s+3)^2 + 1} + 3\frac{1}{(s+3)^2 + 1}\right];$$

transforms (20) and (19)

$$y = \frac{1}{2}(e^{-3t}\cos t + 3e^{-3t}\sin t)$$

$$= \frac{1}{2}e^{-3t}(\cos t + 3\sin t)$$

29. $0.2Di + 10i = 50e^{-100t}, \ i(0) = 0$

$$0.2L(Di) + 10L(i) = 50L(e^{-100t});$$

$$0.2(sL(i) - i(0)) + 10L(i) = 50\frac{1}{s+100};$$

$$\frac{s}{5}L(i) + 10L(i) = \frac{50}{s+100};$$

$$sL(i) + 50L(i) = \frac{250}{s+100}; \ (s+50)L(i) = \frac{250}{s+100}$$

$$L(i) = \frac{250}{(s+50)(s+100)} = \frac{5}{s+50} + \frac{-5}{s+100};$$

$$i = 5e^{-50t} - 5e^{-100t}$$

30. $$L\frac{d^2q}{dt^2} + \frac{q}{C} = E$$

$$10q'' + \frac{q}{10 \times 10^{-6}} = 60$$

$$q'' + 10{,}000q = 6$$

$$L(q'') + 10{,}000L(q) = L(6)$$

$$[s^2L(q) - sq(0) - q'(0)] + 10{,}000L(q) = \frac{6}{s}$$

$$(s^2 + 10{,}000)L(q) = \frac{6}{s}$$

$$L(q) = \frac{6}{s(s^2 + 10{,}000)}$$

$$= \frac{6}{100^2}\frac{100^2}{s(s^2 + 100^2)}$$

$$q = \frac{6}{100^2}(1 - \cos 100t)$$

$$i = \frac{dq}{dt} = \frac{6}{100^2}(100 \sin 100t)$$

$$i = \frac{3}{50} \sin 100t, \text{ current as function of time}$$

31. $LD^2q + RDq + \dfrac{q}{C} = 0 \quad \left(D = \dfrac{d}{dt}\right),$

$q(0) = 400 \ \mu C, q'(0) = 0$
$L = 0.25 \text{ H}, R = 4.0 \ \Omega, C = 100 \ \mu\text{F}$
$0.25D^2q + 4.0Dq + 10^4q = 0$
$D^2q + 16Dq + 4.0 \times 10^4q = 0$
$L(D^2q) + 16L(Dq) + 4.0 \times 10^4L(q) = 0$
$[s^2L(q) - s(4.0 \times 10^{-4}) - 0] + 16[sL(q) - 4.0 \times 10^{-4}]$
$\quad + 4.0 \times 10^4 L(q) = 0$
$(s^2 + 16s + 4.0 \times 10^4)L(q) = 4.0 \times 10^{-4}(s + 16)$

$$L(q) = 4.0 \times 10^{-4}\left[\frac{s + 16}{(s^2 + 16s + 64) + 39{,}936}\right]$$

$$= 4.0 \times 10^{-4}\left[\frac{s + 8}{(s+8)^2 + 200^2} + \frac{8}{(s+8)^2 + 200^2}\right]$$

$$q = 4.0 \times 10^{-4}\left(e^{-8t}\cos 200t + \frac{8}{200}e^{-8t}\sin 200t\right)$$

$$= 10^{-4}e^{-8t}(4.0\cos 200t + 0.16\sin 200t)$$

32. $y^{iv} = k$
$L(y^{iv}) = L(k)$
$s^4L(y) - s^3 \cdot y(0) - s^2 \cdot y'(0) - s \cdot y''(0) - y'''(0) = \dfrac{k}{s}$

$$s^4L(y) - as^2 - b = \frac{k}{s}$$

$$s^4L(y) = as^2 + \frac{k}{s} + b = \frac{as^3 + k + bs}{s}$$

$$L(y) = \frac{as^3 + bs + k}{s^5} = \frac{a}{s^2} + \frac{b}{s^4} + \frac{k}{s^5}$$

$$y = ax + \frac{b}{6}x^3 + \frac{k}{24}x^4$$

$$y' = a + \frac{b}{2}x^2 + \frac{k}{6}x^3$$

$$y'' = bx + \frac{k}{2}x^2$$

$$y''' = b + kx$$

@ $x = L, y = 0 \Rightarrow aL + \dfrac{bL^3}{6} + \dfrac{kL^4}{24} = 0$

@ $x = L, y'' = 0 \Rightarrow bL + \dfrac{k}{2}L^2 = 0 \Rightarrow b = \dfrac{-kL}{2}$

$$aL - \frac{kL}{2} \cdot \frac{L^3}{6} + \frac{kL^4}{24} = 0 \Rightarrow a = \frac{kL^3}{24}$$

$$y = \frac{kL^3}{24}x - \frac{kL}{12}x^3 + \frac{k}{24}x^4 = \frac{k}{24}(L^3x - 2Lx^3 + x^4)$$

$$y = \frac{w}{24EI}(L^3x - 2Lx^3 + x^4)$$

Chapter 16 Review Exercises

1. $y' = x^2 + y^2, 0 \le x \le 0.5, \Delta x = 0.1, (0, 1)$

x	$y_{n+1} = y_n + (x_n^2 - y_n^2)\Delta x$
0	1
0.1	0.9000
0.2	0.8200
0.3	0.7568
0.4	0.7085
0.5	0.6743

2. $y' = \dfrac{1}{1 + x + y}, 0 \le x \le 1, \Delta x = 0.2, (0, 0)$

x	$y_{n+1} = y_n + \dfrac{1}{1 + x_n + y_n}\Delta x$
0	0
0.2	$0 + \dfrac{1}{1 + 0 + 0}(0.2) = 0.2000$
0.4	$0.2000 + \dfrac{1}{1 + 0.2 + 0.2000}(0.2) = 0.3429$
0.6	$0.3429 + \dfrac{1}{1 + 0.4 + 0.3429}(0.2) = 0.4576$
0.8	$0.4576 + \dfrac{1}{1 + 0.6 + 0.4576}(0.2) = 0.5548$
1.0	$0.5548 + \dfrac{1}{1 + 0.8 + 0.5548}(0.2) = 0.6397$

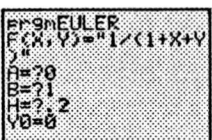

3. $y' = \sqrt{x+y}, 0 \le x \le 2, \Delta x = 0.2, (0,1)$

x	$y_{n+1} = y_n + \sqrt{x_n + y_n}\,\Delta x$
0	1
0.2	1.2000
0.4	1.4366
0.6	1.7077
0.8	2.0115
1.0	2.3469
1.2	2.7128
1.4	3.1084
1.6	3.5330
1.8	3.9861
2.0	4.4672

4. $y' = x^2 + \sin y, 0 \le x \le 1, \Delta x = 0.1, (0,1)$

x	$y_{n+1} = y_n + (x_n^2 + \sin y_n)\Delta x$
0	1
0.1	1.0841
0.2	1.1735
0.3	1.2698
0.4	1.3743
0.5	1.4883
0.6	1.6130
0.7	1.7489
0.8	1.8963
0.9	2.0551
1.0	2.2246

5. $y' = \sin x + \tan y, 0 \le x \le 0.7, \Delta x = 0.1, (0,0.5)$

$y' = x^2 + \sin y, 0 \le x \le 1, \Delta x = 0.1, (0,1)$

x	$y_{n+1} = y_n + (\sin x_n + \tan y_n)\Delta x$
0	0.5
0.1	0.5546
0.2	0.6266
0.3	0.7188
0.4	0.8359
0.5	0.9854
0.6	1.1843
0.7	1.4864

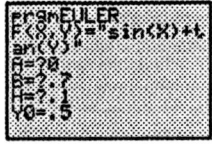

6. $y' = e^{2x} + \sin xy, 0 \le x \le 1, \Delta x = 0.1, (0,0)$

$y' = x^2 + \sin y, 0 \le x \le 1, \Delta x = 0.1, (0,1)$

x	$y_{n+1} = y_n + (e^{2x_n} + \sin(x_n y_n))\Delta x$
0	0
0.1	0.1000
0.2	0.2231
0.3	0.3768
0.4	0.5703
0.5	0.8154
0.6	1.1269
0.7	1.5215
0.8	2.0145
0.9	2.6097
1.0	3.2859

7. $y' = x^2 - y^2, 0 \le x \le 0.5, \Delta x = 0.02, (0,1)$

x	$y_{n+1} = y_n + (x_n^2 - y_n^2)\Delta x$
0	1
0.02	0.9800
0.04	0.9608
0.06	0.9424
0.08	0.9247
0.10	0.9077
0.12	0.8914
0.14	0.8758
0.16	0.8609
0.18	0.8466
0.20	0.8329
0.22	0.8198
0.24	0.8073
0.26	0.7954
0.28	0.7841
0.30	0.7734
0.32	0.7633
0.34	0.7536
0.36	0.7446
0.38	0.7361
0.40	0.7282
0.42	0.7208
0.44	0.7139
0.46	0.7076
0.48	0.7018
0.50	0.6965

8. $y' = \dfrac{1}{1+x+y}, 0 \le x \le 1, \Delta x = 0.05, (0,0)$

	$y_{n+1} = y_n + \frac{1}{1+x_n+y_n}(\Delta x)$
0	0
0.05	0.0500
0.10	0.0955
0.15	0.1373
0.20	0.1761
0.25	0.2125
0.30	0.2466
0.35	0.2790
0.40	0.3097
0.45	0.3389
0.50	0.3669
0.55	0.3936
0.60	0.4194
0.65	0.4441
0.70	0.4680
0.75	0.4911
0.80	0.5134
0.85	0.5350
0.90	0.5560
0.95	0.5763
1.00	0.5961

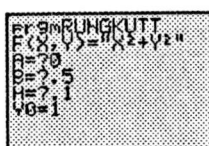

9. $y' = x^2 - y^2, 0 \le x \le 0.5, \Delta x = 0.1, (0,1)$

x	y
0	1
0.1	0.9094
0.2	0.8358
0.3	0.7772
0.4	0.7327
0.5	0.7018

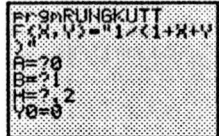

10. $y' = \dfrac{1}{1+x+y}, 0 \le x \le 1, \Delta x = 0.2, (0,0)$

x	y
0	0
0.2	0.1696
0.4	0.3002
0.6	0.4082
0.8	0.5011
1.0	0.5831

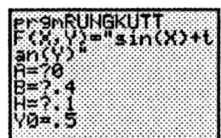

11. $y' = \sin x + \tan y, 0 \le x \le 0.4, \Delta x = 0.1, (0,0.5)$

x	y
0	0.5
0.1	0.5637
0.2	0.6476
0.3	0.7567
0.4	0.8996

12. $y' = e^{2x} + \sin xy, 0 \le x \le 0.8, \Delta x = 0.2, (0,0)$

x	y
0	0
0.2	0.2490
0.4	0.6429
0.6	1.2812
0.8	2.2779

13. $y' = y + x^2, (0,1)$

$dy = (1 + x^2)dx$

$y_1 = x + \dfrac{1}{3}x^3 + c_1,$

$c_1 = 1$

$y_1 = 1 + x + \dfrac{1}{3}x^3$

$dy = \left(1 + x + \dfrac{1}{3}x^3 + x^2\right)dx$

$y_2 = x + \dfrac{1}{2}x^2 + \dfrac{1}{12}x^4 + \dfrac{1}{3}x^3 + c_2,$

$c_2 = 1$

$y_2 = 1 + x + \dfrac{1}{2}x^2 + \dfrac{1}{3}x^3 + \dfrac{1}{12}x^4$

$dy = \left(1 + x + \dfrac{1}{2}x^2 + \dfrac{1}{3}x^3 + \dfrac{1}{12}x^4 + x^2\right)dx$

$y_3 = x + \dfrac{1}{2}x^2 + \dfrac{1}{2}x^3 + \dfrac{1}{12}x^4 + \dfrac{1}{60}x^5 + c_3$

$c_3 = 1$

$y_3 = 1 + x + \dfrac{1}{2}x^2 + \dfrac{1}{2}x^3 + \dfrac{1}{12}x^4 + \dfrac{1}{60}x^5$

14. $y' = \cos x - y, y_3, (0,1)$

$y' = \cos x - 1$

$y_1 = \sin x - x + c$

$1 = \sin 0 - 0 + c,$

$c = 1$

$y_1 = \sin x - x + 1$

$y' = \cos x - \sin x + x - 1$

$y_2 = \sin x + \cos x + \dfrac{x^2}{2} - x + c$

$1 = \sin 0 + \cos 0 + \dfrac{0^2}{2} - 0 + c,$

$c = 0$

$y_2 = \sin x + \cos x + \dfrac{x^2}{2} - x$

$y' = \cos x - \sin x - \cos x - \dfrac{x^2}{2} + x$

$y' = -\sin x - \dfrac{x^2}{2} + x$

$y_3 = \cos x - \dfrac{x^3}{6} + \dfrac{x^2}{2} + c$

$1 = \cos 0 - \dfrac{0^3}{6} + \dfrac{0^2}{2} + c,$

$c = 0$

$y_3 = \cos x + \dfrac{x^2}{2} - \dfrac{x^3}{6}$

15. $y' = 1 + x^2 y, (1,1)$

$dy = (1 + x^2)dx$

$y_1 = x + \dfrac{1}{3}x^3 + c_1,$

$1 = 1 + \dfrac{1}{3} + c_1,$

$c_1 = -\dfrac{1}{3}$

$y_1 = x + \dfrac{1}{3}x^3 - \dfrac{1}{3}$

$dy = \left[1 + x^2\left(x + \dfrac{1}{3}x^3 - \dfrac{1}{3}\right)\right]dx$

$= \left(1 - \dfrac{1}{3}x^2 + x^3 + \dfrac{1}{3}x^5\right)dx$

$y_2 = x - \dfrac{1}{9}x^3 + \dfrac{1}{4}x^4 + \dfrac{1}{18}x^6 + c_2$

$1 = 1 - \dfrac{1}{9} + \dfrac{1}{4} + \dfrac{1}{18} + c_2,$

$c_2 = -\dfrac{7}{36}$

$y_2 = -\dfrac{7}{36} + x - \dfrac{1}{9}x^3 + \dfrac{1}{4}x^4 + \dfrac{1}{18}x^6$

16. $y' = y^2 - \sin x, y_2, (0,2)$

$y' = 2^2 - \sin x = 4 - \sin x$

$y_1 = 4x + \cos x + c$

$2 = 4(0) + \cos 0 + c, c = 1$

$y_1 = 4x + \cos x + 1$

$y' = (4x + \cos x + 1)^2 - \sin x$

$y' = 16x^2 + 8x\cos x + 8x + \cos^2 x$
$\qquad + 2\cos x + 1 - \sin x$

$y_2 = \dfrac{16x^3}{3} + 8\cos x + 8x\sin x + 4x^2$

$\qquad + \dfrac{\sin x \cos x}{2} + \dfrac{x}{2} + 2\sin x + x + \cos x + c$

$2 = \dfrac{16(0)^2}{3} + 8\cos 0 + 8(0)\sin 0 + 4(0)^2$

$\qquad + \dfrac{\sin 0 \cos 0}{2} + \dfrac{0}{2} + 2\sin 0 + 0 + \cos 0 + c$

$2 = 8 + 1 + c$

$c = -7$

$y_2 = \dfrac{16x^3}{3} + 8\cos x + 8x\sin x + 4x^2$

$\qquad + \dfrac{\sin x \cos x}{2} + \dfrac{x}{2} + 2\sin x + x + \cos x - 7$

$y_2 = \dfrac{16x^3}{3} + 4x^2 + \dfrac{3}{2}x - 7 + 9\cos x + 2\sin x$

$\qquad + 8x\cos x + \dfrac{\sin x \cos x}{2}$

17. $4y' - y = 0, y(0) = 1$

$4L(y') - L(y) = L(0)$

$4[sL(y) - 1] - L(y) = 0$

$$L(y) = \frac{4}{4s - 1} = \frac{1}{s - 1/4}$$

$$y = e^{t/4}$$

18. $2y' - y = 4, y(0) = 1$

$$2L(y') - L(y) = L(4)$$

$$2[sL(y) - y(0)] - L(y) = \frac{4}{s}$$

$$2sL(y) - 2 - L(y) = \frac{4}{s}$$

$$(2s - 1)L(y) = 2 + \frac{4}{s} = \frac{2s + 4}{s}$$

$$L(y) = \frac{2(s + 2)}{s(2s - 1)} = \frac{2(s + 2)}{2s(s - \frac{1}{2})} = \frac{s + 2}{s(s - \frac{1}{2})}$$

$$L(y) = \frac{s}{s(s - \frac{1}{2})} + \frac{2}{s(s - \frac{1}{2})}$$

$$L(y) = \frac{1}{s - \frac{1}{2}} - 4\frac{-\frac{1}{2}}{s(s - \frac{1}{2})}$$

$$y = e^{t/2} - 4(1 - e^{t/2})$$

$$y = e^{t/2} - 4 + 4e^{t/2}$$

$$y = 5e^{t/2} - 4$$

19. $y' - 3y = e^t, y(0) = 0$

$L(y') - 3L(y) = L(e^t)$

$$sL(y) - 0 - 3L(y) = \frac{1}{s - 1}$$

$$(s - 3)L(y) = \frac{1}{s - 1}$$

$$L(y) = \frac{1}{(s - 3)(s - 1)} = \frac{1}{2(s - 3)} - \frac{1}{2(s - 1)}$$

$$y = \frac{1}{2}(e^{3t} - e^t)$$

20. $y' + 2y = e^{-2t}, y(0) = 2$

$$L(y') + 2L(y) = L(e^{-2t})$$

$$sL(y) - y(0) + 2L(y) = \frac{1}{s + 2}$$

$$(s + 2)L(y) - 2 = \frac{1}{s + 2}$$

$$(s + 2)L(y) = 2 + \frac{1}{s + 2}$$

$$L(y) = \frac{2}{s + 2} + \frac{1}{(s + 2)^2}$$

$$y = 2e^{-2t} + te^{-2t}$$

$$y = e^{-2t}(t + 2)$$

21. $y'' + y = 0, y(0) = 0, y'(0) = -4$

$L(y'') + L(y) = L(0)$

$s^2 L(y) - s(0) + 4 + L(y) = 0$

$$L(y) = \frac{-4}{s^2 + 1}, y = -4\sin t$$

22. $y'' + 4y' + 5y = 0, y(0) = 1, y'(0) = 1$

$L(y'') + 4L(y') + 5L(y) = L(0)$

$s^2 L(y) - sy(0) - y'(0)$

$+ 4(sL(y) - y(0)) + 5L(y) = 0$

$s^2 L(y) - s - 1 + 4sL(y) - 4 + 5L(y) = 0$

$(s^2 + 4s + 5)L(y) = s + 5$

$$L(y) = \frac{s + 5}{s^2 + 4s + 5} = \frac{s + 5}{s^2 + 4s + 4 + 1}$$

$$= \frac{s + 2 + 3}{(s + 2)^2 + 1}$$

$$L(y) = \frac{s + 2}{(s + 2)^2 + 1} + \frac{3}{(s + 2)^2 + 1}$$

$$y = e^{-2t}\cos t + 3e^{-2t}\sin t$$

$$y = e^{-2t}(\cos t + 3\sin t)$$

23. $y'' + 9y = 3e^t, y(0) = 0, y'(0) = 0$

$L(y'') + 9L(y) = 3L(e^t)$

$$s^2 L(y) - sy(0) - y'(0) + 9L(y) = \frac{3}{s - 1}$$

$$(s^2 + 9)L(y) = \frac{3}{s - 1}$$

$$L(y) = \frac{3}{(s - 1)(s^2 + 9)}$$

$$L(y) = -\frac{3}{10}\frac{s}{s^2 + 9} - \frac{1}{10}\frac{3}{s^2 + 9}$$

$$+ \frac{3}{10}\frac{1}{s - 1}$$

$$y = -\frac{3}{10}\cos 3t - \frac{1}{10}\sin 3t + \frac{3}{10}e^t$$

$$y = \frac{1}{10}(3e^t - \sin 3t - 3\cos 3t)$$

24. $y'' - 2y' + y = e^x + x, y(0) = 0, y'(0) = 1$

$L(y'') - 2L(y') + L(y) = L(e^x) + L(x)$

$s^2 L(y) - sy(0) - y'(0) - 2(sL(y) - y(0))$

$$+ L(y) = \frac{1}{s - 1} + \frac{1}{s^2}$$

$$s^2 L(y) - 1 - 2sL(y) + L(y) = \frac{1}{s - 1} + \frac{1}{s^2}$$

$$(s^2 - 2s + 1)L(y) = 1 + \frac{1}{s - 1} + \frac{1}{s^2}$$

$$(s - 1)^2 L(y) = 1 + \frac{1}{s - 1} + \frac{1}{s^2}$$

$$L(y) = \frac{1}{(s-1)^2} + \frac{1}{(s-1)^3} + \frac{1}{s^2(s-1)^2}$$

$$L(y) = \frac{1}{(s-1)^2} + \frac{1}{(s-1)^3} + \frac{1}{(s-1)^2}$$

$$- \frac{2}{s-1} + \frac{1}{s^2} + \frac{2}{s}$$

$$L(y) = \frac{1}{2}\frac{1}{(s-1)^3} + \frac{1}{(s-1)^2} - \frac{2}{s-1} + \frac{1}{s^2} + \frac{2}{s}$$

$$y = \frac{1}{2}x^2e^x + 2xe^x - 2e^x + x + 2$$

$$y = e^x\left(\frac{x^2}{2} + 2x - 2\right) + x + 2$$

25. $y' = 1 + ye^{-x}, (0,1)$

$dy = (1 + e^{-x})dx$

$y_1 = x - e^{-x} + c_1$

$1 = 0 - 1 + c_1, c_1 = 2$

$y_1 = 2 + x - e^{-x}$

$dy = [1 + (2 + x - e^{-x})e^{-x}]dx$

$\quad = (1 + 2e^{-x} + xe^{-x} - e^{-2x})dx$

$y_2 = x - 2e^{-x} + e^{-x}(-x - 1) + \frac{1}{2}e^{-2x} + c_2$

$1 = 0 - 2 + (-1) + \frac{1}{2} + c_2, c_2 = \frac{7}{2}$

$y = \frac{7}{2} + x - (3 + x)e^{-x} + \frac{1}{2}e^{-2x}$

$y(0.5) = 3.5 + 0.5 - 3.5e^{-0.5} + 0.5e^{-1}$

$\quad\quad = 2.0611$

26. $y' = \dfrac{2x}{x^2 + e^y}, (0,1)$

$y' = \dfrac{2x}{x^2 + e^1}$

$y' = \ln(x^2 + e) + c$

$1 = \ln(0^2 + e) + c, c = 0$

$y_1 = \ln(x^2 + e)$

$y' = \dfrac{2x}{x^2 + e^{\ln(x^2+e)}} = \dfrac{2x}{x^2 + x^2 + e}$

$\quad = \dfrac{2x}{2x^2 + e}$

$y_2 = \dfrac{\ln(2x^2 + e)}{2} + c$

$1 = \dfrac{\ln(2(0)^2 + e)}{2} + c, c = \dfrac{1}{2}$

$y_2 = \dfrac{1}{2}(\ln(2x^2 + e) + 1)$

$y_2(0.5) = \dfrac{1}{2}(\ln(2(0)^2 + e) + 1) = 1.084$

27. $y' = 1 + ye^{-x}, 0 \le x \le 0.5, \Delta x = 0.1, (0,1)$

x	y
0	1
0.1	1.2000
0.2	1.4086
0.3	1.6239
0.4	1.8442
0.5	2.0678

$y = 2.0678$ for $x = 0.5$ as compared with $y_2(0.5) = 2.0611$ from Exercise 25.

28. $y' = 1 + ye^{-x}, 0 \le x \le 0.5, \Delta x = 0.1, (0,1)$

x	y
0	1
0.1	1.2164
0.2	1.4275
0.3	1.7812
0.4	2.1605
0.5	2.6359

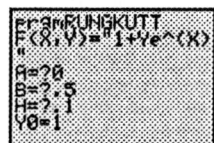

$y = 2.6359$ for $x = 0.5$ as compared with $y_2(0.5) = 2.0611$ from Exercise 25 and $y = 2.0678$ from Exercise 27.

29. $2\dfrac{di}{dt} + i = 12, i(0) = 0$

$$2L(i') + L(i) = L(12)$$

$$2[sL(i) - 0] + L(i) = \frac{12}{s}$$

$$L(i) = \frac{12}{s(2s+1)} = \frac{12(1/2)}{s(s+1/2)}$$

$$i = 12(1 - e^{-t/2})$$

$$i(0.3) = 12(1 - e^{-0.3/2})$$

$$i = 1.67 \text{ A}$$

30. $6i' + 30i = 10 \sin 20t, i(0) = 0$

$$6L(i') + 30L(i) = 10L(\sin 20t)$$

$$6(sL(i) - i(0)) + 30L(i) = 10\frac{20}{s^2 + 20^2}$$

$$(6s + 30)L(i) = \frac{200}{s^2 + 20^2}$$

$$L(i) = \frac{200}{6(s+5)(s^2 + 20^2)} = \frac{100}{3}\left(\frac{1}{(s+5)(s^2 + 400)}\right)$$

$$L(i) = -\frac{4}{51} \cdot \frac{s}{s^2 + 20^2} + \frac{1}{51} \cdot \frac{20}{s^2 + 20^2}$$

$$\frac{4}{51} \cdot \frac{1}{s+5}$$

$$i = -\frac{4}{51}\cos 20t + \frac{1}{51}\sin 20t + \frac{4}{51}e^{-5t}$$

31. $0.25\dfrac{d^2q}{dt^2} + \dfrac{4\,dq}{dt} + \dfrac{q}{10^{-4}} = 0;\ q(0) = 400\ \mu C = 4 \times 10^{-4}\ C$

$i = \dfrac{dq}{dt} = q'0;\ 0.25L(q'') + 4L(q') + 10^4 L(q) = 0;\ L(q'') + 16L(q') + 4 \times 10^4 L(q) = 0$

$s^2 L(q) - 4 \times 10^{-4}s - 0 + 16sL(q) - 4 \times 10^{-4} + 4 \times 10^4 L(q) = 0$

$(s^2 + 16s + 4 \times 10^{-4})L(q) = 4 \times 10^{-4}s + 64 \times 10^{-4}$

$$L(q) = 4 \times 10^{-4}\left[\frac{s + 16}{s^2 + 16s + 64 + 4 \times 10^4}\right] = 4 \times 10^{-4}\left[\frac{s+8}{(s+8)^2 + 200^2} + \frac{8}{(s+8)^2 + 200^2}\right]$$

$$= 4 \times 10^{-4}\left[\frac{s+8}{(s+8)^2 + 200^2} + \frac{8}{200} \times \frac{200}{(s+8)^2 + 200^2}\right]$$

$$q = 4 \times 10^{-4}(e^{-8t}\cos 200t + 0.04e^{-8t}\sin 200t)$$

$$q = 10^{-4}e^{-8t}(4.0\cos 200t + 0.16\sin 200t)$$

32. $0.5q'' + 6q' + 5000q = 0, q(0) = 10 \times 10^{-3}, q'(0) = 0$

$0.5L(q'') + 6L(q') + 5000L(q) = L(0)$

$0.5(s^2 L(q) - sq(0) - q'(0)) + 6(sL(q) - q(0)) + 5000L(q) = L(0)$

$0.5s^2 L(q) - 0.005s + 6sL(q) - 0.06 + 5000L(q) = 0$

$(0.5s^2 + 6s + 5000)L(q) = 0.005s + 0.06$

$0.5(s^2 + 12s + 10{,}000)L(q) = 0.005s + 0.06$

$(s^2 + 12s + 10{,}000)L(q) = 0.01s + 0.12$

$$L(q) = \frac{0.01s + 0.12}{s^2 + 12s + 10{,}000} = \frac{1}{100}\frac{s + 12}{s^2 + 12s + 36 + 9964}$$

$$L(q) = \frac{1}{100}\frac{s + 12}{(s+6)^2 + 99.8^2}$$

$$L(q) = \frac{1}{100}\left[\frac{s+6}{(s+6)^2 + 99.8^2} + \frac{6}{99.8}\frac{99.8}{(s+6)^2 + 99.8^2}\right]$$

$$q = \frac{1}{100}\left[e^{-6t}\cos 99.8t + \frac{6}{99.8}e^{-6t}\sin 99.8t\right]$$

$$q = 0.01e^{-6t}\left[\cos 99.8t + \frac{6}{99.8}\sin 99.8t\right]$$

$$q = 0.01e^{-6t}[\cos 100t + 0.06\sin 100t]$$

33. $m = 0.25$ slug, $F = 8$ lb, $x = 6$ in. $= 0.5$ ft; $8 = 0.5k, k = 16$ lb/ft

$0.25D^2y + 16y = \cos 8t \quad (D = d/dt)$

$D^2y + 64y = 4\cos 8t, y(0) = 0, Dy(0) = 0$

$L(D^2y) + 64L(y) = 4L(\cos 8t); s^2L(y) - s(0) - 0 + 64L(y) = 4\left(\dfrac{s}{s^2 + 64}\right)$

$L(y) = \dfrac{4s}{(s^2 + 64)^2} = \dfrac{1}{4}\left[\dfrac{2(8s)}{(s^2 + 64)^2}\right]; y = 0.25t\sin 8t$

34. $m = 5.00$ kg, $F = 50.0$ N, $x = 1.00$ m; $5.00 = k(1.00), k = 50.0$ N/m

$5.00D^2y + 50.0y = 0, y(0) = 1, y'(0) = 0m; D^2y + 10.0 = 0$

$L(D^2y) + 10.0L(y) = L(0); s^2L(y) - s(1) - 0 + 10.0L(y) = 0$

$L(y) = \dfrac{s}{s^2 + 10}; y = \cos 3.16t$

35. $2i' + i = 12, 0 \le t \le 0.300, \Delta = 0.05, (0,0)$

$i' = \dfrac{12 - i}{2}$

t	i
0.00	0.0000
0.05	0.3000
0.10	0.5988
0.15	0.8963
0.20	1.1925
0.25	1.4875
0.30	1.7813

$i = 1.78$ A for $t = 0.300$ s

36. $2i' + i = 12, i(0) = 0$

$i' = \dfrac{12 - i}{2}$

$i' = \dfrac{12 - 0}{2} = 6$

$i_1 = 6t + c$

$i_1(0) = 0 = 6(0) + c, c = 0$

$i_1 = 6t$

$i' = \dfrac{12 - 6t}{2} = 6 - 3t$

$i_2 = 6t - \dfrac{3}{2}t^2 + c$

$i_2(0) = 0 = 6(0) - \dfrac{3}{2}(0)^2 + c, c = 0$

$i_2 = 6t - \dfrac{3}{2}t^2$

$i' = \dfrac{12 - 6t - \frac{3}{2}t^2}{2} = 6 - 3t - \dfrac{3}{4}t^2$

$i_3 = 6t - \dfrac{3}{2}t^2 - \dfrac{1}{4}t^3 + c$

$i_3(0) = 0 - 6(0) - \dfrac{3}{2}(0)^2 - \dfrac{1}{4}(0)^3 + c, c = 0$

$i_3 = 6t - \dfrac{3}{2}t^2 - \dfrac{t^3}{4}$

$$i_3(0.300) = 6(0.300) - \frac{3}{2}(0.300)^2 - \frac{1}{4}(0.300)^3$$

$i_3 = 1.66$ A as compared with 1.67 A from Exercise 29 and 1.78 A from Exercise 35.

37.
$$0.25y'' + 16y = 0, y(0) = 0.75, y'(0) = 0$$
$$0.25L(y'') + 16L(y) = L(0)$$
$$0.25(s^2 L(y) - sy(0) - y'(0)) + 16L(y) = 0$$
$$0.25s^2 L(y) - 0.75(0.25)s + 16L(y) = 0$$
$$(0.25s^2 + 16)L(y) = 0.75(0.25)s$$

$$L(y) = \frac{0.75(0.25)s}{0.25s^2 + 16} = \frac{0.75(0.25)s}{0.25(s^2 + 64)}$$

$$y = 0.75 \cos 8t$$

38. $5.00y'' + 10y' + 50.0y = 0, y(0) = 1, y'(0) = 0$
$$5.00L(y'') + 10L(y') + 50.0L(y) = L(0)$$
$$5.00(s^2 L(y) - sy(0) - y'(0)) + 10(sL(y) - y(0)) + 50.0L(y) = 0$$
$$5.00(s^2 L(y) - s) + 10(sL(y) - 1) + 50.0L(y) = 0$$
$$(5.00s^2 + 10s + 50.0)L(y) - 5.00s - 10 = 0$$

$$L(y) = \frac{5.00s + 10}{5.00s^2 + 10s + 50.0} = \frac{5.00(s + 2)}{5.00(s^2 + 2s + 10)}$$

$$L(y) = \frac{s + 2}{s^2 + 2s + 1 + 9} = \frac{s + 2}{(s + 1)^2 + 3^2}$$

$$L(y) = \frac{s + 1}{(s + 1)^2 + 3^2} + \frac{1}{3}\frac{3}{(s + 1)^2 + 3^2}$$

$$y = e^{-t}\cos 3t + \frac{1}{3}e^{-t}\sin 3t$$

$$y = e^{-t}\left(\cos 3t + \frac{1}{3}\sin 3t\right)$$

39. $F(s) = \dfrac{b - a}{(s + a)(s + b)} = \dfrac{1}{s + a} - \dfrac{1}{s + b}$

$$L^{-1}(F(s)) = L^{-1}\left(\frac{1}{s + a}\right) - L^{-1}\left(\frac{1}{s + b}\right)$$

$$L^{-1}(F(s)) = e^{-at} - e^{-bt}$$

40. $F(s) = \dfrac{a^2}{s(s^2 + a^2)} = \dfrac{1}{s} - \dfrac{s}{s^2 + a^2}$

$$L^{-1}(F(s)) = L^{-1}\left(\frac{1}{s}\right) - L^{-1}\left(\frac{s}{s^2 + a^2}\right)$$

$$L^{-1}(F(s)) = 1 - \cos at$$

41. Using the methods of Chapter 15 the differential equation is solved first and then the initial conditions are applied to evaluate the arbitrary constants. The Laplace transform usually solves the entire initial-value problem in one step. An exception occurs when the initial conditions are not given at $t = 0$. There are, however, procedures for handling this exception.

SUPPLEMENTARY TOPICS

Exercises A-1 Rotation of Axes

1. $x^2 - y^2 = 25, \theta = 45°; x = x'\cos 45° - y'\sin 45° = \dfrac{x'}{\sqrt{2}} - \dfrac{y'}{\sqrt{2}}; y = x'\sin 45° + y'\cos 45° = \dfrac{x'}{\sqrt{2}} + \dfrac{y'}{\sqrt{2}}$

$x^2 - y^2 = \left(\dfrac{x'}{\sqrt{2}} - \dfrac{y'}{\sqrt{2}}\right)^2 - \left(\dfrac{x'}{\sqrt{2}} + \dfrac{y'}{\sqrt{2}}\right)^2 = 25; \dfrac{x'^2}{2} - \dfrac{2x'y'}{2} + \dfrac{y'^2}{2} - \dfrac{x'^2}{2} - \dfrac{2x'y'}{2} - \dfrac{y'^2}{2} = 25$

$2x'y' + 25 = 0$, hyperbola

2. $x^2 + y^2 = 16, \theta = 60°$

$x = x'\cos 60° - y'\sin 60 = \dfrac{x'}{2} - \dfrac{\sqrt{3}y'}{2}; y = x'\sin 60° + y'\cos 60° = \dfrac{\sqrt{3}x'}{2} + \dfrac{y'}{2}$

$x^2 + y^2 = \left(\dfrac{x'}{2} - \dfrac{\sqrt{3}y'}{2}\right)^2 + \left(\dfrac{\sqrt{3}x'}{2} + \dfrac{y'}{2}\right)^2 = 16; \dfrac{x'^2}{4} - \dfrac{\sqrt{3}x'y'}{2} + \dfrac{3y'^2}{4} + \dfrac{3x'^2}{4} + \dfrac{\sqrt{3}x'y'}{2} + \dfrac{y'^2}{4} = 16$

$x'^2 + y'^2 = 16$, circle

3. $8x^2 - 4xy + 5y^2 = 36, \ \theta = \tan^{-1} 2$

$$\sin\theta = \dfrac{2}{\sqrt{5}}$$

$$\cos\theta = \dfrac{1}{\sqrt{5}}$$

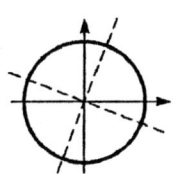

$x = \dfrac{x'}{\sqrt{5}} - \dfrac{2y'}{\sqrt{5}}; y = \dfrac{2x'}{\sqrt{5}} + \dfrac{y'}{\sqrt{5}}$

$8x^2 - 4xy + 5y^2 = 8\left(\dfrac{x'}{\sqrt{5}} - \dfrac{2y'}{\sqrt{5}}\right)^2 - 4\left(\dfrac{x'}{\sqrt{5}} - \dfrac{2y'}{\sqrt{5}}\right)\left(\dfrac{2x'}{\sqrt{5}} + \dfrac{y'}{\sqrt{5}}\right) + 5\left(\dfrac{2x'}{\sqrt{5}} + \dfrac{y'}{\sqrt{5}}\right)^2 = 36$

$8\left(\dfrac{x'^2}{5} - \dfrac{4x'y'}{5} + \dfrac{4y'^2}{5}\right) - 4\left(\dfrac{2x'^2}{5} + \dfrac{x'y'}{5} - \dfrac{4x'y'}{5} - \dfrac{2y'^2}{5}\right) + 5\left(\dfrac{4x'^2}{5} + \dfrac{4x'y'}{5} + \dfrac{4y'^2}{5}\right) = 36$

$\dfrac{8x'^2}{5} - \dfrac{32x'y'}{5} + \dfrac{32y'^2}{5} - \dfrac{8x'^2}{5} - \dfrac{4x'y'}{5} + \dfrac{16x'y'}{5} + \dfrac{8y'^2}{5} + 4x'^2 + 4x'y' + y'^2 = 36$

593

$4x'^2 + 9y'^2 = 36$, ellipse

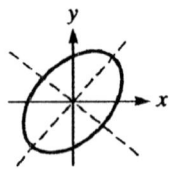

4. $2x^2 + 24xy - 5y^2 = 8, \quad \theta = \tan^{-1}\dfrac{3}{4}$

$$\sin\theta = \frac{3}{5}$$

$$\cos\theta = \frac{4}{5}$$

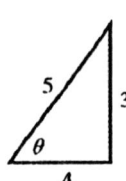

$x = \dfrac{4x'}{5} - \dfrac{3y'}{5}, y = \dfrac{3x'}{5} + \dfrac{4y'}{5}$

$2\left(\dfrac{4x'}{5} - \dfrac{3y'}{5}\right)^2 + 24\left(\dfrac{4x'}{5} - \dfrac{3y'}{5}\right)\left(\dfrac{3x'}{5} + \dfrac{4y'}{5}\right) - 5\left(\dfrac{3x'}{5} + \dfrac{4y'}{5}\right)^2 = 8$

$2\left(\dfrac{16x'^2}{25} - \dfrac{24x'y'}{25} + \dfrac{9y'^2}{25}\right) + 24\left(\dfrac{12x'^2}{25} + \dfrac{16x'y'}{25} - \dfrac{9x'y'}{25} - \dfrac{12y'^2}{25}\right) - 5\left(\dfrac{9x'^2}{25} + \dfrac{24x'y'}{25} + \dfrac{16y'^2}{25}\right) = 8$

$11x'^2 - 14y'^2 = 8$, hyperbola

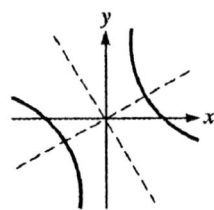

5. $x^2 + 2xy + y^2 - 2x + 2y = 0$

$\tan 2\theta = \dfrac{B}{A-C} = \dfrac{2}{1-1}, A = C, \theta = 45°$

$x = x'\cos 45° - y'\sin 45° = \dfrac{x'}{\sqrt{2}} - \dfrac{y'}{\sqrt{2}}$

$y = x'\sin 45° + y'\cos 45° = \dfrac{x'}{\sqrt{2}} + \dfrac{y'}{\sqrt{2}}$

$\left(\dfrac{x'}{\sqrt{2}} - \dfrac{y'}{\sqrt{2}}\right)^2 + 2\left(\dfrac{x'}{\sqrt{2}} - \dfrac{y'}{\sqrt{2}}\right)\left(\dfrac{x'}{\sqrt{2}} + \dfrac{y'}{\sqrt{2}}\right) + \left(\dfrac{x'}{\sqrt{2}} + \dfrac{y'}{\sqrt{2}}\right)^2 - 2\left(\dfrac{x'}{\sqrt{2}} - \dfrac{y'}{\sqrt{2}}\right) + 2\left(\dfrac{x'}{\sqrt{2}} + \dfrac{y'}{\sqrt{2}}\right) = 0$

$\dfrac{x'^2}{\sqrt{2}} - x'y' + \dfrac{y'^2}{2} + 2\left(\dfrac{x'^2}{2} + \dfrac{x'y'}{2} - \dfrac{x'y'}{2} - \dfrac{y'^2}{2}\right) + \dfrac{x'^2}{2} + x'y' + \dfrac{y'^2}{2} - \dfrac{2x'}{\sqrt{2}} + \dfrac{2y'}{\sqrt{2}} + \dfrac{2x'}{\sqrt{2}} + \dfrac{2y'}{\sqrt{2}} = 0$

$2x'^2 + 2\sqrt{2}y' = 0$

$\quad x'^2 + \sqrt{2}y' = 0$, parabola

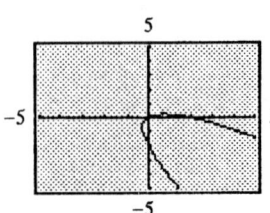

6. $5x^2 - 6xy + 5y^2 = 32;\tan 2\theta = \dfrac{B}{A-C} = \dfrac{-6}{5-5}, A = C, \theta = 45°; x = \dfrac{x'-y'}{\sqrt{2}}, y = \dfrac{x'+y'}{\sqrt{2}}$

$5\left(\dfrac{x'-y'}{\sqrt{2}}\right)^2 - 6\left(\dfrac{x'-y'}{\sqrt{2}}\right)\left(\dfrac{x'+y'}{\sqrt{2}}\right) + \left(\dfrac{x'+y'}{\sqrt{2}}\right)^2 = 32$

$5(x'^2 - 2x'y + y'^2) - 6(x'^2 - y'^2) + 5(x'^2 + 2x'y + y'^2) = 64$

$4x'^2 + 16y'^2 = 64$

$x'^2 + 4y'^2 = 16,$ ellipse

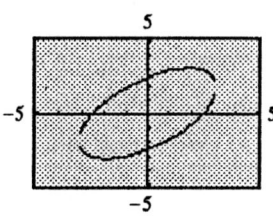

7. $3x^2 + 4xy = 4$

$\tan 2\theta = \dfrac{B}{A-C} = \dfrac{4}{3-0} = \dfrac{4}{3}$

$\cos 2\theta = \dfrac{3}{5}$

$\sin\theta = \sqrt{\dfrac{1-\cos 2\theta}{2}} = \sqrt{\dfrac{1-\frac{3}{5}}{2}} = \dfrac{1}{\sqrt{5}}; \cos\theta = \sqrt{\dfrac{1+\cos 2\theta}{2}} = \sqrt{\dfrac{1+\frac{3}{5}}{2}} = \dfrac{2}{\sqrt{5}}; x = \dfrac{2x'-y'}{\sqrt{5}}, y = \dfrac{x'+2y'}{\sqrt{5}}$

$3\left(\dfrac{2x'-y'}{\sqrt{5}}\right)^2 + 4\left(\dfrac{2x'-y'}{\sqrt{5}}\right)\left(\dfrac{x'+2y'}{\sqrt{5}}\right) = 4; 3(4x'^2 - 4x'y' + y'^2) + 4(2x'^2 + 3x'y' - 2y'^2) = 20$

$20x'^2 - 5y'^2 = 20$

$4x'^2 - y'^2 = 4,$ hyperbola

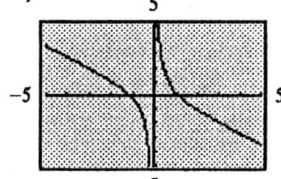

8. $9x^2 - 24xy + 16y^2 - 320x - 240y = 0$

$\tan 2\theta = \dfrac{B}{A-C} = \dfrac{-24}{9-16} = \dfrac{24}{7}; \cos 2\theta = \dfrac{7}{25}$

$\sin\theta = \sqrt{\dfrac{1-\cos 2\theta}{2}} = \sqrt{\dfrac{1-\frac{7}{25}}{2}} = \dfrac{3}{5}; \cos\theta\sqrt{\dfrac{1+\cos 2\theta}{2}} = \sqrt{\dfrac{1+\frac{7}{25}}{2}} = \dfrac{4}{5}$

$x = \dfrac{4x'-3y'}{5}, y = \dfrac{3x'+4y'}{5}$

$9\left(\dfrac{4x'-3y'}{5}\right)^2 - 24\left(\dfrac{4x'-3y'}{5}\right)\left(\dfrac{3x'+4y'}{5}\right) + 16\left(\dfrac{3x'+4y'}{5}\right)^2 - 320\left(\dfrac{4x'-3y'}{5}\right) - 240\left(\dfrac{3x'+4y'}{5}\right) = 0$

$9(16x'^2 - 24x'y' + 9y'^2) - 24(12x'^2 + 16x'y' - 9x'y' - 12y'^2) + 16(9x'^2 + 24x'y' + 16y')$

$\quad - 1600(4x' - 3y') - 1200(3x' + 4y') = 0$

$625y'^2 - 10{,}000x' = 0$

$y'^2 = 16x',$ parabola

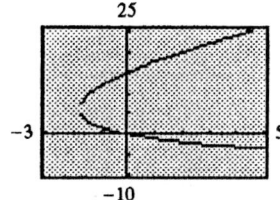

9. $11x^2 - 6xy + 19y^2 = 20$

$$\tan 2\theta = \frac{B}{A-C} = \frac{-6}{11-19} = \frac{6}{8} = \frac{3}{4}; \cos 2\theta = \frac{4}{5}$$

$$\sin\theta = \sqrt{\frac{1-\cos 2\theta}{2}} = \sqrt{\frac{1-\frac{4}{5}}{2}} = \frac{1}{\sqrt{10}}; \cos\theta = \sqrt{\frac{1+\cos 2\theta}{2}} = \sqrt{\frac{1+\frac{4}{5}}{2}} = \frac{3}{\sqrt{10}}$$

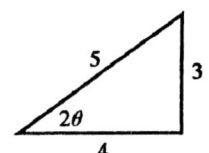

$$x = \frac{3x'-y'}{\sqrt{10}}, y = \frac{x'+3y'}{\sqrt{10}}$$

$$11\left(\frac{3x'-y'}{\sqrt{10}}\right) - 6\left(\frac{3x'-y'}{\sqrt{10}}\right)\left(\frac{x'+3y'}{\sqrt{10}}\right) + 19\left(\frac{x'+3y'}{\sqrt{10}}\right)^2 = 20$$

$$11(9x'^2 - 6x'y' + y'^2) - 6(3x'^2 + 9x'y' - x'y' - 3y'^2) + 19(x'^2 + 6x'y' + 9y'^2) = 200$$

$$100x'^2 + 200y'^2 = 200$$

$$x'^2 + 2y'^2 = 2, \text{ ellipse}$$

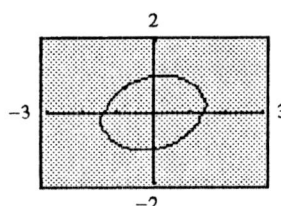

10. $x^2 + 4xy - 2y^2 = 6; \tan 2\theta = \dfrac{B}{A-C} = \dfrac{4}{1-(-2)} = \dfrac{4}{3}$, From Exercise 7

$$\sin\theta = \frac{1}{\sqrt{5}}, \cos\theta = \frac{2}{\sqrt{5}}; x = \frac{2x'-y'}{\sqrt{5}}, y = \frac{x'+2y'}{\sqrt{5}}$$

$$\left(\frac{2x'-y'}{\sqrt{5}}\right)^2 + 4\left(\frac{2x'-y'}{\sqrt{5}}\right)\left(\frac{x'+2y'}{\sqrt{5}}\right) - 2\left(\frac{x'+2y'}{\sqrt{5}}\right)^2 = 6$$

$$4x'^2 - 4x'y' + y'^2 + 4(2x'^2 + 4x'y' - x'y' - 2y'^2) - 2(x'^2 + 4x'y' + 4y'^2) = 30$$

$$10x'^2 - 15y'^2 = 30$$

$$2x'^2 - 3y'^2 = 6, \text{ hyperbola}$$

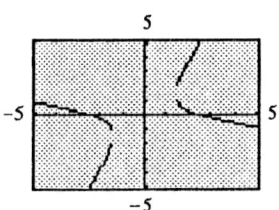

11. $16x^2 - 24xy + 9y^2 - 60x - 80y + 400 = 0$

$$\tan 2\theta = \frac{B}{A-C} = \frac{-24}{16-9} = \frac{-24}{7}$$

$$\cos\theta = \frac{-7}{25}$$

$$\sin\theta = \sqrt{\frac{1-\cos 2\theta}{2}} = \sqrt{\frac{1-\frac{-7}{25}}{2}} = \frac{4}{5}$$

$$\cos\theta = \sqrt{\frac{1+\cos 2\theta}{2}} = \sqrt{\frac{1+\frac{-7}{25}}{2}} = \frac{3}{5}$$

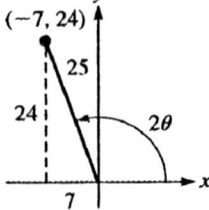

$$x = \frac{3x'-4y'}{5}, y = \frac{4x'+3y'}{5}$$

$$16\left(\frac{3x'-4y'}{5}\right)^2 - 24\left(\frac{3x'-4y'}{5}\right)\left(\frac{4x'+3y'}{5}\right) + 9\left(\frac{4x'+3y'}{5}\right)^2 - 60\left(\frac{3x'-4y'}{5}\right) - 80\left(\frac{4x'+3y'}{5}\right) + 400 = 0$$

$$16(9x'^2 - 24x'y' + 16y'^2) - 24(12x'^2 + 9x'y' - 16x'y' - 12y'^2) + 9(16x'^2 + 24x'y' + 9y'^2)$$
$$- 300(3x' - 4y') - 400(4x' + 3y') + 10{,}000 = 0$$
$$625y'^2 - 2500x' + 10{,}000 = 0$$
$$y'^2 - 4x' + 16 = 0$$
$$(y' - 0)^2 - 4(x' - 4) = 0$$
$$y''^2 - 4x'' = 0$$
$$y''^2 = 4x'', \text{ parabola}$$

12. $73x^2 - 72xy + 52y^2 + 100x - 200y + 100 = 0$

$$\tan 2\theta = \frac{B}{A - C} = \frac{-72}{73 - 52} = \frac{-72}{21} = \frac{-24}{7}$$

From Exercise 11

$$\sin \theta = \frac{4}{5}, \cos \theta = \frac{3}{5}$$

$$x = \frac{3x' - 4y'}{5}, y = \frac{4x' + 3y'}{5}$$

$$73\left(\frac{3x' - 4y'}{5}\right)^2 - 72\left(\frac{3x' - 4y'}{5}\right)\left(\frac{4x' + 3y'}{5}\right) + 52\left(\frac{4x' + 3y'}{5}\right)^2 + 100\left(\frac{3x' - 4y'}{5}\right) - 200\left(\frac{4x' + 3y'}{5}\right) + 100 = 0$$

$$73(9x'^2 - 24x'y' + 16y'^2) - 72(12x'^2 + 9x'y' - 16x'y' - 12y'^2) + 52(16x'^2 + 24x'y' + 9y'^2) + 500(3x' - 4y')$$
$$- 1000(4x' + 3y') + 2500 = 0$$
$$625x'^2 + 2500y'^2 - 2500x' - 5000y' + 2500 = 0$$
$$x'^2 + 4y'^2 - 4x' - 8y' + 4 = 0$$
$$(x'^2 - 4x' + 4) + 4(y'^2 - 2y' + 1) = -4 + 4 + 4$$
$$(x' - 2)^2 + 4(y' - 1)^2 = 4$$
$$x''^2 + 4y''^2 - 4 = 0, \text{ ellipse}$$

Exercises A-2 Regression

1.

x	y	xy	x^2
4	1	4	16
6	4	24	36
8	5	40	64
10	8	80	100
12	9	108	144
40	27	256	360

$n = 5$

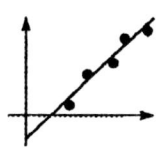

$$m = \frac{5(256) - (40)(27)}{5(360) - 40^2} = 1.0$$

$$b = \frac{360(27) - (256)(40)}{5(360) - 40^2} = -2.6$$

$$y = mx + b; y = 1.0x - 2.6$$

Plot points:

x	y
3	0.4
10	7.4

2.

x	y	xy	x^2
20	160	3 200	400
26	145	3 770	676
30	135	4 050	900
38	120	4 560	1444
48	100	4 800	2304
60	90	5 400	3600
222	750	25 780	9324

$n = 6$

$$m = \frac{6(25\ 780) - (222)(750)}{6(9324) - (222)^2} = -1.77$$

$$b = \frac{(9324)(750) - (25\ 780)(222)}{6(9324) - (222)^2} = 191$$

$y = mx + b;\ y = -1.77x + 191$

Plot points:

x	y
25	147
55	94

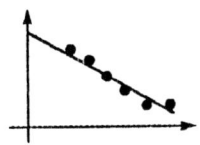

3. $y = m\sqrt{x} + b$

x	y	$\sqrt{x}$	$y\sqrt{x}$	$(\sqrt{x})^2$
0	1	0	0	0
4	9	2	18	4
8	11	$2\sqrt{2}$	$22\sqrt{2}$	8
12	14	$2\sqrt{3}$	$28\sqrt{3}$	12
16	15	4	60	16
	50	12.292	157.610	40

$n = 5$

$$m = \frac{5(157.610) - (12.292)(50)}{5(40) - (12.292)^2} = 3.5$$

$$b = \frac{40(50) - (157.610)(12.292)}{5(40) - (12.292)^2} = 1.3$$

Therefore, $y = 3.5\sqrt{x} + 1.3$

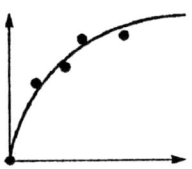

4. $y = m(10^x) + b$

x	y	10^x	yx^{10}	$(10^x)^2$
0.00	6.00	1	6.00	1
0.200	6.60	1.584 893	10.460 295	2.511 886
0.500	8.20	3.162 278	25.930 677	10
0.950	14.0	8.912 509	124.775 131	79.432 823
1.325	26.0	21.134 890	549.507 150	446.683 592
	60.8	35.794 570	716.673 253	539.628 301

$n = 5$

$$m = \frac{5(716.673\ 253) - (35.794\ 570)(60.8)}{5(539.628\ 301) - (35.794\ 570)^2} = 0.99$$

$$b = \frac{(539.628\ 301)(60.8) - (716.673\ 253)(35.794\ 570)}{5(539.628\ 301) - (35.794\ 570)^2} = 5.05$$

$y = 0.99(10^x) + 5.05$

5.

i	V	iV	i^2	
15.0	3.00	45.00	225.00	
10.8	4.10	44.28	116.64	
9.30	5.60	52.08	86.49	$n=5$
3.55	8.00	28.40	12.60	
4.60	10.50	48.30	21.16	
43.25	31.20	218.06	461.89	

$$m = \frac{5(218.06) - (43.25)(31.20)}{5(461.89) - (43.25)^2} = -0.590$$

$$b = \frac{(461.89)(31.20) - (218.06)(43.25)}{5(461.89) - (43.25)^2} = 11.3$$

$$V = mi + b; \ V = -0.590i + 11.3$$

Plot points:

i	V
2	10.1
10	5.3

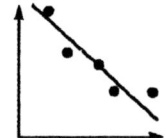

6.

t	T	tT	t^2	
0.0	20.5	0	0	
1.0	20.6	20.6	1.0	
2.0	20.9	41.8	4.0	
3.0	21.3	63.9	9.0	$n=6$
4.0	21.7	86.8	16.0	
5.0	22.0	110.0	25.0	
15.0	127.0	323.1	55.0	

$$m = \frac{6(323.1) - (15.0)(127.0)}{6(55.0) - (15.0)^2} = 0.32$$

$$b = \frac{(55.0)(127.0) - (323.1)(15.0)}{6(55.0) - (15)^2} = 20.4$$

$$T = mt + b; \ T = 0.32t + 20.4$$

Plot points:

t	T
1.5	20.9
4.5	21.8

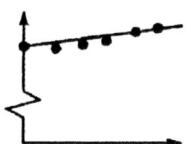

7.

x	p	xp	x^2	
0	650	0	0	
100	630	63,000	10,000	
200	605	121,000	40,000	$n=5$
300	590	177,000	90,000	
400	570	228,000	160,000	
1000	3045	589,000	300,000	

$$m = \frac{5(589,000) - (1000)(3045)}{5(300,000) - (1000)^2} = -0.200$$

$$b = \frac{(300,000)(3045) - (589,000)(1000)}{5(300,000) - (1000)^2} = 649$$

$$p = mx + b; \ p = -0.200x + 649$$

Plot points:

i	V
0	649
350	579

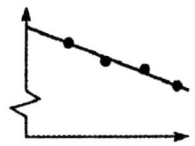

8.

t (in)	L (Btu)	tL	t^2	
3.0	5,900	17,700	9.0	
4.0	4,800	19,200	16.0	
5.0	3,900	19,500	25.0	
6.0	3,100	18,600	36.0	$n=5$
7.0	2,450	17,150	49.0	
25.0	20,150	92,150	135.0	

$$m = \frac{5(92,150) - (25.0)(20,150)}{5(135.0) - (25.0)^2} = -860$$

$$b = \frac{(135.0)(20,150) - (92,150)(25.0)}{5(135.0) - (25.0)^2} = 8330$$

$$L = mt + b; \ L = -860t + 8330$$

Plot points:

x	y
3.5	5320
6.5	2740

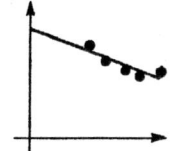

9.

S	$\frac{1}{S}$	P	$\left(\frac{1}{S}\right)P$	$\left(\frac{1}{S}\right)^2$
240	0.00416666667	5.60	0.0233333333	1.736111×10^{-5}
305	0.00327868852	4.40	0.01442622951	1.074980×10^{-5}
420	0.00238095238	3.20	0.00761904762	5.668935×10^{-6}
480	0.0020833333	2.80	0.00583333333	4.340279×10^{-6}
560	0.00178571429	2.40	0.00428571429	3.188776×10^{-6}
2005	0.01369536	18.40	0.05549766	4.13089×10^{-5}

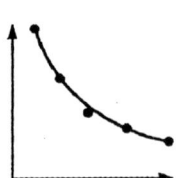

$$m = \frac{5(0.5549766) - (0.01369536)(18.40)}{5(4.13089 \times 10^{-5}) - (0.01369536)^2} = 1343$$

$$b = \frac{(4.13089 \times 10^{-5})(18.40) - (0.05549766)(0.01369536)}{5(4.13089 \times 10^{-5}) - (0.01369536)^2}$$

$$= 1.226612 \times 10^{-3} \text{ or } 0$$

$$P = 1343\left(\frac{1}{S}\right) + 0 = \frac{1343}{S}$$

10. $f = m\left(\dfrac{1}{\sqrt{L}}\right) + b$

L	f	$\frac{1}{\sqrt{L}}$	$f\left(\frac{1}{\sqrt{L}}\right)$	$\left(\frac{1}{\sqrt{L}}\right)^2$
1.0	490	1	490	1
2.0	360	0.707107	254.558441	0.5
4.0	250	0.500000	125	0.25
6.0	200	0.408248	81.649658	0.166667
9.0	170	0.333333	56.666667	0.111111
	1470	2.948688	1007.874766	2.02778

$n = 5$

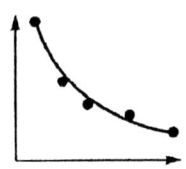

$$m = \frac{5(1007.874766) - (2.948688)(1470)}{5(2.027778) - (2.948688)^2} = 488$$

$$b = \frac{(2.027778)(1470) - (1007.874766)(2.948688)}{5(2.027778) - (2.948688)^2} = 6$$

$$f = \frac{488}{\sqrt{L}} + 6$$

11. $y = me^{-t} + b$

t	y	e^{-t}	e^{-t}	$(e^{-t})^2$
0.0	6.1	1	6.1	1
0.5	3.8	0.606531	2.304817	0.367879
1.0	2.3	0.367879	0.846123	0.135335
1.5	1.3	0.223130	0.290069	0.049871
2.0	0.7	0.135335	0.094735	0.018316
3.0	0.3	0.049787	0.014936	0.002479
	14.5	2.382662	9.65068	1.57388

$n = 6$

$$m = \frac{6(9.65068) - (2.382662)(14.5)}{6(1.57388) - (2.382662)^2} = 6.20$$

$$b = \frac{(1.57388)(14.5) - (9.65068)(2.382662)}{6(1.57388) - (2.382662)^2} = -0.05$$

$$y = 6.20e^{-t} - 0.05$$

12. $T = m \cos\left[\dfrac{\pi}{6}(t - 0.5)\right] + 6$

t	T	$f(t)$	$T \cdot f(t)$	$[f(t)]^2$
0.5	11	1	11	1
1.5	18	0.866025	15.588457	0.75
2.5	29	0.5	14.5	0.25
3.5	46	0	0	0
4.5	57	−0.5	−28.5	0.25
5.5	68	−0.866025	−58.889727	0.75
6.5	73	−1	−73	1
7.5	71	−0.866025	−61.487804	0.75
8.5	61	−0.5	−30.5	0.25
9.5	50	0	0	0
10.5	33	0.5	16.5	0.25
11.5	19	0.866025	16.454483	0.75
	536	0	−178.334591	6

$n = 5$

$m = \dfrac{12(-178.334591) - 0}{12(6) - 0} = -29.7$

$b = \dfrac{6(536) - 0}{12(6) - 0} = 44.7$

$y = -29.7 \cos\left[\dfrac{\pi}{6}(t - 0.5)\right] + 44.7$

THE GRAPHING CALCULATOR

(Most answers have been rounded off to four significant digits.)

1. 56.02   **2.** 1061.8   **3.** 4162.1   **4.** −7.1405   **5.** 18.65   **6.** 24.71   **7.** 0.3954

8. 693.3   **9.** 14.14   **10.** 0.9722   **11.** 0.5251   **12.** 7.769   **13.** 13.35   **14.** 0.001 645

15. 944.6   **16.** 41.92   **17.** 0.7349   **18.** 0.9128   **19.** −0.7594   **20.** −0.4863   **21.** −1.337

22. 0.5954   **23.** 1.015   **24.** −1.133   **25.** 41.35°   **26.** 2.552   **27.** −1.182   **28.** 4.227°

29. 0.5862   **30.** −0.045 18   **31.** 6.695   **32.** 4.256   **33.** 3.508   **34.** 0.8100   **35.** 0.005 685

36. 5.323   **37.** 2.053   **38.** 1.595   **39.** 5.765   **40.** 383.1   **41.** 4.501×10^{10} **42.** 6.313×10^{20}

43. 497.2   **44.** 1.706   **45.** 6.648   **46.** 0.8411   **47.** 401.2   **48.** −1.251   **49.** 8.841

50. 558.2   **51.** 2.523   **52.** −2.088   **53.** 10.08   **54.** 3.658   **55.** 22.36   **56.** 3122

57. 20.3°   **58.** 1.044   **59.** 4729   **60.** 94.82°   **61.** 3.301×10^4 **62.** -3.083×10^{-3} **63.** 1.056

64. 2.781   **65.** 55.5°   **66.** −61.16°   **67.** 3.277   **68.** 0.8210   **69.** 8.125   **70.** 123.5

71. 1.000   **72.** 1.000   **73.** 1.000   **74.** −0.5962   **75.** 12.90   **76.** 5.031   **77.** 8.001

78. 7.001   **79.** 8.053   **80.** 574.2   **81.** 0.042 59   **82.** 8.686   **83.** 0.4219   **84.** −0.052 82

85. 0.7822   **86.** 2.170   **87.** 2.073 646 5   **88.** 0.515 04   **89.** 124.3   **90.** 0.060 83   **91.** 252

92. 0.999 998 9

93.

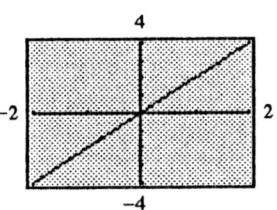

94.

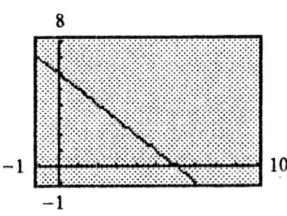

95.

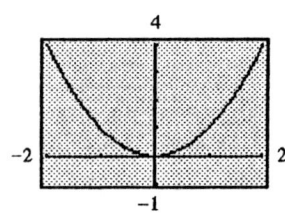

96.

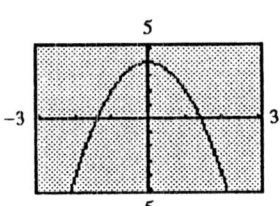

97.

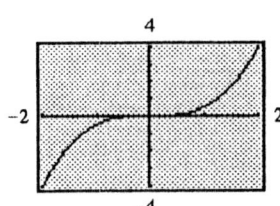

98.

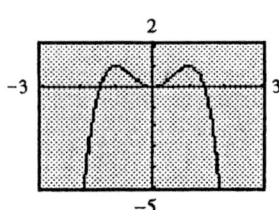

99.

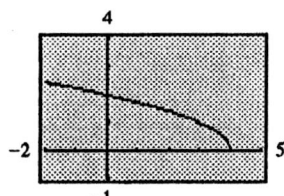

100.

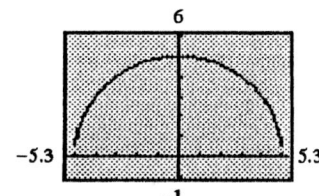

101.

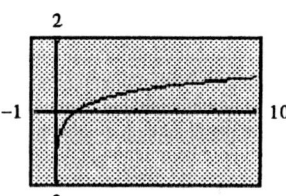

102.

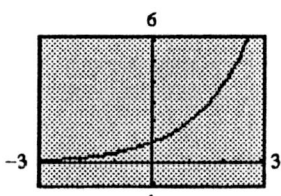

103.

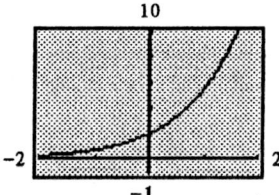

104.

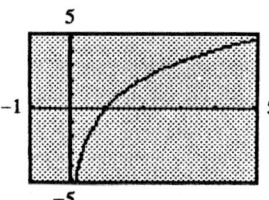

105.

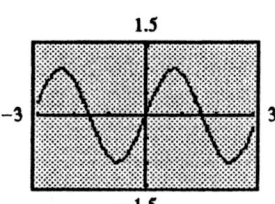

106.

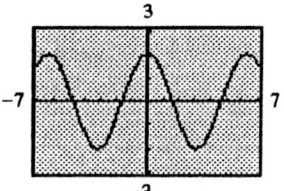

107.

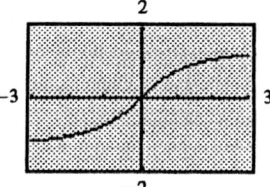

108.

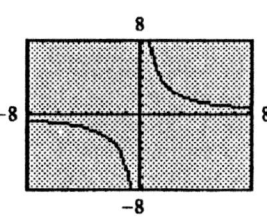